Physics for Engineers and Scientists

Third Edition

Volume 1 (Chapters 1–21)

MOTION, FORCE, AND ENERGY

OSCILLATIONS, WAVES, AND FLUIDS

TEMPERATURE, HEAT, AND THERMODYNAMICS

W · W · NORTON & COMPANY NEW YORK · LONDON

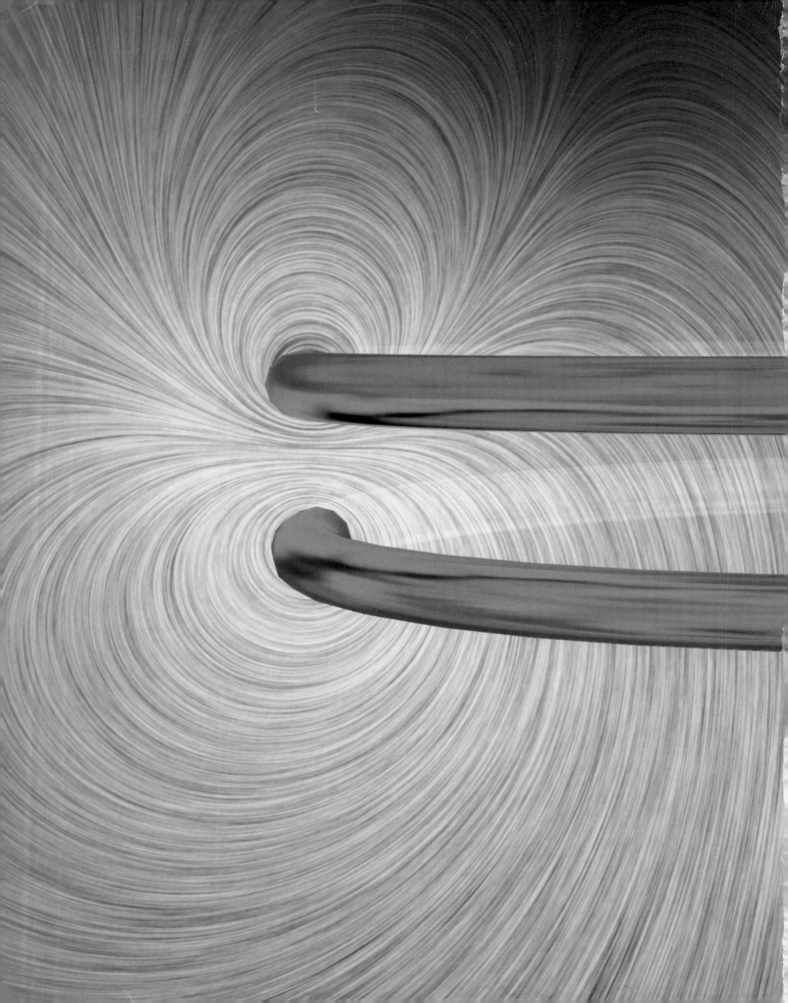

Physics for Engineers and Scientists Third Edition

Volume 1 (Chapters 1–21)

MOTION, FORCE, AND ENERGY

OSCILLATIONS, WAVES, AND FLUIDS

TEMPERATURE, HEAT, AND THERMODYNAMICS

HANS C. OHANIAN,
UNIVERSITY OF VERMONT

JOHN T. MARKERT,
UNIVERSITY OF TEXAS AT AUSTIN

W · W · NORTON & COMPANY      NEW YORK · LONDON

To Susan Ohanian, writer, who gently tried to teach me some of her craft.—H.C.O.

To Frank D. Markert, a printer by trade; to Christiana Park, for her thirst for new knowledge; and to Erin, Ryan, Sean, and Gwen, for their wonder and clarity.—J.T.M.

Copyright © 2007 by W. W. Norton & Company, Inc.

Composition: Techbooks
Manufacturing: RR Donnelley & Sons Company
Editor: Leo A. W. Wiegman
Media Editor: April E. Lange
Director of Manufacturing—College: Roy Tedoff
Senior Project Editor: Christopher Granville
Photo Researcher: Kelly Mitchell
Editorial Assistants: Lisa Rand, Sarah L. Mann
Copy Editor: Richard K. Mickey
Book Designer: Sandy Watanabe
Layout Artist: Paul Lacy
Illustration Studio: JB Woolsey Arts, Inc.
Cover Illustration: John Belcher, inter alia.
Cover Design: Joan Greenfield

The Library of Congress has cataloged the one-volume edition as follows:

Library of Congress Cataloging-in-Publication Data
Ohanian, Hans C.
    Physics for engineers and scientists / Hans C. Ohanian, John T. Markert.—3rd ed.
        p. cm.
    Previously published: Physics. 2nd ed. New York: Norton, c1989.
    "Extended ed., chapters 1-41"
    Includes index.
    ISBN-13: 978-0-393-92631-6
    ISBN-10: 0-393-92631-1
1. Physics. I. Markert, John T. II. Title.
QC21.3.053.2007b
530–dc22   2007061157

Vol. 1 ISBN 978-0-393-93003-0 (pbk.)

W. W. Norton & Company, Inc., 500 Fifth Avenue, New York, N.Y. 10110
www.wwnorton.com

W. W. Norton & Company Ltd., Castle House, 75/76 Wells Street, London W1T 3QT

2 3 4 5 6 7 8 9 0

W. W. Norton & Company has been independent since its founding in 1923, when William Warder Norton and Mary D. Herter Norton first published lectures delivered at the People's Institute, the adult education division of New York City's Cooper Union. The Nortons soon expanded their program beyond the Institute, publishing books by celebrated academics from America and abroad. By mid-century, the two major pillars of Norton's publishing program—trade books and college texts— were firmly established. In the 1950s, the Norton family transferred control of the company to its employees, and today—with a staff of four hundred and a comparable number of trade, college, and professional titles published each year—W. W. Norton & Company stands as the largest and oldest publishing house owned wholly by its employees.

Brief Contents

Chapters 1–21 appear in Volume 1; Chapters 22–36 appear in Volume 2; Chapters 36–41 appear in Volume 3.

PREFACE xiii

OWNER'S MANUAL xxv

PRELUDE: THE WORLD OF PHYSICS xxxv

PART I MOTION, FORCE, AND ENERGY 1

1. SPACE, TIME, AND MASS 2
2. MOTION ALONG A STRAIGHT LINE 28
3. VECTORS 69
4. MOTION IN TWO AND THREE DIMENSIONS 94
5. NEWTON'S LAWS OF MOTION 130
6. FURTHER APPLICATIONS OF NEWTON'S LAWS 173
7. WORK AND ENERGY 204
8. CONSERVATION OF ENERGY 235
9. GRAVITATION 271
10. SYSTEMS OF PARTICLES 305
11. COLLISIONS 338
12. ROTATION OF A RIGID BODY 365
13. DYNAMICS OF A RIGID BODY 394
14. STATICS AND ELASTICITY 429

PART II OSCILLATIONS, WAVES, AND FLUIDS 466

15. OSCILLATIONS 468
16. WAVES 507
17. SOUND 536
18. FLUID MECHANICS 565

PART III TEMPERATURE, HEAT, AND THERMODYNAMICS 600

19. THE IDEAL GAS 602
20. HEAT 628
21. THERMODYNAMICS 661

PART IV ELECTRICITY AND MAGNETISM

22. ELECTRIC FORCE AND ELECTRIC CHARGE
23. THE ELECTRIC FIELD
24. GAUSS' LAW
25. ELECTROSTATIC POTENTIAL AND ENERGY
26. CAPACITORS AND DIELECTRICS
27. CURRENTS AND OHM'S LAW
28. DIRECT CURRENT CIRCUITS
29. MAGNETIC FORCE AND FIELD
30. CHARGES AND CURRENTS IN MAGNETIC FIELDS
31. ELECTROMAGNETIC INDUCTION
32. ALTERNATING CURRENT CIRCUITS

PART V WAVES AND OPTICS

33. ELECTROMAGNETIC WAVES
34. REFLECTION, REFRACTION, AND OPTICS
35. INTERFERENCE AND DIFFRACTION

PART VI RELATIVITY, QUANTA, AND PARTICLES

36. THE THEORY OF SPECIAL RELATIVITY
37. QUANTA OF LIGHT
38. SPECTRAL LINES, BOHR'S THEORY, AND QUANTUM MECHANICS
39. QUANTUM STRUCTURE OF ATOMS, MOLECULES, AND SOLIDS
40. NUCLEI
41. ELEMENTARY PARTICLES AND COSMOLOGY

APPENDICES A-1

Table of Contents

Chapters 1–21 appear in Volume 1; Chapters 22–36 appear in Volume 2; and Chapters 36–41 appear in Volume 3.

PREFACE xiii
OWNER'S MANUAL xxv

PRELUDE: THE WORLD OF PHYSICS xxxv

PART I MOTION, FORCE, AND ENERGY 1

1. SPACE, TIME, AND MASS 2

1.1 Coordinates and Reference Frames 3
1.2 The Unit of Length 5
1.3 The Unit of Time 9
1.4 The Unit of Mass 11
1.5 Derived Units 13
1.6 Significant Figures; Consistency of Units and Conversion of Units 14

SUMMARY / QUESTIONS / PROBLEMS / REVIEW PROBLEMS / ANSWERS TO CHECKUPS 20

2. MOTION ALONG A STRAIGHT LINE 28

2.1 Average Speed 29
2.2 Average Velocity for Motion along a Straight Line 32
2.3 Instantaneous Velocity 35
2.4 Acceleration 39
2.5 Motion with Constant Acceleration 42
2.6 The Acceleration of Free Fall 49
2.7* Integration of the Equations of Motion 54

SUMMARY / QUESTIONS / PROBLEMS / REVIEW PROBLEMS / ANSWERS TO CHECKUPS 57

3. VECTORS 69

3.1 The Displacement Vector and Other Vectors 70
3.2 Vector Addition and Subtraction 72
3.3 The Position Vector; Components of a Vector 76
3.4 Vector Multiplication 81

SUMMARY / QUESTIONS / PROBLEMS / REVIEW PROBLEMS / ANSWERS TO CHECKUPS 87

4. MOTION IN TWO AND THREE DIMENSIONS 94

4.1 Components of Velocity and Acceleration 95
4.2 The Velocity and Acceleration Vectors 98
4.3 Motion with Constant Acceleration 102
4.4 The Motion of Projectiles 104
4.5 Uniform Circular Motion 112
4.6 The Relativity of Motion and the Addition of Velocities 115

SUMMARY / QUESTIONS / PROBLEMS / REVIEW PROBLEMS / ANSWERS TO CHECKUPS 118

5. NEWTON'S LAWS OF MOTION 130

5.1 Newton's First Law 131
5.2 Newton's Second Law 133
5.3 The Combination of Forces 138
5.4 Weight; Contact Force and Normal Force 141
5.5 Newton's Third Law 144
5.6 Motion with a Constant Force 151

SUMMARY / QUESTIONS / PROBLEMS / REVIEW PROBLEMS / ANSWERS TO CHECKUPS 159

6. FURTHER APPLICATIONS OF NEWTON'S LAWS 173

6.1 Friction 174
6.2 Restoring Force of a Spring; Hooke's Law 182
6.3 Force for Uniform Circular Motion 184
6.4* The Four Fundamental Forces 191

SUMMARY / QUESTIONS / PROBLEMS / REVIEW PROBLEMS / ANSWERS TO CHECKUPS 192

7. WORK AND ENERGY 204

7.1 Work 205
7.2 Work for a Variable Force 211
7.3 Kinetic Energy 214
7.4 Gravitational Potential Energy 218

SUMMARY / QUESTIONS / PROBLEMS / REVIEW
PROBLEMS / ANSWERS TO CHECKUPS 224

8. CONSERVATION OF ENERGY 235

8.1 Potential Energy of a Conservative Force 236
8.2 The Curve of Potential Energy 244
8.3 Other Forms of Energy 248
8.4* Mass and Energy 251
8.5 Power 253

SUMMARY / QUESTIONS / PROBLEMS / REVIEW
PROBLEMS / ANSWERS TO CHECKUPS 259

9. GRAVITATION 271

9.1 Newton's Law of Universal Gravitation 272
9.2 The Measurement of G 277
9.3 Circular Orbits 278
9.4 Elliptical Orbits; Kepler's Laws 282
9.5 Energy in Orbital Motion 288

SUMMARY / QUESTIONS / PROBLEMS / REVIEW
PROBLEMS / ANSWERS TO CHECKUPS 293

10. SYSTEMS OF PARTICLES 305

10.1 Momentum 306
10.2 Center of Mass 313
10.3 The Motion of the Center of Mass 323
10.4 Energy of a System of Particles 327

SUMMARY / QUESTIONS / PROBLEMS / REVIEW
PROBLEMS / ANSWERS TO CHECKUPS 328

11. COLLISIONS 338

11.1 Impulsive Forces 339
11.2 Elastic Collisions in One Dimension 344
11.3 Inelastic Collisions in One Dimension 348
11.4* Collisions in Two and Three Dimensions 351

SUMMARY / QUESTIONS / PROBLEMS / REVIEW
PROBLEMS / ANSWERS TO CHECKUPS 354

12. ROTATION OF A RIGID BODY 365

12.1 Motion of a Rigid Body 366

12.2 Rotation about a Fixed Axis 367
12.3 Motion with Constant Angular
Acceleration 374
12.4* Motion with Time-Dependent Angular
Acceleration 376
12.5 Kinetic Energy of Rotation; Moment of
Inertia 378

SUMMARY / QUESTIONS / PROBLEMS / REVIEW
PROBLEMS / ANSWERS TO CHECKUPS 384

13. DYNAMICS OF A RIGID BODY 394

13.1 Work, Energy, and Power in Rotational Motion;
Torque 395
13.2 The Equation of Rotational Motion 399
13.3 Angular Momentum and its Conservation 406
13.4* Torque and Angular Momentum as
Vectors 410

SUMMARY / QUESTIONS / PROBLEMS / REVIEW
PROBLEMS / ANSWERS TO CHECKUPS 417

14. STATICS AND ELASTICITY 429

14.1 Statics of Rigid Bodies 430
14.2 Examples of Static Equilibrium 433
14.3 Levers and Pulleys 441
14.4 Elasticity of Materials 445

SUMMARY / QUESTIONS / PROBLEMS / REVIEW
PROBLEMS / ANSWERS TO CHECKUPS 450

PART II OSCILLATIONS, WAVES, AND FLUIDS 466

15. OSCILLATIONS 468

15.1 Simple Harmonic Motion 469
15.2 The Simple Harmonic Oscillator 476
15.3 Kinetic Energy and Potential Energy 480
15.4 The Simple Pendulum 484
15.5* Damped Oscillations and Forced
Oscillations 488

SUMMARY / QUESTIONS / PROBLEMS / REVIEW
PROBLEMS / ANSWERS TO CHECKUPS 494

16. WAVES 507

16.1 Transverse and Longitudinal Wave Motion 508
16.2 Periodic Waves 509

16.3 The Superposition of Waves 516
16.4 Standing Waves 520

SUMMARY / QUESTIONS / PROBLEMS / REVIEW PROBLEMS / ANSWERS TO CHECKUPS 524

17. SOUND 536

17.1 Sound Waves in Air 538
17.2 Intensity of Sound 540
17.3 The Speed of Sound; Standing Waves 543
17.4 The Doppler Effect 574
17.5* Diffraction 553

SUMMARY / QUESTIONS / PROBLEMS / REVIEW PROBLEMS / ANSWERS TO CHECKUPS 555

18. FLUID MECHANICS 565

18.1 Density and Flow Velocity 567
18.2 Incompressible Steady Flow; Streamlines 569
18.3 Pressure 573
18.4 Pressure in a Static Fluid 575
18.5 Archimedes' Principle 580
18.6 Fluid Dynamics; Bernoulli's Equation 582

SUMMARY / QUESTIONS / PROBLEMS / REVIEW PROBLEMS / ANSWERS TO CHECKUPS 587

PART III TEMPERATURE, HEAT, AND THERMODYNAMICS 600

19. THE IDEAL GAS 602

19.1 The Ideal-Gas Law 603
19.2 The Temperature Scale 609
19.3 Kinetic Pressure 613
19.4 The Internal Energy of an Ideal Gas 616

SUMMARY / QUESTIONS / PROBLEMS / REVIEW PROBLEMS / ANSWERS TO CHECKUPS 619

20. HEAT 628

20.1 Heat as a Form of Energy Transfer 629
20.2 Thermal Expansion of Solids and Liquids 633
20.3 Thermal Conduction 638
20.4 Changes of State 642
20.5 The Specific Heat of a Gas 644
20.6* Adiabatic Expansion of a Gas 647

SUMMARY / QUESTIONS / PROBLEMS / REVIEW PROBLEMS / ANSWERS TO CHECKUPS 650

21. THERMODYNAMICS 661

21.1 The First Law of Thermodynamics 663
21.2 Heat Engines; The Carnot Engine 665
21.3 The Second Law of Thermodynamics 675
21.4 Entropy 677

SUMMARY / QUESTIONS / PROBLEMS / REVIEW PROBLEMS / ANSWERS TO CHECKUPS 681

PART IV ELECTRICITY AND MAGNETISM

22. ELECTRIC FORCE AND ELECTRIC CHARGE

22.1 The Electrostatic Force
22.2 Coulomb's Law
22.3 The Superposition of Electrical Forces
22.4 Charge Quantization and Charge Conservation
22.5 Conductors and Insulators; Charging by Friction or by Induction

SUMMARY / QUESTIONS / PROBLEMS / REVIEW PROBLEMS / ANSWERS TO CHECKUPS

23. THE ELECTRIC FIELD

23.1 The Electric Field of Point Charges
23.2 The Electric Field of Continuous Charge Distributions
23.3 Lines of Electric Field
23.4 Motion in a Uniform Electric Field
23.5 Electric Dipole in an Electric Field

SUMMARY / QUESTIONS / PROBLEMS / REVIEW PROBLEMS / ANSWERS TO CHECKUPS

24. GAUSS' LAW

24.1 Electric Flux
24.2 Gauss' Law
24.3 Applications of Gauss' Law
24.4 Superposition of Electric Fields
24.5 Conductors and Electric Fields

SUMMARY / QUESTIONS / PROBLEMS / REVIEW PROBLEMS / ANSWERS TO CHECKUPS

25. ELECTROSTATIC POTENTIAL AND ENERGY

25.1 The Electrostatic Potential
25.2 Calculation of the Potential from the Field

25.3 Potential in Conductors

25.4 Calculation of the Field from the Potential

25.5 Energy of Systems of Charges

SUMMARY / QUESTIONS / PROBLEMS / REVIEW
PROBLEMS / ANSWERS TO CHECKUPS

26. CAPACITORS AND DIELECTRICS

26.1 Capacitance

26.2 Capacitors in Combination

26.3 Dielectrics

26.4 Energy in Capacitors

SUMMARY / QUESTIONS / PROBLEMS / REVIEW
PROBLEMS / ANSWERS TO CHECKUPS

27. CURRENTS AND OHM'S LAW

27.1 Electric Current

27.2 Resistance and Ohm's Law

27.3 The Resistivity of Materials

27.4 Resistances in Combination

SUMMARY / QUESTIONS / PROBLEMS / REVIEW
PROBLEMS / ANSWERS TO CHECKUPS

28. DIRECT CURRENT CIRCUITS

28.1 Electromotive Force

28.2 Sources of Electromotive Force

28.3 Single-Loop Circuits

28.4 Multi-Loop Circuits

28.5 Energy in Circuits; Joule Heat

28.6* Electrical Measurements

28.7* The *RC* Circuit

28.8* The Hazards of Electric Currents

SUMMARY / QUESTIONS / PROBLEMS / REVIEW
PROBLEMS / ANSWERS TO CHECKUPS

29. MAGNETIC FORCE AND FIELD

29.1 The Magnetic Force

29.2 The Magnetic Field

29.3 Ampère's Law

29.4 Solenoids and Magnets

29.5 The Biot-Savart Law

SUMMARY / QUESTIONS / PROBLEMS / REVIEW
PROBLEMS / ANSWERS TO CHECKUPS

30. CHARGES AND CURRENTS IN MAGNETIC FIELDS

30.1 Circular Motion in a Uniform Magnetic Field

30.2 Force on a Wire

30.3 Torque on a Loop

30.4 Magnetism in Materials

30.5* The Hall Effect

SUMMARY / QUESTIONS / PROBLEMS / REVIEW
PROBLEMS / ANSWERS TO CHECKUPS

31. ELECTROMAGNETIC INDUCTION

31.1 Motional EMF

31.2 Faraday's Law

31.3 Some Examples; Lenz' Law

31.4 Inductance

31.5 Magnetic Energy

31.6* The *RL* Circuit

SUMMARY / QUESTIONS / PROBLEMS / REVIEW
PROBLEMS / ANSWERS TO CHECKUPS

32. ALTERNATING CURRENT CIRCUITS*

32.1 Resistor Circuit

32.2 Capacitor Circuit

32.3 Inductor Circuit

32.4* Freely Oscillating *LC* and *RLC* Circuits

32.5* Series Circuits with Alternating EMF

32.6 Transformers

SUMMARY / QUESTIONS / PROBLEMS / REVIEW
PROBLEMS / ANSWERS TO CHECKUPS

PART V WAVES AND OPTICS

33. ELECTROMAGNETIC WAVES

33.1 Induction of Magnetic Fields; Maxwell's Equations

33.2* The Electromagnetic Wave Pulse

33.3 Plane Waves; Polarization

33.4 The Generation of Electromagnetic Waves

33.5 Energy of a Wave

33.6* The Wave Equation

SUMMARY / QUESTIONS / PROBLEMS / REVIEW
PROBLEMS / ANSWERS TO CHECKUPS

34. REFLECTION, REFRACTION, AND OPTICS

34.1 Huygens' Construction
34.2 Reflection
34.3 Refraction
34.4 Spherical Mirrors
34.5 Thin Lenses
34.6* Optical Instruments

SUMMARY / QUESTIONS / PROBLEMS / REVIEW PROBLEMS / ANSWERS TO CHECKUPS

35. INTERFERENCE AND DIFFRACTION

35.1 Thin Films
35.2* The Michelson Interferometer
35.3 Interference from Two Slits
35.4 Interference from Multiple Slits
35.5 Diffraction by a Single Slit
35.6 Diffraction by a Circular Aperture; Rayleigh's Criterion

SUMMARY / QUESTIONS / PROBLEMS / REVIEW PROBLEMS / ANSWERS TO CHECKUPS

PART VI RELATIVITY, QUANTA, AND PARTICLES

36. THE THEORY OF SPECIAL RELATIVITY

36.1 The Speed of Light; the Ether
36.2 Einstein's Principle of Relativity
36.3 Time Dilation
36.4 Length Contraction
36.5 The Lorentz Transformations and the Combination of Velocities
36.6 Relativistic Momentum and Energy
36.7* Mass and Energy

SUMMARY / QUESTIONS / PROBLEMS / REVIEW PROBLEMS / ANSWERS TO CHECKUPS

37. QUANTA OF LIGHT

37.1 Blackbody Radiation
37.2 Energy Quanta
37.3 Photons and the Photoelectric Effect
37.4 The Compton Effect
37.5 X Rays
37.6 Wave vs. Particle

SUMMARY / QUESTIONS / PROBLEMS / REVIEW PROBLEMS / ANSWERS TO CHECKUPS

38. SPECTRAL LINES, BOHR'S THEORY, AND QUANTUM MECHANICS

38.1 Spectral Lines
38.2 Spectral Series of Hydrogen
38.3 The Nuclear Atom
38.4 Bohr's Theory
38.5 Quantum Mechanics; The Schrödinger Equation

SUMMARY / QUESTIONS / PROBLEMS / REVIEW PROBLEMS / ANSWERS TO CHECKUPS

39. QUANTUM STRUCTURE OF ATOMS, MOLECULES, AND SOLIDS

39.1 Principal, Orbital, and Magnetic Quantum Numbers; Spin
39.2 The Exclusion Principle and the Structure of Atoms
39.3* Energy Levels in Molecules
39.4 Energy Bands in Solids
39.5 Semiconductor Devices

SUMMARY / QUESTIONS / PROBLEMS / REVIEW PROBLEMS / ANSWERS TO CHECKUPS

40. NUCLEI

40.1 Isotopes
40.2 The Strong Force and the Nuclear Binding Energy
40.3 Radioactivity
40.4 The Law of Radioactive Decay
40.5 Fission
40.6* Nuclear Bombs and Nuclear Reactors
40.7 Fusion

SUMMARY / QUESTIONS / PROBLEMS / REVIEW PROBLEMS / ANSWERS TO CHECKUPS

41. ELEMENTARY PARTICLES AND COSMOLOGY

41.1 The Tools of High-Energy Physics
41.2 The Multitude of Particles
41.3 Interactions and Conservation Laws
41.4 Fields and Quanta
41.5 Quarks
41.6 Cosmology

SUMMARY / QUESTIONS / PROBLEMS / REVIEW PROBLEMS / ANSWERS TO CHECKUPS

APPENDICES

APPENDIX 1: GREEK ALPHABET A-1

APPENDIX 2: MATHEMATICS REVIEW A-1

A2.1 Symbols A-1
A2.2 Powers and Roots A-1
A2.3 Arithmetic in Scientific Notation A-2
A2.4 Algebra A-3
A2.5 Equations with Two Unknowns A-5
A2.6 The Quadratic Formula A-5
A2.7 Logarithms and the Exponential Function A-5

APPENDIX 3: GEOMETRY AND TRIGONOMETRY REVIEW A-7

A3.1 Perimeters, Areas, and Volumes A-7
A3.2 Angles A-7
A3.3 The Trigonometric Functions A-8
A3.4 The Trigonometric Identities A-9
A3.5 The Laws of Cosines and Sines A-10

APPENDIX 4: CALCULUS REVIEW A-10

A4.1 Derivatives A-10
A4.2 Important Rules for Differentiation A-11
A4.3 Integrals A-12
A4.4 Important Rules for Integration A-15
A4.5 The Taylor Series A-18
A4.6 Some Approximations A-18

APPENDIX 5: PROPAGATING UNCERTAINTIES A-19

APPENDIX 6: THE INTERNATIONAL SYSTEM OF UNITS (SI) A-21

A6.1 Base Units A-21
A6.2 Derived Units A-23
A6.3 Prefixes A-23

APPENDIX 7: BEST VALUES OF FUNDAMENTAL CONSTANTS A-23

APPENDIX 8: CONVERSION FACTORS A-26

APPENDIX 9: THE PERIODIC TABLE AND CHEMICAL ELEMENTS A-31

APPENDIX 10: FORMULA SHEETS A-33

Chapters 1–21 A-33
Chapters 22–41 A-34

APPENDIX 11: ANSWERS TO ODD-NUMBERED PROBLEMS AND REVIEW PROBLEMS A-35

PHOTO CREDITS A-51

INDEX A-55

Preface

Our aim in *Physics for Engineers and Scientists*, Third Edition, is to present a modern view of classical mechanics and electromagnetism, including some optics and quantum physics. We also want to offer students a glimpse of the practical applications of physics in science, engineering, and everyday life.

The book and its learning package emerged from a collaborative effort that began more than six years ago. We adapted the core of Ohanian's earlier *Physics* (Second Edition, 1989) and combined it with relevant findings from recent physics education research on how students learn most effectively. The result is a text that presents a clear, uncluttered explication of the core concepts in physics, well suited to the needs of undergraduate engineering and science students.

Organization of Topics

The 41 chapters of the book cover the essential topics of introductory physics: mechanics of particles, rigid bodies, and fluids; oscillations, wave motion, heat and thermodynamics; electricity and magnetism; optics; special relativity; and atomic and subatomic physics.

Our arrangement and treatment of topics are fairly traditional with a few deliberate distinctions. We introduce the principle of superposition of forces early in Chapter 5 on Newton's laws of motion, and we give the students considerable exposure to the vector superposition of gravitational forces in Chapter 9. This leaves the students well prepared for the later application of vector superposition of electric and magnetic forces generated by charge or current distributions. We place gravitation in Chapter 9 immediately after the chapters on work and energy, because we regard gravitation as a direct application of these concepts (instructors who prefer to postpone gravitation can, of course, do so). We introduce forces on stationary electric charges in a detailed, complete exposition in Chapter 22, before proceeding to the less obvious concept of the electric field in Chapter 23. We start the study of magnetism in Chapter 29 with the force on a moving charged particle near a current, instead of the more common practice of starting with a postulate about the magnetic field in the abstract. With our approach, the observed magnetic forces on moving charges lead naturally to the magnetic field, and this progression from magnetic force to magnetic field will remind students of the closely parallel progression from electric force to electric field. For efficiency and brevity, we sometimes combine in one chapter closely related topics that other authors elect to spread over more than one chapter. Thus, we cover induction and inductance together in Chapter 31 and interference and diffraction together in Chapter 35.

Concise Writing with Sharp Focus on Core Concepts

Our goal is concise exposition with a sharp focus on core concepts. Brevity is desirable because long chapters with a large number of topics and excessive verbiage are confusing and tedious for the student. In our writing, we obey the admonitions of Strunk and White's *Elements of Style*: use the active voice; make statements in positive form; use definite, specific, concrete language; omit needless words.

We strove for simplicity in organizing the content. Each chapter covers a small set of core topics—rarely more than five or six—and we usually place each core topic in a section of its own. This divides the content into manageable segments and gives the chapter a clear and clean outline. Transitional sentences at the beginning or end of sections spell out the logical connections between each section and the next. Within each section, we strove for a seamless narrative leading from the discussions of concepts to their applications in Example problems. We sought to avoid the patchy, cobbled structure of many texts in which the discussions appear to serve as filler between one equation and the next.

Emphasis on the Atomic Structure of Matter

Throughout the book, we encourage students to keep in mind the atomic structure of matter and to think of the material world as a multitude of restless electrons, protons, and neutrons. For instance, in the mechanics chapters, we emphasize that all macroscopic bodies are systems of particles and that the equations of motion for macroscopic bodies emerge from the equations of motion of the individual particles. We emphasize that macroscopic forces are the result of a superposition of the forces among the particles of the system, and we consider atoms and their bonds in the qualitative discussions of elasticity, thermal expansion, and changes of state. By exposing students to the atomic structure of matter in the first semester, we help them to grasp the nature of the charged particles that play a central role in the treatment of electricity and magnetism in the second semester. Thus, in the electricity chapters, we introduce the concepts of positive and negative charge by referring to protons and electrons, not by referring to the antiquated procedure of rubbing glass rods with silk rags.

We try to make sure that students are always aware of the limitations of the nineteenth-century fiction that matter and electric charge are continua. Blind reliance on this old fiction has often been justified by the claim that, although engineering students need physics as a problem-solving tool, the atomic structure of matter is of little concern to them. This supposition may be adequate for a superficial treatment of mechanical engineering. Yet much of modern engineering—from materials science to electronics—hinges on understanding the atomic structure of matter. For this purpose, engineers need a physicist's view of physics.

Real-World Examples Begin Each Chapter

Each chapter opens with a "Concepts in Context" photograph illustrating a practical application of physics. The caption for this photo explores various core concepts in a concrete real-world context. The questions included in the caption are linked to several solved Examples or discussions later in the chapter. Such revisiting of the chapter-opening application provides layers of learning, as new concepts are carefully built upon a foundation firmly planted in the real world. The emphasis on real-world data is also evident throughout other Examples and in the end-of-chapter problems. By exposing students to realistic data, we give them confidence to apply physics in their later science or engineering courses.

Conceptual Discussions Precede and Motivate the Math

Only after a careful exposition of the conceptual foundations in a qualitative physical context does each section proceed to the mathematical treatment. Thus, we ensure that the mathematical formulas and their consequences and variations are rooted in a firm conceptual foundation. We were very careful to provide clear, thorough, and accurate explanations and derivations of all mathematical statements, to ensure that students acquire a good intuition about why particular equations are applied. Immediately after such derivations, we provide solved Examples to establish a firm connection between theory and concrete practical applications.

Examples Enliven the Text

We devote significant portions of each chapter to carefully selected Examples of solved problems—about 390 altogether or 9 on average per chapter. These Examples are concrete illustrations of the preceding conceptual discussions. They build cumulatively upon each other, from simple to more complicated as the chapter progresses. To enliven the text, we employ realistic data in the Examples, such as students would actually encounter outside the classroom. The solved Examples are designed to cover most variations of possible problems, with solutions that include both general approaches and specific details on how to extract the important information for the given problem. For instance, when such keywords as *initially* or *at rest* occur in a solved Example, we are careful to point out their importance in the problem-solving process. Comments appended to some Examples draw attention to limitations in the solution or to wider implications.

Checkup Questions Implement Active Learning

We conclude each section of a chapter with a series of brief *Checkup* questions. These permit students to test their mastery of core concepts, and they can be of great help in clearing up common misconceptions. Checkup questions include variations and "flip sides" of simple concepts that often occur to students but are rarely addressed. We give detailed answers to each Checkup question at the back of the chapter. The entire book contains roughly 5 Checkup questions per section—comprising a total of about 800 Checkup questions.

The final Checkup question of each section is always in multiple-choice format—specifically designed for interactive teaching. At the University of Texas, instructors use such multiple-choice questions as classroom concept quizzes for welcome breaks in conventional lecturing. When more than one answer is popular, the instructor and class immediately know that more discussion or more examples are needed. Such occasions lend themselves well to peer instruction, in which the students explain to one another their reasoning before responding. This pedagogy implements an active, participatory alternative to the traditional lecture format. In addition, several supplements to the textbook, including the Student Activity Workbook, Online Concept Tutorials, Smartwork online homework, and PhysiQuizzes also implement active learning and a mastery-based approach.

Problem-Solving Techniques

Many chapters have inserts in the form of boxes devoted to Problem-Solving Techniques. These 39 skill boxes summarize the main steps or approaches for the solution of common classes of problems. Often deployed after several seemingly disparate Examples, the Problem-Solving Techniques boxes underscore the unity and generality of the techniques used in the Examples. The boxes list the steps or approaches to be taken, providing a handy reference and review.

Math Help

We have placed a Math Help box wherever students encounter a mathematical concept or technique that may be difficult or unfamiliar. These 6 skill boxes briefly review and summarize such topics as trigonometry, derivatives, integrals, and ellipses. Students can find more detailed help in Appendix 2 on basic algebra, 3 on trigonometry and geometry, 4 on calculus, and 5 on propagation of uncertainties.

Physics in Practice

Many chapters have a short essay on Physics in Practice that illustrates an application of physics in engineering and everyday life. These 27 essay boxes discuss practical topics, such as ultracentrifuges, communication and weather satellites, magnetic levitation, etc. Each of these essays provides a wealth of interesting detail and offers a practical supplement to some of the chapter topics. They have been designed to be engaging, yet sufficiently qualitative to provide some respite from the more analytical discussions, Examples, and Questions.

Figures and Balloon Captions

Over 1,500 figures illustrate the text. We made every effort to assemble a visual narrative as clear as the verbal narrative. Each figure in a sequence carefully builds upon the visual information in the figure that precedes it. Many figures in the text contain a caption in "balloon" that points to important features within the figure. The balloon caption is a concise and informative supplement to the conventional figure caption. The balloons make immediately obvious some details that would require a long, wordy explanation in the conventional caption. Often the balloon captions are arranged so that some cause-effect or other sequential thought process becomes immediately evident. All drawn figures are available to instructors in digital form for use in the course.

End-of-Chapter Summary

Each chapter narrative closes with several support elements, starting with a brief Summary. The Summary contains the essential physical laws, quantities, definitions, and key equations introduced in the chapter. A page reference, key equation number, and often a thumbnail figure accompany these laws, definitions, and equations. The Summary does not include repetition of the detailed explanations of the chapter. The Summary is followed by Questions for Discussion, Problems, Review Problems, and Answers to Checkups.

Questions for Discussion

After the chapter's Summary, we include a large selection of qualitative Questions for Discussion — about 700 in the entire book or roughly 17 per chapter. We intend these qualitative end-of-chapter Questions to stimulate student thinking. Some of these questions are deliberately formulated so as to have no unique answer, which is intended to promote class discussion.

Problems

After the chapter's qualitative Questions, we include computational Problems grouped by chapter section — about 3000 in the entire book, or roughly 73 per chapter. Each problem's level of difficulty is indicated by no asterisk, one asterisk (*), or two asterisks (**). Most no-asterisk Problems are easy and straightforward, only requiring students to "plug in" the correct values to compute answers or to retrace the steps of an Example. One-asterisk Problems are of medium difficulty. They contain a few complications requiring the combination of several concepts or the manipulation of several formulas. Two-asterisk Problems are difficult and challenging. They demand considerable thought and perhaps some insight, and occasionally require appreciable mathematical skill. When an Online Concept Tutorial (see below) is available for help in mastering the concepts in a given section, a dagger footnote (†) tells students where to find the tutorial.

We tried to make the Problems interesting for students by drawing on realistic examples from technology, science, sports, and everyday life. Many of the Problems are based on data extracted from engineering handbooks, car repair manuals, *Jane's Book of Aircraft*, *The Guinness Book of World Records*, newspaper reports, research and industrial instrumentation manuals, etc. Many other Problems deal with atoms and subatomic particles. These Problems are intended to reinforce the atomistic view of the material world. In some cases, experts will perhaps consider the use of classical physics somewhat objectionable in a problem that really ought to be handled by quantum mechanics. But we believe that the advantages of familiarization with atomic quantities and magnitudes outweigh the disadvantages of an occasional naive use of classical mechanics.

Among the Problems are a smaller number of somewhat contrived, artificial Problems that make no pretense of realism (for example, "A block slides on an inclined plane tied by string..."). Such unrealistic Problems are sometimes the best way to bring an important concept into sharp focus. Some Problems are formulated as guided problems, with a series of questions that take the student through an important problem-solving procedure, step by step.

Review Problems

After the Problems section of each chapter, we offer an extra selection of Review Problems — about 600 in the entire book or roughly 15 per chapter. We wrote these Review Problems specifically to help students prepare for examinations. Hence, Review Problems often test comprehension by requiring students to apply concepts from more than one section of the chapter and occasionally from prior, related chapters. Answers to all odd-numbered Problems and Review Problems are given in Appendix 11.

Units and Significant Figures

We use the SI system of units exclusively throughout the text. In the abbreviations for the units we follow the recommendations of the International Committee for Weights and Measures (CIPM), although we retain some traditional units, such as revolution and calorie that have been discontinued by the CIPM. In addition, for the sake of clarity we spell out the name of the unit in full whenever the abbreviation is likely to lead to ambiguity and confusion (for instance, in the case of V for volt, which is easily confused with V for potential; or in the case of C for coulomb, which might be confused with C for capacitance). We try to use realistic numbers of significant figures, with most Examples and Problems using two or three. In cases where it is natural to employ some data with two significant figures and some with three, we have been careful to propagate the appropriate number of significant figures to the result.

For reference purposes, we give the definitions of the British units. Currently only the United States, Bangladesh, and Liberia still adhere to these units. In the United States, automobile manufacturers have already switched to metric units for design and construction. The U. S. Army has also switched to metric units, so soldiers give distances in meters and kilometers (in army slang, the kilometer is called a "klick," a usage that is commendable itself for its brevity). British units are not used in examples or in problems, with the exception of a handful of problems in the early chapters. In the definitions of the British units, the pound (lb) is taken to be the unit of mass, and the pound force (lbf) is taken to be the unit of force. This is in accord with the practice approved by the American National Standards Institute (ANSI), the Institute of Electrical and Electronics Engineers (IEEE), and the U. S. Department of Defense.

Optional Sections and Chapters

We recognize course content varies from institution to institution. Some sections and some chapters can be regarded as optional and can be omitted without loss of continuity. These optional sections are marked by asterisks in the Table of Contents.

Mathematical Prerequisites

In order to accommodate students who are taking an introductory calculus course concurrently, derivatives are used slowly at first (Chapter 2), and routinely later on. Likewise, the use of integrals is postponed as long as possible (Chapter 7), and they come into heavy use only in the second volume (after Chapter 21). For students who need a review of calculus, Appendix 4 contains a concise primer on derivatives and integrals.

Acknowledgments

We have had the benefit of a talented author team for our support resources. In addition to their primary role in the assembly of the learning package, they all have also made substantial contributions to the accuracy and clarity of the text.

Stiliana Antonova, Barnard College
Charles Chiu, University of Texas-Austin
William J. Ellis, University of California-Davis
Mirela Fetea, University of Richmond
Rebecca Grossman, Columbia University
David Harrison, University of Toronto
Prabha Ramakrishnan, North Carolina State University
Krassi Lazarova, Drexel University
Hang Deng-Luzader, Frostburg State University
Stephen Luzader, Frostburg State University
Kevin Martus, William Paterson University
David Marx, Illinois State University
Jason Stevens, Deerfield Academy
Brian Woodahl, Indiana University–Purdue University-Indianapolis
Raymond Zich, Illinois State University
And at Sapling Systems and Science Technologies in Austin, Texas, for content, James Caras, Ph.D.; Jon Harmon, B.S.; Kevin Nelson, Ph.D.; John A. Underwood, Ph.D.; and Jason Vestuto, M.S. and for animation and programming, Jeff Sims and Nathan Wheeler.

Our manuscript was subjected to many rounds of peer review. The reviewers were instrumental in identifying myriad improvements, for which we are grateful:

Yildirim Aktas	University of North Carolina–Charlotte
Patricia E. Allen	Appalachian State University
Steven M. Anlage	University of Maryland
B. Antanaitis	Lafayette College
Laszlo Baksay	Florida Institute of Technology
Marco Battaglia	University of California-Berkeley
Lowell Boone	University of Evansville
Marc Borowczak	Walsh University
Amit Chakrabarti	Kansas State University
D. Cornelison	Northern Arizona University
Corbin Covault	Case Western Reserve University
Kaushik De	University of Texas at Arlington
William E. Dieterle	California University of Pennsylvania
James Dunne	Mississippi State University
R. Eagleton	California Polytechnic University-Pomona
Gregory Earle	University of Texas-Dallas
William Ellis	University of California-Davis
Mark Eriksson	University of Wisconsin-Madison
Morten Eskildsen	University of Notre Dame
Bernard Feldman	University of Missouri–St. Louis
Mirela Fetea	University of Richmond
J. D. Garcia	University of Arizona
U. Garg	University of Notre Dame
Michael Gurvitch	State University of New York at Stony Brook
David Harrison	University of Toronto
John Hernandez	University of North Carolina–Chapel Hill
L. Hodges	Iowa State University
Jean-Pierre Jouas	United Nations International School
Kevin Kimberlin	Bradley University
Sebastian Kuhn	Old Dominion University
Tiffany Landry	Folsom Lake College
Dean Lee	North Carolina State University
Frank Lee	George Washington University
Stephen Luzader	Frostburg State University
Kevin Martus	William Paterson University
M. Matkovich	Oakton Community College
David McIntyre	Oregon State University
Rahul Mehta	University of Central Arkansas
Kenneth Mendelson	Marquette University
Laszlo Mihaly	State University of New York at Stony Brook
Richard Mistrick	Pennsylvania State University
Rabindra Mohapatra	University of Maryland
Philip P. J. Morrison	University of Texas at Austin
Greg Mowry	University of Saint Thomas
David Murdock	Tennessee Technological University
Anthony J. Nicastro	West Chester University
Scott Nutter	Northern Kentucky University
Robert Oerter	George Mason University
Ray H. O'Neal, Jr.	Florida A & M University
Frederick Oho,	Winona State University
Paul Parris	University of Missouri–Rolla
Ashok Puri	University of New Orleans
Michael Richmond	Rochester Institute of Technology
John Rollino	Rutgers University–Newark
David Schaefer	Towson State University
Joseph Serene	Georgetown University
H. Shenton	University of Delaware
Jason Stevens	Deerfield Academy

Jay Strieb	Villanova University
John Swez	Indiana State University
Devki N. Talwar	Indiana University of Pennsylvania
Chin-Che Tin	Auburn University
Tim Usher	California State University-San Bernardino
Andrew Wallace	Angelo State University
Barrett Wells	University of Connecticut
Edward A.P. Whittaker	Stevens Institute of Technology
David Wick	Clarkson University
Don Wieber	Contra Costa College
J. William Gary	University of California-Riverside
Suzanne Willis	Northern Illinois University
Thomas Wilson	Marshall University
William. J. F. Wilson	University of Calgary
Brian Woodahl	Indiana University–Purdue University-Indianapolis
Hai-Sheng Wu	Mankato State University

We thank John Belcher, Michael Danziger, and Mark Bessette of the Massachusetts Institute of Technology for creating the cover image. It illustrates the magnetic field generated by two currents in two copper rings. This is one frame of a continuous animation; at the instant shown, the current in the upper ring is opposite to that in the lower ring and is of smaller magnitude. The magnetic field structure shown in this picture was calculated using a modified intregration technique. This image was created as part of the Technology Enabled Active Learning (TEAL) program in introductory physics at MIT, which teaches physics interactively, combining desktop experiments with visualizations of those experiments to "make the unseen seen."

We thank the several editors that supervised this project: first Stephen Mosberg, then Richard Mixter, John Byram, and finally Leo Wiegman, who had the largest share in the development of the text, and also gave us the benefit of his incisive line-by-line editing of the proofs, catching many slips and suggesting many improvements. We also thank the editorial staff at W. W. Norton & Co., including Chris Granville, April Lange, Roy Tedoff, Rubina Yeh, Rob Bellinger, Kelly Mitchell, Neil Hoos, Lisa Rand, and Sarah Mann, as well as the publishing professionals whom Norton engaged, such as Paul Lacy, Richard K. Mickey, Susan McLaughlin, and John B. Woolsey for their enthusiasm and their patience in dealing with the interminable revisions and corrections of the text and its support package. In addition, JTM is grateful to Robert W. Christy of Dartmouth University for various pointers on textbook writing.

HANS C. OHANIAN
Burlington, Vermont
hohanian@uvm.edu

JOHN T. MARKERT
Austin, Texas
jmarkert@physics.utexas.edu

Publication Formats

Physics for Engineers and Scientists comprises six parts. The text is published in two hardcover versions and several paperback versions.

Hardcover Versions

Third Extended Edition, Parts I–VI, 1450 pages, ISBN 0-393-92631-1
   (Chapters 1–41 including Relativity, Quanta and Particles)
Third Edition, Parts I–V, 1282 pages, ISBN 0-393-97422-7
   (Chapters 1–36, including Special Relativity)

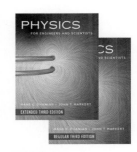

Paperback Versions

Volume 1, (Chapters 1–21) 778 pages, ISBN 0-393-93003-3
   Part I Motion, Force, and Energy (Chapters 1–14)
   Part II Oscillations, Waves, and Fluids (Chapters 15–18)
   Part III Temperature, Heat, and Thermodynamics (Chapter 19–21)
Volume 2, (Chapters 22–36) 568 pages, ISBN 0-393-93004-1
   Part IV Electricity and Magnetism (Chapters 22–32)
   Part V Waves and Optics (Chapters 33–35 and Chapter 36 on Special Relativity)
Volume 3, (Chapters 36–41) 250 pages, ISBN 0-393-92969-8
   Part VI Relativity, Quanta, and Particles

In addition, to explore customized versions, please contact your Norton representative.

Two Norton ebook Options

Physics for Engineers and Scientists is available in a Norton ebook format that retains the content of the print book. The ebook offers a variety of tools for study and review, including sticky notes, highlighters, zoomable images, links to Online Concept Tutorials, and a search function. Purchased together, the SmartWork with integrated ebook bundle makes it easy for students to check text references when completing online homework assignments.

nortonebooks.com

   The ebook may also be purchased as a standalone item. The downloadable PDF version is available for purchase from Powells.com.

Package Options

Each version of the text purchased from Norton—with or without SmartWork—will come with free access to our website at Norton's StudySpace that includes the valuable Online Concept Tutorials. Each version of the text may be purchased as a stand-alone book or as a package that includes—each for a fee—Norton's new SmartWork online homework system or the Student Activity Workbook by David Harrison and William Ellis. Hence, several optional packages are available to instructors:
 • Textbook–StudySpace–Online Concept Tutorials + Student Activity Workbook
 • Textbook–StudySpace–Online Concept Tutorials + SmartWork/ebook
 • Textbook–StudySpace–Online Concept Tutorials + SmartWork/ebook + Student Activity Workbook

The Support Program

To enhance individual learning and also peer instruction, a carefully integrated support program accompanies the text. Each element of the support program has two goals. First, each support resource mirrors the text's emphasis on sharply focused core concepts. Second, treatment of a core concept in a support resource offers a perspective that is different from but compatible with that of the text. If a student needs help beyond the text, he or she would more likely benefit from a fresh presentation on the same concept rather than from one that simply repeats the text presentation.

Hence, the text and its support package offers three or more different approaches to the core concepts. For example, Newton's First and Second Laws are rendered with interactive animations in the Online Concept Tutorial "Forces," with pencil-and-paper exercises in Chapter 5 of the Student Activity Workbook crafted by David Harrison and William Ellis, and with concept test inquiries in PhysiQuiz questions written by Charles Chiu and edited by Jason Stevens.

Both printed and digital resources are offered within the support program. Outstanding web-based resources for both instructors and students include tutorials and a homework system.

SmartWork Online Homework System

www.wwnorton.com/physics

SmartWork—Norton's online homework management system—provides ready-made automatically graded assignments, including guided problems, simple feedback questions, and animated tutorials—all specifically designed to extend the text's emphasis on core concepts and problem-solving skills.

Developed in collaboration with Sapling Systems, SmartWork features an intuitive, easy-to-use interface that offers instructors flexible tools to manage assignments, while making it easy for students to compose mathematical expressions, draw vectors and graphs, and receive helpful and immediate feedback. Two different types of questions expand upon the exposition of concepts in the text:

Simple Feedback Problems present students with problems that anticipate common misconceptions and offer prompts at just the right moment to help them discover the correct solution.

Guided Tutorial Problems addresses more challenging topics. If a student answers a problem incorrectly, SmartWork guides the student through a series of discrete tutorial steps that lead to a general solution. Each step is a simple feedback question that the student answers, with hints if necessary. After completing all of the tutorial steps, the student returns to the original problem ready to apply this newly-obtained knowledge.

SmartWork problems use algorithmic variables so two students are unlikely to see exactly the same problem. Instructors can use the problem sets provided, or can customize these ready-made questions and assignments, or use SmartWork to create their own.

SmartWork is available bundled with the Norton ebook of *Physics for Engineers and Scientists*. Where appropriate, SmartWork prompts students to review relevant sections in the textbook. Links to the **ebook** make it easy for students to consult the text while working through problems online.

Online Concept Tutorials

Online
Concept
Tutorial

www.wwnorton.com/physics

Developed in collaboration with Science Technologies specifically for this course, these 45 tutorials feature interactive animations that reinforce conceptual understanding and develop students' quantitative skills. In-text icons alert students to the availability of a tutorial. All Online Concept Tutorials are available on the free StudySpace web site and are integrated into SmartWork. Tutorials can also be accessed from a CD-ROM that requires no installation, browser tune-ups, or plug-ins.

StudySpace Website

STUDYSPACE www.wwnorton.com/physics. This free and open website is the portal for both public and premium content. Free content at StudySpace includes the Online Concept Tutorials and a Study Plan for each chapter in *Physics for Engineers and Scientists*. Premium content at StudySpace includes links to the online ebook and to SmartWork.

www.wwnorton.com/physics

Additional Instructor Resources

TEST BANK by Mirela Fetea, University of Richmond; Kevin Martus, William Paterson University; and Brian Woodahl, Indiana University-Purdue University-Indianapolis. The Test Bank offers approximately 2000 multiple-choice questions, available in ExamView, WebCT, BlackBoard, rich-text, and printed format.

INSTRUCTOR SOLUTIONS MANUAL by Stephen Luzader and Hang-Deng Luzader, both of Frostburg State University, and David Marx of Illinois State University. The Instructor Solution Manual offers solutions to all end-of-chapter Problems and Review Problems, checked for accuracy and clarity.

PHYSIQUIZ "CLICKER" QUESTIONS by Charles Chiu, University of Texas at Austin, with Jason Stevens, Deerfield Academy. The PhysiQuiz multiple-choice questions are designed for use with classroom response, or "clicker", systems. The 300 PhysiQuiz questions are available as PowerPoint slides, in printed format, and as transparency masters.

NORTON MEDIA LIBRARY INSTRUCTOR CD-ROM The Media Library for instrutors includes selected figures, tables, and equations from the text in JPEG and PowerPoint formats, PhysiQuiz "clicker" questions, and PowerPoint-ready offline versions of the Online Concept Tutorials.

INSTRUCTOR RESOURCE MANUAL offers a guide to the support package with descriptions of the Online Concept Tutorials, information about the SmartWork homework problems available for each chapter, printed PhysiQuiz "clicker" questions, and instructor notes for the workshop activities in the Student Activity Workbook.

TRANSPARENCY ACETATES Approximately 200 printed color acetates of key figures from the text.

BLACKBOARD AND WEBCT COURSE CARTRIDGES Course Cartridges for BlackBoard and WebCT include access to the Online Concept Tutorials, a Study Plan for each chapter, multiple-choice tests, plus links to the premium, password-protected contents of the Norton ebook and SmartWork.

Additional Student Resources

STUDENT ACTIVITY WORKBOOK by David Harrison, University of Toronto, and William Ellis, University of California Davis. The *Student Activity Workbook* is an important part of the learning package. For each chapter of *Physics for Engineers and Scientists*, the Workbook's Activities break down a physical condition into constituent parts. The Activities are pencil and paper exercises well suited to either individual or small group collaboration. The Activities include both conceptual and quantitative exercises. Some Activities are guided problems that pose a question and present a solution scheme via follow up questions. The Workbook is

available in two paperback volumes: Volume 1 comprises Chapters 1–21 and Volume 2 comprises Chapters 22–41.

STUDENT SOLUTIONS MANUAL by Stephen Luzader and Hang-Deng Luzader, both of Frostburg State University, and David Marx of Illinois State University. The Student Solutions Manual contains detailed solutions to approximately 25% of the problems in the book, chosen from the odd-numbered problems whose answers appear in the back of the book. The Manual is available in two paperback volumes: Volume 1 comprises Chapters 1–21 and Volume 2 comprises Chapters 22–41.

ONLINE CONCEPT TUTORIALS CD-ROM The 45 Online Concept Tutorials (see above) can also be accessed from an optional CD-ROM that requires no installation, browser tune-up, or plug-in.

About the Authors

Hans C. Ohanian received his B.S. from the University of California, Berkeley, and his Ph.D from Princeton University, where he worked with John A. Wheeler. He has taught at Rensselaer Polytechnic Institute, Union College, and the University of Vermont. He is the author of several textbooks spanning all undergraduate levels: *Physics, Principles of Physics, Relativity: A Modern Introduction, Modern Physics, Principles of Quantum Mechanics, Classical Electrodynamics*, and, with Remo Ruffini, *Gravitation and Spacetime*. He is also the author of dozens of articles dealing with gravitation, relativity, and quantum theory, including many articles on fundamental physics published in the *American Journal of Physics*, where he served as associate editor for some years. He lives in Vermont. hohanian@uvm.edu

John T. Markert received his B.A. in physics and mathematics from Bowdoin College (1979), and his M.S. (1984) and Ph.D. (1987) in physics from Cornell University, where he was recipient of the *Clark Award for Excellence in Teaching*. After postdoctoral research at the University of California, San Diego, he joined the faculty at the University of Texas at Austin in 1990, where he has received the *College of Natural Sciences Teaching Excellence Award* and is currently Professor of Physics and Department Chair. His introductory physics teaching methods emphasize context-based approaches, interactive techniques, and peer instruction. He is author or coauthor of over 120 journal articles, including experimental condensed-matter physics research in superconductivity, magnetism, and nanoscience. He lives in Austin, Texas, with his spouse and four children. jmarkert@physics.utexas.edu

Owner's Manual for *Physics for Engineers and Scientists*

These pages give a brief tour of the features of *Physics for Engineers and Scientists* and its study resources. Some resources are found within the book. Others are located in accompanying paperback publications or at the StudySpace web portal. Features on the text pages shown here come chiefly from the discussion of friction in Chapter 6, but are common in other chapters.

The learning resources listed below help students study by offering alternative explanations of the core concepts found in the text. These student resources are briefly described at the end of this owner's manual:

- Online Concept Tutorials
- SmartWork Online Homework
- StudySpace
- Student Activity Workbook
- Student Solutions Manual

Each chapter of the textbook starts with a real-world example of a core concept. Chapter 6 opens with the concept of friction and uses automobile tires as an example of friction that is revisited in several different conditions. The opening photograph, it's caption and the caption's closing questions all discuss this example.

CHAPTER
6

Further Applications of Newton's Laws

CONCEPTS IN CONTEXT

Concepts *in* Context

Automobiles rely on the friction between the road and the tires to accelerate and to stop. We will see that one of two types of contact friction, kinetic or static, is involved. To see how these friction forces affect linear and circular motion, we ask:

? In an emergency, an automobile brakes with locked and skidding wheels. What deceleration can be achieved? (Example 1, page 176)

? What is the steepest slope of a street on which an automobile can rest without slipping? (Example 4, page 179)

? When braking without skidding, what maximum deceleration can be achieved? (Example 5, page 180)

? How quickly can a racing car round a curve without skidding sideways? (Example 10, page 186)

? How does a banked curve help to avoid skidding? (Example 11, page 186)

6.1 Friction

6.2 Restoring Force of a Spring; Hooke's Law

6.3 Force for Uniform Circular Motion

6.4 The Four Fundamental Forces

173

Most chapters have six or fewer sections. Most sections are four or five pages in length and cover one major topic.

In this chapter, the rubber tires of an automobile are revisited to explore concepts in friction on pages 176, 179, 180, and 186, as indicated.

174   **CHAPTER 6** Further Applications of Newton's Laws

To find a solution of the equation of motion means to find a force **F** and a corresponding acceleration **a** such that Newton's equation $m\mathbf{a} = \mathbf{F}$ is satisfied. For a physicist, the typical problem involves a known force and an unknown motion; for example, the physicist knows the forces between the planets and the Sun, and she seeks ... for an engineer, the reverse problem with ... often of practical importance; for example, ... a given curve at 60 km/h, and he seeks to ... wheels must withstand. A special problem ... tics; here we know that the body is at rest ... wish to compute the forces that will main- ... depending on the circumstances, we can ... f the equation $m\mathbf{a} = \mathbf{F}$ as an unknown that ... ut the other side.

... ome solutions of the equation of motion ... eight and constant pushes or pulls. In this ... s of the equation of motion, and we will examine other, more complicated forces, such as friction and the forces exerted by springs.

Online Concept Tutorial

LEONARDO da VINCI (1452–1519)
Italian artist, engineer, and scientist. Famous for his brilliant achievements in painting, sculpture, and architecture. Leonardo also made pioneering contributions to science. But Leonardo's investigations of friction were forgotten, and the laws of friction were rediscovered 200 years later by Guillaume Amontons, a French physicist.

6.1 FRICTION

Friction forces, which we have ignored up to now, play an important role in our environment and provide us with many interesting examples of motion with constant force. For instance, if the driver of ... wheels will lock and begin ... an (approximately) constan... the automobile at an (appro... the friction force depends on... the heavy friction of rubber ... of the wheels, which introdu...

For the sake of simplicit... solid block of metal sliding ... in the shape of a brick, slidi... velocity and then let it coast... are the weight **w**, the normal... ward with a magnitude mg. ... upward; the magnitude of th... The friction force **f** exerted ... tabletop, in a direction oppo... contact force which acts ove... 6.1 it is shown as though act...

The friction force arises ... in the block form bonds wi... these bonds are continually r... represents the effort require... scopic level the phenomeno... the resulting friction force ... law, first enunciated by Leo...

6.1 Friction

The magnitude of the friction force between unlubricated, dry surfaces sliding one over the other is proportional to the magnitude of the normal force acting on the surfaces and is independent of the area of contact and of the relative speed.

Friction involving surfaces in relative motion is called **sliding friction**, or **kinetic friction**. According to the above law, the magnitude of the force of kinetic friction can be written mathematically as

$$f_k = \mu_k N \qquad (6.1)$$

where μ_k is the **coefficient of kinetic friction**, a constant characteristic of the material involved. Table 6.1 lists typical friction coefficients for various materials.

Note that Eq. (6.1) states that the magnitudes of the friction force and the normal force are proportional. The *directions* of these forces are, however, quite different: the normal fo... force $\mathbf{f}_k$ is parallel ...

The a... s laws. It is only a... nerely a descriptio... detailed theoretica... m this simple law... devi- ations in many everyday engineering problems in which the speeds are not extreme. The simple friction law is then a reasonably good approximation for a wide range of materials, and it is at its best for metals sliding on metals.

The fact that the friction force is independent of the area of contact means that the friction force of the block sliding on the tabletop is the same whether the block slides on a large face or on one of the small faces (see Fig. 6.2). This may seem surprising at first—we might expect the friction force to be larger when the block slides on the larger face, with more area in contact with the tabletop. However, the normal force is then distributed over a larger area, and is therefore less effective in pressing the atoms together; and the net result is that the friction force is independent of the area of contact.

force of kinetic friction

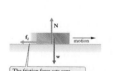

The friction force acts over the bottom surface in a direction opposite to the motion.

FIGURE 6.1 Forces on a block sliding on a plate.

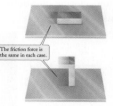

The friction force is the same in each case.

FIGURE 6.2 Steel block on a steel plate, sliding on a large face or on a small face.

TABLE 6.1	KINETIC AND STATIC FRICTION COEFFICIENTS[a]	
MATERIALS	μ_k	μ_s
Steel on steel	0.6	0.7
Steel on lead	0.9	0.9
Steel on copper	0.4	0.5
Copper on cast iron	0.3	1.1
Copper on glass	0.5	0.7
Waxed ski on snow		
at $-10°C$	0.2	—
at $0°C$	0.05	—
Rubber on concrete	≈ 1	≈ 1

[a] The friction coefficient depends on the condition of the surfaces. The values in this table are typical for dry surfaces but not entirely reliable.

Callout annotations (Owner's Manual):

The icon indicates an **Online Concept Tutorial** is available for a key concept. Each such icon includes the identification number of the tutorial—8, in this case. These tutorials offer a visual guide and self-quiz for the concept at hand. Find all the Tutorials at www.wwnorton.com/physics.

In mathematical expressions, such as $m\mathbf{a}=\mathbf{F}$, the **bold type** indicates a **vector** and *italic* indicates variables that are not vectors.

Text in *italic type* indicates major definitions of laws or statements of general principles.

Text in **bold type** highlights the first use of a key term and is generally accompanied by an explanation.

Key concepts or important variants of these concepts have a key-term label in the margin.

Highlighted equations are key equations that express central physics concepts mathematically.

Short **biographical sketches** appear in the margins of this text. Each offer a brief glimpse into the life of some major contributor to our knowledge about the physical world—in this case, Italian artist and engineer Leonardo da Vinci.

Examples are a critical part of each chapter.
• Examples provide concrete illustrations of the concepts being discussed.
• As the chapter unfolds, Examples progress from simple to more complex.

Throughout the text, **figures** often build on each other with a new layer of information.
• **Balloon comments** often point out components of special note in the figure.

The **Concept in Context** icon here indicates the chapter-opening example —automobile tires—is being revisited. In this Example, we explore the slowing down of a skidding automobile with a specific coefficient of kinetic friction for a rubber tire.

---

178   **CHAPTER 6** Further Applications of Newton's Laws

EXAMPLE 3   A man pushes a heavy crate over a floor. The man pushes downward and forward, so his push makes an angle of 30° with the horizontal (Fig. 6.5a). The mass of the crate is 60 kg, and the coefficient of sliding friction is $\mu_k = 0.50$. What force must the man exert to keep the crate moving at uniform velocity?

(a)   (b)

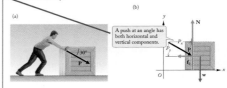

A push at an angle has both horizontal and vertical components.

FIGURE 6.5 (a) Man pushing a crate. (b) "Free-body" diagram for the crate.

SOLUTION: Figure 6.5b is a "free-body" diagram for the crate. The forces on the crate are the push **P** of the man, the weight **w**, the normal force **N**, and the friction force $\mathbf{f}_k$. Note that because the man pushes the crate down against the floor, the magnitude of the normal force is not equal to mg; we will have to treat the magnitude of the normal force as unknown. Taking the x axis horizontal and the y axis vertical, we see from Fig. 6.5b that the x and y components of the forces are (see also Fig. 5.37)

$$P_x = P\cos 30° \qquad\qquad P_y = -P\sin 30°$$
$$w_x = 0 \qquad\qquad w_y = -mg$$
$$N_x = 0 \qquad\qquad N_y = N$$
$$f_{k,x} = -\mu_k N \qquad\qquad f_{k,y} = 0$$

...cceleration of the crate is zero in both the x and the y directions, the ...each of these directions must be zero:

$$P\cos 30° + 0 + 0 - \mu_k N = 0$$
$$-P\sin 30° - mg + N + 0 = 0$$

...wo equations for the two unknowns P and N. By multiplying the second ...μ_k and then adding the resulting equation to the first, we can elimi- ...d we find an equation for P:

$$P\cos 30° - \mu_k P\sin 30° - \mu_k mg = 0$$

...s for P, we find

$$P = \frac{\mu_k mg}{\cos 30° - \mu_k \sin 30°} = \frac{0.50 \times 60\ \text{kg} \times 9.81\ \text{m/s}^2}{\cos 30° - 0.50 \times \sin 30°} \qquad (6.4)$$

$$= 4.8 \times 10^2\ \text{N}$$

Solutions in Examples may cover both general approaches and specific details on how to extract the information from the problem statement.

Comments occasionally close an Example to point out the particular limitations and broader implications of a Solution.

---

176   **CHAPTER 6** Further Applications of Newton's Laws

Concepts —In— Context

A kinetic friction force acts on each wheel, but diagram shows these forces combined in a single force $\mathbf{f}_k$.

Skidding motion is opposed by kinetic friction.

FIGURE 6.3 "Free-body" diagram for an automobile skidding with locked wheels.

EXAMPLE 1   Suppose that the coefficient of kinetic friction of the hard rubber of an automobile tire sliding on the pavement of a street is $\mu_k = 0.8$. What is the deceleration of an automobile on a flat street if the driver brakes sharply, so all the wheels are locked and skidding? (Assume the vehicle is an economy model without an antilock braking system.)

SOLUTION: Figure 6.3 shows the "free-body" diagram with all the forces on the automobile. These forces are the weight **w**, the normal force **N** exerted by the street, and the friction force $\mathbf{f}_k$. The normal force must balance the weight; hence the magnitude of the normal force is the same as the magnitude of the weight, or $N = w = mg$. According to Eq. (6.1), the magnitude of the friction force is then

$$f_k = \mu_k N = 0.8 \times mg$$

Since this friction force is the only horizontal force on the automobile, the deceleration of the automobile along the street is

$$a_x = -\frac{f_k}{m} = -\frac{0.8 \times mg}{m} = -0.8 \times g = -0.8 \times 9.8\ \text{m/s}^2$$

$$= -8\ \text{m/s}^2$$

COMMENT: The normal forces and the friction forces act on all the four wheels of the automobile; but in Fig. 6.3 (and in other "free-body" diagrams in this chapter) these forces have been combined into a net force **N** and a net friction force $\mathbf{f}_k$, which, for convenience, are shown as though acting at the center of the automobile. To the extent that the motion is treated as purely translational motion (that is, particle motion), it makes no difference at what point of the automobile the forces act. Later, in Chapter 13, we will study how forces affect the rotational motion of bodies, and it will then become important to keep track of the exact point at which each force acts.

EXAMPLE 2   A ship is launched toward the water on a slipway making an angle of 5° with the horizontal direction (see Fig. 6.4). The coefficient of kinetic friction between the bottom of the ship and the slipway is $\mu_k = 0.08$. What is the acceleration of the ship along the slipway? What is the speed of the ship after accelerating from rest through a distance of 120 m down the slipway to the water?

SOLUTION: Figure 6.4b is the "free-body" diagram for the ship. The forces shown are the weight **w**, the normal force exerted by the slipway **N**, and the friction force $\mathbf{f}_k$. The magnitude of the weight is $w = mg$.

Since there is no motion in the direction perpendicular to the slipway, we find, as in Eq. (5.36), that the normal force is

$$N = mg\cos\theta$$

and the magnitude of the friction force is

$$f_k = \mu_k N = \mu_k mg\cos\theta \qquad (6.2)$$

With the x axis parallel to the slipway, the x component of the weight is (see Fig. 6.4c)

$$w_x = mg\sin\theta$$

A **Checkup** appears at the end of each section within a chapter.
• Each Checkup is a self-quiz to test the reader's mastery of the concepts in the preceding section.
• Each Checkup has an answer (see below).

Problem-Solving Techniques boxes appear in relevant places throughout the book and offers tips on how to approach problems of a particular kind—in this case, problems involving the use of friction or centripetal force.

Answers to Checkups appear at the very back of each chapter, after the Review Problems.

---

190   CHAPTER 6   Further Applications of Newton's Laws

 Checkup 6.3

QUESTION 1: A stone is being whirled around a circle at the end of a string when the string suddenly breaks. Describe the motion of the stone after the string breaks; ignore gravity.

QUESTION 2: At an intersection, a motorcycle makes a right turn at constant speed. During this turn the motorcycle travels along a 90° arc of a circle. What is the direction of the acceleration of the motorcycle during this turn?

QUESTION 3: A car moves at constant speed along a road leading over a small hill with a spherical top. What is the direction of the acceleration of the car when at the top of this hill?

QUESTION 4: In Example 12, for the aircraft looping the loop, does the chair exert a centripetal or a centrifugal force on the pilot? Does the pilot exert a centripetal or a centrifugal force on the chair? What is the direction of the pilot's apparent, increased weight at the instant the aircraft passes through the bottom of the loop? Does the direction of the apparent weight change as the aircraft climbs up the loop?

QUESTION 5: Two cars travel around a traffic circle in adjacent (outer and inner) lanes. If the two cars travel at the same constant speed, which completes the circle first? Which has the larger acceleration?

(A) Outer; outer.   (B) Inner; outer.
(C) Outer; inner.   (D) Inner; inner.

PROBLEM-SOLVING TECHNIQUES   **FRICTION FORCES AND CENTRIPETAL FORCES**

The problems involving applications of Newton's laws in this chapter can be solved by the techniques discussed in the preceding chapter. In dealing with friction forces and with the centripetal force for uniform circular motion, pay special attention to the directions of the forces.

1 The magnitude of the sliding friction force is proportional to the magnitude of the normal force, but the direction is not the direction of the normal force. Instead, the sliding friction force is always parallel to the sliding surfaces, opposite to the direction of motion.

2 The static friction force is also always parallel to the sliding surfaces, opposite to the direction in which the body tends to move. If you have any doubts about the direction of the static friction force, pretend that the friction is absent, and ask yourself in what direction the body would then move; the static friction force is in the opposite direction.

3 Uniform circular motion requires a force toward the center of the circle, that is, a centripetal force. When preparing a "free-body" diagram for a body in uniform circular motion, include all the pushes and pulls acting on the moving body, but do *not* include a "centripetal mv^2/r force." This would be a mistake, like including an "*ma* force" in the "free-body" diagram for a body with some kind of translational motion. The quantity mv^2/r is not a force; it is merely the product of mass and centripetal acceleration. This acceleration is caused by one force or by the resultant of several forces already included among the pushes and pulls displayed in the "free-body" diagram. For instance, in Example 11 the resultant force is $w \tan \theta$, in Example 12 the resultant force is $N - mg$, and these resultants equal mv^2/r by Newton's Second Law [see Eqs. (6.17) and (6.18)]. To prevent confusion, do not include the resultant in the "free-body" diagram for a body in uniform circular motion. Instead, draw the resultant on a separate diagram (see Fig. 6.21b).

---

202   CHAPTER

*85. Two springs of constants 2.0×10^3 N/m and $3.0 \times$ are connected in tandem, and a mass of 5.0 kg hang from this spring. By what amount does the mass st combined spring? Each individual spring?

*86. A block of mass 1.5 kg is placed on a flat surface, an being pulled horizontally by a spring with a spring 1.2×10^3 N/m (see Fig. 6.48). The coefficient of st between the block and the table is $\mu_s = 0.60$, and cient of sliding friction is $\mu_k = 0.40$.

(a) By what amount must the spring be stretched t block moving?

(b) What is the acceleration of the block if the stre spring is maintained at a constant value equal t required to start the motion?

(c) By what amount must the spring be stretched to keep the mass moving at constant speed?

FIGURE 6.48 Mass pulled by spring.

*87. A block of mass 1.5 kg is placed on a plane inclined at 30°, and it is being pulled upward by a spring with a spring constant 1.2×10^3 N/m (see Fig. 6.49). The direction of pull of the spring is parallel to the inclined plane. The coefficient of static ... lined plane is $\mu_s = 0.60$, ... $\mu_k = 0.40$.

... be stretched to start the

...ck if the stretch of the ... value equal to that

(c) By what amount must the spring be stretched to keep the mass moving at constant speed?

*88. A mass m_1 slides on a smooth, frictionless table. The mass is constrained to move in a circle by a string that passes through a hole in the center of the table and is attached to a second mass m_2 hanging vertically below the table (Fig. 6.50). If the radius of the circular motion of the first mass is r, what must be its speed?

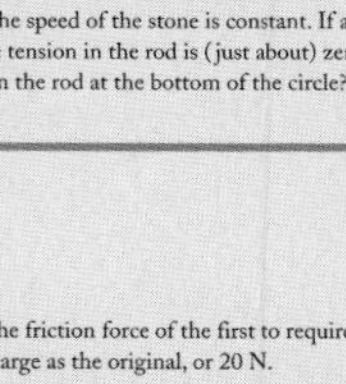

FIGURE 6.50 Mass in circular motion and hanging mass.

89. An automobile enters a curve of radius 45 m at 70 km/h. Will the automobile skid? The curve is not banked, and the coefficient of static friction between the wheels and the road is 0.80.

*90. A stone of 0.90 kg attached to a rod is being whirled around a vertical circle of radius 0.92 m. Assume that during this motion the speed of the stone is constant. If at the top of the circle the tension in the rod is (just about) zero, what is the tension in the rod at the bottom of the circle?

Answers to Checkups

Checkup 6.1

1. The weight of the second book results in a normal force between the first book and the table that is twice as large, so the friction force, and thus the horizontal push to overcome it, will be twice as large, or 20 N. If the first book pushes the second, then the friction force of the second book on the first adds to the friction force of the first to require a push also twice as large as the original, or 20 N.

2. While the block coasts up the incline, the friction, which always opposes the *motion*, is directed down the plane (the corresponding "free-body" diagram would have the weight

MATH HELP   ELLIPSES

An ellipse is defined geometrically by the condition that the sum of the distance from one focus of the ellipse and the distance from the other focus is the same for all points on the ellipse. This geometrical condition leads to a simple method for the construction of an ellipse: Stick pins into the two foci and tie a length of string to these points. Stretch the string taut to the tip of a pencil, and move this pencil around the foci while keeping the string taut (see Fig. 1a).

An ellipse can also be constructed by slicing a cone obliquely (see Fig. 1b). Because of this, an ellipse is said to be a conic section.

The largest diameter of the ellipse is called the major axis, and the smallest diameter is called the minor axis. The semimajor axis and the semiminor axis are one-half of these diameters, respectively (see Fig. 1c).

If the semimajor axis of length a is along the x axis and the semiminor axis of length b is along the y axis, then the x and y coordinates of an ellipse centered on the origin satisfy

$$\frac{x^2}{a^2} + \frac{y^2}{b^2} = 1$$

The foci are on the major axis at a distance f from the origin given by

$$f = \sqrt{a^2 - b^2}$$

The separation between a planet and the Sun is $a - f$ at perihelion and is $a + f$ at aphelion.

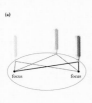

FIGURE 1 (a) Constructing an ellipse. (b) Ellipse as a conic section. (c) Focal distance f, semimajor axis a, and semiminor axis b of an ellipse.

Figure 9.10 illustrates this law. The two colored areas are equal, and the planet takes equal times to move from P to P' and from Q to Q'. According to Fig. 9.10, the speed of the planet is larger when it is near the Sun (at Q) than when it is far from the Sun (at P).

Kepler's Second Law, also called the law of areas, is a direct consequence of the central direction of the gravitational force. We can prove this law by a simple geometrical argument. Consider three successive positions P, P', P'' on the orbit, separated by a relatively small distance. Suppose that the time intervals between P, P' and between P', P'' are equal—say, each of the two intervals is one second. Figure 9.11 shows the positions P, P', P''. Between these positions the curved orbit can be approximated by straight line segments PP' and $P'P''$. Since the time intervals are one unit of time (1 second), the lengths of the segments PP' and $P'P''$ are in proportion to the

---

Math Help boxes offer specific mathematical guidance at the initial location in the text where that technique is most relevant.
• In this case in Chapter 9, ellipses are important in studying orbits.
• Additional math help is available in Appendices 2, 3, 4, and 5 at the back of the textbook.

Throughout the text, **Physics in Practice** boxes offer specific details on a real-world application of the concept under discussion—in this case, forces at work in automobile collisions in Chapter 11.

The text frequently offers **tables of typical values** of physical quantities.
• Such tables usually are labeled "**Some ...**," as in this case, from Chapter 5.
• These tables give some impression of the magnitudes encountered in the real world.

PHYSICS IN PRACTICE   AUTOMOBILE COLLISIONS

Concepts in Context

We can fully appreciate the effects of the secondary impact on the human body if we compare the impact speeds of a human body on the dashboard or the windshield with the speed attained by a body in free fall from some height. The impact of the head on the windshield at 15 m/s is equivalent to falling four floors down from an apartment building and landing headfirst on a hard surface. Our intuition tells us that this is likely to be fatal. Since our intuition about the dangers of heights is much better than our intuition about the dangers of speeds, it is often instructive to compare impact speeds with equivalent heights of fall. The table lists impact speeds and equivalent heights, expressed as the number of floors the body has to fall down to acquire the same speed.

The number of fatalities in automobile collisions has been reduced by the use of air bags. The air bag helps by cushioning the impact over a longer time, reducing the time-average force. To be effective, the air bag must inflate quickly, before the passenger reaches it, typically in about 10 milliseconds. Because of this, a passenger, especially a child, too near an air bag prior to inflation can be injured or killed by the impulse from the inflation. But for a properly seated adult passenger, the inflated air bag cushions the passenger, reducing the severity of injuries.

However, the impact can still be fatal—you wouldn't expect to survive a jump from an 11-floor building onto an air mattress.

For maximum protection, a seat belt should always be worn even in vehicles equipped with air bags. In lateral collisions, in repeated collisions (such as in car pileups), and in rollovers, an air bag is of little help, and a seat belt is essential. The effectiveness of seat belts is well demonstrated by the experiences of race car drivers. Race car drivers wear lap belts and crossed shoulder belts. Even in spectacular crashes at very high speeds (see the figure), the drivers rarely suffer severe injuries.

COMPARISON OF IMPACT SPEEDS AND HEIGHTS OF FALL

SPEED	SPEED	EQUIVALENT HEIGHT (NUMBER OF FLOORS)[a]
15 km/h	9 mi/h	$\frac{1}{3}$
30	19	1
45	28	3
60	37	5
75	47	8
90	56	11
105	65	15

[a]Each floor is 2.9 m.

In a race at the California Speedway in October 2000, a car flips over and breaks in half after a crash, but the driver, Luis Diaz, walks away from the wreck.

SOLUTION: The only horizontal force on the ball is the normal force exerted by the wall; this force reverses the motion of the ball (see Fig. 11.3). Since the wall is very massive, the reaction force of the ball on the wall will not give the wall any appreciable velocity. Hence the kinetic energy of the system, both before and after the collision, is merely the kinetic energy of the ball. Conservation of this

---

TABLE 5.1   SOME FORCES

Gravitational pull of Sun on Earth	3.5×10^{22} N
Thrust of Saturn V rocket engines (a)	3.3×10^{7} N
Pull of large tugboat	1×10^{6} N
Thrust of jet engines (Boeing 747)	7.7×10^{5} N
Pull of large locomotive	5×10^{5} N
Decelerating force on automobile during braking	1×10^{4} N
Force between two protons in a nucleus	$\approx 10^{4}$ N
Accelerating force on automobile	7×10^{3} N
Gravitational pull of Earth on man	7.3×10^{2} N
Maximum upward force exerted by forearm (isometric)	2.7×10^{2} N
Gravitational pull of Earth on apple (b)	2 N
Gravitational pull of Earth on 5¢ coin	5.1×10^{-2} N
Force between electron and nucleus of atom (hydrogen)	8×10^{-8} N
Force on atomic-force microscope tip	10^{-12} N
Smallest force detected (mechanical oscillator)	10^{-19} N

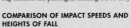

Each chapter closes with a Summary followed by Questions for Discussion, Problems, Review Problems, and Answers to Checkups.

A **Summary** lists the subjects and page references for any special content in this chapter—such as Math Help, Problem-Solving Techniques, or Physics in Practice boxes.
 • Next the Summary lists the chapter's core concepts in the order they are treated. The concept appears on the left in bold.
 • The mathematical expression for the concept appears in the middle column with an equation number on the far right.

About 15 or more **Questions for Discussion** follow the Summary in each chapter.
 • These questions require thought, but not calculation; e.g. "Why are wet streets slippery?"
 • Some of these questions are intended as brain teasers that have no unique answer, but will lead to provocative discussions.

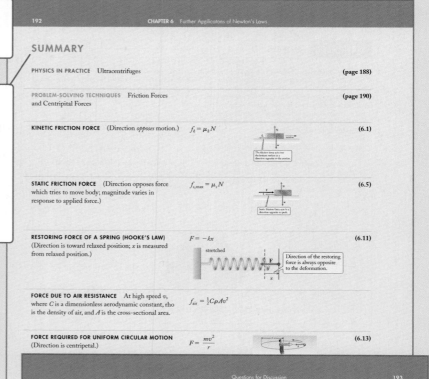

192   CHAPTER 6   Further Applications of Newton's Laws

SUMMARY

PHYSICS IN PRACTICE   Ultracentrifuges   (page 188)

PROBLEM-SOLVING TECHNIQUES   Friction Forces   (page 190)
and Centripetal Forces

KINETIC FRICTION FORCE   (Direction *opposes* motion.) $f_k = \mu_k N$ (6.1)

STATIC FRICTION FORCE   (Direction opposes force which tries to move body; magnitude varies in response to applied force.) $f_{s,max} = \mu_s N$ (6.5)

RESTORING FORCE OF A SPRING (HOOKE'S LAW)   (Direction is toward relaxed position; x is measured from relaxed position.) $F = -kx$ (6.11)

FORCE DUE TO AIR RESISTANCE   At high speed v, where C is a dimensionless aerodynamic constant, rho is the density of air, and A is the cross-sectional area. $f_{air} = \frac{1}{2} C \rho A v^2$

FORCE REQUIRED FOR UNIFORM CIRCULAR MOTION   (Direction is centripetal.) $F = \frac{mv^2}{r}$ (6.13)

Questions for Discussion   193

QUESTIONS FOR DISCUSSION

1. According to the adherents of parapsychology, some people are endowed with the supernormal power of psychokinesis, e.g., spoon-bending-at-a-distance via mysterious psychic forces emanating from the brain. Physicists are confident that the only forces acting between pieces of matter are those listed in Section 6.4, none of which are implicated in psychokinesis. Given that the brain is nothing but a (very complicated) piece of matter, what conclusions can a physicist draw about psychokinesis?

2. If you carry a spring balance from London to Hong Kong, do you have to recalibrate it? If you carry a beam balance?

3. When you stretch a rope horizontally between two fixed points, it always sags a little, no matter how great the tension. Why?

4. What are the forces on a soaring bird? How can the bird gain altitude without flapping its wings?

5. How could you use a pendulum suspended from the roof of your automobile to measure its acceleration?

6. When an airplane flies along a parabolic path similar to that of a projectile, the passengers experience a sensation of weightlessness. How would the airplane have to fly to give the passengers a sensation of enhanced weight?

7. A frictionless chain hangs over two adjoining inclined planes (Fig. 6.24a). Prove the chain is in equilibrium, i.e., the chain will not slip to the left or to the right. [Hint: One method of proof, due to the seventeenth-century engineer and mathematician Simon Stevin, asks you to pretend that an extra piece of chain is hung from the ends of the original chain (Fig. 6.24b). This makes it possible to conclude that the original chain cannot slip.]

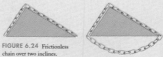

FIGURE 6.24 Frictionless chain over two inclines.

8. Seen from a reference frame moving with the wave, the motion of a surfer is analogous to the motion of a skier down a mountain.[2] If the wave were to last forever, could the surfer ride it forever? In order to stay on the wave as long as possible, in what direction should the surfer ski the wave?

9. Excessive polishing of the surfaces of a block of metal increases its friction. Explain.

10. Some drivers like to spin the wheels of their automobiles for a quick start. Does this give them greater acceleration? (Hint: $\mu_s > \mu_k$.)

[2] There is, however, one complication: surf waves grow higher as they approach the beach. Ignore this complication.

11. Cross-country skiers like to use a ski wax that gives their skis a large coefficient of static friction, but a low coefficient of kinetic friction. Why is this useful? How do "waxless" skis achieve the same effect?

12. Designers of locomotives usually reckon that the maximum force available for moving the train ("tractive force") is one-fourth or one-fifth of the weight resting on the drive wheels of the locomotive. What value of the friction coefficient between the wheels and the track does this implicitly assume?

13. When an automobile with rear-wheel drive accelerates from rest, the maximum acceleration that it can attain is less than the maximum deceleration that it can attain while braking. Why? (Hint: Which wheels of the automobile are involved in acceleration? In braking?)

14. Can you think of some materials with $\mu_s > 1$?

15. For a given initial speed, the stopping distance of a train is much longer than that of a truck. Why?

16. Why does the traction on snow or ice of an automobile with rear-wheel drive improve when you place extra weight over the rear wheels?

17. Why are wet streets slippery?

18. In order to stop an automobile on a slippery street in the shortest distance, it is best to brake as hard as possible without initiating a skid. Why does skidding lengthen the stopping distance? (Hint: $\mu_s > \mu_k$.)

19. Suppose that in a panic stop, a driver locks the wheels of his automobile and leaves skid marks on the pavement. How can you deduce his initial speed from the length of the skid marks?

20. Hot-rod drivers in drag races find it advantageous to spin their wheels very fast at the start so as to burn and melt the rubber on their tires (Fig. 6.25). How does this help them to attain a larger acceleration than expected from the static coefficient of friction?

FIGURE 6.25 Drag racer at the start of the race.

About 70 **Problems** and 15 **Review Problems** follow each chapter's Questions for Discussion.
 • The Problem's statement contains data and conditions upon which a solution will hinge.
 • Problems are grouped by chapter section and proceed from simple to more complex within each section.
 • Many Problems employ real-world data and occasionally may introduce applications beyond those treated in the chapter.

Review Problems are specifically designed to help students prepare for examinations.
 • Review Problems often test comprehension of concepts from more than one section within the chapter.
 • Review Problems often take a guided approach by posing series of questions that build on each other.

---

194   CHAPTER 6   Further Applications of Newton's Laws

21. A curve on a highway consists of a quarter circle connecting two straight segments. If this curve is banked perfectly for motion at some given speed, can it be joined to the straight segments without a bump? How could you design a curve that is banked perfectly along its entire length and merges smoothly into straight segments without any bump?

22. Automobiles with rear engines (such as the old VW "Beetle") tend to oversteer; that is, in a curve the rear end tends to swing toward the outside of the curve, turning the car excessively into the curve. Explain.

23. When rounding a curve in your automobile, you get the impression that a force tries to pull you toward the outside of the curve. Is there such a force?

24. If the Earth were to stop spinning (other things remaining equal), the value of g at all points of the surface except the poles would become slightly larger. Why?

25. (a) If a pilot in a fast aircraft very suddenly pulls out of a dive (Fig. 6.26a), he will suffer blackout caused by loss of blood pressure in the brain. If he suddenly begins a dive while climbing (Fig. 6.26b), he will suffer *redout* caused by excessive blood pressure in the brain. Explain.
    (b) A pilot wearing a G suit—a tightly fitting garment that squeezes the tissues of the legs and abdomen—can tolerate $8g$ while pulling out of a dive (Fig. 6.26c). How does this G suit prevent blackout? A pilot can tolerate no more

than $-2g$ while beginning a dive. Why does the G suit not help against redout?

26. While rounding a curve at high speed, a motorcycle rider leans the motorcycle toward the center of the curve. Why?

FIGURE 6.26 (a) Aircraft pulling out of a dive. (b) Aircraft beginning a dive. (c) Pilot wearing a G suit.

PROBLEMS

6.1 Friction†

1. The ancient Egyptians moved large stones by dragging them across the sand in sleds. How many Egyptians were needed to drag an obelisk of 700 metric tons? Assume that $\mu_k = 0.30$ for the sled on sand and that each Egyptian exerted a horizontal force of 360 N.

2. The base of a winch is bolted to a mounting plate with four bolts. The base and the mounting plate are flat surfaces made of steel; the friction coefficient of these surfaces in contact is $\mu_s = 0.40$. The bolts provide a normal force of 2700 N each. What maximum static friction force will act between the steel surfaces and help oppose lateral slippage of the winch on its base?

3. According to tests performed by the manufacturer, an automobile with an initial speed of 65 km/h has a stopping distance of 20 m on a level road. Assuming that no skidding occurs during braking, what is the value of μ_s between the wheels and the road required to achieve this stopping distance?

4. A crate sits on the load platform of a truck. The coefficient of friction between the crate and the platform is $\mu_s = 0.40$. If the truck stops suddenly, the crate will slide forward and crash into the cab of the truck. What is the maximum braking deceleration that the truck may have if the crate is to stay put?

5. When braking (without skidding) on a dry road, the stopping distance of a sports car with a high initial speed is 38 m. What would have been the stopping distance of the same car with the same initial speed on an icy road? Assume that $\mu_s = 0.85$ for the dry road and $\mu_s = 0.20$ for the icy road.

6. In a remarkable accident on motorway M1 (in England), a Jaguar car initially speeding "in excess of 100 mph" skidded 290 m before coming to a rest. Assuming that the wheels were completely locked during the skid and that the coefficient of kinetic friction between the wheels and the road was 0.80, find the initial speed.

† For help, see Online Concept Tutorial 8 at www.wwnorton.com/physics

---

Review Problems

REVIEW PROBLEMS

76. At liftoff, the Saturn V rocket used for the Apollo missions has a mass of 2.45×10^6 kg.
    (a) What is the minimum thrust that the rocket engines must develop to achieve liftoff?
    (b) The actual thrust that the engines develop is 3.3×10^7 N. What is the vertical acceleration of the rocket at liftoff?
    (c) At burnout, the rocket has spent its fuel, and its remaining mass is 0.75×10^6 kg. What is the acceleration just before burnout? Assume that the motion is still vertical and that the strength of gravity is the same as when the rocket is on the ground.

77. If the coefficient of static friction between the tires of an automobile and the road is $\mu_s = 0.80$, what is the minimum distance the automobile needs in order to stop without skidding from an initial speed of 90 km/h? How long does it take to stop?

78. Suppose that the last car of a train becomes uncoupled while the train is moving upward on a slope of 1:6 at a speed of 48 km/h.
    (a) What is the deceleration of the car? Ignore friction.
    (b) How far does the car coast up the slope before it stops?

79. A 40-kg crate falls off a truck traveling at 80 km/h on a level road. The crate slides along the road and gradually comes to a halt. The coefficient of kinetic friction between the crate and the road is 0.80.
    (a) Draw a "free-body" diagram for the crate sliding on the road.
    (b) What is the normal force the road exerts on the crate?
    (c) What is the friction force the road exerts on the crate?
    (d) What is the weight force on the crate? What is the net force on the crate?
    (e) What is the deceleration of the crate? How far does the crate slide before coming to a halt?

80. A 2.0-kg box rests on an inclined plane which makes an angle of 30° with the horizontal. The coefficient of static friction between the box and the plane is 0.90.
    (a) Draw a "free-body" diagram for the box.
    (b) What is the normal force the inclined plane exerts on the box?
    (c) What is the friction force the inclined plane exerts on the box?
    (d) What is the net force the inclined plane exerts on the box? What is the direction of this force?

81. The body of an automobile is held above the axles of the wheels by means of four springs, one near each wheel. Assume

that the springs are vert[...]
springs are the same. Th[...]
is 1200 kg, and the sprin[...]
N/m. When the autome[...]
far are the springs comp[...]

*82. A block of wood rests o[...]
The coefficient of static[...]
paper is $\mu_s = 0.70$, and [...]
$\mu_s = 0.50$. If you tilt the[...]
begin to move?

*83. Two blocks of masses m_1 and m_2 are connected by a string. One block slides on a table, and the other hangs from the string, which passes over a pulley (see Fig. 6.46). The coefficient of sliding friction between the first block and the table is $\mu_k = 0.20$. What is the acceleration of the blocks?

FIGURE 6.46 Mass on table, pulley, and hanging mass.

*84. A man of mass 75 kg is pushing a heavy box on a flat floor. The coefficient of sliding friction between the floor and the box is 0.20, and the coefficient of static friction between the man's shoes and the floor is 0.80. If the man pushes horizontally (see Fig. 6.47), what is the maximum mass of the box he can move?

FIGURE 6.47 Pushing a box.

The **dagger footnote** (†) that accompanies a Problem heading—in this case, "6.1 Friction"—indicates the availability of an Online Concept Tutorial on this specific topic and states its web address.

Problems and Review Problems are marked by **level of difficulty**:
 • Those without an asterisk are the most common and require very little manipulation of existing equations; or they may merely require retracing the steps of a worked Example.
 • Problems marked with one asterisk (*) are of medium difficulty and may require use of several concepts and manipulation more than one equation to isolate and solve for the unknown variable.
 • Problems marked with two asterisks (**) are challenging, demand considerable thought, may require significant mathematical skill, and are the least common.

Online Concept Tutorials

www.wwnorton.com/physics

An **Online Concept Tutorial** accompanies many central topics in this textbook. When a Tutorial is available, its numbered icon appears at section heading within the chapter and a dagger footnote appears in the end-of-chapter Problems section as reminder. These Tutorials are digitally delivered, either via the Internet or via a CD-ROM for those without Internet access.

Online *Concept* Tutorial

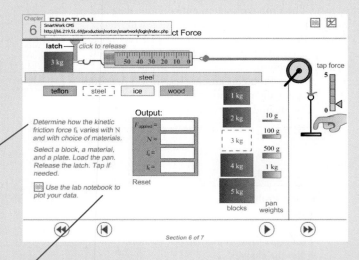

Many Tutorials contain **online experiments**— in this case, determining how the kinetic friction force varies with the normal force and with the choice of materials.

The online experiments allow students to change independent variables—in this case, mass and material.
• Students may collect and display data in a built-in **lab notebook**.
• Each Tutorial includes an interactive **self-quiz**.

The Online Concept Tutorials listed here indicate each textbook section supported by the tutorial (in paratheses).

1 Unit Conversion 1.5, 1.6

2 Significant Digits 1.6

3 Acceleration 2.4, 2.5, 2.6

4 Vector Addition and Vector Components 3.1, 3.2, 3.3

5 Projectile Motion 4.4

6 Forces 5.4

7 "Free-Body" Diagrams 5.3, 5.5, 5.6

8 Friction 6.1

9 Work of a Variable Force 7.1, 7.2, 7.4

10 Conservation of Energy 8.1, 8.2, 8.3

11 Circular Orbits 9.1, 9.3

12 Kepler's Laws 9.4

13 Momentum in Collisions 11.1, 11.3

14 Elastic and Inelastic Collisions 11.2, 11.3

15 Rotation about a Fixed Axis 12.2

16 Oscillations and Simple Harmonic Motion 15.1

17 Simple Pendulum 15.4

18 Wave Superposition 16.3, 16.4

19 Doppler Effect 17.4

20 Fluid Flow 18.1, 18.2, 18.6

21 Ideal-Gas Law 19.1

22 Specific Heat and Changes of State 20.1, 20.4

23 Heat Engines 21.2

24 Coulomb's Law 22.2

25 Electric Charge 22.1, 22.5

26 Electric Force Superposition 22.3

27 Electric Field 23.1, 23.3

28 Electric Flux 24.1

29 Gauss' Law 24.2, 24.3

30 Electrostatic Potential 25.1, 25.2, 25.4

31 Superconductors 27.3, 31.4

32 DC Circuits 28.1, 28.2, 28.3, 28.4, 28.7

33 Motion in a Uniform Magnetic Field 30.1

34 Electromagnetic Induction 31.2, 31.3

35 AC Circuits 32.1, 32.2, 32.3, 32.5

36 Polarization 33.3

37 Huygens' Construction 34.1, 34.2, 34.3

38 Geometric Optics and Lenses 34.4, 34.5

39 Interference and Diffraction 35.3, 35.5

40 X-ray Diffraction 35.4

41 Special Relativity 36.1, 36.2

42 Implications of Special Relativity 36.2, 36.3

43 Bohr Model of the Atom 38.1, 38.2, 38.4

44 Quantum Numbers 39.1, 39.2

45 Radioactive Decay 40.4

www.wwnorton.com/physics

SmartWork is a subscription-based online homework-management system that makes it easy for instructors to assign, collect and grade end-of-chapter problems from *Physics for Engineers and Scientists*. Built-in hinting and feedback address common misperceptions and help students get the maximum benefit from these assignments.

Simple Feedback Problems anticipate common misconceptions and offer prompts at just the right moment to help students reach the correct solution.

Guided Tutorial Problems address challenging topics.
• If a student solves one of these problems incorrectly, she is presented with a series of discrete tutorial steps that lead to a general solution.
• Each step includes hinting and feedback. After working through these remedial steps, the student returns to a restatement of the original problem, ready to apply this newly obtained knowledge.

SmartWork is available as a stand-alone purchase, or with an integrated ebook version of *Physics for Engineers and Scientists*.
• Where appropriate, SmartWork prompts students to review relevant sections of the text.
• Links to the ebook make it easy for students consult the text while working through problems.

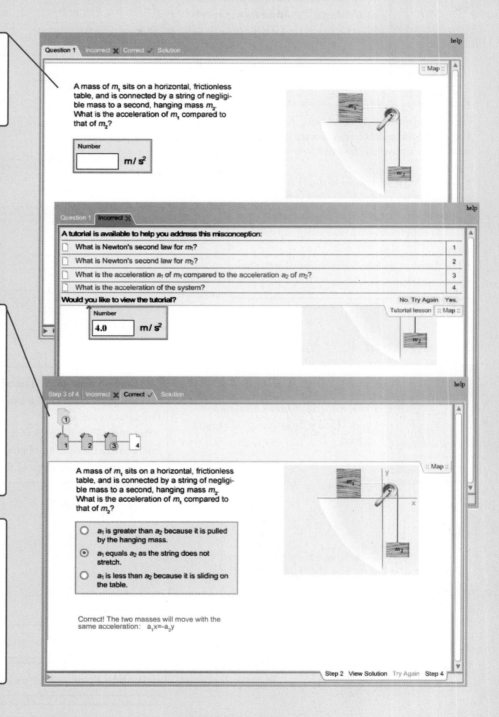

Student Activity Workbook

The Student Activity Workbook is available in two paperback volumes: Volume 1 comprises Chapters 1–21 and Volume 2 comprises Chapters 22–41.

For each chapter of the textbook, the **Student Activity Workbook** offers Activities designed to break down a physical condition into constituent parts.

• The Activities are unique to the Workbook and not found in the textbook.

• The Activities are pencil and paper exercises well suited to either individual or small group collaboration.

• The Activities include both conceptual and quantitative questions.

ACTIVITY 7

Joe is standing on the ground, Pete is standing on a 10m high cliff, and Amanda is at the bottom of a 20m deep pit, as shown. All three are using coordinate systems with the vertical axis directed up.

Joe's coordinate system has the zero of the vertical axis and zero of gravitational potential at ground level.

Peter's coordinate system has the zero of the vertical axis and zero of gravitational potential at the height of the cliff.

Amanda's coordinate system has the zero of the vertical axis and zero of gravitational potential at the bottom of the pit.

A ball of mass m is initially at rest at ground level, Position A.

a) What is the gravitational potential energy of the ball as measured by Joe, by Peter, and by Amanda?

b) The ball is then raised to the height of the cliff, Position B, and is held at rest. What is the gravitational potential energy of the ball at Position B as measured by Joe, by Peter, and by Amanda?

c) The ball is then released from rest and strikes the ground at the bottom of the pit, Position C. What is the gravitational potential energy of the ball at Position C as measured by Joe, by Peter, and by Amanda?

d) What is this change in gravitational potential energy measured by J, P, and A between positions A and C?

Student Solution Manual

The Student Solution Manual is available in two paperback volumes: Volume 1 comprises Chapters 1–21 and Volume 2 comprises Chapters 22–41.

The **Student Solutions Manual** contains worked solutions for about 50% of the odd-numbered Problems and Review Problems in the text.

• Appendix 11 in the back of the textbook contains only the final answer for odd-numbered problems in the chapters, not the intermediate steps of the solutions.

StudySpace www.wwnorton.com/physics

The StudySpace website is the free and open portal through which students access the resources that accompany this text.

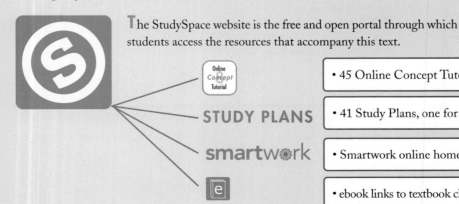

• 45 Online Concept Tutorials—at no additional cost.

• 41 Study Plans, one for each chapter—at no additional cost.

• Smartwork online homework system—a subscription service.

• ebook links to textbook chapters—as part of subscription service.

Prelude

The World of Physics

Physics is the study of matter. In a quite literal sense, physics is the greatest of all natural sciences: it encompasses the smallest particles, such as electrons and quarks; and it also encompasses the largest bodies, such as galaxies and the entire Universe. The smallest particles and the largest bodies differ in size by a factor of more than ten thousand billion billion billion billion! In the pictures on the following pages we will survey the world of physics and attempt to develop some rough feeling for the sizes of things in this world. This preliminary survey sets the stage for our explanations of the mechanisms that make things behave in the way they do. Such explanations are at the heart of physics, and they are the concern of the later chapters of this book.

Since the numbers we will be dealing with in this prelude and in the later chapters are often very large or very small, we will find it convenient to employ the **scientific notation** for these numbers. In this notation, numbers are written with powers of 10; thus, hundred is written as 10^2, thousand is written as 10^3, ten thousand is written as 10^4, and so on. A tenth is written as 10^{-1}, a hundredth is written as 10^{-2}, a thousandth is written as 10^{-3}, and so on. The following table lists some powers of ten:

$10 = 10^1$	$0.1 = 1/10 = 10^{-1}$
$100 = 10^2$	$0.01 = 1/100 = 10^{-2}$
$1000 = 10^3$	$0.001 = 1/1000 = 10^{-3}$
$10000 = 10^4$	$0.0001 = 1/10000 = 10^{-4}$
$100000 = 10^5$	$0.00001 = 1/100000 = 10^{-5}$
$1000000 = 10^6$	$0.000001 = 1/1000000 = 10^{-6}$ etc.

Note that the power of 10, or the exponent on the 10, simply tells us how many zeros follow the 1 in the number (if the power of 10 is positive) or how many zeros follow the 1 in the denominator of the fraction (if the power of 10 is negative).

In scientific notation, a number that does not coincide with one of the powers of 10 is written as a product of a decimal number and a power of 10. For example, in this notation, 1500000000 is written as 1.5×10^9. Alternatively, this number could be written as 15×10^8 or as 0.15×10^{10}; but in scientific notation it is customary to place the decimal point immediately after the first nonzero digit. The same rule applies to numbers smaller than 1; thus, 0.000015 is written as 1.5×10^{-5}.

The pictures on the following pages fall into two sequences. In the first sequence we zoom out: we begin with a picture of a woman's face and proceed step by step to pictures of the entire Earth, the Solar System, the Galaxy, and the Universe. This ascending sequence contains 27 pictures, with the scale decreasing in steps of factors of 10.

Most of our pictures are photographs. Many of these have become available only in recent years; they were taken by high-flying aircraft, Landsat satellites, astronauts, or sophisticated electron microscopes. For some of our pictures no photographs are available and we have to rely, instead, on drawings.

PART I: THE LARGE-SCALE WORLD

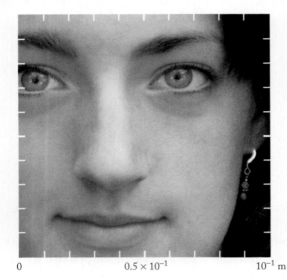

0   0.5×10^{-1} 10^{-1} m

SCALE 1:1.5   This is Erin, an intelligent biped of the planet Earth, Solar System, Orion Spiral Arm, Milky Way Galaxy, Local Group, Local Supercluster. Erin belongs to the phylum Chordata, class Mammalia, order Primates, family Hominidae, genus *Homo*, species *sapiens*. She is made of 5.4×10^{27} atoms, with 1.9×10^{28} electrons, the same number of protons, and 1.5×10^{28} neutrons.

0   0.5×10^{0} 10^{0} m

SCALE 1:1.5 $\times$ 10   Erin has a height of 1.7 meters and a mass of 57 kilograms. Her chemical composition (by mass) is 65% oxygen, 18.5% carbon, 9.5% hydrogen, 3.3% nitrogen, 1.5% calcium, 1% phosphorus, and 1.2% other elements.

The matter in Erin's body and the matter in her immediate environment occur in three states of aggregation: **solid**, **liquid**, and **gas**. All these forms of matter are made of atoms and molecules, but solid, liquid, and gas are qualitatively different because the arrangements of the atomic and molecular building blocks are different.

In a solid, each building block occupies a definite place. When a solid is assembled out of molecular or atomic building blocks, these blocks are locked in place once and for all, and they cannot move or drift about except with great difficulty. This rigidity of the arrangement is what makes the aggregate hard—it makes the solid "solid." In a liquid, the molecular or atomic building blocks are not rigidly connected. They are thrown together at random and they move about fairly freely, but there is enough adhesion between neighboring blocks to prevent the liquid from dispersing. Finally, in a gas, the molecules or atoms are almost completely independent of one another. They are distributed at random over the volume of the gas and are separated by appreciable distances, coming in touch only occasionally during collisions. A gas will disperse spontaneously if it is not held in confinement by a container or by some restraining force.

SCALE 1:1.5 × 10²   The building behind Erin is the New York Public Library, one of the largest libraries on Earth. This library holds 1.4×10^{10} volumes, containing roughly 10% of the total accumulated knowledge of our terrestrial civilization.

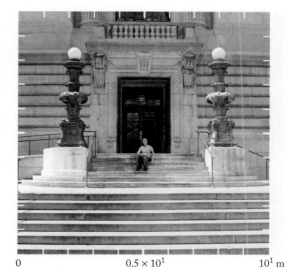

0   0.5×10^1 10^1 m

SCALE 1:1.5 × 10³   The New York Public Library is located at the corner of Fifth Avenue and 42nd Street, in the middle of New York City, with Bryant Park immediately behind it.

0   0.5×10^2 10^2 m

SCALE 1:1.5 × 10⁴   This aerial photograph shows an area of 1 kilometer × 1 kilometer in the vicinity of the New York Public Library. The streets in this part of the city are laid out in a regular rectangular pattern. The library is the building in the park in the middle of the picture. The photograph was taken early in the morning, and the high buildings typical of New York cast long shadows.

The photograph was taken from an airplane flying at an altitude of a few thousand meters. North is at the top of the photograph.

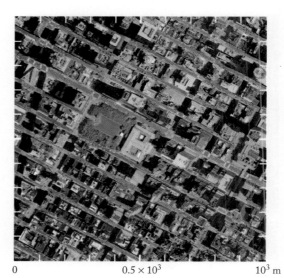

0   0.5×10^3 10^3 m

0   0.5×10^4 10^4 m

SCALE 1:1.5 × 10⁵   This photograph shows a large portion of New York City. We can barely recognize the library and its park as a small rectangular patch slightly above the center of the picture. The central mass of land is the island of Manhattan, with the Hudson River on the left and the East River on the right.

    This photograph and the next two were taken by satellites orbiting the Earth at an altitude of about 700 kilometers.

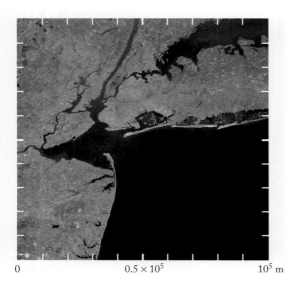

0   0.5×10^5 10^5 m

SCALE 1:1.5 × 10⁶   In this photograph, Manhattan is in the upper middle. On this scale, we can no longer distinguish the pattern of streets in the city. The vast expanse of water in the lower right of the picture is part of the Atlantic Ocean. The mass of land in the upper right is Long Island. Parallel to the south shore of Long Island we can see a string of very narrow islands; they almost look man-made. These are barrier islands; they are heaps of sand piled up by ocean waves in the course of thousands of years.

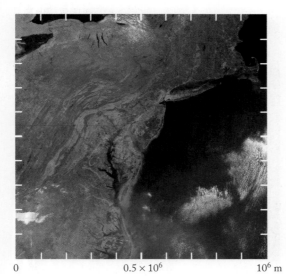

0   0.5×10^6 10^6 m

SCALE 1:1.5 × 10⁷   Here we see the eastern coast of the United States, from Cape Cod to Cape Fear. Cape Cod is the hook near the northern end of the coastline, and Cape Fear is the promontory near the southern end of the coastine. Note that on this scale no signs of human habitation are visible. However, at night the lights of large cities would stand out clearly.

    This photograph was taken in the fall, when leaves had brilliant colors. Streaks of orange trace out the spine of the Appalachian mountains.

SCALE 1:1.5 × 10⁸ In this photograph, taken by the Apollo 16 astronauts during their trip to the Moon, we see a large part of the Earth. Through the gap in the clouds in the lower middle of the picture, we can see the coast of California and Mexico. We can recognize the peninsula of Baja California and the Gulf of California. Erin's location, the East Coast of the United States, is covered by a big system of swirling clouds on the right of the photograph.

Note that a large part of the area visible in this photograph is ocean. About 71% of the surface of the Earth is ocean; only 29% is land. The atmosphere covering this surface is about 100 kilometers thick; on the scale of this photograph, its thickness is about 0.7 millimeter. Seen from a large distance, the predominant colors of the planet Earth are blue (oceans) and white (clouds).

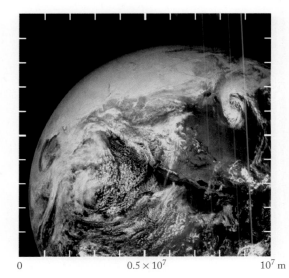

$0 \qquad\qquad 0.5 \times 10^7 \qquad\qquad 10^7 \text{ m}$

SCALE 1:1.5 × 10⁹ This photograph of the Earth was taken by the Apollo 16 astronauts standing on the surface of the Moon. Sunlight is striking the Earth from the top of the picture.

As is obvious from this and from the preceding photograph, the Earth is a sphere. Its radius is 6.37×10^6 meters and its mass is 5.98×10^{24} kilograms.

$0 \qquad\qquad 0.5 \times 10^8 \qquad\qquad 10^8 \text{ m}$

SCALE 1:1.5 × 10¹⁰ In this drawing, the dot at the center represents the Earth, and the solid line indicates the orbit of the Moon around the Earth (many of the pictures on the following pages are also drawings). As in the preceding picture, the Sun is far below the bottom of the picture. The position of the Moon is that of January 1, 2000.

The orbit of the Moon around the Earth is an **ellipse,** but an ellipse that is very close to a circle. The solid red curve in the drawing is the orbit of the Moon, and the dashed green curve is a circle; by comparing these two curves we can see how little the ellipse deviates from a circle centered on the Earth. The point on the ellipse closest to the Earth is called the **perigee,** and the point farthest from the Earth is called the **apogee.** The distance between the Moon and the Earth is roughly 30 times the diameter of the Earth. The Moon takes 27.3 days to travel once around the Earth.

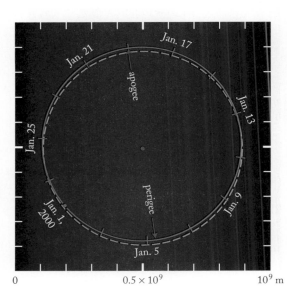

$0 \qquad\qquad 0.5 \times 10^9 \qquad\qquad 10^9 \text{ m}$

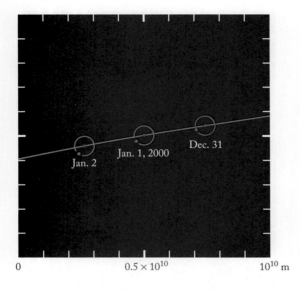

SCALE 1:1.5 × 10¹¹   This picture shows the Earth, the Moon, and portions of their orbits around the Sun. On this scale, both the Earth and the Moon look like small dots. Again, the Sun is far below the bottom of the picture. In the middle, we see the Earth and the Moon in their positions for January 1, 2000. On the right and on the left we see, respectively, their positions for 1 day before and 1 day after this date.

Note that the net motion of the Moon consists of the combination of two simultaneous motions: the Moon orbits around the Earth, which in turn orbits around the Sun.

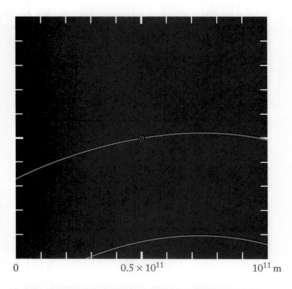

SCALE 1:1.5 × 10¹²   Here we see the orbits of the Earth and of Venus. However, Venus itself is beyond the edge of the picture. The small circle is the orbit of the Moon. The dot representing the Earth is much larger than what it should be, although the artist has drawn it as minuscule as possible. On this scale, even the Sun is quite small; if it were included in this picture, it would be only 1 millimeter across.

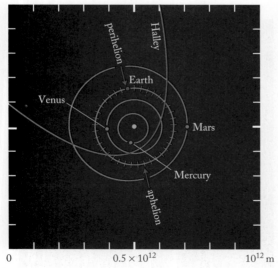

SCALE 1:1.5 × 10¹³   This drawing shows the positions of the Sun and the inner planets: Mercury, Venus, Earth, and Mars. The positions of the planets are those of January 1, 2000. The orbits of all these planets are ellipses, but they are close to circles. The point of the orbit nearest to the Sun is called the **perihelion** and the point farthest from the Sun is called the **aphelion**. The Earth reaches perihelion about January 3 and aphelion about July 6 of each year.

All the planets travel around their orbits in the same direction: counterclockwise in our picture. The marks along the orbit of the Earth indicate the successive positions at intervals of 10 days.

Beyond the orbit of Mars, a large number of asteroids orbit around the Sun; these have been omitted to prevent excessive clutter. Furthermore, a large number of comets orbit around the Sun. Most of these have pronounced elliptical orbits. The comet Halley has been included in our drawing.

The Sun is a sphere of radius 6.96×10^8 meters. On the scale of the picture, the Sun looks like a very small dot, even smaller than the dot drawn here. The mass of the Sun is 1.99×10^{30} kilograms.

The matter in the Sun is in the **plasma** state, sometimes called the fourth state of matter. Plasma is a very hot gas in which violent collisions between the atoms in their random thermal motion have fragmented the atoms, ripping electrons off them. An atom that has lost one or more electrons is called an **ion**. Thus, plasma consists of a mixture of electrons and ions engaging in frequent collisions. These collisions are accompanied by the emission of light, making the plasma luminous.

SCALE 1:1.5 × 10¹⁴   This picture shows the positions of the outer planets of the Solar System: Jupiter, Saturn, Uranus, Neptune, and Pluto. On this scale, the orbits of the inner planets are barely visible. As in our other pictures, the positions of the planets are those of January 1, 2000.

The outer planets move slowly and their orbits are very large; thus they take a long time to go once around their orbit. The extreme case is that of Pluto, which takes 248 years to complete one orbit.

Uranus, Neptune, and Pluto are so far away and so faint that their discovery became possible only through the use of telescopes. Uranus was discovered in 1781, Neptune in 1846, and the tiny Pluto in 1930. Pluto is now known as one of several dwarf planets.

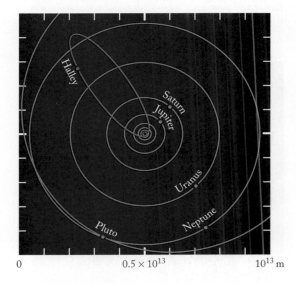

SCALE 1:1.5 × 10¹⁵   We now see that the Solar System is surrounded by a vast expanse of space. Although this space is shown empty in the picture, the Solar System is encircled by a large cloud of millions of comets whose orbits crisscross the sky in all directions. Furthermore, the interstellar space in this picture and in the succeeding pictures contains traces of gas and of dust. The interstellar gas is mainly hydrogen; its density is typically 1 atom per cubic centimeter.

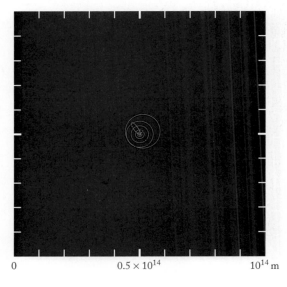

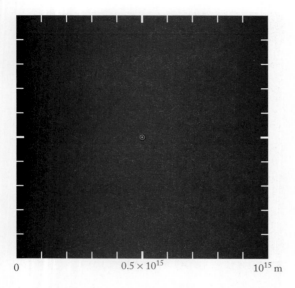

0   0.5×10^{15} 10^{15} m

SCALE 1:1.5 × 10^{16}   More interstellar space. The small circle is the orbit of Pluto.

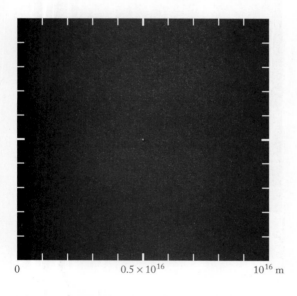

0   0.5×10^{16} 10^{16} m

SCALE 1:1.5 × 10^{17}   And more interstellar space. On this scale, the Solar System looks like a minuscule dot, 0.1 millimeter across.

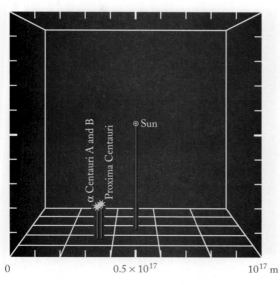

0   0.5×10^{17} 10^{17} m

SCALE 1:1.5 × 10^{18}   Here, at last, we see the stars nearest to the Sun. The picture shows all the stars within a cubical box 10^{17} meters × 10^{17} meters × 10^{17} meters centered on the Sun: Alpha Centauri A, Alpha Centauri B, and Proxima Centauri. All three are in the constellation Centaurus, in the southern sky.

The star closest to the Sun is Proxima Centauri. This is a very faint, reddish star (a "red dwarf"), at a distance of 4.0 × 10^{16} meters from the Sun. Astronomers like to express stellar distances in light-years: Proxima Centauri is 4.2 light-years from the Sun, which means light takes 4.2 years to travel from this star to the Sun.

SCALE 1:1.5 × 10¹⁹   This picture displays the brightest stars within a cubical box 10^{18} meters × 10^{18} meters × 10^{18} meters centered on the Sun. There are many more stars in this box besides those shown—the total number of stars in this box is close to 2000.

Sirius is the brightest of all the stars in the night sky. If it were at the same distance from the Earth as the Sun, it would be 28 times brighter than the Sun.

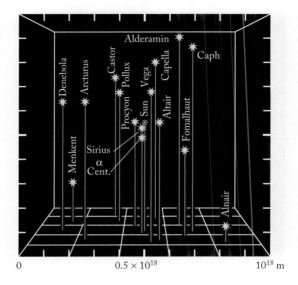

SCALE 1:1.5 × 10²⁰   Here we expand our box to 10^{19} meters × 10^{19} meters × 10^{19} meters, again showing only the brightest stars and omitting many others. The total number of stars within this box is about 2 million. We recognize several clusters of stars in this picture: the Pleiades Cluster, the Hyades Cluster, the Coma Berenices Cluster, and the Perseus Cluster. Each of these has hundreds of stars crowded into a fairly small patch of sky. In this diagram, Starbursts signify single stars, circles with starbursts indicate star clusters, and a circle with a single star indicate a star cluster with its brightest star.

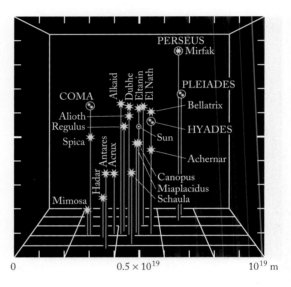

SCALE 1:1.5 × 10²¹   This photograph shows a view of the Milky Way in the direction of the constellation Sagittarius. Now there are so many stars in our field of view that they appear to form clouds of stars. There are about a million stars in this photograph, and there are many more stars too faint to show up distinctly. Although this photograph is not centered on the Sun, it is similar to what we would see if we could look toward the Solar System from very far away.

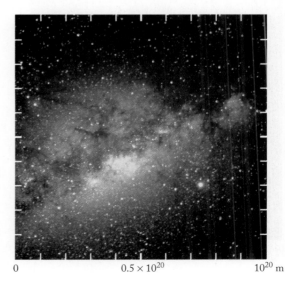

0   0.5×10^{21} 10^{21} m

SCALE 1:1.5 × 10^{22}   This is the spiral galaxy NGC 2997. Its clouds of stars are arranged in spiral arms wound around a central bulge. The bright central bulge is the nucleus of the galaxy; it has a more or less spherical shape. The surrounding region, with the spiral arms, is the disk of the galaxy. This disk is quite thin; it has a thickness of only about 3% of its diameter. The stars making up the disk circle around the galactic center in a clockwise direction.

Our Sun is in a spiral galaxy of roughly similar shape and size: the **Milky Way Galaxy**. The total number of stars in this galaxy is about 10^{11}. The Sun is in one of the spiral arms, roughly one-third inward from the edge of the disk toward the center.

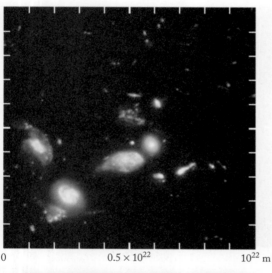

0   0.5×10^{22} 10^{22} m

SCALE 1:1.5 × 10^{23}   Galaxies are often found in clusters of several galaxies. Some of these clusters consist of just a few galaxies, others of hundreds or even thousands. The photograph shows a cluster, or group, of galaxies beyond the constellation Fornax. The group contains an elliptical galaxy like a luminous yellow egg (center), three large spiral galaxies (left), and a spiral with a bar (bottom left).

Our Galaxy is part of a modest cluster, the **Local Group**, consisting of our own Galaxy, the great Andromeda Galaxy, the Triangulum Galaxy, the Large Magellanic Cloud, plus 16 other small galaxies.

According to recent investigations, the dark, apparently empty, space near galaxies contains some form of distributed matter, with a total mass 20 or 30 times as large as the mass in the luminous, visible galaxies. But the composition of this invisible, extragalactic **dark matter** is not known.

0   0.5×10^{23} 10^{23} m

SCALE 1:1.5 × 10^{24}   The Local Group lies on the fringes of a very large cluster of galaxies, called the **Local Supercluster**. This is a cluster of clusters of galaxies. At the center of the Local Supercluster is the **Virgo Cluster** with several thousand galaxies. Seen from a large distance, our supercluster would present a view comparable with this photograph, which shows a multitude of galaxies beyond the constellation Fornax, all at a very large distance from us. The photograph was taken with the Hubble Space Telescope coupled to two very sensitive cameras using an exposure time of almost 300 hours.

All these distant galaxies are moving away from us and away from each other. The very distant galaxies in the photo are moving away from us at speeds almost equal to the speed of light. This motion of recession of the galaxies is analogous to the outward motion of, say, the fragments of a grenade after its explosion. The motion of the galaxies suggests that the Universe began with a big explosion, the **Big Bang**, that launched the galaxies away from each other.

SCALE 1:1.5 × 10²⁵   On this scale a galaxy equal in size to our own Galaxy would look like a fuzzy dot, 0.1 millimeter across. Thus, the galaxies are too small to show up clearly on a photograph. Instead we must rely on a plot of the positions of the galaxies. The plot shows the positions of about 200 galaxies. The dense cluster of galaxies in the lower half of the plot is the Virgo Cluster.

Since we are looking into a volume of space, some of the galaxies are in the foreground, some are in the background; but our plot takes no account of perspective.

The luminous stars in the galaxies constitute only a small fraction of the total mass of the Universe. The space around the galaxies and the clusters of galaxies contains dark matter, and the space between the clusters contains **dark energy**, a strange form of matter that causes an acceleration of the expansion of the Universe.

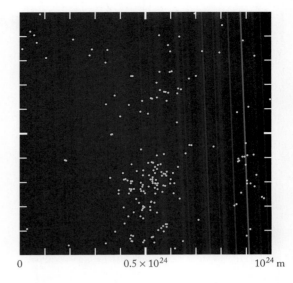

0   0.5×10^{24} 10^{24} m

SCALE 1:1.5 × 10²⁶   This plot shows the positions of about 100,000 galaxies in a patch of the sky at distances of up to 1×10^{9} light years from the Earth. The false color in this image indicates the distance–red for shorter distances, blue for larger distances.

The visible galaxies plotted here contribute only about 5% of the total mass in the universe. The dark matter near the galaxies contribute another 25%. The remaining 70% of the total mass in the universe is in the form of dark energy, which is uniformly distributed over the vast reaches of intergalactic space.

This is the last of our pictures in the ascending series. We have reached the limits of our zoom out. If we wanted to draw another picture, 10 times larger than this, we would need to know the shape and size of the entire Universe. We do not yet know that.

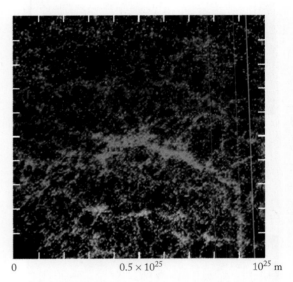

0   0.5×10^{25} 10^{25} m

PART II: THE SMALL-SCALE WORLD

MAGNIFICATION 0.67 ×

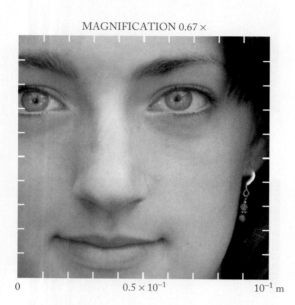

0   0.5 × 10⁻¹   10⁻¹ m

SCALE 1:1.5   We now return to Erin and zoom in on her eye. The surface of her skin appears smooth and firm. But this is an illusion. Matter appears continuous because the number of atoms in each cubic centimeter is extremely large. In a cubic centimeter of human tissue there are about 10^{23} atoms. This large number creates the illusion that matter is continuously distributed—we see only the forest and not the individual trees. The solidity of matter is also an illusion. The atoms in our bodies are mostly vacuum. As we will discover in the following pictures, within each atom the volume actually occupied by subatomic particles is only a very small fraction of the total volume.

MAGNIFICATION 6.7 ×

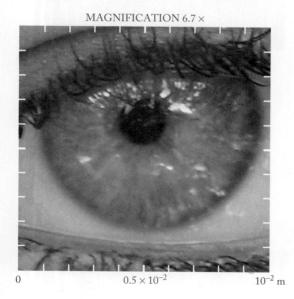

0   0.5 × 10⁻²   10⁻² m

SCALE 1:1.5 × 10⁻¹   Our eyes are very sophisticated sense organs; they collect more information than all our other sense organs taken together. The photograph shows the pupil and the iris of Erin's eye. Annular muscles in the iris change the size of the pupil and thereby control the amount of light that enters the eye. In strong light the pupil automatically shrinks to about 2 millimeters; in very weak light it expands to as much as 7 millimeters.

SCALE 1:1.5 × 10⁻² This false-color photograph shows the delicate network of blood vessels on the front surface of the retina, the light-sensitive membrane lining the interior of the eyeball. The rear surface of the retina is densely packed with two kinds of cells that sense light: cone cells and rod cells. In a human retina there are about 6 million cone cells and 120 million rod cells. The cone cells distinguish colors; the rod cells distinguish only brightness and darkness, but they are more sensitive than the cone cells and therefore give us vision in faint light ("night vision").

This and the following photographs were made with various kinds of **electron microscopes**. An ordinary microscope uses a beam of light to illuminate the object; an electron microscope uses a beam of electrons. Electron microscopes can achieve much sharper contrast and much higher magnification than ordinary microscopes.

MAGNIFICATION $6.7 \times 10 \times$

0   0.5×10^{-3} 10^{-3} m

SCALE 1:1.5 × 10⁻³ Here we have a false-color photograph of rod cells prepared with a scanning electron microscope (SEM). For this photograph, the retina was cut apart and the microscope was aimed at the edge of the cut. In the top half of the picture we see tightly packed rods. Each rod is connected to the main body of a cell containing the nucleus. In the bottom part of the picture we can distinguish tightly packed cell bodies of the cell.

MAGNIFICATION $6.7 \times 10^2 \times$

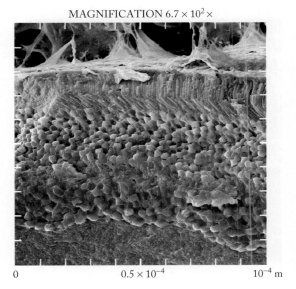

0   0.5×10^{-4} 10^{-4} m

SCALE 1:1.5 × 10⁻⁴ This is a close-up view of a few rods cells. The upper portions of the rods contain a special pigment—visual purple—which is very sensitive to light. The absorption of light by this pigment initiates a chain of chemical reactions that finally trigger nerve pulses from the eye to the brain.

MAGNIFICATION $6.7 \times 10^3 \times$

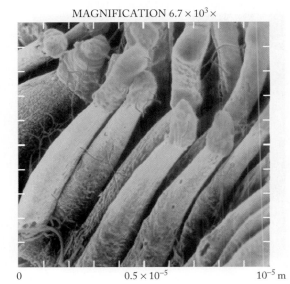

0   0.5×10^{-5} 10^{-5} m

MAGNIFICATION $6.7 \times 10^4 \times$

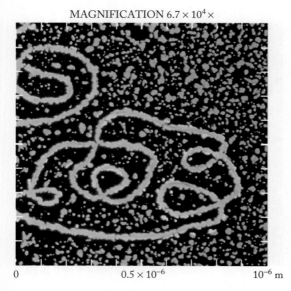

0   0.5×10^{-6} 10^{-6} m

MAGNIFICATION $6.7 \times 10^5 \times$

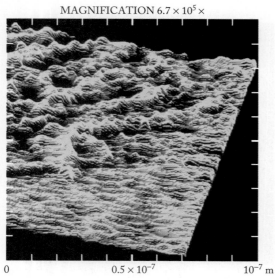

0   0.5×10^{-7} 10^{-7} m

MAGNIFICATION $6.7 \times 10^6 \times$

0   0.5×10^{-8} 10^{-8} m

SCALE 1:1.5 $\times$ 10^{-5}   These are strands of DNA, or deoxyribonucleic acid, as seen with a transmission electron microscope (TEM) at very high magnification. DNA is found in the nuclei of cells. It is a long molecule made by stringing together a large number of nitrogenous base molecules on a backbone of sugar and phosphate molecules. The base molecules are of four kinds, the same in all living organisms. But the sequence in which they are strung together varies from one organism to another. This sequence spells out a message—the base molecules are the "letters" in this message. The message contains all the genetic instructions governing the metabolism, growth, and reproduction of the cell.

The strands of DNA in the photograph are encrusted with a variety of small protein molecules. At intervals, the strands of DNA are wrapped around larger protein molecules that form lumps looking like the beads of a necklace.

SCALE 1:1.5 $\times$ 10^{-6}   The highest magnifications are attained by a newer kind of electron microscope, the scanning tunneling microscope (STM). This picture was prepared with such a microscope. The picture shows strands of DNA deposited on a substrate of graphite. In contrast to the strands of the preceding picture, these strands are uncoated; that is, they are without protein encrustations.

SCALE 1:1.5 $\times$ 10^{-7}   This close-up picture of strands of DNA reveals the helical structure of this molecule. The strand consists of a pair of helical coils wrapped around each other. This picture was generated by a computer from data obtained by illuminating DNA samples with X rays (X-ray scattering).

SCALE 1:1.5 × 10⁻⁸   This picture shows a layer of palladium atoms on surface of graphite as seen with an STM. Here we have visual evidence of the atomic structure of matter. The palladium atoms are arranged in a symmetric, repetitive hexagonal pattern. Materials with such regular arrangements of atoms are called **crystals**.

Each of the palladium atoms is approximately a sphere, about 3×10^{-10} meter across. However, the atom does not have a sharply defined boundary; its surface is somewhat fuzzy. Atoms of other elements are also approximately spheres, with sizes that range from 2×10^{-10} to 4×10^{-10} meter across.

At present we know of more than 100 kinds of atoms or chemical elements. The lightest atom is hydrogen, with a mass of 1.67×10^{-27} kilogram; the heaviest is element 114, ununquadium, with a mass about 289 times as large.

MAGNIFICATION $6.7 \times 10^7 \times$

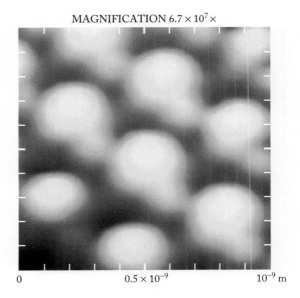

0   0.5×10^{-9} 10^{-9} m

SCALE 1:1.5 × 10⁻⁹   The drawing shows the interior of an atom of neon. This atom consists of 10 electrons orbiting around a nucleus. In the drawing, the electrons have been indicated by small dots, and the nucleus by a slightly larger dot at the center of the picture. These dots have been drawn as small as possible, but even so the size of these dots does not give a correct impression of the actual sizes of the electrons and of the nucleus. The electron is smaller than any other particle we know; maybe the electron is truly pointlike and has no size at all. The nucleus has a finite size, but this size is much too small to show up on the drawing. Note that the electrons tend to cluster near the center of the atom. However, the overall size of the atom depends on the distance to the outermost electron; this electron defines the outer edge of the atom.

The electrons move around the nucleus in a very complicated motion, and so the resulting electron distribution resembles a fuzzy cloud, similar to the STM image of the previous picture. This drawing, however, shows the electrons as they would be seen at one instant of time with a hypothetical microscope that employs gamma rays instead of light rays to illuminate an object; no such microscope has yet been built.

The mass of each electron is 9.11×10^{-31} kilogram, but most of the mass of the atom is in the nucleus; the 10 electrons of the neon atom have only 0.03% of the total mass of the atom.

MAGNIFICATION $6.7 \times 10^8 \times$

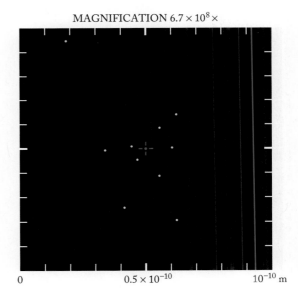

0   0.5×10^{-10} 10^{-10} m

MAGNIFICATION $6.7 \times 10^9 \times$

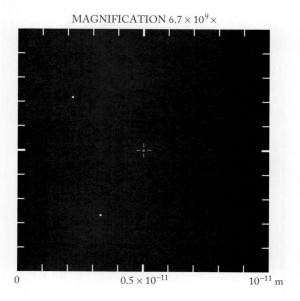

0   0.5×10^{-11} 10^{-11} m

SCALE 1:1.5 × 10⁻¹⁰   Here we are closing in on the nucleus. We are seeing the central part of the atom. Only two electrons are in our field of view; the others are beyond the margin of the drawing. The size of the nucleus is still much smaller than the size of the dot at the center of the drawing.

MAGNIFICATION $6.7 \times 10^{10} \times$

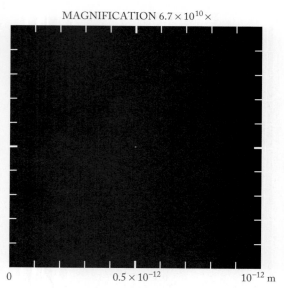

0   0.5×10^{-12} 10^{-12} m

SCALE 1:1.5 × 10⁻¹¹   In this drawing we finally see the nucleus in its true size. At this magnification, the nucleus of the neon atom looks like a small dot, 0.5 millimeter in diameter. Since the nucleus is extremely small and yet contains most of the mass of the atom, the density of the nuclear material is enormous. If we could assemble a drop of pure nuclear material of a volume of 1 cubic centimeter, it would have a mass of 2.3×10^{11} kilograms, or 230 million metric tons!

    Our drawings show clearly that most of the volume within the atom is empty space. The nucleus occupies only a very small fraction of this volume.

MAGNIFICATION $6.7 \times 10^{11} \times$

0   0.5×10^{-13} 10^{-13} m

SCALE 1:1.5 × 10⁻¹²   We can now begin to distinguish the nuclear structure. The nucleus has a nearly spherical shape, but its surface is slightly fuzzy.

SCALE 1:1.5 × 10⁻¹³   At this extreme magnification we can see the details of the nuclear structure. The nucleus of the neon atom is made up of 10 protons (white balls) and 10 neutrons (red balls). Each proton and each neutron is a sphere with a diameter of about 2×10^{-15} meter, and a mass of 1.67×10^{-27} kilogram. In the nucleus, these protons and neutrons are tightly packed together, so tightly that they almost touch. The protons and neutrons move around the volume of the nucleus at high speed in a complicated motion.

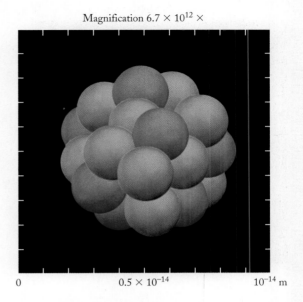

Magnification $6.7 \times 10^{12} \times$

0   0.5×10^{-14} 10^{-14} m

SCALE 1:1.5 × 10⁻¹⁴   This final picture shows three pointlike bodies within a proton. These pointlike bodies are **quarks**—each proton and each neutron is made of three quarks. Recent experiments have told us that the quarks are much smaller than protons, but we do not yet know their precise size. Hence the dots in the drawing probably do not give a fair description of the size of the quarks. The quarks within protons and neutrons are of two kinds, called **up** and **down**. The proton consists of two *up* quarks and one *down* quark joined together; the neutron consists of one *up* quark and two *down* quarks joined together.

This final picture takes us to the limits of our knowledge of the subatomic world. As a next step we would like to zoom in on the quarks and show what they are made of. According to a speculative theory, they are made of small snippets or loops of strings, 10^{-35}m long. But we do not yet have any evidence for this theory.

Magnification $6.7 \times 10^{13} \times$

0   0.5×10^{-15} 10^{-15} m

Motion, Force, and Energy

CONTENTS

CHAPTER 1   Space, Time, and Mass

CHAPTER 2   Motion Along a Straight Line

CHAPTER 3   Vectors

CHAPTER 4   Motion in Two and Three Dimensions

CHAPTER 5   Newton's Laws of Motion

CHAPTER 6   Further Applications of Newton's Laws

CHAPTER 7   Work and Energy

CHAPTER 8   Conservation of Energy

CHAPTER 9   Gravitation

CHAPTER 10   Systems of Particles

CHAPTER 11   Collisions

CHAPTER 12   Rotation of a Rigid Body

CHAPTER 13   Dynamics of a Rigid Body

CHAPTER 14   Statics and Elasticity

At launch, the space shuttle assembly, including the external fuel tank and auxiliary boosters, has a mass of 2×10^6 kg. The thrust of the powerful rocket engines, including the shuttle orbiter's main engines shown here, accelerates the entire launch vehicle to the speed of sound in just 45 seconds.

Space, Time, and Mass

1.1 Coordinates and Reference Frames

1.2 The Unit of Length

1.3 The Unit of Time

1.4 The Unit of Mass

1.5 Derived Units

1.6 Significant Figures; Consistency of Units and Conversion of Units

CONCEPTS IN CONTEXT

Concepts
— in —
Context

A laser-ranging device is used by geologists and surveyors to measure distance. It emits a pulse of laser light toward a mirror placed at some unknown distance and measures the time taken by the pulse to travel to the mirror and back. From this round-trip time and the known speed of light, it then calculates the distance. Since the speed of light is very large, the round-trip time is very small.

With the definitions of the units of distance and time given in this chapter, we can consider the following questions:

❓ How far does light travel in a small fraction of a second? (Example 1, page 10)

❓ How is the precision of the distance determination limited by the precision of the time measurement? (Example 3, page 15)

The investigator of any phenomenon—an earthquake, a flash of lightning, a collision between two ships—must begin by asking, Where and when did it happen? Phenomena happen at points in space and at points in time. A complicated phenomenon, such as a collision between two ships, is spread out over many points of space and time. But no matter how complicated, any phenomenon can be fully described by stating what happened at diverse points of space at successive instants of time. Measurements of positions and times require the use of coordinate grids and reference frames, which we will discuss in the first section of this chapter.

Ships and other macroscopic bodies are made of atoms. Since the sizes of the atoms are extremely small compared with the sizes of macroscopic bodies, we can regard atoms as almost pointlike masses for most practical purposes. *A pointlike mass of no discernible size or internal structure is called an* **ideal particle.** At any given instant of time, the ideal particle occupies a single point of space. Furthermore, the particle has a mass. And that is all: if we know the position of an ideal particle at each instant of time and we know its mass, then we know everything that can be known about the particle. *Position, time, and mass give a complete description of the behavior and the attributes of an ideal particle.*[1] Since every macroscopic body consists of particles, we can describe the behavior and the attributes of such a body by describing the particles within the body. Thus, measurements of position, time, and mass are of fundamental significance in physics. We will discuss the units for these measurements in later sections of this chapter.

1.1 COORDINATES AND REFERENCE FRAMES

If you are lost somewhere on the highways in Canada and you stop at a service station to ask for directions to Moose Jaw, the attendant might instruct you to go 90 kilometers north on Route 6 and then 70 kilometers west on Route 1 (see Fig. 1.1). In giving these instructions, the attendant is taking the service station as **origin**, and he is specifying the position of Moose Jaw relative to this origin. To achieve a precise, quantitative description of the position of a particle, physicists use much the same procedure. They first take some convenient point of space as origin and then specify the position of the particle relative to this origin. For this purpose, they imagine a grid of lines around the origin and give the location of the particle within this grid; that is, they imagine that the ground is covered with graph paper, and they specify the position of the particle by means of coordinates read off this graph paper.

The most common coordinates are **rectangular coordinates** x and y, which rely on a rectangular grid. Figure 1.2 shows such a rectangular grid. The mutually perpendicular lines through the origin O are called the x and y axes. The coordinates of the grid point P, where the particle is located, simply tell us how far we must move parallel to the corresponding axis in order to go from the origin O to the point P. For example, the point P shown in Fig. 1.2a has coordinates $x = 3$ units and $y = 5$ units. If we move from the origin in a direction opposite to that indicated by the arrow on the axis, then the coordinate is negative; thus, the point P shown in Fig. 1.2b has a negative x coordinate, $x = -3$ units.

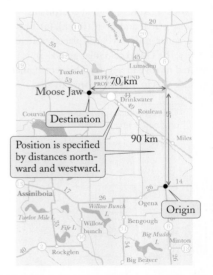

FIGURE 1.1 To reach Moose Jaw, Canada, the automobile has to travel 90 km north and then 70 km west.

---

[1] We will disregard for now the possibility that the particle also has an electric charge. Electricity is the subject of Chapters 22–33.

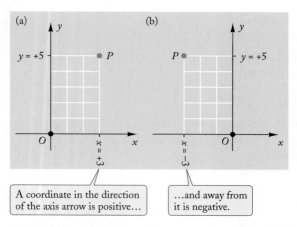

A coordinate in the direction of the axis arrow is positive…

…and away from it is negative.

FIGURE 1.2 Rectangular coordinates x and y of a point P. (a) Both coordinates are positive; (b) the x coordinate is negative.

The two-dimensional grid shown in Fig. 1.2 is adequate when we want to describe the two-dimensional (east–west and north–south) motion of an automobile traveling on flat ground or the motion of a ship on the (nearly) flat surface of the water of a harbor. However if we want to describe the three-dimensional (east–west, north–south, and up–down) motion of an aircraft flying through the air or a submarine diving through the ocean, then we need a three-dimensional grid, with x, y, and z axes. And if we want to describe the motion of an automobile along a straight road, then we need only a one-dimensional grid; that is, we need only the x axis, which we imagine placed along the road.

When we determine the position of a particle by means of a coordinate grid erected around some origin, we perform a *relative* measurement—the coordinates of the point at which the particle is located depend on the choice of origin and on the choice of coordinate grid. The choice of origin of coordinates and the choice of coordinate grid are matters of convenience. For instance, a harbormaster might use a coordinate grid with the origin at the harbor; but a municipal engineer might prefer a displaced coordinate grid with its origin at the center of town (see Fig. 1.3a) or a rotated coordinate grid oriented along the streets of the town (see Fig. 1.3b). The navigator of a ship might find it convenient to place the origin at the midpoint of her ship and to use a coordinate grid erected around this origin; the grid then moves with the ship (see Fig. 1.3c). If the navigator plots the track of a second ship on this grid, she can tell at a glance what the distance of closest approach will be, and whether the other ship is on a collision course (whether it will cross the origin).

For the description of the motion of a particle, we must specify both its position and the time at which it has this position. To determine the time, we use a set of synchronized clocks which we imagine arranged at regular intervals along the coordinate grid. When a particle passes through a grid point P, the coordinates give us the position of the particle in space, and the time registered by the nearby clock gives us the time t. *Such a coordinate grid with an array of synchronized clocks is called a* **reference frame**. Like the choice of origin and the choice of coordinate grid, the choice of reference frame is a matter of convenience. For instance, Fig. 1.4a shows a reference frame erected around the harbor, and Fig. 1.4b shows a reference frame erected around the ship. Reference frames are usually named after the body or point around which they are erected. Thus, we speak of the reference frame of the harbor, the reference frame of the ship, the reference frame of the laboratory, the reference frame of the Earth, etc.

FIGURE 1.3 Rectangular coordinate grids x–y (red) and x'–y' (black). (a) This rectangular grid x'–y' is displaced relative to x–y. (b) This rectangular grid x'–y' is rotated relative to x–y. (c) This rectangular grid x'–y' is in motion relative to x–y.

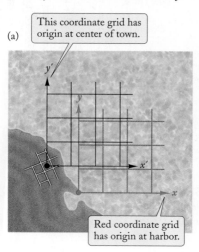

(a) This coordinate grid has origin at center of town.

Red coordinate grid has origin at harbor.

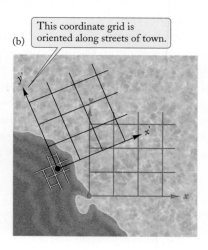

(b) This coordinate grid is oriented along streets of town.

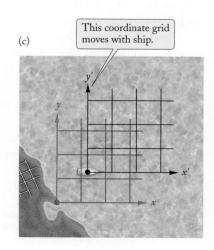

(c) This coordinate grid moves with ship.

✔ Checkup 1.1

QUESTION 1: For an ideal particle, position and mass are the only two measurable quantities (at any given instant). Consider an extended body, such as a bowling ball. What quantities can you measure about the bowling ball, besides position and mass? Do you know the units of any of these quantities?

QUESTION 2: Consider a coordinate grid x'–y' shifted relative to the grid x–y by a fixed amount, as in Fig. 1.3a. Mark a point P on this grid. Is the x' value for the given point P larger or smaller than the x value? What about the y' and y values?

QUESTION 3: What is the distinction between a coordinate grid and a reference frame?

1.2 THE UNIT OF LENGTH : METER

In order to make numerical records of our measurements of position, time, and mass we need to adopt a unit of length, a unit of time, and a unit of mass, so we can express our measurements as numerical multiples or fractions of these units. In this book we will use the **metric system of units**, *which is based on the meter as the unit of length, the second as the unit of time, and the kilogram as the unit of mass.* These units of length, time, and mass, in conjunction with the unit of temperature and the unit of electric charge (to be introduced in later chapters), are sufficient for the measurement of any physical quantity. Scientists and engineers refer to this set of units as the **International System of Units**, or **SI units** (from the French, *Système International*).[2]

Originally, the standard of length that specified the size of one meter was the standard meter bar kept at the International Bureau of Weights and Measures at Sèvres, France. This is a bar made of platinum–iridium alloy with a fine scratch mark near each end (see Fig. 1.5). By definition, the distance between these scratch marks was taken to be exactly one meter. The length of the meter was originally chosen so as to make the polar circumference of the Earth exactly 40 million meters (see Fig. 1.6); however, modern determinations of this circumference show it to be about 0.02% more than 40 million meters.

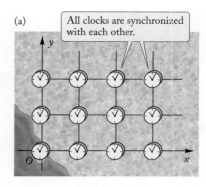

(a) All clocks are synchronized with each other.

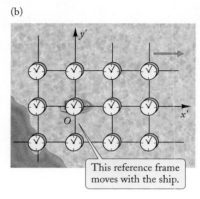

(b) This reference frame moves with the ship.

FIGURE 1.4 (a) A reference frame consists of a coordinate grid and a set of synchronized clocks. (b) A reference frame erected around a ship.

FIGURE 1.5 International standard meter bar.

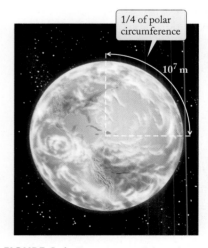

1/4 of polar circumference

10^7 m

FIGURE 1.6 One-quarter of the polar circumference of the Earth equals approximately 10^7 m.

---

[2] SI units are also discussed in Appendix 5; more information may be found at the National Institute of Standards and Technology website, http://www.physics.nist.gov/cuu/Units.

FIGURE 1.7 Some gauge blocks, commonly used by machinists as length standards. The height or thickness of each block serves as a length standard.

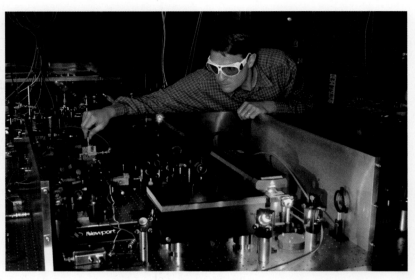

FIGURE 1.8 Stabilized laser at the National Institute of Standards and Technology.

Copies of the prototype standard meter were manufactured in France and distributed to other countries to serve as secondary standards. The length standards used in industry and engineering have been derived from these secondary standards. For instance, Fig. 1.7 shows a set of gauge blocks commonly used as length standards in machine shops.

The precision of the standard meter is limited by the coarseness of the scratch marks at the ends. For the sake of higher precision, physicists developed improved definitions of the standard of length. The most recent improvement emerged from the development of stabilized lasers (see Fig. 1.8). These lasers emit light waves of extreme uniformity which make it possible to determine the speed of light with extreme precision. This led to the adoption of a new definition for the length of the meter in terms of the speed of light: the **meter** (1 m) is the length traveled by a light wave in vacuum in a time interval of 1/299 792 458 second. Note that, since the meter is adjusted so that light travels exactly one meter in 1/299 792 458 second, the speed of light is exactly

$$[\text{speed of light}] = 299\,792\,458 \text{ meters per second} \tag{1.1}$$

Thus, the new definition of the meter amounts to the adoption of the speed of light as a standard of speed.

Table 1.1 lists a few distances and sizes, from the largest to the smallest. Many of the quantities listed in the table have already been mentioned in the Prelude. Quantities indicated with an ≈ (approximately equals) sign are not precisely defined; these quantities are rough approximations.

Table 1.2 lists some multiples and submultiples of the meter and their abbreviations. The **prefixes** used in Table 1.2, and other standard prefixes, are abbreviations for particular powers of ten. Prefixes representing powers of ten that differ by factors of 10^3 are often used with any unit, for convenience or conciseness. These standard prefixes and their abbreviations are listed in Table 1.3.

In the **British system of units**—abandoned by Britain and almost all other countries, but regrettably still in use in the United States—the unit of length is the **foot** (ft), which is exactly 0.3048 m. (For quick mental conversion from feet to meters, multiply by 0.3.) Table 1.4 gives the multiples and submultiples of the foot, but we will have little need for this table because we will hardly ever use British units in this book. Sporadic efforts to adopt metric units in the United States have failed, although most American automobile manufacturers now use metric units, and so does the U.S. Army (note that in U.S. Army slang, a kilometer is called a "klick," a usage commendable for its brevity).

TABLE 1.1	SOME DISTANCES AND SIZES
Distance to boundary of observable Universe	$\approx 1 \times 10^{26}$ m
Distance to Andromeda galaxy (a)	2.1×10^{22} m
Diameter of our Galaxy	7.6×10^{20} m
Distance to nearest star (Proxima Centauri)	4.0×10^{16} m
Earth–Sun distance	1.5×10^{11} m
Radius of Earth	6.4×10^{6} m
Wavelength of radio wave (AM band)	$\approx 3 \times 10^{2}$ m
Length of ship *Queen Elizabeth* (b)	3.1×10^{2} m
Height of average human male	1.8 m
Diameter of 5¢ coin (c)	2.1×10^{-2} m
Diameter of red blood cell (human)	7.5×10^{-6} m
Wavelength of visible light	$\approx 5 \times 10^{-7}$ m
Diameter of smallest virus (potato spindle) (d)	2×10^{-8} m
Diameter of atom	$\approx 1 \times 10^{-10}$ m
Diameter of atomic nucleus (iron)	$\approx 8 \times 10^{-15}$ m
Diameter of proton	$\approx 2 \times 10^{-15}$ m

(a)

(b)

(c)

(d)

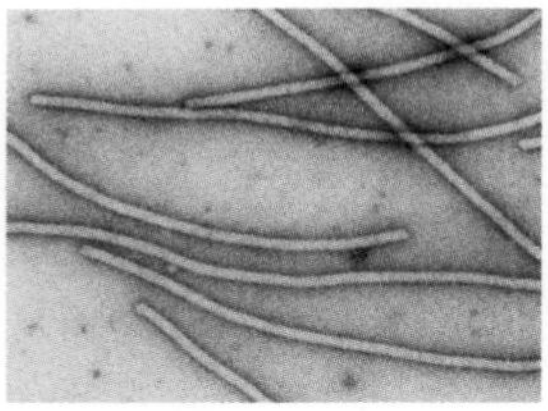

TABLE 1.2	MULTIPLES AND SUBMULTIPLES OF THE METER
kilometer (klick)	$1 \text{ km} = 10^{3}$ m
meter	1 m
centimeter	$1 \text{ cm} = 10^{-2}$ m
millimeter	$1 \text{ mm} = 10^{-3}$ m
micrometer (micron)	$1 \text{ } \mu\text{m} = 10^{-6}$ m
nanometer	$1 \text{ nm} = 10^{-9}$ m
angstrom	$1 \text{ Å} = 10^{-10}$ m
picometer	$1 \text{ pm} = 10^{-12}$ m
femtometer (fermi)	$1 \text{ fm} = 10^{-15}$ m

TABLE 1.3	PREFIXES FOR UNITS	
MULTIPLICATION FACTOR	**PREFIX**	**SYMBOL**
10^{21}	zetta	Z
10^{18}	exa	E
10^{15}	peta	P
10^{12}	tera	T
10^{9}	giga	G
10^{6}	mega	M
10^{3}	kilo	k
10^{-3}	milli	m
10^{-6}	micro	μ
10^{-9}	nano	n
10^{-12}	pico	p
10^{-15}	femto	f
10^{-18}	atto	a
10^{-21}	zepto	z

TABLE 1.4	MULTIPLES AND SUBMULTIPLES OF THE FOOT
mile	1 mi = 5280 ft = 1609.38 m
yard	1 yd = 3 ft = 0.9144 m
foot	1 ft = 0.3048 m
inch	1 in. = $\frac{1}{12}$ ft = 2.540 cm
mil	1 mil = 0.001 in.

 Checkup 1.2

QUESTION 1: How many centimeters are there in one kilometer? How many millimeters in a kilometer?

QUESTION 2: How many microns are there in a fermi?

QUESTION 3: How many microns are there in an angstrom?

  (A) 10^6 (B) 10^4 (C) 10^{-4} (D) 10^{-6}

1.3 THE UNIT OF TIME : SECOND

The unit of time is the second. Originally one second was defined as $1/(60 \times 60 \times 24)$, or 1/86 400, of a mean solar day. The solar day is the time interval required for the Earth to complete one rotation relative to the Sun. The length of the solar day depends on the rate of rotation of the Earth, which is subject to a host of minor variations, both seasonal and long-term, which make the rotation of the Earth an imperfect timekeeper.

To avoid any variation in the unit of time, we now use an atomic standard of time. This standard is the period of one vibration of microwaves emitted by an atom of cesium. The **second** (1 s) is defined as the time needed for 9 192 631 770 vibrations of a cesium atom. Figure 1.9 shows one of the atomic clocks at the National Institute of Standards and Technology in Boulder, Colorado. In this clock, the feeble vibrations of the cesium atoms are amplified to a level that permits them to control the dial of the clock. Good cesium clocks are very, very good—they lose or gain no more than 1 second in 20 million years.

Precise time signals keyed to the cesium atomic clocks of the National Institute of Standards and Technology are continuously transmitted by radio station WWV, Fort Collins, Colorado. These time signals can be picked up worldwide on shortwave receivers tuned to 2.5, 5, 10, 15, or 20 megahertz. Precise time signals are also announced continuously by telephone [in the United States, the telephone number is (303) 499-7111; the precise time is also available online at www.time.gov]. The time announced on the radio and on the telephone is Coordinated Universal Time, or Greenwich time, which is exactly 5 hours ahead of Eastern Standard Time.

Table 1.5 lists some typical time intervals, and Table 1.6 gives multiples and submultiples of the second.

FIGURE 1.9 Cesium atomic clock at the National Institute of Standards and Technology.

TABLE 1.5	SOME TIME INTERVALS
Age of the Universe	$\approx 4 \times 10^{17}$ s
Age of the Solar System	1.4×10^{17} s
Age of oldest known fossils	1.1×10^{17} s
Age of human species	7.9×10^{12} s
Age of the oldest written records (Sumerian)	1.6×10^{11} s
Life span of man (average)	2.2×10^{9} s
Travel time for light from nearest star	1.4×10^{8} s
Revolution of Earth (1 year)	3.2×10^{7} s
Rotation of Earth (1 day)	8.6×10^{4} s
Life span of free neutron (average)	9.2×10^{2} s
Travel time for light from Sun	5×10^{2} s
Travel time for light from Moon	1.3 s
Period of heartbeat (human)	≈ 0.9 s
Period of sound wave (middle C)	3.8×10^{-3} s
Period of radio wave (AM band)	$\approx 1 \times 10^{-6}$ s
Period of light wave	$\approx 2 \times 10^{-15}$ s
Life span of shortest-lived unstable particle	$\approx 10^{-24}$ s

TABLE 1.6	MULTIPLES AND SUBMULTIPLES OF THE SECOND
century	1 century = 100 yr = 3.156×10^9 s
year	1 year = 3.156×10^7 s = 365.25 days
day	1 day = 86 400 s
hour	1 h = 3600 s
minute	1 min = 60 s
millisecond	1 ms = 10^{-3} s
microsecond	1 μs = 10^{-6} s
nanosecond	1 ns = 10^{-9} s
picosecond	1 ps = 10^{-12} s
femtosecond	1 fs = 10^{-15} s

Concepts — in — Context

EXAMPLE 1   The laser-ranging device shown in the chapter photo is capable of measuring the travel time of a light pulse to within better than a billionth of a second. How far does light travel in one billionth of a second (a nanosecond)?

SOLUTION: The distance light travels in a nanosecond is

$$[\text{distance}] = [\text{speed}] \times [\text{time}]$$

$$= \left(2.997\,924\,58 \times 10^8 \, \frac{\text{m}}{\text{s}}\right) \times (1.0 \times 10^{-9} \, \text{s})$$

$$= (2.997\,924\,58 \times 1.0) \times (10^8 \times 10^{-9}) \times \left(\frac{\text{m}}{\cancel{\text{s}}} \times \cancel{\text{s}}\right)$$

$$\approx 3.0 \times (10^{-1}) \times (\text{m})$$

$$= 30 \text{ cm}$$

or, in British units, almost one foot. The ruler drawn diagonally across this page shows the distance light travels in 1 nanosecond.

✔ Checkup 1.3

QUESTION 1: How many milliseconds are there in an hour? How many picoseconds in a microsecond?

QUESTION 2: How many femtoseconds are there in minute?
(A) 3.6×10^{18} (B) 6.0×10^{16} (C) 6.0×10^{15}
(D) 1.7×10^{13} (E) 1.7×10^{14}

1.4 THE UNIT OF MASS: KILOGRAM

The unit of mass is the kilogram. The standard of mass is a cylinder of platinum–iridium alloy kept at the International Bureau of Weights and Measures (see Fig. 1.10). The **kilogram** (1 kg) is defined as exactly equal to the mass of this cylinder. Mass is the only fundamental unit for which we do not, as yet, have an atomic standard. → *why is this necessary?*

Mass is measured with a balance, an instrument that compares the *weight* of an unknown mass with a known force, such as the weight of the standard mass or the pull of a calibrated spring. Weight is directly proportional to mass, and hence equal weights imply equal masses (the precise distinction between mass and weight will be spelled out in Chapter 5). Figure 1.11 shows a watt balance, an extremely accurate balance especially designed by the National Institute of Standards and Technology. The watt balance is effectively a spring balance, but instead of a mechanical spring suspension it uses a magnetic suspension with calibrated magnetic forces.

To relate the mass of an atom to the kilogram mass we need to know **Avogadro's number** N_A, or the number of atoms per mole. One **mole** *of any chemical element (or any chemical compound) is that amount of matter containing as many atoms (or molecules) as there are atoms in exactly 12 grams of carbon-12.* The "atomic mass" of a chemical element (or the "molecular mass" of a compound) is the mass of one mole expressed in grams. Thus, according to the table of atomic masses in Appendix 8, one mole of carbon atoms (C) has a mass of 12.0 grams, one mole of hydrogen atoms (H) has a mass of 1.0 gram, one mole of oxygen atoms (O) has a mass of 16.0 grams, one mole of oxygen molecules (O_2) has a mass of 32.0 grams, one mole of water molecules (H_2O) has a mass of 18.0 grams, and so on.

The available experimental data yield the following value for N_A:

$$N_A = 6.022\,14 \times 10^{23} \text{ atoms or molecules per mole} \tag{1.2}$$

Since there are N_A atoms in one mole, the mass of one atom is the mass of one mole, or the "atomic mass," divided by N_A:

$$[\text{mass of atom}] = \frac{[\text{"atomic mass"}]}{N_A} \tag{1.3}$$

Hence, the mass of, say, a carbon-12 atom is

$$[\text{mass of carbon-12 atom}] = \frac{12 \text{ grams}}{6.022\,14 \times 10^{23}} = 1.992\,65 \times 10^{-23} \text{ gram}$$

$$= 1.992\,65 \times 10^{-26} \text{ kg} \tag{1.4}$$

Masses of atoms are often measured in terms of the **atomic mass unit** (1 u), which is exactly $\frac{1}{12}$ the mass of a carbon-12 atom:

$$1 \text{ atomic mass unit} = 1 \text{ u} = \frac{1.992\,65 \times 10^{-26} \text{ kg}}{12} \tag{1.5}$$

That is,

$$1 \text{ u} = 1.660\,54 \times 10^{-27} \text{ kg} \tag{1.6}$$

Note that with this definition of the atomic mass unit, the "atomic mass," or the number of grams in one mole, necessarily has the same numerical value as the mass of one atom expressed in u. Thus, the carbon atom has a mass of 12 u, the hydrogen atom has a mass of 1.0 u, the oxygen atom has a mass of 16.0 u, and so on.

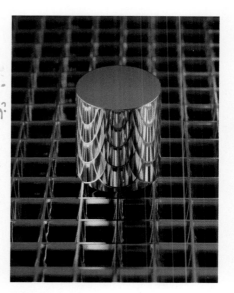

FIGURE 1.10 International standard kilogram.

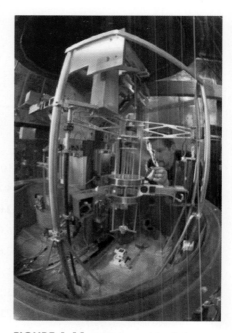

FIGURE 1.11 A high-precision balance.

EXAMPLE 2   How many atoms are there in a 5-cent coin? Assume that the coin is made of nickel and has a mass of 5.2×10^{-3} kg, or 5.2 grams. Atomic masses appear in Appendix 9.

SOLUTION: We recall that the atomic mass is the mass of one atom expressed in u. According to the periodic table of chemical elements in Appendix 9, the atomic mass of nickel is 58.69. Thus, the mass of one nickel atom is 58.69 u, or, $58.69 \times 1.66 \times 10^{-27}$ kg $= 9.74 \times 10^{-26}$ kg. The number of atoms in our 5.2×10^{-3} kg is then

$$\frac{5.2 \times 10^{-3} \text{ kg}}{9.74 \times 10^{-26} \text{ kg/atom}} = 5.3 \times 10^{22} \text{ atoms}$$

Table 1.7 gives some examples of masses expressed in kilograms, and Table 1.8 lists multiples and submultiples of the kilogram. In the British system of units, the unit of mass is the **pound**, which is exactly 0.453 592 37 kg.

For a quick grasp of the rough relationship between British and metric units, remember the following approximate equalities, good to within $\pm 10\%$ (more exact conversion factors are tabulated in Appendix 7):

$$1 \text{ yard} \approx 1 \text{ m}$$
$$1 \text{ mile} \approx 1.6 \text{ km}$$
$$1 \text{ pound} \approx \tfrac{1}{2} \text{ kg}$$
$$1 \text{ quart} \approx 1 \text{ liter}$$
$$1 \text{ gallon} \approx 4 \text{ liters}$$

TABLE 1.7	SOME MASSES	
Observable Universe	$\approx 10^{53}$ kg	
Galaxy	4×10^{41} kg	
Sun	2.0×10^{30} kg	
Earth (a)	6.0×10^{24} kg	
Ship *Queen Elizabeth*	7.6×10^{7} kg	
Jet airliner (Boeing 747, empty)	1.6×10^{5} kg	
Automobile (b)	1.5×10^{3} kg	
Man (average male)	73 kg	
Apple (c)	0.2 kg	(a) (b)
5¢ coin	5.2×10^{-3} kg	
Raindrop	2×10^{-6} kg	
Red blood cell (d)	9×10^{-14} kg	
Smallest virus (potato spindle)	4×10^{-21} kg	
Atom (iron)	9.5×10^{-26} kg	
Proton	1.7×10^{-27} kg	
Electron	9.1×10^{-31} kg	(c) (d)

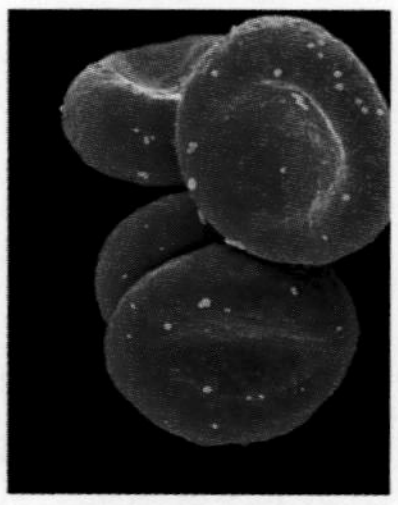

TABLE 1.8	MULTIPLES AND SUBMULTIPLES OF THE KILOGRAM
metric ton (tonne)	$1\text{ t} = 10^3\text{ kg}$
kilogram	1 kg
gram	$1\text{ g} = 10^{-3}\text{ kg}$
milligram	$1\text{ mg} = 10^{-6}\text{ kg}$
microgram	$1\text{ }\mu\text{g} = 10^{-9}\text{ kg}$
atomic mass unit	$1\text{ u} = 1.66 \times 10^{-27}\text{ kg}$
pound	$1\text{ lb} = 0.454\text{ kg}$
ounce ($\frac{1}{16}$ lb)	$1\text{ oz} = 28.3\text{ g}$
ton (2000 lb)	$1\text{ ton} = 907\text{ kg}$

To obtain a better feel for metric units, it helps to know that:

- The height of a person is typically 1.6 to 1.8 m.
- The mass of a person is typically 60 to 75 kg.
- The distance from the centerline of the body to the end of the outstretched arm is about 1 m.

✔ Checkup 1.4

QUESTION 1: How many grams are there in 1 metric ton? How many metric tons in one milligram?

QUESTION 2: How many atomic mass units (u) are there in 1 kilogram?
   (A) 1.66×10^{-26} (B) 1.66×10^{-23} (C) 6.02×10^{23} (D) 6.02×10^{26}

1.5 DERIVED UNITS

Online *Concept* Tutorial

 m, s, kg

The meter, the second, and the kilogram are the fundamental units, or **base units**, of the metric system of units. Any other physical quantity can be measured by introducing a **derived unit** *constructed by some combination of the base units*. For example, **area** can be measured with a derived unit that is the square of the unit of length; thus, in the metric system, the unit of area is the square meter ($1\text{ m} \times 1\text{ m} = 1\text{ m}^2$), which is the area of a square, one meter on a side (Fig. 1.12a). And **volume** can be measured with a derived unit that is the cube of the unit of length; in the metric system, the unit of volume is the cubic meter ($1\text{ m} \times 1\text{ m} \times 1\text{ m} = 1\text{ m}^3$), which is the volume of a cube, one meter on a side (Fig. 1.12b). Tables 1.9 and 1.10 give multiples and submultiples of these units.

Similarly, **density**, or mass per unit volume, can be measured with a derived unit that is the ratio of the unit of mass and the unit of volume. In the metric system, the unit of density is the kilogram per cubic meter (1 kg/m^3). For example, the density of water is 1000 kg/m^3, which means that one cubic meter of water has a mass of 1000 kilograms. We will see in later chapters that other physical quantities, such as speed, acceleration, force, etc., are also measured with derived units.

FIGURE 1.12 (a) One square meter. (b) One cubic meter.

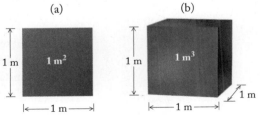

TABLE 1.9	MULTIPLES AND SUBMULTIPLES OF THE SQUARE METER
square meter	1 m^2
square kilometer	$1 \text{ km}^2 = 10^6 \text{ m}^2$
square centimeter	$1 \text{ cm}^2 = 10^{-4} \text{ m}^2$
square millimeter	$1 \text{ mm}^2 = 10^{-6} \text{ m}^2$

TABLE 1.10	MULTIPLES AND SUBMULTIPLES OF THE CUBIC METER
cubic meter	1 m^3
cubic kilometer	$1 \text{ km}^3 = 10^9 \text{ m}^3$
liter	$1 \text{ liter} = 10^{-3} \text{ m}^3 = 10^3 \text{ cm}^3$
cubic centimeter	$1 \text{ cm}^3 = 10^{-6} \text{ m}^3$
cubic millimeter	$1 \text{ mm}^3 = 10^{-9} \text{ m}^3$

The International System of Units, or SI, used in this book is the most widely accepted system of units in science and engineering. It is based on the meter, the second, and the kilogram, plus a special unit for temperature and a special unit for electric current.

 **Checkup 1.5**

QUESTION 1: How many square centimeters are there in a square meter? How many cubic centimeters are there in a cubic meter?

QUESTION 2: How many square millimeters are there in a square kilometer?

QUESTION 3: How many cubic millimeters are in a cubic kilometer?

    (A) 10^{-18} (B) 10^{-6} (C) 10^{6} (D) 10^{18}

Online
Concept
2
Tutorial

1.6 SIGNIFICANT FIGURES; CONSISTENCY OF UNITS AND CONVERSION OF UNITS

Significant Figures

The numbers in Tables 1.1, 1.5, and 1.7 are written in scientific notation, with powers of ten. This not only has the advantage that very large or very small numbers can be written compactly, but it also serves to indicate the precision of the numbers. For instance, a scientist observing the 1998 Berlin marathon at which Ronaldo da Costa set the world record of 2 h 6 min 5.0 s would have reported the running time as 7.5650 ×

10^3 s, or 7.565×10^3 s, or 7.57×10^3 s, or 7.6×10^3 s, depending on whether the measurement of time was made with a stopwatch, or a wristwatch with a second hand (but no stop button), or a wristwatch without a second hand, or a "designer" watch with one of those daft blank faces without any numbers at all. The first of these watches permits measurements to within $\frac{1}{10}$ s, the second to within about 1 s, the third to within 10 or 20 s, and the fourth to within 1 or 2 minutes (if the scientist is good at guessing the position of the hand on the blank face). We will adopt the rule that only as many digits, or **significant figures**, are to be written down as are known to be fairly reliable. In accordance with this rule, the number 7.5650×10^3 s comprises five significant figures, of which the last (0) represents tenths of a second; the number 7.565×10^3 s comprises four significant figures, of which the last (5) represents seconds; and so on. Thus, the scientific notation gives us an immediate indication of the precision to within which the number has been measured.

When numbers in scientific notation are multiplied or divided in calculations, the final result should always be rounded off so that it has no more significant figures than the original numbers, because the final result can be no more accurate than the original numbers on which it is based. Thus, the result of multiplying 7.57×10^3 s by 7.57×10^3 s is 5.73049×10^4 s^2, which should be rounded off to 5.73×10^4 s^2, because we were given only three significant figures in the original numbers. When numbers are added or subtracted, the result should be rounded to the largest decimal place among the last digits of the original numbers. Thus, $89.23 + 5.7 = 94.93$ should be rounded to 94.9, because one of the original numbers is known only to the tenths place.

EXAMPLE 3	A surveyor's laser-ranging device measures a time interval of 1.176×10^{-6} s for a laser light pulse to make a round trip from

a marker. What is the round-trip distance, expressed with the correct number of significant figures?

SOLUTION: The distance light travels in a time interval of 1.176×10^{-6} s is

$$[\text{distance}] = [\text{speed}] \times [\text{time}]$$

$$= \left(2.997\,924\,58 \times 10^8 \, \frac{\text{m}}{\text{s}} \right) \times (1.176 \times 10^{-6} \, \text{s}) \qquad (1.7)$$

$$= 352.6 \text{ m}$$

where the result which appears on a calculator, $352.555\,9306$ m, has been rounded in the last step to 352.6 m, that is, to the same number of digits as the least accurately known factor from which it was calculated. This fraction-of-a-meter accuracy agrees with the distance calculated for a nanosecond time interval in Example 1.

Sometimes even a number known to many significant figures is rounded off to fewer significant figures for the sake of convenience, when high accuracy is not required. For instance, the exact value of the speed of light is $2.997\,924\,58 \times 10^8$ m/s, but for most purposes, it is adequate to round this off to 3.00×10^8 m/s, and we will often employ this approximate value of the speed of light in our calculations.[3]

---

[3] When rounding a number, the following rule is employed (for example, in your handheld calculator): if the first of the digits to be rounded off is from 5 to 9, the prior digit is increased by one ("rounded up"); if the first digit of those to be rounded is 0 to 4, the prior digit is unchanged ("rounded down").

Consistency of Units

In all the equations of physics, the units on the left and the right sides of the equation must be consistent. This consistency is illustrated by the calculations in Example 3, where we see that on the right side of Eq. (1.7) the units of time cancel, and the final result for the right side then has units of length, in agreement with the units for the left side, which is a distance and requires units of length. It is a general rule that in any calculation with the equations of physics, the units can be multiplied and divided as though they were algebraic quantities, and this automatically yields the correct units for the final result. This requirement of consistency of units in the equations of physics can be reformulated in a more general way as a requirement of consistency of dimensions. In this context, the **dimensions** of a physical quantity are said to be length, time, mass, or some product or ratio of these if the units of this physical quantity are those of length, time, mass, or some product or ratio of these. Thus, volume has the dimensions of [length]^3, density has the dimensions of [mass]/[length]^3, speed has the dimensions of [length]/[time], and so on. In any equation of physics, *the dimensions of the two sides of the equation must be the same.* For instance, we can test the consistency of Eq. (1.7) by examining the dimensions of the quantities appearing in this equation:

$$[\text{length}] = \frac{[\text{length}]}{[\text{time}]} \times [\text{time}] \tag{1.8}$$

Dimensions are often used in preliminary tests of the consistency of equations, when there is some suspicion of a mistake in the equation. A test of the consistency of dimensions tells us no more than a test of the consistency of units, but has the advantage that we need not commit ourselves to a particular choice of units, and we need not worry about conversions among multiples and submultiples of the units. Bear in mind that if an equation fails this consistency test, it is proved wrong; but if it passes, it is not proved right.

Dimensions are sometimes used to find relationships between physical quantities. Such a determination of the appropriate proportionality between powers of relevant quantities is called **dimensional analysis**. Such analysis is performed by requiring the consistency of dimensions of units on each side of an equation. Dimensional analysis will prove useful when we have become familiar with more physical quantities and their dimensions.

The disastrous end of the space mission *Mars Climate Orbiter* on December 3, 1999 (see Fig. 1.13), teaches us a lesson on the importance of always attaching units to a number. Engineers at Lockheed Martin provided spacecraft operating data needed for navigation in British units rather than metric units. Flight controllers assumed the data were in metric units, and thus the probe did not behave as intended when the relevant thrusters were fired near Mars. The $155 000 000 mission became a total loss when the spacecraft entered the atmosphere and crashed instead of going into orbit around Mars.

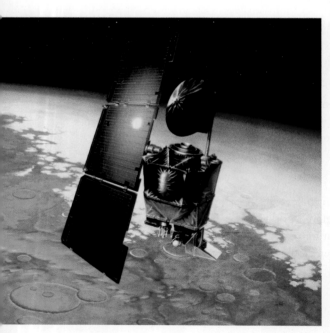

FIGURE 1.13 An artist's conception of the *Mars Climate Orbiter* spacecraft.

Conversion of Units

In many calculations, it is necessary to convert quantities expressed in one set of units to another set of units. Such conversions involve no more than simple substitutions of the equivalent amounts in the two sets of units (a comprehensive list of equivalent amounts in different units can be found in Appendix 7). For example, the density of water is $1.000 \times 10^3 \text{ kg/m}^3$. To express this in g/cm^3, we substitute 1 kg = 1000 g and 1 m = 100 cm, and we find

$$1.000 \times 10^3 \frac{\text{kg}}{\text{m}^3} = 1.000 \times 10^3 \times \frac{1000 \text{ g}}{(100 \text{ cm})^3} = 1.000 \times 10^3 \times \frac{10^3 \text{ g}}{10^6 \text{ cm}^3}$$

$$= 1.000 \frac{\text{g}}{\text{cm}^3}$$

An alternative method for the conversion of units from one set of units to another takes advantage of multiplication by factors that are identically equal to 1. Since 1 kg = 1000 g, we have the identity

$$1 = \frac{1000 \text{ g}}{1 \text{ kg}}$$

and, similarly,

$$1 = \frac{1 \text{ m}}{100 \text{ cm}}$$

This means that any quantity can be multiplied by (1000 g)/(1 kg) or (1 m)/(100 cm) without changing its value. Thus, starting with $1.000 \times 10^3 \text{ kg/m}^3$, we obtain

$$1.000 \times 10^3 \frac{\text{kg}}{\text{m}^3} = 1.000 \times 10^3 \frac{\text{kg}}{\text{m}^3} \times \frac{1000 \text{ g}}{1 \text{ kg}} \times \frac{1 \text{ m}}{100 \text{ cm}} \times \frac{1 \text{ m}}{100 \text{ cm}} \times \frac{1 \text{ m}}{100 \text{ cm}}$$

$$= 1.000 \times 10^3 \times 1000 \times \frac{1}{100} \times \frac{1}{100} \times \frac{1}{100} \times \frac{\text{kg}}{\text{m}^3} \times \frac{\text{g}}{\text{kg}} \times \frac{\text{m}^3}{\text{cm}^3}$$

Multiplying this out, and canceling the kg and m^3, we find

$$1.000 \times 10^3 \frac{\text{kg}}{\text{m}^3} = 1.000 \frac{\text{g}}{\text{cm}^3} \qquad (1.9)$$

Ratios such as (1000 g)/(1 kg) or (1 m)/(100 cm) that are identically equal to 1 are called **conversion factors**. To change the units of a quantity, simply multiply the quantity by one or several conversion factors that will bring about the desired cancellation of the old units.

The ratio of two quantities with identical dimensions or units will have no dimensions at all. For example, the **slope** of a path relative to the horizontal direction is defined as the ratio of the increment of height to the increment of horizontal distance. Since this is the ratio of two lengths, it is a dimensionless quantity. Likewise, the sine of an angle is defined as the ratio of two lengths; in a right triangle, the sine of one of the acute angles is equal to the length of the opposite side divided by the length of the hypotenuse (see Math Help: Trigonometry of the Right Triangle for a review). Thus the slope and the sine, cosine, and tangent of an angle are examples of **dimensionless quantities**.

EXAMPLE 4   Immediately after takeoff, a jet airliner climbs away from the runway at an upward angle of 12° (see Fig. 1.14). What is the slope of the path of ascent of the airliner? What altitude does it reach at a horizontal distance of 2000 m, or 2.0 km, from the point of takeoff?

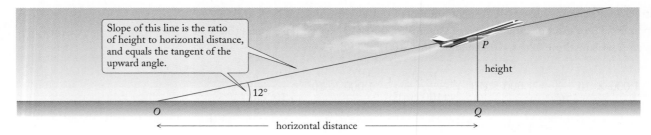

FIGURE 1.14 A jet airliner climbs after takeoff.

SOLUTION: If P is some point on the path at a horizontal distance OQ and a height PQ (see Fig. 1.13), then the slope of the path is

$$[\text{slope}] = \frac{[\text{height}]}{[\text{horizontal distance}]} = \frac{PQ}{OQ} \qquad (1.10)$$

Trigonometry tells us that in the right triangle OPQ, the ratio PQ/OQ is the tangent of the angle θ (see the Math Help box). Therefore

$$[\text{slope}] = \tan\theta$$

With our calculator, we find that the tangent of 12° is 0.21. Hence,

$$[\text{slope}] = 0.21$$

This dimensionless number means that the airliner climbs 0.21 m for every 1 m it advances in the horizontal direction. Slopes are often quoted as ratios; thus a slope of 0.21 can be expressed in the alternative form 21:100.

By proportions, the height reached for 2000 m of horizontal advance must be 2000 times as large as the height for 1 m of horizontal advance; that is,

$$[\text{height}] = 2000 \text{ m} \times 0.21 = 4.2 \times 10^2 \text{ m}$$

PROBLEM-SOLVING TECHNIQUES   UNITS AND SIGNIFICANT FIGURES

In all calculations with the equations of physics, always include the units in your calculations, and multiply and divide them as though they were algebraic quantities. This will automatically yield the correct units for the final result. If it does not, you have made some mistake in the calculation. Thus, it is always worthwhile to keep track of the units in calculations, because this provides some extra protection against costly mistakes. A failure of the expected cancellations is a sure sign of trouble!

If it is necessary to change the units of a quantity, either substitute the old units for equal amounts of new units, or else multiply the old units by whatever conversion factors will bring about the cancellation of the old units.

Always round off your final result to as many significant figures as specified in the given data. For instance, in Example 3, we rounded off the final result to four significant figures, since four significant figures were specified in the time interval measured by the device. Any additional significant figures in the final result would be unreliable and misleading. In fact, even the fourth significant figure in the answer [Eq. (1.7)] is not quite reliable—the calculation from a time specified to 1 nanosecond accuracy would give a distance accurate to only 0.3 m, as in Example 1. It is always wise to doubt the accuracy of the last significant figure, in the final result and (sometimes) also in the initial data.

EXAMPLE 5   We can obtain a rough estimate of the size of a molecule by means of the following simple experiment. Take a droplet of oil and let it spread out on a smooth surface of water. When the oil slick attains its maximum area, it consists of a monomolecular layer; that is, it consists of a single layer of oil molecules which stand on the water surface side by side. Given that an oil droplet of mass 8.4×10^{-7} kg and of density 920 kg/m^3 spreads out into an oil slick of maximum area 0.55 m^2, calculate the length of an oil molecule.

SOLUTION: The volume of the oil droplet is

$$[\text{volume}] = \frac{[\text{mass}]}{[\text{density}]}$$

$$= \frac{8.4 \times 10^{-7} \text{ kg}}{920 \text{ kg/m}^3} = 9.1 \times 10^{-10} \text{ m}^3 \qquad (1.11)$$

The volume of the oil slick must be exactly the same. This latter volume can be expressed in terms of the thickness and the area of the oil slick:

$$[\text{volume}] = [\text{thickness}] \times [\text{area}]$$

Consequently,

$$[\text{thickness}] = \frac{[\text{volume}]}{[\text{area}]}$$

$$= \frac{9.1 \times 10^{-10} \text{ m}^3}{0.55 \text{ m}^2} = 1.7 \times 10^{-9} \text{ m} \qquad (1.12)$$

Since we are told that the oil slick consists of a single layer of molecules standing side by side, the length of a molecule is the same as the calculated thickness, 1.7×10^{-9} m.

MATH HELP   TRIGONOMETRY OF THE RIGHT TRIANGLE

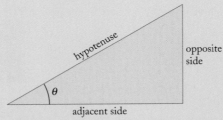

$$\cos \theta = \frac{[\text{adjacent side}]}{[\text{hypotenuse}]}$$

$$\tan \theta = \frac{[\text{opposite side}]}{[\text{adjacent side}]}$$

If θ is one of the acute angles of a right triangle, the side facing this angle is the *opposite side*, the side next to the angle is the *adjacent side*, and the side facing the right angle is the *hypotenuse*.

The Pythagorean theorem states:

$$[\text{hypotenuse}]^2 = [\text{opposite side}]^2 + [\text{adjacent side}]^2$$

This theorem implies that $1 = \sin^2 \theta + \cos^2 \theta$.

In principle, the numerical value of the sine, cosine, or tangent of any angle can be found by laying out a right triangle with this angle, measuring its sides, and evaluating the ratios given in the definitions. In practice, numerical values of tangents, cosines, and sines are obtained from handheld electronic calculators.

The figure shows a right triangle with an angle θ, its opposite and adjacent sides, and the hypotenuse. The sine, cosine, and tangent of θ are defined as follows:

$$\sin \theta = \frac{[\text{opposite side}]}{[\text{hypotenuse}]}$$

Appendix 3 gives a further review of trigonometry.

Checkup 1.6

QUESTION 1: If you use an ordinary ruler, marked in millimeters, to how many significant figures can you measure the length, width, and thickness of your textbook? Accordingly, to how many significant figures can you calculate the volume?

QUESTION 2: How many significant figures are there in the following numbers: 7.3, 1.24, 12.4, 4.85×10^6?

QUESTION 3: You multiply 7.3 and 1.24. How many significant figures are there in the result?

QUESTION 4: You add 73 and 1.2×10^2. What is the result? How many significant figures are there in the result?

QUESTION 5: What is the conversion factor for converting m to km? km to m? cm to km? m^2 to km^2? m^3 to km^3? m/s to mi/h?

QUESTION 6: What is the conversion factor for converting m/s to km/h?

    (A) $(1 \text{ km}/10^3 \text{ m}) \times (3600 \text{ s}/1 \text{ h})$ (B) $(10^3 \text{ km}/1 \text{ m}) \times (3600 \text{ s}/1 \text{ h})$

    (C) $(10^3 \text{ m}/1 \text{ km}) \times (3600 \text{ s}/1 \text{ h})$ (D) $(10^3 \text{ m}/1 \text{ km}) \times (1 \text{ h}/3600 \text{ s})$

    (E) $(10^3 \text{ m}/1 \text{ km}) \times (3600 \text{ h}/1 \text{ s})$

SUMMARY

PROBLEM-SOLVING TECHNIQUES Units and Significant Figures	**(page 18)**
MATH HELP Trigonometry of the Right Triangle	**(page 19)**

IDEAL PARTICLE   A pointlike mass, whose motion can be described completely by giving its position as a function of time

REFERENCE FRAME   A coordinate grid with a set of synchronized clocks

This reference frame moves with the ship.

 SI UNITS OF LENGTH, TIME, AND MASS   Meter, second, kilogram

STANDARDS OF LENGTH, TIME, AND MASS   Speed of light, cesium atomic clock, and standard cylinder of platinum–iridium

AVOGADRO'S NUMBER	$N_A = 6.022 \times 10^{23}$ atoms or molecules per mole	**(1.2)**
ATOMIC MASS UNIT	$1 \text{ u} = 1.66 \times 10^{-27} \text{ kg}$	**(1.5)**

MOLE   That amount of matter containing as many atoms (or molecules) as in exactly 12 g of carbon-12

DERIVED UNIT   A unit constructed from some combination of the base units of length, time, and mass

SIGNIFICANT FIGURES   The digits in a number that are known with certainty (the last such digit is often not entirely reliable). When multiplying or dividing two or more quantities, the result has the same number of significant figures as the least number in the original quantities. When adding or subtracting two or more quantities, the number of significant figures in the result is determined by the largest decimal place among the last digits in the original quantities.

CONSISTENCY OF UNITS   In any equation, the dimensions (the powers of length, time, and mass) on each side of the equation must be the same.

CONVERSION FACTORS   Ratios that are identically equal to 1, used as factors to change the units of a quantity

QUESTIONS FOR DISCUSSION

1. Try to estimate by eye the lengths, in centimeters or meters, of a few objects in your immediate environment. Then measure them with a ruler or meterstick. How good were your estimates?

2. How close is your watch to standard time right now? Roughly how many minutes does your watch gain or lose per month?

3. What is meant by the phrase *a point in time*?

4. Mechanical clocks (with pendulums) were not invented until the tenth century A.D. What clocks were used by the ancient Greeks and Romans?

5. By counting aloud "One Mississippi, two Mississippi, three Mississippi," etc., at a fairly fast rate, you can measure seconds reasonably accurately. Try to measure 30 seconds in this way. How good a timekeeper are you?

6. Pendulum clocks are affected by the temperature and pressure of air. Why?

7. In 1761 an accurate chronometer built by John Harrison (see Fig. 1.15) was tested aboard HMS *Deptford* during a voyage at sea for 5 months. During this voyage, the chronometer accumulated an error of less than 2 minutes. For this achievement, Harrison was ultimately awarded a prize of £20000 that the British government had offered for the discovery of an accurate method for the determination of geographical longitude at sea. Explain how the navigator of a ship uses a chronometer

and observation of the position of the Sun in the sky to find longitude.

FIGURE 1.15 Harrison's chronometer H.4.

8. Captain Lecky's *Wrinkles in Practical Navigation*, a famous nineteenth-century textbook of celestial navigation, recommends that each ship carry three chronometers for accurate timekeeping. What can the navigator do with three chronometers that cannot be done with two?

9. Suppose that by an "act of God" (or by the act of a thief) the standard kilogram at Sèvres were destroyed. Would this destroy the metric system?

10. Estimate the masses, in grams or kilograms, of a few bodies in your environment. Check the masses with a balance if you have one available.

11. Consider the piece of paper on which this sentence is printed. If you had available suitable instruments, what physical quantities could you measure about this piece of paper? Make the

longest list you can and give the units. Are all these units derived from the meter, second, and kilogram?

12. Could we take length, time, and density as the three fundamental units? What could we use as a standard of density?

13. Could we take length, mass, and density as the three fundamental units? Length, mass, and speed?

PROBLEMS

1.2 The Unit of Length

1. What is your height in feet? In meters?

2. With a ruler, measure the thickness of this book, excluding the cover. Deduce the thickness of each of the sheets of paper making up the book.

3. A football field measures 100 yd $\times$ 53$\frac{1}{3}$ yd. Express each of these lengths in meters.

4. If each step you take is 0.60 m, how many steps do you need to cover 1.0 km?

5. The pica is a unit of length used by printers and book designers; 1 pica = $\frac{1}{6}$ in., which is the standard distance between one line of typing produced by a typewriter and the next (single-spaced). How many picas long and wide is a standard sheet of paper, 11 in. long and 8$\frac{1}{2}$ in. wide?

6. Express the last four entries in Table 1.1 in inches.

7. Express the following fractions of an inch in millimeters: $\frac{1}{2}$, $\frac{1}{4}$, $\frac{1}{8}$, $\frac{1}{16}$, $\frac{1}{32}$, and $\frac{1}{64}$ in.

8. Express one mil (one thousandth of an inch) in micrometers (microns). Express one millimeter in mils.

9. Analogies can often help us to imagine the very large or very small distances that occur in astronomy or in atomic physics.

    (a) If the Sun were the size of a grapefruit, how large would the Earth be? How far away would the nearest star be?

    (b) If your head were the size of the Earth, how large would an atom be? How large would a red blood cell be?

10. One of the most distant objects observed by astronomers is the quasar Q1208+1011, at a distance of 12.4 billion light-years from the Earth. If you wanted to plot the position of this quasar on the same scale as the diagram at the top of page xliii of the Prelude, how far from the center of the diagram would you have to place this quasar?

11. On the scale of the second diagram on page xl of the Prelude, what would have to be the size of the central dot if it were to represent the size of the Sun faithfully?

12. An *interferometer* uses the pattern created by mixing laser light waves in order to measure distances extremely accurately. The

wavelength of the laser light waves used is 633 nanometers. A fiber-optic interferometer can measure a distance 10^{-6} times the size of a wavelength. How does such precision compare with the diameter of an atom?

*13. The *thread* of a screw is often described either in terms of the number of complete turns required for the screw to advance one inch (English units) or in terms of the number of millimeters advanced in one complete turn (metric units). For delicate adjustments, scientists often use screws either with a thread of 80 turns per inch or a thread of 0.5 mm per turn. Express each of these in terms of the number of micrometers the screw advances for a partial turn through an angle of 5°.

*14. A nautical mile (nmi) equals 1.151 mi, or 1852 m. Show that the distance of 1 nmi along a meridian of the Earth corresponds to a change in latitude of 1 minute of arc.

**15. A physicist plants a vertical pole at the waterline on the shore of a calm lake. When she stands next to the pole, its top is at eye level, 175 cm above the waterline. She then rows across the lake and walks along the waterline on the opposite shore until she is so far away from the pole that the entire view of it is blocked by the curvature of the surface of the lake; that is, the entire pole is below the horizon (Fig. 1.16). She finds that this happens when her distance from the pole is 9.4 km. From this information, deduce the radius of the Earth.

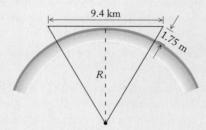

FIGURE 1.16 The distance between the physicist and the pole is 9.4 km.

1.3 The Unit of Time

16. What is your age in days? In seconds?

17. The age of the Earth is 4.5×10^9 years. Express this in seconds.

18. A computer can perform a single calculational step each nanosecond (10^{-9} s). How many steps can be performed in one hour?

19. An Olympic marathon record of 2 h 9 min 21 s was set by Carlos Lopes of Portugal in 1984. Express this time in seconds.

20. Joan Benoit of the United States set the women's Olympic marathon record in 1984 with a time of 2 h 24 min 52 s. Express this time in seconds.

21. The solar day is the interval for the Earth to complete one rotation in relation to the Sun, and the sidereal day is the interval for the Earth to complete one rotation in relation to distant stars. The solar day has exactly 24 hours. How many hours and minutes are there in one sidereal day? (Hint: 1 year is 365.24 solar days, but 366.24 sidereal days. Why?)

22. A mechanical wristwatch ticks 4 times per second. Suppose this watch runs for 10 years. How often does it tick in this time interval?

23. How many days is a million seconds?

24. How many hours are there in a week? How many seconds?

25. Your heart beats 71 times per minute. How often does it beat in a year?

*26. Each day at noon a mechanical wristwatch was compared with WWV time signals. The watch was not reset. It consistently ran late, as follows: June 24, late 4 s; June 25, late 20 s; June 26, late 34 s; June 27, late 51 s.

    (a) For each of the three 24-hour intervals, calculate the rate at which the wristwatch lost time. Express your answer in seconds lost per hour.

    (b) What is the average of the rates of loss found in part (a)?

    (c) When the wristwatch shows 10^h30^m on June 30, what is the correct WWV time? Do this calculation with the average rate of loss of part (b) and also with the largest rates of loss found in part (a). Estimate to within how many seconds the wristwatch can be trusted on June 30 after the correction for rate of loss has been made.

*27. The navigator of a sailing ship seeks to determine his longitude by observing at what time (Coordinated Universal Time) the Sun reaches the zenith at his position (local noon). Suppose that the navigator's chronometer is in error and is late by 1.0 s compared with Coordinated Universal Time. What will be the consequent error of longitude (in minutes of arc)? What will be the error in position (in kilometers) if the ship is on the equator?

1.4 The Unit of Mass

28. What is your mass in pounds? In kilograms? In atomic mass units?

29. What percentage of the mass of the Solar System is in the planets? What percentage is in the Sun? Use the data given in the table printed inside the cover of this book.

30. What is the ratio of the largest to the smallest length listed in Table 1.1? The ratio of the longest to the shortest time in Table 1.5? The ratio of the largest to the smallest mass in Table 1.7? Do you see any coincidences (or near-coincidences) between these numbers? Some physicists have proposed that coincidences between these large numbers must be explained by new cosmological theories.

31. The atom of uranium consists of 92 electrons, each of mass 9.1×10^{-31} kg, and a nucleus. What percentage of the total mass is in the electrons and what percentage is in the nucleus of the atom?

32. A laboratory microbalance can measure a mass of one-tenth of a microgram, a very small speck of matter. How many atoms are there in such a speck of gold, which has 197 grams in one mole?

33. English units use the ordinary pound, also called the *avoirdupois* pound, to specify the mass of most types of things. However, the *troy* pound is often used to measure precious stones, precious metals, and drugs, where 1 troy pound = 0.822 86 avoirdupois pound. If we adopt these different pounds, how many grams are there in a troy pound of gold? In an avoirdupois pound of feathers?

34. Mechanical nano-oscillators can detect a mass change as small as 10^{-21} kg. How many atoms of iron (55.85 g/mole) must be deposited on such an oscillator to produce a measurable mass change?

35. (a) How many molecules of water are there in one cup of water? A cup is about 250 cm^3.

    (b) How many molecules of water are there in the ocean? The total volume of the ocean is 1.3×10^{18} m^3.

    (c) Suppose you pour a cup of water into the ocean, allow it to become thoroughly mixed, and then take a cup of water out of the ocean. On the average, how many molecules originally in the cup will again be in the cup?

*36. How many atoms are there in the Sun? The mass of the Sun is 1.99×10^{30} kg, and its chemical composition (by mass) is approximately 70% hydrogen and 30% helium.

*37. The chemical composition of air is (by mass) 75.5% N_2, 23.2% O_2, and 1.3% Ar. What is the average "molecular mass" of air; that is, what is the mass of 6.02×10^{23} molecules of air?

*38. How many atoms are there in a human body of 73 kg? The chemical composition (by mass) of a human body is 65% oxygen, 18.5% carbon, 9.5% hydrogen, 3.3% nitrogen, 1.5% calcium, 1% phosphorus, and 0.35% other elements (ignore the "other elements" in your calculation).

1.5 Derived Units[†]
1.6 Significant Figures; Conversion of Units[†]

39. As seen from Earth, the Sun has an angular diameter of 0.53°. The distance between the Earth and Sun is 1.5×10^{11} m. From this, calculate the radius of the Sun.

40. The light-year is the distance that light travels in one year. Express the light-year in meters.

41. The distance from our Galaxy to the Andromeda galaxy is 2.2×10^6 light-years. Express this distance in meters.

42. In analogy with the light-year, we can define the light-second as the distance light travels in one second and the light-minute as the distance light travels in one minute. Express the Earth–Sun distance in light-minutes. Express the Earth–Moon distance in light-seconds.

43. Astronomers often use the astronomical unit (AU), the parsec (pc), and the light-year. The AU is the distance from the Earth to the Sun;[4] 1 AU = 1.496×10^{11} m. The pc is the distance at which 1 AU subtends an angle of exactly one second of arc (Fig. 1.17). The light-year is the distance that light travels in one year.

    (a) Express the pc in AU.

    (b) Express the pc in light-years.

    (c) Express the pc in meters.

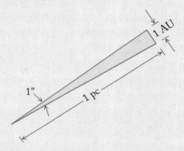

FIGURE 1.17 Geometry relating the astronomical unit (AU) to the parsec (pc).

44. How many square feet are there in a square meter?

45. How many cubic feet are there in a cubic meter?

46. A tennis court measures 78 ft × 27 ft. Calculate the area of this court. Express your result in square meters.

47. The tallest man was Robert Wadlow, who continued to grow throughout his life and attained 8 ft 11.1 in. a few days before his death in 1940. Express his height in meters. How many significant figures are there in your result?

48. A football field measures 100 yd × $53\frac{1}{3}$ yd. Calculate the area of this field; express your result in square meters.

49. The density of copper is 8.9 g/cm^3. Express this in kg/m^3, lb/ft^3, and lb/in.^{3}.

50. What is the volume of an average human body? (Hint: The density of the body is about the same as that of water.)

51. Your heart pumps 92 cm^3 of blood per second (when you are resting). How much blood does it pump per day? Express the answer in m^3.

52. As stated in the preceding problem, your heart pumps 97 cubic centimeters of blood per second. If your total volume of blood is 5.2 liters, what is the average travel time for your blood to complete one trip around your circulatory system?

53. Computers typically use many millions of transistors; each transistor occupies an area of approximately 10^{-6} m × 10^{-6} m = 10^{-12} m^2. How many such transistors can fit on a 1 cm^2 silicon chip? Future-generation computers may exploit a three-dimensional arrangement of transistors. If each layer of transistors is 10^{-7} m thick, how many transistors could fit in a 1-cm^3 silicon cube?

54. Water has a density of 1.00 g/cm^3. Express this in pounds per gallon.

55. Express the results of the following calculations in scientific notation with an appropriate number of significant figures:

    (a) $3.6 \times 10^4 \times 2.049 \times 10^{-2}$.

    (b) $2.581 \times 10^2 - 7.264 \times 10^1$.

    (c) $0.079832 \div 9.43$.

56. Our Sun has a radius of 7.0×10^8 m and a mass of 2.0×10^{30} kg. What is its average density? Express your answer in grams per cubic centimeter.

57. Pulsars, or neutron stars, typically have a radius of 20 km and a mass equal to that of the Sun (2.0×10^{30} kg). What is the average density of such a pulsar? Express your answer in metric tons per cubic centimeter.

58. The total volume of the oceans of the Earth is 1.3×10^{18} m^3. What percentage of the mass of the Earth is in the oceans?

59. A fire hose delivers 300 liters of water per minute. Express this in m^3/s. How many kilograms of water per second does this amount to?

60. Meteorologists usually report the amount of rain in terms of the depth in inches to which the water would accumulate on a flat surface if it did not run off. Suppose that 1 in. of rain falls during a storm. Express this in cubic meters of water per square meter of surface. How many kilograms of water per square meter of surface does this amount to?

61. The nuclei of all atoms have approximately the same density of mass. The nucleus of a copper atom has a mass of 1.06×10^{-25} kg and a radius of 4.8×10^{-15} m. The nucleus of a lead atom has a mass of 3.5×10^{-25} kg; what is its radius? The nucleus of an oxygen atom has a mass of 2.7×10^{-26} kg; what is its radius? Assume that the nuclei are spherical.

62. The table printed inside the book cover gives the masses and radii of the major planets. Calculate the average density of each planet and make a list of the planets in order of decreasing densities. Is there a correlation between the density of a planet and its distance from the Sun?

---

[4] Strictly, it is the semimajor axis of the Earth's orbit.

[†] For help, see Online Concept Tutorial 1 and 2 at www.wwnorton.com/physics

63. The roof of a house has a slope, or pitch, of 1:1 (that is, 45°). The roof has a complex shape, with several gables and dormers (see Fig. 1.18), but all roof surfaces have the same pitch. The area of the ground floor is 250 m². What is the area of the roof surface?

FIGURE 1.18 Roof of a house.

64. The Global Positioning System (GPS) used by navigators of ships and aircraft exploits radio signals from artificial satellites to determine the position of the ship or the aircraft. Portable GPS units for use on yachts (see Fig. 1.19) incorporate a radio receiver and a computer; they give the position to within ±15 m. What error in latitude angle corresponds to a north–south error of 15 m along the surface of the Earth?

FIGURE 1.19 Global Positioning System (GPS) receiver.

65. Some crystals can be polished at an angle to produce sharp "atomic steps" (see Fig. 1.20.). What angle θ should be chosen to produce one step every five atoms? One step every ten atoms? How many atoms will there be along a step if a crystal is polished as flat as is normally practical, usually around 0.10°?

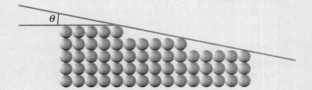

FIGURE 1.20 Atomic steps of a crystal.

66. You want to estimate the height of a skyscraper from the ground. To do so, you walk 150 steps (approximately 75 m) away from a vertical wall and, using a protractor, measure an angle of 78° that the line of sight to the top of the skyscraper makes with the horizontal. How tall is the skyscraper? How many significant figures are in your result?

67. In 1 astronomical year, or "tropical year," of 365.24 days, the Earth moves once around the Sun; that is, it moves 360° along its orbit and returns to the same point of the orbit. How far around the Sun, in degrees, does the Earth move in 1 calendar year of 365 days? How far, in degrees, does the Earth move in 4 consecutive calendar years (one of which is a leap year of 366 days)?

*68. In the Galápagos (on the equator) the small island of Marchena is 60 km west of the small island of Genovesa. If the sun sets at 8:00 P.M. at Genovesa, when will it set at Marchena?

*69. For tall trees, the diameter at the base (or the diameter at any given point of the trunk, such as the midpoint) is roughly proportional to the $\frac{3}{2}$ power of the length. The tallest sequoia in Sequoia National Park in California has a length of 81 m, a diameter of 7.6 m at the base, and a mass of 6100 metric tons. A petrified sequoia found in Nevada has a length of 90 m. Estimate its diameter at the base, and estimate the mass it had when it was still alive.

REVIEW PROBLEMS

70. The Earth is approximately a sphere of radius 6.37×10^6 m. Calculate the distance from the pole to the equator, measured along the surface of the Earth. Calculate the distance from the pole to the equator, measured along a straight line passing through the Earth.

71. The "atomic mass" of fissionable uranium is 235.0 g. What is the mass of a single uranium atom? Express your answer in kilograms and in atomic mass units.

72. How many water molecules are there in 1.0 liter of water? How many oxygen atoms? Hydrogen atoms?

73. How many molecules are there in one cubic centimeter of air? Assume that the density of air is 1.3 kg/m³ and that it consists entirely of nitrogen molecules (N_2). The atomic mass of nitrogen is 14.0 g.

74. Normal human blood contains 5.1×10^6 red blood cells per cubic millimeter. The total volume of blood in a man of 70 kg is 5.2 liters. How may red blood cells does this man have?

75. An epoxy paint used for painting the hull of a ship is supposed to be applied at the rate of 1 liter of paint per 8 square meters of hull surface. How thick will the film of freshly applied paint be?

76. Smokestacks in the United States spew out about 8×10^6 metric tons of fly ash per year. If this stuff settles uniformly over all of the area of the United States (9.4×10^6 km^2), how many kilograms of fly ash will be deposited per square meter per year?

77. For many years, the federal highway speed limit was 55 mi/h. Express this in kilometers per hour, feet per second, and meters per second.

78. The nucleus of an iron atom is spherical and has a radius of 4.6×10^{-15} m; the mass of the nucleus is 9.5×10^{-26} kg. What is the density of the nuclear material? Express your answer in metric tons per cubic centimeter.

79. The longest officially verified human life span was attained by the Japanese man Shigechiyo Izumi, who died in 1986 at an age of 120 years and 237 days. Express this age in seconds. How many significant figures are there in your result?

80. A driveway up a hill has a slope of 1:9. How high does it ascend in 300 m of horizontal distance? What is the corresponding distance measured along the driveway?

81. A small single-engine plane is flying at a height of 5000 m at a (horizontal) distance of 18 km from the San Francisco airport when the engine quits. The pilot knows that, without the engine, the plane will glide downward at an angle of 15°. Can she reach San Francisco?

*82. You are crossing the Atlantic in a sailboat and hoping to make landfall in the Azores. The highest peak in the Azores has a height of 2300 m. From what distance can you see this peak just emerging over the horizon? Assume that your eye is (almost) at the level of the water.

*83. Some engineers have proposed that for long-distance travel between cities we should dig perfectly straight connecting

tunnels through the Earth (see Fig. 1.21). A train running along such a tunnel would initially pick up speed in the first half of the tunnel as if running downhill; it would reach maximum speed at the midpoint of the tunnel; and it would gradually slow down in the second half of the tunnel, as if running uphill. Suppose that such a tunnel were dug between San Francisco and Washington, D.C. The distance between these cities, measured along the Earth's surface, is 3900 km.

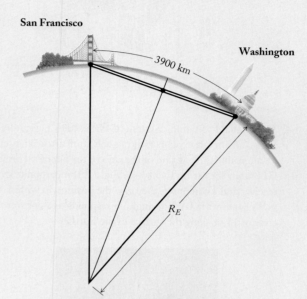

FIGURE 1.21 A proposed tunnel through the Earth.

(a) What is the distance along the straight tunnel?

(b) What is the depth of the tunnel at its midpoint, somewhere below Kansas?

(c) What is the downward slope of the tunnel relative to the horizontal direction at San Francisco?

Answers to Checkups

Checkup 1.1

1. Extended bodies have many characteristics that can be measured; for example, size and shape may be specified by measuring the bowling ball's diameter, surface area, or volume, with units of length, length-squared, and length-cubed, respectively. Other measurable quantities include density, hardness, temperature, color, and chemical composition; we will examine the units of such quantities in later chapters.

2. Since the origin of the $x'-y'$ coordinate grid is shifted from the $x-y$ origin by negative amount in the x direction, and a positive amount in the y direction any point will have larger values of x' compared to x and smaller values of y' compared to y, as in Fig. 1.3a.

3. A coordinate grid is used to specify positions in space. A reference frame includes clocks which specify the time when something occurs at a certain position.

Checkup 1.2

1. There are 100 centimeters in a meter, and there are 1000 meters in a kilometer, so there are $100 \times 1000 = 10^5$ centimeters in a kilometer. Similarly, with 10^3 millimeters in a meter, there are $10^3 \times 10^3 = 10^6$ millimeters in a kilometer.

2. Since there are 10^6 microns in a meter and there are 10^{-15} m in a fermi; there are $10^{-15} \times 10^6$ microns $= 10^{-9}$ microns in a fermi.

3. (C) 10^{-4}. There are 10^6 microns in a meter, and 10^{-10} m in an angstrom, so there are $10^{-10} \times 10^6$ microns $= 10^{-4}$ micron in an angstrom.

Checkup 1.3

1. There are $60 \times 60 = 3600$ seconds in an hour, and there are 1000 milliseconds in a second, so there are $1000 \times 3600 = 3.6 \times 10^6$ milliseconds in an hour. There are 10^{12} picoseconds in a second and 10^6 microseconds in a second, so there are 10^6 picoseconds in a microsecond.

2. (B) 6.0×10^{16}. There are 10^{15} femtoseconds in a second, and 60 seconds in a minute, so there are $60 \times 10^{15} = 6.0 \times 10^{16}$ femtoseconds in a minute.

Checkup 1.4

1. From Table 1.8, there are 10^3 kg in a metric ton. Since there are 10^3 grams in a kilogram, there are $10^3 \times 10^3 = 10^6$ grams in a metric ton. Furthermore, there are 10^9 milligrams in a metric ton, and therefore 10^{-9} ton in one milligram.

2. (D) 6.02×10^{26}. There are 1.66×10^{-27} kilograms in one u, so the number of u per kilogram is the inverse, 1 u/$(1.66 \times 10^{-27}$ kg$) = 6.02 \times 10^{26}$ u/kg.

Checkup 1.5

1. Since 1 m $= 10^2$ cm, then one square meter is $(1\ \text{m})^2 = (10^2\ \text{cm})^2 = 10^4\ \text{cm}^2$. Similarly, a cubic meter is $(1\ \text{m})^3 = (10^2\ \text{cm})^3 = 10^6\ \text{cm}^3$. These values can also be obtained directly from Tables 1.9 and 1.10.

2. Using Table 1.9, 1 km$^2 = 10^6$ m$^2 \times (1\ \text{mm}^2)/(10^{-6}\ \text{m}^2) = 10^{12}$ mm^2.

3. (D) 10^{18}. From Table 1.10, 1 km$^3 = 10^9$ m$^3 \times (1\ \text{mm}^3)/(10^{-9}\ \text{m}^3) = 10^{18}$ mm^3.

Checkup 1.6

1. Since the length and width are around 20 to 30 cm, a measurement of either to the nearest millimeter (1 mm = 0.1 cm) will provide three significant figures. The thickness is only a few centimeters, so its measurement to the nearest 0.1 cm will have two significant figures. The volume is the product of these three lengths, and will have only as many significant figures as the least number in the multiplied quantities; the volume thus has two significant figures.

2. One merely reads off the number of digits; the first number given has two significant figures, and the others have three.

3. The product has only as many significant figures as the least number of the multiplied quantities; here, the product has two significant figures.

4. When we add, only the largest decimal place among the last significant figures of the quantities added is significant in the result. Here, we must round to the nearest 10, since only the 10s digit was known in the second number given. Thus we write 1.9×10^2 for the sum.

5. For the conversion factors, we merely write the ratio of equal quantities that provides the desired units. For the given conversions, the factors are, respectively: 1 km/10^3 m; 10^3 m/1 km; 1 km/10^5 cm; 1 km^2/$(10^3$ m$)^2$; 1 km^3/$(10^3$ m$)^3$; (1 km/10^3 m) $\times$ (3600 s/1 h); (1 mi/1609 m) $\times$ (3600 s/1 h).

6. (A) (1 km/10^3 m) $\times$ (3600 s/1 h). When multiplied by m/s, this both provides the desired units of km/h and contains correct conversion factors.

Motion along a Straight Line

2.1 Average Speed

2.2 Average Velocity for Motion along a Straight Line

2.3 Instantaneous Velocity

2.4 Acceleration

2.5 Motion with Constant Acceleration

2.6 The Acceleration of Free Fall

2.7 Integration of the Equations of Motion

CONCEPTS IN CONTEXT

Aircraft are launched from the deck of an aircraft carrier by a catapult. In conjunction with the jet engines of the aircraft, the catapult quickly accelerates the aircraft to the speed required for takeoff. Typically, during such a "cat shot" the speed of the aircraft increases from zero to 260 km/h (160 mi/h) in just under two seconds. Some pilots find such extreme accelerations exhilarating.

The concepts of velocity and acceleration introduced in this chapter will permit us to answer the following questions:

Concepts
— in —
Context

? What is the magnitude of the acceleration during the launch? (Example 3, page 40)

? How is the distance traveled by the aircraft related to the acceleration and the time? (Example 5, page 44)

? How does the acceleration during launch compare with the acceleration of free fall? (Example 11, page 53)

The branch of physics that studies the motion of bodies is called **mechanics**. In antiquity, the science of mechanics (from the Greek *mechane*, machine) was the study of machines, and this is still what we have in mind when we call an automobile repairperson a mechanic. But physicists soon recognized that the essential aspect of the study of machines is motion, and mechanics thus became the study of motion. Broadly speaking, mechanics is divided into kinematics and dynamics. **Kinematics** deals with the mathematical description of motion in terms of position, velocity, and acceleration; **dynamics** deals with force as the cause of changes in motion.

In this chapter, and in the next nine chapters, we will be concerned only with the **translational motion** of a particle, that is, a change of position of the particle with time. In the case of an ideal particle—a pointlike mass of infinitesimal size—the dependence of position on time provides a complete description of the motion of the particle. In the case of a more complicated body—an automobile, a ship, or a planet— the dependence of position on time does not provide a complete description. Such a complicated body can rotate, or change its orientation in space; furthermore, the body has many internal parts which can move in relation to one another. We will examine such nontranslational motions later in the book. But insofar as we are not interested in the size, shape, orientation, and internal structure of a complicated body, we may find it useful to concentrate on its translational motion and ignore all rotational or internal motions. Under these circumstances, we may pretend that the motion of the complicated body is particle motion.

2.1 AVERAGE SPEED

If your automobile travels 160 km along a highway in 2 h, you would say that your average speed for the trip is 80 kilometers per hour. But this statement relies on the implicit assumption that the motion of the automobile can be regarded as particle motion, and that we ignore the size of the automobile and the complicated behavior of its internal machinery. To make the definition and the measurement of the average speed of the automobile precise and unambiguous, we must select some marker point that we imagine painted on the automobile. For instance, we might select the mid-point of the front bumper as our marker point. We reckon departure and arrival accord-ing to when the marker point crosses the starting line and the finishing line, and we reckon distance according to the path traced out by the marker point (see Fig. 2.1). We can then treat the motion of the automobile as particle motion.

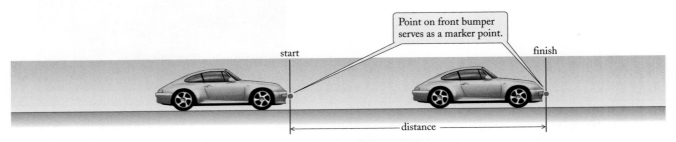

FIGURE 2.1 For a precise measurement, the departure and the arrival of the automobile must be reckoned according to when the front end of the bumper crosses the starting line and the finish line.

For a particle that travels along some path, straight or curved, the total distance traveled and the time taken to travel this distance have a precise, unambiguous meaning. The **average speed** *of the particle is defined as the ratio of this distance and the time taken*. Expressing this as a formula, we can write

average speed

$$[\text{average speed}] = \frac{[\text{total distance traveled}]}{[\text{time taken}]} \tag{2.1}$$

Thus, average speed is the rate of change of the distance, or the change of distance per unit time. We see from Eq. (2.1) that the unit of speed is the unit of length divided by the unit of time. In the SI system, *the unit of speed is the* **meter per second** (m/s). In practice, the speed of automobiles, aircraft, and other everyday objects is often measured in kilometers per hour (km/h):

$$1 \text{ km/h} = \frac{1000 \text{ m}}{3600 \text{ s}} = 0.278 \text{ m/s} \tag{2.2}$$

Table 2.1 gives some examples of typical speeds.

TABLE 2.1	SOME SPEEDS	
Light	3.0×10^8 m/s	
Recession of fastest known quasar	2.8×10^8 m/s	
Electron around nucleus (hydrogen)	2.2×10^6 m/s	
Earth around Sun	3.0×10^4 m/s	
Rifle bullet (muzzle velocity) (a)	$\approx 7 \times 10^2$ m/s	(a)
Random motion of molecules in air (average)	4.5×10^2 m/s	
Sound	3.3×10^2 m/s	
Jet airliner (Boeing 747, maximum airspeed)	2.7×10^2 m/s	
Cheetah (maximum) (b)	28 m/s	
Typical highway speed limit (55 mi/h)	25 m/s	
Human (maximum)	12 m/s	
Human (walking briskly)	1.3 m/s	(b)
Snail	$\approx 10^{-3}$ m/s	
Glacier (c)	$\approx 10^{-6}$ m/s	
Rate of growth of hair (human)	3×10^{-9} m/s	
Continental drift	$\approx 10^{-9}$ m/s	(c)

EXAMPLE 1   The world record set by Asafa Powell in 2005 for the 100-m run was 9.77 s (see Fig. 2.2). What average speed did he attain while setting this record?

SOLUTION: According to Eq. (2.1),

$$[\text{average speed}] = \frac{[\text{total distance traveled}]}{[\text{time taken}]} = \frac{100 \text{ m}}{9.77 \text{ s}} = 10.2 \text{ m/s} \qquad (2.3)$$

Here the final result has been rounded to three significant figures, since the data for the problem have three significant figures.

Motion and speed are relative; the value of the speed depends on the frame of reference with respect to which it is calculated. Example 1 gives the speed of the runner relative to a reference frame attached to the surface of the Earth. However, relative to the reference frame of a bicyclist riding in the same direction as the runner, the speed will be different (Fig. 2.3). For instance, if the ground speed of the bicyclist is also 10.2 m/s, then relative to her reference frame, the runner will have zero speed. Thus, questions regarding speed are meaningless unless the frame of reference is first specified. In everyday language, "speed" often means speed relative the the Earth's surface. If the speed is reckoned relative to some other body, this will usually be clear from the context. For example, in Table 2.1 the speed of the jet airliner is reckoned relative to the air (which may be in motion relative to the Earth). We will be careful to specify the frame of reference whenever it is not clear from the context.

FIGURE 2.2 Asafa Powell in 2005.

FIGURE 2.3 A bicyclist and her reference frame. If both the bicyclist and the runner are moving toward the right at the same speed, then the runner is at rest relative to the reference frame of the bicyclist.

✔ Checkup 2.1

QUESTION 1: A man takes 100 s to walk 50 m along a straight road. What is his average speed? A woman jogs the same distance in 50 s, and then stands at the endpoint for 50 s. What is her average speed over the 100-s interval?

QUESTION 2: Consider a runner running at 10.1 m/s relative to the surface of the Earth and consider a spectator standing at rest on the surface of the Earth. A bicyclist is riding in the same direction as the runner, but at a different speed. Is it possible that in the reference frame of the bicyclist the spectator has a larger speed than the runner? That the spectator and the runner have equal speeds?

QUESTION 3: A car starts at rest ($v = 0$) and increases its speed to 30 m/s in 5.0 s; during this time, the car travels a distance of 60 m. The average speed of the car during this time interval is:

(A) 150 m/s   (B) 60 m/s   (C) 45 m/s   (D) 15 m/s   (E) 12 m/s

Online
Concept
Tutorial

2.2 AVERAGE VELOCITY FOR MOTION ALONG A STRAIGHT LINE

For the rest of this chapter, we will consider the special case of motion along a straight line, that is, motion in one dimension. For convenience, we will assume that the straight line coincides with the *x* axis (Fig. 2.4). We can then give a complete description of the

For motion along a straight line…

…position of automobile can be completely specified by its *x* coordinate.

FIGURE 2.4 An automobile moving along a straight line. The *x* axis coincides with this straight line.

motion of the particle by specifying the *x* coordinate at each instant of time. In mathematical language, this means that *the x coordinate is a function of time*. Graphically, we can represent the motion by means of a plot of the *x* coordinate vs. the time coordinate. For example, Fig. 2.5 is such a plot of the position coordinate *x* vs. the time coordinate *t* for an automobile that starts from rest, accelerates along a straight road for 10 seconds, and then brakes and comes to a full stop 4.3 seconds later (the plot is based on data from an acceleration test of a Maserati sports car). The position is measured from the starting point on the road to a marker point marked on the automobile. The plot shown in this figure gives us a complete description of the (translational) motion of the automobile.

Suppose that at time t_1 the automobile is at position x_1, and at a subsequent time t_2 the automobile is at position x_2 (see Fig. 2.6). Then $x_2 - x_1$ is the change of position that occurs in the time interval $t_2 - t_1$. The **average velocity** *is defined as the ratio of this change of position and the time interval*:

average velocity

$$\overline{v} = \frac{x_2 - x_1}{t_2 - t_1} \tag{2.4}$$

Equation (2.4) can also be written as

$$\overline{v} = \frac{\Delta x}{\Delta t} \tag{2.5}$$

with $\Delta x = x_2 - x_1$ and $\Delta t = t_2 - t_1$ (here, the overbar on the symbol for velocity is a standard notation used in science to indicate an *average* quantity; and Δ, the Greek capital letter *delta*, is a standard notation used to indicate a *change* in a quantity). Thus, the average velocity is the average rate of change of the position, or the average change of position per unit time.

Graphically, *in the plot of position vs. time, the average velocity is the ratio of the vertical separation between the points P_1 and P_2 and the horizontal separation*. If we draw a straight line connecting the points P_1 and P_2 in the plot, the ratio of the vertical separation and the horizontal separation between any two points on this line is the **slope** of the line. Note that this mathematical definition of slope agrees with the everyday notion of slope: a steep line, with a large vertical separation between the points P_1 and P_2, has a large slope; and a nearly horizontal line, with a small vertical separation between the points P_1 and P_2, has a small slope. With this mathematical definition of slope, we can say that the average velocity equals the slope of the straight line connecting the points P_1 and P_2 (see Fig. 2.6). For instance, if $t_1 = 8.0$ s and $t_2 = 14.3$ s,

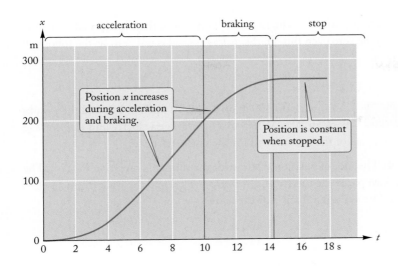

FIGURE 2.5 Plot of position vs. time for an automobile that accelerates (0 s < *t* < 10 s), then brakes (10 s < *t* < 14.3 s), and then stops (*t* > 14.3 s) (based on data from a road test of a Maserati Bora by *Road & Track* magazine).

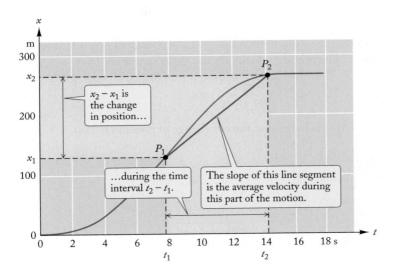

FIGURE 2.6 The average velocity for the interval from $t_1 = 8.0$ s to $t_2 = 14.3$ s is the slope of the straight line P_1P_2.

and $x_1 = 130$ m and $x_2 = 272$ m, then the average velocity or, alternatively, the slope of the straight line connecting the points P_1 and P_2 in Fig. 2.6 is

$$\overline{v} = \frac{x_2 - x_1}{t_2 - t_1} = \frac{272 \text{ m} - 130 \text{ m}}{14.3 \text{ s} - 8.0 \text{ s}} = \frac{142 \text{ m}}{6.3 \text{ s}} = 23 \text{ m/s} \qquad (2.6)$$

A positive or negative slope of the position vs. time plot corresponds, respectively, to a positive or negative sign of the velocity. According to the general formula (2.4), the velocity is positive or negative depending on whether x_2 is larger or smaller than x_1, that is, depending on whether the x coordinate increases or decreases in the time interval $t_2 - t_1$. This means that the sign of the velocity depends on the direction of motion. If the motion is in the positive x direction—as in the example plotted in Fig. 2.5—the velocity is positive; if the motion is in the negative x direction, the velocity is negative. Thus, according to the precise terminology used in physics, speed [defined by Eq. (2.1)] and velocity [defined by Eq. (2.4)] are not the same thing, because *speed is always positive, whereas velocity is positive or negative depending on the direction of motion.* Furthermore, if the motion has one portion in the positive x

direction and another portion in the negative x direction, then it is possible for the average velocity to be zero even though the average speed is not zero, as the following example illustrates.

EXAMPLE 2   A runner runs 100 m on a straight track in 11 s and then walks back in 80 s. What are the average velocity and the average speed for each part of the motion and for the entire motion?

SOLUTION: The plot of position vs. time for the runner is shown in Fig. 2.7. The motion has two parts: the run (from $t = 0$ to $t = 11$ s) and the walk (from $t = 11$ s to $t = 91$ s). The average velocity for the run is

$$\overline{v} = \frac{\Delta x}{\Delta t} = \frac{+100 \text{ m}}{11 \text{ s}} = +9.1 \text{ m/s}$$

The average velocity for the walk is

$$\overline{v} = \frac{-100 \text{ m}}{80 \text{ s}} = -1.3 \text{ m/s}$$

(Here the minus sign in -100 m indicates that the change of position is in the negative direction.) The average velocity for the entire motion is

$$\overline{v} = \frac{0 \text{ m}}{91 \text{ s}} = 0 \text{ m/s}$$

This average velocity is zero because the net change of position is zero.

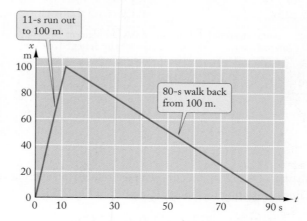

FIGURE 2.7 Plot of position vs. time for a runner.

The average speeds for the run and the walk are, respectively, 9.1 m/s and 1.3 m/s (here, there is no minus sign, since distance is always reckoned as positive). The average speed for the entire motion is the total distance traveled divided by the time taken,

$$[\text{average speed}] = \frac{200 \text{ m}}{91 \text{ s}} = 2.2 \text{ m/s}$$

The average *speed* differs from the average *velocity* because the distance traveled (200 m) differs from the net change of position (zero).

Checkup 2.2

QUESTION 1: You lean out of a window and throw a ball straight down, so it bounces off the sidewalk and returns to your hand 2.0 s later. What is the average velocity of the ball for this motion? Can you calculate the average speed from the information given here?

QUESTION 2: Is it possible that for the trip of an automobile the speed is positive and the average velocity negative? That the speed is positive and the average velocity zero?

    (A) No; no   (B) Yes; no   (C) No; yes   (D) Yes; yes

2.3   INSTANTANEOUS VELOCITY

If, during the road test of the Maserati, the driver was keeping an eye on the speedometer, he would have seen the needle gradually climb from zero to a value well above the legal limit while he was stepping on the accelerator, and then quickly fall to zero when he slammed on the brakes. The speedometer of an automobile displays the magnitude of the instantaneous velocity, that is, the velocity at one instant of time. But although we are all familiar with the speedometers of automobiles, this familiarity does not fully enable us to grasp the mathematical definition of instantaneous velocity, which has to be based on purely kinematic concepts.

Since a moving automobile or a moving particle does not cover any distance in one instant, it is not immediately obvious how we should define the instantaneous velocity. Only in the exceptional case of a particle moving uniformly, with constant velocity, is the instantaneous velocity obvious—it then coincides with the average velocity. We can see this from an examination of the plot of position vs. time. If a particle moves at constant velocity, the plot of position vs. time is a straight line, with a slope equal to the velocity. For example, Fig. 2.8 shows a plot of position vs. time for an automobile moving along a straight road at a constant velocity of 25 m/s. This plot is a straight line of constant slope—the slope in any time interval is equal to the slope in any other time interval. Thus, the average velocity is the same for all time intervals—it is always 25 m/s. Since the velocity is always the same, we may regard the instantaneous velocity for this motion as identical to the average velocity.

If a particle moves with a varying velocity (accelerated motion), the plot of position vs. time is a curve. The plot of position vs. time for the accelerating automobile shown in Fig. 2.5 gives us an example of this: the automobile first accelerates and then decelerates, and the plot of position vs. time is a curve of varying slope. How can we construct a definition of the instantaneous velocity of the automobile on the basis of this plot?

To formulate a definition of the instantaneous velocity, consider the instant $t = 4$ s. We can find an *approximate* value for the velocity at this instant by taking a small time interval of, say, 0.10 s centered on 4 s, that is, a time interval from 3.95 s to 4.05 s. In this time interval the automobile covers some small distance Δx, and we can approximate the actual (curved) plot of position vs. time by a straight line segment connecting the endpoints of the interval (see Fig. 2.9a). According to the discussion at the beginning of this section, the instantaneous velocity associated with a straight plot of position vs. time is simply the slope of the plot; hence the instantaneous velocity at $t = 4$ s can be evaluated

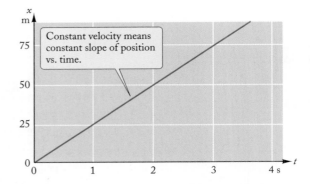

FIGURE 2.8 Plot of position vs. time for an automobile moving at constant velocity.

(a)

(b)

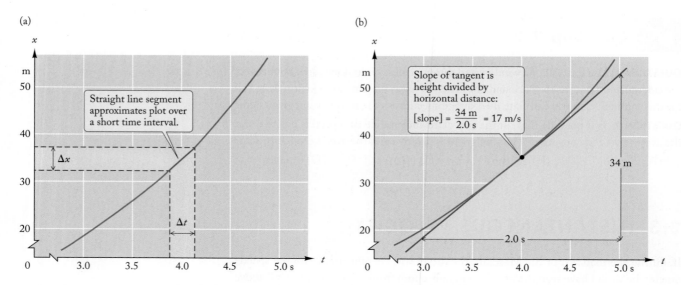

FIGURE 2.9 (a) Plot of position vs. time for an automobile moving with varying velocity. Over a small time interval, we can approximate the plot by a short straight line (blue). (b) The straight line tangent to the plot at 4.0 s rises by 34 m in 2.0 s; thus, the line has a slope of (34 m)/(2.0 s) = 17 m/s.

approximately as the slope of the short line segment shown in Fig. 2.9a. Whether this is a good approximation depends on how closely the straight line segment coincides with the actual curved plot. Obviously, the approximation can be improved by taking a shorter time interval, 0.0010 s or even less. In the limiting case of extremely small time intervals (infinitesimal time intervals), the line segment has the direction of the tangent that touches the plot at the point $t = 4$ s. Hence the **instantaneous velocity** *at any given time equals the slope of the tangent that touches the plot at that time.* For example, drawing the tangent that touches the plot at $t = 4$ s (Fig. 2.9b) and measuring its slope on the graph, we readily find that this slope is 17 m/s; hence the instantaneous velocity at $t = 4$ s is 17 m/s.

By drawing tangents at other points of the plot and measuring their slopes, as illustrated in Fig. 2.10, we can obtain a complete table of values of instantaneous velocities at different times. Figure 2.11 is a plot of the results of such a determination of the instantaneous velocities. The velocity is initially zero (zero slope in Fig. 2.10), then increases to a maximum of 34.9 m/s at $t = 10.0$ s (maximum slope in Fig. 2.10), and finally decreases to zero at $t = 14.3$ s (zero slope in Fig. 2.10).

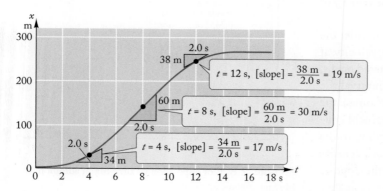

FIGURE 2.10 To find the instantaneous velocities at different times, we draw tangents to the plot at these times and measure their slopes.

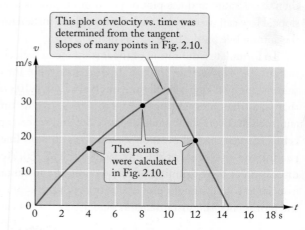

FIGURE 2.11 Instantaneous velocity as a function of time.

The graphical method for determining the instantaneous velocity has the disadvantage that we must first prepare a careful plot of position vs. time. If we have available a complete table of values of the position at different times or an exact mathematical formula for the position as a function of time, then we can calculate the instantaneous velocity by a numerical method, without bothering with a plot. We recall that in the graphical method we began with some small time interval centered on the relevant time, and then took an even smaller time interval, and ultimately considered what happens to the slope in the limiting case of extremely small time intervals. Likewise, in the numerical method, we begin with some small time interval Δt, obtain the change of position Δx, and evaluate the ratio $\Delta x/\Delta t$. Then we take a smaller time interval and repeat the calculation. And then we take an even smaller time interval and repeat the calculation again, and so on. Ultimately, we find that in the limiting case of extremely small time intervals (infinitesimal time intervals), the result of the calculation does not depend on the size of the time interval; that is, we obtain the same result for an extremely small time interval and for an even smaller time interval. This limiting result is the correct value of the instantaneous velocity. Thus, we can write the following formula for the instantaneous velocity:

$$v = \lim_{\Delta t \to 0} \frac{\Delta x}{\Delta t} \qquad (2.7)$$

instantaneous velocity

Here, the notation $\lim \Delta t \to 0$ indicates that $\Delta x/\Delta t$ is to be evaluated in the limiting case of an infinitesimal time interval Δt.

As an example of the application of this equation for the instantaneous velocity, consider the case of the accelerating Maserati whose motion is described by the graph plotted in Fig. 2.5. For a numerical calculation based on Eq. (2.7), we need a formula that gives us x at each instant t, that is, a formula that gives us x as a function of t. The initial, accelerated part of the motion is described by the formula

$$x = 2.376t^2 - 0.042t^3 \qquad \text{(for } t \text{ between 0 s and 10 s)} \qquad (2.8)$$

where x is measured in meters and t in seconds. This formula is merely an alternative way of presenting the information contained in Fig. 2.5 for the interval from 0 s to 10 s. By making a table of values of x at different times t, it is possible to check that the graph plotted in Fig. 2.5 and the formula (2.8) agree—they give the same value of x for any specified time t [in fact, the formula (2.8) was constructed from the data in the graph so as to ensure this agreement].

If we want the instantaneous velocity at $t = 4$ s, we can take $t_1 = 3.9995$ s and $t_2 = 4.0005$ s, for which the formula (2.8) yields

$$x_1 = 2.376 \times (3.9995)^2 - 0.042 \times (3.9995)^3 = 35.3195 \text{ m}$$

and

$$x_2 = 2.376 \times (4.0005)^2 - 0.042 \times (4.0005)^3 = 35.3365 \text{ m}$$

so the instantaneous velocity at $t = 4$ s is approximately

$$v = \frac{\Delta x}{\Delta t} = \frac{x_2 - x_1}{0.001 \text{ s}} = \frac{35.3365 \text{ m} - 35.3195 \text{ m}}{0.001 \text{ s}} = \frac{0.017 \text{ m}}{0.001 \text{ s}} = 17 \text{ m/s}$$

To check the accuracy of this value of the instantaneous velocity, we take a smaller time interval, say, $t_1 = 3.9998$ s and $t_2 = 4.0002$ s. This yields $x_1 = 35.3246$ m and $x_2 = 35.3314$ m, so the instantaneous velocity at $t = 4$ s is approximately

$$v = \frac{x_2 - x_1}{0.0004 \text{ s}} = \frac{35.3314 \text{ m} - 35.3246 \text{ m}}{0.0004 \text{ s}} = 17 \text{ m/s} \qquad (2.9)$$

MATH HELP   DIFFERENTIAL CALCULUS; RULES FOR DERIVATIVES

The derivative of the function $f = f(t) = t^n$:

$$\frac{d}{dt}f = \frac{d}{dt}t^n = nt^{n-1}$$

The derivative of a function f multiplied by a constant c:

$$\frac{d}{dt}(cf) = c\frac{df}{dt}$$

The derivative of the sum of two functions f and g:

$$\frac{d}{dt}(f + g) = \frac{df}{dt} + \frac{dg}{dt}$$

(For more on calculus, see Appendix 4.)

This agrees with the first calculation and confirms that our first choice of time interval was already small enough for an accurate calculation of the instantaneous velocity. Note that the result obtained by the numerical method also agrees with the result obtained by the graphical method, as it should.[1]

In the language of differential calculus, the limit $\Delta x/\Delta t$ for $\Delta t \to 0$ is written as dx/dt, and Eq. (2.7) becomes

instantaneous velocity as derivative of x with respect to t

$$v = \frac{dx}{dt} \tag{2.10}$$

The expression dx/dt is called the **derivative** of x with respect to t. Thus, the *instantaneous velocity is the derivative of the position with respect to time.*

The derivative of a function of t can be evaluated according to the rules of differential calculus. For example, consider a common function, namely, the time t taken to a power n. According to the rules of differential calculus, the derivative of the function t^n with respect to t equals nt^{n-1}. More generally, the derivative of ct^n (where c is some arbitrary constant) is cnt^{n-1}, and the derivative of the sum or difference of several such functions is the sum or difference of their derivatives.

For example, the derivative of $2.376t^2 - 0.042t^3$ with respect to time is $2.376 \times 2t - 0.042 \times 3t^2$, or $4.752t - 0.126t^2$. Accordingly, the instantaneous velocity implied by the formula (2.8) for x is

$$v = \frac{dx}{dt} = 4.752t - 0.126t^2 \quad \text{(for t between 0 s and 10 s)} \tag{2.11}$$

where v is measured in m/s and t in s. With this formula we can readily evaluate the instantaneous velocity at any time. Thus, at $t = 4.0$ s, we obtain

$$v = 4.752 \times 4.0 - 0.126 \times (4.0)^2 = 17 \text{ m/s}$$

which agrees with our previous calculation.

Note that we now have available three methods for the calculation of the instantaneous velocity: the graphical method (based on a determination of the slope in the plot of position vs. time), the numerical method based on small time intervals, and the method of derivatives. These methods are equivalent—they all give the same result. But the method of derivatives is the most convenient and most accurate, and we will hereafter use it whenever we have available an explicit formula for the position as a function of time.

---

[1] It might seem more reasonable to calculate the desired instantaneous velocity by taking a much smaller time interval, say, 10^{-6} s or so. But if we use an ordinary calculator to evaluate x, we reach a point of diminishing returns. The calculator can handle only ten digits, and if the difference between x_2 and x_1 is smaller than the last digit, the calculator will reckon the difference as zero. Thus, we must take care not to use excessively small time intervals when evaluating Eq. (2.7) on such a calculator.

Checkup 2.3

QUESTION 1: What is the meaning of a negative velocity v? Can an aircraft have a negative velocity v of, say, -400 km/h?

QUESTION 2: You are driving an automobile along a straight road, and the speedometer registers 60 km/h (17 m/s). Is the velocity positive or negative? What information do you need to decide this question?

QUESTION 3: Suppose that your average velocity for a 1-hour bike trip on a straight road was 8 m/s. Is it possible that your instantaneous velocity during the entire trip was always larger than 8 m/s?

QUESTION 4: The position x in meters of a body as a function of the time t in seconds is given by $x = \frac{1}{4}t^4$. What is the instantaneous velocity at $t = 2.0$ s?

(A) 16 m/s.   (B) 8.0 m/s.   (C) 6.0 m/s.

(D) 4.0 m/s.   (E) 2.0 m/s.

2.4 ACCELERATION

Online
Concept
Tutorial

Any motion with a change of velocity is accelerated motion. Thus, the motion of an automobile that speeds up is accelerated motion, but so is the motion of an automobile that slows down while braking—in both cases there is a *change* of velocity. If a particle has velocity v_1 at time t_1 and velocity v_2 at time t_2, then the **average acceleration** *for this time interval is defined as the change of velocity divided by the change of time*,

$$\bar{a} = \frac{v_2 - v_1}{t_2 - t_1} \qquad (2.12)$$

average acceleration

or

$$\bar{a} = \frac{\Delta v}{\Delta t} \qquad (2.13)$$

where $\Delta v = v_2 - v_1$ and $\Delta t = t_2 - t_1$. Accordingly, the average acceleration is the average rate of change of the velocity, or the average change of velocity per unit time. The unit of acceleration is the unit of velocity divided by the unit of time. Consequently, in the SI system, *the unit of acceleration is the meter per second per second*, or **meter per second squared** [(m/s)/s, or m/s²]. Table 2.2 gives some typical values of accelerations.

The acceleration can be positive or negative, depending on the sign of the velocity change $v_2 - v_1$. If the velocity is positive and increasing in magnitude, the acceleration is positive; if the velocity is positive and decreasing in magnitude, the acceleration is negative. However, note that if the velocity is *negative* (motion in the negative x direction) and *increasing* in magnitude, that is, becoming *more* negative, the acceleration is *negative*. Thus, an automobile speeding up while moving in the negative x direction has negative acceleration;

TABLE 2.2	SOME ACCELERATIONS
Baseball being struck by bat	3×10^4 m/s²
Soccer ball being struck by foot	3×10^3 m/s²
Rat flea, starting a jump	2×10^3 m/s²
Automobile crash (60 mi/h into fixed barrier)	1×10^3 m/s²
Parachute opening (extreme)	3.2×10^2 m/s²
Free fall on surface of Sun	2.7×10^2 m/s²
Explosive seat ejection from aircraft (extreme)	1.5×10^2 m/s²
Loss of consciousness of human ("blackout")	70 m/s²
Free fall on surface of Earth	9.8 m/s²
Braking of automobile	≈ 8 m/s²
Free fall on surface of Moon	1.7 m/s²

(a)

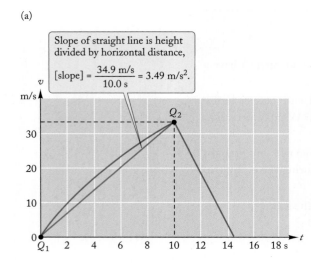

Slope of straight line is height divided by horizontal distance,

$$[\text{slope}] = \frac{34.9 \text{ m/s}}{10.0 \text{ s}} = 3.49 \text{ m/s}^2.$$

(b)

$$t = 4 \text{ s, } [\text{slope}] = \frac{7.4 \text{ m/s}}{2.0 \text{ s}} = 3.7 \text{ m/s}^2$$

FIGURE 2.12 (a) The average acceleration for the interval from $t_1 = 0$ to $t_2 = 10.0$ s is the slope of the straight line Q_1Q_2. (b) The instantaneous acceleration at $t = 4$ s is the slope of the tangent at that point.

conversely, an automobile slowing down or "**decelerating**" while moving in the negative x direction has positive acceleration! These quirks of the formula (2.12) for the acceleration must be kept in mind.

On a plot of velocity vs. time, the average acceleration is the slope of the straight line connecting the points corresponding to t_1 and t_2 on the plot. For instance, in the case of the accelerating Maserati discussed above, the plot of velocity as a function of time is shown in Fig. 2.11, and the average acceleration for the time interval from $t_1 = 0$ to $t_2 = 10.0$ s is the ratio of the vertical separation to the horizontal separation between the two points Q_1 and Q_2 on the plot (see Fig. 2.12a):

$$\bar{a} = \frac{v_2 - v_1}{t_2 - t_1} = \frac{34.9 \text{ m/s} - 0 \text{ m/s}}{10.0 \text{ s} - 0 \text{ s}} = 3.49 \text{ m/s}^2$$

EXAMPLE 3   The chapter photo shows a fighter jet being launched by a catapult from the deck of an aircraft carrier. During this launch, the fighter jet attains a speed of 260 km/h in only 1.8 s. What is the average acceleration of the jet during this time interval?

SOLUTION: We first convert $\Delta v = 260$ km/h to m/s:

$$\Delta v = 260 \text{ km/h} \times \frac{1000 \text{ m}}{1 \text{ km}} \times \frac{1 \text{ h}}{3600 \text{ s}} = 72 \text{ m/s}$$

With $\Delta v = 72$ m/s and $\Delta t = 1.8$ s, the average acceleration is then

$$\bar{a} = \frac{\Delta v}{\Delta t} = \frac{72 \text{ m/s}}{1.8 \text{ s}} = 4.0 \times 10^1 \text{ m/s}^2 = 40 \text{ m/s}^2$$

The **instantaneous acceleration** *at some instant of time is the slope of the tangent drawn on the plot of velocity vs. time.* For example, at $t = 4$ s, we can draw the tangent to the velocity curve in Fig. 2.11 and find that the slope, or the instantaneous acceleration, is 3.7 m/s² (see Fig. 2.12b). By drawing tangents for different times and measuring their slopes, we can prepare a complete table of values of the instantaneous acceleration at different times. Figure 2.13 shows a plot of the values of the instantaneous acceleration obtained from our road test of the Maserati. At the initial instant t = 0, the acceleration is large (large slope in Fig. 2.11); as the automobile gains velocity, the acceleration gradually drops (decreasing slope in Fig. 2.11); at $t = 10$ s, the brakes are applied, leading to a large negative acceleration of -8.1 m/s² (negative slope in Fig. 2.11); this negative acceleration is maintained until the Maserati comes to a halt. Note that when the driver steps on the brakes at time $t = 10$ s, the acceleration suddenly switches from positive to negative; thus, the plot in Fig. 2.13 has a jump at this point.

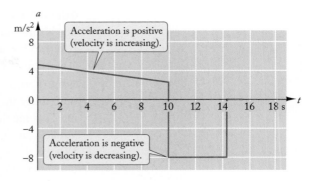

FIGURE 2.13 Instantaneous acceleration as a function of time.

As in the case of the instantaneous velocity, the instantaneous acceleration can also be calculated as the limit of the ratio of small increments:

$$a = \lim_{\Delta t \to 0} \frac{\Delta v}{\Delta t}$$

or

$$a = \frac{dv}{dt} \tag{2.14}$$

instantaneous acceleration as derivative of v with respect to t

This says that the acceleration is the derivative of the velocity with respect to time. Equivalently, we can say that the acceleration is the second derivative of the position x with respect to time, that is

$$a = \frac{d}{dt}\left(\frac{dx}{dt}\right) \quad \text{or} \quad a = \frac{d^2x}{dt^2} \tag{2.15}$$

For example, the acceleration calculated by differentiation of the velocity formula (2.11) is

$$a = \frac{d}{dt}(4.752t - 0.126t^2) = 4.752 - 0.126 \times 2t$$
$$= 4.752 - 0.252t \quad \text{(for } t \text{ between 0 s and 10 s)} \tag{2.16}$$

where the acceleration is measured in m/s². For example, at $t = 4$ s, this relation gives $a = 3.74$ m/s², in agreement with the slope of the tangent to the velocity curve, $a = 3.7$ m/s², mentioned above.

 ## Checkup 2.4

QUESTION 1: What is the meaning of a negative acceleration? Can an "accelerating" automobile starting from rest have a negative acceleration a? Can a braking automobile have a positive acceleration a?

QUESTION 2: Give an example of motion with positive instantaneous velocity and a simultaneous negative acceleration. Give an example of motion with negative instantaneous velocity and a simultaneous positive acceleration.

QUESTION 3: Suppose that at one instant of time the velocity of a body is zero. Can this body have a nonzero acceleration at this instant? Give an example.

QUESTION 4: You drop a tennis ball on a hard floor, and it bounces upward. If the x axis is directed upward, what is the sign of the velocity before the ball hits the floor? After? What is the sign of the acceleration before the ball hits the floor? After? What is the sign of the acceleration while the ball is in contact with the floor?

QUESTION 5: A train strikes an automobile abandoned on a railroad crossing. The train drags the automobile along, and then stops at some distance beyond the crossing. Assume the x axis is along the railroad track, in the direction of travel of the train. What is the sign of the average acceleration of the automobile for the complete motion? What is the sign of the average acceleration of the train? What is the sign of the instantaneous acceleration of the automobile during the impact? After the impact?

QUESTION 6: Two automobiles collide head-on. Compare the signs of the velocities of the automobiles just before the collision. Compare the signs of their accelerations during the collision.

(A) Same; same   (B) Same; opposite
(C) Opposite; same   (D) Opposite; opposite

Online *Concept* Tutorial

2.5 MOTION WITH CONSTANT ACCELERATION

The acceleration of a body may vary as a function of position or time, as in the instantaneous acceleration plot of Fig. 2.13; later, in Section 2.7, we will learn how to use integral calculus to determine the position of a body as a function of time for arbitrary, varying acceleration. However, it is very common for a body to experience a *constant acceleration*, at least for some interval of time; this permits a simpler analysis. Constant acceleration implies a constant slope in the plot of velocity vs. time; thus the plot is a straight line. In this case, the velocity simply increases (or decreases) by equal amounts in each 1-second time interval. For example, in the interval between 10.0 s and 14.3 s, the velocity plotted in Fig. 2.11 decreases by 8.1 m/s in each second while the automobile brakes.

In the case of constant acceleration, there are some simple relations between acceleration, velocity, position, and time that permit us to calculate one of these quantities from the others. Suppose that the initial velocity at time zero is v_0 and that the velocity increases at a constant rate given by the constant acceleration a. After a time t has elapsed, the velocity will have increased by an amount at, and it will have attained the value

$$v = v_0 + at \tag{2.17}$$

Suppose that the initial position is x_0 at time zero. After a time t has elapsed, the position will have changed by an amount equal to the product of the average velocity multiplied by the time; that is, the position will have changed from the initial value x_0 to

$$x = x_0 + \bar{v}t \tag{2.18}$$

Since the velocity increases uniformly with time, the average value of the velocity is simply the average of the initial value and the final values, or

$$\bar{v} = \tfrac{1}{2}(v_0 + v) \tag{2.19}$$

and therefore

$$x = x_0 + \tfrac{1}{2}(v_0 + v)t \tag{2.20}$$

Thus, the change in the position is

$$x - x_0 = \tfrac{1}{2}(v_0 + v)t \tag{2.21}$$

To express this in terms of the acceleration, we substitute Eq. (2.17) into (2.21), and we find

$$x - x_0 = \tfrac{1}{2}(v_0 + v_0 + at)t$$

or

$$x - x_0 = v_0 t + \tfrac{1}{2}at^2 \tag{2.22}$$

The right side of this equation consists of two terms: the term $v_0 t$ represents the change in position that the particle would suffer if moving at constant velocity v_0, and the term $\tfrac{1}{2}at^2$ represents the effect of the acceleration.

Equations (2.17) and (2.22) express velocity and position in terms of time. By eliminating the time t between these two equations, we obtain a direct relation between position and velocity, which is sometimes useful. According to Eq. (2.17),

$$t = \frac{v - v_0}{a} \tag{2.23}$$

and if we substitute this into Eq. (2.22), we obtain

$$
\begin{aligned}
x - x_0 &= v_0\left(\frac{v - v_0}{a}\right) + \tfrac{1}{2}a\left(\frac{v - v_0}{a}\right)^2 \\
&= \frac{v_0 v - v_0^2}{a} + \tfrac{1}{2}\frac{v^2 - 2vv_0 + v_0^2}{a} \\
&= \tfrac{1}{2}\frac{v^2 - v_0^2}{a}
\end{aligned}
\tag{2.24}
$$

which we can rearrange as follows:

$$a(x - x_0) = \tfrac{1}{2}(v^2 - v_0^2) \tag{2.25}$$

Figure 2.14 shows graphs of acceleration, velocity, and position for motion with constant acceleration. For the sake of simplicity, the initial velocity and position have been taken as zero ($v_0 = 0$, $x_0 = 0$). In these graphs, the motion for negative values of t (instants before zero) has also been included with the assumption that a always has the same constant value (this corresponds, for instance, to a subway car that slows down as it reaches a terminal station at the end of the track, stops instantaneously, and then speeds up as it travels away from the station, back in the same direction it came from). Note that the plot in Fig. 2.14b is a straight line; this merely means that the velocity increases in direct proportion to the time, in accord with Eq. (2.17). The plot in Fig. 2.14c is a parabola; this parabola results because the change of position is proportional to the square of the time, as indicated by Eq. (2.22). The parabolic shape of the plot is a distinctive characteristic of motion with constant acceleration.

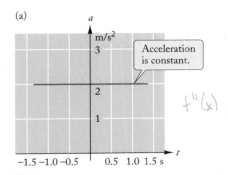

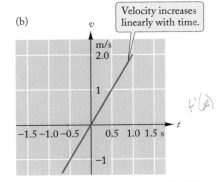

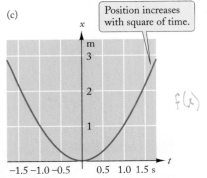

FIGURE 2.14 (a) Acceleration vs. time for motion with constant acceleration; this plot shows a constant value of 2.0 m/s². (b) Velocity vs. time; this plot is a straight line of slope 2.0 m/s². (c) Plot of position vs. time; the plot is a parabola.

EXAMPLE 4   Use the rules of differential calculus to evaluate the second derivative of x with respect to t from Eq. (2.22), and verify that this second derivative equals the acceleration a.

SOLUTION: From Eq. (2.22),

$$x = x_0 + v_0 t + \tfrac{1}{2}at^2$$

In the differentiation, x_0, v_0, and a are constants. Hence the first derivative of x is

$$\frac{dx}{dt} = \frac{d}{dt}\left(x_0 + v_0 t + \tfrac{1}{2}at^2\right) = 0 + v_0 + \tfrac{1}{2}a \times 2t = v_0 + at$$

and the second derivative is

$$\frac{d^2 x}{dt^2} = \frac{d}{dt}(v_0 + at) = a$$

Equations (2.17), (2.21), (2.22), and (2.25) contain the acceleration a, the time t, and the instantaneous and initial positions and velocities, x, x_0, v, and v_0. In a typical problem of motion with constant acceleration, some of these quantities will be known, and the others will be unknown, to be calculated from the equations. Which of these equations are the most useful will depend on the problem.

Concepts
— in —
Context

EXAMPLE 5   In Example 3 we found that during a catapult launch from the deck of an aircraft carrier (see the chapter photo), the average acceleration of a fighter jet is 40 m/s^2 during a time interval of 1.8 s. Assuming the motion proceeds with constant acceleration, how far does the fighter jet travel along the deck during this time interval?

SOLUTION: For motion with constant acceleration, the relevant equations are Eqs. (2.17), (2.21), (2.22), and (2.25). To solve this problem, we must decide which of these equations we need. The unknown quantity is the distance, and the known quantities are the acceleration, the final and initial speeds, and the time. For convenience, we assume that the origin of our coordinates is at the initial position of the aircraft, so $x_0 = 0$ (see Fig. 2.15). Then the known and the unknown quantities are as follows:

UNKNOWN	KNOWN
x	$a = 40$ m/s^2
	$v_0 = 0$
	$x_0 = 0$
	$t = 1.8$ s

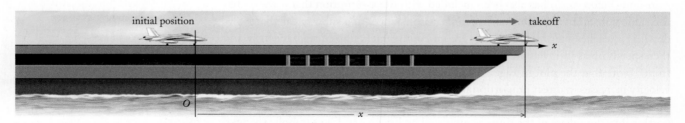

FIGURE 2.15 A catapulted jet. The origin of coordinates is at the initial position of the jet ($x_0 = 0$).

Equations (2.21), (2.22), and (2.25) contain the unknown quantity x, and they contain other quantities, all of which are known. We can use any one of these equations to find x. A simple choice is Eq. (2.22), since it is already solved for x:

$$x = x_0 + v_0 + \tfrac{1}{2}at^2 = 0 + 0 + \tfrac{1}{2} \times (40 \text{ m/s}^2) \times (1.8 \text{ s})^2$$

$$= 65 \text{ m}$$

EXAMPLE 6   An automobile initially traveling at 50 km/h crashes into a stationary, rigid barrier. The front end of the automobile crumples, and the passenger compartment comes to rest after advancing by 0.40 m (see Fig. 2.16). Assuming constant deceleration during the crash, what is the value of the deceleration? How long does it take the passenger compartment to stop?

SOLUTION: The known quantities are the initial velocity ($v_0 = 50$ km/h just before the automobile comes in contact with the barrier), the final velocity ($v = 0$ when the passenger compartment comes to rest), and the change of position of the passenger compartment ($x - x_0 = 0.40$ m; see Fig. 2.16):

UNKNOWN	KNOWN
a	$x - x_0 = 0.40$ m
t	$v_0 = 50$ km/h
	$v = 0$

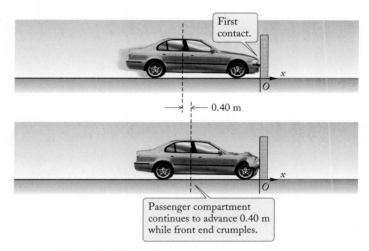

FIGURE 2.16 Deformation of the front end of an automobile crashing into a barrier. The origin of coordinates is at the barrier.

To find the unknown acceleration, it will be best to use Eq. (2.25), since there the acceleration appears as the *only* unknown quantity.

Solving Eq. (2.25) for a, we find

$$a = \frac{v^2 - v_0^2}{2(x - x_0)}$$

Substitution of the known quantities, with $v_0 = 50$ km/h $= 50 \times (1000 \text{ m})/(3600 \text{ s})$ $= 13.9$ m/s, yields

$$a = \frac{v^2 - v_0^2}{2(x - x_0)} = \frac{-(13.9 \text{ m/s})^2}{2 \times 0.40 \text{ m}} = -240 \text{ m/s}^2$$

This is a large deceleration. A passenger involved in such a crash would suffer severe injuries, unless well restrained by a snug seat belt or an air bag.

We can next calculate, from Eq. (2.17), the time the passenger compartment takes to stop:

$$t = \frac{v - v_0}{a} = \frac{-13.9 \text{ m/s}}{-240 \text{ m/s}^2} = 0.058 \text{ s}$$

EXAMPLE 7

An automobile is traveling at 86 km/h on a straight road when the driver spots a wreck ahead and slams on the brakes. The **reaction time** of the driver, that is, the time interval between seeing the wreck and stepping on the brakes, is 0.75 s. Once the brakes are applied, the automobile decelerates at 8.0 m/s². What is the total stopping distance (see Fig. 2.17)?

SOLUTION: The motion has two parts. The first part, before the brakes are applied, is motion at constant velocity; the second part, after the brakes are applied, is motion with constant (negative) acceleration.

The first part of the motion lasts for a time $\Delta t = 0.75$ s, with a constant velocity $v_0 = 86$ km/h, that is,

$$v_0 = 86 \, \frac{\text{km}}{\text{h}} \times \frac{10^3 \, \text{m}}{1 \, \text{km}} \times \frac{1 \, \text{h}}{3600 \, \text{s}} = 24 \, \text{m/s}$$

With this velocity, the automobile travels a distance

$$v_0 \, \Delta t = 24 \, \text{m/s} \times 0.75 \, \text{s} = 18 \, \text{m}$$

The second part of the motion therefore has an initial position $x_0 = 18$ m, an initial velocity $v_0 = 24$ m/s, a final velocity $v = 0$, and an acceleration $a = -8.0$ m/s² (the acceleration is negative since the automobile is decelerating while moving in the positive x direction). The final distance is the unknown:

UNKNOWN	KNOWN
x	$a = -8.0 \, \text{m/s}^2$
	$v = 0$
	$v_0 = 24 \, \text{m/s}$
	$x_0 = 18 \, \text{m}$

The most suitable equation for the solution of this problem is Eq. (2.25), since it contains the unknown quantity and all the other quantities in it are known. Solving this equation for x, we find that the total stopping distance is

$$x = x_0 + \frac{v^2 - v_0^2}{2a}$$

$$= 18 \, \text{m} + \frac{0 - (24 \, \text{m/s})^2}{2 \times (-8.0 \, \text{m/s}^2)} = 18 \, \text{m} + 36 \, \text{m} = 54 \, \text{m}$$

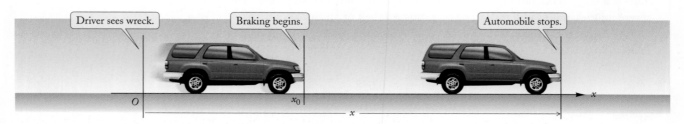

Driver sees wreck.   Braking begins.   Automobile stops.

FIGURE 2.17 A braking automobile. The origin of coordinates is at the point where the driver spots wreck.

PHYSICS IN PRACTICE   STOPPING DISTANCES

The accompanying table lists total stopping distances of an automobile for several initial speeds. These stopping distances were calculated as in Example 7. The reaction time of the driver was assumed to be 0.75 s, and the braking deceleration was assumed to be 8.0 m/s². At all speeds, the reaction time makes a significant contribution to the stopping distance. For an alert driver, the average reaction time is 0.75 s, but for a nonalert driver (such as a driver chatting on a cell phone), the reaction time can be much longer. The braking deceleration of 8 m/s² is the typical deceleration that an automobile with good tires can attain during sharp braking on a dry pavement. If the tires are worn or if the pavement is wet, the attainable deceleration is less than 8 m/s².

The deceleration also depends on the characteristics of the automobile. For instance, a high-performance sports car with a rear engine, such as the Porsche Carrera, can attain decelerations of nearly 11 m/s² (during sharp braking, the nose of a car tends to dive down, placing extra pressure on the front wheels; the large mass of the rear engine distributes the braking effort more equally over the front and rear wheels).

If the driver pushes the brakes too hard, the wheels will lock, and the automobile will skid, which results in a reduced deceleration and a substantial increase in stopping distance. Skidding often leads to loss of directional control, and the automobile might spin around several times and then crash into an obstacle or roll over. Cars equipped with an antilock braking system (ABS) avoid skidding by automatic, rapid, repeated application of the brakes.

AUTOMOBILE STOPPING DISTANCES

v_0	$v_0\,\Delta t$	$-\dfrac{v_0^2}{2a}$	TOTAL STOPPING DISTANCE
15 km/h	3.1 m	1.1 m	4.2 m
30	6.3	4.3	10.6
45	9.4	10	19
60	12	18	30
75	16	27	43
90	19	39	58

EXAMPLE 8 On a foggy day, a minivan is traveling at 80 km/h along a straight road when the driver notices a truck ahead traveling at 25 km/h in the same direction. The driver begins to brake when the truck is 12 m ahead, decelerating the minivan at 8.0 m/s², while the truck continues at a steady 25 km/h. How long after this instant does the minivan collide with the truck? What is the speed of the minivan at the instant of collision?

SOLUTION: We designate the position, velocity, and acceleration of the minivan by x, v, and a and the position, velocity, and acceleration of the truck by x', v', and a'. We reckon the x and x' coordinates from the point at which braking begins (see Fig. 2.18).

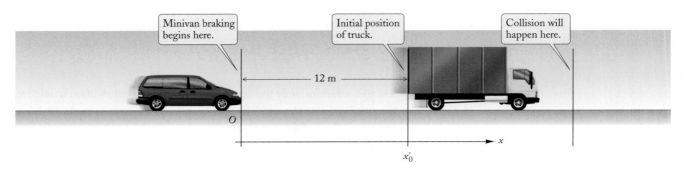

FIGURE 2.18 A braking minivan and a truck traveling at constant velocity. The origin of coordinates is at the point where braking begins. The initial position of the minivan is $x_0 = 0$, and the initial position of the truck (measured to the rear of the truck) is $x'_0 = 12$ m.

The time t of the collision and the positions x and x' at that time are unknown. The initial positions x_0 and x'_0, velocities v_0 and v'_0, and accelerations a and a' are known:

UNKNOWN	KNOWN
t	$x_0 = 0$
x	$x'_0 = 12$ m
x'	$v_0 = 80$ km/h
	$v'_0 = 25$ km/h
	$a = -8.0$ m/s^2
	$a' = 0$

To relate the unknowns x and x' to the known quantities, we use Eq. (2.22) for the minivan:

$$x = x_0 + v_0 t + \tfrac{1}{2} a t^2 = v_0 t + \tfrac{1}{2} a t^2$$

and for the truck:

$$x' = x'_0 + v'_0 t + \tfrac{1}{2} a' t^2 = x'_0 + v'_0 t$$

Here we have two equations but three unknowns (t, x, x'). We can extract the unknown time t from these equations by taking into account that when the vehicles collide, $x = x'$. This condition tells us that

$$v_0 t + \tfrac{1}{2} a t^2 = x'_0 + v'_0 t$$

This is a quadratic equation. Before proceeding with the solution, it is convenient to substitute the known numbers $a = -8.0$ m/s^2, $v_0 = 80$ km/h $= 80 \times 1000$ m/3600 s $=$ 22.2 m/s, $x'_0 = 12$ m, and $v'_0 = 25$ km/h $= 25 \times 1000$ m/3600 s $= 6.9$ m/s. Since the acceleration values are in m/s^2 and the velocity values in m/s, the time will be in seconds. Omitting the units, we obtain

$$22.2t - 4.0t^2 = 12 + 6.9t$$

and if we shift all the terms to the left side, we obtain

$$-4.0t^2 + 15.3t - 12 = 0$$

This has the standard form for a quadratic equation

$$at^2 + bt + c = 0$$

with the two solutions (see Appendix 2)

$$t = \frac{-b \pm \sqrt{b^2 - 4ac}}{2a}$$

$$= \frac{-15.3 \pm \sqrt{(15.3)^2 - 4 \times (-4.0) \times (-12)}}{2 \times (-4.0)} = 1.1 \text{ s or } 2.7 \text{ s}$$

Of these two solutions, only the first is relevant (the second solution would require that the minivan pass through the truck while continuing to brake at 1.1 s and that the truck then again approach the minivan when the minivan has nearly stopped at 2.7 s). Thus, the collision occurs at a time 1.1 s.

The speed of the minivan at this time is

$$v = v_0 + at = 22.2 \text{ m/s} - 8.0 \text{ m/s}^2 \times 1.1 \text{ s} = 13 \text{ m/s}$$

Checkup 2.5

QUESTION 1: According to Example 7, the braking distance for an automobile with an initial speed of 86 km/h is 36 m. What is the braking distance for the same automobile if the initial speed is twice as large?

QUESTION 2: According to Example 6, the deceleration of the passenger compartment of an automobile crashing into a barrier is 240 m/s^2 if its front end crumples 0.40 m. What would be the deceleration of an automobile with the same speed but a stiffer front end, which crumples only 0.20 m?

QUESTION 3: An automobile with initial velocity v_0 brakes to a stop with constant deceleration in a time t. If the initial velocity were twice as large but the constant deceleration were half as large, the time to stop would be:

    (A) $8t$ (B) $4t$ (C) $2t$ (D) t (E) $\frac{1}{2}t$

2.6   THE ACCELERATION OF FREE FALL

Online
Concept
Tutorial

A body released near the surface of the Earth will accelerate downward under the influence of the pull of gravity exerted by the Earth. If the frictional resistance of the air has been eliminated (by placing the body in an evacuated container), then the body is in **free fall**, and the downward motion proceeds with constant acceleration. It is a remarkable fact that the value of this acceleration of free fall is exactly the same for all bodies released at the same location—the value of the acceleration is completely independent of the speeds, masses, sizes, shapes, chemical compositions, etc., of the bodies. Figure 2.19 shows a simple experiment that verifies this equality of the rates of free fall for two bodies of different masses. The universality of the rates of free fall is one of the most precisely and rigorously tested laws of nature; a long series of careful experiments have tested the equality of the rates of free fall of different bodies to within 1 part in 10^{10}, and in some special cases even to within 1 part in 10^{12}.

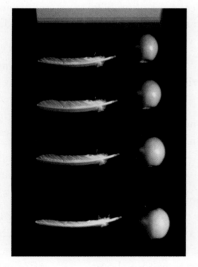

FIGURE 2.19 Stroboscopic photograph of an apple and a feather in free fall in a partially evacuated chamber. The apple and feather were released simultaneously from the trapdoor at the top. The photograph was made by leaving the shutter of the camera open and triggering a flash of light at regular intervals.

The downward acceleration of a freely falling body near the surface of the Earth is usually denoted by g. The numerical value of g is approximately

$$g \approx 9.81 \text{ m/s}^2 \tag{2.26}$$

acceleration of free fall

The exact value of the acceleration of free fall varies somewhat from location to location on the Earth, and it varies with altitude. However, at or near the surface of the Earth this variation amounts to a mere fraction of a percent, and we will neglect it.

For the description of free-fall motion, we can use the formulas for motion with constant acceleration developed in the preceding section. To apply these formulas, we must make a choice for the direction of the x axis. We can take the x axis positive in the upward direction or positive in the downward direction; but once we make one of these choices at the beginning of a problem, we must adhere to it to the end.

For the sake of uniformity, in all the examples in this section, we will take the x axis positive in the upward direction. With this choice of x axis, the acceleration for a freely falling particle is negative, that is, $a = -g$; and Eqs. (2.17), (2.22), and (2.25) become

$$v = v_0 - gt \tag{2.27}$$

free-fall motion

$$x = x_0 + v_0 t - \tfrac{1}{2}gt^2 \tag{2.28}$$

$$-g\,(x - x_0) = \tfrac{1}{2}(v^2 - v_0^2) \tag{2.29}$$

PROBLEM-SOLVING TECHNIQUES   GENERAL GUIDELINES

The solving of problems is an art; there is no simple recipe for obtaining the solutions. Most of the problems in this and the following chapters are applications of the concepts and principles developed in the text. The examples scattered throughout each chapter illustrate typical cases of problem solving. Sometimes you will be able to solve a problem by imitating one of these examples. But if you can't see how to begin the solution, try the following steps:

1   Draw a sketch of the situation described in the problem and label all relevant quantities.

2   If the problem deals with some kind of motion, try to visualize the progress of the motion in time, as though you were watching a movie.

3   For problems that deal with motion, you need to introduce a coordinate axis to describe the motion. Velocities and accelerations must be reckoned as positive when they coincide with the direction of the axis, and negative if they are opposite to the direction of the axis. The origin can be chosen at any convenient location; once chosen, you must use it throughout the problem.

4   Prepare a complete list of the given (known) and sought (unknown) quantities.

5   Ask yourself what physical conditions and principles are applicable to the situation. For instance, does the motion proceed with constant velocity? With constant acceleration? Does the principle of universality of the rate of free fall ($a = -g$) apply?

6   Examine the formulas that are valid under the identified conditions. Then try to spot a formula that permits you to express the unknowns in terms of the known quantities (see Examples 6 and 7). Be discriminating in your selection of formulas—sometimes a formula will tempt you because it displays all the desired quantities, but it will be an invalid formula if the assumptions that went into its derivation are not satisfied in your problem.

7   You will sometimes find that you seem to have too many unknowns and too few equations. Then ask yourself, Are there any special conditions that relate the unknowns (see Example 8)? Are there any quantities that you can calculate from the known quantities? Do these calculated quantities bring you nearer to the answer?

8   It is good technique to solve all the equations by algebraic manipulations and substitute numbers only at the very end; this makes it easier to spot and correct mistakes.

9   When you substitute numbers, also include the units of these numbers. The units in your equations should then combine or cancel in such a way as to give the correct units for the final result. If the units do not combine or cancel in the expected way, something has gone wrong with your algebra.

10   After you have finished your calculations, always check whether the answer is plausible. For instance, if your calculation yields the result that a diver jumping off a cliff hits the water at 3000 km/h, then somebody has made a mistake somewhere!

11   Last, remember to round your final answer to the number of significant figures appropriate for the data given in the problem.

Strictly, these equations are valid only for bodies falling in a vacuum, where there is no frictional air resistance. But they are also good approximations for dense bodies, such as chunks of metal or stone, released in air. For such bodies the frictional resistance offered by the air is unimportant as long as the speed is low (the exact restriction to be imposed on the speed depends on the mass, size, and shape of the body, and on the desired accuracy of the calculation). Unless specifically mentioned, we will ignore the resistance of air, even when the speeds are not all that low.

GALILEO GALILEI (1564–1642)
Italian mathematician, astronomer, and physicist. Galileo demonstrated experimentally that all bodies fall with the same acceleration, and he deduced that the trajectory of a projectile is a parabola. He initiated a new approach to mechanics by recognizing that the natural state of motion of a body, in the absence of forces, is motion with uniform velocity. With a telescope of his own design, he discovered the satellites of Jupiter and sunspots. He vociferously defended the heliocentric system of Copernicus, for which he was condemned by the Inquisition.

EXAMPLE 9   At Acapulco, professional divers amuse tourists by jumping from a 36-m-high cliff into the sea (Fig. 2.20). How long do they fall? What is their impact velocity on the water?

SOLUTION: For this problem, the relevant equations are Eqs. (2.17), (2.22), and (2.25) [or their equivalents with $a = -g$, Eqs. (2.27), (2.28), and (2.29)]. The known quantities are the acceleration of free fall, the change of position, and the initial velocity; the unknown quantities are the time of fall and the final velocity. The initial velocity is $v_0 = 0$ (we assume that the diver merely drops from the initial position, without pushing up or down when jumping off the cliff). The change of position is $x - x_0 = -36$ m. This is negative because the final position is below the initial position; that is, the motion is in the negative x direction (recall that we choose $+x$ upward); see Fig. 2.20. Whether we choose the origin at the top of the cliff ($x_0 = 0$ and $x = -36$ m) or at the bottom ($x = 0$ and $x_0 = 36$ m), the change of position is the same ($x - x_0 = -36$ m).

UNKNOWN	KNOWN
t	$x - x_0 = -36$ m
v	$v_0 = 0$
	$g = 9.81$ m/s^2

To calculate the time from the known quantities, we will use Eq. (2.28), in which the time is the only unknown. With $v_0 = 0$, Eq. (2.28) yields

$$x - x_0 = -\tfrac{1}{2}gt^2$$

which we can solve for t by dividing both sides by $-\tfrac{1}{2}g$ and taking the square root of both sides:

$$t = \sqrt{-\frac{2(x - x_0)}{g}}$$

(2.30)

$$= \sqrt{-\frac{2 \times (-36 \text{ m})}{9.81 \text{ m/s}^2}} = 2.7 \text{ s}$$

From Eq. (2.27), the impact velocity is then

$$v = 0 - gt = -9.81 \text{ m/s}^2 \times 2.7 \text{ s}$$
$$= -26 \text{ m/s}$$

This is about 94 km/h!

COMMENT: Remember that the sign of the velocity or acceleration tells you the direction of the velocity or the acceleration. For instance, the result $v = -26$ m/s means that the motion of the diver is opposite to the direction of the x axis; the x axis is directed upward, and the motion of the diver is directed downward.

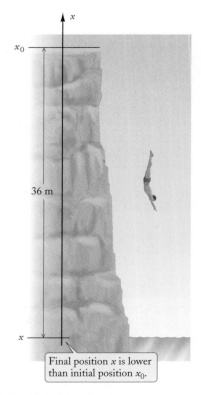

FIGURE 2.20 Jump of a diver. The change of position is negative ($x - x_0 < 0$).

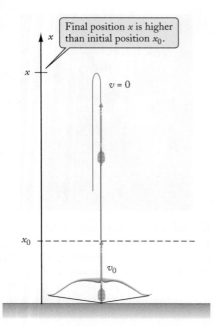

Final position x is higher than initial position x_0.

$v = 0$

v_0

FIGURE 2.21 An ascending arrow. The change of position is positive ($x - x_0 > 0$).

EXAMPLE 10   A powerful bow, like one of those used to establish world records in archery, can launch an arrow at a velocity of 90 m/s. How high will such an arrow rise if aimed vertically upward? How long will it take to return to the ground? What will be its velocity when it hits the ground? For simplicity, ignore air friction and treat the arrow as an ideal particle.

SOLUTION: At the ground, the initial velocity is positive, $v_0 = 90$ m/s (see Fig. 2.21). The arrow moves upward while its velocity decreases at the rate of 9.81 m/s^2. At the highest point of the motion, the arrow ceases to move upward and is momentarily at rest; at this point the instantaneous velocity is zero, $v = 0$. For the upward motion, we can therefore regard the initial and final velocities as known. The height reached and the time are unknown:

UNKNOWN	KNOWN
$x - x_0$	$v_0 = 90$ m/s
t	$v = 0$
	$g = 9.81$ m/s^2

Equation (2.29) relates the height to the known quantities. Dividing this equation by $-g$, we find

$$x - x_0 = \frac{-(v^2 - v_0^2)}{2g} = \frac{-0 + (90 \text{ m/s})^2}{2 \times 9.81 \text{ m/s}^2} = 4.1 \times 10^2 \text{ m}$$

Equation (2.27) relates the time for the upward motion to the known quantities. Solving this equation for t, we find the time for the upward motion:

$$t = \frac{v_0 - v}{g} = \frac{90 \text{ m/s} - 0}{9.81 \text{ m/s}^2} = 9.2 \text{ s} \tag{2.31}$$

The downward motion is simply the reverse of the upward motion—during the downward motion, the arrow accelerates at the rate of 9.81 m/s^2, just as it decelerated at this same rate during the upward motion. The downward motion therefore takes exactly as long as the upward motion, and the total time required for the arrow to complete the up and down motion is twice the time required for the upward motion, that is, 2×9.2 s $= 18.4$ s.

The velocity of the arrow when it hits the ground is simply the reverse of the initial velocity; thus, it is -90 m/s.

COMMENT: Keep in mind that although the instantaneous velocity of the arrow is zero at the highest point of the motion, the acceleration is still the same as that at any other point, $a = -g$. The arrow is momentarily at rest, but it is still accelerating!

Acceleration is sometimes measured in multiples of the "standard" acceleration of gravity; we will call this the **standard** g, where

standard g

$$1 \text{ standard } g = 9.80665 \text{ m/s}^2 \approx 9.81 \text{ m/s}^2 \tag{2.32}$$

Note the distinction between g and standard g: g is the acceleration of gravity at or near the surface of the Earth—its value is approximately 9.81 m/s^2, but its exact value depends on location. The standard g is a unit of acceleration—to three significant figures, its value is 9.81 m/s^2 by definition, and it does not depend on location. Thus, the acceleration of gravity at the Earth's surface is approximately 1 standard g, the acceleration of gravity on the Moon's surface (see Table 2.2) is 0.17 standard g, and so on.

Concepts
— *in* —
Context

EXAMPLE 11   In Example 3, we saw that a catapult launching involved an acceleration of 40 m/s². Express this in standard g's.

SOLUTION: To express this acceleration in standard g's, we multiply by the conversion factor provided by the definition (2.32):

$$a = 40 \text{ m/s}^2 = 40 \text{ m/s}^2 \times \frac{1 \text{ standard } g}{9.81 \text{ m/s}^2} = 4.1 \text{ standard } g\text{'s}$$

It is sometimes said that the pilot is then "pulling 4.1 g's."

Finally, we make some brief comments on the effects of the frictional resistance of air on bodies falling at high speeds. By holding your hand out of the window of a speeding automobile, you can readily feel that air offers a substantial frictional resistance to motion at speeds in excess of a few tens of kilometers per hour. This frictional resistance increases with speed (roughly in proportion to the square of the speed). Hence a falling, accelerating body will experience a larger and larger frictional resistance as its speed increases. Ultimately, this resistance becomes so large that it counterbalances the pull of gravity—the body ceases to accelerate and attains a constant speed. This ultimate speed is called the **terminal speed**, or **terminal velocity**. The precise value of the terminal speed depends on the mass of the body, its size, and its shape; for instance, a skydiver falling with a closed parachute (see Fig. 2.22) attains a terminal speed of about 200 km/h, whereas a skydiver with an open parachute attains a terminal speed of only about 18 km/h. We will reconsider effects of air friction in Section 6.1, after we become familiar with the concept of force.

FIGURE 2.22 Skydivers falling with closed parachutes.

Checkup 2.6

QUESTION 1: You drop an empty wine bottle and a full wine bottle out of a second-floor window. Does the full wine bottle have a larger acceleration? A larger impact speed on the ground?

QUESTION 2: You drop a full cup of water out of a second-floor window. Will the water spill out of the cup while the cup falls?

QUESTION 3: You throw a stone straight up, so it reaches a maximum height and then falls back down. At the instant the stone reaches maximum height, is its velocity positive, negative, or zero? Is its acceleration positive, negative, or zero? Assume the x axis is directed upward.

QUESTION 4: What is the acceleration of a falling skydiver who has attained his terminal velocity?

QUESTION 5: How would Eqs. (2.27) to (2.29) be different if, instead of taking the x axis positive in the upward direction, we took it positive in the downward direction?

QUESTION 6: According to Example 10, an arrow launched upward at 90 m/s reaches a maximum height of 410 m. What maximum height will the arrow reach when launched with a speed half as large?

(A) 820 m   (B) 410 m   (C) 205 m   (D) 103 m   (E) 51 m

2.7 INTEGRATION OF THE EQUATIONS OF MOTION[2]

In the sections above, we dealt with the special case of motion with constant acceleration. However, often the acceleration is not constant, as in Fig. 2.13. In this section we will see how to obtain the position as a function of time when the acceleration is time-dependent. Suppose that the instantaneous acceleration is $a = dv/dt$, where a is a function of time. We rearrange this relation and obtain

$$dv = a\,dt$$

We can integrate this expression directly, for example, from the initial value of the velocity v_0 at time $t = 0$ to some final value v at time t (in the following equations, the integration variables are indicated by primes to distinguish them from the upper limits of integration):

$$\int_{v_0}^{v} dv' = \int_{0}^{t} a\,dt'$$

$$v - v_0 = \int_{0}^{t} a\,dt' \tag{2.33}$$

This gives the velocity as a function of time:

$$v = v_0 + \int_{0}^{t} a\,dt' \tag{2.34}$$

---

[2] Students not familiar with integral calculus should read Section 7.2 before reading this section.

If the acceleration a were constant, we could bring it outside the integral and immediately obtain our previous result, Eq. (2.17). Equation (2.34) enables us to calculate the velocity as a function of time for any acceleration which is a known function of time.

If the velocity $v = dx/dt$ is a known function of time, then the position x can be obtained in a similar manner:

$$dx = v \, dt$$

$$\int_{x_0}^{x} dx' = \int_{0}^{t} v \, dt'$$

$$x - x_0 = \int_{0}^{t} v \, dt' \qquad (2.35)$$

In the special case of constant acceleration a, we could insert Eq. (2.17), $v = v_0 + at$, into the last integral: we then would again obtain Eq. (2.22). When a is not constant, we must first use Eq. (2.34) to find v as a function of time and then insert this function in Eq. (2.35) to find the position as a function of time, as in the following example.

EXAMPLE 12   A motorcycle accelerates quickly from rest, with an acceleration that has an initial value $a_0 = 4.0$ m/s^2 at $t = 0$ and decreases to zero during the interval $0 \le t \le 5.0$ s according to

$$a = a_0 \left(1 - \frac{t}{5.0 \text{ s}} \right)$$

After $t = 5.0$ s, the motorcycle maintains a constant velocity. What is this final velocity? In the process of "getting up to speed," how far does the motorcycle travel?

SOLUTION: The acceleration a is given as an explicit function of time. Since we are beginning from rest, the initial velocity is $v_0 = 0$, so Eq. (2.34) gives v as a function of t:

$$v = v_0 + \int_{0}^{t} a \, dt' = 0 + \int_{0}^{t} a_0 \left(1 - \frac{t'}{5.0 \text{ s}} \right) dt'$$

$$= a_0 \left(\int_{0}^{t} dt' - \frac{1}{5.0 \text{ s}} \int_{0}^{t} t' \, dt' \right) = a_0 \left(t' \Big|_{0}^{t} - \frac{1}{5.0 \text{ s}} \frac{t'^2}{2} \Big|_{0}^{t} \right)$$

$$= a_0 \left(t - \frac{t^2}{10 \text{ s}} \right)$$

where we have used the properties that the integral of the sum is the sum of the integrals and that $\int t^n \, dt = t^{n+1}/(n + 1)$. At $t = 5.0$ s, this velocity reaches its final value of

$$v = 4.0 \text{ m/s}^2 \times \left[5.0 \text{ s} - \frac{(5.0 \text{ s})^2}{10 \text{ s}} \right] = 10 \text{ m/s}$$

To obtain the distance traveled during the acceleration, we must insert the time-dependent velocity into Eq. (2.35):

$$x - x_0 = \int_0^t v \, dt' = \int_0^t a_0 \left(t' - \frac{t'^2}{10 \text{ s}} \right) dt'$$

$$= a_0 \left(\int_0^t t' \, dt' - \frac{1}{10 \text{ s}} \int_0^t t'^2 \, dt' \right) = a_0 \left(\frac{t'^2}{2} \Big|_0^t - \frac{1}{10 \text{ s}} \frac{t'^3}{3} \Big|_0^t \right)$$

$$= a_0 \left(\frac{t^2}{2} - \frac{t^3}{30 \text{ s}} \right)$$

Evaluating this expression at $t = 5.0$ s, we find

$$x - x_0 = 4.0 \text{ m/s}^2 \times \left[\frac{(5.0 \text{ s})^2}{2} - \frac{(5.0 \text{ s})^3}{30 \text{ s}} \right] = 33 \text{ m}$$

Sometimes the acceleration is a known function of the velocity, instead of a known function of time. This is true, for example, for the case of frictional air resistance discussed in Section 2.6 (we will examine this in detail in Chapter 6). Consider one-dimensional motion with an acceleration that is a function of the velocity. In this case, it is possible to integrate the relation $a = dv/dt$ by first rearranging it:

$$dt = \frac{dv}{a}$$

and integrating from time $t = 0$, when $v = v_0$, to some time t:

$$\int_0^t dt' = \int_{v_0}^v \frac{dv'}{a}$$

or simply

$$t = \int_{v_0}^v \frac{dv'}{a} \qquad (2.36)$$

This provides t as a function of v (and v_0), which can sometimes be easily inverted to find v as a function of t.

We have now seen that direct integration of the equations of motion can be applied when the acceleration is known as a function of time or velocity.

 Checkup 2.7

QUESTION 1: Beginning at $t = 0$, a particle accelerates from rest and then moves in one dimension. The acceleration increases from zero in proportion to the time t. By what factor is the particle's speed at $t = 2$ s greater than it was at $t = 1$ s?

QUESTION 2: A water rocket accelerates from rest beginning at $t = 0$ so that its acceleration increases in proportion to the time t. The distance traveled increases with time in proportion to what function of time?

(A) $\ln t$ (B) t^2 (C) t^3 (D) t^4

SUMMARY

MATH HELP   Differential Calculus; Rules for Derivatives   **(page 38)**

PHYSICS IN PRACTICE   Stopping Distances   **(page 47)**

PROBLEM-SOLVING TECHNIQUES   General Guidelines   **(page 50)**

AVERAGE SPEED $\dfrac{[\text{total distance traveled}]}{[\text{time taken}]}$ **(2.1)**

AVERAGE VELOCITY $\bar{v} = \dfrac{\Delta x}{\Delta t}$ **(2.5)**

INSTANTANEOUS VELOCITY $v = \dfrac{dx}{dt}$ **(2.10)**

AVERAGE ACCELERATION $\bar{a} = \dfrac{\Delta v}{\Delta t}$ **(2.13)**

INSTANTANEOUS ACCELERATION $a = \dfrac{dv}{dt}$ **(2.14)**

MOTION WITH CONSTANT ACCELERATION $v = v_0 + at$ **(2.17)**

$x - x_0 = \frac{1}{2}(v_0 + v)t$ **(2.21)**

$x - x_0 = v_0 t + \frac{1}{2}at^2$ **(2.22)**

$a(x - x_0) = \frac{1}{2}(v^2 - v_0^2)$ **(2.23)**

ACCELERATION OF FREE FALL $g \approx 9.81 \text{ m/s}^2$ **(2.26)**

STANDARD g   1 standard $g = 9.81 \text{ m/s}^2$ **(2.32)**

MOTION OF FREE FALL   (x axis is upward) $v = v_0 - gt$ **(2.27)**

$x - x_0 = v_0 t - \frac{1}{2}gt^2$ **(2.28)**

$-g(x - x_0) = \frac{1}{2}(v^2 - v_0^2)$ **(2.29)**

QUESTIONS FOR DISCUSSION

1. The motion of a runner can be regarded as particle motion, but the motion of a gymnast cannot. Explain.

2. A newspaper reports that the world record for speed skiing is 203.160 km/h. This speed was measured on a 100-m "speed trap." The skier took about 1.7 s to cross this trap. In order to calculate speed to six significant figures, we need to measure distance and time to six significant figures. What accuracy in distance and time does this require?

3. Do our sense organs permit us to feel velocity? Acceleration?

4. What is your velocity at this instant? Is this a well-defined question? What is your acceleration at this instant?

5. Does the speedometer on your car give you speed or velocity? Does the speedometer care whether you drive eastward or westward along a road?

6. Suppose at one instant in time the velocity of a body is zero. Can this body have a nonzero acceleration at this instant? Give an example.

7. Give an example of a body in motion with instantaneous velocity and acceleration of the same sign. Give an example of a body in motion with instantaneous velocity and acceleration of opposite signs.

8. Experienced drivers recommend that when driving in traffic you should stay at least 2 s behind the car in front of you. This is equivalent to a distance of about two car lengths for every 15 km/h. Why is it necessary to leave a larger distance between the cars when the speed is larger?

9. Is the average speed equal to the average magnitude of the velocity? Is the average speed equal to the magnitude of the average velocity?

10. In the seventeenth century Galileo Galilei investigated the acceleration of gravity by rolling balls down an inclined plane. Why did he not investigate the acceleration directly by dropping a stone from a tower?

11. Why did astronauts find it easy to jump on the Moon? If an astronaut can jump to a height of 20 cm on the Earth (with his space suit), how high can he jump on the Moon?

12. An elevator is moving upward with a constant velocity of 5 m/s. If a passenger standing in this elevator drops an apple, what will be the acceleration of the apple relative to the elevator?

13. Some people are fond of firing guns into the air when under the influence of drink or patriotic fervor. What happens to the bullets? Is this practice dangerous?

14. A woman riding upward in an elevator drops a penny in the elevator shaft when she is passing the third floor. At the same instant, a man standing at the elevator door at the third floor also drops a penny in the elevator shaft. Which coin hits the bottom first? Which coin hits with the higher speed? Neglect friction.

15. If you take the frictional resistance of air into account, how does this change your answers to Question 14?

16. Suppose that you drop a $\frac{1}{2}$-kg packet of sugar and a $\frac{1}{2}$-kg ball of lead from the top of a building. Taking air friction into account, which will take the shorter time to reach the ground? Suppose you place the sugar and the lead in identical sealed glass jars before dropping them. Which will now take the shorter time?

17. Is air friction important in the falling motion of a raindrop? If a raindrop were to fall without friction from a height of 300 m and hit you, what would it do to you?

18. An archer shoots an arrow straight up. If you consider the effects of the frictional resistance of air, would you expect the arrow to take a longer time to rise or to fall?

PROBLEMS

2.1 Average Speed

1. The speed of nerve pulses in mammals is typically 10^2 m/s. If a shark bites the tail of a 30-m-long whale, roughly how long will it take before the whale knows of this?

2. The world record for the 100-yard run is 9.0 s. What is the corresponding average speed in miles per hour?

3. Glaciers sometimes advance 20 m per year. Express this speed in m/s and in cm/day.

4. In the fall, the monarch butterfly migrates some 3500 km from the Northeastern United States to Mexico. Guess the speed of the butterfly, and estimate how long it takes to make this trip.

5. In 1958 the nuclear-powered submarine *Nautilus* took 6 days and 12 hours to travel submerged 5068 km across the Atlantic from Portland, England, to New York City. What was the average speed (in km/h) for this trip?

6. A galaxy beyond the constellation Corona Borealis is moving directly away from our Galaxy at the rate of 21 600 km/s. This galaxy is now at a distance of 1.4×10^9 light-years from our Galaxy. Assuming that the galaxy has always been moving at a constant speed, how many years ago was it right on top of our Galaxy?

7. On one occasion a tidal wave (tsunami) originating near Java was detected in the English Channel 32 h later. Roughly measure the distance from Java to England by sea (around the Cape of Good Hope) on a map of the world, and calculate the average speed of the tidal wave.

8. In 1971, Francis Chichester, in the yacht *Gypsy Moth V*, attempted to sail the 4000 nautical miles (nmi) from Portuguese Guinea to Nicaragua in no more than 20 days.

    (a) What minimum average speed (in nautical miles per hour) does this require?

    (b) After sailing 13 days, he still had 1720 nmi to go. What minimum average speed did he require to reach his goal in the remaining 7 days? Knowing that his yacht could at best achieve a maximum speed of 10 nmi/h, what could he conclude at this point?

9. The first "marathon" was run in 490 B.C. by the Greek Pheidippides, who ran the 35 km (22 mi) from the battlefield of Marathon to Athens in 2.5 h. What was his average speed?

10. In the open ocean, dolphins sometimes travel 110 km per day, and in short bursts, they can attain speeds of 32 km/h. Express the average travel speed and the burst speed in m/s.

11. The surface of the Earth consists of several large plates which move relative to each other. The Pacific plate (which includes a slice of the coastal part of California) is sliding northward along the continental North American plate at a speed of 4.0 cm per year. How long does it take for this plate to move 1.0 km? 1.0×10^3 km?

12. In setting a world record for speed, the horse Big Racket ran 402 m ($\frac{1}{4}$ mile) in 23.8 s. What was its average speed?

13. The speed-skating record for women is held by Bonnie Blair, who completed a 500-m race in 39.10 s. What is the corresponding average speed? Give the answer in m/s and in km/h.

14. The fastest speed ever measured for a tennis ball served by a player was 263 km/h. At this speed, how long does the tennis ball take to go from one end of the court to the other, a distance of 23.8 m?

15. The distance from New York to Belem, Brazil, is 5280 km. How long does it take you to travel this distance by airliner, at 900 km/h? How long does it take you by ship, at 35 km/h?

16. At the 1988 Olympics, Florence Griffiths-Joyner (Fig. 2.23) ran 100 m in 10.54 s, and she ran 200 m in 21.34 s. What was her average speed in each case?

17. In the case of the closest verdict for a championship ski race, the winner of a cross-country race reached the finish line one hundredth of a second ahead of his closest competitor. If both were moving at a speed of 6 m/s, what was the distance between them at the finish?

FIGURE 2.23 Florence Griffiths-Joyner wins the race.

*18. A hunter shoots an arrow at a deer running directly away from him. When the arrow leaves the bow, the deer is at a distance of 40 m. When the arrow strikes, the deer is at a distance of 50 m. The speed of the arrow is 65 m/s. What must have been the speed of the deer? How long did the arrow take to travel to the deer?

*19. The fastest land animal is the cheetah, which runs at a speed of up to 101 km/h. The second-fastest is the antelope, which runs at a speed of up to 88 km/h.

    (a) Suppose that a cheetah begins to chase an antelope. If the antelope has a head start of 50 m, how long does it take the cheetah to catch the antelope? How far will the cheetah have traveled by this time?

    (b) The cheetah can maintain its top speed for only about 20 s (and then has to rest), whereas the antelope can continue at top speed for a considerably longer time. What is the maximum head start the cheetah can allow the antelope?

20. Carl Lewis set an Olympic record for the 100-m run in 1991 with a time of 9.86 s. What was his average speed?

21. The distance for a marathon is 26 mi 385 yd. In 1984, Joan Benoit set an Olympic record with a time of 2 h 24 min 52 s. What was her average speed in m/s?

*22. On a particular wristwatch, the tips of the minute and second hands are each 0.90 cm from the center, whereas the tip of the hour hand is 0.50 cm from the center. What is the average speed of each of these three tips?

*23. The position of a runner is given by $x = 4.0t - 0.50t^2$, where x is in meters and t is in seconds. What is the average speed between $t = 0$ and $t = 8.0$ s? (Hint: Find the maximum value of x to determine each of the outward and backward distances.)

*24. You make a 100-km trip in an automobile, traveling the first 50 km at 60 km/h and the second 50 km at 80 km/h. What is your average speed for this trip? Explain why the average speed is not 70 km/h.

*25. The table printed inside the book cover gives the radii of the orbits of the planets (mean distance from the Sun) and the times required for moving around the orbit (period of revolution).

   (a) Calculate the speed of motion of each of the nine planets in its orbit around the Sun. Assume that the orbits are circular.

   (b) In a logarithmic graph of speed vs. radius, plot the logarithm of the speed of each planet and the logarithm of its radial distance from the Sun as a point. Draw a curve through the nine points. Can you represent this curve by a simple equation?

2.2 Average Velocity for Motion along a Straight Line[†]

26. Suppose you throw a baseball straight up so that it reaches a maximum height of 8.00 m and returns 2.55 s after you throw it. What is the average speed for this motion of the ball? What is the average velocity?

27. Consider the automobile with the position as a function of time plotted in Fig. 2.5. What is the average velocity for the interval from $t = 0$ to $t = 10.0$ s? From $t = 10.0$ s to $t = 14.3$ s?

28. An elevator travels up from the first floor to the ninth floor in 20 s, down from the ninth floor to the fifth floor in 12 s, and finally up from the fifth floor to the twelfth floor in 18 s. The spacing between any two adjacent floors is 4.0 m. For the entire period of 50 s, what is the average speed? What is the average velocity?

29. A messenger carries a package twelve city blocks north in 14 min 5 s; there, she receives a second package, which she brings six blocks south in 6 min 28 s; finally, she gets a third package, which she brings 3 blocks north in 3 min 40 s. If the city block spacing is 81 m, what is her average speed? What is her average velocity?

30. The position of a person is given by $x = 4.0t - 0.50t^2$, where x is in meters and t is in seconds. What is the average velocity between $t = 0$ and $t = 8.0$ s? Between $t = 8.0$ s and $t = 10.0$ s?

31. A horse trots around a quarter-mile track three complete times in 1 min 40 s. What is the average speed? The average velocity?

32. A dog runs 35 m to catch a Frisbee in 4.5 s, turns quickly and trots 22 m back in 3.6 s, but then stops to guard the Frisbee. What is the dog's average speed? The dog's average velocity?

33. A squirrel walks along a telephone cable, occasionally stopping, even turning back once before proceeding. The squirrel's position as a function of time is shown in Fig. 2.24. What is the average speed for the entire time shown, $0 \leq t \leq 30$ s? The average velocity for the entire time?

34. The position of an automobile as a function of time is plotted in Fig. 2.25. What is the average velocity for $0 \leq t \leq 2.0$ s? For $2.0 \leq t \leq 4.0$ s? Estimate the instantaneous velocity at $t = 1.0$ s and at $t = 3.0$ s.

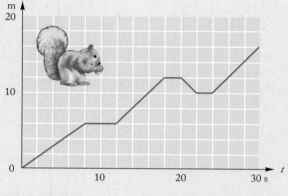

FIGURE 2.24 Position of a squirrel as a function of time.

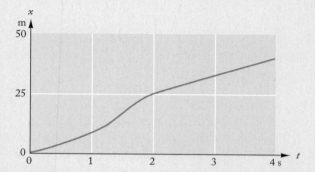

FIGURE 2.25 Automobile position as a function of time.

2.3 Instantaneous Velocity
2.4 Acceleration[†]

35. In an experiment with a water-braked rocket sled, an Air Force volunteer was subjected to an acceleration of 810 m/s^2 for 0.040 s. What was his change of speed in this time interval?

36. A Porsche racing car takes 2.2 s to accelerate from 0 to 96 km/h (60 mi/h). What is the average acceleration?

37. A football player kicks a stationary ball (Fig. 2.26) and sends it flying. Slow-motion photography shows that the ball is in contact with the foot for 8.0×10^{-3} s and leaves with a speed of 27 m/s. What is the average acceleration of the ball while in contact with the foot?

FIGURE 2.26 Kicking a football.

38. The driver of an automobile traveling at 80 km/h suddenly slams on the brakes and stops in 2.8 s. What is the average deceleration during braking?

†39. (a) The blue curve in Fig. 2.27 is a plot of velocity vs. time for a Triumph sports car undergoing an acceleration test. By drawing tangents to the velocity curve, find the accelerations at time $t = 0, 10, 20, 30,$ and 40 s.

(b) The red curve in Fig. 2.27 is a plot of velocity vs. time for the same car when coasting with its gears in neutral. Find the accelerations at time $t = 0, 10, 20, 30,$ and 40 s.

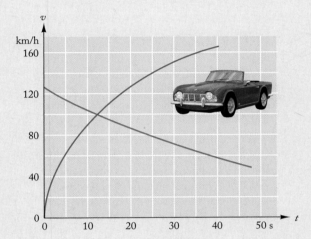

FIGURE 2.27 Instantaneous velocity as a function of time for a Triumph sports car in an acceleration test.

40. The position of a particle in meters is given by $x = 2.5t + 3.1t^2 - 4.5t^3$, where t is the time in seconds. What are the instantaneous velocity and instantaneous acceleration at $t = 0.0$ s? At $t = 2.0$ s? What are the average velocity and average acceleration for the time interval $0 \leq t \leq 2.0$ s?

41. The position of a particle in meters is given by $x = 3.6t^2 - 2.4t^3$, where t is in seconds. Find the two times when the velocity is zero. Calculate the position for each of those times. Sketch x as a function of t for $0 \leq t \leq 2$ s.

42. The velocity of an automobile as a function of time is given by $v = Bt - Ct^2$, where $B = 6.0$ m/s^2 and $C = 2.0$ m/s^3. At what times is the velocity zero? At what time is the acceleration zero? Sketch v as a function of t for $0 \leq t \leq 4.0$ s.

43. The velocity of a bicycle as a function of time is shown in Fig. 2.28. What is the average acceleration for $0 \leq t \leq 5.0$ s? For 5.0 s $\leq t \leq 10$ s? Estimate the instantaneous acceleration at $t = 3.0$ s.

*44. The velocity of a parachutist as a function of time is given by $v = v_f + (v_0 - v_f)e^{[-t/(2.5\text{ s})]}$, where $t = 0$ corresponds to the instant the parachute is opened, $v_0 = 200$ km/h is the velocity before opening of the parachute, and $v_f = 18$ km/h is the final (terminal) velocity. What acceleration does the parachutist experience just after opening the parachute?

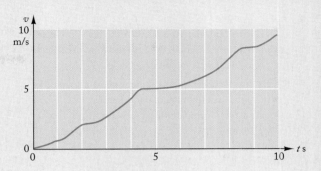

FIGURE 2.28 Bicycle velocity as a function of time.

*45. The velocity of a particle is given by $v = v_0/(1 + At^2)$, where $v_0 = 25$ m/s and $A = 2.0$ s^{-2}. What is the acceleration at $t = 0$? At $t = 2.0$ s? At $t \to \infty$?

*46. Consider the red curve in Fig. 2.27 showing velocity vs. time for a decelerating sports car.

(a) From this curve, estimate the average velocity for the interval $t = 0$ s to $t = 5$ s. (Hint: The average velocity for a small time interval is roughly the average of the initial and final velocities; alternatively, it is roughly the velocity at the midpoint of the time interval.) Estimate how far the car travels in this time interval.

(b) Repeat the calculation of part (a) for every 5-s interval between $t = 5$ s and $t = 45$ s.

(c) What is the *total* distance that the car will have traveled at the end of 45 s?

*47. Table 2.3 gives the horizontal velocity as a function of time for a projectile of 100 lb fired horizontally from a 6-inch naval gun. The velocity decreases with time because of the frictional drag of the air.

TABLE 2.3	EFFECT OF AIR RESISTANCE ON A PROJECTILE		
TIME	**VELOCITY**	**TIME**	**VELOCITY**
0 s	657 m/s	1.80 s	557 m/s
0.30	638	2.10	542
0.60	619	2.40	528
0.90	604	2.70	514
1.20	588	3.00	502
1.50	571		

(a) On a piece of graph paper, make a plot of velocity vs. time and draw a smooth curve through the points of the plot.

(b) Estimate the average velocity for each time interval [the estimated average velocity for the time interval $t = 0$ s to

† Due to the difficulty of accurately drawing tangents, answers for this problem that differ by up to 10% are acceptable.

$t = 0.30$ s is $\frac{1}{2}(657 + 638)$ m/s, etc.]. From the average velocities calculate the distances that the projectile travels in each time interval. What is the total distance that the projectile travels in 3.00 s?

(c) By directly counting the squares on your graph paper, estimate the area (in units of s·m/s) under the curve plotted in part (a), and compare this area with the result of part (b).

48. The instantaneous velocity of the projectile described in Problem 47 can be approximately represented by the formula (valid for 0 s $\leq t \leq$ 3.00 s)

$$v = 655.9 - 61.14t + 3.26t^2$$

where v is measured in meters per second and t is in seconds. Calculate the instantaneous acceleration of the projectile at $t = 0$ s, $t = 1.50$ s, and $t = 3.00$ s.

*49. The position of a particle as a function of time is given by

$$x = A \cos bt$$

where A and b are constants. Assume that $A = 2.0$ m and $b = 1.0$ radian/s.

(a) Roughly plot the position of this particle for the time interval 0 s $\leq t \leq$ 7.0 s.

(b) At what time does the particle pass the origin ($x = 0$)? What are its velocity and acceleration at this instant?

(c) At what time does the particle reach maximum distance from the origin? What are its velocity and acceleration at this instant?

*50. The motion of a rocket burning its fuel at a constant rate while moving through empty interstellar space can be described by

$$x = u_{ex} t + u_{ex} (1/b - t) \ln(1 - bt)$$

where u_{ex} and b are constants (u_{ex} is the exhaust velocity of the gases at the tail of the rocket, and b is proportional to the rate of fuel consumption).

(a) Find a formula for the instantaneous velocity of the rocket.

(b) Find a formula for the instantaneous acceleration.

(c) Suppose that a rocket with $u_{ex} = 3.0 \times 10^3$ m/s and $b = 7.5 \times 10^{-3}$/s takes 120 s to burn all its fuel. What is the instantaneous velocity at $t = 0$ s? At $t = 120$ s?

(d) What is the instantaneous acceleration at $t = 0$ s? At $t = 120$ s?

2.5 Motion with Constant Acceleration[†]

51. The takeoff speed of a jetliner is 360 km/h. If the jetliner is to take off from a runway of length 2100 m, what must be its acceleration along the runway (assumed constant)?

52. A British 6-inch naval gun has a barrel 6.63 m long. The muzzle speed of a projectile fired from the gun is 657 m/s. Assuming that upon detonation of the explosive charge the projectile moves along the barrel with a constant acceleration, what is the magnitude of this acceleration? How long does it take the projectile to travel the full length of the barrel?

53. The nearest star is Proxima Centauri, at a distance of 4.2 light-years from the Sun. Suppose we wanted to send a spaceship to explore this star. To keep the astronauts comfortable, we want the spaceship to travel with a constant acceleration of 9.81 m/s² at all times (this will simulate ordinary gravity within the spaceship). If the spaceship accelerates at 9.81 m/s² until it reaches the midpoint of its trip and then decelerates at 9.81 m/s² until it reaches Proxima Centauri, how long will the one-way trip take? What will be the speed of the spaceship at the midpoint? Do your calculations according to Newtonian physics (actually, the speed is so large that the calculation should be done according to relativistic physics; see Chapter 36).

54. In an accident on motorway M1 in England, a Jaguar sports car made skid marks 290 m long while braking. Assuming that the deceleration was 10 m/s² during this skid (this is approximately the maximum deceleration that a car with rubber wheels can attain on ordinary pavements), calculate the initial speed of the car before braking.

55. The front end of an automobile has been designed so that upon impact it progressively crumples by as much as 0.70 m. Suppose that the automobile crashes into a solid brick wall at 80 km/h. During the collision the passenger compartment decelerates over a distance of 0.70 m. Assume that the deceleration is constant. What is the magnitude of the deceleration? If the passenger is held by a safety harness, is he likely to survive? (Hint: Compare the deceleration with the acceleration listed for a parachutist in Table 2.2.)

56. A jet-powered car racing on the Salt Flats in Utah went out of control and made skid marks 9.6 km long. Assuming that the deceleration during the skid was about 5.0 m/s², what must have been the initial speed of the car? How long did the car take to come to a stop?

57. The operation manual of a passenger automobile states that the stopping distance is 50 m when the brakes are fully applied at 96 km/h. What is the deceleration? What is the stopping time?

58. With an initial speed of 260 km/h, the French TGV (*train à grande vitesse*) takes 1500 m to stop on a level track. Assume that the deceleration is constant. What is the magnitude of the deceleration? What is the time taken for stopping?

59. An automobile accelerates from rest for 20 s with a constant acceleration of 1.5 m/s². What is its final velocity? How far does it travel in that time?

60. At one instant, a body in motion with constant acceleration has velocity 3.0 m/s when its position is $x = 5.0$ m. At a time 4.0 s later, its position is $x = 1.0$ m. What is the acceleration? What is the body's velocity at the later instant?

61. A small airplane accelerates from rest at a constant acceleration of 1.2 m/s² along the runway. The plane rises from the runway after traveling 150 m from its starting point. How long did this take?

62. A Concorde jet traveling at 550 km/h accelerates uniformly at 0.60 m/s² for 90 s. What is its final velocity?

63. A type-A driver guns the engine when a light turns green, and then brakes somewhat quickly for the next red light; the acceleration as a function of time is shown in Fig. 2.29 for $0 \le t \le 10$ s. If the driver started from rest at the origin at $t = 0$, accurately sketch the velocity and position as functions of time.

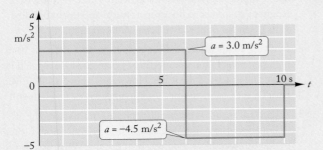

FIGURE 2.29 Automobile acceleration as a function of time.

*64. A truck is initially traveling on a highway at constant speed. The driver wishes to pass an automobile and so begins a 30-s period of constant acceleration at 0.050 m/s². During this period, the truck travels 700 m. What was the truck's initial speed? The truck's final speed?

*65. A train with velocity v_0 accelerates uniformly at 0.50 m/s² for 15 s. It travels 550 m during this time. What is its final velocity? What is v_0?

*66. In a drag race a car starts at rest and attempts to cover 440 yd in the shortest possible time. The world record for a piston-engined car is 5.637 s; while setting this record, the car reached a final speed of 250.69 mi/h at the 440-yd mark.

    (a) What was the average acceleration for the run?

    (b) Prove that the car did not move with constant acceleration.

    (c) What would have been the final speed if the car had moved with constant acceleration so as to reach 440 yd in 5.637 s?

*67. In a large hotel, a fast elevator takes you from the ground floor to the 21st floor. The elevator takes 17 s for this trip: 5 s at constant acceleration, 7 s at constant velocity, and 5 s at constant deceleration. Each floor in the hotel has a height of 2.5 m. Calculate the values of the acceleration and deceleration (assume they are equal). Calculate the maximum speed of the elevator.

*68. (a) In a skyscraper, an elevator takes 55 s to descend from the top floor to ground level, a distance of 400 m. What is the average speed of the elevator for this trip?

    (b) The elevator is at rest at the beginning and at the end of the trip. If you wanted to program the elevator so that it completes the trip in the specified time with a minimum acceleration and a minimum deceleration, how would you have to accelerate and decelerate the elevator? What would be this minimum value of the acceleration and deceleration? What would be the maximum speed during the trip?

*69. For a sleepy or drunk driver, the reaction time is much longer than for an alert driver. Recalculate the table of stopping distances for an automobile in Section 2.5 if the reaction time is 2.0 s instead of 0.75 s.

*70. As discussed in Section 2.5, the total stopping distance for an automobile has two contributions: the reaction-time contribution and the braking contribution. On the basis of the data given in the table on page 47, for what initial speed are these two contributions equal?

*71. In a collision, an automobile initially traveling at 60 km/h decelerates at a constant rate of 200 m/s². A passenger not wearing a seat belt crashes against the dashboard. Before the collision, the distance between the passenger and the dashboard was 0.60 m. With what speed, relative to the automobile, does the passenger crash into the dashboard? Assume that the passenger has no deceleration before contact with the dashboard.

**72. Figure 2.30 (adapted from a diagram in the operation manual of an automobile) describes the automobile's passing ability at low speed. From the data supplied in this figure, calculate the acceleration of the automobile during the pass and the time required for the pass. Assume constant acceleration.

*73. The speed of a body released from rest falling through a viscous medium (for instance, an iron pellet falling in a jar full of oil) is given by the formula

$$v = -g\tau + g\tau e^{-t/\tau}$$

where τ is a constant that depends on the size and shape of the body and on the viscosity of the medium and $e = 2.718 \ldots$ is the basis of the natural logarithms.

    (a) Find the acceleration as a function of time.

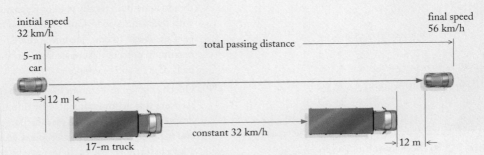

FIGURE 2.30 Diagram from the operation manual of an automobile.

(b) Show that for $t \rightarrow \infty$, the speed approaches the terminal value $-g\tau$.

(c) By differentiation, verify that the equation for the position as a function of time consistent with the above expression for the speed is

$$x = -g\tau t - g\tau^2 e^{-t/\tau} + g\tau^2 + x_0$$

(d) Show that for small values of t ($t \ll \tau$), the equation for x is approximately $x \approx -\frac{1}{2}gt^2 + x_0$.

2.6 The Acceleration of Free Fall†

74. An apple drops from the top of the Empire State Building, 380 m above street level. How long does the apple take to fall? What is its impact velocity on the street? Ignore air resistance.

75. Peregrine falcons dive on their prey with speeds of up to 130 km/h. From what height must a falcon fall freely to achieve this speed? Ignore air resistance.

76. Cats are known for their ability to survive falls from buildings several floors high. If a cat falls three floors down, a distance of 8.7 m, what is its speed when it hits the ground?

77. The world record for a high jump from a standing position is 1.90 m. With what speed must a jumper leave the ground to attain this height? (Treat this as particle motion, although it really is not.)

78. The muzzle speed of a .22-caliber bullet fired from a rifle is 366 m/s. If there were no air resistance, how high would this bullet rise when fired straight up?

79. An engineer standing on a bridge drops a penny toward the water and sees the penny splashing into the water 3.0 s later. How high is the bridge?

80. The volcano Loki on Io, one of the moons of Jupiter, ejects debris to a height of 200 km (Fig. 2.31). What must be the initial ejection velocity of the debris? The acceleration of gravity on Io is 1.80 m/s². There is no atmosphere on Io, hence no air resistance.

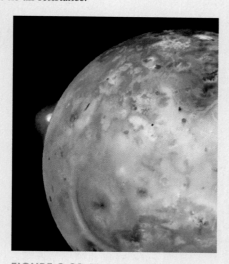

FIGURE 2.31 The volcano Loki on Io.

† For help, see Online Concept Tutorial 3 at www.wwnorton.com/physics

81. The crash of an automobile into a solid barrier can be simulated by dropping the automobile, nose down, from some height onto a hard platform (some of the early crash tests were performed in this manner). From what height must you drop the automobile to simulate a crash at 45 km/h? 75 km/h? 105 km/h? To gain a better appreciation of what these heights mean, express each height as a number of floors up a typical apartment building, with 2.9 m per floor (2.9 m is one floor up, etc.).

82. In diving competitions, a diver jumping from a platform 10 m high performs $1\frac{1}{2}$ somersaults (head up, to down, to up, to down) while falling toward the water. How much time does the diver have for each $\frac{1}{2}$ somersault (head up to down, or down to up)?

83. A hardball is thrown vertically upward, attains a height of 9.5 m above the point of release, falls downward, and is caught by the thrower. How long was it in the air? What was its initial velocity?

84. A Super Ball is thrown to the ground with an initial downward velocity of 1.0 m/s and pops back up to the hand of the thrower with a velocity of the same magnitude as the initial velocity; the total time elapsed is 0.75 s. What height was the Super Ball thrown from?

85. Standing at the edge of a building of height 9.2 m, a woman reaches out over the edge and throws a penny vertically upward. It rises, then falls past her, hitting the ground below 2.5 s after release. What height did the penny attain? What was its initial velocity?

86. A stone is released from rest and falls through a height h. What is the average speed of the stone during this fall?

*87. A rubber ball is thrown vertically downward from a height of 1.5 m and bounces back to the same height with a reversed velocity of the same in a total time of 0.90 s. What was the initial downward velocity? What was the velocity just before hitting the floor?

*88. A particle is initially at rest at some height. If the particle is allowed to fall freely, what distance does it cover in the time from $t = 0$ s to $t = 1$ s? From $t = 1$ s to $t = 2$ s? From $t = 2$ s to $t = 3$ s? Show that these successive distances are in the ratios 1:3:5:7 . . . .

*89. Galileo claimed in his *Dialogues* that "the variation of speed in air between balls of gold, lead, copper, porphyry, and other heavy materials is so slight that in a fall of 100 cubits a ball of gold would surely not outstrip one of copper by as much as four fingers. Having observed this, I came to the conclusion that in a medium totally void of resistance all bodies would fall with the same speed." One cubit is about 46 cm, and four fingers are about 10 cm. According to Galileo's data, what is the maximum percent difference between the accelerations of the balls of gold and of copper?

*90. Quechee gorge in Vermont has a depth of 45 m. If you want to measure this depth within 10% by timing the fall of a stone dropped from the bridge across the gorge, how accurately must you measure the time? Is an ordinary watch with a second hand adequate for this task, or do you need a stopwatch?

*91. A golf ball released from a height of 1.5 m above a concrete floor bounces back to a height of 1.1 m. If the ball is in contact with the floor for 6.2×10^{-4} s, what is the average acceleration of the ball while in contact with the floor?

*92. In 1978 the stuntman A. J. Bakunas died when he jumped from the 23rd floor of a skyscraper and hit the pavement. The air bag that was supposed to cushion his impact ripped.

(a) The height of his jump was 96 m. What was his impact speed?

(b) The air bag was 3.7 m thick. What would have been the man's deceleration had the air bag not ripped? Assume that his deceleration would have been uniform over the 3.7-m interval.

*93. The HARP (High-Altitude Research Project) gun can fire an 84-kg projectile containing scientific instruments straight up to an altitude of 180 km. If we pretend there is no air resistance, what muzzle speed is required to attain this altitude? How long does the projectile remain at a height in excess of 100 km, the height of interest for high-altitude research?

*94. The International Geodesy Association has adopted the following formula from the acceleration of gravity as a function of the latitude Θ (at sea level):

$$g = 978.0318 \text{ cm/s}^2 \times (1 + 53.024 \times 10^{-4} \sin^2 \Theta - 5.9 \times 10^{-6} \sin^2 2\Theta)$$

(a) According to this formula, what is the acceleration of gravity at the equator? At a latitude of 45°? At the pole?

(b) Show that according to this formula, g has a minimum at the equator, a maximum at the pole, and no minima or maxima at intermediate latitudes.

*95. At a height of 1500 m, a dive bomber in a vertical dive at 300 km/h shoots a cannon at a target on the ground. Relative to the bomber the initial speed of the projectile is 700 m/s. What will be the impact speed of the projectile on the ground? How long will it take to get there? Ignore air friction in your calculation.

**96. The elevators in the CN tower in Toronto travel at 370 m/min from ground level to the Skypod, 335 m up. Suppose that when an elevator begins to rise from ground level, you drop a penny from the Skypod down the elevator shaft. At what height does the elevator meet the penny?

**97. You throw a baseball straight up, so it bounces off a ceiling 10 m above your hand and returns to your hand 2.0 s later. What initial speed is required? What is the speed at impact? Assume that the impact on the ceiling reverses the velocity of the baseball but does not change its magnitude.

**98. Suppose you throw a stone straight up with an initial speed of 15.0 m/s.

(a) If you throw a second stone straight up 1.00 s after the first, with what speed must you throw the second stone if it is to hit the first at a height of 11.0 m? (There are two answers. Are both plausible?)

(b) If you throw the second stone 1.30 s after the first, with what speed must you throw the second stone if it is to hit the first at a height of 11.0 m?

***99. Raindrops drip from a spout at the edge of a roof and fall to the ground. Assume that the drops drip at a steady rate of n drops per second (where n is large) and that the height of the roof is h.

(a) How many drops are in the air at one instant?

(b) What is the median height of the drops (i.e., the height above and below which an equal number of drops are found)?

(c) What is the average of the heights of these drops?

2.7 Integration of the Equations of Motion

100. An automobile has an initial velocity $v_0 = 8.0$ m/s. During $0 \le t \le 3.0$ s, it experiences a time-dependent acceleration given by $a = Ct^2$, where $C = 0.25$ m/s^4. What is the instantaneous velocity at $t = 3.0$ s? What is the change in position between $t = 0$ and $t = 1.0$ s?

101. A particle is initially at rest. Beginning at $t = 0$, it begins moving, with an acceleration given by $a = a_0\{1 - [t^2/(4.0 \text{ s}^2)]\}$ for $0 \le t \le 2.0$ s and $a = 0$ thereafter. The initial value is $a_0 = 20$ m/s^2. What is the particle's velocity after 1.0 s? After a long time? How far has the particle traveled after 2.0 s?

102. A particle moves in one dimension and experiences an acceleration which varies with time, given by $a = At + Bt^2$, where $A = 15$ m/s^3 and $B = 25$ m/s^4. The particle is at rest at $t = 0$. What is its speed at $t = 2.0$ s? How far does it travel between $t = 1.0$ s and $t = 2.0$ s?

*103. A parachutist in free fall has a downward speed v_0 when she opens her parachute. Thereafter, she experiences an acceleration proportional to her speed given by $a = g - Av$, where A is a constant (which depends on the size and shape of the parachute, the air density, and the mass of the parachutist). Show that her speed as a function of time is given by

$$v = \frac{g}{A}(1 - e^{-At}) + v_0 e^{-At}$$

After a time $t \gg 1/A$, v approaches the constant value $v = g/A$; this is the terminal velocity.

*104. At moderate to high speeds, an automobile coasting in neutral experiences a deceleration (due to air resistance) of the form $a = -Bv^2$, where $B = 6.1 \times 10^{-4}$ m^{-1} and v is the speed in m/s. The driver of a car traveling horizontally at 120 km/h on a level road shifts into neutral. Calculate the amount of time required for the car to slow to 90 km/h.

REVIEW PROBLEMS

105. In 1993, Noureddine Morcelli set a new record of 3 min 44.39 s for a run of 1 mile, which superseded the previous record of 3 min 46.32 s by Steve Cram. If both runners had run together, at constant speed, how far behind would Cram have been at the finish line?

106. The world record for speed skating set in Salt Lake City in 2002 by Derek Parra was 1 min 43.95 s for 1500 m. What was his average speed?

107. A runner runs 100 m in 10 s, rests 60 s, and returns at a walk in 80 s. What is the average speed for the complete motion? What is the average velocity?

108. You travel in your car for 30 min at 35 km/h, and then for another 30 min at 85 km/h.

   (a) What distance do you cover in the first 30 min? In the second 30 min?

   (b) What is your average speed for the entire trip?

109. The fastest predator in the ocean is the sailfish. When pursuing prey, it swims at 109 km/h. Suppose a sailfish spots a mackerel at a distance of 20 m. The mackerel attempts to escape at 33 km/h, on a straight path. How long does the sailfish take to catch it? How far does the sailfish travel during the pursuit?

*110. The distance from San Francisco to Vancouver is 1286 km by air. A plane leaves San Francisco at 10:00 A.M. heading north toward Vancouver, and another plane leaves Vancouver at 11:00 A.M. heading south toward San Francisco. The first plane travels at 720 km/h, and the second plane, slowed by a headwind, travels at 640 km/h. Where do the planes meet? At what time?

*111. You are traveling on a highway at 80 km/h and you are overtaking a car traveling in the same direction at 50 km/h. How long does it take you to go from 10 m behind this car to 10 m ahead of this car? The length of the car is 4 m.

*112. The highest speed attained by a cyclist on level ground is 105 km/h. To attain this speed, the cyclist used a streamlined recumbent bike. Starting from rest, he gradually built up his speed by pedaling furiously over a distance of 3.2 km. If his acceleration was uniform over this distance, what was the acceleration and how long did the cyclist take to build up his final speed?

*113. A particle moves along the x axis, with the following equation for the position as a function of time:

$$x = 2.0 + 6.0t - 3.0t^2$$

where x is measured in meters and t is measured in seconds.

   (a) What is the position of the particle at $t = 0.50$ s?

   (b) What is the instantaneous velocity at this time?

   (c) What is the instantaneous acceleration at this time?

*114. An automobile speeding at 100 km/h passes a stationary police cruiser. The police officer starts to move her cruiser in pursuit 8.0 s after the automobile passes. She accelerates uniformly to 120 km/h in 10 s, and then continues at uniform speed until she catches the speeder.

   (a) How far ahead is the speeder when the police cruiser starts?

   (b) How far ahead, relative to the cruiser, is the speeder when the cruiser reaches the uniform speed of 120 km/h?

   (c) How long does the police cruiser take to catch the speeder? How far from the initial position?

*115. An automobile is traveling at 90 km/h on a country road when the driver suddenly notices a cow in the road 30 m ahead. The driver attempts to brake the automobile, but the distance is too short. With what velocity does the automobile hit the cow? Assume that, as in the table on page 47, the reaction time of the driver is 0.75 s and that the deceleration of the automobile is 8.0 m/s² when the brakes are applied.

*116. The nozzle of a fire hose discharges water at the rate of 280 liters/min at a speed of 26 m/s. How high will the stream of water rise if the nozzle is aimed straight up? How many liters of water will be in the air at any given instant?

*117. According to an estimate, a man who survived a fall from a 56-m cliff took 0.015 s to stop upon impact on the ground. What was his speed just before impact? What was his average deceleration during impact?

*118. A skydiver jumps out of an airplane at a height of 1000 m. Calculate how long she takes to fall to the ground. Assume that she falls freely with a downward acceleration g until she reaches the terminal speed of 200 km/h, and that she then continues to fall without acceleration until very near the ground (where she opens her parachute a few moments before reaching the ground).

**119. From a window on the fifth floor of a building, 13 m up, you drop two tennis balls with an interval of 1.0 s between the first and the second.

   (a) Where is the first ball when you release the second? Where is the second ball when the first hits the ground?

   (b) What is the instantaneous velocity of the first ball relative to the second just before the first hits the ground?

   (c) What is the instantaneous acceleration of the first ball relative to the second?

Answers to Checkups

Checkup 2.1

1. The average speed is the total distance traveled divided by the time; for the man, this is (50 m)/(100 s), or 0.50 m/s. For the entire 100-s interval, the woman has the same average speed, 0.50 m/s (although her average speed for the first 50 s is twice as large, 1.0 m/s).

2. It is possible that, to the bicyclist, the spectator has a larger speed than the runner; for example, if the bicyclist travels at nearly the same speed as the runner (relative to the Earth), then the runner has a relative speed near zero, and the spectator has a relative speed near 10.1 m/s (but going in the backward direction). It is also possible that both the runner and the spectator have the same speed; this occurs in the special case when the bicyclist travels at half the runner's speed (5.05 m/s) relative to the Earth. In that case, the runner is traveling at 5.05 m/s relative to the bicyclist (in the forward direction), and the spectator is traveling at 5.05 m/s relative to the bicyclist (in the backward direction).

3. (E) 12 m/s. The average speed is the total distance divided by the time taken, (60 m)/(5.0 s) = 12 m/s.

Checkup 2.2

1. Since there is zero net change of position for the round-trip motion of the ball, the average velocity will be zero (the negative and positive parts cancel). We cannot calculate the average speed, since we would need to use the total distance, i.e., twice the distance from the hand to the sidewalk; this was not given (and cannot be calculated from the information given).

2. (D) Yes; yes. As in Example 2, the speed is always positive, but the average velocity can be positive or negative, depending on the relative locations of the starting and ending points for the coordinate system chosen. Also, for any round trip, the net change in position, and thus the average velocity, is zero, whereas, unless there was no motion at any time, the average speed will always be positive.

Checkup 2.3

1. A negative velocity refers to motion in the negative direction for the coordinate axis chosen; in one dimension, this would imply in the negative x direction. An aircraft could have a negative velocity if, for example, the x axis points east but the aircraft flies west.

2. We cannot tell if the velocity is positive or negative until the direction of the x axis is specified.

3. No. If the average velocity was 8 m/s, then the instantaneous velocity must have been sometimes larger than and sometimes smaller than (or always equal to) 8 m/s.

4. (B) 8.0 m/s. The instantaneous velocity is given by the derivative $v = dx/dt = \frac{1}{4} \times 4t^3 = t^3$. Evaluated at $t = 2.0$ s, this is $v = (2.0)^3 = 8.0$ m/s.

Checkup 2.4

1. A negative acceleration means that the velocity in the x direction is decreasing (by becoming either less positive or more negative). An automobile starting from rest can have a negative acceleration by beginning to move in the negative x direction (for example, if the x axis points east and the automobile moves west). A braking automobile can have a positive acceleration (same example: if the x axis points east and an automobile moving west slows down, the acceleration is in the positive x direction).

2. Let the x axis be pointing east; in that case, an automobile moving east and slowing down has a positive velocity and a negative acceleration. Let the x axis point west; in that case, the same automobile (moving east and slowing down) has a negative velocity and a positive acceleration.

3. Yes—for example, a ball bouncing off a wall is instantaneously at rest, but has a nonzero acceleration (since its velocity is changing from one direction to the other).

4. As the ball drops, the velocity is opposite to the chosen axis direction, and so it is negative. After it hits the floor, the velocity is positive (the x axis is upward) until the ball reaches its peak and starts to drop again. Before the ball hits the floor, the acceleration is negative (objects fall increasingly quickly because of gravity); after it hits the floor, the acceleration is still negative (it slows down as it goes back up). During the collision, the acceleration is positive; this acceleration changes the velocity from negative to positive.

5. Since the automobile begins at rest and ends at rest, its average acceleration for the complete motion is zero ($\bar{a} = \Delta v/\Delta t$). The average acceleration of the train is negative, since the x axis is in the direction of travel and the train slows to a stop ($v_2 < v_1$). During the impact, the automobile velocity increases from zero to a positive value, and so it is instantaneously positive. After the impact, the automobile slows with the train, and so it has negative instantaneous acceleration.

6. (D) Opposite; opposite. Just before the collision, the signs of the velocities will be opposite, one positive and one negative, since the velocities are in opposite directions ("head-on"). During the collision, the accelerations will also be opposite; the car with positive velocity will undergo a negative acceleration, and the car with a negative velocity will have a positive acceleration.

Checkup 2.5

1. In Example 7, we saw that the braking distance varies as the square of the velocity ($x - x_0 = -v_0^2/2a$). Thus, twice the initial speed will result in four times the braking distance, or 144 m.

2. In Example 6, we found that the deceleration varies inversely with the crumpling distance [$a = (v^2 - v_0^2)/2(x - x_0)$]. Thus, halving the crumpling distance doubles the deceleration to 480 m/s^2.

3. (B) $4t$. From $v = v_0 + at$, with final velocity $v = 0$ (stopped), we see the time to stop is $t = -v_0/a$. Thus if v_0 is twice as large and a is half as large, t is four times as long.

Checkup 2.6

1. No to both questions. The two bottles have the same acceleration and the same impact speed; the acceleration due to gravity is the same for different masses (neglecting air friction).

2. No. The water accelerates at the same rate as the cup (as usual, we neglect air friction effects).

3. The stone's velocity is zero at the maximum height, since it momentarily stops there. The velocity changes from positive on the way up to negative on the way down, and so it is zero at the top. The acceleration remains negative (and constant, $a = -g$) throughout the motion.

4. Zero. Terminal speed refers to the constant velocity attained under the combined influences of gravity and air resistance. Constant velocity means zero rate of change of velocity, that is, zero acceleration.

5. With the x axis positive down, the acceleration due to gravity becomes positive; that is, the quantity $-g$ in each of Eqs. (2.27) to (2.29) becomes $+g$.

6. (D) 103 m. Under constant acceleration, the change in position is proportional to the change in the square of the velocity [$a(x - x_0) = \frac{1}{2}(v^2 - v_0^2)$]. So if the velocity is half as large, the height will be one-fourth as high, or (410 m)/4 ≈ 103 m.

Checkup 2.7

1. The acceleration is given as proportional to t. Since the velocity, by Eq. (2.34), is the integral of the acceleration over time, the speed will be proportional to t^2. Thus the velocity will be 4 times as large at $t = 2$ s as it was at $t = 1$ s.

2. (C) t^3. If the acceleration a increases in proportion to the time t, then Eq. (2.34) indicates that the velocity $v = \int a\, dt$ increases in proportion to t^2 (for zero initial velocity). The distance then increases with time according to Eq. (2.35), $x - x_0 = \int v\, dt$, which is proportional to t^3, because $\int t^2\, dt = t^3/3$.

Vectors

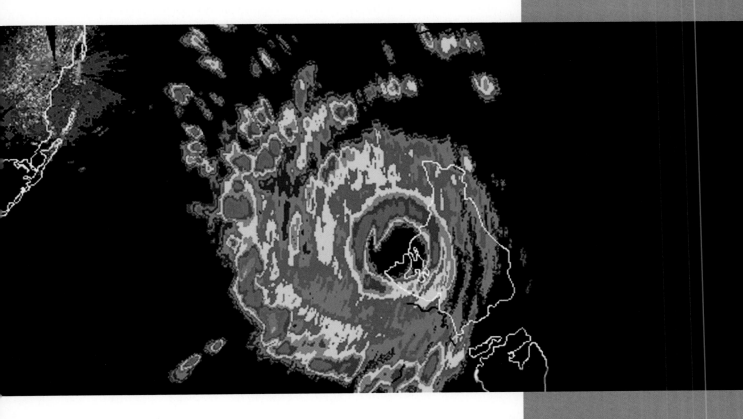

CONCEPTS IN CONTEXT

The colorful circular feature on this radar display is a hurricane off Miami, Florida. The distance and direction from Miami to the hurricane can be represented by a *displacement vector*. Such displacement vectors are commonly used in navigation. Many other types of vectors are used in science and engineering.

While we learn about vectors, we will examine such questions as:

? How are vectors used in navigation? (Physics in Practice: Vectors in Navigation, page 71)

? How can two positions be used to determine relative position? (Example 2, page 78)

? How can vector algebra be used to determine direction? (Example 5, page 83)

3.1 The Displacement Vector and Other Vectors

3.2 Vector Addition and Subtraction

3.3 The Position Vector; Components of a Vector

3.4 Vector Multiplication

The mathematical concept of vector turns out to be very useful for the description of position, velocity, and acceleration in two- or three-dimensional motion. We will see that a vector description of the motion gives precise meaning to the intuitive notion that velocity and acceleration have a *direction* as well as a magnitude: the velocity and the acceleration of a particle can point north, or east, or up, or down, or in any direction in between. And we will see in later chapters that the vector concept is also useful for the description of many other physical quantities, such as force and momentum, which have both a magnitude and a direction. The present chapter is an introduction to vectors and their addition, subtraction, and multiplication. After developing these mathematical tools in this chapter, we will apply vectors to some aspects of motion in two and three dimensions in the next and subsequent chapters.

Online
Concept
4
Tutorial

3.1 THE DISPLACEMENT VECTOR AND OTHER VECTORS

We begin with the concepts of displacement and displacement vector. The displacement of a particle is simply a change of its position. If a particle moves from a point P_1 to a point P_2, we can represent the change of position graphically by an arrow, or directed line segment, from P_1 to P_2. *The directed line segment is the* **displacement vector** *of the particle.* For example, if a ship moves from Liberty Island to the Battery in New York harbor, then the displacement vector is as shown in Fig. 3.1. Note that the displacement vector tells us only where the final position (P_2) is in relation to the initial position (P_1); it does not tell us what path the ship followed between the two positions. Thus, any of the paths shown by the red lines in Fig. 3.2 results in the same final displacement vector.

A displacement vector has a length and a *direction,* which are graphically represented by the length of the arrow and the direction of its tip. Although every vector can be represented by drawing an arrow, not every drawing of an arrow you encounter in everyday life is a vector. For instance, the arrows commonly used in traffic signs are *not* vectors—they tell you only the direction in which you are required to travel, but not the distance (see Fig. 3.3).

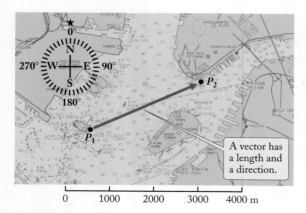

FIGURE 3.1 Displacement vector for a ship moving from Liberty Island to the Battery in New York harbor. The length of this vector is 2790 m; its direction is 65° east of north. (Adapted from National Ocean Survey Chart 12328.)

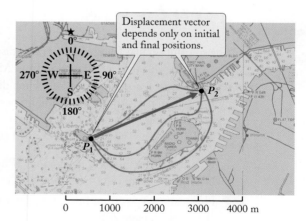

FIGURE 3.2 Several alternative paths from Liberty Island to the Battery (red lines). All of these result in the same displacement vector.

PHYSICS IN PRACTICE   VECTORS IN NAVIGATION

Concepts
— in —
Context

Vectors are widely used by pilots and navigators of ships and aircraft in calculations of positions and directions of travel. For instance, the pilot of an aircraft approaching an airport might ask the airport controller, "Give me a vector," which means that the pilot wants to be told what the displacement vector is from her aircraft to the airport. The controller will consult the radar display showing aircraft positions in relation to the airport, and he might answer, "Heading one five zero, thirty miles," to indicate to the pilot the direction and the distance the aircraft needs to travel to reach the airport. In navigational practice, angles are reckoned clockwise from (magnetic) north; thus, a heading of "one five zero," or 150°, corresponds to 60° south of east.

Incidentally: It is common for the heading of an airport runway to be abbreviated to its first two digits; thus, if the airport controller orders the pilot to land on "runway 24" he means the runway that has a direction 240° clockwise from north, that is, a direction 30° south of west. Runways are used in one of two opposite directions depending on the wind; although only one number is used to specify the active direction, both directions (which always differ by 18, or 180°) appear on airport runway signs (for example, "6/24").

Instead of describing the vector graphically by drawing an arrow, we can describe it numerically by giving the numerical value of its length (in, say, meters) and the numerical value of the angle (in, say, degrees) it makes with some reference direction. For example, we can specify the displacement vector in Fig. 3.1 by stating that it is 2790 m long and points at an angle of 65° east of north. Note that the length, or magnitude, of a vector is always a positive quantity. If we want to construct a vector opposite to the vector shown in Fig. 3.1, we must reverse its direction, but we keep its length positive.

FIGURE 3.3
Arrows painted on a traffic sign.

Since the displacement vector describes a *change* in position, any two line segments of identical length and direction represent equal vectors, regardless of whether the initial points of the line segments are the same. Thus, the parallel directed line segments shown in Fig. 3.4 do not represent different vectors; they both involve the same *change* of position (same distance and same direction), and they both represent equal displacement vectors.

In printed books, vectors are usually indicated by boldface letters, such as **A**, and we will follow this most common convention. In handwritten calculations, an alternative notation consisting of either a small arrow over the letter, such as $\vec{A}$, or a wavy underline, such as $\underset{\sim}{A}$, is usually more convenient. Each means the same as our **A**. We will use an ordinary italic letter, such as *A*, to denote the (positive) length or magnitude of a displacement vector. The magnitude of a vector is also often expressed by placing the vector between vertical lines; that is, $A = |\mathbf{A}|$.

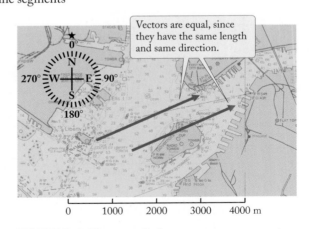

Vectors are equal, since they have the same length and same direction.

FIGURE 3.4 These two displacement vectors are equal.

The displacement vector serves as prototype for all other vectors. To decide whether some quantity endowed with both magnitude and direction is a vector, we compare its mathematical properties with those of the displacement vector. *Any quantity that has magnitude and direction and that behaves mathematically like the displacement vector is a* **vector**. For example, velocity, acceleration, and force are vectors; they can be represented graphically by directed line segments of a length equal to the magnitude of the velocity, acceleration, or force (in some suitable units) and a corresponding direction.

By contrast, any quantity that has a magnitude but *no* direction is called a **scalar**. For example, length, time, mass, area, volume, density, temperature, and energy are scalars; they can be completely specified by their numerical magnitude and units. Note that the *length* of a displacement vector is a quantity that has a magnitude but no direction—that is, the length is a scalar.

✔ Checkup 3.1

QUESTION 1: An airliner flies nonstop from San Francisco to New York. Another airliner flies from San Francisco to New Orleans, and from there to New York. Are the displacement vectors for the two airliners equal? Are the distances traveled equal?

QUESTION 2: Can the magnitude of a vector be negative? Zero?

QUESTION 3: Two aircraft are flying in formation on parallel paths separated by 200 m. The aircraft fly 3000 m due west. Are their displacement vectors equal?

QUESTION 4: The hurricane in the chapter photo is 200 km from Miami, in a direction 30° south of east. Consider a point due south of Miami and due west of the hurricane. How far is this point from Miami?

(A) 71 km	(B) 87 km	(C) 100 km
(D) 141 km	(E) 173 km	

Online
Concept
Tutorial

3.2 VECTOR ADDITION AND SUBTRACTION

Since by definition all vectors have the mathematical properties of displacement vectors, we can investigate all the mathematical operations with vectors by looking at displacement vectors. The most important of these mathematical operations is **vector addition**.

Two displacements carried out in succession result in a net displacement, which is regarded as the vector sum of the individual displacements. For example, Fig. 3.5 shows a displacement vector **A** (from P_1 to P_2) and a displacement vector **B** (from P_2 to P_3). The net displacement vector is the directed line segment from the initial position P_1 to the final position P_3. This net displacement vector is denoted by **C** in Fig. 3.5. This vector **C** can be regarded as the sum of the individual displacements:

$$\mathbf{C} = \mathbf{A} + \mathbf{B} \tag{3.1}$$

The sum of two vectors is usually called the resultant of these vectors. Thus, **C** is called the resultant of **A** and **B**.

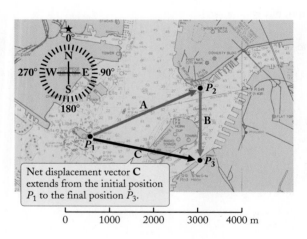

Net displacement vector **C** extends from the initial position P_1 to the final position P_3.

FIGURE 3.5 The displacement **A** is followed by the displacement **B**. The net displacement is **C**.

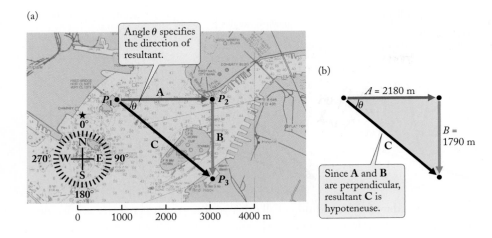

(a)

Angle θ specifies the direction of resultant.

(b)

$A = 2180$ m

$B = 1790$ m

Since **A** and **B** are perpendicular, resultant **C** is hypoteneuse.

FIGURE 3.6 (a) The vectors **A** and **B** represent the displacements of the motorboat moving from Ellis Island (P_1) to the Battery (P_2) and from the Battery to the Atlantic Basin (P_3). The vector **C** is the sum of these two vectors. (b) The vectors **A**, **B**, and **C** form a right triangle. Two of the sides are known, and the hypotenuse is unknown.

EXAMPLE 1 A motorboat moves from Ellis Island in New York harbor to the Battery and from there to the Atlantic Basin (see Fig. 3.6). The first displacement vector is 2180 m due east, and the second is 1790 m due south. What is the resultant?

SOLUTION: The resultant of the two displacement vectors **A** and **B** is the vector **C**, from the tail of **A** to the head of **B**. For a graphical determination of **C**, we can measure the length of **C** directly on the chart using the scale of length marked on the chart, and we can measure the direction of **C** with a protractor (the way the navigator of the ship would solve the problem). This yields a length of about 2800 m and an angle of about 39° for the resultant vector **C**.

For a more precise (numerical) determination of **C**, we note that **A**, **B**, and **C** form a right triangle. We can therefore find **C** by using the standard trigonometric methods for the solution of triangles.[1] The lengths of the two known sides of this triangle are $A = 2180$ m and $B = 1790$ m (see Fig. 3.6). The unknown side is the the hypotenuse of the triangle. By the Pythagorean theorem, the length C of the hypotenuse is

$$C = \sqrt{A^2 + B^2} \tag{3.2}$$

from which

$$C = \sqrt{(2180 \text{ m})^2 + (1790 \text{ m})^2} = 2820 \text{ m}$$

The tangent of the angle θ is the ratio of the opposite side B to the adjacent side A:

$$\tan \theta = \frac{B}{A} = \frac{1790 \text{ m}}{2180 \text{ m}} = 0.821$$

With our calculator, we find that the angle whose tangent is 0.821 is

$$\theta = 39.4°$$

Thus, the resultant vector **C** has a length of 2820 m at an angle 39.4° south of east, which is consistent with values obtained by the graphical method.

---

[1] Appendix 3 gives a review of trigonometry.

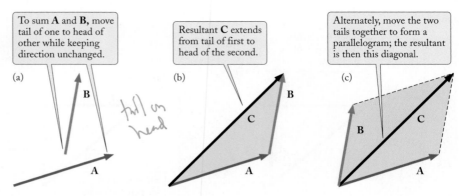

FIGURE 3.7 The vector sum $\mathbf{A} + \mathbf{B}$; the resultant is $\mathbf{C}$. (a) The two vectors $\mathbf{A}$ and $\mathbf{B}$. (b) Addition of $\mathbf{A}$ and $\mathbf{B}$ by the tail-to-head method. (c) Addition of $\mathbf{A}$ and $\mathbf{B}$ by the parallelogram method.

The procedure for the addition of all other vectors—such as velocity, acceleration, and force vectors—mimics that for displacement vectors. All such vectors can be represented by arrows. If $\mathbf{A}$ and $\mathbf{B}$ are two arbitrary vectors (see Fig. 3.7a), then their resultant can be obtained by placing the tail of $\mathbf{B}$ on the head of $\mathbf{A}$ while keeping the magnitude and direction of $\mathbf{B}$ unchanged; the directed line segment connecting the tail of $\mathbf{A}$ to the head of $\mathbf{B}$ is the resultant (see Fig. 3.7b). Alternatively, the resultant can be obtained by placing the tail of $\mathbf{B}$ on the tail of $\mathbf{A}$ and drawing a parallelogram with $\mathbf{A}$ and $\mathbf{B}$ as two of the sides; the diagonal of the parallelogram is then the resultant (see Fig. 3.7c).

Note that the order in which the two vectors are added makes no difference to the final result. Whether we place the tail of $\mathbf{B}$ on the head of $\mathbf{A}$ or the tail of $\mathbf{A}$ on the head of $\mathbf{B}$, the resultant is the same (see Fig. 3.8). Hence

commutative law for vector addition

$$\mathbf{A} + \mathbf{B} = \mathbf{B} + \mathbf{A} \qquad (3.3)$$

This identity is called the **commutative law** for vector addition; it indicates that, just as in ordinary addition of numbers, the order of the terms is irrelevant.

The magnitude of the resultant of two vectors is usually less than the sum of the magnitudes of the vectors. Thus, if

$$\mathbf{C} = \mathbf{A} + \mathbf{B} \qquad (3.4)$$

then

$$C \leq A + B \qquad (3.5)$$

This inequality simply expresses the fact that in a triangle (see Fig. 3.7b) the length of any side is less than the sum of the lengths of the other two sides. Only in the special case where $\mathbf{A}$ and $\mathbf{B}$ are parallel (see Fig. 3.9) will the magnitude of $\mathbf{C}$ equal the sum of the magnitudes of $\mathbf{A}$ and $\mathbf{B}$; it can never exceed this sum.

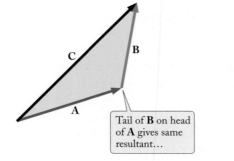

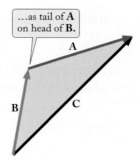

FIGURE 3.8 The resultant for the vector sum $\mathbf{A} + \mathbf{B}$ is the same as for the vector sum $\mathbf{B} + \mathbf{A}$ (compare Fig. 3.7).

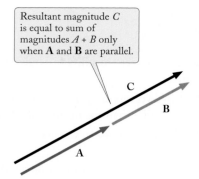

FIGURE 3.9. Parallel vectors **A** and **B**, and their resultant **C**.

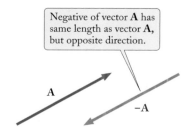

FIGURE 3.10 The vector **A** and its negative −**A**.

The negative of a given vector **A** is a vector of the same magnitude, but opposite direction; this new vector is denoted by −**A** (see Fig. 3.10). Obviously, the sum of a vector and its negative gives a vector of zero magnitude:

$$\mathbf{A} + (-\mathbf{A}) = 0 \qquad (3.6)$$

sum of vector and its negative

The subtraction of two vectors **A** and **B** is defined as the sum of **A** and −**B**; that is,

$$\mathbf{A} - \mathbf{B} = \mathbf{A} + (-\mathbf{B}) \qquad (3.7)$$

Figure 3.11a shows the vector difference **A** − **B** obtained by constructing the vector sum of **A** and −**B** by the parallelogram method. Alternatively, Fig. 3.11b shows how to obtain the vector difference **A** − **B** by drawing the directed line segment from the head of **B** to the head of **A**. Comparison of these two diagrams establishes that this directed line segment from the head of **B** to the head of **A** is equal to the vector difference **A** − **B**.

A vector can be multiplied by any positive or negative number. For instance, if **A** is a given vector, then 3**A** is a vector of the same direction and of a magnitude three times as large (see Fig. 3.12a), whereas −3**A** is a vector of the opposite direction and, again, of a magnitude three times as large (see Fig. 3.12b). In particular, if we multiply a vector by −1, we obtain the negative of that vector:

$$(-1)\mathbf{A} = -\mathbf{A} \qquad (3.8)$$

(a)

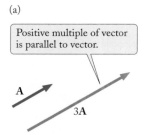

(b)

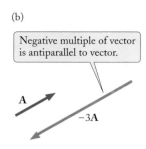

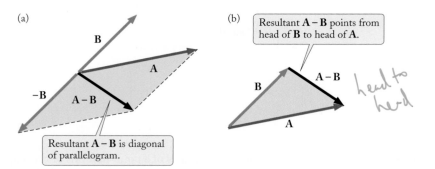

FIGURE 3.11 (a) To obtain the vector difference **A** − **B,** we first draw the vector −**B**, and then construct the vector sum **A** + (−**B**) by the parallelogram method. (b) Alternatively, we can draw the directed line segment from the head of **B** to the head of **A**; this gives the same result as the parallelogram method, and establishes that this directed line segment is equal to the vector difference **A** − **B**.

FIGURE 3.12 (a) The vector 3**A** has the same direction as the vector **A** and is three times as long. (b) The vector −3**A** has a direction opposite to that of **A** and is three times as long.

✔ Checkup 3.2

QUESTION 1: Two vectors have nonzero magnitude. Under what conditions will their sum be zero? Their difference?

QUESTION 2: Suppose that two vectors are perpendicular to each other. Is the magnitude of their sum larger or smaller than the magnitude of each?

QUESTION 3: Is it possible for the sum of two vectors to have the same magnitude as the difference of the two vectors?

QUESTION 4: Three vectors have the same magnitude. Under what conditions will their sum be zero?

QUESTION 5: An automobile drives 3 km south and then drives 4 km west. What is the magnitude of the resultant displacement vector?

    (A) 1 km   (B) 5 km   (C) 7 km   (D) 16 km   (E) 25 km

Online Concept 4 Tutorial

3.3 THE POSITION VECTOR; COMPONENTS OF A VECTOR

In Section 1.1 we saw that to describe the position of a point in space, we must choose an origin and construct a coordinate grid. If the grid is rectangular, then the position of a point will be specified by the three rectangular coordinates x, y, and z. Alternatively, we can describe the position by means of the displacement vector from the origin to the point. For the sake of simplicity, we will usually restrict our discussion to points in a plane, so the position of a point is specified by two rectangular coordinates x and y. Figure 3.13 shows the point, its coordinates x and y, and the displacement vector from the origin to the point. This displacement vector is called the **position vector**, usually denoted by **r**.

As shown in Fig. 3.13, the position vector of a point has a length r, and it makes an angle θ with the x axis. Consider the right triangle shown in Fig. 3.13. The length of the position vector is the hypotenuse of this triangle, and the x and y coordinates form the sides. From the definitions of the cosine and the sine of the angle θ, we find

$$\cos \theta = \frac{x}{r}$$

$$\sin \theta = \frac{y}{r}$$

Hence

$$x = r \cos \theta \tag{3.9}$$

$$y = r \sin \theta \tag{3.10}$$

These two equations express the x and y coordinates of a point in terms of the length r and direction θ of the position vector. Conversely, we can express r and θ in terms of the coordinates x and y. To do this, we apply the Pythagorean theorem to the right triangle in Fig. 3.13:

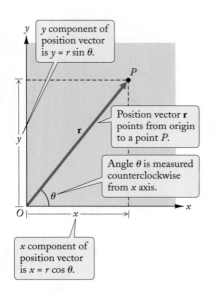

y component of position vector is $y = r \sin \theta$.

Position vector **r** points from origin to a point P.

Angle θ is measured counterclockwise from x axis.

x component of position vector is $x = r \cos \theta$.

FIGURE 3.13 The position vector **r** in two dimensions and its components.

$$r = \sqrt{x^2 + y^2} \qquad (3.11)$$

Furthermore, from the definition of the tangent of the angle θ, we find

$$\tan \theta = \frac{y}{x} \qquad (3.12)$$

These two equations make it easy to calculate r and θ from x and y.

According to the terminology introduced in Chapter 1, x and y are the components of the position. We will now adopt a vectorial terminology according to which *x and y are the* **components** *of the position vector.* Note that graphically the components of the position vector are determined by dropping perpendiculars from the head of the vector to the x and y axes (see Fig. 3.13).

In general, for an arbitrary vector $\mathbf{A}$, the definition of the components is analogous to the definition of the components of the position vector. We place the tail of the vector at the origin, and we drop perpendiculars from the head of the vector to the x and y axes. The intercepts of these perpendiculars with the axes (which may be positive or negative) give us the x and y components of the vector $\mathbf{A}$ (see Fig. 3.14). Designating these components by A_x and A_y, we see from Fig. 3.14 that

$$A_x = A \cos \theta \qquad (3.13)$$

components of a vector

$$A_y = A \sin \theta \qquad (3.14)$$

These equations are analogous to Eqs. (3.9) and (3.10). Furthermore, we see from Fig. 3.14 that

$$A = \sqrt{A_x^2 + A_y^2} \qquad (3.15)$$

magnitude in terms of components

and that

$$\tan \theta = \frac{A_y}{A_x} \qquad (3.16)$$

These two equations are analogous to Eqs. (3.11) and (3.12).

Note that the components uniquely specify the vector—if we know the components, we can find the magnitude and direction of the vector from Eqs. (3.15) and (3.16).

The addition or subtraction of two vectors can be performed by adding or subtracting their components. Thus, if A_x, A_y and B_x, B_y are the components of the vectors $\mathbf{A}$ and $\mathbf{B}$, then the components of the resultant $\mathbf{C} = \mathbf{A} + \mathbf{B}$ of these two vectors are

$$\begin{aligned} C_x &= A_x + B_x \\ C_y &= A_y + B_y \end{aligned} \qquad (3.17)$$

This is sometimes the most convenient method for evaluating sums or differences of vectors.

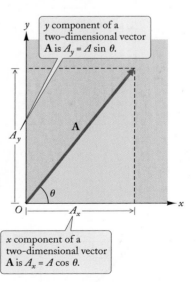

y component of a two-dimensional vector $\mathbf{A}$ is $A_y = A \sin \theta$.

x component of a two-dimensional vector $\mathbf{A}$ is $A_x = A \cos \theta$.

FIGURE 3.14 A vector $\mathbf{A}$ in two dimensions and its components A_x and A_y.

(a)

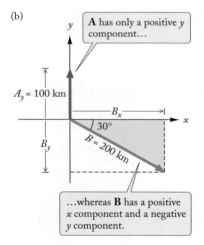

C is the vector difference
B – **A**, pointing from head
of **A** to head of **B**.

(b)

A has only a positive y
component…

…whereas **B** has a positive
x component and a negative
y component.

(c)

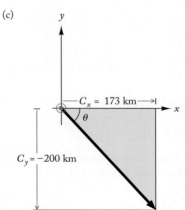

EXAMPLE 2 The eye of the hurricane in the chapter photo is 200 km from Miami on a bearing of 30° south of east. Suppose a reconnaissance aircraft is initially 100 km due north of Miami. What displacement vector will bring the aircraft to the eye of the hurricane?

SOLUTION: In Example 1 we saw how to use a graphical method and a trigonometric method for finding an unknown vector. Here we will see how to use the component method.

For this calculation, we need to make a choice of coordinate system, that is, a choice of origin and of axes. In Fig. 3.15, we have placed the origin on Miami, with the x axis eastward and the y axis northward. In this coordinate system, the airplane has a position vector **A** with components

$$A_x = 0 \text{ km} \qquad \text{and} \qquad A_y = 100 \text{ km}$$

The hurricane has a position vector **B** with components [see Eqs. (3.13) and (3.14)]

$$B_x = 200 \text{ km} \times \cos(-30°) = 173 \text{ km}$$
$$B_y = 200 \text{ km} \times \sin(-30°) = -100 \text{ km}$$

The displacement vector **C** from the airplane to the hurricane is the directed line segment from the head of **A** to the head of **B**. According to our discussion of vector subtraction, this directed line segment is equal to the vector difference **B** − **A**, that is, **C** = **B** − **A** (see Fig. 3.15; you may find it easier to recognize that **A** + **C** = **B**). The vector **C** therefore has components

$$C_x = B_x - A_x = 173 \text{ km} - 0 \text{ km} = 173 \text{ km}$$
$$C_y = B_y - A_y = -100 \text{ km} - 100 \text{ km} = -200 \text{ km}$$

The magnitude of the vector **C** is then, according to Eq. (3.15),

$$C = \sqrt{C_x^2 + C_y^2} = \sqrt{(173 \text{ km})^2 + (-200 \text{ km})^2} = 264 \text{ km}$$

and the angle between **C** and the x axis is given by [see Eq. (3.13)]

$$\cos\theta = \frac{C_x}{C} = \frac{173}{264} = 0.655$$

Our calculator then tells us

$$\theta = 49°$$

Hence a displacement of 264 km at 49° south of east will bring the airplane to the hurricane.

FIGURE 3.15 (a) Position vectors of the aircraft (**A**) and of the hurricane (**B**). The displacement vector from the aircraft to the hurricane is the difference between these vectors, **C** = **B** − **A**. (b) The x and y components of **A** and **B**. (c) The x and y components of **C**. To display these components, the tail of **C** has been shifted to the origin.

PROBLEM-SOLVING TECHNIQUES | VECTOR ADDITION AND SUBTRACTION

When dealing with the addition or subtraction of two or more vectors, it is usually convenient to use a method based on components, as in Example 2.

1 Begin by making some convenient choice of coordinate axes x and y. Selecting a coordinate system such that x or y is parallel to one of the vectors may help.

2 Draw the vectors. If possible, place the tails of the vectors at the origin, since this makes it easier to calculate their components. Try to draw the lengths and angles of the vectors fairly accurately, since this will help you to see roughly what answer to expect.

3 Drop perpendiculars from the heads (and, if necessary, the tails) of each vector, and mark the x and y components of each vector along the axes. The signs of the components will be obvious by inspection: a component is

positive if it extends from the origin along the positive part of the axis, and a component is negative if it extends from the origin to the negative part of the axis.

4 Add or subtract the components as required to obtain the components of the resultant.

5 If the problem asks for the magnitude and the direction of the resultant (instead of components), then calculate these by means of Eq. (3.15), $A = \sqrt{A_x^2 + A_y^2}$, and Eq. (3.16), $\tan \theta = A_y/A_x$, using the components found in step 4.

6 Keep in mind that in any problem that has a vector as answer, you must either state all the components of the vector or else the magnitude *and* the direction of the vector. It is a common mistake to state merely the magnitude, without the direction—this is an incomplete answer.

In the preceding, we have dealt with vectors in only two dimensions, with only x and y components. More generally, a vector in three dimensions has x, y, and z components. To obtain these components, we place the tail of the vector at the origin and drop perpendiculars from the head of the vector to the x–y, x–z, and y–z planes, so as to form a "box" of which the vector is the diagonal (see Fig. 3.16). The sides of this box give us the components A_x, A_y, and A_z of the vector. The generalization of our two-dimensional case to three dimensions gives for the length of the vector

$$A = \sqrt{A_x^2 + A_y^2 + A_z^2}$$

magnitude of a 3D vector

The position vector and other vectors can be expressed in terms of **unit vectors** along the coordinate axes. The unit vectors along the x, y, and z axes are designated by **i**, **j**, and **k**, respectively (see Fig. 3.17). The magnitude of each of these vectors is 1; that is, $|\mathbf{i}| = 1$, $|\mathbf{j}| = 1$, and $|\mathbf{k}| = 1$. Thus, the magnitudes of these *unit vectors have no*

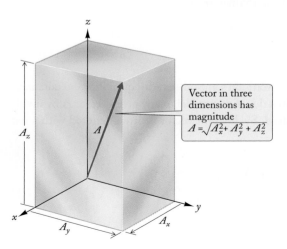

Vector in three dimensions has magnitude $A = \sqrt{A_x^2 + A_y^2 + A_z^2}$

FIGURE 3.16 A vector **A** in three dimensions and its components A_x, A_y, and A_z. The components A_x, A_y, and A_z are represented by the sides of the rectangular box constructed by dropping perpendiculars from the head of the vector to the x–y, x–z, and y–z planes.

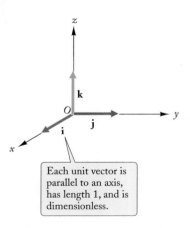

Each unit vector is parallel to an axis, has length 1, and is dimensionless.

FIGURE 3.17 The unit vectors **i**, **j**, and **k**.

dimensions of length, time, or mass—they are pure numbers. Now consider the vector sum $x\mathbf{i} + y\mathbf{j} + z\mathbf{k}$. This vector sum consists of a displacement of magnitude x in the x direction, followed by a displacement of magnitude y in the y direction, followed by a displacement of magnitude z in the z direction. Thus, this vector sum brings us from the origin to the point with coordinates x, y, z. Hence, this vector sum coincides with the position vector:

$$\mathbf{r} = x\mathbf{i} + y\mathbf{j} + z\mathbf{k} \tag{3.18}$$

More generally, for any arbitrary vector with components A_x, A_y, and A_z, we can write

$$\mathbf{A} = A_x\mathbf{i} + A_y\mathbf{j} + A_z\mathbf{k} \tag{3.19}$$

The advantage of expressing vectors in terms of unit vectors is that we can then manipulate them algebraically, as illustrated in the following example and in the calculations of the next section.

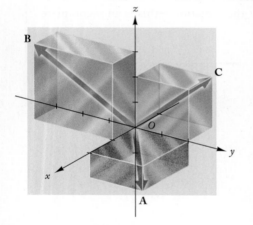

FIGURE 3.18 Three vectors **A**, **B**, and **C** in three dimensions.

EXAMPLE 3   Consider the following three vectors (see Fig. 3.18):

$$\mathbf{A} = 2\mathbf{i} + 2\mathbf{j} - \mathbf{k}$$
$$\mathbf{B} = \mathbf{i} - 3\mathbf{j} + 3\mathbf{k}$$
$$\mathbf{C} = -\mathbf{i} + 2\mathbf{j} + 2\mathbf{k}$$

What is the resultant $\mathbf{D} = \mathbf{A} + \mathbf{B} + \mathbf{C}$ of these three vectors? What is the magnitude of the resultant?

SOLUTION: When the vectors in a sum are expressed in terms of unit vectors, we can manipulate them algebraically, by collecting terms involving the same unit vectors:

$$\mathbf{D} = \mathbf{A} + \mathbf{B} + \mathbf{C} = (2\mathbf{i} + 2\mathbf{j} - \mathbf{k}) + (\mathbf{i} - 3\mathbf{j} + 3\mathbf{k}) + (-\mathbf{i} + 2\mathbf{j} + 2\mathbf{k})$$
$$= (2\mathbf{i} + \mathbf{i} - \mathbf{i}) + (2\mathbf{j} - 3\mathbf{j} + 2\mathbf{j}) + (-\mathbf{k} + 3\mathbf{k} + 2\mathbf{k})$$

Addition of the terms involving the same unit vectors is mathematically equivalent to addition of the components, as in Eq. (3.17). When we perform these additions, we find the resultant

$$\mathbf{D} = 2\mathbf{i} + \mathbf{j} + 4\mathbf{k}$$

From this addition we see that the components of the resultant are $D_x = 2$, $D_y = 1$, and $D_z = 4$. The magnitude D of the resultant is given by our three-dimensional generalization of the Pythagorean theorem:

$$D = \sqrt{D_x^2 + D_y^2 + D_z^2} = \sqrt{2^2 + 1^2 + 4^2} = \sqrt{21} = 4.6$$

We finish the section with a few words on some other vector quantities and notations. We can obtain a unit vector in a direction parallel to any given vector **A** by dividing the vector **A** by its magnitude A. Such a unit vector is often denoted with a hat above the original vector:

$$\hat{\mathbf{A}} = \frac{\mathbf{A}}{A}$$

This can be calculated from our usual explicit forms $\mathbf{A} = A_x\mathbf{i} + A_y\mathbf{j} + A_z\mathbf{k}$ and $A = \sqrt{A_x^2 + A_y^2 + A_z^2}$. From these relations, we see that this form of a unit vector indeed has a magnitude identically equal to 1.

Many books in science and engineering use a compact notation for a vector: the x, y, and z components are placed in order between parentheses. Thus we can write

$$\mathbf{A} = (A_x,\ A_y,\ A_z)$$

For example, the resultant of Example 3 would be written $\mathbf{D} = (2, 1, 4)$. Which of the several notations is most convenient depends on the calculation; we will most often use the unit-vector form.

Checkup 3.3

QUESTION 1: The magnitude of a vector is never smaller than the magnitude of any one component of the vector. Explain.

QUESTION 2: A vector is parallel to the y axis. What is its x component?

QUESTION 3: A vector has equal, positive x and y components and zero z component. What is the angle between this vector and the x axis?

QUESTION 4: A vector $\mathbf{A}$ has an x component $A_x = 3$ and a y component $A_y = -1$. What are the x and y components of the vector $2\mathbf{A}$? The vector $-4\mathbf{A}$?

QUESTION 5: What is the magnitude of the vector $\mathbf{i} + \mathbf{j}$?

   (A) 0   (B) 1   (C) $\sqrt{2}/2$ (D) $\sqrt{2}$ (E) 2

3.4 VECTOR MULTIPLICATION[2]

There are several ways of multiplying vectors. The reason for this diversity is that in forming the "product" of two vectors, we must take into account both their magnitudes and their directions. Depending on how we combine these quantities, we obtain different kinds of products. The two most important kinds of products are the dot product and the cross product.

Dot Product

The **dot product** (also called the **scalar product**) of two vectors $\mathbf{A}$ and $\mathbf{B}$ is denoted by $\mathbf{A} \cdot \mathbf{B}$. This quantity is simply *the product of the magnitudes of the two vectors and the cosine of the angle between them* (Fig. 3.19):

$$\mathbf{A} \cdot \mathbf{B} = AB \cos \phi \qquad (3.20)$$

Thus, the dot product of two vectors simply gives a number, that is, a scalar rather than a vector. The number will be positive if $\phi < 90°$, negative if $\phi > 90°$, and zero if $\phi = 90°$. If the two vectors are perpendicular, then their dot product is zero. Note that the dot product is commutative; as in ordinary multiplication, the order of the factors is irrelevant, that is, $\mathbf{A} \cdot \mathbf{B} = \mathbf{B} \cdot \mathbf{A}$.

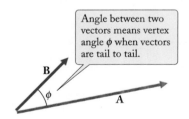

Angle between two vectors means vertex angle ϕ when vectors are tail to tail.

FIGURE 3.19 Two vectors $\mathbf{A}$ and $\mathbf{B}$. The angle between them is ϕ.

dot product in terms of angle

---

[2] The dot product will first be used in Section 7.1 and the cross product will first be used in Section 13.4. Those sections contain brief, self-contained expositions of the dot product and cross products.

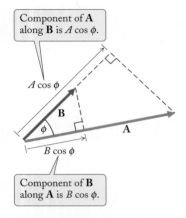

Component of **A** along **B** is $A \cos \phi$.

$A \cos \phi$

B

ϕ

A

$B \cos \phi$

Component of **B** along **A** is $B \cos \phi$.

FIGURE 3.20 The component of **B** along **A** (obtained by dropping a perpendicular from the head of **B** toward **A**) equals $B \cos \phi$. Hence the scalar product $\mathbf{A} \cdot \mathbf{B} = AB \cos \phi$ equals the magnitude of **A** multiplied by the component of **B** along **A**.

We see from Fig. 3.20 that the dot product $\mathbf{A} \cdot \mathbf{B}$ can be regarded as B times the component of **A** along **B**, or as A times the component of **B** along **A**. The special case of the dot product of a vector with itself gives the square of the magnitude of the vector:

$$\mathbf{A} \cdot \mathbf{A} = A A \cos 0° = A^2 \tag{3.21}$$

By means of the the usual rules for the multiplication of a sum of terms, we can derive a useful expression for the dot product in terms of components. Suppose that $\mathbf{A} = A_x \mathbf{i} + A_y \mathbf{j}$ and $\mathbf{B} = B_x \mathbf{i} + B_y \mathbf{j}$; then

$$\mathbf{A} \cdot \mathbf{B} = (A_x \mathbf{i} + A_y \mathbf{j}) \cdot (B_x \mathbf{i} + B_y \mathbf{j})$$

and if we multiply this out term by term, we find

$$\mathbf{A} \cdot \mathbf{B} = A_x B_x \, \mathbf{i} \cdot \mathbf{i} + A_x B_y \, \mathbf{i} \cdot \mathbf{j} + A_y B_x \, \mathbf{j} \cdot \mathbf{i} + A_y B_y \, \mathbf{j} \cdot \mathbf{j} \tag{3.22}$$

Here, $\mathbf{i} \cdot \mathbf{i}$ is the product of a vector with itself, which is the square of the magnitude of the vector:

$$\mathbf{i} \cdot \mathbf{i} = 1^2 = 1$$

Likewise,

$$\mathbf{j} \cdot \mathbf{j} = 1^2 = 1$$

The middle two terms in Eq. (3.22) involve $\mathbf{i} \cdot \mathbf{j}$ and $\mathbf{j} \cdot \mathbf{i}$, both of which are zero, since **i** and **j** are perpendicular to each other (compare Fig. 3.17). Hence Eq. (3.22) reduces to

$$\mathbf{A} \cdot \mathbf{B} = A_x B_x + A_y B_y \tag{3.23}$$

More generally, we can show that in the three-dimensional case, the dot product can be expressed as

dot product in terms of components

$$\mathbf{A} \cdot \mathbf{B} = A_x B_x + A_y B_y + A_z B_z \tag{3.24}$$

Thus the dot product is simply the sum of the products of the x, y, and z components of the two vectors.

Finally, note that the components of a vector are equal to the dot product of the vector and the corresponding unit vector. For instance, if $\mathbf{A} = A_x \mathbf{i} + A_y \mathbf{j}$, then

$$\mathbf{i} \cdot \mathbf{A} = \mathbf{i} \cdot (A_x \mathbf{i} + A_y \mathbf{j})$$
$$= A_x \mathbf{i} \cdot \mathbf{i} + A_y \mathbf{i} \cdot \mathbf{j} = A_x \times 1 + A_y \times 0 = A_x$$

and

$$\mathbf{j} \cdot \mathbf{A} = \mathbf{j} \cdot (A_x \mathbf{i} + A_y \mathbf{j})$$
$$= A_x \mathbf{j} \cdot \mathbf{i} + A_y \mathbf{j} \cdot \mathbf{j} = A_x \times 0 + A_y \times 1 = A_y$$

EXAMPLE 4   Find the dot product of the vectors **A** and **B** of Example 2.

SOLUTION: The vector **A** has a magnitude $A = 100$ km and the vector **B** has a magnitude $B = 200$ km; the angle between the vectors is $\phi = 120°$. Hence

$$\mathbf{A} \cdot \mathbf{B} = AB \cos \phi = 100 \text{ km} \times 200 \text{ km} \times \cos 120°$$
$$= -10\,000 \text{ km}^2 \tag{3.25}$$

Alternatively, this calculation can be done using the components from Example 2:

$$\mathbf{A} \cdot \mathbf{B} = A_x B_x + A_y B_y$$
$$= 0 \text{ km} \times 173 \text{ km} + 100 \text{ km} \times (-100 \text{ km}) = -10\,000 \text{ km}^2 \qquad (3.26)$$

This agrees with Eq. (3.25).

EXAMPLE 5 The displacement from Miami to the hurricane in Example 2 is $\mathbf{B} = (173 \text{ km})\mathbf{i} + (-100 \text{ km})\mathbf{j}$; the displacement from the initial position of the reconnaissance aircraft to the hurricane is $\mathbf{C} = (173 \text{ km})\mathbf{i} + (-200 \text{ km})\mathbf{j}$. Use the dot product to determine the angle between these two vectors (the angle ϕ in Fig. 3.21).

SOLUTION: Since

$$\mathbf{B} \cdot \mathbf{C} = BC \cos \phi$$

the cosine of the angle ϕ between the vectors is

$$\cos \phi = \frac{\mathbf{B} \cdot \mathbf{C}}{BC} \qquad (3.27)$$

Here, we can substitute for $\mathbf{B} \cdot \mathbf{C}$ the component expression (3.23), and for B and C the usual expressions (3.15) for the magnitudes:

$$\cos \phi = \frac{\mathbf{B} \cdot \mathbf{C}}{BC} = \frac{B_x C_x + B_y C_y}{\sqrt{B_x^2 + B_y^2} \sqrt{C_x^2 + C_y^2}}$$

$$= \frac{(173 \text{ km})(173 \text{ km}) + (-100 \text{ km})(-200 \text{ km})}{\sqrt{(173 \text{ km})^2 + (-100 \text{ km})^2} \sqrt{(173 \text{ km})^2 + (-200 \text{ km})^2}}$$

$$= \frac{49\,900}{200 \times 264} = 0.945$$

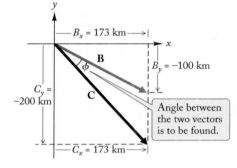

FIGURE 3.21 The two vectors $\mathbf{B}$ and $\mathbf{C}$ from Fig. 3.15.

With our calculator, we take the inverse cosine of 0.945 and find that the angle is

$$\phi = 19.1°$$

COMMENT: This trick for calculating the unknown angle between two vectors is also useful for calculating the angle between two lines in space; simply take any two vectors pointing along the lines and calculate the angle between them.

Cross Product

In contrast to the dot product of two vectors, which is a scalar, the **cross product** (also called the **vector product**) of two vectors is a vector. The cross product of two vectors $\mathbf{A}$ and $\mathbf{B}$ is denoted by $\mathbf{A} \times \mathbf{B}$. The magnitude of this vector is equal to *the product of the magnitudes of the two vectors and the sine of the angle between them.* Thus if we write the vector resulting from the cross product as

$$\mathbf{C} = \mathbf{A} \times \mathbf{B} \qquad (3.28)$$

then the magnitude of this vector is

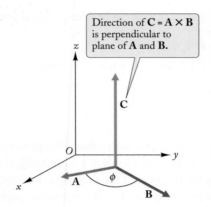

FIGURE 3.22 Two vectors **A** and **B** and their cross product **C** = **A** × **B**.

magnitude of cross product

$$C = AB \sin \phi \qquad (3.29)$$

The direction of the vector **C** is defined to be along the perpendicular to the plane formed by **A** and **B** (Fig. 3.22). The direction of **C** along this perpendicular is given by the **right-hand rule**: put the fingers of your right hand along **A** (Fig. 3.23a), and curl them toward **B** in the direction of the smaller angle from **A** to **B** (Fig. 3.23b); the thumb then points along **C**. Note that the fingers must be curled from the first vector in the product toward the second. Thus, **A** × **B** is not the same as **B** × **A**. For the latter product, the fingers must be curled from **B** toward **A** (rather than vice versa); hence, the direction of the vector **B** × **A** is opposite to that of **A** × **B**:

cross product is anticommutative

$$\mathbf{B} \times \mathbf{A} = -\mathbf{A} \times \mathbf{B} \qquad (3.30)$$

Accordingly, the cross product of two vectors is anticommutative; in contrast to ordinary multiplication, the result depends on the order of the factors and changes sign when the order of factors is reversed.

As we see from Fig. 3.24, the magnitude of **A** × **B** is equal to the magnitude of **B** times the component of **A** perpendicular to **B** (or the magnitude of **A** times the component of **B** perpendicular to **A**). Furthermore, from Fig. 3.24, we see that the magnitude of **A** × **B** is equal to the area of the parallelogram formed out of the vectors **A** and **B**.

If the vectors **A** and **B** are parallel, then their cross product is zero, since for two parallel vectors, $\phi = 0$ in Eq. (3.29). In particular, the cross product of any vector with

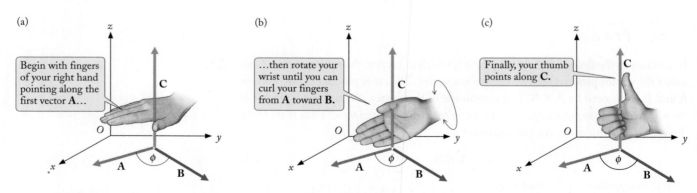

FIGURE 3.23 The right-hand rule for the cross product. If the fingers of the right hand curl from **A** toward **B**, the thumb points along **C**.

FIGURE 3.24 (a) The component of **A** perpendicular to **B** is $A \sin \phi$. (b) The area of the parallelogram that has **A** and **B** as sides equals the length of the base (B) multiplied by the height $A \sin \phi$.

(a)

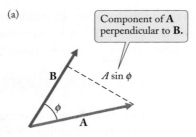

Component of **A** perpendicular to **B**.

(b)

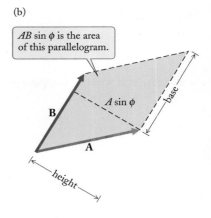

$AB \sin \phi$ is the area of this parallelogram.

itself is zero. In the case of the unit vectors, this tells us $\mathbf{i} \times \mathbf{i} = 0$, $\mathbf{j} \times \mathbf{j} = 0$, and $\mathbf{k} \times \mathbf{k} = 0$. The cross product of two different unit vectors is particularly simple: since each vector has unit magnitude and each pair is at right angles ($\sin 90° = 1$), application of the right-hand rule gives (see Fig. 3.17)

$$\mathbf{i} \times \mathbf{j} = \mathbf{k} \qquad \mathbf{j} \times \mathbf{k} = \mathbf{i} \qquad \mathbf{k} \times \mathbf{i} = \mathbf{j} \tag{3.31}$$

As in Eq. (3.30), reversing the order of multiplication changes the sign of the cross products:

$$\mathbf{j} \times \mathbf{i} = -\mathbf{k} \qquad \mathbf{k} \times \mathbf{j} = -\mathbf{i} \qquad \mathbf{i} \times \mathbf{k} = -\mathbf{j} \tag{3.32}$$

Notice that the unit vectors in these equations are in cyclic order (**ijkijk**) for the positive products, and in reverse order (**kjikji**) for the negative products.

EXAMPLE 6 What is the cross product of the two vectors **A** and **B** of Example 2?

SOLUTION: Figure 3.25 shows the vectors **A** and **B**. The magnitudes of these vectors are 100 km and 200 km, respectively; the angle between them is 120°.

Hence the magnitude of the cross product is

$$C = AB \sin \phi$$
$$= 100 \text{ km} \times 200 \text{ km} \times \sin 120° = 17\,300 \text{ km}^2$$

The direction of the vector **C** is perpendicular to the plane of **A** and **B**; according to the right-hand rule, **C** is along the negative z axis (Fig. 3.25).

Alternatively, we could have used the component notation throughout. From Example 2, we know

$$\mathbf{A} = (100 \text{ km})\mathbf{j}$$
$$\mathbf{B} = (173 \text{ km})\mathbf{i} + (-100 \text{ km})\mathbf{j}$$

To find the cross product, we multiply out the expressions, taking care to maintain the order of the products:

$$\mathbf{C} = \mathbf{A} \times \mathbf{B} = (100 \text{ km})\mathbf{j} \times [(173 \text{ km})\mathbf{i} + (-100 \text{ km})\mathbf{j}]$$
$$= (100 \text{ km})(173 \text{ km})\mathbf{j} \times \mathbf{i} + (100 \text{ km})(-100 \text{ km})\mathbf{j} \times \mathbf{j}$$
$$= (17\,300 \text{ km}^2)(-\mathbf{k}) + 0 = (-17\,300 \text{ km}^2)\mathbf{k}$$

The second term was set to zero, since $\mathbf{j} \times \mathbf{j} = 0$.

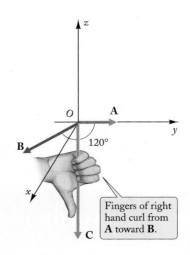

Fingers of right hand curl from **A** toward **B**.

FIGURE 3.25 The two vectors **A** and **B**, and their cross product.

In most situations, one of the techniques of Example 6 will provide the most direct means of evaluating the cross product. The second technique, multiplying out the components, can also be used to write down a general expression for the cross product of two arbitrary vectors. By multiplying out the cross product of $\mathbf{A} = A_x\mathbf{i} + A_y\mathbf{j} + A_z\mathbf{k}$ and $\mathbf{B} = B_x\mathbf{i} + B_y\mathbf{j} + B_z\mathbf{k}$ and evaluating the nine resulting unit-vector cross products, we can obtain the general result (see Problem 56)

cross product in terms of components

$$\mathbf{A} \times \mathbf{B} = (A_yB_z - A_zB_y)\mathbf{i} + (A_zB_x - A_xB_z)\mathbf{j} + (A_xB_y - A_yB_x)\mathbf{k} \qquad (3.33)$$

Those readers familiar with determinants can verify that the right side of Eq. (3.33) can be obtained by expanding a 3×3 determinant:

$$\mathbf{A} \times \mathbf{B} = \begin{vmatrix} \mathbf{i} & \mathbf{j} & \mathbf{k} \\ A_x & A_y & A_z \\ B_x & B_y & B_z \end{vmatrix} = \mathbf{i}\begin{vmatrix} A_y & A_z \\ B_y & B_z \end{vmatrix} - \mathbf{j}\begin{vmatrix} A_x & A_z \\ B_x & B_z \end{vmatrix} + \mathbf{k}\begin{vmatrix} A_x & A_y \\ B_x & B_y \end{vmatrix} \qquad (3.34)$$

where a 2×2 determinant is given by $\begin{vmatrix} a & b \\ c & d \end{vmatrix} = ad - bc$.

As a final note, we point out that since the cross product is a vector, its calculation, like any relation equating vectors, really involves three separate equations, one for each component.

 Checkup 3.4

QUESTION 1: Two vectors have nonzero magnitude. Under what conditions will their dot product be zero? Their cross product?

QUESTION 2: Suppose that $\mathbf{A} \cdot \mathbf{i} = 0$ and $\mathbf{A} \cdot \mathbf{k} = 0$. What can you conclude about the direction of a nonzero vector $\mathbf{A}$?

QUESTION 3: The dot product of two vectors $\mathbf{A}$ and $\mathbf{B}$ is -2. What is the dot product of the two vectors $2\mathbf{A}$ and $5\mathbf{B}$?

QUESTION 4: Suppose that $\mathbf{A} \times \mathbf{j} = \mathbf{0}$. What can you conclude about the direction of a nonzero vector $\mathbf{A}$?

QUESTION 5: Suppose that the vector $\mathbf{A}$ points west and the vector $\mathbf{B}$ points vertically up. What is the direction of $\mathbf{A} \times \mathbf{B}$? The direction of $\mathbf{B} \times \mathbf{A}$?

QUESTION 6: Suppose that $\mathbf{A} \cdot \mathbf{B} > 0$. What can you conclude about the angle ϕ between $\mathbf{A}$ and $\mathbf{B}$?

(A) $0 \le \phi < 90°$ (B) $\phi = 90°$ (C) $90° < \phi < 180°$ (D) $\phi = 180°$

PROBLEM-SOLVING TECHNIQUES   **DOT PRODUCT AND CROSS PRODUCT OF VECTORS**

- We have available two alternative formulas for evaluating the dot product of two vectors, Eq. (3.20) and Eq. (3.24). Which is best depends on how the vectors are specified. If the magnitudes and the directions of the vectors are given, then Eq. (3.20) is best; if the components are given, then Eq. (3.24) is best.

- We also have available two formulas for the cross product, Eq. (3.29) and Eq. (3.33). If we know the magnitudes and the directions of the vectors; then Eq. (3.29) is best. If we

know the components, Eq. (3.33) provides a general expression for the cross product, although it is rather messy. In many cases, it is easier to merely multiply out the components and evaluate the simpler unit-vector cross products of Eqs. (3.31) and (3.32), as we did in Example 6.

- Because the right hand is often busy with the pencil, it is a common mistake to use the left hand while trying to determine the direction of a cross product. Do not make this mistake—it gives the opposite direction.

SUMMARY

PHYSICS IN PRACTICE   Vectors in Navigation   **(page 71)**

PROBLEM-SOLVING TECHNIQUES   Vector Addition and Subtraction   **(page 79)**

PROBLEM-SOLVING TECHNIQUES   Dot Product and Cross Product Vectors   **(page 86)**

VECTOR   Quantity with magnitude and direction;
it behaves like a displacement.

ADDITION OF VECTORS   Use the parallelogram method or
the tail-to-head method; alternatively, add components.

(3.3)

COMPONENTS OF A 2D VECTOR
(θ is measured counterclockwise
from the x axis.)

$A_x = A \cos \theta$
$A_y = A \sin \theta$

(3.13)
(3.14)

DIRECTION OF A 2D VECTOR

$\tan \theta = \dfrac{A_y}{A_x}$

(3.15)

MAGNITUDE OF A 2D VECTOR

$A = \sqrt{A_x^2 + A_y^2}$

(3.16)

MAGNITUDE OF A 3D VECTOR

$A = \sqrt{A_x^2 + A_y^2 + A_z^2}$

(3.18)

UNIT VECTORS $\mathbf{i}$, $\mathbf{j}$, and $\mathbf{k}$ along
the x, y, and z axes, respectively

DOT PRODUCT $\mathbf{A} \cdot \mathbf{B} = AB \cos \phi = A_x B_x + A_y B_y + A_z B_z$

(3.21)

CROSS PRODUCT $\mathbf{C} = \mathbf{A} \times \mathbf{B}$ (direction of $\mathbf{C}$ is perpendicular
to both $\mathbf{A}$ and $\mathbf{B}$, given by right-hand rule: curl the fingers from
$\mathbf{A}$ to $\mathbf{B}$ and the thumb gives the direction of $\mathbf{C}$)

MAGNITUDE OF CROSS PRODUCT	$C = AB \sin \phi$	(3.30)

CROSS PRODUCTS OF UNIT VECTORS		
In cyclic order, positive:	$\mathbf{i} \times \mathbf{j} = \mathbf{k}$ $\mathbf{j} \times \mathbf{k} = \mathbf{i}$ $\mathbf{k} \times \mathbf{i} = \mathbf{j}$	(3.31)
In reverse order, negative:	$\mathbf{j} \times \mathbf{i} = -\mathbf{k}$ $\mathbf{k} \times \mathbf{j} = -\mathbf{i}$ $\mathbf{i} \times \mathbf{k} = -\mathbf{j}$	(3.32)

| CROSS PRODUCT IN TERMS OF COMPONENTS | $\mathbf{A} \times \mathbf{B} = (A_y B_z - A_z B_y)\mathbf{i} + (A_z B_x - A_x B_z)\mathbf{j} + (A_x B_y - A_y B_x)\mathbf{k}$ | (3.33) |

QUESTIONS FOR DISCUSSION

1. A large oil tanker proceeds from Kharg Island (Persian Gulf) to Rotterdam via the Cape of Good Hope. A small oil tanker proceeds from Kharg Island to Rotterdam via the Suez Canal. Are the displacement vectors of the two tankers equal? Are the distances covered equal?

2. An airplane flies from Boston to Houston and back to Boston. Is the displacement zero in the reference frame of the Earth? In the reference frame of the Sun?

3. Does a vector of zero magnitude have a direction? Does it matter?

4. If $\mathbf{A}$ and $\mathbf{B}$ are any arbitrary vectors, then $\mathbf{A} \cdot (\mathbf{B} \times \mathbf{A}) = 0$. Explain.

5. Why is there no vector division? (Hint: If $\mathbf{A}$ and $\mathbf{B}$ are given and if $A = \mathbf{B} \cdot \mathbf{C}$, then there exist several vectors $\mathbf{C}$ that satisfy this equation, and similarly for $\mathbf{A} = \mathbf{B} \times \mathbf{C}$.)

6. Assume that $\mathbf{A}$ is some nonzero vector. If $\mathbf{A} \cdot \mathbf{B} = \mathbf{A} \cdot \mathbf{C}$, can we conclude that $\mathbf{B} = \mathbf{C}$? If $\mathbf{A} \times \mathbf{B} = \mathbf{A} \times \mathbf{C}$, can we conclude that $\mathbf{B} = \mathbf{C}$? What if *both* $\mathbf{A} \cdot \mathbf{B} = \mathbf{A} \cdot \mathbf{C}$ and $\mathbf{A} \times \mathbf{B} = \mathbf{A} \times \mathbf{C}$?

PROBLEMS

3.2 Vector Addition and Subtraction[†]

1. A ship moves from the Golden Gate Bridge in San Francisco Bay to Alcatraz Island and from there to Point Blunt. The first displacement vector is 10.2 km due east, and the second is 5.9 km due north. What is the resultant displacement vector?

2. In midtown Manhattan, the street blocks have a uniform size of 80 m × 280 m, with the shortest side oriented at 29° east of north ("uptown") and the long side oriented at 29° north of west. Suppose you walk three blocks uptown and then two blocks to the left. What is the magnitude and direction of your displacement vector?

3. Figure 3.26 shows the successive displacements of an aircraft flying a search pattern. The initial position of the aircraft is P and the final position is P'. What is the net displacement (magnitude and direction) between P and P'? Find the answer both graphically (by carefully drawing a page-size diagram with protractor and ruler and measuring the resultant) and trigonometrically (by solving triangles).

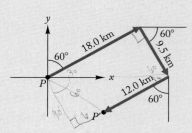

FIGURE 3.26 Successive displacement vectors of an aircraft.

4. A sailboat tacking against the wind moves as follows: 3.2 km at 45° east of north, 4.5 km at 50° west of north, 2.6 km at 45° east of north. What is the net displacement for the entire motion?

5. The recommended route for yachts entering the small harbor of South Bay at Mosquito Island (British Virgin Islands) is as follows: from the northeast side of Mosquito Rock proceed on a course of 135° (magnetic) for 1450 m; then alter course to 180° and proceed for 620 m; then alter course to 285° and proceed for 1190 m; finally alter course to 0° and proceed for

[†] For help, see Online Concept Tutorial 4 at www.wwnorton.com/physics

730 m to reach South Bay harbor. What is the displacement vector from Mosquito Rock to the harbor? Solve this problem graphically, by making an accurate, large diagram showing the successive displacement vectors.

6. A man runs 2.5 km north and then 1.5 km in a direction 30° east of north. A woman walks directly between the same initial and final points. What distance does the woman walk? In what direction?

7. An airplane flies 480 km in a direction 40° east of north to a hub city, and then 370 km in a direction 10° north of west to a final destination. By plotting these individual displacements on graph paper, carefully determine the magnitude and direction of the net displacement.

8. One speedboat travels 14 km in a direction 60° south of east. A second boat has the same displacement, but instead travels due east and then due south. What distance did the second boat travel east? What distance south?

9. The resultant of two displacement vectors has a length of 5.0 m and a direction due north. One of the displacement vectors has a length of 2.2 m and a direction 35° east of north. What is the other displacement vector?

10. Three displacement vectors **A, B,** and **C** are, respectively, 4.0 cm at 30° west of north, 8.0 cm at 30° east of north, and 3.0 cm due north. Carefully draw these vectors on a sheet of paper. Find **A** + **B** + **C** graphically. Find **A** + **B** − **C** graphically.

11. During the maneuvers preceding the Battle of Jutland, the British battle cruiser *Lion* moved as follows (distances are in nautical miles): 1.2 nmi due north, 6.1 nmi at 38° east of south, 2.9 nmi at 59° east of south, 4.0 nmi at 89° east of north, and 6.5 nmi at 31° east of north.

    (a) Draw each of these displacement vectors and draw the net displacement vector.

    (b) Graphically or algebraically find the distance between the initial position and the final position.

*12. The Earth moves around the Sun in a circle of radius 1.50×10^{11} m at (approximately) constant speed.

    (a) Taking today's position of the Earth as origin, draw a diagram showing the position vector 3 months, 6 months, 9 months, and 12 months later.

    (b) Draw the displacement vector between the 0-month and the 3-month positions, the 3-month and the 6-month positions, etc. Calculate the magnitude of the displacement vector for one of these 3-month intervals.

*13. Both Singapore and Quito are (nearly) on the Earth's equator; the longitude of Singapore is 104° east and that of Quito is 78° west. What is the magnitude of the displacement vector between these cities? What is the distance between them measured along the equator?

*14. By a method known as "doubling the angle on the bow," the navigator of a ship can determine his position relative to a fixed point, such as a lighthouse. Figure 3.27 shows the (straight) track of a ship passing by a lighthouse. At the point P, the navigator measures the angle α between the line of

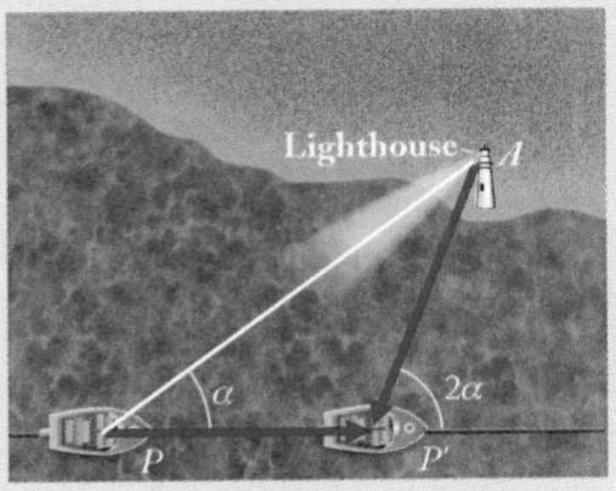

FIGURE 3.27 A lighthouse and two positions of a ship.

sight to the lighthouse and the direction of motion of the ship. He then measures how far the ship advances through the water until the angle between the line of sight and the direction of motion is twice as large as it was initially. Prove that the magnitude of the displacement vector PP' equals the magnitude of the position vector AP' of the ship relative to the lighthouse.

*15. The radar operator of a stationary Coast Guard cutter observes that at 10^h30^m an unidentified ship is at a distance of 9.5 km on a bearing of 60° east of north and at 11^h10^m the unidentified ship is at a distance of 4.2 km on a bearing of 33° east of north. Measured from its position at 10^h30^m, what is the displacement vector of the unidentified ship at 11^h10^m? Assuming that the unidentified ship continues on the same course at the same speed, what will be its displacement vector at 11^h30^m? What will be its distance and bearing from the cutter?

*16. Suppose that two ships proceeding at constant speeds are on converging straight tracks. Prove that the ships will collide if and only if the bearing of each remains constant as seen from the other. This constant-bearing rule is routinely used by mariners to check whether there is danger of collision. (Hint: A convenient method of proof is to draw the displacement vector from one ship to the other at several successive times.)

3.3 The Position Vector; Components of a Vector†

17. A displacement vector has a magnitude of 12.0 km in the direction 40.0° west of north. What is the north component of this vector? The west component?

18. A vector length of 5.0 m is in the *x–y* plane at an angle of 30° with the *x* axis. What is the *x* component of this vector? The *y* component?

19. A displacement vector has a magnitude of 4.0 m and a vertically downward direction. What is the component of this vector along a line sloping upward at 25°? (Hint: Place your *x* axis along the sloping line.)

20. Air traffic controllers usually describe the position of an aircraft relative to the airport by altitude, horizontal distance, and bearing. Suppose an aircraft is at altitude 500 m, distance 15 km, and bearing 35° east of north. What are the *x, y,* and *z* components (in meters) of the position vector? The *x* axis is east, the *y* axis is north, and the *z* axis is vertical.

† For help, see Online Concept Tutorial 4 at www.wwnorton.com/physics

21. The displacement vectors **A** and **B** are in the x–y plane. Their components are $A_x = 3.0$ cm, $A_y = 2.0$ cm, $B_x = -1.0$ cm, $B_y = 3.0$ cm.

    (a) Draw a diagram showing these vectors.

    (b) Calculate the resultant of **A** and **B**. Draw the resultant on your diagram.

22. A vector in the x–y plane has a magnitude of 8.0 units. The angle between the vector and the x axis is 52°. What are the x, y, and z components of this vector?

23. Given that a vector has a magnitude of 6.0 units and makes an angle of 45° and 85° with the x and y axes, respectively, find the x and y components of this vector. Does the given information determine the z component? What can you say about the z component?

24. What is the magnitude of the vector $3\mathbf{i} + 2\mathbf{j} - \mathbf{k}$?

25. Suppose that $\mathbf{A} = -5\mathbf{i} - 3\mathbf{j} + \mathbf{k}$ and $\mathbf{B} = 2\mathbf{i} + \mathbf{j} - 3\mathbf{k}$. Calculate the following:

    (a) $\mathbf{A} + \mathbf{B}$

    (b) $\mathbf{A} - \mathbf{B}$

    (c) $2\mathbf{A} - 3\mathbf{B}$

26. A position vector has x component 6.0 m, y component -8.0 m, and zero z component. What is the magnitude of this vector? What angle does it make with the x axis?

27. A vector of length 14 m points in a direction 135° counterclockwise from the x axis. What is its x component? Its y component?

28. Four vectors are given by $\mathbf{A} = 2.5\mathbf{i} + 3.5\mathbf{j}$, $\mathbf{B} = 1.0\mathbf{i} + 4.5\mathbf{j} + 2.5\mathbf{k}$, $\mathbf{C} = 1.5\mathbf{i} + 2.0\mathbf{j} + 3.0\mathbf{k}$, and $\mathbf{D} = 3.0\mathbf{j} + 1.5\mathbf{k}$. What is the vector $\mathbf{E} = \mathbf{A} + \mathbf{B} + \mathbf{C} + \mathbf{D}$? What is the vector $\mathbf{F} = \mathbf{A} - \mathbf{B} + \mathbf{C} - \mathbf{D}$? What is the magnitude of **E**? The magnitude of **F**?

29. What is the unit vector parallel to the vector $\mathbf{A} = 2.0\mathbf{i} + 4.0\mathbf{j} + 4.0\mathbf{k}$?

*30. Two vectors are given by $\mathbf{A} = 2.0\mathbf{i} + 3.0\mathbf{j} + 1.0\mathbf{k}$ and $\mathbf{B} = -1.0\mathbf{i} + 2.0\mathbf{j} + B_z\mathbf{k}$. The magnitude of the resultant $\mathbf{A} + \mathbf{B}$ is 6.0. What are the two possible values of B_z?

*31. Three vectors are given by $\mathbf{A} = 2.0\mathbf{i} + 3.0\mathbf{j}$, $\mathbf{B} = 1.0\mathbf{i} + 5.0\mathbf{j}$, and $\mathbf{C} = -1.0\mathbf{i} + 3.0\mathbf{j}$. Find constants c_1 and c_2 such that $c_1\mathbf{A} + c_2\mathbf{B} = \mathbf{C}$.

*32. An air traffic controller notices that one aircraft approaching the airport is at an altitude of 2500 m, (horizontal) distance 120 km, and bearing 20° south of east. A second aircraft is at altitude 3500 m, distance 110 km, and bearing 25° south of east. What is the displacement vector from the first aircraft to the second? Express your answer in terms of altitude, (horizontal) distance, and bearing.

*33. A remarkably fast crossing of the Atlantic by sail was achieved in 1916 by the four-masted ship *Lancing*, which sailed from New York (latitude 40°48′ north, longitude 73°58′ west) to Cape Wrath, Scotland (latitude 58°36′ north, longitude 5°1′ west), in $6\frac{3}{4}$ days. What was the magnitude of the displacement vector for this trip?

*34. A vector has components $A_x = 5.0$, $A_y = -3.0$, $A_z = 1.0$. What is the magnitude of this vector? What is the angle between this vector and the x axis? The y axis? The z axis?

*35. Find a vector that has the same direction as $3\mathbf{i} - 6\mathbf{j} + 2\mathbf{k}$ but a magnitude of 2 units.

*36. Given that $\mathbf{A} = 6\mathbf{i} - 2\mathbf{j}$ and $\mathbf{B} = -4\mathbf{i} - 3\mathbf{j} + 8\mathbf{k}$, find a vector **C** such that $3\mathbf{A} - 2\mathbf{C} = 4\mathbf{B}$.

3.4 Vector Multiplication

37. Calculate the dot product of the vectors $5\mathbf{i} - 2\mathbf{j} + \mathbf{k}$ and $2\mathbf{i} - \mathbf{k}$.

38. Calculate the dot product of the vectors **A** and **B** described in Example 1.

39. Find the magnitude of the vector $-2\mathbf{i} + \mathbf{j} + 2\mathbf{k}$. Find the magnitude of the vector $3\mathbf{i} - 6\mathbf{j} + 2\mathbf{k}$. Find the angle between these two vectors.

40. Show that, in three dimensions, the dot product of two vectors can be expressed as [see Eq. (3.24)]

$$\mathbf{A} \cdot \mathbf{B} = A_x B_x + A_y B_y + A_z B_z$$

41. Find the angle between the vector $\mathbf{A} = 3.0\mathbf{i} + 4.0\mathbf{j} + 2.0\mathbf{k}$ and the x axis.

42. For the two vectors $\mathbf{A} = 4.0\mathbf{i} + 3.0\mathbf{j} + 2.0\mathbf{k}$ and $\mathbf{B} = -1.0\mathbf{i} + 2.0\mathbf{j} + 1.0\mathbf{k}$, calculate $\mathbf{A} \cdot \mathbf{B}$ and $\mathbf{A} \times \mathbf{B}$.

43. The dot product of two vectors and the magnitude of the cross product of the same vectors are equal. What is the angle between the two vectors?

44. The displacement vector **A** has a length of 50 m and a direction of 30° east of north; the displacement vector **B** has a length of 35 m and a direction 70° west of north. What are the magnitude and direction of the cross product $\mathbf{A} \times \mathbf{B}$? The cross product $\mathbf{B} \times \mathbf{A}$?

45. Calculate the cross product of the vectors **A** and **B** described in Example 1.

46. Calculate the cross product of the vectors $2\mathbf{i} - 5\mathbf{j} + 3\mathbf{k}$ and $\mathbf{i} - 2\mathbf{k}$.

*47. Suppose that

$$\mathbf{A} = \mathbf{i} \cos \omega t + \mathbf{j} \sin \omega t$$

where ω is a constant. Find $d\mathbf{A}/dt$ (note that **i** and **j** behave as constants in differentiation). Show that $d\mathbf{A}/dt$ is perpendicular to **A**.

*48. The displacement vector **A** has a length of 30 m and a direction 20° south of east. The displacement vector **B** has a length of 40 m and a direction 20° west of north. Find the component of **A** along **B**. Find the component of **B** along **A**.

*49. The two vectors $\mathbf{A} = 5.0\mathbf{i} - 2.0\mathbf{j} + 3.0\mathbf{k}$ and $\mathbf{B} = B_x\mathbf{i} + 3.0\mathbf{j} + B_z\mathbf{k}$ have cross product $\mathbf{C} = \mathbf{A} \times \mathbf{B} = 2.0\mathbf{j} + C_z\mathbf{k}$. Find the values of B_x, B_z, and C_z.

*50. Two vectors **A** and **B** lie in the x–y plane. Show that the tangent of the angle θ between them is given by

$$\tan \theta = \frac{A_x B_y - A_y B_x}{A_x B_x + A_y B_y}$$

*51. A vector **A** has components $A_x = 2, A_y = -1, A_z = -4$. Find a vector (give its components) that has the same direction as **A** but a magnitude of 1 unit.

*52. Find a unit vector that bisects the angle between the vectors $\mathbf{j} + 2\mathbf{k}$ and $3\mathbf{i} - \mathbf{j} + \mathbf{k}$.

*53. Find a unit vector that points toward a position halfway between the two position vectors $4\mathbf{i} + 2\mathbf{j}$ and $-\mathbf{i} + 3\mathbf{j} + 2\mathbf{k}$.

*54. Find the angle between the diagonal of a cube and one of its edges. (Hint: Suppose that the edges of the cube are parallel to the vectors **i**, **j**, and **k**. What vector is then parallel to the diagonal?)

*55. The dot product of two vectors **A** and **B** is zero. The magnitudes of the two vectors are, respectively, $A = 4$ and $B = 6$. What can you say about the cross product of these two vectors?

*56. (a) Figure 3.17 displays the unit vectors **i**, **j**, and **k** along the x, y, and z axes. The positive directions for these axes, and the directions for the unit vectors, have been chosen according to the standard convention for a "right-handed" coordinate system (if, say, the positive direction of the x axis and the direction of **i** were reversed, the coordinate system would become "left-handed"). Use the right-hand rule to establish the following multiplication table for the unit vectors displayed in Fig. 3.17 [see Eqs. (3.31) and (3.32)]:

$$
\begin{array}{lll}
\mathbf{i} \times \mathbf{i} = 0 & \mathbf{j} \times \mathbf{i} = -\mathbf{k} & \mathbf{k} \times \mathbf{i} = \mathbf{j} \\
\mathbf{i} \times \mathbf{j} = \mathbf{k} & \mathbf{j} \times \mathbf{j} = 0 & \mathbf{k} \times \mathbf{j} = -\mathbf{i} \\
\mathbf{i} \times \mathbf{k} = -\mathbf{j} & \mathbf{j} \times \mathbf{k} = \mathbf{i} & \mathbf{k} \times \mathbf{k} = 0
\end{array}
$$

(b) With the above multiplication table, derive Eq. (3.33) for the cross product of two vectors in terms of their components:

$$\mathbf{A} \times \mathbf{B} = (A_y B_z - A_z B_y)\mathbf{i} + (A_z B_x - A_x B_z)\mathbf{j} + (A_x B_y - A_y B_x)\mathbf{k}$$

(c) From this formula, verify that if **A** and **B** are parallel, then $\mathbf{A} \times \mathbf{B} = 0$.

*57. Verify that the evaluation of the determinant in Eq. (3.34) (according to the usual rules for determinants) yields the formula for the cross product given in Eq. (3.33):

$$
\mathbf{A} \times \mathbf{B} = \begin{vmatrix}
\mathbf{i} & \mathbf{j} & \mathbf{k} \\
A_x & A_y & A_z \\
B_x & B_y & B_z
\end{vmatrix}
$$

*58. Evaluate $((\mathbf{i} \times \mathbf{j}) \times \mathbf{i}) \times \mathbf{i}$. Evaluate $(((\mathbf{i} \times \mathbf{j}) \times \mathbf{j}) \times \mathbf{j})$.

*59. Given that $\mathbf{A} = 2\mathbf{i} - 3\mathbf{j} + 2\mathbf{k}$ and $\mathbf{B} = -3\mathbf{i} + 4\mathbf{k}$, calculate the cross product $\mathbf{A} \times \mathbf{B}$.

*60. The vectors **A**, **B**, and **C** have components $A_x = 3, A_y = -2, A_z = 2, B_x = 0, B_y = 0, B_z = 4, C_x = 2, C_y = -3, C_z = 0$. Calculate the following:

(a) $\mathbf{A} \cdot (\mathbf{B} + \mathbf{C})$

(b) $\mathbf{A} \times (\mathbf{B} + \mathbf{C})$

(c) $\mathbf{A} \cdot (\mathbf{B} \times \mathbf{C})$

(d) $\mathbf{A} \times (\mathbf{B} \times \mathbf{C})$

*61. Find a unit vector perpendicular to both $4\mathbf{i} + 3\mathbf{j}$ and $-\mathbf{i} - 3\mathbf{j} + 2\mathbf{k}$.

62. Show that the magnitude of $\mathbf{A} \cdot (\mathbf{B} \times \mathbf{C})$ is the volume of the parallelepiped determined by **A, **B**, and **C**.

63. Show that $\mathbf{A} \times (\mathbf{B} \times \mathbf{C}) = \mathbf{B}(\mathbf{A} \cdot \mathbf{C}) - \mathbf{C}(\mathbf{A} \cdot \mathbf{B})$. (Hint: Choose the orientation of your coordinate axes in such a way that **B is along the x axis and **C** is in the x–y plane.)

*64. In the vicinity of New York City, the direction of magnetic north is 11°55′ west of true north (that is, a magnetic compass needle points 11°55′ west of north). Suppose that an aircraft flies 5.0 km on a bearing of 56° east of magnetic north.

(a) What are the north and east components of this displacement in a coordinate system based on the direction of magnetic north?

(b) What are the north and east components of the displacement in a coordinate system based on the direction of true north?

*65. A vector has components $A_x = 6, A_y = -3, A_z = 0$ in a given rectangular coordinate system. Find a new coordinate system, with new directions of the x and y axes, such that the only nonzero component of the vector is A'_x.

REVIEW PROBLEMS

66. To reach Moose Jaw, Canada, you drive your automobile due north 90 km, and then due west 70 km. What are the magnitude and the direction of your displacement vector? What is the distance you traveled?

67. The displacement vector **A** has a length of 350 m in the direction 45° west of north; the displacement vector **B** has a length of 120 m in the direction 20° east of north. Find the magnitude and direction of the resultant of these vectors.

68. In Chapter 10 we will become acquainted with the center of mass, which for a collection of equal particles is simply the average position of the particles. Suppose that three particles have position vectors $5\mathbf{i} + 3\mathbf{k}$, $-2\mathbf{i} + \mathbf{j} - 3\mathbf{k}$, and $4\mathbf{i} + 2\mathbf{j} + \mathbf{k}$, respectively. What is the average of these position vectors? What is the length of this average position vector?

69. A vector drawn on a wall has a magnitude of 2.0 and makes an angle of 30° with the vertical direction. What are the vertical and horizontal components of this vector?

70. An aircraft flies 250 km in a direction 30° east of south, and then 250 km in a direction 30° west of south. What are the magnitude and the direction of the resultant displacement vector?

71. The vector **A** has a length of 6.2 units in a direction 30° south of east. The vector **B** has a length of 9.6 units in a direction due south. What is the sum **A** + **B** of these vectors? What is the difference **A** − **B**?

*72. A room measures 4 m in the x direction, 5 m in the y direction, and 3 m in the z direction. A lizard crawls along the walls from one corner of the room to the diametrically opposite corner. If the starting point is the origin of coordinates, what is the displacement vector? What is the length of the displacement vector? If the lizard chooses the shortest path along the walls, what is the length of its path?

73. What is the magnitude of the vector $2\mathbf{i} + \mathbf{j} - 4\mathbf{k}$?

74. Suppose that **A** = 4**i** − 2**j** and **B** = −3**i** − 4**j**. Calculate the following:
    (a) **A** + **B**
    (b) **A** − **B**
    (c) 3**A** − **B**

75. The displacement vector **A** has a length of 50 m and direction of 30° east of north; the displacement vector **B** has a length of 35 m and a direction 70° west of north. What is the dot product of these vectors?

76. The displacement vector **A** has a length of 6.0 m in the direction 30° east of north; the displacement vector **B** has a length of 8.0 m in the direction 40° south of east. Find the magnitude and the direction of **A** × **B**.

*77. Suppose that **A** = 3**i** + 4**j** and **B** = **i** + 3**j** − 2**k**. Find the component of **A** along the direction of **B**. Find the component of **B** along the direction of **A**.

*78. Given the vectors **A** = 2**i** + 2**j** − **k** and **B** = 3**i** − **j**, calculate
    (a) The sum **A** + **B**
    (b) The difference **A** − **B**
    (c) The dot product **A** · **B**
    (d) The cross product **A** × **B**

Answers to Checkups

Checkup 3.1

1. A displacement vector depends only on the net change in position, that is, on the beginning and ending points, and so the displacement vectors for the two airliners are equal. The distances traveled refer to the total distances covered, and so are unequal for the two different routes.

2. The magnitude of a vector is its length, which is defined as a positive quantity, and cannot be negative. Only the zero vector has zero length (and it is the only vector with no direction).

3. Yes, the two displacement vectors are equal. Both have the same direction (due west) and the same magnitude (3000 m); only the *change* in position matters, not any particular positions.

4. (C) 100 km. The vector from that point to Miami is one side of a 30°–60°–90° right triangle (the side opposite the 30° angle), a triangle with a 200-km hypotenuse. Thus the distance to Miami is (200 km) × (sin 30°) = 100 km.

Checkup 3.2

1. Their sum will be zero if the two vectors have the same magnitude but point in opposite directions (antiparallel). Their difference will be zero if they are identical vectors, that is, if they have the same magnitude and direction (parallel).

2. For two perpendicular vectors, the magnitude of the sum must be larger than the magnitude of each. This is so because two such vectors form the sides of a right triangle, with their sum forming the hypotenuse.

3. Yes; any two perpendicular vectors have the same magnitude for their sum and for their difference. (Picture adding two perpendicular vectors tail to head for the sum, and then reverse one of the vectors for the difference.)

4. By picturing adding the three vectors tail to head, a zero resultant requires that the head of the third vector close back on the tail of the first vector. Thus one can see that the only way three equal-magnitude vectors can sum to zero is if they form the sides of an equilateral triangle.

5. (B) 5 km. Since the two displacements are in perpendicular directions, the net displacement is given by the hypotenuse
$$C = \sqrt{A^2 + B^2} = \sqrt{(3\text{ km})^2 + (4\text{ km})^2} = 5\text{ km}.$$

Checkup 3.3

1. The magnitude is obtained by squaring the components, adding them, and taking the square root of the result. Since the squares are positive, the sum of the squares will always be greater than or equal to the square of any given component (only equal when the other components are zero). See Eq. (3.15).

2. A vector parallel to the y axis has only a y component, and so its x component must be zero.

3. With equal x and y components, it must point in a direction halfway between the x and y directions, that is, at 45° to the x axis. One can also calculate this from $\tan\theta = y/x = 1$.

4. A constant multiplying a vector multiplies each component; thus $2\mathbf{A}$ has x component $2A_x = 6$ and y component $2A_y = -2$. Similarly, $-4\mathbf{A}$ has x component $-4A_x = -12$ and y component $-4A_y = 4$.

5. (D) $\sqrt{2}$. Each of the two (perpendicular) components has unit magnitude, so the vector has magnitude $\sqrt{1^2 + 1^2} = \sqrt{2}$.

Checkup 3.4

1. The dot product will be zero if the vectors are perpendicular (since $\cos\phi = 0$ when $\phi = 90°$). The cross product will be zero when the vectors are parallel or antiparallel (since $\sin\phi = 0$ when $\phi = 0°$ or $180°$).

2. The vector must be perpendicular to both the $\mathbf{i}$ and $\mathbf{k}$ directions; thus it has only a $\mathbf{j}$ component (along the $\pm y$ direction).

3. Multiplying each vector by a constant mutiplies the product by that constant, so the dot product is $2 \times 5 = 10$ times as large, or equals -20.

4. If the cross product is zero, then $\mathbf{A}$ must have only a $\mathbf{j}$ component, that is, $\mathbf{A}$ is along the $\pm y$ axis. Alternatively, the sine of the angle between $\mathbf{A}$ and $\mathbf{j}$ must be zero ($\phi = 0°$ or $180°$), so that $\mathbf{A}$ must be parallel or antiparallel to $\mathbf{j}$.

5. Pointing the fingers of the right hand along $\mathbf{A}$ (west) and orienting the hand so that the fingers can curl toward $\mathbf{B}$ (upward), we find that the thumb points north (so this is the direction of $\mathbf{A} \times \mathbf{B}$). Similarly, pointing the fingers up and curling them toward the west, the thumb points south, the direction of $\mathbf{B} \times \mathbf{A}$. Also, if we know $\mathbf{A} \times \mathbf{B}$ is north, then $\mathbf{B} \times \mathbf{A} = -\mathbf{A} \times \mathbf{B}$ tells us that $\mathbf{B} \times \mathbf{A}$ is south.

6. (A) $0 \le \phi < 90°$. If the dot product is greater than zero, then the cosine of the angle between the vectors is positive, that is, $\phi < 90°$.

Motion in Two and Three Dimensions

4.1 Components of Velocity and Acceleration

4.2 The Velocity and Acceleration Vectors

4.3 Motion with Constant Acceleration

4.4 The Motion of Projectiles

4.5 Uniform Circular Motion

4.6 The Relativity of Motion and the Addition of Velocities

CONCEPTS IN CONTEXT

Concepts
— *in* —
Context

This time-exposure photograph shows the curved trajectories of incandescent chunks of lava ejected during an eruption of the Stromboli volcano. Such chunks of lava, called "volcanic bombs," are often ejected at speeds of 600 km/h or more, and they can land at distances of several km from the volcano. They sometimes start fires at the point of impact.

The motion of volcanic bombs is an instance of projectile motion, and with the concepts developed in this chapter we can address the following questions:

? Given the initial speed and direction of motion of the projectile, what are the horizontal and vertical components of the velocity and what are the horizontal and vertical motions? (Example 2, page 100)

? How high does the projectile rise vertically? (Example 4, page 105)

? At what time does the projectile reach its maximum height? (Example 4, page 105)

? What is the shape of the trajectory? (Section 4.4, page 108)

In this chapter we will deal with translational motion in a plane, such as the motion of an automobile on the crisscrossing and curving streets of a flat city or the motion of a boat on the surface of a lake. This is two-dimensional motion, and it is a simple generalization of the one-dimensional motion we studied in Chapter 2. In essence, two-dimensional motion consists of two one-dimensional motions occurring simultaneously. Thus, we will have to apply the formulas of Chapter 2 separately to each of these one-dimensional motions. We can further generalize to the case of three-dimensional motion, such as the motion of an aircraft or the motion of an automobile on a mountain road, consisting of three one-dimensional motions. But we will rarely have to deal with three-dimensional motion, because one- or two-dimensional motion adequately describes most of the physical problems to be discussed in later chapters.

We will examine in detail some cases of two-dimensional motion: projectile motion and uniform circular motion. A stone or a ball thrown by hand or hit by a racket, a bomb released from an aircraft, and a volcanic bomb ejected from a volcano (see the chapter photo) are examples of projectile motion. The projectile always remains in the fixed vertical plane defined by its initial vertical and horizontal velocities, and it traces out a two-dimensional curved trajectory in this plane. An automobile traveling at constant speed around a traffic circle, a child riding a merry-go-round, and a communications satellite circling the Earth are examples of uniform circular motion. The moving particle remains in a fixed plane, and it traces out a circular path in this plane.

4.1 COMPONENTS OF VELOCITY AND ACCELERATION

Online Concept Tutorial 5

To describe the translational motion of a particle in a plane, we need two coordinates, say, an x coordinate and a y coordinate. For instance, if the particle is an automobile traveling on the streets of a (flat) city, we can describe the position by choosing an origin at, say, the library, and laying out an x axis in the eastward direction and a y axis in the northward direction (see Fig. 4.1). The x and y coordinates measured with respect to these axes then provide a complete description of the translational motion of the automobile.

For a moving particle, both the x and the y components of the position change with time. Correspondingly, there are x and y components of the velocity. In the case of the average velocity, we can define these x and y components by analogy with Eq. (2.5):

$$\overline{v}_x = \frac{\Delta x}{\Delta t} \tag{4.1}$$

and

$$\overline{v}_y = \frac{\Delta y}{\Delta t} \tag{4.2}$$

where $\Delta x = x_2 - x_1$ and $\Delta y = y_2 - y_1$ are the changes in the x and y components of the position in the time interval $\Delta t = t_2 - t_1$ (see Fig. 4.1). Stated in words: the x component of the velocity is the rate of change of the x coordinate, and the y component of velocity is the rate of change of the y coordinate.

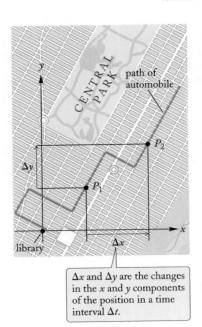

Δx and Δy are the changes in the x and y components of the position in a time interval Δt.

FIGURE 4.1 Path of an automobile along the streets of New York City.

Likewise, in the case of the instantaneous velocity, we can define the x and y components by analogy with Eq. (2.10):

$$v_x = \frac{dx}{dt} \tag{4.3}$$

and

$$v_y = \frac{dy}{dt} \tag{4.4}$$

As we saw in Chapter 2, for one-dimensional motion the instantaneous speed (such as the speed indicated by the speedometer of an automobile) equals the magnitude of the instantaneous velocity. For two-dimensional motion, the velocity has two components and the relationship between speed and velocity is not that obvious. To discover what the relationship is, consider the motion of a particle in a small time interval dt. In this time interval the particle travels a distance P_1P_2 from the point P_1 to the point P_2, and the changes in its x and y coordinates are dx and dy. As indicated in Fig. 4.2, dx and dy form the sides of a small right triangle, and P_1P_2 is the hypotenuse of this triangle. According to the Pythagorean theorem,

$$[\text{distance traveled}] = P_1P_2 = \sqrt{(dx)^2 + (dy)^2} \tag{4.5}$$

Hence the instantaneous speed is

$$[\text{instantaneous speed}] = \frac{[\text{distance traveled}]}{[\text{time taken}]} = \frac{P_1P_2}{dt} \tag{4.6}$$

$$= \frac{\sqrt{(dx)^2 + (dy)^2}}{dt} = \sqrt{\left(\frac{dx}{dt}\right)^2 + \left(\frac{dy}{dt}\right)^2}$$

$$= \sqrt{v_x^2 + v_y^2} \tag{4.7}$$

(a)

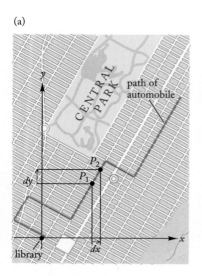

(b)

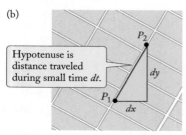

Hypotenuse is distance traveled during small time dt.

FIGURE 4.2 (a) In a small time interval dt, the displacement is P_1P_2, and the changes in the x and y coordinates are dx and dy. (b) the displacements dx and dy are the sides of a right triangle.

Thus, the instantaneous speed is the square root of the sum of the squares of the components of the instantaneous velocity. If we use the letter v (without any subscript) to represent the instantaneous speed in two-dimensional motion, we can write Eq. (4.7) as

$$v = \sqrt{v_x^2 + v_y^2} \tag{4.8}$$

The definitions of the components of the acceleration in two dimensions can be formulated in much the same way. For the average acceleration we have

$$\bar{a}_x = \frac{\Delta v_x}{\Delta t} \tag{4.9}$$

and

$$\bar{a}_y = \frac{\Delta v_y}{\Delta t} \tag{4.10}$$

where Δv_x and Δv_y are the changes in the x and the y components of the velocity. For the instantaneous acceleration we have

$$a_x = \frac{dv_x}{dt} \qquad (4.11)$$

components of the
instantaneous acceleration

and

$$a_y = \frac{dv_y}{dt} \qquad (4.12)$$

It is an important consequence of these definitions that *there is an acceleration when-ever any of the components of the velocity change.* This means that not only is there an acceleration when an automobile increases or decreases its speed, but there is also an acceleration when the automobile travels around a curve at constant speed, as we will see in the following example.

EXAMPLE 1   An automobile, traveling at a constant speed of 25 m/s, enters a 90°curve and emerges from this curve 6.0 s later. What are the components of the average acceleration for this time interval?

SOLUTION: Figure 4.3 shows the path of the automobile and the orientation of the axes. The initial velocity has an x component (v_x = 25 m/s) and no y compo-nent. The final velocity has a y component ($v_y = -25$ m/s) and no x component. Hence the changes in the velocity components are

$$\Delta v_x = [\text{final } x \text{ velocity}] - [\text{initial } x \text{ velocity}] = 0 - 25 \text{ m/s} = -25 \text{ m/s}$$

and

$$\Delta v_y = [\text{final } y \text{ velocity}] - [\text{initial } y \text{ velocity}] = -25 - 0 \text{ m/s} = -25 \text{ m/s}$$

The components of the average acceleration are then

$$\bar{a}_x = \frac{\Delta v_x}{\Delta t} = \frac{-25 \text{ m/s}}{6.0 \text{ s}} = -4.2 \text{ m/s}^2$$

and

$$\bar{a}_y = \frac{\Delta v_y}{\Delta t} = \frac{-25 \text{ m/s}}{6.0 \text{ s}} = -4.2 \text{ m/s}^2$$

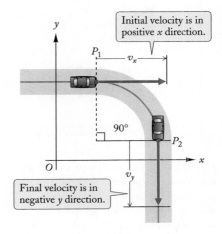

FIGURE 4.3 Automobile rounding a curve. The arrows indicate the directions of the motion.

✔ Checkup 4.1

QUESTION 1: Consider a ship moving on the sea. The x axis is eastward and the y axis is northward. For each of the following cases, state whether the x and y components of the velocity are positive, negative, or zero $(+, -, 0)$: (a) ship is moving northwest; (b) ship is moving south; (c) ship is moving southeast.

QUESTION 2: The speedometer of your automobile shows that you are proceeding at a steady speed of 80 km/h. Is it nevertheless possible that your automobile is in accel-erated motion?

QUESTION 3: A particle travels once around a circle at uniform speed. What are the average velocity and the average acceleration for this motion?

QUESTION 4: A motorcycle is traveling counterclockwise around a traffic circle at a steady speed of 50 km/h. The x axis is eastward and the y axis is northward. What are the x and y components of the velocity at the instant the motorcycle is at the eastern point of the traffic circle?

(A) $v_x = 50$ km/h; $v_y = 0$ (B) $v_x = -50$ km/h; $v_y = 0$

(C) $v_x = 0$; $v_y = 50$ km/h   (D) $v_x = 0$; $v_y = -50$ km/h

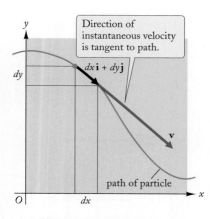

FIGURE 4.4 In a small time interval dt, the changes in the x and y coordinates are dx and dy. The displacement vector is $dx\mathbf{i} + dy\mathbf{j}$. This vector is tangent to the path of the particle, and so is the velocity vector $\mathbf{v} = (dx\mathbf{i} + dy\mathbf{j})/dt$.

4.2 THE VELOCITY AND ACCELERATION VECTORS

In the preceding section we described two-dimensional velocity and acceleration by components. Now we will see how to describe these quantities more concisely by vectors.

The x coordinate and the y coordinate of a particle can be regarded as the x component and the y component of the position vector:

$$\mathbf{r} = x\mathbf{i} + y\mathbf{j} \tag{4.13}$$

Likewise, the x and y components of the velocity introduced in Eqs. (4.3) and (4.4) can be regarded as the x and y components of the velocity vector:

$$\mathbf{v} = v_x\mathbf{i} + v_y\mathbf{j} \tag{4.14}$$

or

$$\mathbf{v} = \frac{dx}{dt}\,\mathbf{i} + \frac{dy}{dt}\,\mathbf{j} \tag{4.15}$$

The velocity vector is the rate of change of the position vector. Note that the velocity vector equals the small displacement $dx\mathbf{i} + dy\mathbf{j}$ divided by dt. Thus, the direction of the velocity vector is the direction of the small displacement $dx\mathbf{i} + dy\mathbf{j}$, that is, the direction of the instantaneous velocity vector is the direction of the instantaneous motion, which is tangent to the path of the particle (see Fig. 4.4). For example, Fig. 4.5 shows the velocity vectors at different instants for a projectile launched upward at some angle; at each instant, the velocity is tangent to the path. The lengths of the velocity

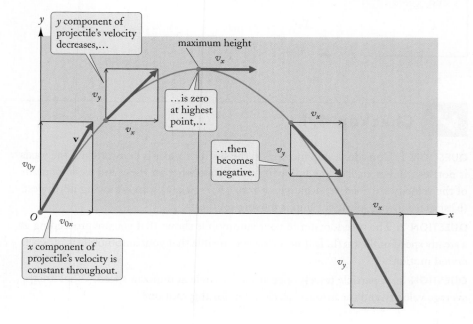

FIGURE 4.5 Velocity vectors of a projectile at different instants.

vectors indicate the magnitude of the velocity. In Fig. 4.5, we can see that the magnitude of the velocity is largest at the start of the motion, and it is smallest at the apex. In order to prepare this drawing of velocity vectors, the artist had to make a choice of scale for these vectors; for instance, a centimeter of length for a velocity vector in Fig. 4.5 might represent 10 m/s. Often, we will be mainly interested in how the relative magnitudes of various vectors look at different points, and for such comparisons we do not need to know the scale.

According to the usual equations for the components of a vector [Eqs. (3.13) and (3.14)],

$$v_x = v \cos \theta \tag{4.16}$$

$$v_y = v \sin \theta \tag{4.17}$$

where θ is the angle between the velocity vector and the x axis (see Fig. 4.6) and v is the magnitude of the velocity vector; that is, v is the speed. These relations permit us to calculate the x and y components of the velocity if we know the speed and the direction of motion.

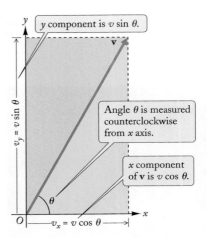

FIGURE 4.6 The x and y components of the velocity vector **v**.

PHYSICS IN PRACTICE   VELOCITY VECTORS

Charts of velocity vectors are used to visualize the flow of bodies of water, such as the tidal flow in harbors, and the flow of air masses, such as the updrafts and downdrafts in thunderstorm cells or the flow of air around obstacles. For instance, Figure 1 displays the tidal flow in Tampa Bay. The velocity vectors indicate the direction and the magnitude of the flow of water at one particular time. Information on tidal flow is important to navigators of ships and also to engineers concerned with the construction of harbors and bridges or the dispersion of pollutants. Figure 2 shows the velocity vectors associated with the flow of air in one of the diffuser channels in the centrifugal compressor of a jet engine. The complex velocity vector pattern indicates a pressure surge and a compressor stall, with a reversal of the normal direction of flow. This results in loss of power and engine flameout. By understanding the details of the flow, engineers can optimize compressor design and provide protection against potentially catastrophic engine failures.

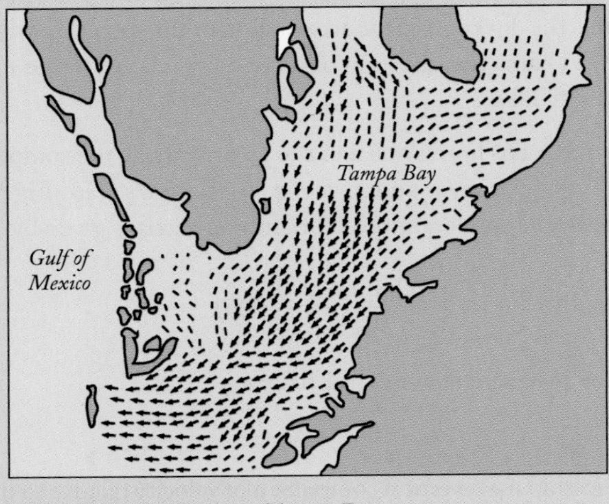

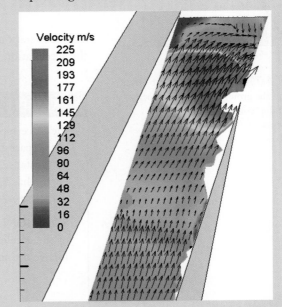

FIGURE 2 The airflow in one of the diffuser channels of a centrifugal compressor. The velocity vectors were measured by Digital Particle Imaging Velocimetry (DPIV). The airflow is seeded with a sprinkling of shiny particles illuminated with quick bursts of laser light. A high-speed digital camera records the positions of the particles, and velocities are calculated from changes in the positions.

FIGURE 1 Tidal flow in Tampa Bay.

Concepts
— in —
Context

(a)

(b)

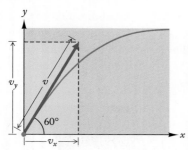

FIGURE 4.7 (a) A "volcanic bomb" after impact on the ground. (b) Initial velocity vector of the "volcanic bomb" and its components.

EXAMPLE 2   Volcanos sometime eject large "volcanic bombs," with masses of several tons (see the chapter photo and Figure 4.7a). Suppose that such a bomb is ejected eastward by a volcano with an initial speed of 330 km/h at an upward angle of 60.0° with the horizontal (see Fig. 4.7b). What are the components of the initial velocity of the bomb in the horizontal and vertical directions? Assume the x axis is horizontal and eastward and the y axis is vertical and upward.

SOLUTION: A speed of 330 km/h corresponds to

$$330\frac{\text{km}}{\text{h}} \times \frac{1000 \text{ m}}{1 \text{ km}} \times \frac{1 \text{ h}}{3600 \text{ s}} = 91.7 \text{ m/s}$$

With $v = 91.7$ m/s and $\theta = 60.0°$, Eqs. (4.16) and (4.17) immediately give the components of the velocity:

$$v_x = v \cos \theta = 91.7 \text{ m/s} \times \cos 60.0° = 91.7 \text{ m/s} \times 0.500$$
$$= 45.8 \text{ m/s}$$
$$v_y = v \sin \theta = 91.7 \text{ m/s} \times \sin 60.0° = 91.7 \text{ m/s} \times 0.866$$
$$= 79.4 \text{ m/s}$$

The x and y components of the acceleration can be regarded as the components of the acceleration vector:

$$\mathbf{a} = a_x\mathbf{i} + a_y\mathbf{j} \tag{4.18}$$

or

$$\mathbf{a} = \frac{dv_x}{dt}\mathbf{i} + \frac{dv_y}{dt}\mathbf{j} \tag{4.19}$$

The acceleration vector is the rate of change of the velocity vector. The direction of the acceleration vector is the direction of the change of the velocity vector.

EXAMPLE 3   An aircraft in level flight releases a rocket spaceship at time $t = 0$. The engine of the spaceship ignites 2.0 s after release and gives the spaceship a horizontal acceleration of 6.0 m/s². Assume that the vertical motion of the spaceship is free-fall motion during the first few seconds after its release. Ignore air resistance. (a) What is the direction of the acceleration of the spaceship when the engine has ignited? (b) What is the direction of its velocity relative to the moving aircraft at $t = 2.0$ s? At $t = 3.0$ s?

SOLUTION: (a) With the x axis horizontal and the y axis vertical, the components of the acceleration are $a_x = 6.0$ m/s² and $a_y = -g = -9.8$ m/s² (see Fig. 4.8a). Hence the angle between the acceleration vector and the x axis is given by

$$\tan \theta = \frac{a_y}{a_x} = \frac{-9.8 \text{ m/s}^2}{6.0 \text{ m/s}^2} = -1.63$$

With our calculator, the inverse tangent of -1.63 gives

$$\theta = -58°$$

(b) At $t = 2.0$ s, the spaceship has a vertical component of velocity relative to the aircraft, but no horizontal component of velocity (until its engine ignites, the spaceship

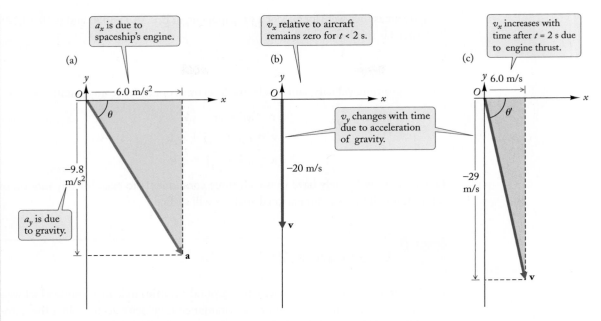

FIGURE 4.8 (a) The x and y components of the acceleration of the spaceship. (b) The components of the velocity at $t = 2.0$ s. (c) At $t = 3.0$ s.

moves forward with the same horizontal motion as the aircraft; that is, it has no horizontal velocity relative to the aircraft). The vertical component of velocity is simply that acquired by free fall in the interval 2.0 s:

$$v_y = a_y t = -gt = -9.8 \text{ m/s}^2 \times 2.0 \text{ s} = -20 \text{ m/s}$$

The direction of the velocity (relative to the aircraft) at this time is vertically down (see Fig. 4.8b).

At $t = 3.0$ s, the vertical component of the velocity is

$$v_y = -9.8 \text{ m/s}^2 \times 3.0 \text{ s} = -29 \text{ m/s}$$

The horizontal component of velocity of the spaceship (relative to the horizontal velocity of the aircraft) arises from the horizontal acceleration of 6.0 m/s^2 provided by the engine. This acceleration begins at $t = 2.0$ s, and in the time interval Δt from $t = 2.0$ s to 3.0 s it contributes a horizontal component

$$v_x = a_x \Delta t = 6.0 \text{ m/s}^2 \times 1.0 \text{ s} = 6.0 \text{ m/s}$$

The angle between the velocity vector and the x axis is then given by

$$\tan \theta' = \frac{v_y}{v_x} = \frac{-29 \text{ m/s}}{6.0 \text{ m/s}} = -4.8$$

from which our calculator finds

$$\theta' = -78°$$

This means that the direction of the velocity vector is 78° below the horizontal (see Fig. 4.8c).

Our various definitions and results are easily extended to three dimensions. The average and instantaneous velocities and accelerations are defined the same way for a z

component as they were for an x or a y component, in Eqs. (4.1)–(4.4) and (4.9)–(4.12). The speed is similarly

$$v = \sqrt{v_x^2 + v_y^2 + v_z^2}$$

and the position, velocity, and acceleration vectors contain a z component:

$$\mathbf{r} = x\mathbf{i} + y\mathbf{j} + z\mathbf{k}$$

$$\mathbf{v} = v_x\mathbf{i} + v_y\mathbf{j} + v_z\mathbf{k}$$

$$\mathbf{a} = a_x\mathbf{i} + a_y\mathbf{j} + a_z\mathbf{k}$$

However, we will rarely have to use all three components to examine any motion of interest; usually, a two-dimensional analysis will suffice.

Checkup 4.2

QUESTION 1: An automobile is traveling around a traffic circle in a counterclockwise direction. What is the direction of its instantaneous velocity vector when the automobile is at the east, north, west, and south extremes of the circle?

QUESTION 2: The x component of the velocity of an aircraft is 150 km/h, and the y component of the velocity is also 150 km/h. What is the direction of the velocity vector relative to the x and y axes?

QUESTION 3: A tennis ball, initially traveling horizontally, collides with a wall and bounces back horizontally. What is the direction of the acceleration vector during the collision?

QUESTION 4: A skier accelerates along a slope at a rate of 6.0 m/s². The horizontal component of the skier's acceleration is 3.0 m/s². What is the angle of the slope with respect to horizontal?

(A) 27°　　　(B) 30°　　　(C) 45°　　　(D) 60°　　　(E) 63°

4.3 MOTION WITH CONSTANT ACCELERATION

For a particle moving in two dimensions with constant acceleration, we can derive equations relating acceleration, velocity, position, and time analogous to the equations that hold in one dimension (see Section 2.5). If the x component of the acceleration is a_x and the y component a_y, then, by analogy with Eq. (2.17), we see that the components of the velocity at time t will be

$$v_x = v_{0x} + a_x t \tag{4.20}$$

$$v_y = v_{0y} + a_y t \tag{4.21}$$

These two equations can be regarded as components of the following vector equation:

$$\mathbf{v} = \mathbf{v}_0 + \mathbf{a}t \tag{4.22}$$

where, as always, the subscript 0 (zero) indicates the values at the initial time zero. Furthermore, a mathematical argument analogous to that used in the one-dimensional case [see Eqs. (2.18)–(2.22)] leads us to the following expressions for the change in the position:

$$x - x_0 = v_{0x}t + \tfrac{1}{2}a_x t^2 \qquad (4.23)$$

$$y - y_0 = v_{0y}t + \tfrac{1}{2}a_y t^2 \qquad (4.24)$$

These equations can be regarded as components of the following vector equation:

$$\mathbf{r} - \mathbf{r}_0 = \mathbf{v}_0 t + \tfrac{1}{2}\mathbf{a}t^2 \qquad (4.25)$$

Equations (4.20)–(4.25) state that *the x and y components of the motion evolve completely independently of one another.* Thus, the x acceleration affects only the x velocity, and the change in the x position is completely determined by the x acceleration and the initial x velocity. Figure 4.9 shows an experimental demonstration of this independence between the x and y components of the motion. Two balls were released simultaneously from a platform; one was merely dropped from rest, the other was launched with an initial horizontal velocity. According to our discussion of the acceleration of gravity in Chapter 2, the vertical (downward) free-fall accelerations of the two balls are the same. According to Eq. (4.24), the vertical motions (y motions) of the two balls should then be the same, even though their horizontal motions (x motions) differ. Furthermore, according to Eq. (4.23), the horizontal motion of the second ball should simply proceed with uniform horizontal velocity, even though this ball has a vertical acceleration.

The stroboscopic images of the balls recorded on the photograph at equal intervals of time confirm these predictions. The balls indeed fall downward in unison, reaching equal heights at the same instants of time. The red grid lines drawn in Fig. 4.9b permit us to verify that the vertical components of the positions of the two balls are always exactly the same, even though the horizontal components differ. Furthermore, the blue grid lines permit us to verify that the horizontal component of the velocity of the second ball is constant.

(a)

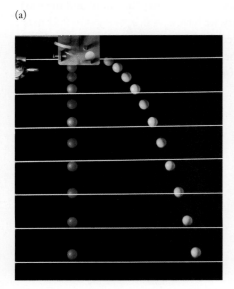

(b)

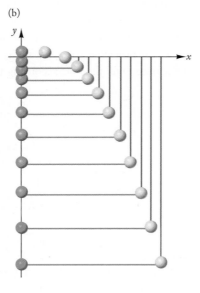

FIGURE 4.9 (a) Stroboscopic photograph showing multiple exposures of two balls that have been released simultaneously from a platform; one ball has a horizontal velocity, the other does not. The time interval between the exposures is 1/40 s. (b) For the analysis of the photograph, we draw horizontal (red) and vertical (blue) grid lines through the positions of the balls. The red grid lines verify that the vertical components of the positions of the two balls always coincide. The blue grid lines, by their uniform spacing, verify that the horizontal component of the velocity of the second ball remains constant.

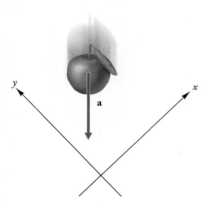

FIGURE 4.10 Apple falling in tilted coordinate system.

Online
Concept
Tutorial

 Checkup 4.3

QUESTION 1: Consider a motorboat maneuvering on the surface of a lake, with x and y axes along the surface. Is it possible for the boat to have accelerated motion in the x direction, and unaccelerated motion in the y direction? Is it possible for the boat to have accelerated motion in both the x and the y directions?

QUESTION 2: Suppose the direction of motion of a particle lies between the x and y axes. You know that the x motion is accelerated, and you see that the particle moves along a straight line. Is the y motion accelerated or unaccelerated?

QUESTION 3: Consider the motion of a falling apple in a coordinate system whose axes are arranged so the upward direction is at 45° between the x and the y axes (see Fig. 4.10). What are the accelerations a_x and a_y of the apple?

(A) $a_x = 0$; $a_y = 9.8 \text{ m/s}^2$ (B) $a_x = 6.9 \text{ m/s}^2$; $a_y = 6.9 \text{ m/s}^2$
(C) $a_x = 6.9 \text{ m/s}^2$; $a_y = -6.9 \text{ m/s}^2$ (D) $a_x = -6.9 \text{ m/s}^2$; $a_y = 6.9 \text{ m/s}^2$
(E) $a_x = -6.9 \text{ m/s}^2$; $a_y = -6.9 \text{ m/s}^2$

4.4 THE MOTION OF PROJECTILES

We know that near the surface of the Earth, the pull of gravity gives a freely falling body a downward acceleration of about 9.81 m/s². If we ignore air resistance, this is the only acceleration that the body experiences when launched from some initial position with some initial velocity. Thus, the motion of a baseball thrown by hand (or hit by a bat) is motion with constant vertical acceleration and zero horizontal acceleration. This kind of motion is called **projectile motion**, or ballistic motion. In most cases of projectile motion, we deliberately launch the body with an initial upward component of velocity, so that it flies farther before striking the ground. But we can also launch the projectile horizontally, without any initial upward component of velocity (as in the case of the ball whose motion we examined in Fig. 4.9), and we can even launch the projectile with an initial downward component of velocity.

In any case, the initial velocity of the projectile can be characterized by its vertical and horizontal components. If we take the y axis in the upward direction[1] and the x axis in the direction of the initial horizontal velocity, we have $a_x = 0$, $a_y = -g = -9.81 \text{ m/s}^2$. Furthermore, let us assume that the origin of coordinates coincides with the initial position of the projectile, so $x_0 = 0$ and $y_0 = 0$. The components of the velocity and position at the time t will then be, according to Eqs. (4.20), (4.21), (4.23), and (4.24):

horizontal motion of projectile

$$v_x = v_{0x} \tag{4.26}$$

$$x = v_{0x}t \qquad (x_0 = 0) \tag{4.27}$$

$$v_y = v_{0y} - gt \tag{4.28}$$

vertical motion of projectile

$$y = v_{0y}t - \tfrac{1}{2}gt^2 \qquad (y_0 = 0) \tag{4.29}$$

[1] Note that in Section 2.6 we took the x axis in the upward direction, whereas now we are taking the y axis in the upward direction.

These equations represent a motion with constant velocity in the x direction and a simultaneous motion with constant (downward) acceleration in the y direction. According to Eq. (2.29), the vertical position and velocity also obey the relation

$$-gy = \tfrac{1}{2}(v_y^2 - v_{0y}^2) \tag{4.30}$$

If the motion is initially upward, the projectile will ascend to some maximum height and then begin its descent. At the instant of maximum height, the vertical component of the velocity is zero, since at this one instant the projectile has ceased to move upward and has not yet begun to move downward. We can find the instant of maximum height and the value of the maximum height by inserting the condition $v_y = 0$ into Eqs. (4.28) and (4.30), respectively.

| EXAMPLE 4 | Consider the "volcanic bomb" described in Example 2. At what time does this projectile reach its maximum height? What is |

this maximum height?

SOLUTION: The known quantities are the components of the initial velocity; according to Example 2, these components are $v_{0x} = 45.8$ m/s and $v_{0y} = 79.4$ m/s. Furthermore, we know that when the projectile reaches its maximum height, $v_y = 0$ (see Fig. 4.11). Since the problem asks no questions about the horizontal motion, we can ignore this motion altogether, and we can concentrate on the vertical motion. For this vertical motion, Eqs. (4.28) and (4.29) are applicable. The known and unknown quantities are

UNKNOWN	KNOWN
t	$v_{0y} = 79.4$ m/s
y	$v_y = 0$
	$g = 9.81$ m/s^2

From these known values of v_{0y} and v_y, Eq. (4.28) determines the instant of maximum height:

$$0 = v_{0y} - gt$$

or, solving this for t,

$$t = \frac{v_{0y}}{g}$$

$$= \frac{79.4 \text{ m/s}}{9.81 \text{ m/s}^2} = 8.09 \text{ s}$$

With this value of the time, we can then calculate the maximum height from Eq. (4.29):

$$y = v_{0y}t - \tfrac{1}{2}gt^2$$

$$= 79.4 \text{ m/s} \times 8.09 \text{ s} - \tfrac{1}{2} \times 9.81 \text{ m/s}^2 \times (8.09 \text{ s})^2$$

$$= 321 \text{ m}$$

COMMENT: Note that we could instead have used Eq. (4.30), with $v_y = 0$, to calculate the same maximum height directly. Solving Eq. (4.30) for y we have

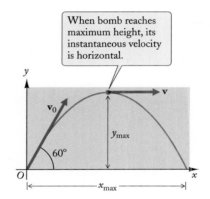

FIGURE 4.11 Trajectory of a volcanic bomb launched with an initial velocity $\mathbf{v}_0$.

$$y = \frac{\frac{1}{2}(-v_{0y}^2)}{-g} = \frac{v_{0y}^2}{2g}$$

$$= \frac{(79.4 \text{ m/s})^2}{2 \times 9.81 \text{ m/s}^2} = 321 \text{ m}$$

In this case, we did not need to determine the time t, because Eq. (4.30) was obtained by eliminating the variable t from Eqs. (4.28) and (4.29) (see Section 2.5).

In our calculation we have ignored the effects of air resistance. For a dense projectile of large mass, such as a large volcanic bomb, this is a good approximation.

EXAMPLE 5   You throw a baseball toward a wall, 10 m away. The initial direction of motion of the ball is horizontal, and the initial speed is 20 m/s. How far below its initial height does this projectile hit the wall?

SOLUTION: It is convenient to place the origin of coordinates at the initial height of the ball (see Fig. 4.12). We first consider the horizontal motion. This horizontal motion proceeds at uniform velocity, with $v_{0x} = 20$ m/s. Hence the time the ball takes to reach the wall is

$$t = \frac{x}{v_{0x}} = \frac{10 \text{ m}}{20 \text{ m/s}} = 0.50 \text{ s}$$

Next, we consider the vertical motion, which proceeds with uniform acceleration, according to Eqs. (4.28)–(4.30). The initial velocity is $v_{0y} = 0$, since the ball was thrown horizontally. The final time is now known, and the final value of y is unknown:

UNKNOWN	KNOWN
y	$v_{0y} = 0$
	$t = 0.50$ s
	$g = 9.81$ m/s^2

From these known values, Eq. (4.29) immediately gives us y:

$$y = v_{0y}t - \tfrac{1}{2}gt^2 = 0 - \tfrac{1}{2} \times 9.81 \text{ m/s}^2 \times (0.50 \text{ s})^2$$

$$= -1.2 \text{ m}$$

The negative sign means that the ball has descended 1.2 m below its initial height by the time it reaches the wall.

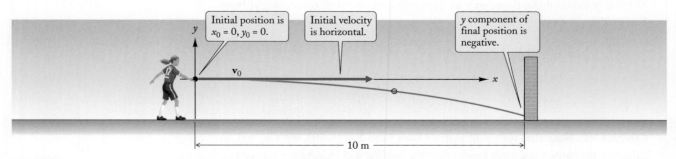

FIGURE 4.12 Trajectory of a baseball, thrown with an initial horizontal velocity.

EXAMPLE 6   In low-level bombing (at "smokestack level"), a bomber releases a bomb at a height of 50 m above the surface of the sea while in horizontal flight at a constant speed of 300 km/h. How long does the bomb take to fall to the surface of the sea? How far ahead (horizontally) of the point of release is the point of impact?

SOLUTION: It is convenient to place the origin of coordinates at the point of release, 50 m above the level of the sea, with the x axis along the horizontal path of the bomber (see Fig. 4.13). The initial velocity of the bomb is the same as that of the bomber:

$$v_{0x} = 300 \text{ km/h} = 300 \frac{\text{km}}{\text{h}} \times \frac{1000 \text{ m}}{1 \text{ km}} \times \frac{1 \text{ h}}{3600 \text{ s}} = 83.3 \text{ m/s}$$

$$v_{0y} = 0$$

When the bomb reaches the level of the sea, its vertical position is $y = -50$ m (i.e., the bomb is 50 m below our origin of coordinates).

We begin with the vertical motion, described by Eqs. (4.28) and (4.29). The initial value of the velocity and the final value of y are known:

UNKNOWN	KNOWN
t	$y = -50 \text{ m}$
	$v_{0y} = 0$
	$g = 9.81 \text{ m/s}^2$

With $v_{0y} = 0$, Eq. (4.29) then determines the time of impact:

$$y = -\tfrac{1}{2}gt^2$$

To solve for t, we divide both sides of this equation by $-\tfrac{1}{2}g$ and take the square root of both sides:

$$t = \sqrt{-\frac{2y}{g}}$$

$$= \sqrt{\frac{2 \times 50 \text{ m}}{g}} = \sqrt{\frac{2 \times 50 \text{ m}}{9.81 \text{ m/s}^2}} = 3.2 \text{ s}$$

FIGURE 4.13 Trajectory of a bomb dropped by a bomber. The initial vertical component of the velocity is zero, and the initial horizontal component is the same as that of the bomber.

Next, we consider the horizontal motion. The bomb moves with a constant horizontal velocity $v_{0x} = 83.3$ m/s. Hence at the time 3.2 s, the horizontal position of the bomb is

$$x = v_{0x}t = 83.3 \text{ m/s} \times 3.2 \text{ s} = 270 \text{ m}$$

Note that the bomber moves exactly the same *horizontal* distance in this time; that is, the bomb always remains directly below the bomber because both have exactly the same horizontal velocity $v_{0x} = 83.3$ m/s. Figure 4.14 shows bombs released by a bomber at successive instants of time, and demonstrates that they remain directly below the bomber.

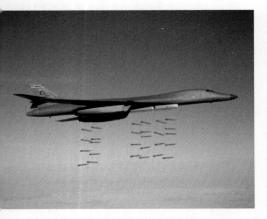

FIGURE 4.14 "Strings" of bombs released from a bomber. The bombs continue to move forward with the same horizontal velocity as that of the bomber.

The path of a baseball, a golf ball, a bomb, or any other projectile is a *parabola*. The mathematical proof of this statement rests on Eqs. (4.27) and (4.29). Suppose that time is reckoned from the instant the projectile reaches its maximum height (so t is negative before the instant of maximum height, and positive after). Then $v_{0y} = 0$ and Eqs. (4.27) and (4.29) become

$$x = v_{0x}t$$
$$y = -\tfrac{1}{2}gt^2$$

If we square both sides of the first equation and divide it into the second equation, we obtain

$$\frac{y}{x^2} = -\frac{1}{2}\frac{gt^2}{v_{0x}^2 t^2} \tag{4.31}$$

and, canceling the factors of t^2 and multiplying by x^2, we obtain

$$y = -\left(\frac{1}{2}\frac{g}{v_{0x}^2}\right)x^2 \tag{4.32}$$

Concepts — *in* — Context

This equation says that y is proportional to x^2, which is the equation for a parabola with apex at $y = 0$. The parabolic shape of the trajectory can be seen clearly in a multiple-exposure photograph, such as Fig. 4.15. It can also be seen clearly in the time-exposure photograph of incandescent volcanic bombs in the chapter photo.

PROBLEM-SOLVING TECHNIQUES   PROJECTILE MOTION

The solution of problems of projectile motion exploits the independence of the two components of the motion.

1 First solve for one of the components of the motion, and then the other. Whether you should begin with the y motion (vertical) or the x motion (horizontal) depends on circumstances. If the motion is limited vertically (by the ground), as in Examples 4 and 6, then you must first deal with the y motion. If the motion is limited horizontally (by a wall), as in Example 5, then you must first deal with the x motion.

2 Each of the separate components of motion can be treated as one-dimensional motion. The x motion proceeds with constant velocity; the y motion proceeds with constant acceleration. The solution of these two parts of the problem relies on the techniques we used for one-dimensional problems with constant velocity or with constant acceleration in Chapter 2.

If air resistance is negligible, the path of a projectile launched with some horizontal and vertical velocity is always some portion of a parabola. In the case of a bomb dropped from an airplane, the relevant portion of the parabola begins at the apex and descends to the level of the target. In the case of a ball or stone launched by hand, a bomb launched by a volcano, or a shot or bullet fired from a gun, the relevant portion begins at the level of the launch point or muzzle, rises to the apex, and then descends to the target. In the latter case, it is often important to calculate the **maximum height** reached, the **time of flight** (time between the instants of launch and impact), and the **range** (the horizontal distance between the points of launch and impact). This kind of calculation is illustrated in the following example.

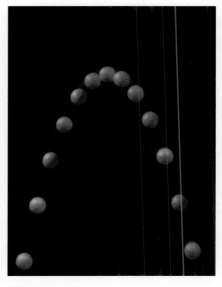

FIGURE 4.15 Stroboscopic photograph showing the path of a projectile (a golf ball) launched upward at an angle. The path is a parabola. At the apex of the parabola (the point of maximum height), the vertical component of the velocity is instantaneously zero.

EXAMPLE 7 A champion discus thrower (Fig. 4.16) throws a discus with an initial speed of 26 m/s. If the discus is thrown upward at an angle of 45°, what maximum height and what range will the discus attain? What is its time of flight? For the sake of simplicity, ignore the height of the hand over the ground, so the launch point and the impact point are at the same height.

SOLUTION: We take the origin of coordinates at the launch point. Equations (4.26)–(4.29) are then applicable to this problem. However, to extract the answers we want, we must manipulate these equations somewhat.

Since Eqs. (4.26)–(4.29) depend on the components v_{0x} and v_{0y} of the initial velocity, we begin by calculating these components. From Fig. 4.17 we see that

$$v_{0x} = v_0 \cos \theta$$
$$= v_0 \cos 45° = 26 \text{ m/s} \times \cos 45° = 26 \text{ m/s} \times 0.707 \qquad (4.33)$$
$$= 18.4 \text{ m/s}$$

and

$$v_{0y} = v_0 \sin \theta$$
$$= v_0 \sin 45° = 26 \text{ m/s} \times \sin 45° = 26 \text{ m/s} \times 0.707 \qquad (4.34)$$
$$= 18.4 \text{ m/s}$$

FIGURE 4.16 A champion discus thrower.

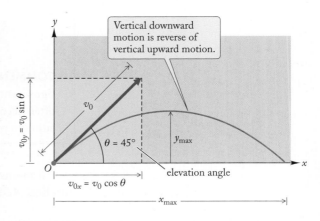

FIGURE 4.17 Trajectory of a discus thrown upward. The arrow indicates the initial velocity vector of the discus. The angle θ between this direction and the horizontal is the elevation angle.

As a next step, we calculate the time at which the discus reaches maximum height. From this we can immediately find the time of impact, or the time of flight, because this time is twice the time required to reach maximum height; we can recognize this if we concentrate on the y coordinate, for which the downward motion, after maximum height, is merely the reverse of the upward motion. Thus, the time of flight (which includes the upward and the downward motions) must be twice the time for the upward motion:

$$t_{\text{flight}} = 2t_{\text{height}} \tag{4.35}$$

The discus reaches maximum height when $v_y = 0$. Equation (4.28) then implies

$$0 = v_{0y} - gt_{\text{height}} \tag{4.36}$$

and, solving this for t_{height}, following the method we used in Example 4, we find

$$
\begin{aligned}
t_{\text{height}} &= \frac{v_{0y}}{g} \\
&= \frac{18.4 \text{ m/s}}{9.81 \text{ m/s}^2} = 1.87 \text{ s}
\end{aligned}
\tag{4.37}
$$

Accordingly,

$$
\begin{aligned}
t_{\text{flight}} &= 2t_{\text{height}} = 2\frac{v_{0y}}{g} \\
&= 2 \times 1.87 \text{ s} = 3.7 \text{ s}
\end{aligned}
\tag{4.38}
$$

At the time t_{height}, the height reached by the discus is, by Eq. (4.29),

$$y_{\max} = v_{0y}t_{\text{height}} - \tfrac{1}{2}g(t_{\text{height}})^2 = v_{0y}\left(\frac{v_{0y}}{g}\right) - \tfrac{1}{2}g\left(\frac{v_{0y}}{g}\right)^2 = \tfrac{1}{2}\frac{v_{0y}^2}{g} \tag{4.39}$$

$$= \frac{1}{2}\frac{(18.4 \text{ m/s})^2}{9.81 \text{ m/s}^2} = 17 \text{ m}$$

The horizontal distance from the launch point attained at the time $t = t_{\text{flight}}$ is equal to the range:

$$x_{\max} = v_{0x}t_{\text{flight}} = v_{0x} \times 2t_{\text{height}} = v_{0x} \times 2\frac{v_{0y}}{g} = \frac{2v_{0x}v_{0y}}{g} \tag{4.40}$$

$$= \frac{2 \times 18.4 \text{ m/s} \times 18.4 \text{ m/s}}{9.81 \text{ m/s}^2} = 69 \text{ m}$$

COMMENT: Note that in the intermediate stages of this calculation, we kept three significant figures, even though the data contain only two significant figures. It is often necessary to carry extra significant figures in intermediate stages of a calculation to prevent accumulation of errors from round-off.

An alternative method of solution is to substitute $v_{0x} = v_0 \cos\theta$ and $v_{0y} = v_0 \sin\theta$ into Eqs. (4.38), (4.39), and (4.40). These substitutions yield

time of flight

$$t_{\text{flight}} = \frac{2v_0 \sin\theta}{g} \tag{4.41}$$

$$y_{max} = \frac{v_0^2 \sin^2\theta}{2g}$$

(4.42)   **maximum height**

and, for the horizontal range,

$$x_{max} = \frac{2v_0^2 \sin\theta \cos\theta}{g}$$

(4.43)   **range**

These equations express parameters of interest explicitly in terms of quantities that are usually known: the initial speed v_0 (the launch speed) and the angle θ between the horizontal and the direction of the initial motion (the elevation angle). Thus, these equations permit a direct calculation of the answers to our problem, without further intermediate steps.

Equations (4.41)–(4.43) are often handy for the solution of other problems of projectile motion; for instance, these equations permit us to calculate the launch speed and the elevation angle required to hit a target at a known range. However, keep in mind that these equations do not give us a complete description of the motion. The complete description of the motion is contained in Eqs. (4.26)–(4.29), and we can always extract anything we want to know about the motion from these equations.

Figure 4.18 shows the trajectories of projectiles of the same launch speed as a function of the elevation angle. Note that *the range is maximum for an elevation angle of 45°.* We can see this by recalling that at a maximum, the derivative of a function is zero. If we take the derivative (as a function of the angle θ) of the angular part of the range function, Eq. (4.43), we have, using the product rule,

$$\frac{d}{d\theta}(\sin\theta \cos\theta) = \left(\frac{d}{d\theta}\sin\theta\right) \times \cos\theta + \sin\theta \times \left(\frac{d}{d\theta}\cos\theta\right)$$

$$= \cos\theta \times \cos\theta + \sin\theta \times (-\sin\theta) = \cos^2\theta - \sin^2\theta$$

This is zero for $\cos^2\theta = \sin^2\theta$, or $\cos\theta = \sin\theta$, which indeed corresponds to an angle of 45°. This angle of 45° represents a compromise: an angle smaller than 45° gives the projectile a larger x velocity but reduces the time it spends in flight, whereas an angle larger than 45° increases the time it spends in flight but gives it a smaller x velocity.

By inspection of Fig. 4.18 we see that any two angles that are equal amounts above or below 45° yield equal ranges; for instance, 60° (15° above 45°) and 30° (15° below 45°) yield equal ranges. This is evident from Eq. (4.43), since $\sin\theta \cos\theta = \sin\theta \sin(90° - \theta)$, which is the same product for any two complementary angles, which differ from 45° by the same amount.

In Fig. 4.18, as in all calculations of this section, air resistance has been neglected. For a high-speed projectile, such as a rifle bullet, air resistance is quite important, and the parabolic trajectory is distorted into a trajectory of more complicated shape, called a **ballistic curve**. If we attempt to use the simple formulas of this section in the calculation of the motion of a high-speed projectile, our results will bear only a vague resemblance to reality.

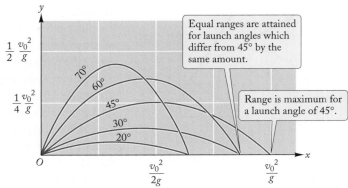

FIGURE 4.18 Trajectories of projectiles of the same launch speed, but different elevation angles. In this plot, the horizontal and the vertical coordinates are measured in fractions of v_0^2/g, which is the maximum range the projectile can achieve, when it is launched with an elevation angle of 45°.

Checkup 4.4

QUESTION 1: What is the instantaneous acceleration of a projectile when it reaches the top of its trajectory?

QUESTION 2: For a given, fixed launch speed, how should you launch a projectile to achieve the longest range? The longest time of flight? The largest height?

QUESTION 3: Consider a projectile launched with some elevation angle larger than 0° and smaller than 90°. Are the velocity and the acceleration of this projectile ever parallel? Perpendicular?

QUESTION 4: You launch two projectiles with the same elevation angle, but different launch speeds. If one projectile has twice the launch speed as the other, how much farther will it go?

QUESTION 5: Consider the trajectories for several projectiles with the same launch speed, but different elevation angles. Which of these projectiles will return to the ground in the shortest time?

QUESTION 6: You launch six projectiles with the same launch speed, but different elevation angles. The first projectile has an elevation angle of 20°. Which of the other projectiles has a shorter range than the first? The other projectiles have elevation angles of

(A) 40°   (B) 50°   (C) 60°   (D) 70°   (E) 80°

4.5 UNIFORM CIRCULAR MOTION

Uniform circular motion is motion with constant speed along a circular path, such as the motion of an automobile traveling around a traffic circle. Figure 4.19 shows the positions at different times for a particle in uniform circular motion. The velocity vector at any instant is tangent to the path, or tangent to the circle. All the velocity vectors shown have the same magnitude (same speed), but they differ in direction. Because of this change of direction, *uniform circular motion is accelerated motion.*

Suppose that the constant speed of the particle is v and the radius of the circle is r. To find the value of the instantaneous acceleration, we must look at the velocity change in a very short time interval Δt. We choose a convenient origin at the center of the circle. Figure 4.20a shows the particle at two positions $\mathbf{r}_1$ and $\mathbf{r}_2$ a short time apart; the difference between these vectors is $\Delta \mathbf{r} = \mathbf{r}_2 - \mathbf{r}_1$. The figure also shows the two velocity vectors $\mathbf{v}_1$ and $\mathbf{v}_2$; both velocity vectors have the same magnitude v; Fig. 4.20b shows the two velocity vectors tail to tail and their difference $\Delta \mathbf{v} = \mathbf{v}_2 - \mathbf{v}_1$. The position vectors make a small angle $\Delta \theta$ with each other; the velocity vectors are always perpendicular to the position vectors, and the angle between the velocity vectors is therefore the same as the angle between the position vectors. Since the angles $\Delta \theta$ in Figs. 4.20a and b are equal, the triangles formed by the position vectors and by the velocity vectors are similar—hence the ratio of the short sides of these triangles equals the ratio of their long sides; that is, the ratio of the magnitudes of the vectors $\Delta \mathbf{v}$ and $\Delta \mathbf{r}$ equals the ratio of v and r:

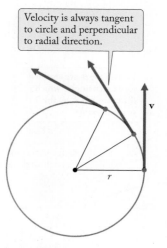

Velocity is always tangent to circle and perpendicular to radial direction.

FIGURE 4.19 Instantaneous velocity vectors for a particle in uniform circular motion.

$$\frac{[\text{magnitude of } \Delta \mathbf{v}]}{[\text{magnitude of } \Delta \mathbf{r}]} = \frac{v}{r} \qquad (4.44)$$

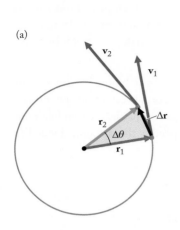

(a)

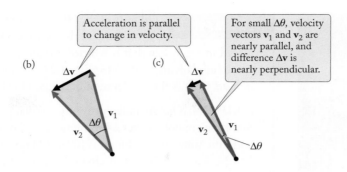

(b) Acceleration is parallel to change in velocity.

(c) For small $\Delta\theta$, velocity vectors $\mathbf{v}_1$ and $\mathbf{v}_2$ are nearly parallel, and difference $\Delta\mathbf{v}$ is nearly perpendicular.

FIGURE 4.20 (a) Position vectors $\mathbf{r}_1$ and $\mathbf{r}_2$ at two times with an interval Δt. The difference between these two position vectors is $\Delta\mathbf{r}$. (b) Velocity vectors $\mathbf{v}_1$ and $\mathbf{v}_2$ at the two times. The difference between these velocity vectors is $\Delta\mathbf{v}$. The colored triangle is similar to the colored triangle in part (a). (c) Velocity vectors $\mathbf{v}_1$ and $\mathbf{v}_2$ for two times with a very small time interval Δt and a very small angle $\Delta\theta$.

We can rewrite this as

$$[\text{magnitude of } \Delta\mathbf{v}] = \frac{v}{r} \times [\text{magnitude of } \Delta\mathbf{r}] \qquad (4.45)$$

If Δt is very small, then $\Delta\theta$ will also be very small, and the straight line segment $\Delta\mathbf{r}$ will approximately coincide with the circular arc from the tip of $\mathbf{r}_1$ to the tip of $\mathbf{r}_2$. The latter length is simply the distance traveled by the particle in the time Δt:

$$[\text{magnitude of } \Delta\mathbf{v}] \approx \frac{v}{r} \times [\text{distance traveled in time } \Delta t] \qquad (4.46)$$

The magnitude of the acceleration is the magnitude of $\Delta\mathbf{v}$ divided by the time Δt:

$$a = \frac{[\text{magnitude of } \Delta\mathbf{v}]}{\Delta t} \approx \frac{v}{r} \times \frac{[\text{distance traveled in time } \Delta t]}{\Delta t} \qquad (4.47)$$

But the distance traveled divided by the time Δt is the speed v, so

$$a \approx \frac{v}{r} \times v \qquad (4.48)$$

This equation becomes exact if the time Δt is extremely small, giving us the result

$$a = \frac{v^2}{r} \qquad (4.49)$$

The *direction* of this acceleration remains to be determined. From Fig. 4.20c it is clear that, for the case of very small Δt, the direction of $\Delta\mathbf{v}$ will be perpendicular to the velocity vectors $\mathbf{v}_1$ and $\mathbf{v}_2$ (which will be nearly parallel in this limiting case). Hence the instantaneous acceleration is perpendicular to the instantaneous velocity. Since the velocity vector corresponding to circular motion is tangential to the circle, *the acceleration vector points inward along the radius, toward the center of the circle.* Figure 4.21 shows the velocity and acceleration vectors at several positions along the circular path. The acceleration of uniform circular motion, with a magnitude given by Eq. (4.49), is called **centripetal acceleration**, because it is directed toward the center of the circle.

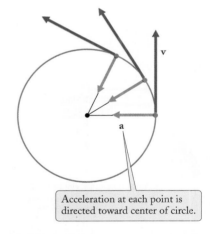

Acceleration at each point is directed toward center of circle.

FIGURE 4.21 Instantaneous acceleration vectors for a particle in uniform circular motion.

centripetal acceleration

FIGURE 4.22 Centrifuge at the NASA Spacecraft Center in Houston.

EXAMPLE 8   In tests of the effects of high acceleration on the human body, astronauts at the NASA Spacecraft Center in Houston are placed in a gondola that is whirled around a circular path of radius 15 m at the end of a revolving girder (see Fig. 4.22). If the girder makes 24 revolutions per minute, what is the acceleration of the gondola?

SOLUTION: The circumference of the circular path is $2\pi \times$ [radius] $= 2\pi \times 15$ m. Since the gondola makes 24 revolutions per minute, or 24 revolutions per 60 seconds, the time per revolution is (60 s)/(24 revolutions); that is, the gondola takes (60/24) s to go around this circumference. The speed is the distance divided by the time; therefore

$$v = \frac{2\pi \times 15 \text{ m}}{(60/24) \text{ s}} = 38 \text{ m/s}$$

From Eq. (4.49), the centripetal acceleration is then

$$a = \frac{v^2}{r} = \frac{(38 \text{ m/s})^2}{15 \text{ m}} = 95 \text{ m/s}^2$$

This is almost 10 times the acceleration due to gravity (almost 10 standard g's); this is near the limit of human tolerance.

EXAMPLE 9   An automobile drives around a traffic circle of radius 30 m. If the wheels of the automobile can withstand a maximum transverse acceleration of 8.0 m/s^2 without skidding, what is the maximum permissible speed?

SOLUTION: With a radius of 30 m and a centripetal acceleration of 8.0 m/s^2, Eq. (4.49) gives

$$v^2 = ar = 8.0 \text{ m/s}^2 \times 30 \text{ m} = 240 \text{ (m/s)}^2$$

Hence

$$v = \sqrt{240 \text{ (m/s)}^2} = 15 \text{ m/s}$$

This is the same as 15 m/s $\times$ (1 km)/(1000 m) $\times$ (3600 s)/(1 h) = 56 km/h. If the driver tries to go around the circle faster than this, the automobile will skid out of the circle.

(a)

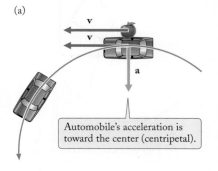

Automobile's acceleration is toward the center (centripetal).

(b)

In automobile reference frame, released apple accelerates outward (centrifugal).

FIGURE 4.23 (a) Automobile in uniform circular motion. The driver has just released an apple, which now moves with constant velocity **v** relative to the ground. The automobile accelerates away from the apple with the centripetal acceleration **a**. (b) In the reference frame of the automobile, the apple has a centrifugal acceleration $-\mathbf{a}$.

Note that the acceleration of a particle, or an automobile, in uniform circular motion is always centripetal (toward the center), never centrifugal (away from the center). However, with respect to the reference frame of the accelerating automobile, any body free to move in a horizontal direction—such as an apple that the driver of the automobile has placed on the (frictionless) dashboard—slides away and accelerates toward the outer side of the automobile with a centrifugal acceleration of magnitude equal to the centripetal acceleration of the automobile with respect to the ground (see Fig. 4.23). This centrifugal acceleration of the apple exists only in the reference frame of the automobile; it does not exist in the reference frame of the ground, where the apple merely continues its horizontal motion with constant velocity while the dashboard of the automobile accelerates sideways away from it.

✔ Checkup 4.5

QUESTION 1: A truck drives along a curved road with an S-shaped curve. Describe the direction of the acceleration vector at different points of the S.

QUESTION 2: A particle travels once around a circle with uniform circular motion. What are the average velocity and the average acceleration for this motion?

QUESTION 3: A girl sits in a swing, which is swinging back and forth along an arc of a circle. What are the directions of the velocity vector and the acceleration vector when the girl passes through the lowest point of the circle?

QUESTION 4: A road leads over the crest of a hill. Near the crest, the vertical cross section of the hill is an arc of circle. If an automobile drives along the road at constant speed, what are the directions of the velocity vector and the acceleration vector at the crest of the hill?

(A) **v** forward, **a** upward   (B) **v** forward, **a** downward

(C) **v** upward, **a** upward   (D) **v** upward, **a** downward

4.6 THE RELATIVITY OF MOTION AND THE ADDITION OF VELOCITIES

Motion is relative—the values of the position, velocity, and acceleration of a particle depend on the frame of reference in which these quantities are measured. For example, consider one reference frame attached to the shore and a second reference frame attached to a ship moving away from the shore due east at a constant velocity of 5 m/s. Suppose that observers in both reference frames measure and plot the position vector of a fast motorboat passing by. The observers will then find different results for the position and the velocity of the motorboat. If the velocity of the motorboat is 12 m/s due east in the reference frame of the shore, it will be 7 m/s due east in the reference frame of the ship. The velocities of the motorboat in the two reference frames are related by a simple addition rule: the velocity relative to the shore is the velocity relative to the ship plus the velocity of the ship relative to the shore, that is, 12 m/s = 7 m/s + 5 m/s. This simple addition rule seems intuitively obvious, but to see where it comes from, we need to examine the position vectors of the motorboat in the two reference frames.

The position vector measured in the first reference frame (shore) will be denoted by **r**; that measured in the second reference frame (ship) will be denoted by **r′**. These two position vectors are different (see Fig. 4.24, where the position vector **r** is green and the position vector **r′** is blue). The velocity of the ship relative to the shore will be denoted by $\mathbf{V}_O$. For the sake of simplicity, let us assume that the velocity of the ship relative to the shore is constant,

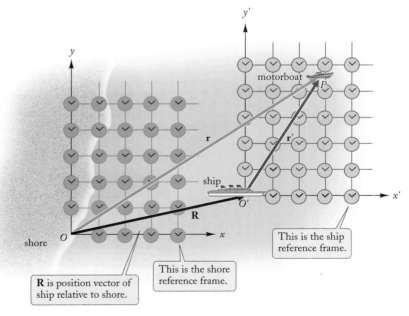

FIGURE 4.24 The coordinate grid x'–y' (blue) of the ship moves relative to the coordinate grid x–y (green) of the shore.

and that the ship started at the shore at time $t = 0$. The position vector of the ship relative to the shore is then $\mathbf{R} = \mathbf{V}_O t$. The vectors $\mathbf{r}$, $\mathbf{r}'$, and $\mathbf{R}$ form a vector triangle (see Fig. 4.24), and by inspection of this triangle we recognize that

$$\mathbf{r} = \mathbf{r}' + \mathbf{R}$$

or

$$\mathbf{r} = \mathbf{r}' + \mathbf{V}_O t \tag{4.50}$$

This equation merely says that the position vector of the motorboat relative to the shore is the position vector relative to the ship plus the position vector of the ship relative to the shore.

To extract the addition rule for velocities from Eq. (4.50), we contemplate what happens in a small time interval Δt. In such a time interval, the position vectors $\mathbf{r}$ and $\mathbf{r}'$ change by $\Delta \mathbf{r}$ and $\Delta \mathbf{r}'$. According to Eq. (4.50), these small changes are related by

$$\Delta \mathbf{r} = \Delta \mathbf{r}' + \mathbf{V}_O \Delta t \tag{4.51}$$

Dividing this by Δt, we find

$$\frac{\Delta \mathbf{r}}{\Delta t} = \frac{\Delta \mathbf{r}'}{\Delta t} + \mathbf{V}_O \tag{4.52}$$

In the limit $\Delta t \to 0$, $\Delta \mathbf{r}/\Delta t$ is the instantaneous velocity of the motorboat relative to the shore, and $\Delta \mathbf{r}'/\Delta t$ is the instantaneous velocity of the motorboat relative to the ship. Designating these velocites by $\mathbf{v}$ and $\mathbf{v}'$, respectively, we obtain

addition rule for velocities

$$\mathbf{v} = \mathbf{v}' + \mathbf{V}_O \tag{4.53}$$

This is the **addition rule for velocities**, also called the *Galilean velocity transformation,* because Galileo was the first to investigate the relativity of motion.

Note that since the velocities in the two reference frames differ by only a constant quantity $\mathbf{V}_O$, the *changes* in these velocities are the same in the two reference frames. Thus, the accelerations in the two reference frames are the same,

$$\mathbf{a}' = \mathbf{a} \tag{4.54}$$

If the motorboat is accelerating at, say, 3 m/s^2 relative to the shore, it will be accelerating at the same rate relative to the ship. This means that *for reference frames in uniform motion relative to one another, acceleration is an absolute quantity.*

EXAMPLE 10   Off the coast of Miami, the Gulf Stream current has a velocity of 4.8 km/h in a direction due north. The captain of a motorboat wants to travel from Miami to North Bimini island, due east of Miami. His boat has a speed of 18 km/h relative to the water. (a) If he heads his boat due east, what will be his actual course relative to the shore? (b) To attain a straight course, due east, from Miami to Bimini, in what direction must he head his boat? What is his speed relative to the shore?

SOLUTION: (a) Figure 4.25a shows the boat heading due east. The Gulf Stream current carries the boat northward, resulting in a northeast course. Figure 4.25b shows the velocity vector $\mathbf{V}_O$ of the water relative to the shore and the velocity

(a)

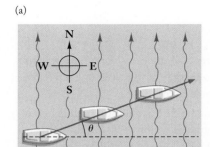

(b)

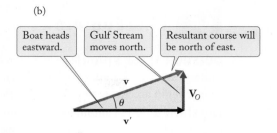

FIGURE 4.25   (a) The boat is heading due east, but the current carries the boat northward, resulting in a northeast course. (b) The velocity $\mathbf{V}_O$ of the water relative to the shore is northward, and the velocity $\mathbf{v}'$ of the boat relative to the water is eastward. The vector sum $\mathbf{v}' + \mathbf{V}_O$ is northeastward.

vector $\mathbf{v}'$ of the boat relative to the water. According to Eq. (4.53), the velocity vector $\mathbf{v}$ of the boat relative to the shore is the sum of $\mathbf{v}'$ and $\mathbf{V}_O$; that is,

$$\mathbf{v} = \mathbf{v}' + \mathbf{V}_O \qquad (4.55)$$

From Fig. 4.25b we see that the tangent of the angle θ between $\mathbf{v}$ and the eastward direction is

$$\tan\theta = \frac{V_O}{v'}$$
$$= \frac{4.8 \text{ km/h}}{18 \text{ km/h}} = 0.27$$

With a calculator, we find that the inverse tangent of 0.27, and thus the direction of the actual course of the motorboat, is

$$\theta = 15°$$

(b) Figure 4.26a shows the boat on a due east course. To achieve this course, the captain must head his boat southeastward at an angle ϕ, to compensate for the Gulf Stream current. Figure 4.26b shows the velocity vector $\mathbf{V}_O$ of the water relative to the shore and the velocity vector $\mathbf{v}'$ of the boat relative to the water. By hypothesis, the velocity vector $\mathbf{v}$ relative to the shore points due east. Since the vector triangle is a right triangle, the sine of the angle ϕ between $\mathbf{v}'$ and the eastward direction is

$$\sin\phi = \frac{V_O}{v'} \qquad (4.56)$$
$$= \frac{4.8 \text{ km/h}}{18 \text{ km/h}} = 0.27$$

With our calculator, we find that if the sine is 0.27, the angle is

$$\phi = 16°$$

Thus, the boat must head 16° south of east.

The speed of the boat relative to the shore can be calculated from the right triangle of Fig. 4.26b. It is

$$v = v'\cos\phi = 18 \text{ km/h} \times \cos 16° = 17 \text{ km/h}$$

(a)

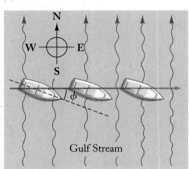

(b)

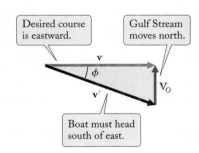

FIGURE 4.26   (a) The course of the boat is due east, but the heading of the boat is south of east. (b) The velocity $\mathbf{V}_O$ of the water relative to the shore is northward, and the velocity $\mathbf{v}'$ of the boat relative to the water is southeastward. The angle ϕ has been selected so that the vector sum $\mathbf{v}' + \mathbf{V}_O$ is due east.

Checkup 4.6

QUESTION 1: Can a particle be at rest in one reference frame, but be in motion with uniform velocity in another reference frame? Can a particle be at rest in one reference frame, but in accelerated motion in another reference frame?

QUESTION 2: A red automobile traveling along a straight road at 80 km/h is being overtaken by a black automobile traveling in the same direction at 90 km/h. What is the velocity of the black automobile relative to the red? The red relative to the black?

QUESTION 3: A wind is blowing at 30 km/h in the southward direction. An airplane is flying northward with an airspeed (relative to the air) of 200 km/h. What is the ground speed (relative to the ground) of the airplane?

QUESTION 4: To swim across a flowing river in the shortest possible time, in what direction should you swim? To reach a point on the shore directly opposite your starting point, how should you swim?

QUESTION 5: Relative to particle A, particle B is in uniform motion. Relative to particle B, particle C is in accelerated motion. Relative to particle A, is the motion of particle C accelerated or unaccelerated?

QUESTION 6: The wind on Lake Champlain is blowing from the north. A motorboat heads across the lake, in the west direction. Relative to the motorboat, the wind seems to blow from:

(A) East of north   (B) West of north   (C) East of south
(D) West of south   (E) North

SUMMARY

Here, we review the relations for two dimensions. As discussed at the end of Section 4.2, the extension to three dimensions is straightforward, but rarely needed.

PHYSICS IN PRACTICE Velocity Vectors		**(page 99)**
PROBLEM-SOLVING TECHNIQUES Projectile Motion		**(page 108)**
AVERAGE VELOCITY COMPONENTS	$\bar{v}_x = \dfrac{\Delta x}{\Delta t} \qquad \bar{v}_y = \dfrac{\Delta y}{\Delta t}$	**(4.1; 4.2)**
INSTANTANEOUS VELOCITY COMPONENTS	$v_x = \dfrac{dx}{dt} \qquad v_y = \dfrac{dy}{dt}$	**(4.3; 4.4)**
AVERAGE ACCELERATION COMPONENTS	$\bar{a}_x = \dfrac{\Delta v_x}{\Delta t} \qquad \bar{a}_y = \dfrac{\Delta v_y}{\Delta t}$	**(4.9; 4.10)**

| **INSTANTANEOUS ACCELERATION COMPONENTS** | $a_x = \dfrac{dv_x}{dt} \qquad a_y = \dfrac{dv_y}{dt}$ | (4.11; 4.12) |

| **POSITION VECTOR** | $\mathbf{r} = x\mathbf{i} + y\mathbf{j}$ | (4.13) |

VELOCITY VECTOR

$$\mathbf{v} = v_x\mathbf{i} + v_y\mathbf{j} = \frac{dx}{dt}\mathbf{i} + \frac{dy}{dt}\mathbf{j}$$

(4.14; 4.15)

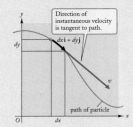

ACCELERATION VECTOR

$$\mathbf{a} = a_x\mathbf{i} + a_y\mathbf{j} = \frac{dv_x}{dt}\mathbf{i} + \frac{dv_y}{dt}\mathbf{j}$$

(4.18; 4.19)

MOTION WITH CONSTANT ACCELERATION

$$v_x = v_{0x} + a_x t \qquad (4.20)$$
$$v_y = v_{0y} + a_y t \qquad (4.21)$$
$$x - x_0 = v_{0x}t + \tfrac{1}{2}a_x t^2 \qquad (4.23)$$
$$y - y_0 = v_{0y}t + \tfrac{1}{2}a_y t^2 \qquad (4.24)$$

PROJECTILE MOTION (With initial position $x_0 = 0$, $y_0 = 0$; $g \approx 9.81\ \text{m/s}^2$, and elevation angle θ measured from the horizontal.)

$$v_x = v_{0x} = v_0 \cos\theta \qquad (4.26)$$
$$v_y = v_{0y} - gt = v_0 \sin\theta - gt \qquad (4.28)$$
$$x = v_{0x}t \qquad (4.27)$$
$$y = v_{0y}t - \tfrac{1}{2}gt^2 \qquad (4.29)$$

RANGE $x_{\max}$, MAXIMUM HEIGHT $y_{\max}$, AND TIME OF FLIGHT t_{flight} (Over flat ground)

$$x_{\max} = \frac{2v_0^2 \sin\theta \cos\theta}{g} \qquad (4.43)$$

$$y_{\max} = \frac{v_0^2 \sin^2\theta}{2g} \qquad (4.42)$$

$$t_{\text{flight}} = \frac{2v_0 \sin\theta}{g} \qquad (4.41)$$

CENTRIPETAL ACCELERATION IN UNIFORM CIRCULAR MOTION (The direction of $\mathbf{a}$ is toward the center of the circle.)

$$a = \frac{v^2}{r}$$

(4.49)

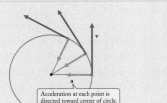

| **ADDITION RULE FOR VELOCITIES** (Galilean velocity transformation) | $\mathbf{v} = \mathbf{v}' + \mathbf{V}_O$ | (4.53) |

| **ACCELERATION IS ABSOLUTE** (For uniform relative motion of reference frames.) | $\mathbf{a} = \mathbf{a}'$ | (4.54) |

QUESTIONS FOR DISCUSSION

1. Can an automobile have eastward instantaneous velocity and northward instantaneous acceleration? Give an example.

2. Consider an automobile that is rounding a curve and braking at the same time. Draw a diagram showing the relative directions of the instantaneous velocity and acceleration.

3. A projectile is launched over level ground. Its initial velocity has a horizontal component v_{0x} and a vertical component v_{0y}. What is the average velocity of the projectile between the instants of launch and of impact?

4. If you throw a crumpled piece of paper, its trajectory is not a parabola. How does it differ from a parabola and why?

5. If a projectile is subject to air resistance, then the elevation angle for maximum range is not 45°. Do you expect the angle to be larger or smaller than 45°?

6. Baseball pitchers are fond of throwing curveballs. How does the trajectory of such a ball differ from the simple parabolic trajectory we studied in this chapter? What accounts for the difference?

7. A pendulum is swinging back and forth. Is this uniform circular motion? Draw a diagram showing the directions of the velocity and the acceleration at the top of the swing. Draw a similar diagram at the bottom of the swing.

8. Why do raindrops fall down at a pronounced angle with the vertical when seen from the window of a speeding train? Is this angle necessarily the same as that of the path of a water drop sliding down the outside surface of the window?

9. When a sailboat is sailing to windward ("beating"; see Fig. 4.27), the wind feels much stronger than when the sailboat is sailing downwind ("running"). Why?

10. Rain is falling vertically. If you run through the rain, at what angle should you hold your umbrella? If you don't have an umbrella, should you bend forward while running?

11. In the reference frame of the ground, the path of a sailboat beating to windward makes an angle of 45° with the direction of the wind. In the reference frame of the sailboat, the angle is somewhat smaller. Explain.

12. According to a theory proposed by Galileo, the tides on the oceans are caused by the Earth's rotational motion about its axis combined with its translational motion around the Sun. At midnight these motions are in the same direction; at noon they are in opposite directions (Fig. 4.28). Thus, any point on the Earth alternately speeds up and slows down. Galileo was of the opinion that the speeding up and slowing down of the ocean basins would make the water slosh back and forth, thus giving rise to tides. What is wrong with this theory? (Hint: What is the acceleration of a point on the Earth? Does this acceleration depend on the translational motion?)

FIGURE 4.27 Sailboat beating to windward.

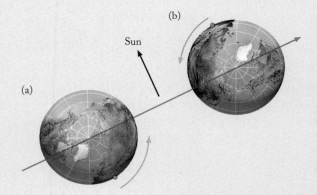

FIGURE 4.28 Rotational and translational motions of the Earth (a) at midnight and (b) at noon.

PROBLEMS

4.1 Components of Velocity and Acceleration[†]
4.2 The Velocity and Acceleration Vectors[†]

1. A sailboat tacking against the wind moves as follows: 3.2 km at 45° east of north, 4.5 km at 50° west of north, and 2.6 km at 45° east of north. The entire motion takes 1 h 15 min.
   (a) What is the total displacement for this motion?
   (b) What is the average velocity for this motion?
   (c) What is the speed if it is assumed to be constant?

2. In one-half year, the Earth moves halfway around its orbit, a circle of radius 1.50×10^{11} m centered on the Sun. What is the average speed, and what is the magnitude of the average velocity for this time interval?

3. The fastest bird is the spine-tailed swift, which reaches speeds of 171 km/h. Suppose that you wish to shoot such a bird with a .22-caliber rifle that fires a bullet with a speed of 366 m/s. If you fire at the instant when the bird is 30 m directly overhead, how many meters ahead of the bird must you aim the rifle? Ignore gravity in this problem.

4. An automobile with a drunken driver at the wheel travels round and round a traffic circle at 30 km/h. The automobile takes 80 s to go once around the circle. At $t = 0$, the automobile is at the east of the traffic circle; at $t = 20$ s it is at the north; at $t = 40$ s it is at the west; etc. What are the components of the velocity of the automobile at $t = 0$, $t = 10$ s, $t = 20$ s, $t = 30$ s, and $t = 40$ s? The x axis points eastward and the y axis points northward.

5. Suppose that a particle moving in three dimensions has a position vector
$$\mathbf{r} = (4 + 2t)\mathbf{i} + (3 + 5t + 4t^2)\mathbf{j} + (2 - 2t - 3t^2)\mathbf{k}$$
where distance is measured in meters and time in seconds.
   (a) Find the instantaneous velocity vector.
   (b) Find the instantaneous acceleration vector. What are the magnitude and the direction of the acceleration?

6. A particle is moving in the x–y plane; the components of its position are
$$x = A \cos bt \qquad y = A \sin bt$$
where A and b are constants.
   (a) What are the components of the instantaneous velocity vector? The instantaneous acceleration vector?
   (b) What is the magnitude of the instantaneous velocity? The instantaneous acceleration?

7. For the motion of the cruise missile described in Example 3, calculate the displacement of the missile relative to the aircraft at $t = 2.0$ s and at $t = 3.0$ s. What are the magnitude and direction of the displacement vector at each of these times?

8. The components of the position of a body as a function of time are given by:
$$x = 5t + 4t^2 \qquad y = 3t^2 + 2t^3 \qquad z = 0$$
where x and y are in meters and t is in seconds. What is the velocity vector as a function of time? What is the acceleration vector as a function of time? What is the speed at $t = 2.0$ s?

9. An airplane traveling at a constant speed of 300 km/h flies 30° north of east for 0.50 h and then flies 30° west of south for 1.00 h. What is the average velocity vector for the entire flight? What is the average acceleration vector for the entire flight?

10. The components of the position vector of a particle moving in the x–y plane are
$$x = A \cos bt \qquad y = Bt$$
where A, b, and B are constants. What are the components of the instantaneous velocity vector? The instantaneous acceleration vector? What is the speed of the particle?

11. As an aircraft approaches landing, the components of its position are given by
$$x = 90t \qquad y = 500 - 15t$$
where x and y are in meters and t is in seconds. What is the velocity vector of the aircraft during this descent? What is the value of its speed during the descent? What angle does the velocity vector make with the horizontal?

*12. Two football players are initially 20 m apart. The first player (a receiver) runs perpendicularly to the initial line joining the two players at a constant speed of 7.0 m/s. After two seconds, the second player (the quarterback) throws the ball at a horizontal speed of 15 m/s (ignore any vertical motion). In what horizontal direction should the quarterback aim so that the ball reaches the same spot the receiver will be? At what time will the ball be caught?

4.3 Motion with Constant Acceleration
4.4 The Motion of Projectiles[†]

13. Suppose that the acceleration vector of a particle moving in the x–y plane is
$$\mathbf{a} = 3\mathbf{i} + 2\mathbf{j}$$
where the acceleration is measured in m/s². The velocity vector and the position vector are zero at $t = 0$.
   (a) What is the velocity vector of this particle as a function of time?
   (b) What is the position vector as a function of time?

14. The fastest recorded speed of a baseball thrown by a pitcher is 162.3 km/h (100.9 mi/h), achieved by Nolan Ryan in 1974 at

[†] For help, see Online Concept Tutorial 5 at www.wwnorton.com/physics

Anaheim Stadium. If the baseball leaves the pitcher's hand with a horizontal velocity of this magnitude, how far will the ball have fallen vertically by the time it has traveled 20 m horizontally?

15. At Acapulco, professional divers jump from a 36-m-high cliff into the sea (compare Example 9 in Chapter 2). At the base of the cliff, a rocky ledge sticks out for a horizontal distance of 6.4 m. With what minimum horizontal velocity must the divers jump off if they are to clear this ledge?

16. Consider the bomb dropped from the bomber described in Example 6.

    (a) What are the final horizontal and vertical components of the velocity of the bomb when it strikes the surface of the sea?

    (b) What is the final speed of the bomb? Compare this with the initial speed of the bomb.

17. A stunt driver wants to make his car jump over 10 cars parked side by side below a horizontal ramp (Fig. 4.29). With what minimum speed must he drive off the ramp? The vertical height of the ramp is 2.0 m, and the horizontal distance he must clear is 24 m.

18. A particle has an initial position vector $\mathbf{r} = 0$ and an initial velocity $\mathbf{v}_0 = 3\mathbf{i} + 2\mathbf{j}$ (where distance is measured in meters and velocity in meters per second). The particle moves with a constant acceleration $\mathbf{a} = \mathbf{i} - 4\mathbf{j}$ (measured in m/s^2). At what time does the particle reach a maximum y coordinate? What is the position vector of the particle at that time?

19. According to a reliable report, in 1795 a member of the Turkish embassy in England shot an arrow to a distance of 441 m. According to a less reliable report, a few years later the Turkish Sultan Selim shot an arrow to 889 m. In each of these cases calculate what must have been the minimum initial speed of the arrow.

20. A golfer claims that a golf ball launched with an elevation angle of 12° can reach a horizontal range of 250 m. Ignoring air friction, what would the initial speed of such a golf ball have to be? What maximum height would it reach?

21. A gunner wants to fire a gun at a target at a horizontal distance of 12 500 m from his position.

    (a) If his gun fires with a muzzle speed of 700 m/s and if $g = 9.81$ m/s^2, what is the correct elevation angle? Pretend that there is no air resistance.

    (b) If the gunner mistakenly assumes $g = 9.80$ m/s^2, by how many meters will he miss the target?

22. According to the *Guinness Book of World Records,* during a catastrophic explosion in Halifax on December 6, 1917, William Becker was thrown through the air for some 1500 m and was found, still alive, in a tree. Assume that Becker left the ground and returned to the ground (ignore the height of the tree) at an angle of 45°. With what speed did he leave the ground? How high did he rise? How long did he stay in flight?

23. In a circus act at the Ringling Bros. and Barnum & Bailey Circus, a "human cannonball" was fired from a large cannon with a muzzle speed of 87 km/h. Assume that the firing angle was 45° from the horizontal. How many seconds did the human cannonball take to reach maximum height? How high did he rise? How far from the cannon did he land?

24. The world record for the javelin throw by a woman established in 1976 by Ruth Fuchs in Berlin was 69.11 m (226 ft 9 in.). If Fuchs had thrown her javelin with the same initial velocity in Buenos Aires rather than in Berlin, how much farther would it have gone? The acceleration of gravity is 9.8128 m/s^2 in Berlin and 9.7967 m/s^2 in Buenos Aires. Pretend that air resistance plays no role in this problem.

25. The motion of an ICBM can be regarded as the motion of a projectile, because along the greatest part of its trajectory the missile is in free fall, outside of the atmosphere. Suppose that the missile is to strike a target 1000 km away. What minimum speed must the missile have at the beginning of its trajectory? What maximum height does it reach when launched with this minimum speed? How long does it take to reach its target? For these calculations assume that $g = 9.8$ m/s^2 everywhere along the trajectory and ignore the (short) portions of the trajectory inside the atmosphere.

26. The natives of the South American Andes throw stones by means of slings which they whirl (see Fig. 4.30). They can accurately throw a 0.20-kg stone to a distance of 50 m.

    (a) What is the minimum speed with which the stone must leave the sling to reach this distance?

    (b) Just before the release, the stone is being whirled around a circle of radius 1.0 m with the speed calculated in part (a). How many revolutions per second does the stone make?

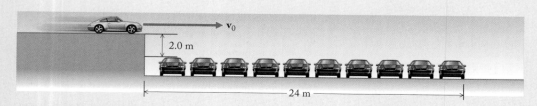

FIGURE 4.29 A stunt.

FIGURE 4.30 Whirling a stone before slinging.

27. The nozzle of a fire hose ejects 280 liters of water per minute at a speed of 26 m/s. How far away will the stream of water land if the nozzle is aimed at an angle of 35° with the horizontal? How many liters of water are in the air at any given instant?

28. According to an ancient Greek source, a stone-throwing machine on one occasion achieved a range of 730 m. If this is true, what must have been the minimum initial speed of the stone as it was ejected from the engine? When thrown with this speed, how long would the stone have taken to reach its target?

29. For what launch angle will the height and range of a projectile be equal?

30. A juggler tosses and catches balls at waist level; the balls are tossed at launch angles of 60°. If a ball attains a height 60 cm above waist level, how long is a ball in the air?

31. At $t = 0$, a small particle begins at the origin with initial velocity components $v_{0x} = -10$ m/s and $v_{0y} = 25$ m/s. Throughout its motion, the particle experiences an acceleration $\mathbf{a} = (2.0\mathbf{i} - 4.5\mathbf{j})$ m/s^2. Find the speed of the particle at $t = 3.0$ s. Find the position vector of the particle at $t = 3.0$ s.

32. A baseball is popped up, remaining aloft for 6.0 s before being caught at a horizontal distance of 75 m from the starting point. What was the launch angle?

33. An errant speeding bus launches from an unfinished highway ramp angled 10° upward. To complete the jump across a horizontal roadway gap of 15 m, what minimum initial speed must the bus have?

34. A child rolls a ball horizontally off the edge of a table. For what initial speed will the ball strike the floor a horizontal distance away from the table edge equal to the table height? In that case, what is the velocity of the ball just before it hits the floor?

*35. A boy stands at the edge of a cliff and launches a rock upward at an angle of 45.0°. The rock comes back down to the elevation where it was released 2.25 s later, then continues until it is seen to splash into the lake below 4.00 s after release. How far below the point of release is the lake surface? What horizontal distance from the point of release is the splash?

*36. A rock is thrown from a bridge at an upward launch angle of 30° with an initial speed of 25 m/s. The bridge is 30 m above the river. How much time elapses before the rock hits the water?

*37. A hockey player 25 m from the goal hits the hockey puck toward the goal, imparting a launch speed of 65 m/s at a launch angle of 10°. If the goal is 1.5 m high, does the shot score? At what vertical height does the puck pass the goal? How long does the puck take to reach the goal?

*38. (a) A golfer wants to drive a ball to a distance of 240 m. If he launches the ball with an elevation angle of 14.0°, what is the appropriate initial speed? Ignore air resistance.

  (b) If the speed is too great by 0.6 m/s, how much farther will the ball travel when launched at the same angle?

  (c) If the elevation angle is 0.5° larger than 14.0°, how much farther will the ball travel if launched with the speed calculated in part (a)?

*39. Show that for a projectile launched with an elevation angle of 45°, the maximum height reached is one-quarter of the range.

*40. During a famous jump in Richmond, Virginia, in 1903, the horse Heatherbloom with its rider jumped over an obstacle 8 ft 8 in. high while covering a horizontal distance of 37 ft. At what angle and with what speed did the horse leave the ground? Make the (somewhat doubtful) assumption that the motion of the horse is particle motion.

*41. With what elevation angle must you launch a projectile if its range is to equal twice its maximum height?

*42. In a baseball game, the batter hits the ball and launches it upward at an angle of 52° with a speed of 38 m/s. At the same instant, the center fielder starts to run toward the (expected) point of impact of the ball from a distance of 45 m. If he runs at 8.0 m/s, can he reach the point of impact before the ball?

*43. The gun of a coastal battery is emplaced on a hill 50 m above the water level. It fires a shot with a muzzle speed of 600 m/s at a ship at a horizontal distance of 12 000 m. What elevation angle must the gun have if the shot is to hit the ship? Pretend there is no air resistance.

*44. In a flying ski jump, the skier acquires a speed of 110 km/h by racing down a steep hill and then lifts off into the air from a horizontal ramp. Beyond this ramp, the ground slopes downward at an angle of 45°.

  (a) Assuming that the skier is in a free-fall motion after he leaves the ramp, at what distance down the slope will he land?

  (b) In actual jumps, skiers reach distances of up to 165 m. Why does this not agree with the result you obtained in part (a)?

*45. Olympic target archers shoot arrows at a bull's-eye 12 cm across from a distance of 90.00 m. If the initial speed of the arrow is 70.00 m/s, what must be the elevation angle? If the archer misaims the arrow by 0.03° in the vertical direction, will it hit the bull's-eye? If the archer misaims the arrow by 0.03° in the horizontal direction, will it hit the bull's-eye? Assume that the height of the bull's-eye above the ground is the same as the initial arrow height of the bow and ignore air resistance.

*46. The muzzle speed for a Lee−Enfield rifle is 630 m/s. Suppose you fire this rifle at a target 700 m away and at the same level as the rifle.

  (a) In order to hit the target, you must aim the barrel at a point above the target. How many meters above the target must you aim? Pretend there is no air resistance.

  (b) What will be the maximum height that the bullet reaches along its trajectory?

  (c) How long does the bullet take to reach the target?

*47. In artillery, it is standard practice to fire a sequence of trial shots at a target before commencing to fire "for effect." The artillerist first fires a shot short of the target, then a shot beyond the target, and then makes the necessary adjustment in elevation so that the third shot is exactly on target. Suppose that the first shot fired from a gun aimed with an elevation angle of 7°20′ lands 180 m short of the target; the second shot fired with an elevation of 7°35′ lands 120 m beyond the target. What is the correct elevation angle to hit the target?

*48. A hay-baling machine throws each finished bundle of hay 2.5 m up in the air so it can land on a trailer waiting 5.0 m behind the machine. What must be the speed with which the bundles are launched? What must be the angle of launch?

*49. Consider the trajectories for projectiles with the same launch speed, but different elevation angles. If you launch a large number of such projectiles simultaneously, will any of them ever collide while in flight? Explain carefully.

*50. Suppose that at the top of its parabolic trajectory a projectile has a horizontal speed v_{0x}. The segment at the top of the parabola can be approximated by a circle, called the osculating circle (Fig. 4.31). What is the radius of this circle? (Hint: The projectile is instantaneously in uniform circular motion at the top of the parabola.)

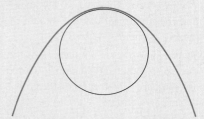

FIGURE 4.31 The osculating circle.

*51. A battleship steaming at 45 km/h fires a gun at right angles to the longitudinal axis of the ship. The elevation angle of the gun is 30°, and the muzzle velocity of the shot is 720 m/s; the gravitational acceleration is 9.8 m/s². What is the range of this shot in the reference frame of the ground? Pretend that there is no air resistance.

**52. The maximum speed with which you can throw a stone is about 25 m/s (a professional baseball pitcher can do much better than this). Can you hit a window 50 m away and 13 m up from the point where the stone leaves your hand? What is the maximum height of a window you can hit at this distance?

**53. A gun standing on sloping ground (see Fig. 4.32) fires up the slope. Show that the slant range of the gun (measured along the slope) is

$$l = \frac{2v_0^2 \cos^2 \theta}{g \cos \alpha} (\tan \theta - \tan \alpha)$$

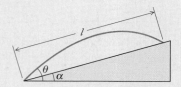

FIGURE 4.32 Projectile motion up a slope.

where α is the angle of the slope and the other symbols have their usual meaning. For what value of θ is this range a maximum?

**54. Two football players are initially 15 m apart. The first player (a receiver) runs perpendicular to the line joining the two players at a constant speed of 8.0 m/s. After two seconds, the second player (the quarterback) throws the ball with a horizontal component velocity of 20 m/s. In what horizontal direction and with what vertical launch angle should the quarterback throw so that the ball reaches the same spot the receiver will be? At what time will the ball be caught?

**55. When a tractor leaves a muddy field and drives on the highway, clumps of mud will sometimes come off the rear wheels and be launched into the air (see Fig. 4.33). In terms of the speed u of the tractor and the radius R of the wheel, find the maximum possible height that a clump of dirt can reach. In your calculation be careful to take into account both the initial velocity of the clump and the initial height at which it comes off the wheel. Evaluate numerically for $u = 30$ km/h and $R = 0.80$ m. (Hint: Solve this problem in the reference frame of the tractor.)

FIGURE 4.33 Tractor wheel flinging mud.

**56. A gun on the shore (at sea level) fires a shot at a ship which is heading directly toward the gun at a speed of 40 km/h. At the instant of firing, the distance to the ship is 15 000 m. The muzzle speed of the shot is 700 m/s. Pretend that there is no air resistance.

(a) What is the required elevation angle for the gun? Assume $g = 9.8$ m/s².

(b) What is the time interval between firing and impact?

**57. A ship is steaming at 30 km/h on a course parallel to a straight shore at a distance of 17 000 m. A gun emplaced on the shore (at sea level) fires a shot with a muzzle speed of 700 m/s when the ship is at the point of closest approach. If the shot is to hit the ship, what must be the elevation angle of the gun? How far ahead of the ship must the gun be aimed? Give the answer to the latter question both in meters and in minutes of arc. Pretend that there is no air resistance. (Hint: Solve this problem by the following method of successive approximations. First calculate the time of flight of the shot, neglecting the motion of the ship; then calculate how far the ship moves in this time; and then calculate the elevation angle and the aiming angle required to hit the ship at this new position.)

4.5 Uniform Circular Motion

58. An audio compact disk (CD) player is rotating at an angular velocity of 32.5 radians per second when playing a track at a radius of 4.0 cm. What is the linear speed at that radius? What is the rotation rate in revolutions per minute?

59. In science fiction movies, large, ring-shaped space stations rotate so that astronauts experience an acceleration, which feels the same as gravity. If the station is 200 m in radius, how many revolutions per minute are required to provide an acceleration of 9.81 m/s²?

60. When drilling metals, excess heat is avoided by staying below a recommended linear cutting speed. A 3.0-mm-diameter hole and a 25-mm-diameter hole need to be drilled. At what maximum number of revolutions per minute can the drill bit rotate so that a point on its perimeter does not exceed the material's linear cutting speed limit of 3.0 m/s?

61. The Space Shuttle orbits the Earth on a circle of radius 6500 km every 87 minutes. What is the centripetal acceleration of the Space Shuttle in this orbit?

62. A mechanical pitcher hurls baseballs for batting practice. The arm of the pitcher is 0.80 m long and is rotating at 45 radians/second at the instant of release. What is the speed of the pitched ball?

63. An ultracentrifuge spins a small test tube in a circle of radius 10 cm at 1000 revolutions per second. What is the centripetal acceleration of the test tube? How many standard *g*'s does this amount to?

64. The blade of a circular saw has a diameter of 20 cm. If this blade rotates at 7000 revolutions per minute (its maximum safe speed), what are the speed and the centripetal acceleration of a point on the rim?

65. At the Fermilab accelerator (one of the world's largest atom smashers), protons are forced to travel in an evacuated tube in a circular orbit of diameter 2.0 km (Fig. 4.34). The protons have a speed nearly equal to the speed of light (99.999 95% of the speed of light). What is the centripetal acceleration of these protons? Express your answer in m/s² and in standard *g*'s.

66. A phonograph record rotates at $33\frac{1}{3}$ revolutions per minute. The radius of the record is 15 cm. What is the speed of a point at its rim?

67. The Earth moves around the Sun in a circular path of radius 1.50×10^{11} m at uniform speed. What is the magnitude of the centripetal acceleration of the Earth toward the Sun?

68. An automobile has wheels of diameter 64 cm. What is the centripetal acceleration of a point on the rim of this wheel when the automobile is traveling at 95 km/h?

*69. The Earth rotates about its axis once in one sidereal day of 23 h 56 min. Calculate the centripetal acceleration of a point located on the equator. Calculate the centripetal acceleration of a point located at a latitude of 45°.

*70. When looping the loop, the Blue Angels stunt pilots of the U.S. Navy fly their jet aircraft along a vertical circle of diameter 1000 m (Fig. 4.35). At the top of the circle, the speed is 350 km/h; at the bottom of the circle, the speed is 620 km/h. What is the centripetal acceleration at the top? At the bottom? In the reference frame of one of these aircraft, what is the acceleration that the pilot feels at the top and at the bottom; i.e., what is the acceleration relative to the aircraft of a small body, such as a coin, released by the pilot?

71. The table inside the book cover lists the radii of the orbits of the planets around the Sun and the time taken to complete an orbit ("period of revolution"). Assume that the planets move along circles at constant speed. Calculate the centripetal acceleration for each of the first three planets (Mercury, Venus, Earth). Verify that the centripetal accelerations are in proportion to the inverses of the squares of the orbital radii.

FIGURE 4.34 The main accelerator ring at Fermilab.

FIGURE 4.35 Blue Angels looping the loop.

4.6 The Relativity of Motion and the Addition of Velocities

72. On a rainy day, a steady wind is blowing at 30 km/h. In the reference frame of the *air*, the raindrops are falling vertically with a speed of 10 m/s. What are the magnitude and the direction of the velocity of the raindrops in the reference frame of the ground?

73. In an airport, a moving walkway has a speed of 1.5 m/s relative to the ground. What is the speed, relative to the ground, of a passenger running forward on this walkway at 4.0 m/s? What is the speed, relative to the ground, of a passenger running backward on this walkway at 4.0 m/s?

74. On a rainy day, raindrops are falling with a vertical velocity of 10 m/s. If an automobile drives through the rain at 25 m/s, what is the velocity (magnitude and direction) of the raindrops relative to the automobile?

75. A battleship steaming at 13 m/s toward the shore fires a shot in the forward direction. The elevation angle of the gun is 20°, and the muzzle speed of the shot is 660 m/s. What is the velocity vector of the shot relative to the shore?

76. A wind of 30 m/s is blowing from the west. What will be the speed, relative to the ground, of a sound signal traveling due north? The speed of sound, relative to air, is 330 m/s.

77. On a windy day, a hot-air balloon is ascending at a rate of 1.5 m/s relative to the air. Simultaneously, the air is moving with a horizontal velocity of 12.0 m/s. What is the velocity (magnitude and direction) of the balloon relative to the ground?

78. You can paddle your kayak at a speed of 3.5 km/h relative to the water. If a river is flowing at 2.5 km/h, how far can you paddle downstream in 40 minutes? How long will it take you to paddle back upstream from there?

79. As a train rolls by at 5.00 m/s, you see a cat on one of the flatcars. The cat is walking toward the back of the train at a speed of 0.50 m/s relative to the car. On the cat is a flea which is walking from the cat's neck to its tail at a speed of 0.10 m/s relative to the cat. How fast is the flea moving relative to you?

80. A boat with maximum speed v (relative to the water) is on one shore of a river of width d. The river is flowing at speed V. Traveling in a straight line, how long does it take to get to a point directly opposite? What is the fastest crossing time to any point?

81. Each step on an up escalator is 20 cm high and 30 cm deep. The escalator advances 1.5 step per second. If you also walk up the escalator stairs at a rate of 1.0 step per second, what is your velocity (magnitude and direction) relative to a fixed observer?

*82. A villain in a car traveling at 30 m/s fires a projectile along the direction of motion toward the front of the car with a launch speed of 50 m/s relative to the car. A hero standing nearby observes the projectile to travel straight up. What was the launch angle as viewed by the villain? What height does the projectile attain?

*83. A blimp is motoring at constant altitude. The airspeed indicator on the blimp shows that its speed relative to the air is 20 km/h, and the compass shows that the heading of the blimp is 10° east of north. If the air is moving over the ground with a velocity of 15 km/h due east, what is the velocity (magnitude and direction) of the blimp relative to the ground? For an observer on the ground, what is the angle between the longitudinal axis of the blimp and the direction of motion?

*84. A sailboat is moving in a direction 50° east of north at a speed of 14 km/h. The wind measured by an instrument aboard the sailboat has an apparent (relative to the sailboat) speed of 32 km/h coming from an apparent direction of 10° east of north. Find the true (relative to ground) speed and direction of the wind.

*85. (a) In still air, a high-performance sailplane has a rate of descent (or sinking rate) of 0.50 m/s at a forward speed (or airspeed) of 60 km/h. Suppose the plane is at an initial altitude of 1500 m. How far can it travel horizontally in still air before it reaches the ground?

    (b) Suppose the plane is in a (horizontal) wind of 20 km/h. With the same initial conditions, how far can it travel in the downwind direction? In the upwind direction?

*86. A wind is blowing at 50 km/h from a direction 45° west of north. The pilot of an airplane wishes to fly on a route due north from an airport. The airspeed of the airplane is 250 km/h.

    (a) In what direction must the pilot point the nose of the airplane?

    (b) What will be the airplane's speed relative to the ground?

*87. At the entrance of Ambrose Channel at New York harbor, the tidal current at one time of the day has a velocity of 4.2 km/h in a direction 20° south of east. Consider a ship in this current; suppose that the ship has a speed of 16 km/h relative to the water. If the helmsman keeps the bow of the ship aimed due north, what will be the actual velocity (magnitude and direction) of the ship relative to the ground?

*88. A white automobile is traveling at a constant speed of 90 km/h on a highway. The driver notices a red automobile 1.0 km behind, traveling in the same direction. Two minutes later, the red automobile passes the white automobile.

    (a) What is the average speed of the red automobile relative to the white?

    (b) What is the speed of the red automobile relative to the ground?

*89. Two automobiles travel at equal speeds in opposite directions on two separate lanes of a highway. The automobiles move at constant speed v_0 on straight parallel tracks separated by a distance h. Find a formula for the rate of change of the distance between the automobiles as a function of time; take the instant of closest approach as $t = 0$. Plot v vs. t for $v_0 = 60$ km/h, $h = 50$ m.

*90. A ferryboat on a river has a speed v relative to the water. The water of the river flows with speed V relative to the ground. The width of the river is d.

  (a) Show that the ferryboat takes a time $2d/\sqrt{v^2 - V^2}$ to travel across the river and back.

  (b) Show that the ferryboat takes a time $2dv/(v^2 - V^2)$ to travel a distance d up the river and back. Which trip takes a shorter time?

**91. An AWACS aircraft is flying at high altitude in a wind of 150 km/h from due west. Relative to the air, the heading of the aircraft is due north and its speed is 750 km/h. A radar operator on the aircraft spots an unidentified target approaching from northeast; relative to the AWACS aircraft, the bearing of the target is 45° east of north, and its speed is 950 km/h. What is the speed of the unidentified target relative to the ground?

REVIEW PROBLEMS

92. In the fastest-ever descent from Mt. Everest, two climbers slid down the side of the mountain on the seats of their pants, using their ice picks as brakes. They descended a height of 2340 m in 3.5 h. From these data, can you calculate the average of the vertical component of the velocity? The average of the horizontal component of the velocity? The average speed? What extra information do you need to calculate these quantities?

93. With its engine cut off, a small airplane glides downward at an angle of 15° below the horizontal at a speed of 240 km/h.

  (a) What are the horizontal and the vertical components of its velocity?

  (b) If the airplane is initially at a height of 2000 m above the ground, how long does it take to crash into the ground?

94. An automobile enters a 180° curve at a constant speed of 25 m/s and emerges from this curve 12 s later. What are the components of the average acceleration for this time interval?

95. At the entrance to Ambrose Channel at New York harbor, the maximum tidal current has a velocity of 4.2 km/h in a direction 20° south of east. What is the component of this velocity in the east direction? In the north direction?

96. A blimp motoring at a constant altitude has a velocity component of 15 km/h in the north direction and a velocity component of 15 km/h in the east direction. What is the speed of the blimp? What is the direction of motion of the blimp?

97. Suppose that the position vector of a particle is given by the following function of time:

$$\mathbf{r} = (6.0 + 2.0t^2)\mathbf{i} + (3.0 - 2.0t + 3.0t^2)\mathbf{j}$$

where distance is measured in meters and time in seconds.

  (a) What is the instantaneous velocity vector at $t = 2.0$ s? What is the magnitude of this vector?

  (b) What is the instantaneous acceleration vector? What are the magnitude and direction of this vector?

98. An archer shoots an arrow over level ground. The arrow leaves the bow at a height of 1.5 m with an initial velocity of 60 m/s in a horizontal direction.

  (a) How long does this arrow take to fall to the ground?

  (b) At what horizontal distance does this arrow strike the ground?

99. Volcanos on the Earth eject rocks at speeds of up to 700 m/s. Assume that the rocks are ejected in all directions; ignore the height of the volcano and ignore air friction.

  (a) What is the maximum height reached by the rocks?

  (b) What is the maximum horizontal distance reached by the rocks?

  (c) Is it reasonable to ignore air friction in these calculations?

100. A large stone-throwing engine designed by Archimedes could throw a 77-kg stone over a range of 180 m. What must have been the initial speed of the stone if thrown at an initial angle of 45° with the horizontal?

101. The world record for the discus throw set by M. Wilkins in 1976 was 70.87 m. What is the initial speed of the discus required to achieve this range? Assume that the discus was launched with an elevation angle of 45°, and that the height of the hand over the ground was 2.0 m at the instant of launch.

*102. When you hold the nozzle of a water hose horizontally, at a height of 1.0 m above the ground, the stream of water lands 4.0 m from you. If you now aim the nozzle straight up, how high (above the nozzle) will the stream of water rise?

*103. An automobile travels at a steady 90 km/h along a road leading over a small hill. The top of the hill is rounded so, in the vertical plane, the road approximately follows an arc of a circle of radius 70 m. What is the centripetal acceleration of the automobile at the top of the hill?

*104. A lump of concrete falls off a crumbling overpass and strikes an automobile traveling on a highway below. The lump of concrete falls 5.0 m before impact, and the automobile has a speed of 90 km/h.

  (a) What is the speed of impact of the lump in the reference frame of the automobile?

  (b) What is the angle of impact?

*105. You are driving an automobile at a steady speed of 90 km/h along a straight highway. Ahead of you is a 10-m-long truck traveling at a steady speed of 60 km/h. You decide to pass this truck, and you switch into the passing lane when 40 m behind the truck.

(a) What is your speed in the reference frame of the truck?

(b) How long do you take to pass the truck, starting 40 m behind the truck and ending 40 m ahead of the truck? (Hint: Calculate this time in the reference frame of the truck.)

Answers to Checkups

Checkup 4.1

1. With the x axis eastward and the y axis northward, we must have: (a) for motion northwest, the x component is $-$ and the y component is $+$; (b) for southward motion, the x component is 0 and the y component is $-$; and (c) for motion southeast, the x component is $+$ and the y component is $-$.

2. Yes—if the direction of the velocity is changing, then the motion is accelerated. When traveling at constant speed, you experience acceleration when the path is curved.

3. Since the initial and final positions are the same ($\Delta x = 0$ and $\Delta y = 0$), the average velocity is zero. Since the initial and final velocities are the same ($\Delta v_x = 0$ and $\Delta v_y = 0$), the average acceleration is zero.

4. (C) $v_x = 0$; $v_y = 50$ km/h. We assume the motorcycle travels counterclockwise (looking from above) around the traffic circle. In that case, at the eastern point, the motorcycle is traveling northward, and so has $v_x = 0$ and $v_y = 50$ km/h.

Checkup 4.2

1. In general, the instantaneous velocity vector will be tangent to the circle wherever the automobile is heading on the circle, and so will be in the north, west, south, and east directions, respectively.

2. Since the x and y components of the velocity are equal, the direction of the aircraft is halfway between the x and y axes, that is, at an angle of 45° with respect to either axis. We can also see that $\tan \theta = v_y/v_x = 1$ implies $\theta = 45°$.

3. Since the change in velocity ($v_2 - v_1$) is opposite to the original motion, the acceleration vector during the collision is also opposite to the initial motion.

4. (D) 60°. The magnitude of the acceleration is $a = 6.0$ m/s²; the x component is $a_x = 3.0$ m/s². But $a_x = a \cos \theta$, where θ is the angle with respect to the horizontal, so $\cos \theta = a_x/a = 3/6 = 1/2$. The inverse cosine of 1/2 is 60°.

Checkup 4.3

1. It is possible to have accelerated motion in the x direction only; for example, if the boat moves in the x direction while speeding up. It is possible to have accelerated motion in both the x and y directions; for example, if the boat moves in a direction between the x and y directions while speeding up.

2. If the y motion were unaccelerated, the particle would move equal distance in equal times in the y direction, while moving greater and greater distances in equal times in the (accelerated) x direction. Its path would thus not be a straight line. Since the particle moves along a straight line, and we know the x motion is accelerated, then the y motion must be accelerated also.

3. (E) $a_x = -6.9$ m/s²; $a_y = -6.9$ m/s². Since the coordinate axes are tilted by 45° with respect to vertical, both the x and y directions will have equal accelerations; these will both be equal to the component of the acceleration due to gravity along the axes. For the x direction, we will have $a_x = -g \times \sin 45° = -9.8$ m/s² $\times \sin 45° = -6.9$ m/s²; for the y direction, we will have $a_y = -g \times \cos 45° = -9.8$ m/s² $\times \cos 45° = -6.9$ m/s².

Checkup 4.4

1. At all points of its trajectory, the acceleration of a projectile is the same; it is always $a_y = -g = -9.81$ m/s².

2. As we saw graphically and by maximizing the range function, the maximum range is achieved by launching at an angle of 45° (for a given fixed launch speed). From Eqs. (4.41) and (4.42), we see that both the maximum time of flight and the maximum height are achieved by launching the projectile vertically ($\sin \theta = 1$ when $\theta = 90°$).

3. The acceleration is constant and always downward ($a_y = -g$), but the velocity is never downward (since we are told that the projectile was launched at an angle *smaller* than 90°), and so is never parallel to the acceleration. At the top of its trajectory (y_{max}), the velocity is horizontal, and so is perpendicular to the acceleration there.

4. Since the range varies as the square of the launch speed [Eq. (4.43)], the projectile launched at twice the launch speed will go four times as far (for the same elevation angle).

5. Since the time of flight varies proportionally to $\sin \theta$, the sine of the launch angle, the projectile with the smallest launch angle will return to the ground in the shortest time.

6. (E) 80°. The range is largest at 45°, and drops off symmetrically above and below 45°. Thus, compared with 20° (which

differs from 45° by 25°), elevation angles of 30°, 40°, 50°, and 60° will result in larger ranges (since they differ from 45° by smaller angles); the projectile launched at 70° will have the same range as the 20° one. However, the projectile launched at 80° will have a shorter range (since that elevation angle differs from 45° by the larger angle of 35°).

Checkup 4.5

1. By an "S-shaped curve" we usually mean one that turns first one way and then the other way before returning parallel to the original direction. For example, if one enters the S from below, the acceleration is first toward the left and then briefly zero as the curvature changes to the opposite direction, and then the acceleration is toward the right at the top of the S.

2. Since the velocity traces out all directions uniformly as the particle travels once around the circle, the average velocity is zero. Similarly, the acceleration points toward the center of the circle, and rotates around including all directions equally as the particle travels once around the circle, so the average acceleration is also zero.

3. At the lowest point, the instantaneous velocity is in the direction the girl is moving: straight ahead (or straight back) horizontally. For the acceleration, we note that although the motion is not uniform circular motion, at the bottom the speed goes through a maximum: she switches from speeding up on the way down to slowing down on the way back up. Since there is no change in speed at the bottom point, the acceleration is purely centripetal there, and thus is toward the center of the circle, or vertically up.

4. (B) **v** forward, **a** downward. At the moment the automobile crests the hill, its velocity is horizontal, straight ahead. At constant speed along the arc of a circle, the acceleration is purely centripetal, and so is toward the center of the circle, or vertical and downward.

Checkup 4.6

1. Yes, a particle at rest in one reference frame is in motion with uniform velocity in any reference frame that moves with uniform velocity with respect to the reference frame where the particle was at rest. Yes, a particle at rest in one reference frame is in accelerated motion in any reference frame that is in accelerated motion relative to the reference frame where the particle was at rest.

2. The black automobile is moving forward at 10 km/h relative to the red [see Eq. (4.53)]. The red automobile is moving backward relative to the black automobile, and so the red is moving at −10 km/h relative to the black.

3. Since the wind opposes the motion, the speed relative to the ground is $v = v' - V_O = 200$ km/h − 30 km/h = 170 km/h.

4. To swim across in the minimum time, you would swim perpendicular to the river, so that all your swimming velocity was directed at crossing the river; however, you would wind up somewhat downstream. To reach a point on the shore directly opposite your starting point, you would have to swim somewhat upstream, as with the motorboat in Example 10.

5. For reference frames in uniform motion with respect to one another, acceleration is an absolute quantity; thus, relative to particle A, the motion of particle C is accelerated, and it is exactly the same acceleration as the motion of particle C relative to particle B.

6. (B) West of north. The westward motion of the motorboat makes the wind seem to have a component coming from the west (looking forward, you feel a breeze on your face); added to the actual wind from the north, the wind seems to come from west of north.

Newton's Laws of Motion

5.1   Newton's First Law

5.2   Newton's Second Law

5.3   The Combination of Forces

5.4   Weight; Contact Force and Normal Force

5.5   Newton's Third law

5.6   Motion with a Constant Force

CONCEPTS IN CONTEXT

Concepts *in* Context

Elevators remind us of familiar sensations triggered by vertical accelerated motion: When accelerating upward we feel heavy; the floor seems to push harder on our feet. As the upward motion slows toward a stop, we may feel the "butterflies in the stomach" sensation associated with free fall.

With the concepts of this chapter we will describe forces and predict accelerations, and we can ask:

❓ What force must the floor of an elevator apply to your feet to accelerate you upward? (Checkup 5.4, question 6, page 143; and Example 6, page 147)

❓ How do an elevator and its counterweight accelerate if the cable connecting them is permitted to run freely? (Example 10, page 154)

❓ What devices are installed on elevators to ensure their safety? (Physics in Practice: Elevators, page 157)

So far we have dealt only with the mathematical description of motion—the definitions of position, velocity, and acceleration and the relationships between these quantities. We did not inquire what causes a body to accelerate. In this chapter we will see that *the cause of acceleration is a force exerted on the body by some external agent.* The fundamental properties of force and the relationship between force and acceleration are contained in Newton's three laws of motion. The first of these laws describes the natural state of motion of a free body on which no net external force is acting, whereas the other two laws deal with the behavior of bodies under the influence of external forces.

SIR ISAAC NEWTON (1642–1727)
English mathematician and physicist, widely regarded as the greatest scientist of all time. His brilliant discoveries in mechanics were published in 1687 in his book Principia Mathematica, *one of the glories of the Age of Reason. In this book, Newton laid down the laws of motion and the Law of Universal Gravitation, and he demonstrated that planets in the sky as well as bodies on the Earth obey the same mathematical equations. For over 200 years, Newton's laws stood as the unchallenged basis of all our attempts at a scientific explanation of the physical world.*

The first law was actually discovered by Galileo Galilei early in the seventeenth century, but it remained for Isaac Newton, in the second half of the seventeenth century, to formulate a coherent theory of forces and to lay down a complete set of equations from which the motion of bodies under the influence of arbitrary forces can be calculated. The study of forces and their effects on the motion of bodies is called **dynamics**, and Newton's laws of motion are sometimes called the laws of dynamics.

5.1 NEWTON'S FIRST LAW

Everyday experience seems to suggest that a force—a push or a pull—is needed to keep a body moving at constant velocity. For example, if the wind pushing a sailboat suddenly ceases, the boat will coast along for some distance, but it will gradually slow down, stop, and remain stopped until a new gust of wind comes along. However, everyday experience misleads us: what actually slows down the sailboat is not the *absence* of a propulsive force but, rather, the *presence* of friction forces exerted by the water and the air, which oppose the motion. *Under ideal frictionless conditions, a body in motion would continue to move forever.* Experiments with pucks or gliders riding on a cushion of air on a low-friction air table or air track give a clear indication of the persistence of motion (see Fig. 5.1); but in order to eliminate friction entirely, it is best to use bodies moving in a vacuum, without even air against which to rub. Experiments with particles moving in vacuum show that a body left to itself, on which no net external force is acting, persists indefinitely in its state of uniform motion.

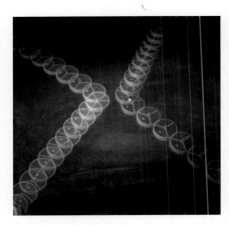

FIGURE 5.1 Multiple-exposure photograph of two pucks moving on an air table. The pucks move along straight lines with uniform velocity, except when they collide.

In this context, an external force is any force exerted on the body by some *other* body. By contrast, internal forces are those exerted by some part of a body on another part of the *same* body. For instance, the forces that the screws or bolts in the sailboat exert on its planks are internal forces; such internal forces do not affect the motion of the boat.

Newton's First Law summarizes experiments and observations on the motion of bodies on which no net external force is acting:

> *In the absence of a net external force, a body at rest remains at rest, and a body in motion continues to move at constant velocity.*

Newton's First Law

(a)

(b)

Ball accelerates
toward rear of truck.

accelerating truck

FIGURE 5.2 (a) In the absence of a net external force, a ball at rest on a level street remains at rest. (b) But a ball at rest on the platform of an accelerating truck acquires a "spontaneous" acceleration toward the rear of the truck (in the reference frame of the truck).

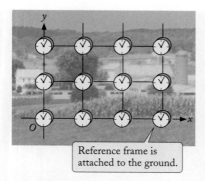

FIGURE 5.3 A reference frame with origin fixed at a point on the surface of the Earth.

The tendency of a body to continue in its initial state of motion (a state of rest or a state of uniform velocity) is called its **inertia**. Accordingly, the First Law is often called the Law of Inertia.

A crucial restriction on Newton's First Law concerns the choice of reference frame: the law is not valid in all reference frames but only in certain special frames. If this law is valid in one given reference frame, then it cannot be valid in a second reference frame that has an accelerated motion relative to the first. For example, in the reference frame of the ground, a bowling ball at rest on a level street remains at rest, but in the reference frame of an accelerating truck, a bowling ball initially at rest on the platform of the truck acquires a "spontaneous" acceleration toward the rear of the truck, in contradiction to Newton's First Law (see Fig. 5.2). Those special reference frames in which the law is valid are called **inertial reference frames**. Thus, the reference frame of the ground is an inertial reference frame, but that of the accelerating truck is not.

Note that *if some first reference frame is inertial, any other reference frame in uniform translational motion relative to the first will also be inertial, and any other reference frame in accelerated motion relative to the first will not be inertial.* Thus, any two inertial reference frames can differ only by some constant relative velocity; they cannot differ by an acceleration. This implies that, as measured with respect to inertial reference frames, *acceleration is absolute:* when a particle has some acceleration in one inertial reference frame, then the particle will have exactly the same acceleration in any other [see Eq. (4.54)]. By contrast, the velocity of the particle is relative; the velocities are related by the addition rule for velocities [see Eq. (4.53)].

Finally we must address an important question: Which of the reference frames in practical use for everyday measurements are inertial? For the description of everyday phenomena, the most commonly used reference frame is one attached to the ground, with the origin of coordinates fixed at some point on the surface of the Earth (see Fig. 5.3). Although crude experiments indicate that this reference frame is inertial (for example, a ball placed on a level street remains at rest), more precise experiments show that this reference frame is not inertial. The Earth rotates about its axis, and this rotational motion gives points on the ground a centripetal acceleration; thus, a reference frame attached to the ground is an accelerated, noninertial reference frame. However, the numerical value of the centripetal acceleration of points on the surface of the Earth is fairly small—about 0.034 m/s^2 at the equator—and it can be neglected for most purposes. Our additional centripetal acceleration due to the motion of the Earth around the Sun is even smaller, about 0.002 m/s^2. Hereafter, unless otherwise stated, we will take it for granted that the reference frames in which we express the laws of physics are inertial reference frames, either exactly inertial or at least so nearly inertial that no appreciable deviation from Newton's First Law occurs within the region of space and time in which we are interested.

✔ Checkup 5.1

QUESTION 1: To keep a stalled car moving steadily along a level street you have to keep pushing it. Does this contradict Newton's First Law?

QUESTION 2: When you roll a bowling ball on a level surface, you find it gradually slows down. Does this contradict Newton's First Law?

QUESTION 3: A car is traveling at constant speed along a straight, level road. Is the reference frame of this car an inertial reference frame? What if the car rounds a curve at constant speed? What if the car brakes?

QUESTION 4: An elevator is descending at constant speed. Is the reference frame of this elevator an inertial reference frame?

QUESTION 5: A diver is in free fall after jumping off a diving board. Is the reference frame of this diver an inertial reference frame?

QUESTION 6: Which of the following represents an inertial reference frame?

(A) The reference frame of an elevator in free fall (constant acceleration).
(B) The reference frame of a bird descending at constant velocity.
(C) The reference frame of a particle in uniform circular motion.
(D) The reference frame of a car slowing down while coasting uphill.

5.2 NEWTON'S SECOND LAW

Online
Concept
Tutorial

6

Newton's Second Law of motion establishes the relationship between the force acting on a body and the acceleration caused by this force. This law summarizes experiments and observations on bodies moving under the action of external forces. Qualitatively, a **force** is any push or pull exerted on a body, such as the push of the wind on a sailboat, or the pull of your hand on a doorknob. It is intuitively obvious that such a push or pull has a direction as well as a magnitude—in fact, force is a vector quantity, and it can be represented graphically by an arrow (see Fig. 5.4). For the sake of simplicity, we assume for now that only one force is acting on the body, but we will eliminate this assumption in the next section.

Newton's Second Law states:

An external force acting on a body gives it an acceleration that is in the direction of the force and has a magnitude directly proportional to the magnitude of the force and inversely proportional to the mass of the body:

$$\mathbf{a} = \frac{\mathbf{F}}{m} \tag{5.1}$$

or

$$m\mathbf{a} = \mathbf{F} \tag{5.2}$$

Magnitude and direction of force are represented by the arrow.

FIGURE 5.4 Man pushing an automobile. The force has a magnitude and a direction.

Newton's Second Law, for single force

According to Eq. (5.1) or (5.2), the acceleration vector is equal to the force vector divided by the mass; thus, this equation specifies both the magnitude and the direction of the acceleration, as asserted by the verbal statement of the law.

The Second Law is subject to the same restrictions as the First Law: it is valid only in inertial reference frames.[1]

---

[1] The validity of the Second Law requires that the clocks of the inertial reference frame be correctly synchronized. Such a correct synchronization can be achieved by slowly transporting a calibrating clock from place to place in the reference frame or by using light signals and making an allowance for light travel time (this will be discussed in Chapter 36).

Before we deal with applications of the Second Law, we must give the precise definitions of mass and of force. These definitions are contained in the Second Law itself; that is, the Second Law plays a dual role as a law of physics and as a definition of mass and force.

The definition of mass hinges on comparing the unknown mass with a standard mass, which is assumed known. To compare the two masses, we exert forces of identical magnitudes on each, and we measure the accelerations that these forces produce on each. For instance, we might attach identical rubber bands or springs to the unknown mass and to the standard mass, and stretch these rubber bands or springs by identical amounts, thereby producing forces of identical magnitudes. According to the Second Law, *if two bodies of different masses are subjected to forces of identical magnitudes, the accelerations will be in the inverse ratio of the masses*. If we designate the acceleration of the unknown mass by a and that of the standard mass by a_s and their masses by m and m_s, we can express this inverse ratio of masses and accelerations as

definition of mass

$$\frac{m}{m_s} = \frac{a_s}{a} \tag{5.3}$$

This relation serves to define the unknown mass m in terms of the standard mass m_s. The relation says that the unknown mass is large if its acceleration is small. This is of course quite reasonable. A large mass is hard to accelerate—it has a large inertia. If we pull a baseball bat with our rubber band, it will accelerate readily; but if we pull a supertanker, it will hardly accelerate at all. The precise definition given by Eq. (5.3) expresses the intuitive notion that mass is a measure of the resistance that the body offers to changes in its velocity.

As already mentioned in Section 1.4, the unit of mass in the SI system of units is the **kilogram** (1 kg), and the standard of mass is the standard kilogram, a cylinder of platinum–iridium alloy kept at the International Bureau of Weights and Measures. Table 1.7 gives some examples of masses expressed in kilograms, and Table 1.8 lists multiples and submultiples of the kilogram. Among these submultiples is the **pound** (1 lb = 0.4536 kg), which is a unit of mass in the British system of units.

kilogram (kg)

pound (lb)

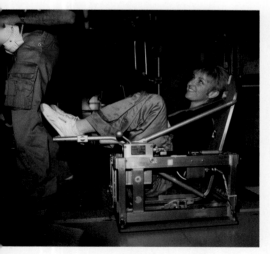

FIGURE 5.5 An astronaut trains on the body-mass measurement device (BMMD).

EXAMPLE 1   During the Skylab mission, three astronauts were kept for about 2 months in weightless conditions. Since an ordinary balance will not work under these conditions, scientists who wanted a daily record of the astronauts' masses had to invent a new mass measurement device. Figure 5.5 shows the device used aboard Skylab. It consisted of a small chair that could be accelerated back and forth by a spring attached to it. Instruments connected to the chair measured the acceleration. With a known standard mass of 66.9 kg placed in the chair, the bent spring produced an acceleration a_s. With the standard mass removed and with astronaut J. R. Lousma sitting in the chair, the bent spring (with the same amount of bending) produced an acceleration a, which was smaller by a factor of 0.779. Deduce the mass of Lousma. Ignore the mass of the chair.

SOLUTION: The bent spring provides the same force F when the standard mass is placed in the chair and when the astronaut is placed in the chair. Consequently, the accelerations must be in the inverse ratio of the masses, as in Eq. (5.3):

$$\frac{m}{m_s} = \frac{a_s}{a}$$

from which

$$m = \frac{a_s}{a} m_s$$

Since the ratio of the measured acceleration is $a_s/a = 1/0.779$, we then find

$$m = \frac{a_s}{a} m_s = \frac{1}{0.779} \times 66.9 \text{ kg} = 85.9 \text{ kg} \qquad (5.4)$$

for the mass of Lousma.

The quantitative definition of force also relies on the Second Law. To measure a given force—say, the force generated by a spring that has been stretched a certain amount—we apply this force to the standard kilogram. If the resulting acceleration of the standard kilogram is a_s, then the force has a magnitude

$$F = m_s a_s = 1 \text{ kg} \times a_s \qquad (5.5)$$

After the standard mass has been used to measure the force, any other masses to which this same force is applied will be found to obey the Second Law. In regard to these other masses, the Second Law is an assertion about the physical world that can be verified by experiments—it is a law of physics.

In the SI system of units, the unit of force is the **newton** (N); this is the force that will give a mass of 1 kg an acceleration of 1 m/s^2:

$$\boxed{1 \text{ newton} = 1 \text{ N} = 1 \text{ kg·m/s}^2} \qquad (5.6)$$

Table 5.1 lists the magnitudes of some typical forces.

TABLE 5.1	SOME FORCES	
Gravitational pull of Sun on Earth	3.5×10^{22} N	
Thrust of Saturn V rocket engines (a)	3.3×10^{7} N	
Pull of large tugboat	1×10^{6} N	
Thrust of jet engines (Boeing 747)	7.7×10^{5} N	
Pull of large locomotive	5×10^{5} N	
Decelerating force on automobile during braking	1×10^{4} N	
Force between two protons in a nucleus	$\approx 10^{4}$ N	(a) 
Accelerating force on automobile	7×10^{3} N	
Gravitational pull of Earth on man	7.3×10^{2} N	
Maximum upward force exerted by forearm (isometric)	2.7×10^{2} N	
Gravitational pull of Earth on apple (b)	2 N	
Gravitational pull of Earth on 5¢ coin	5.1×10^{-2} N	
Force between electron and nucleus of atom (hydrogen)	8×10^{-8} N	
Force on atomic-force microscope tip	10^{-12} N	
Smallest force detected (mechanical oscillator)	10^{-19} N	(b) 

FIGURE 5.6 *Spirit of America* on the Salt Flats of Utah.

EXAMPLE 2 The racing car *Spirit of America* (see Fig. 5.6), which set a world record for speed on the Salt Flats of Utah, had a mass of 4100 kg, and its jet engine could develop up to 68000 N of thrust. What acceleration could this car achieve?

SOLUTION: According to Newton's Second Law, a horizontal force of magnitude 68000 N produces an acceleration

$$a = \frac{F}{m} = \frac{68\,000\ \text{N}}{4100\ \text{kg}} = 17\ \text{m/s}^2 \tag{5.7}$$

EXAMPLE 3 Some small animals—locusts, beetles, and fleas—attain very large accelerations while starting a jump. The rat flea attains an acceleration of about 2.0×10^3 m/s². Calculate what force the hind legs of the flea must exert on the body while pushing it off with this acceleration. The mass of the flea is about 6.0×10^{-11} kg; neglect the mass of the legs.

SOLUTION: According to Newton's Second Law, the magnitude of the force is

$$F = ma = 6.0 \times 10^{-11}\ \text{kg} \times 2.0 \times 10^3\ \text{m/s}^2 = 1.2 \times 10^{-7}\ \text{N}$$

In the British system of units, the unit of force is the **pound-force** (lbf), which equals 4.4482 N. In everyday usage, the pound-force is often simply called pound, but we must be careful not to confuse the pound-force (a unit of force) with the pound (a unit of mass). The widespread confusion between the two kinds of pounds stems from their close relationship—the pound-force is the weight of a pound-mass.[2] Confusion is displayed on labels on grocery packages, which typically state "weight 1 lb" when they should state "mass 1 lb." Also, labels on tire gauges state "pressure lb/in²" when they should state "pressure lbf/in²."

We now turn to the question of the practical measurement of force and mass. Measurements of force can be conveniently performed with a spring balance (see Fig. 5.7), by matching the unknown force with a known force supplied by a stretched, calibrated spring. Alternatively, measurements of force can be performed by comparing the unknown force with a known weight. Weight is the downward pull that the gravity of the Earth exerts on a body. The weight of a body is proportional to its mass, and standard sets of weights are usually constructed by taking multiples and submultiples of the standard of mass (Fig. 5.8).

(a)

(b)

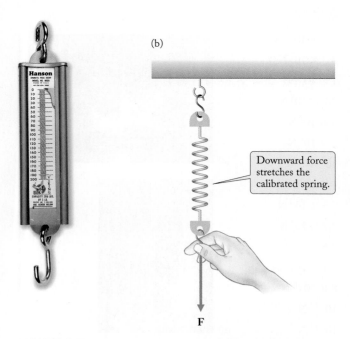

Downward force stretches the calibrated spring.

F

FIGURE 5.7 (a) Spring balance. (b) A spring balance, used to measure an unknown force **F**.

[2] Note that 1 lbf is *not* equal to 1 lb $\times$ 1 ft/s²; instead, 1 lbf = 1 lb $\times g$ = 1 lb $\times$ 32.2 ft/s². If you use British units in Newton's Second Law, you must take into account this extra conversion between lb·ft/s² and lbf. An alternative British unit of mass is the **slug**; 1 slug = (1 lbf)/(1 ft/s²) = 32.2 lb. With this unit of mass, Newton's Second Law automatically delivers the correct units, without any need for extra conversions. But the slug is hardly ever used by practicing engineers.

FIGURE 5.8
A set of standard weights.

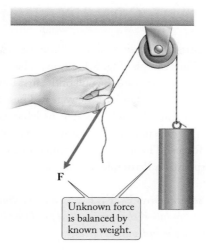

Unknown force is balanced by known weight.

FIGURE 5.9 The unknown force **F** pulling on one end of the string is measured by balancing it with a known weight acting on the other end.

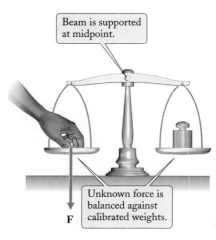

Beam is supported at midpoint.

Unknown force is balanced against calibrated weights.

FIGURE 5.10 In a beam balance, the unknown force **F** pushing down on one pan is measured by balancing it with known weights placed in the other pan.

We will discuss the meaning of "weight" in more detail in Section 5.4. A simple comparison of an unknown force and a known weight can be performed by letting the force act on one end of a string while the weight acts on the other end (see Fig. 5.9). A more precise comparison can be achieved with a beam balance (see Fig. 5.10), by letting the force push down on one of the balance pans while a known weight is placed in the other balance pan.

Measurements of mass are commonly carried out with beam balances that compare the weights of the masses. Since the weight of a body is proportional to its mass, measurements of mass via weight give results consistent with those obtained by the primary procedure based on Eq. (5.3).

The masses of electrons and protons and the masses of ions (atoms with missing electrons or added electrons) are too small to be measured by their weight. Instead, they are measured with a procedure based on Eq. (5.2), by applying a known force to the particle, measuring the resulting acceleration, and then calculating the mass. Table 5.2 lists the masses of the electron, the proton, and the neutron.

 Checkup 5.2

QUESTION 1: Two cardboard boxes rest on a smooth, frictionless table. How can you determine which box has more mass without lifting them off the table?

QUESTION 2: To get your stalled automobile moving, you can either push against the rear end, or pull on the front end. What is the direction of the force in each case?

QUESTION 3: Tired of waiting for the wind, a sailor decides to stand up in his sailboat and push on the mast. Will this push accelerate the sailboat?

QUESTION 4: A bobsled slides on flat ice, without friction. A man pushing the empty bobsled as hard as possible gives it an acceleration of 4 m/s². What will be the acceleration of the sled if two men push on it equally? What will be the acceleration of a loaded bobsled, of twice the mass of the empty sled, if one man pushes? If two men push? Choose among the following respective quantities in units of m/s²:

(A) 8, 1, 2   (B) 8, 1, 4   (C) 8, 2, 4   (D) 16, 1, 2   (E) 16, 1, 4

TABLE 5.2	
THE MASSES OF ELECTRONS, PROTONS, AND NEUTRONS	
PARTICLE	**MASS**
Electron	9.11×10^{-31} kg
Proton	1.673×10^{-27} kg
Neutron	1.675×10^{-27} kg

5.3 THE COMBINATION OF FORCES

More often than not, a body will be subjected to the simultaneous action of several forces. For example, Fig. 5.11 shows a barge under tow by two tugboats. The forces acting on the barge are the pull of the first towrope, the pull of the second towrope, and the frictional resistance of the water.[3] These forces are indicated by the arrows in Fig. 5.11. Newton's Second Law tells us what each of these forces would do if acting by itself. The question now is, How can we calculate the simultaneous effect of two or more forces? The answer is supplied by an addition principle for forces, called the **Superposition Principle** for forces:

If several forces $\mathbf{F}_1, \mathbf{F}_2, \mathbf{F}_3, \ldots$ act simultaneously on a body, then the acceleration they produce is the same as that produced by a single force $\mathbf{F}_{net}$ given by the vector sum of the individual forces,

Superposition Principle

$$\mathbf{F}_{net} = \mathbf{F}_1 + \mathbf{F}_2 + \mathbf{F}_3 + \cdots \tag{5.8}$$

The single force $\mathbf{F}_{net}$ that has the same effect as the combination of the individual forces is called the **net force**, or the **resultant force**. The net force then determines the acceleration, and Newton's Second Law takes the form

Newton's Second Law, for net force

$$m\mathbf{a} = \mathbf{F}_{net} \tag{5.9}$$

We must emphasize that this Superposition Principle is a law of physics, which has the same status as Newton's laws. Crude tests of this principle can be performed in laboratory experiments by pulling on a body with known forces in known directions. But the most precise empirical test of this principle emerges from the study of planetary motion; there it is found that the net force on a planet is indeed the vector sum of all the gravitational pulls exerted by the Sun and by the other planets.

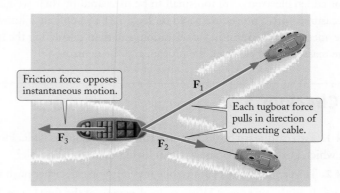

Friction force opposes instantaneous motion.

Each tugboat force pulls in direction of connecting cable.

$\mathbf{F}_1$

$\mathbf{F}_2$

$\mathbf{F}_3$

FIGURE 5.11 A barge under tow by two tugboats. $\mathbf{F}_1$ and $\mathbf{F}_2$ are the forces exerted by the tugboats; $\mathbf{F}_3$ is the frictional resistance of the water.

---

[3] These are the horizontal forces. There are also vertical forces: the downward pull of gravity (the weight) and the upward pressure of the water (the buoyancy). The vertical forces can be ignored, since they cancel each other, and do not contribute to the net force.

EXAMPLE 4   Suppose that the two towropes in Fig. 5.11 pull with horizontal forces of 2.5×10^5 N and 1.0×10^5 N, respectively, and that these forces make angles of 30° and 15° with the long axis of the barge (see Fig. 5.12). Suppose that the friction force is zero. What are the magnitude and direction of the net horizontal force the towropes exert on the barge?

SOLUTION: The net force is the vector sum

$$\mathbf{F}_{net} = \mathbf{F}_1 + \mathbf{F}_2 \tag{5.10}$$

where $\mathbf{F}_1$ is the force of the first towrope and $\mathbf{F}_2$ that of the second. The net force is shown in Fig. 5.12a. With the x and y axes arranged as in Fig. 5.12a, the forces can be resolved into x and y components. The x component of the net force is the sum of the x components of the individual forces (see Fig. 5.12b),

$$
\begin{aligned}
F_{net,x} &= F_{1,x} + F_{2,x} \\
&= 2.5 \times 10^5 \, \text{N} \times \cos 30° + 1.0 \times 10^5 \, \text{N} \times \cos 15° \\
&= 2.5 \times 10^5 \, \text{N} \times 0.866 + 1.0 \times 10^5 \, \text{N} \times 0.966 \\
&= 3.1 \times 10^5 \, \text{N}
\end{aligned}
\tag{5.11}
$$

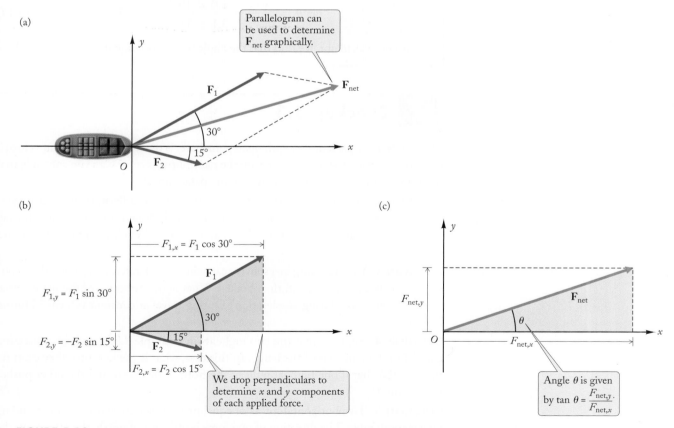

FIGURE 5.12 (a) One tugboat pulls with a force $\mathbf{F}_1$, and the other pulls with a force $\mathbf{F}_2$. The magnitudes of these forces are $F_1 = 2.5 \times 10^5$ N and $F_2 = 1.0 \times 10^5$ N, respectively. The net force $\mathbf{F}_{net}$ is the vector sum of the two forces $\mathbf{F}_1$ and $\mathbf{F}_2$. (b) The forces $\mathbf{F}_1$ and $\mathbf{F}_2$ and their x and y components. (c) The net force $\mathbf{F}_{net}$ and its x and y components.

Likewise, the y component of the net force is the sum of the y components of the individual forces,

$$
\begin{aligned}
F_{\text{net},y} &= F_{1,y} + F_{2,y} \\
&= 2.5 \times 10^5 \, \text{N} \times \sin 30° - 1.0 \times 10^5 \, \text{N} \times \sin 15° \\
&= 2.5 \times 10^5 \, \text{N} \times 0.500 - 1.0 \times 10^5 \, \text{N} \times 0.259 \\
&= 1.0 \times 10^5 \, \text{N}
\end{aligned}
\tag{5.12}
$$

The y components of the individual forces are of opposite sign because one tugboat pulls the barge to the left (up in Fig. 5.12) and the other to the right (down in Fig. 5.12).

The components $F_{\text{net},x}$ and $F_{\text{net},y}$ uniquely specify the net force, and we could end our calculation of the net force with these components. However, the problem asks for the magnitude and the direction of the net force, and we therefore have to take our calculation a step further. According to Eq. (3.15), the magnitude of the net force is the square root of the sum of squares of the components:

$$
\begin{aligned}
F_{\text{net}} &= \sqrt{(F_{\text{net},x})^2 + (F_{\text{net},y})^2} \\
&= \sqrt{(3.1 \times 10^5 \, \text{N})^2 + (1.0 \times 10^5 \, \text{N})^2} = 3.3 \times 10^5 \, \text{N}
\end{aligned}
\tag{5.13}
$$

The direction of the net force makes an angle θ with the x axis (see Fig. 5.12c). According to Eq. (3.16), this angle is given by

$$
\tan \theta = \frac{F_{\text{net},y}}{F_{\text{net},x}} = \frac{1.0 \times 10^5 \, \text{N}}{3.1 \times 10^5 \, \text{N}} = 0.32
\tag{5.14}
$$

With our calculator, we find that the angle with this tangent is 18°.

 Checkup 5.3

QUESTION 1: A parachutist, with open parachute, is descending at uniform velocity. Can you conclude that the net force on the parachutist is zero? Can you conclude that there are no forces whatsoever acting on the parachutist?

QUESTION 2: An elevator is initially at rest at the ground floor. It then accelerates briefly, and then continues to ascend at constant speed. What is the direction of the net force on the elevator when at rest? When accelerating? When ascending at constant speed?

QUESTION 3: You are riding in a subway car, which accelerates, then proceeds at constant velocity for a while, and then brakes. What is the direction of the net external force on your body during acceleration? During travel at constant velocity? During braking?

QUESTION 4: Suppose that the two tugboats in Example 4 both pull in a direction parallel to the long axis of the barge. In this case, what is the net force they exert on the barge? What if one tugboat pulls at the front of the barge and the other pushes from behind?

QUESTION 5: Two horizontal forces of equal magnitudes are acting on a box sliding on a smooth table. The direction of one force is 30° west of north, the other is in the west direction. What is the direction of the acceleration of the box?

(A) 15° north of west   (B) 30° north of west   (C) Directly northwest
(D) 30° west of north   (E) 15° west of north

5.4 WEIGHT; CONTACT FORCE AND NORMAL FORCE

Online
Concept
Tutorial

The gravity of the Earth is the most familiar of all forces. When you hold a body, say, an apple, in your hands, you can feel the downward pull of gravity on the apple; and if you release the apple, you can see it accelerating under the influence of this pull. In the terminology of physics, the pull of gravity on a body is called the **weight** of the body. Thus, weight is a force; it is a vector quantity—it has a direction (downward) as well as a magnitude. The unit of weight is the unit of force, that is, the newton (N).

The magnitude of the weight force is directly proportional to the mass of the body. To understand this, consider a body of mass m in free fall near the surface of the Earth, say, an apple you have released from your hand (see Fig. 5.13). The body has a downward acceleration g. Since we attribute this acceleration to the weight force, Newton's Second Law tells us that the magnitude of the weight force acting on the body must be

$$F = ma = mg \qquad (5.15)$$

We will denote the weight by the vector symbol **w**. According to Eq. (5.15), the magnitude of the weight is

$$w = mg \qquad (5.16)$$

If the body is not in free fall but is held in a stationary position by some support, then the weight is of course still the same as that given by Eq. (5.16); however, the support balances the downward weight force and prevents it from producing a downward motion.

weight

| EXAMPLE 5 | What is the weight of a 54-kg woman? Assume that $g = 9.81 \text{ m/s}^2$. |

SOLUTION: By Eq. (5.16), the magnitude of the weight is

$$w = mg = 54 \text{ kg} \times 9.81 \text{ m/s}^2 = 530 \text{ N}$$

and its direction is downward.

COMMENT: Since the value of g depends on location, the weight of a body also depends on its location. For example, if the 54-kg woman travels from London ($g = 9.81 \text{ m/s}^2$) to Hong Kong ($g = 9.79 \text{ m/s}^2$), her weight will decrease from 530 N to 529 N, a difference of 1 N. And if this woman were to travel to the Moon ($g = 1.62 \text{ m/s}^2$), her weight would decrease to 87 N!

The preceding example illustrates an essential distinction between mass and weight. *Mass is an intrinsic property of a body,* measuring the inertial resistance with which the body opposes changes in its motion. The definition of mass is formulated in such a way that a given body has the same mass regardless of its location in the universe. *Weight is an extrinsic property of a body,* measuring the pull of gravity on the body. It depends on the (gravitational) environment in which the body is located, and it therefore depends on location.

A body deep in intergalactic space, far from the gravitational pull of any star or planet, will experience hardly any gravitational pull—the weight of the body will be nearly zero; that is, the body will be weightless. Although such a condition of true

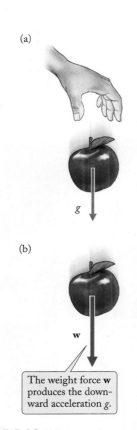

(a)

(b)

The weight force **w** produces the downward acceleration g.

FIGURE 5.13 (a) An apple in free fall has a downward acceleration g. (b) The force on the apple is also downward, and it has a magnitude $w = mg$.

FIGURE 5.14 This soccer player has jumped into the air. Both he and the ball are in free fall.

FIGURE 5.15 Astronauts training in an airplane.

weightlessness is impossible at any location on or near the Earth, *a condition of apparent weightlessness can be simulated on or near the Earth by means of a freely falling reference frame.* Consider an observer in free fall, such as the soccer player in Fig. 5.14, who has jumped into the air to kick a ball. The player and the ball both accelerate downward at the same rate; thus, the ball does not accelerate relative to the player. In the reference frame of the player (a freely falling reference frame accelerating downward with the acceleration g), the freely falling ball, or any other freely falling body, continues to move with constant velocity, as though there were no force acting on it. This means that in such a reference frame, the gravitational pull is *apparently* zero; the weight is *apparently* zero. Of course, this simulated weightlessness arises from the accelerated motion of the reference frame—in the unaccelerated, inertial reference frame of the ground, the weight of the ball is certainly not zero. Nevertheless, if the player insists on looking at things from his own reference frame, he will judge the weight of the ball, and the weight of his own body, as zero. This condition of weightlessness is also simulated within an airplane flying along a parabola, imitating the motion of a (frictionless) projectile (see Fig. 5.15); and it is also simulated in a spacecraft orbiting the Earth (see Fig. 5.16). Both of these motions are free-fall motions.

The gravity of the Earth reaches from the Earth to any other body, even a body placed high above the surface. Gravity bridges empty space and requires no perceptible medium for its transmission. In contrast, most of the other forces familiar from everyday experience require direct contact between the bodies. You cannot exert a push on a box unless your hand is in contact with the box; and the box cannot exert a push on the floor unless it is in contact with the floor. The push that the surface of a body exerts on the adjacent surface of another body is called a **contact force**. If the two bodies are solid, the contact forces between their

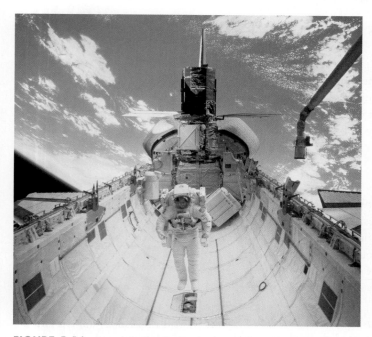

FIGURE 5.16 Astronaut floating in the cargo bay of the Space Shuttle.

adjacent surfaces are of two kinds: the compressional force, or normal force, that arises when the surfaces are pressed together perpendicularly, and the friction force that arises when the surfaces are sliding, or trying to slide, past each other. We will deal with frictional forces in the next chapter.

The **normal force** represents the resistance that solid bodies offer to interpenetration. When you try to push two bodies together, their surfaces begin to repel as soon as they come into contact. You can feel this kind of repulsive contact force when you push with your hand or your foot against any solid surface. For instance, if you push with your hand horizontally against a wall, you can feel the wall pushing against your hand, stopping your hand from penetrating the wall (see Fig. 5.17). This push of the wall is called a normal force, because it is "normal," meaning perpendicular, to the wall. This normal force arises from the contact between the atoms of your hand and the atoms of the wall; the atoms of your hand and the atoms on the surface of the wall exert repulsive forces on each other, which oppose their interpenetration.

How does the wall succeed in preventing your hand from penetrating the wall, regardless of how hard you push? The resistance offered by the wall results from a slight compression of the material of the wall. The atoms in the material in the wall behave like an array of miniature springs; these atomic springs compress slightly when you push your hand against them, and the force that these springs exert on your hand increases with the amount of compression. Hence your hand compresses the wall until the increasing force of the atomic springs stops your hand. If the material of the wall is hard—for example, concrete—the amount of compression is so slight as to be unnoticeable, and the wall seems impenetrable.

FIGURE 5.17 When you push against a wall with your hand, the wall pushes back against your hand and resists penetration.

 ## Checkup 5.4

QUESTION 1: You throw a 1.0-kg stone straight up. What is the force of gravity on the stone while it is traveling upward? When it is instantaneously at rest at the top of its trajectory? When it is traveling back down?

QUESTION 2: A star deep in intergalactic space, far from the gravitational pull of any other star or planet, is weightless. Is it also massless?

QUESTION 3: The accelerations due to gravity on the surface of the Earth, the Moon, and Jupiter are 9.81 m/s^2, 1.62 m/s^2, and 24.8 m/s^2, respectively. Where would your weight be largest? Smallest?

QUESTION 4: An astronaut and her spacecraft are initially at rest on the launchpad. Then the rocket engines fire, and the spacecraft lifts off and ascends. After some minutes, the rocket engines cut off, and the spacecraft coasts through empty space. At what point will the astronaut begin to experience (apparent) weightlessness?

QUESTION 5: A book with a weight of 50 N lies on a table. What is the normal force that the table exerts on the book? If we place a second, identical book on top of the first, what is the normal force that the table exerts on the first book? What is the normal force that the first book exerts on the second?

QUESTION 6: An elevator traveling upward decelerates to stop at a floor. Is the normal force on the feet of a passenger during the deceleration larger or smaller than her weight? Another elevator traveling downward decelerates to stop at a floor. Is the normal force on the feet of a passenger during the deceleration larger or smaller than his weight?

Concepts
— in —
Context

(A) Larger; larger
(B) Larger; smaller
(C) Smaller; smaller
(D) Smaller; larger

Online
Concept
Tutorial

7

5.5 NEWTON'S THIRD LAW

When you push with your hand against a body, such as a wall, the body pushes back at you. Thus, the mutual interaction of your hand and the wall involves two normal forces: the "action" force of the hand on the wall and the "reaction" force of the wall on the hand (see Fig. 5.18). These forces are said to form an **action–reaction pair**. Which of the forces is regarded as "action" and which as "reaction" is irrelevant. It may seem reasonable to regard the push of the hand as an action; then the push of the wall is a reaction. However, it is equally valid to regard the push of the wall on the hand as an action, and then the push of the hand on the wall is a reaction. At the microscopic level, both the hand and the wall consist of atoms, and when two atoms exert forces on each other, it is equally valid to regard the first atom as "acting" and the second as "reacting" or vice versa. The important point is that forces always occur in pairs; each of them cannot exist without the other. This is true not only for normal forces, as in the example of the hand and the wall, but for all forces.

Newton's Third Law gives the quantitative relationship between the action force and the reaction force:

Newton's Third Law

Whenever a body exerts a force on another body, the latter exerts a force of equal magnitude and opposite direction on the former.

For instance, if the push of your hand on a wall has a magnitude of 60 N and is perpendicular to the wall, then the push of the wall on your hand also has a magnitude of 60 N and is also perpendicular to the wall, but in the opposite direction (both your push and the push of the wall are normal forces). But if the push of your stationary hand on the wall makes an angle with the wall (see Fig. 5.19), then the push of the wall on your hand makes a corresponding angle (both your push and the push of the wall are then a combination of normal and friction forces).

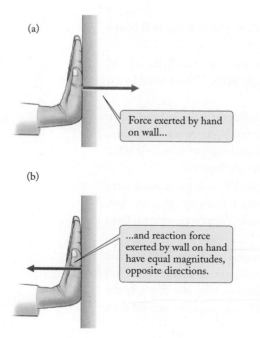

(a)

Force exerted by hand on wall...

(b)

...and reaction force exerted by wall on hand have equal magnitudes, opposite directions.

FIGURE 5.18 (a) Hand pushes on wall; (b) wall pushes on hand.

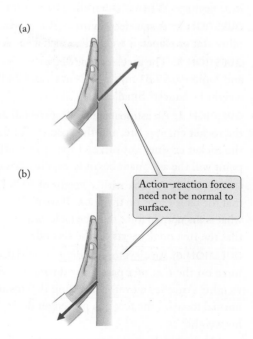

(a)

(b)

Action–reaction forces need not be normal to surface.

FIGURE 5.19 (a) Hand pushes on wall at an angle; (b) wall pushes on hand at an angle.

The equality of the magnitudes of action and reaction is valid even if the body you push against is not held in a fixed position (like a wall) but is free to move. For example, if you push on a cart with a force of 60 N, the cart will push back on you with a force of 60 N (see Fig. 5.20), even while the cart accelerates away from you. Note that although these action and reaction forces are of equal magnitudes, they act on different bodies and their effects are quite different: the first force gives an acceleration to the cart (if there is no other force acting on the cart), whereas the second force merely slows your hand and prevents it from accelerating as much as it would if the cart were not there. *Thus, although action and reaction are forces of equal magnitudes and of opposite directions, their effects do not cancel because they act on different bodies.*

We can express Newton's Third Law mathematically simply by equating the force exerted by a first body on a second body with the negative of the force exerted by the second body on the first body:

$$\mathbf{F}_{1 \text{ on } 2} = -\mathbf{F}_{2 \text{ on } 1} \tag{5.17}$$

action and reaction forces

Reaction forces play a crucial role in all animals and machines that produce loco-motion by pushing against the ground, water, or air. For example, a man walks by push-ing backward on the ground; the reaction of the ground then pushes the man forward (see Fig. 5.21). An automobile moves by pushing backward on the ground with its wheels; the reaction of the ground then pushes the automobile forward (see Fig. 5.22). A tugboat moves by pushing backward against the water with its propeller; the reaction of the water on the propeller then pushes the tugboat forward (see Fig. 5.23). Even the propulsion of a jet aircraft or a rocket relies on reaction forces. The rocket engine expels exhaust gases; the reaction of the exhaust gases then pushes the engine and the rocket forward (see Fig. 5.24). The atmosphere is of no help in rocket propulsion; rather it is a hindrance, since it exerts a frictional resistance on the rocket.

Push exerted by hand accelerates cart; reaction exerted by cart slows hand.

FIGURE 5.20 Woman pushes on cart; cart pushes on woman.

Reaction force pushes the man forward.

FIGURE 5.21 Man pushes against ground; ground pushes against man.

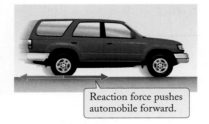

Reaction force pushes automobile forward.

FIGURE 5.22 Automobile pushes against ground; ground pushes against automobile.

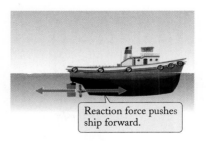

Reaction force pushes ship forward.

FIGURE 5.23 Propeller pushes against water; water pushes against propeller.

FIGURE 5.24 Rocket pushes against exhaust gases; exhaust gases push against rocket.

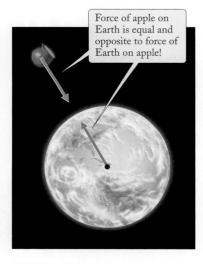

FIGURE 5.25 Earth pulls on apple; apple pulls on Earth.

Reaction forces exist even if the two interacting bodies are not in direct contact, so the forces between them must bridge the intervening empty space. For instance, consider an apple in free fall at some height above the ground. The Earth pulls on the apple by means of gravity. If this pull has a magnitude of, say, 2 N, then Newton's Third Law requires that the apple pull on the Earth with an opposite force of 2 N (see Fig. 5.25). This reaction force is also a form of gravity—it is the gravity that the apple exerts on the Earth. However, the effect of the apple on the motion of the Earth is insignificant because the mass of the Earth is so large that a force of only 2 N produces only a negligible acceleration of the Earth.

Keep in mind that although the two forces in an action–reaction pair are always of equal magnitudes and opposite directions, two forces of equal magnitudes and opposite directions are not always an action–reaction pair. For instance, consider a box of mass m sitting on the floor (see Fig. 5.26). There are two forces acting on the box: the weight **w** of the box pointing downward in the vertical direction, and the normal force of the floor **N** pointing upward in the vertical direction. Figure 5.27 shows the box and these two forces **w** and **N** acting on it. Since the box is supposed to remain at rest, the net force on the box must be zero, which requires that the two forces **w** and **N** have equal magnitudes. The magnitude of the weight is mg, and therefore the magnitude of the normal force must also be mg. However, although **w** and **N** are of equal magnitudes, they are *not* an action–reaction pair. Instead, the normal force **N** exerted by the floor upward on the box forms an action–reaction pair with the normal force **N'** exerted by the box downward on the floor; and the weight **w**, or the gravitational pull exerted by the Earth on the box, forms an action–reaction pair with the gravitational pull **w'** exerted by the box on the Earth.

A diagram such as Fig. 5.27 that shows a body and all the external forces acting on the body, but not the reaction forces that the body exerts on its environment, is called a **"free-body" diagram**. (In this context *free* does not mean free of forces; it means that the body is shown free of its environment, and this environment is represented by the forces it exerts.) Thus, the floor on which the box rests is not shown in Fig. 5.27—the effects of the floor are entirely contained in the normal force **N**. The "free-body" diagram eliminates clutter and helps us to focus on the body and on the forces that we need to formulate the equation of motion of the body.

Note that Figs. 5.20–5.25 are not "free-body" diagrams; for example, Figs. 5.21, 5.22, and 5.25 show the forces on the body (man, or car, or apple) and also the reaction forces on the ground in the same diagram.

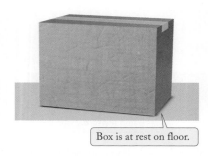

Box is at rest on floor.

FIGURE 5.26 A box resting on a floor.

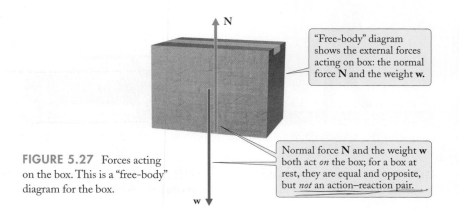

N

"Free-body" diagram shows the external forces acting on box: the normal force **N** and the weight **w.**

Normal force **N** and the weight **w** both act *on* the box; for a box at rest, they are equal and opposite, but *not* an action–reaction pair.

FIGURE 5.27 Forces acting on the box. This is a "free-body" diagram for the box.

w

EXAMPLE 6   A man of mass 75.0 kg is standing in an elevator which is accelerating upward at 2.00 m/s² (see Fig. 5.28). What is the normal force that the floor of the elevator exerts on the man? What is the normal force that the man exerts on the floor?

Concepts
— *in* —
Context

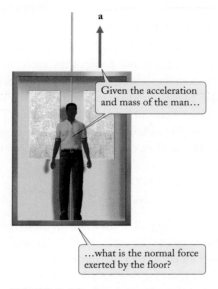

Given the acceleration and mass of the man...

...what is the normal force exerted by the floor?

FIGURE 5.28   A man standing in an elevator accelerating upward.

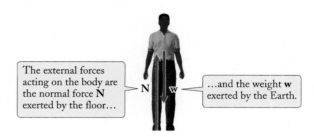

The external forces acting on the body are the normal force **N** exerted by the floor...

...and the weight **w** exerted by the Earth.

FIGURE 5.29   "Free-body" diagram for the man.

SOLUTION: The two forces on the man are his weight and the normal force; thus we need only consider vertical forces. These forces are shown in the "free-body" diagram of Fig. 5.29. The net force on the man is

$$F_{\text{net}} = N - mg$$

where the forces are regarded as positive when directed upward. Since the net force F_{net} gives the man an acceleration a, Newton's Second Law tells us that

$$ma = N - mg$$

or

$$N = ma + mg$$

Hence

$$N = 75.0 \text{ kg} \times 2.00 \text{ m/s}^2 + 75.0 \text{ kg} \times 9.81 \text{ m/s}^2 = 150 \text{ N} + 736 \text{ N} = 886 \text{ N}$$

Thus, the normal force on the man is *larger* than his weight by 150 N.

According to Newton's Third Law, the normal reaction force that the man exerts downward on the floor of the elevator has the same magnitude, also 886 N, since the normal forces on the elevator and on the man form an action–reaction pair. If the man were standing on a spring balance (bathroom scale), the balance would register this larger "weight" of 886 N, as though gravity had increased, and the spring balance would indicate a reading of (886 N)/(9.81 m/s²) = 90 kg, instead of 75 kg. Note that the direction of the velocity of the elevator is irrelevant; only the direction of the acceleration matters. If the elevator were descending and braking (again, a positive acceleration!), the result would be the same.

EXAMPLE 7   A tugboat tows a barge of mass 50 000 kg by means of a cable (see Fig. 5.30a). If the tugboat exerts a horizontal pull of 6000 N on the cable, what is the acceleration of the cable and the barge? What is the magnitude of the pull that the cable exerts on the barge? Assume that the mass of the cable can be neglected (i.e., assume the cable is practically massless), and ignore the friction of the water on the barge.

SOLUTION: Before drawing the "free-body" diagram, we must decide what is our "body." We could take the barge as our body, or the cable, or both jointly. Since the barge and the cable accelerate jointly, it will be best to take the barge and cable jointly as our body. The "free-body" diagram for this body is shown in Fig. 5.30b (only horizontal forces have been included in this diagram). The force of 6000 N exerted by the tugboat accelerates both the cable and the barge; that is, it accelerates a total mass of 50 000 kg. Hence, the resulting acceleration is

$$a = \frac{F}{m} = \frac{6000 \text{ N}}{50\,000 \text{ kg}} = 0.12 \text{ m/s}^2$$

By Newton's Third Law, the pull of the cable on the barge has the same magnitude as the pull of the barge on the cable. To find this magnitude we can examine the "free-body" diagram either for the barge or for the cable. Let us choose the cable; the "free-body" diagram for this body is shown in Fig. 5.31. The tugboat pulls at the forward end of the cable with a force of 6000 N, and the barge pulls at the rearward end. For a cable of zero mass, the net force on the cable must be zero (for a body of zero mass, $F = ma = 0 \times a = 0$). Hence the pull of the barge on the rearward end of the cable must match the pull of the tugboat at the forward end—both pulls must be 6000 N. Newton's Third Law then requires that the cable pull on the barge with a force of 6000 N.

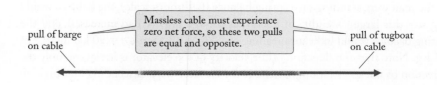

FIGURE 5.30(a)   A tugboat tows a barge.

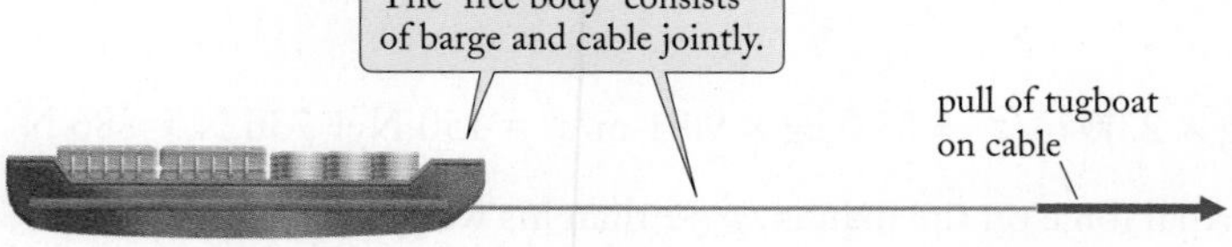

FIGURE 5.30(b)   "Free-body" diagram for barge and cable.

The "free body" consists of barge and cable jointly.

pull of tugboat on cable

FIGURE 5.31   "Free body" diagram for the forces acting on the cable, showing both the external pulls exerted by the tugboat and the barge.

pull of barge on cable

Massless cable must experience zero net force, so these two pulls are equal and opposite.

pull of tugboat on cable

COMMENT: Note that although the pulls at the forward and rearward ends of the cable are two forces of equal magnitudes and opposite directions, these two forces are *not* an action–reaction pair. The pull at the forward end forms an action–reaction pair with the pull exerted by the cable on the tugboat, and the pull at the backward end forms an action–reaction pair with the pull exerted by the cable on the barge. These pulls of the cable are shown in red in Fig. 5.32.

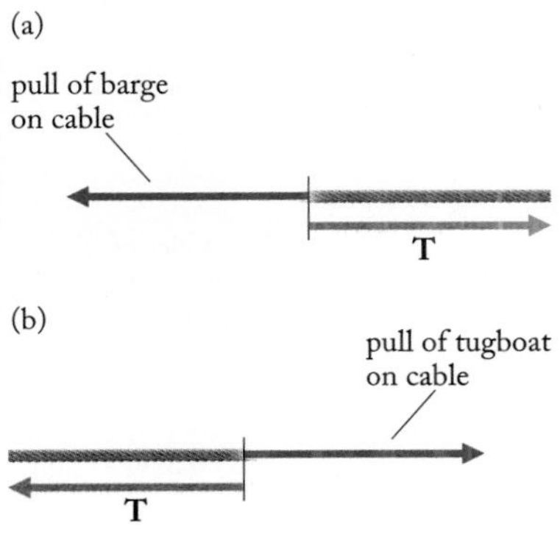

The force with which a cable pulls on what is attached to it is called the **tension T.** *The direction of the tension is along the cable.* Thus, in Fig. 5.32, the magnitude of the tension of the cable is $T = 6000$ N at its forward end and also $T = 6000$ N at its rearward end. This equality of the magnitudes of the tensions at the forward and the rearward ends of the cable is a consequence of neglecting the mass of the cable. If we were to take into account the mass of the cable, then a net force would be needed to accelerate the cable. This means that the force pulling on the forward end of the cable would have to be *larger* than the force pulling on the rearward end—by Newton's Third Law, the tensions would be of unequal magnitudes.

For practical purposes, the mass of a cable, rope, string, wire, or chain can often be neglected compared with the mass of the body to which it is attached. Under these conditions, *the cable transmits the magnitude of the tension without change.* In subsequent problems we will always neglect the mass of the cable unless we explicitly state otherwise. The transmission of the magnitude of the tension without change occurs even if the cable is led around (frictionless and massless) pulleys, which change only the direction of the pull, as illustrated in the following example. We examine a frictional, locked pulley later in Example 10. We will examine pulleys that are not massless much later, in Chapter 13.

EXAMPLE 8   Figure 5.33 shows a traction apparatus used in hospitals to exert a steady pull on a broken leg, in order to keep the bones aligned. The middle pulley is attached to the cast, and the other two pulleys are attached to the bed or the wall. A flexible wire passes over these pulleys, and a brick hanging from this wire provides a tension. The upper and the lower portions of the wire are oriented, respectively, upward and downward from the middle pulley at angles of 35° with respect to the horizontal. If the horizontal pull on the leg is to be 50 N, what tension must the brick provide at the end of the wire?

SOLUTION: As discussed above, the tension is constant along the entire wire. If the magnitude of the tension at the lower end of the wire is T, the magnitude of the tension at all other points of the wire must also be T. Under static conditions, the upper and the lower portions of the wire may be regarded as attached to the middle pulley at the points of first contact. Thus the upper portion of the wire pulls upward at an angle of 35° with a force $\mathbf{T}_1$ of magnitude T, and the lower portion pulls downward at an angle of 35° with a force $\mathbf{T}_2$ of the same magnitude.

Figure 5.34a shows these forces $\mathbf{T}_1$ and $\mathbf{T}_2$ that the wire exerts on the middle pulley. The x axis is horizontal, and the y axis vertical. The vertical components of these forces cancel, since they have opposite signs and equal magnitudes. The horizontal components add, since they both have positive signs. The resultant force exerted by the wire on the middle pulley is therefore in the horizontal direction, that is, the x direction.

(a)

pull of barge
on cable

T

(b)

pull of tugboat
on cable

T

FIGURE 5.32 Action-reaction pairs. (a) Barge pulls on cable; cable pulls on barge with tension **T**. (b) Tugboat pulls on cable; cable pulls on tugboat with tension **T**.

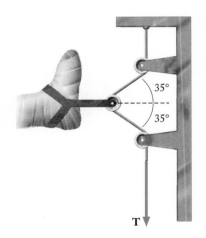

35°

35°

T

FIGURE 5.33 Traction apparatus.

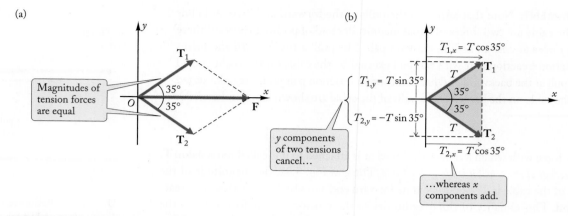

FIGURE 5.34 (a) Forces exerted by the wire on the middle pulley. The upper portion of the wire pulls with a tension $\mathbf{T}_1$ and the lower portion pulls with a tension $\mathbf{T}_2$. The magnitudes of these forces are equal: $T_1 = T_2 = T$. The resultant $\mathbf{F}$ is in the x direction. (b) Components of these forces.

The components of $\mathbf{T}_1$ and $\mathbf{T}_2$ in the x direction are $T \cos 35°$ (see Fig. 5.34b). Hence the x component of the resultant force is

$$F_x = T_{1,x} + T_{2,x} = T \cos 35° + T \cos 35° = 2T \cos 35° \qquad (5.18)$$

Since the magnitude of F_x is supposed to be 50 N, we obtain

$$T = \frac{F_x}{2 \cos 35°} = \frac{50 \text{ N}}{2 \times 0.819} = 31 \text{ N} \qquad (5.19)$$

Therefore the brick must provide a tension of 31 N.

 Checkup 5.5

QUESTION 1: Draw a "free-body" diagram for an apple in free fall. Draw a "free-body" diagram for an apple at rest on a table.

QUESTION 2: An apple of weight 2 N hangs from the branch of a tree. The two forces on the apple are the weight and the upward pull exerted by the branch. If these two forces are regarded as actions, what are the reactions?

QUESTION 3: While driving a car, you accelerate, then proceed at constant velocity for a while, and then brake. What is the direction of the net external force on the car during acceleration? During travel at constant velocity? During braking? What external body exerts the force on the car during acceleration and during braking?

QUESTION 4: While sitting at the edge of a dock, you push with your feet against a supertanker with a force of 400 N. What is the force with which the supertanker pushes against your feet? If, instead, you push against a rowboat with a force of 400 N, what is the force with which the rowboat pushes against your feet? Is there any difference in the behaviors of the supertanker and the rowboat?

QUESTION 5: A book of weight 50 N lies on a table, and a second book lies on top of the first. Draw the "free-body" diagram for each book. List all the forces acting on the books. Which of these forces are action–reaction pairs?

QUESTION 6: A man pulls with a force of 150 N on one end of a rope, and a woman with a force of 150 N on the other end. What is the tension in the rope? If the woman now ties her end of the rope to a tree and walks away, while the man continues pulling, what will be the tension in the rope?

QUESTION 7: A mass of 10 kg hangs on a rope attached to a spring scale which hangs from the ceiling by a second rope (Fig. 5.35). Assume that the masses of the ropes and of the spring scale can be neglected. What is the tension in the first rope? What is the tension in the second rope? What is the weight (in N) on the spring scale? Choose among the following respective quantities:

(A) 0 N; 98 N; 98 N   (B) 98 N; 0 N; 98 N
(C) 98 N; 98 N; 0 N   (D) 98 N; 98 N; 98 N

FIGURE 5.35 A 10-kg mass hanging on a spring balance.

Online
Concept
Tutorial

5.6 MOTION WITH A CONSTANT FORCE

Newton's Second Law is often called the **equation of motion**. If the force on a particle is known, then *the Second Law determines the acceleration, and from this the position of the particle at any time can be calculated.* Thus, in principle, the motion of the particle is completely predictable.

If the force acting on a particle is constant, then the acceleration is also constant. The motion is then given by the formulas we developed for motion with constant acceleration in Chapter 4 [Eqs. (4.20)–(4.24)]. If the acceleration is not constant, then formulas for the motion can be obtained by calculations using integration; we will learn to calculate integrals in Chapter 7, and we discuss their application to the equations of motion in Sections 2.7 and 12.4.

As a simple example of motion with a constant force, consider a rectangular box of mass m being pushed along a smooth, frictionless floor. Figure 5.36 shows the box and the man pushing it. The push can be represented by a vector **P** of magnitude P pointing at an angle θ with the horizontal direction. Besides this push, there are two other forces acting on the box: the weight **w** of the box pointing downward in the vertical direction, and the normal force of the floor **N** pointing upward in the vertical direction. Figure 5.37 is a "free-body" diagram showing these forces.

equation of motion

Push on box is both forward and downward.

FIGURE 5.36 Man pushing a box.

(a)

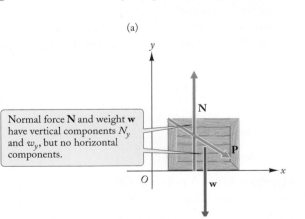

Normal force **N** and weight **w** have vertical components N_y and w_y, but no horizontal components.

(b)

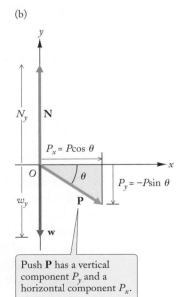

FIGURE 5.37 (a) "Free-body" diagram for the box. The three forces acting on the box are **P**, **w**, and **N**. The magnitudes of these forces are P, mg, and N. (b) The x and y components of the forces **P**, **w**, and **N**.

Push **P** has a vertical component P_y and a horizontal component P_x.

The net force on our box is the vector sum of **P**, **w**, and **N**; and the acceleration of the box is related to this net force by Newton's Second Law. Although the motion of the box is one-dimensional (horizontal, along the floor), the forces acting on it are not one-dimensional, and we must consider both the horizontal and the vertical components of these forces. For this purpose, it is convenient to take the x axis in the horizontal direction and the y axis in the vertical direction. Looking at Fig. 5.37b, we then find that the x and the y components of the individual forces are

$$P_x = P \cos \theta \qquad P_y = -P \sin \theta \tag{5.20}$$

$$w_x = 0 \qquad w_y = -mg \tag{5.21}$$

$$N_x = 0 \qquad N_y = N \tag{5.22}$$

Here N is the (unknown) magnitude of the normal force.

The components of the net force are

$$F_x = P_x + w_x + N_x = P \cos \theta + 0 + 0$$
$$= P \cos \theta \tag{5.23}$$

$$F_y = P_y + w_y + N_y$$
$$= -P \sin \theta - mg + N \tag{5.24}$$

The x and y components of the equation of motion for the box are $ma_x = F_x$ and $ma_y = F_y$, from which we obtain the components of the acceleration:

$$a_x = \frac{F_x}{m} = \frac{P \cos \theta}{m} \tag{5.25}$$

$$a_y = \frac{F_y}{m} = \frac{-P \sin \theta - mg + N}{m} \tag{5.26}$$

Equation (5.25) says that the acceleration of the box along the floor is $(P \cos \theta)/m$. This determines the motion of the box along the floor, since the magnitude and direction (P and θ) of the push exerted by the man are assumed known. To find the velocity and the position at any time we need only substitute the acceleration a_x into our old equations [(4.20)–(4.24)] for uniformly accelerated motion.

Equation (5.26) can be used to evaluate the magnitude of the normal force. Since the motion is necessarily along the floor, the acceleration a_y in the direction perpendicular to the floor is zero; hence

$$0 = \frac{-P \sin \theta - mg + N}{m} \tag{5.27}$$

We can solve this for N, with the result

$$N = mg + P \sin \theta \tag{5.28}$$

Note that if the push is horizontal ($\theta = 0$ and $\sin \theta = 0$) or if the push is absent ($P = 0$), then the normal force is $N = mg$. Thus, under these conditions, the normal force simply balances the weight; this is, of course, exactly what we would expect for a box sitting on a floor when there is no push. Furthermore, note that if the push is vertically downward ($\theta = 90°$ and $\sin \theta = 1$, $\cos \theta = 0$), the normal force is $N = mg + P$ (the normal force balances the sum of weight and downward push), and the horizontal acceleration is zero. Finally, if the push has an upward component, then $P_y = P \sin \theta$

has become positive (the angle θ is above the x axis in Fig. 5.37). If this upward push is so large that it exceeds the weight (when $P\sin\theta > mg$), then the box no longer contacts the floor ($N = 0$) and it accelerates upward with $a_y = (P\sin\theta - mg)/m$.

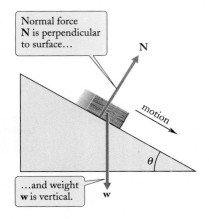

FIGURE 5.38 Block sliding down an inclined plane.

EXAMPLE 9   Figure 5.38 shows a block of mass m sliding down a smooth, frictionless plane, or ramp, inclined at an angle θ with respect to the horizontal direction. Find the acceleration of the block along the inclined plane. Find the magnitude of the normal force that the plane exerts on the block.

SOLUTION: There are two forces acting on the block: the weight **w** pointing vertically downward and the normal force **N** pointing in a direction perpendicular to the inclined plane. Figure 5.39a shows these two forces on a "free-body" diagram. The net force acting on the block is the vector sum of **w** and **N**. For the calculation of the components of these forces, it is convenient to take the x axis parallel to the inclined plane and the y axis perpendicular to it; this simplifies the calculation of the motion, since the velocity and the acceleration are then entirely along the chosen x axis. With this choice of axes, we find that the components of the two forces are (see Fig. 5.39b)

$$N_x = 0 \qquad\qquad N_y = N \qquad\qquad (5.29)$$

$$w_x = mg\sin\theta \qquad w_y = -mg\cos\theta \qquad (5.30)$$

and the components of the net force are

$$F_x = N_x + w_x = mg\sin\theta \qquad\qquad (5.31)$$

$$F_y = N_y + w_y = N - mg\cos\theta \qquad\qquad (5.32)$$

The equation of motion then gives us the corresponding components of the acceleration of the block:

$$a_x = \frac{F_x}{m} = \frac{mg\sin\theta}{m} = g\sin\theta \qquad\qquad (5.33)$$

$$a_y = \frac{F_y}{m} = \frac{N - mg\cos\theta}{m} \qquad\qquad (5.34)$$

Equation (5.33) tells us that the acceleration of the block along the inclined plane is $g\sin\theta$. In the case of a horizontal plane ($\theta = 0$ and $\sin\theta = 0$), there is no acceleration. In the case of a vertical plane ($\theta = 90°$ and $\sin\theta = 1$), the acceleration is $a_x = g$, which is the acceleration of free fall (with the x axis directed downward). Both of these extreme cases are as we would expect them to be.

Equation (5.34) can be used to evaluate the normal force. Since the motion is along the plane, the acceleration perpendicular to the plane must be identically zero, and thus

$$0 = \frac{N - mg\cos\theta}{m} \qquad\qquad (5.35)$$

From this we find

$$N = mg\cos\theta \qquad\qquad (5.36)$$

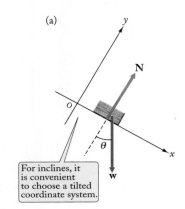

(a)

(b)

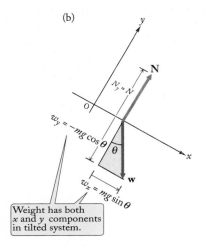

FIGURE 5.39 (a) "Free-body" diagram showing the forces **N** and **w**. (b) The x and the y components of the forces **N** and **w**.

COMMENT: Note that in the case of a horizontal plane ($\theta = 0$), the normal force has a magnitude mg, that is, the magnitude of the normal force matches the weight; in the case of a vertical plane ($\theta = 90°$), the normal force vanishes. This is reasonable, and we might have expected it.

Concepts
— in —
Context

EXAMPLE 10   A passenger elevator consists of an elevator cage of 1000 kg (empty) and a counterweight of 1100 kg connected by a cable running over a large pulley (see Fig. 5.40). Neglect the masses of the cable and of the pulley. (a) What is the upward acceleration of the elevator cage if the pulley is permitted to run freely, without friction? (b) What is the tension in the cable? (c) What are the tensions in the cable if the pulley is locked (by means of a brake) so that the elevator remains stationary?

SOLUTION: (a) Since the elevator cage and the counterweight are linked by the cable, it is necessary to solve the equations of motion for these two bodies simultaneously. For the free pulley, the cable merely transmits the tension from one body to the other, without change of its magnitude; consequently, the upward tension forces exerted by the ends of the cable on each body are exactly equal.

Figure 5.41 shows "free-body" diagrams for the elevator cage and the counterweight. The masses of these two bodies are designated by m_1 and m_2, respectively. For a system consisting of several bodies, such as the system of two bodies we are dealing with here, the "free-body" diagrams are especially helpful, since they permit us to view each body in isolation, and they give us a clear picture of what happens to each individual body. The vector **T** in Fig. 5.41 represents the tension force, and $\mathbf{w}_1$ and $\mathbf{w}_2$ represent the weights. Only vertical forces are present. With the y axis in the vertical direction, as indicated in Fig. 5.41, the y component of the force acting on m_1 is $F_1 = T - w_1$, and the y component of the force on m_2 is $F_2 = T - w_2$. Hence, the equation for the vertical motion of each mass is

$$m_1 a_1 = F_1 = T - w_1 \tag{5.37}$$

$$m_2 a_2 = F_2 = T - w_2 \tag{5.38}$$

where the forces and the accelerations are regarded as positive when directed upward. Since the two masses are tied together by a fixed length of cable, their accelerations a_1 and a_2 are always of the same magnitudes and in opposite directions; that is,

$$a_1 = -a_2 \tag{5.39}$$

With this equation and with $w_1 = m_1 g$ and $w_2 = m_2 g$, we obtain from Eqs. (5.37) and (5.38)

$$m_1 a_1 = T - m_1 g \tag{5.40}$$

$$-m_2 a_1 = T - m_2 g \tag{5.41}$$

These are two simultaneous equations for the two unknowns a_1 and T. To solve these equations first for a_1, we can eliminate T by subtracting each side of the second equation from each side of the first equation:

$$m_1 a_1 - (-m_2 a_1) = T - m_1 g - (T - m_2 g) \tag{5.42}$$

So the unknown T cancels out, leaving us with an equation for a_1:

$$m_1 a_1 + m_2 a_1 = -m_1 g + m_2 g \tag{5.43}$$

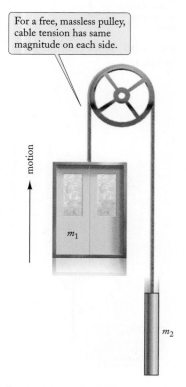

For a free, massless pulley, cable tension has same magnitude on each side.

motion

m_1

m_2

FIGURE 5.40 Elevator with counterweight.

We can solve this equation for a_1 by rearranging:

$$a_1 = \frac{m_2 - m_1}{m_1 + m_2}g \qquad (5.44)$$

This equation tells us the acceleration. With $m_1 = 1000$ kg and $m_2 = 1100$ kg, we find

$$a_1 = \frac{1100 \text{ kg} - 1000 \text{ kg}}{1100 \text{ kg} + 1000 \text{ kg}}g = \frac{100}{2100}g$$

$$= 0.0476g = 0.0476 \times 9.81 \text{ m/s}^2 = 0.467 \text{ m/s}^2 \qquad (5.45)$$

The positive value of this acceleration indicates that the elevator cage accelerates upward, as we might have expected, since it has the smaller mass.

(b) Next, we must find the tension in the cable. Substituting the result for a_1 into Eq. (5.40), we obtain an equation for T:

$$m_1 \frac{m_2 - m_1}{m_1 + m_2}g = T - m_1 g \qquad (5.46)$$

which leads to

$$T = m_1 \frac{m_2 - m_1}{m_1 + m_2}g + m_1 g = m_1 g \times \left(\frac{m_2 - m_1}{m_1 + m_2} + 1 \right)$$

$$= \frac{2m_1 m_2 g}{m_1 + m_2} \qquad (5.47)$$

This tells us the tension in the cable; numerically,

$$T = \frac{2m_1 m_2 g}{m_1 + m_2} = \frac{2 \times 1000 \text{ kg} \times 1100 \text{ kg} \times 9.81 \text{ m/s}^2}{1000 \text{ kg} + 1100 \text{ kg}}$$

$$= 1.03 \times 10^4 \text{ N} \qquad (5.48)$$

(c) If the pulley is locked and the elevator is stationary, the tension in the cable on either side of the pulley must match the weight hanging on that side. Thus

$$T_1 = w_1 = m_1 g = 1000 \text{ kg} \times 9.81 \text{ m/s}^2 = 9.81 \times 10^3 \text{ N} \qquad (5.49)$$

and

$$T_2 = w_2 = m_2 g = 1100 \text{ kg} \times 9.81 \text{ m/s}^2 = 1.08 \times 10^4 \text{ N} \qquad (5.50)$$

COMMENT: As we might have expected, Eq. (5.44) shows that if the two masses m_1 and m_2 are equal, the acceleration is zero—the two masses are then in equilibrium, and they either remain at rest or move with uniform velocity (until the cable runs out).

Note that when the two masses are unequal and the pulley is locked, the tensions in the two parts of the cable are *not* equal. This is because the locked pulley exerts extra friction forces on the cable, and the two portions of the cable now behave as though they were suspended independently.

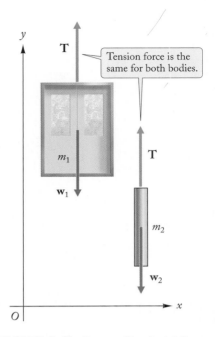

FIGURE 5.41 Separate "free-body" diagrams for the elevator and the counterweight.

EXAMPLE 11 The acceleration of a large block of mass m_1 down a friction-less ramp is to be kept small by using a hanging counterweight of mass m_2; the two are connected over a light, frictionless pulley by a light rope as shown in Fig. 5.42a. The angle of the incline is $\theta = 15°$. If the desired acceleration is to be one-hundredth of a standard g, what should the mass m_2 of the counter-weight be?

SOLUTION: The "free-body" diagrams for the two masses are shown in Fig. 5.42b. It is convenient to use tilted coordinate axes for the mass m_1 on the incline, as in Example 9. The forces on the mass m_1 are the same as in Example 9, except that there is, additionally, the tension $\mathbf{T}$ from the rope. Accordingly, the x and the y components of the equation of motion of the mass m_1 are

$$m_1 a_{1,x} = m_1 g \sin\theta - T \tag{5.51}$$
$$m_1 a_{1,y} = N - m_1 g \cos\theta$$

The question does not ask about the normal force, so we will not need the second equation. For the hanging mass m_2, we use an upward $+y$ axis, and we have only vertical forces, so

$$m_2 a_{2,y} = T - m_2 g \tag{5.52}$$

As in Example 10, note that the linked accelerations must have the same magnitude, since they are connected by a taught rope. For our axis directions, this gives

$$a_{1,x} = a_{2,y} \tag{5.53}$$

Note that the pulley has linked motion in two different directions. The sign in Eq. (5.53) is chosen as positive, since motion of m_1 along the positive x axis results in motion of m_2 along the positive y axis. Using Eq. (5.53) in Eq. (5.52), and then adding Eq. (5.52) and Eq. (5.51), we can eliminate the tension T:

$$m_1 a_{1,x} + m_2 a_{1,x} = m_1 g \sin\theta - m_2 g$$

Collecting the m_2 terms on the left side, we obtain

$$m_2 (a_{1,x} + g) = m_1 g \sin\theta - m_1 a_{1,x}$$

which we can solve for m_2:

$$m_2 = m_1 \frac{g \sin\theta - a_{1,x}}{a_{1,x} + g}$$

Finally, we substitute the desired value $a_{1,x} = 0.01g$, cancel the common factor of g, and evaluate the result for $\theta = 15°$:

$$m_2 = m_1 \frac{g \sin\theta - 0.01g}{0.01g + g}$$
$$= m_1 \frac{\sin\theta - 0.01}{1.01}$$
$$= m_1 \frac{\sin 15° - 0.01}{1.01} = m_1 \frac{0.26 - 0.01}{1.01}$$
$$= 0.25 m_1$$

(a)

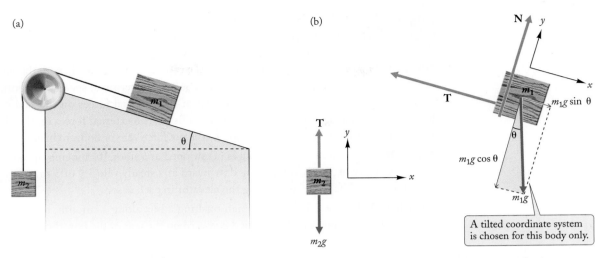

(b)

A tilted coordinate system
is chosen for this body only.

FIGURE 5.42   (a) Slowly accelerating mass m_1 on incline with counterweight m_2. (b) "Free-body" diagrams for the two masses.

PHYSICS IN PRACTICE   ELEVATORS

Concepts
— *in* —
Context

The counterweight of an elevator usually consists of a concrete block, sliding within guide rails attached to the side of the elevator shaft. The mass of the counterweight is chosen to equal the mass of the elevator plus one-half of the mass of the expected average payload. With this choice of mass, the elevator–counterweight system will be nearly in equilibrium most of the time, and the extra force that the motor has to supply to move the elevator up (or down) will be minimized.

In Example 10, we neglected the mass of the elevator cable. For the sake of safety, modern elevators use several cables strung in parallel. In the elevators of skyscrapers, the total mass of the cables is considerable, and it often exceeds the mass of the elevator cage. If we take the mass of the cable into account, the acceleration of a freely running elevator depends on how much cable hangs on the elevator side and how much on the counterweight side. Since the lengths of these two segments of cable change as a function of the position of the elevator, the acceleration also changes with position.

Elevators have braking systems that would prevent their fall even if the cables were to break. The braking system consists of powerful jaws with braking pads that grip the guide rails (Figure 1). The jaws are triggered automatically if the speed of the elevator exceeds a critical value. Some elevators also have an air cushion at the bottom of the shaft. This is a downward extension of the shaft, which fits tightly around the sides of the elevator. If the elevator falls into this tightly fitting shaft, it will be cushioned by the air that it traps and compresses in the shaft. To prevent a bounce of the elevator, small vents allow the compressed air to leak out gradually.

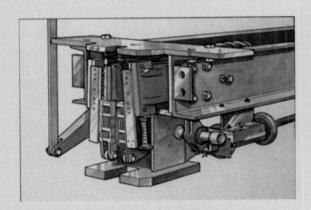

FIGURE 1   Elevator brake assembly.

Checkup 5.6

QUESTION 1: A skydiver jumps out of a helicopter. At first, she falls freely, with the standard downward acceleration g. After a while, she reaches terminal speed and continues to fall with uniform velocity. Make a list of the external forces on the skydiver during the first part of the motion, and make a list of the external forces on the skydiver during the second part of the motion. Is the net force zero or nonzero in each case?

QUESTION 2: An elevator of 1000 kg is (a) stopped at a floor, (b) moving upward at a steady velocity of 5 m/s, (c) moving downward at a steady velocity of 5 m/s. What is the tension in the cable supporting the elevator in each case?

QUESTION 3: Figure 5.36 shows a man pushing a box along a smooth, frictionless floor. Suppose that instead of pushing downward on the rear of the box, the man pulls

PROBLEM-SOLVING TECHNIQUES   "FREE-BODY" DIAGRAMS

From the examples analyzed in this section, we see that the solution of a problem of motion with several forces acting on a body often proceeds in a sequence of steps.

1 The first step is always a careful enumeration of all the forces. Make a complete list of these forces, and label each with a vector symbol.

2 Identify the body whose motion or whose equilibrium is to be investigated and draw a separate "free-body" diagram for this body in which each force is represented by an arrow labeled by a vector symbol. Remember that the only forces to be included in the "free-body" diagram are the forces that act *on* the body, not the forces exerted *by* the body. If there are several separate bodies in the problem (as in Examples 10 and 11), then you need to draw a separate "free-body" diagram for each. When drawing the arrows for the forces, try to draw the lengths of the arrows in proportion to the magnitudes of the forces; this will help you to see what the direction of the resultant is. (Do *not* include an "*ma*" force in the "free-body" diagram; the acceleration is caused by the resultant of several forces already included among the pushes and pulls displayed in the diagram.)

3 Then draw coordinate axes on each diagram, preferably placing one of the axes along the direction of motion. If the motion proceeds along an inclined plane, it is convenient to use tilted coordinate axes, with one axis along the plane, as in Fig. 5.39b. It is even sometimes convenient to use a different orientation of the axes for different bodies, as in Example 11, where one body was on an incline and attached by a rope to one that was hanging straight down.

4 Next, examine the components of the individual forces and the components of the net force. Remember that the signs of the components correspond to the directions of the force vectors. Remember that for freely moving massless pulleys and massless cables, the magnitude (but not necessarily the direction) of the tension in the cable is the same at every point.

5 Then apply Newton's Second Law for the components of the net force ($F_{\text{net},x} = ma_x$, $F_{\text{net},y} = ma_y$), and calculate the components of the acceleration (if the acceleration is unknown), or the components of some force (if some force is unknown). If there are several separate bodies in the problem, you need to apply Newton's Law separately for each. Another relation is available for bodies attached by a cable: the magnitude of the acceleration of each is the same (the direction may be opposite, as in Eq. (5.39) in Example 10, or even along different axes, as in Eq. (5.53) of Example 11).

6 As in the kinematics problems of Chapters 2 and 3, it is a good idea to solve for the unknowns algebraically, and substitute numbers only as a last step. This makes it easier to spot and correct mistakes, and it also makes it possible to check whether the final result behaves as you might have expected in various limiting cases (for instance, see the comments attached to Examples 9 and 10).

7 When substituting numbers, also substitute the units. Use the conversion 1 N = 1 kg·m/s^2 where necessary. This should automatically result in the correct units for the final result.

upward at the front of the box, at the same angle θ and with the same magnitude of force. Does this change the acceleration? The normal force?

QUESTION 4: Consider a block placed on a smooth (frictionless) inclined plane. With your hands, you push against this block horizontally with a force of 100 N, and this barely holds the block stationary. (a) What must be the magnitude of your push if you want to keep the block moving at constant speed up the plane? Down the plane? (b) If instead of pushing horizontally, you push parallel to the surface of the plane, will your push have to be larger or smaller? To calculate how much larger or smaller, what do you need to know?

QUESTION 5: Which of the "free-body" diagrams in Fig. 5.43 respectively corresponds to each of the following bodies: (1) a book lying on a flat table, (2) a box on the floor with another box on top, and (3) a lamp hanging by a cord from a ceiling?

(A) a, b, c   (B) a, c, b   (C) b, a, c   (D) b, c, a   (E) c, a, b

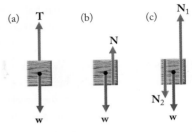

FIGURE 5.43 Three "free-body" diagrams.

SUMMARY

PHYSICS IN PRACTICE Elevators	**(page 157)**

PROBLEM-SOLVING TECHNIQUES "Free-Body" Diagrams	**(page 158)**

NEWTON'S FIRST LAW   In an inertial reference frame, a body at rest remains at rest and a body in motion continues to move at constant velocity unless acted upon by a net external force.

NEWTON'S SECOND LAW $m\mathbf{a} = \mathbf{F}_{net}$

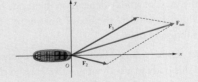

(5.2)

UNIT OF FORCE $1 \text{ newton} = 1 \text{ N} = 1 \text{ kg·m/s}^2$ (5.6)

DEFINITION OF MASS $\dfrac{m}{m_s} = \dfrac{a_s}{a}$ (5.3)

SUPERPOSITION OF FORCES $\mathbf{F}_{net} = \mathbf{F}_1 + \mathbf{F}_2 + \mathbf{F}_3 + \cdots$

(5.8)

WEIGHT   Magnitude: $w = mg$
Direction: downward   (5.16)

NORMAL FORCE   The force provided by a surface that prevents interpenetration of solid bodies. The normal force has a direction perpendicularly outward from the surface, and, if the surface is not accelerating, a magnitude that balances the net force pushing perpendicularly toward the surface.

NEWTON'S THIRD LAW   Whenever a body exerts a force on another body, the latter exerts a force of equal magnitude and opposite direction on the former. These *action–reaction pairs* of forces do not "cancel," since they *act on different bodies.*

$$\mathbf{F}_{1 \text{ on } 2} = -\mathbf{F}_{2 \text{ on } 1}$$

(5.17)

"FREE-BODY" DIAGRAM   A diagram showing a body in isolation and each of the external force vectors acting on the body.

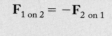

TENSION   The force with which a string, rope, or cable pulls on what is attached to it. For a massless string, rope, or cable, the magnitude of the tension is the same at each point, even when passing around a massless, freely moving pulley.

QUESTIONS FOR DISCUSSION

1. If a glass stands on a table on top of a sheet of paper, you can remove the paper without touching the glass by jerking the paper away very sharply. Explain why the glass more or less stays put.

2. Make a critical assessment of the following statement: An automobile is a device for pushing the air out of the way of the passenger so that his body can continue to its destination in its natural state of motion at uniform velocity.

3. Does the mass of a body depend on the frame of reference from which we observe the body? Answer this by appealing to the definition of mass.

4. Suppose that a (strange) body has negative mass. Suppose that you tie this body to a body of positive mass of the same magnitude by means of a stretched rubber band. Describe the motion of the two bodies.

5. Does the magnitude or the direction of a force depend on the frame of reference?

6. A fisherman wants to reel in a large dead shark hooked on a thin fishing line. If he jerks the line, it will break; but if he reels it in very gradually and smoothly, it will hold. Explain.

7. If a body crashes into a water surface at high speed, the impact is almost as hard as on a solid surface. Explain.

8. Moving downwind, a sailboat can go no faster than the wind. Moving across the wind, a sailboat can go faster than the wind. How is this possible? (Hint: What are the horizontal forces on the sail and on the keel of a sailboat?)

9. A boy and girl are engaged in a tug-of-war. Draw a diagram showing the horizontal forces (a) on the boy, (b) on the girl, and (c) on the rope. Which of these forces are action–reaction pairs?

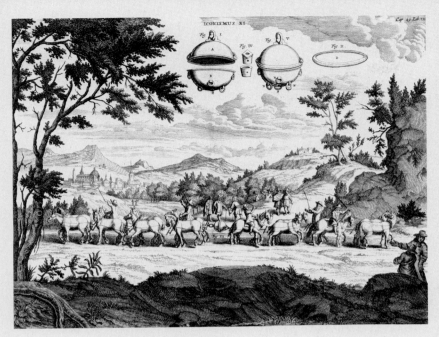

FIGURE 5.44 The Magdeburg hemispheres.

10. In an experiment performed in 1654, Otto von Guericke, mayor of Magdeburg and inventor of the air pump, gave a demonstration of air pressure before Emperor Ferdinand. He had two teams of 15 horses each pull in opposite directions on two evacuated hemispheres held together by nothing but air pressure. The horses failed to pull these hemispheres apart (Fig. 5.44). If each horse exerted a pull of 3000 N, what was the tension in the harness attached to each hemisphere? If the harness attached to one of the hemispheres had simply been tied to a stout tree, what would have been the tension exerted by a single team of horses hitched to the other harness? What would have been the tension exerted by the two teams of horses hitched in series to the other harness? Can you guess why von Guericke hitched up his horses in the way he did?

11. In a tug-of-war, two teams of children pull on a rope (Fig. 5.45). Is the tension constant along the entire length of the rope? If not, along what portion of the rope is it constant?

FIGURE 5.45 Two teams of children in a tug-of-war.

12. If you carry a spring balance from London to Hong Kong, do you have to recalibrate it? If you carry a beam balance?

13. When you stretch a rope horizontally between two fixed points, it always sags a little, no matter how great the tension. Why?

14. What are the forces on a soaring bird? How can the bird gain altitude without flapping its wings?

15. An automobile is parked on a street. (a) Draw a "free-body" diagram showing the forces acting on the automobile. What is the net force? (b) Draw a "free-body" diagram showing the forces that the automobile exerts on the Earth. Which of the forces in the diagrams (a) and (b) are action–reaction pairs?

16. A ship sits in calm water. What are the forces acting on the ship? Draw a "free-body" diagram for the ship.

17. Some old-time roofers claim that when walking on a rotten roof, it is important to "walk with a light step so that your full weight doesn't rest on the roof." Can you walk on a roof with less than your full weight? What is the advantage of a light step?

18. In a tug-of-war on sloping ground the party on the low side has the advantage. Why?

19. The label on a package of sugar claims that the weight of the contents is "1 lb or 454 g." What is wrong with this statement?

20. A physicist stands on a bathroom scale in an elevator. When the elevator is stationary, the scale reads 73 kg. Describe qualitatively how the reading of the scale will fluctuate while the elevator makes a trip to a higher floor.

21. How could you use a pendulum suspended from the roof of your automobile to measure its acceleration?

22. When you are standing on the Earth, your feet exert a force (push) against the surface. Why does the Earth not accelerate away from you?

23. When an automobile accelerates on a level road, the force that produces this acceleration is the push of the road on the wheels. If so, why does the automobile need an engine?

24. You are in a small boat in the middle of a calm lake. You have no oars, and you cannot put your hands in the water because the lake is full of piranhas. The boat carries a large load of coconuts. How can you get to the shore?

25. On a windless day, a sailor puts an electric fan, powered by a battery, on the stern of his boat and blows a stream of air into the sail. Will the boat move forward?

26. You are inside a ship that is trying to make headway against the strong current of a river. Without looking at the shore or other outside markers, is there any way you can tell whether the ship is making any progress?

PROBLEMS

5.2 Newton's Second Law†

1. According to the *Guinness Book of Records,* the heaviest man ever had a confirmed mass of 975 pounds. Express this in kilograms.

2. The hydrogen atom consists of one proton and one electron. What is the mass of one hydrogen atom? (See Table 5.2.) How many hydrogen atoms are there in 1.0 kg of hydrogen gas?

3. The oxygen atom consists of 8 protons, 8 neutrons, and 8 electrons. What is the mass of one oxygen atom? How many oxygen atoms are there in 1.0 kg of oxygen gas?

4. A boy and a girl are engaged in a tug-of-war while standing on the slippery (frictionless) surface of a sheet of ice (see Fig. 5.46). While they are pulling on the rope, the instantaneous acceleration of the boy is 7.0 m/s² toward the girl, and the instantaneous acceleration of the girl is 8.2 m/s² toward the boy. The mass of the boy is 50 kg. What is the mass of the girl?

FIGURE 5.46 A boy and a girl in a tug-of-war on ice.

5. On a flat road, a Maserati sports car can accelerate from 0 to 80 km/h (0 to 50 mi/h) in 5.8 s. The mass of the car is 1620 kg. What are the average acceleration and the average horizontal force on the car?

6. The Grumman F-14B fighter plane has a mass of 16 000 kg, and its engines develop a thrust of 2.7×10^5 N when at full power. What is the maximum horizontal acceleration that this plane can achieve? Ignore friction.

7. A woman of 57 kg is held firmly in the seat of her automobile by a lap-and-shoulder seat belt. During a collision, the automobile decelerates from 50 to 0 km/h in 0.12 s. What is the average horizontal force that the seat belt exerts on the woman? Compare the force with the weight of the woman.

8. A heavy freight train has a total mass of 16 000 metric tons. The locomotive exerts a pull of 670 000 N on this train. What is the acceleration? How long does it take to increase the speed from 0 to 50 km/h?

9. With brakes fully applied, a 1500-kg automobile decelerates at the rate of 8.0 m/s² on a flat road. What is the braking force acting on the automobile? Draw a "free-body" diagram showing the direction of motion of the automobile and the direction of the braking force.

10. Consider the impact of an automobile on a barrier. The initial speed is 50 km/h and the automobile comes to rest within a distance of 0.40 m, with constant deceleration. If the mass of the automobile is 1400 kg, what is the force acting on the automobile during the deceleration?

11. In a crash at the Silverstone circuit in England, a race-car driver suffered more than 30 fractures and dislocations and several heart stoppages after a deceleration from 174 km/h to 0 km/h within a distance of about 66 cm. If the deceleration was constant during the crash and the mass of the driver was 75 kg, what were the deceleration and force on the driver?

12. The projectile fired by the gun described in Problem 52 in Chapter 2 has a mass of 45 kg. What is the force on this projectile as it moves along the barrel?

13. A 70-g racquetball is accelerated from 0 to 30 m/s during an impact lasting 0.060 s. What is the average force experienced by the ball?

14. A Roman candle (a pyrotechnic device) of mass 35 g accelerates vertically from 0 to 22 m/s in 0.20 s. What is the average acceleration? What is the average force on the firework during this time? Neglect any loss in mass.

15. When the engine of a 240-kg motorboat is shut off, the boat slows from 15.0 m/s to 10.0 m/s in 1.2 s, and then from 10.0 m/s to 5.0 m/s in 2.1 s. What is the average acceleration during each of these intervals? What average frictional force does the water provide during each interval?

16. A tennis ball of mass 57 g is initially at rest. While being hit, it experiences an average force of 55 N during a 0.13-s interval. What is its final velocity?

17. An astronaut (with spacesuit) of mass 95 kg is tethered to a 750-kg satellite. By pulling on the tether, she accelerates toward the satellite at 0.50 m/s². What is the acceleration of the satellite toward the astronaut?

††*18. Figure 2.27 shows the plot of velocity vs. time for a Triumph sports car coasting along with its gears in neutral. The mass of the car is 1160 kg. From the values of the deceleration at the times $t = 0, 10, 20, 30,$ and 40 s [see Problem 39(b) in Chapter 2], calculate the friction force that the car experiences at these times. Make a plot of friction force vs. velocity.

*19. Table 2.3 gives the velocity of a projectile as a function of time. The projectile slows down because of the friction force exerted by the air. For the first 0.30-s time interval and for the last 0.20-s time interval, calculate the average friction force.

*20. A proton moving in an electric field has an equation of motion

$$\mathbf{r} = (5.0 \times 10^4 t)\mathbf{i} + (2.0 \times 10^4 t - 2.0 \times 10^5 t^2)\mathbf{j}$$
$$- (4.0 \times 10^5 t^2)\mathbf{k}$$

where distance is measured in meters and time in seconds. The proton has a mass of 1.7×10^{-27} kg. What are the components of the force acting on this proton? What is the magnitude of the force?

---

† For help, see Online Concept Tutorial 6 at www.wwnorton.com/physics

†† Due to the difficulty of accurately drawing tangents, answers for this problem that differ by up to 10% are acceptable.

*21. You launch a stone of mass 40 g horizontally with a slingshot. During the launch, the position of the stone is given by

$$x = x_0 [1 - \cos(bt)]$$

where $x_0 = 30$ cm is the position of the stone at the end of the launch and $b = 4\pi$ s^{-1} is a constant. The launch begins at time $t = 0$ and ends at $t = 0.125$ s. Ignore any vertical motion. During the launch, what is the velocity of the stone as a function of time? What are the acceleration and the force as functions of time? Rewrite the force in terms of the position x.

5.3 The Combination of Forces[†]

22. While braking, an automobile of mass 1200 kg decelerates along a level road at 7.8 m/s^2. What is the horizontal force that the road exerts on each wheel of the automobile? Assume all wheels contribute equally to the braking. Ignore the friction of the air.

23. In 1978, in an accident at a school in Harrisburg, Pennsylvania, several children lost parts of their fingers when a nylon rope suddenly snapped during a giant tug-of-war among 2300 children. The rope was known to have a breaking tension of 58000 N. Each child can exert a pull of approximately 130 N. Was it safe to employ this rope in this tug-of-war?

24. An 1800-kg barge is pulled via cables by two donkeys on opposite riverbanks. The (horizontal) cables each make an angle of 30° with the direction of motion of the barge. At low speed, ignore friction and determine the acceleration of the barge when each donkey exerts a force of 460 N on a cable.

25. Three rescuers are pulling horizontally on a safety net to keep it taut. One pulls northward with a force of 270 N; the second pulls in a direction 30° south of west with a force of 240 N. In which direction and with what force must the third pull to keep the net stationary?

26. Two soccer players kick a ball at the same instant. During the kick, one applies a force of 25 N in a direction 30° east of north and the other a force of 35 N in a direction 30° east of south. What are the magnitude and direction of the net force?

27. An ocean current applies a force of 2500 N to a 1400-kg sailboat in a direction 15° east of north. The wind applies a force of 3200 N in a direction 30° east of north. What are the magnitude and direction of the resulting acceleration?

28. A 5.0-kg mass has the following forces on it:

$$\mathbf{F}_1 = (4.0 \text{ N})\mathbf{i} + (3.0 \text{ N})\mathbf{j}$$

$$\mathbf{F}_2 = (2.0 \text{ N})\mathbf{i} - (5.0 \text{ N})\mathbf{j}$$

Find the magnitude and direction of the acceleration of this mass.

*29. The Earth exerts a gravitational pull of 2.0×10^{20} N on the Moon; the Sun exerts a gravitational pull of 4.3×10^{20} N on the Moon. What is the net force on the Moon when the angular separation between the Earth and the Sun is 90° as seen from the Moon?

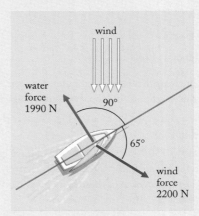

FIGURE 5.47 Forces on a sailboat. The angles are measured relative to the line of motion.

*30. A sailboat is propelled through the water by the combined action of two forces: the push ("lift") of the wind on the sail and the push of the water on the keel. Figure 5.47 shows the magnitudes and the directions of these forces acting on a medium-sized sailboat (this oversimplified diagram does not include the drag of wind and water). What is the resultant of the forces in Fig. 5.47?

*31. A boat is tied to a dock by four (horizontal) ropes. Two ropes, with a tension of 260 N each, are at right angles to the dock. Two other ropes, with a tension of 360 N each, are at an angle of 20° with the dock (Fig. 5.48). What is the resultant of these forces?

FIGURE 5.48 Ropes holding a boat at a dock.

**32. In a tug-of-war, a jeep of mass 1400 kg and a tractor of mass 2000 kg pull on a horizontal rope in opposite directions. At one instant, the tractor pulls on the rope with a force of 1.50×10^4 N while its wheels push horizontally against the ground with a force of 1.60×10^4 N. Calculate the instantaneous accelerations of the tractor and of the jeep; calculate the horizontal push of the wheels of the jeep. Assume the rope does not stretch or break.

[†] For help, see Online Concept Tutorial 7 at www.wwnorton.com/physics

5.4 Weight; Contact Force and Normal Force[†]
5.5 Newton's Third Law[†]

33. What is the mass of a laptop computer that weighs 25 N? What is the weight of a dictionary of mass 3.5 kg?

34. The surface gravity on Pluto is $0.045g$. What is the weight of a 60-kg woman on Pluto?

35. A man weighs 750 N on the Earth. The surface gravities of Mars and Jupiter are $0.38g$ and $2.53g$, respectively. What is the mass of the man on each of these planets? What is his weight on each planet?

36. What is the weight (in pounds-force) of a 1-lb bag of sugar in New York ($g = 9.803$ m/s^2)? In Hong Kong ($g = 9.788$ m/s^2)? In Quito ($g = 9.780$ m/s^2)?

37. A bar of gold of mass 500.00 g is transported from Paris ($g = 9.8094$ m/s^2) to San Francisco ($g = 9.7996$ m/s^2).

    (a) What is the decrease of the weight of the gold? Express your answer as a fraction of the initial weight.

    (b) Does the decrease of weight mean that the bar of gold is worth less in San Francisco?

38. A woman stands on a chair. Her mass is 60 kg, and the mass of the chair is 20 kg. What is the force that the chair exerts on the woman? What is the force that the floor exerts on the chair?

39. A chandelier of 10 kg hangs from a cord attached to the ceiling, and a second chandelier of 3.0 kg hangs from a cord below the first (see Fig. 5.49). Draw the "free-body" diagram for the first chandelier and the "free-body" diagram for the second chandelier. Find the tension in each of the cords.

FIGURE 5.49 Two chandeliers.

40. A mass $m = 200$ kg hangs from a horizontal ceiling by two cables, one of length 3.0 m and the other 4.0 m; the two cables subtend an angle of 90° at the mass. What is the tension in each cable?

41. You hold a cable from which hangs a first mass m_1. A second cable below it connects to a second hanging mass m_2, and a third cable below the second mass connects to a third hanging mass m_3. If you hold the system stationary, what upward force F do you apply? What is the tension in the first cable? What is the tension in the second cable? In the third? Assume massless cables.

42. A horizontal force F holds a block on a frictionless inclined plane in equilibrium. If the block has a mass $m = 5.0$ kg and the incline makes an angle of 50° with the horizontal, what is the value of F? What is the normal force exerted by the incline on the block?

43. In a bosun's chair, a cable from a seat runs up over a pulley and back down. If a sailor of mass M sits on the seat of the otherwise massless system, with what force must a second sailor pull downward on the free end of the cable to get the first sailor moving? If, instead, the seated sailor pulls, what force must he apply to pull himself upward?

44. A locomotive pulls a train consisting of three equal boxcars with constant acceleration along a straight, frictionless track. Suppose that the tension in the coupling between the locomotive and the first boxcar is 12 000 N. What is the tension in the coupling between the first and the second boxcar? The second and the third? Does the answer depend on the absence of friction?

*45. A long, thick cable of diameter d and density ρ is hanging vertically down the side of a building. The length of the cable is l. What tension does the weight of the cable produce at its upper end? At its midpoint?

46. A small truck of 2800 kg collides with an initially stationary automobile of 1200 kg. The acceleration of the truck during the collision is -500 m/s^2, and the collision lasts for 0.20 s. What is the acceleration of the automobile? What is the speed of the automobile after the collision? Assume that the frictional force due to the road can be neglected during the collision.

*47. Two heavy boxes of masses 20 kg and 30 kg sit on a smooth, frictionless surface. The boxes are in contact, and a horizontal force of 60 N pushes horizontally against the smaller box (Fig. 5.50). What is the acceleration of the two boxes? What is the force that the smaller box exerts on the larger box? What is the force that the larger box exerts on the smaller box?

FIGURE 5.50 Two boxes in contact.

*48. An archer pulls the string of her bow back with her hand with a force of 180 N. If the two halves of the string above and below her hand make an angle of 120° with each other, what is the tension in each half of the string?

[†] For help, see Online Concept Tutorial 6 and 7 at www.wwnorton.com/physics

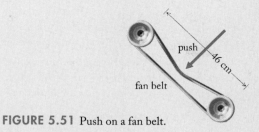

FIGURE 5.51 Push on a fan belt.

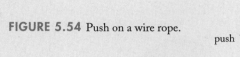

FIGURE 5.54 Push on a wire rope.

*49. A mechanic tests the tension in a fan belt by pushing against it with his thumb (Fig. 5.51). The force of the push is 130 N, and it is applied to the midpoint of a segment of belt 46 cm long. The lateral displacement of the belt is 2.5 cm. What is the tension in the belt (while the mechanic is pushing)?

*50. On a sailboat, a rope holding the foresail passes through a block (a pulley) and is made fast on the other side to a cleat (Fig. 5.52). The two parts of the rope make an angle of 140° with each other. The sail pulls on the rope with a force of 1.2×10^4 N. What is the force that the rope exerts on the block?

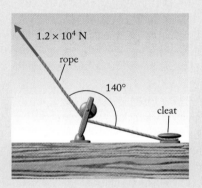

FIGURE 5.52 The left end of the rope is attached to the sail, and the right end is attached to a cleat.

*51. In order to pull an automobile out of the mud in which it is stuck, the driver stretches a rope tautly from the front end of the automobile to a stout tree. He then pushes sideways against the rope at the midpoint (see Fig. 5.53). When he pushes with a force of 900 N, the angle between the two halves of the rope on his right and his left is 170° . What is the tension in the rope under these conditions?

**52. A sailor tests the tension in a wire rope holding up a mast by pushing against the rope with his hand at a distance l from the lower end of the rope. When he exerts a transverse push N, the wire rope suffers a transverse displacement s (Fig. 5.54).

(a) Show that for $s \ll l$ the tension in the wire rope is given approximately by the formula

$$T = Nl/s$$

In your calculation, assume that the distance to the upper end of the rope is effectively infinite; i.e., the total length of the rope is much larger than l.

(b) What is the tension in the rope when it suffers a transverse displacement of 2.0 cm under a force of 150 N applied at a distance of 1.5 m from the lower end?

*53. On a windy day, a small tethered balloon is held by a long string making an angle of 70° with the ground. The vertical buoyant force on the balloon (exerted by the air) is 67 N. During a sudden gust of wind, the (horizontal) force of the wind is 200 N; the tension in the string is 130 N. What are the magnitude and direction of the force on the balloon?

*54. A horse, walking along the bank of a canal, pulls a barge. The horse exerts a pull of 300 N on the barge at an angle of 30° (Fig. 5.55). The bargeman relies on the rudder to steer the barge on a straight course parallel to the bank. What transverse force (perpendicular to the bank) must the rudder exert on the barge?

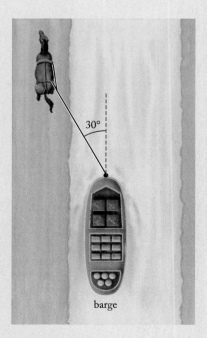

FIGURE 5.55 Horse pulling a barge.

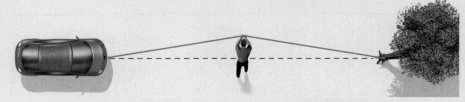

FIGURE 5.53 The rope is stretched between the automobile and a tree. The man pushes at the midpoint.

**55. A flexible massless rope is placed over a cylinder of radius R. A tension T is applied to each end of the rope, which remains stationary (see Fig. 5.56). Show that each small segment $d\theta$ of the rope in contact with the cylinder pushes against the cylinder with a force $T\,d\theta$ in the radial direction. By integration of the forces exerted by all the small segments, show that the net vertical force on the cylinder is $2T$ and the net horizontal force is zero.

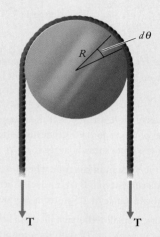

FIGURE 5.56 Rope and cylinder.

5.6 Motion with a Constant Force†

56. An elevator accelerates upward at 1.8 m/s^2. What is the normal force on the feet of an 80-kg passenger standing in the elevator? By how much does this force exceed his weight?

57. A parachutist of mass 80 kg approaches the ground at 5.0 m/s. Suppose that when he hits the ground, he decelerates at a constant rate (while his legs buckle under him) over a distance of 1.0 m. What is the force the ground exerts on his feet during the deceleration?

58. If the elevator described in Example 10 carries four passengers of 70 kg each, what speed will the elevator attain running down freely from a height of 10 m, starting from rest?

59. You lift a cable attached to a first mass m_1. A second cable below it connects to a second mass m_2, and a third cable below the second mass connects to a third mass m_3. If you apply an upward force F, what is the tension in the first cable? What is the tension in the second cable? In the third? What is the acceleration of the system? Assume massless cables.

60. Two adjacent blocks of mass $m_1 = 3.0 \text{ kg}$ and $m_2 = 4.0 \text{ kg}$ are on a frictionless surface. A force of 6.0 N is applied to m_1, and a force of 4.0 N is applied to m_2; these antiparallel forces squeeze the blocks together. What is the force due to m_1 on m_2? Due to m_2 on m_1? What is the acceleration of the system?

61. A 220-lb man stands on a scale in an elevator. What does the scale read when the elevator accelerates upward at 1.6 m/s^2? What does it read when accelerating downward at the same rate?

62. You are on an elevator holding your backpack by a single loose strap. The backpack and its contents have a mass of 9.5 kg. What is the tension in the strap when the elevator accelerates upward at 1.9 m/s^2?

63. Each of a pair of dice hangs from the rearview mirror of a car by a string. When the car accelerates forward at 2.5 m/s^2, the strings make an angle with the vertical. What is the angle? If each die has a mass of 25 g, what is the tension in the string?

64. A horizontal force $F = 25.0 \text{ N}$ attempts to push a block up a frictionless inclined plane. If the block has a mass $m = 3.50 \text{ kg}$ and the incline makes an angle of $50.0°$ with the horizontal, does the force succeed? What is the acceleration of the block along the incline?

65. A boy on a skateboard rolls down a hill of slope 1:5. What is his acceleration? What speed will he reach after rolling for 50 m? Ignore friction.

66. A skier of mass 75 kg is sliding down a frictionless hillside inclined at $35°$ to the horizontal.

  (a) Draw a "free-body" diagram showing all the forces acting on the skier (regarded as a particle); draw a separate diagram showing the resultant of these forces.

  (b) What is the magnitude of each force? What is the magnitude of the resultant?

  (c) What is the acceleration of the skier?

67. A bobsled slides down an icy track making an angle of $30°$ with the horizontal. How far must the bobsled slide in order to attain a speed of 90 km/h if initially at rest? When will it attain this speed? Assume that the motion is frictionless.

68. A man carrying a 20-kg sack on his shoulder rides in an elevator. What is the force the sack exerts on his shoulder when the elevator is accelerating upward at 2.0 m/s^2?

*69. Figure 5.57 shows a spherical ball hanging on a string on a smooth, frictionless wall. The mass of the ball is m, its radius is R, and the length of the string is l. Draw a "free-body" diagram with all the forces acting on the ball. Find the normal force between the ball and the wall. Show that $N \to 0$ as $l \to \infty$.

FIGURE 5.57
Ball on string against wall.

70. Figure 5.58 shows two masses hanging from a string running over a pulley. Such a device can be used to measure the acceleration of gravity; it is then called Atwood's machine. If the masses are nearly equal, then the acceleration a of the masses will be much smaller than g; that makes it convenient to measure a and then to calculate g by means of Eq. (5.44). Suppose that an experimenter using masses $m_1 = 400.0$ g and $m_2 = 402.0$ g finds that the masses move a distance of 0.50 m in 6.4 s starting from rest. What value of g does this imply? Assume the pulley is massless.

(a)

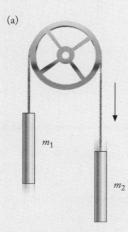

FIGURE 5.58 Two masses and a pulley.

*71. A woman pushes horizontally on a wooden box of mass 60 kg sitting on a frictionless ramp inclined at an angle of 30° (see Fig. 5.59).

(a) Draw the "free-body" diagram for the box.

(b) Calculate the magnitudes of all the forces acting on the box under the assumption that the box is at rest or in uniform motion along the ramp.

30°

FIGURE 5.59 Woman pushing box.

*72. During takeoff, a jetliner is accelerating along the runway at 1.2 m/s². In the cabin, a passenger holds a pocket watch by a chain (a plumb). Draw a "free-body" diagram with the forces acting on the watch. What angle will the chain make with the vertical during this acceleration?

*73. In a closed subway car, a girl holds a helium-filled balloon by a string. While the car is traveling at constant velocity, the string of the balloon is exactly vertical. (Hint: For constant velocity, a buoyant force acts upward, but for constant deceleration, the buoyant force becomes tilted, but remains antiparallel to the tension.)

(a) While the subway car is braking, will the string be inclined forward or backward relative to the car?

(b) Suppose that the string is inclined at an angle of 20° with the vertical and remains there. What is the acceleration of the car?

**74. A string passes over a frictionless, massless pulley attached to the ceiling (see Fig. 5.60). A mass m_1 hangs from one end of this string, and a second massless, frictionless pulley hangs from the other end. A second string passes over the second pulley, and a mass m_2 hangs from one end of the string, whereas the other end is attached firmly to the ground. Draw separate "free-body" diagrams for the mass m_1, the second pulley, and the mass m_2. Find the accelerations of the mass m_1, the second pulley, and the mass m_2.

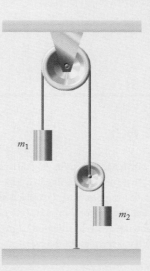

FIGURE 5.60 Two masses and two pulleys.

**75. A particle sliding down a frictionless ramp is to attain a given *horizontal* displacement Δx in a minimum amount of time. What is the best angle for the ramp? What is the minimum time?

**76. A mass m_1 hangs from one end of a string passing over a frictionless, massless pulley. A second frictionless, massless pulley hangs from the other end of the string (Fig. 5.61). Masses m_2 and m_3 hang from a second string passing over this second pulley. Find the acceleration of the three masses, and find the tensions in the two strings.

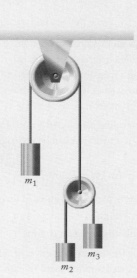

FIGURE 5.61 Three masses and two pulleys.

**77. A flexible cable of length *l* and mass *m* hangs over a small pulley. Initially, the cable is at rest, the length of the cable hanging on one side is more than $l/2$ by x_0, and the length of the cable hanging on the other side is less than $l/2$ by x_0 (see Fig. 5.62). What is the acceleration of the cable as a function of the distance *x* measured from the position of equal lengths? What is the position of the end of the cable as a function of time? (Hint: The differential equation $d^2x/dt^2 = kx$ has the solution $x = Ae^{\sqrt{k}t}$.)

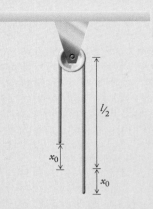

FIGURE 5.62 Massive cable and pulley.

REVIEW PROBLEMS

78. A 2000-kg truck, equipped with a winch, is trying to pull a car across a slippery (frictionless) sheet of ice by reeling in a cable attached to the car (see Fig. 5.63). Suppose that the truck is also on the sheet of ice, so the truck and the car both slip and accelerate toward each other with instantaneous accelerations of 1.2 m/s² and 2.5 m/s², respectively. What is the mass of the car?

FIGURE 5.63 Car and tow truck on frictionless surface.

79. Pushing with both hands, a sailor standing on a pier exerts a horizontal force of 270 N on a destroyer of 3400 metric tons. Assuming that the mooring ropes do not interfere and that the water offers no resistance, what is the acceleration of the ship? How far does the ship move in 60 s?

80. The speed of a projectile traveling horizontally and slowing down under the influence of air friction can be approximately represented by

$$v = 655.9 - 61.14t + 3.260t^2$$

where *v* is measured in meters per second and *t* in seconds; the mass of the projectile is 45.36 kg.

(a) What is the acceleration as a function of time?

(b) What is the force of air friction as a function of time? Can these formulas for the speed and the force remain valid when the speed becomes low ($v \approx 0$)?

81. Two forces $\mathbf{F}_1$ and $\mathbf{F}_2$ act on a particle of mass 6.0 kg. The forces are

$$\mathbf{F}_1 = 2\mathbf{i} - 5\mathbf{j} + 3\mathbf{k}$$
$$\mathbf{F}_2 = -4\mathbf{i} + 8\mathbf{j} + \mathbf{k}$$

where the force is measured in newtons.

(a) What is the net force vector?

(b) What is the acceleration vector of the particle, and what is the magnitude of the acceleration?

82. A diver of mass 75 kg is in free fall after jumping off a high platform.

(a) What is the force that the Earth exerts on the diver? What is the force that the diver exerts on the Earth?

(b) What is the acceleration of the diver? What is the acceleration of the Earth?

83. During a storm, a 2500-kg sailboat is anchored in a 10-m-deep harbor. The wind pushes against the boat with a horizontal force of 7000 N. The anchor rope that holds the boat in place is 50 m long and is stretched straight between the anchor and the boat (see Fig. 5.64).

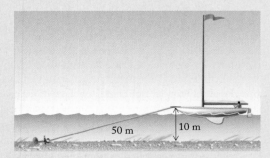

FIGURE 5.64 Anchored boat.

(a) Draw the "free-body" diagram for the boat. Be sure to include the upward force (buoyant force) that the water exerts on the boat, keeping it afloat.

(b) Calculate the tension in the anchor rope.

(c) Calculate the upward force (buoyant force) exerted by the water.

84. A box of mass 25 kg sits on a smooth, frictionless table. You push down on the box at an angle of 30° with a force of 80 N (see Fig. 5.65).

(a) Draw a "free-body" diagram for the box; include all the forces that act on the box.

(b) What is the acceleration of the box?

(c) What is the normal force that the table exerts on the box?

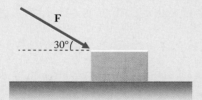

FIGURE 5.65 Pushing on a box.

85. A boy and a girl are engaged in a tug-of-war. Suppose that they are evenly matched, and neither moves. The boy pulls toward the left on the rope with a force of 250 N, and the girl pulls toward the right with a force of 250 N.

(a) Draw separate "free-body" diagrams for the boy, the girl, and the rope. In each of these diagrams include all the appropriate *horizontal* forces (ignore the vertical forces).

(b) What is the force that the ground exerts on the boy? On the girl?

(c) What is the tension in the rope?

(d) The girl ties her end of the rope to a stout tree and walks away, while the boy continues pulling as before. What is the tension in the rope now?

86. A long freight train consists of 250 cars each of mass 64 metric tons. The pull of the locomotive accelerates this train at the rate of 0.043 m/s² along a level track. What is the tension in the coupling that holds the first car to the locomotive? What is the tension in the coupling that holds the last car to the next-to-last car? Ignore friction.

87. The world's steepest railroad track, found in Guatemala, has a slope of 1:11 (see Fig. 5.66). A boxcar of 20 metric tons is being pulled up this track.

(a) Draw a "free-body" diagram for the boxcar.

(b) What force (along the track) is required to move the boxcar up the track at constant speed? Ignore friction and treat the motion as particle motion.

(c) What force is required to move the boxcar down the track at constant speed?

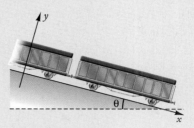

FIGURE 5.66 Boxcar on a steep railroad track.

88. To drag a heavy log of mass 500 kg up a ramp inclined at 30° to the horizontal, you attach the log to a cable that runs over a pulley at the top of the ramp, and you attach a 300-kg counterweight at the other end (see Fig. 5.67). Assume the log moves without friction.

(a) What is the acceleration of the log up the ramp?

(b) Could you use a smaller counterweight to move the log? How much smaller?

FIGURE 5.67 A log on a ramp with pulley and counterweight.

89. A 60-kg woman stands on a bathroom scale placed on the floor of an elevator.

(a) What does the scale read when the elevator is at rest?

(b) What does the scale read when the elevator is accelerating upward at 1.8 m/s²?

(c) What does the scale read when the elevator is moving upward with constant velocity?

(d) What does the scale read if the cable of the elevator is cut (and the brakes have not yet engaged), so that the elevator is in free fall?

90. A crate of mass 2000 kg is hanging from a crane at the end of a cable 12 m long. If we attach a horizontal rope to this crate and gradually apply a pull of 1800 N, what angle will the cable finally make with the vertical?

*91. While a train is moving up a track of slope 1:10 at 50 km/h, the last car of the train suddenly becomes uncoupled. The car continues to roll up the slope for a while, then stops and rolls back.

(a) What is the deceleration of the car while it continues to roll up the slope? Assume there is no friction.

(b) How far along the track will the car roll before it stops and begins to roll back?

(c) What speed does the car attain when it has rolled back to the place where it first decoupled?

*92. An elevator consists of an elevator cage and a counterweight attached to the ends of a cable which runs over a pulley (see Fig. 5.68). The mass of the cage (with its load) is 1200 kg, and the mass of the counterweight is 1000 kg. Suppose that the elevator is moving upward at 1.5 m/s when its motor fails. The elevator then continues to coast upward until it stops and begins to fall down. Assume there is no friction, and assume the emergency brake of the elevator does not engage until the elevator begins to fall down. The brake locks the elevator cage to its guide rails.

(a) What is the deceleration of the elevator? How long does it take to stop?

(b) What is the tension in the cable when the elevator is coasting up?

(c) What is the tension in the cable when the emergency brake has engaged and the elevator is stopped?

FIGURE 5.68 Elevator with counterweight.

93. A mass m_1 sits on a horizontal, frictionless table, and is connected by a light string to a second, hanging mass m_2, as shown in Fig 5.69. (a) Find the acceleration of the system. (b) Find the tension in the string.

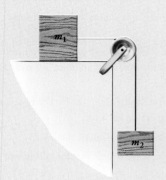

FIGURE 5.69 Mass m_1 on table with pulley and counterweight m_2.

94. Two masses m_1 and m_2 sit on a horizontal, frictionless table and are connected by a light string. Another light string connects the mass m_2 to a third, hanging mass m_3, as shown in Fig. 5.70.

(a) Find the acceleration of the system.

(b) Find the tension in each of the two strings.

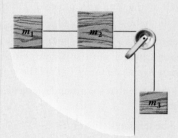

FIGURE 5.70 Two masses on table with pulley and counterweight.

Answers to Checkups

Checkup 5.1

1. No; the street exerts a friction force on the car's tires (the tires are both a source of friction and transfer friction forces from the axle to the road), and so an external pushing force is needed to counter it to maintain zero net external force.

2. No; the surface exerts friction forces on the ball that both roll and slow the ball. Thus, the ball does not have zero net external force on it, and so Newton's First Law does not apply.

3. Yes, the reference frame of a car traveling at constant speed along a straight, level road is an inertial reference frame (as

usual, we neglect the Earth's rotation). A car rounding a curve is accelerating centripetally, and so is not an inertial reference frame; similarly, a car braking on a level road is decelerating, and so is not an inertial reference frame.

4. Yes—while descending at constant speed, the elevator is inertial, since it moves at constant speed relative to an inertial reference frame, the Earth.

5. No—the diver is accelerating downward with $a_y = -g$, and so the reference frame of the diver is not an inertial reference frame.

6. (B) The reference frame of a bird descending at constant velocity. The other reference frames all involve accelerated motion; only the bird moves at constant velocity.

Checkup 5.2

1. You could push or pull each one horizontally with the same force; the one that accelerates more slowly has the greater mass.

2. In each case, the force is in the same direction, in the automobile's forward direction. The forces are merely applied to different points in each case.

3. No, the force is not an external force if the sailor is standing in the sailboat. The sailor's push on the mast is balanced by the push of his feet on the boat; wholly internal forces always cancel.

4. (C) 8, 2, 4. We assume that by "push on it equally" we mean each of the two men applies the same force. If two men push on the bobsled, the force is twice as large. By Newton's Second Law, $m\mathbf{a} = \mathbf{F}$, if the force is twice as large and the mass is the same, the acceleration will be twice as large, or 8 m/s². If the mass is twice as large, then the acceleration provided by a single man's force will be half as large, or 2 m/s²; if both men push the doubled mass, then the doubling of both the force and the mass will leave the acceleration the same at 4 m/s².

Checkup 5.3

1. You can conclude that the net force is zero, since constant velocity means zero acceleration, and zero acceleration requires zero *net* force (Newton's Second Law). However, you cannot conclude that there are no forces: in this case of the parachutist, the downward force of gravity is balanced by the upward force of air friction.

2. When the elevator is at rest, there is zero net force on it. Only when the elevator is accelerating is there a net force; when accelerating upward, the net force is upward. When the elevator ascends at constant speed (zero acceleration), there is again zero net force on it.

3. When the subway car accelerates (and causes you to do so), the net external force on you is forward (even though you feel yourself being pushed backward in the reference frame of the

car, this is merely your inertia trying to keep you at rest). When the car travels at constant velocity (zero acceleration), the net force on you is then zero ($ma = F$). During braking (deceleration), a negative acceleration means the net force on you is toward the rear of the car (even though you feel yourself lurch forward in the reference frame of the car, this is your inertia trying to maintain a constant velocity).

4. If both tugboats pull directly forward, then the magnitudes of their forces simply add (the two vectors are parallel), so that the net force is 2.5×10^5 N $+ 1.0 \times 10^5$ N $= 3.5 \times 10^5$ N. If one pulls from the front and the other pushes from the rear, the two force vectors are still parallel, so the net force is again 3.5×10^5 N.

5. (B) 30° north of west. Since the two forces are of equal magnitude, the net force, and thus the acceleration, will be in a direction halfway between the two. Halfway between 30° west of north and west (which is 90° west of north) is 60° west of north, the same as 30° north of west.

Checkup 5.4

1. The force due to gravity is the stone's weight and, whether the stone is traveling up, or momentarily at rest, or traveling down, is always equal to $F = w = mg = 1.0 \text{ kg} \times 9.8 \text{ m/s}^2 = 9.8$ N in the downward direction.

2. No, the star has mass (inertia), and so will offer resistance to any attempt to accelerate it. That it is weightless reflects the fact that there are no external bodies exerting any measurable gravitational force on it.

3. Since weight is $w = mg$, your weight would be largest where the local value of the acceleration due to gravity, g, is largest, namely, Jupiter; your weight is smallest where g is smallest, on the Moon.

4. Apparent weightlessness will begin when the engines cut off (no external force from the engines' thrust); this is when the spacecraft begins coasting, that is, when free fall begins.

5. For the first book alone, the table exerts a normal force that balances the weight of the book, and so is 50 N upward. If a second, identical book is placed on top of the first, then the first book experiences two downward forces: its own weight (50 N) and a normal force on its top surface from the second book (50 N). Thus, the table exerts a normal force that balances both of these forces, and so is now 100 N upward.

6. (D) Smaller; larger. As the upward-traveling elevator decelerates, the passenger experiences a downward acceleration. Since her weight (a downward force) does not change, the normal force of the floor on her feet (an upward force) must be smaller than her weight to have a net downward force (and thus a net downward acceleration). For the downward-traveling elevator, the deceleration implies an upward acceleration. Thus the normal force (upward) is now larger than the passenger's weight.

Checkup 5.5

1. For an apple in free fall, your "free-body" diagram should show only the apple with its weight **w** as the only force acting on it (downward, acting at its center of mass). For an apple resting on a table, your diagram should show two forces: the apple's weight **w** acting downward at its center of mass, and the normal force of the table **N**, acting upward on the bottom of the apple; these two forces have the same magnitude (the two force vectors have the same length).

2. The reaction force to the apple's weight is the pull of gravity exerted by the apple on the Earth. The reaction force to the upward pull of the tree branch is the downward pull exerted by the apple on the tree branch.

3. During acceleration, the external force acts in the forward direction; while traveling at constant velocity, there is zero net external force on the car (no direction); while braking, the external force acts rearward on the car. The ground (the road) exerts these external forces on the tires of the car.

4. In both the cases of the supertanker and the rowboat, the vessels push backward against your feet with the same reaction force of 400 N. Because the supertanker has a much larger mass, your push of 400 N causes only a negligible acceleration, while for the much less massive rowboat, the same push of 400 N will cause a noticeable acceleration ($a = F/m$).

5. The top book has two forces acting upon it: its weight $\mathbf{w}_1$ downward and a normal force $\mathbf{N}_1$ of the lower book acting upward on the upper book; these two forces have equal magnitudes, since the book is not accelerating. The lower book has three forces acting on it: its downward weight $\mathbf{w}_2$, a downward normal force from the top book $\mathbf{N}_2$, and an upward normal force from the table $\mathbf{N}_3$; the two downward forces sum to the same magnitude as the upward force. Only the two normal forces of each book on the other are an action–reaction pair; that is, $\mathbf{N}_2 = -\mathbf{N}_1$.

6. If the man pulls with a force of 150 N on the rope, the rope pulls back on the man with the same force, and this is the tension in the rope, 150 N; the same tension acts at the woman's end. If the woman ties her end to a tree, the tension continues to be the same, since it still forms an action–reaction pair with the man's pull.

7. (D) 98 N; 98 N; 98 N. The tension T in the first, lower rope forms an action–reaction pair with the weight of the 10-kg mass, and so is $T = -w = -(-mg) = 10 \text{ kg} \times 9.8 \text{ m/s}^2 = 98 \text{ N}$. Since the spring scale and the ropes are assumed massless, this tension is transmitted in full to the second (upper) rope, which then also has a tension of 98 N. The spring scale is set up to measure weight; since there is no net acceleration, it measures the actual weight of 98 N (we will examine the behavior of springs in detail in the next chapter).

Checkup 5.6

1. In the first part of the motion (low speed), friction is negligible ("she falls freely"), so the only force on the skydiver is the downward force due to her weight, $w = mg$. This is then the net force, and is nonzero. Later, reaching a uniform, terminal velocity implies zero acceleration; thus there is also zero net force. At that time, the force due to her weight (downward) is balanced by the force of air friction (upward).

2. In each case (zero or uniform velocity), there is zero acceleration, so there is zero net force. Thus, the upward tension exactly balances the weight, and the magnitude of the tension is $T = mg = 1000 \text{ kg} \times 9.81 \text{ m/s}^2 = 9810 \text{ N}$ in each case.

3. The acceleration may be changed, and the normal force is certainly changed. By pulling upward on the front, the vertical component of the force from the man is now opposite to its direction when pushing downward. If the vertical component of the pull is less than the weight, the normal force is reduced to a value equal to the difference between the weight and the vertical pull, and the vertical acceleration remains zero; if the vertical pull is greater than the weight, the normal force is zero, and the box accelerates vertically off the floor. In either case, the horizontal acceleration remains the same.

4. (a) Moving at constant speed up or down the plane implies no acceleration, so the force remains zero, and the horizontal push needed is the same as when the block was stationary, 100 N in each case. (b) If you push along the incline instead of horizontally, then more of your total push acts parallel to the motion (none merely opposes the normal force), so your push will have to be smaller to maintain zero net force. To calculate the value of push needed to balance the component of the weight along the incline, you would need to know the mass of the block and the angle θ of the incline [see Eq. (5.33)].

5. (D) b, c, a. For (1), the book diagram should have a weight vector downward, acting at the center of mass, and the normal force vector upward, on the surface in contact with the table; these two vectors are equal and opposite, as in (b). For (2), the box should also have a weight downward at the center of mass and an upward normal force on the bottom surface; in addition, there is another normal force from the second box downward on the top surface, as in (c). For (3), the lamp diagram should have a weight vector downward, acting at the center of mass, and a tension vector upward, acting at the point where the cord attaches to the lamp; the two vectors are equal and opposite, as in (a).

Further Applications of Newton's Laws

CONCEPTS IN CONTEXT

Concepts — in — Context

Automobiles rely on the friction between the road and the tires to accelerate and to stop. We will see that one of two types of contact friction, kinetic or static, is involved. To see how these friction forces affect linear and circular motion, we ask:

? In an emergency, an automobile brakes with locked and skidding wheels. What deceleration can be achieved? (Example 1, page 176)

? What is the steepest slope of a street on which an automobile can rest without slipping? (Example 4, page 179)

? When braking without skidding, what maximum deceleration can be achieved? (Example 5, page 180)

? How quickly can a racing car round a curve without skidding sideways? (Example 10, page 186)

? How does a banked curve help to avoid skidding? (Example 11, page 186)

6.1   Friction

6.2   Restoring Force of a Spring; Hooke's Law

6.3   Force for Uniform Circular Motion

6.4   The Four Fundamental Forces

To find a solution of the equation of motion means to find a force **F** and a corresponding acceleration **a** such that Newton's equation $m\mathbf{a} = \mathbf{F}$ is satisfied. For a physicist, the typical problem involves a known force and an unknown motion; for example, the physicist knows the forces between the planets and the Sun, and she seeks to calculate the motion of these bodies. But for an engineer, the reverse problem with a known motion and an unknown force is often of practical importance; for example, the engineer knows that a train is to round a given curve at 60 km/h, and he seeks to calculate the forces that the track and the wheels must withstand. A special problem with a known motion is the problem of statics; here we know that the body is at rest (zero velocity and zero acceleration), and we wish to compute the forces that will maintain this condition of equilibrium. Thus, depending on the circumstances, we can regard either the right side or the left side of the equation $m\mathbf{a} = \mathbf{F}$ as an unknown that is to be calculated from what we know about the other side.

In the preceding chapter we found some solutions of the equation of motion with simple, constant forces, such as the weight and constant pushes or pulls. In this chapter we will examine further solutions of the equation of motion, and we will examine other, more complicated forces, such as friction and the forces exerted by springs.

6.1 FRICTION

Friction forces, which we have ignored up to now, play an important role in our environment and provide us with many interesting examples of motion with constant force. For instance, if the driver of a moving automobile suddenly slams on the brakes, the wheels will lock and begin to skid on the pavement. The skidding wheels experience an (approximately) constant friction force that opposes the motion and decelerates the automobile at an (approximately) constant rate of, say, 8 m/s². The magnitude of the friction force depends on the characteristics of the tires and the pavement; besides, the heavy friction of rubber wheels on a typical pavement is accompanied by abrasion of the wheels, which introduces additional complications.

For the sake of simplicity, let us focus on an idealized case of friction, involving a solid block of metal sliding on a flat surface of metal. Figure 6.1 shows a block of steel, in the shape of a brick, sliding on a tabletop of steel. If we give the block some initial velocity and then let it coast, friction will decelerate it. The forces acting on the block are the weight **w**, the normal force **N**, and the friction force **f**. The weight **w** acts downward with a magnitude mg. The normal force **N** exerted by the table on the block acts upward; the magnitude of this normal force must be mg, so that it balances the weight. The friction force **f** exerted by the table on the block acts horizontally, parallel to the tabletop, in a direction opposite to the motion. This force, like the normal force, is a contact force which acts over the entire bottom surface of the block; however, in Fig. 6.1 it is shown as though acting at the center of the surface.

The friction force arises from adhesion between the two pieces of metal: the atoms in the block form bonds with the atoms in the tabletop, and when the block slides, these bonds are continually ruptured and formed again. The macroscopic friction force represents the effort required to rupture the microscopic bonds. Although at the microscopic level the phenomenon of friction is very complicated, at the macroscopic level the resulting friction force can often be described adequately by a simple empirical law, first enunciated by Leonardo da Vinci:

Online
Concept
Tutorial

LEONARDO da VINCI (1452–1519)
Italian artist, engineer, and scientist. Famous for his brilliant achievements in painting, sculpture, and architecture. Leonardo also made pioneering contributions to science. But Leonardo's investigations of friction were forgotten, and the laws of friction were rediscovered 200 years later by Guillaume Amontons, a French physicist.

The magnitude of the friction force between unlubricated, dry surfaces sliding one over the other is proportional to the magnitude of the normal force acting on the surfaces and is independent of the area of contact and of the relative speed.

Friction involving surfaces in relative motion is called **sliding friction**, or **kinetic friction**. According to the above law, the magnitude of the force of kinetic friction can be written mathematically as

$$f_k = \mu_k N \tag{6.1}$$

where μ_k is the **coefficient of kinetic friction**, a constant characteristic of the material involved. Table 6.1 lists typical friction coefficients for various materials.

Note that Eq. (6.1) states that the magnitudes of the friction force and the normal force are proportional. The *directions* of these forces are, however, quite different: the normal force **N** is perpendicular to the surface of contact, whereas the friction force $\mathbf{f}_k$ is parallel to this surface, in a direction opposite to that of the motion.

The above simple "law" of friction lacks the general validity of, say, Newton's laws. It is only approximately valid, and it is phenomenological, which means that it is merely a descriptive summary of empirical observations which does not rest on any detailed theoretical understanding of the mechanism that causes friction. Deviations from this simple law occur at high speeds and at low speeds. However, we can ignore these deviations in many everyday engineering problems in which the speeds are not extreme. The simple friction law is then a reasonably good approximation for a wide range of materials, and it is at its best for metals sliding on metals.

The fact that the friction force is independent of the area of contact means that the friction force of the block sliding on the tabletop is the same whether the block slides on a large face or on one of the small faces (see Fig. 6.2). This may seem surprising at first—we might expect the friction force to be larger when the block slides on the larger face, with more area in contact with the tabletop. However, the normal force is then distributed over a larger area, and is therefore less effective in pressing the atoms together; and the net result is that the friction force is independent of the area of contact.

force of kinetic friction

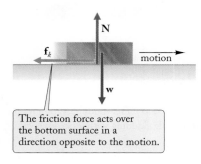

The friction force acts over the bottom surface in a direction opposite to the motion.

FIGURE 6.1 Forces on a block sliding on a plate.

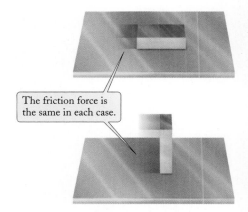

The friction force is the same in each case.

FIGURE 6.2 Steel block on a steel plate, sliding on a large face or on a small face.

TABLE 6.1	KINETIC AND STATIC FRICTION COEFFICIENTS[a]	
MATERIALS	μ_k	μ_s
Steel on steel	0.6	0.7
Steel on lead	0.9	0.9
Steel on copper	0.4	0.5
Copper on cast iron	0.3	1.1
Copper on glass	0.5	0.7
Waxed ski on snow		
at $-10°C$	0.2	—
at $0°C$	0.05	—
Rubber on concrete	≈ 1	≈ 1

[a] The friction coefficient depends on the condition of the surfaces. The values in this table are typical for dry surfaces but not entirely reliable.

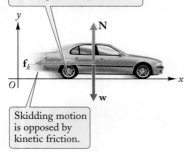

A kinetic friction force acts on each wheel, but diagram shows these forces combined in a single force $\mathbf{f}_k$.

Skidding motion is opposed by kinetic friction.

FIGURE 6.3 "Free-body" diagram for an automobile skidding with locked wheels.

Concepts
—in—
Context

EXAMPLE 1   Suppose that the coefficient of kinetic friction of the hard rubber of an automobile tire sliding on the pavement of a street is $\mu_k = 0.8$. What is the deceleration of an automobile on a flat street if the driver brakes sharply, so all the wheels are locked and skidding? (Assume the vehicle is an economy model without an antilock braking system.)

SOLUTION: Figure 6.3 shows the "free-body" diagram with all the forces on the automobile. These forces are the weight $\mathbf{w}$, the normal force $\mathbf{N}$ exerted by the street, and the friction force $\mathbf{f}_k$. The normal force must balance the weight; hence the magnitude of the normal force is the same as the magnitude of the weight, or $N = w = mg$. According to Eq. (6.1), the magnitude of the friction force is then

$$f_k = \mu_k N = 0.8 \times mg$$

Since this friction force is the only horizontal force on the automobile, the deceleration of the automobile along the street is

$$a_x = -\frac{f_k}{m} = -\frac{0.8 \times mg}{m} = -0.8 \times g = -0.8 \times 9.8 \text{ m/s}^2$$
$$= -8 \text{ m/s}^2$$

COMMENT: The normal forces and the friction forces act on all the four wheels of the automobile; but in Fig. 6.3 (and in other "free-body" diagrams in this chapter) these forces have been combined into a net force $\mathbf{N}$ and a net friction force f_k, which, for convenience, are shown as though acting at the center of the automobile. To the extent that the motion is treated as purely translational motion (that is, particle motion), it makes no difference at what point of the automobile the forces act. Later, in Chapter 13, we will study how forces affect the rotational motion of bodies, and it will then become important to keep track of the exact point at which each force acts.

EXAMPLE 2   A ship is launched toward the water on a slipway making an angle of 5° with the horizontal direction (see Fig. 6.4). The coefficient of kinetic friction between the bottom of the ship and the slipway is $\mu_k = 0.08$. What is the acceleration of the ship along the slipway? What is the speed of the ship after accelerating from rest through a distance of 120 m down the slipway to the water?

SOLUTION: Figure 6.4b is the "free-body" diagram for the ship. The forces shown are the weight $\mathbf{w}$, the normal force exerted by the slipway $\mathbf{N}$, and the friction force $\mathbf{f}_k$. The magnitude of the weight is $w = mg$.

Since there is no motion in the direction perpendicular to the slipway, we find, as in Eq. (5.36), that the normal force is

$$N = mg \cos \theta$$

and the magnitude of the friction force is

$$f_k = \mu_k N = \mu_k mg \cos \theta \tag{6.2}$$

With the x axis parallel to the slipway, the x component of the weight is (see Fig. 6.4c)
$$w_x = mg \sin \theta$$

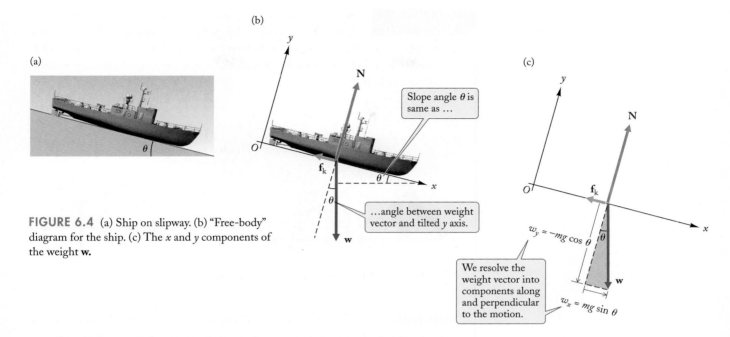

FIGURE 6.4 (a) Ship on slipway. (b) "Free-body" diagram for the ship. (c) The x and y components of the weight **w**.

The x component of the net force is then

$$F_x = w_x - f_k = mg \sin\theta - \mu_k mg \cos\theta$$

Hence the acceleration of the ship along the slipway is

$$a_x = \frac{F_x}{m} = \frac{mg \sin\theta - \mu_k mg \cos\theta}{m} = (\sin\theta - \mu_k \cos\theta)g \qquad (6.3)$$

Note that in this equation the mass has canceled—the acceleration is the same for a large ship and a small ship. With $\theta = 5°$ and $\mu_k = 0.08$, Eq. (6.3) gives

$$a_x = (\sin 5° - 0.08 \times \cos 5°) \times 9.81 \text{ m/s}^2 = 0.07 \text{ m/s}^2$$

From kinematics, Eqs. (4.20)–(4.24), we know that the velocity and displacement after constant acceleration from rest ($v_{0x} = 0$) will be

$$v_x = a_x t$$

$$x - x_0 = \tfrac{1}{2}a_x t^2$$

We can solve for the time t in the second equation:

$$t = \sqrt{\frac{2(x - x_0)}{a_x}}$$

and substitute it into the first:

$$v_x = a_x \times \sqrt{\frac{2(x - x_0)}{a_x}} = \sqrt{2(x - x_0)a_x}$$

$$= \sqrt{2 \times 120 \text{ m} \times 0.07 \text{ m/s}^2} = 4 \text{ m/s}$$

This is the speed of the ship as it enters the water.

EXAMPLE 3 A man pushes a heavy crate over a floor. The man pushes downward and forward, so his push makes an angle of 30° with the horizontal (Fig. 6.5a). The mass of the crate is 60 kg, and the coefficient of sliding friction is $\mu_k = 0.50$. What force must the man exert to keep the crate moving at uniform velocity?

(a)

(b)

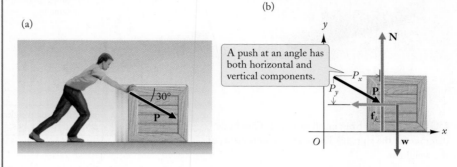

FIGURE 6.5 (a) Man pushing a crate. (b) "Free-body" diagram for the crate.

SOLUTION: Figure 6.5b is a "free-body" diagram for the crate. The forces on the crate are the push **P** of the man, the weight **w**, the normal force **N**, and the friction force $\mathbf{f}_k$. Note that because the man pushes the crate down against the floor, the magnitude of the normal force is not equal to *mg;* we will have to treat the magnitude of the normal force as unknown. Taking the *x* axis horizontal and the *y* axis vertical, we see from Fig. 6.5b that the *x* and *y* components of the forces are (see also Fig. 5.37)

$$P_x = P \cos 30° \qquad P_y = -P \sin 30°$$
$$w_x = 0 \qquad w_y = -mg$$
$$N_x = 0 \qquad N_y = N$$
$$f_{k,x} = -\mu_k N \qquad f_{k,y} = 0$$

Since the acceleration of the crate is zero in both the *x* and the *y* directions, the net force in each of these directions must be zero:

$$P \cos 30° + 0 + 0 - \mu_k N = 0$$
$$-P \sin 30° - mg + N + 0 = 0$$

These are two equations for the two unknowns *P* and *N*. By multiplying the second equation by μ_k and then adding the resulting equation to the first, we can eliminate *N*, and we find an equation for *P*:

$$P \cos 30° - \mu_k P \sin 30° - \mu_k mg = 0$$

Solving this for *P*, we find

$$P = \frac{\mu_k mg}{\cos 30° - \mu_k \sin 30°} = \frac{0.50 \times 60 \text{ kg} \times 9.81 \text{ m/s}^2}{\cos 30° - 0.50 \times \sin 30°} \qquad (6.4)$$

$$= 4.8 \times 10^2 \text{ N}$$

Friction forces also act between two surfaces at rest. If we exert a force against the side of, say, a steel block initially at rest on a steel tabletop, the block will not move unless the force is sufficiently large to overcome the friction that holds it in place. Friction between surfaces at rest is called **static friction**. The maximum magnitude of the static friction force, that is, the magnitude that this force attains when the lateral push is just about to start the motion, can be described by an empirical law quite similar to that for the kinetic friction force:

The magnitude of the maximum static friction force between unlubricated, dry surfaces at rest with respect to each other is proportional to the magnitude of the normal force and independent of the area of contact.

Mathematically,

$$f_{s,\text{max}} = \mu_s N \tag{6.5}$$

force of static friction

Here μ_s is a constant of proportionality, called the **coefficient of static friction**, which depends on the material. The direction of the static friction force is parallel to the surface, so as to oppose the total lateral push that tries to move the body (like the force **F** in Fig. 6.6).

The force in Eq. (6.5) is labeled with the subscript "max" because it represents the largest friction force that the surfaces can support without beginning to slide; in other words, $f_{s,\text{max}}$ is the force at the "breakaway" point, when the lateral push is just about to start the motion. Of course, *if the lateral push is less than this critical value, then the static friction force f_s is less than $f_{s,\text{max}}$ and its magnitude exactly matches the magnitude of the total lateral push.* This makes the net force on the block zero, as required if the block is to remain at rest.

Table 6.1 includes some typical values of the coefficient of static friction. *For most materials μ_s is larger than μ_k,* and therefore the maximum static friction force is larger than the kinetic friction force. This implies that if the lateral push applied to the block is large enough to overcome the static friction and start the block moving, it will more than compensate for the subsequent, smaller kinetic friction, and it will therefore accelerate the block continuously.

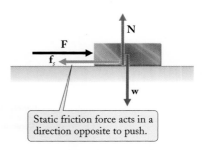

Static friction force acts in a direction opposite to push.

FIGURE 6.6 Forces on a steel block at rest on a steel plate. The friction force $\mathbf{f}_s$ has the same magnitude as the force **F**.

EXAMPLE 4   The coefficient of static friction of the rubber of an automobile tire on a street surface is $\mu_s = 0.90$. What is the steepest slope of a street on which an automobile with such tires (and locked wheels) can rest without slipping?

Concepts — *in* — Context

SOLUTION: The "free-body" diagram is shown in Fig. 6.7. The angle θ is assumed to be at its maximum value, so that the friction force has its maximum value $f_{s,\text{max}} = \mu_s N$. As in Example 2, $N = mg \cos \theta$, and hence $f_{s,\text{max}} = \mu_s mg \cos \theta$. With the x axis parallel to the street surface, we then find that the x component of the net force is

$$F_x = w_x - f_{s,\text{max}} = mg \sin \theta - \mu_s mg \cos \theta \tag{6.6}$$

This component of the force determines the motion along the street. If the automobile is to remain stationary, F_x must be zero:

$$0 = mg \sin \theta - \mu_s mg \cos \theta$$

Hence, dividing by mg,

$$\sin \theta = \mu_s \cos \theta$$

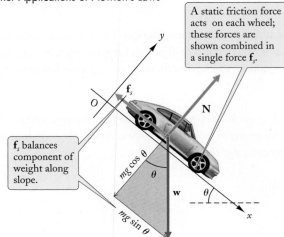

A static friction force acts on each wheel; these forces are shown combined in a single force **f**$_s$.

f$_s$ balances component of weight along slope.

FIGURE 6.7 "Free-body" diagram for an automobile parked on a very steep street.

or, dividing this equation by $\cos\theta$,

$$\tan\theta = \mu_s \tag{6.7}$$

With $\mu_s = 0.90$, this gives $\tan\theta = 0.90$. Thus, the slope of the street is 0.90, or 9:10. (With a calculator, we find that the inverse tangent of 0.90 gives $\theta = 42°$ for the angle of the incline.)

Concepts
— in —
Context

EXAMPLE 5   An automobile is braking on a level road. What is the maximum deceleration that the automobile can achieve when it brakes without skidding? As in the preceding example, assume that the tires of the automobile have a coefficient of static friction $\mu_s = 0.9$.

SOLUTION: Figure 6.8 shows the "free-body" diagram. If wheels are rolling without skidding, their rubber surface does *not* slide on the street surface; that is, the point of contact between the rolling wheel and the street is instantaneously at rest on the street (you can easily convince yourself of this by rolling any round object on a tabletop). Since there is no sliding, the relevant friction force is the *static* friction force. The maximum value of this force is

$$f_{s,\max} = \mu_s N \tag{6.8}$$

Here the magnitude of the normal force N is simply mg, since the normal force must balance the weight. The deceleration is then given by

$$a_x = -\frac{f_{s,\max}}{m} = -\frac{\mu_s mg}{m} = -\mu_s g \tag{6.9}$$

which yields

$$a_x = -0.9 \times 9.8 \text{ m/s}^2 = -9 \text{ m/s}^2$$

COMMENTS: Note that this is a larger deceleration than in Example 1, where the automobile skidded, because the coefficient of static friction is larger than that of kinetic friction. A car equipped with an antilock braking system (ABS) can exploit this difference and avoid skidding by rapid, repeated application of the brakes. The ABS also permits the driver to maintain directional control during rapid braking.

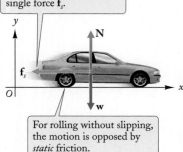

A static friction force acts on each wheel; these forces are shown combined in a single force **f**$_s$.

For rolling without slipping, the motion is opposed by *static* friction.

FIGURE 6.8 "Free-body" diagram for an automobile braking without skidding.

Note that throughout this section we have dealt only with the friction forces that act between solid surfaces in contact. There are also other friction forces that act when a solid body moves through a liquid or a gas, for instance, the friction experienced by an automobile moving through air. The magnitude of these **drag forces** can depend in a complicated way on the shape of the body, the speed of its motion, and the properties of the liquid or gas.

At low speeds in a liquid or gas, the friction force which opposes motion through the medium is very nearly proportional to the speed; this low-speed friction is due to the viscocity, or stickiness, of the medium; we explore such **viscous forces** in Problems 25 and 26. For automobiles or projectiles, the force due to **air resistance** at moderate to high speeds varies instead in direct proportion to the *square* of speed. The magnitude of the force can be written

$$f_{air} = \tfrac{1}{2}C\rho A v^2 \qquad (6.10)$$

force due to air resistance

Here, C is a dimensionless constant related to the shape of the body (its "aerodynamic design"), ρ is the density of air ($\rho = 1.3$ kg/m^3 for dry air), and A is the cross-sectional area, perpendicular to the motion, of the automobile. For modern automobiles, values of C are in the range 0.3–0.5.

EXAMPLE 6

A manufacturer quotes an aerodynamic constant $C = 0.30$ for an automobile of mass 900 kg and cross-sectional area $A = 2.8$ m^2. If the driver were to coast (in neutral) down a long hill with a slope of 8.0° (see Fig. 6.9a), what would be the terminal velocity? Recall from Section 2.6 that terminal velocity is the constant final velocity of a body moving under the combined influence of gravity and air resistance. Assume that air resistance is the only source of friction.

SOLUTION: We need only consider the components of the forces parallel to the motion. From Fig. 6.9b and Eq. (6.10), these are

$$w_x = mg\sin\theta \quad \text{and} \quad f_{air,x} = -\tfrac{1}{2}C\rho A v^2$$

When terminal (constant) velocity is reached, the two forces balance and the acceleration is zero. Thus,

$$\tfrac{1}{2}C\rho A v^2 = mg\sin\theta$$

Solving for the velocity, we find

$$v = \sqrt{\frac{2\,mg\sin\theta}{C\rho A}}$$

Inserting the values given in the problem, plus the air density $\rho = 1.3$ kg/m^3 from above, we have

$$v = \sqrt{\frac{2\times 900 \text{ kg} \times 9.81 \text{ m/s}^2 \times \sin 8.0°}{0.30 \times 1.3 \text{ kg/m}^3 \times 2.8 \text{ m}^2}}$$

$$= 47 \text{ m/s}$$

This is the same as

$$v = 47 \text{ m/s} \times \frac{1 \text{ km}}{1000 \text{ m}} \times \frac{3600 \text{ s}}{1 \text{ h}} = 170 \text{ km/h}$$

On such a slope, considerable speed can be attained before air resistance limits the motion.

(a)

(b)

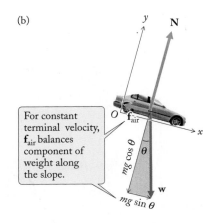

For constant terminal velocity, $\mathbf{f}_{air}$ balances component of weight along the slope.

FIGURE 6.9 (a) An automobile coasting down a hill. (b) "Free-body" diagram for the automobile, including air resistance.

Checkup 6.1

QUESTION 1: To overcome friction and to keep a book moving at constant speed on a flat table, you must push with a horizontal force of 10 N. If another, equal book is piled on the first, what horizontal force will be required to keep the books moving at constant speed? What if instead of piling the second book on top of the first you place it on the table in front of the first, so the first book pushes the second?

QUESTION 2: A block is sliding on an inclined plane, with friction. The block coasts up the plane, stops, and then slides down. What is the direction of the friction force while the block is coasting up the plane? While the block is sliding down the plane? Draw "free-body" diagrams for the block in these two cases.

QUESTION 3: You exert a lateral push on a brick resting on a surface with friction. If the brick remains at rest, you can conclude that the friction force is (a) equal to $\mu_s N$, (b) equal to $2\mu_s N$, or (c) smaller than or equal to $\mu_s N$.

QUESTION 4: A box sits on the floor of a delivery truck. If the truck brakes sharply, the box slides forward, but if the truck brakes gently, the box does not slide. Explain.

QUESTION 5: If the automobile of Example 6 had an extremely optimized aerodynamic constant of $C = 0.15$, by what factor would its terminal velocity increase?

QUESTION 6: A box is sliding down a ramp. Consider the mass m of the box, the friction coefficient μ_k, and the angle θ of the ramp. On which of these does the acceleration of the box depend?

(A) m and μ_k only   (B) m and θ only

(C) μ_k and θ only   (D) m, μ_k, and θ

6.2 RESTORING FORCE OF A SPRING; HOOKE'S LAW

A body is said to be **elastic** if it suffers a deformation when subjected to a stretching or compressing force, and if it returns to its original shape when the force is removed. For example, suitable forces can stretch a coil spring or a rubber band, and they can bend a flexible rod or a beam of metal or wood. Even bodies normally regarded as rigid, such as the balls of a ball bearing made of hardened steel, are somewhat elastic, and they experience slight deformations; but these deformations can be neglected unless the force is extremely large.

The force with which a body resists deformation is called its **restoring force**. If we stretch a spring by pulling with a hand at one end (see Fig. 6.10), we can feel the restoring force opposing our pull.

Under static conditions, the restoring force with which an elastic body opposes whatever pulls on it often obeys a simple empirical law known as **Hooke's Law:**

Hooke's Law

The magnitude of the restoring force is directly proportional to the deformation.

This law, like the law for friction, is not a general law of physics—the exact restoring force produced by the deformation of an elastic body depends in a complicated way on the shape of the body and on the detailed properties of the material of the body. Hooke's Law is only an approximate description of the restoring force. However, it is often a quite good approximation, provided the restoring force and the deformation are small.

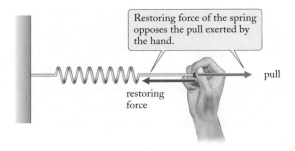

FIGURE 6.10 Restoring force of a stretched spring. The more the spring is stretched, the stronger the restoring force.

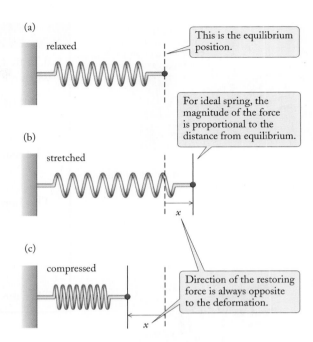

FIGURE 6.11 (a) Spring, relaxed. (b) Spring, stretched by a length x. (c) Spring, compressed by a length x.

As an example, consider a coil spring. Figure 6.11a shows such a spring in its relaxed state; it is loosely coiled and can therefore be compressed as well as stretched. Suppose we attach the left end of the spring to a wall and we apply a stretching or compressing force to the right end. Under the influence of this force, the spring will settle into a new equilibrium configuration such that the restoring force exactly balances the externally applied force. We can measure the deformation of the spring by the displacement that the right end undergoes relative to its initial position. In Fig. 6.11b, this displacement is denoted by x. Clearly, x is simply the change in the length of the spring. A positive value of x corresponds to an elongation, or stretching, of the spring, and a negative value of x corresponds to a compression.

Expressed mathematically, Hooke's Law then says that the restoring force opposes and is directly proportional to the displacement x:

$$F = -kx \qquad\qquad (6.11)$$

restoring force of a spring

The constant of proportionality k is the **spring constant**; it is a positive number characteristic of the spring. The spring constant is a measure of the stiffness of the spring—a stiff spring has a high value of k, and a soft spring has a low value of k. The units for the spring constant are newtons per meter (N/m). The negative sign in Eq. (6.11) indicates that the restoring force opposes the deformation; if the spring in Fig. 6.11 is elongated (positive x), then the restoring force is negative and opposes the external stretching force; if the spring is compressed (negative x), then the restoring force is positive and opposes the external compressing force.

EXAMPLE 7   The manufacturer's specifications for the coil spring for the front suspension of a Triumph sports car call for a spring with a relaxed length of 0.316 m, and a length of 0.205 m when under a load of 399 kg. What is the spring constant?

SOLUTION: The weight of 399 kg is $w = mg = 399 \text{ kg} \times 9.81 \text{ m/s}^2 = 3.91 \times 10^3$ N. The magnitude of the restoring force that will balance this weight must then also be 3.91×10^3 N. For the given relaxed and compressed lengths, the corresponding change of length is $x = 0.205 \text{ m} - 0.316 \text{ m} = -0.111$ m. Hence, Eq. (6.11) gives us

$$k = -\frac{F}{x} = -\frac{3.91 \times 10^3 \text{ N}}{-0.111 \text{ m}} = 3.53 \times 10^4 \text{ N/m}$$

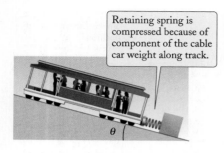

FIGURE 6.12 A cable car and retaining spring at a terminal station.

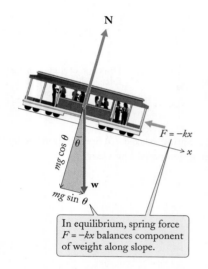

FIGURE 6.13 "Free-body" diagram for cable car parked against a retaining spring.

EXAMPLE 8   A cable car of mass 1200 kg is parked on a slope of 20°, compressing a giant retaining spring (see Fig. 6.12) to a length of 0.75 m. If the spring constant is 2.0×10^4 N/m, what is the length of the spring when relaxed? Neglect friction.

SOLUTION: A "free-body" diagram for the cable car is shown in Fig. 6.13, with the x axis parallel to the incline. The spring force $F = -kx$ balances the component of the cable car's weight that is parallel to the slope, $w_x = mg \sin \theta$. For the stationary cable car, the x component of the net force must be zero:

$$0 = mg \sin \theta - kx$$

This implies the length has been compressed by

$$x = \frac{mg \sin \theta}{k} = \frac{1200 \text{ kg} \times 9.81 \text{ m/s}^2 \times \sin 20°}{2.0 \times 10^4 \text{ N/m}}$$
$$= 0.20 \text{ m}$$

Thus the relaxed length of the spring is 0.75 m + 0.20 m = 0.95 m.

✔ Checkup 6.2

QUESTION 1: Nylon strings and ropes are elastic when stretched. Suppose that a mountain climber hanging from a long nylon rope stretches it by 20 cm. If two mountain climbers of equal mass hang from this rope, by how much will the rope stretch?

QUESTION 2: A spring is attached horizontally to a mass $m = 1.0$ kg, which sits on a table. The spring stretches an amount $x = 0.10$ m before the mass starts to move. If the spring constant is $k = 50$ N/m, what is the coefficient of static friction between the mass and the table?

QUESTION 3: A force F compresses a first spring by an amount x_1. If a second spring has twice the spring constant and a force $\frac{1}{2}F$ compresses it by an amount x_2, then the ratio x_2/x_1 is:

(A) 4   (B) 2   (C) 1   (D) $\frac{1}{2}$ (E) $\frac{1}{4}$

6.3 FORCE FOR UNIFORM CIRCULAR MOTION

All the examples of applications of Newton's laws we have examined so far involved particles moving along straight lines. But Newton's laws are also valid for motion along curved paths, for instance, motion with uniform speed along a circular path. As we saw in Section 4.5, such uniform circular motion is accelerated motion with a centripetal acceleration. If the motion proceeds with speed v around a circle of radius r, Eq. (4.49) tells us that the magnitude of the centripetal acceleration is

$$a = \frac{v^2}{r} \tag{6.12}$$

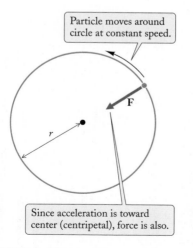

FIGURE 6.14 A particle in uniform circular motion. The force that acts on the particle is directed toward the center of the circle.

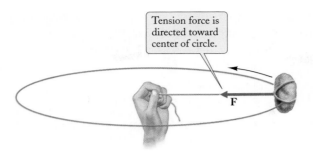

FIGURE 6.15 A stone being whirled around a circle. The string must exert a pull toward the center of the circle to produce a centripetal acceleration and to keep the stone in uniform circular motion.

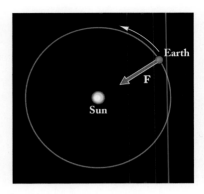

FIGURE 6.16 The Earth in its circular orbit around the Sun. The gravitational pull of the Sun produces the centripetal acceleration.

According to Newton's Second Law, this acceleration must be caused by a net force having the same direction as that of the acceleration; that is, *the direction of the net force must be centripetal, toward the center of the circle* (see Fig. 6.14). Such a force directed toward the center is called a **centripetal force**. For instance, the centripetal acceleration of a stone being whirled around a circle at the end of a string is caused by the pull of the string toward the center of the circle (see Fig. 6.15), and the centripetal acceleration of the Earth moving in its orbit around the Sun is caused by the gravitational pull toward the Sun (see Fig. 6.16).

The magnitude of the centripetal force required to maintain uniform circular motion is

$$F = ma = \frac{mv^2}{r} \tag{6.13}$$

centripetal force for circular motion

Note that this equation does not tell us how the force is produced. It is not a law of force, such as the law of friction or Hooke's Law, that tells us how to relate the force to the characteristics of the materials involved in producing the force. Instead, Eq. (6.13) merely tells us what magnitude of force we must produce, somehow, to keep the body in circular motion. For instance, to keep a stone whirling around a circle at the end of a string, we must exert this force with the string. If the string suddenly breaks, the stone will fly off in the direction of its instantaneous velocity, that is, in the direction of a tangent to the circle.

Equation (6.13) can be used to calculate the magnitude of the centripetal force required if the speed of the motion is known, or it can be used to calculate the speed if the force is known. The following examples illustrate such calculations with different kinds of forces.

EXAMPLE 9 In the hammer throw, an athlete launches a "hammer" consisting of a heavy metal ball attached to a handle by a steel cable (see Fig. 6.17). Just before launching the hammer, the athlete swings it around several times in a circle. The mass of the ball is 7.3 kg, and the distance from the hammer to the center of its circular motion is 1.9 m (including some length from the athlete's arms; see Fig. 6.18). The speed of the hammer is 27 m/s. What is the centripetal force that the athlete must exert with his arms to keep the hammer moving in its circle?

FIGURE 6.17 Hammer throw.

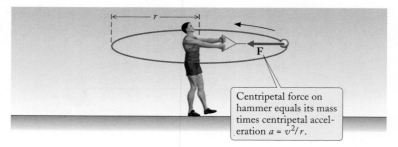

FIGURE 6.18 Circular motion of the hammer.

SOLUTION: According to Eq. (6.13), the magnitude of the force must be

$$F = ma = \frac{mv^2}{r} = \frac{7.3 \text{ kg} \times (27 \text{ m/s})^2}{1.9 \text{ m}} = 2.8 \times 10^3 \text{ N}$$

This is a rather large force! Hammer throwing requires great physical strength, and hammer throwers must be of hefty build.

EXAMPLE 10 What is the maximum speed with which an automobile can round a curve of radius 100 m without skidding sideways? Assume that the road is flat and that the coefficient of static friction between the tires and road surface is $\mu_s = 0.80$.

SOLUTION: The "free-body" diagram for the automobile is given in Fig. 6.19. The forces on the automobile are the weight **w**, the normal force **N**, and the friction force **f**$_s$. The weight balances the normal force; that is, $N = mg$. The horizontal friction force must provide the centripetal force; hence the magnitude of the friction force must be

$$f_s = ma = \frac{mv^2}{r} \tag{6.14}$$

The friction is *static* because, by assumption, there is no lateral slippage. At the maximum possible speed, the friction force has its maximum value $f_s = f_{s,\text{max}} = \mu_s N = \mu_s mg$, and consequently

$$\mu_s mg = \frac{mv^2}{r} \tag{6.15}$$

We can cancel the masses on both sides of this equation and then multiply both sides by r. Taking the square root of both sides then yields

$$v = \sqrt{\mu_s gr}$$
$$= \sqrt{0.80 \times 9.81 \text{ m/s}^2 \times 100 \text{ m}} = 28 \text{ m/s} \tag{6.16}$$

This is about 100 km/h.

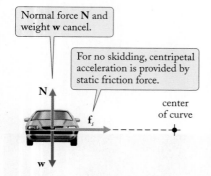

Normal force **N** and weight **w** cancel.

For no skidding, centripetal acceleration is provided by static friction force.

center of curve

FIGURE 6.19 "Free-body" diagram for an automobile rounding a curve.

EXAMPLE 11 At a speedway in Texas, a curve of radius 500 m is banked at an angle of 22° (see Fig. 6.20). If the driver of a racing car does not wish to rely on lateral friction, at what speed should he take this curve?

SOLUTION: The "free-body" diagram for the car is shown in Fig. 6.21a. Lateral (sideways) friction is assumed absent, and hence the normal force **N** and the weight **w** are the only forces acting on the racing car perpendicular to its motion.[1] The resultant of these forces must play the role of centripetal force. Hence the resultant must be horizontal, as in Fig. 6.21b. From this figure we see that the magnitude of the resultant is $F = w \tan\theta$, which must coincide with the magnitude of the centripetal force:

$$w \tan\theta = ma = \frac{mv^2}{r} \tag{6.17}$$

or

$$mg \tan\theta = \frac{mv^2}{r}$$

which yields

$$v = \sqrt{rg \tan\theta}$$
$$= \sqrt{500 \text{ m} \times 9.81 \text{ m/s}^2 \times \tan 22°} = 45 \text{ m/s}$$

This is 160 km/h. If the car goes faster than this, it will tend to skid up the embankment; if it goes slower than this, it will tend to skid down the embankment unless friction holds it there.

FIGURE 6.20 Racing car on a banked curve.

(a)

(b)

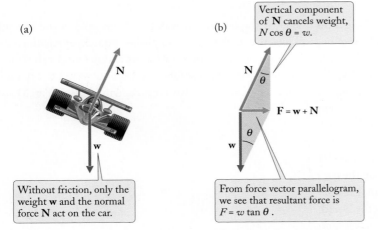

Vertical component of **N** cancels weight, $N \cos\theta = w$.

$F = w + N$

Without friction, only the weight **w** and the normal force **N** act on the car.

From force vector parallelogram, we see that resultant force is $F = w \tan\theta$.

FIGURE 6.21 (a) "Free-body" diagram for the car rounding a banked curve. (b) The resultant **F** of the forces **N** and **w**.

EXAMPLE 12 A pilot in a fast jet aircraft loops the loop (see Fig. 6.22). The radius of the loop is 400 m, and the aircraft has a speed of 150 m/s when it passes through the bottom of the loop. What is the apparent weight that the pilot feels; in other words, what is the force with which she presses against her chair? Express the answer as a multiple of her normal weight.

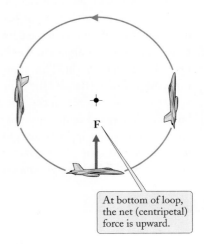

At bottom of loop, the net (centripetal) force is upward.

FIGURE 6.22 Jet aircraft looping the loop.

---

[1] Air resistance also acts on the car, but it is compensated by the propulsive force that the wheels produce, by reaction, with the *forward* friction force from the road.

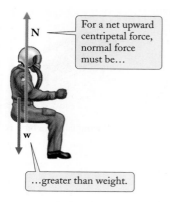

N

For a net upward centripetal force, normal force must be…

w

…greater than weight.

FIGURE 6.23 "Free-body" diagram for jet pilot at the bottom of the loop. The forces acting on the pilot are her weight **w** and the normal force **N** exerted by the chair.

SOLUTION: Figure 6.23 shows a "free-body" diagram for the pilot at the bottom of the loop. The forces acting on her are the (true) weight **w** and the normal force **N** exerted by the chair. The net vertical upward force is $N - mg$, and this must provide the centripetal acceleration:

$$N - mg = ma = \frac{mv^2}{r} \qquad (6.18)$$

Note that here we used the the familiar formula v^2/r for the centripetal acceleration, even though the speed is not constant (the speed of the aircraft increases somewhat as it goes down the loop, and the speed decreases as it goes up the loop). Such a change of speed along the loop implies that there can be an extra acceleration *along* the loop. But this extra acceleration does not affect the centripetal acceleration—the two accelerations are at right angles, and they are independent. For the purposes of this problem, we do not need to pay any attention to the extra tangential acceleration along the loop at any point.

Solving Eq. (6.18) for N, we obtain

PHYSICS IN PRACTICE   ULTRACENTRIFUGES

The operation of centrifuges and ultracentrifuges hinges on the effective increase of weight associated with circular motion. For instance, consider a test tube with some liquid that is being spun in a horizontal circle in a centrifuge (see Fig. 1). Suppose the liquid contains some particles in suspension; for instance, the liquid might be blood, consisting of a suspension of red blood cells (and other corpuscles) in liquid blood plasma. The radial force required to keep a corpuscle of mass m suspended in a fixed position relative to the test tube is mv^2/r in the centripetal direction. This centripetal force on the corpuscle has to be exerted by the liquid. By reaction, the corpuscle exerts a force of equal magnitude mv^2/r on the liquid, in the outward, or centrifugal, direction. These action and reaction forces are as though the particle were at rest but had an apparent weight mv^2/r in the outward, or centrifugal, direction. Note that the apparent weight is proportional to the mass, like a true weight. This apparent weight is called the **centrifugal force**. (Besides the apparent weight mv^2/r, there is also a true weight mg, in the vertical downward direction; but in a high-speed centrifuge, the true weight is negligible compared with the apparent weight.) Since the corpuscle is more dense than the plasma, its large apparent weight will enable it to shoulder the plasma aside, and to settle quickly

against the outermost wall (the "bottom") of the test tube. Thus, the centrifuge accomplishes a quick segregation of the contents of the test tube into layers of different density (Fig. 2), just as though the test tube had been subjected to a manifold enhancement of gravity. Ultracentrifuges (see Fig. 3) spin-

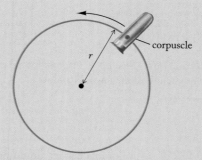

r

corpuscle

FIGURE 1 The test tube rotates around a circle of radius r. A corpuscle of mass m is suspended in the liquid in the test tube. A centripetal force mv^2/r is required to keep this particle moving on the circular path.

$$N = mg + \frac{mv^2}{r} = mg\left(1 + \frac{v^2}{gr}\right) \tag{6.19}$$

With $v = 150$ m/s and $r = 400$ m, we then find

$$N = mg\left(1 + \frac{(150 \text{ m/s})^2}{9.81 \text{ m/s}^2 \times 400 \text{ m}}\right) = mg \times 6.7$$

This is the force with which the chair presses against the pilot, and it is therefore the apparent weight that the pilot feels. The apparent weight equals the true weight multiplied by a factor of 6.7. In these circumstances, the pilot would say she is "pulling 6.7 g's," because she feels as though gravity had been magnified by a factor of 6.7.

COMMENT: This example shows that *the centripetal acceleration can generate an effective increase of weight*—the apparent weight in the reference frame of a body in circular motion can be much larger than its normal weight.

ning at up to 100 000 revolutions per minute generate apparent centrifugal weights of up to 500 000 times the true weight; they are used in chemical and biochemical research. Special ultracentrifuges can separate even different isotopes of chemical elements, such as the different isotopes of uranium.

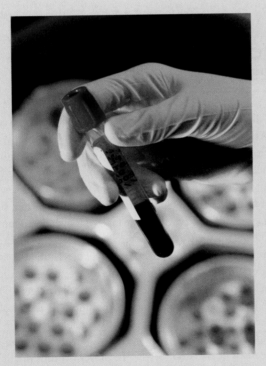

FIGURE 2 Test tube with a blood sample after centrifuging.

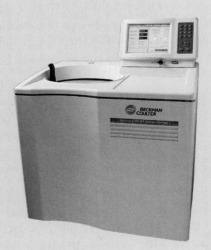

FIGURE 3 An ultracentrifuge.

Checkup 6.3

QUESTION 1: A stone is being whirled around a circle at the end of a string when the string suddenly breaks. Describe the motion of the stone after the string breaks; ignore gravity.

QUESTION 2: At an intersection, a motorcycle makes a right turn at constant speed. During this turn the motorcycle travels along a 90° arc of a circle. What is the direction of the acceleration of the motorcycle during this turn?

QUESTION 3: A car moves at constant speed along a road leading over a small hill with a spherical top. What is the direction of the acceleration of the car when at the top of this hill?

QUESTION 4: In Example 12, for the aircraft looping the loop, does the chair exert a centripetal or a centrifugal force on the pilot? Does the pilot exert a centripetal or a centrifugal force on the chair? What is the direction of the pilot's apparent, increased weight at the instant the aircraft passes through the bottom of the loop? Does the direction of the apparent weight change as the aircraft climbs up the loop?

QUESTION 5: Two cars travel around a traffic circle in adjacent (outer and inner) lanes. If the two cars travel at the same constant speed, which completes the circle first? Which has the larger acceleration?

(A) Outer; outer   (B) Inner; outer

(C) Outer; inner   (D) Inner; inner

PROBLEM-SOLVING TECHNIQUES

FRICTION FORCES AND CENTRIPETAL FORCES

The problems involving applications of Newton's laws in this chapter can be solved by the techniques discussed in the preceding chapter. In dealing with friction forces and with the centripetal force for uniform circular motion, pay special attention to the directions of the forces.

1   The magnitude of the sliding friction force is proportional to the magnitude of the normal force, but the direction is not the direction of the normal force. Instead, the sliding friction force is always parallel to the sliding surfaces, opposite to the direction of motion.

2   The static friction force is also always parallel to the sliding surfaces, opposite to the direction in which the body tends to move. If you have any doubts about the direction of the static friction force, pretend that the friction is absent, and ask yourself in what direction the body would then move; the static friction force is in the opposite direction.

3   Uniform circular motion requires a force toward the center of the circle, that is, a centripetal force. When preparing a "free-body" diagram for a body in uniform circular motion, include all the pushes and pulls acting on the moving body, but do *not* include a "centripetal mv^2/r force." This would be a mistake, like including an "*ma* force" in the "free-body" diagram for a body with some kind of translational motion. The quantity mv^2/r is not a force; it is merely the product of mass and centripetal acceleration. This acceleration is caused by one force or by the resultant of several forces already included among the pushes and pulls displayed in the "free-body" diagram. For instance, in Example 11 the resultant force is $w \tan\theta$, in Example 12 the resultant force is $N - mg$, and these resultants equal mv^2/r by Newton's Second Law [see Eqs. (6.17) and (6.18)]. To prevent confusion, do not include the resultant in the "free-body" diagram for a body in uniform circular motion. Instead, draw the resultant on a separate diagram (see Fig. 6.21b).

6.4 THE FOUR FUNDAMENTAL FORCES

In everyday experience we encounter an enormous variety of forces: the gravity of the Earth that pulls all bodies downward, contact forces between rigid bodies that resist their interpenetration, friction forces that resist the motion of a surface sliding over another surface, elastic forces that oppose the deformation of springs and beams, pressure forces exerted by air or water on bodies immersed in them, adhesive forces exerted by a layer of glue bonding two surfaces, electrostatic forces between two electrified bodies, magnetic forces between the poles of magnets, and so on.

Besides these forces that act in the macroscopic world of everyday experience, there are many others that act in the microscopic world of atomic and nuclear physics. There are intermolecular forces that attract or repel molecules to or from each other, interatomic forces that bind atoms into molecules or repel them if they come too close to each other, atomic forces within the atom that hold its parts together, nuclear forces that act on the parts of the nucleus, and even more esoteric forces which act during radioactive decay or act only for a brief instant when subnuclear particles are made to suffer violent collisions in high-energy experiments performed in accelerator laboratories.

Yet, at the fundamental level, this bewildering variety of forces involves only four different kinds of forces. The four fundamental forces are the gravitational force, the electromagnetic force, the "strong" force, and the "weak" force.

The **gravitational force** is a mutual attraction between all masses. Gravitation is the weakest of the four forces. The gravitational attraction between two masses of, say, 1 kg placed next to each other is so small that it is detectable only with extremely sensitive equipment. On the surface of the Earth, we feel the force of gravity only because the mass of the Earth is very large. We discuss gravity further in Chapter 9.

gravitational force

The **electromagnetic force** is an attraction or repulsion between electric charges. The electric and the magnetic forces, once considered to be separate, are now grouped together because they are closely related: the magnetic force is nothing but an extra electric force that acts whenever charges are in motion. Of all the forces, the electric force plays the largest role in our lives. With the exception of the Earth's gravity, every force in our immediate macroscopic environment is electric. Contact forces between rigid bodies, elastic forces, pressure forces, adhesive forces, friction forces, etc., are nothing but electric forces between charged particles in the atoms of one body and those in the atoms of another. Electricity and magnetism are the subject of Chapters 22–33.

electromagnetic force

The **"strong" force** acts mainly within the nuclei of atoms. It plays the role of a nuclear glue that prevents the protons and neutrons of the nucleus from flying apart. This nuclear force is called "strong" because it is the strongest of the four forces. It can be either attractive or repulsive: the strong force will push protons and neutrons apart if they come too near each other, and it will pull them together if they begin to drift too far apart. We will examine nuclei in Chapter 40.

"strong" force

Finally, the **"weak" force** manifests itself only in certain reactions among elementary particles. Most of the reactions caused by the weak force are radioactive-decay reactions; they involve the spontaneous breakup of a particle into several other particles (we will discuss this in Chapters 40 and 41). This force is called "weak" because it is weak compared with the "strong" force and the electromagnetic force.

"weak" force

SUMMARY

PHYSICS IN PRACTICE   Ultracentrifuges   **(page 188)**

PROBLEM-SOLVING TECHNIQUES   Friction Forces   **(page 190)**
and Centripetal Forces

KINETIC FRICTION FORCE   (Direction *opposes* motion.) $f_k = \mu_k N$ **(6.1)**

The friction force acts over the bottom surface in a direction opposite to the motion.

STATIC FRICTION FORCE   (Direction opposes force $f_{s,\mathrm{max}} = \mu_s N$ **(6.5)**
which tries to move body; magnitude varies in
response to applied force.)

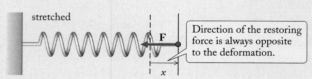

Static friction force acts in a direction opposite to push.

RESTORING FORCE OF A SPRING (HOOKE'S LAW) $F = -kx$ **(6.11)**
(Direction is toward relaxed position; x is measured
from relaxed position.)

stretched

Direction of the restoring force is always opposite to the deformation.

FORCE DUE TO AIR RESISTANCE   At high speed v, $f_{\mathrm{air}} = \frac{1}{2}C\rho A v^2$
where C is a dimensionless aerodynamic constant, ρ is
the density of air, and A is the cross-sectional area.

FORCE REQUIRED FOR UNIFORM CIRCULAR MOTION   **(6.13)**
(Direction is centripetal.) $F = \dfrac{mv^2}{r}$

THE FOUR FUNDAMENTAL FORCES
Gravitational, "weak," electromagnetic, "strong"

QUESTIONS FOR DISCUSSION

1. According to the adherents of parapsychology, some people are endowed with the supernormal power of psychokinesis, e.g., spoon-bending-at-a-distance via mysterious psychic forces emanating from the brain. Physicists are confident that the only forces acting between pieces of matter are those listed in Section 6.4, none of which are implicated in psychokinesis. Given that the brain is nothing but a (very complicated) piece of matter, what conclusions can a physicist draw about psychokinesis?

2. If you carry a spring balance from London to Hong Kong, do you have to recalibrate it? If you carry a beam balance?

3. When you stretch a rope horizontally between two fixed points, it always sags a little, no matter how great the tension. Why?

4. What are the forces on a soaring bird? How can the bird gain altitude without flapping its wings?

5. How could you use a pendulum suspended from the roof of your automobile to measure its acceleration?

6. When an airplane flies along a parabolic path similar to that of a projectile, the passengers experience a sensation of weightlessness. How would the airplane have to fly to give the passengers a sensation of enhanced weight?

7. A frictionless chain hangs over two adjoining inclined planes (Fig. 6.24a). Prove the chain is in equilibrium, i.e., the chain will not slip to the left or to the right. [Hint: One method of proof, due to the seventeenth-century engineer and mathematician Simon Stevin, asks you to pretend that an extra piece of chain is hung from the ends of the original chain (Fig. 6.24b). This makes it possible to conclude that the original chain cannot slip.]

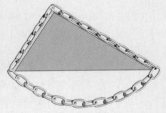
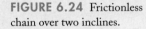

FIGURE 6.24 Frictionless chain over two inclines.

8. Seen from a reference frame moving with the wave, the motion of a surfer is analogous to the motion of a skier down a mountain.[2] If the wave were to last forever, could the surfer ride it forever? In order to stay on the wave as long as possible, in what direction should the surfer ski the wave?

9. Excessive polishing of the surfaces of a block of metal increases its friction. Explain.

10. Some drivers like to spin the wheels of their automobiles for a quick start. Does this give them greater acceleration? (Hint: $\mu_s > \mu_k$.)

[2] There is, however, one complication: surf waves grow higher as they approach the beach. Ignore this complication.

11. Cross-country skiers like to use a ski wax that gives their skis a large coefficient of static friction, but a low coefficient of kinetic friction. Why is this useful? How do "waxless" skis achieve the same effect?

12. Designers of locomotives usually reckon that the maximum force available for moving the train ("tractive force") is one-fourth or one-fifth of the weight resting on the drive wheels of the locomotive. What value of the friction coefficient between the wheels and the track does this implicitly assume?

13. When an automobile with rear-wheel drive accelerates from rest, the maximum acceleration that it can attain is less than the maximum deceleration that it can attain while braking. Why? (Hint: Which wheels of the automobile are involved in acceleration? In braking?)

14. Can you think of some materials with $\mu_k > 1$?

15. For a given initial speed, the stopping distance of a train is much longer than that of a truck. Why?

16. Why does the traction on snow or ice of an automobile with rear-wheel drive improve when you place extra weight over the rear wheels?

17. Why are wet streets slippery?

18. In order to stop an automobile on a slippery street in the shortest distance, it is best to brake as hard as possible without initiating a skid. Why does skidding lengthen the stopping distance? (Hint: $\mu_s > \mu_k$.)

19. Suppose that in a panic stop, a driver locks the wheels of his automobile and leaves skid marks on the pavement. How can you deduce his initial speed from the length of the skid marks?

20. Hot-rod drivers in drag races find it advantageous to spin their wheels very fast at the start so as to burn and melt the rubber on their tires (Fig. 6.25). How does this help them to attain a larger acceleration than expected from the static coefficient of friction?

FIGURE 6.25 Drag racer at the start of the race.

21. A curve on a highway consists of a quarter circle connecting two straight segments. If this curve is banked perfectly for motion at some given speed, can it be joined to the straight segments without a bump? How could you design a curve that is banked perfectly along its entire length and merges smoothly into straight segments without any bump?

22. Automobiles with rear engines (such as the old VW "Beetle") tend to oversteer; that is, in a curve the rear end tends to swing toward the outside of the curve, turning the car excessively into the curve. Explain.

23. When rounding a curve in your automobile, you get the impression that a force tries to pull you toward the outside of the curve. Is there such a force?

24. If the Earth were to stop spinning (other things remaining equal), the value of g at all points of the surface except the poles would become slightly larger. Why?

25. (a) If a pilot in a fast aircraft very suddenly pulls out of a dive (Fig. 6.26a), he will suffer blackout caused by loss of blood pressure in the brain. If he suddenly begins a dive while climbing (Fig. 6.26b), he will suffer *redout* caused by excessive blood pressure in the brain. Explain.

    (b) A pilot wearing a G suit—a tightly fitting garment that squeezes the tissues of the legs and abdomen—can tolerate $8g$ while pulling out of a dive (Fig. 6.26c). How does this G suit prevent blackout? A pilot can tolerate no more than $-2g$ while beginning a dive. Why does the G suit not help against redout?

26. While rounding a curve at high speed, a motorcycle rider leans the motorcycle toward the center of the curve. Why?

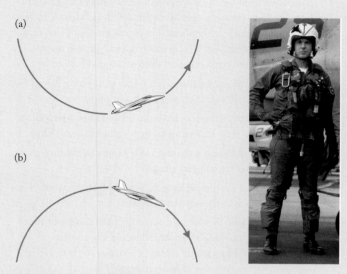

(a)

(b)

FIGURE 6.26 (a) Aircraft pulling out of a dive. (b) Aircraft beginning a dive. (c) Pilot wearing a G suit.

PROBLEMS

6.1 Friction[†]

1. The ancient Egyptians moved large stones by dragging them across the sand in sleds. How many Egyptians were needed to drag an obelisk of 700 metric tons? Assume that $\mu_k = 0.30$ for the sled on sand and that each Egyptian exerted a horizontal force of 360 N.

2. The base of a winch is bolted to a mounting plate with four bolts. The base and the mounting plate are flat surfaces made of steel; the friction coefficient of these surfaces in contact is $\mu_s = 0.40$. The bolts provide a normal force of 2700 N each. What maximum static friction force will act between the steel surfaces and help oppose lateral slippage of the winch on its base?

3. According to tests performed by the manufacturer, an automobile with an initial speed of 65 km/h has a stopping distance of 20 m on a level road. Assuming that no skidding occurs during braking, what is the value of μ_s between the wheels and the road required to achieve this stopping distance?

4. A crate sits on the load platform of a truck. The coefficient of friction between the crate and the platform is $\mu_s = 0.40$. If the truck stops suddenly, the crate will slide forward and crash into the cab of the truck. What is the maximum braking deceleration that the truck may have if the crate is to stay put?

5. When braking (without skidding) on a dry road, the stopping distance of a sports car with a high initial speed is 38 m. What would have been the stopping distance of the same car with the same initial speed on an icy road? Assume that $\mu_s = 0.85$ for the dry road and $\mu_s = 0.20$ for the icy road.

6. In a remarkable accident on motorway M1 (in England), a Jaguar car initially speeding "in excess of 100 mph" skidded 290 m before coming to a rest. Assuming that the wheels were completely locked during the skid and that the coefficient of kinetic friction between the wheels and the road was 0.80, find the initial speed.

[†] For help, see Online Concept Tutorial 8 at www.wwnorton.com/physics

7. Because of a failure of its landing gear, an airplane has to make a belly landing on the runway of an airport. The landing speed of the airplane is 90 km/h, and the coefficient of kinetic friction between the belly of the airplane and the runway is $\mu_k = 0.60$. How far will the airplane slide along the runway?

8. A child slides down a playground slide; the coefficient of kinetic friction is $\mu_k = 0.15$, and the angle that the slide makes with the horizontal is 30°. She begins from rest and slides through a vertical height of 3.5 m. With what speed does she exit the slide?

9. A baseball player sprinting at 4.5 m/s begins to slide with his body flat on the ground when 2.8 m from home base. The coefficient of friction between the player's uniform and the ground is 0.30. Does he make it home? If so, what is his speed as he reaches home?

10. For microscopic objects, friction can be overwhelming. For example, tiny silicon microstructures (see Fig. 6.27) can become stuck when in contact. If a silicon cube 10 μm on each side with density 2.33 g/cm^3 requires a horizontal force of 0.50×10^{-9} N to begin sliding on a horizontal silicon surface, what is the effective coefficient of static friction?

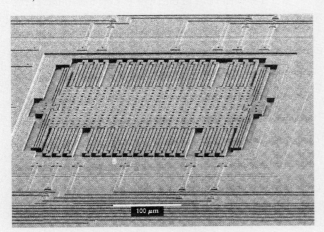

FIGURE 6.27 Micromachined silicon structures (an automobile accelerometer).

11. While braking in an emergency a train traveling at 35 m/s screeches to a halt with all wheels locked. The constant deceleration takes 7.5 s to complete. What is the coefficient of kinetic friction between the train and the tracks?

12. An automobile has an aerodynamic constant $C = 0.35$ and a cross-sectional area of 3.4 m^2. To balance air resistance, what force must be provided when traveling at 20 m/s? At 40 m/s?

13. The driver of the automobile in Example 6 is traveling on a flat road at 25 m/s. Considering only air resistance, what forward friction force must the road provide if the driver wants to begin accelerating at 2.0 m/s^2?

14. A Ping-Pong ball has an aerodynamic constant $C = 0.51$, a mass of 2.5 g, and a radius of 1.6 cm. What is its terminal velocity when dropped?

15. A falling golf ball (mass 45 g, radius 20 mm) reaches a high terminal speed of 45 m/s. What is the value of the aerodynamic constant C for this dimpled sphere?

16. A sky surfer (see Fig. 6.28) has a mass of 70 kg; in the position shown, the product of his aerodynamic constant and cross-sectional area is $CA = 0.42$ m^2. What is this surfer's terminal speed?

FIGURE 6.28 A sky surfer.

*17. A girl pulls a sled along a level dirt road by means of a rope attached to the front of the sled (Fig. 6.29). The mass of the sled is 40 kg, the coefficient of kinetic friction is $\mu_k = 0.60$, and the angle between the rope and the road is 30°. What pull must the girl exert to move the sled at constant velocity?

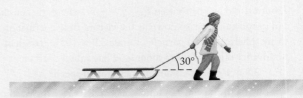

FIGURE 6.29 Pulling a sled.

*18. The "Texas" locomotives of the old T&P railways had a mass of 200 000 kg, of which 136 000 kg rested on the driving wheels. What maximum acceleration could such a locomotive attain (without slipping) when pulling a train of 100 boxcars of mass 18 000 kg each on a level track? Assume that the coefficient of static friction between the driving wheels and the track is 0.25.

*19. During braking, a truck has a steady deceleration of 7.0 m/s^2. A box sits on the platform of this truck. The box begins to slide when the braking begins and, after sliding a distance of 2.0 m (relative to the truck), it hits the cab of the truck. With what speed (relative to the truck) does the box hit? The coefficient of kinetic friction for the box is $\mu_k = 0.50$.

*20. The Schleicher ASW-22 is a high-performance sailplane of a wingspan of 24 m and a mass of 750 kg (including the pilot). At a forward speed (airspeed) of 35 knots, the sink rate, or the rate of descent, of this sailplane is 0.46 m/s. Draw a "free-body" diagram showing the forces on the plane. What is the friction force (antiparallel to the direction of motion) exerted on the plane by air resistance under these conditions? What is the lift force (perpendicular to the direction of motion) generated by air streaming past the wings?

*21. The friction force (including air friction and rolling friction) acting on an automobile traveling at 65 km/h amounts to 500 N. What slope must a road have if the automobile is to roll down this road at a constant speed of 65 km/h (with its gears in neutral)? The mass of the automobile is 1.5×10^3 kg.

*22. In a downhill race, a skier slides down a 40° slope. Starting from rest, how far must he slide down the slope in order to reach a speed of 130 km/h? How many seconds does it take him to reach this speed? The friction coefficient between his skis and the snow is $\mu_k = 0.10$. Ignore the resistance offered by the air.

*23. To measure the coefficient of static friction of a block of plastic on a plate of steel, an experimenter places the block on the plate and then gradually tilts the plate. The block suddenly begins to slide when the plate makes an angle of 38° with the horizontal. What is the value of μ_s?

*24. A solid steel ball bearing of radius 0.25 cm falling in air has a terminal speed of 88 m/s. What is the terminal speed of a solid steel ball of radius 5.0 cm?

*25. At very low speeds, the resistance to motion offered by a liquid or gas is nearly proportional to velocity (instead of the square of the velocity); such a *viscous drag force* opposes the motion and can be written $\mathbf{f}_{\text{viscous}} = -b\mathbf{v}$, where the constant of proportionality is known as the viscous drag coefficient. A tiny spherical metal particle of mass 3.9×10^{-6} g (neglect buoyancy effects) falling though oil has a drag coefficient $b = 2.8 \times 10^{-5}$ kg/s. What is its terminal speed?

**26. Show that the speed as a function of time of a particle falling from rest under the influence of gravity and a viscous force of the form $\mathbf{f}_{\text{viscous}} = -b\mathbf{v}$ (see Problem 25) is given by

$$v = \frac{mg}{b}(1 - e^{-bt/m})$$

[Hint: Integrate Newton's law in the form $m(dv/dt) = mg - bv$.] What is the value of the characteristic time $t = m/b$ for the particle in Problem 25?

**27. On a level road, the stopping distance for an automobile is 35 m for an initial speed of 90 km/h. What is the stopping distance of the same automobile on a similar road with a downhill slope of 1:10?

**28. Two masses, of 2.0 kg each, connected by a string slide down a ramp making an angle of 50° with the horizontal (Fig. 6.30). The mass m_1 has a coefficient of kinetic friction 0.60, and the mass m_2 has a coefficient of kinetic friction 0.40. Find the acceleration of the masses and the tension in the string.

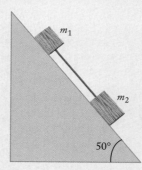

FIGURE 6.30 Two masses connected by a string sliding down a ramp.

**29. You are holding a book against a wall by pushing with your hand. Your push makes an angle of θ with the wall (see Fig. 6.31). The mass of the book is m, and the coefficient of static friction between the book and the wall is μ_s.

(a) Draw the "free-body" diagram for the book.

(b) Calculate the magnitude of the push you must exert to (barely) hold the book stationary.

(c) For what value of the angle θ is the magnitude of the required push as small as possible? What is the magnitude of the smallest possible push?

(d) If you push at an angle larger than 90°, you must push very hard to hold the book in place. For what value of the angle will it become impossible to hold the book in place?

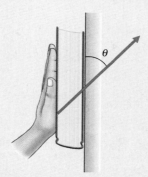

FIGURE 6.31 Pushing a book.

**30. A box is being pulled along a level floor at constant velocity by means of a rope attached to the front end of the box. The rope makes an angle θ with the horizontal. Show that for a given mass m of the box and a given coefficient of kinetic friction μ_k, the tension required in the rope is minimum if $\tan\theta = \mu_k$. What is the tension in the rope when at this optimum angle?

**31. Consider the man pushing the crate described in Example 3. Assume that instead of pushing down at an angle of 30°, he pushes down at an angle θ. Show that he will not be able to keep the crate moving if θ is larger than $\tan^{-1}(1/\mu_k)$.

32. A block of mass m_1 sits on top of a larger block of mass m_2 which sits on a flat surface (Fig. 6.32). The coefficient of kinetic friction between the upper and lower blocks is μ_1, and that between the lower block and the flat surface is μ_2. A horizontal force **F pushes against the upper block, causing it to slide; the friction force between the blocks then causes the lower block to slide also. Find the acceleration of the upper block and the acceleration of the lower block.

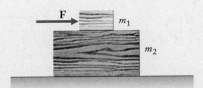

FIGURE 6.32 Block on block on surface.

**33. Two masses $m_1 = 1.5$ kg and $m_2 = 3.0$ kg are connected by a thin string running over a massless pulley. One of the masses hangs from the string; the other mass slides on a 35° ramp with a coefficient of kinetic friction $\mu_k = 0.40$ (Fig. 6.33). What is the acceleration of the masses?

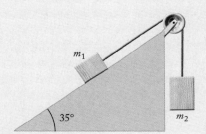

FIGURE 6.33 Two masses, an incline, and a pulley.

**34. A man pulls a sled up a ramp by means of a rope attached to the front of the sled (Fig. 6.34). The mass of the sled is 80 kg, the coefficient of kinetic friction between the sled and the ramp is $\mu_k = 0.70$, the angle between the ramp and the horizontal is 25°, and the angle between the rope and the ramp is 35°. What pull must the man exert to keep the sled moving at constant velocity?

FIGURE 6.34 Pulling a sled up an incline.

**35. Two blocks of masses m_1 and m_2 are sliding down an inclined plane making an angle θ with the horizontal. The leading block has a coefficient of kinetic friction μ_k; the trailing block has a coefficient of kinetic friction $2\mu_k$. A string connects the two blocks; this string makes an angle ϕ with the ramp (Fig. 6.35). Find the tension in the string.

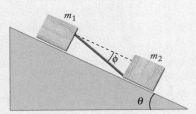

FIGURE 6.35 Two blocks connected by a slanted string sliding down an inclined plane.

**36. A man of 75 kg is pushing a heavy box along a flat floor. The coefficient of sliding friction between the floor and the box is 0.20, and the coefficient of static friction between the man's shoes and the floor is 0.80.

(a) If the man pushes downward on the box at an angle of 30° (see Fig. 6.36a), what is the maximum mass of the box he can move?

(b) If the man pushes upward on the box at an angle of 30° (see Fig. 6.36b), what is the maximum mass of the box he can move?

(a)

(b)

FIGURE 6.36 Pushing a box.

6.2 Restoring Force of a Spring; Hooke's Law

37. A spring with a force constant $k = 150$ N/m has a relaxed length of 0.15 m. What force must you exert to stretch this spring to twice its length? What force must you exert to compress this spring to one-half its length?

38. Attempting to measure the force constant of a spring, an experimenter clamps the upper end of the spring in a vise and suspends a mass of 1.5 kg from the lower end. This stretches the spring by 0.20 m. What is the force constant of the spring?

39. A rubber band of relaxed length 6.3 cm stretches to 10.2 cm under a force of 1.0 N, and to 15.5 cm under 2.0 N. Does this rubber band obey Hooke's Law?

40. A cantilever (such as a diving board) can be regarded as a spring. When a 70-kg diver stands on the edge of the board, it deflects downward by 16 cm. What is the effective spring constant of this diving board?

41. When pulled back to launch a ball in a pinball machine, a spring is compressed by 7.0 cm. This requires a pull of 3.5 N. What is the spring constant of the pinball spring?

42. Atomic-force microscopes (AFMs) use tiny cantilevers which act like springs; one manufacturer quotes a value of $k = 4.8 \times 10^{-2}$ N/m for the cantilever's spring constant. When the "spring" moves over an atom, it is compressed a distance of 2.0×10^{-11} m. What is the value of the atomic force in this case?

43. Retractable ballpoint pens contain a spring. When a mass of 250 g is placed on top of such a spring, the spring compresses by 2.8 mm. What is the spring constant of this spring?

44. When a 75-kg bungee jumper hangs from a bungee cord, it stretches by 2.9 m. What is the spring constant of the bungee cord?

45. Solid materials can act much like springs. Consider a steel cable with radius 2.0 cm and length 20 m, which has a spring constant of 1.4×10^7 N/m. If a 1500-kg elevator car is hung from this cable, how much does the cable stretch?

*46. Suppose that a uniform spring with a constant $k = 120$ N/m is cut into two pieces, one twice as long as the other. What are the spring constants of the two pieces?

*47. Show that if two springs, of constants k_1 and k_2, are connected in parallel (Fig. 6.37), the net spring constant k of the combination is given by

$$k = k_1 + k_2$$

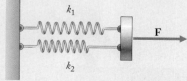

FIGURE 6.37 Springs acting in parallel.

*48. Show that if two springs, of constants k_1 and k_2, are connected in series (Fig. 6.38), the net spring constant k of the combination is given by

$$\frac{1}{k} = \frac{1}{k_1} + \frac{1}{k_2}$$

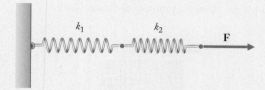

FIGURE 6.38 Springs acting in series.

6.3 Force for Uniform Circular Motion

49. A man of 80 kg is standing in the cabin of a Ferris wheel of radius 30 m rotating at 1.0 rev/min. What is the force that the feet of the man exert on the floor of the cabin when he reaches the highest point? The lowest point?

50. The Moon moves around the Earth in a circular orbit of radius 3.8×10^8 m in 27 days. The mass of the Moon is 7.3×10^{22} kg. From these data, calculate the magnitude of the force required to keep the Moon in its orbit.

51. A swing consists of a seat supported by a pair of ropes 5.0 m long. A 60-kg woman sits in the swing. Suppose that the speed of the woman is 5.0 m/s at the instant the swing goes through its lowest point. What is the tension in each of the two ropes? Ignore the masses of the seat and of the ropes.

52. A few copper coins are lying on the (flat) dashboard of an automobile. The coefficient of static friction between the copper and the dashboard is 0.50. Suppose the automobile rounds a curve of radius 90 m. At what speed of the automobile will the coins begin to slide? The curve is *not* banked.

53. A curve of radius 400 m has been designed with a banking angle such that an automobile moving at 75 km/h does not have to rely on friction to stay in the curve. What is the banking angle?

54. In an amusement park ride called "Drop Out," riders are spun in a horizontal circle of radius 6.0 m, which pins their backs against an outer wall. When they are spinning quickly enough, the floor drops out, and they are suspended by friction. If the coefficient of static friction between the riders and the wall is as small as 0.25, how many revolutions per second must the ride achieve before the floor is allowed to drop out?

55. An ant walks from the center toward the edge of a turntable of radius 15 cm. If the coefficient of friction between the ant's feet and the turntable is 0.30, at what radius does the ant begin to slide when the turntable rotates at 45 revolutions per minute?

56. Public skateboarding parks often include a well in the shape of a half cylinder (or "half-pipe"). The skateboarder's path traces out a semicircular arc, with the midpoint of the arc at the lowest point (Fig. 6.39). When starting from rest at one of the upper edges of the arc (at height h), the horizontal speed attained at the bottom can be shown to be the same as for vertical free fall, $v = \sqrt{2gh}$. At the bottom point, by what factor does the normal force from the skateboard on the skateboarder's feet exceed his weight?

FIGURE 6.39 Skateboard half-pipe.

57. A geosynchronous satellite orbits the Earth once per day; this requires an orbital radius of 4.23×10^4 km. From these data, deduce the weight of a (stationary) 1-kg mass at this distance.

58. A jet traveling at 140 m/s makes a turn of radius 6.0 km. What bank angle should the pilot use for the turn so that a passenger does not feel any lateral force?

*59. A rider of a swing carousel initially sits on a seat suspended vertically by a 7.0-m cable from a point 3.0 m from the center axis. When the carousel rotates, the seat swings outward to its equilibrium angle. If the seat speed during this rotation is 6.0 m/s, what is the angle that the cable makes with the vertical?

*60. Two identical automobiles enter a curve side by side, one traveling on the inside lane, the other on the outside. The curve is an arc of a circle, and it is unbanked. Each automobile travels through the curve at the maximum speed tolerated without skidding. Which automobile has a higher speed? Which automobile emerges from the curve first? Prove your answer.

*61. The highest part of a road over the top of a hill follows an arc of a vertical circle of radius 50 m. With what minimum speed must you drive an automobile along this road if its wheels are to lose contact with the road at the top of the hill?

*62. A woman holds a pail full of water by the handle and whirls it around a vertical circle at constant speed. The radius of this circle is 0.90 m. What is the minimum speed that the pail must have at the top of its circular motion if the water is not to spill out of the upside-down pail?

*63. In ice speedway races, motorcycles run at a high speed on an ice-covered track and are kept from skidding by long spikes on their wheels. Suppose that a motorcycle runs around a curve of radius 30 m at a speed of 96 km/h. What is the angle of inclination of the force exerted by the track on the wheels?

*64. An automobile traveling at speed v on a level surface approaches a brick wall (Fig. 6.40). When the automobile is at a distance d from the wall, the driver suddenly realizes that he must either brake or turn. If the coefficient of static friction between the tires and the surface is μ_s, what is the minimum distance that the driver needs to stop (without turning)? What is the minimum distance that the driver needs to complete a 90° turn (without braking)? What is the safest tactic for the driver?

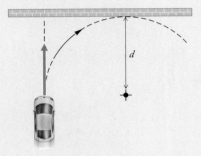

FIGURE 6.40 Automobile approaching a brick wall.

*65. A mass is attached to the lower end of a string of length l; the upper end of the string is held fixed. Suppose that the string initially makes an angle θ with the vertical. With what horizontal velocity must we launch the mass so that it continues to travel at constant speed along a horizontal circular path under the influence of the combined forces of the tension of the string and gravity? This device is called a **conical pendulum** (Fig. 6.41).

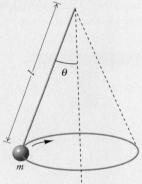

FIGURE 6.41 Mass suspended from a string swinging around a circle (conical pendulum).

*66. An automobile of mass 1200 kg rounds a curve at a speed of 25 m/s. The radius of the curve is 400 m, and its banking angle is 6.0°. What is the magnitude of the normal force on the automobile? The friction force?

*67. An airplane flies in a horizontal circular path at 320 km/h. Looking at the horizon, a passenger notices that the angle of bank of the airplane is 30°. What radius of the circular path can the passenger deduce from these data? [Hint: The force exerted by the air on the wings (lift) is perpendicular to the wings.]

*68. A mass m_1 slides on a smooth, frictionless table. The mass is constrained to move in a circle by a string that passes through a hole in the center of the table and is attached to a second mass m_2 hanging below the table. The second mass swings in a circle, so the string makes an angle θ with the vertical (see Fig. 6.42). The two masses move around their circles in unison, so they are always at diametrically opposite points from the hole. If the radius of the circular motion of the first mass is r_1, what must be the radius of the circular motion of the second mass?

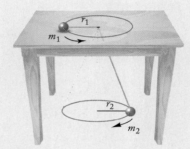

FIGURE 6.42 Mass on table and hanging mass are both in circular motion.

**69. Two masses m_1 and m_2 hang at the ends of a string that passes over a small pulley. The masses swing along circular arcs of equal radii and $m_1 > m_2$. Find a relation between m_1, m_2, and the initial angles θ_1 and θ_2 of swing, where the masses are at rest, if both ends of the string are to be under the same tension, so that the string is in equilibrium. Assume that the two masses later reach the bottom in equal times, so that the mass m_1 is moving faster there. Will the string be in equilibrium just before the masses reach the bottom?

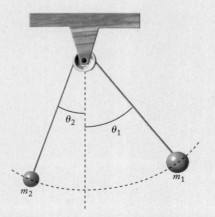

FIGURE 6.43 Masses hanging over pulley.

**70. A flexible drive belt runs over a flywheel turning freely on a frictionless axle (see Fig. 6.44). The mass per unit length of the drive belt is σ, and the tension in the drive belt is T. The speed of the drive belt is v. Show that each small segment $d\theta$

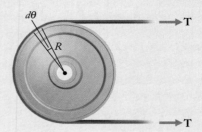

FIGURE 6.44 Belt over flywheel.

of the drive belt exerts a radial force $(T - \sigma v^2)\,d\theta$ on the flywheel. For what value of v is this force zero?

**71. A circle of rope of mass m and radius r is spinning about its center so each point of the rope has a speed v. Calculate the tension in the rope.

**72. The rotor of a helicopter consists of two blades 180° apart. Each blade has a mass of 140 kg and a length of 3.6 m. What is the tension in each blade at the hub when rotating at 320 rev/min? Pretend that each blade is a uniform thin rod.

**73. Assume that the Earth is a sphere and that the force of gravity (mg) points precisely toward the center of the Earth. Taking into account the rotation of the Earth about its axis, calculate the angle between the direction of a plumb line and the direction of the Earth's radius as a function of latitude. What is this deviation angle at a latitude of 45°?

**74. A curve of radius 120 m is banked at an angle of 10°. If an automobile with wheels with $\mu_s = 0.90$ is to round this curve without skidding, what is the maximum permissible speed?

**75. Figure 6.45 shows a pendulum hanging from the edge of a horizontal disk which rotates around its axis at a constant rate. The angle α that the rotating pendulum makes with the vertical increases with the speed of rotation, and can therefore be used as an indicator of this speed. Find a formula for the speed v_0 of the edge of the disk in terms of the angle α, the radius R of the disk, and the length l of the pendulum. If $R = 0.20$ m and $l = 0.30$ m, what is the speed when $\alpha = 45°$?

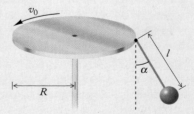

FIGURE 6.45 Pendulum hanging from edge of disk.

REVIEW PROBLEMS

76. At liftoff, the Saturn V rocket used for the Apollo missions has a mass of 2.45×10^6 kg.

    (a) What is the minimum thrust that the rocket engines must develop to achieve liftoff?

    (b) The actual thrust that the engines develop is 3.3×10^7 N. What is the vertical acceleration of the rocket at liftoff?

    (c) At burnout, the rocket has spent its fuel, and its remaining mass is 0.75×10^6 kg. What is the acceleration just before burnout? Assume that the motion is still vertical and that the strength of gravity is the same as when the rocket is on the ground.

77. If the coefficient of static friction between the tires of an automobile and the road is $\mu_s = 0.80$, what is the minimum distance the automobile needs in order to stop without skidding from an initial speed of 90 km/h? How long does it take to stop?

78. Suppose that the last car of a train becomes uncoupled while the train is moving upward on a slope of 1:6 at a speed of 48 km/h.

    (a) What is the deceleration of the car? Ignore friction.

    (b) How far does the car coast up the slope before it stops?

79. A 40-kg crate falls off a truck traveling at 80 km/h on a level road. The crate slides along the road and gradually comes to a halt. The coefficient of kinetic friction between the crate and the road is 0.80.

    (a) Draw a "free-body" diagram for the crate sliding on the road.

    (b) What is the normal force the road exerts on the crate?

    (c) What is the friction force the road exerts on the crate?

    (d) What is the weight force on the crate? What is the net force on the crate?

    (e) What is the deceleration of the crate? How far does the crate slide before coming to a halt?

80. A 2.0-kg box rests on an inclined plane which makes an angle of 30° with the horizontal. The coefficient of static friction between the box and the plane is 0.90.

    (a) Draw a "free-body" diagram for the box.

    (b) What is the normal force the inclined plane exerts on the box?

    (c) What is the friction force the inclined plane exerts on the box?

    (d) What is the net force the inclined plane exerts on the box? What is the direction of this force?

81. The body of an automobile is held above the axles of the wheels by means of four springs, one near each wheel. Assume that the springs are vertical and that the forces on all the springs are the same. The mass of the body of the automobile is 1200 kg, and the spring constant of each spring is 2.0×10^4 N/m. When the automobile is stationary on a level road, how far are the springs compressed from their relaxed length?

*82. A block of wood rests on a sheet of paper lying on a table. The coefficient of static friction between the block and the paper is $\mu_s = 0.70$, and that between the paper and the table is $\mu_s = 0.50$. If you tilt the table, at what angle will the block begin to move?

*83. Two blocks of masses m_1 and m_2 are connected by a string. One block slides on a table, and the other hangs from the string, which passes over a pulley (see Fig. 6.46). The coefficient of sliding friction between the first block and the table is $\mu_k = 0.20$. What is the acceleration of the blocks?

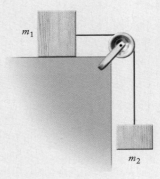

FIGURE 6.46 Mass on table, pulley, and hanging mass.

*84. A man of mass 75 kg is pushing a heavy box on a flat floor. The coefficient of sliding friction between the floor and the box is 0.20, and the coefficient of static friction between the man's shoes and the floor is 0.80. If the man pushes horizontally (see Fig. 6.47), what is the maximum mass of the box he can move?

FIGURE 6.47 Pushing a box.

*85. Two springs of constants 2.0×10^3 N/m and 3.0×10^3 N/m are connected in tandem, and a mass of 5.0 kg hangs vertically from the bottom of the lower spring. By what amount does the mass stretch the combined spring? Each individual spring?

*86. A block of mass 1.5 kg is placed on a flat surface, and it is being pulled horizontally by a spring with a spring constant 1.2×10^3 N/m (see Fig. 6.48). The coefficient of static friction between the block and the table is $\mu_s = 0.60$, and the coefficient of sliding friction is $\mu_k = 0.40$.

(a) By what amount must the spring be stretched to start the block moving?

(b) What is the acceleration of the block if the stretch of the spring is maintained at a constant value equal to that required to start the motion?

(c) By what amount must the spring be stretched to keep the mass moving at constant speed?

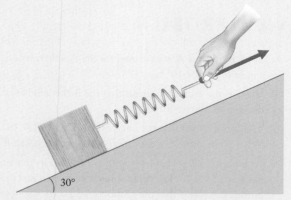

FIGURE 6.49 Block on incline pulled by spring.

*88. A mass m_1 slides on a smooth, frictionless table. The mass is constrained to move in a circle by a string that passes through a hole in the center of the table and is attached to a second mass m_2 hanging vertically below the table (Fig. 6.50). If the radius of the circular motion of the first mass is r, what must be its speed?

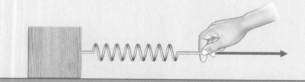

FIGURE 6.48 Mass pulled by spring.

*87. A block of mass 1.5 kg is placed on a plane inclined at 30°, and it is being pulled upward by a spring with a spring constant 1.2×10^3 N/m (see Fig. 6.49). The direction of pull of the spring is parallel to the inclined plane. The coefficient of static friction between the block and the inclined plane is $\mu_s = 0.60$, and the coefficient of sliding friction is $\mu_k = 0.40$.

(a) By what amount must the spring be stretched to start the block moving?

(b) What is the acceleration of the block if the stretch of the spring is maintained at a constant value equal to that required to start the motion?

(c) By what amount must the spring be stretched to keep the mass moving at constant speed?

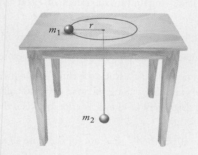

FIGURE 6.50 Mass in circular motion and hanging mass.

89. An automobile enters a curve of radius 45 m at 70 km/h. Will the automobile skid? The curve is not banked, and the coefficient of static friction between the wheels and the road is 0.80.

*90. A stone of 0.90 kg attached to a rod is being whirled around a vertical circle of radius 0.92 m. Assume that during this motion the speed of the stone is constant. If at the top of the circle the tension in the rod is (just about) zero, what is the tension in the rod at the bottom of the circle?

Answers to Checkups

Checkup 6.1

1. The weight of the second book results in a normal force between the first book and the table that is twice as large, so the friction force, and thus the horizontal push to overcome it, will be twice as large, or 20 N. If the first book pushes the second, then the friction force of the second book on the first

adds to the friction force of the first to require a push also twice as large as the original, or 20 N.

2. While the block coasts up the incline, the friction, which always opposes the *motion*, is directed down the plane (the corresponding "free-body" diagram would have the weight

component $mg \sin \theta$ and the friction $\mathbf{f}_k$ both pointing down the incline, the normal force $\mathbf{N}$ perpendicularly out from the incline, and the weight component $mg \cos \theta$ balancing the normal force). When the block slides down the incline, the friction force now points up the incline, again opposing the motion (except for this one reversed vector, the "free-body" diagram is the same).

3. For static friction, $\mu_s N$ is the *maximum* friction force, beyond which a body will begin to move. Since the brick remains at rest, the friction force can be any value less than or equal to $\mu_s N$, and so the correct answer is (c).

4. The static friction force can be no larger than $f_{s,\text{max}} = \mu_s N$; thus the box will not slide if the deceleration is less than is $f_{s,\text{max}}/m$, and will slide if the deceleration is larger than that value.

5. For automotive speeds, the friction force due to air resistance is proportional to the aerodynamic constant C and varies as the square of the speed [Eq. (6.10)], so for the same force (same road slope), the speed varies inversely with the square root of C. Thus, since C is reduced by a factor of 2 (from $C = 0.30$ to $C = 0.15$), the speed will increase by a factor of $\sqrt{2} \approx 1.4$.

6. (C) μ_k and θ only. The motion of the box down the ramp is the same as that of the ship down the slipway in Example 2; from Eq. (6.3), we see $a_x = (\sin \theta - \mu_k \cos \theta)g$. Thus, the acceleration does not depend on the mass, but depends on μ_k and θ. However, the friction force $f_k = \mu_k N = \mu_k mg \cos \theta$ does depend on all three quantities.

Checkup 6.2

1. The weight of the climbers has doubled, so the spring force will double to balance the weight. Since $F = -kx$, where k is constant, the amount of stretch x will double to 40 cm.

2. The mass starts to move when the magnitude of the spring force equals the maximum static friction force, or $kx = \mu_s N = \mu_s mg$. Thus, $\mu_s = kx/mg = (50 \text{ N/m} \times 0.10 \text{ m})/(1.0 \text{ kg} \times 9.8 \text{ m/s}^2) = 0.51$.

3. (E) $\frac{1}{4}$. The relationship between force, spring constant, and distance is $F = -kx$. So for the first spring, $x_1 = -F/k$. If k is twice as large and F half as large, then for the second spring $x_2 = -(F/2)/(2k) = -\frac{1}{4}F/k$. Comparing, we have $x_2/x_1 = \frac{1}{4}$.

Checkup 6.3

1. Once the string breaks, the stone is in free motion; by Newton's First Law, the stone will continue with uniform velocity in a straight line in the direction of its motion when the centripetal force from the string was removed. That direction is tangent to the circle.

2. The acceleration is centripetal, and so is always to the right of the motorcyclist, perpendicular to the instantaneous direction of motion.

3. At the top, the centripetal acceleration points straight downward, toward the center of the circle of motion.

4. The chair exerts a force upward on the pilot, toward the center of the circle, and so that force is centripetal. The pilot exerts a force downward on the chair, away from the center of the circle, and so that force is centrifugal. At the bottom of the loop, the apparent weight is downward (see Example 12). The apparent weight (if defined as the perpendicular force on the seat of the chair) changes direction as the chair swings around, varying in magnitude from $mv^2/r + mg$ at the bottom to $mv^2/r - mg$ at the top. There is an additional force parallel to the seat of the chair at other points (a tangential force), when the weight has a component parallel to the seat of the chair; there is also a tangential force on the chair if the speed v changes around the loop.

5. (D) Inner; inner. Since both cars have the same constant speed, the driver on the inner lane (the circle of smaller radius r) will have less distance ($2\pi r$) to travel, and so will finish first. That inner car also has the larger acceleration, since it has the smaller r, and $a = v^2/r$.

Work and Energy

7.1 Work

7.2 Work for a Variable Force

7.3 Kinetic Energy

7.4 Gravitational Potential
       Energy

CONCEPTS IN CONTEXT

Concepts
— in —
Context

The high-speed and high-acceleration thrills of a roller coaster are made possible by the force of gravity. We will see that gravity does work on the roller-coaster car while it descends, increasing its kinetic energy.

　　To see how energy considerations provide powerful approaches for understanding and predicting motion, we will ask:

? What is the work done by gravity when the roller-coaster car descends along an incline? (Example 3, page 209)

? As a roller-coaster car travels up to a peak, over it, and then down again, does gravity do work? Does the normal force? (Checkup 7.1, question 1, page 210)

? For a complex, curving descent, how can the final speed be determined in a simple way? (Example 8, page 222; and Checkup 7.4, question 1, page 224)

Conservation laws play an important role in physics. Such laws assert that some quantity is conserved, which means that the quantity remains constant even when particles or bodies suffer drastic changes involving motions, collisions, and reactions. One familiar example of a conservation law is the conservation of mass. Expressed in its simplest form, this law asserts that the mass of a given particle remains constant, regardless of how the particle moves and interacts with other particles or other bodies. In the preceding two chapters we took this conservation law for granted, and we treated the particle mass appearing in Newton's Second Law ($m\mathbf{a} = \mathbf{F}$) as a constant, time-independent quantity. More generally, the sum of all the masses of the particles or bodies in a system remains constant, even when the bodies suffer transformations and reactions. In everyday life and in commercial and industrial operations, we always rely implicitly on the conservation of mass. For instance, in the chemical plants that reprocess the uranium fuel for nuclear reactors, the batches of uranium compounds are carefully weighed at several checkpoints during the reprocessing operation to ensure that none of the uranium is diverted for nefarious purposes. This procedure would make no sense if mass were not conserved, if the net mass of a batch could increase or decrease spontaneously.

This chapter and the next deal with the conservation of energy. This conservation law is one of the most fundamental laws of nature. Although we will derive this law from Newton's laws, it is actually much more general than Newton's laws, and it remains valid even when we step outside of the realm of Newtonian physics and enter the realm of relativistic physics or atomic physics, where Newton's laws fail. No violation of the law of conservation of energy has ever been discovered.

In mechanics, *we can use the conservation law for energy to deduce some features of the motion of a particle or of a system of particles* when it is undesirable or too difficult to calculate the full details of the motion from Newton's Second Law. This is especially helpful in those cases where the forces are not known exactly; we will see some examples of this kind in Chapter 11.

But before we can deal with energy and its conservation, we must introduce the concept of work. Energy and work are closely related. We will see that the work done by the net force on a body is equal to the change of the kinetic energy (the energy of motion) of the body.

7.1 WORK

Online
Concept
Tutorial

To introduce the definition of work done by a force, we begin with the simple case of motion along a straight line, with the force along the line of motion, and then we will generalize to the case of motion along some arbitrary curved path, with the force in some arbitrary direction at each point. Consider a particle moving along such a straight line, say, the x axis, and suppose that a constant force F_x, directed along the same straight line, acts on the particle. Then the **work done by the force** F_x *on the particle as it moves some given distance is defined as the product of the force and the displacement* Δx:

$$W = F_x \, \Delta x \qquad\qquad (7.1)$$

work done by one constant force

This rigorous definition of work is consistent with our intuitive notion of what constitutes "work." For example, the particle might be a stalled automobile that you are pushing along a road (see Fig. 7.1). Then the work that you perform is proportional to the magnitude of the force you have to exert, and it is also proportional to the distance you move the automobile.

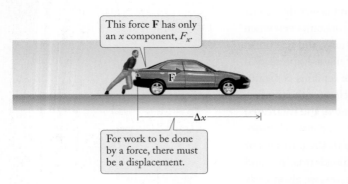

This force **F** has only an x component, F_x.

For work to be done by a force, there must be a displacement.

FIGURE 7.1 You do work while pushing an automobile along a road with a horizontal force **F**.

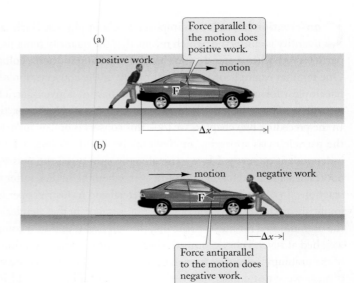

(a)

Force parallel to the motion does positive work.

positive work

motion

(b)

motion   negative work

Force antiparallel to the motion does negative work.

FIGURE 7.2 (a) The work you do on the automobile is positive if you push in the direction of motion. (b) The work you do on the automobile is negative if you push in the direction opposite to the motion.

Note that in Eq. (7.1), F_x is reckoned as positive if the force is in the positive x direction and negative if in the negative x direction. The subscript x on the force helps us to remember that F_x has a magnitude and a sign; in fact, F_x is the x component of the force, and this x component can be positive or negative. According to Eq. (7.1), *the work is positive if the force and the displacement are in the same direction* (both positive, or both negative), *and the work is negative if they are in opposite directions* (one positive, the other negative). When pushing the automobile, you do positive work on the automobile if you push in the direction of the motion, so your push tends to accelerate the automobile (Fig. 7.2a); but you do negative work on the automobile (it does work on you) if you push in the direction opposite to the motion, so your push tends to decelerate the automobile (Fig. 7.2b).

Equation (7.1) gives the work done by one of the forces acting on the particle. If several forces act, then Eq. (7.1) can be used to calculate the work done by each force. If we add the amounts of work done by all the forces acting on the particle, we obtain the net amount of work done by all these forces together. This net amount of work can be directly calculated from the net force:

$$W = F_{\mathrm{net},x}\, \Delta x$$

In the SI system, *the unit of work is the* **joule** (J), which is the work done by a force of 1 N during a displacement of 1 m. Thus,

$$1 \text{ joule} = 1 \text{ J} = 1 \text{ N·m}$$

EXAMPLE 1   Suppose you push your stalled automobile along a straight road (see Fig. 7.1). If the force required to overcome friction and to keep the automobile moving at constant speed is 500 N, how much work must you do to push the automobile 30 m?

SOLUTION: With $F_x = 500$ N and $\Delta x = 30$ m, Eq. (7.1) gives

$$W = F_x\,\Delta x = 500\text{ N} \times 30\text{ m} = 15\,000\text{ J} \qquad (7.2)$$

EXAMPLE 2 A 1000-kg elevator cage descends 400 m within a skyscraper. (a) What is the work done by gravity on the elevator cage during this displacement? (b) Assuming that the elevator cage descends at constant velocity, what is the work done by the tension of the suspension cable?

SOLUTION: (a) With the x axis arranged vertically upward (see Fig. 7.3), the displacement is negative, $\Delta x = -400$ m; and the x component of the weight is also negative, $w_x = -mg = -1000\text{ kg} \times 9.81\text{ m/s}^2 = -9810$ N. Hence by the definition (7.1), the work done by the weight is

$$W = w_x\,\Delta x = (-9810\text{ N}) \times (-400\text{ m}) = 3.92 \times 10^6\text{ J} \qquad (7.3)$$

(b) For motion at constant velocity, the tension force must exactly balance the weight, so the net force $F_{\text{net},x}$ is zero. Therefore, the tension force of the cable has the same magnitude as the weight, but the opposite direction:

$$T_x = +mg = 9810\text{ N}$$

The work done by this force is then

$$W = T_x\,\Delta x = 9810\text{ N} \times (-400\text{ m}) = -3.92 \times 10^6\text{ J} \qquad (7.4)$$

This work is negative because the tension force and the displacement are in opposite directions. Gravity does work on the elevator cage, and the elevator cage does work on the cable.

COMMENTS: (a) Note that the work done by gravity is completely independent of the details of the motion; the work depends on the total vertical displacement and on the weight, but not on the velocity or the acceleration of the motion. (b) Note that the work done by the tension is exactly the negative of the work done by gravity, and thus the net work done by both forces together is zero (we can also see this by examining the work done by the net force; since the net force $F_{\text{net},x} = w_x + T_x$ is zero, the net work $W = F_{\text{net},x}\,\Delta x$ is zero). However, the result (7.4) for the work done by the tension depends implicitly on the assumptions made about the motion. Only for unaccelerated motion does the tension force remain constant at 9810 N. For instance, if the elevator cage were allowed to fall freely with the acceleration of gravity, then the tension would be zero; the work done by the tension would then also be zero, whereas the work done by gravity would still be 3.92×10^6 J.

Although the rigorous definition of work given in Eq. (7.1) agrees to some extent with our intuitive notion of what constitutes "work," the rigorous definition clashes with our

JAMES PRESCOTT JOULE
(1818–1889) *English physicist. He established experimentally that heat is a form of mechanical energy, and he made the first direct measurement of the mechanical equivalent of heat. By a series of meticulous mechanical, thermal, and electrical experiments, Joule provided empirical proof of the general law of conservation of energy.*

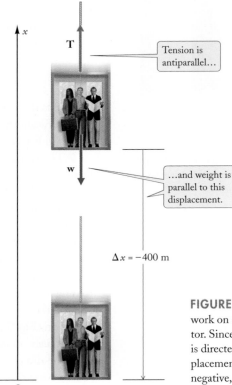

FIGURE 7.3 Gravity does work on a descending elevator. Since the positive x axis is directed upward, the displacement of the elevator is negative, $\Delta x = -400$ m.

No work is done on a stationary ball.

FIGURE 7.4 Man holding a ball. The displacement of the ball is zero; hence the work done on the ball is zero.

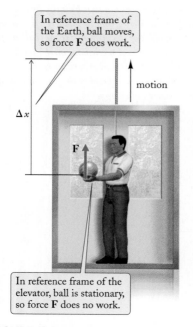

In reference frame of the Earth, ball moves, so force **F** does work.

motion

Δx

F

In reference frame of the elevator, ball is stationary, so force **F** does no work.

FIGURE 7.5 The man holding the ball rides in an elevator. The work done depends on the reference frame.

intuition in some instances. For example, consider a man holding a bowling ball in a fixed position in his outstretched hand (see Fig. 7.4). Our intuition suggests that the man does work—yet Eq. (7.1) indicates that no work is done on the ball, since the ball does not move and the displacement Δx is zero. The resolution of this conflict hinges on the observation that, although the man does no work *on the ball*, he does work *within his own muscles* and, consequently, grows tired of holding the ball. A contracted muscle is never in a state of complete rest; within it, atoms, cells, and muscle fibers engage in complicated chemical and mechanical processes that involve motion and work. This means that work is done, and wasted, internally within the muscle, while no work is done externally on the bone to which the muscle is attached or on the bowling ball supported by the bone.

Another conflict between our intuition and the rigorous definition of work arises when we consider a body in motion. Suppose that the man with the bowling ball in his hand rides in an elevator moving upward at constant velocity (Fig. 7.5). In this case, the displacement is not zero, and the force (push) exerted by the hand on the ball does work—the displacement and the force are in the same direction, and consequently the man continuously does positive work on the ball. Nevertheless, to the man the ball feels no different when riding in the elevator than when standing on the ground. This example illustrates that *the amount of work done on a body depends on the reference frame*. In the reference frame of the ground, the ball is moving upward and work is done on it; in the reference frame of the elevator, the ball is at rest, and no work is done on it. The lesson we learn from this is that before proceeding with a calculation of work, we must be careful to specify the reference frame.

If the motion of the particle and the force are not along the same line, then the simple definition of work given in Eq. (7.1) must be generalized. Consider a particle moving along some arbitrary curved path, and suppose that the force that acts on the particle is constant (we will consider forces that are not constant in the next section). The force can then be represented by a vector **F** (see Fig. 7.6a) that is constant in magnitude and direction. *The work done by this constant force during a (vector) displacement* **s** *is defined as*

$$W = Fs\cos\theta \tag{7.5}$$

where F is the magnitude of the force, s is the length of the displacement, and θ is the angle between the direction of the force and the direction of the displacement. Both F and s in Eq. (7.5) are positive; the correct sign for the work is provided by the factor $\cos\theta$. The work done by the force **F** is positive if the angle between the force and the displacement is less than 90°, and it is negative if this angle is more than 90°.

As shown in Fig. 7.6b, the expression (7.5) can be regarded as the magnitude of the displacement (s) multiplied by the component of the force along the direction of the displacement ($F\cos\theta$). If the force is parallel to the direction of the displacement ($\theta = 0$ and $\cos\theta = 1$), then the work is simply Fs; this coincides with the case of motion along a straight line [see Eq. (7.1)]. If the force is perpendicular to the direction of the displacement ($\theta = 90°$ and $\cos\theta = 0$), then the work vanishes. For instance, if a woman holding a bowling ball walks along a level road at constant speed, she does not do any work on the ball, since the force she exerts on the ball is perpendicular to the direction of motion (Fig. 7.7a). However, if the woman climbs up some stairs while holding the ball, then she does work on the ball, since now the force she exerts has a component along the direction of motion (Fig. 7.7b).

For two arbitrary vectors **A** and **B**, the product of their magnitudes and the cosine of the angle between them is called the **dot product** (or **scalar product**) of the vec-

(a)

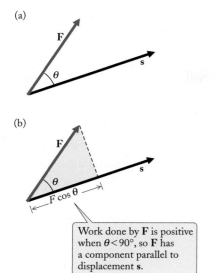

(b)

Work done by **F** is positive when $\theta < 90°$, so **F** has a component parallel to displacement **s**.

FIGURE 7.6 (a) A constant force **F** acts during a displacement **s**. The force makes an angle θ with the displacement. (b) The component of the force along the direction of the displacement is $F \cos \theta$.

(a)

Zero work is done when $\theta = 90°$.

(b)

Positive work is done for $\theta < 90°$.

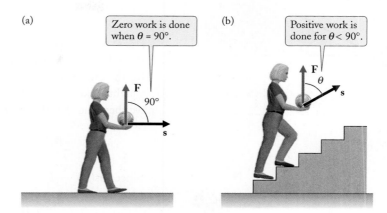

FIGURE 7.7 (a) The force exerted by the woman is perpendicular to the displacement. (b) The force exerted by the woman is now not perpendicular to the displacement.

tors (see Section 3.4). The standard notation for the dot product consists of the two vector symbols separated by a dot:

$$\mathbf{A} \cdot \mathbf{B} = AB \cos \theta \qquad (7.6)$$

dot product (scalar product)

Accordingly, the expression (7.5) for the work can be written as the dot product of the force vector **F** and displacement vector **s**,

$$W = \mathbf{F} \cdot \mathbf{s} \qquad (7.7)$$

In Section 3.4, we found that the dot product is also equal to the sum of the products of the corresponding components of the two vectors, or

$$\mathbf{A} \cdot \mathbf{B} = A_x B_x + A_y B_y + A_z B_z \qquad (7.8)$$

If the components of **F** are F_x, F_y, and F_z and those of **s** are Δx, Δy, and Δz, then the second version of the dot product means that the work can be written

$$W = F_x \, \Delta x + F_y \, \Delta y + F_z \, \Delta z \qquad (7.9)$$

Note that although this equation expresses the work as a sum of contributions from the *x, y,* and *z* components of the force and the displacement, the work does not have separate components. The three terms on the right are merely three terms in a sum. Work is a single-component, scalar quantity, not a vector quantity.

EXAMPLE 3   A roller-coaster car of mass m glides down to the bottom of a straight section of inclined track from a height h. (a) What is the work done by gravity on the car? (b) What is the work done by the normal force? Treat the motion as particle motion.

Concepts —*in*— Context

(a)

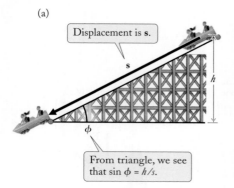

Displacement is **s**.

s

h

φ

From triangle, we see that sin φ = h/s.

(b)

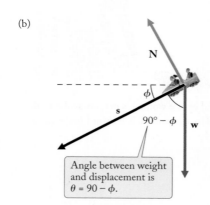

N

φ

s

90° − φ   **w**

Angle between weight and displacement is θ = 90 − φ.

FIGURE 7.8 (a) A roller-coaster car undergoing a displacement along an inclined plane. (b) "Free-body" diagram showing the weight, the normal force, and the displacement of the car.

SOLUTION: (a) Figure 7.8a shows the inclined track. The roller-coaster car moves down the full length of this track. By inspection of the right triangle formed by the incline and the ground, we see that the displacement of the car has a magnitude

$$s = \frac{h}{\sin \phi} \tag{7.10}$$

[Here we use the label ϕ (Greek phi) for the angle of the incline to distinguish it from the angle θ appearing in Eq. (7.5).] Figure 7.8b shows a "free-body" diagram for the car; the forces acting on it are the normal force **N** and the weight **w**. The weight makes an angle $\theta = 90° - \phi$ with the displacement. According to Eq. (7.5), we then find that the work W done by the weight **w** is

$$W = ws \cos \theta = mg \times \frac{h}{\sin \phi} \times \cos(90° - \phi)$$

Since $\cos(90° - \phi) = \sin \phi$, the work is

$$W = mg \times \frac{h}{\sin \phi} \times \sin \phi = mgh \tag{7.11}$$

Alternatively, we can use components to calculate the work. For example, if we choose the x axis horizontal and the y axis vertical, the motion is two-dimensional, and we need to consider x and y components. The components of the weight are $w_x = 0$ and $w_y = -mg$. According to Eq. (7.9), the work done by the weight is then

$$W = w_x \Delta x + w_y \Delta y = 0 \times \Delta x + (-mg) \times \Delta y = 0 + (-mg) \times (-h) = mgh$$

Of course, this alternative calculation agrees with Eq. (7.11).

(b) The work done by the normal force is zero, since this force makes an angle of 90° with the displacement.

COMMENTS: (a) Note that the result (7.11) for the work done by the weight is independent of the angle of the incline—it depends only on the change of height, not on the angle or the length of the inclined plane. (b) Note that the result of zero work for the normal force is quite general. The normal force **N** acting on any body rolling or sliding on any kind of fixed surface never does work on the body, since this force is always perpendicular to the displacement.

✔ Checkup 7.1

Concepts
— *in* —
Context

QUESTION 1: Consider a frictionless roller-coaster car traveling up to, over, and down from a peak. The forces on the car are its weight and the normal force of the tracks. Does the normal force of the tracks perform work on the car? Does the weight?

QUESTION 2: While cutting a log with a saw, you push the saw forward, then pull backward, etc. Do you do positive or negative work on the saw while pushing it forward? While pulling it backward?

QUESTION 3: While walking her large dog on a leash, a woman holds the dog back to a steady pace. Does the dog's pull do positive or negative work on the woman? Does the woman's pull do positive or negative work on the dog?

QUESTION 4: You are trying to stop a moving cart by pushing against its front end. Do you do positive or negative work on the cart? What if you pull on the rear end?

QUESTION 5: You are whirling a stone tied to a string around a circle. Does the tension of the string do any work on the stone?

QUESTION 6: Figure 7.9 shows several equal-magnitude forces **F** and displacements **s**. For which of these is the work positive? Negative? Zero? For which of these is the work largest?

QUESTION 7: To calculate the work performed by a known constant force **F** acting on a particle, which two of the following do you need to know? (1) The mass of the particle; (2) the acceleration; (3) the speed; (4) the displacement; (5) the angle between the force and the displacement.

(A) 1 and 2   (B) 1 and 5   (C) 2 and 3
(D) 3 and 5   (E) 4 and 5

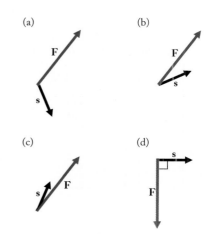

FIGURE 7.9 Several equal-magnitude forces and displacements.

7.2 WORK FOR A VARIABLE FORCE

Online *Concept* Tutorial

The definition of work in the preceding section assumed that the force was constant (in magnitude and in direction). But many forces are not constant, and we need to refine our definition of work so we can deal with such forces. For example, suppose that you push a stalled automobile along a straight road, and suppose that the force you exert is not constant—as you move along the road, you sometimes push harder and sometimes less hard. Figure 7.10 shows how the force might vary with position. (The reason why you sometimes push harder is irrelevant—maybe the automobile passes through a muddy portion of the road and requires more of a push, or maybe you get impatient and want to hurry the automobile along; all that is relevant for the calculation of the work is the value of the force at different positions, as shown in the plot.)

Such a variable force can be expressed as a function of position:

$$F_x = F_x(x)$$

(here the subscript indicates the x component of the force, and the x in parentheses indicates that this component is a function of x; that is, it varies with x, as shown in the diagram). To evaluate the work done by this variable force on the automobile, or on a particle, during a displacement from $x = a$ to $x = b$, we divide the total displacement into a large number of small intervals, each of length Δx (see Fig. 7.11). The beginnings and ends of these intervals are located at $x_0, x_1, x_2, \ldots, x_n$, where the first location x_0 coincides with a and the last location x_n coincides with b. Within each of the small intervals, the force can be regarded as approximately constant—within the interval x_{i-1} to x_i (where $i = 1$, or 2, or 3, . . . , or n), the force is approximately $F_x(x_i)$. This approximation is at its best if we select Δx to be very small. The work done by this force as the particle moves from x_{i-1} to x_i is then

$$W_i = F_x(x_i)\,\Delta x \tag{7.12}$$

and the total work done as the particle moves from a to b is simply the sum of all the small amounts of work associated with the small intervals:

$$W = \sum_{i=1}^{n} W_i = \sum_{i=1}^{n} F_x(x_i)\,\Delta x \tag{7.13}$$

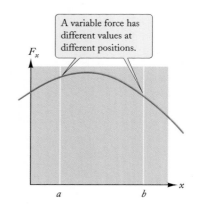

A variable force has different values at different positions.

FIGURE 7.10 Plot of F_x vs. x for a force that varies with position.

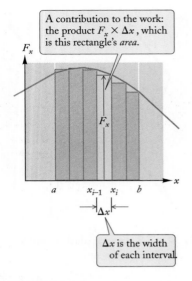

A contribution to the work: the product $F_x \times \Delta x$, which is this rectangle's *area*.

Δx is the width of each interval.

FIGURE 7.11 The curved plot of F_x vs. x has been approximated by a series of horizontal and vertical steps. This is a good approximation if Δx is very small.

Note that each of the terms $F_x(x_i)\, \Delta x$ in the sum is the area of a rectangle of height $F_x(x_i)$ and width Δx, highlighted in color in Fig. 7.11. Thus, Eq. (7.13) gives the sum of all the rectangular areas shown in Fig. 7.11.

Equation (7.13) is only an approximation for the work. In order to improve this approximation, we must use a smaller interval Δx. In the limiting case $\Delta x \to 0$ (and $n \to \infty$), the width of each rectangle approaches zero and the number of rectangles approaches infinity, so we obtain an exact expression for the work. Thus, the exact definition for the work done by a variable force is

$$W = \lim_{\Delta x \to 0} \sum_{i=1}^{n} F_x(x_i)\, \Delta x$$

This expression is called the **integral** of the function $F_x(x)$ between the limits a and b. The usual notation for this integral is

$$W = \int_a^b F_x(x)\, dx \tag{7.14}$$

where the symbol $\int$ is called the integral sign and the function $F_x(x)$ is called the integrand. The quantity (7.14) is equal to *the area bounded by the curve representing* $F_x(x)$, *the x axis, and the vertical lines $x = a$ and $x = b$* in Fig. 7.12. More generally, for a curve that has some portions above the x axis and some portions below, the quantity (7.14) is the net area bounded by the curve above and below the x axis, with areas above the x axis being reckoned as positive and areas below the x axis as negative.

We will also need to consider arbitrarily small contributions to the work. From Eq. (7.12), the infinitesimal work dW done by the force $F_x(x)$ when acting over an infinitesimal displacement dx is

$$dW = F_x(x)\, dx \tag{7.15}$$

We will see later that the form (7.15) is useful for calculations of particular quantities, such as power or torque.

Finally, if the force is variable and the motion is in more than one dimension, the work can be obtained by generalizing Eq. (7.7):

$$W = \int \mathbf{F} \cdot d\mathbf{s} \tag{7.16}$$

work done by a variable force

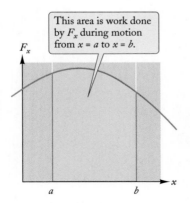

This area is work done by F_x during motion from $x = a$ to $x = b$.

FIGURE 7.12 The integral $\int_a^b F_x(x)\,dx$ is the area (colored) under the curve representing $F_x(x)$ between $x = a$ and $x = b$.

To evaluate Eq. (7.16), it is often easiest to express the integral as the sum of three integrals, similar to the form of Eq. (7.9). For now, we consider the use of Eq. (7.14) to determine the total work done by a variable force as it acts over some distance in one dimension.

EXAMPLE 4   A spring exerts a restoring force $F_x(x) = -kx$ on a particle attached to it (compare Section 6.2). What is the work done by the spring on the particle when it moves from $x = a$ to $x = b$?

SOLUTION: By Eq. (7.14), the work is the integral

$$W = \int_a^b F_x(x)\, dx = \int_a^b (-kx)\, dx$$

To evaluate this integral, we rely on a result from calculus (see the Math Help box on integrals) which states that the integral between a and b of the function x is the difference between the values of $\frac{1}{2}x^2$ at $x = b$ and $x = a$:

$$\int_a^b x \, dx = \frac{1}{2}x^2 \Big|_a^b = \frac{1}{2}(b^2 - a^2)$$

where the vertical line | means that we evaluate the preceding function at the upper limit and then subtract its value at the lower limit. Since the constant $-k$ is just a multiplicative factor, we may pull it outside the integral and obtain for the work

$$W = \int_a^b (-kx) \, dx = -k \int_a^b x \, dx = -\frac{1}{2}k(b^2 - a^2) \qquad (7.17)$$

This result can also be obtained by calculating the area in a plot of force vs. position. Figure 7.13 shows the force $F(x) = -kx$ as a function of x. The area of the quadrilateral $aQPb$ that represents the work W is the difference between the areas of the two triangles OPb and OQa. The triangular area above the $F_x(x)$ curve between the origin and $x = b$ is $\frac{1}{2}$ [base] $\times$ [height] $= \frac{1}{2}b \times kb = \frac{1}{2}kb^2$. Likewise, the triangular area between the origin and $x = a$ is $\frac{1}{2}ka^2$. The difference between these areas is $\frac{1}{2}k(b^2 - a^2)$. Taking into account that areas below the x axis must be reckoned as negative, we see that the work W is $W = -\frac{1}{2}k(b^2 - a^2)$, in agreement with Eq. (7.17).

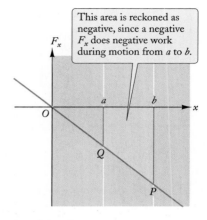

This area is reckoned as negative, since a negative F_x does negative work during motion from a to b.

FIGURE 7.13 The plot of the force $F = -kx$ is a straight line. The work done by the force as the particle moves from a to b equals the (colored) quadrilateral area $aQPb$ under this plot.

MATH HELP   INTEGRALS

The following are some theorems for integrals that we will frequently use.

The integral of a constant times a function is the constant times the integral of the function:

$$\int_a^b cf(x) \, dx = c \int_a^b f(x) \, dx$$

The integral of the sum of two functions is the sum of the integrals:

$$\int_a^b [f(x) + g(x)] \, dx = \int_a^b f(x) \, dx + \int_a^b g(x) \, dx$$

The integral of the function x^n (for $n \neq -1$) is

$$\int_a^b x^n \, dx = \frac{1}{n+1} x^{n+1} \Big|_a^b = \frac{1}{n+1}(b^{n+1} - a^{n+1})$$

In tables of integrals, this is usually written in the compact notation

$$\int x^n \, dx = \frac{x^{n+1}}{n+1} \qquad \text{(for } n \neq -1)$$

where it is understood that the right side is to be evaluated at the upper and at the lower limits of integration and then subtracted.

In a similar compact notation, here are a few more integrals of widely used functions (the quantity k is any constant):

$$\int \frac{1}{x} \, dx = \ln x$$

$$\int e^{kx} \, dx = \frac{1}{k} e^{kx}$$

$$\int \sin(kx) \, dx = -\frac{1}{k} \cos(kx)$$

$$\int \cos(kx) \, dx = \frac{1}{k} \sin(kx)$$

Appendix 4 gives more information on integrals.

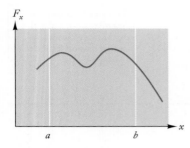

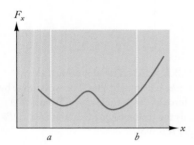

FIGURE 7.14 Two examples of plots of variable forces.

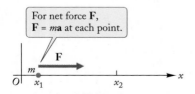

FIGURE 7.15 A particle moves on a straight line from x_1 to x_2 while a force **F** acts on it.

*For net force **F**,*
$\mathbf{F} = m\mathbf{a}$ at each point.

✔ Checkup 7.2

QUESTION 1: Figure 7.14 shows two plots of variable forces acting on two particles. Which of these forces will perform more work during a displacement from a to b?

QUESTION 2: Suppose that a spring exerts a force $F_x(x) = -kx$ on a particle. What is the work done by the spring as the particle moves from $x = -b$ to $x = +b$?

QUESTION 3: What is the work that *you* must do to pull the end of the spring described in Example 4 from $x = a$ to $x = b$?

QUESTION 4: An amount of work W is performed to stretch a spring by a distance d from equilibrium. How much work is performed to further stretch the spring from d to $2d$?

(A) $\frac{1}{2}W$ (B) W (C) $2W$ (D) $3W$ (E) $4W$

7.3 KINETIC ENERGY

In everyday language, energy means a capacity for vigorous activities and hard work. Likewise, in the language of physics, energy is a capacity for performing work. Energy is "stored" work, or latent work, which can be converted into actual work under suitable conditions. *A body in motion has energy of motion, or kinetic energy.* For instance, a speeding arrow has kinetic energy that will be converted into work when the arrow strikes a target, such as a the trunk of a tree. The tip of the arrow then performs work on the wood, prying apart and cutting the wood fibers. The arrow continues to perform work and to penetrate the wood for a few centimeters, until all of its kinetic energy has been exhausted. A high-speed arrow has a deeper penetration and delivers a larger amount of work to the target than a low-speed arrow. Thus, we see that the kinetic energy of the arrow, or the kinetic energy of any kind of particle, must be larger if the speed is larger.

We now examine how work performed by or on a particle is related to changes of the speed of the particle. For clarity, we consider the work done *on* a particle by the net external force F_{net} acting on it (rather than the work done *by* the particle). When the force F_{net} acts on the particle, it accelerates the particle; if the acceleration has a component along the direction of motion of the particle, it will result in a change of the speed of the particle. The force does work on the particle and "stores" this work in the particle; or, if this force decelerates the particle, it does negative work on the particle and removes "stored" work.

We can establish an important identity between the work done by the net force and the change of speed it produces. Let us do this for the simple case of a particle moving along a straight line (see Fig. 7.15). If this straight line coincides with the x axis, then the work done by the net force $F_{\text{net},x}$ during a displacement from x_1 to x_2 is

$$W = \int_{x_1}^{x_2} F_{\text{net},x}\, dx \tag{7.18}$$

By Newton's Second Law, the net force equals the mass m times the acceleration $a = dv/dt$, and therefore the integral equals

$$\int_{x_1}^{x_2} F_{\text{net},x}\, dx = \int_{x_1}^{x_2} ma\, dx = m \int_{x_1}^{x_2} \frac{dv}{dt}\, dx \tag{7.19}$$

The velocity v is a function of time; but in the integral (7.19) it is better to regard the velocity as a function of x, and to rewrite the integrand as follows:

$$\frac{dv}{dt} = \frac{dv}{dx}\frac{dx}{dt} = \frac{dv}{dx}v = v\frac{dv}{dx} \tag{7.20}$$

Consequently, the work becomes

$$m\int_{x_1}^{x_2}\frac{dv}{dt}\,dx = m\int_{x_1}^{x_2}v\frac{dv}{dx}\,dx = m\int_{v_1}^{v_2}v\,dv = m\left.\frac{1}{2}v^2\right|_{v_1}^{v_2} \tag{7.21}$$

$$= \tfrac{1}{2}mv_2^2 - \tfrac{1}{2}mv_1^2$$

or

$$W = \tfrac{1}{2}mv_2^2 - \tfrac{1}{2}mv_1^2 \tag{7.22}$$

This shows that the change in the square of the speed is proportional to the work done by the force.

Although we have here obtained the result (7.22) for the simple case of motion along a straight line, it can be shown that the same result is valid for motion along a curve, in three dimensions.

According to Eq. (7.22), whenever we perform positive work on the particle, we increase the "amount of $\frac{1}{2}mv^2$" in the particle; and whenever we perform negative work on the particle (that is, when we let the particle perform work on us), we decrease the "amount of $\frac{1}{2}mv^2$" in the particle. Thus, *the quantity $\frac{1}{2}mv^2$ is the amount of work stored in the particle, or the* **kinetic energy** *of the particle.* We represent the kinetic energy by the symbol K:

$$K = \tfrac{1}{2}mv^2 \tag{7.23}$$

kinetic energy

With this notation, Eq. (7.22) states that *the change of kinetic energy equals the net work done on the particle;* that is,

$$K_2 - K_1 = W \tag{7.24}$$

or

$$\Delta K = W \tag{7.25}$$

work–energy theorem

This result is called the **work–energy theorem**. Keep in mind that the work in Eqs. (7.22), (7.24), and (7.25) must be evaluated with the *net* force; that is, all the forces that do work on the particle must be included in the calculation.

When a force does positive work on a particle initially at rest, the kinetic energy of the particle increases. The particle then has a capacity to do work: if the moving particle subsequently is allowed to push against some obstacle, then this obstacle does negative work on the particle and simultaneously the particle does positive work on the obstacle. When the particle does work, its kinetic energy decreases. The total amount of work the particle can deliver to the obstacle is equal to its kinetic energy. Thus, *the kinetic energy represents the capacity of a particle to do work by virtue of its speed.*

The acquisition of kinetic energy through work and the subsequent production of work by this kinetic energy are neatly illustrated in the operation of a waterwheel driven by falling water. In a flour mill of an old Spanish Colonial design, the water runs down from a reservoir in a steep, open channel (see Fig. 7.16). The motion of the water particles is essentially that of particles sliding down an inclined plane. If we

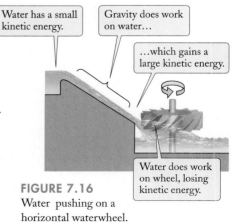

FIGURE 7.16
Water pushing on a horizontal waterwheel.

ignore friction, then the only force that does work on the water particles is gravity. This work is positive, so the kinetic energy of the water increases and it attains a maximum value at the lower end of the channel (where its speed is greatest). The stream of water emerges from this channel with high kinetic energy and hits the blades of the waterwheel. The water pushes on the wheel, turns it, and gives up its kinetic energy while doing work—and the wheel runs the millstones and does useful work on them. Thus, the work that gravity does on the descending water is ultimately converted into useful work, with the kinetic energy playing an intermediate role in this process.

The unit of kinetic energy is the joule, the same as the unit of work. Table 7.1 lists some typical kinetic energies.

FIGURE 7.17 Pitcher throwing a ball. The ball leaves his hand with a speed of 30 m/s.

EXAMPLE 5 During a baseball game, the pitcher throws the ball with a speed of 30 m/s (Fig. 7.17). The mass of the ball is 0.15 kg. What is the kinetic energy of the ball when it leaves his hand? How much work did his hand do on the ball during the throw?

SOLUTION: The final speed of the ball, when it leaves the hand at the end of the throwing motion, is $v_2 = 30$ m/s. The final kinetic energy of the ball is

$$K_2 = \tfrac{1}{2}mv_2^2 = \tfrac{1}{2} \times 0.15 \text{ kg} \times (30 \text{ m/s})^2 = 68 \text{ J} \qquad (7.26)$$

According to the work–energy theorem [Eq. (7.25)], the work done by the hand on the ball equals the change of kinetic energy. Since the initial kinetic energy at the beginning of the throwing motion is zero ($v_1 = 0$), the change of kinetic energy equals the final kinetic energy, and the work is

$$W = K_2 - K_1 = 68 \text{ J} - 0 = 68 \text{ J}$$

Note that for this calculation of the work we did not need to know the (complicated) details of how the force varies during the throwing motion. The work–energy theorem gives us the answer directly.

TABLE 7.1	SOME KINETIC ENERGIES
Orbital motion of Earth	2.6×10^{33} J
Ship *Queen Elizabeth* (at cruising speed)	9×10^{9} J
Jet airliner (Boeing 747 at maximum speed)	7×10^{9} J
Automobile (at 90 km/h)	5×10^{5} J
Rifle bullet	4×10^{3} J
Person walking	60 J
Falling raindrop	4×10^{-5} J
Proton from large accelerator (Fermilab)	1.6×10^{-7} J
Electron in atom (hydrogen)	2.2×10^{-18} J
Air molecule (at room temperature)	6.2×10^{-21} J

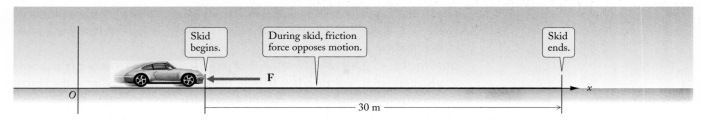

FIGURE 7.18 Automobile skidding on a street.

EXAMPLE 6   While trying to stop his automobile on a flat street, a drunk driver steps too hard on the brake pedal and begins to skid. He skids for 30 m with all wheels locked, leaving skid marks on the pavement, before he releases the brake pedal and permits the wheels to resume rolling (see Fig. 7.18). How much kinetic energy does the automobile lose to friction during this skid? If you find skid marks of 30 m on the pavement, what can you conclude about the initial speed of the automobile? The mass of the automobile is 1100 kg, and the coefficient of sliding friction between the wheels and the street is $\mu_k = 0.90$.

SOLUTION: The magnitude of the sliding friction force is $f_k = \mu_k N = \mu_k mg$. With the x axis along the direction of motion, the x component of this friction force is negative:

$$F_x = -\mu_k mg$$

Since the force is constant, the work done by this force is

$$W = F_x \, \Delta x = -\mu_k mg \times \Delta x$$

$$= -0.90 \times 1100 \text{ kg} \times 9.81 \text{ m/s}^2 \times 30 \text{ m} = -2.9 \times 10^5 \text{ J}$$

According to the work–energy theorem, this work equals the change of kinetic energy:

$$\Delta K = W = -2.9 \times 10^5 \text{ J}$$

Since the kinetic energy of the automobile decreases by 2.9×10^5 J, its initial kinetic energy must have been at least 2.9×10^5 J. Hence its initial speed must have been at least large enough to provide this kinetic energy; that is,

$$\tfrac{1}{2}mv_1^2 \geq 2.9 \times 10^5 \text{ J}$$

and so

$$v_1 \geq \sqrt{\frac{2 \times 2.9 \times 10^5 \text{ J}}{m}} = \sqrt{\frac{2 \times 2.9 \times 10^5 \text{ J}}{1100 \text{ kg}}} = 23 \text{ m/s} = 83 \text{ km/h}$$

 Checkup 7.3

QUESTION 1: Two automobiles of equal masses travel in opposite directions. Can they have equal kinetic energies?

QUESTION 2: A car is traveling at 80 km/h on a highway, and a truck is traveling at 40 km/h. Can these vehicles have the same kinetic energy? If so, what must be the ratio of their masses?

QUESTION 3: Consider a golf ball launched into the air. The ball rises from the ground to a highest point, and then falls back to the ground. At what point is the kinetic energy largest? Smallest? Is the kinetic energy ever zero?

QUESTION 4: A horse is dragging a sled at steady speed along a rough surface, with friction. The horse does work on the sled, but the kinetic energy of the sled does not increase. Does this contradict the work–energy theorem?

QUESTION 5: If you increase the speed of your car by a factor of 3, from 20 km/h to 60 km/h, by what factor do you change the kinetic energy?

(A) $\frac{1}{9}$ (B) $\frac{1}{3}$ (C) 1   (D) 3   (E) 9

PROBLEM-SOLVING TECHNIQUES   CALCULATION OF WORK

In calculations of the work done by a force acting on a body, keep in mind that

- A force that has a component in the direction of the displacement does positive work; a force that has a component in the direction opposite to the displacement does negative work.

- A force perpendicular to the displacement does no work [examples: the normal force acting on a body sliding on a surface, the centripetal force acting on a body in circular motion (uniform or not)].

- For a constant force, the work can be calculated either from Eq. (7.5) or from Eq. (7.9); use the former if you

know the magnitude of the force and the angle, and use the latter if you know the components.

- For a variable force, the calculation of the work involves integration along the path [Eq. (7.14)]; also, Eq. (7.15) can be used for the work during an infinitesimal displacement.

- The work–energy theorem is valid only if the work is calculated with the *net* force. When two of the three quantities (work done, initial kinetic energy, and final kinetic energy) are known, the theorem can be applied to determine the third: $W = K_2 - K_1$.

Online
Concept
Tutorial

7.4 GRAVITATIONAL POTENTIAL ENERGY

As we saw in the preceding section, the kinetic energy represents the capacity of a particle to do work by virtue of its speed. We will now become acquainted with another form of energy that represents the capacity of the particle to do work by virtue of its position in space. This is the **potential energy**. In this section, we will examine the special case of gravitational potential energy for a particle moving under the influence of the constant gravitational force near the surface of the Earth, and we will formulate a law of conservation of energy for such a particle. In the next chapter we will examine other cases of potential energy and formulate the General Law of Conservation of Energy.

The gravitational potential energy represents the capacity of the particle to do work by virtue of its height above the surface of the Earth. When we lift a particle to some height above the surface, we have to do work against gravity, and we thereby store work in the particle. Thus, a particle high above the surface is endowed with a large amount of latent work, which can be exploited and converted into actual work by allowing the particle to push against some obstacle as it descends. A good example of such an exploitation of gravitational potential energy is found in a grandfather clock, where a weight hanging on a cord drives the wheel of the clock (Fig. 7.19). The weight does work on the wheel, and gradually converts all of its gravitational potential energy into work as it descends (in a typical grandfather clock, the weight takes about a week to sink down from the top to the bottom, and you must then rewind the clock, by lifting the weight).

To obtain a general expression for the gravitational potential energy of a particle moving on a straight or a curving path, we first consider a particle moving on an inclined plane. According to Eq. (7.11), when a particle of mass m descends a distance h along an inclined plane, the work done by gravity is

$$W = mgh \qquad (7.27)$$

As already remarked on in Example 3, this result is independent of the angle of inclination of the plane—it depends only on the change of height. More generally, for a curved path, the result is independent of the shape of the path that the particle follows from its starting point to its endpoint. For instance, the curved path and the straight sloping path in Fig. 7.20a lead to exactly the same result (7.27) for the work done by gravity. To recognize this, we simply approximate the curved path by small straight segments (see Fig. 7.20b). Each such small segment can be regarded as a small inclined plane, and therefore the work is mg times the small change of height. The net amount of work for all the small segments taken together is then mg times the net change of height, in agreement with Eq. (7.27).

If the vertical coordinate of the starting point is y_1 and the vertical coordinate of the endpoint is y_2 (see Fig. 7.20), then $h = y_1 - y_2$ and Eq. (7.27) becomes

$$W = mg(y_1 - y_2) \qquad \text{or} \qquad W = -(mgy_2 - mgy_1) \qquad (7.28)$$

According to Eq. (7.28), whenever gravity performs positive work on the particle ($y_1 > y_2$, a descending particle), the "amount of mgy" of the particle decreases; and

FIGURE 7.19 The descending weights of the grandfather clock pull on the cords and do work on the wheel of the clock.

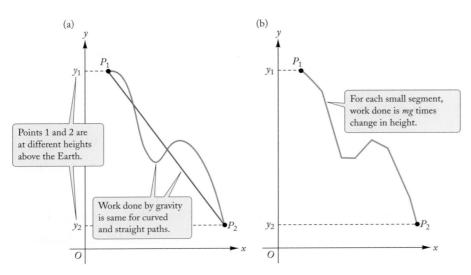

FIGURE 7.20 (a) A curved path (red) and a straight path (blue) from point P_1 to point P_2. (b) The curved path can be approximated by short straight segments.

whenever gravity performs negative work on the particle ($y_1 < y_2$, an ascending particle), the "amount of mgy" increases. Thus, *the quantity mgy represents the amount of stored, or latent, gravitational work; that is, it represents the gravitational potential energy.* We will adopt the notation U for the **gravitational potential energy**:

gravitational potential energy

$$U = mgy \tag{7.29}$$

This potential energy is directly proportional to the height y, and it has been chosen to be zero at $y = 0$ (see Fig. 7.21).

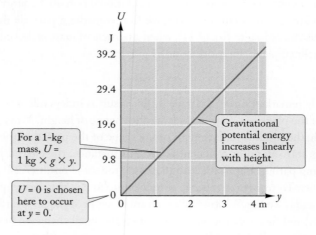

FIGURE 7.21 Plot of the gravitational potential energy of a mass of 1 kg as a function of height y.

In terms of the gravitational potential energy, Eq. (7.28) for the work done by gravity becomes

$$W = -U_2 + U_1 \tag{7.30}$$

Since $\Delta U = U_2 - U_1$ is the change in potential energy, Eq. (7.30) says that the work equals the negative of the change in potential energy,

$$W = -\Delta U \tag{7.31}$$

EXAMPLE 7 What is the kinetic energy and what is the gravitational potential energy (relative to the ground) of a jet airliner of mass 73 000 kg cruising at 240 m/s at an altitude of 9000 m?

SOLUTION: The kinetic energy is

$$K = \tfrac{1}{2}mv^2 = \tfrac{1}{2} \times 7.3 \times 10^4 \text{ kg} \times (240 \text{ m/s})^2 = 2.1 \times 10^9 \text{ J}$$

The gravitational potential energy is $U = mgy$. If we measure the y coordinate from the ground level, then $y = 9000$ m for our airliner, and

$$U = mgy = 7.3 \times 10^4 \text{ kg} \times 9.81 \text{ m/s}^2 \times 9.0 \times 10^3 \text{ m} = 6.4 \times 10^9 \text{ J}$$

We see that the airliner has about three times more potential energy than kinetic energy.

If we let the particle push or pull on some obstacle (such as the wheel of the grandfather clock) during its descent from y_1 to y_2, then the total amount of work that we can extract during this descent is equal to the work done by gravity; that is, it is equal to $-U_2 + U_1 = -(U_2 - U_1) = -\Delta U$, or the negative of the change of potential energy. Of course, the work extracted in this way really arises from the Earth's gravity—the particle can do work on the obstacle because gravity is doing work on the particle. Hence *the gravitational potential energy is really a joint property of the particle and the Earth; it is a property of the configuration of the particle–Earth system.*

If the only force acting on the particle is gravity, then by combining Eqs. (7.24) and (7.30) we can obtain a relation between potential energy and kinetic energy. According to Eq. (7.24), the change in kinetic energy equals the work, or $K_2 - K_1 = W$; and according to Eq. (7.30), the negative of the change in potential energy also equals the work: $W = -U_2 + U_1$. Hence the change in kinetic energy must equal the negative of the change in potential energy:

$$K_2 - K_1 = -U_2 + U_1$$

We can rewrite this as follows:

$$K_2 + U_2 = K_1 + U_1 \tag{7.32}$$

This equality indicates that the quantity $K + U$ is a constant of the motion; that is, it has the same value at the endpoint as it had at the starting point. We can express this as

$$K + U = [\text{constant}] \tag{7.33}$$

The sum of the kinetic and potential energies is called the **mechanical energy** *of the particle.* It is usually designated by the symbol E:

$$E = K + U \tag{7.34}$$

This energy represents the total capacity of the particle to do work by virtue of both its speed and its position.

Equation (7.33) shows that if the only force acting on the particle is gravity, then the mechanical energy remains constant:

$$E = K + U = [\text{constant}] \tag{7.35}$$

This is the **Law of Conservation of Mechanical Energy**.

Since the sum of the potential and kinetic energies must remain constant during the motion, an increase in one must be compensated by a decrease in the other; this means that *during the motion, kinetic energy is converted into potential energy and vice versa.* For instance, if we throw a baseball straight upward from ground level ($y = 0$), the initial kinetic energy is large and the initial potential energy is zero. As the baseball rises, its potential energy increases and, correspondingly, its kinetic energy decreases, so as to keep the sum of the kinetic and potential energies constant. When the baseball reaches its maximum height, its potential energy has the largest value, and the kinetic energy is (instantaneously) zero. As the baseball falls, its potential energy decreases, and its kinetic energy increases (see Fig. 7.22).

Apart from its practical significance in terms of work, the mechanical energy is very helpful in the study of the motion of a particle. If we make use of the formulas for K and U, Eq. (7.35) becomes

$$E = \tfrac{1}{2}mv^2 + mg\,y = [\text{constant}] \tag{7.36}$$

CHRISTIAAN HUYGENS (1629–1695)
Dutch mathematician and physicist. He invented the pendulum clock, made improvements in the manufacture of telescope lenses, and discovered the rings of Saturn. Huygens investigated the theory of collisions of elastic bodies and the theory of oscillations of the pendulum, and he stated the Law of Conservation of Mechanical Energy for motion under the influence of gravity.

mechanical energy

Law of Conservation of
Mechanical Energy

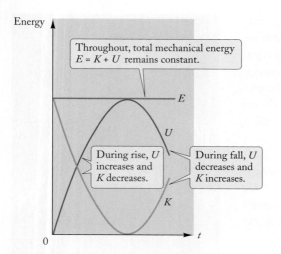

FIGURE 7.22 Kinetic energy K, potential energy U, and mechanical energy $E = K + U$ as functions of time during the upward and downward motions of a baseball.

This shows explicitly how the baseball, or any other particle moving under the influence of gravity, trades speed for height during the motion: whenever y increases, v must decrease (and conversely) so as to keep the sum of the two terms on the left side of Eq. (7.36) constant.

If we consider the vertical positions (y_1 and y_2) and speeds (v_1 and v_2) at two different times, we can equate the total mechanical energy at those two times:

$$\tfrac{1}{2}mv_1^2 + mgy_1 = \tfrac{1}{2}mv_2^2 + mgy_2$$

Rearranging, we immediately obtain

$$-g\,\Delta y = \tfrac{1}{2}(v_2^2 - v_1^2) \tag{7.37}$$

where $\Delta y = y_2 - y_1$. We recognize Eq. (7.37) as the same form that we obtained when studying the equations of motion [see Eq. (2.29)]. Here, however, the result follows directly from conservation of mechanical energy; we did not need to determine the detailed time dependence of the motion.

An important aspect of Eq. (7.36) is that it is valid not only for a particle in free fall (a projectile), but also for a particle sliding on a surface or a track of arbitrary shape, provided that there is no friction. Of course, under these conditions, besides the gravitational force there also acts the normal force; but this force does no work, and hence does not affect Eq. (7.28), or any of the equations following after it. The next example illustrates how these results can be applied to simplify the study of fairly complicated motions, which would be extremely difficult to investigate by direct calculation with Newton's Second Law. This example gives us a glimpse of the elegance and power of the Law of Conservation of Mechanical Energy.

EXAMPLE 8   A roller-coaster car descends 38 m from its highest point to its lowest. Suppose that the car, initially at rest at the highest point, rolls down this track without friction. What speed will the car attain at the lowest point? Treat the motion as particle motion.

SOLUTION: The coordinates of the highest and the lowest points are $y_1 = 38$ m and $y_2 = 0$, respectively (see Fig. 7.23). According to Eq. (7.36), the energy at the start of the motion for a car initially at rest is

$$E = \tfrac{1}{2}mv_1^2 + mgy_1 = 0 + mgy_1 \tag{7.38}$$

and the energy at the end of the motion is

$$E = \tfrac{1}{2}mv_2^2 + mgy_2 = \tfrac{1}{2}mv_2^2 + 0 \tag{7.39}$$

The conservation of energy implies that the right sides of Eqs. (7.38) and (7.39) are equal:

$$\tfrac{1}{2}mv_2^2 = mgy_1 \tag{7.40}$$

Solving this for v_2, we find

$$v_2 = \sqrt{2gy_1} \tag{7.41}$$

(a)

(b)

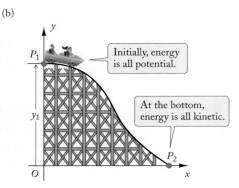

FIGURE 7.23 (a) A roller coaster. (b) Profile of a roller coaster. The roller-coaster car descends from P_1 to P_2.

which gives

$$v_2 = \sqrt{2 \times 9.81 \text{ m/s}^2 \times 38 \text{ m}} = 27 \text{ m/s}$$

Note that according to Eq. (7.41) the final velocity is independent of the mass of the car; since both the kinetic energy and the gravitational potential energy are proportional to mass, the mass cancels in this calculation.

COMMENT: This example illustrates how energy conservation can be exploited to answer a question about motion. To obtain the final speed by direct computation of forces and accelerations would have been extremely difficult—it would have required detailed knowledge of the shape of the path down the hill. With the Law of Conservation of Energy we can bypass these complications.

PROBLEM-SOLVING TECHNIQUES ENERGY CONSERVATION IN ANALYSIS OF MOTION

As illustrated by the preceding example, the use of energy conservation in a problem of motion typically involves three steps:

1 First write an expression for the energy at one point of the motion [Eq. (7.38)].

2 Then write an expression for the energy at another point [Eq. (7.39)].

3 And then rely on energy conservation to equate the two expressions [Eq. (7.40)]. This yields one equation, which can be solved for the unknown final speed or the unknown final position (if the final speed is known).

Note that the value of the gravitational potential energy $U = mgy$ depends on the level from which you measure the y coordinate. However, the *change* in the potential energy does not depend on the choice of this level, and therefore any choice will lead to the same result for the change of kinetic energy. Thus, you can make any choice of zero level, but you must continue to use this choice throughout the entire calculation. You will usually find it convenient to place the zero level for the y coordinate either at the final position of the particle (as in the preceding example), or at the initial position, or at some other distinctive height, such as the bottom of a hill or the ground floor of a building. And always remember that the formula $U = mgy$ for the gravitational potential energy assumes that the y axis is directed vertically upward.

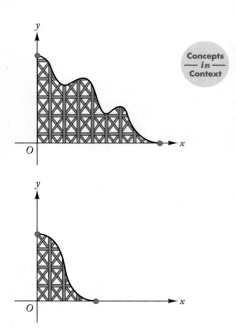

FIGURE 7.24 Two roller-coaster profiles. The two plots have the same vertical scale.

Concepts
— *in* —
Context

✓ Checkup 7.4

QUESTION 1: Figure 7.24 shows two roller coasters in profile. Cars are released at the top of each, from rest. Which, if either, of these roller coasters gives the car a larger speed at the bottom? Neglect friction.

QUESTION 2: A piano is being moved from the second floor of one house to the second floor of another, nearby house. Describe the changes in the gravitational potential energy of the piano during this move.

QUESTION 3: A skidding truck slides down a mountain road, at constant speed. Is the mechanical energy $E = K + U$ conserved?

QUESTION 4: At an amusement park, a girl jumps off a high tower and lands in a pool. Meanwhile, a boy slides down a (frictionless) water slide that also takes him from the tower into the pool. Who reaches the pool with the higher speed? Who reaches the pool first?

QUESTION 5: A bicyclist rolls down a hill without braking, starting at the top, from rest. A second bicyclist rolls down the same hill, starting at one-half the height, from rest. By what factor will the speed of the first bicyclist be larger than that of the second, at the bottom? Ignore friction.

(A) $\sqrt{2}$ (B) 2   (C) $2\sqrt{2}$ (D) 4

SUMMARY

MATH HELP Integrals		**(page 213)**
PROBLEM-SOLVING TECHNIQUES Calculation of Work		**(page 218)**
PROBLEM-SOLVING TECHNIQUES Energy Conservation in Analysis of Motion		**(page 223)**
SI UNIT OF WORK (Unit of energy)	joule $= J = N\cdot m$	

WORK DONE BY A CONSTANT FORCE

Parallel to a displacement Δx $W = F_x \, \Delta x$ **(7.1)**

Not parallel to a displacement **s** $W = Fs\cos\theta = \mathbf{F} \cdot \mathbf{s}$ **(7.6; 7.5)**

DOT PRODUCT (OR SCALAR PRODUCT) OF TWO VECTORS	$\mathbf{A} \cdot \mathbf{B} = AB \cos\theta = A_x B_x + A_y B_y + A_z B_z$	**(7.6; 7.8)**

WORK DONE BY CONSTANT GRAVITATIONAL FORCE (Descending from a height h)	$W = mgh$	**(7.11)**

WORK DONE BY A VARIABLE FORCE In one dimension	$W = \displaystyle\int_a^b F_x(x)\,dx$ 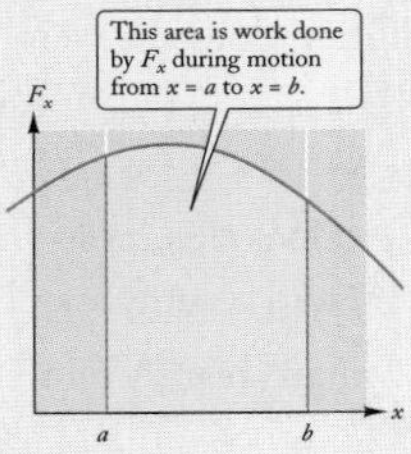	**(7.15)**

This area is work done by F_x during motion from $x = a$ to $x = b$.

In two or three dimensions	$W = \displaystyle\int \mathbf{F} \cdot d\mathbf{s}$	**(7.16)**

WORK DONE BY A SPRING (Moving from $x = a$ to $x = b$)	$W = -\frac{1}{2}k(b^2 - a^2)$	**(7.17)**

KINETIC ENERGY	$K = \frac{1}{2}mv^2$	**(7.23)**

WORK–ENERGY THEOREM	$\Delta K = W$	**(7.25)**

GRAVITATIONAL POTENTIAL ENERGY	$U = mgy$	**(7.29)**

RELATION BETWEEN WORK AND CHANGE IN POTENTIAL ENERGY	$W = -\Delta U$	**(7.31)**

MECHANICAL ENERGY	$E = K + U$	**(7.34)**

CONSERVATION OF MECHANICAL ENERGY	$E = K + U = [\text{constant}]$	**(7.35)**

CONSERVATION OF MECHANICAL ENERGY AT TWO POINTS	$\frac{1}{2}mv_1^2 + mgy_1 = \frac{1}{2}mv_2^2 + mgy_2$ 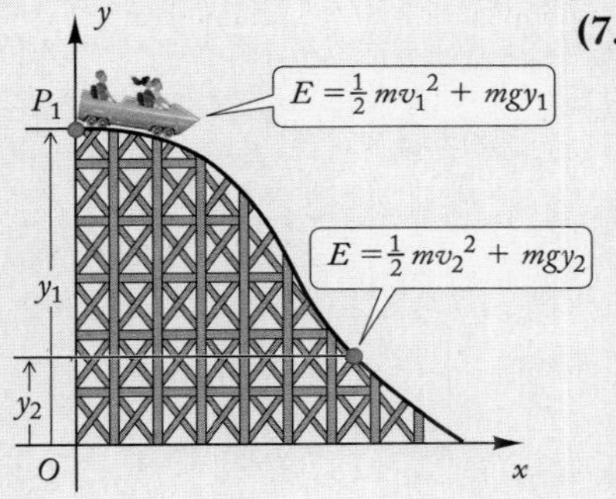	**(7.37)**

$E = \frac{1}{2}mv_1^2 + mgy_1$

$E = \frac{1}{2}mv_2^2 + mgy_2$

QUESTIONS FOR DISCUSSION

1. Does the work of a force on a body depend on the frame of reference in which it is calculated? Give some examples.

2. Does your body do work (external or internal) when standing at rest? When walking steadily along a level road?

3. Consider a pendulum swinging back and forth. During what part of the motion does the weight do positive work? Negative work?

4. Since $v^2 = v_x^2 + v_y^2 + v_z^2$, Eq.(7.23) implies $K = \frac{1}{2}mv_x^2 + \frac{1}{2}mv_y^2 + \frac{1}{2}mv_z^2$. Does this mean that the kinetic energy has x, y, and z components?

5. Consider a woman steadily climbing a flight of stairs. The external forces on the woman are her weight and the normal force of the stairs against her feet. During the climb, the weight does negative work, while the normal force does no work. Under these conditions how can the kinetic energy of the woman remain constant? (Hint: The entire woman cannot be regarded as a particle, since her legs are not rigid; but the upper part of her body can be regarded as a particle, since it is rigid. What is the force of her legs against the upper part of her body? Does this force do work?)

6. An automobile increases its speed from 80 to 88 km/h. What is the percentage of increase in kinetic energy? What is the percentage of reduction of travel time for a given distance?

7. Two blocks in contact slide past one another and exert friction forces on one another. Can the friction force *increase* the kinetic energy of one block? Of both? Does there exist a reference frame in which the friction force decreases the kinetic energy of both blocks?

8. When an automobile with rear-wheel drive is accelerating on, say, a level road, the horizontal force of the road on the rear wheels does not give the automobile any energy because the point of application of this force (point of contact of wheel on ground) is instantaneously at rest if the wheel is not slipping. What force gives the body of the automobile energy? Where does this energy come from? (Hint: Consider the force that the rear axle exerts against its bearings.)

9. Why do elevators have counterweights? (See Fig. 5.40.)

10. A parachutist jumps out of an airplane, opens a parachute, and lands safely on the ground. Is the mechanical energy for this motion conserved?

11. If you release a tennis ball at some height above a hard floor, it will bounce up and down several times, with a gradually decreasing amplitude. Where does the ball suffer a loss of mechanical energy?

12. Two ramps, one steeper than the other, lead from the floor to a loading platform (Fig. 7.25). It takes more force to push a (frictionless) box up the steeper ramp. Does this mean it takes more work to raise the box from the floor to the platform?

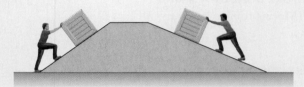

FIGURE 7.25 Two ramps of different steepness.

13. Consider the two ramps described in the preceding question. Taking friction into account, which ramp requires less work for raising a box from the floor to the platform?

14. A stone is tied to a string. Can you whirl this stone in a vertical circle with constant speed? Can you whirl this stone with constant energy? For each of these two cases, describe how you must move your hand.

PROBLEMS

7.1 Work[†]

1. If it takes a horizontal force of 300 N to push a stalled automobile along a level road at constant speed, how much work must you do to push this automobile a distance of 5.0 m?

2. In an overhead lift, a champion weight lifter raises 254 kg from the floor to a height of 1.98 m. How much work does he do?

3. Suppose that the force required to push a saw back and forth through a piece of wood is 35 N. If you push this saw back and forth 30 times, moving it forward 12 cm and back 12 cm each time, how much work do you do?

4. It requires 2200 J of work to lift a 15-kg bucket of water from the bottom of a well to the top. How deep is the well?

5. A child drags a 20-kg box at constant speed across a lawn for 10 m and along a sidewalk for 30 m; the coefficient of friction is 0.25 for the first part of the trip and 0.55 for the second. If the child always pulls horizontally, how much work does the child do on the box?

6. A man moves a vacuum cleaner 1.0 m forward and 1.0 m back 300 times while cleaning a floor, applying a force of 40 N during each motion. The pushes and pulls make an angle of 60° with the horizontal. How much work does the man do on the vacuum cleaner?

[†] For help, see Online Concept Tutorial 9 at www.wwnorton.com/physics

7. A record for stair climbing was achieved by a man who raced up the 1600 steps of the Empire State Building to a height of 320 m in 10 min 59 s. If his average mass was 75 kg, how much work did he do against gravity? At what average rate (in J/s) did he do this work?

8. Suppose you push on a block sliding on a table. Your push has a magnitude of 50 N and makes a downward angle of 60° with the direction of motion. What is the work you do on the block while the block moves a distance of 1.6 m?

9. Consider the barge being pulled by two tugboats, as described in Example 4 of Chapter 5. The pull of the first tugboat is 2.5×10^5 N at 30° to the left, and the pull of the second tugboat is 1.0×10^5 N at 15° to the right (see Fig. 7.26). What is the work done by each tugboat on the barge while the barge moves 100 m forward (in the direction of the x axis in Fig. 7.26)? What is the total work done by both tugboats on the barge?

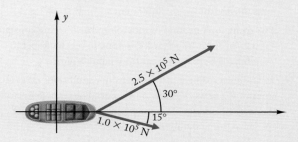

FIGURE 7.26 A barge pulled by two tugboats.

10. A 2.0-kg stone thrown upward reaches a height of 4.0 m at a horizontal distance of 6.0 m from the point of launch. What is the work done by gravity during this displacement?

*11. A man pushes a heavy box up an inclined ramp making an angle of 30° with the horizontal. The mass of the box is 60 kg, and the coefficient of kinetic friction between the box and the ramp is 0.45. How much work must the man do to push the box to a height of 2.5 m at constant speed? Assume that the man pushes on the box in a direction parallel to the surface of the ramp.

12. The driver of a 1200-kg automobile notices that, with its gears in neutral, it will roll downhill at a constant speed of 110 km/h on a road of slope 1:20. Draw a "free-body" diagram for the automobile, showing the force of gravity, the normal force (exerted by the road), and the friction force (exerted by the road and by air resistance). What is the magnitude of the friction force on the automobile under these conditions? What is the work done by the friction force while the automobile travels 1.0 km down the road?

13. Driving an automobile down a slippery, steep hill, a driver brakes and skids at constant speed for 10 m. If the automobile mass is 1700 kg and the angle of slope of the hill is 25°, how much work does gravity do on the car during the skid? How much work does friction do on the car?

14. The automobile in Example 6 of Chapter 6 is traveling on a flat road. For a trip of length 250 km, what is the total work done against air friction when traveling at 20 m/s? At 30 m/s?

15. A constant force of 25 N is applied to a body while it moves along a straight path for 12 m. The force does 175 J of work on the body. What is the angle between the force and the path of the body?

*16. A strong, steady wind provides a force of 150 N in a direction 30° east of north on a pedestrian. If the pedestrian walks first 100 m north and then 200 m east, what is the total work done by the wind?

*17. A man pulls a cart along a level road by means of a short rope stretched over his shoulder and attached to the front end of the cart. The friction force that opposes the motion of the cart is 250 N.

   (a) If the rope is attached to the cart at shoulder height, how much work must the man do to pull the cart 50 m at constant speed?

   (b) If the rope is attached to the cart below shoulder height so it makes an angle of 30° with the horizontal, what is the tension in the rope? How much work must the man now do to pull the cart 50 m? Assume that enough mass was added so the friction force is unchanged.

*18. A particle moves in the $x-y$ plane from the origin $x = 0$, $y = 0$ to the point $x = 2$, $y = -1$ while under the influence of a force $\mathbf{F} = 3\mathbf{i} + 2\mathbf{j}$. How much work does this force do on the particle during this motion? The distances are measured in meters and the force in newtons.

*19. An elevator consists of an elevator cage and a counterweight attached to the ends of a cable that runs over a pulley (Fig. 7.27). The mass of the cage (with its load) is 1200 kg, and the mass of the counterweight is 1000 kg. The elevator is driven by an electric motor attached to the pulley. Suppose that the elevator is initially at rest on the first floor of the building and the motor makes the elevator accelerate upward at the rate of 1.5 m/s².

   (a) What is the tension in the part of the cable attached to the elevator cage? What is the tension in the part of the cable attached to the counterweight?

   (b) The acceleration lasts exactly 1.0 s. How much work has the electric motor done in this interval? Ignore friction forces and ignore the mass of the pulley.

   (c) After the acceleration interval of 1.0 s, the motor pulls the elevator upward at constant speed until it reaches the third floor, exactly 10.0 m above the first floor. What is the total amount of work that the motor has done up to this point?

1200 kg

1000 kg

FIGURE 7.27
Elevator cage and counterweight.

*20. By means of a towrope, a girl pulls a sled loaded with firewood along a level, icy road. The coefficient of friction between the sled and the road is $\mu_k = 0.10$, and the mass of the sled plus its load is 150 kg. The towrope is attached to the front end of the sled and makes an angle of 30° with the horizontal. How much work must the girl do on the sled to pull it 1.0 km at constant speed?

*21. During a storm, a sailboat is anchored in a 10-m-deep harbor. The wind pushes against the boat with a steady horizontal force of 7000 N.

(a) The anchor rope that holds the boat in place is 50 m long and is stretched straight between the anchor and the boat (Fig. 7.28a). What is the tension in the rope?

(b) How much work must the crew of the sailboat do to pull in 30 m of the anchor rope, bringing the boat nearer to the anchor (Fig. 7.28b)? What is the tension in the rope when the boat is in this new position?

(a)

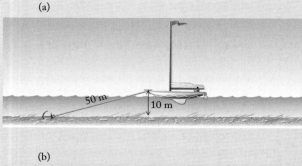

(b)

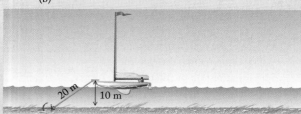

FIGURE 7.28   A sailboat at anchor.

7.2 Work for a Variable Force†

22. The spring used in the front suspension of a Triumph sports car has a spring constant $k = 3.5 \times 10^4$ N/m. How much work must you do to compress this spring by 0.10 m from its relaxed condition? How much more work must you do to compress the spring a further 0.10 m?

23. A particle moving along the x axis is subjected to a force F_x that depends on position as shown in the plot in Fig. 7.29. From this plot, find the work done by the force as the particle moves from $x = 0$ to $x = 8.0$ m.

24. A 250-g object is hung from a vertical spring, stretching it 18 cm below its original equilibrium position. How much work was done by gravity on the object? By the spring?

† For help, see Online Concept Tutorial 9 at www.wwnorton.com/physics

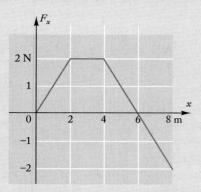

FIGURE 7.29   Position-dependent force.

25. When an ideal, horizontal spring is at equilibrium, a mass attached to its end is at $x = 0$. If the spring constant is 440 N/m, how much work does the spring do on the mass if the mass moves from $x = -0.20$ m to $x = +0.40$ m?

26. The spring on one kind of mousetrap has a spring constant of 4500 N/m. How much work is done to set the trap, by stretching the spring 2.7 cm from equilibrium?

*27. To stretch a spring a distance d from equilibrium takes an amount W_0 of work. How much work does it take to stretch the spring from d to $2d$ from equilibrium? From Nd to $(N + 1)d$ from equilibrium?

*28. A particular spring is not ideal; for a distance x from equilibrium, the spring exerts a force $F_x = -6x - 2x^3$, where x is in meters and F_x is in newtons. Compared with an ideal spring with a spring constant $k = 6.0$ N/m, by what factor does the work done by the nonideal spring exceed that done by the ideal spring when moving from $x = 0$ to $x = 0.50$ m? From $x = 1.0$ m to $x = 1.5$ m? From $x = 2.0$ m to $x = 2.5$ m?

*29. The ends of a relaxed spring of length l and force constant k are attached to two points on two walls separated by a distance l.

(a) How much work must you do to push the midpoint of the spring up or down a distance y (see Fig. 7.30)?

(b) How much force must you exert to hold the spring in this configuration?

*30. A particle moves along the x axis from $x = 0$ to $x = 2.0$ m. A force $F_x(x) = 2x^2 + 8x$ acts on the particle (the distance x is measured in meters, and the force in newtons). Calculate the work done by the force $F_x(x)$ during this motion.

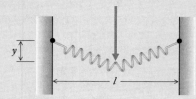

FIGURE 7.30   The midpoint of the spring has been pushed down a distance y. When the spring is relaxed, its length matches the distance l between the walls.

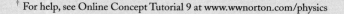

*31. Suppose that the force acting on a particle is a function of position; the force has components $F_x = 4x^2 + 1$, $F_y = 2x$, $F_z = 0$, where the force is measured in newtons and distance in meters. What is the work done by the force if the particle moves on a straight line from $x = 0$, $y = 0$, $z = 0$ to $x = 2.0$ m, $y = 2.0$ m, $z = 0$?

*32. A horse pulls a sled along a snow-covered curved ramp. Seen from the side, the surface of the ramp follows an arc of a circle of radius R (Fig. 7.31). The pull of the horse is always parallel to this surface. The mass of the sled is m, and the coefficient of sliding friction between the sled and the surface is μ_k. How much work must the horse do on the sled to pull it to a height $(1 - \sqrt{2}/2)R$, corresponding to an angle of 45° along the circle (Fig. 7.31)? How does this compare with the amount of work required to pull the sled from the same starting point to the same height along a straight ramp inclined at 22.5°?

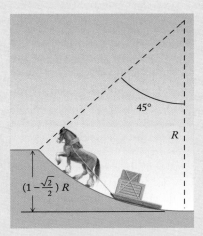

FIGURE 7.31 A horse pulling a sled along a curved ramp.

**33. The force between two inert gas atoms is often described by a function of the form

$$F_x = Ax^{-13} - Bx^{-7}$$

where A and B are positive constants and x is the distance between the atoms. Answer in terms of A and B.

(a) What is the equilibrium separation?

(b) What is the work done if the atoms are moved from their equilibrium separation to a very large distance apart?

7.3 Kinetic Energy

34. In a serve, a champion tennis player sends the ball flying at 160 km/h. The mass of the ball is 60 g. What is the kinetic energy of the ball?

35. Calculate the kinetic energy that the Earth has owing to its motion around the Sun.

36. The electron in a hydrogen atom has a speed of 2.2×10^6 m/s. What is the kinetic energy of this electron?

37. The fastest skier is Graham Wilkie, who attained 212.52 km/h on a steep slope at Les Arcs, France. The fastest runner is Robert Hayes, who briefly attained 44.88 km/h on a level track. Assume that the skier and the runner each have a mass of 75 kg. What is the kinetic energy of each? By what factor is the kinetic energy of the skier larger than that of the runner?

38. The Skylab satellite disintegrated when it reentered the atmosphere. Among the pieces that crashed down on the surface of the Earth, one of the heaviest was a lead-lined film vault of 1770 kg that had an estimated impact speed of 120 m/s on the surface. What was its kinetic energy? How many kilograms of TNT would we have to explode to release the same amount of energy? (One kilogram of TNT releases 4.6×10^6 J.)

39. An automobile of mass 1600 kg is traveling along a straight road at 80 km/h.

(a) What is the kinetic energy of this automobile in the reference frame of the ground?

(b) What is the kinetic energy in the reference frame of a motorcycle traveling in the same direction at 60 km/h?

(c) What is the kinetic energy in the reference frame of a truck traveling in the opposite direction at 60 km/h?

40. According to statistical data, the probability that an occupant of an automobile suffers lethal injury when involved in a crash is proportional to the square of the speed of the automobile. At a speed of 80 km/h, the probability is approximately 3%. What are the probabilities at 95 km/h, 110 km/h, and 125 km/h?

41. For the projectile described in Problem 47 of Chapter 2, calculate the initial kinetic energy ($t = 0$) and calculate the final kinetic energy ($t = 3.0$ s). How much energy does the projectile lose to friction in 3.0 s?

42. Compare the kinetic energy of a 15-g bullet fired at 630 m/s with that of a 15-kg bowling ball released at 6.3 m/s.

43. Compare the kinetic energy of a golf ball ($m = 45$ g) falling at a terminal velocity of 45 m/s with that of a person (75 kg) walking at 1.0 m/s.

44. A child's toy horizontally launches a 20-g ball using a spring that was originally compressed 8.0 cm. The spring constant is 30 N/m. What is the work done by the spring moving the ball from its compressed point to its relaxed position, where the ball is released? What is the kinetic energy of the ball at launch? What is the speed of the ball?

45. A mass of 150 g is held by a horizontal spring of spring constant 20 N/m. It is displaced from its equilibrium position and released from rest. As it passes through equilibrium, its speed is 5.0 m/s. For the motion from the release position to the equilibrium position, what is the work done by the spring? What was the initial displacement?

46. A 60-kg hockey player gets moving by pushing on the rink wall with a force of 500 N. The force is in effect while the skater extends his arms 0.50 m. What is the player's kinetic energy after the push? The player's speed?

47. A 1300-kg communication satellite has a speed of 3.1 km/s. What is its kinetic energy?

48. Suppose you throw a stone straight up so it reaches a maximum height h. At what height does the stone have one-half its initial kinetic energy?

49. The velocity of small bullets can be roughly measured with ballistic putty. When the bullet strikes a slab of putty, it penetrates a distance that is roughly proportional to the kinetic energy. Suppose that a bullet of velocity 160 m/s penetrates 0.80 cm into the putty and a second, identical bullet fired from a more powerful gun penetrates 1.2 cm. What is the velocity of the second bullet?

50. A particle moving along the x axis is subject to a force

$$F_x = -ax + bx^3$$

where a and b are constants.

  (a) How much work does this force do as the particle moves from x_1 to x_2?

  (b) If this is the only force acting on the particle, what is the change of kinetic energy during this motion?

*51. In the "tapping mode" used in atomic-force microscopes, a tip on a cantilever taps against the atoms of a surface to be studied. The cantilever acts as a spring of spring constant 2.5×10^{-2} N/m. The tip is initially displaced away from equilibrium by 3.0×10^{-8} m; it accelerates toward the surface, passes through the relaxed spring position, begins to slow down, and strikes the surface as the displacement approaches 2.5×10^{-8} m. What kinetic energy does the tip have just before striking the surface?

*52. With the brakes fully applied, a 1500-kg automobile decelerates at the rate of 8.0 m/s².

  (a) What is the braking force acting on the automobile?

  (b) If the initial speed is 90 km/h, what is the stopping distance?

  (c) What is the work done by the braking force in bringing the automobile to a stop from 90 km/h?

  (d) What is the change in the kinetic energy of the automobile?

*53. A box of mass 40 kg is initially at rest on a flat floor. The coefficient of kinetic friction between the box and the floor is $\mu_k = 0.60$. A woman pushes horizontally against the box with a force of 250 N until the box attains a speed of 2.0 m/s.

  (a) What is the change of kinetic energy of the box?

  (b) What is the work done by the friction force on the box?

  (c) What is the work done by the woman on the box?

7.4 Gravitational Potential Energy†

54. It has been reported that at Cherbourg, France, waves smashing on the coast lifted a boulder of 3200 kg over a 6.0-m wall. What minimum energy must the waves have given to the boulder?

55. A 75-kg man walks up the stairs from the first to the third floor of a building, a height of 10 m. How much work does he do against gravity? Compare your answer with the food energy he acquires by eating an apple (see Table 8.1).

56. What is the kinetic energy and what is the gravitational potential energy (relative to the ground) of a goose of mass 6.0 kg soaring at 30 km/h at a height of 90 m?

57. Surplus energy from an electric power plant can be temporarily stored as gravitational energy by using this surplus energy to pump water from a river into a reservoir at some altitude above the level of the river. If the reservoir is 250 m above the level of the river, how much water (in cubic meters) must we pump in order to store 2.0×10^{13} J?

58. The track of a cable car on Telegraph Hill in San Francisco rises more than 60 m from its lowest point. Suppose that a car is ascending at 13 km/h along the track when it breaks away from its cable at a height of exactly 60 m. It will then coast up the hill some extra distance, stop, and begin to race down the hill. What speed does the car attain at the lowest point of the track? Ignore friction.

59. In pole vaulting, the jumper achieves great height by converting her kinetic energy of running into gravitational potential energy (Fig. 7.32). The pole plays an intermediate role in this process. When the jumper leaves the ground, part of her translational kinetic energy has been converted into kinetic energy of rotation (with the foot of the pole as the center of rotation) and part has been converted into elastic potential energy of deformation of the pole. When the jumper reaches her highest point, all of this energy has been converted into gravitational potential energy. Suppose that a jumper runs at a speed of 10 m/s. If the jumper converts all of the corresponding kinetic energy into gravitational potential energy, how high will her center of mass rise? The actual height reached by pole vaulters is 5.7 m (measured from the ground). Is this consistent with your calculation?

FIGURE 7.32 A pole vaulter.

60. Because of brake failure, a bicycle with its rider careens down a steep hill 45 m high. If the bicycle starts from rest and there is no friction, what is the final speed attained at the bottom of the hill?

61. Under suitable conditions, an avalanche can reach extremely great speeds because the snow rides down the mountain on a cushion of trapped air that makes the sliding motion nearly frictionless. Suppose that a mass of 2.0×10^7 kg of snow breaks loose from a mountain and slides down into a valley 500 m below the starting point. What is the speed of the snow when it hits the valley? What is its kinetic energy? The explo-

---

† For help, see Online Concept Tutorial 9 at www.wwnorton.com/physics

sion of 1 short ton (2000 lb) of TNT releases 4.2×10^9 J. How many tons of TNT release the same energy as the avalanche?

62. A parachutist of mass 60 kg jumps out of an airplane at an altitude of 800 m. Her parachute opens and she lands on the ground with a speed of 5.0 m/s. How much energy has been lost to air friction in this jump?

63. A block released from rest slides down to the bottom of a plane of incline 15° from a height of 1.5 m; the block attains a speed of 3.5 m/s at the bottom. By considering the work done by gravity and the frictional force, determine the coefficient of friction.

64. A bobsled run leading down a hill at Lake Placid, New York, descends 148 m from its highest to its lowest point. Suppose that a bobsled slides down this hill without friction. What speed will the bobsled attain at the lowest point?

65. A 2.5-g Ping-Pong ball is dropped from a window and strikes the ground 20 m below with a speed of 9.0 m/s. What fraction of its initial potential energy was lost to air friction?

66. A roller coaster begins at rest from a first peak, descends a vertical distance of 45 m, and then ascends a second peak, cresting the peak with a speed of 15 m/s. How high is the second peak? Ignore friction.

67. A skateboarder starts from rest and descends a ramp through a vertical distance of 5.5 m; he then ascends a hill through a vertical distance of 2.5 m and subsequently coasts on a level surface. What is his coasting speed? Ignore friction.

*68. In some barge canals built in the nineteenth century, barges were slowly lifted from a low level of the canal to a higher level by means of wheeled carriages. In a French canal, barges of 70 metric tons were placed on a carriage of 35 tons that was pulled, by a wire rope, to a height of 12 m along an inclined track 500 m long.

(a) What was the tension in the wire rope?

(b) How much work was done to lift the barge and carriage?

(c) If the cable had broken just as the carriage reached the top, what would have been the final speed of the carriage when it crashed at the bottom?

*69. A wrecking ball of mass 600 kg hangs from a crane by a cable of length 10 m. If this wrecking ball is released from an angle of 35°, what will be its kinetic energy when it swings through the lowest point of its arc?

*70. Consider a stone thrown vertically upward. If we take air friction into account, we see that $\frac{1}{2}mv^2 + mgy$ must decrease as a function of time. From this, prove that the stone will take longer for the downward motion than for the upward motion.

*71. A stone of mass 0.90 kg attached to a string swings around a vertical circle of radius 0.92 m. Assume that during this motion the energy (kinetic plus potential) of the stone is constant. If, at the top of the circle, the tension in the string is (just about) zero, what is the tension in the string at the bottom of the circle?

*72. A center fielder throws a baseball of mass 0.17 kg with an initial speed of 28 m/s and an elevation angle of 30°. What is the kinetic energy and what is the potential energy of the

baseball when it reaches the highest point of its trajectory? Ignore friction.

*73. A jet aircraft looping the loop (see Problem 70 in Chapter 4) flies along a vertical circle of diameter 1000 m with a speed of 620 km/h at the bottom of the circle and a speed of 350 km/h at the top of the circle. The change of speed is due mainly to the downward pull of gravity. For the given speed at the bottom of the circle, what speed would you expect at the top of the circle if the thrust of the aircraft's engine exactly balances the friction force of air (as in the case for level flight)?

*74. A pendulum consists of a mass hanging from a string of length 1.0 m attached to the ceiling. Suppose that this pendulum is initially held at an angle of 30° with the vertical (see Fig. 7.33) and then released. What is the speed with which the mass swings through its lowest point? For the same initial angle, at what angle with the vertical will the mass have one-half of this speed?

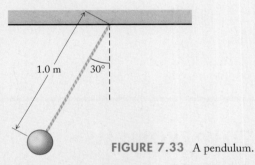

1.0 m   30°

FIGURE 7.33 A pendulum.

**75. A stone is tied to a string of length R. A man whirls this stone in a vertical circle. Assume that the energy of the stone remains constant as it moves around the circle. Show that if the string is to remain taut at the top of the circle, the speed of the stone at the bottom of circle must be at least $\sqrt{5gR}$.

**76. In a loop coaster at an amusement park, cars roll along a track that is bent in a full vertical loop (Fig. 7.34). If the upper portion of the track is an arc of a circle of radius $R = 10$ m, what is the minimum speed that a car must have at the top of the loop if it is not to fall off? If the highest point of the loop has a height $h = 40$ m, what is the minimum speed with which the car must enter the loop at its bottom? Ignore friction.

FIGURE 7.34 A roller coaster with a full loop.

**77. You are to design a roller coaster in which cars start from rest at a height $h = 30$ m, roll down into a valley, and then up a mountain (Fig. 7.35). Ignore friction.

(a) What is the speed of the cars at the bottom of the valley?

(b) If the passengers are to feel $8g$ at the bottom of the valley, what must be the radius R of the arc of the circle that fits the bottom of the valley?

(c) The top of the next mountain is an arc of a circle of the same radius R. If the passengers are to feel $0g$ at the top of this mountain, what must be its height h'?

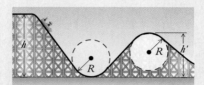

FIGURE 7.35 Profile of a roller coaster.

**78. One portion of the track of a toy roller coaster is bent into a full vertical circle of radius R. A small cart rolling on the track enters the bottom of the circle with a speed $2\sqrt{gR}$. Show that this cart will fall off the track before it reaches the top of the circle, and find the (angular) position at which the cart loses contact with the track.

**79. A particle initially sits on top of a large, smooth sphere of radius R (Fig. 7.36). The particle begins to slide down the sphere, without friction. At what angular position θ will the particle lose contact with the surface of the sphere? Where will the particle land on the ground?

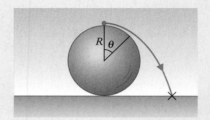

FIGURE 7.36 Particle sliding down a sphere.

REVIEW PROBLEMS

80. An apple falls down 35 m from the fifth floor of an apartment building to the street. The mass of the apple is 0.20 kg. How much work does gravity do on the apple during this fall?

81. A woman pulls a sled by a rope. The rope makes an upward angle of 45° with the ground, and the woman exerts a pull of 150 N on the rope. How much work does the woman do if she pulls this sled 20 m?

82. A man pushes a crate along a flat concrete floor. The mass of the crate is 120 kg, and the coefficient of friction between the crate and the floor is $\mu_k = 0.50$. How much work does the man do if, pushing horizontally, he moves the crate 15 m at constant speed?

83. A 1500-kg automobile is traveling at 20 m/s on a level road. How much work must be done on the automobile to accelerate it from 20 m/s to 25 m/s? From 25 m/s to 30 m/s?

84. A woman slowly lifts a 20-kg package of books from the floor to a height of 1.8 m, and then slowly returns it to the floor. How much work does she do on the package while lifting? How much work does she do on the package while lowering? What is the total work she does on the package? For the information given, can you tell how much work she expends internally in her muscles, that is, how many calories she expends?

85. An automobile of 1200 kg is traveling at 25 m/s when the driver suddenly applies the brakes so as to lock the wheels and cause the automobile to skid to a stop. The coefficient of sliding friction between the tires and the road is 0.90.

(a) What is the deceleration of the automobile, and what is the stopping distance?

(b) What is the friction force of the road on the wheels, and what is the amount of work that this friction force does during the stopping process?

*86. A golf ball of mass 50 g released from a a height of 1.5 m above a concrete floor bounces back to a height of 1.0 m.

(a) What is the kinetic energy of the ball just before contact with the floor begins? Ignore air friction.

(b) What is the kinetic energy of the ball just after contact with the floor ends?

(c) What is the loss of energy during contact with the floor?

87. A small aircraft of mass 1200 kg is cruising at 250 km/h at an altitude of 2000 m.

(a) What is the gravitational potential energy (relative to the ground), and what is the kinetic energy of the aircraft?

(b) If the pilot puts the aircraft into a dive, what will be the gravitational potential energy, what will be the kinetic energy, and what will be the speed when the aircraft reaches an altitude of 1500 m? Assume that the engine of the aircraft compensates the friction force of air, so the aircraft is effectively in free fall.

88. In a roller coaster, a car starts from rest on the top of a 30-m-high mountain. It rolls down into a valley, and then up a 20-m-high mountain. What is the speed of the car at the bottom of the valley, at ground level? What is the speed of the car at the top of the second mountain?

*89. In a compound bow (see Fig. 7.37), the pull of the limbs of the bow is communicated to the arrow by an arrangement of strings and pulleys that ensures that the force of the string against the arrow remains roughly constant while you pull the arrow back in preparation for letting it fly (in an ordinary bow, the force of the string increases as you pull back, which makes it difficult to continue pulling). A typical compound bow provides a steady force of 300 N. Suppose you pull an arrow of 0.020 kg back 0.50 m against this force.

(a) What is the work you do?

(b) When you release the arrow, what is the kinetic energy with which it leaves the bow?

(c) What is the speed of the arrow?

(d) How far will this arrow fly when launched with an elevation angle of 45°? Ignore friction and assume that the heights of the launch and impact points are the same.

(e) With what speed will it hit the target?

FIGURE 7.37   A compound bow.

*90. A large stone-throwing engine designed by Archimedes could throw a 77-kg stone over a range of 180 m. Assume that the stone is thrown at an initial angle of 45° with the horizontal.

(a) Calculate the initial kinetic energy of this stone.

(b) Calculate the kinetic energy of the stone at the highest point of its trajectory.

*91. The luge track at Lillehammer, the site of the 1994 Olympics, starts at a height of 350 m and finishes at 240 m. Suppose that a luger of 95 kg, including the sled starts from rest and reaches the finish at 130 km/h. How much energy has been lost to friction against the ice and the air?

*92. A pendulum consists of a mass m tied to one end of a string of length l. The other end of the string is attached to a fixed point on the ceiling. Suppose that the pendulum is initially held at an angle of 90° with the vertical. If the pendulum is released from this point, what will be the speed of the pendulum at the instant it passes through its lowest point? What will be the tension in the string at this instant?

*93. A roller coaster near St. Louis is 34 m high at its highest point.

(a) What is the maximum speed that the car can attain by rolling down from the highest point if initially at rest? Ignore friction.

(b) Some people claim that cars reach a maximum speed of 100 km/h. If this is true, what must be the initial speed of the car at the highest point?

*94. At a swimming pool, a water slide starts at a height of 6.0 m and ends at a height of 1.0 m above the water level with a short horizontal segment (see Fig. 7.38). A girl slides down the water slide.

(a) What is her speed at the bottom of the slide?

(b) How far from the slide does she land in the water?

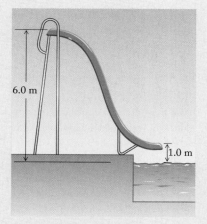

FIGURE 7.38   A water slide.

Answers to Checkups

Checkup 7.1

1. The normal force, which is perpendicular to the motion, does no work. The weight of the roller-coaster car does negative work as the car travels upward, and positive work as the car moves downward, since it has a component against or along the motion, respectively. At the peak, the work done by the weight is zero, since the weight is then perpendicular to the displacement.

2. For both the pushing and the pulling, the force is in the same direction as the displacement (you push when the saw moves forward and pull when the saw moves backward); thus, the work is positive in both cases.

3. The dog's pull is in the same direction as the displacement, and thus does positive work on the woman. The woman's pull is in the opposite direction to the displacement, and thus does negative work on the dog.

4. In each case, the force is opposite to the displacement (whether pushing against the front or pulling on the rear, the force is rearward), and so negative work is done on the cart in both cases.

5. No. The tension provides a centripetal acceleration, which is perpendicular to the (tangential) motion, and thus does no work.

6. The work is positive in (b) and (c), where the angle between the force and displacement is less than 90°; the work is negative in (a), where the angle is greater than 90°. The work is zero in (d), where the force is perpendicular to the displacement. The work is largest when the force is most nearly parallel to the displacement; for force vectors (and displacement vectors) of equal magnitude, this occurs in (c).

7. (E) 4 and 5. To calculate the work done by a constant force, $W = Fs \cos \theta$, you do not need to know the mass, acceleration, or speed. You do need to know the force, the displacement, and the angle between the two.

Checkup 7.2

1. The work done by a variable force is equal to the area under the $F(x)$ vs. x curve. Assuming the two plots are drawn to the same vertical scale, for a displacement from a to b, the upper plot clearly has a greater area between the $F(x)$ curve and the x axis.

2. If we consider a plot such as Fig. 7.13 and imagine extending the curve to the left to $x = -b$ [where $F(x) = +kb$], then we see that *positive* work is done on the particle as it moves from $x = -b$ to $x = 0$ [where the area between the $F(x)$ curve and the x axis is *above* the x axis]. *Negative* work is done on the particle as it moves from $x = 0$ to $x = +b$ [where the area between the $F(x)$ curve and the x axis is *below* the origin]. Thus the net work is zero.

3. The work you must do on the spring is the opposite of what the spring does on you, since the forces involved are an action–reaction pair. Thus the work you do is the negative of the result of Example 4, or $W = +\frac{1}{2}k(b^2 - a^2)$.

4. (D) $3W$. The work to stretch from equilibrium is $\frac{1}{2}kx^2$, so the first stretch requires $W = \frac{1}{2}kd^2$. The second stretch requires work $W' = \frac{1}{2}kx^2 \big|_d^{2d} = \frac{1}{2}k(2d)^2 - \frac{1}{2}kd^2 = 4W - W = 3W$.

Checkup 7.3

1. Yes—the kinetic energy, $K = \frac{1}{2}mv^2$, depends only on the square of the speed, and not on the direction of the velocity. Thus if the two equal masses have the same speed, they have the same kinetic energy.

2. Yes, the kinetic energies can be equal. Since the kinetic energy is proportional to mass and proportional to the square of the speed ($K = \frac{1}{2}mv^2$), if the car has twice the speed of the truck (a factor of 4 contribution to the kinetic energy), then the kinetic energies can be equal if the truck has 4 times the mass of the car.

3. The kinetic energy of the golf ball is largest at the beginning (and end, if we neglect air resistance) of the trajectory; at higher points, the force of gravity has slowed the ball down. The kinetic energy is smallest at the top of the trajectory, where there is only a horizontal contribution to the speed ($v = \sqrt{v_x^2 + v_y^2}$). The kinetic energy is not zero while the ball is in the air (unless the ball was accidentally launched vertically; in that case, the kinetic energy would be zero at the top of the trajectory).

4. No. For the work–energy theorem to apply, one must consider the *net* external force on the sled. If traveling at constant velocity (zero acceleration), the total force must be zero (the horse's pull does positive work and is canceled by the friction force, which does negative work), and so the total work done on the sled is zero. Thus there is no change in kinetic energy.

5. (E) 9. The kinetic energy, $K = \frac{1}{2}mv^2$, is proportional to the square of the speed; thus increasing the speed by a factor of 3 increases the kinetic energy by a factor of 9.

Checkup 7.4

1. As in Example 8, the velocity at the bottom depends only on the height of release (the cars do not even have to have the same mass!); thus, the upper roller coaster will provide the larger speed at the bottom, since Δy is greater.

2. The gravitational potential energy U decreases as the piano is brought to street level from the first house; U remains constant during the trip to the nearby house (assuming travel over flat ground); then, the gravitational potential energy increases back to its original value as the piano is brought up to the second floor of the second house (assuming similar houses).

3. No. At constant speed, K is constant; since U decreases as the truck moves down, $E = K + U$ decreases also, and so is not conserved.

4. Since both the girl and the boy change height by the same amount, they both reach the pool with the same speed (at *any* vertical height, they have the same speed, but the boy's velocity has a horizontal component, so his vertical velocity is slower than that of the girl). Since the girl's velocity is all vertical, a larger vertical velocity implies that she reaches the pool first.

5. (A) $\sqrt{2}$. As in Example 8, the speed at the bottom (starting from rest) is proportional to the square root of the initial height. Thus, for twice the height, the speed of the first bicyclist will be $\sqrt{2}$ times as large at the bottom.

Conservation of Energy

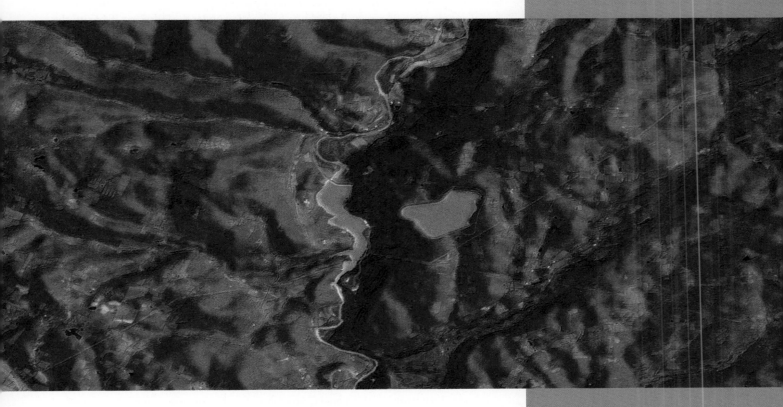

Concepts
in
Context

CONCEPTS IN CONTEXT

The two orange areas in the middle of this satellite image are the reservoirs of the hydroelectric pumped-storage plant on Brown Mountain in New York State. When full, the upper reservoir (at right) holds 19 million cubic meters of water. This reservoir is linked to the lower reservoir at the base, part of the Schoharie Creek, by a 320-m vertical shaft bored through the mountain. The water flowing out of this shaft drives four large turbines that generate electric power. During periods of low demand, the turbines are operated in reverse, so they pump water back into the upper reservoir.

With the concepts developed in this chapter we can address questions such as:

? How do pumped-storage power plants complement other power plants? (Physics in Practice: Hydroelectric Pumped Storage, page 242)

? What is the speed of water spurting out of the shaft at the bottom? (Example 3, page 242)

8.1 Potential Energy of a Conservative Force

8.2 The Curve of Potential Energy

8.3 Other Forms of Energy

8.4 Mass and Energy

8.5 Power

235

? How much gravitational potential energy is stored in the upper reservoir, and how much available electric energy does this represent? (Example 5, page 249)

? When generating power at its maximum capacity, at what rate does the power plant remove water from the upper reservoir? How many hours can it run? (Example 10, page 257)

In the preceding chapter we found how to formulate a law of conservation of mechanical energy for a particle moving under the influence of the Earth's gravity. Now we will seek to formulate the law of conservation of mechanical energy when other forces act on the particle—such as the force exerted by a spring—and we will state the general law of conservation of energy. As in the case of motion under the influence of gravity, the conservation law permits us to deduce some features of the motion without having to deal with Newton's Second Law.

Online *Concept* Tutorial

JOSEPH LOUIS, COMTE LAGRANGE (1736–1813) *French mathematician and theoretical astronomer. In his elegant mathematical treatise* Analytical Mechanics, *Lagrange formulated Newtonian mechanics in the language of advanced mathematics and introduced the general definition of the potential-energy function. Lagrange is also known for his calculations of the motion of planets and for his influential role in securing the adoption of the metric system of units.*

8.1 POTENTIAL ENERGY OF A CONSERVATIVE FORCE

To formulate the law of conservation of energy for a particle moving under the influence of gravity, we began with the work–energy theorem [see Eq. (7.24)],

$$K_2 - K_1 = W \tag{8.1}$$

We then expressed the work W as a difference of two potential energies [see Eq. (7.30)],

$$W = -U_2 + U_1 \tag{8.2}$$

This gave us

$$K_2 - K_1 = -U_2 + U_1$$

from which we immediately found the conservation law for the sum of the kinetic and potential energies, $K_2 + U_2 = K_1 + U_1$, or

$$E = K + U = [\text{constant}] \tag{8.3}$$

As an illustration of this general procedure for the construction of the conservation law for mechanical energy, let us deal with the case of a particle moving under the influence of the elastic force exerted by a spring attached to the particle. If the particle moves along the x axis and the spring lies along this axis, the force has only an x component F_x, which is a function of position:

$$F_x(x) = -kx \tag{8.4}$$

Here, as in Section 6.2, the displacement x is measured from the relaxed position of the spring. The crucial step in the construction of the conservation law is to express the work W as a difference of two potential energies. For this purpose, we take advantage of the result established in Section 7.2 [see Eq. (7.17)], according to which the work done by the spring force during a displacement from x_1 to x_2 is

$$W = \tfrac{1}{2}kx_1^2 - \tfrac{1}{2}kx_2^2 \tag{8.5}$$

This shows that if we identify the **elastic potential energy of the spring** as

potential energy of spring

$$U = \tfrac{1}{2}kx^2 \tag{8.6}$$

then the work is, indeed, the difference between two potential energies $U_1 = \frac{1}{2}kx_1^2$ and $U_2 = \frac{1}{2}kx_2^2$. According to Eq. (8.6), the potential energy of the spring is proportional to the square of the displacement. Figure 8.1 gives a plot of this elastic potential energy.

The potential energy of the spring represents the capacity of the spring to do work by virtue of its deformation. When we compress a spring, we store latent work in it, which we can recover at a later time by letting the spring push against something. An old-fashioned watch, operated by a wound spring, illustrates this storage of energy in a spring (however, the springs in watches are not coil springs, but spiral springs, which are compressed by turning the knob of the watch).

As in the case of the particle moving under the influence of gravity, we conclude that for the particle moving under the influence of the spring force, the sum of the kinetic and elastic potential energies is constant,

$$E = K + U = \tfrac{1}{2}mv^2 + \tfrac{1}{2}kx^2 = [\text{constant}] \qquad (8.7)$$

This equation gives us some information about the general features of the motion; it shows how the particle trades speed for an increase in the distance from the relaxed position of the spring. For instance, an increase of the magnitude of x requires a decrease of the speed v so as to keep the sum $\frac{1}{2}mv^2 + \frac{1}{2}kx^2$ constant.

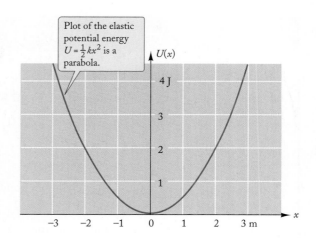

Plot of the elastic potential energy $U = \frac{1}{2}kx^2$ is a parabola.

FIGURE 8.1 Plot of the potential energy of a spring as a function of the displacement x. In this plot, the spring constant is $k = 1$ N/m.

EXAMPLE 1   A child's toy gun shoots a dart by means of a compressed spring. The constant of the spring is $k = 320$ N/m, and the mass of the dart is 8.0 g. Before shooting, the spring is compressed by 6.0 cm, and the dart is placed in contact with the spring (see Fig. 8.2). The spring is then released. What will be the speed of the dart when the spring reaches its relaxed position?

SOLUTION: The dart can be regarded as a particle moving under the influence of a force $F_x = -kx$, with a potential energy $U = \frac{1}{2}kx^2$. Taking the positive x axis along the direction of motion, the initial value of x is negative ($x_1 = -6.0$ cm); also, the initial speed is zero. According to Eq. (8.7), the initial energy is

$$E = \tfrac{1}{2}mv_1^2 + \tfrac{1}{2}kx_1^2 = 0 + \tfrac{1}{2}kx_1^2 \qquad (8.8)$$

When the spring reaches its relaxed position ($x_2 = 0$), the energy will be

$$E = \tfrac{1}{2}mv_2^2 + \tfrac{1}{2}kx_2^2 = \tfrac{1}{2}mv_2^2 + 0 \qquad (8.9)$$

Conservation of energy demands that the right sides of Eqs. (8.8) and (8.9) be equal:

$$\tfrac{1}{2}mv_2^2 = \tfrac{1}{2}kx_1^2 \qquad (8.10)$$

If we cancel the factors of $\frac{1}{2}$ in this equation, divide both sides by m, and take the square root of both sides, we find that the speed of the dart as it leaves the spring at $x_2 = 0$ is

$$v_2 = \sqrt{\frac{k}{m}x_1^2}$$

$$= \sqrt{\frac{320 \text{ N/m}}{0.0080 \text{ kg}} \times (-0.060 \text{ m})^2} = 12 \text{ m/s}$$

$$(8.11)$$

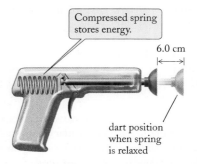

Compressed spring stores energy.

6.0 cm

dart position when spring is relaxed

FIGURE 8.2 A toy gun. The spring is initially compressed 6.0 cm.

PROBLEM-SOLVING TECHNIQUES | ENERGY CONSERVATION

To obtain an expression for the total mechanical energy, you must include terms for the different kinds of energy that are present:

1 Begin with an expression for the energy at one point [Eq. (8.8)].

2 And an expression for the energy at another point [Eq. (8.9)].

3 Then use energy conservation to equate these expressions [Eq. (8.10)].

With the appropriate expression for the mechanical energy, you can apply energy conservation to solve some problems of motion. As illustrated in the preceding example, this involves the three steps outlined in Section 7.4 and 8.1.

CONTRIBUTIONS TO THE MECHANICAL ENERGY

KIND OF ENERGY	APPLICABLE IF	CONTRIBUTION TO TOTAL MECHANICAL ENERGY
Kinetic energy	Particle is in motion	$K = \frac{1}{2}mv^2$
Gravitational potential energy	Particle is moving up or down near the Earth's surface	$U = mgy$
Elastic potential energy	Particle is subject to a spring force	$U = \frac{1}{2}kx^2$

To formulate the law of conservation of mechanical energy for a particle moving under the influence of some other force, we want to imitate the above construction. We will be able to do this if, and only if, the work performed by this force can be expressed as a difference between two potential energies, that is,

$$W = -U_2 + U_1 \qquad (8.12)$$

If the force meets this requirement (and therefore permits the construction of a conservation law), the force is called **conservative.** Thus, the force of gravity and the force of a spring are conservative forces. Note that for any such force, the work done when the particle starts at the point x_1 and *returns* to the same point is necessarily zero, since, with $x_2 = x_1$, Eq. (8.12) implies

$$W = -U_1 + U_1 = 0 \qquad (8.13)$$

This simply means that for a round trip that starts and ends at x_1, the work the force does during the outward portion of the trip is exactly the negative of the work the force does during the return portion of the trip, and therefore the net work for the round trip is zero (see Fig. 8.3). Thus, the energy supplied by the force is recoverable: the energy supplied by the force during motion in one direction is restored during the return motion in the opposite direction. For instance, when a particle moves downward from some starting point, gravity performs positive work; and when the particle moves upward, returning to its starting point, gravity performs negative work of a magnitude exactly equal to that of the positive work.

The requirement of zero work for a round trip can be used to discriminate between conservative and nonconservative forces. Friction is an example of a nonconservative force. If we slide a metal block through some distance along a table and then slide the block back to its starting point, the net work is not zero. The work performed by the friction force during the outward portion of the motion is negative, and the work performed by the friction force during the return portion of the trip is also negative—the friction

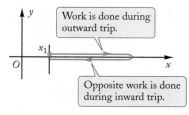

FIGURE 8.3 A particle starts at a point x_1 and returns to the point x_1 after completing some round trip. If the force is conservative, the work done is zero, because the work for the outward portion of the trip is opposite to that for the inward portion.

force always opposes the motion, and the work done by the friction force is always negative. Thus, the work done by the friction force cannot be expressed as a difference between two potential energies, and we cannot formulate a law of conservation of mechanical energy if friction forces are acting. However, as we will see in Section 8.3, we can formulate a more general law of conservation of energy, involving kinds of energy other than mechanical, which remains valid even when there is friction.

In the case of one-dimensional motion, a force is conservative whenever it can be expressed as an explicit function of position, $F_x = F_x(x)$. (Note that the friction force does *not* fit this criterion; the sign of the friction force depends on the direction of motion, and therefore the friction force is not uniquely determined by the position x.) For any such force $F_x(x)$, we can construct the potential energy function by integration. We take a point x_0 as reference point at which the potential energy is zero. The potential energy at any other point x is constructed by evaluating an integral (in the following equations, the integration variables are indicated by primes to distinguish them from the upper limits of integration):

$$U(x) = -\int_{x_0}^{x} F_x(x')\, dx' \qquad (8.14)$$

potential energy as integral of force

To check that this construction agrees with Eq. (8.12), we examine $U_1 - U_2$:

$$U_1 - U_2 = U(x_1) - U(x_2) = -\int_{x_0}^{x_1} F_x(x')\, dx' + \int_{x_0}^{x_2} F_x(x')\, dx'$$

By one of the basic rules for integrals (see Appendix 4), the integral changes sign when we reverse the limits of integration. Hence

$$U_1 - U_2 = \int_{x_1}^{x_0} F_x(x')\, dx' + \int_{x_0}^{x_2} F_x(x')\, dx'$$

And by another basic rule, the sum of an integral from x_1 to x_0 and an integral from x_0 to x_2 is equal to a single integral from x_1 to x_2. Thus

$$U_1 - U_2 = \int_{x_1}^{x_2} F_x(x')\, dx' \qquad (8.15)$$

Here the right side is exactly the work done by the force as the particle moves from x_1 to x_2, in agreement with Eq. (8.12). This confirms that our construction of the potential energy is correct.

In the special case of the spring force $F_x(x) = -kx$, our general construction (8.14) of the potential energy immediately yields the result (8.6), provided we take $x_0 = 0$.

For a particle moving under the influence of any conservative force, the total mechanical energy is the sum of the kinetic energy and the potential energy; as before, this total mechanical energy is conserved:

$$E = K + U = [\text{constant}] \qquad (8.16)$$

or

$$E = \tfrac{1}{2}mv^2 + U = [\text{constant}] \qquad (8.17)$$

conservation of mechanical energy

EXAMPLE 2   As we will see in later chapters, the **inverse-square force** plays a large role in physics—gravitational forces are inverse square, and electric forces are inverse square. If we consider a particle that can move in only one dimension along the positive x axis, this force has the form

$$F_x(x) = \frac{A}{x^2} \tag{8.18}$$

where A is a constant. The point $x = 0$ is called the **center of force**. If A is positive, the force is repulsive (F_x is positive, and the force therefore pushes a particle on the positive x axis away from the center of force); if A is negative, the force is attractive (F_x is negative, and the force pulls the particle toward the center of force). The magnitude of the force is very large near $x = 0$, and it decreases as the distance from this point increases (Figs. 8.4a and b). What is the potential energy for this force?

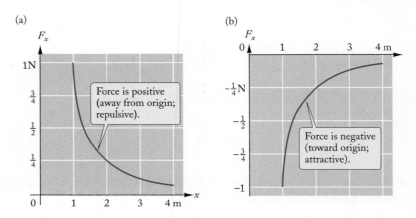

FIGURE 8.4   The inverse-square force A/x^2 as a function of x, (a) for a positive value of A (repulsive force; $A = 1$ N·m^2) and (b) for a negative value of A (attractive force; $A = -1$ N·m^2).

SOLUTION: According to Eq. (8.14),

$$U(x) = -\int_{x_0}^{x} \frac{A}{x'^2}\, dx'$$

In the compact notation of tables of integrals, $\int (1/x'^2)\, dx' = -1/x'$. Hence

$$U(x) = -\left[-\frac{A}{x'} \right]_{x_0}^{x} = -\left[-\frac{A}{x} - \left(-\frac{A}{x_0} \right) \right] = \frac{A}{x} - \frac{A}{x_0}$$

It is usually convenient to take $x_0 = \infty$ as the reference point, with $U_0 = 0$ at $x = \infty$. With this choice,

$$U(x) = \frac{A}{x} \tag{8.19}$$

COMMENT: Note that for a repulsive force ($A > 0$), the potential energy decreases with x (see Fig. 8.5a), and for an attractive force ($A < 0$), the potential energy increases with x (the potential energy is large and negative near $x = 0$, and it increases toward zero as x increases; see Fig. 8.5b).

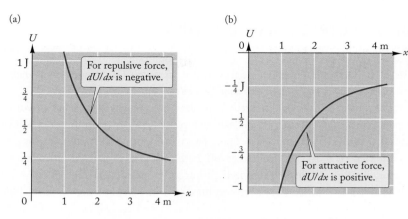

FIGURE 8.5   The potential A/x as a function of x, (a) for a positive value of A and (b) for a negative value of A.

For both the spring force and the inverse-square force, the force can be expressed in terms of the potential energy as $F_x(x) = -dU/dx$; that is, the force is the negative of the derivative of the potential energy. This relationship holds generally, for any kind of conservative force. We can see this by examining the change in potential energy produced by a small displacement dx. From Eq. (8.12) we see that if the points x_1 and x_2 are separated by a small distance $dx = x_2 - x_1$, then [see also Eq. (7.15)]

$$dU = U_2 - U_1 = -dW = -F_x\,dx \qquad (8.20)$$

and if we divide this by dx, we obtain

$$F_x = -\frac{dU}{dx} \qquad (8.21)$$

force as derivative of potential

This relation gives us a quick way to calculate the force if the potential energy is known.

From Eq. (8.21) we see that the force F_x is positive wherever the potential is a decreasing function of x, that is, wherever the derivative dU/dx is negative. Conversely, the force F_x is negative wherever the potential is an increasing function of x, that is, wherever the derivative dU/dx is positive. This is in agreement with the result we found for repulsive and attractive forces in Example 2.

Although in this section we have focused on one-dimensional motion, the criterion of zero work for a round trip is also valid for conservative forces in two or three dimensions. In one dimension, the path for a round trip is necessarily back and forth along a straight line; in two or three dimensions, the path can be of any shape, provided it forms a closed loop that starts and ends at the same point.

Furthermore, the law of conservation of mechanical energy is valid not only for the motion of a single particle, but also for the motion of more general systems, such as systems consisting of solids, liquids, or gases. When applying the conservation law to the kinetic and potential energies of such bodies, it may be necessary to take into account other forms of energy, such as the heat produced by friction and stored in the bodies (see Section 8.3). However, if such other forms of energy stored in the bodies are constant, then we can ignore them in our examination of the motion, as illustrated in the following example of the motion of water in a pipe.

PHYSICS IN PRACTICE   HYDROELECTRIC PUMPED STORAGE

Concepts
— in —
Context

The demand for electric energy by industrial and commercial users is high during working hours, but low during nights and on weekends. For maximum efficiency, electric power companies prefer to run their large nuclear or coal-fired power plants at a steady, full output for 24 hours a day, 7 days a week. Thus electric power companies often have a surplus of electric energy available at night and on weekends, and they often have a deficit of energy during peak-demand times, which requires them to purchase energy from neighboring power companies. Hydroelectric pumped-storage plants help to deal with this mismatch between a fluctuating demand and a steady supply. A hydroelectric pumped-storage plant is similar to an ordinary hydroelectric power plant. It consists of an upper water reservoir and a lower water reservoir, typically separated by a few hundred meters in height. Large pipes (penstocks) connect the upper reservoir to turbines placed at the level of the lower reservoir. The water spurting out of the pipes drives the turbines, which drive electric generators. However, in contrast to an ordinary hydroelectric plant, the pumped-storage plant can be operated in reverse. The electric generators then act as electric motors which drive the turbines in reverse, and thereby pump water from the lower reservoir into the upper reservoir. At peak-demand times the hydroelectric storage plant is used for the generation of electric energy—it converts the gravitational potential energy of the water into electric energy. At low-demand times, the hydroelectric storage plant is used to absorb electric energy—it converts surplus electric energy into grav-

itational potential energy of the water. This gravitational potential energy can then be held in storage until needed.

The chapter photo shows the reservoirs of a large hydroelectric pumped-storage plant on Brown Mountain in New York State. The upper reservoir on top of the mountain is linked to the lower reservoir at the base by a vertical shaft of more than 320 m bored through the mountain. Each of the four reversible pump/turbines (see Fig. 8.17) and motor/generators in the powerhouse at the base (see Fig. 1) is capable of generating 260 MW of electric power. The upper reservoir holds 1.9×10^7 m^3 of water, which is enough to run the generators at full power for about half a day.

FIGURE 1 Powerhouse at the lower reservoir of the Brown Mountain hydroelectric pumped-storage plant.

Concepts
— in —
Context

EXAMPLE 3

At the Brown Mountain hydroelectric storage plant, water from the upper reservoir flows down a pipe in a long vertical shaft (Fig. 8.6). The pipe ends 330 m below the water level of the (full) upper reservoir. Calculate the speed with which the water emerges from the bottom of the pipe. Consider two cases: (a) the bottom of the pipe is wide open, so the pipe does not impede the downward motion of the water; and (b) the bottom of the pipe is closed except for a small hole through which water spurts out. Ignore frictional losses in the motion of the water.

SOLUTION: (a) If the pipe is wide open at the bottom, any parcel of water simply falls freely along the full length of the pipe. Thus, the pipe plays no role at all in the motion of the water, and the speed attained by the water is the same as for a reservoir suspended in midair with water spilling out and falling freely through a height $h = 330$ m. For such free-fall motion, the final speed v can be obtained either from the equations for uniformly accelerated motion [from Eq. (2.29)] or from energy conservation [see Eq. (7.41)]. The result is

$$v = \sqrt{2gh} = \sqrt{2 \times 9.81 \text{ m/s}^2 \times 330 \text{ m}} = 80 \text{ m/s}$$

(b) For a closed pipe with a small hole, the motion of a parcel of water from the top of the upper reservoir to the hole at the bottom of the pipe is complicated and unknown. However, we can find the final speed of the water by relying on the law of energy conservation as applied to the system consisting of the entire volume of water in the reservoir and the pipe. For this purpose, we must examine the kinetic and the potential energy of the water. The water spurting out at the bottom has a large kinetic energy but a low potential energy. In contrast, the water at the top of the upper reservoir has a high potential energy, but next to no kinetic energy (while the water spurts out at the bottom, the water level in the reservoir gradually decreases; but the speed of this downward motion of the water level is very small if the reservoir is large, and this speed can be ignored compared with the large speed of the spurting water).

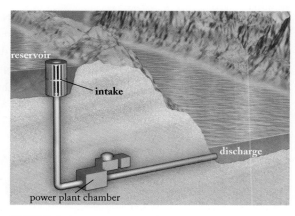

FIGURE 8.6 Cross-sectional view of hydroelectric pumped-storage power plant.

Consider, then, the energy changes that occur when a mass m of water, say, 1 kg of water, spurts out at the bottom of the pipe while, simultaneously, the water level of the upper reservoir decreases slightly. As concerns the energy balance, this effectively amounts to the removal of the potential energy of 1 kg from the top of the reservoir and the addition of the kinetic energy of 1 kg at the bottom of the pipe. All the water at intermediate locations, in the pipe and the reservoir, has the same energy it had before. Thus, energy conservation demands that the kinetic energy of the mass m of water emerging at the bottom be equal to the potential energy of a mass m at the top:

$$\tfrac{1}{2}mv^2 = mgh$$

This again gives

$$v = \sqrt{2gh} = 80 \text{ m/s}$$

that is, the same result as in part (a).

COMMENT: Note that the way the water acquires the final speed of 80 m/s in the cases (a) and (b) is quite different. In case (a), the water accelerates down the pipe with the uniform free-fall acceleration g. In case (b), the water flows down the pipe at a slow and nearly constant speed, and accelerates (strongly) only at the last moment, as it approaches the hole at the bottom. However, energy conservation demands that the result for the final speed of the emerging water be the same in both cases.

 Checkup 8.1

QUESTION 1: The potential energy corresponding to the spring force $F = -kx$ is $U = \tfrac{1}{2}kx^2$. Suppose that some new kind of force has a potential energy $U = -\tfrac{1}{2}kx^2$. How does this new kind of force differ from the spring force?

QUESTION 2: A particle moves along the positive x axis under the influence of a conservative force. Suppose that the potential energy of this force is as shown in Fig. 8.5a. Is the force directed along the positive x direction or the negative x direction?

QUESTION 3: Suppose that the force acting on a particle is given by the function $F_x = ax^3 + bx^2$, where a and b are constants. How do we know that the work done by this force during a round trip from, say, $x = 1$ back to $x = 1$ is zero?

QUESTION 4: Is the equation $W = U_1 - U_2$ valid for the work done by every kind of force? Is the equation $W = K_2 - K_1$ valid for the work done by each individual force acting on a particle?

(A) Yes; yes   (B) Yes; no   (C) No; yes   (D) No; no

8.2 THE CURVE OF POTENTIAL ENERGY

If a particle of some given energy is moving in one dimension under the influence of a conservative force, then Eq. (8.17) permits us to calculate the speed of the particle as a function of position. Suppose that the potential energy is some known function $U = U(x)$; then Eq. (8.17) states

$$E = \tfrac{1}{2}mv^2 + U(x) \tag{8.22}$$

or, rearranging,

$$v^2 = \frac{2}{m}[E - U(x)] \tag{8.23}$$

Since the left side of this equation is never negative, we can immediately conclude that the particle must always remain within a range of values of x for which $U(x) \leq E$. If $U(x)$ is increasing and the particle reaches a point at which $U(x) = E$, then $v = 0$; that is, the particle will stop at this point, and its motion will reverse. Such a point is called a **turning point** of the motion.

According to Eq. (8.23), v^2 is directly proportional to $E - U(x)$; thus, v^2 is large wherever the difference between E and $U(x)$ is large. We can therefore gain some insights into the qualitative features of the motion by drawing a graph of potential energy as a function of x on which it is possible to display the difference between E and $U(x)$. Such a graph of $U(x)$ vs. x is called the **curve of potential energy**. For example, Fig. 8.7 shows the curve of potential energy for an atom in a diatomic molecule. Treating the atom as a particle, we can indicate the value of the energy of the particle by a horizontal line in the graph (the red line in Fig. 8.7). We call this horizontal line the **energy level** of the particle. At any point x, we can then see the difference between E and $U(x)$ at a glance; according to Eq. (8.23), this tells us v^2. For instance, suppose that a particle has an energy $E = E_1$. Figure 8.7 shows this energy level. The particle has maximum speed at the point $x = x_0$, where the separation between the energy level and the potential-energy curve is maximum. The speed gradually decreases as the particle moves, say, toward the right. The potential-energy curve intersects the energy level at $x = a$; at this point the speed of the particle will reach zero, so this point is a turning point of the motion. The particle then moves toward the left, again attaining the same greatest speed at $x = x_0$. The speed gradually decreases as the particle continues to move toward the left, and the speed reaches zero at $x = a'$, the second turning point of the motion. Here the particle begins to move toward the right, and so on. Thus the particle continues to move back and forth between the two turning points—the particle is confined between the two turning points. The regions $x > a$ and $x < a'$ are forbidden regions;

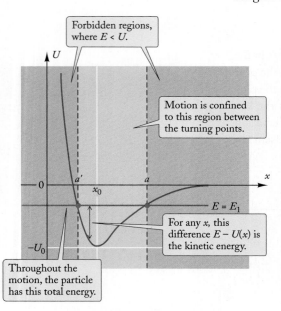

Forbidden regions, where $E < U$.

Motion is confined to this region between the turning points.

For any x, this difference $E - U(x)$ is the kinetic energy.

Throughout the motion, the particle has this total energy.

FIGURE 8.7 Potential-energy curve for an atom in a diatomic molecule. The horizontal line (red) is the energy level. The turning points are at $x = a$ and at $x = a'$.

only the region $a' \leq x \leq a$ is permitted. The particle is said to be in a **bound orbit.** The motion is periodic, that is, repeats again and again whenever the particle returns to its starting point.

The location of the turning points depends on the energy. For a particle with a lower energy level, the turning points are closer together. The lowest possible energy level intersects the potential-energy curve at its minimum (see $E = -U_0$ in Fig. 8.8); the two turning points then merge into the single point $x = x_0$. A particle with this lowest possible energy cannot move at all—it remains stationary at $x = x_0$. Note that the potential-energy curve has zero slope at $x = x_0$; this corresponds to zero force, $F_x = -dU/dx = 0$. A point such as $x = x_0$, where the force is zero, is called an **equilibrium point.** The point $x = x_0$ in Fig. 8.8 is a **stable** equilibrium point, since, after a small displacement, the force pushes the particle back toward that point. In contrast, at an **unstable** equilibrium point, after a small displacement, the force pushes the particle away from the point (see the point x_1 for the potential-energy curve shown in Fig. 8.9); and at a **neutral** equilibrium point no force acts nearby (see the point x_2 in Fig. 8.9). Equivalently, since the force is zero at an equilibrium point, the stable, unstable, and neutral equilibrium points correspond to negative, positive, or zero changes in the force with increasing x, that is, to negative, positive, or zero values of dF_x/dx. But $dF_x/dx = -d^2U/dx^2$, so the stable and unstable equilibrium points respectively correspond to positive and negative second derivatives of the function $U(x)$; in the former case the plot of $U(x)$ curves upward, and in the latter, downward (see Fig. 8.9).

In Fig. 8.7, the right side of the potential-energy curve never rises above $U = 0$. Consequently, if the energy level is above this value (for instance, $E = E_2$; see Fig. 8.10), then there is only one single turning point on the left, and no turning point on the right. A particle with energy E_2 will continue to move toward the right forever; it is not confined. Such a particle is said to be in an **unbound orbit.**

The above qualitative analysis based on the curve of potential energy cannot tell us the details of the motion such as, say, the travel time from one point to another. But the qualitative analysis is useful because it gives us a quick survey of the types of motion that are possible for different values of the energy.

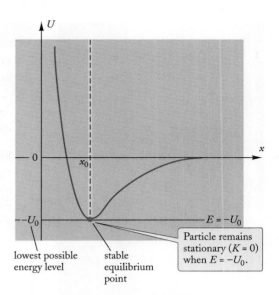

FIGURE 8.8 The energy level (red) coincides with the minimum of the potential-energy curve.

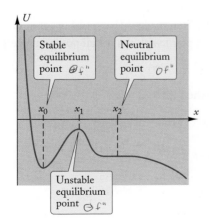

FIGURE 8.9 Types of equilibrium points. At the stable, unstable, and neutral equilibrium points, respectively, the potential-energy curve has a minimum, has a maximum, or is flat.

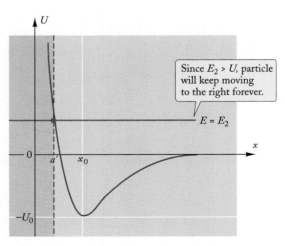

FIGURE 8.10 The energy level (red) is above the maximum height the potential-energy curve attains at its right. There is only one turning point, at $x = a'$.

FIGURE 8.11 Bungee jumping.

EXAMPLE 4   Some fanatics, in search of dangerous thrills, jump off high bridges or towers with bungee cords (long rubber cords) tied to their ankles (Fig. 8.11). Consider a jumper of mass 70 kg, with a 9.0-m cord tied to his ankles. When stretched, this cord may be treated as a spring, of spring constant 150 N/m. Plot the potential-energy curve for the jumper, and from this curve estimate the turning point of the motion, that is, the point at which the stretched cord stops the downward motion of the jumper.

SOLUTION: It is convenient to arrange the x axis vertically upward, with the origin at the point where the rubber cord becomes taut, that is, 9.0 m below the jump-off point (see Fig. 8.12a). The potential-energy function then consists of two pieces. For $x > 0$, the rubber cord is slack, and the potential energy is purely gravitational:

$$U = mgx \quad \text{for } x > 0$$

For $x < 0$, the rubber cord is stretched, and the potential energy is a sum of gravitational and elastic potential energies:

$$U = mgx + \tfrac{1}{2}kx^2 \quad \text{for } x < 0$$

With the numbers specified for this problem,

$$U = 70 \text{ kg} \times 9.81 \text{ m/s}^2 \times x$$
$$= 687x \quad \text{for } x > 0 \tag{8.24}$$

and

$$U = 70 \text{ kg} \times 9.81 \text{ m/s}^2 \times x + \tfrac{1}{2} \times 150 \text{ N/m} \times x^2$$
$$= 687x + 75x^2 \quad \text{for } x < 0 \tag{8.25}$$

where x is in meters and U in joules. Figure 8.12b gives the plot of the curve of potential energy, according to Eqs. (8.24) and (8.25).

At the jump-off point $x = +9.0$ m, the potential energy is $U = 687x = 687 \times 9.0 \text{ J} = 6180 \text{ J}$. The red line in Fig. 8.12b indicates this energy level. The left intersection of the red line with the curve indicates the turning point at the lower end of the motion. By inspection of the plot, we see that this turning point is at $x \approx -15$ m. Thus, the jumper falls a total distance of 9.0 m + 15 m = 24 m before his downward motion is arrested.

We can accurately calculate the position of the lower turning point ($x < 0$) by equating the potential energy at that point with the initial potential energy:

$$687x + 75x^2 = 6180 \text{ J}$$

This provides a quadratic equation of the form $ax^2 + bx + c = 0$:

$$75x^2 + 687x - 6180 = 0$$

This has the standard solution $x = (-b \pm \sqrt{b^2 - 4ac})/2a$, or

$$x = \frac{-687 \pm \sqrt{(687)^2 + 4 \times 75 \times 6180}}{2 \times 75}$$
$$= -14.7 \text{ m} \approx -15 \text{ m}$$

in agreement with our graphical result. Here we have chosen the negative solution, since we are solving for x at the lower turning point using the form (8.25), which is valid only for $x < 0$.

(a)   (b)

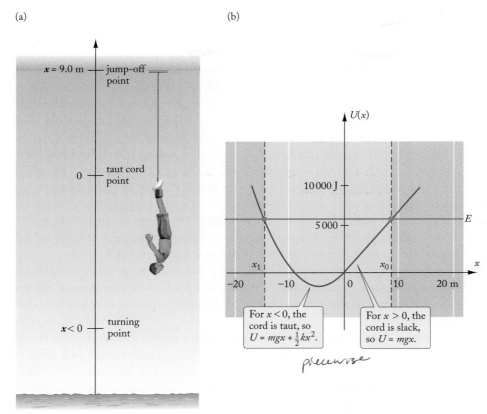

FIGURE 8.12 (a) The origin for the x coordinate is at the point where the rubber cord becomes taut. The jump-off point is at $x = 9.0$ m, and the turning point is at some negative value of x. (b) Curve of potential energy for the bungee jumper. The red line indicates the energy level. This line intersects the curve at approximately $x = -15$ m. This is the turning point for the jumper.

COMMENTS: If there were no friction, the motion would reverse, and the jumper would ascend to the bridge and bang against it. However, like a bouncing ball, the rubber cord has some energy loss due to friction within the material, and the jumper will not bounce back as high as the starting point.

Bungee jumping is a dangerous stunt. The human body has poor tolerance to deceleration in the head-down position. The pooling of blood in the head can lead to loss of consciousness ("redout"), rupture of blood vessels, eye damage, and temporary blindness. And in several instances, jumpers were killed by smashing their heads into the ground or by becoming entangled in their cords during the fall.

 Checkup 8.2

QUESTION 1: A particle moving in one dimension under the influence of a given conservative force has either no turning point, one turning point, or two turning points, depending on the energy. Does the number of turning points increase or decrease with the energy? Is there any conceivable value of the energy that will result in three turning points?

QUESTION 2: By examining the curve of potential energy in Fig. 8.12, estimate at what points the bungee jumper attains his maximum downward speed and his maximum acceleration.

QUESTION 3: A particle moving under the influence of the spring force has a positive energy $E = 50$ J. How many turning points are there for this particle?

(A) 1   (B) 2   (C) 3   (D) 0

8.3 OTHER FORMS OF ENERGY

If the forces acting on a particle are conservative, then the mechanical energy of the particle is conserved. But if some of the forces acting on the particle are not conservative, then the mechanical energy of the particle—consisting of the sum of the kinetic energy and the net potential energy of all the conservative forces acting on the particle—will not remain constant. For instance, if friction forces are acting, they do negative work and thereby decrease the mechanical energy of the particle.

However, it is a remarkable fact about our physical universe that *whenever mechanical energy is lost by a particle or some other body, this energy never disappears—it is merely changed into other forms of energy.* Thus, in the case of friction, the mechanical energy lost by the body is transformed into kinetic and potential energy of the atoms in the body and in the surface against which it is rubbing. The energy that the atoms acquire in the rubbing process is disorderly kinetic and potential energy—it is spread out among the atoms in an irregular, random fashion. At the macroscopic level, we perceive the increase of the disorderly kinetic and potential energy of the rubbed surfaces as an increase of temperature. Thus, friction produces **heat** or **thermal energy.** (You can easily convince yourself of this by vigorously rubbing your hands against each other.)

Heat is a form of energy, but whether it is to be regarded as a new form of energy or not depends on what point of view we adopt. Taking a macroscopic point of view, we ignore the atomic motions; then heat is to be regarded as distinct from mechanical energy. Taking a microscopic point of view, we recognize heat as kinetic and potential energy of the atoms; then heat is to be regarded as mechanical energy. (We will further discuss heat in Chapter 20.)

Chemical energy and nuclear energy are two other forms of energy. The former is kinetic and potential energy of the electrons within the atoms; the latter is kinetic and potential energy of the protons and neutrons within the nuclei of atoms. As in the case of heat, whether these are to be regarded as new forms of energy depends on the point of view.

Electric and magnetic energy are forms of energy associated with electric charges and with light and radio waves. (We will examine these forms of energy in Chapters 25 and 31.)

Table 8.1 lists some examples of different forms of energy. All the energies in Table 8.1 are expressed in joules, the SI unit of energy. However, for reasons of tradition and convenience, some other energy units are often used in specialized areas of physics and engineering.

The energy of atomic and subatomic particles is usually measured in **electron-volts** (eV), where

$$1 \text{ electron-volt} = 1 \text{ eV} = 1.60 \times 10^{-19} \text{ J} \qquad (8.26)$$

Electrons in atoms typically have kinetic and potential energies of a few eV.

The energy supplied by electric power plants is usually measured in **kilowatt-hours** (kW·h), where

$$1 \text{ kilowatt-hour} = 1 \text{ kW·h} = 3.60 \times 10^{6} \text{ J} \qquad (8.27)$$

The electric energy used by appliances such as vacuum cleaners, hair dryers, or toasters during one hour of operation is typically 1 kilowatt-hour.

And the thermal energy supplied by the combustion of fuels is often expressed in **kilocalories** (kcal):

HERMANN von **HELMHOLTZ**
(1821–1894) *Prussian surgeon, biologist, mathematician, and physicist. His scientific contributions ranged from the invention of the ophthalmoscope and studies of the physiology and physics of vision and hearing to the measurement of the speed of light and studies in theoretical mechanics. Helmholtz formulated the general Law of Conservation of Energy, treating it as a consequence of the basic laws of mechanics and electricity.*

TABLE 8.1	SOME ENERGIES	
Nuclear fuel in the Sun	1×10^{45} J	
Explosion of a supernova	1×10^{44} J	
Fossil fuel available on Earth	2.0×10^{23} J	
Yearly energy expenditure of the United States (a)	8×10^{19} J	
Volcanic explosion (Krakatoa)	6×10^{18} J	
Annihilation of 1 kg of matter–antimatter	9.0×10^{16} J	(a)
Explosion of thermonuclear bomb (1 megaton)	4.2×10^{15} J	
Gravitational potential energy of airliner (Boeing 747 at 10 000 m)	2×10^{10} J	
Combustion of 1 gal of gasoline (b)	1.3×10^{8} J	
Daily food intake of man (3000 kcal)	1.3×10^{7} J	(b)
Explosion of 1 kg of TNT	4.6×10^{6} J	
Metabolization of one apple (110 kcal)	4.6×10^{5} J	
One push-up (c)	3×10^{2} J	
Fission of one uranium nucleus	3.2×10^{-11} J	
Energy of ionization of hydrogen atom	2.2×10^{-18} J	(c)

$$1 \text{ kilocalorie} = 1 \text{ kcal} = 4.187 \times 10^3 \text{ J} \qquad (8.28)$$

or in **British thermal units** (Btu):

$$1 \text{ Btu} = 1.055 \times 10^3 \text{ J} \qquad (8.29)$$

We will learn more about these units in later chapters.

All these forms of energy can be transformed into one another. For example, in an internal combustion engine, chemical energy of the fuel is transformed into heat and kinetic energy; in a hydroelectric power station, gravitational potential energy of the water is transformed into electric energy; in a nuclear reactor, nuclear energy is transformed into heat, light, kinetic energy, etc. However, in any such transformation process, the sum of all the energies of all the pieces of matter involved in the process remains constant: *the form of the energy changes, but the total amount of energy does not change.* This is the general **Law of Conservation of Energy**.

law of conservation of energy

EXAMPLE 5 At the Brown Mountain hydroelectric pumped-storage plant, the average height of the water in the upper reservoir is 320 m above the lower reservoir, and the upper reservoir holds 1.9×10^7 m^3 of water. Expressed in kW·h, what is the gravitational potential energy available for conversion into electric energy?

Concepts — *in* — Context

SOLUTION: A cubic meter of water has a mass of 1000 kg. Hence the total mass of water is 1.9×10^{10} kg, and the gravitational potential energy is

$$U = mgh = 1.9 \times 10^{10} \text{ kg} \times 9.81 \text{ m/s}^2 \times 320 \text{ m} = 6.0 \times 10^{13} \text{ J}$$

Expressed in kW·h , this amounts to

$$6.0 \times 10^{13} \text{ J} \times \frac{1 \text{ kW·h}}{3.6 \times 10^6 \text{ J}} = 1.7 \times 10^7 \text{ kW·h}$$

(The actual electric energy that can be generated is about 30% less than that, because of frictional losses during the conversion from one form of energy to the other. These frictional losses result in the generation of heat.)

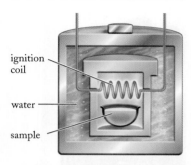

ignition
coil

water

sample

FIGURE 8.13 A bomb calorimeter. The sample is ignited electrically, by a glowing wire.

EXAMPLE 6 The "calorie" used by dietitians to express the energy equivalents of different foods is actually a kilocalorie, or a "large" calorie. To measure the energy equivalent of some kind of food—for instance, sugar—a sample is placed in a bomb calorimeter, a closed vessel filled with oxygen at high pressure (see Fig. 8.13). The sample is ignited and burned completely (complete oxidation). The number of calories released in this chemical reaction—for instance, 4.1 kcal for 1.0 g of sugar—tells us the maximum amount of energy that can be extracted from this food. The human body does not necessarily "burn" food quite as completely, and the muscles do not convert all of the available chemical energy into mechanical energy. However, energy conservation tells us that from one gram of sugar the body cannot produce more than 4.1 kcal of mechanical work.

If you eat one spoonful (4.0 g) of sugar, what is the maximum height to which this permits you to climb stairs? Assume your mass is 70 kg.

SOLUTION: Since 1.0 g of sugar releases 4.1 kcal of energy, the energy equivalent of 4.0 g of sugar is

$$4.0 \times 4.1 \text{ kcal} = 16.4 \text{ kcal} = 16.4 \text{ kcal} \times 4.18 \times 10^3 \text{ J/kcal} = 6.9 \times 10^4 \text{ J}$$

When you climb the stairs to a height y, this energy becomes gravitational potential energy:

$$mgy = 6.9 \times 10^4 \text{ J}$$

from which

$$y = \frac{6.9 \times 10^4 \text{ J}}{mg} = \frac{6.9 \times 10^4 \text{ J}}{70 \text{ kg} \times 9.81 \text{ m/s}^2} = 100 \text{ m}$$

In practice, because of the limited efficiency of your body, only about 20% of the chemical energy of food is converted into mechanical energy; thus, the actual height you can climb is only about 20 m. (Because of the strong musculature of the human leg, stair climbing is one of your most efficient activities; other physical activities are considerably less efficient in converting chemical energy into mechanical energy.)

 Checkup 8.3

QUESTION 1: A parachutist descends at uniform speed. Is the mechanical energy conserved? What happens to the lost mechanical energy?

QUESTION 2: You fire a bullet from a rifle. The increase of kinetic energy of the bullet upon firing must be accompanied by a decrease of some other kind of energy. What energy decreases?

QUESTION 3: A truck travels at constant speed down a road leading from a mountain peak to a valley. What happens to the gravitational potential energy of the truck? How is it dissipated?

QUESTION 4: When you apply the brakes and stop a moving automobile, what happens to the kinetic energy?

    (A) Kinetic energy is converted to gravitational potential energy.
    (B) Kinetic energy is converted to elastic potential energy.
    (C) Kinetic energy is converted to heat due to frictional forces.
    (D) Kinetic energy is converted to chemical energy.

8.4 MASS AND ENERGY

One of the great discoveries made by Albert Einstein early in the twentieth century is that energy can be transformed into mass, and mass can be transformed into energy. Thus, *mass is a form of energy*. The amount of energy contained in an amount m of mass is given by Einstein's famous formula

$$E = mc^2 \qquad (8.30)$$

energy–mass relation

where c is the speed of light, $c = 3.00 \times 10^8$ m/s. This formula is a consequence of Einstein's relativistic physics. It cannot be obtained from Newton's physics, and its theoretical justification will have to wait until we study the theory of relativity in Chapter 36.

    The most spectacular demonstration of Einstein's mass–energy formula is found in the annihilation of matter and antimatter (as we will see in Chapter 41, particles of antimatter are similar to the particles of ordinary matter, except that they have opposite electric charge). If a proton collides with an antiproton, or an electron with an antielectron, the two colliding particles react violently, and they annihilate each other in an explosion that generates an intense flash of very energetic light. According to Eq. (8.30), the annihilation of just 1000 kg of matter and antimatter (500 kg of each) would release an amount of energy

$$E = mc^2 = 1000 \text{ kg} \times (3.00 \times 10^8 \text{ m/s})^2 = 9.0 \times 10^{19} \text{ J} \qquad (8.31)$$

This is enough energy to satisfy the requirements of the United States for a full year. Unfortunately, antimatter is not readily available in large amounts. On Earth, antiparticles can be obtained only from reactions induced by the impact of beams of high-energy particles on a target. These collisions occasionally result in the creation of a particle–antiparticle pair. Such pair creation is the reverse of pair annihilation. The creation process transforms some of the kinetic energy of the collision into mass, and a subsequent annihilation merely gives back the original energy.

    But the relationship between energy and mass in Eq. (8.30) also has another aspect. *Energy has mass.* Whenever the energy of a body is changed, its mass (and weight) are changed. The change in mass that accompanies a given change of energy is

mass–energy relation

$$\Delta m = \frac{\Delta E}{c^2} \tag{8.32}$$

For instance, if the kinetic energy of a body increases, its mass (and weight) increase. At speeds small compared with the speed of light, the mass increment is not noticeable. But when a body approaches the speed of light, the mass increase becomes very large. The high-energy electrons produced at the Stanford Linear Accelerator provide an extreme example of this effect: these electrons have a speed of 99.999 999 97% of the speed of light, and their mass is 44 000 times the mass of electrons at rest!

The fact that energy has mass indicates that energy is a form of mass. Conversely, as we have seen above, mass is a form of energy. Hence mass and energy must be regarded as two aspects of the same thing. The laws of conservation of mass and conservation of energy are therefore not two independent laws—each implies the other. For example, consider the fission reaction of uranium inside the reactor vessel of a nuclear power plant. The complete fission of 1.0 kg of uranium yields an energy of 8.2×10^{13} J. The reaction conserves energy—it merely transforms nuclear energy into heat, light, and kinetic energy, but does not change the total amount of energy. The reaction also conserves mass—if the reactor vessel is hermetically sealed and thermally insulated from its environment, then the reaction does not change the mass of the contents of the vessel. However, if we open the vessel during or after the reaction and let some of the heat and light escape, then the mass of the residue will not match the mass of the original amount of uranium. The mass of the residues will be about 0.1% smaller than the original mass of the uranium. This mass defect represents the mass carried away by the energy that escapes. Thus, the nuclear fission reactions merely transform energy into new forms of energy and mass into new forms of mass. In this regard, a nuclear reaction is not fundamentally different from a chemical reaction. The mass of the residues in a chemical reaction that releases heat (exothermic reaction) is slightly less than the original mass. The heat released in such a chemical reaction carries away some mass, but, in contrast to a nuclear reaction, this amount of mass is so small as to be quite immeasurable.

EXAMPLE 7   As an example of the small mass loss in a chemical reaction, consider the binding energy of the electron in the hydrogen atom (one proton and one electron), which is 13.6 eV. What is the fractional mass loss when an electron is captured by a proton and the binding energy is allowed to escape?

SOLUTION: In joules, the binding energy is 13.6 eV $\times$ 1.60 $\times$ 10^{-19} J/eV = 2.18×10^{-18} J. The mass loss corresponding to this binding energy is

$$\Delta m = \frac{\Delta E}{c^2} = \frac{2.18 \times 10^{-18} \text{ J}}{(3.00 \times 10^8 \text{ m/s})^2} = 2.42 \times 10^{-35} \text{ kg}$$

Since the mass of a proton and electron together is 1.67×10^{-27} kg (see Table 5.2), the fractional mass loss is

$$\frac{\Delta m}{m} = \frac{2.42 \times 10^{-35} \text{ kg}}{1.67 \times 10^{-27} \text{ kg}} = 1.45 \times 10^{-8}$$

This is about a millionth of one percent.

Checkup 8.4

QUESTION 1: The Sun radiates heat and light. Does the Sun consequently suffer a loss of mass?

QUESTION 2: In the annihilation of matter and antimatter, a particle and an antiparticle—such as a proton and an antiproton, or an electron and an antielectron—disappear explosively upon contact, giving rise to an intense flash of light. Is energy conserved in this reaction? Is mass conserved?

QUESTION 3: You heat a potful of water to the boiling point. If the pot is sealed so no water molecules can escape, then, compared with the cold water, the mass of the boiling water will:

    (A) Increase   (B) Decrease   (C) Remain the same

8.5 POWER

When we use an automobile engine to move a car up a hill or when we use an electric motor to lift an elevator cage, the important characteristic of the engine is not how much force it can exert, but rather how much work it can perform in a given amount of time. The force is only of secondary importance, because by shifting to a low gear we can make sure that even a "weak" engine exerts enough force on the wheels to propel the automobile uphill. But the work performed in a given amount of time, or the rate of work, is crucial, since it determines how fast the engine can propel the car up the hill. While the car moves uphill, the gravitational force takes energy from the car; that is, it performs negative work on the car. To keep the car moving, the engine must perform an equal amount of positive work. If the engine is able to perform this work at a fast rate, it can propel the car uphill at a fast speed.

The rate at which a force does work on a body is called the **power** _delivered by the force._ If the force does an amount of work W in an interval of time Δt, then the **average power** is the ratio of W and Δt:

$$\overline{P} = \frac{W}{\Delta t} \tag{8.33}$$

average power

The **instantaneous power** is defined by a procedure analogous to that involved in the definition of the instantaneous velocity. We consider the small amount of work dW done in the small interval of time dt and take the ratio of these small quantities:

$$P = \frac{dW}{dt} \tag{8.34}$$

instantaneous power

According to these definitions, the engine of your automobile delivers high power if it performs a large amount of work on the wheels (or, rather, the driveshaft) in a short time. The maximum power delivered by the engine determines the maximum speed of which this automobile is capable, since at high speed the automobile loses energy to air resistance at a prodigious rate, and this loss has to be made good by the engine. You might also expect that the power of the engine determines the maximum acceleration of which the automobile is capable. But the acceleration is determined by the maximum force exerted by the engine on the wheels, and this is not directly related to the power as defined above.

The SI unit of power is the **watt (W)**, which is the rate of work of one joule per second:

$$1 \text{ watt} = 1 \text{ W} = 1 \text{ J/s}$$

In engineering practice, power is often measured in **horsepower** (hp) units, where

$$1 \text{ horsepower} = 1 \text{ hp} = 746 \text{ W} \tag{8.35}$$

This is roughly the rate at which a (very strong) horse can do work.

Note that multiplication of a unit of power by a unit of time gives a unit of energy. An example of this is the kilowatt-hour (kW·h), already mentioned in Section 8.3:

$$1 \text{ kilowatt-hour} = 1 \text{ kW·h} = 1 \text{ kW} \times 1 \text{ h} = 1000 \text{ W} \times 3600 \text{ s}$$
$$= 3.6 \times 10^6 \text{ J} \tag{8.36}$$

This unit is commonly used to measure the electric energy delivered to homes and factories.

For a constant (or average) power P delivered to a body during a time Δt, the work W delivered is the rate times the time [see Eq. (8.33)]:

$$W = P \Delta t \tag{8.37}$$

If the rate of doing work P varies with time, then the total work W done between a time t_1 and another time t_2 is the sum of the infinitesimal $P\Delta t$ contributions; that is, the work done is the integral of the power over time:

$$W = \int dW = \int_{t_1}^{t_2} P \, dt \tag{8.38}$$

JAMES WATT (1736–1819) *Scottish inventor and engineer. He modified and improved an earlier steam engine and founded the first factory constructing steam engines. Watt introduced the* horsepower *as a unit of mechanical power.*

EXAMPLE 8   An elevator cage has a mass of 1000 kg. How many horsepower must the motor deliver to the elevator if it is to raise the elevator cage at the rate of 2.0 m/s? The elevator has no counterweight (see Fig. 8.14).

SOLUTION: The weight of the elevator is $w = mg = 1000 \text{ kg} \times 9.81 \text{ m/s}^2 \approx 9800 \text{ N}$. By means of the elevator cable, the motor must exert an upward force equal to the weight to raise the elevator at a steady speed. If the elevator moves up a distance Δy, the work done by the force is

$$W = F \Delta y \tag{8.39}$$

To obtain the power, or the rate of work, we must divide this by the time interval Δt:

$$P = \frac{W}{\Delta t} = \frac{F \Delta y}{\Delta t} = F \frac{\Delta y}{\Delta t} = Fv \tag{8.40}$$

where $v = \Delta y/\Delta t$ is the speed of the elevator. With $F = 9800$ N and $v = 2.0$ m/s, we find

$$P = Fv = 9800 \text{ N} \times 2.0 \text{ m/s} = 2.0 \times 10^4 \text{ W}$$

Since 1 hp = 746 W [see Eq. (8.35)], this equals

$$P = 2.0 \times 10^4 \text{ W} \times \frac{1 \text{ hp}}{746 \text{ W}} = 27 \text{ hp}$$

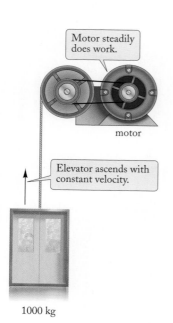

Motor steadily does work.

motor

Elevator ascends with constant velocity.

1000 kg

FIGURE 8.14 Elevator cage and motor.

Equation (8.40) is a special instance of a simple formula, which expresses the instantaneous power as the scalar product of force and velocity. To see this, consider that when a body suffers a small displacement $d\mathbf{s}$, the force $\mathbf{F}$ acting on the body will perform an amount of work

$$dW = \mathbf{F} \cdot d\mathbf{s} \qquad (8.41)$$

or

$$dW = F\,ds\,\cos\theta$$

where θ is the angle between the direction of the force and the direction of the displacement (see Fig. 8.15). The instantaneous power delivered by this force is then

$$P = \frac{dW}{dt} = F\frac{ds}{dt}\cos\theta \qquad (8.42)$$

Since ds/dt is the speed v, this expression for the power equals

$$P = Fv\cos\theta \qquad (8.43)$$

or

$$P = \mathbf{F} \cdot \mathbf{v} \qquad (8.44)$$

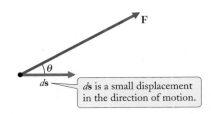

FIGURE 8.15 The force $\mathbf{F}$ makes an angle θ with the displacement $d\mathbf{s}$.

ds is a small displacement in the direction of motion.

power delivered by a force

EXAMPLE 9 A horse pulls a sled up a steep snow-covered street of slope 1:7 (see Fig. 8.16a). The sled has a mass of 300 kg, and the coefficient of sliding friction between the sled and the snow is 0.12. If the horse pulls parallel to the surface of the street and delivers a power of 1.0 hp, what is the maximum (constant) speed with which the horse can pull the sled uphill? What fraction of the horse's power is expended against gravity? What fraction against friction?

SOLUTION: Figure 8.16b is a "free-body" diagram for the sled, showing the weight ($w = mg$), the normal force ($N = mg\cos\phi$), the friction force ($f_k = \mu_k N$), and the pull of the horse (T). With the x axis along the street and the y axis at right angles to the street, the components of these forces are

$$w_x = -mg\sin\phi \qquad w_y = -mg\cos\phi$$
$$N_x = 0 \qquad N_y = mg\cos\phi$$
$$f_{k,x} = -\mu_k mg\cos\phi \qquad f_{k,y} = 0$$
$$T_x = T \qquad T_y = 0$$

Since the acceleration along the street is zero (constant speed), the sum of the x components of these forces must be zero:

$$-mg\sin\phi + 0 - \mu_k mg\cos\phi + T = 0 \qquad (8.45)$$

We can solve this equation for the pull of the horse:

$$T = mg\sin\phi + \mu_k mg\cos\phi \qquad (8.46)$$

This simply says that the pull of the horse must balance the component of the weight along the street plus the friction force. The direction of this pull is parallel to the direction of motion of the sled. Hence, in Eq. (8.43), $\theta = 0$, and the power delivered by the horse is

$$P = Tv = (mg\sin\phi + \mu_k mg\cos\phi)v \qquad (8.47)$$

Solving this equation for v, we find

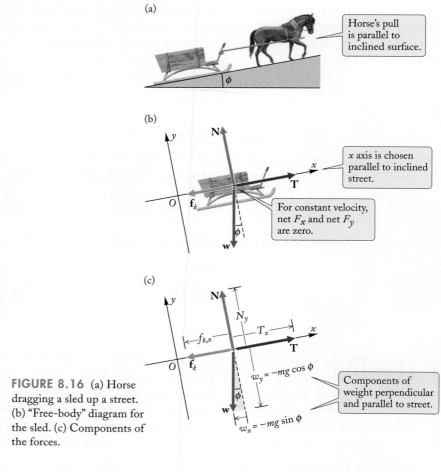

FIGURE 8.16 (a) Horse dragging a sled up a street. (b) "Free-body" diagram for the sled. (c) Components of the forces.

$$v = \frac{P}{mg \sin \phi + \mu_k mg \cos \phi} = \frac{P}{mg (\sin \phi + \mu_k \cos \phi)} \tag{8.48}$$

For a slope of 1:7, the tangent of the angle of inclination is $\tan \phi = 1/7$, and, using a calculator, the inverse tangent of 1/7 gives $\phi = 8.1°$. Hence

$$v = \frac{746 \text{ W}}{300 \text{ kg} \times 9.81 \text{ m/s}^2 \times (\sin 8.1° + 0.12 \cos 8.1°)}$$
$$= 0.98 \text{ m/s}$$

The weight of the sled makes an angle of $90.0° + 8.1° = 98.1°$ with the direction of motion (see Fig. 8.16b). The power exerted by the weight of the sled is given by Eq. (8.43), with $F = mg$ and $\cos \theta = \cos 98.1°$:

$$P_{\text{weight}} = mgv \cos 98.1° = 300 \text{ kg} \times 9.81 \text{ m/s}^2 \times 0.98 \text{ m/s} \times \cos 98.1°$$
$$= -406 \text{ W} = -0.54 \text{ hp}$$

Since the total power is 1.0 hp, this says that 54% of the horse's power is expended against gravity and, consequently, the remaining 46% against friction. The friction portion can also be calculated directly. The friction force acts opposite to the velocity ($\cos \theta = -1$), and so the power exerted is negative:

$$P_{\text{friction}} = -f_k v = -\mu_k mg \cos \phi \, v = -0.12 \times 300 \text{ kg} \times 9.81 \text{ m/s}^2 \times \cos 8.1° \times 0.98 \text{ m/s}$$
$$= -343 \text{ W} = -0.46 \text{ hp}$$

The above equations all refer to *mechanical* power. In general, *power is the rate at which energy is transferred from one form of energy to another or the rate at which energy is transported from one place to another.* For instance, an automobile engine converts chemical energy of fuel into mechanical energy and thermal energy. A nuclear power plant converts nuclear energy into electric energy and thermal energy. And a high-voltage power line transports electric energy from one place to another. Table 8.2 gives some examples of different kinds of power.

TABLE 8.2	SOME POWERS	
Light and heat emitted by the Sun		3.9×10^{26} W
Mechanical power generated by hurricane	(a)	2×10^{13} W
Total power used in United States (average)		2×10^{12} W
Large electric power plant		$\approx 10^{9}$ W
Jet airliner engines (Boeing 747)	(b)	2.1×10^{8} W
Automobile engine		1.5×10^{5} W
Solar light and heat per square meter at Earth		1.4×10^{3} W
Electricity used by toaster		1×10^{3} W
Work output of man (athlete at maximum)		2×10^{2} W
Electricity used by light bulb		1×10^{2} W
Basal metabolic rate for man (average)		88 W
Heat and work output of bumblebee (in flight)	(c)	2×10^{-2} W
Atom radiating light		$\approx 10^{-10}$ W

(a)   (b)   (c)

In the previous example, part of the horse's work was converted into heat by the friction between the sled and the snow, and part was converted into gravitational potential energy. In the following example, gravitational potential energy is converted into electric energy.

EXAMPLE 10   Each of the four generators (Fig. 8.17) of the Brown Mountain hydroelectric plant generates 260 MW of electric power. When generating this power, at what rate does the power plant take water from the upper reservoir? How long does a full reservoir last? See the data in Example 5.

Concepts
— *in* —
Context

SOLUTION: We will assume that all of the potential energy of the water in the upper reservoir, at a height of 320 m, is converted into electric energy. The electric

FIGURE 8.17 Turbine generator at Brown Mountain hydroelectric plant, shown during installation.

power $P = 4 \times 260 \times 10^6\,\text{W} = 1.0 \times 10^9\,\text{W}$ must then equal the negative of the rate of change of the potential energy (see Eq. 7.31):

$$P = -\frac{dU}{dt} = -\frac{dm}{dt}gh$$

from which we obtain the rate of change of mass,

$$\frac{dm}{dt} = -\frac{P}{gh} = -\frac{1.0 \times 10^9\,\text{W}}{9.81\,\text{m/s}^2 \times 320\,\text{m}} = -3.3 \times 10^5\,\text{kg/s}$$

Expressed as a volume of water, this amounts to an outflow of 330 m³ per second. At this rate, the 1.9×10^7 m³ of water in the reservoir will last for

$$\frac{1.9 \times 10^7 \text{m}^3}{330\ \text{m}^3/\text{s}} = 5.7 \times 10^4\,\text{s} = 16\,\text{h}$$

As mentioned in Example 5, there are also some frictional losses. As a result, the reservoir will actually be depleted about 30% faster than this, that is, in a bit less than half a day.

✔ Checkup 8.5

QUESTION 1: (a) You trot along a flat road carrying a backpack. Do you deliver power to the pack? (b) You trot uphill. Do you deliver power to the pack? (c) You trot downhill. Do you deliver power to the pack? Does the pack deliver power to you?

QUESTION 2: To reach a mountaintop, you have a choice between a short, steep road or a longer, less steep road. Apart from frictional losses, is the energy you have to expend in walking up these two roads the same? Why does the steeper road require more of an effort?

QUESTION 3: In order to keep a 26-m motor yacht moving at 88 km/h, its engines must supply about 5000 hp. What happens to this power?

QUESTION 4: Two cars are traveling up a sloping road, each at a constant speed. The second car has twice the mass and twice the speed of the first car. What is the ratio of the power delivered by the second car engine to that delivered by the first? Ignore friction and other losses.

(A) 1   (B) 2   (C) 4   (D) 8   (E) 16

SUMMARY

PROBLEM-SOLVING TECHNIQUES   Energy Conservation   (page 238)

PHYSICS IN PRACTICE   Hydroelectric Pumped Storage   (page 242)

CONSERVATIVE FORCE   The work done by the force is zero for any round trip.

WORK DONE BY A CONSERVATIVE FORCE $W = -U_2 + U_1 = -\Delta U$ (8.2)

POTENTIAL ENERGY OF A SPRING $U = \frac{1}{2}kx^2$ (8.6)

CONTRIBUTIONS TO THE MECHANICAL ENERGY

Kinetic energy $K = \frac{1}{2}mv^2$ (for motion)

Gravitational potential energy $U = mgy$ (near Earth's surface)

Elastic potential energy $U = \frac{1}{2}kx^2$ (for a spring)

POTENTIAL ENERGY AS INTEGRAL OF FORCE $U(x) = -\int_{x_0}^{x} F_x(x')\,dx'$ (8.14)

POTENTIAL OF INVERSE-SQUARE FORCE   If $F(x) = \dfrac{A}{x^2}$ then $U(x) = \dfrac{A}{x}$ (8.19)

(for $x > 0$; attractive for $A < 0$, repulsive for $A > 0$.)

FORCE AS DERIVATIVE OF POTENTIAL ENERGY $F_x = -\dfrac{dU}{dx}$ (8.21)

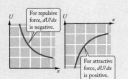

CONSERVATION OF MECHANICAL ENERGY $E = \frac{1}{2}mv^2 + U = [\text{constant}]$ (8.17)

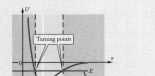

MASS IS A FORM OF ENERGY $E = mc^2$ (8.30)

ENERGY HAS MASS $\Delta m = \dfrac{\Delta E}{c^2}$ (8.32)

SI UNIT OF POWER	$1 \text{ watt} = 1 \text{ W} = 1 \text{ J/s}$	
AVERAGE POWER	$\overline{P} = \dfrac{W}{\Delta t}$	(8.33)
INSTANTANEOUS POWER	$P = \dfrac{dW}{dt}$	(8.34)
MECHANICAL POWER DELIVERED BY A FORCE	$P = \mathbf{F} \cdot \mathbf{v} = Fv \cos \theta$	(8.43; 8.44)
WORK DONE AT CONSTANT POWER	$W = P \Delta t$	(8.37)
WORK DONE WITH TIME-DEPENDENT POWER	$W = \displaystyle\int_{t_1}^{t_2} P \, dt$	(8.38)

QUESTIONS FOR DISCUSSION

1. A body slides on a smooth horizontal plane. Is the normal force of the plane on the body a conservative force? Can we define a potential energy for this force according to the recipe in Section 8.1?

2. If you stretch a spring so far that it suffers a permanent deformation, is the force exerted by the spring during this operation conservative?

3. Is there any frictional dissipation of mechanical energy in the motion of the planets of the Solar System or in the motion of their satellites? (Hint: Consider the tides.)

4. What happens to the kinetic energy of an automobile during braking without skidding? With skidding?

5. An automobile travels down a road leading from a mountain peak to a valley. What happens to the gravitational potential energy of the automobile? How is it dissipated?

6. Suppose you wind up a watch and then place it into a beaker full of nitric acid and let it dissolve. What happens to the potential energy stored in the spring of the watch?

7. News reporters commonly speak of "energy consumption." Is it accurate to say that energy is *consumed*? Would it be more accurate to say that energy is *dissipated*?

8. The explosive yield of thermonuclear bombs (Fig. 8.18) is usually reported in kilotons or megatons of TNT. Would the explosion of a 1-megaton hydrogen bomb really produce the same effects as the explosion of 1 megaton of TNT (a mountain of TNT more than a hundred meters high)?

9. When you heat a potful of water, does its mass increase?

10. Since mass is a form of energy, why don't we measure mass in the same units as energy? How could we do this?

11. In order to travel at 130 km/h, an automobile of average size needs an engine delivering about 40 hp to overcome the effects of air friction, road friction, and internal friction (in the transmission and drive train). Why do most drivers think they need an engine of 150 or 200 hp?

FIGURE 8.18 A thermonuclear explosion.

PROBLEMS

8.1 Potential Energy of a Conservative Force[†]

1. The spring used in the front suspension of a Triumph sports car has a spring constant $k = 3.5 \times 10^4$ N/m. How far must we compress this spring to store a potential energy of 100 J?

2. A particle moves along the x axis under the influence of a variable force $F_x = 2x^3 + 1$ (where force is measured in newtons and distance in meters). Show that this force is conservative; that is, show that for any back-and-forth motion that starts and ends at the same place (round trip), the work done by the force is zero.

3. Consider a force that is a function of the velocity of the particle (and is not perpendicular to the velocity). Show that the work for a round trip along a closed path can then be different from zero.

4. The force acting on a particle moving along the x axis is given by the formula $F_x = K/x^4$, where K is a constant. Find the corresponding potential-energy function. Assume that $U(x) = 0$ for $x = \infty$.

5. A 50-g particle moving along the x axis experiences a force $F_x = -Ax^3$, where $A = 50$ N/m^3. Find the corresponding potential-energy function. If the particle is released from rest at $x = 0.50$ m, what is its speed as it passes the origin?

6. The force on a particle confined to move along the positive x axis is constant, $F_x = -F_0$, where $F_0 = 25$ N. Find the corresponding potential-energy function. Assume $U(x) = 0$ at $x = 0$.

7. A particular spring is not ideal; for a distance x from equilibrium, the spring exerts a force $F_x = -2x - x^3$, where x is in meters and F_x is in newtons. What is the potential-energy function for this spring? Assume $U(x) = 0$ at $x = 0$. How much energy is stored in the spring when it is stretched 1.0 m? 2.0 m? 3.0 m?

8. The force on a particle moving along the x axis is given by

$$F_x = \begin{cases} F_0 & x \le -a \\ 0 & -a < x < a \\ -F_0 & x \ge a \end{cases}$$

where F_0 is a constant. What is the potential-energy function for this force? Assume $U(x) = 0$ for $x = 0$.

9. Consider a particle moving in a region where the potential energy is given by $U = 2x^2 + x^4$, where U is in joules and x is in meters. What is the position-dependent force on this particle?

10. The force on an electron in a particular region of space is given by $\mathbf{F} = F_0 \sin(ax)\mathbf{i}$, where F_0 and a are constants (this force is achieved with two oppositely directed laser beams). What is the corresponding potential-energy function?

*11. A bow may be regarded mathematically as a spring. The archer stretches this "spring" and then suddenly releases it so that the bowstring pushes against the arrow. Suppose that when the archer stretches the "spring" 0.52 m, he must exert a force of 160 N to hold the arrow in this position. If he now releases the arrow, what will be the speed of the arrow when the "spring" reaches its equilibrium position? The mass of the arrow is 0.020 kg. Pretend that the "spring" is massless.

*12. A mass m hangs on a vertical spring of a spring constant k.
    (a) How far will this hanging mass have stretched the spring from its relaxed length?
    (b) If you now push up on the mass and lift it until the spring reaches its relaxed length, how much work will you have done against gravity? Against the spring?

*13. A particle moving in the x–y plane experiences a conservative force

$$\mathbf{F} = by\mathbf{i} + bx\mathbf{j}$$

where b is a constant.
    (a) What is the work done by this force as the particle moves from $x_1 = 0, y_1 = 0$ to $x_2 = x, y_2 = y$? (Hint: Use a path from the origin to the point x_2, y_2 consisting of a segment parallel to the x axis and a segment parallel to the y axis.)
    (b) What is the potential energy associated with this force? Assume that the potential energy is zero when the particle is at the origin.

*14. The four wheels of an automobile of mass 1200 kg are suspended below the body by vertical springs of constant $k = 7.0 \times 10^4$ N/m. If the forces on all wheels are the same, what will be the maximum instantaneous deformation of the springs if the automobile is lifted by a crane and dropped on the street from a height of 0.80 m?

*15. A rope can be regarded as a long spring; when under tension, it stretches and stores elastic potential energy. Consider a nylon rope similar to that which snapped during a giant tug-of-war at a school in Harrisburg, Pennsylvania (see Problem 23 of Chapter 5). Under a tension of 58 000 N (applied at its ends), the rope of initial length 300 m stretches to 390 m. What is the elastic energy stored in the rope at this tension? What happens to this energy when the rope breaks?

*16. Among the safety features on elevator cages are spring-loaded brake pads which grip the guide rail if the elevator cable should break. Suppose that an elevator cage of 2000 kg has two such brake pads, arranged to press against opposite sides of the guide rail, each with a force of 1.0×10^5 N. The friction coefficient for the brake pads sliding on the guide rail is 0.15. Assume that the elevator cage is falling freely with an initial speed of 10 m/s when the brake pads come into action. How long will the elevator cage take to stop? How far will it travel? How much energy is dissipated by friction?

[†] For help, see Online Concept Tutorial 10 at www.ww norton.com/physics

*17. The force between two inert-gas atoms is often described by a function of the form

$$F_x = Ax^{-13} - Bx^{-7}$$

where A and B are positive constants and x is the distance between the atoms. What is the corresponding potential-energy function, called the **Lennard–Jones potential**?

*18. A particle moving in three dimensions is confined by a force $\mathbf{F} = -k(x\mathbf{i} + y\mathbf{j} + z\mathbf{k})$, where k is a constant. What is the work required to move the particle from the origin to a point $\mathbf{r} = x\mathbf{i} + y\mathbf{j} + z\mathbf{k}$? What is the potential-energy function?

**19. Mountain climbers use nylon safety rope whose elasticity plays an important role in cushioning the sharp jerk if a climber falls and is suddenly stopped by the rope.

(a) Suppose that a climber of 80 kg attached to a 10-m rope falls freely from a height of 10 m above to a height of 10 m below the point at which the rope is anchored to a vertical wall of rock. Treating the rope as a spring with $k = 4.9 \times 10^3$ N/m (which is the appropriate value for a braided nylon rope of 9.2 mm diameter), calculate the maximum force that the rope exerts on the climber during stopping.

(b) Repeat the calculations for a rope of 5.0 m and an initial height of 5.0 m. Assume that this second rope is made of the same material as the first, and remember to take into account the change in the spring constant due to the change in length. Compare your results for (a) and (b) and comment on the advantages and disadvantages of long ropes vs. short ropes.

**20. A package is dropped on a horizontal conveyor belt (Fig. 8.19). The mass of the package is m, the speed of the conveyor belt is v, and the coefficient of kinetic friction for the package on the belt is μ_k. For what length of time will the package slide on the belt? How far will it move in this time? How much energy is dissipated by friction? How much energy does the belt supply to the package (including the energy dissipated by friction)?

FIGURE 8.19
Package dropped on a conveyor belt

*21. The potential energy of a particle moving in the x–y plane is $U = a/(x^2 + y^2)^{1/2}$, where a is a constant. What is the force on the particle? Draw a diagram showing the particle at the position x, y and the force vector.

22. The potential energy of a particle moving along the x axis is $U(x) = K/x^2$, where K is a constant. What is the corresponding force acting on the particle?

23. According to theoretical calculations, the potential energy of two quarks (see the Prelude) separated by a distance r is $U = \eta r$, where $\eta = 1.18 \times 10^{24}$ eV/m. What is the force between the two quarks? Express your answer in newtons.

8.2 The Curve of Potential Energy

24. The potential energy of a particle moving along the x axis is $U(x) = 2x^4 - x^2$, where x is measured in meters and the energy is measured in joules.

(a) Plot the potential energy as a function of x.

(b) Where are the possible equilibrium points?

(c) Suppose that $E = -0.050$ J. What are the turning points of the motion?

(d) Suppose that $E = 1.0$ J. What are the turning points of the motion?

25. In Example 4, we determined the turning point for a bungee jump graphically and numerically. Use the data given in this example for the following calculations.

(a) At what point does the jumper attain maximum speed? Calculate this maximum speed.

(b) At what point does the jumper attain maximum acceleration? Calculate this maximum acceleration.

26. The potential energy of one of the atoms in the hydrogen molecule is

$$U(x) = U_0 \left[e^{-2(x-x_0)/b} - 2e^{-(x-x_0)/b} \right]$$

with $U_0 = 2.36$ eV, $x_0 = 0.037$ nm, and b = 0.034 nm.[2] Under the influence of the force corresponding to this potential, the atom moves back and forth along the x axis within certain limits. If the energy of the atom is $E = -1.15$ eV, what will be the turning points of the motion; i.e., at what positions x will the kinetic energy be zero? [Hint: Solve this problem graphically by making a careful plot of $U(x)$; from your plot find the values of x that yield $U(x) = -1.15$ eV.]

27. Suppose that the potential energy of a particle moving along the x axis is

$$U(x) = \frac{b}{x^2} - \frac{2c}{x}$$

where b and c are positive constants.

(a) Plot $U(x)$ as a function of x; assume $b = c = 1$ for this purpose. Where is the equilibrium point?

(b) Suppose the energy of the particle is $E = -\frac{1}{2}c^2/b$. Find the turning points of the motion.

(c) Suppose that the energy of the particle is $E = \frac{1}{2}c^2/b$. Find the turning points of the motion. How many turning points are there in this case?

---

[2] These values of U_0, x_0, and b are half as large as those usually quoted, because we are looking at the motion of *one* atom relative to the center of the molecule.

28. A particle moves along the x axis under the influence of a conservative force with a potential energy $U(x)$. Figure 8.20 shows the plot of $U(x)$ vs. x. Figure 8.20 shows several alternative energy levels for the particle: $E = E_1$, $E = E_2$, and $E = E_3$. Assume that the particle is initially at $x = 1$ m. For each of the three alternative energies, describe the motion qualitatively, answering the following questions:

(a) Roughly, where are the turning points (right and left)?

(b) Where is the speed of the particle maximum? Where is the speed minimum?

(c) Is the orbit bound or unbound?

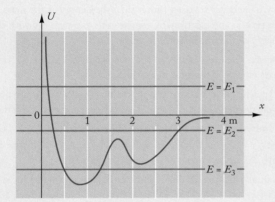

FIGURE 8.20 Plot of $U(x)$ vs. x.

29. A particle moving along the x axis experiences a potential of the form $U(x) = A|x|$, where A is a constant. A particle of mass m has speed v at the origin. Where are the turning points of its motion?

30. A particle initially at the origin moves in a potential of the form $U(x) = -U_0 \cos(ax)$, where U_0 and a are constants. What is the lowest energy the particle may have? If the energy of the particle is $E = 0$ and the particle is initially at $x = 0$, what are the turning points of the motion? For what energies is the particle motion unbound?

31. The potential energy of a particle moving along the x axis is $U(x) = -U_0/[1 + (x/a)^2]$, where $U_0 = 2.0$ J and $a = 1.0$ m. Sketch this function for -3 m $\le x \le 3$ m. What are the turning points for a particle with energy $E = -1.0$ J? For what energies is the particle unbound?

*32. Consider a particle moving in a region where the potential energy is given by $U = 2x^2 + x^4$, where U is in joules and x is in meters. Where are the turning points for a particle with total mechanical energy $E = 1.0$ J? with $E = 2.0$ J?

*33. The potential-energy function (Lennard–Jones potential) for two argon atoms as a function of their separation x is given by $U(x) = Cx^{-12} - Dx^{-6}$, where $C = 1.59 \times 10^{-24}$ J·(nm)^{12} and $D = 1.03 \times 10^{-21}$ J·(nm)^6. (Recall that 1 nm $= 10^{-9}$ m.)

(a) What is their equilibrium separation in nanometers (nm)?

(b) What is the lowest possible energy?

(c) What are the turning points for a particle with energy $E = -2.0 \times 10^{-21}$ J?

8.3 Other Forms of Energy[†]

34. Express the last two entries in Table 8.1 in electron-volts.

35. The chemical formula for TNT is $CH_3C_5H_2(NO_2)_3$. The explosion of 1 kg of TNT releases 4.6×10^6 J. Calculate the energy released per molecule of TNT. Express your answer in electron-volts.

36. Using the data of Table 8.1, calculate the amount of gasoline that would be required if all the energy requirements of the United States were to be met by the consumption of gasoline. How many gallons per day would have to be consumed?

37. The following table lists the fuel consumption and the passenger capacity of several vehicles. Assume that the energy content of the fuel is that of gasoline (see Table 8.1). Calculate the amount of energy used by each vehicle per passenger per mile. Which is the most energy-efficient vehicle? The least energy-efficient?

VEHICLE	PASSENGER CAPACITY	FUEL CONSUMPTION
Motorcycle	1	60 mi/gal
Snowmobile	1	12
Automobile	4	12
Intercity bus	45	5
Concorde SST	110	0.12
Jetliner	360	0.1

38. The energy released by the metabolization of fat is about 9000 kcal per kg of fat. While jogging on a level road, you use 750 kcal/h. How long do you need to jog to eliminate 1.0 kg of fat?

39. A 12-ounce can of soda typically contains 150 kcal of food energy (150 food "calories"). If your body uses one-fifth of this to climb stairs, how high does one soda enable you to climb?

40. A large household may use as much as 3000 kilowatt-hours of energy during a hot summer month. Express this amount of energy in joules.

41. On food labels in Europe, energy content is typically listed in kilojoules (kJ) instead of kcal (food "calories"). Express a daily intake of 2500 kcal in kJ.

*42. When a humpback whale breaches, or jumps out of the water (see Fig. 8.21), it typically leaves the water at an angle of about 70° at high speed and sometimes attains a height of 3 m,

FIGURE 8.21 A whale breaching.

measured from the water surface to the center of the whale. For a rough estimate of the energy requirements for such a breach, we can treat the translational motion of the whale as that of a particle moving from the surface of the water upward to a height of 3.0 m (for a more accurate calculation, we would have to take into account the buoyancy of the whale, which assists it in getting out of the water, but let us ignore this). What is the initial speed of the whale when it emerges from the water? What is the initial kinetic energy of a whale of 33 metric tons? Express the energy in kilocalories.

*43. The following table gives the rate of energy dissipation by a man engaged in diverse activities; the energies are given per kilogram of body mass:

RATE OF ENERGY DISSIPATION OF A MAN (PER kg OF BODY MASS)

Standing	1.3 kcal/(kg·h)
Walking (5 km/h)	3.3
Running (8 km/h)	8.2
Running (16 km/h)	15.2

Suppose the man wants to travel a distance of 2.5 km in one-half hour. He can walk this distance in exactly half an hour, or run slow and then stand still until the half hour is up, or run fast and then stand still until the half hour is up. What is the energy per kg of body mass dissipated in each case? Which program uses the most energy? Which the least?

8.4 Mass and Energy

44. The atomic bomb dropped on Hiroshima had an explosive energy equivalent to that of 20 000 tons of TNT, or 8.4×10^{13} J. How many kilograms of mass must have been converted into energy in this explosion?

45. How much energy is released by the annihilation of one proton and one antiproton (both initially at rest)? Express your answer in electron-volts.

46. How much energy is released by the annihilation of one electron and one antielectron (both initially at rest)? Express your answer in electron-volts.

47. The mass of the Sun is 2×10^{30} kg. The thermal energy in the Sun is about 2×10^{41} J. How much does the thermal energy contribute to the mass of the Sun?

48. The masses of the proton, electron, and neutron are $1.672\,623 \times 10^{-27}$ kg, 9.11×10^{-31} kg, and $1.674\,929 \times 10^{-27}$ kg, respectively. If a neutron decays into a proton and an electron, how much energy is released (other than the energy of the mass of the proton and electron)? Compare this extra energy with the energy of the mass of the electron.

49. Express the mass energy of the electron in keV. Express the mass energy of the proton in MeV.

50. A typical household may use approximately 1000 kilowatt-hours of energy per month. What is the equivalent amount of rest mass?

51. Combustion of one gallon of gasoline releases 1.3×10^8 J of energy. How much mass is converted to energy? Compare this with 2.8 kg, the mass of one gallon of gasoline.

52. A small silicon particle of diameter 0.20 micrometers has a mass of 9.8×10^{-18} kg. What is the mass energy of such a "nanoparticle" (in J)?

*53. In a high-speed collision between an electron and an antielectron, the two particles can annihilate and create a proton and an antiproton. The reaction

$$e + \bar{e} \rightarrow p + \bar{p}$$

converts the mass energy and kinetic energy of the electron and antielectron into the mass energy of the proton and the antiproton. Assume that the electron and the antielectron collide head-on with opposite velocities of equal magnitudes and that the proton and the antiproton are at rest immediately after the reaction. Calculate the kinetic energy of the electron required for this reaction; express your answer in electron-volts.

8.5 Power[†]

54. For an automobile traveling at a steady speed of 65 km/h, the friction of the air and the rolling friction of the ground on the wheels provide a total external friction force of 500 N. What power must the engine supply to keep the automobile moving? At what rate does the friction force remove momentum from the automobile?

55. In 1979, B. Allen flew a very lightweight propeller airplane across the English Channel. His legs, pushing bicycle pedals, supplied the power to turn the propeller. To keep the airplane flying, he had to supply about 0.30 hp. How much energy did he supply for the full flight lasting 2 h 49 min? Express your answer in kilocalories.

[†] For help, see Online Concept Tutorial 10 at www.ww norton.com/physics

56. The ancient Egyptians and Romans relied on slaves as a source of mechanical power. One slave, working desperately by turning a crank, could deliver about 200 W of mechanical power (at this power the slave would not last long). How many slaves would be needed to match the output of a modern automobile engine (150 hp)? How many slaves would an ancient Egyptian have to own in order to command the same amount of power as the average per capita power used by residents of the United States (14 kW)?

57. An electric clock uses 2.0 W of electric power. How much electric energy (in kilowatt-hours) does this clock use in 1 year? What happens to this electric energy?

58. While an automobile is cruising at a steady speed of 65 km/h, its engine delivers a mechanical power of 20 hp. How much energy does the engine deliver per hour?

59. A large windmill delivers 10 kW of mechanical power. How much energy does the windmill deliver in a working day of 8 hours?

60. The heating unit of a medium-sized house produces 170 000 Btu/h. Is this larger or smaller than the power produced by a typical automobile engine of 150 hp?

61. The heart of a resting person delivers a mechanical power of about 1.1 W for pumping blood. Express this power in hp. How much work does the heart do on the blood per day? Express this work in kcal.

62. The lasers to be used for controlled fusion experiments at the National Ignition Facility at the Lawrence Livermore Laboratory will deliver a power of 2.0×10^{15} W, a thousand times the output of all the power stations in the United States, in a brief pulse lasting 1.0×10^{-9} s. What is the energy in this laser pulse? How does it compare with the energy output of all the power stations in the United States in one day?

63. During the seven months of the cold season in the Northeastern United States, a medium-sized house requires about 1.0×10^8 Btu of heat to keep warm. A typical furnace delivers 1.3×10^5 Btu of heat per gallon of fuel oil.

    (a) How many gallons of fuel oil does the house consume during the cold season?

    (b) What is the average power delivered by the furnace?

64. Experiments on animal muscle tissue indicate that it can produce up to 100 watts of power per kilogram. A 600-kg horse has about 180 kg of muscle tissue attached to the legs in such a way that it contributes to the external work the horse performs while pulling a load. Accordingly, what is the theoretical prediction for the maximum power delivered by a horse? In trials, the actual maximum power that a horse can deliver in a short spurt was found to be about 12 hp. How does this compare with the theoretical prediction?

65. If a 60-W light bulb is left on for 24 hours each day, how many kilowatt-hours of electric power does it use in one year? If the electric energy costs you 15 cents per kilowatt-hour, what is your cost for one year?

66. Nineteenth-century English engineers reckoned that a laborer turning a crank can do steady work at the rate of 5000 ft·lbf/min. Suppose that four laborers working a manual crane attempt to lift a load of 9.0 short tons (1 short ton = 2000 lb). If there is no friction, what is the rate at which they can lift this load? How long will it take them to lift the load 15 ft?

67. The driver of an automobile traveling on a straight road at 80 km/h pushes forward with his hands on the steering wheel with a force of 50 N. What is the rate at which his hands do work on the steering wheel in the reference frame of the ground? In the reference frame of the automobile?

68. An automobile with a 100-hp engine has a top speed of 160 km/h. When at this top speed, what is the friction force (from air and road) acting on the automobile?

69. A horse walks along the bank of a canal and pulls a barge by means of a long horizontal towrope making an angle of 35° with the bank. The horse walks at the rate of 5.0 km/h, and the tension in the rope is 400 N. What horsepower does the horse deliver?

70. A 900-kg automobile accelerates from 0 to 80 km/h in 7.6 s. What are the initial and the final translational kinetic energies of the automobile? What is the average power delivered by the engine in this time interval? Express your answer in horsepower.

71. A six-cylinder internal combustion engine, such as used in an automobile, delivers an average power of 150 hp while running at 3000 rev/min. Each of the cylinders fires once every two revolutions. How much energy does each cylinder deliver each time it fires?

72. In Chapter 6, we saw that an automobile must overcome the force of air resistance, $f_{air} = \frac{1}{2} \rho C A v^2$. For the automobile of Example 6 of Chapter 6 ($C = 0.30$, $A = 2.8$ m^2, and $\rho = 1.3$ kg/m^3), calculate the power dissipation due to air resistance when traveling at 30 km/h and when traveling at 90 km/h. What is the difference in the total energy supplied to overcome air friction for a 300-km trip at 30 km/h? For a 300-km trip at 90 km/h?

73. A constant force of 40 N is applied to a body as the body moves uniformly at a speed of 3.5 m/s. The force does work on the body at a rate of 90 W. What is the angle between the force and the direction of motion of the body?

74. An electric motor takes 1.0 s to get up to speed; during this time, the power supplied by the motor varies with time according to $P = P_1 + (P_0 - P_1)(t - 1)^2$, where t is in seconds, $P_0 = 1.50$ kW, and $P_1 = 0.75$ kW. What is the total energy supplied for the time period $0 \le t \le 1$ s?

75. A constant force $\mathbf{F} = (6.0 \text{ N})\mathbf{i} + (8.0 \text{ N})\mathbf{j}$ acts on a particle. At what instantaneous rate is this force doing work on a particle with velocity $\mathbf{v} = (3.0 \text{ m/s})\mathbf{i} - (2.5 \text{ m/s})\mathbf{j}$?

76. An automobile engine typically has an efficiency of about 25%; i.e., it converts about 25% of the chemical energy available in gasoline into mechanical energy. Suppose that an automobile engine has a mechanical output of 110 hp. At what

rate (in gallons per hour) will this engine consume gasoline? See Table 8.1 for the energy content in gasoline.

77. The takeoff speed of a DC-3 airplane is 100 km/h. Starting from rest, the airplane takes 10 s to reach this speed. The mass of the (loaded) airplane is 11 000 kg. What is the average power delivered by the engines to the airplane during takeoff?

78. The Sun emits energy in the form of radiant heat and light at the rate of 3.9×10^{26} W. At what rate does this energy carry away mass from the Sun? How much mass does this amount to in 1 year?

79. The energy of sunlight arriving at the surface of the Earth amounts to about 1.0 kW per square meter of surface (facing the Sun). If all of the energy incident on a collector of sunlight could be converted into useful energy, how many square meters of collector area would we need to satisfy all of the energy demands in the United States? See Table 8.1 for the energy expenditure of the United States.

80. Equations (2.11) and (2.16) give the velocity and the acceleration of an accelerating Maserati sports car as a function of time. The mass of this automobile is 1770 kg. What is the instantaneous power delivered by the engine to the automobile? Plot the instantaneous power as a function of time in the time interval from 0 to 10 s. At what time is the power maximum?

81. The ship *Globtik Tokyo*, a supertanker, has a mass of 650 000 metric tons when fully loaded.

    (a) What is the kinetic energy of the ship when her speed is 26 km/h?

    (b) The engines of the ship deliver a power of 44 000 hp. According to the energy requirements, how long a time does it take the ship to reach a speed of 26 km/h, starting from rest? Make the assumption that 50% of the engine power goes into friction or into stirring up the water and 50% remains available for the translational motion of the ship.

    (c) How long a time does it take the ship to stop from an initial speed of 26 km/h if her engines are put in reverse? Estimate roughly how far the ship will travel during this time.

82. At Niagara Falls, 6200 m³ per second of water falls down a height of 49 m.

    (a) What is the rate (in watts) at which gravitational potential energy is dissipated by the falling water?

    (b) What is the amount of energy (in kilowatt-hours) wasted in 1 year?

    (c) Power companies get paid about 5 cents per kilowatt-hour of electric energy. If all the gravitational potential energy wasted in Niagara Falls could be converted into electric energy, how much money would this be worth?

83. The movement of a grandfather clock is driven by a 5.0-kg weight which drops a distance of 1.5 m in the course of a week. What is the power delivered by the weight to the movement?

84. A 27 000-kg truck has a 550-hp engine. What is the maximum speed with which this truck can move up a 10° slope?

*85. Consider a "windmill ship," which extracts mechanical energy from the wind by means of a large windmill mounted on the deck (see Fig. 8.22). The windmill generates electric power, which is fed into a large electric motor, which propels the ship. The mechanical efficiency of the windmill is 70% (that is, it removes 70% of the kinetic energy of the wind and transforms it into rotational energy of its blades). The efficiency of the electric generator attached to the windmill is 90%, and the efficiency of the electric motor connected to the generator is also 90%. We want the electric motor to deliver 20 000 hp in a (relative) wind of 40 km/h. What size windmill do we need? The density of air is 1.29 kg/m³.

FIGURE 8.22
A "windmill ship."

*86. An electric water pump is rated at 15 hp. If this water pump is to lift water to a height of 30 m, how many kilograms of water can it lift per second? How many liters? Neglect the kinetic energy of the water.

*87. The engines of the Sikorski Blackhawk helicopter generate 3080 hp of mechanical power, and the maximum takeoff mass of this helicopter is 7400 kg. Suppose that this helicopter is climbing vertically at a steady rate of 5.0 m/s.

    (a) What is the power that the engines deliver to the body of the helicopter?

    (b) What is the power that the engines deliver to the air (by friction and by the work that the rotors of the helicopter perform on the air)?

*88. In order to overcome air friction and other mechanical friction, an automobile of mass 1500 kg requires a power of 20 hp from its engine to travel at 64 km/h on a level road. Assuming the friction remains the same, what power does the same automobile require to travel uphill on an incline of slope 1:10 at the same speed? Downhill on the same incline at the same speed?

*89. With the gears in neutral, an automobile rolling down a long incline of slope 1:10 reaches a terminal speed of 95 km/h. At this speed the rate of decrease of the gravitational potential energy matches the power required to overcome air friction and other mechanical friction. What power (in horsepower) must the engine of this automobile deliver to drive it at 95 km/h on a level road? The mass of the automobile is 1500 kg.

*90. The power supplied to an electric circuit decreases exponentially with time according to $P = P_0 e^{-t/\tau}$, where $P_0 = 2.0$ W and $\tau = 5.0$ s are constants. What is the total energy supplied to the circuit during the time interval $0 \le t \le 5.0$ s? During $0 \le t \le \infty$?

*91. Each of the two Wright Cyclone engines on a DC-3 airplane generates a power of 850 hp. The mass of the loaded plane is 10 900 kg. The plane can climb at the rate of 260 m/min. When the plane is climbing at this rate, what percentage of the engine power is used to do work against gravity?

*92. A fountain shoots a stream of water 10 m up in the air. The base of the stream is 10 cm across. What power is expended to send the water to this height?

*93. The record of 203.1 km/h for speed skiing set by Franz Weber at Velocity Peak in Colorado was achieved on a mountain slope inclined downward at 51°. At this speed, the force of friction (air and sliding friction) balances the pull of gravity along the slope, so the motion proceeds at constant velocity.

   (a) What is the rate at which gravity does work on the skier? Assume that the mass of the skier is 75 kg.

   (b) What is the rate at which sliding friction does work? Assume that the coefficient of friction is $\mu_k = 0.03$.

   (c) What is the rate at which air friction does work?

*94. A windmill for the generation of electric power has a propeller of diameter 1.8 m. In a wind of 40 km/h, this windmill delivers 200 W of electric power.

   (a) At this wind speed, what is the rate at which the air carries kinetic energy through the circular area swept out by the propeller? The density of air is 1.29 kg/m^3.

   (b) What percentage of the kinetic energy of the air passing through this area is converted into electric energy?

*95. A small electric kitchen fan blows 8.5 m^3/min of air at a speed of 5.0 m/s out of the kitchen. The density of air is 1.3 kg/m^3. What electric power must the fan consume to give the ejected air the required kinetic energy?

*96. The final portion of the Tennessee River has a downward slope of 0.074 m per kilometer. The rate of flow of water in the river is 280 m^3/s. Assume that the speed of the water is constant along the river. How much power is dissipated by friction of the water against the riverbed per kilometer?

*97. Off the coast of Florida, the Gulf Stream has a speed of 4.6 km/h and a rate of flow of 2.2×10^3 km^3/day. At what rate is kinetic energy flowing past the coast? If all this kinetic energy could be converted into electric power, how many kilowatts would it amount to?

*98. Figure 8.23 shows an overshot waterwheel, in which water flowing onto the top of the wheel fills buckets whose weight causes the wheel to turn. The water descends in the buckets to the bottom, and there it is spilled out, so the ascending buckets are always empty. If in a waterwheel of diameter 10 m the amount of water carried down by the wheel is 20 liters per second (or 20 kg per second), what is the mechanical power

that the descending water delivers to the wheel? Assume that the water flowing onto the top of the wheel has roughly the same speed as the wheel and exerts no horizontal push on the wheel. [Hint: The kinetic energy of the water is the same when the water enters the bucket and when it spills out (since the speed of the bucket is constant); hence the kinetic energy of the water does not affect the answer.]

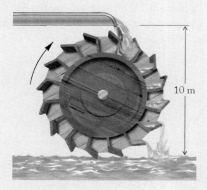

FIGURE 8.23 An overshot waterwheel.

*99. Suppose that in the undershot waterwheel shown in Fig. 8.24, the stream of water against the blades of the wheel has a speed of 15 m/s, and the amount of water is 30 liters per second (or 30 kg per second). If the water gives all of its kinetic energy to the blades (and then drips away with zero horizontal speed), how much mechanical power does the water deliver to the wheel?

FIGURE 8.24 An undershot waterwheel.

*100. (a) With its engines switched off, a small two-engine airplane of mass 1100 kg glides downward at an angle of 13° at a speed of 90 knots. Under these conditions, the weight of the plane, the lift force (perpendicular to the direction of motion) generated by air flowing over the wings, and the frictional force (opposite to the direction of motion) exerted by the air are in balance. Draw a "free-body" diagram for these forces, and calculate their magnitudes.

   (b) Suppose that with its engine switched on, the plane climbs at an upward angle of 13° at a speed of 90 knots. Draw a "free-body" diagram for the forces acting on the airplane under these conditions; include the push that the air exerts on the propeller. Calculate the magnitudes of all the forces.

(c) Calculate the power that the engine must deliver to compensate for the rate of increase of the potential energy of the plane and the power lost to friction. For a typical small plane of 1100 kg, the actual engine power required for such a climb of 13° is about 400 hp. Explain the discrepancy between your result and the actual engine power. (Hint: What does the propeller do to the air?)

*101. The reaction that supplies the Sun with energy is

$$H + H + H + H \rightarrow He + [energy]$$

(The reaction involves several intermediate steps, but this need not concern us now.) The mass of the hydrogen (H) atom is 1.008 13 u, and that of the helium (He) atom is 4.003 88 u.

(a) How much energy is released in the reaction of four hydrogen atoms (by the conversion of mass into energy)?

(b) How much energy is released in the reaction of 1.0 kg of hydrogen atoms?

(c) The Sun releases energy at the rate of 3.9×10^{26} W. At what rate (in kg/s) does the Sun consume hydrogen?

(d) The Sun contains about 1.5×10^{30} kg of hydrogen. If it continues to consume hydrogen at the same rate, how long will the hydrogen last?

REVIEW PROBLEMS

102. A particle moves along the x axis under the influence of a variable force $F_x = 5x^2 + 3x$ (where force is measured in newtons and distance in meters).

(a) What is the potential energy associated with this force? Assume that $U(x) = 0$ at $x = 0$.

(b) How much work does the force do on a particle that moves from $x = 0$ to $x = 2.0$ m?

*103. A particle is subjected to a force that depends on position as follows:

$$\mathbf{F} = 4\mathbf{i} + 2x\mathbf{j}$$

where the force is measured in newtons and the distance in meters.

(a) Calculate the work done by this force as the particle moves from the origin to the point $x = 1.0$ m, $y = 1.0$ m along the straight path I shown in Fig. 8.25.

(b) Calculate the work done by this force if the particle returns from the point $x = 1.0$ m, $y = 1.0$ m to the origin along the

path II consisting of a horizontal and a vertical segment (see Fig. 8.25). Is the force conservative?

*104. A 3.0-kg block sliding on a horizontal surface is accelerated by a compressed spring. At first, the block slides without friction. But after leaving the spring, the block travels over a new portion of the surface, with a coefficient of friction 0.20, for a distance of 8.0 m before coming to rest (see Fig. 8.26). The force constant of the spring is 120 N/m.

(a) What was the maximum kinetic energy of the block?

(b) How far was the spring compressed before being released?

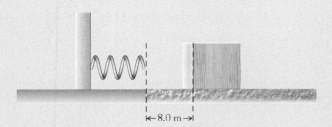

FIGURE 8.26 Block released from a spring.

105. The ancient Egyptians moved large stones by dragging them across the sand in sleds (Fig. 8.27). Suppose that 6000 Egyptians are dragging a sled with a coefficient of sliding friction $\mu_k = 0.30$ along a level surface of sand. Each Egyptian can exert a force of 360 N, and each can deliver a mechanical power of 0.20 hp.

(a) What is the maximum weight they can move at constant speed?

(b) What is the maximum speed with which they can move this weight?

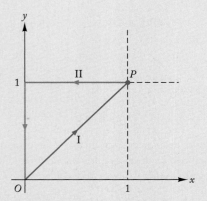

FIGURE 8.25 Outward and return paths of a particle.

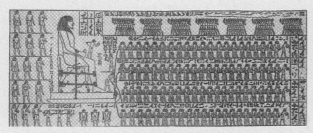

FIGURE 8.27   Ancient Egyptian wall mural from 1900 B.C.

106. In a braking test, a 990-kg automobile takes 2.1 s to come to a full stop from an initial speed of 60 km/h. What is the amount of energy dissipated in the brakes? What is the average power dissipated in the brakes? Ignore external friction in your calculation and express the power in horsepower.

107. In a waterfall on the Alto Paraná river (between Brazil and Paraguay), the height of fall is 33 m and the average rate of flow is 13 000 m^3 of water per second. What is the power dissipated by this waterfall?

108. When jogging at 12 km/h on a level road, a 70-kg man uses 750 kcal/h. How many kilocalories per hour does he require when jogging up a 1:10 incline at the same speed? Assume that the frictional losses are independent of the value of the slope.

109. Consider a projectile traveling horizontally and slowing down under the influence of air resistance, as described in Problems 47 and 48 of Chapter 2. The mass of this projectile is 45.36 kg, and the speed as a function of time is

$$v = 655.9 - 61.1t + 3.26t^2$$

where speed is in m/s and time in seconds.

(a) What is the instantaneous power removed from the projectile by air resistance?

(b) What is the kinetic energy at time $t = 0$? At time $t = 3.00$ s?

(c) What is the average power for the time interval from 0 to 3.00 s?

110. A woman exercising on a rowing machine pulls the oars back once per second. During such a pull, each hand moves 0.50 m while exerting an average force of 100 N.

(a) What is the work the woman does during each stroke (with both hands)?

(b) What is the average power the woman delivers to the oars?

111. The world's tallest staircase, of 2570 steps, is located in the CN tower in Toronto. It reaches a height of 457 m. Estimate how long it would take an athlete to climb this staircase. The athlete has a mass of 75 kg, and his leg muscle can deliver a power of 200 W.

112. A pump placed on the shore of a pond takes in 0.80 kg of water per second and squirts it out of a nozzle at 50 m/s. What mechanical power does this pump supply to the water?

113. The hydroelectric pumped-storage plant in Northfield, Massachusetts, has a reservoir holding 2.2×10^7 m^3 of water on top of a mountain. The water flows 270 m vertically down the mountain in pipes and drives turbines connected to electric generators.

(a) How much electric energy, in kW·h, can this storage plant generate with the water available in the reservoir?

(b) In order to generate 1000 MW of electric energy, at what rate, in m^3/s, must this storage plant withdraw water from the reservoir?

114. A 50-kg circus clown is launched vertically from a spring-loaded cannon using a spring with spring constant 3500 N/m. The clown attains a height of 4.0 m above the initial position (when the spring was compressed).

(a) How far was the spring compressed before launch?

(b) What was the maximum acceleration of the clown during launch?

(c) What was the maximum speed of the clown?

Answers to Checkups

Checkup 8.1

1. The force can be obtained from $F_x = -dU/dx = +kx$. Thus, the force is positive for positive x and negative for negative x; that is, the new force is repulsive (it pushes a particle away from $x = 0$), whereas the spring force is attractive.

2. The potential energy shown in Fig. 8.5a has a negative slope as a function of x. By Eq. (8.21), the force is the negative of the slope of the potential; the negative of a negative is positive, and so the force is directed along the positive x direction.

3. A force is always conservative if the force is an explicit function of position x. In that case, a potential-energy function can always be constructed by integration of the force according to Eq. (8.14).

4. (D) No; no. The work done is equal to the negative of the change in potential energy only for conservative forces. The work done is equal to the change in kinetic energy only for the *net* force acting on a particle.

Checkup 8.2

1. The number of turning points must decrease with increasing energy [we do not consider a stationary point of stable equilibrium (Fig. 8.8), since the particle is moving]. Consider the potential of Fig. 8.7: for small energies, the particle will move back and forth (two turning points); for somewhat higher energy, the particle will move back from the left end but escape from the right end (one turning point). Unless $U = \infty$, for sufficiently high energy the particle could escape from the left end also (no turning point). In one dimension, there cannot be more than two turning points, although the two turning points will of course be different for different energies.

2. The maximum speed corresponds to the deepest part of the curve (maximum kinetic energy, $K = E - U$); from the figure, this occurs at $x \approx -6$ m. The maximum acceleration and force ($F = -dU/dx$) occurs where the slope is largest; for the bungee jumper, this is at $x \approx -15$ m.

3. (B) 2. The potential-energy curve of the spring force is a simple parabola (Fig. 8.1), so there are two turning points for any positive energy.

Checkup 8.3

1. No. Gravitational potential energy is lost as the parachutist descends (at uniform speed, there is no change in kinetic energy). From a macroscopic viewpoint, the energy lost due to friction with the air is converted into heat.

2. The energy comes from a decrease in the chemical energy of the exploding gunpowder; microscopically, such chemical energy comes from changes in the kinetic and potential energy of electrons in the atoms and bonds of the elements involved.

3. The energy is converted to heat due to frictional forces; these may include friction in the engine, brakes, tires, and road, as well as air friction.

4. (C) The kinetic energy is converted into heat due to frictional forces, mostly in the brakes (brake pads rub against drums or disks), partly where the tires contact the road, and some from air friction. All the heat is eventually transferred to the air as the brakes, tires, and road cool.

Checkup 8.4

1. Yes; the Sun continually loses mass in the form of heat and light, as well as by emitting particles with mass.

2. Energy and mass are both conserved; the original rest mass is converted to the energy of the light (electromagnetic radiation), and this light carries away mass as well as energy.

3. (A) Increase. The mass of the water will increase by the usual $\Delta m = \Delta E/c^2$, where ΔE is the increase in thermal energy of the water.

Checkup 8.5

1. (a) No; there is no force parallel to the motion, so there is no work done and no power expended. (b) Trotting uphill, you deliver power at a rate $P = Fv = mg \sin \phi\, v$, where m is the mass of the pack, ϕ is the angle of the incline, and v is the speed along the incline. (c) Trotting downhill, the component of $\mathbf{F}$ along $\mathbf{v}$ is negative, so you do negative work on the backpack; that is, the backpack delivers power to you.

2. Yes, the energy you have to expend is mgh, whichever slope of road you take. The steeper road requires more of an effort, since, for example, for the same walking speed, the force is more nearly parallel to the velocity, and so the power expended, $P = \mathbf{F} \cdot \mathbf{v}$, is greater.

3. Some of the power is lost as heat, due to the friction force between the boat and the water; some of the energy is converted into a more macroscopic kinetic energy of the water, by the generation of water waves.

4. (C) 4. The power is equal to the force times the speed. At the same speed, a car with twice the mass will require twice the power to move against gravity; if that car is also traveling at twice the speed, it will then require four times as much power (ignoring other losses).

Gravitation

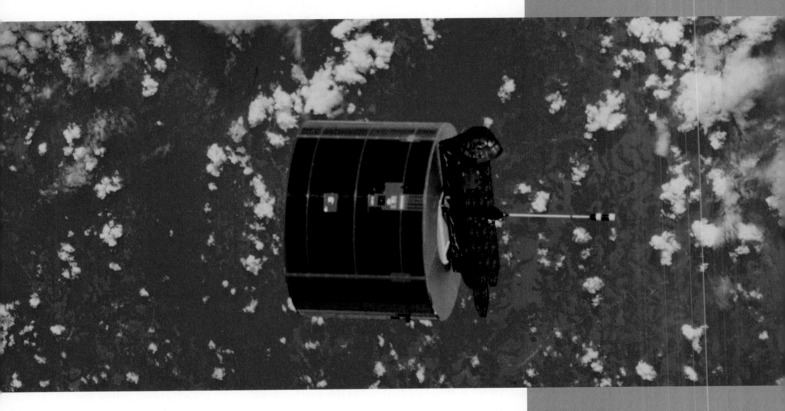

CONCEPTS IN CONTEXT

Hundreds of artificial satellites have been placed in orbit around the Earth, such as this Syncom communications satellite shown just after launch from the Space Shuttle.

With the concepts we will develop in this chapter, we can answer various questions about artificial satellites:

? Communications satellites and weather satellites are placed in high-altitude "geosynchronous" orbits that permit them to keep in step with the rotation of the Earth, so the satellite always remains at a fixed point above the equator. What is the radius of such a geosynchronous orbit? (Example 6, page 279; and Physics in Practice: Communications Satellites and Weather Satellites, page 281)

? Surveillance satellites and spacecraft such as the Space Shuttle usually operate in low-altitude orbits, just above the Earth's atmosphere. How quickly does such a satellite circle the Earth? (Example 7, page 280)

9.1 Newton's Law of Universal Gravitation

9.2 The Measurement of G

9.3 Circular Orbits

9.4 Elliptical Orbits; Kepler's Laws

9.5 Energy in Orbital Motion

? The Syncom satellite was carried by the Space Shuttle to a low-altitude orbit, and then it used its own booster rocket to lift itself to the high-altitude geosynchronous orbit. What is the increase of mechanical energy (kinetic and gravitational) of the satellite during this transfer from one orbit to another? (Example 9, page 290)

Within the Solar System, planets orbit around the Sun, and satellites orbit around the planets. These circular, or nearly circular, motions require a centripetal force pulling the planets toward the Sun and the satellites toward the planets. It was Newton's great discovery that this interplanetary force holding the celestial bodies in their orbits is of the same kind as the force of gravity that causes apples, and other things, to fall downward near the surface of the Earth. Newton found that a single formula, his Law of Universal Gravitation, encompasses both the gravitational forces acting between celestial bodies and the gravitational force acting on bodies near the surface of the Earth.

By the nineteenth century, Newton's theory of gravitation had proved itself so trustworthy that when astronomers noticed an irregularity in the motion of Uranus, they could not bring themselves to believe that the theory was at fault. Instead, they suspected that a new, unknown planet caused these irregularities by its gravitational pull on Uranus. The astronomers J. C. Adams and U. J. J. Leverrier proceeded to calculate the expected position of this hypothetical planet—and the new planet was immediately found at just about the expected position. This discovery of a new planet, later named Neptune, was a spectacular success of Newton's theory of gravitation. Newton's theory remains one of the most accurate and successful theories in all of physics, and in all of science.

In this chapter, we will examine Newton's Law of Universal Gravitation; we will see how it includes the familiar gravitational force near the Earth's surface. We will also examine circular and elliptical orbits of planets and satellites, and we will become familiar with Kepler's laws describing these orbits. Finally, we will discuss gravitational potential energy and apply energy conservation to orbital motion.

Online
Concept
Tutorial
11

9.1 NEWTON'S LAW OF UNIVERSAL GRAVITATION

Newton proposed that just as the Earth gravitationally attracts bodies placed near its surface and causes them to fall downward, the Earth also attracts more distant bodies, such as the Moon, or the Sun, or other planets. In turn, the Earth is gravitationally attracted by all these bodies. More generally, *every* body in the Universe attracts *every* other body with a gravitational force that depends on their masses and on their distances. The gravitational force that two bodies exert on each other is large if their masses are large, and small if their masses are small. The gravitational force decreases if we increase the distance between the bodies. The **Law of Universal Gravitation** formulated by Newton can be stated most easily for the case of particles:

Every particle attracts every other particle with a force directly proportional to the product of their masses and inversely proportional to the square of the distance between them.

Expressed mathematically, the magnitude of the gravitational force that two particles of masses M and m separated by a distance r exert on each other is

Law of Universal Gravitation

$$F = \frac{GMm}{r^2} \tag{9.1}$$

where G is a universal constant of proportionality, the same for all pairs of particles.

The direction of the force on each particle is directly toward the other particle. Figure 9.1 shows the directions of the forces on each particle. Note that the two forces are of equal magnitudes and opposite directions; they form an action–reaction pair, as required by Newton's Third Law.

The constant G is known as the **gravitational constant**. In SI units its value is approximately given by

$$G = 6.67 \times 10^{-11}\,\text{N·m}^2/\text{kg}^2 \qquad (9.2)$$

gravitational constant

The gravitational force of Eq. (9.1) is an inverse-square force: it decreases by a factor of 4 when the distance increases by a factor of 2, it decreases by a factor of 9 when the distance increases by a factor of 3, and so on. Figure 9.2 is a plot of the magnitude of the gravitational force as a function of the distance. Although the force decreases with distance, it never quite reaches zero. Thus, every particle in the universe continually attracts every other particle at least a little bit, even if the distance between the particles is very, very large.

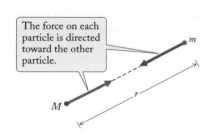

The force on each particle is directed toward the other particle.

FIGURE 9.1 Two particles attract each other gravitationally. The forces are of equal magnitudes and of opposite directions.

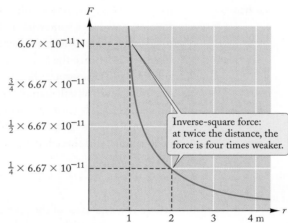

Inverse-square force: at twice the distance, the force is four times weaker.

FIGURE 9.2   Magnitude of the gravitational force exerted by a particle of mass 1 kg on another particle of mass 1 kg.

EXAMPLE 1

What is the gravitational force between a 70-kg man and a 70-kg woman separated by a distance of 10 m? Treat both masses as particles.

SOLUTION: From Eq. (9.1),

$$F = \frac{GMm}{r^2}$$

$$= \frac{6.67 \times 10^{-11}\,\text{N·m}^2/\text{kg}^2 \times 70\ \text{kg} \times 70\ \text{kg}}{(10\ \text{m})^2}$$

$$= 3.3 \times 10^{-9}\,\text{N}$$

This is a very small force, but as we will see in the next section, the measurement of such small forces is not beyond the reach of sensitive instruments.

The gravitational force does not require any contact between the interacting particles. In reaching from one remote particle to another, the gravitational force somehow bridges the empty space between the particles. This is called **action-at-a-distance**.

It is also quite remarkable that the gravitational force between two particles is unaffected by the presence of intervening masses. For example, a particle in Washington attracts a particle in Beijing with exactly the force given by Eq. (9.1), even though all of the bulk of the Earth lies between Washington and Beijing. This means that it is impossible to shield a particle from the gravitational attraction of another particle.

Since the gravitational attraction between two particles is completely independent of the presence of other particles, it follows that the net gravitational force between two bodies (e.g., the Earth and the Moon or the Earth and an apple) is merely the vector sum of the individual forces between all the particles making up the bodies—that is, the gravitational force obeys the principle of linear superposition of forces (see Section 5.3). As a consequence of this simple vector summation of the gravitational forces of the individual particles in a body, it can be shown that *the net gravitational force between two spherical bodies acts just as though each body were concentrated at the center of its respective sphere.* This result is known as **Newton's theorem**. The proof of Newton's theorem involves a somewhat tedious summation. Later, in the context of electrostatic force, we provide a much simpler derivation of Newton's theorem using Gauss' Law (see Chapter 24). Since the Sun, the planets, and most of their satellites are almost exactly spherical, this important theorem permits us to treat all these celestial bodies as pointlike particles in all calculations concerning their gravitational attractions. For instance, since the Earth is (nearly) spherical, the gravitational force exerted by the Earth on a particle above its surface is as though the mass of the Earth were concentrated at its center; thus, this force has a magnitude

$$F = \frac{GM_E m}{r^2} \tag{9.3}$$

where m is the mass of the particle, M_E is the mass of the Earth, and r is the distance from the center of the Earth (see Fig. 9.3).

If the particle is at the surface of the Earth, at a radius $r = R_E$, then Eq. (9.3) gives a force

$$F = \frac{GM_E m}{R_E^2} \tag{9.4}$$

The corresponding acceleration of the mass m is

$$a = \frac{F}{m} = \frac{GM_E}{R_E^2} \tag{9.5}$$

But this acceleration is what we usually call the acceleration of free fall; and usually designate by g. Thus, g is related to the mass and the radius of the Earth,

$$g = \frac{GM_E}{R_E^2} \tag{9.6}$$

This equation establishes the connection between the ordinary force of gravity we experience at the surface of the Earth and Newton's Law of Universal Gravitation. Notice that g is only approximately constant. Small changes in height near the Earth's surface have little effect on the value given by Eq. (9.6), since $R_E \approx 6.4 \times 10^6$ m is so large. But for a large altitude h above the Earth's surface, we must replace R_E with $R_E + h$ in Eq. (9.6), and appreciable changes in g can occur.

Note that an equation analogous to Eq. (9.6) relates the acceleration of free fall at the surface of any (spherical) celestial body to the mass and the radius of that body.

Newton's theorem

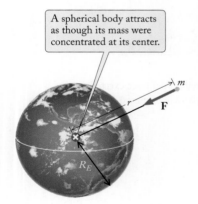

A spherical body attracts as though its mass were concentrated at its center.

m

r F

R_E

FIGURE 9.3 The gravitational force exerted by the Earth on a particle is directed toward the center of the Earth.

For example, we can calculate the acceleration of free fall on the surface of the Moon from its mass and radius.

EXAMPLE 2 The mass of the Moon is 7.35×10^{22} kg, and its radius is 1.74×10^{6} m. Calculate the acceleration of free fall on the Moon and compare with acceleration of free fall on the Earth.

SOLUTION: For the Moon, the formula analogous to Eq. (9.6) is

$$g_{\text{Moon}} = \frac{GM_{\text{Moon}}}{R_{\text{Moon}}^2} = \frac{6.67 \times 10^{-11}\,\text{N·m}^2/\text{kg}^2 \times 7.35 \times 10^{22}\,\text{kg}}{(1.74 \times 10^{6}\,\text{m})^2}$$

$$= 1.62 \text{ m/s}^2$$

This is about 1/6 the acceleration of free fall on the surface of the Earth ($g = 9.81$ m/s^2). If you can jump upward to a height of one-half meter on the Earth, then this same jump will take you to a height of 3 meters on the Moon!

EXAMPLE 3 The masses of the Sun, Earth, and Moon are 1.99×10^{30} kg, 5.98×10^{24} kg, and 7.35×10^{22} kg, respectively. Assume that the location of the Moon is such that the angle subtended by the lines from the Moon to the Sun and from the Moon to the Earth is 45.0°, as shown in Fig. 9.4a. What is the net force on the Moon due to the gravitational forces of the Sun and Earth? The Moon is 1.50×10^{11} m from the Sun and 3.84×10^{8} m from the Earth.

SOLUTION: Before finding the resultant force, we first find the magnitudes of the individual forces. The magnitude of the force due to the Sun on the Moon is

$$F_{SM} = \frac{GM_S M_M}{R_{SM}^2}$$

$$= \frac{6.67 \times 10^{-11}\,\text{N·m}^2/\text{kg}^2 \times 1.99 \times 10^{30}\,\text{kg} \times 7.35 \times 10^{22}\,\text{kg}}{(1.50 \times 10^{11}\,\text{m})^2}$$

$$= 4.34 \times 10^{20}\,\text{N}$$

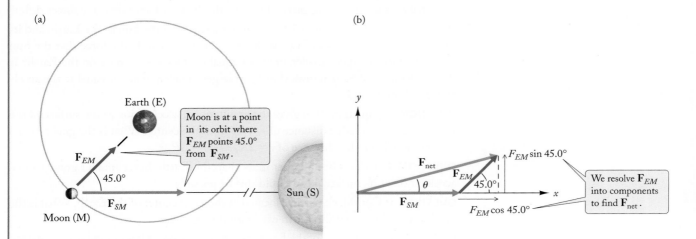

FIGURE 9.4 (a) Each of the gravitational forces on the Moon is directed toward the body producing the force. (b) Vector addition of the two forces.

The magnitude of the force due to the Earth on the Moon is

$$F_{EM} = \frac{GM_E M_M}{R_{EM}^2}$$

$$= \frac{6.67 \times 10^{-11}\,\text{N·m}^2/\text{kg}^2 \times 5.98 \times 10^{24}\,\text{kg} \times 7.35 \times 10^{22}\,\text{kg}}{(3.84 \times 10^8\,\text{m})^2}$$

$$= 1.99 \times 10^{20}\,\text{N}$$

The direction of each force on the Moon is toward the body producing that force, as indicated in Fig. 9.4a. We choose the x axis along the Moon–Sun direction and add the two forces vectorially as shown in Fig. 9.4b. By resolving $\mathbf{F}_{EM}$ into components, we see that the resultant force $\mathbf{F}_{\text{net}}$ has x component

$$F_{\text{net},x} = F_{SM} + F_{EM}\cos 45.0°$$

$$= 4.34 \times 10^{20}\,\text{N} + 1.99 \times 10^{20}\,\text{N} \times \cos 45.0° = 5.75 \times 10^{20}\,\text{N}$$

and y component

$$F_{\text{net},y} = F_{EM}\sin 45.0° = 1.99 \times 10^{20}\,\text{N} \times \sin 45.0° = 1.41 \times 10^{20}\,\text{N}$$

Thus the resultant force has magnitude

$$F_{\text{net}} = \sqrt{F_{\text{net},x}^2 + F_{\text{net},y}^2}$$

$$= \sqrt{(5.75 \times 10^{20}\,\text{N})^2 + (1.41 \times 10^{20}\,\text{N})^2} = 5.92 \times 10^{20}\,\text{N}$$

The direction of $\mathbf{F}_{\text{net}}$ is given by $\tan\theta = \dfrac{F_{\text{net},y}}{F_{\text{net},x}} = \dfrac{1.41 \times 10^{20}\,\text{N}}{5.75 \times 10^{20}\,\text{N}} = 0.245.$

With a calculator, we find that the inverse tangent of 0.238 is

$$\theta = 13.8°$$

 Checkup 9.1

QUESTION 1: Neptune is about 30 times as far away from the Sun as the Earth. Compare the gravitational force that the Sun exerts on a 1-kg piece of Neptune with the force it exerts on a 1-kg piece of the Earth. By what factor do these forces differ?

QUESTION 2: Saturn is about 10 times as far away from the Sun as the Earth, and its mass is about 100 times as large as the mass of the Earth. Is the force that the Sun exerts on Saturn larger, smaller, or about equal to the force it exerts on the Earth? Is the acceleration of Saturn toward the Sun larger, smaller, or about equal to the acceleration of the Earth?

QUESTION 3: Equation (9.6) gives the gravitational acceleration at the surface of the Earth, that is, at a radial distance of $r = R_E$ from the center. What is the gravitational acceleration at a radial distance of $r = 2R_E$? At $r = 3R_E$?

QUESTION 4: Uranus has a larger mass than the Earth, but a smaller gravitational acceleration at its surface. How could this be possible?

QUESTION 5: Consider a particle located at the exact center of the Earth. What is the gravitational force that the Earth exerts on this particle?

QUESTION 6: If the radius of the Earth were twice as large as it is but the mass remained unchanged, what would be the gravitational acceleration at its surface?

(A) $\frac{1}{8}g$ (B) $\frac{1}{4}g$ (C) g (D) $4g$ (E) $8g$

9.2 THE MEASUREMENT OF G

The gravitational constant G is rather difficult to measure with precision. The trouble is that gravitational forces between masses of laboratory size are extremely small, and thus a very delicate apparatus is needed to detect these forces. Measurements of G are usually done with a **Cavendish torsion balance** (see Fig. 9.5). Two equal, small spherical masses m and m' are attached to a lightweight horizontal beam which is suspended at its middle by a thin vertical fiber. When the beam is left undisturbed, it will settle into an equilibrium position such that the fiber is completely untwisted. If two equal, large masses M and M' are brought near the small masses, the gravitational attraction between each small mass and the neighboring large mass tends to rotate the beam clockwise (as seen from above). The twist of the fiber opposes this rotation, and the net result is that the beam settles into a new equilibrium position in which the force on the beam generated by the gravitational attraction between the masses is exactly balanced by the force exerted by the twisted fiber. The gravitational constant can then be calculated from the measured values of the angular displacement between the two equilibrium positions, the values of the masses, their distances, and the characteristics of the fiber.

Note that the mass of the Earth can be calculated from Eq. (9.6) using the known values of G, R_E, and g:

$$M_E = \frac{R_E^2 g}{G} = \frac{(6.38 \times 10^6 \text{ m})^2 \times 9.81 \text{ m/s}^2}{6.67 \times 10^{-11} \text{ N} \cdot \text{m}^2/\text{kg}^2}$$

$$= 5.98 \times 10^{24} \text{ kg}$$

This calculation would seem to be a rather roundabout way to arrive at the mass of the Earth, but there is no direct route, since we cannot place the Earth on a balance. Because the calculation requires a prior measurement of the value of G, the Cavendish experiment has often been described figuratively as "weighing the Earth."

HENRY CAVENDISH (1731–1810)
English experimental physicist and chemist. His torsion balance for the absolute measurement of the gravitational force was based on an earlier design used by Coulomb for the measurement of the electric force.

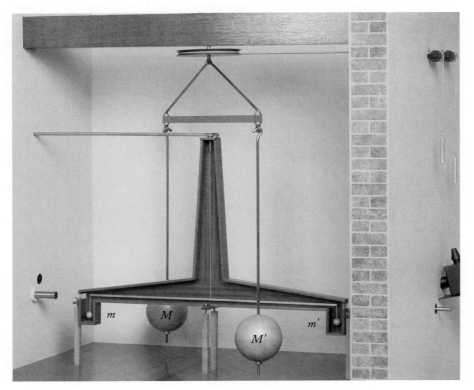

FIGURE 9.5 Model of large torsion balance used by Cavendish. The small masses m, m' hang from the ends of a horizontal beam which is suspended at its middle by a thin vertical fiber.

QUESTION 1: Why don't we determine G by measuring the (fairly large) force between the Earth and a mass of, say, 1 kg?

QUESTION 2: Large mountains produce a (small) deflection of a plumb bob suspended nearby. Could we use this effect to determine G?

    (A) Yes   (B) No

Online
Concept
Tutorial

9.3 CIRCULAR ORBITS

The gravitational force is responsible for holding the Solar System together; it makes the planets orbit around the Sun, and it makes the satellites orbit around the planets. Although the mutual gravitational forces of the Sun on a planet and of the planet on the Sun are of equal magnitudes, the mass of the Sun is more than a thousand times as large as the mass of even the largest planet, and hence its acceleration is much smaller. It is therefore an excellent approximation to regard the Sun as fixed and immovable, and it then only remains to investigate the motion of the planet. If we designate the masses of the Sun and the planet by M_S and m, respectively, and their center-to-center separation by r, then the magnitude of the gravitational force on the planet is

$$F = \frac{GM_S m}{r^2} \tag{9.7}$$

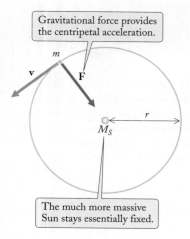

FIGURE 9.6 Circular orbit of a planet around the Sun.

(figure labels: Gravitational force provides the centripetal acceleration. — The much more massive Sun stays essentially fixed.)

This force points toward the center of the Sun; that is, the center of the Sun is the center of force (see Fig. 9.6). For a particle moving under the influence of such a central force, the simplest orbital motion is uniform circular motion, with the gravitational force acting as centripetal force. The motion of the planets in our Solar System is somewhat more complicated than that—as we will see in the next section, the planets move along ellipses, instead of circles. However, none of these planetary ellipses deviates very much from a circle, and as a first approximation we can pretend that the planetary orbits are circles.

By combining the expression (9.7) for the centripetal force with Newton's Second Law we can find a relation between the radius of the circular orbit and the speed. If the speed of the planet is v, then the centripetal acceleration is v^2/r [see Eq. (4.49)], and the equation of motion, $ma = F$, becomes

$$\frac{mv^2}{r} = F \tag{9.8}$$

Consequently,

$$\frac{mv^2}{r} = \frac{GM_S m}{r^2} \tag{9.9}$$

We can cancel a factor of m and a factor of $1/r$, in this equation, and we obtain

$$v^2 = \frac{GM_S}{r}$$

or

speed for circular orbit

$$v = \sqrt{\frac{GM_S}{r}} \tag{9.10}$$

| EXAMPLE 4 | The mass of the Sun is 1.99×10^{30} kg, and the radius of the Earth's orbit around the Sun is 1.5×10^{11} m. From this, |

calculate the orbital speed of the Earth.

SOLUTION: According to Eq. (9.10), the orbital speed is

$$v = \sqrt{\frac{GM_S}{r}} = \sqrt{\frac{6.67 \times 10^{-11}\, \text{N·m}^2/\text{kg}^2 \times 1.99 \times 10^{30}\, \text{kg}}{1.5 \times 10^{11}\, \text{m}}}$$

$$= 3.0 \times 10^4\, \text{m/s} = 30\, \text{km/s}$$

NICHOLAS COPERNICUS (1473–1543)
Polish astronomer. In his book De Revolutionibus Orbium Coelestium *he formulated the helio-centric system for the description of the motion of the planets, according to which the Sun is immovable and the planets orbit around it.*

The time a planet takes to travel once around the Sun, or the time for one revolution, is called the **period** *of the planet.* We will designate the period by T. The speed of the planet is equal to the circumference $2\pi r$ of the orbit divided by the time T:

$$v = \frac{2\pi r}{T} \tag{9.11}$$

With this expression for the speed, the square of Eq. (9.10) becomes

$$\frac{4\pi^2 r^2}{T^2} = \frac{GM_S}{r} \tag{9.12}$$

which can be rearranged to read

$$T^2 = \frac{4\pi^2}{GM_S} r^3 \tag{9.13}$$

period for circular orbit

This says that *the square of the period is proportional to the cube of the radius of the orbit,* with a constant of proportionality depending on the mass of the central body.

| EXAMPLE 5 | Both Venus and the Earth have approximately circular orbits around the Sun. The period of the orbital motion of Venus is |

0.615 year, and the period of the Earth is 1 year. According to Eq. (9.13), by what factor do the sizes of the two orbits differ?

SOLUTION: If we take the cube root of both sides of Eq. (9.13), we see that the orbital radius is proportional to the 2/3 power of the period. Hence we can set up the following proportion for the orbital radii of the Earth and Venus:

$$\frac{r_E}{r_V} = \frac{T_E^{2/3}}{T_V^{2/3}}$$

$$= \frac{(1\ \text{year})^{2/3}}{(0.615\ \text{year})^{2/3}} = 1.38 \tag{9.14}$$

An equation analogous to Eq. (9.13) also applies to the circular motion of a moon or artificial satellite around a planet. In this case, the planet plays the role of the central body and, in Eq. (9.13), its mass replaces the mass of the Sun.

| EXAMPLE 6 | A communications satellite is in a circular orbit around the Earth, in the equatorial plane. The period of the orbit of such |

a satellite is exactly 1 day, so that the satellite always hovers in a fixed position relative to the rotating Earth. What must be the radius of such a "geosynchronous," or "geostationary," orbit?

Concepts — in — Context

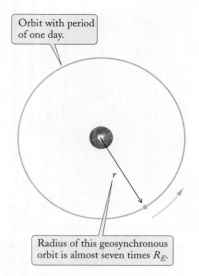

Orbit with period of one day.

r

Radius of this geosynchronous orbit is almost seven times R_E.

FIGURE 9.7 Orbit of a "geostationary" satellite around the Earth.

SOLUTION: Since the central body is the Earth, the equation analogous to Eq. (9.13) is

$$T^2 = \frac{4\pi^2}{GM_E} r^3 \tag{9.15}$$

or

$$r^3 = \frac{GM_E T^2}{4\pi^2} \tag{9.16}$$

Taking the cube root of both sides of this equation, we find

$$r = \left(\frac{GM_E T^2}{4\pi^2}\right)^{1/3}$$

$$= \left(\frac{6.67 \times 10^{-11}\,\text{N·m}^2/\text{kg}^2 \times 5.98 \times 10^{24}\,\text{kg} \times (24 \times 60 \times 60\,\text{s})^2}{4\pi^2}\right)^{1/3}$$

$$= 4.23 \times 10^7\,\text{m} \tag{9.17}$$

The orbit is shown in Fig. 9.7, which is drawn to scale. The radius of the orbit is about 6.6 times the radius of the Earth.

Concepts —in— Context

EXAMPLE 7   Surveillance satellites and spacecraft such as the Space Shuttle (Fig. 9.8) often operate in low-altitude orbits quite near the Earth, just above the atmosphere. Such orbits can have a radius as small as $r_{\text{low}} = 6.6 \times 10^6$ m; this is less than one-sixth of the geostationary orbit radius $r_{\text{geo}} = 4.23 \times 10^7$ m. Calculate how often the low-altitude satellites and spacecraft circle the Earth.

SOLUTION: Taking the square root of both sides of Eq. (9.13), we see that the period is proportional to the 3/2 power of the orbital radius. Hence we can set up the following proportion for the orbital periods:

$$\frac{T_{\text{low}}}{T_{\text{geo}}} = \left(\frac{r_{\text{low}}}{r_{\text{geo}}}\right)^{3/2} = \left(\frac{6.6 \times 10^6\,\text{m}}{4.23 \times 10^7\,\text{m}}\right)^{3/2}$$

$$= 0.062$$

or, since the geostationary period T_{geo} is one day, or 24 h,

$$T_{\text{low}} = 0.062 \times 24\,\text{h} = 1.5\,\text{h}$$

Thus such "fly-bys" occur quite frequently: 16 times per day.

FIGURE 9.8   The Space Shuttle in orbit with its cargo bay open.

✓ Checkup 9.3

QUESTION 1: The orbit of the geostationary satellite illustrated in Fig. 9.7 is in the equatorial plane, and the satellite is stationary above a point on the Earth's equator. Why can't we keep a satellite stationary above a point that is not on the equator, say, above San Francisco?

PHYSICS IN PRACTICE

COMMUNICATIONS SATELLITES AND WEATHER SATELLITES

Concepts —in— Context More than a hundred communications and weather satellites have been placed in geostationary orbits. The communications satellites use radio signals to relay telephone and TV signals from one point on the Earth to another. The weather satellites capture pictures of the cloud patterns and measure the heights of clouds, wind speeds, atmospheric and ground temperatures, and moisture in the atmosphere. These observations are especially useful for monitoring weather conditions over the oceans, where there are few observation stations at ground level. Data collected by weather satellites permit early detection of dangerous tropical storms (hurricanes, typhoons) and forecasting of the tracks and the strengths of these storms.

The launch vehicle for these satellites usually consists of a two-stage rocket, which carries the satellite to a low-altitude orbit. A small rocket motor attached to the satellite is then used to lift the satellite from the low-altitude orbit to the high-altitude geostationary orbit. Alternatively, the satellite can be ferried to the low-altitude orbit by the Space Shuttle.

At the high altitude of the geostationary orbit there is no atmospheric drag, and a satellite placed in such an orbit will continue to orbit the Earth indefinitely. However, the orbital motion of the satellite is disturbed by the Moon and the Sun, and it is also affected by the nonspherical shape of the Earth, which produces deviations from the ideal uniform centripetal force. These disturbances cause the satellite to drift from its geostationary position. This requires an adjustment of the orbit every few weeks, which is done with small control nozzles on the satellite. Typically, a satellite carries enough propellant to operate its control nozzles for 10 years, by which time other components in the satellite will also have worn out, or will have been superseded by new technology, so it becomes desirable to switch the satellite off, and replace it by a new model.

Communications satellites contain a radio receiver and a transmitter connected to dish antennas aimed at radio stations on the ground. The signal received from one station on the ground is amplified by the satellite, and then this amplified signal is retransmitted to the other station (the satellite acts as a transponder).

FIGURE 1   Astronauts handle an INTELSAT communications satellite.

Figure 1 shows a recent model of the INTELSAT series of communications satellites. This satellite has a length of 5.2 m, a diameter of 3.6 m, and a mass of 2240 kg. It is powered by solar panels that convert the energy of sunlight into electricity, delivering 2300 watts of power. It contains 50 transponders and is capable of handling 40 000 telephone circuits simultaneously.

For intercontinental communications, three groups of INTELSAT satellites are deployed at geostationary positions over the Atlantic, Pacific, and Indian Oceans. But communications satellites are also cost-effective for communications over shorter ranges, when there is a shortage of telephone cables. Many countries have launched communications satellites to handle telephone traffic within their borders. Communications satellites also relay TV transmissions. A small dish antenna connected to an amplifier permits home television sets to pick up a multitude of TV channels from these satellites.

QUESTION 2: The period of the orbital motion of the Moon around the Earth is 27 days. If the orbit of the Moon were twice as large as it is, what would be the period of its motion?

QUESTION 3: The mass of a planet can be determined by observing the period of a moon in a circular orbit around the planet. For such a mass determination, which of the following do we need: the period, the radius of the moon's orbit, the mass of the moon, the radius of the planet?

QUESTION 4: The radius of the orbit of Saturn around the Sun is about 10 times the radius of the orbit of the Earth. Accordingly, what must be the approximate period of its orbital motion?

(A) 1000 yr   (B) 100 yr   (C) 30 yr   (D) 10 yr   (E) 3 yr

Online *Concept* Tutorial

9.4 ELLIPTICAL ORBITS; KEPLER'S LAWS

Although the orbits of the planets around the Sun are approximately circular, none of these orbits are *exactly* circular. We will not attempt the general solution of the equation of motion for such noncircular orbits. A complete calculation shows that with the inverse-square force of Eq. (9.1), the planetary orbits are ellipses. This is **Kepler's First Law**:

Kepler's First Law

The orbits of the planets are ellipses with the Sun at one focus.

Figure 9.9 shows an elliptical planetary orbit (for the sake of clarity, the elongation of this ellipse has been exaggerated; actual planetary orbits have only very small elongations). The point closest to the Sun is called the **perihelion**; the point farthest from the Sun is called the **aphelion**. The sum of the perihelion and the aphelion distances is the **major axis** of the ellipse. The distance from the center of the ellipse to the perihelion (or aphelion) is the **semimajor axis**; this distance equals the average of the perihelion and aphelion distances.

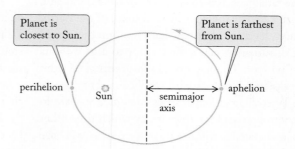

FIGURE 9.9 Orbit of a planet around the Sun. The orbit is an ellipse, with the Sun at one focus.

Kepler originally discovered his First Law and his other two laws (see below) early in the seventeenth century, by direct analysis of the available observational data on planetary motions. Hence, Kepler's laws were originally purely phenomenological statements; that is, they described the phenomenon of planetary motion but did not explain its causes. The explanation came only later, when Newton laid down his laws of motion and his Law of Universal Gravitation and deduced the features of planetary motion from these fundamental laws.

Kepler's Second Law describes the variation in the speed of the motion:

Kepler's Second Law

The radial line segment from the Sun to the planet sweeps out equal areas in equal times.

MATH HELP　　ELLIPSES

An ellipse is defined geometrically by the condition that the sum of the distance from one focus of the ellipse and the distance from the other focus is the same for all points on the ellipse. This geometrical condition leads to a simple method for the construction of an ellipse: Stick pins into the two foci and tie a length of string to these points. Stretch the string taut to the tip of a pencil, and move this pencil around the foci while keeping the string taut (see Fig. 1a).

An ellipse can also be constructed by slicing a cone obliquely (see Fig. 1b). Because of this, an ellipse is said to be a conic section.

The largest diameter of the ellipse is called the major axis, and the smallest diameter is called the minor axis. The semimajor axis and the semiminor axis are one-half of these diameters, respectively (see Fig. 1c).

If the semimajor axis of length a is along the x axis and the semiminor axis of length b is along the y axis, then the x and y coordinates of an ellipse centered on the origin satisfy

$$\frac{x^2}{a^2} + \frac{y^2}{b^2} = 1$$

The foci are on the major axis at a distance f from the origin given by

$$f = \sqrt{a^2 - b^2}$$

The separation between a planet and the Sun is $a - f$ at perihelion and is $a + f$ at aphelion.

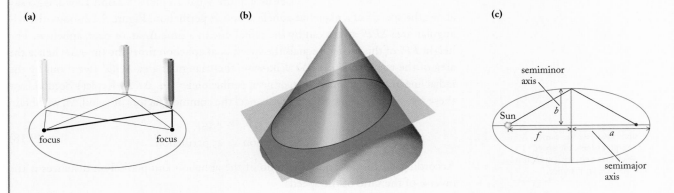

(a)　　**(b)**　　**(c)**

FIGURE 1 (a) Constructing an ellipse. (b) Ellipse as a conic section. (c) Focal distance f, semimajor axis a, and semiminor axis b of an ellipse.

Figure 9.10 illustrates this law. The two colored areas are equal, and the planet takes equal times to move from P to P' and from Q to Q'. According to Fig. 9.10, the speed of the planet is larger when it is near the Sun (at Q) than when it is far from the Sun (at P).

Kepler's Second Law, also called the law of areas, is a direct consequence of the central direction of the gravitational force. We can prove this law by a simple geometrical argument. Consider three successive positions P, P', P'' on the orbit, separated by a relatively small distance. Suppose that the time intervals between P, P' and between P', P'' are equal—say, each of the two intervals is one second. Figure 9.11 shows the positions P, P', P''. Between these positions the curved orbit can be approximated by straight line segments PP' and $P'P''$. Since the time intervals are one unit of time (1 second), the lengths of the segments PP' and $P'P''$ are in proportion to the

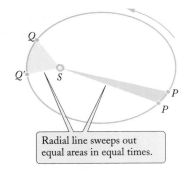

Radial line sweeps out equal areas in equal times.

FIGURE 9.10 For equal time intervals, the areas SQQ' and SPP' are equal. The distance QQ' is larger than the distance PP'.

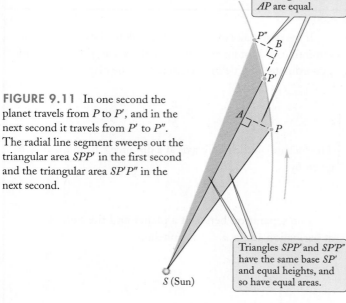

For a central force, the heights BP'' and AP are equal.

Triangles SPP' and $SP'P''$ have the same base SP' and equal heights, and so have equal areas.

S (Sun)

FIGURE 9.11 In one second the planet travels from P to P', and in the next second it travels from P' to P''. The radial line segment sweeps out the triangular area SPP' in the first second and the triangular area $SP'P''$ in the next second.

average velocities in the two time intervals. The velocities differ because the gravitational force causes an acceleration. However, since the direction of the force is toward the center, parallel to the radius, the component of the velocity perpendicular to the radius cannot change. The component of the velocity perpendicular to the radius is represented by the line segment PA for the first time interval, and it is represented by BP'' for the second time interval. These line segments perpendicular to the radius are, respectively, the heights of the triangles SPP' and $SP'P''$ (see Fig. 9.11). Since these heights are equal and since both triangles have the same base SP', their areas must be equal. Thus, the areas swept out by the radial line in the two time intervals must be equal, as asserted by Kepler's Second Law. Note that this geometrical argument depends only on the fact that the force is directed toward a center; it does not depend on the magnitude of the force. This means that Kepler's Second Law is valid not only for planetary motion, but also for motion with any kind of central force.

Let us explore what Kepler's Second Law has to say about the speeds of a planet at aphelion and at perihelion. Figure 9.12 shows the triangular area SPP' swept out by the radial line in a time Δt at, or near, aphelion. The height PP' of this triangle equals the speed v_1 at aphelion times the time Δt; hence the area of the triangle is $\frac{1}{2}r_1 v_1 \Delta t$. Likewise, the triangular area SQQ' swept out by the radial line in an equal time Δt at, or near, perihelion is $\frac{1}{2}r_2 v_2 \Delta t$. By Kepler's Second Law these two areas must be equal; if we cancel the common factors of $\frac{1}{2}$ and Δt, we obtain

$$r_1 v_1 = r_2 v_2$$
$$\text{at aphelion} \qquad \text{at perihelion} \tag{9.18}$$

According to this equation, the ratio of the aphelion and perihelion distances is the inverse of the ratio of the speeds.

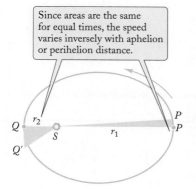

Since areas are the same for equal times, the speed varies inversely with aphelion or perihelion distance.

FIGURE 9.12 Triangular area SPP' swept out in one interval Δt after aphelion, and triangular area SQQ' swept out in an identical interval Δt after perihelion.

EXAMPLE 8   The perihelion and aphelion distances for Mercury are 45.9×10^9 m and 69.8×10^9 m, respectively. The speed of Mercury at aphelion is 3.88×10^4 m/s. What is the speed at perihelion?

SOLUTION: From Eq. (9.18),

$$v_2 = \frac{r_1}{r_2} v_1 = \frac{69.8 \times 10^9 \text{ m}}{45.9 \times 10^9 \text{ m}} \times 3.88 \times 10^4 \text{ m/s}$$
$$= 5.90 \times 10^4 \text{ m/s}$$

In Chapter 13 we will become acquainted with the **angular momentum** L, which, for a planet at aphelion or perihelion, is equal to the product rmv. By multiplying both sides of Eq. (9.18) by the mass of the planet m, we see that $r_1 mv_1 = r_2 mv_2$; that is, the angular momentum at aphelion equals the angular momentum at perihelion. Thus, Kepler's Second Law can be regarded as a consequence of a conservation law for angular momentum. We will see that angular momentum is conserved when a particle is under the influence of any central force.

Kepler's Third Law relates the period of the orbit to the size of the orbit:

The square of the period is proportional to the cube of the semimajor axis of the planetary orbit.

This Third Law, or law of periods, is nothing but a generalization of Eq. (9.13) to elliptical orbits.

Table 9.1 lists the orbital data for the planets of the Solar System. The mean distance listed in this table is defined as the average of the perihelion and aphelion distances; that is, it is the semimajor axis of the ellipse. The difference between the perihelion and aphelion distances gives an indication of the elongation of the ellipse.

TABLE 9.1	THE PLANETS				
PLANET (a)	MASS	MEAN DISTANCE FROM SUN (SEMIMAJOR AXIS)	PERIHELION DISTANCE	APHELION DISTANCE	PERIOD
Mercury	3.30×10^{23} kg	57.9×10^{6} km	45.9×10^{6} km	69.8×10^{6} km	0.241 yr
Venus	4.87×10^{24}	108	107	109	0.615
Earth	5.98×10^{24}	150	147	152	1.00
Mars	6.42×10^{23}	228	207	249	1.88
Jupiter	1.90×10^{27}	778	740	816	11.9
Saturn	5.67×10^{26}	1430	1350	1510	29.5
Uranus	8.70×10^{25}	2870	2730	3010	84.0
Neptune	1.03×10^{26}	4500	4460	4540	165
Pluto	1.50×10^{22}	5890	4410	7360	248

(a) A photomontage of the planets in sequence from Mercury (top left, partly hidden) to Pluto (bottom left).

JOHANNES KEPLER (1571–1630) *German astronomer and mathematician. Kepler relied on the theoretical framework of the Copernican system, and he extracted his three laws by a meticulous analysis of the observational data on planetary motions collected by the great Danish astronomer Tycho Brahe.*

(a)

(b)

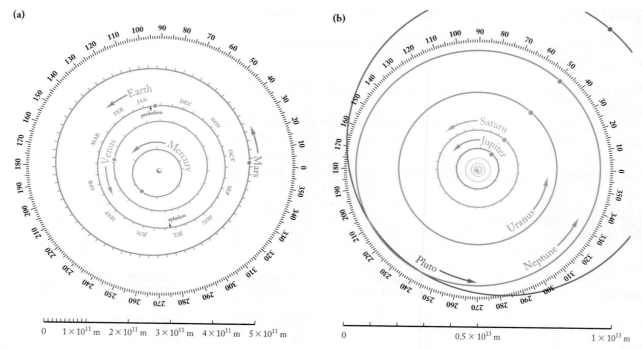

FIGURE 9.13 (a) Orbits of Mercury, Venus, Earth, and Mars. Elliptical orbits can appear quite circular, even when the focus is noticeably off-center, as with Mercury and Mars. The colored dots indicate the positions of these planets on January 1, 2000. The tick marks indicate the positions at intervals of 10 days. (b) Orbits of Jupiter, Saturn, Uranus, and Neptune, and a portion of the orbit of Pluto. The tick marks for Jupiter and Saturn indicate the positions at intervals of 1 year.

Figure 9.13a shows the orbits of the planets Mercury, Venus, Earth, and Mars on scale diagrams. The orbits of Saturn, Jupiter, Uranus, and Neptune and part of the orbit of Pluto are shown in Fig. 9.13b. Inspection of these diagrams reveals that the orbits of Mercury, Mars, and Pluto are noticeably different from circles.[1]

Kepler's three laws apply not only to planets, but also to satellites and to comets. For example, Fig. 9.14 shows the orbits of a few of the many artificial Earth satellites. All these orbits are ellipses. For Earth orbits, the point closest to the Earth is called **perigee**; the point farthest from Earth is called **apogee**. The early artificial satellites were quite small, with masses below 100 kg (see Fig. 9.15). Nowadays, satellites with masses

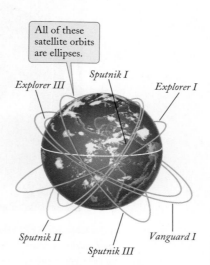

FIGURE 9.14 Orbits of the first artificial Earth satellites. See Table 9.3 for more data.

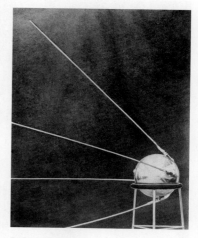

FIGURE 9.15 *Sputnik I*, the first artificial Earth satellite. This satellite had a mass of 83 kg.

[1] Pluto has recently been reclassified by the International Astronomical Union as a dwarf planet, in the same category as Ceres and 2003 UB₃₁₃ (X ena).

of several tons are not unusual. All of the early artificial satellites burned up in the atmosphere after a few months or a few years because they were not sufficiently far from the Earth to avoid the effects of residual atmospheric friction.

Kepler's laws also apply to the motion of a projectile near the Earth. For instance, Fig. 9.16 shows the trajectory of an intercontinental ballistic missile (ICBM). During most of its trajectory, the only force acting on the missile is the gravity of the Earth; the thrust of the engines and the friction of the atmosphere act only during the relatively short initial and final segments of the trajectory (on the scale of Fig. 9.16, these initial and final segments of the trajectory are too small to be noticed). The trajectory is a portion of an elliptical orbit cut short by the impact on the Earth. Likewise, the motion of an ordinary low-altitude projectile, such as a cannonball, is also a portion of an elliptical orbit (if we ignore atmospheric friction). In Chapter 4 we made the near-Earth approximation that gravity was constant in magnitude and direction; with these approximations we found that the orbit of a projectile was a parabola. Although the exact orbit of a projectile is an ellipse, the parabola approximates this ellipse quite well over the relatively short distance involved in ordinary projectile motion; deviations do become noticeable for long-range trajectories (see Fig. 9.17).

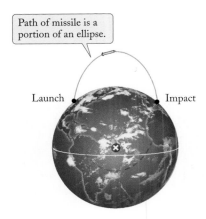

FIGURE 9.16 Orbit of an intercontinental ballistic missile (ICBM). The elongation of the ellipse and the height of the orbit are exaggerated.

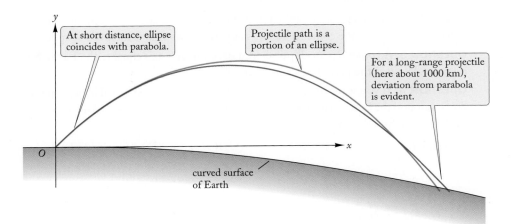

FIGURE 9.17 The parabola (blue curve) approximates the ellipse (red curve) for short distances.

The connection between projectile motion and orbital motion was neatly illustrated by Newton by means of an imaginary experiment, or what today we would call a *Gedankenexperiment*.[1] Newton proposed to fire a cannonball horizontally from a gun emplaced on a high mountain (see Fig. 9.18). If the muzzle velocity is fairly low, the cannonball will arc toward the Earth and strike near the base of the mountain. The trajectory is a segment of a parabola, or, more precisely, a segment of an ellipse. If we increase the muzzle velocity, the cannonball will describe larger and larger arcs. Finally, if the muzzle velocity is just large enough, the rate at which the trajectory curves downward is precisely matched by the curvature of the surface of the Earth—the cannonball never hits the Earth and keeps on falling forever while moving in a circular orbit. This example makes it very clear that orbital motion is free-fall motion.

FIGURE 9.18 This drawing from Newton's *Principia* illustrates an imaginary experiment with a cannonball fired from a gun on a high mountain. For a sufficiently large muzzle velocity, the trajectory of the cannonball is a circular orbit.

---

[1] *Gedankenexperiment* is German for "thought experiment." This word is used by physicists for an imaginary experiment that can be done in principle, but that has never been done in practice, and whose outcome can be discovered by thought.

Finally, we note that in our mathematical description of planetary motion we have neglected the gravitational forces that the planets exert on one another. These forces are much smaller than the force exerted by the Sun, but in a precise calculation the vector sum of all the forces must be taken into account. The net force on any planet then depends on the positions of all the other planets. This means that the motions of all the planets are coupled together, and the calculation of the motion of one planet requires the simultaneous calculation of the motions of all the other planets. This makes the precise mathematical treatment of planetary motion extremely complicated. Kepler's simple laws neglect the complications introduced by the interplanetary forces; these laws therefore do not provide an exact description of planetary motions, but only a very good first approximation.

✔ Checkup 9.4

QUESTION 1: Suppose that the gravitational force were an inverse-cube force, instead of an inverse-square force. Would Kepler's Second Law remain valid? Would Kepler's Third Law remain valid?

QUESTION 2: A comet has an aphelion distance twice as large as its perihelion distance. If the speed of the comet is 40 km/s at perihelion, what is its speed at aphelion?

QUESTION 3: A comet has an elliptical orbit of semimajor axis equal to the Earth–Sun distance. What is the period of such a comet?

QUESTION 4: If you want to place an artificial satellite in an elliptical orbit of period 8 years around the Sun, what must be the semimajor axis of this ellipse? (Answer in units of the Earth–Sun distance.)

(A) 64   (B) $16\sqrt{2}$ (C) 8   (D) 4   (E) 2

9.5 ENERGY IN ORBITAL MOTION

The gravitational force is a conservative force; that is, the work done by this force on a particle moving from some point P_1 to some other point P_2 can be expressed as a difference between two potential energies, and the work done on any round trip starting and ending at some given point is zero. To construct the potential energy, we proceed as in Section 8.1: we calculate the work done by the gravitational force as the particle moves from point P_1 to point P_2, and we seek to express this work as a difference of two terms. In Fig. 9.19, the points P_1 and P_2 are at distances r_1 and r_2, respectively, from the central mass. To calculate the work, we must take into account that the force is a function of the distance; that is, the force is variable. From Section 7.2, we know that for such a variable force, the work is the integral of the force over the distance. If we place the x axis along the line connecting P_1 and P_2 (see Fig. 9.19), then the force can be expressed as

$$F_x = -\frac{GMm}{x^2}$$

and the work is

$$W = \int_{P_1}^{P_2} F_x(x)\, dx = \int_{r_1}^{r_2} \left(-\frac{GMm}{x^2}\right) dx$$

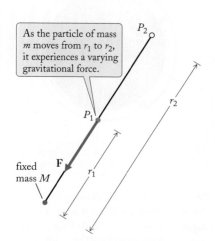

As the particle of mass m moves from r_1 to r_2, it experiences a varying gravitational force.

FIGURE 9.19 Two points P_1 and P_2 at distances r_1 and r_2 from the central mass.

We already have evaluated this kind of integral in Example 2 of Chapter 8 (in the case of the gravitational force, the constant A in that example is $A = -GMm$). The result of the integration is

$$W = \frac{GMm}{x}\bigg|_{r_1}^{r_2} = \frac{GMm}{r_2} - \frac{GMm}{r_1} \qquad (9.19)$$

As expected, this result shows that the work is the difference between two potential energies. Accordingly, we can identify the **gravitational potential energy** as

$$U = -\frac{GMm}{r} \qquad (9.20)$$

gravitational potential energy

Note that in this calculation of the gravitational potential energy we assumed that the points P_1 and P_2 lie on the same radius (see Fig. 9.19). However, Eq. (9.19) is valid in general, even if P_1 and P_2 do not lie on the same radial line. We can see this by introducing an intermediate point Q, which is on the radial line of P_1 but at the radial distance of P_2 (see Fig. 9.20). To move the particle from P_1 to P_2, we first move it from P_1 to Q along the radial line; this takes the amount of work given by Eq. (9.19). We then move the particle from Q to P_2, along the circular arc of radius r_2; this costs no work, since such a displacement is perpendicular to the force. Any more general path can be constructed from small radial segments and small arcs of circles, and so Eq. (9.19) is true in general.

The potential energy (9.20) is always negative, and its magnitude is inversely proportional to r. Figure 9.21 gives a plot of this potential energy as a function of distance. If the distance r is small, the potential energy is low (the potential energy is much below zero); if the distance r is large, the potential energy is higher (the potential energy is still negative, but not so much below zero). Thus, the potential energy *increases* with distance; it increases from a large negative value to a smaller negative value or to zero. Such an increase of potential energy with distance is characteristic of an attractive force. For instance, if we want to lift a communications satellite from a low initial orbit (just above the Earth's atmosphere) into a high final orbit (such as the geostationary orbit described in Example 6), we must do work on this satellite (by means of a rocket engine). The work we do while lifting the satellite increases the gravitational potential energy from a large negative value (much below zero) to a smaller negative value (not so much below zero).

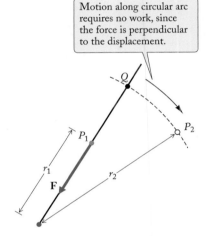

FIGURE 9.20 Two points P_1 and P_2 at distances r_1 and r_2 in different directions. The particle moves from P_1 to Q and then from Q to P_2.

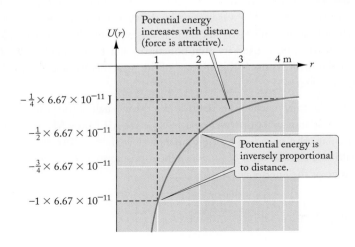

FIGURE 9.21 Gravitational potential energy for a particle of mass 1 kg gravitationally attracted by another particle of mass 1 kg.

The total mechanical energy is the sum of the potential energy and the kinetic energy. Since we are assuming that the mass M is stationary, the kinetic energy is entirely due to the motion of the mass m, and the Law of Conservation of Energy takes the form

Law of Conservation of Energy

$$E = K + U = \tfrac{1}{2}mv^2 - \frac{GMm}{r} = [\text{constant}] \tag{9.21}$$

If the only force acting on the body is the gravitational force (no rocket engine or other external force!), then this total energy remains constant during the motion. For instance, the energy (9.21) is constant for a planet orbiting the Sun, and for a satellite or a spacecraft (with rocket engines shut off) orbiting the Earth. As we saw in Chapter 8, examination of the energy reveals some general features of the motion. Equation (9.21) shows how the orbiting body trades distance ("height") for speed; it implies that if r decreases, v must increase, so that the sum of the two terms $\tfrac{1}{2}mv^2$ and $-GMm/r$ remains constant. Conversely, if r increases, v must decrease.

Let us now investigate the possible orbits around, say, the Sun from the point of view of their energy. For a circular orbit, we saw in Eq. (9.10) that the orbital speed is

$$v = \sqrt{\frac{GM_S}{r}} \tag{9.22}$$

and so the kinetic energy is

$$K = \tfrac{1}{2}mv^2 = \frac{GM_S m}{2r} \tag{9.23}$$

Hence the total energy is

$$E = K + U = \tfrac{1}{2}mv^2 - \frac{GM_S m}{r} = \frac{GM_S m}{2r} - \frac{GM_S m}{r}$$

or

$$E = -\frac{GM_S m}{2r} \tag{9.24}$$

Consequently, the total energy for a circular orbit is negative and is exactly one-half of the potential energy.

 Concepts *in* Context

EXAMPLE 9 The 1300-kg Syncom communications satellite was placed in its high-altitude geosynchronous orbit of radius 4.23×10^7 m in two steps. First the satellite was carried by the Space Shuttle to a low-altitude circular orbit of radius 6.65×10^6 m; there it was released from the cargo bay of the Space Shuttle, and it used its own booster rocket to lift itself to the high-altitude circular orbit. What is the increase of the total mechanical energy during this change of orbit?

SOLUTION: The total mechanical energy is exactly one-half of the potential energy [Eq. (9.24)]. For an Earth orbit, we replace M_S in Eq. (9.24) by M_E. For the low-altitude circular orbit of radius r_1, the total energy is $E_1 = -GM_E m/2r_1$, and for the high-altitude circular orbit of radius r_2, the total energy is $E_2 = -GM_E m/2r_2$. So the change of the energy is

$$E_2 - E_1 = -\frac{GM_E m}{2}\left(\frac{1}{r_2} - \frac{1}{r_1}\right)$$

$$= -\frac{6.67 \times 10^{-11}\,\text{N·m}^2/\text{kg}^2 \times 5.98 \times 10^{24}\,\text{kg} \times 1300\,\text{kg}}{2}$$

$$\times \left(\frac{1}{4.23 \times 10^7\,\text{m}} - \frac{1}{6.65 \times 10^6\,\text{m}}\right)$$

$$= 3.29 \times 10^{10}\,\text{J}$$

This energy was supplied by the booster rocket of the satellite.

For an elliptical orbit, the total energy is also negative. It can be demonstrated that the energy can still be written in the form of Eq. (9.24), but the quantity r must be taken equal to the semimajor axis of the ellipse. The total energy of the orbit does not depend on the shape of the ellipse, but only on its larger overall dimension. Figure 9.22 shows several orbits of different shapes but with exactly the same total energy.

From Eq. (9.24) we see that if the energy is nearly zero, then the size of the orbit is very large (note that $E \to 0$ as $r \to \infty$). Such orbits are characteristic of comets, many of which have elliptical orbits that extend far beyond the edge of the Solar System (see Fig. 9.23). If the energy is exactly zero, then the "ellipse" extends all the way to infinity and never closes; such an "open ellipse" is actually a parabola (see Fig. 9.24).

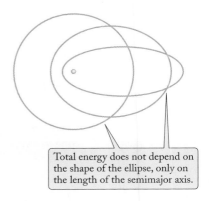

Total energy does not depend on the shape of the ellipse, only on the length of the semimajor axis.

FIGURE 9.22   Orbits of the same total energy. All these orbits have the same semimajor axis.

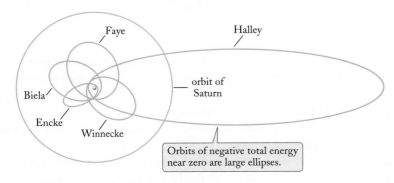

Faye

Halley

Biela

orbit of Saturn

Encke

Winnecke

Orbits of negative total energy near zero are large ellipses.

FIGURE 9.23   Orbits of some periodic comets.

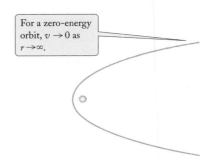

For a zero-energy orbit, $v \to 0$ as $r \to \infty$.

FIGURE 9.24   Orbit of zero energy—a parabola.

Equation (9.21) indicates that if the energy is zero, the comet will reach infinite distance with zero velocity (if $r = \infty$, then $v = 0$). By considering the reverse of this motion, we see that a comet of zero energy, initially at very large distance from the Sun, will fall along this type of parabolic orbit.

If the energy is positive, then the orbit again extends all the way to infinity and again fails to close; such an open orbit is a hyperbola. The comet will then reach infinite distance with some nonzero velocity and continue moving along a straight line (see Fig. 9.25).

EXAMPLE 10   A meteoroid (a chunk of rock) is initially at rest in interplanetary space at a large distance from the Sun. Under the influence of gravity, the meteoroid begins to fall toward the Sun along a straight radial line. With what speed does it strike the Sun? The radius of the Sun is 6.96×10^8 m.

SOLUTION: The energy of the meteoroid is

$$E = K + U = \tfrac{1}{2}mv^2 - \frac{GM_S m}{r} = [\text{constant}] \tag{9.25}$$

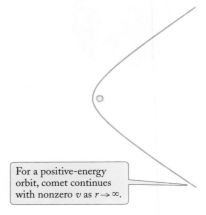

For a positive-energy orbit, comet continues with nonzero v as $r \to \infty$.

FIGURE 9.25   Orbit of positive energy—a hyperbola.

Initially, both the kinetic and potential energies are zero ($v = 0$ and $r \approx \infty$). Hence at any later time

$$\tfrac{1}{2}mv^2 - \frac{GM_S m}{r} = 0$$

or

$$\tfrac{1}{2}mv^2 = \frac{GM_S m}{r} \qquad (9.26)$$

If we cancel a factor of m and multiply by 2 on both sides of this equation, take the square root of both sides, and substitute $r = R_S$ for the impact on the Sun's surface, we find the speed at the moment of impact:

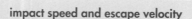

$$v = \sqrt{\frac{2GM_S}{R_S}} \qquad (9.27)$$

With $M_S = 1.99 \times 10^{30}$ kg (see Example 4) and $R_S = 6.96 \times 10^8$ m, we obtain

$$v = \sqrt{\frac{2 \times 6.67 \times 10^{-11}\,\text{N·m}^2/\text{kg}^2 \times 1.99 \times 10^{30}\,\text{kg}}{6.96 \times 10^8\,\text{m}}}$$
$$= 6.18 \times 10^5\ \text{m/s} = 618\ \text{km/s}$$

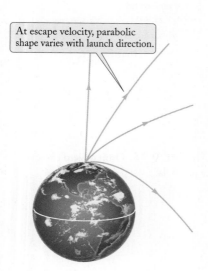

At escape velocity, parabolic shape varies with launch direction.

FIGURE 9.26 Different orbits with the same starting point and initial speed. All these orbits are segments of parabolas.

The quantity given by Eq. (9.27) is called the Sun's **escape velocity** because it is the minimum initial velocity with which a body must be launched upward from the surface of the Sun if it is to escape and never fall back. We can recognize this by looking at the motion of the meteoroid in Example 10 in reverse: it starts with a velocity of 618 km/s at the surface of the Sun and gradually slows as it rises, but never quite stops until it reaches a very large distance ($r \approx \infty$).

The escape velocity for a body launched from the surface of the Earth can be calculated from a formula analogous to Eq. (9.27), provided that we ignore atmospheric friction and the pull of the Sun on the body. Atmospheric friction will be absent if we launch the body from just above the atmosphere. The pull of the Sun has only a small effect on the velocity of escape from the Earth if we contemplate a body that "escapes" to a distance of, say, $r = 100R_E$ or $200R_E$ rather than $r = \infty$, where we would also have to consider escape from the Sun. For such a motion, the displacement relative to the Sun can be neglected, and the escape velocity v is approximately $\sqrt{2GM_E/R_E} \approx 11.2$ km/s.

Note that the direction in which the escaping body is launched is immaterial—the body will succeed in its escape whenever the direction of launch is above the horizon. Of course, the path that the body takes will depend on the direction of launch (see Fig. 9.26).

✔ Checkup 9.5

QUESTION 1: An artificial satellite is initially in a circular orbit of fairly low altitude around the Earth. Because of friction with the residual atmosphere, the satellite loses some energy and enters a circular orbit of smaller radius. The speed of the satellite will then be *larger* in the new orbit. How can friction result in an increase of kinetic energy?

QUESTION 2: Does Kepler's Second Law apply to parabolic and hyperbolic orbits?

QUESTION 3: Suppose that we launch a body horizontally from the surface of the Earth, with a velocity exactly equal to the escape velocity of 11.2 km/s. What kind of orbit will this body have? Ignore atmospheric friction.

QUESTION 4: Uranus has a smaller gravitational acceleration at its surface than the Earth. Can you conclude that the escape velocity from its surface is smaller than from the Earth's surface?

QUESTION 5: Suppose that three comets, I, II, and III, approach the Sun. At the instant they cross the Earth's orbit, comet I has a speed of 42 km/s, comet II has a larger speed, and comet III a smaller speed. Given that the orbit of comet I is parabolic, what are the kinds of orbit for comets II and III, respectively?

(A) Elliptical; hyperbolic (B) Elliptical; parabolic (C) Hyperbolic; elliptical
(D) Hyperbolic; parabolic (E) Parabolic; elliptical

SUMMARY

PHYSICS IN PRACTICE Communication Satellites and Weather Satellites | **(page 281)**

MATH HELP Ellipses | **(page 283)**

LAW OF UNIVERSAL GRAVITATION (9.1)

Magnitude: $F = \dfrac{GMm}{r^2}$

Direction: The force on each mass is directed toward the other mass.

The force on each particle is directed toward the other particle.

GRAVITATIONAL CONSTANT $G = 6.67 \times 10^{-11}\,\text{N·m}^2/\text{kg}^2$ (9.2)

ACCELERATION OF FREE FALL ON EARTH $g = \dfrac{GM_E}{R_E^2}$ (9.6)

SPEED FOR CIRCULAR ORBIT AROUND SUN $v = \sqrt{\dfrac{GM_S}{r}}$ (9.10)

PERIOD OF ORBIT AROUND SUN $T^2 = \dfrac{4\pi^2}{GM_S}r^3$ (9.13)

KEPLER'S FIRST LAW The orbits of the planets are ellipses with the Sun at one focus.

KEPLER'S SECOND LAW The radial line segment from the Sun to a planet sweeps out equal areas in equal times.

Radial line sweeps out equal areas in equal times.

KEPLER'S THIRD LAW The square of the period is proportional to the cube of the semimajor axis of a planetary orbit.

GRAVITATIONAL POTENTIAL ENERGY

$$U = -\frac{GMm}{r}$$

(9.20)

CONSERVATION OF ENERGY

$$E = K + U = \tfrac{1}{2}mv^2 - \frac{GMm}{r} = [\text{constant}]$$

(9.21)

ENERGY FOR A CIRCULAR ORBIT AROUND THE SUN (Also the energy for an elliptical orbit with semimajor axis r.)

$$E = -\frac{GM_S m}{2r}$$

(9.24)

SHAPES OF ORBITS For total mechanical energy E,

$E < 0$ elliptical orbit
$E = 0$ parabolic trajectory
$E > 0$ hyperbolic trajectory

ESCAPE VELOCITY FROM EARTH

$$v = \sqrt{\frac{2GM_E}{R_E}}$$

(9.27)

QUESTIONS FOR DISCUSSION

1. Can you directly feel the gravitational pull of the Earth with your sense organs? (Hint: Would you feel anything if you were in free fall?)

2. According to a tale told by Professor R. Lichtenstein, some apple trees growing in the mountains of Tibet produce apples of negative mass. In what direction would such an apple fall if it fell off its tree? How would such an apple hang on the tree?

3. Eclipses of the Moon can occur only at full Moon. Eclipses of the Sun can occur only at new Moon. Why?

4. Explain why the sidereal day (the time of rotation of the Earth relative to the stars, or 23 h 56 min 4 s) is shorter than the mean solar day (the time between successive passages of the Sun over a given meridian, or 24 h). (Hint: The rotation of the Earth around its axis and the revolution of the Earth around the Sun are in the same direction.)

5. Suppose that an airplane flies around the Earth along the equator. If this airplane flew *very* fast, it would not need wings to support itself. Why not?

6. The mass of Pluto was not known until 1978 when a moon of Pluto was finally discovered. How did the discovery of this moon help?

7. It is easier to launch an Earth satellite into an eastward orbit than into a westward orbit. Why?

8. Would it be advantageous to launch rockets into space from a pad at very high altitude on a mountain? Why has this not been done?

9. Describe how you would play squash on a small, round asteroid (with no enclosing wall). What rules of the game would you want to lay down?

10. According to an NBC news report of April 5, 1983, a communications satellite launched from the Space Shuttle went into an orbit as shown in Fig. 9.27. Is this believable?

FIGURE 9.27 Proposed orbit for a communications satellite.

11. Does the radial line from the Sun to Mars sweep out area at the same rate as the radial line from the Sun to the Earth?

12. Why were the Apollo astronauts able to jump much higher on the Moon than on the Earth (Fig. 9.28)? If they had landed on a small asteroid, could they have launched themselves into a parabolic or hyperbolic orbit by a jump?

FIGURE 9.28 The jump of the astronaut.

13. The Earth reaches perihelion on January 3 and aphelion on July 6. Why is it not warmer in January than in July?

14. When the Apollo astronauts were orbiting around the Moon at low altitude, they detected several mass concentrations ("mascons") below the lunar surface. What is the effect of a mascon on the orbital motion?

15. An astronaut in a circular orbit above the Earth wants to take his spacecraft into a new circular orbit of larger radius. Give him instructions on how to do this.

16. A Russian and an American astronaut are in two separate spacecraft in the same circular orbit around the Earth. The Russian is slightly behind the American, and he wants to overtake him. The Russian fires his thrusters in the *forward* direction, braking for a brief instant. This changes his orbit into an ellipse. One orbital period later, the astronauts return to the vicinity of their initial positions, but the Russian is now ahead of the American. He then fires his thrusters in the *backward* direction. This restores his orbit to the original circle. Carefully explain the steps of this maneuver, drawing diagrams of the orbits.

17. The gravitational force that a hollow spherical shell of uniformly distributed mass exerts on a particle in its interior is zero. Does this mean that such a shell acts as a gravity shield?

18. Consider an astronaut launched in a rocket from the surface of the Earth and then placed in a circular orbit around the Earth. Describe the astronaut's weight (measured in an inertial reference frame) at different times during this trip. Describe the astronaut's *apparent* weight (measured in his own reference frame) at different times.

19. Several of our astronauts suffered severe motion sickness while under conditions of apparent weightlessness. Since the astronauts were not being tossed about (as in an airplane or a ship in a storm), what caused this motion sickness? What other difficulties does an astronaut face in daily life under conditions of weightlessness?

20. An astronaut on the International Space Station lights a candle. Will the candle burn like a candle on Earth?

21. Astrology is an ancient superstition according to which the planets influence phenomena on the Earth. The only force that can reach over the large distances between the planets and act on pieces of matter on the Earth is gravitation (planets do not have electric charge, and they therefore do not exert electric forces; some planets do have magnetism, but their magnetic forces are too weak to reach the Earth). Given that the Earth is in free fall under the action of the net gravitational force of the planets and the Sun, is there any way that the gravitational forces of the planets can affect what happens on the Earth?

PROBLEMS

9.1 Newton's Law of Universal Gravitation
9.2 The Measurement of G[†]

1. Two supertankers, each with a mass of 700 000 metric tons, are separated by a distance of 2.0 km. What is the gravitational force that each exerts on the other? Treat them as particles.

2. What is the gravitational force between two protons separated by a distance equal to their diameter, 2.0×10^{-15} m?

3. Somewhere between the Earth and the Moon there is a point where the gravitational pull of the Earth on a particle exactly balances that of the Moon. At what distance from the Earth is this point?

4. Calculate the value of the acceleration of gravity at the surface of Venus, Mercury, and Mars. Use the data on planetary masses and radii given in the table printed inside the book cover.

5. What is the magnitude of the gravitational force that the Sun exerts on you? What is the magnitude of the gravitational force that the Moon exerts on you? The masses of the Sun and the Moon and their distances are given inside the book cover; assume that your mass is 70 kg. Compare these forces with

[†] For help, see Online Concept Tutorial 11 at www.wwnorton.com/physics

your weight. Why don't you feel these forces? (Hint: You and the Earth are in free fall toward the Sun and the Moon.)

6. Calculate the gravitational force between our Galaxy and the Andromeda galaxy. Their masses are 2.0×10^{11} and 3.0×10^{11} times the mass of the Sun, respectively, and their separation is 2.2×10^6 light-years. Treat both galaxies as point masses.

7. The nearest star is Alpha Centauri, at a distance of 4.4 light-years from us. The mass of this star is 2.0×10^{30} kg. Compare the gravitational force exerted by Alpha Centauri on the Sun with the gravitational force that the Earth exerts on the Sun. Which force is stronger?

8. What is the magnitude of the gravitational attraction the Sun exerts on the Moon? What is the magnitude of the gravitational attraction the Earth exerts on the Moon? Suppose that the three bodies are aligned, with the Earth between the Sun and the Moon (at full moon). What is the direction of the net force acting on the Moon? Suppose that the three bodies are aligned with the Moon between the Earth and the Sun (at new moon). What is the direction of the net force acting on the Moon?

9. Calculate the value of the acceleration due to gravity at the surfaces of Jupiter, Saturn, and Uranus. Use the values of the planetary masses and radii given in the table printed inside the book cover.

10. Somewhere between the Earth and the Sun is a point where the gravitational attraction of the Earth exactly balances that of the Sun. At what fraction of the Earth–Sun distance does this occur?

11. Compare the weight of a 1-kg mass at the Earth's surface with the gravitational force between our Sun and another star of the same mass located at the far end of our galaxy, about 5×10^{20} m away.

12. Each of two adjacent 1.5-kg spheres hangs from a ceiling by a string. The center-to-center distance of the spheres is 8.0 cm. What (small) angle does each string make with the vertical?

13. A 7.0-kg mass is on the x axis at $x = 3.0$ m, and a 4.0-kg mass is on the y axis at $y = 2.0$ m. What is the resultant gravitational force (magnitude and direction) due to these two masses on a third mass of 3.0 kg located at the origin?

14. Three equal masses m are located at the vertices of an equilateral triangle of side a. What is the magnitude of the net gravitational force on each mass due to the other two?

15. Find the acceleration of the Moon due to the pull of the Earth. Express your result in units of the standard g.

16. If a "tower to the sky" of height 2000 km above the Earth's surface could be built, what would be your weight when standing at the top? Assume the tower is located at the South Pole. Express your answer in terms of your weight at the Earth's surface.

17. It has been suggested that strong tidal forces on Io, a moon of Jupiter, could be responsible for the dramatic volcanic activity observed there by Voyager spacecraft. Compare the difference in gravitational accelerations on the near and far surfaces of Io (due to Jupiter) with the difference in accelerations on the near and far side of the Earth (due to the Moon), both as absolute accelerations and as a fraction of the surface g. Io has a mass of 8.9×10^{22} kg and a radius of 1820 km, and is 422×10^3 km from the center of Jupiter.

*18. Suppose that the Earth, Sun, and Moon are located at the vertices of a right triangle, with the Moon located at the right angle (at first or last quarter moon; see Fig. 9.29). Find the magnitude and direction of the sum of the gravitational forces exerted by the Earth and the Sun on the Moon.

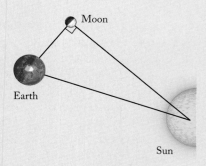

FIGURE 9.29   Earth, Moon, and Sun.

*19. Mimas, a small moon of Saturn, has a mass of 3.8×10^{19} kg and a diameter of 500 km. What is the maximum equatorial velocity with which we can make this moon rotate about its axis if pieces of loose rock sitting on its surface at its equator are not to fly off?

9.3 Circular Orbits[†]

20. The *Midas II* spy satellite was launched into a circular orbit at a height of 500 km above the surface of the Earth. Calculate the orbital period and the orbital speed of this satellite.

21. Consider the communications satellite described in Example 6. What is the speed of this satellite?

22. Calculate the orbital speed of Venus from the data given in Example 5.

23. The Sun is moving in a circular orbit around the center of our Galaxy. The radius of this orbit is 3×10^4 light-years. Calculate the period of the orbital motion and calculate the orbital speed of the Sun. The mass of our Galaxy is 4×10^{41} kg, and all of this mass can be regarded as concentrated at the center of the Galaxy.

24. Table 9.2 lists some of the moons of Saturn. Their orbits are circular.

    (a) From the information given, calculate the periods and orbital speeds of all these moons.

    (b) Calculate the mass of Saturn.

† For help, see Online Concept Tutorial 11 at www.wwnorton.com/physics

TABLE 9.2	SOME MOONS OF SATURN		
MOON	**DISTANCE FROM SATURN**	**PERIOD**	**ORBITAL SPEED**
Tethys (Fig. 9.30)	2.95×10^5 km	1.89 days	—
Dione	3.77	—	—
Rhea	5.27	—	—
Titan	12.22	—	—
Iapetus	35.60	—	—

FIGURE 9.30 Tethys, one of the moons of Saturn.

25. Before clocks with long-term accuracy were constructed, it was proposed that navigators at sea should use the motion of the moons of Jupiter as a clock. The moons Io, Europa, and Ganymede have orbital radii of 422×10^3, 671×10^3, and 1070×10^3 km, respectively. What are the periods of the orbits of these moons? The mass of Jupiter is 1.90×10^{27} kg.

26. A satellite is to be put into an equatorial orbit with an orbital period of 12 hours. What is the radius of the orbit? What is the orbital speed? How many times a day will the satellite be over the same point on the equator if the satellite orbits in the same direction as the Earth's rotation? If it orbits in the opposite direction?

27. An asteroid is in a circular orbit at a distance of two solar diameters from the center of the Sun. What is its orbital period in days?

28. The Sun rotates approximately every 26 days. What is the radius of a "heliosynchronous" orbit, that is, an orbit that stays over the same spot of the Sun?

29. The Apollo command module orbited the Moon while the lunar excursion module visited the surface. If the orbit had a radius of 2.0×10^6 m, how many times per (Earth) day did the command module fly over the excursion module?

30. A Jupiter-sized planet orbits the star 55 Cancri with an orbital radius of 8.2×10^{11} m (see Fig. 9.31). The orbital period of this planet is 13 yr. What is the mass of the star 55 Cancri? How does this compare with the mass of the Sun?

*31. The *Discoverer II* satellite had an approximately circular orbit passing over both poles of the Earth. The radius of the orbit was about 6.67×10^3 km. Taking the rotation of the Earth into account, if the satellite passed over New York City at one instant, over what point of the United States would it pass after completing one more orbit?

*32. The binary star system PSR 1913+16 consists of two neutron stars orbiting with a period of 7.75 h about their center of mass, which is at the midpoint between the stars. Assume that the stars have equal masses and that their orbits are circular with a radius of 8.67×10^8 m.

(a) What are the masses of the stars?

(b) What are their speeds?

*33. Figure 9.32 shows two stars orbiting about their common center of mass in the binary system Krüger 60. The center of mass is at a point between the stars such that the distances of the stars from this point are in the inverse ratio of their masses. Measure the sizes of their orbits and determine the ratio of their masses.

FIGURE 9.32 The orbits of the two stars in the binary system Krüger 60. Each ellipse has its focus at the center of mass.

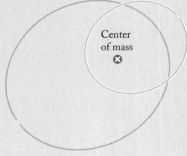

Center of mass

FIGURE 9.31 (a) The Solar System and (b) the 55 Cancri system.

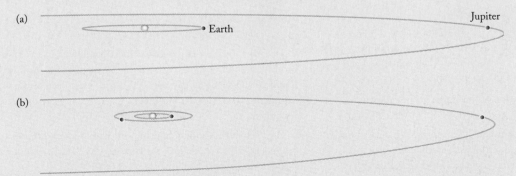

(a)
Earth
Jupiter

(b)

**34. A binary star system consists of two stars of masses m_1 and m_2 orbiting about each other. Suppose that the orbits of the stars are circles of radii r_1 and r_2 centered on the center of mass (Fig. 9.33). The center of mass is a point between the stars such that the radii r_1 and r_2 are in the ratio $r_1/r_2 = m_2/m_1$. Show that the period of the orbital motion is given by

$$T^2 = \frac{4\pi^2}{G(m_1 + m_2)}(r_1 + r_2)^3$$

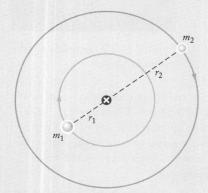

FIGURE 9.33   A binary star system. The orbits are circles about the center of mass.

*35. The binary system Cygnus X-1 consists of two stars orbiting about their center of mass under the influence of their mutual gravitational forces. The orbital period of the motion is 5.6 days. One of the stars is a supergiant with a mass 25 times the mass of the Sun. The other star is believed to be a black hole with a mass about 10 times the mass of the Sun. From the information given, determine the distance between the stars; assume that the orbits of both stars are circular. (Hint: See Problem 34.)

**36. A hypothetical triple star system consists of three stars orbiting about each other. For the sake of simplicity, assume that all three stars have equal masses and that they move along a common circular orbit maintaining an angular separation of 120° (Fig. 9.34). In terms of the mass M of each star and the orbital radius R, what is the period of the motion?

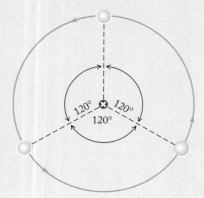

FIGURE 9.34   Three identical stars orbiting about their center of mass.

**37. Take into account the rotation of the Earth in the following problem:

(a) Cape Canaveral is at a latitude of 28° north. What eastward speed (relative to the ground) must a satellite be given if it is to achieve a low-altitude circular orbit (Fig. 9.35)? What

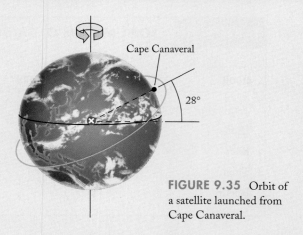

FIGURE 9.35   Orbit of a satellite launched from Cape Canaveral.

westward speed must the satellite be given if it is to travel along the same orbit in the opposite direction? For the purpose of this problem, pretend that "low altitude" means essentially "zero altitude."

(b) Suppose that the satellite has a mass of 14.0 kg. What kinetic energy must the launch vehicle give to the satellite for an eastward orbit? For a westward orbit?

9.4 Elliptical Orbits; Kepler's Laws[†]

38. Halley's comet (Fig. 9.36) orbits the Sun in an elliptical orbit (the comet reached perihelion in 1986). When the comet is at perihelion, its distance from the Sun is 8.78×10^{10} m, and its speed is 5.45×10^4 m/s. When the comet is at aphelion, its distance is 5.28×10^{12} m. What is the speed at aphelion?

FIGURE 9.36   Halley's comet photographed in 1986.

39. *Explorer I,* the first American artificial satellite, had an elliptical orbit around the Earth with a perigee distance of 6.74×10^6 m and an apogee distance of 8.91×10^6 m. The speed of this satellite was 6.21×10^3 m/s at apogee. Calculate the speed at perigee.

40. The *Explorer X* satellite had an orbit with perigee 175 km and apogee 181,200 km above the surface of the Earth. What was the period of this satellite?

[†] For help, see Online Concept Tutorial 12 at www.wwnorton.com/physics

TABLE 9.3 THE FIRST ARTIFICIAL EARTH SATELLITES

SATELLITE	MASS	MEAN DISTANCE FROM CENTER OF EARTH (SEMIMAJOR AXIS)	PERIGEE DISTANCE	APOGEE DISTANCE	PERIOD
Sputnik I	83 kg	6.97×10^3 km	6.60×10^3 km	7.33×10^3 km	96.2 min
Sputnik II	3000	7.33	6.61	8.05	104
Explorer I	14	7.83	6.74	8.91	115
Vanguard I	1.5	8.68	7.02	10.3	134
Explorer III	14	7.91	6.65	9.17	116
Sputnik III	1320	7.42	6.59	8.25	106

41. Calculate the orbital periods of *Sputnik I* and *Explorer I* from their apogee and perigee distances in Table 9.3.

42. The aphelion distance for Saturn is 1510×10^6 km; its perihelion distance is 1350×10^6 km. By Kepler's First Law, the Sun is at one focus of this ellipse. How far from the Sun is the other focus? How does this compare with the orbital radius of Mercury?

43. The comet Hale–Bopp was spectacularly visible in the spring of 1997 (see Fig. 9.37) and may be the most viewed comet in history. Its perihelion distance was 137×10^6 km, and its orbital period is 2380 yr. What is its aphelion distance? How does this compare with the mean distance of Pluto from the Sun?

FIGURE 9.37 Comet Hale–Bopp photographed in 1997.

44. The orbit of the Earth deviates slightly from circular: at aphelion, the Earth–Sun distance is 1.52×10^8 km, and at perihelion it is 1.47×10^8 km. By what factor is the speed of the Earth at perihelion greater than the speed at aphelion?

9.5 Energy in Orbital Motion

45. The *Voskhod I* satellite, which carried Yuri Gagarin into space in 1961, had a mass of 4.7×10^3 kg. The radius of the orbit was (approximately) 6.6×10^3 km. What were the orbital speed and the orbital energy of this satellite?

46. What is the kinetic energy and what is the gravitational potential energy for the orbital motion of the Earth around the Sun? What is the total energy?

47. Compare the escape velocity given by Eq. (9.27) with the velocity required for a circular orbit of radius R_S, according to Eq. (9.10). By what factor is the escape velocity larger than the velocity for the circular orbit?

48. In July of 1994, fragments of the comet Shoemaker–Levy struck Jupiter.

    (a) What is the impact speed (equal to the escape speed) for a fragment falling on the surface of Jupiter?

    (b) What is the kinetic energy at impact for a fragment of 1.0×10^{10} kg? Express this energy as an equivalent number of short tons of TNT (the explosion of 1 short ton, or 2000 lb, of TNT releases 4.2×10^9 J).

49. A 1.0-kg mass is in the same orbit around the Earth as the Moon (but far from the Moon). What is the kinetic energy for this orbit? The gravitational potential energy? The total energy?

50. The boosters on a satellite in geosynchronous orbit accidentally fire for a prolonged period. At the instant this "burn" ends, the velocity is parallel to the original tangential direction, but the satellite has been slowed to one-half of its original speed. The satellite is thus at apogee of its new orbit. What is the perigee distance for such an orbit? What happens to the satellite?

51. A black hole is so dense that even light cannot escape its gravitational pull. Assume that all of the mass of the Earth is compressed in a sphere of radius R. How small must R be so the escape speed is the speed of light?

52. The spectacular comet Hale–Bopp (Fig. 9.37), most visible in 1997, entered the Solar System in an elliptical orbit with period 4206 yr. However, after a close encounter with Jupiter on its inbound path, it continues on a new elliptical orbit with a period of 2380 yr. By what fraction did the encounter with Jupiter change the energy of Hale–Bopp's orbit?

53. The typical speed of nitrogen molecules at a temperature of 117°C, the temperature of the Moon's surface at "noon," is 600 m/s; some molecules move slower, others faster. What fraction of the escape velocity from the Moon is this? Can you guess why the Moon has not retained an atmosphere?

*54. The Andromeda galaxy is at a distance of 2.1×10^{22} m from our Galaxy. The mass of the Andromeda is 6.0×10^{41} kg, and the mass of our Galaxy is 4.0×10^{41} kg.

   (a) Gravity accelerates the galaxies toward each other. As reckoned in an inertial reference frame, what is the acceleration of the Andromeda galaxy? What is the acceleration of our Galaxy? Treat both galaxies as point particles.

   (b) The speed of the Andromeda galaxy *relative to our Galaxy* is 266 km/s. What is the speed of the Andromeda and what is the speed of our Galaxy *relative to the center of mass* of the two galaxies? The center of mass is at a point between the galaxies such that the distances of the galaxies from this point are in the inverse ratios of their masses.

   (c) What is the kinetic energy of each galaxy relative to the center of mass? What is the total energy (kinetic and potential) of the system of the two galaxies? Will the two galaxies eventually escape from each other?

*55. Neglect the gravity of the Moon, neglect atmospheric friction, and neglect the rotational velocity of the Earth in the following problem. A long time ago, Jules Verne, in his book *From Earth to the Moon* (1865), suggested sending an expedition to the Moon by means of a projectile fired from a gigantic gun.

   (a) With what muzzle speed must a projectile be fired vertically from a gun on the surface of the Earth if it is to (barely) reach the distance of the Moon?

   (b) Suppose that the projectile has a mass of 2000 kg. What energy must the gun deliver to the projectile? The explosion of 1 short ton (2000 lb) of TNT releases 4.2×10^9 J. How many tons of TNT are required for firing this gun?

   (c) If the gun barrel is 500 m long, what must be the average acceleration of the projectile during firing?

*56. An artificial satellite of 1300 kg made of aluminum is in a circular orbit at a height of 100 km above the surface of the Earth. Atmospheric friction removes energy from the satellite and causes it to spiral downward so that it ultimately crashes into the ground.

   (a) What is the initial orbital energy (gravitational plus kinetic) of the satellite? What is the final energy when the satellite comes to rest on the ground? What is the energy change?

   (b) Suppose that all of this energy is absorbed in the form of heat by the material of the satellite. Is this enough heat to melt the material of the satellite? To vaporize it? The heats of fusion and of vaporization of aluminum are given in Table 20.4.

*57. According to one theory, glassy meteorites (tektites) found on the surface of the Earth originate in volcanic eruptions on the Moon. With what minimum speed must a volcano on the Moon eject a stone if it is to reach the Earth? With what speed will this stone strike the surface of the Earth? In this problem ignore the orbital motion of the Moon around the Earth; use the data for the Earth–Moon system listed in the tables printed inside the book cover. (Hint: When the rock reaches the intermediate point where the gravitational pulls of the Moon and the Earth cancel out, it must have zero velocity.)

*58. A spacecraft is launched with some initial velocity toward the Moon from 300 km above the surface of the Earth.

   (a) What is the minimum initial speed required if the spacecraft is to coast all the way to the Moon without using its rocket motors? For this problem pretend that the Moon does not move relative to the Earth. The masses and radii of the Earth and the Moon and their distance are listed in the tables printed inside the book cover. (Hint: When the spacecraft reaches the point in space where the gravitational pulls of the Earth and the Moon cancel, it must have zero velocity.)

   (b) With what speed will the spacecraft strike the Moon?

59. The Pons–Brooks comet had a speed of 47.30 km/s when it reached its perihelion point, 1.160×10^8 km from the Sun. Is the orbit of this comet elliptical, parabolic, or hyperbolic?

*60. At a radial distance of 2.00×10^7 m from the center of the Earth, three artificial satellites (I, II, III) are ejected from a rocket. The three satellites I, II, III are given initial speeds of 5.47 km/s, 4.47 km/s, and 3.47 km/s, respectively; the initial velocities are all in the tangential direction.

   (a) Which of the satellites I, II, III will have a circular orbit? Which will have elliptical orbits? Explain your answer.

   (b) Draw the circular orbit. Also, superimposed on the same diagram, draw the elliptical orbits of the other satellites; label the orbits with the names of the satellites. (Note: You need not calculate the exact sizes of the ellipses, but your diagram should show where the ellipses are larger or smaller than the circle.)

*61. (a) Since the Moon (*our* Moon) has no atmosphere, it is possible to place an artificial satellite in a circular orbit that skims along the surface of the Moon (provided that the satellite does not hit any mountains!). Suppose that such a satellite is to be launched from the *surface* of the Moon by means of a gun that shoots the satellite in a horizontal direction. With what velocity must the satellite be shot out from the gun? How long does the satellite take to go once around the Moon?

   (b) Suppose that a satellite is shot from the gun with a horizontal velocity of 2.00 km/s. Make a rough sketch showing the Moon and the shape of the satellite's orbit; indicate the position of the gun on your sketch.

(c) Suppose that a satellite is shot from the gun with a horizontal velocity of 3.00 km/s. Make a rough sketch showing the Moon and the shape of the satellite's orbit. Is this a closed orbit?

*62. According to an estimate, a large crater on Wilkes Land, Antarctica, was produced by the impact of a 1.2×10^{13}–kg meteoroid incident on the surface of the Earth at 70 000 km/h. What was the speed of this meteoroid relative to the Earth when it was at a "large" distance from the Earth?

*63. An experienced baseball player can throw a ball with a speed of 140 km/h. Suppose that an astronaut standing on Mimas, a small moon of Saturn of mass 3.76×10^{19} kg and radius 195 km, throws a ball with this speed.

(a) If the astronaut throws the ball horizontally, will it orbit around Mimas?

(b) If the astronaut throws the ball vertically, how high will it rise?

*64. An electromagnetic launcher, or rail gun, accelerates a projectile by means of magnetic fields. According to some calculations, it may be possible to attain muzzle speeds as large as 15 km/s with such a device. Suppose that a projectile is launched upward from the surface of the Earth with this speed; ignore air resistance.

(a) Will the projectile escape permanently from the Earth?

(b) Can the projectile escape permanently from the Solar System? (Hint: Take into account the speed of 30 km/s of the Earth around the Sun.)

*65. *Sputnik I,* the first Russian satellite (1957), had a mass of 83.5 kg; its orbit reached perigee at a height of 225 km and apogee at 959 km. *Explorer I,* the first American satellite (1958), had a mass of 14.1 kg; its orbit reached perigee at a height of 368 km and apogee at 2540 km. What was the orbital energy of these satellites?

*66. The orbits of most meteoroids around the Sun are nearly parabolic.

(a) With what speed will a meteoroid reach a distance from the Sun equal to the distance of the Earth from the Sun? (Hint: In a parabolic orbit the speed at any radius equals the escape velocity at the radius. Why?)

(b) Taking into account the Earth's orbital speed, what will be the speed of the meteoroid *relative to the Earth* in a head-on collision with the Earth? In an overtaking collision? Ignore the effect of the gravitational pull of the Earth on the meteoroid.

*67. Calculate the perihelion and the aphelion speeds of Encke's comet. The perihelion and aphelion distances of this comet are 5.06×10^7 km and 61.25×10^7 km. (Hint: Consider the total energy of the orbit.)

*68. The *Explorer XII* satellite was given a tangential velocity of 10.39 km/s when at perigee at a height of 457 km above the Earth. Calculate the height of the apogee. (Hint: Consider the total energy of the orbit.)

**69. Prove that the orbital energy of a planet or a comet in an elliptical orbit around the Sun can be expressed as

$$E = -\frac{GM_S m}{r_1 + r_2}$$

where r_1 and r_2 are, respectively, the perihelion and aphelion distances. [Hint: Use the conservation of energy and the conservation of angular momentum $(r_1 v_1 = r_2 v_2)$ at perihelion and at aphelion to solve for v_1^2 and v_2^2 in terms of r_1 and r_2.]

*70. Suppose that a comet is originally at rest at a distance r_1 from the Sun. Under the influence of the gravitational pull, the comet falls radially toward the Sun. Show that the time it takes to reach a radius r_2 is

$$t = -\int_{r_1}^{r_2} \frac{dr}{\sqrt{2GM_S/r - 2GM_S/r_1}}$$

*71. Suppose that a projectile is fired horizontally from the surface of the Moon with an initial speed of 2.0 km/s. Roughly sketch the orbit of the projectile. What maximum height will this projectile reach? What will be its speed when it reaches maximum height?

**72. The Earth has an orbit of radius 1.50×10^8 km around the Sun; Mars has an orbit of radius 2.28×10^8 km. In order to send a spacecraft from the Earth to Mars, it is convenient to launch the spacecraft into an elliptical orbit whose perihelion coincides with the orbit of the Earth and whose aphelion coincides with the orbit of Mars (Fig. 9.38); this orbit requires the least amount of energy for a trip to Mars.

(a) To achieve such an orbit, with what speed (relative to the Earth) must the spacecraft be launched? Ignore the pull of the gravity of the Earth and Mars on the spacecraft.

(b) With what speed (relative to Mars) does the spacecraft approach Mars at the aphelion point? Assume that Mars actually is at the aphelion point when the spacecraft arrives.

(c) How long does the trip from Earth to Mars take?

(d) Where must Mars be (in relation to the Earth) at the instant the spacecraft is launched? Where will the Earth be when the spacecraft arrives at its destination? Draw a diagram showing the relative positions of Earth and Mars at these two times.

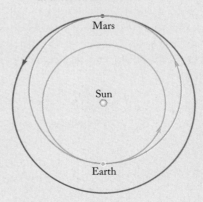

FIGURE 9.38
Orbit for a spacecraft on a trip to Mars.

**73. Repeat the calculations of Problem 72 for the case of a spacecraft launched on a trip to Venus. The orbit of Venus has a radius of 1.08×10^8 km.

**74. If a spacecraft, or some other body, approaches a moving planet on a hyperbolic orbit, it can gain some energy from the motion of the planet and emerge with a larger speed than it had initially. This slingshot effect has been used to boost the speeds of the two Voyager spacecraft as they passed near Jupiter. Suppose that the line of approach of the satellite makes an angle θ with the line of motion of the planet and the line of recession of the spacecraft is parallel to the line of motion of the planet (Fig. 9.39; the planet can be regarded as moving on a straight line during the time interval in question). The speed of the planet is u, and the initial speed of the spacecraft is v (in the reference frame of the Sun).

(a) Show that the final speed of the spacecraft is

$$v' = u + \sqrt{v^2 + u^2 - 2uv\cos\theta}$$

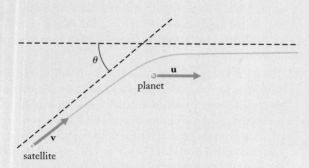

FIGURE 9.39 Trajectory of a spacecraft passing by a planet.

(b) Show that the spacecraft will not gain any speed in this encounter if $\theta = 0$, and show that the spacecraft will gain maximum speed if $\theta = 180°$.

(c) If a spacecraft with $v = 3.0$ km/s approaches Jupiter at an angle of $\theta = 20°$, what will be its final speed?

75. According to one design studied by NASA, a large space colony in orbit around the Earth would consist of a torus of diameter 1.8 km, looking somewhat like a gigantic wheel (see Fig. 9.40). In order to generate artificial gravity of $1g$, how fast must this space colony rotate about its axis?

FIGURE 9.40 A rotating space station.

REVIEW PROBLEMS

76. Calculate the gravitational force that the Earth exerts on an astronaut of mass 75 kg in a space capsule at a height of 100 km above the surface of the Earth. Compare with the gravitational force that this astronaut would experience if on the surface of the Earth.

77. The masses used in the Cavendish experiment typically are a few kilograms for the large masses and a few tens of grams for the small masses. Suppose that a "large" spherical mass of 8.0 kg is at a center-to-center distance of 10 cm from a "small" spherical mass of 30 g. What is the magnitude of the gravitational force?

78. The asteroid Ceres has a diameter of 1100 km and a mass of (approximately) 7×10^{20} kg. What is the value of the acceleration of gravity at its surface? On the surface of this asteroid, what would be the weight (in lbf) of a man whose weight on the surface of the Earth is 170 lbf?

79. The asteroid belt of the Solar System consists of chunks of rock orbiting around the Sun in approximately circular orbits.

The mean distance of the asteroid belt from the Sun is about 2.9 times the distance of the Earth. What is the mean period of the orbital motion of the asteroids?

80. Imagine that somewhere in interstellar space a small pebble is in a circular orbit around a spherical asteroid of mass 1000 kg. If the radius of the circular orbit is 1.0 km, what is the period of the motion?

81. Europa (Fig. 9.41) is a moon of Jupiter. Astronomical observations show that this moon is in a circular orbit of radius 6.71×10^8 m with a period of 3.55 days. From these data deduce the mass of Jupiter.

82. Observations with the Hubble Space Telescope have revealed that at the center of the galaxy M87, gas orbits around a very massive compact object, believed to be a black hole. The measurements show that gas clouds in a circular orbit of radius 250 light-years have an orbital speed of 530 km/s. From this information, deduce the mass of the black hole.

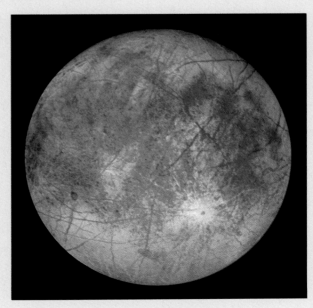

FIGURE 9.41 Europa, one of the moons of Jupiter.

83. Consider a space station in a circular orbit at an altitude of 400 km around the Earth and a piece of debris, left over from, say, the disintegration of a rocket, in an orbit of the same radius but of opposite direction.

    (a) What is the speed of the debris relative to the space station when they pass?

    (b) If the debris hit the spacecraft, it would penetrate the space station with catastrophic consequences for the crew. Penetration depends on the kinetic energy of the debris. What must be the mass of a piece of debris if it is to have an impact energy of 4.6×10^5 J, which corresponds to the explosion of 100 g of TNT?

84. *Vanguard I*, the second American artificial satellite (Fig. 9.42), moved in an elliptical orbit around the Earth with a perigee distance of 7.02×10^6 m and an apogee distance of 10.3×10^6 m. At perigee, the speed of this satellite was 8.22×10^3 m/s. What was the speed at apogee?

*85. The motor of a Scout rocket uses up all its fuel and stops when the rocket is at an altitude of 200 km above the surface of the Earth and is moving vertically at 8.50 km/s.

How high will this rocket rise? Neglect any residual atmospheric friction.

*86. An astronaut in a spacecraft in a circular orbit around the Earth wants to get rid of a defective solar panel that he has detached from the spacecraft. He hits the panel with a blast from the steering rocket of the spacecraft, giving it an increment of velocity. This sends the solar panel into an elliptical orbit.

    (a) Sketch the circular orbit of the spacecraft and the elliptical orbit of the solar panel if the velocity increment is parallel to the velocity of the spacecraft and if it is antiparallel.

    (b) If the ratio of the semimajor axis of the ellipse to the radius of the circle has a special value, it is possible for the panel to meet with the spacecraft again after several orbits. What are these special values of the ratio?

*87. A communications satellite of mass 700 kg is placed in a circular orbit of radius 4.23×10^7 m around the Earth.

    (a) What is the total orbital energy of this satellite?

    (b) How much extra energy would we have to give this satellite to put it into a parabolic orbit that permits it to escape to infinite distance from the Earth?

88. What is the escape velocity for a projectile launched from the surface of our Moon?

FIGURE 9.42 The *Vanguard I* satellite.

Answers to Checkups

Checkup 9.1

1. The gravitational force varies inversely with the square of the distance, so the force will be $(30)^2 = 900$ times weaker for a 1-kg piece of Neptune than for a 1-kg piece of the Earth.

2. The gravitational force varies in proportion to the mass and in inverse proportion to the square of the distance, so the 100-times-larger mass for Saturn cancels the 10-times-larger

distance; thus, the gravitational force that the Sun exerts on Saturn is about equal to that on the Earth. The acceleration is $a = F/m$, and so is about 100 times smaller for Saturn.

3. The acceleration varies inversely with the square of the distance, and so is $\frac{1}{4}g$ at $r = 2R_E$, and is $\frac{1}{9}g$ at $r = 3R_E$.

4. Since the acceleration at a planet's surface is $a = GM/R^2$, a larger mass M and a smaller gravitational acceleration a are possible only because the radius R of Uranus is sufficiently larger than that of the Earth.

5. At the exact center of the Earth, a particle would be equally attracted in all directions, and so would experience zero net force.

6. (B) $\frac{1}{4}g$. The acceleration at the surface is $a = GM_E/R_E^2$, so a doubled radius would result in an acceleration one-fourth as large, or $\frac{1}{4}g$.

Checkup 9.2

1. To determine G by measuring the force between the Earth and some known mass, we would also have to know the mass of the Earth; we have no independent way of determining the mass of the Earth.

2. (A) Yes. If we knew the mass of the mountain (and the spatial distribution of such mass), then we could determine the gravitational force from the plumb bob's deflection, and thus G.

Checkup 9.3

1. An orbit that is a circle at the latitude of San Francisco is impossible, since the center of every orbit must coincide with the center of the Earth.

2. The period is proportional to the 3/2 power of the radius of the orbit, so for a doubled radius, the period of the Moon would become $2^{3/2} \times 27$ days ≈ 76 days.

3. As in Eq. (9.13), we need only know the period and radius of the moon's orbit to determine the mass of the planet.

4. (C) 30 yr. The period is proportional to the 3/2 power of the radius of the orbit, so the period of Saturn's motion is $10^{3/2} \times 1$ yr ≈ 30 yr.

Checkup 9.4

1. Kepler's Second Law would remain valid, since it depends only on the central nature of the force, and otherwise not on any particular form of the force. Kepler's Third Law, however, like the law of periods, Eq. (9.13), depends on the inverse-square nature of the force. If we were to perform a similar derivation to that preceding Eq. (9.13) for an inverse-cube force, we would find that the period was proportional to the square of the radius.

2. As in Eq. (9.18), the speeds vary inversely with the distances, so for an aphelion distance twice as large as the perihelion distance, the speed at aphelion will be half as large as the speed at perihelion, or will be 20 km/s.

3. According to Kepler's Third Law, the period must be exactly one year. This is so because both the Earth's orbit (nearly circular; the semimajor axis of a circle is its radius) and the comet's orbit have the same semimajor axis, and both orbit the same central body, the Sun.

4. (D) 4. Kepler's Third Law states that the square of the period is proportional to the cube of the semimajor axis of the orbit, so to make the period 8 times as large as the Earth's period would make the cube of the semimajor axis 64 times as large; thus the semimajor axis would be $64^{1/3} = 4$ times as large as the Earth–Sun distance.

Checkup 9.5

1. For a circular orbit, we found that the magnitude of the (negative) potential energy is twice the size of the kinetic energy. Thus the potential energy decreases so much for the lower orbit (it becomes more negative) that the kinetic energy can increase and energy can be lost to friction.

2. Yes—our derivation of the law depended only on the central nature of the force, not on any particular type of orbit (or even any particular form of the central force).

3. If we ignore air friction (and the body does not encounter any obstacles), then the body will escape the Earth's influence in a parabolic "orbit," since the escape velocity provides for zero net energy. The orbit would be similarly parabolic if we launched the body at any angle (except straight up, although that resulting linear path can be considered a special case of the parabola). Ultimately, far from the Earth's influence, the path would be modified by the Sun.

4. No. The gravitational acceleration is $g = GM/R^2$, whereas the escape velocity depends on the gravitational potential energy, which is proportional to M/R. For example, a body with twice the mass and twice the radius of the Earth would have half the gravitational acceleration at the surface, but would have the same escape velocity.

5. (C) Hyperbolic; elliptical. Recall that a parabolic orbit is a zero-energy orbit, where the comet can just barely escape to infinity. The energy of comet II must be positive, since it has a larger speed (a greater kinetic energy, but the same potential energy as it crosses the Earth's orbit); we found that a positive-energy orbit is a hyperbola. Similarly, the energy of comet III must be negative, since it has a smaller speed; negative-energy orbits are ellipses, with a semimajor axis given by Eq. (9.24).

Systems of Particles

CONCEPTS IN CONTEXT

While this high jumper is passing over the bar, he bends backward and keeps his extremities below the level of the bar. This means that the average height of his body parts is less than if he were to keep his body straight, and he requires less energy to pass over the bar.

The concepts introduced in this chapter permit us to examine in detail several aspects of the motion of the jumper:

? The body of the jumper is a system of particles. Where is the average position of the mass of this system of particles when the body is in a straight configuration? How does this change when the jumper reconfigures his extremities? (Example 8, part (a), page 322)

? What is the gravitational potential energy of a system of particles, and how much does the jumper reduce his potential energy by bending his body? (Page 321 in Section 10.2 and Example 8, part (b), page 322)

10.1 Momentum

10.2 Center of Mass

10.3 The Motion of the Center of Mass

10.4 Energy of a System of Particles

? What is equation of motion of a system of particles, and to what extent does the translational motion of a jumper resemble projectile motion? (Page 324 in Section 10.3)

So far we have dealt almost exclusively with the motion of a single particle. Now we will begin to study systems of particles interacting with each other via some forces. This means we must examine, and solve, the equations of motion of all these particles simultaneously.

Since chunks of ordinary matter are made of particles—electrons, protons, and neutrons—all the macroscopic bodies that we encounter in our everyday environment are in fact many-particle systems containing a very large number of particles. However, for most practical purposes, it is not desirable to adopt such an extreme microscopic point of view, and in the preceding chapters we treated the motion of a macroscopic body, such as an automobile, as motion of a particle. Likewise, in dealing with a system consisting of several macroscopic bodies, we will often find it convenient to treat each of these bodies as a particle and ignore the internal structure of the bodies. For example, when investigating a collision between two automobiles, we may find it convenient to pretend that each of the automobiles is a particle—we then regard the colliding automobiles as a system of two particles which exert forces on each other when in contact. And when investigating the Solar System, we may find it convenient to pretend that each planet and each satellite is a particle—we then regard the Solar System as a system of such planet and satellite particles loosely held together by gravitation and orbiting around the Sun and around each other.

The equations of motion of a system of several particles are often hard, and sometimes impossible, to solve. It is therefore necessary to make the most of any information that can be extracted from the general conservation laws. In the following sections we will become familiar with the *momentum* vector, and we will see how the laws of conservation of momentum and of energy apply to a system of particles.

10.1 MOMENTUM

Newton's laws can be expressed very neatly in terms of **momentum,** a vector quantity of great importance in physics. *The momentum of a single particle is defined as the product of the mass and the velocity of the particle:*[1]

momentum of a particle

$$\mathbf{p} = m\mathbf{v} \tag{10.1}$$

Thus, the momentum **p** is a vector that has the same direction as the velocity vector, but a magnitude that is m times the magnitude of the velocity. The SI unit of momentum is kg·m/s; this is the momentum of a mass of 1 kg when moving at 1 m/s.

The mathematical definition of momentum is consistent with our intuitive, everyday notion of "momentum." If two cars have equal masses but one has twice the velocity of the other, it has twice the momentum. And if a truck has three times the mass of a car and the same velocity, it has three times the momentum. During the nineteenth century physicists argued whether momentum or kinetic energy was the best measure of the "amount of motion" in a body. They finally decided that the answer

---

[1] The momentum $\mathbf{p} = m\mathbf{v}$ is sometimes referred to as *linear momentum* to distinguish it from *angular momentum,* discussed in Chapter 13.

depends on the context—as we will see in the examples in this chapter and the next, sometimes momentum is the most relevant quantity, sometimes energy is, and sometimes both are relevant.

Newton's First Law states that, in the absence of external forces, the velocity of a particle remains constant. Expressed in terms of momentum, *the First Law therefore states that the momentum remains constant:*

$$\mathbf{p} = [\text{constant}] \qquad \text{(no external forces)} \qquad (10.2)$$

First Law in terms of momentum

Thus, we can say that the momentum of the particle is conserved. Of course, we could equally well say that the velocity of this particle is conserved; but the deeper significance of momentum will emerge when we study the motion of a system of several particles exerting forces on one another. We will find that the total momentum of such a system is conserved—any momentum lost by one particle is compensated by a momentum gain of some other particle or particles.

To express the Second Law in terms of momentum, we note that since the mass is constant, the time derivative of Eq. (10.1) is

$$\frac{d\mathbf{p}}{dt} = m\frac{d\mathbf{v}}{dt}$$

or

$$\frac{d\mathbf{p}}{dt} = m\mathbf{a}$$

But, according to Newton's Second Law, $m\mathbf{a}$ equals the force; hence, *the rate of change of the momentum with respect to time equals the force:*

$$\frac{d\mathbf{p}}{dt} = \mathbf{F} \qquad (10.3)$$

Second Law in terms of momentum

This equation gives the Second Law a concise and elegant form.

EXAMPLE 1   A tennis player smashes a ball of mass 0.060 kg at a vertical wall. The ball hits the wall perpendicularly with a speed of 40 m/s and bounces straight back with the same speed. What is the change of momentum of the ball during the impact?

SOLUTION: Take the positive x axis along the direction of the initial motion of the ball (see Fig. 10.1a). The momentum of the ball before impact is then in the positive direction, and the x component of the momentum is

$$p_x = mv_x = 0.060 \text{ kg} \times 40 \text{ m/s} = 2.4 \text{ kg·m/s}$$

The momentum of the ball after impact has the same magnitude but the opposite direction:

$$p'_x = -2.4 \text{ kg·m/s}$$

(Throughout this chapter, the primes on mathematical quantities indicate that these quantities are evaluated *after* the collision.) The change of momentum is

$$\Delta p_x = p'_x - p_x = -2.4 \text{ kg·m/s} - 2.4 \text{ kg·m/s} = -4.8 \text{ kg·m/s}$$

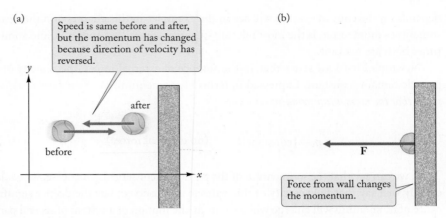

FIGURE 10.1 (a) A tennis ball bounces off a wall. (b) At the instant of impact, the wall exerts a large force on the ball.

This change of momentum is produced by the (large) force that acts on the ball during impact on the wall (see Fig. 10.1b). The change of momentum is negative because the force is negative (the force is in the negative x direction, opposite to the direction of the initial motion).

We can also express Newton's Third Law in terms of momentum. Since the action force is exactly opposite to the reaction force, the rate of change of momentum generated by the action force on one body is exactly opposite to the rate of change of momentum generated by the reaction force on the other body. Hence, we can state the Third Law as follows:

Third Law in terms of momentum

> *Whenever two bodies exert forces on each other, the resulting changes of momentum are of equal magnitudes and of opposite directions.*

This balance in the changes of momentum leads us to a general law of conservation of the total momentum for a system of particles.

The total momentum of a system of n particles is simply the (vector) sum of all the individual momenta of all the particles. Thus, if $\mathbf{p}_1 = m_1\mathbf{v}_1$, $\mathbf{p}_2 = m_2\mathbf{v}_2$, ..., and $\mathbf{p}_n = m_n\mathbf{v}_n$ are the individual momenta of the particles, then the total momentum is

momentum of a system of particles

$$\mathbf{P} = \mathbf{p}_1 + \mathbf{p}_2 + \cdots + \mathbf{p}_n \tag{10.4}$$

The simplest of all many-particle systems consists of just two particles exerting some mutual forces on one another (see Fig. 10.2). Let us assume that the two particles are isolated from the rest of the Universe so that, apart from their mutual forces, they experience no extra forces of any kind. According to the above formulation of the Third Law, the rates of change of $\mathbf{p}_1$ and $\mathbf{p}_2$ are then exactly opposite:

$$\frac{d\mathbf{p}_1}{dt} = -\frac{d\mathbf{p}_2}{dt}$$

The rate of change of the sum $\mathbf{p}_1 + \mathbf{p}_2$ is therefore zero, since the rate of change of the first term in this sum is canceled by the rate of change of the second term:

$$\frac{d(\mathbf{p}_1 + \mathbf{p}_2)}{dt} = 0$$

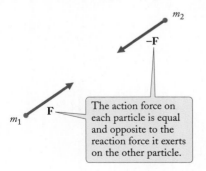

FIGURE 10.2 Two particles exerting mutual forces on each other. The net change of momentum of the isolated particle pair is zero.

This means that the sum $\mathbf{p}_1 + \mathbf{p}_2$ is a constant of the motion:

$$\mathbf{p}_1 + \mathbf{p}_2 = [\text{constant}] \tag{10.5}$$

momentum conservation for two particles

This is the **Law of Conservation of Momentum.** Note that Newton's Third Law is an essential ingredient for establishing the conservation of momentum: the total momentum is constant because the equality of action and reaction keeps the momentum changes of the two particles exactly equal in magnitude but opposite in direction—the particles merely exchange some momentum by means of their mutual forces. Thus, for our particles, the total momentum $\mathbf{P}$ at some instant equals the total momentum $\mathbf{P}'$ at some other instant, so

$$\mathbf{P} = \mathbf{P}'$$

Conservation of momentum is a powerful tool which permits us to calculate some general features of the motion even when we are ignorant of the detailed properties of the interparticle forces. The following examples illustrate how we can use conservation of momentum to solve some problems of motion.

(a)

EXAMPLE 2   A gun used onboard an eighteenth-century warship is mounted on a carriage which allows the gun to roll back each time it is fired (Fig. 10.3). The mass of the gun, including the carriage, is 2200 kg. The gun fires a 6.0-kg shot horizontally with a velocity of 500 m/s. What is the recoil velocity of the gun?

SOLUTION: The total momentum of the shot plus the gun must be the same before the firing and just after the firing. Before, the total momentum is zero (Fig. 10.3a):

$$\mathbf{P} = 0$$

(b)

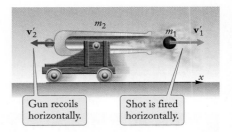

Gun recoils horizontally.   Shot is fired horizontally.

FIGURE 10.3 (a) Initially, the gun and the shot are at rest. (b) After the firing, the gun recoils toward the left (the velocity $\mathbf{v}'_2$ of the gun is negative).

After, the (horizontal) velocity of the shot is $\mathbf{v}'_1$, and the velocity of the gun is $\mathbf{v}'_2$ (as above, the primes on mathematical quantities indicate that these are evaluated *after* the firing); hence the total momentum is

$$\mathbf{P}' = m_1\mathbf{v}'_1 + m_2\mathbf{v}'_2$$

where $m_1 = 6.0$ kg is the mass of the shot and $m_2 = 2200$ kg is the mass of the gun (including the carriage). Thus, momentum conservation tells us

$$0 = m_1\mathbf{v}'_1 + m_2\mathbf{v}'_2$$

or

$$\mathbf{v}'_2 = -\frac{m_1}{m_2}\,\mathbf{v}'_1$$

The negative sign indicates that $\mathbf{v}'_2$, the recoil velocity of the gun, is opposite to the velocity of the shot and has a magnitude

$$v'_2 = \frac{m_1}{m_2}\,v'_1$$

$$= \frac{6.0 \text{ kg}}{2200 \text{ kg}} \times 500 \text{ m/s} = 1.4 \text{ m/s}$$

COMMENTS: Note that the final velocities are in the inverse ratio of the masses: the shot emerges with a large velocity, and the gun rolls back with a low velocity.

This is a direct consequence of the equality of the magnitudes of the action and reaction forces that act on the shot and the gun during the firing. The force gives the shot (of small mass) a large acceleration, and the reaction force gives the gun (of large mass) a small acceleration.

In this calculation we neglected the mass and momentum of the gases released in the explosion of the gunpowder. This extra momentum increases the recoil velocity somewhat.

EXAMPLE 3   An automobile of mass 1500 kg traveling at 24 m/s crashes into a similar parked automobile. The two automobiles remain joined together after the collision. What is the velocity of the wreck immediately after the collision? Neglect friction against the road, since this force is insignificant compared with the large mutual forces that the automobiles exert on each other.

SOLUTION: Under the assumptions of the problem, the only horizontal forces are the mutual forces of one automobile on the other. Thus, momentum conservation applies to the horizontal component of the momentum: the value of this component must be the same before and after the collision. Before the collision, the (horizontal) velocity of the moving automobile is $v_1 = 24$ m/s and that of the other is $v_2 = 0$. With the x axis along the direction of motion (see Fig. 10.4), the total momentum is therefore

$$P_x = m_1v_1 + m_2v_2 = m_1v_1$$

After the collision, both automobiles have the same velocity (see Fig. 10.4b). We will designate the velocities of the automobiles after the collision by v_1' and v_2', respectively. We can write $v_1' = v_2' = v'$ (the automobiles have a common v', since they remain joined), so the total momentum is

$$P_x' = m_1v_1' + m_2v_2' = (m_1 + m_2)v'$$

PROBLEM-SOLVING TECHNIQUES   **CONSERVATION OF MOMENTUM**

Note that the solution of these examples involves three steps similar to those we used in examples of energy conservation:

1 First write an expression for the total momentum **P** before the firing of the gun or the collision of the automobiles.

2 Then write an expression for the total momentum **P′** after the firing or the collision.

3 And then use momentum conservation to equate these expressions.

However, in contrast to energy conservation, you must keep in mind that momentum conservation applies to the components of the momentum—*the x, y, and z components of the momentum are conserved separately.* Thus, before writing the expressions for the momentum, you need to select coordinate axes and decide which components of the momentum you want to examine. If the motion is one-dimensional, place one axis along the direction of motion, such as the x axis in the above examples. It then suffices to examine the x component of the momentum. However, sometimes it is necessary to examine two components of the momentum (or, rarely, three); then two (or three) equations result. When writing the components of the momentum, pay attention to the signs; the component is positive if the motion is along the direction of the axis, negative if the motion is opposite to the direction of the axis.

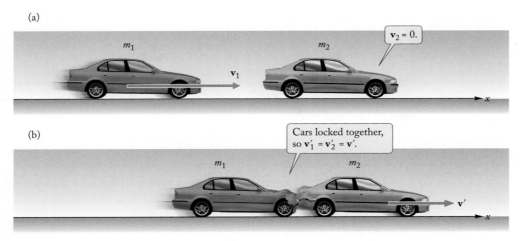

(a)

$v_2 = 0.$

m_1

m_2

v_1

x

(b)

Cars locked together, so $v_1' = v_2' = v'.$

m_1

m_2

v'

x

FIGURE 10.4 (a) Initially, the red automobile has a speed of 24 m/s, and the blue automobile is at rest. (b) After the collision, both automobiles are in motion with velocity v'.

By momentum conservation, the momenta P_x and P_x' before and after the collision must be equal:

$$m_1 v_1 = (m_1 + m_2)v' \tag{10.6}$$

When we solve this for the velocity of the wreck v', we find

$$v' = \frac{m_1 v_1}{m_1 + m_2}$$
$$= \frac{1500\,\text{kg} \times 24\,\text{m/s}}{1500\,\text{kg} + 1500\,\text{kg}} = 12\,\text{m/s} \tag{10.7}$$

The forces acting during the firing of the gun or the collision of the automobiles are quite complicated, but momentum conservation permits us to bypass these complications and directly obtain the answer for the final velocities. Incidentally: It is easy to check that kinetic energy is *not* conserved in these examples. During the firing of the gun, kinetic energy is supplied to the shot and the gun by the explosive combustion of the gunpowder, and during the collision of the automobiles, some kinetic energy is used up to produce changes in the shapes of the automobiles.

The conservation law for momentum depends on the absence of "extra" forces. If the particles are not isolated from the rest of the Universe, then besides the mutual forces exerted by one particle on the other, there are also forces exerted by other bodies not belonging to the particle system. The former forces are called **internal forces** of the system and the latter **external forces.** For instance, for the colliding automobiles of Example 3 the gravity of the Earth, the normal force of the road, and the friction of the road are external forces. In Example 3 we ignored these external forces, because gravity and the normal force cancel each other, and the friction force can be neglected in comparison with the much larger impact force that the automobiles exert on each other. But if the external forces are significant, we must take them into account, and we must modify Eq. (10.5). If the internal force on particle 1 is $\mathbf{F}_{1,\text{int}}$ and the external force is $\mathbf{F}_{1,\text{ext}}$, then the total force on particle 1 is $\mathbf{F}_{1,\text{int}} + \mathbf{F}_{1,\text{ext}}$ and its equation of motion will be

$$\frac{d\mathbf{p}_1}{dt} = \mathbf{F}_{1,\text{int}} + \mathbf{F}_{1,\text{ext}} \tag{10.8}$$

internal forces and external forces

Likewise

$$\frac{d\mathbf{p}_2}{dt} = \mathbf{F}_{2,\text{int}} + \mathbf{F}_{2,\text{ext}} \tag{10.9}$$

If we add the left sides of these equations and the right sides, the contributions from the internal forces cancel (that is, $\mathbf{F}_{1,\text{int}} + \mathbf{F}_{2,\text{int}} = 0$), since they are action–reaction pairs. What remains is

$$\frac{d\mathbf{p}_1}{dt} + \frac{d\mathbf{p}_2}{dt} = \mathbf{F}_{1,\text{ext}} + \mathbf{F}_{2,\text{ext}} \tag{10.10}$$

The sum of the rates of change of the momenta is the same as the rate of change of the sum of the momenta; hence,

$$\frac{d(\mathbf{p}_1 + \mathbf{p}_2)}{dt} = \mathbf{F}_{1,\text{ext}} + \mathbf{F}_{2,\text{ext}} \tag{10.11}$$

The sum $\mathbf{P} = \mathbf{p}_1 + \mathbf{p}_2$ is the total momentum, and the sum $\mathbf{F}_{1,\text{ext}} + \mathbf{F}_{2,\text{ext}}$ is the total external force on the particle system. Thus, Eq. (10.11) states that the rate of change of the total momentum of the two-particle system equals the total *external* force.

For a system containing more than two particles, we can obtain similar results. If the system is isolated so that there are no external forces, then the mutual interparticle forces acting between pairs of particles merely transfer momentum from one particle of the pair to the other, just as in the case of two particles. Since all the internal forces necessarily arise from such forces between pairs of particles, these internal forces cannot change the total momentum. For example, Fig. 10.5 shows three isolated particles exerting forces on one another. Consider particle 1; the mutual forces between particles 1 and 2 exchange momentum between these two, while the mutual forces between particles 1 and 3 exhange momentum between those two. But none of these momentum transfers will change the total momentum. The same holds for particles 2 and 3. Consequently, the total momentum is constant. More generally, for an isolated system of *n* particles, the total momentum $\mathbf{P} = \mathbf{p}_1 + \mathbf{p}_2 + \cdots + \mathbf{p}_n$ obeys the conservation law

$$\mathbf{P} = [\text{constant}] \qquad (\text{no external forces}) \tag{10.12}$$

If, besides the internal forces, there are external forces, then the latter will change the momentum. The rate of change can be calculated in essentially the same way as for the two-particle system, and again, the rate of change of the total momentum is equal to the total external force. We can write this as

$$\frac{d\mathbf{P}}{dt} = \mathbf{F}_{\text{ext}} \tag{10.13}$$

where $\mathbf{F}_{\text{ext}} = \mathbf{F}_{1,\text{ext}} + \mathbf{F}_{2,\text{ext}} + \cdots + \mathbf{F}_{n,\text{ext}}$ is the total external force acting on the system.

Equations (10.12) and (10.13) have exactly the same mathematical form as Eqs. (10.2) and (10.3), and they may be regarded as the generalizations for a system of particles of Newton's First and Second Laws. As we will see in Section 10.3, Eq. (10.13) is an equation of motion for the system of particles—it determines the overall translational motion of the system.

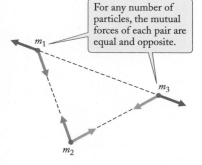

For any number of particles, the mutual forces of each pair are equal and opposite.

FIGURE 10.5 Three particles exerting forces on each other. As in the case of two particles, the mutual forces between pairs of particles merely exchange momentum between them.

momentum conservation for a system of particles

Second Law for a system of particles

Checkup 10.1

QUESTION 1: An automobile and a truck have equal momenta. Which has the larger speed? Which has the larger kinetic energy?

QUESTION 2: An automobile and a truck are traveling along a street in opposite directions. Can they have the same momentum? The same kinetic energy?

QUESTION 3: A rubber ball, dropped on a concrete floor, bounces up with reversed velocity. Is the momentum before the impact the same as after the impact?

QUESTION 4: Is the net momentum of the Sun and all the planets and moons of the Solar System constant? Is the net kinetic energy constant?

QUESTION 5: Consider two automobiles of equal masses m and equal speeds v. (a) If both automobiles are moving southward on a street, what are the total kinetic energy and the total momentum of this system of two automobiles? (b) If one automobile is moving southward and one northward? (c) If one automobile is moving southward and one eastward?

QUESTION 6: An automobile and a truck have equal kinetic energies. Which has the larger speed? Which has the larger momentum? Assume that the truck has the larger mass.

(A) Truck; truck   (B) Truck; automobile

(C) Automobile; truck   (D) Automobile; automobile

10.2 CENTER OF MASS

In our study of kinematics and dynamics in the preceding chapters we always ignored the size of the bodies; even when analyzing the motion of a large body—an automobile or a ship—we pretended that the motion could be treated as particle motion, position being described by means of some reference point marked on the body. In reality, large bodies are systems of particles, and their motion obeys Eq. (10.13) for a system of particles. This equation can be converted into an equation of motion containing just one acceleration rather than the rate of change of momentum of the entire system, by taking as reference point the **center of mass** of the body. The equation that describes the motion of this special point has the same mathematical form as the equation of motion of a particle; that is, the motion of the center of mass mimics particle motion (see, for example, Fig. 10.6).

center of mass

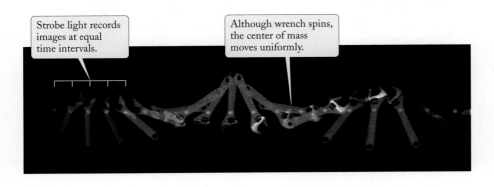

FIGURE 10.6 A wrench moving freely in the absence of external forces. The center of mass, marked with a dot, moves with uniform velocity, along a straight line (you can check this by laying a ruler along the dots).

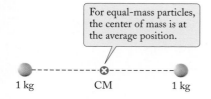

For equal-mass particles, the center of mass is at the average position.

1 kg   CM   1 kg

FIGURE 10.7 Two particles of equal masses, and their center of mass.

The position of the center of mass is merely the <u>average position of the mass of the system</u>. For instance, if the system consists of two particles, each of mass 1 kg, then the center of mass is halfway between them (see Fig. 10.7). In any system consisting of *n* particles of equal masses—such as a piece of pure metal with atoms of only one kind—the *x* coordinate of the center of mass is simply the sum of the *x* coordinates of all the particles divided by the number of particles,

$$x_{CM} = \frac{x_1 + x_2 + \cdots + x_n}{n} \quad \text{(for equal-mass particles)} \quad (10.14)$$

Similar equations apply to the *y* and the *z* coordinates, if the particles of the system are distributed over a three-dimensional region. The three coordinate equations can be expressed concisely in terms of position vectors:

$$\mathbf{r}_{CM} = \frac{\mathbf{r}_1 + \mathbf{r}_2 + \cdots + \mathbf{r}_n}{n} \quad \text{(for equal-mass particles)} \quad (10.15)$$

If the system consists of particles of unequal mass, then the position of the center of mass can be calculated by first subdividing the particles into fragments of equal mass. For instance, if the system consists of two particles, the first of mass 2 kg and the second of 1 kg, then we can pretend that we have *three* particles of equal masses 1 kg, two of which are located at the same position. The coordinate of the center of mass is then

$$x_{CM} = \frac{x_1 + x_1 + x_2}{3}$$

We can also write this in the equivalent form

$$x_{CM} = \frac{m_1 x_1 + m_2 x_2}{m_1 + m_2} \quad (10.16)$$

where $m_1 = 2$ kg and $m_2 = 1$ kg. The formula (10.16) is actually valid for any values of the masses m_1 and m_2. The formula simply asserts that in the average position, the position of particle 1 is included m_1 times and the position of particle 2 is included m_2 times—that is, the number of times each particle is included in the average is directly proportional to its mass.

EXAMPLE 4   A 50-kg woman and an 80-kg man sit on the two ends of a seesaw of length 3.00 m (see Fig. 10.8). Treating them as particles, and ignoring the mass of the seesaw, find the center of mass of this system.

SOLUTION: In Fig. 10.8, the origin of coordinates is at the center of the seesaw; hence the woman has a negative *x* coordinate ($x = -1.50$ m) and the man a positive *x* coordinate ($x = +1.50$ m). According to Eq. (10.16), the coordinate of the center of mass is

$$x_{CM} = \frac{m_1 x_1 + m_2 x_2}{m_1 + m_2} = \frac{50 \, kg \times (-1.50 \, m) + 80 \, kg \times 1.50 \, m}{50 \, kg + 80 \, kg}$$

$$= 0.35 \, m$$

COMMENT: Note that the distance of the woman from the center of mass is 1.50 m + 0.35 m = 1.85 m, and the distance of the man from the center of mass is 1.50 m

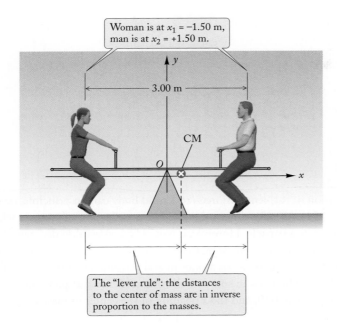

FIGURE 10.8 A woman and a man on a seesaw.

− 0.35 m = 1.15 m. The ratio of these distances is 1.6, which coincides with the inverse of the ratio of the masses, 50/80 = 1/1.6. This "lever rule" is quite general: the position of the center of mass of two particles divides the line segment connecting them in the ratio $m_1:m_2$, with the smaller length segment nearer to the larger mass.

If the system consists of n particles of different masses $m_1, m_2, \ldots, m_n$, then we apply the same prescription: the number of times each particle is included in the average is in direct proportion to its mass; the exact factor by which each particle's coordinate is multiplied is that particle's fraction of the total mass. This gives the following general expression for the coordinate of the center of mass:

$$x_{CM} = \frac{m_1 x_1 + m_2 x_2 + \cdots + m_n x_n}{m_1 + m_2 + \cdots + m_n} \tag{10.17}$$

or

$$x_{CM} = \frac{m_1 x_1 + m_2 x_2 + \cdots + m_n x_n}{M} \tag{10.18}$$

where M is the total mass of the system, $M = m_1 + m_2 + \cdots + m_n$. Similar formulas apply to the y and the z coordinates, if the particles of the system are distributed over a three-dimensional region:

$$y_{CM} = \frac{m_1 y_1 + m_2 y_2 + \cdots + m_n y_n}{M} \tag{10.19}$$

$$z_{CM} = \frac{m_1 z_1 + m_2 z_2 + \cdots + m_n z_n}{M} \tag{10.20}$$

By introducing the standard notation Σ for a summation of n terms, we can express these formulas more concisely as

$$x_{CM} = \frac{1}{M} \sum_{i=1}^{n} m_i x_i \qquad (10.21)$$

$$y_{CM} = \frac{1}{M} \sum_{i=1}^{n} m_i y_i \qquad (10.22)$$

$$z_{CM} = \frac{1}{M} \sum_{i=1}^{n} m_i z_i \qquad (10.23)$$

The position of the center of mass of a solid body can, in principle, be calculated from Eqs. (10.21)–(10.23), since a solid body is a collection of atoms, each of which can be regarded as a particle. However, it would be awkward to deal with the 10^{23} or so atoms that make up a chunk of matter the size of, say, a coin. It is more convenient to pretend that matter in bulk has a smooth and continuous distribution of mass over its entire volume. The mass in some small volume element at position x_i in the body is then Δm_i (see Fig. 10.9), and the x position of the center of mass is

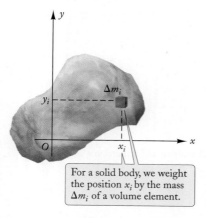

For a solid body, we weight the position x_i by the mass Δm_i of a volume element.

FIGURE 10.9 A small volume element of the body at position x_i has a mass Δm_i.

$$x_{CM} = \frac{1}{M} \sum_{i=1}^{n} x_i \Delta m_i \qquad (10.24)$$

In the limiting case of $\Delta m_i \rightarrow 0$ (and $n \rightarrow \infty$), this sum becomes an integral:

$$x_{CM} = \frac{1}{M} \int x \, dm \qquad (10.25)$$

Similar expressions are valid for the y and z positions of the center of mass:

$$y_{CM} = \frac{1}{M} \int y \, dm \qquad (10.26)$$

$$z_{CM} = \frac{1}{M} \int z \, dm \qquad (10.27)$$

Thus, the position of the center of mass is the average position of all the mass elements making up the body.

For a body of uniform density, the amount of mass dm in any given volume element dV is directly proportional to the amount of volume. *For a uniform-density body, the position of the center of mass is simply the average position of all the volume elements of the body* (in mathematics, this is called the **centroid** of the volume). If the body has a symmetric shape, this average position will often be obvious by inspection. For instance, a sphere of uniform density, or a ring, or a circular plate, or a cylinder, or a parallelepiped will have its center of mass at the geometrical center (see Fig. 10.10). But for a less symmetric body, the center of mass must often be calculated, either by considering parts of the body (as in the next example) or by integrating over the entire body (as in the two subsequent examples).

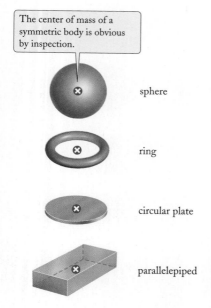

The center of mass of a symmetric body is obvious by inspection.

sphere

ring

circular plate

parallelepiped

FIGURE 10.10 Several bodies for which the center of mass coincides with the geometrical center.

EXAMPLE 5 A meterstick of aluminum is bent at its midpoint so that the two halves are at right angles (see Fig. 10.11). Where is the center of mass of this bent stick?

SOLUTION: We can regard the bent stick as consisting of two straight pieces, each of 0.500 m. The centers of mass of these straight pieces are at their midpoints,

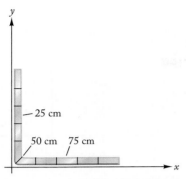

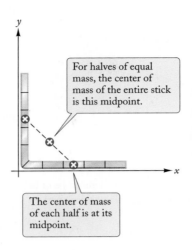

For halves of equal mass, the center of mass of the entire stick is this midpoint.

The center of mass of each half is at its midpoint.

FIGURE 10.11 A meterstick, bent through 90° at its midpoint.

FIGURE 10.12 The center of mass of the bent meterstick is at the midpoint of the line connecting the centers of the halves. The coordinates x_{CM} and y_{CM} of this midpoint are one-half of the distances to the centers of mass of the horizontal and vertical sides—that is, 0.125 m each.

0.250 m from their ends (see Fig. 10.12). The center of mass of the entire stick is the average position of the centers of mass of the two halves. With the coordinate axes arranged as in Fig. 10.12, the x coordinate of the center of mass is, according to Eq. (10.14),

$$x_{CM} = \frac{0.250 \text{ m} + 0}{2} = 0.125 \text{ m} \qquad (10.28)$$

Likewise, the y coordinate is

$$y_{CM} = \frac{0.250 \text{ m} + 0}{2} = 0.125 \text{ m}$$

Note that the center of mass of this bent stick is *outside* the stick; that is, it is not in the volume of the stick (see Fig. 10.12).

EXAMPLE 6   Figure 10.13 shows a mobile by Alexander Calder, which contains a uniform sheet of steel, in the shape of a triangle, suspended at its center of mass. Where is the center of mass of a right triangle of perpendicular sides a and b?

SOLUTION: Figure 10.14 shows the triangle positioned with a vertex at the origin and its right angle at a distance b along the x axis. To calculate the x coordinate of the center of mass, we need to sum mass contributions dm at each value of x; one such contribution is the vertical strip in Fig. 10.14, which has a height $y = (a/b)x$ and a width dx. Since the sheet is uniform, the strip has a fraction of the total mass M equal to the strip's area $y\,dx = (a/b)x\,dx$ divided by the total area $\frac{1}{2}ab$:

$$\frac{dm}{M} = \frac{(a/b)x\,dx}{\frac{1}{2}ab}$$

or

$$dm = M\frac{2x}{b^2}\,dx$$

FIGURE 10.13 This mobile by Alexander Calder contains a triangle suspended above its center of mass.

(a)

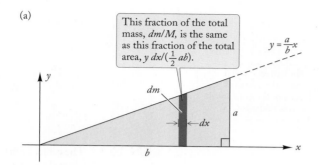

This fraction of the total mass, dm/M, is the same as this fraction of the total area, $y\,dx/(\frac{1}{2}\,ab)$.

$y = \dfrac{a}{b}x$

dm

dx

a

b

(b)

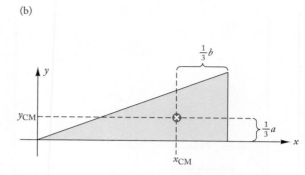

$\frac{1}{3}b$

y_{CM}

$\frac{1}{3}a$

x_{CM}

FIGURE 10.14 (a) A right triangle, with mass element dm of height y and width dx. (b) The center of mass is one-third of the distance from the right angle along sides a and b.

We integrate this in Eq. (10.25) for x_{CM} and sum the contributions from $x = 0$ to $x = b$:

$$x_{\mathrm{CM}} = \frac{1}{M}\int x\,dm = \frac{1}{M}\int_0^b xM\,\frac{2x}{b^2}\,dx$$

$$= \frac{2}{b^2}\int_0^b x^2\,dx = \frac{2}{b^2}\frac{1}{3}x^3\Big|_0^b$$

$$= \frac{2}{b^2}\frac{1}{3}(b^3 - 0) = \frac{2}{3}b$$

So the center of mass is two-thirds of the distance toward the right angle. Performing a similar calculation for y_{CM} yields $y_{\mathrm{CM}} = \frac{1}{3}a$. Thus each of x_{CM} and y_{CM} is a distance *away* from the right angle equal to one-third of the length of the corresponding side (see Fig. 10.14b).

EXAMPLE 7   The Great Pyramid at Giza (see Fig. 10.15) has a height of 147 m and a square base. Assuming that the entire volume is completely filled with stone of uniform density, find its center of mass.

SOLUTION: Because of symmetry, the center of mass must be on the vertical line through the apex. For convenience, we place the y axis along this line, and we arrange this axis downward, with origin at the apex. We must then find where the center of mass is on this y axis. Figure 10.16a shows a cross section through the pyramid, looking parallel to two sides. The half-angle at the apex is ϕ. By examination of the colored triangle, we see that at a height y (measured from the apex) the half-width is $x = y\tan\phi$ and the full width is $2x = 2y\tan\phi$. A horizontal slice through the pyramid at this height is a square measuring $2x \times 2x$ (see Fig. 10.16b). The volume of a horizontal slab of thickness dy at this height y is therefore $dV = (2x)^2\,dy = (2y\tan\phi)^2\,dy$. If we represent the uniform density of the stone by ρ (the Greek letter *rho*), the proportionality between mass and volume can be written

$$dm = \rho\,dV$$

FIGURE 10.15 The Great Pyramid.

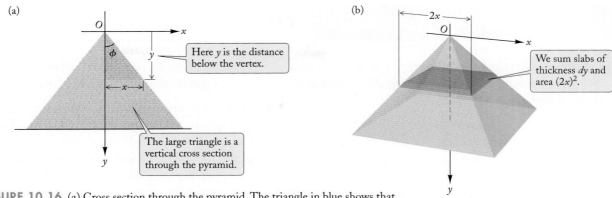

FIGURE 10.16 (a) Cross section through the pyramid. The triangle in blue shows that at a height y measured from the apex, the half-width of the pyramid is $x = y \tan \phi$. (b) The thin horizontal slab indicated in red is a square measuring $2x \times 2x$ with a thickness dy.

Thus the mass of the slab of thickness dy at this height y is

$$dm = \rho \, dV = \rho (2y \tan \phi)^2 \, dy = 4\rho (\tan^2 \phi) y^2 \, dy$$

Equation (10.26) then gives us the y coordinate of the center of mass:

$$y_{CM} = \frac{1}{M} \int y \, dm = \frac{1}{M} \int 4\rho (\tan^2 \phi) y^3 \, dy \tag{10.29}$$

The total mass is

$$M = \int dm = \int 4\rho (\tan^2 \phi) y^2 \, dy \tag{10.30}$$

When we substitute Eq. (10.30) into Eq. (10.29), the common factor $4\rho \tan^2 \phi$ cancels, leaving

$$y_{CM} = \frac{\displaystyle\int y^3 \, dy}{\displaystyle\int y^2 \, dy} \tag{10.31}$$

As we sum the square slabs of thickness dy in both of these integrals, the integration runs from $y = 0$ at the top of the pyramid to $y = h$ at the bottom, where h is the height of the pyramid. Evaluation of these integrals yields

$$\int_0^h y^3 \, dy = \frac{y^4}{4} \bigg|_0^h = \frac{h^4}{4}$$

$$\int_0^h y^2 \, dy = \frac{y^3}{3} \bigg|_0^h = \frac{h^3}{3}$$

The y coordinate of the center of mass is therefore

$$y_{CM} = \frac{h^4/4}{h^3/3} = \frac{3}{4} h$$

This means that the center of mass is $3/4 \times 147$ m below the apex; that is, it is $1/4 \times 147$ m $= 37$ m above the ground.

PROBLEM-SOLVING TECHNIQUES   CENTER OF MASS

Calculations of the position of the center of mass of a body can often be simplified by exploiting the shape or the symmetry of the body.

• Sometimes it is profitable to treat the body as consisting of several parts and to begin by calculating the positions of the centers of mass of these parts (as in the example of the bent meterstick). Each part can then be treated as a particle located at its center of mass, and the center of mass of the entire body is then the center of mass of this system of particles, which can be calculated by the sums, Eqs. (10.18)–(10.20).

• If the body or some part of it has symmetry, the position of the center of mass will often be obvious by inspection.

For instance, in the example of the bent meterstick, it is obvious that the center of mass of each half is at its center.

• Geometrical arguments can sometimes replace algebraic calculations of the coordinates of the center of mass. For instance, in the example of the bent meterstick, instead of the algebraic calculations of the coordinates [such as for x_{CM} in Eq. (10.28)], the coordinates can be obtained by regarding the stick as consisting of two straight pieces with known centers of mass; then the coordinates of the overall center of mass can be found from the geometry of a diagram, such as Fig. 10.12.

PHYSICS IN PRACTICE   CENTER OF MASS AND STABILITY

In the design of ships, engineers need to ensure that the position of the center of mass is low in the ship, to enhance the stability. If the center of mass is high, the ship is top-heavy and liable to tip over. Ships often carry ballast at the bottom of the hull to lower the center of mass. Many ships have been lost because of insufficient ballast or because of an unexpected shifting of the ballast. For instance, in 1628, the Swedish ship *Vasa* (see Fig. 1), the pride and joy of the Swedish navy and King Gustavus II Adolphus, capsized and sank on its maiden

voyage when struck by a gust of wind, just barely out of harbor. It carried an excessive number of heavy guns on its upper decks, which made it top-heavy; and it should have carried more ballast to lower its center of mass.

The position of the center of mass is also crucial in the design of automobiles. A top-heavy automobile, such as an SUV, will tend to roll over when speeding around a sharp curve. High-performance automobiles, such as the Lamborghini shown in Fig. 2, have a very low profile, with the engine and transmission slung low in the body, so the center of mass is as low as possible and the automobile hugs the ground.

FIG. 1 The Swedish ship *Vasa*.

FIG. 2 A Lamborghini Gallardo sports car.

The position of the center of mass enters into the calculation of the gravitational potential energy of an extended body located near the surface of the Earth. According to Eq. (7.29), the potential energy of a single particle of mass m at a height y above the ground is mgy. For a system of particles, the total gravitational potential energy is then

$$U = m_1 g y_1 + m_2 g y_2 + \cdots + m_n g y_n$$
$$= (m_1 y_1 + m_2 y_2 + \cdots + m_n y_n)g \qquad (10.32)$$

Comparison with Eq. (10.19) shows that the quantity in parentheses is My_{CM}. Hence, Eq. (10.32) becomes

$$U = Mgy_{CM} \qquad (10.33)$$

potential energy in terms of height of center of mass

This expression for the gravitational potential energy of a system near the Earth's surface has the same mathematical form as for a single particle—it is as though the entire mass of the system were located at the center of mass.

For a human body standing upright, the position of the center of mass is in the middle of the trunk, at about the height of the navel. This is therefore the height to be used in the calculation of the gravitational potential energy of the body. However, if the body adopts any bent position, the center of mass shifts.

Concepts
— in —
Context

(a)

(b)

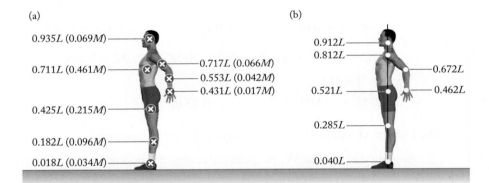

FIGURE 10.17 (a) Centers of mass of the body segments of an average male of mass M and height L standing upright. The numbers give the heights of the centers of mass of the body segments from the floor and (in parentheses) the masses of the body segments; right and left limbs are shown combined. (b) Hinge points of the body. The numbers give the heights of the joints from the floor.

Figure 10.17a gives the centers of mass of the body segments of a man of average proportions standing upright. Figure 10.17b shows the hinge points at which these body segments are joined. From the data in this figure, we can calculate the location of the center of mass when the body adopts any other position, and we can calculate the work done against gravity to change the position of any segment. For instance, if the body is bent in a tight backward arc, the center of mass shifts to a location just outside the body, about 10 cm below the middle of the trunk. Olympic jumpers (see Fig. 10.18) take advantage of this shift of the center of mass to make the most of the gravitational potential energy they can supply for a high jump. By adopting a bent position as they pass over the bar, they raise their trunk above the center of mass, so the trunk passes over the bar while the center of mass can pass *below* the bar. By this trick, the jumper raises the center of her trunk by about 10 cm relative to the center of mass, and she gains extra height without expending extra energy.

FIGURE 10.18 High jumper passing over the bar.

Concepts
— in —
Context

EXAMPLE 8

Suppose a man of average proportions performs a high jump, while arching his back (see the chapter opening photo). At the peak of his jump, his torso is approximately horizontal; his thighs, arms, and head make an angle of 45° with the horizontal; and his lower legs are vertical, as shown in Fig. 10.19b. (a) How much is his center of mass shifted downward compared with a man who goes over the pole horizontally (Fig. 10.19a)? (b) How much is his potential energy reduced? Assume the mass of the jumper is $M = 73$ kg and his height $L = 1.75$ m.

SOLUTION: (a) In Fig. 10.19a, the center of mass of the horizontal body is at $y = 0$, since each segment is essentially at $y = 0$. In Fig. 10.20, we have used the relative locations of the hinge points and centers of mass from Fig. 10.17 to determine the vertical position of each body segment in the arched-back position. For example, the center of mass of the thigh is at a distance $0.521L - 0.425L = 0.096L$ from the hip joint, and so is at a vertical distance $0.096L \times \sin 45° = 0.068L$ below $y = 0$. Similarly, we can determine that the centers of mass of the lower legs, the feet, the head, the upper arms, the forearms, and the hands are at $y = -0.270L$, $-0.434L$, $-0.016L$, $-0.067L$, $-0.183L$, and $-0.269L$, respectively. From Fig. 10.17, the masses of all seven segments are $0.215M$, $0.096M$, $0.034M$, $0.069M$, $0.066M$, $0.042M$, and $0.017M$, respectively. The torso, of mass $0.461M$, is again at $y = 0$. Thus, using Eq. (10.19) or (10.22), the arched-back center of mass is at

$$y_{CM} = \frac{1}{M} \sum_{i=1}^{n} m_i y_i$$

$$= -\frac{1}{M}(0.215 \times 0.068 + 0.096 \times 0.270 + 0.034 \times 0.434 + 0.069$$

$$\times 0.016 + 0.066 \times 0.067 + 0.042 \times 0.183 + 0.017 \times 0.269$$

$$+ 0 \times 0.461)ML$$

$$= -0.073L = -0.073 \times 1.75 \text{ m} = -0.13 \text{ m}$$

Thus a height advantage of 13 cm is gained in this arched position.

(b) According to Eq. (10.33), the potential energy is changed by

$$\Delta U = Mg\,\Delta y_{CM}$$

$$= 73 \text{ kg} \times 9.81 \text{ m/s}^2 \times (-0.13 \text{ m}) \tag{10.34}$$

$$= -93 \text{ J}$$

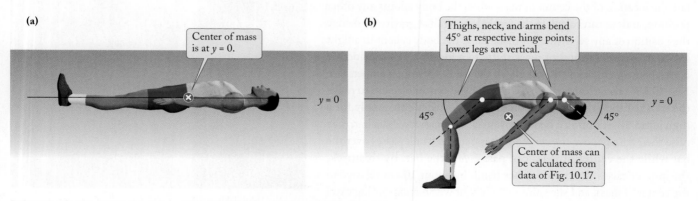

(a)

Center of mass is at $y = 0$.

$y = 0$

(b)

Thighs, neck, and arms bend 45° at respective hinge points; lower legs are vertical.

$y = 0$

45°   45°

Center of mass can be calculated from data of Fig. 10.17.

FIGURE 10.19 (a) Horizontal position. (b) High jumper in arched-back position.

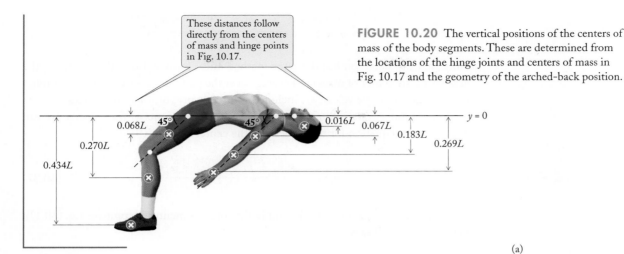

These distances follow directly from the centers of mass and hinge points in Fig. 10.17.

FIGURE 10.20 The vertical positions of the centers of mass of the body segments. These are determined from the locations of the hinge joints and centers of mass in Fig. 10.17 and the geometry of the arched-back position.

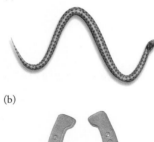

(a)

(b)

FIGURE 10.21
(a) A snake. (b) A horseshoe.

✔ Checkup 10.2

QUESTION 1: Roughly where is the center of mass of the snake shown in Fig. 10.21a?

QUESTION 2: Roughly where is the center of mass of the horseshoe shown in Fig. 10.21b?

QUESTION 3: Is it possible for the center of mass of a body to be above the highest part of the body?

QUESTION 4: A sailboat has a keel with a heavy lead bulb at the bottom. If the bulb falls off, the center of mass of the sailboat:

(A) Remains at the same position   (B) Shifts downward
(C) Shifts upward

10.3 THE MOTION OF THE CENTER OF MASS

When the particles in a system move, often so does the center of mass. We will now obtain an equation for the motion of the center of mass, an equation which relates the acceleration of the center of mass to the external force. This equation will permit us to calculate the overall translational motion of a system of particles.

According to Eq. (10.18), if the x components of positions of the respective particles change by $dx_1, dx_2, \ldots, dx_n$, then the position of the center of mass changes by

$$dx_{CM} = \frac{1}{M}(m_1 dx_1 + m_2 dx_2 + \cdots + m_n dx_n) \tag{10.35}$$

Dividing this by the time dt taken for these changes of position, we obtain

$$\frac{dx_{CM}}{dt} = \frac{1}{M}\left(m_1 \frac{dx_1}{dt} + m_2 \frac{dx_2}{dt} + \cdots + m_n \frac{dx_n}{dt}\right) \tag{10.36}$$

The left side of this equation is the x component of the velocity of the center of mass, and the rates of change on the right side are the x components of the velocities of the individual particles; thus

$$v_{x,\text{CM}} = \frac{m_1 v_{x,1} + m_2 v_{x,2} + \cdots + m_n v_{x,n}}{M}$$

Note that this equation has the same mathematical form as Eq. (10.18); that is, the velocity of the center of mass is an average over the particle velocities, and the number of times each particle velocity is included is directly proportional to its mass.

Since similar equations apply to the y and z components of the velocity, we can write a vector equation for the velocity of the center of mass:

velocity of the center of mass

$$\mathbf{v}_{\text{CM}} = \frac{m_1 \mathbf{v}_1 + m_2 \mathbf{v}_2 + \cdots + m_n \mathbf{v}_n}{M} \tag{10.37}$$

The quantity in the numerator is simply the total momentum [compare Eq. (10.1)]; hence Eq. (10.37) says

$$\mathbf{v}_{\text{CM}} = \frac{\mathbf{P}}{M} \tag{10.38}$$

or

momentum in terms of velocity of CM

$$\mathbf{P} = M\mathbf{v}_{\text{CM}} \tag{10.39}$$

This equation expresses the total momentum of a system of particles as the product of the total mass and the velocity of the center of mass. Obviously, this equation is analogous to the familiar equation $\mathbf{p} = m\mathbf{v}$ for the momentum of a single particle.

We know, from Eq. (10.13), that the rate of change of the total momentum equals the net external force on the system,

$$\frac{d\mathbf{P}}{dt} = \mathbf{F}_{\text{ext}}$$

If we substitute $\mathbf{P} = M\mathbf{v}_{\text{CM}}$ and take into account that the mass is constant, we find

$$\frac{d\mathbf{P}}{dt} = \frac{d}{dt}(M\mathbf{v}_{\text{CM}}) = M\frac{d\mathbf{v}_{\text{CM}}}{dt} = M\mathbf{a}_{\text{CM}}$$

and consequently

motion of center of mass

$$M\mathbf{a}_{\text{CM}} = \mathbf{F}_{\text{ext}} \tag{10.40}$$

This equation for a system of particles is the analog of Newton's equation for motion for a single particle. The equation asserts that the center of mass moves as though it were a particle of mass M under the influence of a force $\mathbf{F}_{\text{ext}}$.

Concepts
— in —
Context

This result justifies some of the approximations we made in previous chapters. For instance, in Example 9 of Chapter 2 we treated a diver falling from a cliff as a particle. Equation (10.40) shows that this treatment is legitimate: the center of mass of the diver, under the influence of the external force (gravity), moves with a downward acceleration g, just as though it were a freely falling particle. Likewise, after a high jumper leaves the ground, his center of mass moves along a parabolic trajectory, as though it were a projectile, and the shape and height of this parabolic trajectory is unaffected by any contortions the high jumper might perform while in flight. From Chapter 4, we know that the initial vertical velocity v_y determines the maximum height h of the center of mass; that is, $v_y = \sqrt{2gh}$. The contortions of the jumper enable his body to pass over a bar roughly 10 cm above the maximum height of the center of mass.

If the net external force vanishes, then the acceleration of the center of mass vanishes; hence the center of mass remains at rest or it moves with uniform velocity.

EXAMPLE 9   During a "space walk," an astronaut floats in space 8.0 m from his spacecraft orbiting the Earth. He is tethered to the spacecraft by a long umbilical cord (see Fig. 10.22); to return, he pulls himself in by this cord. How far does the spacecraft move toward him? The mass of the spacecraft is 3500 kg, and the mass of the astronaut, including his space suit, is 110 kg.

SOLUTION: In the reference frame of the orbiting (freely falling) astronaut and spacecraft, each is effectively weightless; that is, the external force on the system is effectively zero. The only forces in the system are the forces exerted when the astronaut pulls on the cord; these forces are internal. The forces exerted by the cord on the spacecraft and on the astronaut during the pulling in are of equal magnitudes and opposite directions; the astronaut is pulled toward the spacecraft, and the spacecraft is pulled toward the astronaut. In the absence of external forces, the center of mass of the astronaut–spacecraft system remains at rest. Thus, the spacecraft and the astronaut both move toward the center of mass, and there they meet.

FIGURE 10.22 Astronaut on a "space walk" during the *Gemini 4* mission.

With the x axis as in Fig. 10.23, the x coordinate of the center of mass is

$$x_{CM} = \frac{m_1 x_1 + m_2 x_2}{m_1 + m_2} \tag{10.41}$$

where $m_1 = 3500$ kg is the mass of the spacecraft and $m_2 = 110$ kg is the mass of the astronaut. Strictly, the coordinates x_1 and x_2 of the spacecraft and of the astronaut should correspond to the centers of mass of these bodies, but, for the sake of simplicity, we neglect their size and treat both as particles. The initial values of the coordinates are $x_1 = 0$ and $x_2 = 8.0$ m; hence

$$x_{CM} = \frac{0 + 110 \text{ kg} \times 8.0 \text{ m}}{3500 \text{ kg} + 110 \text{ kg}} = 0.24 \text{ m}$$

During the pulling in, the spacecraft will move from $x_1 = 0$ to $x_1 = 0.24$ m; simultaneously, the astronaut will move from $x_2 = 8.0$ m to $x_2 = 0.24$ m.

(a)

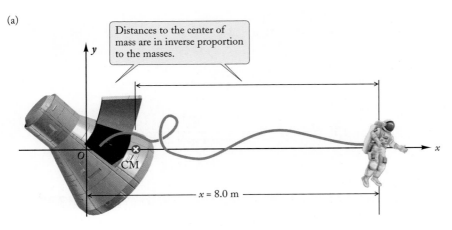

(b)

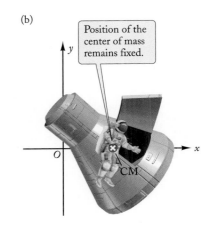

FIGURE 10.23 (a) Initial position of the astronaut and the spacecraft. The center of mass is between them. (b) Final position of the astronaut and the spacecraft. They are both at the center of mass.

COMMENT: The distances moved by the astronaut and by the spacecraft are in the inverse ratio of their masses. The astronaut (of small mass) moves a large distance, and the spacecraft (of large mass) moves a smaller distance. This is the result of the accelerations that the pull of the cord gives to these bodies: with forces of equal magnitudes, the accelerations of the astronaut and spacecraft are in the inverse ratio of their masses. However, our method of calculation based on the fixed position of the center of mass gives us the final positions directly, without any need to examine accelerations.

EXAMPLE 10   A projectile is launched at some angle θ with respect to the horizontal, $0° < \theta < 90°$. Just as it reaches its peak, it explodes into two pieces. The explosion causes a first, rear piece to come to a momentary stop, and it simply drops, striking the ground directly below the peak position. The explosion also causes the speed of the second piece to increase, and it hits the ground a distance five times further from the launch point than the first piece (see Fig. 10.24). If the original projectile had a mass of 12.0 kg, what are the masses of the pieces?

SOLUTION: Because the explosion does not produce external forces, the center of mass continues on its original path, a parabolic trajectory which strikes the ground at the range x_{max}, given by Eq. (4.43). The peak of the parabolic trajectory occurs at half this distance; thus the first piece, of some mass m_1, hits the ground a distance $\frac{1}{2}x_{max}$ from the launch point. We are also told that the second piece, of mass m_2, hits the ground a distance $5 \times \frac{1}{2}x_{max}$ from the launch point. The two pieces will reach the ground at the same instant, since this explosion affected only each piece's horizontal momentum. If we take our origin at the launch point, the x component of the center of mass is thus

$$x_{CM} = x_{max} = \frac{m_1 x_1 + m_2 x_2}{m_1 + m_2} = \frac{m_1 x_{max}/2 + 5 m_2 x_{max}/2}{m_1 + m_2}$$

We can divide both sides of this equation by x_{max} and rearrange to obtain

$$m_1 = 3m_2$$

Since we know the total mass is $m_1 + m_2 = 12.0$ kg, or $4m_2 = 12.0$ kg, we obtain

$$m_1 = 9.0 \text{ kg} \qquad \text{and} \qquad m_2 = 3.0 \text{ kg}$$

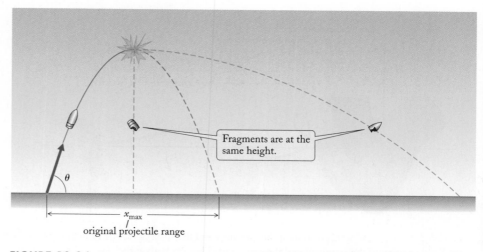

FIGURE 10.24 A projectile explodes at its apex. The rear fragment simply drops, and the forward piece lands five times further from the launch point.

COMMENT: Note that to relate both points of impact to the center of mass, we had to know that the impacts occurred at the *same instant*; we must always use the coordinates of a system of particles at a particular instant when calculating the center of mass.

 Checkup 10.3

QUESTION 1: When you crawl from the rear end of a canoe to the front end, the boat moves backward relative to the water. Explain.

QUESTION 2: You are locked inside a boxcar placed on frictionless wheels on railroad tracks. If you walk from the rear end of the boxcar to the front end, the boxcar rolls backward. Is it possible for you to make the boxcar roll a distance longer than its length?

QUESTION 3: You drop a handful of marbles on a smooth floor, and they bang into each other and roll away in all directions. What can you say about the motion of the center of mass of the marbles after the impact on the floor?

QUESTION 4: An automobile is traveling north at 25 m/s. A truck with twice the mass of the automobile is heading south at 20 m/s. What is the velocity of the center of mass of the two vehicles?

(A) 0   (B) 5 m/s south   (C) 5 m/s north
(D) 10 m/s south   (E) 10 m/s north

10.4 ENERGY OF A SYSTEM OF PARTICLES

The total kinetic energy of a system of particles is simply the sum of the individual kinetic energies of all the particles,

$$K = \tfrac{1}{2} m_1 v_1^2 + \tfrac{1}{2} m_2 v_2^2 + \cdots + \tfrac{1}{2} m_n v_n^2 \tag{10.42}$$

kinetic energy of a system of particles

Since Eq. (10.39) for the momentum of a system of particles resembles the expression for the momentum of a single particle, we might be tempted to guess that the equation for the kinetic energy for a system of particles also can be expressed in the form of the translational kinetic energy of the center of mass $\tfrac{1}{2} M v_{CM}^2$, resembling the kinetic energy of a single particle. But this is wrong! The total kinetic energy of a system of particles is usually larger than $\tfrac{1}{2} M v_{CM}^2$. We can see this in the following simple example: Consider two automobiles of equal masses moving toward each other at equal speeds. The velocity of the center of mass is then zero, and consequently $\tfrac{1}{2} M v_{CM}^2 = 0$. However, since each automobile has a positive kinetic energy, the total kinetic energy is *not* zero.

If the internal and external forces acting on a system of particles are conservative, then the system will have a potential energy. We saw above that for the specific example of the gravitational potential energy near the Earth's surface, the potential energy of the system took the same form as for a single particle, $U = Mgy_{CM}$ [see Eq. (10.33)]. But this form is a result of the particular force (uniform and proportional to mass); in general, the potential energy for a system does not have the same form as for a single particle. Unless we specify all of the forces, we cannot write down an explicit formula for the potential energy; but in any case, this potential energy will be some function of the positions of all the particles. The total mechanical energy is the sum of the total kinetic energy [Eq. (10.42)] and the total potential energy. This total energy will be conserved during the motion of the system of particles. Note that in reckoning the total potential energy of the system, we must include the potential energy of both the external forces and the internal forces. We know that the internal forces do not contribute to the changes of total momentum of the system, but these internal forces, and their potential energies, contribute to the total energy. For instance, if two particles are falling toward each other under the influence of their mutual gravitational attraction, the momentum gained by one particle is balanced by momentum lost by the other, but the kinetic energy gained by one particle is *not* balanced by kinetic energy lost by the other—both particles gain kinetic energy. In this example the gravitational attraction plays the role of an internal force in the system, and the gain of kinetic energy is due to a loss of mutual gravitational potential energy.

 Checkup 10.4

QUESTION 1: Consider a system consisting of two automobiles of equal mass. Initially, the automobiles have velocities of equal magnitudes in opposite directions. Suppose the automobiles collide head-on. Is the kinetic energy conserved?

QUESTION 2: The Solar System consists of the Sun, nine planets, and their moons. Is the total energy of this system conserved? Is the kinetic energy conserved? Is the potential energy conserved?

QUESTION 3: Two equal masses on a frictionless horizontal surface are connected by a spring. Each is given a brief push in a different direction. During the subsequent motion, which of the following remain(s) constant? (P = total momentum; K = total kinetic energy; U = total potential energy.)

(A) P only   (B) P and K (C) P and U
(D) K and U (E) P, K, and U

SUMMARY

PROBLEM-SOLVING TECHNIQUES Conservation of Momentum	(page 310)
PROBLEM-SOLVING TECHNIQUES Center of Mass	(page 320)
PHYSICS IN PRACTICE Center of Mass and Stability	(page 320)
MOMENTUM OF A PARTICLE	$\mathbf{p} = m\mathbf{v}$	(10.1)
MOMENTUM OF A SYSTEM OF PARTICLES	$\mathbf{P} = \mathbf{p}_1 + \mathbf{p}_2 + \cdots + \mathbf{p}_n$	(10.4)
RATE OF CHANGE OF MOMENTUM	$\dfrac{d\mathbf{P}}{dt} = \mathbf{F}_{\text{ext}}$	(10.13)
CONSERVATION OF MOMENTUM (in the absence of external forces)	$\mathbf{P} = [\text{constant}]$	(10.12)

CENTER OF MASS
(Using $M = m_1 + m_2 + \cdots + m_n$)

$$x_{CM} = \frac{1}{M}(m_1 x_1 + m_2 x_2 + \cdots + m_n x_n) \qquad (10.18)$$

$$y_{CM} = \frac{1}{M}(m_1 y_1 + m_2 y_2 + \cdots + m_n y_n) \qquad (10.19)$$

$$z_{CM} = \frac{1}{M}(m_1 z_1 + m_2 z_2 + \cdots + m_n z_n) \qquad (10.20)$$

CENTER OF MASS OF CONTINUOUS DISTRIBUTION OF MASS

where $dm = \rho\, dV$ (ρ is density and dV is a volume element).

sphere   circular plate

ring   parallelepiped

$$x_{CM} = \frac{1}{M}\int x\, dm \qquad (10.25)$$

$$y_{CM} = \frac{1}{M}\int y\, dm \qquad (10.26)$$

$$z_{CM} = \frac{1}{M}\int z\, dm \qquad (10.27)$$

VELOCITY OF THE CENTER OF MASS

$$\mathbf{v}_{CM} = \frac{m_1 \mathbf{v}_1 + m_2 \mathbf{v}_2 + \cdots + m_n \mathbf{v}_n}{M} \qquad (10.37)$$

MOMENTUM OF A SYSTEM OF PARTICLES

$$\mathbf{P} = M\mathbf{v}_{CM} \qquad (10.39)$$

MOTION OF THE CENTER OF MASS

$$M\mathbf{a}_{CM} = \mathbf{F}_{ext} \qquad (10.40)$$

GRAVITATIONAL POTENTIAL ENERGY OF A SYSTEM OF PARTICLES (near the Earth's surface)

$$U = Mgy_{CM} \qquad (10.33)$$

KINETIC ENERGY OF A SYSTEM OF PARTICLES

$$K = \tfrac{1}{2}m_1 v_1^2 + \tfrac{1}{2}m_2 v_2^2 + \cdots + \tfrac{1}{2}m_n v_n^2 \qquad (10.42)$$

QUESTIONS FOR DISCUSSION

1. When the nozzle of a fire hose discharges a large amount of water at high speed, several strong firefighters are needed to hold the nozzle steady. Explain.

2. When firing a shotgun, a hunter always presses it tightly against his shoulder. Why?

3. As described in Example 2, guns onboard eighteenth-century warships were often mounted on carriages (see Fig. 10.3). What was the advantage of this arrangement?

4. Hollywood movies often show a man being knocked over by the impact of a bullet while the man who shot the bullet remains standing, quite undisturbed. Is this reasonable?

5. Where is the center of mass of this book when it is closed? Mark the center of mass with a cross.

6. Roughly, where is the center of mass of this book when it is open, as it is at this moment?

7. A fountain shoots a stream of water up into the air (Fig. 10.25). Roughly, where is the center of mass of the water that is in the air at one instant? Is the center of mass higher or lower than the middle height?

FIGURE 10.25 Stream of water from a fountain.

8. Consider the moving wrench shown in Fig. 10.6. If the center of mass on this wrench had not been marked, how could you have found it by inspection of this photograph?

9. Is it possible to propel a sailboat by mounting a fan on the deck and blowing air on the sail? Is it better to mount the fan on the stern and blow air toward the rear?

10. Cyrano de Bergerac's sixth method for propelling himself to the Moon was as follows: "Seated on an iron plate, to hurl a magnet in the air—the iron follows—I catch the magnet—throw again—and so proceed indefinitely." What is wrong with this method (other than the magnet's insufficient pull)?

11. Within the Mexican jumping bean, a small insect larva jumps up and down. How does this lift the bean off the table?

12. Answer the following question, sent by a reader to the *New York Times:*

A state trooper pulls a truck driver into the weigh station to see if he's overloaded. As the vehicle rolls onto the scales, the driver jumps out and starts beating on the truck box with a club. A bystander asks what he's doing. The trucker says: "I've got five tons of canaries in here. I know I'm overloaded. But if I can keep them flying I'll be OK." If the canaries are flying in that enclosed box, will the truck really weigh any less than if they're on the perch?

13. An elephant jumps off a cliff. Does the Earth move upward while the elephant falls?

14. A juggler stands on a balance, juggling five balls (Fig. 10.26). On the average, will the balance register the weight of the juggler plus the weight of the five balls? More than that? Less?

FIGURE 10.26 Juggler on a balance.

15. Suppose you fill a rubber balloon with air and then release it so that the air spurts out of the nozzle. The balloon will fly across the room. Explain.

16. The combustion chamber of a rocket engine is closed at the front and at the sides, but it is open at the rear (Fig. 10.27). Explain how the pressure of the gas on the walls of this combustion chamber gives a net forward force that propels the rocket.

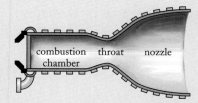

combustion   throat   nozzle
chamber

FIGURE 10.27 Combustion chamber of a rocket engine.

PROBLEMS

10.1 Momentum

1. What is the momentum of a rifle bullet of mass 15 g and speed 600 m/s? An arrow of mass 40 g and speed 80 m/s?

2. What is the momentum of an automobile of mass 900 kg moving at 65 km/h? If a truck of mass 7200 kg is to have the same momentum as the automobile, what must be its speed?

3. Using the entries listed in Tables 1.7 and 2.1, find the magnitude of the momentum for each of the following: Earth moving around the Sun, jet airliner at maximum airspeed, automobile at 55 mi/h, man walking, electron moving around a nucleus.

4. The push that a bullet exerts during impact on a target depends on the momentum of the bullet. A Remington .244

rifle, used for hunting deer, fires a bullet of 90 grains (1 grain is $\frac{1}{7000}$ lb) with a speed of 975 m/s. A Remington .35 rifle fires a bullet of 200 grains with a speed of 674 m/s. What is the momentum of each bullet?

5. An electron, of mass 9.1×10^{-31} kg, is moving in the $x-y$ plane; its speed is 2.0×10^5 m/s, and its direction of motion makes an angle of $25°$ with the x axis. What are the components of the momentum of the electron?

6. A skydiver of mass 75 kg is in free fall. What is the rate of change of his momentum? Ignore friction.

7. A soccer player kicks a ball and sends it flying with an initial speed of 26 m/s at an upward angle of $30°$. The mass of the ball is 0.43 kg. Ignore friction.

    (a) What is the initial momentum of the ball?

    (b) What is the momentum when the ball reaches maximum height on its trajectory?

    (c) What is the momentum when the ball returns to the ground? Is this final momentum the same as the initial momentum?

8. The Earth moves around the Sun in a circle of radius 1.5×10^{11} m at a speed of 3.0×10^4 m/s. The mass of the Earth is 6.0×10^{24} kg. Calculate the magnitude of the rate of change of the momentum of the Earth from these data. (Hint: The magnitude of the momentum does not change, but the direction does.)

9. A 1.0-kg mass is released from rest and falls freely. How much momentum does it acquire after one second? After ten seconds?

10. A 55-kg woman in a 20-kg rowboat throws a 3.0-kg life preserver with a horizontal velocity of 5.0 m/s. What is the recoil velocity of the woman and rowboat?

11. A 90-kg man dives from a 20-kg boat with an initial horizontal velocity of 2.0 m/s (relative to the water). What is the initial recoil velocity of the boat? (Neglect water friction.)

12. A hydrogen atom (mass 1.67×10^{-27} kg) at rest can emit a photon (a particle of light) with maximum momentum 7.25×10^{-27} kg·m/s. What is the maximum recoil velocity of the hydrogen atom?

13. Calculate the change of the kinetic energy in the collision between the two automobiles described in Example 3.

14. A rifle of 10 kg lying on a smooth table discharges accidentally and fires a bullet of mass 15 g with a muzzle speed of 650 m/s. What is the recoil velocity of the rifle? What is the kinetic energy of the bullet, and what is the recoil kinetic energy of the rifle?

15. A typical warship built around 1800 (such as the USS *Constitution*) carried 15 long guns on each side. The guns fired a shot of 11 kg with a muzzle speed of about 490 m/s. The mass of the ship was about 4000 metric tons. Suppose that all of the 15 guns on one side of the ship are fired (almost) simultaneously in a horizontal direction at right angle to the ship. What is the recoil velocity of the ship? Ignore the resistance offered by the water.

16. Two automobiles, moving at 65 km/h in opposite directions, collide head-on. One automobile has a mass of 700 kg; the other, a mass of 1500 kg. After the collision, both remain joined together. What is the velocity of the wreck? What is the change of the velocity of each automobile during the collision?

17. The nucleus of an atom of radium (mass 3.77×10^{-25} kg) suddenly ejects an alpha particle (mass 6.68×10^{-27} kg) of an energy of 7.26×10^{-16} J. What is the velocity of the recoil of the nucleus? What is the kinetic energy of the recoil?

18. A lion of mass 120 kg leaps at a hunter with a horizontal velocity of 12 m/s. The hunter has an automatic rifle firing bullets of mass 15 g with a muzzle speed of 630 m/s, and he attempts to stop the lion in midair. How many bullets would the hunter have to fire into the lion to stop its horizontal motion? Assume the bullets stick inside the lion.

*19. Find the recoil velocity for the gun described in Example 2 if the gun is fired with an elevation angle of $20°$.

*20. Consider the collision between the moving and the initially stationary automobiles described in Example 3. In this example we neglected effects of the friction force exerted by the road during the collision. Suppose that the collision lasts for 0.020 s, and suppose that during this time interval the joined automobiles are sliding with locked wheels on the pavement with a coefficient of friction $\mu_k = 0.90$. What change of momentum and what change of speed does the friction force produce in the joined automobiles in the interval of 0.020 s? Is this change of speed significant?

*21. A Maxim machine gun fires 450 bullets per minute. Each bullet has a mass of 14 g and a velocity of 630 m/s.

    (a) What is the average force that the impact of these bullets exerts on a target? Assume that the bullets penetrate the target and remain embedded in it.

    (b) What is the average rate at which the bullets deliver their energy to the target?

*22. An owl flies parallel to the ground and grabs a stationary mouse with its talons. The mass of the owl is 250 g, and that of the mouse is 50 g. If the owl's speed was 4.0 m/s before grabbing the mouse, what is its speed just after the capture?

*23. A particle moves along the x axis under the influence of a time-dependent force of the form $F_x = 2.0t + 3.0t^2$, where F_x is in newtons and t is in seconds. What is the change in momentum of the particle between $t = 0$ and $t = 5.0$ s? [Hint: Rewrite Eq. (10.3) as $dp_x = F_x \, dt$ and integrate.]

*24. A vase falls off a table and hits a smooth floor, shattering into three fragments of equal mass which move away horizontally along the floor. Two of the fragments leave the point of impact with velocities of equal magnitudes v at right angles. What are the magnitude and direction of the horizontal velocity of the third fragment? (Hint: The x and y components of the momentum are conserved separately.)

*25. The nucleus of an atom of radioactive copper undergoing beta decay simultaneously emits an electron and a neutrino. The momentum of the electron is 2.64×10^{-22} kg·m/s, that of the neutrino is 1.97×10^{-22} kg·m/s, and the angle between their directions of motion is 30.0°. The mass of the residual nucleus is 63.9 u. What is the recoil velocity of the nucleus? (Hint: The x and y components of the momentum are conserved separately.)

*26. The solar wind sweeping past the Earth consists of a stream of particles, mainly hydrogen ions of mass 1.7×10^{-27} kg. There are about 1.0×10^{7} ions per cubic meter, and their speed is 4.0×10^{5} m/s. What force does the impact of the solar wind exert on an artificial Earth satellite that has an area of 1.0 m² facing the wind? Assume that upon impact the ions at first stick to the surface of the satellite.

*27. The record for the heaviest rainfall is held by Unionville, Maryland, where 3.12 cm of rain (1.23 in.) fell in an interval of 1.0 min. Assuming that the impact velocity of the raindrops on the ground was 10 m/s, what must have been the average impact force on each square meter of ground during this rainfall?

*28. An automobile is traveling at a speed of 80 km/h through heavy rain. The raindrops are falling vertically at 10 m/s, and there are 7.0×10^{-4} kg of raindrops in each cubic meter of air. For the following calculation assume that the automobile has the shape of a rectangular box 2.0 m wide, 1.5 m high, and 4.0 m long.

   (a) At what rate (in kg/s) do the raindrops strike the front and top of the automobile?

   (b) Assume that when a raindrop hits, it initially sticks to the automobile, although it falls off later. At what rate does the automobile give momentum to the raindrops? What is the horizontal drag force that the impact of the raindrops exerts on the automobile?

*29. A spaceship of frontal area 25 m² passes through a cloud of interstellar dust at a speed of 1.0×10^{6} m/s. The density of dust is 2.0×10^{-18} kg/m³. If all the particles of dust that impact on the spaceship stick to it, find the average decelerating force that the impact of the dust exerts on the spaceship.

**30. A basketball player jumps straight up to launch a long jump shot at an angle of 45° with the horizontal and a speed of 15 m/s. The 75-kg player is momentarily at rest at the top of his jump just before the shot is released, with his feet 0.80 m above the floor. (a) What is the player's velocity immediately after the shot is released? (b) How far from his original position does he land? Treat the player as a point particle. The mass of a basketball is 0.62 kg.

**31. A gun mounted on a cart fires bullets of mass m in the backward direction with a horizontal muzzle velocity u. The initial mass of the cart, including the mass of the gun and the mass of the ammunition, is M, and the initial velocity of the cart is zero. What is the velocity of the cart after firing n bullets? Assume that the cart moves without friction, and ignore the mass of the gunpowder.

10.2 Center of Mass

32. A penny coin lies on a table at a distance of 20 cm from a stack of three penny coins. Where is the center of mass of the system of four coins?

33. A 59-kg woman and a 73-kg man sit on a seesaw, 3.5 m long. Where is their center of mass? Neglect the mass of the seesaw.

34. Consider the system Earth–Moon; use the data in the table printed inside the book cover. How far from the center of the Earth is the center of mass of this system?

35. Consider the Sun and the planet Jupiter as a two-particle system. How far from the center of the Sun is the center of mass of this system? Express your result as a multiple of the radius of the Sun. (Use the data inside the cover of this book.)

36. Two bricks are adjacent, and a third brick is positioned symmetrically above them, as shown in Fig. 10.28. Where is the center of mass of the three bricks?

FIGURE 10.28 Three bricks.

*37. Where is the center of mass of a uniform sheet in the shape of an isosceles triangle? Assume that the height of the triangle is h when the unequal side is the base.

*38. Consider a pyramid with height h and a triangular base. Where is its center of mass?

*39. In order to balance the wheel of an automobile, a mechanic attaches a piece of lead alloy to the rim of the wheel. The mechanic finds that if he attaches a piece of 40 g at a distance of 20 cm from the center of the wheel of 30 kg, the wheel is perfectly balanced; that is, the center of the wheel coincides with the center of mass. How far from the center of the wheel was the center of mass before the mechanic balanced the wheel?

*40. The distance between the oxygen and each of the hydrogen atoms in a water (H_2O) molecule is 0.0958 nm; the angle between the two oxygen–hydrogen bonds is 105° (Fig. 10.29). Treating the atoms as particles, find the center of mass.

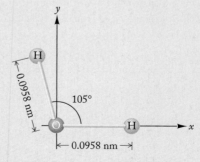

FIGURE 10.29 Atoms in a water molecule.

*41. Figure 10.30 shows the shape of a nitric acid (HNO_3) molecule and its dimensions. Treating the atoms as particles, find the center of mass of this molecule.

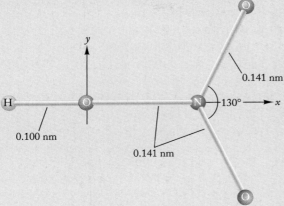

FIGURE 10.30 Atoms in a nitric acid molecule.

*42. Figure 9.13a shows the positions of the three inner planets (Mercury, Venus, and Earth) on January 1, 2000. Measure angles and distances off this figure and find the center of mass of the system of these planets (ignore the Sun). The masses of the planets are listed in Table 9.1.

*43. The Local Group of galaxies consists of our Galaxy and its nearest neighbors. The masses of the most important members of the Local Group are as follows (in multiples of the mass of the Sun): our Galaxy, 2×10^{11}; the Andromeda galaxy, 3×10^{11}; the Large Magellanic Cloud, 2.5×10^{10}; and NGC598, 8×10^9. The x, y, z coordinates of these galaxies are, respectively, as follows (in thousands of light-years): (0, 0, 0); (1640, 290, 1440), (8.5, 56.7, −149), and (1830, 766, 1170). Find the coordinates of the center of mass of the Local Group. Treat all the galaxies as point masses.

*44. A thin, uniform rod is bent in the shape of a semicircle of radius R (see Fig. 10.31). Where is the center of mass of this rod?

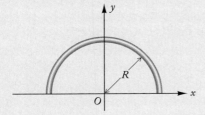

FIGURE 10.31 A rod bent in a semicircle.

*45. Three uniform square pieces of sheet metal are joined along their edges so as to form three of the sides of a cube (Fig. 10.32). The dimensions of the squares are $L \times L$. Where is the center of mass of the joined squares?

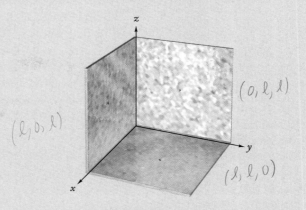

FIGURE 10.32 Three square pieces of sheet metal joined together at their edges.

*46. A box made of plywood has the shape of a cube measuring $L \times L \times L$. The top of the box is missing. Where is the center of mass of the open box?

*47. A cube of iron has dimensions $L \times L \times L$. A hole of radius $\frac{1}{4}L$ has been drilled all the way through the cube so that one side of the hole is tangent to the middle of one face along its entire length (Fig. 10.33). Where is the center of mass of the drilled cube?

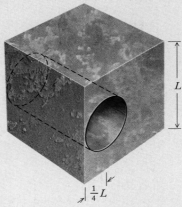

FIGURE 10.33 Iron cube with a hole.

*48. A semicircle of uniform sheet metal has radius R (Fig. 10.34). Find the center of mass.

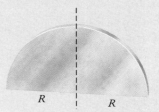

FIGURE 10.34 Semicircle of sheet metal.

*49. Mount Fuji has approximately the shape of a cone. The half-angle at the apex of this cone is 65°, and the height of the apex is 3800 m. At what height is the center of mass? Assume that the material in Mount Fuji has uniform density.

*50. Show that the center of mass of a uniform flat triangular plate is at the point of intersection of the lines drawn from the vertices to the midpoints of the opposite sides.

*51. Consider a man of mass 80 kg and height 1.70 m with the mass distribution described in Fig. 10.17. How much work does this man do to raise his arms from a hanging position to a horizontal position? To a vertically raised position?

*52. Suppose that a man of mass 75 kg and height 1.75 m runs in place, raising his legs high, as in Fig. 10.35. If he runs at the rate of 80 steps per minutes for each leg (160 total per minute), what power does he expend in raising his legs?

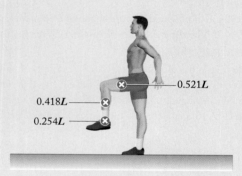

FIGURE 10.35 Man with raised leg.

*53. A lock on the Champlain Canal is 73 m long and 9.2 m wide; the lock has a lift of 3.7 m—that is, the difference between the water levels of the canal on one side of the lock and on the other side is 3.7 m. How much gravitational potential energy is wasted each time the lock goes through one cycle (involving the filling of the lock with water from the high level and then the spilling of this water to the low level)?

*54. The Great Pyramid at Giza has a mass of 6.6×10^6 metric tons and a height of 147 m (see Example 7). Assume that the mass is uniformly distributed over the volume of the pyramid.

  (a) How much work must the ancient Egyptian laborers have done against gravity to pile up the stones in the pyramid?

  (b) If each laborer delivered work at an average rate of 4.0×10^5 J/h, how many person-hours of work have been stored in this pyramid?

**55. A thin hemispherical shell of uniform thickness is suspended from a point above its center of mass as shown in Fig. 10.36. Where is that center of mass?

**56. Suppose that water drops are released from a point at the edge of a roof with a constant time interval Δt between one water drop and the next. The drops fall a distance l to the ground. If Δt is very short (so the number of drops falling though the air at any given instant is very large), show that the center of mass of the falling drops is at a height of $\frac{2}{3}l$ above the ground.

FIGURE 10.36 A hemispherical shell used as a gong.

From this, deduce that the time-average height of a projectile released from the ground and returning to the ground is $\frac{2}{3}$ of its maximum height. (This theorem is useful in the calculation of the average air pressure and air resistance encountered by a projectile.)

10.3 The Motion of the Center of Mass

57. A proton of kinetic energy 1.6×10^{-13} J is moving toward a proton at rest. What is the velocity of the center of mass of the system?

58. In a molecule, the atoms usually execute a rapid vibrational motion about their equilibrium position. Suppose that in an isolated potassium bromide (KBr) molecule the speed of the potassium atom is 5.0×10^3 m/s at one instant (relative to the center of mass). What is the speed of the bromine atom at the same instant?

59. A fisherman in a boat catches a great white shark with a harpoon. The shark struggles for a while and then becomes limp when at a distance of 300 m from the boat. The fisherman pulls the shark by the rope attached to the harpoon. During this operation, the boat (initially at rest) moves 45 m in the direction of the shark. The mass of the boat is 5400 kg. What is the mass of the shark? Pretend that the water exerts no friction.

60. A 75-kg man climbs the stairs from the ground floor to the fourth floor of a building, a height of 15 m. How far does the Earth recoil in the opposite direction as the man climbs?

61. A 6000-kg truck stands on the deck of an 80000-kg ferryboat. Initially the ferry is at rest and the truck is located at its front end. If the truck now drives 15 m along the deck toward the rear of the ferry, how far will the ferry move forward relative to the water? Pretend that the water has no effect on the motion.

62. While moving horizontally at 5.0×10^3 m/s at an altitude of 2.5×10^4 m, a ballistic missile explodes and breaks apart into

two fragments of equal mass which fall freely. One of the fragments has zero speed immediately after the explosion and lands on the ground directly below the point of the explosion. Where does the other fragment land? Ignore the friction of air.

63. A 15-g bullet moving at 260 m/s is fired at a 2.5-kg block of wood. What is the velocity of the center of mass of the bullet–block system?

64. A 60-kg woman and a 90-kg man walk toward each other, each moving with speed v relative to the ground. What is the velocity of their center of mass?

65. A projectile of mass M reaches the peak of its motion a horizontal distance D from the launch point. At its peak, it explodes into three equal fragments. One fragment returns directly to the launch point, and one lands a distance $2D$ from the launch point, at a point in the same plane as the initial motion. Where does the third fragment land?

*66. A projectile is launched with speed v_0 at an angle of θ with respect to the horizontal. At the peak of its motion, it explodes into two pieces of equal mass, which continue to move in the original plane of motion. One piece strikes the ground a horizontal distance D further from the launch point than the point directly below the explosion at a time $t < v_0 \sin\theta/g$ after the explosion. How high does the other piece go? Where does the other piece land? Answer in terms of v_0, θ, D, and t.

**67. Figure 9.13a shows the positions of the three inner planets (Mercury, Venus, Earth) on January 1, 2000. Measuring angles off this figure and using the data on masses, orbital radii, and periods given in Table 9.1, find the velocity of the center of mass of this system of three planets.

10.4 Energy of a System of Particles

68. Two automobiles, each of mass 1500 kg, travel in the same direction along a straight road. The speed of one automobile is 25 m/s, and the speed of the other automobile is 15 m/s. If we regard these automobiles as a system of two particles, what is the translational kinetic energy of the center of mass? What is the total kinetic energy?

69. Repeat the calculation of Problem 68 if the two automobiles travel in *opposite* directions.

70. A projectile of 45 kg fired from a gun has a speed of 640 m/s. The projectile explodes in flight, breaking apart into a fragment of 32 kg and a fragment of 13 kg (we assume that no mass is dispersed in the explosion). Both fragments move along the original direction of motion. The speed of the first fragment is 450 m/s and that of the second is 1050 m/s.

(a) Calculate the translational kinetic energy of the center of mass motion before the explosion.

(b) Calculate the translational kinetic energy of the center of mass motion after the explosion. Calculate the total kinetic energy. Where does the extra kinetic energy come from?

71. Consider the automobile collision described in Problem 16. What is the translational kinetic energy of the center of mass motion before the collision? What is the total kinetic energy before and after the collision?

72. Two isolated point masses m_1 and m_2 are connected by a spring. The masses attain their maximum speeds at the same instant. A short time later both masses are stationary. The maximum speed of the first mass is v_1. What is the maximum speed of the second mass? When the masses are stationary, what is the energy stored in the spring?

73. The typical speed of a helium atom in helium gas at room temperature is 1.4 km/s; that of an oxygen molecule (O_2) in oxygen gas is close to 500 m/s. Find the total kinetic energy of one mole of helium atoms and that of one mole of oxygen molecules.

*74. Two automobiles, each of mass $M/2$ and speed v, drive around a one-lane traffic circle. What is the total kinetic energy of the two-car system? What is the quantity $\frac{1}{2}Mv_{CM}^2$ if the automobiles are (a) on opposite sides of the traffic circle, (b) one-quarter of the circle apart, and (c) locked together?

*75. Consider the Sun and Jupiter to be a two-particle system, orbiting around the center of mass. Find the ratio of the kinetic energy of the Sun to that of Jupiter. (Use the data inside the book cover.)

*76. The typical speed of the vibrational motion of the iron atoms in a piece of iron at room temperature is 360 m/s. What is the total kinetic energy of a 1.0-kg chunk of iron?

REVIEW PROBLEMS

77. A hunter on skates on a smooth sheet of ice shoots 10 bullets at a target at the shore. Each bullet has a mass of 15 g and a speed of 600 m/s. The hunter has a mass of 80 kg. What recoil speed does he acquire?

78. Grain is being loaded into an almost full railroad car from an overhead chute (see Fig. 10.37). If 500 kg per second falls freely from a height of 4.0 m to the top of the car, what downward push does the impact of the grain exert on the car?

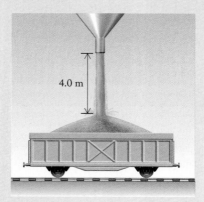

FIGURE 10.37 Grain from a chute falls into a railroad car.

4.0 m

79. A boy and a girl are engaged in a tug-of-war on smooth, frictionless ice. The mass of the boy is 40 kg, and that of the girl is 30 kg; their separation is initially 4.0 m. Each pulls with a force of 200 N on the rope. What is the acceleration of each? If they keep pulling, where will they meet?

80. An automobile of 1200 kg and an automobile of 1500 kg are traveling in the same direction on a straight road. The speeds of the two automobiles are 60 km/h and 80 km/h, respectively. What is the velocity of the center of mass of the two-automobile system?

81. An automobile traveling 40 km/h collides head-on with a truck which has 5 times the mass of the automobile. The wreck remains at rest after the collision. Deduce the speed of the truck.

82. The nozzle of a fire hose ejects 800 liters of water per minute at a speed of 26 m/s. Estimate the recoil force on the nozzle. By yourself, can you hold this nozzle steady in your hands?

83. The distance between the centers of the atoms of potassium and bromine in the potassium bromide (KBr) molecule is 0.282 nm (Fig. 10.38). Treating the atoms as particles, find the center of mass.

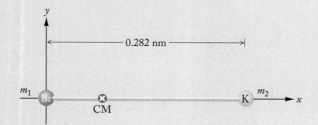

FIGURE 10.38 Atoms in a potassium bromide molecule.

84. A tugboat of mass 400 metric tons and a ship of 28000 metric tons are joined by a long towrope of 400 m. Both vessels are initially at rest in the water. If the tugboat reels in 200 m of towrope, how far does the ship move relative to the water? The tugboat? Ignore the resistance that the water offers to the motion.

85. A cat stands on a plank of balsa wood floating in water. The mass of the cat is 3.5 kg, and the mass of the balsa is 5.0 kg. If the cat walks 1.0 m along the plank, how far does she move in relation to the water?

86. Three firefighters of equal masses are climbing a long ladder. When the first firefighter is 20 m up the ladder, the second is 15 m up, and the third is 5 m up. Where is the center of mass of the three firefighters?

87. Four identical books are arranged on the vertices of an equilateral triangle of side 1.0 m. Two of the books are together at one vertex of the triangle, and the other two are at the other two vertices. Where is the center of mass of this arrangement?

88. Three identical metersticks are arranged to form the letter U. Where is the center of mass of this system?

89. Two uniform squares of sheet metal of dimensions $L \times L$ are joined at right angles along one edge (see Fig. 10.39). One of the squares has twice the mass of the other. Find the center of mass of the combined squares.

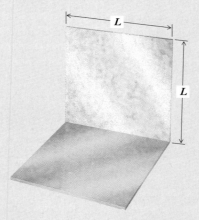

FIGURE 10.39 Two square pieces of sheet metal joined along one edge.

*90. Find the center of mass of a uniform solid hemisphere of radius R.

Answers to Checkups

Checkup 10.1

1. We assume the usual case that the truck has a larger mass than the automobile. Then the equality of their momenta (mv) implies that the automobile has a larger speed (the ratio of the velocities will be the inverse of the ratio of the masses). Since the kinetic energy is $\frac{1}{2}mv^2 = \frac{1}{2} \times mv \times v$ and the momenta (mv) are equal, then the vehicle with the larger speed, the automobile, will also have the larger kinetic energy.

2. They cannot have the same momentum, since the signs of their momenta will be opposite. They can have the same kinetic energy, since that depends only on the speed: $K = \frac{1}{2}mv^2$.

3. No—the momentum, like the velocity, is also reversed, and it has the opposite sign after the impact.

4. Yes—for practical purposes, the Solar System is essentially an isolated system, and so the net momentum is constant. The net kinetic energy is not constant, since during motion, kinetic energy is converted to potential energy and vice versa.

5. (a) For any directions, the total kinetic energy is $\frac{1}{2}mv^2 + \frac{1}{2}mv^2 = mv^2$. For parallel motion (both southward), the total momentum is $mv + mv = 2mv$ southward. (b) The total kinetic energy is again mv^2, while for antiparallel motion, the total momentum is $mv - mv = 0$ (no direction). (c) The total kinetic energy is again mv^2, while for perpendicular motion, the total momentum has magnitude $\sqrt{(mv)^2 + (mv)^2} = \sqrt{2}mv$ and is directed 45° south of east.

6. (C) Automobile; truck. The truck has a larger mass M than the automobile mass m. Let the truck speed be V and the automobile speed be v. The equal kinetic energies ($\frac{1}{2}MV^2 = \frac{1}{2}mv^2$) then imply that the automobile will have the larger speed $v = (M/m)^{1/2}V$. If we substitute one power of this v into the kinetic energy equality and cancel a factor of $\frac{1}{2}V$, we find $MV = (M/m)^{1/2}mv$; thus, the truck momentum is larger.

Checkup 10.2

1. Consider the average position of the mass distribution. For the curved snake shown in the figure, the center of mass is at a point in the space below the top arc, perhaps slightly below center (because of the *two* bottom arcs) and slightly to the right of center (because of the head).

2. Consider the average position of the mass distribution. For the horseshoe, the center of mass is below center in the space in the middle of the arc, along the vertical line of symmetry, at a point well away from the open end.

3. No. The center of mass is a weighted average of position; such an average can never be greater than all of the positions averaged.

4. (C) Shifts upward. If the heavy mass at the bottom falls off, the center of mass is higher. When the center of mass of the sailboat is too high, it is top-heavy, and prone to tip over.

Checkup 10.3

1. For the (isolated) system of person plus canoe, with no initial motion, the center of mass stays at the same fixed position as you begin and continue your crawl. Thus as your mass moves from the rear to the front, the boat moves backward a sufficient distance to keep the center of mass of the combined system fixed.

2. No. In the extreme case where the boxcar has zero mass, you remain fixed relative to the ground, and the boxcar rolls a distance equal to its length as you walk from the rear to the front. If the boxcar has appreciable mass, you will move toward your common center of mass, which will be a distance less than the length of the boxcar.

3. If the marbles were dropped vertically, then the center of mass remains fixed at the point of impact with the floor, even though the marbles scatter in all directions.

4. (B) 5 m/s south. Using Eq. (10.37) with positive velocity northward, $v_{CM} = (M \times 25 \text{ m/s} - 2M \times 20 \text{ m/s})/(3M)$ $= (-15 \text{ m/s})/3 = -5 \text{ m/s}$.

Checkup 10.4

1. No. The initial kinetic energy is large, and the final kinetic energy is small or zero; the energy is transformed into other forms: elastic energy (deformation of automobile parts), friction, sound, and heat.

2. The total energy is conserved, if we consider only gravitational potential energy and kinetic energy (in actuality, some other energy is lost, for example, as the Sun's light is radiated away into space). Neither kinetic nor potential energy is separately conserved; these two are traded back and forth, for example, as the planets move in their elliptical orbits.

3. (A) **P** only. Since there is no net external force, the total momentum of such an isolated system is always simply conserved. However, the spring will stretch and compress during the motion, trading kinetic for potential energy, so K and U will not remain constant.

11.1 Impulsive Forces

11.2 Elastic Collisions in One Dimension

11.3 Inelastic Collisions in One Dimension

11.4 Collisions in Two and Three Dimensions

CONCEPTS IN CONTEXT

**Concepts
— in —
Context**

In this crash test, the automobile was towed at a speed of 56 km/h (35 mi/h) and then crashed into a rigid concrete barrier. Anthropomorphic dummies that simulate human bodies are used for evaluation of injuries that would be sustained by driver and passengers. Accelerometers installed on the body of the automobile and the bodies of the dummies permit calculation of impact forces.

With the concepts of this chapter we can answer questions such as:

? What is the average force on the front of an automobile during impact? (Example 1, page 341)

? What is force on the head of a dummy during a collision with the windshield or the steering wheel? (Example 2, page 341)

? How do seat belts and air bags protect occupants of an automobile in a crash? (Physics in Practice: Automobile Collisions, page 343)

? How does the stiffness of the front end of an automobile affect the safety of its occupants in a collision? (Checkup 11.1, question 1, page 344)

? In a two-car collision, how are the initial velocities related to the final direction of motion? (Example 7, page 351)

The collision between two bodies—an automobile and a solid wall, a ship and an iceberg, a molecule of oxygen and a molecule of nitrogen—involves a violent change of the motion, a change brought about by very strong forces that begin to act suddenly when the bodies come into contact, last a short time, and then cease just as suddenly when the bodies separate. The forces that act during a collision are usually rather complicated, so their complete theoretical description is impossible (e.g., in an automobile collision) or at least very difficult (e.g., in a collision between subatomic particles). However, even without exact knowledge of the details of the forces, we can make some predictions about the collision by taking advantage of the general laws of conservation of momentum and energy we studied in the preceding chapters. In the following sections we will see what constraints these laws impose on the motion of the colliding bodies.

The study of collisions is an important tool in engineering and physics. In automobile collision and safety studies, engineers routinely subject vehicles to crash tests. Collisions are also essential for the experimental investigation of atoms, nuclei, and elementary particles. All subatomic bodies are too small to be made visible with any kind of microscope. Just as you might use a stick to feel your way around a dark cave, a physicist who cannot see the interior of an atom uses probes to "feel" for subatomic structures. The probe used by physicists in the exploration of subatomic structures is simply a stream of fast-moving particles—electrons, protons, alpha particles (helium nuclei), or others. These projectiles are aimed at a target containing a sample of the atoms, nuclei, or elementary particles under investigation. From the manner in which the projectiles collide and react with the target, physicists can deduce some of the properties of the subatomic structures in the target. Similarly, materials scientists, chemists, and engineers deduce the structure and composition of solids and liquids by bombarding such materials with particles and examining the results of such collisions.

11.1 IMPULSIVE FORCES

Online
Concept
Tutorial

The force that two colliding bodies exert on one another acts for only a short time, giving a brief but strong push. Such a force that acts for only a short time is called an **impulsive force.** During the collision, the impulsive force is much stronger than any other forces that may be present; consequently the impulsive force produces a large change in the motion while the other forces produce only small and insignificant changes. For instance, during the automobile collision shown in Fig. 11.1, the only important force on the automobile is the push of the wall on its front end; the effects produced by gravity and by the friction force of the road during the collision are insignificant.

Suppose the collision lasts some short time Δt, say, from $t = 0$ to $t = \Delta t$, and that during this time an impulsive force **F** acts on one of the colliding bodies. This force is zero before $t = 0$ and it is zero after $t = \Delta t$, but it is large between these times. For example, Fig. 11.2 shows a plot of the force experienced by an automobile in a collision with a solid wall lasting 0.120 s. The force is zero before $t = 0$ and after $t = 0.120$ s, and varies in a complicated way between these times.

*The **impulse** delivered by such a force to the body is defined as the integral of the force over time:*

$$\mathbf{I} = \int_0^{\Delta t} \mathbf{F}\, dt \qquad\qquad (11.1)$$

impulse

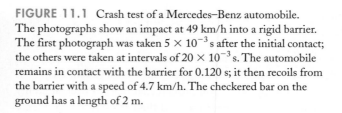

FIGURE 11.1 Crash test of a Mercedes–Benz automobile. The photographs show an impact at 49 km/h into a rigid barrier. The first photograph was taken 5×10^{-3} s after the initial contact; the others were taken at intervals of 20×10^{-3} s. The automobile remains in contact with the barrier for 0.120 s; it then recoils from the barrier with a speed of 4.7 km/h. The checkered bar on the ground has a length of 2 m.

According to this equation, the x component of the impulse for the force plotted in Fig. 11.2 is the area between the curve $F_x(t)$ and the t axis.

The SI units of impulse are N·s, or kg·m/s; these units are the same as those for momentum.

By means of the equation of motion, $\mathbf{F} = d\mathbf{p}/dt$, we can transform Eq. (11.1) into

$$\mathbf{I} = \int_0^{\Delta t} \mathbf{F}\, dt = \int_0^{\Delta t} \frac{d\mathbf{p}}{dt}\, dt = \int d\mathbf{p} = \mathbf{p}' - \mathbf{p} \qquad (11.2)$$

where $\mathbf{p}$ is the momentum of the body before the collision (at time 0) and $\mathbf{p}'$ is the momentum after the collision (at time $t = \Delta t$). Thus, the impulse of a force is simply equal to the momentum change produced by this force. This equality of impulse and momentum change is sometimes referred to as the *impulse–momentum relation*. However, since the force acting during a collision is usually not known in detail, Eq. (11.2) is not very helpful for calculating momentum changes. It is often best to apply Eq. (11.2) in reverse, for calculating the time-average force from the known momentum change. This time-average force is defined by

$$\overline{\mathbf{F}} = \frac{1}{\Delta t} \int_0^{\Delta t} \mathbf{F}\, dt \qquad (11.3)$$

In a plot of force vs. time, such as shown in Fig. 11.2, the time-average force simply represents the mean height of the function above the t axis; this mean height is shown

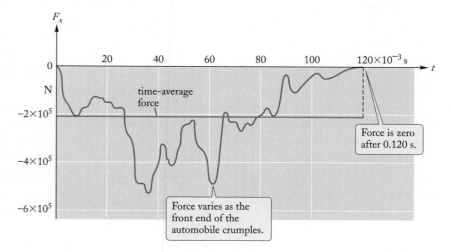

FIGURE 11.2 Force on the automobile as a function of time during the impact shown in Fig. 11.1. The colored horizontal line indicates the time-average force. (Calculated from data supplied by Mercedes–Benz of North America, Inc.)

by the red horizontal line in Fig. 11.2. According to Eq. (11.2), we can write the time-average force as

$$\overline{\mathbf{F}} = \frac{1}{\Delta t}\mathbf{I} = \frac{1}{\Delta t}(\mathbf{p}' - \mathbf{p}) \qquad (11.4)$$

average force in collision

This relation gives a quick estimate of the average magnitude of the impulsive force acting on the body if the duration of the collision and the momentum change are known.

EXAMPLE 1   The collision between the automobile and the barrier shown in Fig. 11.1 lasts 0.120 s. The mass of the automobile is 1700 kg, and the initial and final velocities in the horizontal direction are $v_x = 13.6$ m/s and $v'_x = -1.3$ m/s, respectively (the final velocity is negative because the automobile recoils, or bounces back from the barrier). From these data, evaluate the average force that acts on the automobile during the collision. Evaluate the average force that acts on the barrier.

SOLUTION: With the x axis along the direction of the initial motion, the change of momentum is

$$p'_x - p_x = mv'_x - mv_x$$

$$= 1700 \text{ kg} \times (-1.3 \text{ m/s}) - 1700 \text{ kg} \times 13.6 \text{ m/s}$$

$$= -2.53 \times 10^4 \text{ kg·m/s}$$

According to Eq. (11.4), the average force is then

$$\overline{F}_x = \frac{p'_x - p_x}{\Delta t}$$

$$= \frac{-2.53 \times 10^4 \text{ kg·m/s}}{0.120 \text{ s}} = -2.11 \times 10^5 \text{ N} \qquad (11.5)$$

Since the mutual forces on two bodies engaged in a collision are an action–reaction pair, the forces on the automobile and on the barrier are of equal magnitudes and of opposite directions. Thus, the average force on the barrier is $F_x = +2.11 \times 10^5$ N. This is quite a large force—it equals the weight of about 2×10^4 kg, or 20 tons.

EXAMPLE 2   When an automobile collides with an obstacle and suddenly stops, a passenger not restrained by a seat belt will not stop simultaneously with the automobile, but instead will continue traveling at nearly constant speed until he or she hits the dashboard and the windshield. The collision of the passenger's head with the windshield often results in severe or fatal injuries. In crash tests, dummies with masses, shapes, and joints simulating human bodies are used to determine likely injuries. Consider a dummy head striking a windshield at 15 m/s (54 km/h) and stopping in a time of 0.015 s (this time is considerably shorter than the time of about 0.12 s for stopping the automobile because the front end of the automobile crumples gradually and cushions the collision to some extent; there is no such cushioning for the head striking the windshield). What is the average force on the head during impact on the windshield? What is the average deceleration? Treat the head as a body of mass 5.0 kg, moving independently of the neck and trunk.

SOLUTION: The initial momentum of the head is

$$p_x = mv_x = 5.0 \text{ kg} \times 15 \text{ m/s} = 75 \text{ kg·m/s}$$

When the head stops, the final momentum is zero. Hence the average force is

$$\overline{F}_x = \frac{p'_x - p_x}{\Delta t} = -\frac{p_x}{\Delta t}$$

$$= -\frac{75 \text{ kg·m/s}}{0.015 \text{ s}} = -5.0 \times 10^3 \text{ N}$$

The average acceleration is

$$\overline{a}_x = \frac{\overline{F}_x}{m} = -\frac{5.0 \times 10^3 \text{ N}}{5.0 \text{ kg}} = -1.0 \times 10^3 \text{ m/s}^2$$

which is about 100 standard g's!

Often it is not possible to calculate the motion of the colliding bodies by direct solution of Newton's equation of motion because the impulsive forces that act during the collision are not known in sufficient detail. We must then glean whatever information we can from the general laws of conservation of momentum and energy, which do not depend on the details of these forces. In some simple instances, these general laws permit the deduction of the motion after the collision from what is known about the motion before the collision.

In all collisions between two or more particles, the total momentum of the system is conserved. Whether or not the mechanical energy is conserved depends on the character of the forces that act between the particles. *A collision in which the total kinetic energy before and after the collision is the same is called* **elastic.** (This usage of the word *elastic* is consistent with the usage we encountered previously when discussing the restoring force of a deformable body in Section 6.2. For example, if the colliding bodies exert a force on each other by means of a massless elastic spring placed between them, then the kinetic energy before and after the collision will indeed be the same—that is, the collision will be elastic.) Collisions between macroscopic bodies are usually not elastic—during the collision some of the kinetic energy is transformed into heat by the internal friction forces and some is used up in doing work to change the internal configuration of the bodies. For example, the automobile collision shown in Fig. 11.1 is highly *inelastic;* almost the entire initial kinetic energy is used up in doing work on the automobile parts, changing their shape. On the other hand, the collision of a "Super Ball" and a hard wall or the collision of two billiard balls comes pretty close to being elastic—that is, the kinetic energies before and after the collision are almost the same.

Collisions between "elementary" particles—such as electrons, protons, and neutrons—are often elastic. These particles have no internal friction forces which could dissipate kinetic energy. A collision between such particles can be inelastic only if it involves the creation of new particles; such new particles may arise either by conversion of some of the available kinetic energy into mass or else by transmutation of the old particles by means of a change of their internal structure.

elastic collision

EXAMPLE 3   A Super Ball, made of a rubberlike plastic, is thrown against a hard, smooth wall. The ball strikes the wall from a perpendicular direction with speed v. Assuming that the collision is elastic, find the speed of the ball after the collision.

PHYSICS IN PRACTICE   AUTOMOBILE COLLISIONS

Concepts
— in —
Context

We can fully appreciate the effects of the secondary impact on the human body if we compare the impact speeds of a human body on the dashboard or the windshield with the speed attained by a body in free fall from some height. The impact of the head on the windshield at 15 m/s is equivalent to falling four floors down from an apartment building and landing headfirst on a hard surface. Our intuition tells us that this is likely to be fatal. Since our intuition about the dangers of heights is much better than our intuition about the dangers of speeds, it is often instructive to compare impact speeds with equivalent heights of fall. The table lists impact speeds and equivalent heights, expressed as the number of floors the body has to fall down to acquire the same speed.

The number of fatalities in automobile collisions has been reduced by the use of air bags. The air bag helps by cushioning the impact over a longer time, reducing the time-average force. To be effective, the air bag must inflate quickly, before the passenger reaches it, typically in about 10 milliseconds. Because of this, a passenger, especially a child, too near an air bag prior to inflation can be injured or killed by the impulse from the inflation. But for a properly seated adult passenger, the inflated air bag cushions the passenger, reducing the severity of injuries.

However, the impact can still be fatal—you wouldn't expect to survive a jump from an 11-floor building onto an air mattress.

For maximum protection, a seat belt should always be worn even in vehicles equipped with air bags. In lateral collisions, in repeated collisions (such as in car pileups), and in rollovers, an air bag is of little help, and a seat belt is essential. The effectiveness of seat belts is well demonstrated by the experiences of race car drivers. Race car drivers wear lap belts and crossed shoulder belts. Even in spectacular crashes at very high speeds (see the figure), the drivers rarely suffer severe injuries.

COMPARISON OF IMPACT SPEEDS AND HEIGHTS OF FALL

SPEED	SPEED	EQUIVALENT HEIGHT (NUMBER OF FLOORS)[a]
15 km/h	9 mi/h	$\frac{1}{3}$
30	19	1
45	28	3
60	37	5
75	47	8
90	56	11
105	65	· 15

[a]Each floor is 2.9 m.

In a race at the California Speedway in October 2000, a car flips over and breaks in half after a crash, but the driver, Luis Diaz, walks away from the wreck.

SOLUTION: The only horizontal force on the ball is the normal force exerted by the wall; this force reverses the motion of the ball (see Fig. 11.3). Since the wall is very massive, the reaction force of the ball on the wall will not give the wall any appreciable velocity. Hence the kinetic energy of the system, both before and after the collision, is merely the kinetic energy of the ball. Conservation of this

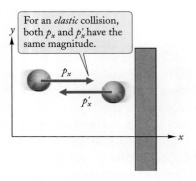

For an *elastic* collision, both p_x and p'_x have the same magnitude.

FIGURE 11.3 The initial momentum p_x of the ball is positive; the final momentum p'_x is negative.

kinetic energy then requires that the ball rebound with a speed v equal to the incident speed.

Note that although the kinetic energy of the ball is the same before and after the collision, the momentum is not the same before and after (see also Example 1 of Chapter 10). If the x axis is in the direction of the initial motion, then the momentum of the ball before the collision is $p_x = mv$, and after the collision it is $p'_x = -mv$. Hence the change of momentum is $p'_x - p_x = -2mv$. The wall suffers an equal and opposite momentum change of $+2mv$, so that the total momentum of the system is conserved. The wall can acquire the momentum $2mv$ without acquiring any appreciable velocity because its mass is large and it is attached to a building of even larger mass.

Concepts
— *in* —
Context

✓ Checkup 11.1

QUESTION 1: In order to protect the occupants of an automobile in a collision, is it better to make the front end of the automobile very hard (a solid block of steel) or fairly soft and crushable?

QUESTION 2: If a golf ball and a steel ball of the same mass strike a concrete floor with equal speeds, which will exert the larger average force on the floor?

QUESTION 3: You drop a Super Ball on a hard, smooth floor from a height of 1 m. If the collision is elastic, how high will the ball bounce up?

QUESTION 4: A child throws a wad of chewing gum against a wall, and it sticks. Is this an elastic collision?

QUESTION 5: A 3000-kg truck collides with a 1000-kg car. During this collision the average force exerted by the truck on the car is 3×10^6 N in an eastward direction. What is the magnitude of the average force exerted by the car on the truck?

  (A) 0   (B) 1×10^6 N   (C) 3×10^6 N   (D) 9×10^6 N

Online
Concept
Tutorial

Online
Concept
Tutorial

11.2 ELASTIC COLLISIONS IN ONE DIMENSION

The collision of two boxcars on a railroad track is an example of a collision on a straight line. More generally, the collision of any two bodies that approach head-on and recoil along their original line of motion is a collision along a straight line. Such collisions will occur only under exceptional circumstances; nevertheless, we find it instructive to study such collisons because they display in a simple way some of the broad features of more complicated collisions.

In an elastic collision of two particles moving along a straight line, the laws of conservation of momentum and energy completely determine the final velocities in terms of the initial velocities. In the following calculations, we will assume that one particle (the "projectile") is initially in motion and the other (the "target") is initially at rest.

Figure 11.4a shows the particles before the collision, and Fig. 11.4b shows them after; the x axis is along the direction of motion. We will designate the x components of the velocity of particle 1 and particle 2 before the collision by v_1 and v_2, respectively. We will designate the x components of these velocities after the collision by v'_1 and v'_2.

Particle 2 is the target, initially at rest, so $v_2 = 0$. Particle 1 is the projectile. The initial momentum is therefore simply the momentum $m_1 v_1$ of particle 1. The final momentum, after the collision, is $m_1 v_1' + m_2 v_2'$. Conservation of momentum then tells us that

$$m_1 v_1 = m_1 v_1' + m_2 v_2' \tag{11.6}$$

The initial kinetic energy is $\frac{1}{2} m_1 v_1^2$, and the final kinetic energy is $\frac{1}{2} m_1 v_1'^2 + \frac{1}{2} m_2 v_2'^2$. Since this collision is *elastic*, conservation of kinetic energy[1] tell us that

$$\tfrac{1}{2} m_1 v_1^2 = \tfrac{1}{2} m_1 v_1'^2 + \tfrac{1}{2} m_2 v_2'^2 \tag{11.7}$$

In these equations, we can regard the initial velocities v_1 and v_2 as known, and the final velocities v_1' and v_2' as unknown. We therefore want to solve these equations for the unknown quantities. For this purpose, it is convenient to rearrange the two equations somewhat. If we subtract $m_1 v_1'$ from both sides of Eq. (11.6), we obtain

$$m_1(v_1 - v_1') = m_2 v_2' \tag{11.8}$$

If we multiply both sides of Eq. (11.7) by 2 and subtract $m_1 v_1'^2$ from both sides, we obtain

$$m_1(v_1^2 - v_1'^2) = m_2 v_2'^2 \tag{11.9}$$

With the identity $v_1^2 - v_1'^2 = (v_1 - v_1')(v_1 + v_1')$, this becomes

$$m_1(v_1 - v_1')(v_1 + v_1') = m_2 v_2'^2 \tag{11.10}$$

Now divide Eq. (11.10) by Eq. (11.8)—that is, divide the left side of Eq. (11.10) by the left side of Eq. (11.8) and the right side of Eq. (11.10) by the right side of Eq. (11.8). The result is

$$v_1 + v_1' = v_2' \tag{11.11}$$

This trick gets rid of the bothersome squares in Eq. (11.7) and leaves us with two equations—Eqs. (11.8) and (11.11)—without squares. To complete the solution for our unknowns, we take the value $v_2' = v_1 + v_1'$ given by Eq. (11.11) and substitute it into the right side of Eq. (11.8):

$$m_1(v_1 - v_1') = m_2(v_1 + v_1') \tag{11.12}$$

We can solve this immediately for the unknown v_1', with the result

$$v_1' = \frac{m_1 - m_2}{m_1 + m_2} v_1 \tag{11.13}$$

final projectile velocity in elastic collision

Finally, we substitute this value of v_1' into the expression from Eq. (11.11), $v_2' = v_1 + v_1'$, and we find

$$v_2' = v_1 + \frac{m_1 - m_2}{m_1 + m_2} v_1 = \frac{(m_1 + m_2)v_1 + (m_1 - m_2)v_1}{m_1 + m_2}$$

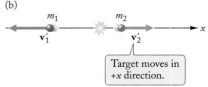

Particle 1 is the moving projectile. Particle 2 is a stationary target.

(a)

m_1 m_2

$\mathbf{v}_1$

(b)

m_1 m_2

$\mathbf{v}_1'$ $\mathbf{v}_2'$

Target moves in $+x$ direction.

FIGURE 11.4 (a) Before the collision, particle 2 is at rest, and particle 1 has velocity $\mathbf{v}_1$. (b) After the collision, particle 1 has velocity $\mathbf{v}_1'$, and particle 2 has velocity $\mathbf{v}_2'$.

---

[1] In the context of elastic collisions, "conservation of kinetic energy" is taken to mean that the kinetic energy is the same before and after the collision; during the collision, when the particles are interacting, what is conserved is not the kinetic energy itself, but the sum of kinetic and potential energies.

or

$$v_2' = \frac{2m_1}{m_1 + m_2} v_1 \qquad (11.14)$$

Equations (11.13) and (11.14) give us the final velocities v_1' and v_2' in terms of the initial velocity v_1.

EXAMPLE 4   An empty boxcar of mass $m_1 = 20$ metric tons rolling on a straight track at 5.0 m/s collides with a loaded stationary boxcar of mass $m_2 = 65$ metric tons (see Fig. 11.5). Assuming that the cars bounce off each other elastically, find the velocities after the collision.

SOLUTION: With $m_1 = 20$ tons and $m_2 = 65$ tons, Eqs. (11.13) and (11.14) yield

$$v_1' = \frac{20 \text{ tons} - 65 \text{ tons}}{20 \text{ tons} + 65 \text{ tons}} \times 5.0 \text{ m/s} = -2.6 \text{ m/s}$$

$$v_2' = \frac{2 \times 20 \text{ tons}}{20 \text{ tons} + 65 \text{ tons}} \times 5.0 \text{ m/s} = 2.4 \text{ m/s}$$

Thus, boxcar 2 acquires a speed of 2.4 m/s, and boxcar 1 recoils with a speed of 2.6 m/s (note the negative sign of v_1').

(a)

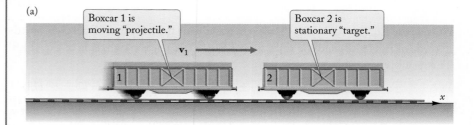

Boxcar 1 is moving "projectile."

Boxcar 2 is stationary "target."

(b)

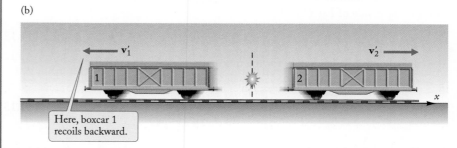

Here, boxcar 1 recoils backward.

FIGURE 11.5 (a) Initially, boxcar 1 is moving toward the right, and boxcar 2 is stationary. (b) After the collision, boxcar 1 is moving toward the left, and boxcar 2 is moving toward the right.

Note that if the mass of the target is much larger than the mass of the projectile, then m_1 can be neglected compared with m_2. Equation (11.13) then becomes

$$v_1' \approx -\frac{m_2}{m_2} v_1 = -v_1 \qquad (11.15)$$

and Eq. (11.14) becomes

$$v_2' \approx \frac{2m_1}{m_2} v_1 \approx 0 \qquad (11.16)$$

This means the projectile bounces off the target with a reversed velocity and the target remains nearly stationary (as in the case of the Super Ball bouncing off the wall; see Example 3).

Conversely, if the mass of the projectile is much larger than the mass of the target, then m_2 can be neglected compared with m_1, and Eqs. (11.13) and (11.14) become

$$v_1' \approx \frac{m_1}{m_1} v_1 = v_1 \qquad (11.17)$$

and

$$v_2' \approx \frac{2m_1}{m_1} v_1 = 2v_1 \qquad (11.18)$$

This means that the projectile plows along with unchanged velocity and the target bounces off with *twice* the speed of the incident projectile. For example, when a (heavy) golf club strikes a golf ball, the ball bounces away at twice the speed of the club (see Fig. 11.6).

Also, if the two masses are equal, Eqs. (11.13) and (11.14) give

$$v_1' = 0 \qquad \text{and} \qquad v_2' = v_1$$

Thus, the projectile stops and the target moves off with the projectile's initial speed. This is common in a head-on collision in billiards, and is also realized in certain pendulum toys (see Discussion Question 9 at the end of the chapter).

Finally, if both particles involved in a one-dimensional elastic collision are initially moving ($v_1 \neq 0$ and $v_2 \neq 0$), conservation of the total momentum and the total kinetic energy can again be applied to uniquely determine the final velocities. The results are more complicated, but they are obtained in the same manner as in the stationary target case above.

Exposures are at equal time intervals.

When struck by a larger mass, the smaller mass moves off at higher speed.

FIGURE 11.6 Impact of club on golf ball. By inspection of this multiple-exposure photograph, we see that the speed of the ball is larger than the initial speed of the club.

 ## Checkup 11.2

In the following questions assume that a projectile traveling in the direction of the positive x axis strikes a stationary target head-on and the collision is elastic.

QUESTION 1: Under what conditions will the velocity of the projectile be positive after the collision? Negative?

QUESTION 2: Can the speed of recoil of the target ever exceed twice the speed of the incident projectile?

QUESTION 3: For an elastic collision, the kinetic energies before and after the collision are the same. Is the kinetic energy *during* the collision also the same?

QUESTION 4: A marble with velocity v_1 strikes a stationary, identical marble elastically and head-on. The final velocities of the shot and struck marbles are, respectively:

(A) $\frac{1}{2}v_1$; $\frac{1}{2}v_1$ (B) v_1; $2v_1$ (C) $-v_1$; 0

(D) $-v_1$; $2v_1$ (E) 0; v_1

Online
Concept
Tutorial
13

Online
Concept
Tutorial
14

11.3 INELASTIC COLLISIONS IN ONE DIMENSION

If the collision is inelastic, kinetic energy is not conserved, and then the only conservation law that is applicable is the conservation of momentum. This, by itself, is insufficient to calculate the velocities of both particles after the collision. Thus, for most inelastic collisions, one of the final velocities must be measured in order for momentum conservation to provide the other. Alternatively, we must have some independent knowledge of the amount of kinetic energy lost. However, *if the collision is* **totally inelastic,** *so a maximum amount of kinetic energy is lost, then the common velocity of both particles after the collision can be calculated.*

totally inelastic collision

In a totally inelastic collision, the particles do not bounce off each other at all; instead, *the particles stick together,* like two automobiles that form a single mass of interlocking wreckage after a collision, or two railroad boxcars that couple together. Under these conditions, the velocities of both particles must coincide with the velocity of the center of mass. But the velocity of the center of mass *after* the collision is the same as the velocity of the center of mass *before* the collision, because there are no external forces and the acceleration of the center of mass is zero [see Eq. (10.40)]. We again consider a stationary target, so that before the collision the velocity of the target particle is zero ($v_2 = 0$) and the general equation [Eq. (10.37)] for the velocity of the center of mass yields

$$v_{CM} = \frac{m_1 v_1}{m_1 + m_2} \tag{11.19}$$

This must then be the final velocity of both particles after a totally inelastic collision:

final velocities in totally inelastic collision with stationary target

$$v'_1 = v'_2 = v_{CM} = \frac{m_1 v_1}{m_1 + m_2} \tag{11.20}$$

We have already come across an instance of this formula in Example 3 of Chapter 10.

EXAMPLE 5 Suppose that the two boxcars of Example 4 couple during the collision and remain locked together (see Fig. 11.7). What is the velocity of the combination after the collision? How much kinetic energy is dissipated during the collision?

SOLUTION: Since the boxcars remain locked together, this is a totally inelastic collision. With $m_1 = 20$ tons, $m_2 = 65$ tons, and $v_1 = 5.0$ m/s, Eq. (11.19) gives us the velocity of the center of mass:

$$v_{CM} = \frac{m_1 v_1}{m_1 + m_2} = \frac{20 \text{ tons} \times 5.0 \text{ m/s}}{20 \text{ tons} + 65 \text{ tons}} = 1.2 \text{ m/s}$$

and this must be the velocity of the coupled cars after the collision.

The kinetic energy before the collision is that of the moving boxcar,

$$\tfrac{1}{2}m_1 v_1^2 = \tfrac{1}{2} \times 20\,000 \text{ kg} \times (5.0 \text{ m/s})^2 = 2.5 \times 10^5 \text{ J}$$

and the kinetic energy after the collision is that of the two coupled boxcars,

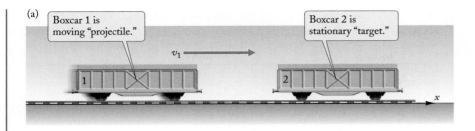

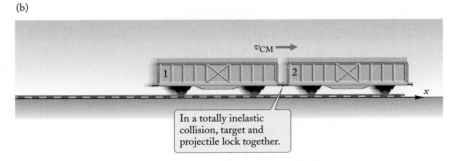

FIGURE 11.7 (a) Initially, boxcar 1 is moving toward the right, and boxcar 2 is stationary, as in Fig. 11.5. (b) After the collision, the boxcars remain locked together. Their common velocity must be the velocity of the center of mass.

$$\tfrac{1}{2}m_1 v_{CM}^2 + \tfrac{1}{2}m_2 v_{CM}^2 = \tfrac{1}{2}(m_1 + m_2)v_{CM}^2$$

$$= \tfrac{1}{2}(20\,000 \text{ kg} + 65\,000 \text{ kg}) \times (1.2 \text{ m/s})^2 = 0.61 \times 10^5 \text{ J}$$

Thus, the loss of kinetic energy is

$$2.5 \times 10^5 \text{ J} - 0.61 \times 10^5 \text{ J} = 1.9 \times 10^5 \text{ J} \tag{11.21}$$

This energy is absorbed by friction in the bumpers during the coupling of the boxcars.

EXAMPLE 6   Figure 11.8a shows a **ballistic pendulum,** a device once commonly used to measure the speeds of bullets. The pendulum consists of a large block of wood of mass m_2 suspended from thin wires. Initially, the pendulum is at rest. The bullet, of mass m_1, strikes the block horizontally and remains stuck in it. The impact of the bullet puts the block in motion, causing it to swing upward to a height h (see Fig. 11.8b), where it momentarily stops. In a test of a Springfield rifle firing a bullet of 9.7 g, a ballistic pendulum of 4.0 kg swings up to a height of 19 cm. What was the speed of the bullet before impact?

SOLUTION: The collision of the bullet with the wood is totally inelastic. Hence, immediately after the collision, bullet and block move horizontally with the velocity of the center of mass:

$$v_{CM} = \frac{m_1 v_1}{m_1 + m_2} \tag{11.22}$$

(a)

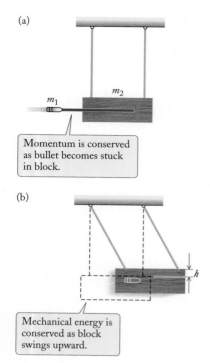

Momentum is conserved as bullet becomes stuck in block.

(b)

Mechanical energy is conserved as block swings upward.

FIGURE 11.8 (a) Before the bullet strikes, the block of wood is at rest. (b) After the bullet strikes, the block, with the embedded bullet, moves toward the right and swings upward to a height h.

After the collision is over, during the subsequent swinging motion of the pendulum, the total mechanical energy (kinetic plus potential) is conserved. At the bottom of the swing, the energy is kinetic, $\frac{1}{2}(m_1 + m_2)v_{CM}^2$; and at the top of the swing at height h, it is potential, $(m_1 + m_2)gh$. Hence, conservation of the total mechanical energy tells us that

$$\tfrac{1}{2}(m_1 + m_2)v_{CM}^2 = (m_1 + m_2)gh \tag{11.23}$$

If we divide this by $(m_1 + m_2)$ and take the square root of both sides, we find

$$v_{CM} = \sqrt{2gh} \tag{11.24}$$

Substitution of this into Eq. (11.22) yields

$$\sqrt{2gh} = \frac{m_1 v_1}{m_1 + m_2} \tag{11.25}$$

which we can solve for v_1, with the result

$$v_1 = \frac{m_1 + m_2}{m_1}\sqrt{2gh}$$

$$= \frac{0.0097 \text{ kg} + 4.0 \text{ kg}}{0.0097 \text{ kg}}\sqrt{2 \times 9.81 \text{ m/s}^2 \times 0.19 \text{ m}} \tag{11.26}$$

$$= 800 \text{ m/s}$$

COMMENT: Note that during the collision, momentum is conserved but not kinetic energy (the collision is totally inelastic); and that during the swinging motion, the total mechanical energy is conserved, but not momentum (the swinging motion proceeds under the influence of the "external" forces of gravity and the tensions in the wires).

✔ Checkup 11.3

QUESTION 1: In a totally inelastic collision, do both particles lose kinetic energy?

QUESTION 2: Consider a collision between two particles of equal masses and of opposite velocities. What is the velocity after this collision if the collision is totally inelastic? If the collision is elastic?

QUESTION 3: Under what conditions is the velocity of the particles after a totally inelastic collision equal to one-half the velocity of the incident projectile? (Assume a stationary target.)

QUESTION 4: Does the length of the suspension wires affect the operation of the ballistic pendulum described in Example 6?

QUESTION 5: A particle is traveling in the positive x direction with speed v. A second particle with one-half the mass of the first is traveling in the opposite direction with the same speed. The two experience a totally inelastic collision. The final x component of the velocity is:

(A) 0   (B) $\frac{1}{3}v$ (C) $\frac{1}{2}v$ (D) $\frac{2}{3}v$ (F) v

11.4 COLLISIONS IN TWO AND THREE DIMENSIONS

In the previous sections, we have focused on collisions on a straight line, in one dimension. Collisions in two or three dimensions are more difficult to analyze, because the conservation laws for momentum and energy do not provide sufficient information to determine the final velocities completely in terms of the initial velocities. Momentum is always conserved during a collision, and this conservation provides one equation for each of the x, y, and z directions. If it is known that the collision is totally elastic, then conservation of the total kinetic energy provides another equation. However, these are not enough to determine the three final velocity components for each and every particle. Some information concerning the final velocities must also be known or measured.

The case of totally inelastic collisions is an exception: in this case, the conservation of momentum determines the outcome completely, even in two or three dimensions. The particles stick together, and their final velocities coincide with the velocity of the center of mass, as illustrated by the following example. The subsequent example explores a case where the solution exploits some knowledge of the final velocities.

EXAMPLE 7 A red automobile of mass 1100 kg and a green automobile of mass 1300 kg collide at an intersection. Just before this collision, the red automobile was traveling due east at 34 m/s, and the green automobile was traveling due north at 15 m/s (see Fig. 11.9). After the collision, the wrecked automobiles remain joined together, and they skid on the pavement with locked wheels. What is the direction of the skid?

Concepts
— *in* —
Context

SOLUTION: The final velocity of the wreck coincides with the final velocity of the center of mass, which is the same as the initial velocity of the center of mass. According to Eq. (10.37), this velocity is

$$\mathbf{v}_{CM} = \frac{m_1\mathbf{v}_1 + m_2\mathbf{v}_2}{m_1 + m_2} \qquad (11.27)$$

final velocity in totally inelastic collision

With the x axis eastward and the y axis northward, the initial velocity $\mathbf{v}_1$ of the red automobile has an x component but no y component, and the initial velocity $\mathbf{v}_2$ of the green automobile has a y component but no x component. Hence the x component of $\mathbf{v}_{CM}$ is

$$v_{CM,x} = \frac{m_1 v_1}{m_1 + m_2} = \frac{1100 \text{ kg} \times 34 \text{ m/s}}{1100 \text{ kg} + 1300 \text{ kg}} = 16 \text{ m/s}$$

and the y component of $\mathbf{v}_{CM}$ is

$$v_{CM,y} = \frac{m_2 v_2}{m_1 + m_2} = \frac{1300 \text{ kg} \times 15 \text{ m/s}}{1100 \text{ kg} + 1300 \text{ kg}} = 8.1 \text{ m/s}$$

The angle between the direction of this velocity and the x axis is given by

$$\tan\theta = \frac{v_{CM,y}}{v_{CM,x}} = \frac{8.1 \text{ m/s}}{16 \text{ m/s}} = 0.51$$

from which

$$\theta = 27°$$

Since the x axis is eastward, this is 27° north of east.

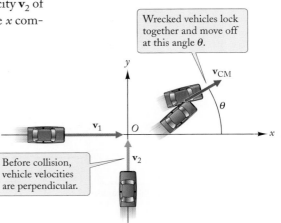

FIGURE 11.9 An automobile collision. Before the collision, the velocities of the automobiles were $\mathbf{v}_1$ and $\mathbf{v}_2$. After the collision, both velocities are $\mathbf{v}_{CM}$.

EXAMPLE 8   In an atomic collision experiment, or "scattering" experiment, a helium ion of mass $m_1 = 4.0$ u with speed $v_1 = 1200$ m/s strikes an oxygen (O_2) molecule of mass $m_2 = 32$ u which is initially at rest (see Fig. 11.10a). The helium ion exits the collision at 90° from its incident direction with one-fourth of its original kinetic energy. What is the recoil speed of the oxygen molecule? What fraction of the total kinetic energy is lost during the collision? [This energy is lost to the internal (vibrational and rotational) motions of the oxygen molecule.]

SOLUTION: In the absence of external forces, momentum is always conserved. If we choose the direction of incident motion along the x axis, then for 90° scattering, we can choose the direction in which the helium ion exits (the direction of $\mathbf{v}_1'$) to be along the y axis (see Fig. 11.10b). Conservation of momentum in the two directions then requires

$$\text{for } x \text{ direction: } m_1 v_1 = m_2 v_{2x}'$$

$$\text{for } y \text{ direction: } 0 = m_1 v_1' + m_2 v_{2y}'$$

Since the helium ion exits with one-fourth of its initial kinetic energy,

$$\tfrac{1}{2} m_1 v_1'^2 = \tfrac{1}{4} \times \tfrac{1}{2} m_1 v_1^2$$

or

$$v_1' = \tfrac{1}{2} v_1$$

Substituting this v_1' and the given $m_2 = 8m_1$ into the x and y components of the momentum gives for the velocity of the oxygen molecule:

$$v_{2x}' = \frac{m_1}{m_2} v_1 = \frac{1}{8} v_1 = \frac{1}{8} \times 1200 \text{ m/s} = 150 \text{ m/s}$$

$$v_{2y}' = -\frac{m_1}{m_2} v_1' = -\frac{1}{8} \times \frac{1}{2} v_1 = -\frac{1}{16} \times 1200 \text{ m/s} = -75 \text{ m/s}$$

The speed of recoil of the oxygen molecule is thus

$$v_2' = \sqrt{v_{2x}'^2 + v_{2y}'^2} = \sqrt{(150 \text{ m/s})^2 + (-75 \text{ m/s})^2} = 170 \text{ m/s}$$

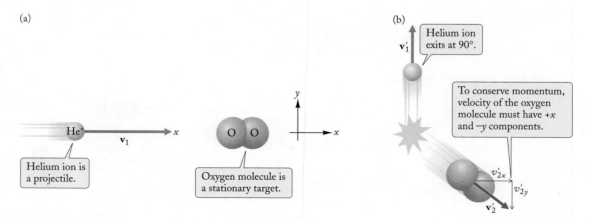

(a)

(b)

Helium ion exits at 90°.

To conserve momentum, velocity of the oxygen molecule must have $+x$ and $-y$ components.

Helium ion is a projectile.

Oxygen molecule is a stationary target.

FIGURE 11.10   (a) A helium ion with velocity $\mathbf{v}_1 = v_{1x}\mathbf{i}$ is moving toward a stationary oxygen molecule. (b) After the collision, the helium ion exits perpendicular to its incident direction with velocity $\mathbf{v}_1'$, while the oxygen molecule acquires a velocity $\mathbf{v}_2' = v_{2x}'\mathbf{i} + v_{2y}'\mathbf{j}$.

The fraction of kinetic energy lost is the amount of kinetic energy lost divided by the original kinetic energy:

$$[\text{fraction lost}] = \frac{K - K'}{K} = \frac{\frac{1}{2}m_1v_1^2 - \left(\frac{1}{4} \times \frac{1}{2}m_1v_1^2 + \frac{1}{2}m_2v_2'^2\right)}{\frac{1}{2}m_1v_1^2}$$

$$= 1 - \frac{1}{4} - \frac{m_2}{m_1}\frac{v_2'^2}{v_1^2} = \frac{3}{4} - \frac{32}{4.0} \times \frac{(170 \text{ m/s})^2}{(1200 \text{ m/s})^2}$$

$$= 0.59$$

Thus, about 59% of the helium ion's initial kinetic energy is lost to internal motions of the oxygen molecule during the collision.

 Checkup 11.4

QUESTION 1: A car traveling south collides with and becomes entangled with a car of the same mass and speed heading west. In what direction does the wreckage emerge from the collision?

QUESTION 2: An object at rest explodes into three pieces; one travels due west and another due north. In which quadrant of directions does the third piece travel?

  (A) Northeast   (B) Southeast   (C) Southwest   (D) Northwest

PROBLEM-SOLVING TECHNIQUES	CONSERVATION OF ENERGY AND MOMENTUM IN COLLISIONS

For solving problems involving collisions, it is essential to know what conservation laws are applicable. The following table summarizes the conservation laws applicable for different collisions:

TYPE OF COLLISION	CONSERVATION OF KINETIC ENERGY	CONSERVATION OF MOMENTUM	COMMENTS
Elastic	Yes	Yes	For a one-dimensional collision, energy and momentum conservation determine the final velocities in terms of the initial velocities. For a 2- or 3-dimensional collision, there is not enough information in the initial velocities alone to determine the final velocities uniquely.
Totally inelastic	No	Yes	The two colliding bodies stick together, and momentum conservation determines the final velocities (in 1, 2, and 3 dimensions).
Inelastic	No	Yes	If the collision is not totally inelastic, there is not enough information in the initial velocities alone to determine the final velocities. Some information about the energy loss and final velocities must also be known.

SUMMARY

PHYSICS IN PRACTICE   Automobile Collisions   **(page 343)**

PROBLEM-SOLVING TECHNIQUES   **(page 353)**
Conservation of Energy and Momentum in Collisions

IMPULSE

$$\mathbf{I} = \int_0^{\Delta t} \mathbf{F}\, dt = \mathbf{p}' - \mathbf{p}$$

(11.1)

AVERAGE FORCE IN COLLISION

$$\overline{\mathbf{F}} = \frac{\mathbf{p}' - \mathbf{p}}{\Delta t}$$

(11.4)

ALL COLLISIONS
The total momentum is conserved.

$$m_1\mathbf{v}_1 + m_2\mathbf{v}_2 + \cdots = m_1\mathbf{v}'_1 + m_2\mathbf{v}'_2 + \cdots$$

(11.6)

ELASTIC COLLISION
The total kinetic energy is conserved.

$$\tfrac{1}{2}m_1 v_1^2 + \tfrac{1}{2}m_2 v_2^2 + \cdots = \tfrac{1}{2}m_1 v_1'^2 + \tfrac{1}{2}m_2 v_2'^2 + \cdots$$

(11.7)

**VELOCITIES IN ONE-DIMENSIONAL ELASTIC
COLLISION WITH STATIONARY TARGET**

Before: $v_1 \neq 0, \quad v_2 = 0$

After: $v'_1 = \dfrac{m_1 - m_2}{m_1 + m_2} v_1$ **(11.13)**

$$v'_2 = \frac{2m_1}{m_1 + m_2} v_1$$

(11.14)

INELASTIC COLLISION
Kinetic energy is not conserved.

TOTALLY INELASTIC COLLISION
The colliding particles stick together.

VELOCITIES IN TOTALLY INELASTIC COLLISION
(1, 2, or 3 dimensions).

Before: $\mathbf{v}_1$ and $\mathbf{v}_2$

After: $\mathbf{v}'_1 = \mathbf{v}'_2 = \mathbf{v}_{\mathrm{CM}} = \dfrac{m_1\mathbf{v}_1 + m_2\mathbf{v}_2}{m_1 + m_2}$ **(11.27)**

QUESTIONS FOR DISCUSSION

1. According to the data given in Example 1, what percentage of the initial kinetic energy does the automobile have after the collision?

2. A (foolish) stuntman wants to jump out of an airplane at high altitude without a parachute. He plans to jump while tightly encased in a strong safe which can withstand the impact on the ground. How would you convince the stuntman to abandon this project?

3. In the crash test shown in the photographs of Fig. 11.1, anthropomorphic dummies were riding in the automobile. These dummies were (partially) restrained by seat belts, which limited their motion relative to the automobile. How would the motion of the dummies have differed from that shown in these photographs if they had not been restrained by seat belts?

4. For the sake of safety, would it be desirable to design automobiles so that their collisions are elastic or inelastic?

5. Two automobiles have collided at a north–south east–west intersection. The skid marks their tires made after the collision point roughly northwest. One driver claims he was traveling west; the other driver claims he was traveling south. Who is lying?

6. Statistics show that, on the average, the occupants of a heavy ("full-size") automobile are more likely to survive a crash than those of a light ("compact") automobile. Why would you expect this to be true?

7. In Joseph Conrad's tale "Gaspar Ruiz", the hero ties a cannon to his back and, hugging the ground on all fours, fires several shots at the gate of a fort. How does the momentum absorbed by Ruiz compare with that absorbed by the gate? How does the energy absorbed by Ruiz compare with that absorbed by the gate?

8. Give an example of a collision between two bodies in which *all* of the kinetic energy is lost to inelastic processes.

9. Explain the operation of the five-pendulum toy, called Newton's cradle, shown in Fig. 11.11.

10. In order to split a log with a small ax, you need a greater impact speed than you would need with a large ax. Why? If the energy required to split the log is the same in both cases, why is it more tiring to use the small ax? (Hint: Think about the kinetic energy of your arms.)

11. If you throw an (elastic) baseball at an approaching train, the ball will bounce back at you with an increased speed. Explain.

12. You are investigating the collision of two automobiles at an intersection. The automobiles remained joined together after this collision, and their wheels made measurable skid marks on the pavement before they came to rest. Assume that during skidding all the wheels remained locked so that the deceleration was entirely due to sliding friction. You know the direction of motion of the automobiles before the collision (drivers are likely to be honest about this), but you do not know the speeds (drivers are likely to be dishonest about this). What do you have to measure at the scene of the accident to calculate the speeds of both the automobiles before the collision?

13. You are sitting in your car, stopped at an intersection. You notice another car approaching from behind, and you notice this car is not slowing down and is going to ram you. Because the time to impact is short, you have only two choices: push hard on your brake, or take your foot off the brake and give your car freedom to roll. Which of these tactics will minimize damage to yourself? Which will minimize damage to your car? Which will minimize damage to the other car?

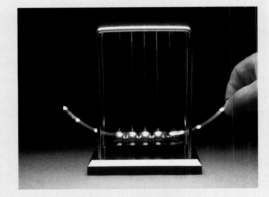

FIGURE 11.11 Newton's cradle.

PROBLEMS

11.1 Impulsive Forces†

1. A stuntman of mass 77 kg "belly-flops" on a shallow pool of water from a height of 11 m. When he hits the pool, he comes to rest in about 0.050 s. What is the impulse that the water and the bottom of the pool deliver to his body during this time interval? What is the time-average force?

2. A large ship of 7.0×10^5 metric tons steaming at 20 km/h runs aground on a reef, which brings it to a halt in 5.0 s. What is the impulse delivered to the ship? What is the average force on the ship? What is the average deceleration?

† For help, see Online Concept Tutorial 13 at www.wwnorton.com/physics

3. The photographs in Fig. 11.1 show the impact of an automobile on a rigid wall.

   (a) Measure the positions of the automobile on these photographs and calculate the average velocity for each of the 20×10^{-3}-s intervals between one photograph and the next; calculate the average acceleration for each time interval from the change between one average velocity and the next.

   (b) The mass of this automobile is 1700 kg. Calculate the average force for each time interval.

   (c) Make a plot of this force as a function of time and find the impulse by estimating the area under this curve.

4. The "land divers" of Pentecost Island (Vanuatu) jump from platforms 21 m high. Long liana vines tied to their ankles jerk them to a halt just short of the ground. If the pull of the liana takes 0.020 s to halt the diver, what is the average acceleration of the diver during this time interval? If the mass of the diver is 64 kg, what is the corresponding average force on his ankles?

5. A shotgun fires a slug of lead of mass 28 g with a muzzle velocity of 450 m/s. The slug acquires this velocity while it accelerates along the barrel of the shotgun, which is 70 cm long.

   (a) What is the impulse the shotgun gives the slug?

   (b) Estimate the average impulsive force; assume constant acceleration of the slug along the barrel.

6. A rule of thumb for automobile collisions against a rigid barrier is that the collision lasts about 0.11 s, for any initial speed and for any model of automobile (for instance, the collision illustrated in Fig. 11.1 lasted 0.120 s, in rough agreement with this rule of thumb). Accordingly, the deceleration experienced by an automobile during a collision is directly proportional to the change of velocity Δv (with a constant factor of proportionality), and therefore Δv can be regarded as a measure of the severity of the collision.

   (a) If the collision lasts 0.11 s, what is the average deceleration experienced by an automobile in an impact on a rigid barrier at 55 km/h? 65 km/h? 75 km/h?

   (b) For each of these speeds, what is the crush distance of the front end of the automobile? Assume constant deceleration for this calculation.

   (c) For each of these speeds, what is the average force the seat belt must exert to hold a driver of 75 kg in his seat during the impact?

7. Suppose that a seat-belted mother riding in an automobile holds a 10-kg baby in her arms. The automobile crashes and decelerates from 50 km/h to 0 in 0.10 s. What average force would the mother have to exert on the baby to hold it? Do you think she can do this?

8. In a test, an air force volunteer belted in a chair placed on a rocket sled was decelerated from 143 km/h to 0 in a distance of 5.5 m. Assume that the mass of the volunteer was 75 kg, and assume that the deceleration was uniform. What was the deceleration? What impulse did the seat belt deliver to the volunteer? What time-average force did the seat belt exert?

9. Assume that the Super Ball of Example 3 has a mass of 60 g and is initially traveling with speed 15 m/s. For simplicity, assume that the acceleration is constant while the ball is in contact with the wall. After touching the wall, the center of mass of the Super Ball moves 0.50 cm toward the wall, and then moves the same distance away to complete the bounce. What is the impulse delivered by the wall? What is the time-average force?

10. A 0.50-kg hammerhead moving at 2.0 m/s strikes a board and stops in 0.020 s. What is the impulse delivered to the board? What is the time-average force?

11. A soccer player applies an average force of 180 N during a kick. The kick accelerates a 0.45-kg soccer ball from rest to a speed of 18 m/s. What is the impulse imparted to the ball? What is the collision time?

12. When an egg ($m = 50$ g) strikes a hard surface, the collision lasts about 0.020 s. The egg will break when the average force during impact exceeds 3.0 N. From what minimum height will a dropped egg break?

*13. The net force on a body varies with time according to $F_x = 3.0t + 0.5t^2$, where F_x is in newtons and t is in seconds. What is the impulse imparted to the body during the time interval $0 \le t \le 3.0$ s?

*14. Suppose that in a baseball game, the batter succeeds in hitting the baseball thrown toward him by the pitcher. Suppose that just before the bat hits, the ball is moving toward the batter horizontally with a speed of 35 m/s; and that after the bat has hit, the ball is moving away from the batter and upward at an angle of 50° and finally lands on the ground 110 m away. The mass of the ball is 0.15 kg. From this information, calculate the magnitude and direction of the impulse the ball receives in the collision with the bat. Neglect air friction and neglect the initial height of the ball above the ground.

*15. Bobsleds racing down a bobsled run often suffer glancing collisions with the vertical walls enclosing the run. Suppose that a bobsled of 600 kg traveling at 120 km/h approaches a wall at an angle of 3.0° and bounces off at the same angle. Subsequent inspection of the wall shows that the side of the bobsled made a scratch mark of length 2.5 m along the wall. From these data, calculate the time interval the bobsled was in contact with the wall, and calculate the average magnitude of the force that acted on the side of the bobsled during the collision.

11.2 Elastic Collisions in One Dimension[†]

16. A particle moving at 10 m/s along the x axis collides elastically with another particle moving at 5.0 m/s in the *same* direction along the x axis. The particles have equal masses. What are their speeds after this collision?

17. In a lecture demonstration, two masses collide elastically on a a frictionless air track. The moving mass (projectile) is 60 g,

[†] For help, see Online Concept Tutorial 13 and 14 at www.wwnorton.com/physics

and the initially stationary mass (target) is 120 g. The initial velocity of the projectile is 0.80 m/s.

   (a) What is the velocity of each mass after the collision?

   (b) What is the kinetic energy of each mass before the collision? After the collision?

18. A target sometimes used for target shooting with small bullets consists of a steel disk hanging on a rod which is free to swing on a pivot (in essence, a pendulum). The collision of the bullet with the steel disk is not elastic and not totally inelastic, but somewhere between these extremes. Suppose that a .22-caliber bullet of 15 g and initial speed 600 m/s strikes such a target of mass 40 g. With what velocity would this bullet bounce back (ricochet) if the collision were elastic? Assume that the disk acts like a free particle during the collision.

19. The impact of the head of a golf club on a golf ball can be approximately regarded as an elastic collision. The mass of the head of the golf club is 0.15 kg, and that of the ball is 0.045 kg. If the ball is to acquire a speed of 60 m/s in the collision, what must be the speed of the club before impact?

20. Suppose that a neutron in a nuclear reactor initially has an energy of 4.8×10^{-13} J. How many head-on collisions with carbon nuclei at rest must this neutron make before its energy is reduced to 1.6×10^{-19} J? The collisions are elastic.

21. The impact of a hammer on a nail can be regarded as an elastic collision between the head of the hammer and the nail. Suppose that the mass of the head of the hammer is 0.50 kg and it strikes a nail of mass 12 g with an impact speed of 5.0 m/s. How much energy does the nail acquire in this collision?

22. Consider two coins: a quarter of mass 5.6 g and a dime of mass 2.3 g. If one is sliding at 2.0 m/s on a frictionless surface and hits the other head-on, find the final velocities when either (a) the quarter or (b) the dime is the stationary target. Assume the collision is elastic.

23. Using a straw, a child shoots a series of small balls of mass 1.0 g with speed v at a block of mass 40 g on a frictionless surface. If the small balls elastically collide head-on with the block, how fast will the block be moving after five strikes?

24. A projectile of unknown mass and speed strikes a ball of mass $m = 0.15$ kg initially at rest. The collision is head-on and elastic. The ball moves off at 1.50 m/s, and the projectile continues in its original direction at 0.50 m/s. What is the mass of the projectile? What was its original speed?

25. A marble of unknown mass m is shot at a larger marble of known mass M, initially at rest in the center of a circle. The collision is head-on and elastic. The smaller marble bounces backward and exits the circle in one-third of the time that it takes the larger marble to do so. What is the mass of the smaller marble? Neglect any rolling motion.

26. In materials science, **Rutherford backscattering** is used to determine the composition of materials. In such an experiment, alpha particles (helium nuclei, mass 4.0 u) of typical kinetic energy 1.6×10^{-13} J strike target nuclei at rest. The collisions

are elastic and head-on. What is the recoil kinetic energy of the alpha particle when the target is (a) silicon ($m = 28$ u) and (b) copper ($m = 63$ u)?

*27. An automobile traveling at 60 km/h bumps into the rear of another automobile traveling at 55 km/h in the *same* direction. The mass of the first automobile is 1200 kg, and the mass of the second automobile is 1000 kg. If the collision is elastic, find the velocities of both automobiles immediately after this collision. (Hint: Solve this problem in a reference frame moving with a velocity equal to the initial velocity of one of the automobiles.)

*28. A projectile of 45 kg has a muzzle speed of 656.6 m/s when fired horizontally from a gun held in a rigid support (no recoil). What will be the muzzle speed (relative to the ground) of the same projectile when fired from a gun that is free to recoil? The mass of the gun is 6.6×10^3 kg. (Hint: The kinetic energy of the gun–projectile system is the same in both cases.)

*29. On a smooth, frictionless table, a billiard ball of velocity v is moving toward two other aligned billiard balls in contact (Fig. 11.12). What will be the velocity of each ball after impact? Assume that all balls have the same mass and that the collisions are elastic. Ignore any rotation of the balls. (Hint: Treat this as two successive collisions.)

FIGURE 11.12 Three billiard balls along a line.

*30. Repeat Problem 29 but assume that the middle ball has twice the mass of each of the others.

*31. Two small balls are suspended side by side from two strings of length l so that they touch when in their equilibrium position (see Fig. 11.13). Their masses are m and $2m$, respectively. If the left ball (of mass m) is pulled aside and released from a height h, it will swing down and collide with the right ball (of mass $2m$) at the lowest point. Assume the collision is elastic.

   (a) How high will each ball swing after the collision?

   (b) Both balls again swing down, and they collide once more at the lowest point. How high will each swing after this second collision?

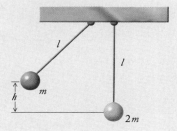

FIGURE 11.13 Two balls suspended from strings.

*32. If a spacecraft, or some other body, approaches a planet at fairly high speed at a suitable angle, it will whip around the planet and recede in a direction almost opposite to the initial direction of motion (Fig. 11.14). This can be regarded approximately as a one-dimensional "collision" between the satellite and the planet; the collision is elastic. In such a collision the satellite will gain kinetic energy from the planet, provided that it approaches the planet along a direction opposite to the direction of the planet's motion. This slingshot effect has been used to boost the speed of both Voyager spacecraft as they passed near Jupiter. Consider the head-on "collision" of a satellite of initial speed 10 km/s with the planet Jupiter, which has a speed of 13 km/s. (The speeds are measured in the reference frame of the Sun.) What is the maximum gain of speed that the satellite can achieve?

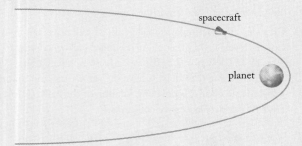

FIGURE 11.14 Spacecraft "colliding" with planet.

**33. A turbine wheel with curved blades is driven by a high-velocity stream of water that impinges on the blades and bounces off (Fig. 11.15). Under ideal conditions the velocity of the water particles after the collision with the blade is exactly zero, so that all of the kinetic energy of the water is transferred to the turbine wheel. If the speed of the water particles is 27 m/s, what is the ideal speed of the turbine blade? (Hint: Treat the collision of a water particle and the blade as a one-dimensional elastic collision.)

FIGURE 11.15 An undershot turbine wheel.

*34. A nuclear reactor designed and built in Canada (CANDU) contains heavy water (D_2O). In this reactor, the fast neutrons are slowed down by elastic collisions with the deuterium nuclei of the heavy-water molecule.

(a) By what factor will the speed of the neutron be reduced in a head-on collision with a deuterium nucleus? The mass of this nucleus is 2.01 u.

(b) After how many head-on collisions with deuterium nuclei will the speed be reduced by the same factor as in a single head-on collision with a proton?

**35. Because of brake failure, an automobile parked on a hill of slope 1:10 rolls 12 m downhill and strikes a parked automobile. The mass of the first automobile is 1400 kg, and the mass of the second automobile is 800 kg. Assume that the first automobile rolls without friction and that the collision is elastic.

(a) What are the velocities of both automobiles immediately after the collision?

(b) After the collision, the first automobile continues to roll downhill, with acceleration, and the second automobile skids downhill, with deceleration. Assume that the second automobile skids with all its wheels locked, with a coefficient of sliding friction 0.90. At what time after the first collision will the automobiles have another collision, and how far from the initial collision?

**36. (a) Show that for an elastic one-dimensional collision the relative velocity reverses during the collision; that is, show that $v_1' - v_2' = -v_1$ (for $v_2 = 0$).

(b) For a partially inelastic collision the relative velocity after the collision will have a smaller magnitude than the relative velocity before the collision. We can express this mathematically as $v_1' - v_2' = -ev_1$, where $e < 1$ is called the **coefficient of restitution.** For some kinds of bodies, the coefficient e is a constant, independent of v_1 and v_2. Show that in this case the final kinetic energy of the motion relative to the center of mass is less than the initial kinetic energy of this motion by a factor of e^2, that is, that $K' = e^2K$.

(c) Derive formulas analogous to Eqs. (11.13) and (11.14) for the velocities v_1' and v_2' in terms of v_1.

11.3 Inelastic Collisions in One Dimension†

37. In karate, the fighter makes the hand collide at high speed with the target; this collision is inelastic, and a large portion of the kinetic energy of the hand becomes available to do damage in the target. According to a crude estimate, the energy required to break a concrete block (28 cm × 15 cm × 1.9 cm supported only at its short edges) is of the order of 10 J. Suppose the fighter delivers a downward hammer-fist strike with a speed of 12 m/s to such a concrete block. In principle, is there enough energy to break the block? Assume that the fist has a mass of 0.4 kg.

38. According to a tall tale told by Baron Münchhausen, on one occasion, while cannon shots were being exchanged between a

† For help, see Online Concept Tutorial 13 and 14 at www.wwnorton.com/physics

besieged city and the enemy camp, he jumped on a cannonball as it was being fired from the city, rode the cannonball toward the enemy camp, and then, in midair, jumped onto an enemy cannonball and rode back to the city. The collision of Münchhausen and the enemy cannonball must have been inelastic, since he held on to it. Suppose that his speed just before hitting the enemy cannonball was 150 m/s southward and the speed of the enemy cannonball was 300 m/s northward. The mass of Münchhausen was 90 kg, and the mass of the enemy cannonball was 20 kg. What must have been the speed just after the collision? Do you think he made it back to the city?

39. As described in Problem 6, the change of velocity Δv of an automobile during a collision is a measure of the severity of the collision. Suppose that an automobile moving with an initial speed of 15 m/s collides with (a) an automobile of equal mass initially at rest, (b) an automobile of equal mass initially moving in the opposite direction at 15 m/s, or (c) a stationary rigid barrier. Assume that the collision is totally inelastic. What is Δv in each case?

40. A 25-kg boy on a 10-kg sled is coasting at 3.0 m/s on level ice toward his 30-kg sister. The girl jumps vertically and lands on her brother's back. What is the final speed of the siblings and sled? Neglect friction.

41. A 75-kg woman and a 65-kg man face each other on a frictionless ice pond. The woman holds a 5.0-kg "medicine ball." The woman throws the ball to the man with a horizontal velocity of 2.5 m/s relative to the ice. What is her recoil velocity? What is the man's velocity after catching the ball? The man then throws the ball horizontally to the woman at 3.0 m/s relative to himself at the instant before release. What is his final velocity? What is the woman's final velocity after catching it?

42. A 16-u oxygen atom traveling at 600 m/s collides head-on with another oxygen atom at rest. The two join and form an oxygen molecule. With what speed does the molecule move? What fraction of the original translational kinetic energy is transferred to internal energy of the molecule?

*43. A circus clown in a cannon is shot vertically upward with an initial speed of 12 m/s. After ascending 3.5 m, she collides with and grabs a performer sitting still on a trapeze. They ascend together and then fall. What is their speed when they reach the original launch height? The clown and trapeze artist have the same mass.

*44. As described in Problem 6, the change in velocity Δv of an automobile during a collision is a measure of the severity of the collision. For a collision between two automobiles of equal masses, Δv has the same magnitude for each automobile. But for a collision between automobiles of different masses, Δv is larger for the automobile of smaller mass. Suppose that an automobile of 800 kg moving with an initial speed of 15 m/s collides with (a) an automobile of 1400 kg initially at rest, (b) an automobile of 1400 kg initially moving in the opposite direction at 15 m/s, or (c) a stationary rigid barrier. Assume that the collision is totally inelastic. What is Δv in each case for each participating automobile?

*45. Two automobiles of 540 and 1400 kg collide head-on while moving at 80 km/h in opposite directions. After the collision the automobiles remain locked together.

   (a) Find the velocity of the wreck immediately after the collision.

   (b) Find the kinetic energy of the two-automobile system before and after the collision.

   (c) The front end of each automobile crumples by 0.60 m during the collision. Find the acceleration (relative to the ground) of the passenger compartment of each automobile; make the assumption that these accelerations are constant during the collision.

*46. A speeding automobile strikes the rear of a parked automobile. After the impact the two automobiles remain locked together, and they skid along the pavement with all their wheels locked. An investigation of this accident establishes that the length of the skid marks made by the automobiles after the impact was 18 m; the mass of the moving automobile was 2200 kg and that of the parked automobile was 1400 kg, and the coefficient of sliding friction between the wheels and the pavement was 0.95.

   (a) What was the speed of the two automobiles immediately after impact?

   (b) What was the speed of the moving automobiles before impact?

*47. A proton of energy 8.0×10^{-13} J collides head-on with a proton at rest. How much energy is available for inelastic reactions between these protons?

*48. According to test procedures laid down by the National Highway Traffic Safety Administration, a stationary barrier (of very large mass) and a towed automobile are used for tests of front impacts (Fig. 11.16a), but a moving barrier of 1800 kg and a stationary, unbraked automobile are used for tests of rear impacts (Fig. 11.16b). Explain how this test with the moving barrier and the stationary automobile could be replaced by an equivalent test with a stationary barrier and an automobile towed *backward* at some appropriate speed. If the automobile has a mass of 1400 kg and the moving barrier has a speed of 8 km/h, what is the appropriate equivalent speed of the moving automobile towed backward to the stationary barrier? Assume the collision is inelastic.

(a)

(b)

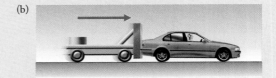

FIGURE 11.16 (a) Test procedure for front impact. (b) Test procedure for rear impact.

*49. Regard the two automobiles described in Example 7 as a system of two particles.

  (a) What is the translational kinetic energy of the center of mass before the collision? After the collision?

  (b) What is the total kinetic energy before the collision? After the collision?

*50. A cat crouches on the floor, at a distance of 1.2 m from a desk chair of height 0.45 m. The cat jumps onto the chair, landing with zero vertical velocity (this is standard procedure for cat jumps). The desk chair has frictionless coasters and rolls away when the cat lands. The mass of the cat is 4.5 kg, and the mass of the chair is 12 kg. What is the speed of recoil of the chair and cat?

*51. A crude but simple method for measuring the speed of a bullet is to shoot the bullet horizontally into a block of wood resting on a table. The block of wood will then slide until its kinetic energy is expended against the friction of the surface of the table. Suppose that a 3.0-kg block of wood slides a distance of 6.0 cm after it is struck by a bullet of 12 g. If the coefficient of sliding friction for the wood on the table is 0.60, what impact speed can you deduce for the bullet?

*52. Another way (not recommended) to measure the speed of bullets with a ballistic pendulum is to shoot a steady stream of bullets into the pendulum, which will push it aside and hold it in a rough equilibrium position at some angle. The speed can be calculated from this equilibrium angle. Suppose that you shoot .22-caliber bullets of mass 15 g into a 4.0-kg ballistic pendulum at the rate of 2 per second. You find that the equilibrium angle is 24°. What is the speed of the bullets?

*53. You shoot a .22-caliber bullet through a piece of wood sitting on a table. The piece of wood acquires a speed of 8.0 m/s, and the bullet emerges with a reduced speed. The mass of the bullet is 15 g, and its initial speed is 600 m/s; the mass of the piece of wood is 300 g.

  (a) What is the change of velocity of the bullet?

  (b) What is the change of kinetic energy of the bullet?

  (c) What is the change of kinetic energy of the wood?

  (d) Account for the missing kinetic energy.

*54. An automobile traveling at 50 km/h strikes the rear of a parked automobile. After the collision, the two automobiles remain joined together. The parked automobile skids with all its wheels locked, but the other automobile rolls with negligible friction. The mass of each automobile is 1300 kg, and the coefficient of sliding friction between the locked wheels and the pavement is 0.90. How far do the joined automobiles move before they stop? How long do they take to stop?

*55. You can make a fairly accurate measurement of the speed of a bullet by shooting it horizontally into a block of wood sitting on a fence. The collision of the bullet and the block is inelastic, and the block will fall off the fence and land on the ground at some distance from the bottom of the fence. The speed of the bullet is proportional to this distance. Suppose that a bullet of mass 15 g fired into a block of 4.0 kg sitting on a 1.8-m fence causes the block to land 1.4 m from the bottom of the fence. Calculate the speed of the bullet.

11.4 Collisions in Two and Three Dimensions

56. A cheetah intercepts a gazelle on the run, and grabs it (a totally inelastic collision). Just before this collision, the gazelle was running due north at 20 m/s, and the cheetah was running on an intercepting course of 45° east of north at 22 m/s. The mass of the cheetah is 60 kg, and the mass of the gazelle is 50 kg. What are the magnitude and the direction of the velocity of the entangled animals at the instant after the collision?

57. Two hydrogen atoms ($m = 1.0$ u) with equal speeds, initially traveling in perpendicular directions, collide and join together to form a hydrogen molecule. If 6.1×10^{-22} J of the initial kinetic energy is transferred to internal energy in the collision, what was the initial speed of the atoms?

*58. Two automobiles of equal masses collide at an intersection. One was traveling eastward and the other northward. After the collision, they remain joined together and skid, with locked wheels, before coming to rest. The length of the skid marks is 18 m, and the coefficient of friction between the locked wheels and the pavement is 0.80. Each driver claims his speed was less than 14 m/s (50 km/h) before the collision. Prove that at least one driver is lying.

*59. Two hockey players (see Fig. 11.17) of mass 80 kg collide while skating at 7.0 m/s. The angle between their initial directions of motion is 130°.

  (a) Suppose that the players remain entangled and that the collision is totally inelastic. What is their velocity immediately after collision?

  (b) Suppose that the collision lasts 0.080 s. What is the magnitude of the average acceleration of each player during the collision?

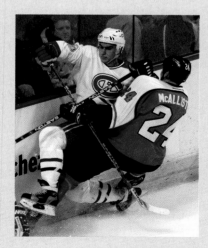

FIGURE 11.17 Collision of two hockey players.

*60. On July 27, 1956, the ships *Andrea Doria* (40 000 metric tons) and *Stockholm* (20 000 metric tons) collided in the fog south of Nantucket and remained locked together (for a while). Immediately before the collision the velocity of the *Andrea Doria* was 22 knots at 15° east of south and that of the *Stockholm* was 19 knots at 48° east of south (1 knot = 1 nmi/h = 1.85 km/h).

(a) Calculate the velocity (magnitude and direction) of the combined wreck immediately after the collision.

(b) Find the amount of kinetic energy that was converted into other forms of energy by inelastic processes during the collision.

(c) The large amount of energy absorbed by inelastic processes accounts for the heavy damage to both ships. How many kilograms of TNT would have to be exploded to obtain the same amount of energy as was absorbed by inelastic processes in the collision? The explosion of 1 kg of TNT releases 4.6×10^6 J.

*61. Your automobile of mass $m_1 = 900$ kg collides at a traffic circle with another automobile of mass $m_2 = 1200$ kg. Just before the collision, your automobile was moving due east and the other automobile was moving 40° south of east. After the collision the two automobiles remain entangled while they skid, with locked wheels, until coming to rest. Your speed before the collision was 14 m/s. The length of the skid marks is 17.4 m, and the coefficient of kinetic friction between the tires and the pavement is 0.85. Calculate the speed of the other automobile before the collision.

*62. Two billiard balls are placed in contact on a smooth, frictionless table. A third ball moves toward this pair with velocity v in the direction shown in Fig. 11.18. What will be the velocity (magnitude and direction) of the three balls after the collision? The balls are identical and the collisions are elastic.

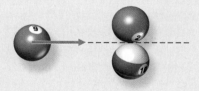

FIGURE 11.18 Three billiard balls.

*63. A billiard ball of mass m and radius R moving with speed v on a smooth, frictionless table collides elastically with an identical stationary billiard ball glued firmly to the surface of the table.

(a) Find a formula for the angular deflection suffered by the moving billiard ball as a function of the impact parameter b (defined in Fig. 11.19). Assume the billiard balls are very smooth so that the force during contact is entirely along the center-to-center line of the balls.

(b) Find a formula for the magnitude of the momentum change suffered by the billiard ball.

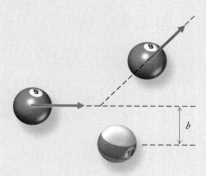

FIGURE 11.19 Two billiard balls.

*64. A coin of mass m slides along a table with speed v and elastically collides with a second, identical coin at rest. The first coin is deflected 60° from its original direction. What are the speeds of each of the two coins after the collision? At what angle does the second coin exit the collision?

*65. In a head-on elastic collision between a projectile and a stationary target of equal mass, we saw that the projectile stops. Show that if such a collision is not head-on, then the projectile and target final velocities are perpendicular (see Fig. 11.20). (Hint: Square the conservation of momentum equation, using $p^2 = \mathbf{p} \cdot \mathbf{p}$, and compare the resulting equation with the energy conservation equation.)

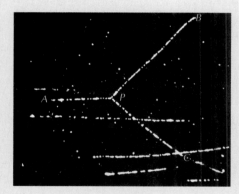

FIGURE 11.20 Elastic collision between two protons. The final velocities of the protons are perpendicular.

*66. In an elastic collision in two dimensions, the projectile has twice the mass of the stationary target. After the collision, the target moves off with three times the final speed of the projectile. Find the angle between the two final directions of motion.

REVIEW PROBLEMS

67. High-speed photography reveals that when a golf club hits a golf ball, the club and the ball typically remain in contact for 1.0×10^{-3} s and the ball acquires a speed of 70 m/s. The mass of the ball is 45 g. What is the impulse the club delivers to the ball? What is the time-average force?

68. In a remarkable incident, a 52-kg woman jumped from the 10th floor of an apartment building, fell 28 m, and landed on her side on soft earth in a freshly dug garden. She fractured her wrist and rib, but remained conscious and fully alert, and recovered completely after some time in a hospital. The earth was depressed 15 cm by her impact.

    (a) What was her impact speed?

    (b) Assuming constant deceleration upon contact with the ground, what was her deceleration?

    (c) What was the force of the ground on her body during deceleration?

69. An automobile approaching an intersection at 10 km/h bumps into the rear of another automobile standing at the intersection with its brakes off and its gears in neutral. The mass of the moving automobile is 1200 kg, and that of the stationary automobile is 700 kg. If the collision is elastic, find the velocities of both automobiles after the collision.

70. It has been reported (fallaciously) that the deer botfly can attain a maximum airspeed of 1318 km/h, that is, 366 m/s. Suppose that such a fly, buzzing along at this speed, strikes a stationary hummingbird and remains stuck in it. What will be the recoil velocity of the hummingbird? The mass of the fly is 2 g; the mass of the hummingbird is 50 g.

*71. A proton of energy 8.0×10^{-13} J collides head-on with a proton of energy 4.0×10^{-13} J moving in the opposite direction. How much energy is available for inelastic reactions between these protons?

*72. When a baseball bat strikes a ball, the impact can be approximately regarded as an elastic collision (the hands of the hitter have little effect on the short time the bat and the ball are in contact). Suppose that a bat of 0.85 kg moving horizontally at 30 m/s encounters a ball of 0.15 kg moving at 40 m/s in the opposite direction. We cannot directly apply the results of Section 11.2 to this collision, since *both* particles are in motion before collision ($v_1 = 40$ m/s and $v_2 = -30$ m/s). However, we can apply these results if we use a reference frame that moves at a velocity $V_0 = -30$ m/s in the direction of the initial motion of the bat; in this reference frame, the initial velocity of the bat is zero ($v_2 = 0$)

    (a) What is the initial velocity of the ball in this reference frame?

    (b) What are the final velocities of the ball and the bat, just after the collision?

    (c) What are these final velocities in the reference frame of the ground?

*73. A boy throws a baseball at another baseball sitting on a 1.5-m-high fence. The collision of the balls is elastic. The thrown ball moves horizontally at 20 m/s just before the head-on collision.

    (a) What are the velocities of the two balls just after the collision?

    (b) Where do the two balls land on the ground?

74. An automobile of 1200 kg traveling at 45 km/h strikes a moose of 400 kg standing on the road. Assume that the collision is totally inelastic (the moose remains draped over the front end of the automobile). What is the speed of the automobile immediately after this collision?

75. A ship of 3.0×10^4 metric tons steaming at 40 km/h strikes an iceberg of 8.0×10^5 metric tons. If the collision is totally inelastic, what fraction of the initial kinetic energy of the ship is converted into inelastic energy? What fraction remains as kinetic energy of the ship–iceberg system? Ignore the effect of the water on the motion of the ship and iceberg.

*76. When William Tell shot the apple off his son's head, the arrow remained stuck in the apple, which means the collision between the arrow and apple was totally inelastic. Suppose that the velocity of the arrow was horizontal at 80 m/s before it hit, the mass of the arrow was 40 g, and the mass of the apple was 200 g. Suppose Tell's son was 1.40 m high.

    (a) Calculate the velocity of the apple and arrow immediately after the collision.

    (b) Calculate how far behind the son the apple and arrow landed on the ground.

*77. Meteor Crater in Arizona (Fig. 11.21), a hole 180 m deep and 1300 m across, was gouged in the surface of the Earth by the impact of a large meteorite. The mass and speed of this meteorite have been estimated at 2.0×10^9 kg and 10 km/s, respectively, before impact.

    (a) What recoil velocity did the Earth acquire during this (inelastic) collision?

FIGURE 11.21 Meteor Crater in Arizona.

(b) How much kinetic energy was released for inelastic processes during the collision? Express this energy in the equivalent of tons of TNT; 1 ton of TNT releases 4.2×10^9 J upon explosion.

(c) Estimate the magnitude of the impulsive force.

*78. A black automobile smashes into the rear of a white automobile stopped at a stop sign. You investigate this collision and find that before the collision, the black automobile made skid marks 5.0 m long; after the collision the black automobile made skid marks 1.0 m long (in the same direction as the initial direction of motion), and the white automobile made skid marks 2.0 m long. Both automobiles made these skid marks with all their wheels. The mass of the black automobile is 1400 kg, and the mass of the white automobile is 800 kg. The coefficient of sliding friction between the wheels and the pavement is 0.90. From these data, deduce the speed of the black automobile just before the collision, and the speed before it started to brake.

*79. (a) Two identical small steel balls are suspended from strings of length l so they touch when hanging straight down, in their equilibrium position (Fig. 11.22). If we pull one of the balls back until its string makes an angle θ with the

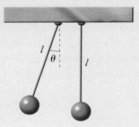

FIGURE 11.22 Two balls suspended from strings.

vertical and then let it go, it will collide elastically with the other ball. How high will the other ball rise?

(b) Suppose that instead of steel balls we use putty balls. They will then collide inelastically and remain stuck together. How high will the balls rise?

*80. While in flight, a peregrine falcon spots a pigeon flying 40 m below. The falcon closes its wings and, in free fall, dives on the pigeon and grabs it (a totally inelastic collision). The mass of the falcon is 1.5 kg, and the mass of the pigeon is 0.40 kg. Suppose that the velocity of the pigeon before this collision is horizontal, at 15 m/s, and the velocity of the falcon is vertical, equal to the free-fall velocity. What is the velocity (magnitude and direction) of both birds after the collision?

*81. On a freeway, a truck of 3500 kg collides with an automobile of 1500 kg that is trying to cut diagonally across the path of the truck. Just before the collision, the truck was traveling due north at 70 km/h, and the automobile was traveling at 30° west of north at 100 km/h. After the collision, the vehicles remain joined together.

(a) What is the velocity (magnitude and direction) of the joined vehicles immediately after the collision?

(b) How much kinetic energy is lost during the collision?

*82. Two asteroids of 1.0×10^7 kg and 8.0×10^7 kg, respectively, are initially at rest in interstellar space separated by a large distance. Their mutual gravitational attraction then causes them to fall toward each other on a straight line. Assume the asteroids are spheres of radius 100 m and 200 m, respectively.

(a) What is the velocity of each asteroid just before they hit? What is the kinetic energy of each? What is the total kinetic energy?

(b) The collision is totally inelastic. What is the velocity of the joined asteroids after they hit?

Answers to Checkups

Checkup 11.1

1. The front end should be soft and crushable to protect automobile occupants in a collision; this will spread the momentum change over a longer time, lowering the force experienced by the occupants.

2. The steel ball will exert a larger force, because it is less deformable than the golf ball. Thus, although the change in momentum (the impulse $\mathbf{I}$) can be the same, the steel ball is in contact for a shorter time Δt, and so exerts a greater average force during that time ($\overline{\mathbf{F}} = \mathbf{I}/\Delta t$).

3. Because the collision is elastic, the ball will rebound with the same kinetic energy; as this energy gets converted to potential energy, the ball will rise up to the same height, 1 m, from which it was dropped before stopping.

4. No. Since the wad of gum stopped, kinetic energy was lost, and thus the collision was not elastic. We will later refer to such a collision (when the bodies stick together) as totally *in*elastic.

5. (C) 3×10^6 N. In a collision, each vehicle exerts an equal-magnitude, but opposite-direction, force on the other (an action–reaction pair), so the force exerted by the car on the truck is 3×10^6 N westward.

Checkup 11.2

1. As in the cases just discussed, and as in Eq. (11.13), where the projectile's final velocity is proportional to $m_1 - m_2$, the velocity of the projectile will be positive when it is more massive than the target ($m_1 > m_2$), and it will be negative when it is less massive than the target ($m_1 < m_2$).

2. No. As we saw in the cases just discussed, the speed of recoil of a massive target is very small; in the limit of a very light target, the speed approaches twice the speed of the projectile. For any values of m_1 and m_2, the final speed of the target (v_2'), given by Eq. (11.14), cannot exceed twice the projectile speed (v_1).

3. No; for instance, in the collision of the Super Ball and the wall, the ball is instantaneously at rest before it bounces back. The kinetic energy is transformed into elastic energy momentarily, and then converted back into kinetic energy.

4. (E) 0; v_1. As discussed above, when the masses of the target and projectile are identical, the speed of the projectile is zero after the collision [since we have $m_1 - m_2 = 0$ in Eq. (11.13)]. For identical masses, the target speed is equal to the initial speed of the projectile, v_1 [since we have $m_1 = m_2$ in Eq. (11.14)].

Checkup 11.3

1. No; for example, if the target is initially at rest, it gains kinetic energy.

2. Two particles of equal mass and opposite velocities have zero net momentum. Thus, in a totally inelastic collision, the composite particle has zero momentum, and thus zero velocity. In an elastic collision, the total kinetic energy is unchanged; since the net momentum is zero, the particles must again have opposite velocities. If we ignore the possibility that the particles might have passed through each other, then this means that their velocities were reversed by the collision.

3. The velocity of the joined particles after a totally inelastic collision is the velocity of the center of mass, $v_{CM} = m_1 v_1 / (m_1 + m_2)$; this is equal to one-half of the velocity of the incident projectile when the masses of the target and projectile are equal, or $m_1 = m_2$.

4. No, assuming the wires are long enough to permit the upward motion of the pendulum to the maximum height h.

5. (B) $\frac{1}{3}v$. Momentum is conserved, so equating the initial and final momenta, we have $mv - (m/2)v = (\frac{3}{2}m)v'$, which implies $v' = \frac{1}{3}v$.

Checkup 11.4

1. Because the cars have equal mass and speed, the total momentum before and after this totally inelastic collision is directed due southwest.

2. (B) Southeast. This explosion is like a three-particle totally inelastic collision in reverse. Since the total momentum before the "collision" (explosion) is zero, so must it be afterward: the third particle must have momentum components which cancel the northward and westward momentum contributions of the other two particles; thus, the third particle travels in the southeast quadrant of directions.

Rotation of a Rigid Body

12.1 Motion of a Rigid Body

12.2 Rotation about a Fixed Axis

12.3 Motion with Constant Angular Acceleration

12.4 Motion with Time-Dependent Angular Acceleration

12.5 Kinetic Energy of Rotation; Moment of Inertia

**Concepts
— in —
Context**

CONCEPTS IN CONTEXT

This large centrifuge at the Sandia National Laboratory is used for testing the behavior of components of rockets, satellites, and reentry vehicles when subjected to high accelerations. The components to be tested are placed in a compartment in one arm of this centrifuge; the opposite arm holds a counterweight. The arms rotate at up to 175 revolutions per minute, and they generate a centripetal acceleration of up to 300g.

The concepts of this chapter permit us to answer several questions about this centrifuge:

? How are the speed and the centripetal acceleration at the end of an arm related to the rate of rotation? (Example 5, page 373)

? How do we determine the resistance that the centrifuge offers to changes in its rotational motion? (Example 12, part (a), page 383)

? How is the kinetic energy of the centrifuge arms related to the rate of rotation? (Example 12, part (b), page 383)

365

A *body is* **rigid** *if the particles in the body do not move relative to one another.* Thus, the body has a fixed shape, and all its parts have a fixed position relative to one another. A hammer is a rigid body, and so is a baseball bat. A baseball is not rigid—when struck a blow by the bat, the ball suffers a substantial deformation; that is, different parts of the ball move relative to one another. However, the baseball can be regarded as a rigid body while it flies through the air—the air resistance is not sufficiently large to produce an appreciable deformation of the ball. This example indicates that whether a body can be regarded as rigid depends on the circumstances. No body is absolutely rigid; when subjected to a sufficiently large force, any body will suffer some deformation or perhaps even break into several pieces. In this chapter, we will ignore such deformations produced by the forces acting on bodies. We will examine the motion of bodies under the assumption that rigidity is a good approximation.

12.1 MOTION OF A RIGID BODY

A rigid body can simultaneously have two kinds of motion: it can change its position in space, and it can change its orientation in space. Change of position is translational motion; as we saw in Chapter 10, this motion can be conveniently described as motion of the center of mass. Change in orientation is rotational motion; that is, it is rotation about some axis.

As an example, consider the motion of a hammer thrown upward (see Fig. 12.1). The orientation of the hammer changes relative to fixed coordinates attached to the ground. Instantaneously, the hammer rotates about a horizontal axis, say, a horizontal axis that passes through the center of mass. In Fig. 12.1, this horizontal axis sticks out of the plane of the page and moves upward with the center of mass. The complete motion can then be described as a rotation of the hammer about this axis and a simultaneous translation of the axis along a parabolic path.

In this example of the thrown hammer, the axis of rotation always remains horizontal, out of the plane of the page. In the general case of motion of a rigid body, the axis of rotation can have any direction and can also change its direction. To describe such complicated motion, it is convenient to separate the rotation into three components along three perpendicular axes. The three components of rotation are illustrated by the motion of an aircraft (see Fig. 12.2): the aircraft can turn left or right (yaw), it can tilt to the left or the right (roll), and it can tilt its nose up or down (pitch). However, in the following sections we will usually not deal with this general case of rotation with three components; we will mostly deal only with the simple case of rotation about a fixed axis, such as the rotational motion of a fan, a roulette wheel, a compact disc, a swinging door, or a merry-go-round (see Fig. 12.3).

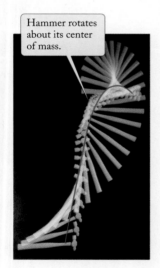

Hammer rotates about its center of mass.

FIGURE 12.1 A hammer in free fall under the influence of gravity. The center of mass of the hammer moves with constant vertical acceleration *g*, just like a particle in free fall.

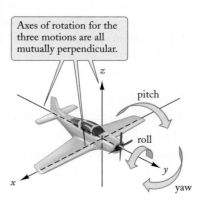

Axes of rotation for the three motions are all mutually perpendicular.

FIGURE 12.2 Pitch, roll, and yaw motions of an aircraft.

(a)

(b)

(c)

(d)

(e)

Checkup 12.1

QUESTION 1: Characterize the following motions as translational, rotational, or both: swinging motion of door, motion of wheel of train, motion of propeller of airplane while in level flight.

QUESTION 2: Suppose that instead of selecting an axis through the center of mass of the hammer in Fig. 12.1, we select a parallel axis through the end of the handle. Can the motion still be described as rotation about this axis and a simultaneous translation of the axis along some path? Is this path parabolic?

QUESTION 3: Under what conditions will the passenger compartment of an automobile exhibit (limited) rolling, pitching, and turning motions?

QUESTION 4: Which of the rotating bodies in Fig. 12.3 does *not* rotate about an axis through its center of mass?

  (A) Fan   (B) Roulette wheel   (C) Compact disc
  (D) Swinging door   (E) Merry-go-round

FIGURE 12.3 Some examples of rotational motion with a fixed axis (a) fan, (b) roulette wheel, (c) compact disc, (d) swinging door, (e) merry-go-round.

12.2 ROTATION ABOUT A FIXED AXIS

Online
Concept
Tutorial

Figure 12.4 shows a rigid body rotating about a fixed axis, which coincides with the z axis. During this rotational motion, each point of the body remains at a given distance from this axis and moves along a circle centered on the axis. To describe the orientation of the body at any instant, we select one particle in the body and use it as a reference point; any particle can serve as reference point, provided that it is not on the axis of rotation. The circular motion of this reference particle (labeled P in Fig. 12.4) is then representative of the rotational motion of the entire body, and the angular position of this particle is representative of the angular orientation of the entire body.

  Figure 12.5 shows the rotating rigid body as seen from along the axis of rotation. The coordinates in Fig. 12.5 have been chosen so the z axis coincides with the axis of rotation, whereas the x and y axes are in the plane of the circle traced out by the motion of the reference particle. The angular position of the reference particle—and hence the angular orientation of the entire rigid body—can be described by the position angle ϕ (the Greek

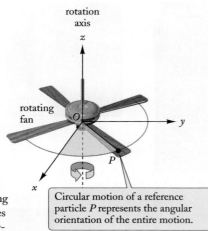

FIGURE 12.4 The four blades of this fan are a rigid body rotating about a fixed axis, which coincides with the *z* axis. The reference particle *P* in this rigid body moves along a circle around this axis.

Circular motion of a reference particle *P* represents the angular orientation of the entire motion.

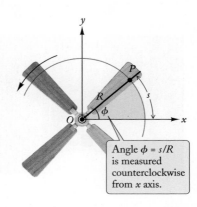

Angle $\phi = s/R$ is measured counterclockwise from *x* axis.

FIGURE 12.5 Motion of a reference particle *P* in the rigid body rotating about a fixed axis. The axis is indicated by the circled dot *O*. The radius of the circle traced out by the motion of the reference particle is *R*.

letter *phi*) between the radial line *OP* and the *x* axis. Conventionally, the angle ϕ is taken as positive when reckoned in a counterclockwise direction (as in Fig. 12.5). We will usually measure this position angle in radians, rather than degrees. By definition, *the **angle ϕ in radians** is the length s of the circular arc divided by the radius R*, or

angle in radians

$$\phi = \frac{s}{R} \qquad (12.1)$$

In Fig. 12.5, the length *s* is the distance traveled by the reference particle from the *x* axis to the point *P*. Note that if the length *s* is the circumference of a full circle, then $s = 2\pi R$, and $\phi = s/R = 2\pi R/R = 2\pi$. Thus, there are 2π radians in a full circle; that is, there are 2π radians in 360°:

$$2\pi \text{ radians} = 360°$$

Accordingly, 1 radian equals 360°/2π, or

$$1 \text{ radian} = 57.3°$$

EXAMPLE 1   The accuracy of the guidance system of the Hubble Space Telescope is such that if the telescope were sitting in New York, the guidance system could aim at a dime placed on top of the Washington Monument, at a distance of 320 km. The width of a dime is 1.8 cm. What angle does the dime subtend when seen from New York?

SOLUTION: Figure 12.6 shows the circular arc subtended by the dime. The radius of the circle is 320 km. For a small angle, such as in this figure, the length *s* of the arc from one side of the dime to the other is approximately the same as the length of the straight line from one side to the other, which is the width of the dime. Hence the angle in radians is

$$\phi = \frac{s}{R} = \frac{1.8 \times 10^{-2}\,\text{m}}{3.2 \times 10^{5}\,\text{m}} = 5.6 \times 10^{-8}\text{ radian}$$

Expressed in degrees, this becomes

$$\phi = 5.6 \times 10^{-8}\text{ radian} \times \frac{360°}{2\pi\text{ radians}} = 3.2 \times 10^{-6}\text{ degree}$$

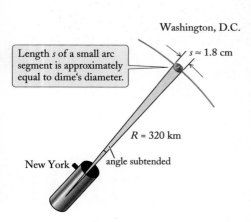

Length *s* of a small arc segment is approximately equal to dime's diameter.

Washington, D.C.

$s \approx 1.8$ cm

$R = 320$ km

New York   angle subtended

FIGURE 12.6 A dime placed at a distance of 320 km from the telescope. The length $s = 1.8$ cm is the diameter of the dime.

When a rigid body rotates, the position angle ϕ changes in time. The body then has an **angular velocity** ω (the Greek letter *omega*). The definition of the angular velocity for rotational motion is mathematically analogous to the definition of velocity for translational motion (see Sections 2.2 and 2.3). The **average angular velocity** $\overline{\omega}$ is defined as

$$\overline{\omega} = \frac{\Delta\phi}{\Delta t} \qquad (12.2)$$

average angular velocity

where $\Delta\phi$ is the change in the angular position and Δt the corresponding change in time. The **instantaneous angular velocity** is defined as

$$\omega = \frac{d\phi}{dt} \qquad (12.3)$$

instantaneous angular velocity

According to these definitions, the angular velocity is the rate of change of the angle with time. The unit of angular velocity is the radian per second (1 radian/s). The radian is the ratio of two lengths [compare Eq. (12.1)], and hence it is a pure number; thus, 1 radian/s is the same thing as 1/s. However, to prevent confusion, it is often useful to retain the vacuous label *radian* as a reminder that angular motion is involved. Table 12.1 gives some examples of angular velocities.

If the body rotates with constant angular velocity, then we can also measure the rate of rotation in terms of the ordinary **frequency** f, or the number of revolutions per second. Since each complete revolution involves a change of ϕ by 2π radians, the frequency of revolution is smaller than the angular velocity by a factor of 2π:

$$f = \frac{\omega}{2\pi} \qquad (12.4)$$

frequency

This expresses the frequency in terms of the angular velocity. The unit of rotational frequency is the revolution per second (1 rev/s). Like the radian, the revolution is a pure number, and hence 1 rev/s is the same thing as 1/s. But we will keep the label *rev* to prevent confusion between rev/s and radian/s.

As in the case of planetary motion, the time per revolution is called the **period** of the motion. If the number of revolutions per second is f, then the time per revolution is $1/f$, that is,

$$T = \frac{1}{f} \qquad (12.5)$$

period of motion

TABLE 12.1   SOME ANGULAR VELOCITIES

Computer hard disk	8×10^2 radians/s	Helicopter rotor	40 radians/s
Circular saw	7×10^2 radians/s	Compact disc (outer track)	22 radians/s
Electric blender blades	5×10^2 radians/s	Phonograph turntable	3.5 radians/s
Jet engine	4×10^2 radians/s	Neutron star (pulsar) rotation	0.1 radian/s
Airplane propeller	3×10^2 radians/s	Earth rotation	7.3×10^{-5} radian/s
Automobile engine	2×10^2 radians/s	Earth revolution about Sun	2.0×10^{-7} radian/s
Small fan	60 radians/s		

EXAMPLE 2   The rotational frequency of machinery is often expressed in revolutions per minute, or rpm. A typical ceiling fan on medium speed rotates at 150 rpm. What is the frequency of revolution? What is the angular velocity? What is the period of the motion?

SOLUTION: Each minute is 60.0 s; hence 150 revolutions per minute amounts to 150 revolutions in 60.0 s; so

$$f = \frac{150 \text{ rev}}{60.0 \text{ s}} = 2.50 \text{ rev/s}$$

Since each revolution comprises 2π radians, the angular velocity is

$$\omega = 2\pi f = 2\pi \times 2.50 \text{ rev/s} = 15.7 \text{ radians/s}$$

Note that here we have dropped a label *rev* in the third step and inserted a label *radians;* as remarked above, these labels merely serve to prevent confusion, and they can be inserted and dropped at will once they have served their purpose.

The period of the motion is

$$T = \frac{1}{f} = \frac{1}{2.50 \text{ rev/s}} = 0.400 \text{ s}$$

One complete revolution takes two-fifths of a second.

If the angular velocity of a rigid body is changing, the body has an **angular acceleration** α (the Greek letter *alpha*). The rotational motion of a ceiling fan that is gradually building up speed immediately after being turned on is an example of accelerated rotational motion. The mathematical definition of the **average angular acceleration** is, again, analogous to the definition of acceleration for translational motion. If the angular velocity changes by $\Delta\omega$ in a time Δt, then the average angular acceleration is

average angular acceleration

$$\overline{\alpha} = \frac{\Delta\omega}{\Delta t} \tag{12.6}$$

and the **instantaneous angular acceleration** is

instantaneous angular acceleration

$$\alpha = \frac{d\omega}{dt} \tag{12.7}$$

Thus, the angular acceleration is the rate of change of the angular velocity. The unit of angular acceleration is the radian per second per second, or radian per second squared (1 radian/s^2).

Since the angular velocity ω is the rate of change of the angular position ϕ [see Eq. (12.3)], the angular acceleration given by Eq. (12.7) can also be written

$$\alpha = \frac{d^2\phi}{dt^2} \tag{12.8}$$

Equations (12.3) and (12.7) give the angular velocity and acceleration of the rigid body; that is, they give the angular velocity and acceleration of every particle in the body. It is interesting to focus on one of the particles and evaluate its *translational* speed and acceleration as it moves along its circular path around the axis of rotation of the rigid body. If the particle is at a distance R from the axis of rotation (see Fig. 12.7), then the length along the circular path of the particle is, according to the definition of angle, Eq. (12.1),

$$s = \phi R \qquad (12.9)$$

Since R is a constant, the rate of change of s is entirely due to the rate of change of ϕ, so

$$\frac{ds}{dt} = \frac{d\phi}{dt} R \qquad (12.10)$$

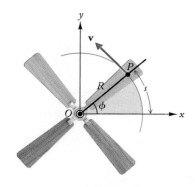

FIGURE 12.7 The instantaneous translational velocity of a particle in a rotating rigid body is tangent to the circular path.

Here ds/dt is the translational speed v with which the particle moves along its circular path, and $d\phi/dt$ is the angular velocity ω; hence Eq. (12.10) is equivalent to

$$v = \omega R \qquad (12.11)$$

translational speed in circular motion

This shows that the translational speed of the particle along its circular path around the axis is directly proportional to the radius: the farther a particle in the rigid body is from the axis, the faster it moves. We can understand this by comparing the motions of two particles, one on a circle of large radius R_1, and the other on a circle of smaller radius R_2 (see Fig. 12.8). For each revolution of the rigid body, both of these particles complete one trip around their circles. But the particle on the larger circle has to travel a larger distance, and hence must move with a larger speed.

For a particle at a given R, the translational speed is constant if the angular velocity is constant. This speed is the distance around the circular path (the circumference) divided by the time for one revolution (the period), or

$$v = \frac{2\pi R}{T} \qquad \text{(constant speed)} \qquad (12.12)$$

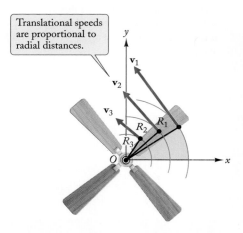

Translational speeds are proportional to radial distances.

FIGURE 12.8 Several particles in a rigid body rotating about a fixed axis and their velocities.

Since $2\pi/T = 2\pi f = \omega$, Eq. (12.12) can be obtained from Eq. (12.11).

If v is changing, it also follows from Eq. (12.11) that the rate of change of v is proportional to the rate of change of ω:

$$\frac{dv}{dt} = \frac{d\omega}{dt} R$$

A rate of change of the speed along the circle implies that the particle has an acceleration along the circle, called a **tangential acceleration.** According to the last equation, this tangential acceleration is

$$a_{\text{tangential}} = \alpha R \qquad (12.13)$$

tangential acceleration

Note that, besides this tangential acceleration directed along the circle, the particle also has a **centripetal acceleration** directed toward the center of the circle. From Section 4.5, we know that the centripetal acceleration for uniform circular motion is

$$a_{\text{centripetal}} = \frac{v^2}{R} \qquad (12.14)$$

centripetal acceleration

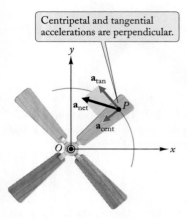

Centripetal and tangential accelerations are perpendicular.

FIGURE 12.9 A particle in a rotating rigid body with an angular acceleration has both a centripetal acceleration $a_{centripetal}$ and a tangential acceleration $a_{tangential}$. The net instantaneous translational acceleration a_{net} is then the vector sum of $a_{centripetal}$ and $a_{tangential}$.

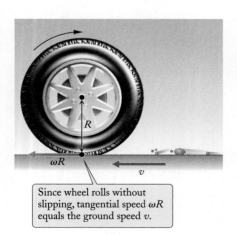

Since wheel rolls without slipping, tangential speed ωR equals the ground speed v.

FIGURE 12.10 Rotating wheel of the automobile as viewed in the reference frame of the automobile. The ground moves toward the left at speed v.

With $v = \omega R$, this becomes

$$a_{centripetal} = \omega^2 R \tag{12.15}$$

The net translational acceleration of the particle is the vector sum of the tangential and the centripetal accelerations, which are perpendicular (see Fig. 12.9); thus, the magnitude of the net acceleration is

$$a_{net} = \sqrt{a_{tangential}^2 + a_{centripetal}^2} \tag{12.16}$$

Although we have here introduced the concept of tangential acceleration in the context of the rotational motion of a rigid body, this concept is also applicable to the translational motion of a particle along a circular path or any curved path. For instance, consider an automobile (regarded as a particle) traveling around a curve. If the driver steps on the accelerator (or on the brake), the automobile will suffer a change of speed as it travels around the curve. It will then have both a tangential and a centripetal acceleration.

EXAMPLE 3 The blade of a circular saw is initially rotating at 7000 revolutions per minute. Then the motor is switched off, and the blade coasts to a stop in 8.0 s. What is the average angular acceleration?

SOLUTION: In radians per second, 7000 rev/min corresponds to an initial angular velocity $\omega_1 = 7000 \times 2\pi$ radians/min, or

$$\omega_1 = \frac{7000 \times 2\pi \text{ radians}}{60 \text{ s}} = 7.3 \times 10^2 \text{ radians/s}$$

The final angular velocity is $\omega_2 = 0$. Hence the average angular acceleration is

$$\bar{\alpha} = \frac{\Delta\omega}{\Delta t} = \frac{\omega_2 - \omega_1}{t_2 - t_1} = \frac{0 - 7.3 \times 10^2 \text{ radians/s}}{8.0 \text{ s} - 0}$$

$$= -91 \text{ radians/s}^2$$

EXAMPLE 4 An automobile accelerates uniformly from 0 to 80 km/h in 6.0 s. The wheels of the automobile have a radius of 0.30 m. What is the angular acceleration of the wheels? Assume that the wheels roll without slipping.

SOLUTION: The translational acceleration of the automobile is

$$a = \frac{v - v_0}{t} = \frac{80 \text{ km/h}}{6.0 \text{ s}} = \frac{(80 \text{ km/h}) \times (1000 \text{ m/1 km}) \times (1 \text{ h/3600 s})}{6.0 \text{ s}}$$

$$= 3.7 \text{ m/s}^2$$

The angular acceleration of the wheel is related to this translational acceleration by $a = \alpha R$, the same relation as Eq. (12.13). We can establish this relationship most conveniently by viewing the motion of the wheel in the reference frame of the automobile (see Fig. 12.10). In this reference frame, the ground is moving backward at speed v, and the bottom point of the rotating wheel is moving backward at the tangential speed ωR. Since the wheel is supposed to move without slipping, the speed v of the ground must match the tangential speed of the bottom point of the wheel; that is, $v = \omega R$. This proportionality of v and ω implies the same proportionality of the accelerations a and α, and therefore establishes the relationship $a = \alpha R$.

The angular acceleration of the wheel is then

$$\alpha = \frac{a}{R} = \frac{3.7 \text{ m/s}^2}{0.30 \text{ m}} = 12 \text{ radians/s}^2$$

EXAMPLE 5   The large centrifuge shown in the chapter photo has an arm of length 8.8 m. When rotating at 175 revolutions per minute, what is the speed of the end of this arm, and what is the centripetal acceleration?

Concepts
— in —
Context

SOLUTION: 175 rpm amounts to $175/60 = 2.9$ revolutions per second. The corresponding angular velocity is

$$\omega = 2\pi f = 2\pi \times 2.9 \text{ rev/s} = 18 \text{ radians/s}$$

According to Eq. (12.11), the speed at a radius $R = 8.8$ m is

$$v = \omega R = 18 \text{ radians/s} \times 8.8 \text{ m} = 1.6 \times 10^2 \text{ m/s}$$

and according to Eq. (12.15), the centripetal acceleration is

$$a_{\text{centripetal}} = \omega^2 R = (18 \text{ radians/s})^2 \times 8.8 \text{ m} = 2.9 \times 10^3 \text{ m/s}^2$$

This is almost 300 standard g's!

 Checkup 12.2

QUESTION 1: Consider a point P on the rim of a rotating, accelerating flywheel and a point Q near the center. Which point has the larger instantaneous speed? The larger instantaneous angular velocity? The larger angular acceleration? The larger tangential acceleration? The larger centripetal acceleration?

QUESTION 2: The Earth rotates steadily around its axis once per day. Do all points on the surface of the Earth have the same radius R for their circular motion? Do they all have the same angular velocity ω? The same speed v around the axis? The same centripetal acceleration $a_{\text{centripetal}}$? If not, which points have the largest R, ω, v, and $a_{\text{centripetal}}$?

QUESTION 3: A short segment of the track of a roller coaster can be approximated by a circle of suitable radius. If a (frictionless) roller-coaster car is passing through the highest point of the track, is there a centripetal acceleration? A tangential acceleration? What if the the roller coaster is some distance beyond the highest point?

QUESTION 4: Consider the motion of the hammer shown in Fig. 12.1. Taking into account only the rotational motion, which end of the hammer has the larger speed v around the axis? The larger centripetal acceleration $a_{\text{centripetal}}$?

(A) Head end; head end   (B) Head end; handle end
(C) Handle end; head end   (D) Handle end; handle end

12.3 MOTION WITH CONSTANT ANGULAR ACCELERATION

We will now examine the kinematic equations describing rotational motion for the special case of *constant* angular acceleration; these are mathematically analogous to the equations describing translational motion with constant acceleration (see Section 2.5), and they can be derived by the same methods. In the next section, we will develop an alternative method, based on integration, for obtaining the kinematic equations describing either angular or translational motion for the general case of accelerations with arbitrary time dependence.

If the rigid body rotates with a constant angular acceleration α, then the angular velocity increases at a constant rate, and after a time t has elapsed, the angular velocity will attain the value

$$\omega = \omega_0 + \alpha t \tag{12.17}$$

constant angular acceleration: ω, α, and t

where ω_0 is the initial value of the angular velocity at $t = 0$.

The angular position can be calculated from this angular velocity by the arguments used in Section 2.5 to calculate x from v [see Eqs. (2.17), (2.22), and (2.25)]. The result is

$$\phi = \phi_0 + \omega_0 t + \tfrac{1}{2}\alpha t^2 \tag{12.18}$$

constant angular acceleration: ϕ, α, and t

Furthermore, the arguments of Section 2.5 lead to an identity between acceleration, position, and velocity [see Eqs. (2.20)–(2.22)]:

$$\alpha(\phi - \phi_0) = \tfrac{1}{2}(\omega^2 - \omega_0^2) \tag{12.19}$$

constant angular acceleration: α, ϕ, and ω

Note that all these equations have exactly the same mathematical form as the equations of Section 2.5, with the angular position ϕ taking the place of the position x, the angular velocity ω taking the place of v, and the angular acceleration α taking the place of a. This analogy between rotational and translational quantities can serve as a useful mnemonic for remembering the equations for rotational motion. Table 12.2 displays analogous equations.

TABLE 12.2	ANALOGIES BETWEEN TRANSLATIONAL AND ROTATIONAL QUANTITIES
$v = \dfrac{dx}{dt}$	$\rightarrow$ $\quad$ $\omega = \dfrac{d\phi}{dt}$
$a = \dfrac{dv}{dt}$	$\rightarrow$ $\quad$ $\alpha = \dfrac{d\omega}{dt}$
$v = v_0 + at$	$\rightarrow$ $\quad$ $\omega = \omega_0 + \alpha t$
$x = x_0 + v_0 t + \tfrac{1}{2}at^2$	$\rightarrow$ $\quad$ $\phi = \phi_0 + \omega_0 t + \tfrac{1}{2}\alpha t^2$
$a(x - x_0) = \tfrac{1}{2}(v^2 - v_0^2)$	$\rightarrow$ $\quad$ $\alpha(\phi - \phi_0) = \tfrac{1}{2}(\omega^2 - \omega_0^2)$

EXAMPLE 6   The cable supporting an elevator runs over a wheel of radius 0.36 m (see Fig. 12.11). If the elevator begins from rest and ascends with an upward acceleration of 0.60 m/s², what is the angular acceleration of the wheel? How many turns does the wheel make if this accelerated motion lasts 5.0 s? Assume that the cable runs over the wheel without slipping.

SOLUTION: If there is no slipping, the speed of the cable must always coincide with the tangential speed of a point on the rim of the wheel. The acceleration $a = 0.60$ m/s² of the cable must then coincide with the tangential acceleration of a point on the rim of the wheel:

$$a = a_{\text{tangential}} = \alpha R \qquad\qquad (12.20)$$

where $R = 0.36$ m is the radius of the wheel. Hence

$$\alpha = \frac{a}{R} = \frac{0.60 \text{ m/s}^2}{0.36 \text{ m}} = 1.7 \text{ radians/s}^2$$

According to Eq. (12.18), the angular displacement in 5.0 s is

$$\phi - \phi_0 = \omega_0 t + \tfrac{1}{2}\alpha t^2$$
$$= 0 + \tfrac{1}{2} \times 1.7 \text{ radians/s}^2 \times (5.0 \text{ s})^2$$
$$= 21 \text{ radians}$$

Each revolution comprises 2π radians; thus, the number of turns the wheel makes is

$$[\text{number of turns}] = \frac{\phi - \phi_0}{2\pi} = \frac{21 \text{ radians}}{2\pi} = 3.3 \text{ revolutions}$$

FIGURE 12.11 Elevator supported by a cable that runs over a rotating wheel.

PROBLEM-SOLVING TECHNIQUES   ANGULAR MOTION

The solution of kinematic problems about angular velocity and angular acceleration involves the same techniques as the problems about translational velocity and translational acceleration in Chapter 2. You might find it useful to review the procedures suggested on page 50.

Sometimes a problem contains a link between a rotational motion and a translational motion, such as the link between the rotational and translational motions of the wheels of an automobile (see Example 4) or the link between the translational motion of the elevator cable and the rotational motion of the wheel over which it runs (Example 6). If the body in contact with the rim of the wheel does not slip, the translational speed of this body equals the tangential speed of the contact point at the rim of the wheel; that is, $v = \omega R$ and $a = \alpha R$.

Keep in mind that although some of the equations in this chapter remain valid if the angular quantities are expressed in degrees, any equation that contains both angular quantities and distances (e.g., $v = \alpha R$) is valid only if the angular quantity is expressed in radians. To prevent mistakes, it is safest to express all angular quantities in radians; if degrees are required in the answer, convert from radians to degrees after completing your calculations.

✔ Checkup 12.3

QUESTION 1: Consider a point on the rim of the wheel shown in Fig. 12.11, (instantaneously) at the top of the wheel. What is the direction of the centripetal acceleration of this point? The tangential acceleration?

QUESTION 2: The wheel of a bicycle rolls on a flat road. Is the angular velocity constant if the translational velocity of the bicycle is constant? Is the angular acceleration constant if the translational acceleration of the bicycle is constant?

QUESTION 3: A grinding wheel accelerates uniformly for 3 seconds after being turned on. In the first second of motion, the wheel rotates 5 times. In the first two seconds of motion, the total number of revolutions is:

(A) 6   (B) 10   (C) 15   (D) 20   (E) 25

12.4 MOTION WITH TIME-DEPENDENT ANGULAR ACCELERATION

The equations of angular motion for the general case when the angular acceleration is a function of time are analogous to the corresponding equations of translational motion discussed in Section 2.7. Such equations are solved by integration. Integral calculus was discussed in detail in Chapter 7, and we now revisit the technique of integration of the equations of motion for the case of angular motion. To see how we can obtain kinematic solutions for nonconstant accelerations, consider the angular acceleration $\alpha = d\omega/dt$. We rearrange this relation and obtain

$$d\omega = \alpha\, dt$$

We can integrate this expression directly, for example, from the initial value of the angular velocity ω_0 at time $t = 0$, to some final value ω at time t (the integration variables are indicated by primes to distinguish them from the upper limits of integration):

$$\int_{\omega_0}^{\omega} d\omega' = \int_0^t \alpha\, dt'$$

$$\omega - \omega_0 = \int_0^t \alpha\, dt' \tag{12.21}$$

This gives the angular velocity as a function of time:

angular velocity for time-dependent angular acceleration

$$\omega = \omega_0 + \int_0^t \alpha\, dt' \tag{12.22}$$

Equation (12.22) enables us to calculate the angular velocity as a function of time for any angular acceleration that is a known function of time.

The angular position ϕ can be obtained in a similar manner:

$$d\phi = \omega\, dt$$

$$\int_{\phi_0}^{\phi} d\phi' = \int_0^t \omega\, dt'$$

angular position for time-dependent angular velocity

$$\phi - \phi_0 = \int_0^t \omega\, dt' \tag{12.23}$$

In the special case of constant angular acceleration α, Eq. (12.22) gives us $\omega = \omega_0 + \alpha t$, which agrees with our previous result, Eq. (12.17). If we insert this into Eq. (12.23), we obtain $\phi - \phi_0 = \int_0^t (\omega_0 + \alpha t)dt = \omega_0 t + \frac{1}{2}\alpha t^2$, which agrees with our previous Eq. (12.18).

In the general case of a time-dependent angular acceleration α, we proceed in the same way: first, use Eq. (12.22) to find ω as a function of time, and then insert this function into Eq. (12.23) to find the angular position as a function of time, as in the following example.

EXAMPLE 7   When turned on, a motor rotates a circular saw wheel, beginning from rest, with an angular acceleration that has an initial value $\alpha_0 = 60$ radians/s^2 at $t = 0$ and decreases to zero acceleration during the interval $0 \le t \le 3.0$ s according to

$$\alpha = \alpha_0\left(1 - \frac{t}{3.0\,\text{s}}\right)$$

After $t = 3.0$ s, the motor maintains the wheel's angular velocity at a constant value. What is this final angular velocity? In the process of "getting up to speed," how many revolutions occur?

SOLUTION: The angular acceleration α is given as an explicit function of time. Since we are beginning from rest, the initial angular velocity is $\omega_0 = 0$, so Eq. (12.22) gives ω as a function of t:

$$\omega = \omega_0 + \int_0^t \alpha\, dt' = 0 + \int_0^t \alpha_0\left(1 - \frac{t'}{3.0\,\text{s}}\right)dt'$$

$$= \alpha_0\left(\int_0^t dt' - \frac{1}{3.0\,\text{s}}\int_0^t t'\, dt'\right) = \alpha_0\left(t'\Big|_0^t - \frac{1}{3.0\,\text{s}}\frac{t'^2}{2}\Big|_0^t\right)$$

$$= \alpha_0\left(t - \frac{t^2}{6.0\,\text{s}}\right) \tag{12.24}$$

where we have used the property that the integral of the sum is the sum of the integrals, and that $\int t^n\, dt = t^{n+1}/(n+1)$. At $t = 3.0$ s, this angular velocity reaches its final value of

$$\omega = 60\ \text{radians/s}^2 \times \left(3.0\,\text{s} - \frac{(3.0\,\text{s})^2}{6.0\,\text{s}}\right) = 90\ \text{radians/s}$$

To obtain the number of revolutions during the time of acceleration, we can calculate the change in angular position and divide by 2π. To do so, we must insert the time-dependent angular velocity obtained in Eq. (12.24) into Eq. (12.23):

$$\phi - \phi_0 = \int_0^t \omega\, dt' = \int_0^t \alpha_0\left(t' - \frac{t'^2}{6.0\,\text{s}}\right)dt'$$

$$= \alpha_0\left(\int_0^t t'\, dt' - \frac{1}{6.0\,\text{s}}\int_0^t t'^2\, dt'\right) = \alpha_0\left(\frac{t'^2}{2}\Big|_0^t - \frac{1}{6.0\,\text{s}}\frac{t'^3}{3}\Big|_0^t\right)$$

$$= \alpha_0\left(\frac{t^2}{2} - \frac{t^3}{18\,\text{s}}\right)$$

Evaluating this expression at $t = 3.0$ s, we find

$$\phi - \phi_0 = 60 \text{ radians/s}^2 \times \left(\frac{(3.0\,\text{s})^2}{2} - \frac{(3.0\ \text{s})^3}{18\ \text{s}} \right) = 180 \text{ radians}$$

Hence the number of revolutions during the acceleration is

$$[\text{number of revolutions}] = \frac{\phi - \phi_0}{2\pi} = \frac{180 \text{ radians}}{2\pi} = 29 \text{ revolutions} \quad (12.25)$$

As discussed in Section 2.7, similar integration techniques can be applied to determine any component of the translational velocity and the position when the time-dependent net force and, thus, the time-dependent translational acceleration are known. In Section 2.7 we also examined the case when the acceleration is a known function of the velocity; in that case, integration provides t as a function of v (and v_0), which can sometimes be inverted to find v as a function of t.

We saw in Chapters 7–9 that a conservation-of-energy approach is often the easiest way to determine the motion when the forces are known as a function of position. Now we have seen that direct integration of the equations of motion can be applied when the translational or angular acceleration is known as a function of time or of velocity.

 Checkup 12.4

QUESTION 1: Beginning from rest at $t = 0$, the angular velocity of a merry-go-round increases in proportion to the square root of the time t. By what factor is the angular position of the merry-go-round at $t = 4$ s greater than it was at $t = 1$ s?

QUESTION 2: A car on a circular roadway accelerates from rest beginning at $t = 0$, so that its angular acceleration increases in proportion to the time t. With what power of time does its centripetal acceleration increase?

(A) t (B) t^2 (C) t^3 (D) t^4 (E) t^5

12.5 KINETIC ENERGY OF ROTATION; MOMENT OF INERTIA

A rigid body is a system of particles, and as for any system of particles, the total kinetic energy of a rotating rigid body is simply the sum of the individual kinetic energies of all the particles (see Section 10.4). If the particles in the rigid body have masses m_1, m_2, m_3, . . . and speeds v_1, v_2, v_3, . . . , then the kinetic energy is

$$K = \tfrac{1}{2}m_1 v_1^2 + \tfrac{1}{2}m_2 v_2^2 + \tfrac{1}{2}m_3 v_3^2 + \cdots \quad (12.26)$$

In a rigid body rotating about a given axis, all the particles move with the same angular velocity ω along circular paths. By Eq. (12.11), the speeds of the particles along their paths are proportional to their radial distances:

$$v_1 = R_1\omega, \quad v_2 = R_2\omega, \quad v_3 = R_3\omega, \ \cdots \quad (12.27)$$

and hence the total kinetic energy is

$$K = \tfrac{1}{2}m_1 R_1^2 \omega^2 + \tfrac{1}{2}m_2 R_2^2 \omega^2 + \tfrac{1}{2}m_3 R_3^2 \omega^2 + \cdots$$

We can write this as

$$K = \tfrac{1}{2}I\omega^2 \qquad (12.28)$$

where the quantity

$$I = m_1 R_1^2 + m_2 R_2^2 + m_3 R_3^2 + \cdots \qquad (12.29)$$

is called the **moment of inertia** of the rotating body about the given axis. The SI unit of moment of inertia is $kg \cdot m^2$.

Note that Eq. (12.28) has a mathematical form reminiscent of the familiar expression $\tfrac{1}{2}mv^2$ for the kinetic energy of a single particle—the moment of inertia replaces the mass, and the angular velocity replaces the translational velocity. As we will see in the next chapter, this analogy between moment of inertia and mass is of general validity. *The moment of inertia is a measure of the resistance that a body offers to changes in its rotational motion*, just as mass is a measure of the resistance that a body offers to changes in its translational motion.

Equation (12.29) shows that the moment of inertia—and consequently the kinetic energy for a given value of ω—is large if most of the mass of the body is at a large distance from the axis of rotation. This is very reasonable: for a given value of ω, particles at large distance from the axis move with high speeds, and therefore have large kinetic energies.

EXAMPLE 8   A 50-kg woman and an 80-kg man sit on a massless seesaw separated by 3.00 m (see Fig. 12.12). The seesaw rotates about a fulcrum (the point of support) placed at the center of mass of the system; the center of mass is 1.85 m from the woman and 1.15 m from the man, as obtained in Example 4 of Chapter 10. If the (instantaneous) angular velocity of the seesaw is 0.40 radian/s, calculate the kinetic energy. Treat both masses as particles.

SOLUTION: The moment of inertia for particles rotating about an axis depends only on the masses and their distances from the axis:

$$I = m_1 R_1^2 + m_2 R_2^2$$
$$= 50 \text{ kg} \times (1.85 \text{ m})^2 + 80 \text{ kg} \times (1.15 \text{ m})^2 = 280 \text{ kg} \cdot \text{m}^2 \qquad (12.30)$$

The kinetic energy for the rotational motion is

$$K = \tfrac{1}{2}I\omega^2$$
$$= \tfrac{1}{2} \times 280 \text{ kg} \cdot \text{m}^2 \times (0.40 \text{ radian/s})^2 = 22 \text{ J} \qquad (12.31)$$

This kinetic energy could equally well have been obtained by first calculating the individual speeds of the woman and the man ($v_1 = R_1\omega$, $v_2 = R_2\omega$) and then adding the corresponding individual kinetic energies.

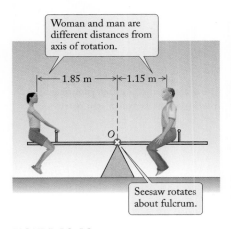

Woman and man are different distances from axis of rotation.

← 1.85 m →|← 1.15 m →|

Seesaw rotates about fulcrum.

FIGURE 12.12 Woman and man on a seesaw.

If we regard the mass of a solid body as continuously distributed throughout its volume, then we can calculate the moment of inertia by the same method we used for the calculation of the center of mass: we subdivide the body into small mass elements and add the moments of inertia contributed by all these small amounts of mass. This leads to an approximation for the moment of inertia,

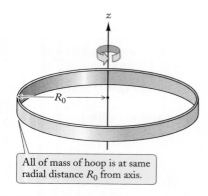

FIGURE 12.13 A thin hoop rotating about its axis of symmetry.

All of mass of hoop is at same radial distance R_0 from axis.

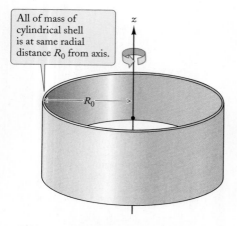

All of mass of cylindrical shell is at same radial distance R_0 from axis.

FIGURE 12.14 A thin cylindrical shell rotating about its axis of symmetry.

$$I \approx \sum_i R_i^2 \, \Delta m_i \qquad (12.32)$$

where R_i is the radial distance of the mass element Δm_i from the axis of rotation. In the limit $\Delta m_i \to 0$, this approximation becomes exact, and the sum becomes an integral:

$$I = \int R^2 \, dm \qquad (12.33)$$

In general, the calculation of the moment of inertia requires the evaluation of the integral (12.33). However, in a few exceptionally simple cases, it is possible to find the moment of inertia without performing this integration. For example, if the rigid body is a thin hoop (see Fig. 12.13) or a thin cylindrical shell (see Fig. 12.14) of radius R_0 rotating about its axis of symmetry, then *all* of the mass of the body is at the same distance from the axis of rotation—the moment of inertia is then simply the total mass M of the hoop or shell multiplied by its radius R_0 squared,

$$I = MR_0^2$$

If all of the mass is *not* at the same distance from the axis of rotation, then we must perform the integration (12.33); when summing the individual contributions, we usually write the small mass contribution as a mass per unit length times a small length, or as a mass per unit area times a small area, as in the following examples.

EXAMPLE 9   Find the moment of inertia of a uniform thin rod of length l and mass M rotating about an axis perpendicular to the rod and through its center.

SOLUTION: Figure 12.15 shows the rod lying along the x axis; the axis of rotation is the z axis. The rod extends from $x = -l/2$ to $x = +l/2$. Consider a small slice dx of the rod. The amount of mass within this slice is proportional to the length dx, and so is equal to the mass per unit length times this length:

$$dm = \frac{M}{l} \, dx$$

The square of the distance of the slice from the axis of rotation is $R^2 = x^2$, so Eq. (12.33) becomes

$$I = \int R^2 \, dm = \int_{-l/2}^{+l/2} x^2 \frac{M}{l} \, dx = \frac{M}{l} \left(\frac{x^3}{3} \right) \Big|_{-l/2}^{+l/2}$$

$$= \frac{M}{l} \times \frac{2(l/2)^3}{3} = \frac{1}{12} M l^2 \qquad (12.34)$$

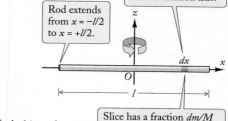

A slice of width dx is located at distance x from rotation axis.

Rod extends from $x = -l/2$ to $x = +l/2$.

Slice has a fraction dm/M of total mass equal to its fraction dx/l of total length.

FIGURE 12.15 A thin rod rotating about its center.

EXAMPLE 10   Repeat the calculation of the preceding example for an axis of rotation through one end of the rod.

SOLUTION: Figure 12.16 shows the rod and the axis of rotation. The rod extends from $x = 0$ to $x = l$. Hence, instead of Eq. (12.34) we now obtain

$$I = \int_0^l x^2 \frac{M}{l}\,dx = \frac{M}{l}\left(\frac{x^3}{3}\right)\Bigg|_0^l = \frac{M}{l} \times \frac{l^3}{3} = \frac{1}{3}Ml^2 \qquad (12.35)$$

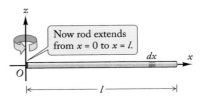

FIGURE 12.16 A thin rod rotating about its end.

EXAMPLE 11   Find the moment of inertia of a wide ring, or annulus, made of sheet metal of inner radius R_1, outer radius R_2, and mass M rotating about its axis of symmetry (see Fig. 12.17).

SOLUTION: The annulus can be regarded as made of a large number of thin concentric hoops fitting around one another. Figure 12.17 shows one such hoop, of radius R and width dR. All of the mass dm of this hoop is at the same radius R from the axis of rotation; hence the moment of inertia of the hoop is

$$dI = R^2\,dm$$

The area dA of the hoop is the product of its length (the perimeter $2\pi R$) and its width dR, so $dA = 2\pi R\,dR$. The mass dm of the hoop equals the product of this area and the mass per unit area of the sheet metal. Since the total area of the annulus is $\pi R_2^2 - \pi R_1^2$, the mass per unit area is $M/\pi(R_2^2 - R_1^2)$. The mass contributed by each hoop is the mass per unit area times its area:

$$dm = \frac{M}{\pi(R_2^2 - R_1^2)} \times 2\pi R\,dR = \frac{2M}{R_2^2 - R_1^2}R\,dR \qquad (12.36)$$

We sum the contributions dI from $R = R_1$ to $R = R_2$; hence

$$I = \int R^2\,dm = \frac{2M}{R_2^2 - R_1^2}\int_{R_1}^{R_2} R^3\,dR$$

$$= \frac{2M}{R_2^2 - R_1^2}\left(\frac{R^4}{4}\right)\Bigg|_{R_1}^{R_2} = \frac{M}{2(R_2^2 - R_1^2)} \times (R_2^4 - R_1^4)$$

$$= \frac{M}{2(R_2^2 - R_1^2)} \times (R_2^2 + R_1^2)(R_2^2 - R_1^2) = \frac{M}{2}(R_2^2 + R_1^2) \qquad (12.37)$$

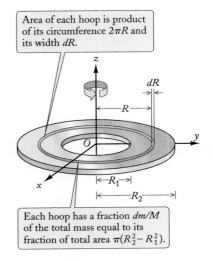

FIGURE 12.17 An annulus of sheet metal rotating about its axis of symmetry. The annulus can be regarded as made of a large number of concentric hoops. The hoop shown in the figure has radius R and width dR.

COMMENT: Note that for $R_1 = 0$, this becomes $I = MR_2^2/2$, which is the moment of inertia of a disk (see Table 12.3). And for $R_1 = R_2$, it becomes $I = MR_1^2$, which is the moment of inertia of a hoop. Note that the result (12.37) for a sheet also applies to a thick annulus or a thick cylindrical shell (rotating about the axis of symmetry).

Comparison of Eqs. (12.34) and (12.35) for the moment of inertia of a rod makes it clear that the value of the moment of inertia depends on the location of the axis of rotation. The moment of inertia is small if the axis passes through the center of mass, and large if it passes through the end of the rod. In the latter case, more of the mass of the rod is at a larger distance from the axis of rotation, which leads to a larger moment of inertia.

TABLE 12.3	SOME MOMENTS OF INERTIA

BODY		MOMENT OF INERTIA
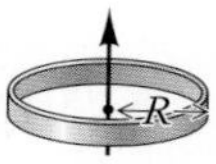	Thin hoop about symmetry axis	MR^2
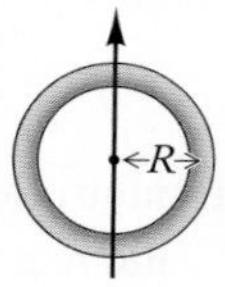	Thin hoop about diameter	$\frac{1}{2}MR^2$
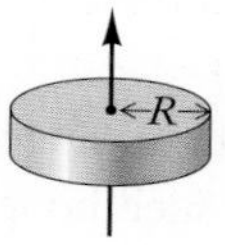	Disk or cylinder about symmetery axis	$\frac{1}{2}MR^2$
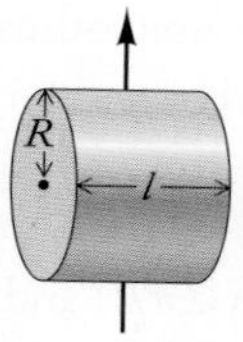	Cylinder about diameter through center	$\frac{1}{4}MR^2 + \frac{1}{12}Ml^2$
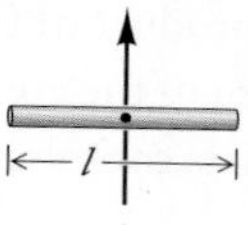	Thin rod about perpendicular axis through center	$\frac{1}{12}Ml^2$
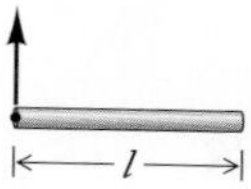	Thin rod about perpendicular axis through end	$\frac{1}{3}Ml^2$
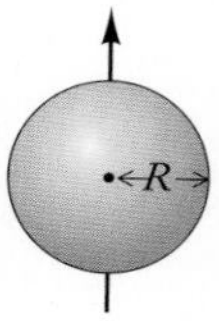	Sphere about diameter	$\frac{2}{5}MR^2$
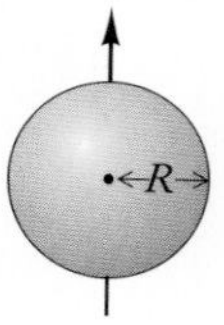	Thin spherical shell about diameter	$\frac{2}{3}MR^2$

It is possible to prove a theorem that relates the moment of inertia I_{CM} about an axis through the center of mass to the moment of inertia I about a parallel axis through some other point. This theorem, called the **parallel-axis theorem,** asserts that

parallel-axis theorem

$$I = I_{CM} + Md^2 \qquad (12.38)$$

where M is the total mass of the body and d the distance between the two axes. We will not give the proof, but merely check that the theorem is consistent with our results for the moments of inertia of the rod rotating about an axis through the center $[I_{CM} = \frac{1}{12}Ml^2$; see Eq. (12.34)] and an axis through an end $[I = \frac{1}{3}Ml^2$; see Eq. (12.35)]. In this case, $d = l/2$, and the parallel-axis theorem asserts

$$\frac{1}{3}Ml^2 = \frac{1}{12}Ml^2 + M\left(\frac{l}{2}\right)^2 \qquad\qquad (12.39)$$

which is identically true.

Note that it is a corollary of Eq. (12.38) that the moment of inertia about an axis passing through the center of mass is always smaller than that about any other parallel axis.

Table 12.3 lists the moments of inertia of a variety of rigid bodies about an axis through their center of mass; all the bodies are assumed to have uniform density.

EXAMPLE 12   The large centrifuge shown in the chapter photo carries the payload in a chamber in one arm and counterweights at the end of the opposite arm. The mass distribution depends on the choice of payload and the choice of counterweights. Figure 12.18 is a schematic diagram of the mass distribution attained with a particular choice of payload and counterweights. The payload arm (including the payload) has a mass of 1.8×10^3 kg uniformly distributed over a length of 8.8 m. The counterweight arm has a mass of 1.1×10^3 kg uniformly distributed over a length of 5.5 m, and it carries a counterweight of 8.6×10^3 kg at its end. (a) What is the moment of inertia of the centrifuge for this mass distribution? (b) What is the rotational kinetic energy when the centrifuge is rotating at 175 revolutions per minute?

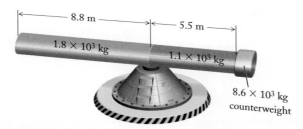

FIGURE 12.18 Centrifuge mass distribution.

SOLUTION: (a) The total moment of inertia is the sum of the moments of inertia of a rod of mass $m_1 = 1.8 \times 10^3$ kg, length $l_1 = 8.8$ m rotating about its end; a second rod of mass $m_2 = 1.1 \times 10^3$ kg, length $l_2 = 5.5$ m also rotating about its end; and a mass of $m = 8.6 \times 10^3$ kg at a radial distance of $R = 5.5$ m. The moments of inertia of the rods are given by Eq. (12.35), and the moment of inertia of the counterweight is mR^2. So the total moment of inertia is

$$I = \tfrac{1}{3}m_1 l_1^2 + \tfrac{1}{3}m_2 l_2^2 + mR^2$$
$$= \tfrac{1}{3} \times 1.8 \times 10^3 \text{ kg} \times (8.8 \text{ m})^2 + \tfrac{1}{3} \times 1.1 \times 10^3 \text{ kg} \times (5.5 \text{ m})^2$$
$$\quad + 8.6 \times 10^3 \text{ kg} \times (5.5 \text{ m})^2$$
$$= 3.2 \times 10^5 \text{ kg}\cdot\text{m}^2$$

(b) At 175 revolutions per minute, the angular velocity is $\omega = 18$ radians/s (see Example 5), and the rotational kinetic energy is

$$K = \tfrac{1}{2}I\omega^2$$
$$= \tfrac{1}{2} \times 3.2 \times 10^5 \text{ kg}\cdot\text{m}^2 \times (18 \text{ radians/s})^2$$
$$= 5.2 \times 10^7 \text{ J}$$

(a)

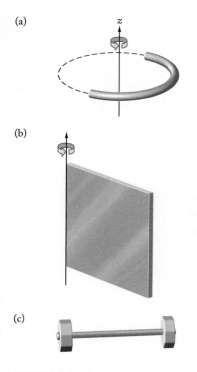

(b)

(c)

FIGURE 12.19 (a) A rod bent into an arc of a circle of radius R, rotating about its center of curvature. (b) A square plate rotating about an axis along one edge. (c) A dumbbell.

✓ Checkup 12.5

QUESTION 1: What is the moment of inertia of a rod of mass M bent into an arc of a circle of radius R when rotating about an axis through the center and perpendicular to the circle (see Fig. 12.19a)?

QUESTION 2: Consider a rod rotating about (a) an axis along the rod, (b) an axis perpendicular to the rod through its center, and (c) an axis perpendicular to the rod through its end. For which axis is the moment of inertia largest? Smallest?

QUESTION 3: What is the moment of inertia of a square plate of mass M and dimension $L \times L$ rotating about an axis along one of its edges (see Fig. 12.19b)? What is the moment of inertia if this square plate rotates about an axis through its center parallel to an edge?

QUESTION 4: A dumbbell consists of two particles of mass m each attached to the ends of a rigid, massless rod of length l (Fig. 12.19c). Assume the particles are point particles. What is the moment of inertia of this rigid body when rotating about an axis through the center and perpendicular to the rod? When rotating about a parallel axis through one end? Are these moments of inertia consistent with the parallel-axis theorem?

QUESTION 5: According to Table 12.3, the moment of inertia of a hoop about its symmetry axis is $I_{CM} = MR^2$. What is the moment of inertia if you twirl a large hoop around your finger, so that in essence it rotates about a point on the hoop, about an axis parallel to the symmetry axis?

    (A) $5MR^2$ (B) $2MR^2$ (C) $\frac{3}{2}MR^2$.

    (D) MR^2 (E) $\frac{1}{2}MR^2$.

SUMMARY

PROBLEM-SOLVING TECHNIQUES   Angular Motion   **(page 375)**

DEFINITION OF ANGLE   (in radians)

$$\phi = \frac{[\text{arc length}]}{[\text{radius}]} = \frac{s}{R}$$

                                                 (12.1)

ANGLE CONVERSIONS

1 revolution $= 2\pi$ radians $= 360°$

AVERAGE ANGULAR VELOCITY

$$\overline{\omega} = \frac{\Delta\phi}{\Delta t}$$

                                                 (12.2)

INSTANTANEOUS ANGULAR VELOCITY

$$\omega = \frac{d\phi}{dt}$$

                                                 (12.3)

FREQUENCY

$$f = \frac{\omega}{2\pi} \qquad (12.4)$$

PERIOD OF MOTION

$$T = \frac{1}{f} = \frac{2\pi}{\omega} \qquad (12.5)$$

AVERAGE ANGULAR ACCELERATION

$$\bar{\alpha} = \frac{\Delta\omega}{\Delta t} \qquad (12.6)$$

INSTANTANEOUS ANGULAR ACCELERATION

$$\alpha = \frac{d\omega}{dt} \qquad (12.7)$$

SPEED OF PARTICLE ON ROTATING BODY

$$v = \omega R \qquad (12.11)$$

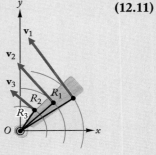

ACCELERATION OF PARTICLE ON ROTATING BODY

$$a_{\text{tangential}} = \alpha R, \qquad (12.13)$$

$$a_{\text{centripetal}} = \omega^2 R \qquad (12.15)$$

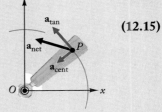

MOTION WITH CONSTANT ANGULAR ACCELERATION

$$\omega = \omega_0 + \alpha t \qquad (12.17)$$

$$\phi = \phi_0 + \omega_0 t + \tfrac{1}{2}\alpha t^2 \qquad (12.18)$$

$$\alpha(\phi - \phi_0) = \tfrac{1}{2}(\omega^2 - \omega_0^2) \qquad (12.19)$$

MOTION WITH TIME-DEPENDENT ANGULAR ACCELERATION

$$\omega = \omega_0 + \int_0^t \alpha\, dt' \qquad (12.22)$$

$$\phi = \phi_0 + \int_0^t \omega\, dt' \qquad (12.23)$$

MOMENT OF INERTIA

where R_i is the radial distance of m_i from the axis of rotation.

$$I = m_1 R_1^2 + m_2 R_2^2 + m_3 R_3^2 + \cdots \qquad (12.29)$$

MOMENT OF INERTIA OF RIGID BODY
(see also Table 12.3)
where R is the radial distance of the mass
element dm from the axis of rotation; for
uniformly distributed mass, dm is given by

$$I = \int R^2 dm$$

$$dm = \frac{M}{[\text{length}]} dx$$

$$dm = \frac{M}{[\text{area}]} dA,$$ (12.33)

$$dm = \frac{M}{[\text{volume}]} dV$$

PARALLEL-AXIS THEOREM
where M is the total mass and d is the distance from CM axis.

$$I = I_{CM} + Md^2$$ (12.38)

KINETIC ENERGY OF ROTATION

$$K = \tfrac{1}{2} I \omega^2$$ (12.28)

QUESTIONS FOR DISCUSSION

1. A spinning flywheel in the shape of a disk suddenly shatters into many small fragments. Draw the trajectories of a few of these small fragments; assume that the fragments do not interfere with each other.

2. You may have noticed that in some old movies the wheels of moving carriages or stagecoaches seem to rotate backwards. How does this come about?

3. Relative to an inertial reference frame, what is your angular velocity right now about an axis passing through your center of mass?

4. Consider the wheel of an accelerating automobile. Draw the instantaneous acceleration vectors for a few points on the rim of the wheel.

5. The hands of a watch are small rectangles with a common axis passing through one end. The minute hand is long and thin; the hour hand is short and thicker. Assume both hands have the same mass. Which has the greater moment of inertia? Which has the greater kinetic energy and angular momentum?

6. What configuration and what axis would you choose to give your body the smallest possible moment of inertia? The greatest?

7. About what axis through the center of mass is the moment of inertia of this book largest? Smallest? (Assume the book is closed.)

8. A circular hoop made of thin wire has a radius R and mass M. About what axis perpendicular to the plane of the hoop must you rotate this hoop to obtain the minimum moment of inertia? What is the value of this minimum?

9. Automobile engines and other internal combustion engines have flywheels attached to their crankshafts. What is the purpose of these flywheels? (Hint: Each explosive combustion in one of the cylinders of such an engine gives a sudden push to the crankshaft. How would the crankshaft respond to this push if it had no flywheel?)

10. Suppose you pump a mass M of seawater into a pond on a hill at the equator. How does this change the moment of inertia of the Earth?

PROBLEMS

12.2 Rotation About A Fixed Axis[†]

1. The minute hand of a wall clock has a length of 20 cm. What is the angular velocity of this hand? What is the speed of the tip of this hand?

[†] For help, see Online Concept Tutorial 15 at www.wwnorton.com/physics

2. Quito is on the Earth's equator; New York is at latitude 41° north. What is the angular velocity of each city about the Earth's axis of rotation? What is the linear speed of each?

3. An automobile has wheels with a radius of 30 cm. What are the angular velocity (in radians per second) and the frequency (in revolutions per second) of the wheels when the automobile is traveling at 88 km/h?

4. In an experiment at the Oak Ridge Laboratory, a carbon fiber disk of 0.70 m in diameter was set spinning at 37 000 rev/min. What was the speed at the edge of this disk?

5. The rim of a phonograph record is at a distance of 15 cm from the center, and the rim of the paper label on the record is at a distance of 5.0 cm from the center.

   (a) When this record is rotating at $33\frac{1}{3}$ rev/min, what is the translational speed of a point on the rim of the record? The translational speed of a point on the rim of the paper label?

   (b) What are the centripetal accelerations of these points?

6. An electric drill rotates at 5000 rev/min. What is the frequency of rotation (in rev/s)? What is the time for one revolution? What is the angular velocity (in radians/s)?

7. An audio compact disk (CD) rotates at 210 rev/min when playing an outer track of radius 5.8 cm. What is the angular velocity in radians/s? What is the tangential speed of a point on the outer track? Because the CD has the same linear density of bits on each track, the drive maintains a constant tangential speed. What is the angular velocity (in radians/s) when playing an inner track of radius 2.3 cm? What is the corresponding rotational frequency (in rev/s)?

8. An automobile travels one-fourth of the way around a traffic circle in 4.5 s. The diameter of the traffic circle is 50 m. The automobile travels at constant speed. What is that speed? What is the angular velocity in radians/s?

9. When a pottery wheel motor is switched on, the wheel accelerates from rest to 90 rev/min in 5.0 s. What is its angular velocity at $t = 5.0$ s (in radians/s)? What is the linear speed of a piece of clay 10 cm from the center of the wheel at $t = 5.0$ s? What is its average angular acceleration during the acceleration?

10. A grinding wheel of radius 6.5 cm accelerates from rest to its operating speed of 3450 rev/min in 1.6 s. When up to speed, what is its angular velocity in radians/s? What is the linear speed at the edge of the wheel? What is its average angular acceleration during this 1.6 s? When turned off, it decelerates to a stop in 35 s. What is its average angular acceleration during this time?

11. When drilling holes, manufacturers stay close to a recommended linear cutting speed in order to maintain efficiency while avoiding overheating. The rotational speed of the drill thus depends on the diameter of the hole. For example, recommended linear cutting speeds are typically 20 m/min for steel and 100 m/min for aluminum. What is the corresponding rotational rate (in rev/s) when drilling a 3.0-mm-diameter hole in aluminum? When drilling a 2.5-cm-diameter hole in steel?

12. An electric blender accelerates from rest to 500 radians/s in 0.80 s. What is the average angular acceleration? What is the corresponding average tangential acceleration for a point on the tip of a blender blade a distance 3.0 cm from the axis? If this point has that tangential acceleration when the blender's angular velocity is 50 radians/s, what is the corresponding total acceleration of the point?

13. The angular position of a ceiling fan during the first two seconds after start-up is given by $\phi = C[t^2 - (t^3/4 \text{ s})]$, where $C = 20/\text{s}^2$ and t is in seconds. What are the angular position, angular frequency, and angular acceleration at $t = 0$ s? At $t = 1.0$ s? At $t = 2.0$ s?

*14. An aircraft passes directly over you with a speed of 900 km/h at an altitude of 10 000 m. What is the angular velocity of the aircraft (relative to you) when directly overhead? Three minutes later?

*15. The outer edge of the grooved area of a long-playing record is at a radial distance of 14.6 cm from the center; the inner edge is at a radial distance of 6.35 cm. The record rotates at $33\frac{1}{3}$ rev/min. The needle of the pickup arm takes 25 min to play the record, and in that time interval it moves uniformly and radially from the outer edge to the inner edge. What is the radial speed of the needle? What is the speed of the outer edge relative to the needle? What is the speed of the inner edge relative to the needle?

*16. Consider the phonograph record described in Problem 15. What is the total length of the groove in which the needle travels?

12.3 Motion with Constant Angular Acceleration

17. The blade of a circular saw of diameter 20 cm accelerates uniformly from rest to 7000 rev/min in 1.2 s. What is the angular acceleration? How many revolutions will the blade have made by the time it reaches full speed?

18. A large ceiling fan has blades of radius 60 cm. When you switch this fan on, it takes 20 s to attain its final steady speed of 1.0 rev/s. Assume a constant angular acceleration.

   (a) What is the angular acceleration of the fan?

   (b) How many revolutions does it make in the first 20 s?

   (c) What is the distance covered by the tip of one blade in the first 20 s?

19. When you switch on a PC computer, the disk in the disk drive takes 5.0 s to reach its final steady speed of 7200 rev/min. What is the average angular acceleration?

20. When you turn off the motor, a phonograph turntable initially rotating at $33\frac{1}{3}$ rev/min makes 25 revolutions before it stops. Calculate the angular deceleration of this turntable; assume it is constant.

21. A large merry-go-round rotates at one revolution each 9.0 seconds. When shut off, it decelerates uniformly to a stop in 16 s. What is the angular acceleration? How many revolutions does the merry-go-round make during the deceleration?

22. A cat swipes at a spool of thread, which then rolls across the floor with an initial speed of 1.0 m/s. The spool decelerates uniformly to a stop 3.0 m from its initial position. The spool has a radius of 1.5 cm and rolls without slipping. What is the

initial angular velocity? Through what total angle does the spool rotate while slowing to a stop? What is the angular acceleration during this motion?

23. If you lift the lid of a washing machine during the rapid spin–dry cycle, the cycle stops (for safety), typically after 5.0 revolutions. If the clothes are spinning at 6.0 rev/s initially, what is their constant angular acceleration during the slowing motion? How long do they take to come to a stop?

24. A toy top initially spinning at 30 rev/s slows uniformly to a stop in 25 seconds. What is the angular acceleration during this motion? Through how many revolutions does the top turn while slowing to a stop?

*25. The rotation of the Earth is slowing down. In 1977, the Earth took 1.01 s longer to complete 365 rotations than in 1900. What was the average angular deceleration of the Earth in the time interval from 1900 to 1977?

*26. An automobile engine accelerates at a constant rate from 200 rev/min to 3000 rev/min in 7.0 s and then runs at constant speed.

    (a) Find the angular velocity and the angular acceleration at $t = 0$ (just after acceleration begins) and at $t = 7.0$ s (just before acceleration ends).

    (b) A flywheel with a radius of 18 cm is attached to the shaft of the engine. Calculate the tangential and the centripetal acceleration of a point on the rim of the flywheel at the times given above.

    (c) What angle does the net acceleration vector make with the radius at $t = 0$ and at $t = 7.0$ s? Draw diagrams showing the wheel and the acceleration vector at these times.

12.4 Motion with Time-Dependent Angular Acceleration

27. A disk has an initial angular velocity of $\omega_0 = 8.0$ radians/s. At $t = 0$, it experiences a time-dependent angular acceleration given by $\alpha = Ct^2$, where $C = 0.25$ radian/s^4. What is the instantaneous angular velocity at $t = 3.0$ s? What is the change in angular position between $t = 0$ and $t = 1.0$ s?

28. A rigid body is initially at rest. Beginning at $t = 0$, it begins rotating, with an angular acceleration given by $\alpha = \alpha_0 \{1 - [t^2/(4 \text{ s}^2)]\}$ for $0 \leq t \leq 2.0$ s and $\alpha = 0$ thereafter. The initial value is $\alpha_0 = 20$ radians/s^2. What is the body's angular velocity after 1.0 s? After a long time? How many revolutions have occurred after 1.0 s?

*29. A sphere is initially rotating with angular velocity ω_0 in a viscous liquid. Friction causes an angular deceleration that is proportional to the instantaneous angular velocity, $\alpha = -A\omega$, where A is a constant. Show that the angular velocity as a function of time is given by

$$\omega = \omega_0 e^{-At}$$

12.5 Kinetic Energy of Rotation; Moment of Inertia

30. Find the moment of inertia of an orange of mass 300 g and diameter 9.0 cm. Treat the orange as a uniform sphere.

31. The original Ferris wheel built by George Ferris (see Fig. 12.20) had a radius of 38 m and a mass of 1.9×10^6 kg. Assume that all of the mass of the wheel was uniformly distributed along its rim. If the wheel was rotating at 0.050 rev/min, what was its kinetic energy?

FIGURE 12.20 The original Ferris wheel.

32. What is the moment of inertia of a broomstick of mass 0.50 kg, length 1.5 m, and diameter 2.5 cm about its longitudinal axis? About an axis at right angles to the broomstick, passing through its center?

33. According to spectroscopic measurements, the moment of inertia of an oxygen molecule about an axis through the center of mass and perpendicular to the line joining the atoms is 1.95×10^{-46} kg·m^2. The mass of an oxygen atom is 2.66×10^{-26} kg. What is the distance between the atoms? Treat the atoms as pointlike particles.

34. The moment of inertia of the Earth about its polar axis is $0.331 M_E R_E^2$, where M_E is the mass and R_E the equatorial radius. Why is the moment of inertia smaller than that of a sphere of uniform density? What would the radius of a sphere of uniform density have to be if its mass and moment of inertia are to coincide with those of the Earth?

35. Problem 41 in Chapter 10 gives the dimensions of a molecule of nitric acid (HNO$_3$). What is the moment of inertia of this molecule when rotating about the symmetry axis passing through the H, O, and N atoms? Treat the atoms as pointlike particles.

36. The water molecule has a shape shown in Fig. 12.21. The distance between the oxygen and the hydrogen atoms is d, and the angle between the hydrogen atoms is θ. From spectroscopic investigations it is known that the moment of inertia of the molecule is 1.93×10^{-47} kg·m² for rotation about the axis AA' and 1.14×10^{-47} kg·m² for rotation about the axis BB'. From this information and the known values of the masses of the atoms, determine the values of d and θ. Treat the atoms as pointlike.

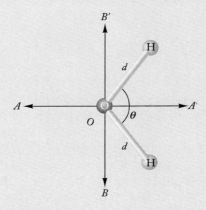

FIGURE 12.21 Atoms in a water molecule.

37. What is the moment of inertia (about the axis of symmetry) of a bicycle wheel of mass 4.0 kg, radius 0.33 m? Neglect the mass of the spokes.

38. An airplane propeller consists of three radial blades, each of length 1.2 m and mass 6.0 kg. What is the kinetic energy of this propeller when rotating at 2500 rev/min? Assume that each blade is (approximately) a uniform rod.

39. Estimate the moment of inertia of a human body spinning rigidly about its longitudinal axis. Treat the body as a uniform cylinder of mass 70 kg, length 1.7 m, and average diameter 23 cm.

40. Use the parallel-axis theorem to determine the moment of inertia of a solid disk or cylinder of mass M and radius R rotating about an axis parallel to its symmetry axis but tangent to its surface.

41. The moment of inertia of the Earth is approximately $0.331 M_E R_E^2$ (see also Problem 34). Calculate the rotational kinetic energy of the Earth.

42. Assume that a potter's kickwheel is a disk of radius 60 cm and mass 120 kg. What is its moment of inertia? What is its rotational kinetic energy when revolving at 2.0 rev/s?

43. A flywheel energy-storage system designed for the International Space Station has a maximum rotational rate of 53 000 rev/min. The cylindrical flywheel has a mass of 75 kg and a radius of 16 cm. For simplicity, assume the cylinder is solid and uniform. What is the moment of inertia of the flywheel? What is the maximum rotational kinetic energy stored in the flywheel?

*44. An empty beer can has a mass of 15 g, a length of 12 cm, and a radius of 3.3 cm. Find the moment of inertia of the can about its axis of symmetry. Assume that the can is a perfect cylinder of sheet metal with no ridges, indentations, or holes.

*45. Suppose that a supertanker transports 4.4×10^8 kg of oil from a storage tank in Venezuela (latitude 10° north) to a storage tank in Holland (latitude 53° north). What is the change of the moment of inertia of the Earth–oil system?

*46. A dumbbell consists of two uniform spheres of mass M and radius R joined by a thin rod of mass m. The distance between the centers of the sphers is l (Fig. 12.22). What is the moment of inertia of this device about an axis through the center of the rod perpendicular to the rod? About an axis along the rod?

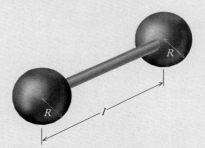

FIGURE 12.22 A dumbbell.

*47. Suppose that the Earth consists of a spherical core of mass $0.22M_E$ and radius $0.54R_E$ and a surrounding mantle (a spherical shell) of mass $0.78M_E$ and outer radius R_E. Suppose that the core is of uniform density and the mantle is also of uniform density. According to this simple model, what is the moment of inertia of the Earth? Express your answer as a multiple of $M_E R_E^2$.

*48. In order to increase her moment of inertia about a vertical axis, a spinning figure skater stretches her arms out horizontally; in order to reduce her moment of inertia, she brings her arms down vertically along her sides. Calculate the change of moment of inertia between these two configurations of the arms. Assume that each arm is a thin, uniform rod of length 0.60 m and mass 2.8 kg hinged at the shoulder a distance of 0.20 m from the axis of rotation.

*49. Find the moment of inertia of a thin rod of mass M and length L about an axis through the center inclined at an angle θ with respect to the rod.

*50. Given that the moment of inertia of a sphere about a diameter is $\frac{2}{5} MR^2$, show that the moment of inertia about an axis tangent to the surface is $\frac{7}{5} MR^2$.

*51. Find a formula for the moment of inertia of a uniform thin square plate (mass m, dimension $l \times l$) rotating about an axis that coincides with one of its edges.

*52. A conical shell has mass M, height h, and base radius R. Assume it is made from a thin sheet of uniform thickness. What is its moment of inertia about its symmetry axis?

*53. Suppose a peach of radius R and mass M consists of a spherical pit of radius $0.50R$ and mass $0.050M$ surrounded by a spherical shell of fruit of mass $0.95M$. What is the moment of inertia of the peach?

*54. Find the moment of inertia of the flywheel shown in Fig. 12.23 rotating about its axis. The flywheel is made of material of uniform thickness; its mass is M.

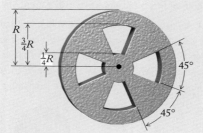

FIGURE 12.23 A flywheel.

*55. A solid cylinder capped with two solid hemispheres rotates about its axis of symmetry (Fig. 12.24). The radius of the cylinder is R, its height is h, and the total mass (hemispheres included) is M. What is the moment of inertia?

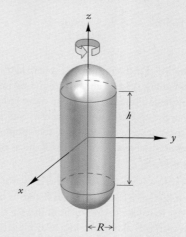

FIGURE 12.24 A solid cylinder capped with two solid hemispheres.

*56. A hole of radius r has been drilled in a circular, flat plate of radius R (Fig. 12.25). The center of the hole is at a distance d from the center of the circle. The mass of this body is M. Find the moment of inertia for rotation about an axis through the center of the circle, perpendicular to the plate.

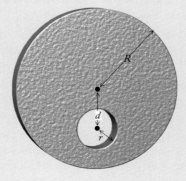

FIGURE 12.25 Circular plate with a hole.

*57. Derive a formula for the moment of inertia of a uniform spherical shell of mass M, inner radius R_1, outer radius R_2, rotating about a diameter.

*58. Find the moment of inertia of a flywheel of mass M made by cutting four large holes of radius r out of a uniform disk of radius R (Fig. 12.26). The holes are centered at a distance $R/2$ from the center of the flywheel.

FIGURE 12.26 Disk with four holes.

*59. Show that the moment of inertia of a long, very thin cone (Fig. 12.27) about an axis through the apex and perpendicular to the centerline is $\frac{3}{5}Ml^2$, where M is the mass and l the height of the cone.

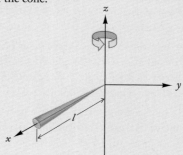

FIGURE 12.27 A long, thin cone rotating about its apex.

*60. The mass distribution within the Earth can be roughly approximated by several concentric spherical shells, each of constant density. The following table gives the outer and the inner radius of each shell and its mass (expressed as a fraction of the Earth's mass):

SHELL	OUTER RADIUS	INNER RADIUS	FRACTION OF MASS
1	6400 km	5400 km	0.28
2	5400	4400	0.25
3	4400	3400	0.16
4	3400	2400	0.20
5	2400	0	0.11

Use these data to calculate the moment of inertia of the Earth about its axis.

61. The drilling pipe of an oil rig is 2.0 km long and 15 cm in diameter, and it has a mass of 20 kg per meter of length. Assume that the wall of the pipe is very thin.

    (a) What is the moment of inertia of this pipe rotating about its longitudinal axis?

    (b) What is the kinetic energy when rotating at 1.0 rev/s?

62. Engineers have proposed that large flywheels be used for the temporary storage of surplus energy generated by electric power plants. A suitable flywheel would have a diameter of 3.6 m and a mass of 300 metric tons and would spin at 3000 rev/min. What is the kinetic energy of rotation of this flywheel? Give the answer in both joules and kilowatt-hours. Assume that the moment of inertia of the flywheel is that of a uniform disk.

63. An automobile of mass 1360 kg has wheels 76.2 cm in diameter of mass 27.2 kg each. Taking into account the rotational kinetic energy of the wheels about their axles, what is the total kinetic energy of the automobile when traveling at 80.0 km/h? What percentage of the kinetic energy belongs to the rotational motion of the wheels about their axles? Pretend that each wheel has a mass distribution equivalent to that of a uniform disk.

*64. The Oerlikon Electrogyro bus uses a flywheel to store energy for propelling the bus. At each bus stop, the bus is briefly connected to an electric power line, so that an electric motor on the bus can spin up the flywheel to 3000 rev/min. If the flywheel is a disk of radius 0.60 m and mass 1500 kg, and if the bus requires an average of 40 hp for propulsion at an average speed of 20 km/h, how far can it move with the energy stored in the rotating flywheel?

*65. Pulsars are rotating stars made almost entirely of neutrons closely packed together. The rate of rotation of most pulsars gradually decreases because rotational kinetic energy is gradually converted into other forms of energy by a variety of complicated "frictional" processes. Suppose that a pulsar of mass 1.5×10^{30} kg and radius 20 km is spinning at the rate of 2.1 rev/s and is slowing down at the rate of 1.0×10^{-15} rev/s^2. What is the rate (in joules per second, or watts) at which the rotational energy is decreasing? If this rate of decrease of the energy remains constant, how long will it take the pulsar to come to a stop? Treat the pulsar as a sphere of uniform density.

66. For the sake of directional stability, the bullet fired from a rifle is given a spin angular velocity about its axis by means of spiral grooves ("rifling") cut into the barrel. The bullet fired by a Lee–Enfield rifle is (approximately) a uniform cylinder of length 3.18 cm, diameter 0.790 cm, and mass 13.9 g. The bullet emerges form the muzzle with a translational velocity of 628 m/s and a spin angular velocity of 2.47×10^3 rev/s. What is the translational kinetic energy of the bullet? What is the rotational kinetic energy? What fraction of the total kinetic energy is rotational?

*67. Find a formula for the moment of inertia of a thin disk of mass M and radius R rotating about a diameter.

*68. Derive the formula for the moment of inertia of a thin hoop of mass M and radius R rotating about a diameter.

*69. Find a formula for the moment of inertia of a uniform thin square plate (mass M, dimension $l \times l$) rotating about an axis through the center and perpendicular to the plate.

*70. Find the moment of inertia of a uniform cube of mass M and edge l. Assume the axis of rotation passes through the center of the cube and is perpendicular to two of the faces.

*71. What is the moment of inertia of a thin, flat plate in the shape of a semicircle rotating about the straight side (Fig. 12.28)? The mass of the plate is M and the radius is R.

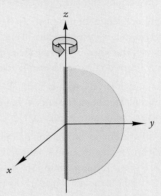

FIGURE 12.28 A semicircle rotating about its straight edge.

**72. Find the moment of inertia of the thin disk with two semicircular cutouts shown in Fig. 12.29 rotating about its axis. The disk is made of material of uniform thickness; its mass is M.

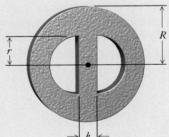

FIGURE 12.29 Disk with two semicircular cutouts.

**73. A cone of mass M has a height h and a base diameter R. Find its moment of inertia about its axis of symmetry.

**74. Derive the formula given in Table 12.3 for the moment of inertia of a sphere.

REVIEW PROBLEMS

75. An automobile has wheels of diameter 0.63 m. If the automobile is traveling at 80 km/h, what is the instantaneous velocity vector (relative to the ground) of a point at the top of the wheel? At the bottom? At the front?

76. The propeller of an airplane is turning at 2500 rev/min while the airplane is cruising at 200 km/h. The blades of the propeller are 1.5 m long. Taking into account both the rotational motion of the propeller and the translational motion of the aircraft, what is the velocity (magnitude and direction) of the tip of the propeller?

77. An automobile accelerates uniformly from 0 to 80 km/h in 6.0 s. The automobile has wheels of radius 30 cm. What is the angular acceleration of the wheels? What is their final angular velocity? How many turns do they make during the 6.0-s interval?

78. The minute hand of a wall clock is a rod of mass 5.0 g and length 15 cm rotating about one end. What is the rotational kinetic energy of the minute hand?

79. What is the kinetic energy of rotation of a phonograph record of mass 170 g and radius 15.2 cm rotating at $33\frac{1}{3}$ revolutions per minute? To give this phonograph record a translational kinetic energy of the same magnitude, how fast would you have to throw it?

80. The wheel of a wagon consists of a rim of mass 20 kg and eight spokes in the shape of rods of length 0.50 m and mass 0.80 kg each.

    (a) What is the moment of inertia of this wheel about its axle?

    (b) What is the kinetic energy of this wheel when rotating at 1.0 rev/s?

*81. A solid body consists of two uniform solid spheres of mass M and radius R welded together where they touch (see Fig. 12.30). What is the moment of inertia of this rigid body about the longitudinal axis through the center of the spheres? About the transverse axis through the point of contact?

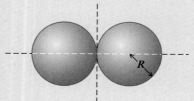

FIGURE 12.30 Two connected solid spheres.

*82. A .22-caliber bullet is a solid cylinder of length 7.0 mm and radius 2.7 mm capped at its front with a hemisphere of the same radius. The mass of the bullet is 15 g.

    (a) What is the moment of inertia of this bullet when rotating about its axis of symmetry?

    (b) What is the rotational kinetic energy of the bullet when rotating at 1.2×10^3 rev/s?

*83. Find the moment of inertia of the wheel shown in Fig. 12.31 rotating about its axis. The wheel is made of material of uniform thickness, its mass is M, and its radius is R. Treat the

spokes approximately as thin rods of length $R/2$ and width $R/12$. Ignore the overlap of the spokes at the center and ignore the curvature of the spokes at their outer edges.

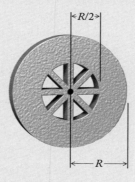

FIGURE 12.31 A wheel.

*84. The total kinetic energy of a rolling body is the sum of its translational kinetic energy $\frac{1}{2}Mv^2$ and its rotational kinetic energy $\frac{1}{2}I\omega^2$. Suppose that a cylinder, a sphere, and a pipe (a cylindrical shell) of equal masses 2.0 kg are rolling with equal speeds of 1.0 m/s. What is the total kinetic energy of each?

*85. A uniform solid cylinder is initially at rest at the top of a ramp of height 1.5 m. If the cylinder rolls down this ramp without slipping, what will be its speed at the bottom? (Hint: Use energy conservation. The kinetic energy of the cylinder at the bottom of the ramp is the sum of its translational kinetic energy $\frac{1}{2}Mv^2$ and its rotational kinetic energy $\frac{1}{2}I\omega^2$.)

**86. An airplane propeller (Fig. 12.32) is rotating at 3000 rev/min when one of the blades breaks off at the hub. Treat the blade as a rod, of length 1.2 m. The blade is horizontal and swinging upward at the instant it breaks.

    (a) What is the velocity (magnitude and direction) of the motion of the center of mass of the blade immediately after this instant?

    (b) What is the angular velocity of the rotational motion of the blade about its own center of mass?

    (c) Suppose that this happens while the aircraft is on the ground, with the hub of the propeller 2.4 m above the ground. How high above the ground does the center of mass of the broken propeller blade rise? Neglect air resistance.

FIGURE 12.32 An airplane propeller.

Answers to Checkups

Checkup 12.1

1. The swinging door executes only rotational motion about its (fixed) hinges. The motions of the wheel of a train and of the propeller of an airplane involve both rotational and translational motion; the wheel and propeller rotate as the vehicle moves through space.

2. Yes, the motion is describable as rotation about an axis and simultaneous translational motion. The rotational motion is rotation about an axis through the end of the hammer; the translational motion, however, is not along a parabolic path, but involves more complicated looping motion (see Fig. 12.1).

3. An automobile exhibits roll motion when driving on a banked surface; the auto is then tilted. Pitch motion can occur during sudden braking, when the front of the auto dives downward. Turning motion occurs whenever the auto is being driven around a curve (compare Fig. 12.2).

4. (D) Swinging door. The axis of rotation is through the hinges, along the edge of the door.

Checkup 12.2

1. The point P has the larger instantaneous speed (it travels through a greater distance per unit time). Both points have the same instantaneous angular velocity ω and the same angular acceleration α (as do all points on the same rigid body). Hence the point P has the larger tangential acceleration ($a_{\text{tangential}} = \alpha R$) and also the larger centripetal acceleration ($a_{\text{centripetal}} = \omega^2 R$).

2. The radius R for circular motion is the perpendicular distance from the axis of rotation, and so is equal to the Earth's radius only at the equator, and is increasingly smaller as one moves toward the poles; at a pole, R is zero. All points have the same angular velocity ω, as for any rigid body. The velocity is not the same for all points; since $v = \omega R$, v is largest at the equator. All points do not have the same centripetal acceleration; since $a_{\text{centripetal}} = \omega^2 R$, the centripetal acceleration is largest at the equator.

3. There is a centripetal acceleration; at the top of the arc, this is directed downward ($a_{\text{centripetal}} = v^2/R$). There is no tangential acceleration at the top (no forces act in this direction). Some distance beyond the highest point, there will be both a centripetal acceleration (since the car still moves along an arc) and a tangential acceleration (since now a component of the gravitational force is tangent to the path).

4. (D) Handle end; handle end. Since the rotation is about an axis through the center of mass (near the hammer head), the end of the handle is furthest from the axis. Thus both the speed $v = \omega R$ and the centripetal acceleration $a_{\text{centripetal}} = \omega^2 R$ are largest at the end of the handle, since R is largest there (and ω is a constant for all points on a rigid body).

Checkup 12.3

1. The centripetal acceleration always points toward the center of curvature of the circular arc of the problem; here, this is verti-

cally down. The tangential acceleration points perpendicular to a radius at any point; since the elevator accelerates upward, the tangential acceleration at the top of the wheel points horizontally toward the left.

2. Yes to both. As long as there is no slipping, we have $\omega = v/R$ and $\alpha = a/R$, so the behavior of an angular quantity is the same as the corresponding translational quantity.

3. (D) 20. For constant acceleration and starting from rest, the angular position is $\phi = \frac{1}{2}\alpha t^2$. Since this is proportional to t^2, the angular position will be four times greater in twice the time. Thus the total number of revolutions in the first two seconds is $4 \times 5 = 20$.

Checkup 12.4

1. Since the angular velocity is proportional to $t^{1/2}$, the angular position, which is the integral of the angular velocity over time [Eq. (12.23)], will be proportional to $t^{1/2+1} = t^{3/2}$. Thus the angular position will be $4^{3/2} = 8$ times as large at $t = 4$ s as it was at $t = 1$ s.

2. (D) t^4. If the angular acceleration α increases in proportion to the time t, then the angular velocity $\omega = \int \alpha\, dt$ increases in proportion to t^2. The centripetal acceleration is given by $a_{\text{centripetal}} = v^2/R = \omega^2 R$, and so increases in proportion to the fourth power of the time.

Checkup 12.5

1. Since all of the mass M is at the same distance from the axis of rotation, the moment of inertia is simply $I = MR^2$.

2. Rotation about an axis perpendicular to the rod through its end gives the largest moment of inertia, since more mass is located at a greater distance from the axis of rotation. Rotation about an axis along the rod must give the smallest moment of inertia, since in this case all of the mass is very close to the axis.

3. About an axis along one edge or through its center parallel to one edge, the distribution of mass (relative to the axis of rotation) in each case is the same as for the corresponding rod (imagine viewing Fig. 12.19b from above, that is, along the axis of rotation). Thus the moment of inertia of the square about an axis along one edge is $I = \frac{1}{3}ML^2$; about an axis through its center parallel to one edge, it is $\frac{1}{12}ML^2$.

4. About an axis through the center, each particle is a distance $l/2$ from the axis, and so the moment of inertia is $I = m(l/2)^2 + m(l/2)^2 = \frac{1}{2}ml^2$. About an axis through one particle, one particle is a distance l from the axis and the other is at zero distance, so $I = ml^2 + 0 = ml^2$. Since we have shifted the axis by $d = l/2$ in the second case, we indeed have $I = I_{\text{CM}} + Md^2 = \frac{1}{2}ml^2 + (2m)(l/2)^2 = ml^2$, so the parallel-axis theorem is satisfied (notice we must use the total mass $M = 2m$).

5. (B) $2MR^2$. Since the axis is shifted by a distance $d = R$, the parallel-axis theorem gives $I = I_{\text{CM}} + Md^2 = MR^2 + MR^2 = 2MR^2$ for rotation about a point on the hoop.

Dynamics of a Rigid Body

13.1 Work, Energy, and Power in Rotational Motion; Torque

13.2 The Equation of Rotational Motion

13.3 Angular Momentum and Its Conservation

13.4 Torque and Angular Momentum as Vectors

CONCEPTS IN CONTEXT

Concepts
— *in* —
Context

The *Gravity Probe B* satellite, containing four high-precision gyroscopes, was recently placed in orbit by a rocket. These gyroscopes are used for a delicate test of Einstein's theory of General Relativity. The rotor of one of these gyroscopes is shown here. It consists of a nearly perfect sphere of quartz, 3.8 cm in diameter, suspended electrically and spinning at 10 000 revolutions per minute.

Some of the questions we can address with the concepts developed in this chapter are:

? When initially placed in orbit, the rotor is at rest. What torque and what force are needed to spin up this gyroscope with a given angular acceleration? (Example 4, page 401)

? A rotating body, such as this rotor, has not only kinetic energy, but also an angular momentum, which is the rotational analog of the linear momentum introduced in Chapter 10. How is the angular

momentum of the gyroscope expressed in terms of its angular velocity?
(Example 8, page 406)

? The gyroscope is used like a compass, to establish a reference direction in space.
How does a gyroscope maintain a fixed reference direction? (Physics in Practice:
The Gyrocompass, page 414)

A s we saw in Chapter 5, Newton's Second Law is the equation that determines the
translational motion of a body. In this chapter, we will derive an equation that
determines the rotational motion of a rigid body. Just as Newton's equation of motion
gives us the translational acceleration and permits us to calculate the change in velocity and position, the analogous equation for rotational motion gives us the angular
acceleration and permits us to calculate the change in angular velocity and angular position. The equation for rotational motion is not a new law of physics, distinct from
Newton's three laws. Rather, it is a consequence of these laws.

13.1 WORK, ENERGY, AND POWER IN ROTATIONAL MOTION; TORQUE

We begin with a calculation of the work done by an external force on a
rigid body constrained to rotate about a fixed axis. Figure 13.1 shows the
body, with the axis of rotation perpendicular to the page. The force is applied
at some point of the body at a distance R from the axis of rotation. For a
start, we will assume that the force has no component parallel to the axis;
any such component is of no interest in the present context since the body
does not move in the direction parallel to the axis, and so a force parallel to
the axis can do no work. In Fig. 13.1, the force is shown entirely in the
plane of the page. The work done by this force during a small displacement of the point at which the force acts is the product of the force F, the
displacement ds, and the cosine of the angle between the force and the displacement [see Eq. (7.5)]. The cosine of this angle is equal to the sine of the angle θ
between the force and the radial line (see Fig. 13.1). Hence, we can write the work as

$$dW = F\,ds\,\sin\theta$$

If the body rotates through a small angle $d\phi$, the displacement is $ds = R\,d\phi$, and
therefore

$$dW = FR\,d\phi\,\sin\theta \tag{13.1}$$

The product $FR\sin\theta$ is called the **torque** of the force F, usually designated by the
symbol τ (the Greek letter *tau*):

$$\tau = FR\sin\theta \tag{13.2}$$

With this notation, the work done by the force, or *the work done by the torque,* is simply

$$dW = \tau\,d\phi \tag{13.3}$$

This is the rotational analog of the familiar equation $dW = F\,dx$ for work done in
translational motion. The torque τ is analogous to the force F, and the angular displacement $d\phi$ is analogous to the translational displacement dx. The analogy between
torque and force extends beyond the equation for the work. As we will see in the next
section, a torque applied to a rigid body causes angular acceleration, just as a force
applied to a particle causes translational acceleration.

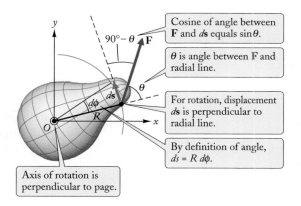

Cosine of angle between **F** and $d\mathbf{s}$ equals $\sin\theta$.

θ is angle between **F** and radial line.

For rotation, displacement $d\mathbf{s}$ is perpendicular to radial line.

By definition of angle, $ds = R\,d\phi$.

Axis of rotation is perpendicular to page.

FIGURE 13.1 Force applied to a rigid
body rotating about a fixed axis. As in
Chapter 12, the axis of rotation is indicated
by a circled dot. The force makes an angle θ
with the radial line and an angle $90° − \theta$
with the instantaneous displacement $d\mathbf{s}$.

torque

work done by torque

According to Eq. (13.3), each contribution to the work is the product of the torque τ and the small angular displacement $d\phi$. Thus the total work done in rotating a body from an initial angle ϕ_1 to a final angle ϕ_2 is

$$W = \int dW = \int_{\phi_1}^{\phi_2} \tau \, d\phi \tag{13.4}$$

In the special case of a constant torque, the torque may be brought outside the integral to obtain

$$W = \tau \int_{\phi_1}^{\phi_2} d\phi = \tau(\phi_2 - \phi_1)$$

or simply

work done by constant torque

$$W = \tau \, \Delta\phi \qquad \text{(for } \tau = \text{constant)} \tag{13.5}$$

where $\Delta\phi = \phi_2 - \phi_1$ is the change in angular position during the time that the torque is applied. Equation (13.5) is analogous to the equation for the work done by a constant force on a body in one-dimensional translational motion, $W = F \, \Delta x$.

From Eq. (13.2), we see that the unit of torque is the unit of force multiplied by the unit of distance; *this SI unit of torque is the newton-meter* (N·m).

Note that according to Eq. (13.2), for a force of given magnitude, the torque is largest if the force acts at right angles to the radial line ($\theta = 90°$) and if the force acts at a large distance from the axis of rotation (large R). This dependence of the torque (and of the work) on the distance from the axis of rotation and on the angle of the push agrees with our everyday experience in pushing doors open or shut. A door is a rigid body, which rotates about a vertical axis through the hinges. If you push perpendicularly against the door, near the edge farthest from the hinge (largest R; see Fig. 13.2a), you produce a large torque, which does work on the door, increases its kinetic energy, and swings the door quickly on its hinges. If you push at a point near the hinge (small R; see Fig. 13.2b), the door responds more sluggishly. You produce a smaller torque, and you have to push harder to do the same amount of work and attain the same amount of kinetic energy and the same final angular velocity. Finally, if you push in a direction that is not perpendicular to the door (small θ; see Fig. 13.2c), the door again responds sluggishly, because the torque is small.

FIGURE 13.2 (a) A push against the door far from the hinge produces a large angular acceleration. (b) The same push near the hinge produces a small angular acceleration. (c) A push against the door at a small angle also produces a small angular acceleration.

(a) Large torque

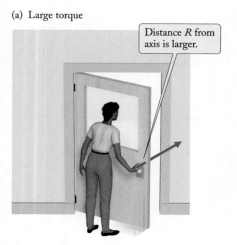

Distance R from axis is larger.

(b) Small torque

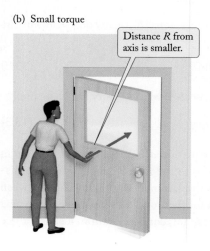

Distance R from axis is smaller.

(c) Small torque

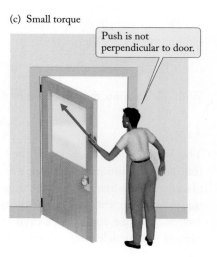

Push is not perpendicular to door.

EXAMPLE 1

Suppose that while opening a 1.0-m-wide door, you push against the edge farthest from the hinge, applying a force with a steady magnitude of 0.90 N at right angles to the surface of the door. How much work do you do on the door during an angular displacement of 30°?

SOLUTION: For a constant torque, the work is given by Eq. (13.5), $W = \tau \, \Delta\phi$. The definition of torque, Eq. (13.2), with $F = 0.90$ N, $R = 1.0$ m, and $\theta = 90°$, gives

$$\tau = FR \sin 90° = 0.90 \text{ N} \times 1.0 \text{ m} \times 1 = 0.90 \text{ N·m}$$

To evaluate the work, the angular displacement must be expressed in radians; $\Delta\phi = 30° \times (2\pi \text{ radians}/360°) = 0.52$ radian. Then

$$W = \tau \, \Delta\phi = 0.90 \text{ N·m} \times 0.52 \text{ radian}$$

$$= 0.47 \text{ J}$$

The equation for the power in rotational motion and the equations that express the work–energy theorem and the conservation law for energy in rotational motion are analogous to the equations we formulated for translational motion in Chapters 7 and 8. If we divide both sides of Eq. (13.3) by dt, we find *the instantaneous power delivered by the torque:*

$$P = \frac{dW}{dt} = \tau \frac{d\phi}{dt}$$

or

$$P = \tau\omega \qquad (13.6)$$

power delivered by torque

where $\omega = d\phi/dt$ is the angular velocity. Obviously, this equation is analogous to the equation $P = Fv$ obtained in Section 8.5 for the power in one-dimensional translational motion.

The work done by the torque changes the rotational kinetic energy of the body. Like the work–energy theorem for translational motion, the work–energy theorem for rotational motion says that the work done on the body by the external torque equals the change in rotational kinetic energy (the internal forces and torques in a rigid body do no net work):

$$W = K_2 - K_1 = \tfrac{1}{2}I\omega_2^2 - \tfrac{1}{2}I\omega_1^2 \qquad (13.7)$$

If the force acting on the body is conservative—such as the force of gravity or the force of a spring—then the work equals the negative of the change in potential energy, and Eq. (13.7) becomes

$$-U_2 + U_1 = \tfrac{1}{2}I\omega_2^2 - \tfrac{1}{2}I\omega_1^2 \qquad (13.8)$$

or

$$\tfrac{1}{2}I\omega_1^2 + U_1 = \tfrac{1}{2}I\omega_2^2 + U_2 \qquad (13.9)$$

This expresses the **conservation of energy in rotational motion**: the sum of the kinetic and potential energies is constant, that is,

$$E = \tfrac{1}{2}I\omega^2 + U = [\text{constant}] \qquad (13.10)$$

conservation of energy
in rotational motion

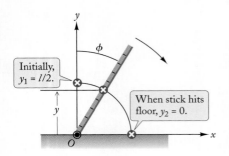

FIGURE 13.3 Meterstick rotating about its lower end.

EXAMPLE 2 A meterstick is initially standing vertically on the floor. If the meterstick falls over, with what angular velocity will it hit the floor? Assume that the end in contact with the floor does not slip.

SOLUTION: The motion of the meterstick is rotation about a fixed axis passing through the point of contact with the floor (see Fig. 13.3). The stick is a uniform rod of mass M and length $l = 1.0$ m. Its moment of inertia about the end is $Ml^2/3$ (see Table 12.3), and its rotational kinetic energy is therefore $\frac{1}{2}I\omega^2 = Ml^2\omega^2/6$. The gravitational potential energy is Mgy, where y is the height of the center of mass above the floor. When the meterstick is standing vertically, the initial angular velocity is $\omega_1 = 0$ and $y_1 = l/2$, so the total energy is

$$E = \tfrac{1}{6}Ml^2\omega_1^2 + Mgy_1 = 0 + Mgl/2 \qquad (13.11)$$

Just before the meterstick hits the floor, the angular velocity is ω_2 and $y_2 = 0$. The energy is

$$E = \tfrac{1}{6}Ml^2\omega_2^2 + Mgy_2 = \tfrac{1}{6}Ml^2\omega_2^2 + 0 \qquad (13.12)$$

Conservation of energy therefore implies

$$\tfrac{1}{6}Ml^2\omega_2^2 = Mgl/2$$

from which we obtain

$$\omega_2^2 = \frac{3g}{l} \qquad (13.13)$$

Taking the square root of both sides, we find

$$\omega_2 = \sqrt{\frac{3g}{l}} = \sqrt{\frac{3 \times 9.81\,\text{m/s}^2}{1.0\ \text{m}}} = 5.4\ \text{radians/s}$$

EXAMPLE 3 At what instantaneous rate is gravity delivering energy to the meterstick of Example 2 just before it hits the floor? The mass of the meterstick is 0.15 kg.

SOLUTION: The rate of energy delivery is the power,

$$P = \tau\omega$$

From Example 2, we know $\omega = 5.4$ radians/s just before the stick hits the floor. At that instant, gravity acts perpendicular to the stick at the center of mass (in the next chapter we will see that the weight acts as if concentrated at the center of mass), a distance $R = l/2 = 0.50$ m from the end. So the torque exerted by gravity is

$$\tau = FR\sin\theta = mg\frac{l}{2}\sin 90° = 0.15\ \text{kg} \times 9.81\ \text{m/s}^2 \times 0.50\ \text{m} \times 1$$

$$= 0.74\ \text{N·m}$$

Thus the instantaneous power delivered by the torque due to gravity is

$$P = \tau\omega = 0.74\ \text{N·m} \times 5.4\ \text{radians/s} = 4.0\ \text{W}$$

Checkup 13.1

QUESTION 1: You are trying to tighten a bolt with a wrench. Where along the handle should you place your hand so you can exert maximum torque? In what direction should you push?

QUESTION 2: A force is being exerted against the rim of a freely rotating wheel, but the work done by this force is zero. What can you conclude about the direction of the force? What is the torque of the force?

QUESTION 3: Consider the meterstick falling over, as in Example 2. What is the torque that the weight exerts on the meterstick when it is in the upright, initial position? After the stick begins to fall over, the torque increases. When is the torque maximum?

QUESTION 4: Suppose you first push a door at its outer edge at right angles to the surface of the door with a force of magnitude F. Next you push the door at its center, again at right angles to the surface, with a force of magnitude $F/2$. In both cases you push the door as it moves through 30°. The ratio of the work done by the second push to the work done by the first push is:

(A) $\frac{1}{4}$ (B) $\frac{1}{2}$ (C) 1   (D) 2   (E) 4

13.2 THE EQUATION OF ROTATIONAL MOTION

Our intuition tells us that a torque acting on a wheel or some other body free to rotate about an axis will produce an angular acceleration. For instance, the push of your hand against a crank on a wheel (see Fig. 13.4) exerts a torque or "twist" that starts the wheel turning. The angular acceleration depends on the magnitude of your push on the crank and also on its direction (as well as on the inertia of the wheel). Your push will be most effective if exerted tangentially, at right angles to the radius (at $\theta = 90°$; see Fig. 13.4a). It will be less effective if exerted at a smaller or larger angle (see Fig. 13.4b). And it will be completely ineffective if exerted parallel to the radius (at $\theta = 0$ or 180°; see Fig. 13.4c)—such a push in the radial direction produces no rotation at all. These qualitative considerations are in agreement with the definition of torque,

$$\tau = FR \sin \theta \tag{13.14}$$

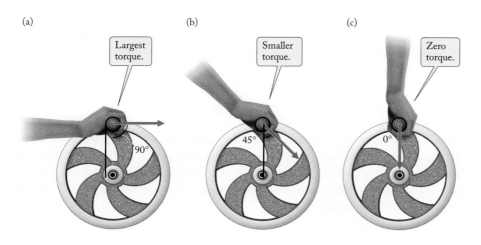

(a) Largest torque. 90°

(b) Smaller torque. 45°

(c) Zero torque. 0°

FIGURE 13.4 (a) A push at right angles to the radius is most effective in producing rotation. (b) A push at 45° is less effective. (c) A push parallel to the radius produces no rotation.

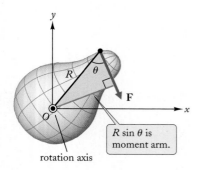

FIGURE 13.5 The distance between the center of rotation and the point of application of the force is R. The perpendicular distance between the center of rotation and the line of action of the force is $R \sin \theta$.

According to this equation, the torque provided by a force of a given magnitude F is maximum if the force is at right angles to the radius ($\theta = 90°$), and it is zero if the force is parallel to the radius ($\theta = 0°$ or $180°$).

The quantity $R \sin \theta$ appearing in Eq. (13.14) has a simple geometric interpretation: it is the perpendicular distance between the line of action of the force and the axis of rotation (see Fig. 13.5); this perpendicular distance is called the **moment arm of the force**. Hence, Eq. (13.14) states that the torque equals the magnitude of the force multiplied by the moment arm.

To find a quantitative relationship between torque and angular acceleration, we recall from Eq. (13.6) that the power delivered by a torque acting on a body is

$$\frac{dW}{dt} = \tau \omega \tag{13.15}$$

The work–energy theorem tells us that the work dW equals the change of kinetic energy in the small time interval dt. The small change in the kinetic energy is $dK = d(\frac{1}{2}I\omega^2) = \frac{1}{2}I \times 2\omega \, d\omega = I\omega \, d\omega$. Thus,

$$dW = I\omega \, d\omega \tag{13.16}$$

Inserting this into the left side of Eq. (13.15), we find

$$\frac{I\omega \, d\omega}{dt} = \tau \omega \tag{13.17}$$

Canceling the factor of ω on both sides of the equation, we obtain

$$I \frac{d\omega}{dt} = \tau \tag{13.18}$$

But $d\omega/dt$ is the angular acceleration α; hence

equation of rotational motion

$$\boxed{I\alpha = \tau} \tag{13.19}$$

This is the equation for rotational motion. As we might have expected, this equation says that *the angular acceleration is directly proportional to the torque*. Equation (13.19) is mathematically analogous to Newton's Second Law, $m\mathbf{a} = \mathbf{F}$, for the translational motion of a particle; the moment of inertia takes the place of the mass, the angular acceleration the place of the acceleration, and the torque the place of the force.

In our derivation of Eq. (13.19) we assumed that only one external force is acting on the rigid body. If several forces act, then each produces its own torque. If an individual torque would produce an angular acceleration in the rotational direction chosen as positive, it is reckoned as positive, and if a torque would produce an angular acceleration in the opposite direction, it is reckoned as negative. The net torque is the sum of these individual torques, and the angular acceleration is proportional to this net torque:

equation of rotational motion for net torque

$$\boxed{I\alpha = \tau_{\text{net}}} \tag{13.20}$$

In the evaluation of the net torque, we need to take into account all the external forces acting on the rigid body, but we can ignore the internal forces that particles in the body exert on other particles also in the body. The torques of such internal forces cancel (this is an instance of the general result mentioned in Section 10.4: for a rigid body, the work of internal forces cancels).

EXAMPLE 4   The rotor of the gyroscope of the *Gravity Probe B* experiment (see the chapter photo and Fig. 13.6) is a quartz sphere of diameter 3.8 cm and mass 7.61×10^{-2} kg. To start this sphere spinning, a stream of helium gas flowing in an equatorial channel in the surface of the housing is blown tangentially against the rotor. What torque must this stream of gas exert on the rotor to accelerate it uniformly from 0 to 10 000 rpm (revolutions per minute) in 30 minutes? What force must it exert on the equator of the sphere?

Concepts
— in —
Context

SOLUTION: The final angular velocity is $2\pi \times 10\,000$ radians $/60$ s $= 1.05 \times 10^3$ radians/s, and therefore the angular acceleration is

$$\alpha = \frac{\omega_2 - \omega_1}{t_2 - t_1}$$

$$= \frac{1.05 \times 10^3 \text{ radians/s} - 0}{30 \times 60 \text{ s} - 0} = 0.582 \text{ radian/s}^2$$

The moment of inertia of the rotor is that of a sphere (see Table 12.3):

$$I = \tfrac{2}{5}MR^2$$

$$= \tfrac{2}{5} \times 7.61 \times 10^{-2} \text{ kg} \times (0.019 \text{ m})^2 = 1.1 \times 10^{-5} \text{ kg·m}^2$$

Hence the required torque is, according to Eq. (13.19),

$$\tau = I\alpha = 1.1 \times 10^{-5} \text{ kg·m}^2 \times 0.582 \text{ radian/s}^2$$

$$= 6.4 \times 10^{-6} \text{ N·m}$$

The driving force is along the equator of the rotor—that is, it is perpendicular to the radius—so $\sin\theta = 1$ and Eq. (13.2) reduces to $\tau = FR$, which yields

$$F = \frac{\tau}{R} = \frac{6.4 \times 10^{-6} \text{ N·m}}{0.019 \text{ m}} = 3.4 \times 10^{-4} \text{ N}$$

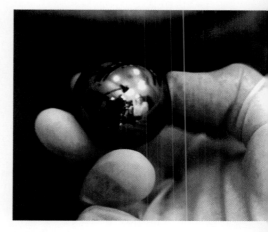

FIGURE 13.6 A gyroscope sphere for *Gravity Probe B*.

EXAMPLE 5   Two masses m_1 and m_2 are suspended from a string that runs, without slipping, over a pulley (see Fig. 13.7a). The pulley has a radius R and a moment of inertia I about its axle, and it rotates without friction. Find the accelerations of the masses.

SOLUTION: We have already found the motion of this system in Example 10 of Chapter 5, where the two masses were an elevator and its counterweight, and where we neglected the inertia of the pulley. Now we will take this inertia into account.

Figure 13.7c shows the "free-body" diagrams for the masses m_1 and m_2. In these diagrams, T_1 and T_2 are the tensions in the two parts of the string attached to the two masses. (Note that now T_1 and T_2 are not equal. For a pulley of zero moment of inertia, these tensions would be equal; but for a pulley of nonzero moment of inertia, a difference between T_1 and T_2 is required to produce the angular acceleration of the pulley.) If the acceleration of mass m_1 is a (reckoned as positive if upward), then the acceleration of mass m_2 is $-a$, and the equations of motion of the two masses are

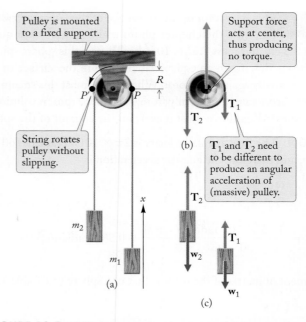

Pulley is mounted to a fixed support.

Support force acts at center, thus producing no torque.

String rotates pulley without slipping.

(b)

$\mathbf{T}_1$ and $\mathbf{T}_2$ need to be different to produce an angular acceleration of (massive) pulley.

(a)

(c)

FIGURE 13.7 (a) Two masses m_1 and m_2 suspended from a string that runs over a pulley. (b) "Free-body" diagram for the pulley. (c) "Free-body" diagrams for the masses m_1 and m_2.

$$m_1 a = T_1 - m_1 g \tag{13.21}$$

$$-m_2 a = T_2 - m_2 g \tag{13.22}$$

Figure 13.7b shows the "free-body" diagram for the pulley. The tension forces act at the ends of the horizontal diameter (since the string does not slip, it behaves as though instantaneously attached to the pulley at the point of first contact; see points P and P' in Fig. 13.7a). The upward supporting force of the axle acts at the center of the pulley, and it generates no torque about the center of the pulley. The tensions act perpendicular to the radial direction, so $\sin \theta = 1$ in Eq. (13.2). Taking the positive direction of rotation as counterclockwise (to match the positive direction for the motion of mass m_1), we see that the tension forces T_1 and T_2 generate torques $-RT_1$ and RT_2 about the center. The equation of rotational motion of the pulley is

$$I\alpha = \tau_{\text{net}} = -RT_1 + RT_2 \tag{13.23}$$

The translational acceleration of each hanging portion of the string must match the instantaneous translational acceleration of the point of first contact (for the given condition of no slipping). Hence the translational acceleration a of the masses is related to the angular acceleration α by $a = \alpha R$, or $\alpha = a/R$ [see Eq. (12.13)]. Furthermore, according to Eqs. (13.21) and (13.22), $T_1 = m_1 g + m_1 a$ and $T_2 = m_2 g - m_2 a$. With these substitutions, Eq. (13.23) becomes

$$I(a/R) = -R(m_1 g + m_1 a) + R(m_2 g - m_2 a)$$

Solving this for a, we find

$$a = \frac{m_2 - m_1}{m_1 + m_2 + (I/R^2)} g \tag{13.24}$$

COMMENT: If the mass of the pulley is small, then I/R^2 can be neglected; with this approximation, Eq. (13.24) reduces to Eq. (5.44), which was obtained without taking into account the inertia of the pulley.

A device of this kind, called **Atwood's machine,** can be used to determine the value of g. For this purpose, it is best to use masses m_1 and m_2 that are nearly equal. Then a is much smaller than g and easier to measure; the value of g can be calculated from the measured value of a according to Eq. (13.24).

Atwood's machine

In some cases—for instance, the rolling motion of a wheel—the axis of rotation is in motion, perhaps accelerated motion, and is *not a fixed axis.* For such problems, some further arguments can be used to demonstrate that Eq. (13.19) remains valid for rotation about an axis in accelerated translational motion, *provided the axis passes through the center of mass of the rotating body.* When this condition is met, we can use the equation of rotational motion (13.19) as in the following examples.

(a)

| **EXAMPLE 6** | An automobile with rear-wheel drive is accelerating at 4.0 m/s^2 |

An automobile with rear-wheel drive is accelerating at 4.0 m/s^2 along a straight road. Consider one of the front wheels of this automobile (see Fig. 13.8a). The axle pushes the wheel forward, providing an acceleration of 4.0 m/s^2. Simultaneously, the friction force of the road pushes the bottom of the wheel backward, providing a torque that gives the wheel an angular acceleration. The wheel has a radius of 0.38 m and a mass of 25 kg. Assume that the wheel is (approximately) a uniform disk, and assume it rolls without slipping. Find the backward force that the friction force exerts on the wheel, and find the forward force that the axle exerts on the wheel.

SOLUTION: Figure 13.8b shows a "free-body" diagram of the wheel, with the horizontal forces acting on it (besides these horizontal forces, there are also a vertical downward push exerted by the axle and a vertical upward normal force exerted by the road; these forces exert no torque and cancel, so they need not concern us here). The forward push of the axle is P, and the rearward push of the ground is f. The force P, acting at the center of the wheel, exerts no torque; the force f, acting at the rim, exerts a torque Rf. Thus, the equation for the rotational motion of the wheel is

$$I\alpha = Rf$$

or, since $I = \tfrac{1}{2}MR^2$ for a uniform disk (see Table 12.3),

$$\tfrac{1}{2}MR\alpha = f$$

As we have seen in Example 4 of Chapter 12, the angular acceleration of a rolling wheel is related to the translational acceleration by $\alpha = a/R$. Hence

$$\tfrac{1}{2}Ma = f$$

from which

$$f = \tfrac{1}{2}Ma = \tfrac{1}{2} \times 25 \text{ kg} \times 4.0 \text{ m/s}^2$$
$$= 50 \text{ N}$$

(b)

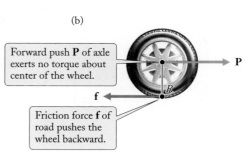

Forward push **P** of axle exerts no torque about center of the wheel.

Friction force **f** of road pushes the wheel backward.

FIGURE 13.8 (a) Front wheel of an automobile. (b) "Free-body" diagram for the wheel. The friction force of the road pushes the wheel backward. The axle pushes the wheel forward.

To find the force P, we need to examine the equation for the translational motion. The net horizontal force is $F_{net} = P - f$. Hence the equation for the translational motion of the wheel is

$$Ma = P - f$$

from which

$$P = Ma + f = 25 \text{ kg} \times 4.0 \text{ m/s}^2 + 50 \text{ N}$$

$$= 150 \text{ N}$$

Thus, the force required to accelerate a rolling wheel is larger than the force required for a wheel that slips on a frictionless surface without rolling—for such a wheel the force would be only $Ma = 25 \text{ kg} \times 4.0 \text{ m/s}^2 = 100 \text{ N}$. Here, the additional rotational inertia $\frac{1}{2}MR^2$ adds an additional amount $f = \frac{1}{2}Ma$ to the required force, so the total required force is $\frac{3}{2}$ that for sliding without rolling.

(a)

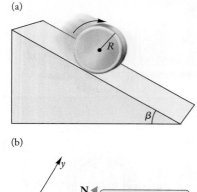

(b)

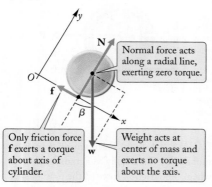

Normal force acts along a radial line, exerting zero torque.

Only friction force **f** exerts a torque about axis of cylinder.

Weight acts at center of mass and exerts no torque about the axis.

FIGURE 13.9 (a) A cylinder rolling down an inclined plane. (b) "Free-body" diagram for the cylinder.

EXAMPLE 7   A solid cylinder of mass M and radius R rolls down a sloping ramp that makes an angle β with the ground (see Fig. 13.9a). What is the acceleration of the cylinder? Assume that the cylinder is uniform and rolls without slipping.

SOLUTION: Figure 13.9b shows the "free-body" diagram for the cylinder. The forces on the cylinder are the normal force **N** exerted by the ramp, the friction force **f** exerted by the ramp, and the weight **w**. The friction force is exerted on the rim of the cylinder, and the weight is effectively exerted at the center of the cylinder (in the next chapter we will see that the weight can always be regarded as concentrated at the center of mass). As axis of rotation, we take the axis that passes through the center of the cylinder. The weight exerts no torque about this axis, and neither does the normal force (zero moment arm). Hence, the only force that exerts a torque is the friction force, and so

$$\tau = Rf$$

The equation of rotational motion is then

$$I\alpha = Rf$$

The moment of inertia of a uniform cylinder is the same as that of a disk, $I = \frac{1}{2}MR^2$. Furthermore, for rolling motion without slipping, $\alpha = a/R$. Hence

$$\tfrac{1}{2}MRa = Rf$$

or

$$a = \frac{2f}{M} \qquad (13.25)$$

To evaluate the acceleration, we need to eliminate the friction force f from this equation. We can do this by appealing to the equation for the component of the translational motion along the ramp (the motion along the x direction in Fig. 13.9b).

The components of the forces along the ramp are $-f$ for the friction force and $Mg \sin \beta$ for the weight. Hence

$$Ma = Mg \sin \beta - f$$

or

$$f = Mg \sin \beta - Ma$$

Substituting this into Eq. (13.25), we find

$$a = 2g \sin \beta - 2a$$

which we can immediately solve for a:

$$a = \tfrac{2}{3} g \sin \beta$$

COMMENT: Note that the force $Mg \sin \beta$ along the ramp here produces an acceleration that is two-thirds of the acceleration that the cylinder would have if it were to slip down a frictionless ramp without rolling. This is consistent with the last example, where we saw that a force $\tfrac{3}{2}$ as large was required to produce a given acceleration. The same factor occurs in both cases, because both the disk and the cylinder have the same moment of inertia, $\tfrac{1}{2} MR^2$.

PROBLEM-SOLVING TECHNIQUES   TORQUES AND ROTATIONAL MOTION

The general techniques for the solution of problems of rotational motion are similar to the techniques we learned in Chapters 5 and 6 for translational motion.

1   The first step is always a careful enumeration of all the forces. Make a complete list of these forces, and label each with a vector symbol.

2   Identify the body whose motion or whose equilibrium is to be investigated and draw the "free-body" diagram showing the forces acting on this body. If there are several distinct bodies in the problem (as in Example 5), then you need to draw a separate "free-body" diagram for each. When drawing the arrows for the forces acting on a rotating body, be sure to draw the head or the tail of the arrow at the actual point of the body where the force acts, since this will be important for the calculation of the torque. Note that the weight acts at the center of mass (we will establish this in the next chapter).

3   Select which direction of rotation will be regarded as positive (for instance, in Example 5, we selected the counterclockwise direction of rotation as positive). If the problem involves joint rotational and translational motions, select coordinate axes for the translational motion, preferably placing one of the axes along the direction of motion.

4   Select an axis for the rotation of the rigid body, either an axis through the center of mass, or else a fixed axis (such as an axle or a pivot mounted on a support) about which the body is constrained to rotate. Calculate the torque of each force acting on the body about this center. Remember that the sign of the torque is positive or negative depending on whether it produces an angular acceleration in the positive or the negative direction of rotation.

5   Then apply the equation of rotational motion, $I\alpha = \tau$, to each rotating body, where τ is the net torque on a given body.

6   If the rigid body has a translational motion besides the rotational motion, apply Newton's Second Law, $\mathbf{F} = m\mathbf{a}$, for the translational motion (see Examples 5 and 6). For rolling without slipping, the translational and the rotational motions are related by $v = \omega R$ and $a = \alpha R$.

7   If there are several distinct bodies in the problem, you need to apply the equation of rotational motion or Newton's Second Law separately for each (see Example 5).

Checkup 13.2

QUESTION 1: Consider a meterstick falling over, as in Example 2. At what instant is the angular acceleration produced by the weight force maximum?

QUESTION 2: A rolling cylinder has both rotational kinetic energy (reckoned about its center of mass) and translational kinetic energy. Which is larger?

QUESTION 3: Consider the rolling cylinder of Example 7. When this cylinder reaches the bottom of the ramp, is its kinetic energy larger, smaller, or the same as that of a similar cylinder that slips down a frictionless ramp without rolling?

QUESTION 4: A sphere and a cylinder of equal masses roll down an inclined plane without slipping. Will they have equal kinetic energies when they reach the bottom? Which will get to the bottom first?

QUESTION 5: A thin hoop and a solid cylinder roll down an inclined plane without slipping. When they reach the bottom, the translational speed of the hoop is

(A) Less than that of the cylinder
(B) Greater than that of the cylinder
(C) Equal to that of the cylinder

13.3 ANGULAR MOMENTUM AND ITS CONSERVATION

In Chapter 10 we saw how to express the equation for the translational motion in terms of the momentum: the rate of change of the momentum equals the force ($dp_x/dt = F_x$). Likewise, we can express the equation for rotational motion in terms of **angular momentum.** *The angular momentum of a body rotating about a fixed axis is defined as the product of the moment of inertia and the angular velocity,*

angular momentum

$$L = I\omega \tag{13.26}$$

This equation for angular momentum is analogous to the equation $p = mv$ for translational momentum. *The SI unit of angular momentum is* $\text{kg·m}^2/\text{s}$, which can also be written in the alternative form J·s. Table 13.1 gives some examples of typical values of angular momenta.

Concepts
— in —
Context

EXAMPLE 8

According to the data given in Example 4, what is the angular momentum of the rotor of the *Gravity Probe B* gyroscope when spinning at 10 000 revolutions per minute?

SOLUTION: From Example 4, the angular velocity is $\omega = 1.05 \times 10^3$ radians/s, and the moment of inertia is $I = 1.1 \times 10^{-5}$ kg·m². So

$$L = I\omega = 1.1 \times 10^{-5} \text{ kg·m}^2 \times 1.05 \times 10^3 \text{ radians/s}$$

$$= 1.2 \times 10^{-2} \text{ kg·m}^2/\text{s}$$

To express the equation for rotational motion in terms of angular momentum, we proceed as we did in the translational case. We note that if the change of angular velocity is $d\omega$, then $dL = I\, d\omega$. Dividing both sides of this relation by dt, we see

$$\frac{dL}{dt} = I\frac{d\omega}{dt}$$

If we compare this with Eq. (13.18), we see that the right side can be expressed as the torque, so

$$\frac{dL}{dt} = \tau \qquad\qquad (13.27)$$

equation of rotational motion in terms of angular momentum

This says that *the rate of change of angular momentum equals the torque.* Obviously, this equation is analogous to the equation $dp_x/dt = F_x$ for translational motion.

We now see that the analogy between rotational and translational quantities mentioned in Section 12.3 can be extended to angular momentum and momentum. Table 13.2 lists analogous quantities, including the quantities for work, power, and kinetic energy.

If there is no torque acting on the rotating body, $\tau = 0$ and therefore $dL/dt = 0$, which means that the angular momentum does not change:

$$L = [\text{constant}] \qquad (\text{when } \tau = 0) \qquad (13.28)$$

conservation of angular momentum

This is the **Law of Conservation of Angular Momentum.** Since $L = I\omega$, we can also write this law as

$$I\omega = [\text{constant}] \qquad\qquad (13.29)$$

TABLE 13.1	SOME ANGULAR MOMENTA
Orbital motion of Earth	2.7×10^{40} J·s
Rotation of Earth	5.8×10^{33} J·s
Helicopter rotor (320 rev/min)	5×10^{4} J·s
Automobile wheel (90 km/h)	1×10^{2} J·s
Electric fan	1 J·s
Frisbee	1×10^{-1} J·s
Toy gyroscope	1×10^{-2} J·s
Phonograph record (33.3 rev/min)	6×10^{-3} J·s
Compact disc (plating outer track)	2×10^{-3} J·s
Bullet fired from rifle	2×10^{-3} J·s
Orbital motion of electron in atom	1.05×10^{-34} J·s
Spin of electron	0.53×10^{-34} J·s

TABLE 13.2

FURTHER ANALOGIES BETWEEN 1D TRANSLATIONAL AND ROTATIONAL QUANTITIES

$dW = F\,dx$	$\rightarrow$	$dW = \tau\, d\phi$
$P = Fv$	$\rightarrow$	$P = \tau\omega$
$K = \frac{1}{2}mv^2$	$\rightarrow$	$K = \frac{1}{2}I\omega^2$
$ma = F$	$\rightarrow$	$I\alpha = \tau$
$p = mv$	$\rightarrow$	$L = I\omega$
$\dfrac{dp}{dt} = F$	$\rightarrow$	$\dfrac{dL}{dt} = \tau$

(a)

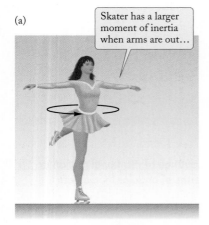

Skater has a larger moment of inertia when arms are out...

(b)

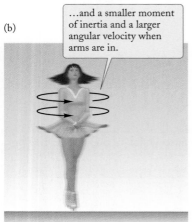

...and a smaller moment of inertia and a larger angular velocity when arms are in.

FIGURE 13.10 Figure skater performing a pirouette. (a) Arms extended. (b) Arms folded against body.

FIGURE 13.11 A figure skater whirling at high speed.

A pirouette performed by a figure skater on ice provides a nice illustration of the conservation of angular momentum. The skater begins the pirouette by spinning about her vertical axis with her arms extended horizontally (see Fig. 13.10a); in this configuration, the arms have a large moment of inertia. She then brings her arms close to her body (see Fig. 13.10b), suddenly decreasing her moment of inertia. Since the ice is nearly frictionless, the external torque on the skater is nearly zero, and therefore the angular momentum is conserved. According to Eq. (13.26), a decrease of I requires an increase of ω to keep the angular momentum constant. Thus, the change of configuration of her arms causes the skater to whirl around her vertical axis with a dramatic increase of angular velocity (see Fig. 13.11).

Like the law of conservation of translational momentum, the Law of Conservation of Angular Momentum is often useful in the solutions of problems in which the forces are not known in detail.

EXAMPLE 9   Suppose that a pottery wheel is spinning (with the motor disengaged) at 80 rev/min when a 6.0-kg ball of clay is suddenly dropped down on the center of the wheel (see Fig. 13.12). What is the angular velocity after the drop? Treat the ball of clay as a uniform sphere of radius 8.0 cm. The pottery wheel has a moment of inertia $I = 7.5 \times 10^{-2}$ kg·m^2. Ignore the (small) friction force in the axle of the turntable.

SOLUTION: Since there is no external torque on the system of pottery wheel and clay, the angular momentum of this system is conserved. The angular momentum before the drop is

$$L = I\omega \tag{13.30}$$

where ω is the initial angular velocity and I the moment of inertia of the pottery wheel. The angular momentum after the drop is

$$L' = I'\omega' \tag{13.31}$$

where ω' is the final angular velocity and I' the moment of inertia of pottery wheel and clay combined. Hence

$$I\omega = I'\omega' \tag{13.32}$$

from which we find

$$\omega' = \frac{I}{I'}\omega \tag{13.33}$$

The wheel is initially rotating with angular velocity

$$\omega = 2\pi \times f = 2\pi \times \frac{80 \text{ rev}}{60 \text{ s}} = 8.4 \text{ radians/s}$$

The moment of inertia of the pottery wheel is given,

$$I = 7.5 \times 10^{-2} \text{ kg·m}^2$$

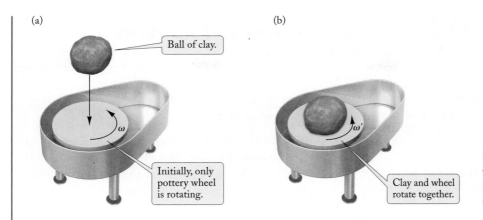

(a)

Ball of clay.

Initially, only
pottery wheel
is rotating.

(b)

Clay and wheel
rotate together.

FIGURE 13.12 (a) A pottery wheel rotates
with angular velocity ω; (b) when a ball of clay
is dropped on the wheel, the angular velocity
slows to ω'.

and the moment of inertia of the clay is that of a uniform sphere (see Table 12.3):

$$I_{\text{clay}} = \tfrac{2}{5}MR^2 = \tfrac{2}{5} \times 6.0 \text{ kg} \times (0.080 \text{ m})^2$$

$$= 1.5 \times 10^{-2} \text{ kg·m}^2$$

Accordingly,

$$\omega' = \frac{I}{I'}\omega = \frac{I}{I + I_{\text{clay}}}\omega$$

$$= \frac{7.5 \times 10^{-2} \text{ kg·m}^2}{7.5 \times 10^{-2} \text{ kg·m}^2 + 1.5 \times 10^{-2} \text{ kg·m}^2} \times 8.4 \text{ radians/s}$$

$$= 7.0 \text{ radians/s}$$

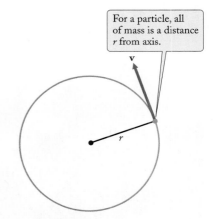

For a particle, all
of mass is a distance
r from axis.

v

r

FIGURE 13.13 A particle moving with
speed v along a circle of radius r. The moment
of inertia of this particle with respect to the
center of the circle is $I = mr^2$.

As already mentioned in Chapter 9, the Law of Conservation of Angular
Momentum also applies to a single particle moving in an orbit under the influence of
a central force. Such a force is always directed along the radial line, and it therefore
exerts no torque. If the particle is moving along a circle of radius r with velocity v (see
Fig. 13.13), its moment of inertia is mr^2 and its angular velocity is $\omega = v/r$. Hence $I\omega$
$= mr^2 \times v/r = mvr$, and the angular momentum of the particle is

$$L = mvr \qquad \text{(circular orbit)} \qquad (13.34)$$

angular momentum for circular orbit

This formula is valid not only for a circular orbit, but also for the perihelion and
aphelion points of an elliptical orbit, where the instantaneous velocity is perpendicular to the radius. In Chapter 9 we took advantage of the conservation of the angular momentum $L = mvr$ to compare the speeds of a planet at perihelion and at
aphelion.

The angular momentum defined by Eq. (13.34) is called the **orbital angular
momentum** to distinguish it from **spin angular momentum** of a body rotating about
its own axis. For instance, the Earth has both an orbital angular momentum (due to its
motion around the Sun) and a spin angular momentum (due to its rotation about its
own axis). Table 13.1 includes examples of both kinds of angular momentum.

**orbital angular momentum
and spin angular momentum**

PROBLEM-SOLVING TECHNIQUES   CONSERVATION OF ANGULAR MOMENTUM

The use of conservation of angular momentum in a problem involving rotational motion involves the familiar three steps we used with conservation of momentum or of energy in translational motion:

1 First write an expression for the angular momentum at one instant of the motion [Eq. (13.30)].

2 Then write an expression for the angular momentum at another instant [Eq. (13.31)].

3 And then rely on conservation of angular momentum to equate the two expressions [Eq. (13.32)]. This yields one equation, which can be solved for an unknown quantity, such as the final angular speed.

✔ **Checkup 13.3**

QUESTION 1: A hoop and a uniform disk have equal radii and equal masses. Both are spinning with equal angular speeds. Which has the larger angular momentum? By what factor?

QUESTION 2: Two automobiles of equal masses are traveling around a traffic circle side by side, with equal angular velocities. Which has the larger angular momentum?

QUESTION 3: You sit on a spinning stool with your legs tucked under the seat. You then stretch your legs outward. How does your angular velocity change?

QUESTION 4: Consider the spinning skater described in Fig. 13.10. While she brings her arms close to her body, does the rotational kinetic energy remain constant?

QUESTION 5: Three children sit on a tire swing (see Fig. 13.14), leaning backward as the wheel rotates about a vertical axis. What happens to the rotational frequency if the children sit up straight?

(A) Frequency increases   (B) Frequency decreases
(C) Frequency remains constant

FIGURE 13.14 Children on a spinning tire swing.

13.4 TORQUE AND ANGULAR MOMENTUM AS VECTORS

The rotational motion of a rigid body about a fixed axis is analogous to one-dimensional translational motion. More generally, if the axis of rotation is not fixed but changes in direction, the motion becomes three-dimensional. A wobbling, spinning top provides an example of such a three-dimensional rotational motion. In this case, the torque and the angular momentum must be treated as vectors, analogous to the force vector and the momentum vector. The definitions of the torque vector and the angular-momentum vector involve the vector cross product that we introduced in Section 3.4. When a force **F** acts at some point with position vector **r**, *the resulting* **torque vector** *is the cross product of the position vector and the force vector:*

torque vector $$\boldsymbol{\tau} = \mathbf{r} \times \mathbf{F}$$ (13.35)

According to the definition of the cross product, the magnitude of $\boldsymbol{\tau}$ is

$$\tau = rF \sin \theta \qquad (13.36)$$

and the direction of $\boldsymbol{\tau}$ is perpendicular to the force vector and the position vector, as specified by the right-hand rule (see Fig. 13.15). Note that since the position vector depends on the choice of origin, *the torque* also *depends on the choice of origin.* We will usually place the origin on some axis or some pivot, and the torque (13.35) is then reckoned in relation to this pivot. For instance, for rotation about a fixed axis, we place the origin on that axis, so $\mathbf{r}$ is in the plane of the circular motion of the point at which the force acts; then $r = R$, and Eq. (13.36) agrees with Eq. (13.2).

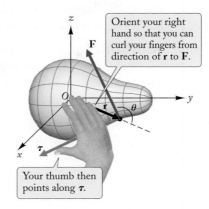

FIGURE 13.15 The torque vector $\boldsymbol{\tau}$ is perpendicular to the force $\mathbf{F}$ and the position vector $\mathbf{r}$, in the direction specified by the right-hand rule: place the fingers of your right hand along the direction of $\mathbf{r}$ and curl toward $\mathbf{F}$ along the smaller angle between these vectors; your thumb will then point in the direction of $\mathbf{r} \times \mathbf{F}$.

The definition of the angular-momentum vector of a rigid body is based on the definition of the angular-momentum vector for a single particle. If a particle has translational momentum $\mathbf{p}$ at position $\mathbf{r}$, then *its* **angular-momentum vector** *is defined as the cross product of the position vector and the momentum vector:*

$$\mathbf{L} = \mathbf{r} \times \mathbf{p} \qquad (13.37)$$

angular-momentum vector

As in the case of the torque, *the angular momentum vector depends on the choice of origin.* For instance, if the particle is moving along a circle, we place the origin at the center of the circle, so $\mathbf{r}$ and $\mathbf{p}$ are in the plane of the circular motion. Since the vectors $\mathbf{r}$ and $\mathbf{p}$ are perpendicular, the magnitude of their cross product is then $L = rp \sin 90° = rp = rmv$. By the right-hand rule, the direction of $\mathbf{r} \times \mathbf{p}$ is perpendicular to the plane of the circular motion, parallel to the axis of rotation. (see Fig. 13.16).

For a rigid body rotating about some (instantaneous) axis, the angular-momentum vector is defined as the sum of the angular-momentum vectors of all the particles in the body,

$$\mathbf{L} = \mathbf{r}_1 \times \mathbf{p}_1 + \mathbf{r}_2 \times \mathbf{p}_2 + \cdots \qquad (13.38)$$

As in the case of a single particle, the value of the angular momentum obtained from this formula depends on the choice of the origin of coordinates. For the calculation of the angular momentum of a rigid body rotating about a fixed axis, it is usually convenient to choose an origin on the axis of rotation.

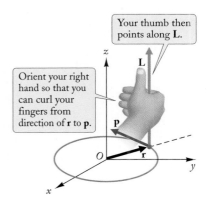

FIGURE 13.16 Angular-momentum vector for a particle.

EXAMPLE 10 Figure 13.17 shows a dumbbell, a rigid body consisting of two particles of mass m attached to the ends of a massless rigid rod of length $2r$. The body rotates with angular velocity ω about a perpendicular axis through the center of the rod. Find the angular momentum about this center.

SOLUTION: Each particle executes circular motion with speed $v = r\omega$. Hence the angular momentum of each has a magnitude $L = rmv = mr^2\omega$ (compare the case of a single particle, illustrated in Fig. 13.13). The direction of each angular-momentum vector is parallel to the axis of rotation (see Fig. 13.16). Thus the direction of the vector sum of the two angular-momentum vectors is also parallel to the axis of rotation, and its magnitude is

$$L = mr^2\omega + mr^2\omega = 2mr^2\omega$$

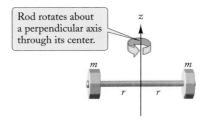

FIGURE 13.17 A rotating dumbbell.

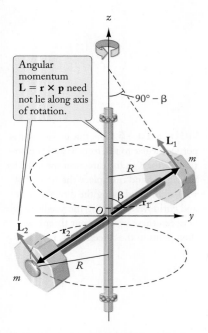

Angular momentum $\mathbf{L} = \mathbf{r} \times \mathbf{p}$ need not lie along axis of rotation.

FIGURE 13.18 A rotating dumbbell oriented at an angle β with the axis of rotation.

EXAMPLE 11 Suppose that the rod of the dumbbell described in the preceding example is welded to an axle inclined at an angle β with respect to the rod. The dumbbell rotates with angular velocity ω about this axis, which is supported by fixed bearings (see Fig. 13.18). Find the angular momentum about an origin on the axis, at the center of mass.

SOLUTION: Each particle executes circular motion, but since the origin is not at the center of the circle, the angular momentum is not the same as in Example 10. The distance between each particle and the axis of rotation is

$$R = r \sin \beta$$

and the magnitude of the velocity of each particle is

$$v = \omega R = \omega r \sin \beta$$

The direction of the velocity is perpendicular to the position vector. Hence the angular-momentum vector of each mass has a magnitude

$$|\mathbf{L}_1| = |\mathbf{L}_2| = m|\mathbf{r} \times \mathbf{v}| = mrv = m\omega r^2 \sin \beta \qquad (13.39)$$

The direction of the angular-momentum vector of each mass is perpendicular to both the velocity and the position vectors, as specified by the right-hand rule. The angular-momentum vector of each mass is shown in Fig. 13.18; these vectors are parallel to each other, they are in the plane of the axis and the rod, and they make an angle of $90° - \beta$ with the axis. The total angular momentum is the vector sum of these individual angular momenta. This vector is in the same direction as the individual angular-momentum vectors, and it has a magnitude twice as large as either of those in Eq. (13.39):

$$L = 2m\omega r^2 \sin \beta \qquad (13.40)$$

As the body rotates, so does the angular-momentum vector, remaining in the plane of the axis and the rod. If at one instant the angular momentum lies in the z–y plane, a quarter of a cycle later it will lie in the z–x plane, etc.

COMMENT: Note that the z component of the angular momentum is

$$L_z = L \cos(90° - \beta) = 2m\omega r^2 \sin \beta \cos(90° - \beta) = 2m\omega r^2 \sin^2\beta$$

This can also be written as

$$L_z = 2m\omega R^2 \qquad (13.41)$$

where $R = r \sin \beta$ is the perpendicular distance between each mass and the axis of rotation. Since $2mR^2$ is simply the moment of inertia of the two particles about the z axis, Eq. (13.41) is the same as

$$L_z = I\omega \qquad (13.42)$$

As we will see below, this formula is of general validity for rotation around a fixed axis.

The preceding example shows that *the angular-momentum vector of a rotating body need not always lie along the axis of rotation.* However, if the body is symmetric about the axis of rotation, then the angular-momentum vector will lie along this axis. In such a symmetric body, each particle on one side of the axis has a counterpart on the other side of the axis, and when we add the angular-momentum vectors contributed by these two particles (or any other pair of particles), the resultant lies along the axis of rotation (see Fig. 13.19).

Since Newton's Second Law for translational motion states that the rate of change of the momentum equals the force, the analogy between the equations for translational and rotational motion suggests that the rate of change of the angular momentum should equal the torque. It is easy to verify this for the case of a single particle. With the usual rule for the differentiation of a product,

$$\frac{d}{dt}\mathbf{L} = \frac{d}{dt}(\mathbf{r} \times \mathbf{p})$$

$$= \frac{d\mathbf{r}}{dt} \times \mathbf{p} + \mathbf{r} \times \frac{d\mathbf{p}}{dt} \tag{13.43}$$

The first term on the right side is

$$\frac{d\mathbf{r}}{dt} \times \mathbf{p} = \mathbf{v} \times (m\mathbf{v}) = m(\mathbf{v} \times \mathbf{v}) = 0 \tag{13.44}$$

This is zero because the cross product of a vector with itself is always zero. According to Newton's Second Law, the second term on the right side of Eq. (13.43) is

$$\mathbf{r} \times \frac{d\mathbf{p}}{dt} = \mathbf{r} \times \mathbf{F} \tag{13.45}$$

where **F** is the force acting on the particle. Therefore, Eq. (13.43) becomes

$$\frac{d\mathbf{L}}{dt} = \mathbf{r} \times \mathbf{F} = \boldsymbol{\tau} \tag{13.46}$$

In the case of a rigid body, the angular momentum is the sum of all the angular momenta of the particles in the body, and the rate of change of this total angular momentum can be shown to equal the net external torque:

$$\frac{d\mathbf{L}}{dt} = \boldsymbol{\tau} \tag{13.47}$$

equation of rotational motion for vector angular momentum

This equation for the rate of change of the angular momentum of a rigid body is analogous to the equation $d\mathbf{p}/dt = \mathbf{F}$ for the rate of change of the translational momentum of a particle.

To compare the vector equation (13.47) with our earlier equation $I\alpha = \tau$, we must focus our attention on the component of the angular momentum along the axis of rotation, that is, the z axis. Figure 13.20 shows an arbitrary rigid body rotating about a fixed axis, which coincides with the z axis. As in Example 11, the angular-momentum vector of this body makes an angle with the axis of rotation. However, as we discussed in Example 11, the z component of the angular momentum of each particle in the rotating body is simply equal to its moment of inertia about the z axis multiplied by the angular velocity [see Eq. (13.42)]. Hence, when we sum the contributions of all the particles in the rotating body, we find that the z component of the net angular momentum of the entire rotating body equals the net moment of inertia of the entire body multiplied by the angular velocity. This establishes that the equation

$$L_z = I\omega \tag{13.48}$$

is of general validity.

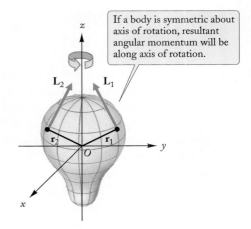

If a body is symmetric about axis of rotation, resultant angular momentum will be along axis of rotation.

FIGURE 13.19 For a rotating symmetric body, the angular momentum is always along the axis of rotation.

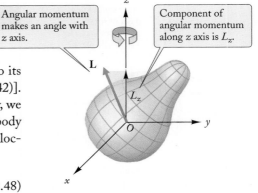

Angular momentum makes an angle with z axis.

Component of angular momentum along z axis is L_z.

FIGURE 13.20 A body rotating about the z axis.

PHYSICS IN PRACTICE   THE GYROCOMPASS

Concepts
— *in* —
Context

A gyroscope is a flywheel suspended in gimbals (pivoted rings; see Fig. 1). The angular-momentum vector of the flywheel lies along its axis of rotation. Since there are no torques on this flywheel, except for the very small and negligible frictional torques in the pivots of the gimbals, the angular-momentum vector remains constant in both magnitude and direction. Hence the direction of the axis of spin remains fixed in space—the gyroscope can be carried about, its base can be twisted and turned in any way, and yet the axis always continues to point in its original direction. Thus, the gyroscope serves as a compass. High-precision gyroscopes are used in the inertial-guidance systems for

ships, aircraft, rockets, and spacecraft (see Fig. 2). They provide an absolute reference direction relative to which the orientation of the vehicle can be established. In such applications, three gyroscopes aimed along mutually perpendicular axes define the absolute orientation of an x, y, z coordinate grid.

The best available high-precision gyroscopes, such as those used in the inertial-guidance system of the Hubble Space Telescope, are capable of maintaining a fixed reference direction with a deviation, or drift, of no more than 10 arcseconds per hour. The special gyroscopes developed for the *Gravity Probe B* experiment are even better than that; their drift is less than 1 milliarcsecond per year!

FIGURE 1 Gyroscope mounted in gimbals.

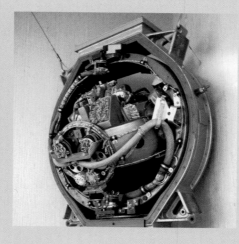

FIGURE 2 Internal-guidance system for an Atlas rocket. This system contains gyroscopes to sense the orientation of the rocket and accelerometers to measure the instantaneous acceleration. From these measurements, computers calculate the position of the rocket and guide it along the intended flight path.

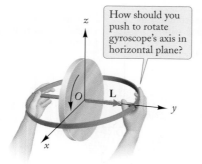

How should you push to rotate gyroscope's axis in horizontal plane?

FIGURE 13.21 A gyroscope held in both hands. The axis of the gyroscope is horizontal, and the hands twist this axis sideways through an angle in the x–y plane.

EXAMPLE 12   You grasp the gimbals of a spinning gyroscope with both hands and you forcibly twist the axis of the gyroscope through an angle in the horizontal plane (see Fig. 13.21). If the angular momentum of the gyroscope spinning about its axis is 3.0×10^{-2} J·s, what are the magnitude and the direction of the torque you need to exert to twist the axis of the gyroscope at a constant rate through 90° in the horizontal plane in 1.0s?

SOLUTION: Figure 13.22a shows the angular-momentum vector **L** of the spinning gyroscope at an initial time and the new angular-momentum vector **L** + d**L** after you have turned the gyroscope through a small angle $d\beta$. From the figure, we see that d**L** is approximately perpendicular to **L**, and that the magnitude of d**L** is

$$dL = L\,d\beta$$

Hence

$$\frac{dL}{dt} = L\frac{d\beta}{dt} \qquad (13.49)$$

According to Eq. (13.49), the magnitude of the torque is

$$\tau = \frac{dL}{dt} = L\frac{d\beta}{dt}$$

With $L = 3.0 \times 10^{-2}$ J·s and $d\beta/dt = (90°)/(1.0 \text{ s}) = \pi/2$ radians/s,

$$\tau = 3.0 \times 10^{-2} \text{ J·s} \times \frac{\pi}{2} \text{ radians/s} = 4.7 \times 10^{-2} \text{ N·m}$$

Since $\boldsymbol{\tau} = d\mathbf{L}/dt$, the direction of the torque vector $\boldsymbol{\tau}$ must be the direction of $d\mathbf{L}$; that is, the torque vector must be perpendicular to $\mathbf{L}$, or initially into the plane of the page (see Fig. 13.22b). To produce such a torque, your left hand must push up, and your right hand must pull down. This is contrary to intuition, which would suggest that to twist the axis in the horizontal plane, you should push forward with your right hand and pull back with your left! This surprising behavior also explains why a downward gravitational force causes the slow precession of a spinning top, as considered in the next example.

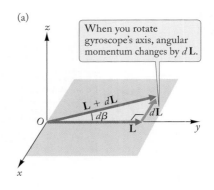

(a)

When you rotate gyroscope's axis, angular momentum changes by $d\mathbf{L}$.

$\mathbf{L} + d\mathbf{L}$ $d\beta$ $d\mathbf{L}$ $\mathbf{L}$

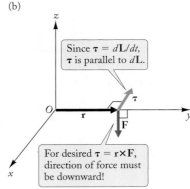

(b)

Since $\boldsymbol{\tau} = d\mathbf{L}/dt$, $\boldsymbol{\tau}$ is parallel to $d\mathbf{L}$.

$\boldsymbol{\tau}$ $\mathbf{r}$ $\mathbf{F}$

For desired $\boldsymbol{\tau} = \mathbf{r} \times \mathbf{F}$, direction of force must be downward!

FIGURE 13.22 (a) $d\mathbf{L}$ is approximately perpendicular to $\mathbf{L}$, in the x–y plane. (b) The torque $\boldsymbol{\tau}$ is parallel to $d\mathbf{L}$, also in the x–y plane.

EXAMPLE 13

A toy top spins with angular momentum of magnitude L; the axis of rotation is inclined at an angle θ with respect to the vertical (see Fig. 13.23). The spinning top has mass M; its point of contact with the ground remains fixed, and its center of mass is a distance r from the point of contact. The top *precesses;* that is, its angular-momentum vector rotates about the vertical. Find the angular velocity Ω_p of this precessional motion. If a top has $r = 4.0$ cm and moment of inertia $I = MR^2/4$, where $R = 3.0$ cm, find the period of the precessional motion when the top is spinning at 250 radians/s.

SOLUTION: From Fig. 13.24a, we see that the weight, Mg, acting at the center of mass, produces a torque τ of magnitude

$$\tau = rMg\sin\theta \qquad (13.50)$$

As in Example 12, the change in angular momentum $d\mathbf{L}$ will be parallel to the torque, since $\boldsymbol{\tau} = d\mathbf{L}/dt$. In a time dt, the top will precess though an angle $d\beta$ given by (see Fig. 13.24b)

$$d\beta = \frac{dL}{L\sin\theta}$$

Using $dL = \tau\, dt = rMg\sin\theta\, dt$, we thus have

$$d\beta = \frac{rMg\sin\theta\, dt}{L\sin\theta} = \frac{rMg}{L}\, dt$$

The precessional angular velocity is the rate of change of this angle:

$$\Omega_p = \frac{d\beta}{dt} = \frac{rMg}{L} \qquad (13.51)$$

Thus the angular velocity of precession is independent of the tilt angle θ.

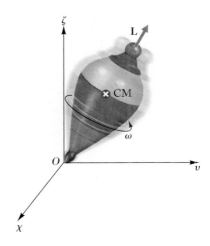

FIGURE 13.23 A tilted top spinning with angular velocity ω.

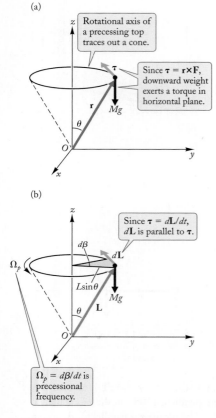

FIGURE 13.24 (a) The weight of the top, acting at the center of mass (a distance r from the point of contact), produces a torque perpendicular to $\mathbf{r}$ and to the weight. (b) The torque is parallel to $d\mathbf{L}$, which results in a slow precession around a vertical axis at an angular velocity Ω_p.

The period of the precession is related to the precessional angular velocity by

$$T = \frac{2\pi}{\Omega_p} = \frac{2\pi L}{rMg}$$

For the particular top described, we insert the angular momentum

$$L = I\omega = \frac{MR^2}{4}\omega$$

and obtain

$$T = \frac{2\pi MR^2\omega}{4rMg} = \frac{\pi R^2\omega}{2rg}$$

$$= \frac{\pi \times (0.030\,\text{m})^2 \times 250\,\text{radians/s}}{2 \times 0.040\,\text{m} \times 9.81\,\text{m/s}^2}$$

$$= 0.90\,\text{s}$$

Since this precessional period is proportional to ω, we see that as the spinning of the top slows down, the top will precess with a shorter period, that is, more quickly.

 Checkup 13.4

QUESTION 1: A particle has a nonzero position vector $\mathbf{r}$ and a nonzero momentum $\mathbf{p}$. Can the angular momentum of this particle be zero?

QUESTION 2: What is the angle between the momentum vector $\mathbf{p}$ and the angular-momentum vector $\mathbf{L}$ of a particle?

QUESTION 3: Suppose that instead of calculating the angular momentum of the dumbbell shown in Fig. 13.17 about the center, we calculate it about an origin on the z axis at some distance below the center. What are the directions of the individual angular-momentum vectors of the two masses m in this case? What is the direction of the total angular momentum?

QUESTION 4: Is a torque required to keep the dumbbell in Fig. 13.18 rotating around the z axis at constant angular velocity?

QUESTION 5: What is the direction of the angular-momentum vector of the rotating minute hand on your watch (calculated with respect to an origin at the center of the watch face)?

(A) In the direction that the minute hand points

(B) Antiparallel to the direction that the minute hand points

(C) In the plane of the watch face, but perpendicular to the minute hand

(D) Perpendicularly out of the face of the watch

(E) Perpendicularly into the face of the watch

SUMMARY

PROBLEM-SOLVING TECHNIQUES Torques and Rotational Motion   **(page 405)**

PROBLEM-SOLVING TECHNIQUES Conservation of Angular Momentum   **(page 410)**

PHYSICS IN PRACTICE The Gyrocompass   **(page 414)**

TORQUE
where θ is the angle between the force **F** and the radial line of length R.

$$\tau = FR\sin\theta \qquad (13.3)$$

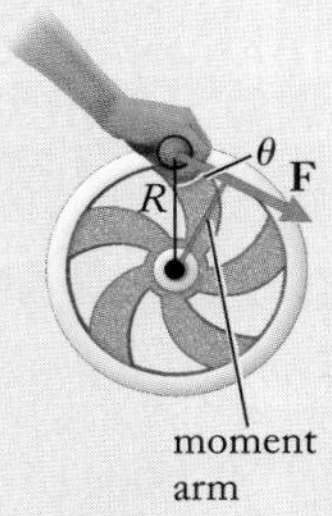

moment arm

WORK DONE BY TORQUE

$$W = \int \tau\, d\phi \qquad (13.4)$$

WORK DONE BY A CONSTANT TORQUE

$$W = \tau\,\Delta\phi \qquad (13.5)$$

POWER DELIVERED BY TORQUE where ω is the angular velocity.

$$P = \tau\omega \qquad (13.6)$$

CONSERVATION OF ENERGY IN ROTATIONAL MOTION

$$E = \tfrac{1}{2}I\omega^2 + U = [\text{constant}] \qquad (13.10)$$

EQUATION OF ROTATIONAL MOTION (Fixed axis)
where I is the moment of inertia and α is the angular acceleration.

$$I\alpha = \tau \qquad (13.19)$$

ANGULAR MOMENTUM OF ROTATION

$$L = I\omega \qquad (13.26)$$

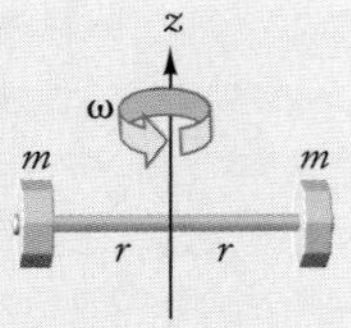

CONSERVATION OF ANGULAR MOMENTUM

$$I\omega = [\text{constant}] \qquad (13.29)$$

ANGULAR MOMENTUM OF PARTICLE (In circular orbit)

$$L = mvr \qquad (13.34)$$

ANGULAR MOMENTUM VECTOR

$$\mathbf{L} = \mathbf{r} \times \mathbf{p} \qquad (13.37)$$

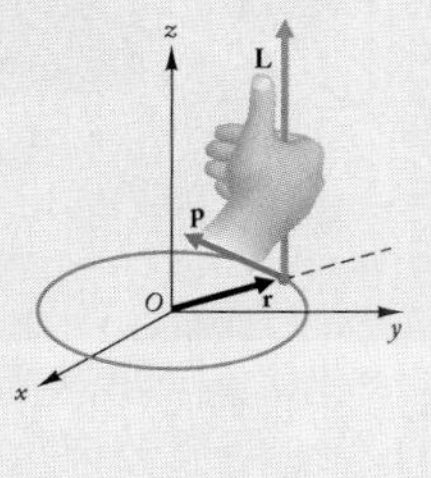

TORQUE VECTOR

$$\boldsymbol{\tau} = \mathbf{r} \times \mathbf{F}$$ (13.35)

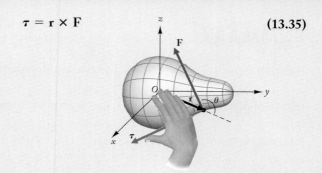

EQUATION OF ROTATIONAL MOTION FOR VECTOR ANGULAR MOMENTUM

$$\frac{d\mathbf{L}}{dt} = \boldsymbol{\tau}$$ (13.47)

GYROSCOPIC PRECESSION ANGULAR VELOCITY

where r is the distance from the point of contact to the center of mass.

$$\Omega_p = \frac{rMg}{L}$$ (13.51)

QUESTIONS FOR DISCUSSION

1. Suppose you push down on the rim of a stationary phonograph turntable. What is the direction of the torque you exert about the center of the turntable?

2. Many farmers have been injured when their tractors suddenly flipped over backward while pulling a heavy piece of farm equipment. Can you explain how this happens?

3. Rifle bullets are given a spin about their axis by spiral grooves ("rifling") in the barrel of the gun. What is the advantage of this?

4. You are standing on a frictionless turntable (like a phonograph turntable, but sturdier). How can you turn 180° without leaving the turntable or pushing against any exterior body?

5. If you give a hard-boiled egg resting on a table a twist with your fingers, it will continue to spin. If you try doing the same with a raw egg, it will not. Why?

6. A tightrope walker uses a balancing pole to keep steady (Fig. 13.25). How does this help?

7. Why do helicopters need a small vertical propeller on their tail?

8. The rate of rotation of the Earth is subject to small seasonal variations. Does this mean that angular momentum is not conserved?

9. Why does the front end of an automobile dip down when the automobile is braking sharply?

10. The friction of the tides against the ocean coasts and the ocean shallows is gradually slowing down the rotation of the Earth. What happens to the lost angular momentum?

11. An automobile is traveling on a straight road at 90 km/h. What is the speed, relative to the ground, of the lowermost point on one of its wheels? The topmost point? The midpoint?

12. A sphere and a hoop of equal masses roll down an inclined plane without slipping. Which will get to the bottom first? Will they have equal kinetic energy when they reach the bottom?

13. A yo-yo rests on a table (Fig. 13.26). If you pull the string horizontally, which way will it move? If you pull vertically?

FIGURE 13.25 A tightrope walker.

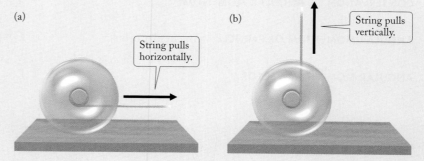

(a) String pulls horizontally.

(b) String pulls vertically.

FIGURE 13.26 Yo-yo resting on a table. (a) String pulls horizontally. (b) String pulls vertically.

14. Stand a pencil vertically on its point on a table and let go. The pencil will topple over.

    (a) If the table is very smooth, the point of the pencil will slip in the direction opposite to that of the toppling. Why?

    (b) If the table is somewhat rough, or covered with a piece of paper, the point of the pencil will jump in the direction of the toppling. Why? (Hint: During the early stages of the toppling, friction holds the point of the pencil fixed; thus the pencil acquires horizontal momentum.)

15. An automobile travels at constant speed along a road consisting of two straight segments connected by a curve in the form of an arc of a circle. Taking the center of the circle as origin, what is the direction of the angular momentum of the automobile? Is the angular momentum constant as the automobile travels along this road?

16. Is the angular momentum of the orbital motion of a planet constant if we choose an origin of coordinates on the Sun?

17. A pendulum is swinging back and forth. Is the angular momentum of the pendulum bob constant?

18. What is the direction of the angular-momentum vector of the rotation of the Earth?

19. A bicycle is traveling east along a level road. What are the directions of the angular-momentum vectors of its wheels?

PROBLEMS

13.1 Work, Energy, and Power in Rotational Motion; Torque

1. The operating instructions for a small crane specify that when the boom is at an angle of 20° above the horizontal (Fig. 13.27), the maximum safe load for the crane is 500 kg. Assuming that this maximum load is determined by the maximum torque that the pivot can withstand, what is the maximum torque for 20° in terms of length R of the boom? What is the maximum safe load for 40°? For 60°?

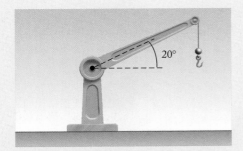

FIGURE 13.27 Small crane.

2. A simple manual winch consists of a drum of radius 4.0 cm to which is attached a handle of radius 25 cm (Fig. 13.28). When you turn the handle, the rope winds up on the drum and pulls the load. Suppose that the load carried by the rope is 2500 N. What force must you exert on the handle to hold this load?

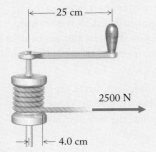

FIGURE 13.28 Manual winch.

3. The repair handbook for an automobile specifies that the cylinder-head bolts are to be tightened to a torque of 62 N·m. If a mechanic uses a wrench of length 20 cm on such a bolt, what perpendicular force must he exert on the end of this wrench to achieve the correct torque?

4. A 2.0-kg trout hangs from one end of a 2.0-m-long stiff fishing pole that the fisherman holds with one hand by the other end. If the pole is horizontal, what is the torque that the weight of the trout exerts about the end the fisherman holds? If the pole is tilted upward at an angle of 60°?

5. You hold a 10-kg book in your hand with your arm extended horizontally in front of you. What is the torque that the weight of this book exerts about your shoulder joint, at a distance of 0.60 m from the book?

6. If you bend over, so your trunk is horizontal, the weight of your trunk exerts a rather strong torque about the sacrum, where your backbone is pivoted on your pelvis. Assume that the mass of your trunk (including arms and head) is 48 kg, and that the weight effectively acts at a distance of 0.40 m from the sacrum. What is the torque that this weight exerts?

7. The engine of an automobile delivers a maximum torque of 203 N·m when running at 4600 rev/min, and it delivers a maximum power of 142 hp when running at 5750 rev/min. What power does the engine deliver when running at maximum torque? What torque does it deliver when running at maximum power?

8. The flywheel of a motor is connected to the flywheel of a pump by a drive belt (Fig. 13.29). The first flywheel has a radius R_1, and the second a radius R_2. While the motor wheel is rotating at a constant angular velocity ω_1, the tensions in the upper and the lower portions of the drive belt are T and T', respectively. Assume that the drive belt is massless.

    (a) What is the angular velocity of the pump wheel?

    (b) What is the torque of the drive belt on each wheel?

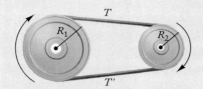

FIGURE 13.29 Motor and pump wheels connected by a drive belt.

(c) By taking the product of torque and angular velocity, calculate the power delivered by the motor to the drive belt, and the power removed by the pump from the drive belt. Are these powers equal?

9. The Wright Cyclone engine on a DC-3 airplane delivers a power of 850 hp with the propeller revolving steadily at 2100 rev/min. What is the torque exerted by air resistance on the propeller?

10. A woman on an exercise bicycle has to exert an (average) tangential push of 35 N on each pedal to keep the wheel turning at constant speed. Each pedal has a radial length of 0.18 m. If she pedals at the rate of 60 rev/min, what is the power she expends against the exercise bicycle? Express your answer in watts and in kilocalories per minute.

11. With what translational speed does the upper end of the meterstick in Example 2 hit the floor? If, instead of a 1.0-m stick, we use a 2.0-m stick, with what translational speed does it hit?

12. A ceiling fan uses 0.050 hp to maintain a rotational frequency of 150 rev/min. What torque does the motor exert?

13. The motor of a grinding wheel exerts a torque of 0.65 N·m to maintain an operating speed of 3450 rev/min. What power does the motor deliver?

14. From the human-body data of Fig. 10.17, calculate (a) the torque about the shoulder for an arm held horizontally and (b) the torque about the hip for a leg held horizontally.

15. A large grinding table is used to thin large batches of silicon wafers in the final stage of semiconductor manufacturing, a process called *backlap*. If the driving motor exerts a torque of 250 N·m while rotating the table 1200 times for one batch of wafers, how much work does the motor do?

16. Recently, a microfabricated torque sensor measured a torque as small as 7.5×10^{-24} N·m. If the torque is produced by a force applied perpendicular to the sensor at a distance of 25 μm from the axis of rotation, what is the smallest force that the sensor can detect?

*17. The angular position of a ceiling fan during the first two seconds after start-up is given by $\phi = Ct^2$, where $C = 7.5$ radians/s^2 and t is in seconds. If the fan motor exerts a torque of 2.5 N·m, how much work has the motor done after $t = 1.0$ s? After $t = 2.0$ s?

*18. While braking, a 1500-kg automobile decelerates at the rate of 8.0 m/s^2. What is the magnitude of the braking force that the road exerts on the automobile? What torque does this

force generate about the center of mass of the automobile? Will this torque tend to lift the front end of the automobile or tend to depress it? Assume that the center of mass of the automobile is 60 cm above the surface of the road.

*19. A tractor of mass 4500 kg has rear wheels of radius 0.80 m. What torque and what power must the engine supply to the rear axle to move the tractor up a road of slope 1:3 at a constant speed of 4.0 m/s?

*20. A bicycle and its rider have a mass of 90 kg. While accelerating from rest to 12 km/h, the rider turns the pedals through three full revolutions. What torque must the rider exert on the pedals? Assume that the torque is constant during the acceleration and ignore friction within the bicycle mechanism.

*21. A meterstick is held to a wall by a nail passing through the 60-cm mark (Fig. 13.30). The meterstick is free to swing about this nail, without friction. If the meterstick is released from an initial horizontal position, what angular velocity will it attain when it swings through the vertical position?

FIGURE 13.30 A meterstick.

*22. A uniform solid sphere of mass M and radius R hangs from a string of length $R/2$. Suppose the sphere is released from an initial position making an angle of 45° with the vertical (Fig. 13.31).

(a) Calculate the angular velocity of the sphere when it swings through the vertical position.

(b) Calculate the tension in the string at this instant.

FIGURE 13.31 A hanging sphere.

*23. The maximum (positive) acceleration an automobile can achieve on a level road depends on the maximum torque the engine can deliver to the wheels.

(a) The engine of a Maserati sports car delivers a maximum torque of 441 N·m to the gearbox. The gearbox steps down the rate of revolution by a factor of 2.58; that is, whenever the engine makes 2.58 revolutions, the wheels make 1 revolution. What is the torque delivered to the wheels? Ignore frictional losses in the gearbox.

(b) The mass of the car (including fuel, driver, etc.) is 1770 kg, and the radius of its wheels is 0.30 m. What is the maximum acceleration? Ignore the moment of inertia of the wheels and frictional losses.

*24. An automobile of mass 1200 kg has four brake drums of diameter 25 cm. The brake drums are rigidly attached to the wheels of diameter 60 cm. The braking mechanism presses brake pads against the rim of each drum, and the friction between the pad and the rim generates a torque that slows the rotation of the wheel. Assume that all four wheels contribute equally to the braking. What torque must the brake pads exert on each drum in order to decelerate the automobile at 7.8 m/s²? If the coefficient of friction between the pad and the drum is $\mu_k = 0.60$, what normal force must the brake pad exert on the rim of the drum? Ignore the masses of the wheels.

*25. In one of the cylinders of an automobile engine, the gas released by internal combustion pushes on the piston, which, in turn, pushes on the crankshaft by means of a piston rod (Fig. 13.32). If the crankshaft experiences a torque of 31 N·m and if the dimensions of the crankshaft and piston rod are as in Fig. 13.32, what must be the force of the gas on the piston when the crankshaft is in the horizontal position as in Fig. 13.32? Ignore friction, and ignore the masses of the piston and rod.

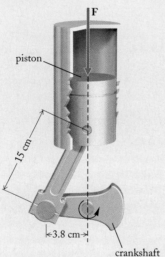

FIGURE 13.32 Automobile piston and crankshaft.

13.2 The Equation of Rotational Motion

26. While starting up a roulette wheel, the croupier exerts a torque of 100 N·m with his hand on the spokes of the wheel. What angular acceleration does this produce? Treat the wheel as a disk of mass 30 kg and radius 0.25 m.

27. The center span of a revolving drawbridge consists of a uniform steel girder of mass 300 metric tons and length 25 m. This girder can be regarded as a uniform thin rod. The bridge opens by rotating about a vertical axis through its center. What torque is required to open this bridge in 60 s? Assume that the bridge first accelerates uniformly through an angular interval of 45° and then the torque is reversed, so the bridge decelerates uniformly through an angular interval of 45° and comes to rest after rotating by 90°.

28. The original Ferris wheel, built by George Ferris, had a radius of 38 m and a mass of 1.9×10^6 kg. Assume that all of its mass was uniformly distributed along the rim of the wheel. If the wheel was initially rotating at 0.050 rev/min, what constant torque had to be applied to bring it to a full stop in 30 s? What force exerted on the rim of the wheel would have given such a torque?

29. The pulley of an Atwood machine for the measurement of g is a brass disk of mass 120 g. When using masses $m_1 = 0.4500$ kg and $m_2 = 0.4550$ kg, an experimenter finds that the larger mass descends 1.6 m in 8.0 s, starting from rest. What is the value of g?

30. A hula hoop rolls down a slope of 1:10 without slipping. What is the (linear) acceleration of the hoop?

31. A uniform cylinder rolls down a plane inclined at an angle θ with the horizontal. Show that if the cylinder rolls without slipping, the acceleration is $a = \frac{2}{3}g \sin \theta$.

32. The spare wheel of a truck, accidentally released on a straight road leading down a steep hill, rolls down the hill without slipping. The mass of the wheel is 60 kg, and its radius is 0.40 m; the mass distribution of the wheel is approximately that of a uniform disk. At the bottom of the hill, at a vertical distance of 120 m below the point of release, the wheel slams into a telephone booth. What is the total kinetic energy of the wheel just before impact? How much of this kinetic energy is translational energy of the center of mass of the wheel? How much is rotational kinetic energy about the center of mass? What is the speed of the wheel?

33. Galileo measured the acceleration of a sphere rolling down an inclined plane. Suppose that, starting from rest, the sphere takes 1.6 s to roll a distance of 3.00 m down a 20° inclined plane. What value of g can you deduce from this?

34. A yo-yo consists of a uniform disk with a string wound around the rim. The upper end of the string is held fixed. The yo-yo unwinds as it drops. What is its downward acceleration?

35. A man is trying to roll a barrel along a level street by pushing forward along its top rim. At the same time another man is pushing backward at the middle, with a force of equal magnitude F (see Fig. 13.33). The barrel rolls without slipping. Which way will the barrel roll? Find the magnitude and direction of the friction force at the point of contact with the street. The barrel is a uniform cylinder of mass M and radius R.

FIGURE 13.33 One man pushes horizontally at a cylinder's top; another pushes with equal force in the opposite direction at its middle. Which way does it roll?

36. An electric blender uniformly accelerates from rest beginning at $t = 0$; at $t = 0.50$ s, the blender has reached 250 radians/s and continues accelerating. If the rotating components have a moment of inertia of 2.0×10^{-4} kg·m^2, at what instantaneous rate is the motor delivering energy at $t = 0.50$ s?

37. A basketball is released from rest on a 15° incline. How many revolutions will the basketball undergo in 4.0 s? Assume the basketball is a thin spherical shell with a diameter of 23 cm, and that it rolls without slipping.

38. A 25-cm length of thin string is wound on the axle of a toy gyroscope that rotates in fixed bearings; the radius of the winding is 2.0 mm. If the string is pulled with a steady force of 5.0 N until completely unwound, how long does it take to complete the pull? What is the final angular velocity? The moment of inertia of the gyroscope (including axle) is 5.0×10^{-5} kg·m^2.

39. A phonograph turntable driven by an electric motor accelerates at a constant rate from 0 to 33.3 revolutions per minute in a time of 2.0 s. The turntable is a uniform disk of metal, of mass 1.2 kg and radius 15 cm. What torque is required to produce this acceleration? If the driving wheel makes contact with the turntable at its outer rim, what force must it exert?

*40. A bowling ball sits on the smooth floor of a subway car. If the car has a horizontal acceleration a, what is the acceleration of the ball? Assume that the ball rolls without slipping.

*41. A hoop rolls down an inclined ramp. The coefficient of static friction between the hoop and the ramp is μ_s. If the ramp is very steep, the hoop will slip while rolling. Show that the critical angle of inclination at which the hoop begins to slip is given by $\tan \theta = 2\mu_s$.

*42. A solid cylinder rolls down an inclined plane. The angle of inclination θ of the plane is large so that the cylinder slips while rolling. The coefficient of kinetic friction between the cylinder and the plane is μ_k. Find the rotational and translational accelerations of the cylinder. Show that the translational acceleration is the same as that of a block sliding down the plane.

**43. Suppose that a tow truck applies a horizontal force of 4000 N to the front end of an automobile similar to that described in Problem 63 of Chapter 12. Taking into account the rotational inertia of the wheels and ignoring frictional losses, what is the acceleration of the automobile? What is the percentage difference between this value of the acceleration and the value calculated by neglecting the rotational inertia of the wheels?

**44. A cart consists of a body and four wheels on frictionless axles. The body has a mass m. The wheels are uniform disks of mass M and radius R. Taking into account the moment of inertia of the wheels, find the acceleration of this cart if it rolls without slipping down an inclined plane making an angle θ with the horizontal.

**45. When the wheels of a landing airliner touch the runway, they are not rotating initially. The wheels first slide on the runway (and produce clouds of smoke and burn marks on the runway, which you may have noticed; see Fig. 13.34), until the sliding friction force has accelerated the wheels to the rotational speed required for rolling without slipping. From the following data, calculate how far the wheel of an airliner slips before it begins to roll without slipping: the wheel has a radius of 0.60 m and a mass of 160 kg, the normal force acting on the wheel is 2.0×10^5 N, the speed of the airliner is 200 km/h, and the coefficient of sliding friction for the wheel on the runway is 0.80. Treat the wheel as a uniform disk.

FIGURE 13.34 A landing airliner.

13.3 Angular Momentum and its Conservation

46. You spin a hard-boiled egg on a table, at 5.0 rev/s. What is the angular momentum of the egg? Treat the egg as a sphere of mass 70 g and mean diameter 5.0 cm.

47. The Moon moves around the Earth in an (approximately) circular orbit of radius 3.8×10^8 m in a time of 27.3 days. Calculate the magnitude of the orbital angular momentum of the Moon. Assume that the origin of coordinates is centered on the Earth.

48. At the Fermilab accelerator, protons of momentum 5.2×10^{-16} kg·m/s travel around a circular path of diameter 2.0 km. What is the orbital angular momentum of one of these protons? Assume that the origin is at the center of the circle.

49. Prior to launching a stone from a sling, a Bolivian native whirls the stone at 3.0 rev/s around a circle of radius 0.75 m. The mass of the stone is 0.15 kg. What is the angular momentum of the stone relative to the center of the circle?

50. A communications satellite of mass 100 kg is in a circular orbit of radius 4.22×10^7 m around the Earth. The orbit is in the equatorial plane of the Earth, and the satellite moves along it from west to east with a speed of 4.90×10^2 m/s. What is the magnitude of the angular momentum of this satellite?

51. According to Bohr's (oversimplified) theory, the electron in the hydrogen atom moves in one or another of several possible circular orbits around the nucleus. The radii and the orbital

velocities of the three smallest orbits are, respectively, 0.529×10^{-10} m, 2.18×10^6 m/s; 2.12×10^{-10} m, 1.09×10^6 m/s; and 4.76×10^{-10} m, 7.27×10^5 m/s. For each of these orbits calculate the orbital angular momentum of the electron, with the origin at the center. How do these angular momenta compare?

52. A high-speed meteoroid moves past the Earth along an (almost) straight line. The mass of the meteoroid is 150 kg, its speed relative to the Earth is 60 km/s, and its distance of closest approach to the center of the Earth is 1.2×10^4 km.

   (a) What is the angular momentum of the meteoroid in the reference frame of the Earth (origin at the center of the Earth)?

   (b) What is the angular momentum of the Earth in the reference frame of the meteoroid (origin at the center of the meteoroid)?

53. A train of mass 1500 metric tons runs along a straight track at 85 km/h. What is the angular momentum of the train about a point 50 m to the side of the track, left of the train? About a point on the track?

54. The electron in a hydrogen atom moves around the nucleus under the influence of the electric force of attraction, a central force pulling the electron toward the nucleus. According to the Bohr theory, one of the possible orbits of the electron is an ellipse of angular momentum $2\hbar$ with a distance of closest approach $(1 - 2\sqrt{2}/3)a_0$ and a distance of farthest recession $(1 + 2\sqrt{2}/3)a_0$, where $\hbar$ and a_0 are two atomic constants with the numerical values 1.05×10^{-34} kg·m²/s ("Planck's constant") and 5.3×10^{-11} m ("Bohr radius"), respectively. In terms of $\hbar$ and a_0, find the speed of the electron at the points of closest approach and farthest recession; then evaluate numerically.

55. According to a simple (but erroneous) model, the proton is a uniform rigid sphere of mass 1.67×10^{-27} kg and radius 1.0×10^{-15} m. The spin angular momentum of the proton is 5.3×10^{-35} J·s. According to this model, what is the angular velocity of rotation of the proton? What is the linear velocity of a point on its equator? What is the rotational kinetic energy? How does this rotational energy compare with the rest-mass energy mc^2?

56. What is the angular momentum of a Frisbee spinning at 20 rev/s about its axis of symmetry? Treat the Frisbee as a uniform disk of mass 200 g and radius 15 cm.

57. A phonograph turntable is a uniform disk of radius 15 cm and mass 1.4 kg. If this turntable accelerates from 0 rev/min to 78 rev/min in 2.5 s, what is the average rate of change of the angular momentum in this time interval?

58. The propeller shaft of a cargo ship has a diameter of 8.8 cm, a length of 27 m, and a mass of 1200 kg. What is the rotational kinetic energy of this propeller shaft when it is rotating at 200 rev/min? What is the angular momentum?

59. The Sun rotates about its axis with a period of about 25 days. Its moment of inertia is $0.20 M_S R_S^2$, where M_S is its mass and

R_S its radius. Calculate the angular momentum of rotation of the Sun. Calculate the total orbital angular momentum of all the planets; make the assumption that each planet moves in a circular orbit of radius equal to its mean distance from the Sun listed in Table 9.1. What percentage of the angular momentum of the Solar System is in the rotational motion of the Sun?

60. Suppose we measure the speed v_1 and the radial distance r_1 of a comet when it reaches perihelion. Use conservation of angular momentum and conservation of energy to determine the speed and the radius at aphelion.

61. A playground merry-go-round is rotating at 2.0 radians/s. Consider the merry-go-round to be a uniform disk of mass 20 kg and radius 1.5 m. A 25-kg child, moving along a radial line, jumps onto the edge of the merry-go-round. What is its new angular velocity? The child then kicks the ground until the merry-go-round (with the child) again rotates at 2.0 radians/s. If the child then walks radially inward, what will the angular velocity be when the child is 0.50 m from the center?

62. The moment of inertia of the Earth is approximately $0.331 M_E R_E^2$. If an asteroid of mass 5.0×10^{18} kg moving at 150 km/s struck (and stuck in) the Earth's surface, by how long would the length of the day change? Assume the asteroid was traveling westward in the equatorial plane and struck the Earth's surface at 45°.

63. In a popular demonstration, a professor rotates on a stool at 0.50 rev/s, holding two 10-kg masses, each 1.0 m from the axis of rotation. If she pulls the weights inward until they are 10 cm from the axis, what is the new rotational frequency? Without the weights, the professor and stool have a moment of inertia of 6.0 kg·m² with arms extended and 4.0 kg·m² with arms pulled in.

64. In a demonstration, a bicycle wheel with moment of inertia 0.48 kg·m² is spun up to 18 radians/s, rotating about a vertical axis. A student holds the wheel while sitting on a rotatable stool. The student and stool are initially stationary and have a moment of inertia of 3.0 kg·m². If the student turns the bicycle wheel over so its axis points in the opposite direction, with what angular velocity will the student and stool rotate? For simplicity, assume the wheel is held overhead, so that the student, wheel, and stool all have the same axis of rotation.

65. A very heavy freight train made up of 250 cars has a total mass of 7700 metric tons. Suppose that such a train accelerates from 0 to 65 km/h on a track running exactly east from Quito, Ecuador (on the equator). The force that the engine exerts on the Earth will slow down the rotational motion of the Earth. By how much will the angular velocity of the Earth have decreased when the train reaches its final speed? Express your answer in revolutions per day. The moment of inertia of the Earth is $0.33 M_E R_E^2$.

66. There are 1.1×10^8 automobiles in the United States, each of an average mass of 2000 kg. Suppose that one morning all these automobiles simultaneously start to move in an eastward direction and accelerate to a speed of 80 km/h.

(a) What total angular momentum about the axis of the Earth do all these automobiles contribute together? Assume that the automobiles travel at an average latitude of 40°.

(b) How much will the rate of rotation of the Earth change because of the action of these automobiles? Assume that the axis of the Earth remains fixed. The moment of inertia of the Earth is 8.1×10^{37} kg·m^2.

*67. Two artificial satellites of equal masses are in circular orbits of radii r_1 and r_2 around the Earth. The second has an orbit of larger radius than the first ($r_2 > r_1$). What is the speed of each? What is the angular momentum of each? Which has the larger speed? Which has the larger angular momentum?

*68. Consider the motion of the Earth around the Sun. Take as origin the point at which the Earth is today and treat the Earth as a particle.

(a) What is the angular momentum of the Earth about this origin today?

(b) What will be the angular momentum of the Earth about the same origin three months from now? Six months from now? Nine months from now? Is the angular momentum conserved?

*69. The friction of the tides on the coastal shallows and the ocean floors gradually slows down the rotation of the Earth. The period of rotation (length of a sidereal day) is gradually increasing by 0.0016 s per century. What is the angular deceleration (in radians/s^2) of the Earth? What is the rate of decrease of the rotational angular momentum? What is the rate of decrease of the rotational kinetic energy? The moment of inertia of the Earth about its axis is $0.331 M_E R_E^2$, where M_E is the mass of the Earth and R_E its equatorial radius.

*70. Phobos is a small moon of Mars. For the purposes of the following problem, assume that Phobos has a mass of 5.8×10^{15} kg and that it has a shape of a uniform sphere of radius 7.5×10^3 m. Suppose that a meteoroid strikes Phobos 5.0×10^3 m off center (Fig. 13.35) and remains stuck. If the momentum of the meteoroid was 3×10^{13} kg·m/s before impact and the mass of the meteoroid is negligible compared with the mass of Phobos, what is the change in the rotational angular velocity of Phobos?

FIGURE 13.35   A meteoroid strikes Phobos.

*71. A woman stands in the middle of a small rowboat. The rowboat is floating freely and experiences no friction against the water. The woman is initially facing east. If she turns around 180° so that she faces west, through what angle will the rowboat turn? Assume that the woman performs her turning movement at constant angular velocity and that her moment

of inertia remains constant during this movement. The moment of inertia of the rowboat about the vertical axis is 20 kg·m^2 and that of the woman is 0.80 kg·m^2.

*72. Two automobiles both of 1200 kg and both traveling at 30 km/h collide on a frictionless icy road. They were initially moving on parallel paths in opposite directions, with a center-to-center distance of 1.0 m (Fig. 13.36). In the collision, the automobiles lock together, forming a single body of wreckage; the moment of inertia of this body about its center of mass is 2.5×10^3 kg·m^2.

(a) Calculate the angular velocity of the wreck.

(b) Calculate the kinetic energy before the collision and after the collision. What is the change of kinetic energy?

FIGURE 13.36   Two automobiles collide.

*73. In one experiment performed under weightless conditions in Skylab, the three astronauts ran around a path on the inside wall of the spacecraft so as to generate artificial gravity for their bodies (Fig. 13.37). Assume that the center of mass of each astronaut moves around a circle of radius 2.5 m; treat the astronauts as particles.

(a) With what speed must each astronaut run if the average normal force on his feet is to equal his normal weight (mg)?

(b) Suppose that before the astronauts begin to run, Skylab is floating in its orbit without rotating. When the astronauts begin to run clockwise, Skylab will begin to rotate counterclockwise. What will be the angular velocity of Skylab when the astronauts are running steadily with the speed calculated above? Assume that the mass of each astronaut is 70 kg and that the moment of inertia of Skylab about its longitudinal axis is 3×10^5 kg·m^2.

(c) How often must the astronauts run around the inside if they want Skylab to rotate through an angle of 30°?

FIGURE 13.37   Three astronauts about to start running around inside Skylab.

*74. A flywheel rotating freely on a fixed shaft is suddenly coupled by means of a drive belt to a second flywheel sitting on a parallel fixed shaft (Fig. 13.38). The initial angular velocity of the first flywheel is ω; that of the second is zero. The flywheels are uniform disks of masses M_1, M_2 and of radii R_1, R_2, respectively. The drive belt is massless and the shafts are frictionless.

(a) Calculate the final angular velocity of each flywheel. (Hint: Since the shafts are fixed, they provide external torques that prevent the flywheels from rotating about each other, that is, angular momentum is not conserved.)

(b) Calculate the kinetic energy lost during the coupling process. What happens to this energy?

FIGURE 13.38 Two flywheels coupled by a drive belt.

*75. A thin rod of mass M and length l hangs from a pivot at its upper end. A ball of clay of mass m and of horizontal velocity v strikes the lower end at right angles and remains stuck (a totally inelastic collision). How high will the rod swing after this collision?

**76. If the melting of the polar ice caps were to raise the water level on the Earth by 10 m, by how much would the day be lengthened? Assume that the moment of inertia of the ice in the polar ice caps is negligible (they are very near the axis), and assume that the extra water spreads out uniformly over the entire surface of the Earth (that is, neglect the area of the continents compared with the area of the oceans). The moment of inertia of the Earth (now) is 8.1×10^{37} kg·m².

77. Consider a projectile of mass m launched with a speed v_0 at an elevation angle of 45°. If the launch point is the origin of coordinates, what is the angular momentum of the projectile at the instant of launch? At the instant it reaches maximum height? At the instant it strikes the ground? Is the angular momentum conserved in this motion with this choice of origin?

13.4 Torque and Angular Momentum as Vectors

78. Show that for a flat plate rotating about an axis perpendicular to the plate, the angular-momentum vector lies along the axis of rotation, even if the body is not symmetric.

79. A child's toy top consists of a uniform thin disk of radius 5.0 cm and mass 0.15 kg with a thin spike passing through its center. The lower part of the spike protrudes 6.0 cm from the disk. If you stand this top on its spike and start it spinning at 200 rev/s, what will be its precession frequency?

80. Suppose that the flywheel of a gyroscope is a uniform disk of mass 250 g and radius 3.0 cm. The distance of the center of this flywheel from the point of support is 4.0 cm. What is the precession frequency if the flywheel is spinning at 120 rev/s?

81. If a bicycle in forward motion begins to tilt to one side, the torque exerted by gravity will tend to turn the bicycle. Draw a diagram showing the angular momentum of a (slightly tilted) front wheel, the weight of the wheel, and the resulting torque. In which direction is the instantaneous change in angular momentum? Will this change make the tilt worse or better?

82. Slow precession can be used to determine a much more rapid rotational frequency. Consider a top made by inserting a small pin radially into a ball (a uniform sphere) of radius $R = 6.0$ cm. The pin extends 1.0 cm from the surface of the ball and supports the top. When set spinning, the top is observed to precess with a period of 0.75 s. What is the rotational frequency of the top?

*83. The wheel of an automobile has a mass of 25 kg and a diameter of 70 cm. Assume that the wheel can be regarded as a uniform disk.

(a) What is the angular momentum of the wheel when the automobile is traveling at 25 m/s (90 km/h) on a straight road?

(b) What is the rate of change of the angular momentum of the wheel when the automobile is traveling at the same speed along a curve of radius 80 m?

(c) For this rate of change of the angular momentum, what must be the torque on the wheel? Draw a diagram showing the path of the automobile, the angular-momentum vector of the wheel, and the torque vector.

*84. Consider the airplane propeller described in Problem 38 in Chapter 12. If the airplane is flying around a curve of radius 500 m at a speed of 360 km/h, what is the rate of change of the angular momentum of the propeller? What torque is required to change the angular momentum at this rate? Draw a diagram showing $\mathbf{L}$, $d\mathbf{L}/dt$, and $\boldsymbol{\tau}$.

*85. A large flywheel designed for energy storage at a power plant has a moment of inertia of 5×10^5 kg·m² and spins at 3000 rev/min. Suppose that this flywheel is mounted on a horizontal axle oriented in the east–west direction. What are the magnitude and direction of its angular momentum? What is the rate of change of this angular momentum due to the rotational motion of the Earth and the consequent motion of the axle of the flywheel? What is the torque that the axle of the flywheel exerts against the bearings supporting it? If the bearings are at a distance of 0.60 m from the center of the flywheel on each side, what are the forces associated with this torque?

REVIEW PROBLEMS

86. A door is 0.80 m wide. What is the torque you exert about the axis passing through the hinges if you push against this door with a perpendicular force of 200 N at its middle? What is it if you push at the edge? A wind is blowing against the other side of the door and trying to push it open. Where should you push to keep the door closed?

87. An elevator of mass 900 kg is being lifted at constant speed by a cable wrapped around a wheel (see Fig. 13.39). The radius of the wheel is 0.35 m. What torque does the cable exert on the wheel?

0.35 m

FIGURE 13.39
Elevator cable
attached to a wheel.

88. Each of the two fuel turbopumps in the Space Shuttle delivers a power of 700 hp. The rotor of this pump rotates at 37 000 rev/min. What is the torque that the rotor exerts while pushing against the fuel?

89. A manual winch has a crank of length (radius) 0.25 m. If a laborer pushes against its handle tangentially with a force of 200 N, how much work does the laborer do while turning the crank through 10 revolutions?

90. A meterstick is initially standing vertically on the floor. If the meterstick falls over, with what angular velocity will it hit the floor? Assume that the end in contact with the floor experiences no friction and slips freely.

*91. A heavy hatch on a ship is made of a uniform plate of steel that measures 1.2 m × 1.2 m and has a mass of 400 kg. The hatch is hinged along one side; it is horizontal when closed, and it opens upward. A torsional spring assists in the opening of the hatch. The spring exerts a torque of 2.00×10^3 N·m when the hatch is horizontal and a torque of 0.30×10^3 N·m when the hatch is vertical; in the range of angles between horizontal and vertical, the torque decreases linearly (e.g., the torque is 1.15×10^3 N·m when the hatch is at 45°).

   (a) At what angle will the hatch be in equilibrium so the spring exactly compensates the torque due to the weight?

   (b) What minimum push must a sailor exert on the hatch to open it from the closed position? To close it from the

open position? Assume that the sailor pushes perpendicularly on the hatch at the edge that is farthest from the hinge.

92. With your bicycle upside down on the ground, and the wheel free to rotate, you grasp the front wheel at the top and give it a horizontal push of 20 N. What is the instantaneous angular acceleration of the wheel? The wheel is a hoop of mass 4.0 kg and radius 0.33 m; ignore the mass of the spokes.

93. A toy top consists of a disk of radius 4.0 cm with a reinforced rim (a ring). The mass of the disk is 20 g, and the mass of the rim is 15 g. The mass of the pivot of this top is negligible.

   (a) What is the moment of inertia of this top?

   (b) When you give this top a twist and start it rotating at 100 rev/min on the floor, friction slows the top to a stop in 1.5 min. Assuming that the angular deceleration is uniform, what is the angular deceleration?

   (c) What is the frictional torque on the top?

   (d) What is the work done by the frictional torque?

94. The turntable of a record player is a uniform disk of radius 0.15 m and mass 1.2 kg. When in operation, it spins at $33\frac{1}{3}$ rev/min. If you switch the record player off, you find that the turntable coasts to a stop in 45 s.

   (a) Calculate the frictional torque that acts on the turntable. Assume the torque is constant, that is, independent of the angular speed.

   (b) Calculate the power that the motor of the record player must supply to keep the turntable in operation at $33\frac{1}{3}$ rev/min.

*95. A barrel of mass 200 kg and radius 0.50 m rolls down a 40° ramp without slipping. What is the value of the friction force acting at the point of contact between the barrel and the ramp? Treat the barrel as a cylinder of uniform density.

*96. A disk of mass M is free to rotate about a fixed horizontal axis. A string is wrapped around the rim of the disk, and a mass m is attached to this string (see Fig. 13.40). What is the downward acceleration of the mass?

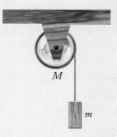

M

m

FIGURE 13.40   A mass m hanging from a disk.

*97. A hoop of mass M and radius R rolls down a sloping ramp that makes an angle of 30° with the ground. What is the acceleration of the hoop if it rolls without slipping?

*98. An automobile has the arrangement of the wheels shown in Fig. 13.41. The mass of this automobile is 1800 kg, the center of mass is at the midpoint of the rectangle formed by the wheels, and the moment of inertia about a vertical axis through the center of mass is 2200 kg·m². Suppose that during braking in an emergency, the left front and rear wheels lock and begin to skid while the right wheels continue to rotate just short of skidding. The coefficient of static friction between the wheels and the road is $\mu_s = 0.90$, and the coefficient of kinetic friction is $\mu_k = 0.50$. Calculate the instantaneous angular acceleration of the automobile about the vertical axis through the center of mass.

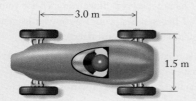

FIGURE 13.41 An automobile.

*99. Neutron stars, or pulsars, spin very quickly about their axes. Their high rate of spin is the result of the conservation of angular momentum during the formation of the neutron star by the gradual contraction (shrinking) of an initially normal star.

  (a) Suppose that the initial star is similar to the Sun, with a radius of 7.0×10^8 m and a rate of rotation of 1.0 revolution per month. If this star contracts to a radius of 1.0×10^4 m, by what factor does the moment of inertia increase? Assume that the relative distribution of mass in the initial and the final stars is roughly the same.

  (b) By what factor does the angular velocity increase? What is the final angular velocity?

*100. A rod of mass M and length l is lying on a flat, frictionless surface. A ball of putty of mass m and initial velocity v at right angles to the rod strikes the rod at a distance $l/4$ from the center (Fig. 13.42). The collision is inelastic, and the putty adheres to the rod.

  (a) Where is the center of mass of the rod with adhering putty?

  (b) What is the velocity of this center of mass after the collision?

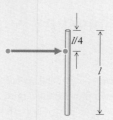

FIGURE 13.42 A ball of putty strikes a rod.

  (c) What is the angular momentum about this center of mass? What is the moment of inertia, and what is the angular velocity?

*101. A communications satellite of mass 1000 kg is in a circular orbit of radius 4.22×10^7 m around the Earth. The orbit is in the equatorial plane of the Earth, and the satellite moves along it from west to east. What are the magnitude and the direction of the angular-momentum vector of this satellite?

*102. The spin angular momentum of the Earth has a magnitude of 5.9×10^{33} kg·m²/s. Because of forces exerted by the Sun and the Moon, the spin angular momentum gradually changes direction, describing a cone of half-angle 23.5° (Fig. 13.43). The angular-momentum vector takes 26 000 years to swing once around this cone. What is the magnitude of the rate of change of the angular-momentum vector; that is, what is the value of $|d\mathbf{L}/dt|$?

FIGURE 13.43 The precessing Earth.

Answers to Checkups

Checkup 13.1

1. You should place your hand at the end of the handle farthest from the bolt; this will provide the largest R in Eq. (13.2) and maximize the torque. Similarly, your push should be perpendicular to the wrench handle, in order to maximize $\sin \theta$ to the value $\sin 90° = 1$ in Eq. (13.2).

2. The direction must be toward the axis (along a radius), so that $\sin \theta = \sin 0° = 0$; thus both the torque and the work done will be zero.

3. Initially, when the stick is upright, the weight acts downward, along the radial direction, and so the torque is zero. As the stick falls, the weight (mg) and the point at which it acts ($R = y_{CM} = l/2$) remain constant. Only the angle between the force and the radial line changes; the sine of this angle is maximum just as the meterstick hits the floor (when $\sin \theta = \sin 90° = 1$), so the torque is maximum then.

4. (A) $\frac{1}{4}$. The work done is $W = \tau \Delta\phi = FR \sin \theta \, \Delta\phi$. In both cases, pushing at right angles implies $\sin \theta = 1$, and both angular displacements $\Delta\phi$ are the same. But with half the force applied at half the radius for the second push, the work will be one-fourth of that for the first push.

Checkup 13.2

1. The angular acceleration results from the torque exerted by gravity at the center of mass; this is maximum when the weight is perpendicular to the radial direction (when $\sin \theta = 1$). That occurs when the meterstick is horizontal, just before it hits the floor.

2. The translational kinetic energy is twice as large for a (uniform) rotating cylinder, because the rotational kinetic energy is $\frac{1}{2}I\omega^2 = \frac{1}{2} \times \frac{1}{2}MR^2 \times (v/R)^2 = \frac{1}{2} \times \frac{1}{2}Mv^2$.

3. The rolling cylinder's total kinetic energy is the same as for a slipping cylinder; in each case, it is equal to the change in potential energy Mgh. For the rolling cylinder, one-third of the total kinetic energy is rotational kinetic energy, and two-thirds is translational kinetic energy; thus, the rolling cylinder's translational speed is smaller when it reaches the bottom than that of a slipping cylinder (by a factor of $\sqrt{2/3}$).

4. The sphere and cylinder must have equal kinetic energies when they reach the bottom; each kinetic energy is equal to the change in potential energy Mgh. The sphere's moment of inertia is only $\frac{2}{5}MR^2$, compared with $\frac{1}{2}MR^2$ for the cylinder, so the sphere will achieve a higher speed and get to the bottom first.

5. (A) Less than that of the cylinder. For the thin hoop ($I = MR^2$), only one-half of its kinetic energy is translational; for the cylinder ($I = \frac{1}{2}Mr^2$), two-thirds of its kinetic energy will be translational. Since the total kinetic energy in each case will equal the change in potential energy (Mgh), the speed of the hoop will be smaller.

Checkup 13.3

1. Since the angular momentum is $L = I\omega$ and the angular speeds (ω) are equal, the hoop (with moment of inertia $I = MR^2$; see Table 12.3) has a larger angular momentum by a factor of 2 compared with the uniform disk (which has $I = \frac{1}{2}MR^2$).

2. Since the angular velocities are equal and the angular momentum is $L = I\omega$, the car with the larger moment of inertia $I = MR^2$ has the greater angular momentum. Since the masses are equal, this is the car on the outside, with the greater value of R.

3. Since there are no external torques on you, angular momentum $L = I\omega$ is conserved. Since you increase your moment of inertia I by stretching your legs outward (increasing R^2), your angular velocity ω must decrease.

4. No. Since angular momentum $L = I\omega$ is conserved and she decreases her moment of inertia I, her angular velocity ω increases. But her rotational kinetic energy is $K = \frac{1}{2}I\omega^2 = \frac{1}{2}I\omega \times \omega$. Since $I\omega$ is constant and ω increases, the kinetic energy increases. Thus the skater must do work to bring her arms close to her body.

5. (A) Frequency increases. The moment of inertia decreases when the children sit up, since more of their mass is closer to the axis. Since the angular momentum $L = I\omega$ is conserved, a smaller moment of inertia requires a larger angular frequency.

Checkup 13.4

1. Yes; since the angular-momentum vector is $\mathbf{L} = \mathbf{r} \times \mathbf{p}$, it will be zero when $\mathbf{r}$ and $\mathbf{p}$ are parallel (or antiparallel).

2. Since the angular-momentum vector is $\mathbf{L} = \mathbf{r} \times \mathbf{p}$, $\mathbf{L}$ is always perpendicular to $\mathbf{p}$; the angle between them is $90°$.

3. The individual angular-momentum vectors will be inclined at an angle with respect to the z axis; each, however, will point toward the z axis, like the angular-momentum vectors $\mathbf{L}_1$ and $\mathbf{L}_2$ in Fig. 13.19. In this case, the horizontal components of the two angular-momentum vectors will cancel, and the total angular-momentum vector will point along the z axis.

4. Yes; the total angular momentum is changing as the dumbbell rotates about the z axis (because the direction of $\mathbf{L}$ is changing), so a torque is required to produce that change in angular momentum.

5. (E) Perpendicularly into the face of the watch. By the right-hand rule, with $\mathbf{r}$ pointing along the minute hand and $\mathbf{p}$ in the direction of motion, the clockwise rotation implies that the angular-momentum vector $\mathbf{L} = \mathbf{r} \times \mathbf{p}$ is perpendicularly *into* the face of the watch.

Statics and Elasticity

CONCEPTS IN CONTEXT

Tower cranes are widely used at construction sites. The K-10000 tower crane shown here is the largest commercially available tower crane. Its central tower is 110 m high, and its long horizontal arm reaches out to 84 m. It can lift 120 tons at the end of the long arm, and more than twice as much at the middle of the long arm. The short arm holds a fixed counterweight of 100 tons (at the end, above the arm) and two additional mobile counterweights (below the arm). For the lift of a small load, the mobile counterweights are parked in the inboard position, near the central tower. For the lift of a large load, the mobile counterweights are moved outward to keep the crane in balance.

The concepts discussed in this chapter permit us to examine many aspects of the operation of such a crane:

? Where must the mobile counterweights be placed to keep the crane in balance for a given load? (Example 2, page 435)

14.1 Statics of Rigid Bodies

14.2 Examples of Static Equilibrium

14.3 Levers and Pulleys

14.4 Elasticity of Materials

? What is the tension in the tie-rod (stretched diagonally from the top of the tower to the end of the arm) that holds the short arm in place? (Example 3, page 435)

? What is the elongation of the lifting cable when subjected to a given load? (Example 8, page 448)

Engineers and architects concerned with the design of bridges, buildings, and other structures need to know under what conditions a body will remain at rest, even when forces act on it. For instance, the designer of a railroad bridge must make sure that the bridge will not tip over or break when a heavy train passes over it. *A body that remains at rest, even though several forces act on it, is said to be in equilibrium.* The branch of physics that studies the conditions for the equilibrium of a body is called **statics.** Statics is the oldest branch of physics. The ancient Egyptians, Greeks, and Romans had a good grasp of the basic principles of statics, as is evident from their construction of elegant arches for doorways and bridges. The oldest surviving physics textbook is a treatise on the statics of ships by Archimedes.

In the first three sections of this chapter, we will rely on the assumption that the "rigid" structural members—such as beams and columns—indeed remain rigid; that is, they do not deform. In essence, this means that we assume that the forces are not so large as to produce a significant bending or compression of the beams or columns. However, in the last section, we will take a brief look at the phenomenon of the elastic deformation of solid bodies when subjected to the action of large forces.

14.1 STATICS OF RIGID BODIES

If a rigid body is to remain at rest, its translational and rotational accelerations must be zero. Hence, the condition for the **static equilibrium** of a rigid body is that *the sum of external forces and the sum of external torques on the body must be zero.* This means that the forces and the torques are in balance; each force is compensated by some other force or forces, and each torque is compensated by some other torque or torques. For example, when a baseball bat rests in your hands (Fig. 14.1), the external forces on the bat are its (downward) weight $\mathbf{w}$ and the (upward) pushes $\mathbf{N}_1$ and $\mathbf{N}_2$ of your hands. If the bat is to remain at rest, the sum of these external forces must be zero—that is, $\mathbf{w} + \mathbf{N}_1 + \mathbf{N}_2 = 0$, or, in terms of magnitudes, $-w + N_1 + N_2 = 0$. Likewise, the sum of the torques of the external forces must be zero. Since the angular acceleration of the bat is zero about any axis of rotation whatsoever that we might choose in Fig. 14.1, the sum of torques must be zero about any such axis. For example, we might choose a horizontal axis of rotation through the center of mass of the bat, out of the plane of the page, as in Fig. 14.1a. With this choice of axis, the force $\mathbf{N}_2$ produces a counterclockwise torque $r_2 N_2$ and the force $\mathbf{N}_1$ produces a clockwise torque $r_1 N_1$, whereas the weight $\mathbf{w}$ (acting at the axis) produces no torque. The equilibrium condition for the torque is then $r_2 N_2 - r_1 N_1 = 0$. Alternatively, we might choose a horizontal axis of rotation through, say, the left hand, out of the plane of the page, as in Fig. 14.1b. With this choice, the force $\mathbf{N}_1$ produces a clockwise torque $(r_1 + r_2)N_1$, the weight produces a counterclockwise torque $r_2 w$, and the force $\mathbf{N}_2$ produces no torque. The equilibrium condition for the torques is then $-(r_1 + r_2)N_1 + r_2 w = 0$. With other choices of axis of rotation, we can generate many more equations than there are unknown forces or torques in a static equilibrium problem. However, the equations obtained with different choices of axis of rotation are related, and they can always be shown to be consistent.

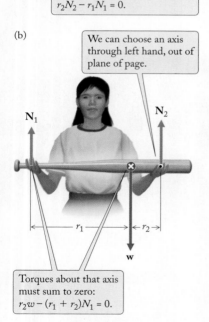

(a)

We can choose an axis through center of mass, out of plane of page.

N_1 N_2

r_1 r_2

$\mathbf{w}$

Bat is at rest, so torques about that axis must sum to zero: $r_2 N_2 - r_1 N_1 = 0$.

(b)

We can choose an axis through left hand, out of plane of page.

N_1 N_2

r_1 r_2

$\mathbf{w}$

Torques about that axis must sum to zero: $r_2 w - (r_1 + r_2)N_1 = 0$.

FIGURE 14.1 A baseball bat at rest in your hands. The external forces are the downward weight $\mathbf{w}$ and the upward pushes $\mathbf{N}_1$ and $\mathbf{N}_2$ of the right and left hands, respectively. These external forces add to zero. The external torques about any axis also add to zero. (a) Axis is through center of mass. (b) Axis is through left hand.

From this discussion, we conclude that for the purposes of static equilibrium, *any line through the body or any line passing at some distance from the body can be thought of as a conceivable axis of rotation, and the torque about every such axis must be zero.* This means we have complete freedom in the choice of the axis of rotation, and we can make whatever choice seems convenient. With some practice, one learns to recognize which choice of axis will be most useful for the solution of a problem in statics.

The force of gravity plays an important role in many problems of statics. The force of gravity on a body is distributed over all parts of the body, each part being subjected to a force proportional to its mass. However, for the calculation of the torque exerted by gravity on a rigid body, *the entire gravitational force may be regarded as acting on the center of mass.* We relied on this rule in Fig. 14.1, where we assumed that the weight acts at the center of mass of the bat. The proof of this rule is easy: Suppose that we release some arbitrary rigid body and permit it to fall freely from an initial condition of rest. Since all the particles in the body fall at the same rate, the body will not change its orientation as it falls. If we consider an axis through the center of mass, the absence of angular acceleration implies that gravity does not generate any torque about the center of mass. Hence, if we want to simulate gravity by a single force acting at one point of the rigid body, that point will have to be the center of mass.

Given that in a rigid body the force of gravity effectively acts on the center of mass, we see that a rigid body supported by a single force acting at its center of mass or acting on the vertical line through its center of mass is in equilibrium, since the support force is then collinear with the effective force of gravity, and such collinear forces of equal magnitudes and opposite directions exert no net torque. This provides us with a simple method for the experimental determination of the center of mass of a body of complicated shape: Suspend the body from a string attached to a point on its surface (Fig. 14.2); the body will then settle into an equilibrium position such that the center of mass is on the vertical downward prolongation of the string (this vertical prolongation is marked dashed in Fig. 14.2). Next, suspend the body from a string attached at another point of its surface, and mark a new vertical downward prolongation of the string. The center of mass is then at the intersection of the new and the old prolongations of the string.

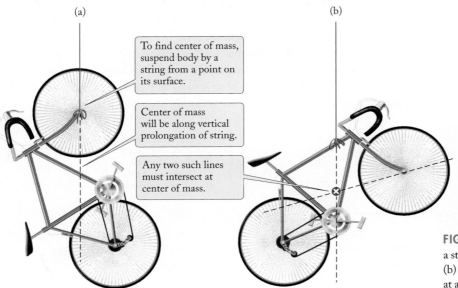

(a)

(b)

To find center of mass, suspend body by a string from a point on its surface.

Center of mass will be along vertical prolongation of string.

Any two such lines must intersect at center of mass.

FIGURE 14.2 (a) Bicycle suspended by a string attached at a point on its "surface." (b) Bicycle suspended by a string attached at a different point.

(a) stable equilibrium

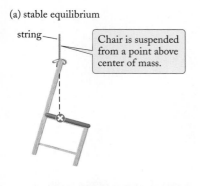

(b) unstable equilibrium

(c) neutral equilibrium

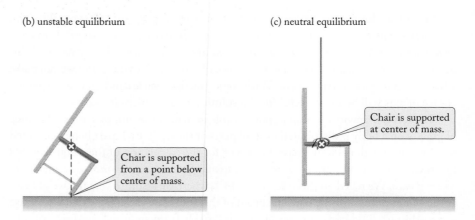

FIGURE 14.3 A body (a) in stable equilibrium; (b) in unstable equilibrium; (c) in neutral equilibrium.

A body suspended from a point above its center of mass, as in Fig. 14.3a, is in **stable equilibrium** (see also Section 8.2). If we turn this body through some angle, so the center of mass is no longer vertically below the point of support, the force of gravity and the supporting force will produce a torque that tends to return the body to the equilibrium position. In contrast, *if this body is supported by a single force applied at a point below the center of mass, as in Fig. 14.3b, the body is in* **unstable equilibrium.** If we turn the body ever so slightly, the force of gravity and the supporting force will produce a torque that tends to turn the body farther away from the equilibrium position—the body tends to topple over. Finally, *a body supported by a single force at its center of mass, as in Fig. 14.3c, is in* **neutral equilibrium.** If we turn such a body, it remains in equilibrium in its new position, and exhibits no tendency to return to its original position or to turn farther away.

Similar stability criteria apply to the translational motion of a body moving on a surface. A body is in stable equilibrium if it resists small disturbances and tends to return to its original position when the disturbance ceases. A car resting at the bottom of a dip in the road is an example of this kind of equilibrium; if we displace the car forward and then let go, the car rolls back to its original position. A body is in unstable equilibrium if it tends to move away from its original position when disturbed. A car resting on the top of a hill is an example of this second kind of equilibrium. If we displace the car forward, it continues to roll down the hill. A car resting on a flat street is in neutral equilibrium with respect to translational displacements. If we displace the car along the street, it merely remains at the new position, without any tendency to return to its original position or to move away from it (see Fig. 14.4).

The first four examples of the next section involve stable or neutral equilibrium; the next two examples involve unstable equilibrium. Engineers take great care to avoid unstable equilibrium in the design of structures and machinery, since an unstable configuration will collapse or come apart at the slightest provocation.

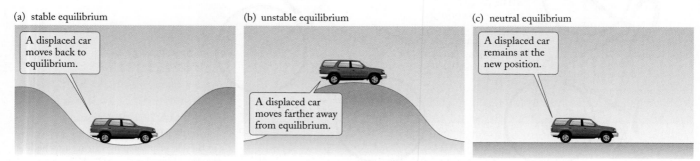

FIGURE 14.4 Stationary automobile in (a) stable, (b) unstable, and (c) neutral equilibrium.

✓ Checkup 14.1

QUESTION 1: Is a cyclist balanced on an upright bicycle in stable or unstable equilibrium? Assume the cyclist sits rigidly, and makes no effort to avoid whatever might befall (see Fig. 14.5).

QUESTION 2: You sit in a swing, with your knees bent. If you now extend your legs fully, how will this change the equilibrium position of the swing and your body?

QUESTION 3: (a) You hold a fishing pole with both hands and point it straight up. Is the support force aligned with the weight? (b) You point the fishing pole horizontally. Is the support force aligned with the weight? Is there a single support force?

QUESTION 4: Consider a cone on a table (a) lying flat on its curved side, (b) standing on its base, (c) standing on its apex. Respectively, the equilibrium of each position is

(A) Stable, unstable, neutral   (B) Stable, neutral, unstable
(C) Unstable, stable, neutral   (D) Neutral, stable, unstable
(E) Neutral, unstable, stable

FIGURE 14.5 Is an upright bicycle in unstable equilibrium?

14.2 EXAMPLES OF STATIC EQUILIBRIUM

The following are some examples of solutions of problems in statics. In these examples, the conditions of a zero sum of external forces,

$$\mathbf{F}_1 + \mathbf{F}_2 + \mathbf{F}_3 + \cdots = 0 \tag{14.1}$$

and a zero sum of external torques,

$$\boldsymbol{\tau}_1 + \boldsymbol{\tau}_2 + \boldsymbol{\tau}_3 + \cdots = 0 \tag{14.2}$$

are used either to find the magnitudes of the forces that hold the body in equilibrium, or to find whether the body can achieve equilibrium at all.

EXAMPLE 1   A locomotive of mass 90 000 kg is one-third of the way across a bridge 90 m long. The bridge consists of a uniform iron girder of mass 900 000 kg, which is supported by two piers (see Fig. 14.6a). What is the load on each pier?

SOLUTION: The body whose equilibrium we want to investigate is the bridge. Figure 14.6b is a "free-body" diagram for the bridge, showing all the forces acting on it: the weight of the bridge, the downward push exerted by the locomotive, and the upward thrust exerted by each pier. The weight of the bridge can be regarded as acting at its center of mass. The bridge is static, and hence the net torque on the bridge reckoned about any point must be zero.

Let us first consider the torques about the point P_2, at the right pier. These torques are generated by the weight of the bridge acting at a distance of 45 m, the downward push of the locomotive acting at a distance of 30 m, and the upward thrust $\mathbf{F}_1$ of the pier at P_1 acting at a distance of 90 m (the upward thrust $\mathbf{F}_2$ has zero moment arm and generates no torque about P_2). The weight of the bridge is $m_{\mathrm{bridge}}\, g = 9.0 \times 10^5$ kg $\times g$, and the downward push exerted by the locomotive equals its weight, $m_{\mathrm{loc}}\, g = 9.0 \times 10^4$ kg $\times g$. Since each of the forces

(a)

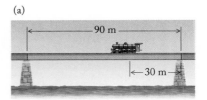

(b)

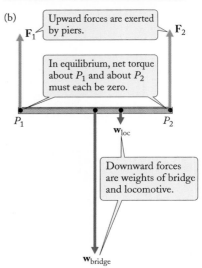

FIGURE 14.6 (a) Bridge with a locomotive on it. (b) "Free-body" diagram for the bridge.

acts at right angles to the (horizontal) line from P_2 to the point of application of the force, the magnitude of the torque $\tau = rF \sin 90°$ for each force is simply the product of the distance and the force, $\tau = rF$. According to the equilibrium condition, we must set the sum of the three torques equal to zero:

$$\tau_{\text{bridge}} + \tau_{\text{loc}} + \tau_{\text{pier}} = 0 \tag{14.3}$$

$$45 \text{ m} \times 9.0 \times 10^5 \text{ kg} \times g + 30 \text{ m} \times 9.0 \times 10^4 \text{ kg} \times g - 90 \text{ m} \times F_1 = 0$$

$$\tag{14.4}$$

Here, we have chosen to reckon the first two torques as positive, since they tend to produce counterclockwise rotation about P_2, and the last torque must then be reckoned as negative, since it tends to produce clockwise rotation. Equation (14.4) contains only the single unknown force F_1. Note that we were able to isolate this unknown force by evaluating the torques about P_2: the other unknown force F_2 is absent because it produces no torque about P_2. Solving this equation for the unknown F_1, we find

$$F_1 = \frac{(45 \text{ m} \times 9.0 \times 10^5 \text{ kg} + 30 \text{ m} \times 9.0 \times 10^4 \text{ kg}) \times g}{90 \text{ m}}$$

$$= 4.8 \times 10^5 \text{ kg} \times g$$

$$= 4.8 \times 10^5 \text{ kg} \times 9.81 \text{ m/s}^2 = 4.7 \times 10^6 \text{ N}$$

Next, consider the torques about the point P_1. These torques are generated by the weight of the bridge, the weight of the locomotive, and the upward thrust $\mathbf{F}_2$ at point P_2 (the upward thrust of $\mathbf{F}_1$ has zero moment arm and generates no torque about P_1). Setting the sum of these three torques about the point P_1 equal to zero, we obtain

$$-45 \text{ m} \times 9.0 \times 10^5 \text{ kg} \times g - 60 \text{ m} \times 9.0 \times 10^4 \text{ kg} \times g + 90 \text{ m} \times F_2 = 0$$

This equation contains only the single unknown force F_2 (the force F_1 is absent because it produces no torque about P_1). Solving for the unknown F_2, we find

$$F_2 = 5.0 \times 10^6 \text{ N}$$

The loads on the piers (the downward pushes of the bridge on the piers) are opposite to the forces $\mathbf{F}_1$ and $\mathbf{F}_2$ (these downward pushes of the bridge on the piers are the reaction forces corresponding to the upward thrusts of the piers on the bridge). Thus, the magnitudes of the loads are 4.7×10^6 N and 5.0×10^6 N, respectively.

COMMENT: Note that the net vertical upward force exerted by the piers is $F_1 + F_2 = 9.7 \times 10^6$ N. It is easy to check that this matches the sum of the weights of the bridge and the locomotive; thus, the condition for zero net vertical force, as required for translational static equilibrium, is automatically satisfied. This automatic result for the equilibrium of vertical forces came about because we used the condition for rotational equilibrium twice. Instead, we could have used the condition for rotational equilibrium once [Eq. (14.4)] and then evaluated F_2 by means of the condition for translational equilibrium [Eq. (14.1)]. The result for zero net torque about the point P_1 would then have emerged automatically.

Also note that instead of taking the bridge as the body whose equilibrium is to be investigated, we could have taken the bridge plus locomotive as a combined body. The downward push of the locomotive on the bridge would then not be an external force, and would not be included in the "free-body" diagram. Instead, the weight of the locomotive would be one of the external forces acting on the combined body and would have to be included in the "free-body" diagram. The vectors in Fig. 14.6b would therefore remain unchanged.

EXAMPLE 2   A large tower crane has a fixed counterweight of 100 tons at the end of its short arm, and it also has a mobile counterweight of 120 tons. The length of the short arm is 56 m, and the length of the long arm is 84 m; the total mass of both arms is 100 tons, and this mass is uniformly distributed along their combined length. The crane is lifting a load of 80 tons hanging at the end of the long arm. Where should the crane operator position the mobile counterweight to achieve a perfect balance of the crane, that is, a condition of zero (external) torque?

SOLUTION: To find the position of the counterweight, we consider the equilibrium condition for the entire crane (alternatively, we could consider the upper part of the crane, that is, the arms and the tie-rods that hold them rigid). Figure 14.7 is a "free-body" diagram for the crane. The external forces are the support force of the base and the weights of the load, the tower, the horizontal arms, the fixed counterweight, and the mobile counterweight. The weight of the arms acts at the center of mass of the combined arms. The total length of these arms is 84 m + 56 m = 140 m, and the center of mass is at the midpoint, 70 m from each end, that is, 14 m from the centerline of the tower.

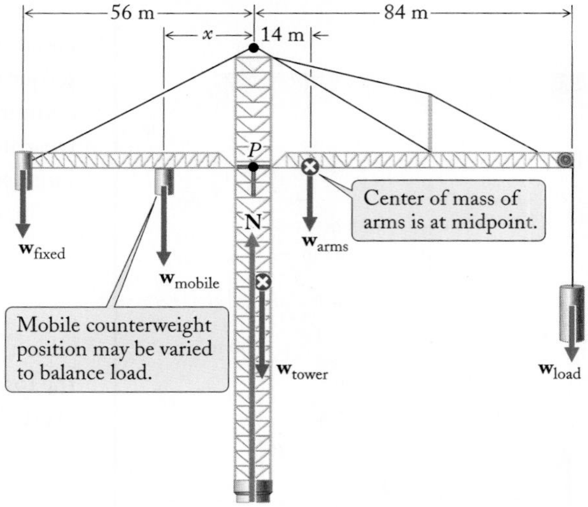

To examine the balance of torques, it is convenient to select the point P at the intersection of the arms and the midline of the tower. All the forces then act at right angles to the line from P to the point of application of the force, and the torque for each is simply the product of the distance and the force. The weight of the tower and the support force of the base do not generate any torques, since they act at zero distance. The equilibrium condition for the sum of the torques generated by the weights of the load, the arms, the fixed counterweight, and the mobile counterweight is

$$\boldsymbol{\tau}_{load} + \boldsymbol{\tau}_{arms} + \boldsymbol{\tau}_{fixed} + \boldsymbol{\tau}_{mobile} = 0 \qquad (14.5)$$

Inserting the values of the weights and moment arms, we have

$$-84 \text{ m} \times 80 \text{ t} \times g - 14 \text{ m} \times 100 \text{ t} \times g + 56 \text{ m} \times 100 \text{ t} \times g$$
$$+ x \times 120 \text{ t} \times g = 0$$

where we have again chosen to reckon counterclockwise torques as positive and clockwise torques as negative. When we solve this equation for x, we obtain

$$x = \frac{84 \text{ m} \times 80 \text{ t} + 14 \text{ m} \times 100 \text{ t} - 56 \text{ m} \times 100 \text{ t}}{120 \text{ t}}$$
$$= 21 \text{ m}$$

FIGURE 14.7 "Free-body" diagram of a tower crane. The crane is balanced, so that no torque is exerted by the base.

EXAMPLE 3   The short arm of the tower crane is held in place by a steel tie-rod stretched diagonally from the top of the tower to the end of the arm, as shown in Fig. 14.8a. The top part of the tower is 30 m high, and the short arm has a length of 56 m and a mass of 40 metric tons. The joint of the arm and the tower is somewhat flexible, so the joint acts as a pivot. Suppose

(a)

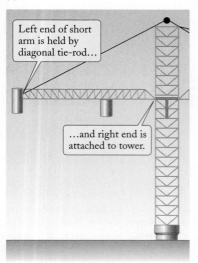

Left end of short arm is held by diagonal tie-rod...

...and right end is attached to tower.

(b)

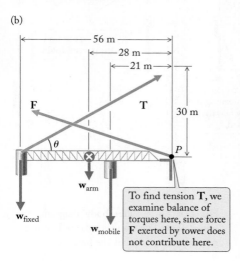

56 m
28 m
21 m

F   **T**

30 m

θ

P

w_{arm}

w_{fixed}

w_{mobile}

To find tension **T**, we examine balance of torques here, since force **F** exerted by tower does not contribute here.

(c)

We resolve the tension **T** into horizontal and vertical components...

y

x **T** w_{fixed}

$T \sin \theta$ w_{arm}

θ $T \cos \theta$

F_x

F_y w_{mobile}

F

...and require that all forces sum to zero (translational equilibrium) to determine **F**.

FIGURE 14.8 (a) Steel tie-rod supporting the short tower crane arm. (b) "Free-body" diagram for the short tower crane arm. (c) The x and y components of the forces.

that the counterweights are placed on the short arm as in the preceding example: the fixed counterweight of 100 metric tons is at the end of the arm, and the mobile counterweight of 120 metric tons is at a distance of 21 m from the centerline. (a) What is the tension in the tie-rod? (b) What is the force that the short arm exerts against the tower at the joint?

SOLUTION: Figure 14.8b is a "free-body" diagram of the short arm, displaying all the external forces acting on it. These forces are the weight w_{arm} of the arm, the weights of the counterweights w_{fixed} and w_{mobile}, the tension **T** of the tie-rod, and the force **F** exerted by the tower at the joint. The force **F** is equal and opposite to the force that the short arm exerts against the tower. The weight of the arm acts at its center of mass, at a distance of 28 m from the centerline; the mobile counterweight acts at a distance of 21 m; and the fixed counterweight and the tension act at the end of the short arm, at a distance of 56 m.

(a) To find the tension **T**, it is convenient to examine the balance of torques about a point P that coincides with the joint. The force **F** does not generate any torque about this point, and hence the condition for the balance of the torques will contain **T** as the sole unknown. The weight of the short arm and the counterweights act at right angles to the line from P to the point of application of the force, so the torque for each is the product of the distance and the force. From Fig. 14.8b, we see that the tension acts at an angle θ, given by

$$\tan \theta = \frac{30 \text{ m}}{56 \text{ m}} = 0.54$$

which corresponds to $\theta = 28°$. With the same sign convention for the direction of the torques as in the preceding example, the equilibrium condition for the torques exerted by the weight of the arm, the counterweights, and the tension is then

$$28 \text{ m} \times 40 \text{ t} \times g + 21 \text{ m} \times 120 \text{ t} \times g + 56 \text{ m} \times 100 \text{ t} \times g$$
$$- 56 \text{ m} \times T \times \sin 28° = 0$$

We can solve this equation for T, with the result

$$T = \frac{28 \text{ m} \times 40 \text{ t} \times g + 21 \text{ m} \times 120 \text{ t} \times g + 56 \text{ m} \times 100 \text{ t} \times g}{56 \text{ m} \times \sin 28°}$$

$$= 351 \text{ t} \times g$$

$$= 351 \times 1000 \text{ kg} \times 9.8 \text{ m/s}^2 = 3.4 \times 10^6 \text{ N}$$

(b) To find the components of the force **F** (Fig. 14.8c), we simply use the conditions for translational equilibrium: the sum of the horizontal components of all the forces and the sum of the vertical components of all the forces must each be zero. The weights of the short arm and the counterweights have vertical components, but no horizontal components. The tension force has a horizontal component $T \cos \theta$ and a vertical component $T \sin \theta$. Hence

$$3.4 \times 10^6 \text{ N} \times \cos 28° + F_x = 0$$

and

$$3.4 \times 10^6 \text{ N} \times \sin 28° - 40 \text{ t} \times g - 120 \text{ t} \times g - 100 \text{ t} \times g + F_y = 0$$

When we solve these equations for F_x and F_y, we find

$$F_x = -3.0 \times 10^6 \text{ N}$$

PROBLEM-SOLVING TECHNIQUES   STATIC EQUILIBRIUM

From the preceding examples we see that the steps in the solution of a problem of statics resemble the steps we employed in Chapter 5.

1 The first step is the selection of the body that is to obey the equilibrium conditions. The body may consist of a genuine rigid body (for instance, the bridge in Example 1), or it may consist of several pieces that act as a single rigid body for the purposes of the problem (for instance, the bridge plus the locomotive in Example 1). It is often helpful to mark the boundary of the selected rigid body with a distinctive color or with a heavy line; this makes it easier to recognize which forces are external and which internal.

2 Next, list all the external forces that act on this body, and display these forces on a "free-body" diagram.

3 If the forces have different directions, it is usually best to draw coordinate axes on the diagram and to resolve the forces into x and y components.

4 For each component, apply the static equilibrium condition for forces: the sum of forces is zero.

5 Make a choice of axis of rotation, calculate the torque of each force about this axis ($\tau = RF \sin \theta$), and apply the static equilibrium condition for torques: the sum of torques is zero. Establish and maintain a sign convention for torques; for example, for an axis pointing into the plane of the paper, counterclockwise torques to be positive and clockwise torques to be negative.

6 As mentioned in Section 14.1, any line can be thought of as an axis of rotation; and the torque about every such axis must be zero. You can make an unknown force disappear from the equation if you place the axis of rotation at the point of action or on the line of action of this force, so that this force has zero moment arm. Furthermore, as illustrated in Example 1, sometimes it is convenient to consider two different axes of rotation, and to examine the separate equilibrium conditions of the torques for each of these axes.

7 As recommended in Chapter 2, it is usually best to solve the equations algebraically for the unknown quantities, and to substitute numbers for the known quantities as a last step. But if the equations are messy, with a clutter of algebraic symbols, it may be convenient to substitute some of the numbers before proceeding with the solution of the equations.

and

$$F_y = 9.5 \times 10^5 \, \text{N}$$

The x and y components of the force exerted by the short arm on the tower are therefore $+3.0 \times 10^6$ N and -9.5×10^5 N, respectively.

EXAMPLE 4   The bottom of a ladder rests on the floor, and the top rests against a wall (see Fig. 14.9a). If the coefficient of static friction between the ladder and the floor is $\mu_s = 0.40$ and the wall is frictionless, what is the maximum angle that the ladder can make with the wall without slipping?

SOLUTION: Figure 14.9b shows the "free-body" diagram for the ladder, with all the forces. The weight of the ladder acts downward at the center of mass. If the ladder is about to slip, the friction force at the floor has the maximum magnitude for a static friction force, that is,

$$f = \mu_s N_1 \tag{14.6}$$

If we reckon the torques about the point of contact with the floor, the normal force $\mathbf{N}_1$ and the friction force $\mathbf{f}$ exert no torques about this point, since their moment

(a)   (b)

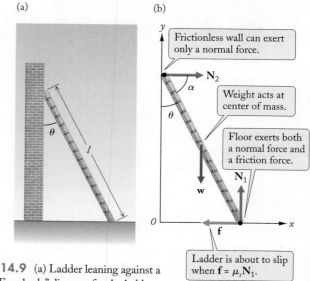

FIGURE 14.9 (a) Ladder leaning against a wall. (b) "Free-body" diagram for the ladder.

arms are zero. The weight $w = mg$ acting at the center of mass exerts a counter-clockwise torque of magnitude $(l/2) \times mg \times \sin \theta$, and the normal force $\mathbf{N}_2$ of the wall exerts a clockwise torque of magnitude $l \times N_2 \times \sin \alpha$, where α is the angle between the ladder and the normal force (see Fig. 14.9b); since $\alpha = 90° - \theta$, the sine of α equals the cosine of θ, and the torque equals $l \times N_2 \times \cos \theta$. For equilibrium, the sum of these torques must be zero,

$$+ \frac{l}{2} mg \sin \theta - l N_2 \cos \theta = 0 \tag{14.7}$$

or, equivalently,

$$\frac{1}{2} mg \sin \theta = N_2 \cos \theta \tag{14.8}$$

We collect the factors that depend on θ by dividing both sides of this equation by $\frac{1}{2} mg \cos \theta$, so

$$\frac{\sin \theta}{\cos \theta} = \frac{2N_2}{mg}$$

or, since $\sin \theta / \cos \theta = \tan \theta$,

$$\tan \theta = \frac{2N_2}{mg} \tag{14.9}$$

To evaluate the angle θ we still need to determine the unknown N_2. For this, we use the condition for translational equilibrium: the net vertical and the net horizontal forces must be zero, or

$$N_1 - mg = 0 \tag{14.10}$$

$$N_2 - \mu_s N_1 = 0 \tag{14.11}$$

From the first of these equations, $N_1 = mg$; therefore, from the second equation, $N_2 = \mu_s mg$. Inserting this into our expression (14.9) for the tangent of the angle θ, we obtain the final result

$$\tan \theta = \frac{2\,\mu_s mg}{mg} = 2\,\mu_s \tag{14.12}$$

With $\mu_s = 0.40$, this yields $\tan \theta = 0.80$. With a calculator, we find that the angle with this tangent is

$$\theta = 39°$$

For any angle larger than this, equilibrium is impossible, because the maximum frictional force is not large enough to prevent slipping of the ladder.

EXAMPLE 5   A uniform rectangular box 2.0 m high, 1.0 m wide, and 1.0 m deep stands on a flat floor. You push the upper end of the box to one side and then release it (see Fig. 14.10a). At what angle of release will the box topple over on its side?

SOLUTION: The forces on the box when it has been released are as shown in the "free-body" diagram in Fig. 14.10b. Both the normal force **N** and the friction force **f** act at the bottom corner, which is the only point of contact of the box with the floor. The weight acts at the center of mass, which is at the center of the box.

Since the box rotates about the bottom corner, let us consider the torque about this point. The only force that produces a torque about the bottom corner is the weight. The weight acts at the center of mass; for a uniform box, this is at the center of the box. The torque exerted by the weight can be expressed as $d \times Mg$, where d is the perpendicular distance from the bottom corner to the vertical line through the center of mass (see Fig. 14.10b). This torque produces counterclockwise rotation if the center of mass is to the left of the bottom corner, and it produces clockwise rotation if the center of mass is to the right of the bottom corner. This means that in the former case, the box returns to its initial position, and in the latter case it topples over on its side. Thus, the critical angle beyond which the box will tip over corresponds to vertical alignment of the bottom corner and the center of the box (see Fig. 14.10c). This critical angle equals the angle between the side of the box and the diagonal. The tangent of this angle is the ratio of the width and the height of the box,

$$\tan \theta = \frac{0.50 \text{ m}}{1.0 \text{ m}} = 0.50$$

With our calculator we find that the critical angle is then

$$\theta = 27°$$

COMMENT: In this example we found that the box begins to topple over if its inclination is such that the center of mass is vertically aligned with the bottom corner. This is a special instance of the general rule that a rigid body resting on a surface (flat or otherwise) becomes unstable when its center of mass is vertically above the outermost point of support.

EXAMPLE 6   A uniform rectangular box 2.0 m high, 1.0 m wide, and 1.0 m deep stands on the platform of a truck (Fig. 14.11a). What is the maximum forward acceleration of the truck that the box can withstand without toppling over? Assume that the coefficient of static friction is large enough that the box will topple over before it starts sliding.

(a)

Box is tilted and does not slip.

(b)

Weight acts at center of mass.

Consider net torque about this corner, where friction and normal forces exert no torque.

(c)

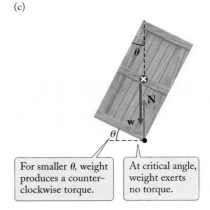

For smaller θ, weight produces a counter-clockwise torque.

At critical angle, weight exerts no torque.

FIGURE 14.10 (a) Box standing on edge. (b) "Free-body" diagram for the box. (c) "Free-body" diagram if the box is tilted at the critical angle. The center of mass is directly above the edge.

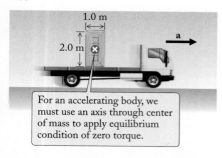

(a)

1.0 m

2.0 m

a

For an accelerating body, we must use an axis through center of mass to apply equilibrium condition of zero torque.

(b)

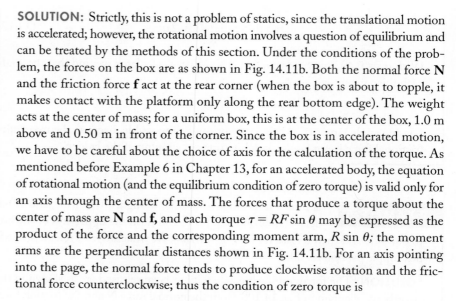

moment arm for **f**

1.0 m

f

When box starts to topple, friction and normal forces act at the rear corner.

N

w

0.50 m

moment arm for **N**

FIGURE 14.11 (a) Box on an accelerating truck. (b) "Free-body" diagram for the box.

SOLUTION: Strictly, this is not a problem of statics, since the translational motion is accelerated; however, the rotational motion involves a question of equilibrium and can be treated by the methods of this section. Under the conditions of the problem, the forces on the box are as shown in Fig. 14.11b. Both the normal force **N** and the friction force **f** act at the rear corner (when the box is about to topple, it makes contact with the platform only along the rear bottom edge). The weight acts at the center of mass; for a uniform box, this is at the center of the box, 1.0 m above and 0.50 m in front of the corner. Since the box is in accelerated motion, we have to be careful about the choice of axis for the calculation of the torque. As mentioned before Example 6 in Chapter 13, for an accelerated body, the equation of rotational motion (and the equilibrium condition of zero torque) is valid only for an axis through the center of mass. The forces that produce a torque about the center of mass are **N** and **f**, and each torque $\tau = RF \sin \theta$ may be expressed as the product of the force and the corresponding moment arm, $R \sin \theta$; the moment arms are the perpendicular distances shown in Fig. 14.11b. For an axis pointing into the page, the normal force tends to produce clockwise rotation and the frictional force counterclockwise; thus the condition of zero torque is

$$-0.50 \text{ m} \times N + 1.0 \text{ m} \times f = 0 \qquad (14.13)$$

We can obtain expressions for f and N from the equations for the horizontal and vertical translational motions. The horizontal acceleration is a and the vertical acceleration is zero; accordingly, the horizontal and vertical components of Newton's Second Law are

$$f = ma$$
$$N - mg = 0$$

Inserting these expressions for f and for N into Eq. (14.13), we obtain

$$0.50 \text{ m} \times mg - 1.0 \text{ m} \times ma = 0$$

from which

$$a = 0.50g = 4.9 \text{ m/s}^2$$

If the acceleration exceeds this value, rotational equilibrium fails, and the box topples.

✔ Checkup 14.2

QUESTION 1: Why is it dangerous to climb a ladder that is leaning against a building at a large angle with the vertical? Why is it dangerous to climb a ladder that is leaning against a building at a small angle with the vertical?

QUESTION 2: Suppose that in Example 5 all the mass of the box is concentrated at the midpoint of the bottom surface, so the center of mass is at this midpoint. What is the critical angle at which such a box topples over on its side?

QUESTION 3: Two heavy pieces of lumber lean against each other, forming an A-frame (see Fig. 14.12). Qualitatively, how does the force that one piece of lumber exerts on the other at the tip of the A vary with the angle?

QUESTION 4: You hold a fishing pole steady, with one hand forward, pushing upward to support the pole, and the other hand further back, pushing downward to maintain

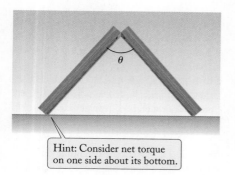

θ

Hint: Consider net torque on one side about its bottom.

FIGURE 14.12 Two pieces of lumber forming an A-frame.

zero net torque. If a fish starts to pull downward on the far end of the pole, then to maintain equilibrium you must

  (A) Increase the upward push and decrease the downward push
  (B) Increase the upward push and increase the downward push
  (C) Increase the upward push and keep the downward push the same

14.3 LEVERS AND PULLEYS

A lever consists of a rigid bar swinging on a pivot (see Fig. 14.13). If we apply a force at the long end, the short end of the bar pushes against a load with a larger force. Thus, the lever permits us to lift a larger load than we could with our bare hands. The relationship between the magnitudes of the forces at the ends follows from the condition for static equilibrium for the lever. Figure 14.13 shows the forces acting on the lever: the force **F** that we exert at one end, the force **F**′ exerted by the load at the other end, and the support force **S** exerted by the pivot point P. The net torque about the pivot point P must be zero. Since, for the arrangement shown in Fig. 14.13, the forces at the ends are at right angles to the distances l and l', the condition on the net torque is

$$Fl - F'l' = 0 \qquad (14.14)$$

from which we find

$$\frac{F'}{F} = \frac{l}{l'} \qquad (14.15)$$

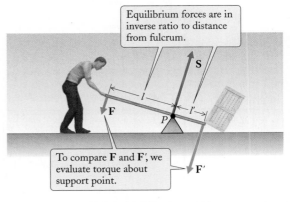

Equilibrium forces are in inverse ratio to distance from fulcrum.

To compare **F** and **F**′, we evaluate torque about support point.

FIGURE 14.13 A lever. The vectors show the forces acting on the lever; **F** is our push, **F**′ is the push of the load, and **S** is the supporting force of the pivot. The force that the lever exerts on the load is of the same magnitude as **F**′, but of opposite direction.

mechanical advantage of lever

By Newton's Third Law, the force that the load exerts on the lever is equal in magnitude to the force that the lever exerts on the load (and of opposite direction). Hence Eq. (14.15) tells us the ratio of the magnitudes of the forces we exert and the lever exerts. *These forces are in the inverse ratio of the distances from the pivot point.* For a powerful lever, we must make the lever arm l as long as possible and the lever arm l' as short as possible. The ratio F'/F of the magnitudes of the force delivered by the lever and the force we must supply is called the **mechanical advantage.**

Apart from its application in the lifting of heavy loads, the principle of the lever finds application in many hand tools, such as pliers and bolt cutters. The handles of these tools are long, and the working ends are short, yielding an enhancement of the force exerted by the hand (see Fig. 14.14). A simple manual winch also relies on the principle of the lever. The handle of the winch is long, and the drum of the winch, which acts as the short lever arm, is small (see Fig. 14.15). The force the winch delivers to the rope attached to the drum is then larger than the force exerted by the hand pushing on the handle. Compound winches, used for trimming sails on sailboats, have internal sets of gears that provide a larger mechanical advantage; in essence, such compound winches stagger one winch within another, so the force ratio generated by one winch is further multiplied by the force ratio of the other.

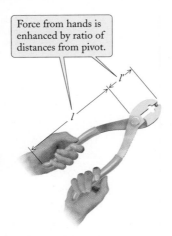

Force from hands is enhanced by ratio of distances from pivot.

FIGURE 14.14 A pair of pliers serves as levers.

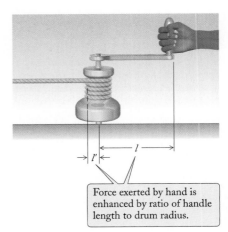

Force exerted by hand is enhanced by ratio of handle length to drum radius.

FIGURE 14.15 A manual winch.

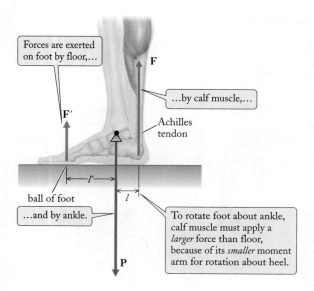

Forces are exerted on foot by floor,...

F'

F

...by calf muscle,...

Achilles tendon

ball of foot

...and by ankle.

l'

l

P

To rotate foot about ankle, calf muscle must apply a *larger* force than floor, because of its *smaller* moment arm for rotation about heel.

FIGURE 14.16 Bones of the foot acting as a lever.

In the human body, many bones play the role of levers that permit muscles or groups of muscles to support or to move the body. For example, Fig. 14.16 shows the bones of the foot; these act as a lever, hinged at the ankle. The rear end of this lever, at the heel, is tied to the muscles of the calf by the Achilles tendon, and the front end of the lever is in contact with the ground, at the ball of the foot. When the muscle contracts, it rotates the heel about the ankle and presses the ball of the foot against the ground, thereby lifting the entire body on tiptoe. Note that the muscle is attached to the short end of this lever—the muscle must provide a larger force than the force generated at the ball of the foot. At first sight, it would seem advantageous to install a longer projecting spur at the heel of the foot and attach the Achilles tendon to the end of this spur; but this would require that the contracting muscle move through a longer distance. Muscle is good at producing large forces, but not so good at contracting over long distances, and the attachment of the Achilles tendon represents the best compromise. In most of the levers found in the human skeleton, the muscle is attached to the short end of the lever.

Equation (14.15) is valid only if the forces are applied at right angles to the lever. A similar equation is valid if the forces are applied at some other angle, but instead of the lengths *l* and *l'* of the lever, we must substitute the lengths of the moment arms of the forces, that is, the perpendicular distances between the pivot point and the lines of action of the forces. These moment arms play the role of effective lengths of the lever.

EXAMPLE 7   When you bend over to pick up something from the floor, your backbone acts as a lever pivoted at the sacrum (see Fig. 14.17). The weight of the trunk pulls downward on this lever, and the muscles attached along the upper part of the backbone pull upward. The actual arrangement of the muscles is rather complicated, but for a simple mechanical model we can pretend that the muscles are equivalent to a string attached to the backbone at an angle of about 12° at a point beyond the center of mass (the other end of the "string" is attached to the pelvis). Assume that the mass of the trunk, including head and arms, is

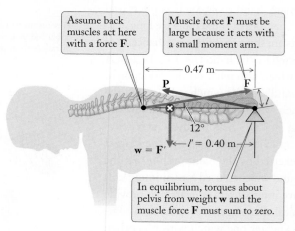

Assume back muscles act here with a force **F**.

Muscle force **F** must be large because it acts with a small moment arm.

0.47 m

P

F

l

12°

w = F'

l' = 0.40 m

In equilibrium, torques about pelvis from weight **w** and the muscle force **F** must sum to zero.

FIGURE 14.17 "Free-body" diagram for the backbone acting as lever. The forces on the backbone are the weight **w** of the trunk (including the weight of the backbone), the pull **F** of the muscles, and the thrust **P** of the pelvis acting as pivot.

48 kg, and that the dimensions are as shown in the diagram. What force must the muscles exert to balance the weight of the trunk when bent over horizontally?

SOLUTION: Figure 14.17 shows a "free-body" diagram for the backbone, with all the forces acting on it. Since the weight **w** of the trunk acts at right angles to the backbone, the lever arm for this weight is equal to the distance $l' = 0.40$ m between the pivot and the center of mass of the trunk. The lever arm for the muscle is the (small) distance l, which equals $l = 0.47$ m $\times \sin 12° = 0.10$ m. According to Eq. (14.15), the force **F** exerted by the muscles then has magnitude

$$F = \frac{l'}{l}F' = \frac{l'}{l}w = \frac{l'}{l}Mg = \frac{0.40 \text{ m}}{0.10 \text{ m}} \times Mg = 4.0 \times Mg$$

$$= 4.0 \times 48 \text{ kg} \times 9.81 \text{ m/s}^2 = 1.9 \times 10^3 \text{ N}$$

This is a quite large force, 4.0 times larger than the weight of the trunk.

COMMENT: Bending over horizontally puts a severe stress on the muscles of the back. Furthermore, it puts an almost equally large compressional stress on the backbone, pulling it hard against the sacrum. The stresses are even larger if you try to lift a load from the floor while your body is bent over in this position. To avoid damage to the muscles and to the lumbosacral disk, it is best to lift by bending the knees, keeping the backbone vertical.

Often, a force is applied to a load by means of a flexible rope, or a string. A pulley is then sometimes used to change the direction of the string or rope and the direction of the force exerted on the body. If the pulley is frictionless, the tension at each point of a flexible rope passing over the pulley is the same. For instance, if we want to lift a load with a rope passing over a single pulley attached to the ceiling (see Fig. 14.18), the force we must exert on the rope has the same magnitude as the weight of the load. Thus, there is no gain of mechanical advantage in such an arrangement of a single pulley; the only benefit is that it permits us to pull more comfortably than if we attempted to lift the load directly.

However, an arrangement of several pulleys linked together, called **block and tackle,** can provide a large gain of mechanical advantage. For example, consider the arrangement of three pulleys shown in Fig. 14.19a; the axles of the two upper pulleys are bolted together, and they are linked to each other and to the third pulley by a single rope. If the rope segments linking the pulleys are parallel and there is no friction, then the mechanical advantage of this arrangement is 3; that is, the magnitudes of the forces F and F' are in the ratio of 1 to 3. This can be most easily understood by drawing the "free-body" diagram for the lower portion of the pulley system, including the load (Fig. 14.19b). In this diagram, the three ropes leading upward have been cut off and replaced by the forces exerted on them by the external (upper) portions of the ropes. Since the tension is the same everywhere along the rope, the forces pulling upward on each of the three rope ends shown in the "free-body" diagram all have the same magnitude F, and thus the net upward force is $3F$.

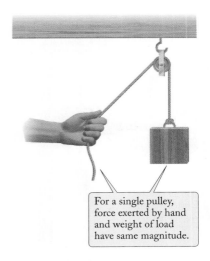

For a single pulley, force exerted by hand and weight of load have same magnitude.

FIGURE 14.18 A single pulley.

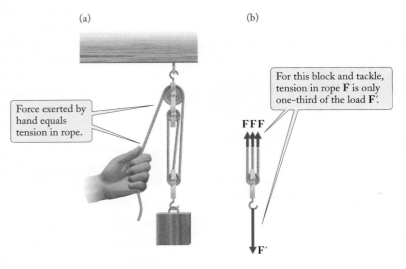

(a)

Force exerted by hand equals tension in rope.

(b)

For this block and tackle, tension in rope **F** is only one-third of the load **F'**.

FFF

F'

FIGURE 14.19 (a) Block and tackle. (b) "Free-body" diagram for the lower portion of the pulley system.

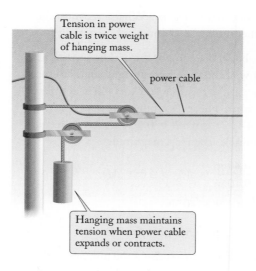

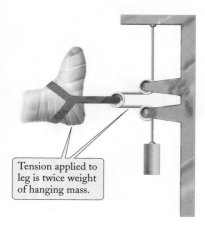

FIGURE 14.20 Block and tackle used for tensioning power line.

FIGURE 14.21 Block and tackle in traction apparatus for fractured leg.

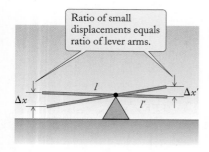

FIGURE 14.22 Rotation of lever by a small angle produces displacements Δx and $\Delta x'$ of the ends.

Block-and-tackle arrangements have many practical applications. For instance, they are used to provide the proper tension in overhead power cables for electric trains and trams (see Fig. 14.20); without such an arrangement, the cables would sag on warm days when thermal expansion increases their length, and they would stretch excessively tight and perhaps snap on cold days, when they contract. One common cause of power failures on cold winter nights is the snapping of power lines lacking such compensating pulleys.

Another practical application of block and tackle is found in the traction devices used in hospitals to immobilize and align fractured bones, especially leg bones. A typical arrangement is shown in Fig. 14.21; here the pull applied to the leg is twice as large as the magnitude of the weight attached on the lower end to the rope. Also, as in the case of the power line, the tension remains constant even if the leg moves.

The mechanical advantage provided by levers, arrangements of pulleys, or other devices can be calculated in a general and elegant way by appealing to the Law of Conservation of Energy. A lever merely transmits the work we supply at one end to the load at the other end. We can express this equality of work input and work output by

$$F' \Delta x' = F \Delta x \tag{14.16}$$

where Δx is the displacement of our hand and $\Delta x'$ the displacement of the load. According to this equation, the forces F' and F are in the inverse ratio of the displacements,

$$\frac{F'}{F} = \frac{\Delta x}{\Delta x'} \tag{14.17}$$

Consider, now, the rotation of the lever by a small angle (see Fig. 14.22). Since the two triangles included between the initial and final positions of the lever are similar, the distances Δx and $\Delta x'$ are in the same ratio as the lever arms l and l'; thus, we immediately recognize from Eq. (14.17) that the mechanical advantage of the lever is l/l'.

Likewise, we immediately recognize from Eq. (14.17) that the mechanical advantage of the arrangement of pulleys shown in Fig. 14.19 is 3, since whenever our hand pulls a length Δx of rope out of the upper pulley, the load moves upward by a distance of only $\Delta x/3$.

✔ Checkup 14.3

QUESTION 1: Figure 14.23 shows two ways of using a lever. Which has the larger mechanical advantage?

QUESTION 2: Is Eq. (14.15) for the ratio of the forces F and F' on a lever valid if one or both of these forces are not perpendicular to the lever?

QUESTION 3: Suppose that the pulleys in a block and tackle are of different sizes. Does this affect the mechanical advantage?

(a)

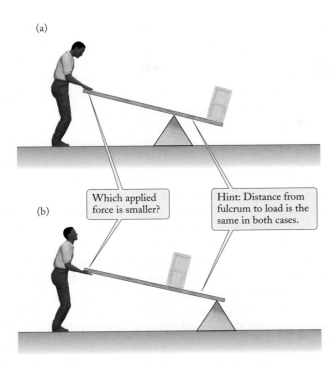

(b)

Which applied
force is smaller?

Hint: Distance from
fulcrum to load is the
same in both cases.

FIGURE 14.23
Two ways of using a lever.

QUESTION 4: A lever is used to lift a 100-kg rock. The distance from the rock to the
fulcrum is roughly one-tenth of the distance from the fulcrum to the handle. If the
rock has a mass of 100 kg, the downward force at the handle necessary to lift the rock
is approximately:

(A) 1 N   (B) 10 N   (C) 100 N   (D) 1000 N

14.4 ELASTICITY OF MATERIALS

In our examples of bridges, tower cranes, etc., we assumed that the bodies
on which the forces act are rigid; that is, they do not deform. Although
solid bodies, such as bars or blocks of steel, are nearly rigid, they are not
exactly rigid, and they will deform by a noticeable amount if a large
enough force is applied to them. A solid bar may be thought of as a very
stiff spring. If the force is fairly small, this "spring" will suffer only an
insignificant deformation, but if the force is large, it will suffer a noticeable deformation. Provided that the force and the deformation remain
within some limits, the deformation of a solid body is **elastic,** *which means
that the body returns to its original shape once the force ceases to act.* Such
elastic deformations of a solid body usually obey Hooke's Law: the deformation is proportional to the force. But the constant of proportionality
is small, giving a small deformation unless the force is large. The corresponding spring constant is thus very large, meaning that an appreciable deformation requires a large force.

A solid block of material can suffer several kinds of deformation, depending on
how the force is applied. If one end of the body is held fixed and the force pulls on the
other end, the deformation is a simple **elongation** of the body (see Fig. 14.24). If one
side of the body is held fixed and the force pushes tangentially along the other side,
then the deformation is a **shear,** which changes the shape of the body from a rectangular

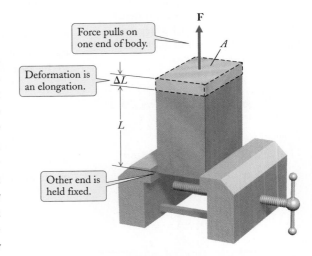

Force pulls on
one end of body.

Deformation is
an elongation.

Other end is
held fixed.

FIGURE 14.24 Tension applied to the end of a block of
material causes elongation.

parallelepiped to a rhomboidal parallelepiped (see Fig. 14.25a). During this deformation, the parallel layers of the body slide with respect to one another just as the pages of a book slide with respect to one another when we push along its cover (see Fig. 14.25b). If the force is applied from all sides simultaneously, by subjecting the body to the pressure of a fluid in which the body is immersed, then the deformation is a **compression** of the volume of the body, without any change of the geometrical shape (see Fig. 14.26).

In all of these cases, *the fractional deformation, or the percent deformation, is directly proportional to the applied force and inversely proportional to the area over which the force is distributed.* For instance, if a given force produces an elongation of 1% when pulling on the end of a block, then the same force pulling on the end of a block of, say, twice the cross-sectional area will produce an elongation of $\frac{1}{2}$%. This can be readily understood if we think of the block as consisting of parallel rows of atoms linked by springs, which represent the interatomic forces that hold the atoms in their places (see Fig. 14.27). When we pull on the end of the block with a given force, we stretch the interatomic springs by some amount; and when we pull on a block of twice the cross-sectional area, we have to stretch twice as many springs, and therefore the force acting on each spring is only half as large and produces only half the elongation in each spring. Furthermore, since the force applied to the end of a row of atoms is communicated to all the interatomic springs in that row, a given force produces a given elongation in each spring in a row. The net elongation of the block is the sum of the elongations of all the interatomic springs in the row, and hence the fractional elongation of the block is the same as the fractional elongation of each spring, regardless of the overall length of the block. For instance, if a block elongates by 0.1 mm when subjected to a given force, then a block of, say, twice the length will elongate by 0.2 mm when subjected to the same force.

To express the relationships among elongation, force, and area mathematically, consider a block of initial length L and cross-sectional area A. If a force F pulls on the end of this block, the elongation is ΔL, and the fractional elongation is $\Delta L / L$. This fractional elongation is directly proportional to the force and inversely proportional to the area A:

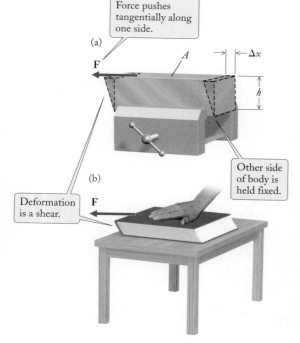

FIGURE 14.25 (a) Tangential force applied to the side of a block of material causes shear. (b) When such a tangential force is applied to the cover of a book, the pages slide past one another.

Force pushes tangentially along one side.

Deformation is a shear.

Other side of body is held fixed.

elongation and Young's modulus

$$\frac{\Delta L}{L} = \frac{1}{Y}\frac{F}{A} \tag{14.18}$$

Here the quantity Y is the constant of proportionality. In Eq. (14.18) this constant written as $1/Y$, so it divides the right side, instead of multiplying it (this is analogous to writing Hooke's Law for a spring as $\Delta x = (1/k)\, F$, where Δx is the elongation

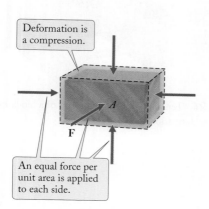

Deformation is a compression.

An equal force per unit area is applied to each side.

FIGURE 14.26 Pressure applied to all sides of a block of material causes compression.

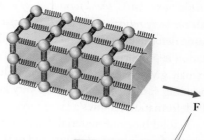

FIGURE 14.27 Microscopically, a block of solid material may be thought of as rows of atoms linked by springs. The springs stretch when a tension is applied to the block.

If force pulling on end is distributed over larger area, more springs need to be stretched, and smaller deformation will result.

TABLE 14.1	ELASTIC MODULI OF SOME MATERIALS		
MATERIAL	**YOUNG'S MODULUS**	**SHEAR MODULUS**	**BULK MODULUS**
Steel	$22 \times 10^{10} \text{ N/m}^2$	$8.3 \times 10^{10} \text{ N/m}^2$	$16 \times 10^{10} \text{ N/m}^2$
Cast iron	15	6.0	11
Brass	9.0	3.5	6.0
Aluminum	7.0	2.5	7.8
Bone (long)	3.2	1.2	3.1
Concrete	2	—	—
Lead	1.6	0.6	4.1
Nylon	0.36	0.12	0.59
Glycol	—	—	0.27
Water	—	—	0.22
Quartz	9.7(max)	3.1	3.6

produced by an applied force F). Thus, a stiff material, such as steel, that elongates by only a small amount has a large value of Y. The constant Y is called **Young's modulus.** Table 14.1 lists values of Young's moduli for a few solid materials. Note that if, instead of exerting a pull on the end of the block, we exert a push, then F in Eq. (14.18) must be reckoned as negative, and the change ΔL of length will then likewise be negative—the block becomes shorter.

In engineering language, *the fractional deformation is usually called the* **strain,** *and the force per unit area is called the* **stress.** In this terminology, Eq. (14.18) simply states that the strain is proportional to the stress.

This proportionality of strain and stress is also valid for shearing deformations and compressional deformations, provided we adopt a suitable definition of strain, or fractional deformation, for these cases. For shear, the fractional deformation is defined as the ratio of the sideways displacement Δx of the edge of the block to the height h of the block (see Fig. 14.25a). This fractional deformation is directly proportional to the force F and inversely proportional to the area A (note that the relevant area A is now the top area of the block, where the force is applied):

$$\frac{\Delta x}{h} = \frac{1}{S}\frac{F}{A} \tag{14.19}$$

shear and shear modulus

Here, the constant of proportionality S is called the **shear modulus.** Table 14.1 includes values of shear moduli of solids.

For compression, the fractional deformation is defined as the ratio of the change ΔV of the volume to the initial volume, and this fractional deformation is, again, proportional to the force F pressing on each face of the block and inversely proportional to the area A of that face:

$$\frac{\Delta V}{V} = -\frac{1}{B}\frac{F}{A} \tag{14.20}$$

compression and bulk modulus

In this equation, the minus sign indicates that ΔV is negative; that is, the volume decreases. The constant of proportionality B in the equation is called the **bulk modulus.**

Table 14.1 includes values of bulk moduli for solids. This table also includes values of bulk moduli for some liquids. The force per unit area, F/A, is also known as the **pressure:**

$$[\text{pressure}] = \frac{F}{A} \qquad (14.21)$$

The formula (14.20) is equally valid for solids and for liquids—when we squeeze a liquid from all sides, it will suffer a compression. Note that Table 14.1 does not include values of Young's moduli and of shear moduli for liquids. *Elongation and shear stress are not supported by a liquid*—we can elongate or shear a "block" of liquid as much as we please without having to exert any significant force.

Concepts *in* **Context**

EXAMPLE 8   The lifting cable of a tower crane is made of steel, with a diameter of 5.0 cm. The length of this cable, from the ground to the horizontal arm, across the horizontal arm, and down to the load, is 160 m (Fig. 14.28). By how much does this cable stretch in excess of its initial length when carrying a load of 60 tons?

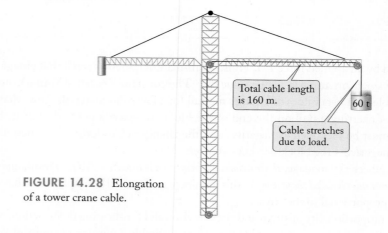

Total cable length is 160 m.

60 t

Cable stretches due to load.

FIGURE 14.28 Elongation of a tower crane cable.

SOLUTION: The cross-sectional area of the cable is

$$A = \pi r^2 = \pi \times (0.025 \text{ m})^2 = 2.0 \times 10^{-3} \text{ m}^2$$

and the force per unit area is

$$\frac{F}{A} = \frac{(60\,000 \text{ kg} \times 9.81 \text{ m/s}^2)}{2.0 \times 10^{-3} \text{ m}^2} = 2.9 \times 10^8 \text{ N/m}^2$$

Since we are dealing with an elongation, the relevant elastic modulus is the Young's modulus. According to Table 14.1, the Young's modulus of steel is $22 \times 10^{10} \text{ N/m}^2$. Hence Eq. (14.18) yields

$$\frac{\Delta L}{L} = \frac{1}{Y}\frac{F}{A} = \frac{1}{22 \times 10^{10} \text{ N/m}^2} \times 2.9 \times 10^8 \text{ N/m}^2$$

$$= 1.3 \times 10^{-3}$$

The change of length is therefore

$$\Delta L = 1.3 \times 10^{-3} \times L = 1.3 \times 10^{-3} \times 160 \text{ m}$$

$$= 0.21 \text{ m}$$

EXAMPLE 9   What pressure must you exert on a sample of water if you want to compress its volume by 0.10%?

SOLUTION: For volume compression, the relevant elastic modulus is the bulk modulus B. By Eq. (14.20), the pressure, or the force per unit area, is

$$\frac{F}{A} = -B \frac{\Delta V}{V}$$

For 0.10% compression, we want to achieve a fractional change of volume of $\Delta V/V = -0.0010$. Since the bulk modulus of water is 0.22×10^{10} N/m^2, the required pressure is

$$\frac{F}{A} = 0.22 \times 10^{10} \text{ N/m}^2 \times 0.0010 = 2.2 \times 10^6 \text{ N/m}^2$$

The simple uniform deformations of elongation, shear, and compression described above require a rather special arrangement of forces. In general, the forces applied to a solid body will produce nonuniform elongation, shear, and compression. For instance, a beam supported at its ends and sagging in the middle because of its own weight or the weight of a load placed on it will elongate along its lower edge, and compress along its upper edge.

Finally, note that the formulas (14.18)–(14.20) are valid only as long as the deformation is reasonably small—a fraction of a percent or so. If the deformation is excessive, the material will be deformed beyond its elastic limit; that is, the material will suffer a permanent deformation and will *not* return to its original size and shape when the force ceases. If the deformation is even larger, the material will break apart or crumble. For instance, steel will break apart (see Fig. 14.29) if the tensile stress exceeds 5×10^8 N/m^2, or if the shearing stress exceeds 2.5×10^8 N/m^2, and it will crumble if the compressive stress exceeds 5×10^8 N/m^2.

FIGURE 14.29 These rods of steel broke apart when a large tension was applied.

 Checkup 14.4

QUESTION 1: When a tension of 70 N is applied to a piano wire of length 1.8 m, it stretches by 2.0 mm. If the same tension is applied to a similar piano wire of length 3.6 m, by how much will it stretch?

QUESTION 2: Is it conceivable that a long cable hanging vertically might snap under its own weight? If so, does the critical length of the cable depend on its diameter?

QUESTION 3: The bulk modulus of copper is about twice that of aluminum. Suppose that a copper and an aluminum sphere have exactly equal volumes at normal atmospheric pressure. Suppose that when subjected to a high pressure, the volume of the aluminum sphere shrinks by 0.01%. By what percentage will the copper sphere shrink at the same pressure?

QUESTION 4: While lifting a load, the steel cable of a crane stretches by 1 cm. If you want the cable to stretch by only 0.5 cm, by what factor must you increase its diameter?

(A) $\sqrt{2}$ (B) 2   (C) $2\sqrt{2}$ (D) 4

SUMMARY

STATIC EQUILIBRIUM   The sums of the external forces and of the external torques on a rigid body are zero.

$$\mathbf{F}_1 + \mathbf{F}_2 + \mathbf{F}_3 + \cdots = 0 \tag{14.1}$$

$$\boldsymbol{\tau}_1 + \boldsymbol{\tau}_2 + \boldsymbol{\tau}_3 + \cdots = 0 \tag{14.2}$$

STATICS CALCULATION TECHNIQUE   To eliminate an unknown force, evaluate torques about the point where that force acts (or about another point where the force has zero moment arm).

TORQUE DUE TO GRAVITY   Gravity effectively acts at the center of mass.

MECHANICAL ADVANTAGE OF LEVER

$$\frac{F'}{F} = \frac{l}{l'} \tag{14.15}$$

BLOCK AND TACKLE   An arrangement of several pulleys that provides a mechanical advantage (equal to the ratio of the distance moved where the force is applied to the distance moved by the load).

PRESSURE

$$[\text{pressure}] = \frac{F}{A} \tag{14.21}$$

DEFORMATIONS OF ELASTIC MATERIAL

A is cross-sectional area.

Elongation: $\dfrac{\Delta L}{L} = \dfrac{1}{Y}\dfrac{F}{A}$ (Y = Young's modulus)   **(14.18)**

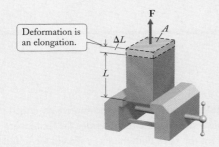

Deformation is an elongation.

Shear: $\dfrac{\Delta x}{h} = \dfrac{1}{S}\dfrac{F}{A}$ (S = shear modulus)   **(14.19)**

Deformation is a shear.

Compression: $\dfrac{\Delta V}{V} = -\dfrac{1}{B}\dfrac{F}{A}$ (B = bulk modulus)   **(14.20)**

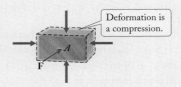

Deformation is a compression.

QUESTIONS FOR DISCUSSION

1. If the legs of a table are exactly the same length and if the floor is exactly flat, then the weight of the table will be equally distributed over all four legs. But if there are small deviations from exactness, then the weight will not be equally distributed. Is it possible for all of the weight to rest on three legs? On two?

2. List as many examples as you can of joints in the human skeleton that act as pivots for levers. Do any of these levers in the human skeleton have a mechanical advantage larger than 1?

3. Design a block and tackle with a mechanical advantage of 4, and another with a mechanical advantage of 5. If you connect these two arrangements in tandem, what mechanical advantage do you get?

4. Figure 14.30 shows a differential windlass consisting of two rigidly joined drums around which a rope is wound. A pulley

holding a load hangs from this rope. Explain why this device gives a very large mechanical advantage if the radii of the two drums are nearly equal.

5. The collapse of several skywalks at the Hyatt Regency hotel in Kansas City on July 17, 1982, with the loss of 114 lives, was due to a defective design of the suspension system. Instead of suspending the beams of the skywalks directly from single, long steel rods anchored at the top of the building, some incompetent engineers decided to use several short steel rods joining the beams of each skywalk to those of the skywalk above (Fig. 14.31). Criticize this design, keeping in mind that the beams are made of a much weaker material than the rods.

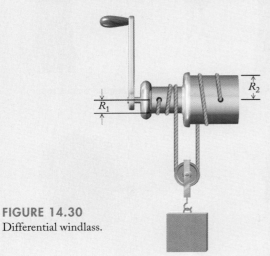

FIGURE 14.30
Differential windlass.

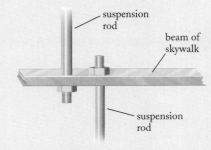

FIGURE 14.31 Beam of skywalk and suspension rods.

6. A steel rod is much less flexible than a woven steel rope of the same strength. Explain this.

7. A carpenter wants to support the (flat) roof of a building with horizontal beams of wood of rectangular cross section. To achieve maximum strength of the roof (least sag), should he install the beams with their narrow side up or with their wide side up?

8. The long bones in the limbs of vertebrates have the shape of hollow pipes. If the same amount of bone tissue had been assembled in a solid rod (of correspondingly smaller cross section), would the limb have been more rigid or less rigid?

PROBLEMS

4.2 Examples of Static Equilibrium

1. At a construction site, a laborer pushes horizontally against a large bucket full of concrete of total mass 600 kg suspended from a crane by a 20-m cable (see Fig. 14.32). What is the force the laborer has to exert to hold the bucket at a distance of 2.0 m from the vertical?

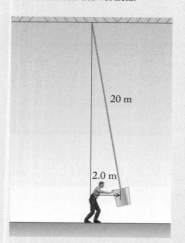

FIGURE 14.32 Bucket hanging from a cable.

2. You are holding a meterstick of 0.20 kg horizontally in one hand. Assume that your hand is wrapped around the last 10 cm of the stick (see Fig. 14.33), so the front edge of your hand exerts an upward force and the rear edge of your hand exerts a downward force. Calculate these forces.

FIGURE 14.33 A meterstick held in a hand.

3. Consider the bridge with the locomotive described in Example 1 and suppose that, besides the first locomotive at 30 m from the right end, there is a second locomotive, also of 90 000 kg, at 80 m from the right end. What is the load on each pier in this case?

4. Repeat the calculations of Example 1 assuming that the bridge has a slope of 1:7, with the left end higher than the right.

5. In order to pull an automobile out of the mud in which it is stuck, the driver stretches a rope taut from the front end of the automobile to a stout tree. He then pushes sideways against the rope at the midpoint (see Fig. 14.34). When he pushes with a force of 900 N, the angle between the two halves of the rope on his right and left is 170°. What is the tension in the rope under these conditions?

6. A mountaineer is trying to cross a crevasse by means of a rope stretched from one side to the other (see Fig. 14.35). The mass of the mountaineer is 90 kg. If the two parts of the rope make angles of 40° and 20° with the horizontal, what are the tensions in the two parts?

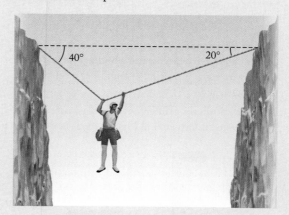

FIGURE 14.35 Mountaineer suspended from a rope.

FIGURE 14.34 The rope is stretched between the automobile and a tree. The driver is pushing at the midpoint.

7. The plant of the foot of an average male is 26 cm, and the height of his center of mass above the floor is 1.03 m. When he is standing upright, the center of mass is vertically aligned with the ankle, 18 cm from the tip of the foot (see Fig. 14.36). Without losing his equilibrium, how far can the man lean forward or backward while keeping his body straight and his feet stiff and immobile?

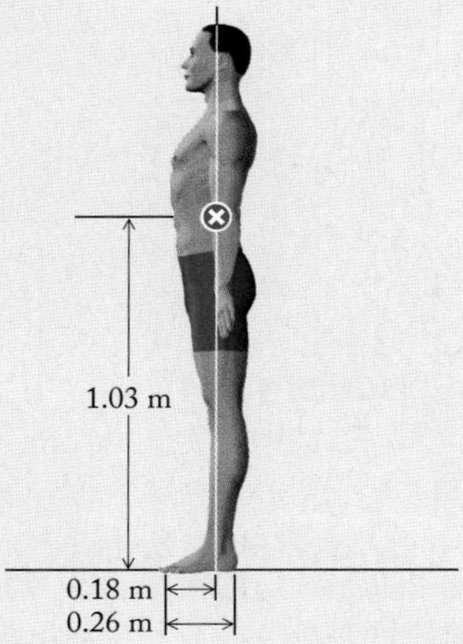

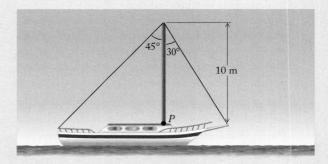

FIGURE 14.36 Man standing on stiff feet.

1.03 m

0.18 m

0.26 m

8. A 50-kg log of uniform thickness lies horizontally on the ground.

(a) What vertical force must you exert on one end of the log to barely lift this end off the ground?

(b) If you continue to exert a purely vertical force on the end of the log, what is the magnitude of the force required to hold the log at an angle of 30° to the ground? At an angle of 60°? At an angle of 85°?

(c) If instead you exert a force at right angles to the length of the log, what is the magnitude of the force required to hold the log at an angle of 30° to the ground? At an angle of 60°? At an angle of 85°?

9. In an unequal-arm balance, the beam is pivoted at a point near one end. With such a balance, large loads can be balanced with small standard weights. Figure 14.37 shows such a balance with an arm of 50 cm swinging on a pivot 1.0 cm from one end. When a package of sugar is deposited in the balance pan, equilibrium is attained with a standard mass of 0.12 kg in the other pan. What is the mass of the sugar? Neglect the masses of the pans.

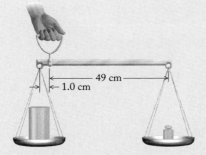

49 cm

1.0 cm

FIGURE 14.37 Unequal-arm balance.

10. One end of a uniform beam of mass 50 kg and length 3.0 m rests on the ground; the other end is held above the ground by a pivot placed 1.0 m from that end (see Fig. 14.38). An 80-kg man walks along the beam, from the low end toward the high end. How far beyond the pivot can the man walk before the high end of the beam swings down?

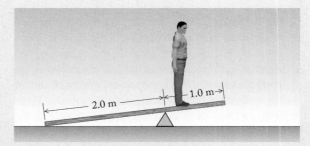

2.0 m

1.0 m

FIGURE 14.38 Man standing on a beam.

11. The mast of a sailboat is held by two steel cables attached as shown in Fig. 14.39. The front cable has a tension of 5.0×10^3 N. The mast is 10 m high. What is the tension in the rear cable? What force does the foot of the mast exert on the sailboat? Assume that the weight of the mast can be neglected and that the foot of the mast is hinged (and therefore exerts no torque).

45°

30°

10 m

P

FIGURE 14.39 Steel cables staying a mast.

12. The center of mass of a 45-kg sofa is 0.30 m above its bottom, at its lateral midpoint. You lift one end of the 2.0-m-long sofa to a height of 1.0 m by applying a vertical force at the bottom of one end; the other end stays on the floor without slipping. What force do you apply? Compare this with the force you apply when a friend lifts the other end, also to 1.0 m, so that the weight is shared equally. Based on this, who has the easier task when a short and a tall person share a bulky load?

13. Suppose that you lift the lid of a chest. The lid is a uniform sheet of mass 12 kg, hinged at the rear. What is the smallest force you can apply at the front of the lid to hold it at an angle of 30° with the horizontal? At 60°?

14. A pole-vaulter holds a 4.5-m pole horizontally with her right hand at one end and her left hand 1.5 m from the same end.

The left hand applies an upward force and the right hand a downward force. If the mass of the pole is 3.0 kg, find those two forces.

15. A 50-kg diving board is 3.0 m long; it is a uniform beam, bolted down at one end and supported from below a distance 1.0 m from the same end. A 60-kg diver stands at the other end. Calculate the downward force at the bolted end and the upward support force.

16. A window washer's scaffolding is 12 m long; it is suspended by a cable at each end. Assume that the scaffolding is a horizontal uniform rod of mass 110 kg. The window washer (with gear) has a mass of 90 kg and stands 2.0 m from one end of the scaffolding. Find the tension in each cable.

17. A pencil is placed on an incline, and the angle of the incline is slowly increased. At what angle will the pencil start to roll? Assume the pencil has an exactly hexagonal cross section and does not slip.

18. A 10-kg ladder is 5.0 m long and rests against a frictionless wall, making an angle of 30° with the vertical. The coefficient of friction between the ladder and the ground is 0.35. A 60-kg painter begins to climb the ladder, standing vertically on each rung. How far up the ladder has the painter climbed when the ladder begins to slip?

19. Figure 14.40 shows the arrangement of wheels on a passenger engine of the Caledonian Railway. The numbers give the distances between the wheels in feet and the downward forces that each wheel exerts on the track in short tons (1 short ton = 2000 lbf; the numbers for the forces include both the right and left wheels). From the information given, find how far the center of mass of the engine is behind the front wheel.

20. A door made of a uniform piece of wood measures 1.0 m by 2.0 m and has a mass of 18 kg. The door is entirely supported by two hinges, one at the bottom corner and one at the top corner. Find the force (magnitude and direction) that the door exerts on each hinge. Assume that the *vertical* force on each hinge is the same.

21. You want to pick up a nearly massless rectangular cardboard box by grabbing its top and side between your forefinger and thumb (see Fig. 14.41). Show that this is impossible unless the coefficient of friction between your fingers and the box is at least 1.

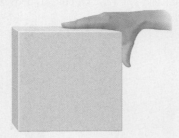

FIGURE 14.41 Box held in a hand.

22. A meterstick of wood of 0.40 kg is nailed to the wall at the 75-cm mark. If the stick is free to rotate about the nail, what horizontal force must you exert at the upper (short) end to deflect the stick 30° to one side?

*23. A wheel of mass M and radius R is to be pulled over a step of height h, where $R > h$. Assume that the pulling force is applied at the axis of the wheel. If the pull is horizontal, what force must be applied to barely begin moving? If the pull at the axis is instead in the direction that requires the least force to begin moving, what force must be applied? What is the new direction? (Hint: Consider the torques about the point of contact with the step.)

*24. Consider a heavy cable of diameter d and density ρ from which hangs a load of mass M. What is the tension in the cable as a function of the distance from the lower end?

*25. Figure 14.42 shows two methods for supporting the mast of a sailboat against the lateral force exerted by the pull of the sail. In Fig. 14.42a, the shrouds (wire ropes) are led directly to the top of the mast; in Fig. 14.42b, the shrouds are led around a rigid pair of spreaders. Suppose that the dimensions of the mast

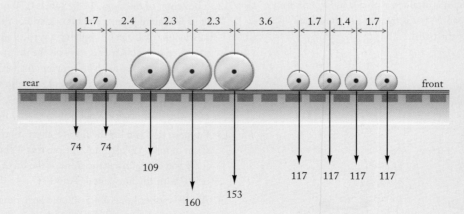

FIGURE 14.40 Wheels of a locomotive.

(a)   (b)

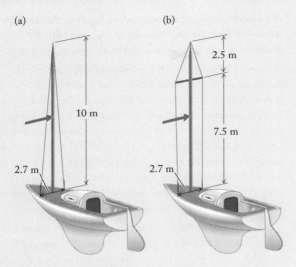

10 m

2.5 m

7.5 m

2.7 m   2.7 m

FIGURE 14.42 Two methods for supporting the mast of a boat.

and the boat are as indicated in this figure, and that the pull of the sail is equivalent to a horizontal force of 2400 N acting from the left at half the height of the mast. The foot of the mast permits the mast to tilt, so the only lateral support of the mast is that provided by the shrouds. What is the excess tension in the left shroud supporting the mast in case (a)? In case (b)? Which arrangement is preferable?

*26. A bowling ball of mass 10 kg rests in a groove with smooth, perpendicular walls, inclined at angles of 30° and 60° with the vertical, as shown in Fig. 14.43. Calculate the magnitudes of the normal forces at the points of contact.

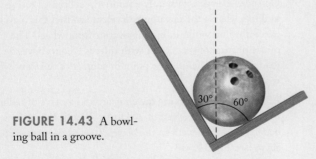

FIGURE 14.43 A bowling ball in a groove.

30°   60°

*27. A tetrahedral tripod consists of three massless legs (see Fig. 14.44). A mass M hangs from the apex of the (regular) tetrahedron. What are the compressional forces in the three legs?

FIGURE 14.44 A tripod.

*28. A sailor is being transferred from one ship to another by means of a bosun's chair (see Fig. 14.45). The chair hangs from a roller riding on a rope strung between the two ships. The distance between the ships is d, and the rope has a length $1.2d$. The mass of the sailor plus the chair is m. If the sailor is at a (horizontal) distance of $0.25d$ from one ship, find the force that must be exerted on the pull rope to keep the sailor in equilibrium. Also find the tension in the long rope. Ignore the masses of the ropes.

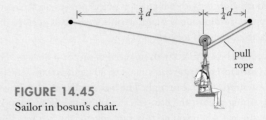

$\frac{3}{4}d$ $\frac{1}{4}d$

pull rope

FIGURE 14.45 Sailor in bosun's chair.

*29. A uniform solid disk of mass M and radius R hangs from a string of length l attached to a smooth vertical wall (see Fig. 14.46). Calculate the tension in the string and the normal force acting at the point of contact of disk and wall.

l

R

FIGURE 14.46 Disk hanging from string.

*30. Three traffic lamps of equal masses of 20 kg hang from a wire stretched between two telephone poles, 15 m apart (Fig. 14.47). The horizontal spacing of the traffic lamps is uniform. At each pole, the wire makes a downward angle of 10° with the horizontal line. Find the tensions in all the segments of wire, and find the distance of each lamp below the horizontal line.

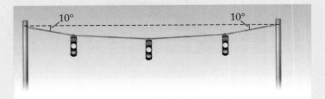

10°   10°

FIGURE 14.47 Three traffic lamps.

*31. Consider the ladder leaning against a wall described in Example 4. If the ladder makes an angle of 30° with the wall, how hard can you push down vertically on the top of the ladder with your hand before slipping begins?

*32. An automobile with a wheelbase (distance from the front wheels to the rear wheels) of 3.0 m has its center of mass at a point midway between the wheels at a height of 0.65 m above the road. When the automobile is on a level road, the force with which each wheel presses on the road is 3100 N. What is the normal force with which each wheel presses on the road when the automobile is standing on a steep road of slope 3:10 with all the wheels locked?

*33. A wooden box is filled with material of uniform density. The box (with its contents) has a mass of 80 kg; it is 0.60 m wide, 0.60 m deep, and 1.2 m high. The box stands on a level floor. By pushing against the box, you can tilt it over (Fig. 14.48). Assume that when you do this, one edge of the box remains in contact with the floor without sliding.

(a) Plot the gravitational potential energy of the box as a function of the angle θ between the bottom of the box and the floor.

(b) What is the critical angle beyond which the box will topple over when released?

(c) How much work must you do to push the box to this critical angle?

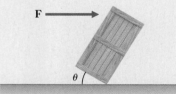

FIGURE 14.48
Tilted box.

*34. A meterstick of mass M hangs from a 1.5-m string tied to the meterstick at the 80-cm mark. If you push the bottom end of the meterstick to one side with a horizontal push of magnitude $Mg/2$, what will be the equilibrium angles of the meterstick and the string?

*35. Five identical books are to be stacked one on top of the other. Each book is to be shifted sideways by some variable amount, so as to form a curved leaning tower with maximum protrusion (see Fig. 14.49). How much must each book be shifted? What is the maximum protrusion? If you had an infinite

number of books, what would be the limiting maximum protrusion? (Hint: Try this experimentally; start with the top book, and insert the others underneath, one by one.)

**36. A wooden box, filled with a material of uniform density, stands on a concrete floor. The box has a mass of 75 kg and is 0.50 m wide, 0.50 m long, and 1.5 m high. The coefficient of friction between the box and the floor is $\mu_s = 0.80$. If you exert a (sufficiently strong) horizontal push against the side of the box, it will either topple over or start sliding without toppling over, depending on how high above the level of the floor you push. What is the maximum height at which you can push if you want the box to slide? What is the magnitude of the force you must exert to start the sliding?

*37. The left and right wheels of an automobile are separated by a transverse distance of $l = 1.5$ m. The center of mass of this automobile is $h = 0.60$ m above the ground. If the automobile is driven around a flat (no banking) curve of radius $R = 25$ m with an excessive speed, it will topple over sideways. What is the speed at which it will begin to topple? Express your answer in terms of l, h, and R; then evaluate numerically. Assume that the wheels do not skid.

*38. An automobile has a wheelbase (distance from front wheels to rear wheels) of 3.0 m. The center of mass of this automobile is at a height of 0.60 m above the ground. Suppose that this automobile has rear-wheel drive and that it is accelerating along a level road at 6.0 m/s². When the automobile is parked, 50% of its weight rests on the front wheels and 50% on the rear wheels. What is the weight distribution when it is accelerating? Pretend that the body of the automobile remains parallel to the road at all times.

*39. Consider a bicycle with only a front-wheel brake. During braking, what is the maximum deceleration that this bicycle can withstand without flipping over its front wheel? The center of mass of the bicycle with rider is 95 cm above the road and 70 cm behind the point of contact of the front wheel with the ground.

*40. A bicycle and its rider are traveling around a curve of radius 6.0 m at a constant speed of 20 km/h. What is the angle at which the rider must lean the bicycle toward the center of the curve (see Fig. 14.50)?

FIGURE 14.49 A stack of books.

FIGURE 14.50 Bicycle traveling around curve.

**41. An automobile is braking on a flat, dry road with a coefficient of static friction of 0.90 between its wheels and the road. The wheelbase (the distance between the front and the rear wheels) is 3.0 m, and the center of mass is midway between the wheels, at a height of 0.60 m above the road.

(a) What is the deceleration if all four wheels are braked with the maximum force that avoids skidding?

(b) What is the deceleration if the rear-wheel brakes are disabled? Take into account that during braking, the normal force on the front wheels is larger than that on the rear wheels.

(c) What is the deceleration if the front-wheel brakes are disabled?

*42. A square framework of steel hangs from a crane by means of cables attached to the upper corners making an angle of 60° with each other (see Fig. 14.51). The framework is made of beams of uniform thickness joined (loosely) by pins at the corners, and its total mass is M. Find the tensions in the cables and the tensional and compressional forces in each beam at each of its two ends.

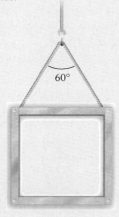

FIGURE 14.51
Hanging framework of beams.

**43. Two smooth balls of steel of mass m and radius R are sitting inside a tube of radius $1.5R$. The balls are in contact with the bottom of the tube and with the wall (at two points; see Fig. 14.52). Find the contact force at the bottom and at the two points on the wall.

FIGURE 14.52 Two balls in a tube.

**44. One end of a uniform beam of length L rests against a smooth, frictionless vertical wall, and the other end is held by a string of length $l = \frac{3}{2}L$ attached to the wall (see Fig. 14.53). What must be the angle of the beam with the wall if it is to remain at rest without slipping?

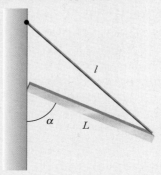

FIGURE 14.53 Beam, string, and wall.

**45. Two playing cards stand on a table leaning against each other so as to form an A-frame "roof." The frictional coefficient between the bottoms of the cards and the table is μ_s. What is the maximum angle that the cards can make with the vertical without slipping?

**46. A rope is draped over the round branch of a tree, and unequal masses m_1 and m_2 are attached to its ends. The coefficient of sliding friction for the rope on the branch is μ_k. What is the acceleration of the masses? Assume that the rope is massless. (Hint: For each small segment of the rope in contact with the branch, the small change in tension across the segment is equal to the friction force.)

**47. The flywheel of a motor is connected to the flywheel of an electric generator by a drive belt (Fig. 14.54). The flywheels are of equal size, each of radius R. While the flywheels are rotating, the tensions in the upper and the lower portions of the drive belt are T_1 and T_2, respectively, so the drive belt exerts a torque $\tau = (T_2 - T_1)R$ on the generator. The coefficient of static friction between each flywheel and the drive belt is μ_s. Assume that the tension in the drive belt is as low as possible with no slipping, and that the drive belt is massless. Show that under these conditions

$$T_1 = \frac{\tau}{R} \frac{1}{e^{\mu_s \pi} - 1}$$

$$T_2 = \frac{\tau}{R} \frac{1}{1 - e^{-\mu_s \pi}}$$

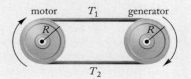

FIGURE 14.54 A drive belt connecting flywheels of a motor and a generator.

**48. A power brake invented by Lord Kelvin consists of a strong flexible belt wrapped once around a spinning flywheel (Fig. 14.55). One end of the belt is fixed to an overhead support; the other end carries a weight w. The coefficient of kinetic friction between the belt and the wheel is μ_k. The radius of the wheel is R, and its angular velocity is ω.

(a) Show that the tension in the belt is

$$T = we^{-\mu_k\theta}$$

as a function of the angle of contact (Fig. 14.55).

(b) Show that the net frictional torque the belt exerts on the flywheel is

$$\tau = wR\left(1 - e^{-2\pi\mu_k}\right)$$

(c) Show that the power dissipated by friction is

$$P = wR\omega\left(1 - e^{-2\pi\mu_k}\right)$$

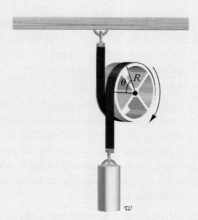

FIGURE 14.55 Belt and flywheel.

14.3 Levers and Pulleys

49. The human forearm (including the hand) can be regarded as a lever pivoted at the joint of the elbow and pulled upward by the tendon of the biceps (Fig. 14.56a). The dimensions of this lever are given in Fig. 14.56b. Suppose that a load of 25 kg rests in the hand. What upward force must the biceps exert to keep the forearm horizontal? What is the downward force at the elbow joint? Neglect the weight of the forearm.

50. Repeat the preceding problem if, instead of being vertical, the upper arm is tilted, so as to make an angle of 135° with the (horizontal) forearm.

51. A simple manual winch consists of a drum of radius 4.0 cm to which is attached a handle of radius 25 cm (Fig. 14.57). When you turn the handle, the rope winds up on the drum and pulls the load. Suppose that the load carried by the rope is 2500 N. What force must you exert on the handle to hold this load?

52. The handle of a crowbar is 60 cm long; the short end is 4.0 cm from a bend, which acts as the fulcrum. If a 75-kg man

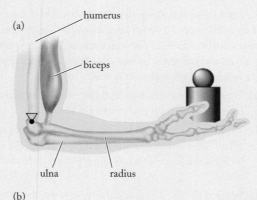

(a) humerus

biceps

ulna   radius

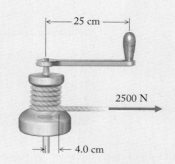

(b)

5.5 cm   30 cm

FIGURE 14.56 Forearm as lever.

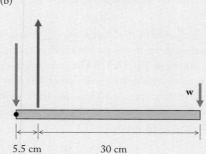

25 cm

2500 N

4.0 cm

FIGURE 14.57 Manual winch.

leans on the handle with all his weight, how much mass can he lift at the short end?

53. A 60-kg woman sits 80 cm from the fulcrum of a 4.0-m-long seesaw. The woman's daughter pulls down on the other end of the seesaw. What minimum force must the child apply to hold her mother's end of the seesaw off the ground?

54. The fingers apply a force of 30 N at the handle of a pair of scissors, 4.0 cm from the hinge point. What force is available for cutting when the object to be cut is placed at the far end of the scissors, 12 cm from the hinge point? When the object is placed as close to the hinge point as possible, at a distance of 1.0 cm?

55. A laboratory microbalance has two weighing pans, one hanging 10 times farther away from the fulcrum than the other.

When an unknown mass is placed on the inner pan, the microbalance can measure changes in mass as small as 100 nanograms (1.0×10^{-7} g) and can measure masses up to 2.0 milligrams. What would you expect that the resolution and maximum load for the outer pan might be?

56. A man of 73 kg stands on one foot, resting all of his weight on the ball of the foot. As described in Section 14.3, the bones of the foot play the role of a lever. The short end of the lever (to the heel) measures 5.0 cm and the long end (to the ball of the foot) 14 cm. Calculate the force exerted by the Achilles tendon and the force at the ankle.

57. A rope hoist consists of four pulleys assembled in two pairs with rigid straps, with a rope wrapped around as shown in Fig. 14.58. A load of 300 kg hangs from the lower pair of pulleys. What tension must you apply to the rope to hold the load steady? Treat the pulleys and the rope as massless, and ignore any friction in the pulleys.

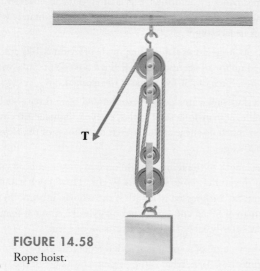

FIGURE 14.58
Rope hoist.

58. A parbuckle is a simple device used by laborers for raising or lowering a barrel or some other cylindrical object along a ramp. It consists of a loop of rope wrapped around the barrel (see Fig. 14.59). One end of the rope is tied to the top of the ramp, and the laborer pulls on the other end. Suppose that the laborer exerts a pull of 500 N on the rope, parallel to the ramp. What is the force that the rope exerts on the barrel? What is the mechanical advantage of the parbuckle?

FIGURE 14.59 Parbuckle used to move a barrel up a ramp.

59. Consider the differential windlass illustrated in Fig. 14.30. Calculate what clockwise torque must be applied to the handle to lift a load of mass m. What tangential force must be exerted on the handle? What is the mechanical advantage of this windlass?

60. Design a block and tackle with a mechanical advantage of 4, and another with a mechanical advantage of 5. If you connect these two arrangements in tandem, what mechanical advantage do you get?

*61. Figure 14.60 shows a compound bolt cutter. If the dimensions are as indicated in this figure, what is the mechanical advantage?

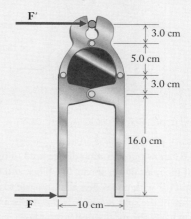

FIGURE 14.60 Compound bolt cutter.

*62. The drum of a winch is rigidly attached to a concentric large gear, which is driven by a small gear attached to a crank. The dimensions of the drum, the gears, and the crank are given in Fig. 14.61. What is the mechanical advantage of this geared winch?

FIGURE 14.61 Geared winch.

*63. The screw of a vise has a *pitch* of 4.0 mm; that is, it advances 4.0 mm when given one full turn. The handle of the vise is 25 cm long, measured from the screw to the end of the handle. What is the mechanical advantage when you push perpendicularly on the end of the handle?

*64. A scissors jack has the dimensions shown in Fig. 14.62. The screw of the jack has a pitch of 5.0 mm (as stated in the previous problem, this is the distance the screw advances when given one full turn). Suppose the scissors jack is partially extended, with an angle of 55° between its upper sides. What is the mechanical advantage provided by the jack?

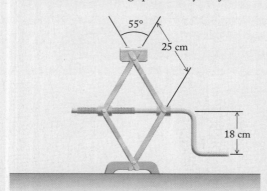

FIGURE 14.62 Scissors jack.

**65. Figure 14.63 shows a tensioning device used to tighten the rear stay of the mast of a sailboat. The block and tackle pulls down a rigid bar with two rollers that squeeze together the two branches of the split rear stay. If the angles are as given in the figure, what is the mechanical advantage?

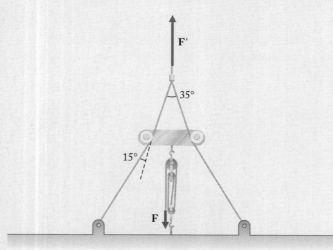

FIGURE 14.63 Tensioning device.

14.4 Elasticity of Materials

66. The anchor rope of a sailboat is a nylon rope of length 60 m and diameter 1.3 cm. While anchored during a storm, the sailboat momentarily pulls on this rope with a force of 1.8×10^4 N. How much does the rope stretch?

67. A piano wire of steel of length 1.8 m and radius 0.30 mm is subjected to a tension of 70 N by a weight attached to its

lower end. By how much does this wire stretch in excess of its initial length?

68. An elastic cord is 5.0 m long and 1.0 cm in diameter and acts as a spring with spring constant 70 N/m. What is the Young's modulus for this material?

69. The piano wire described in Problem 67 can be regarded as a spring. What is the effective spring constant of this spring?

70. A simple hand-operated hydraulic press can generate a pressure of 6.0×10^9 N/m^2. If the system is used to compress a small volume of steel, what fraction of the original volume does the final volume of steel occupy?

71. A 10-m length of 1.0-mm-radius copper wire is stretched by holding one end fixed and pulling on the other end with a force of 150 N. What is the change of length? By briefly increasing the force to exceed the limit of elastic behavior (a fractional elongation of approximately 1.0%), the wire may be permanently deformed; this is often done in order to straighten out bends or kinks in a wire. Approximately what force is necessary?

72. A 0.50-mm-radius fishing line made of nylon is 100 m long when no forces are applied. A fish is hooked and pulls with a tension force of 250 N. What is the elongation?

73. In a skyscraper, an elevator is suspended from three equal, parallel 300-m-long steel cables, each of diameter 1.0 cm. How much do these cables stretch if the mass of the elevator is 1000 kg?

74. The length of the femur (thighbone) of a woman is 38 cm, and the average cross section is 10 cm^2. How much will the femur be compressed in length if the woman lifts another woman of 68 kg and carries her piggyback? Assume that, momentarily, all of the weight rests on one leg.

75. If the volume of a sphere subjected to an external pressure shrinks by 0.10%, what is the percent shrinkage of the radius? In general, show that the percent shrinkage of the volume equals three times the percent shrinkage of the radius, provided the shrinkage is small.

76. At the bottom of the Marianas Trench in the Pacific Ocean, at a depth of 10 900 m, the pressure is 1.24×10^8 N/m^2. What is the percent increase of the density of water at this depth as compared with the density at the surface?

77. A slab of stone of mass 1200 kg is attached to the wall of a building by two bolts of iron of diameter 1.5 cm (see Fig. 14.64). The distance between the wall and the slab of stone is 1.0 cm. Calculate by how much the bolts will sag downward because of the shear stress they are subjected to.

78. According to (somewhat oversimplified) theoretical considerations, the Young's modulus, the shear modulus, and the bulk modulus are related by

$$Y = \frac{9BS}{3B + S}$$

Check this for the first four materials listed in Table 14.1.

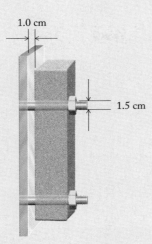

FIGURE 14.64 A slab of stone held by bolts.

79. A nylon rope of diameter 1.3 cm is to be spliced to a steel rope. If the steel rope is to have the same ultimate breaking strength as the nylon, what diameter should it have? The ultimate tensile strength is 2.0×10^9 N/m² for the steel and 3.2×10^8 N/m² for nylon.

*80. A rod of aluminum has a diameter of 1.000 002 cm. A ring of cast steel has an inner diameter of 1.000 000 cm. If the rod and the ring are placed in a liquid under high pressure, at what value of the pressure will the aluminum rod fit inside the steel ring?

*81. A heavy uniform beam of mass 8000 kg and length 2.0 m is suspended at one end by a nylon rope of diameter 2.5 cm and at the other end by a steel rope of diameter 0.64 cm. The ropes are tied together above the beam (see Fig. 14.65). The unstretched lengths of the ropes are 3.0 m each. What angle will the beam make with the horizontal?

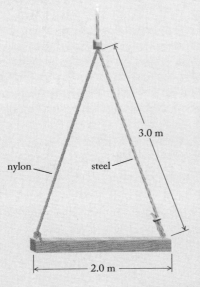

FIGURE 14.65 Beam hanging from two types of rope.

*82. A rod of cast iron is soldered to the upper edges of a plate of copper whose lower edge is held in a vise (see Fig. 14.66). The rod has a diameter of 4.0 cm and a length of 2.0 m. The copper plate measures 6.0 cm × 6.0 cm × 1.0 cm. If we pull the free end of the iron rod forward by 3.0 mm, what is the shear strain ($\Delta x/h$) of the copper plate?

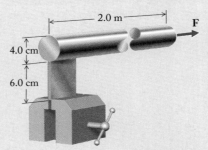

FIGURE 14.66 Iron rod and copper plate.

*83. When a bar of steel is heated, it expands in length by 0.0012% for each degree Celsius of temperature increase. If the length of the heated bar is to be reduced to its original value, a compressive stress must be applied to it. The compressive stress required to cancel the thermal expansion is called **thermal stress.** What is its value for a cylindrical bar of cast steel of cross section 4.0 cm² heated by 150°C?

*84. A power cable of copper is stretched straight between two fixed towers. If the temperature decreases, the cable tends to contract (compare Problem 83). The amount of contraction for a free copper cable or rod is 0.0017% per degree Celsius. Estimate what temperature decrease will cause the cable to snap. Pretend that the cable obeys Eq. (14.18) until it reaches its breaking point, which for copper occurs at a tensile stress of 2.4×10^8 N/m². Ignore the weight of the cable and the sag and stress produced by the weight.

**85. A meterstick of steel, of density 7.8×10^3 kg/m³, is made to rotate about a perpendicular axis passing through its middle. What is the maximum angular velocity with which the stick can rotate if its center is to hold? Mild steel will break when the tensile stress exceeds 3.8×10^8 N/m².

**86. The wall of a pipe of diameter 60 cm is constructed of a sheet of steel of thickness 0.30 cm. The pipe is filled with water under high pressure. What is the maximum pressure, that is, force per unit area, that the pipe can withstand? See Problem 85 for data on mild steel.

**87. A hoop of aluminum of radius 40 cm is made to spin about its axis of symmetry at high speed. The density of aluminum is 2.7×10^3 kg/m³, and the ultimate tensile breaking strength is 7.8×10^7 N/m². At what angular velocity will the hoop begin to break apart?

**88. A pipe of steel with a wall 0.40 cm thick and a diameter of 50 cm contains a liquid at a pressure of 2.0×10^4 N/m². How much will the diameter of the pipe expand due to this pressure?

REVIEW PROBLEMS

89. A traffic lamp of mass 25 kg hangs from a wire stretched between two posts. The traffic lamp hangs at the middle of the wire, and the two halves of the wire sag downward at an angle of 20° (see Fig. 14.67). What is the tension in the wire? Assume the wire is massless.

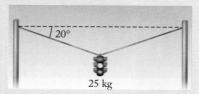

FIGURE 14.67 A traffic lamp.

90. A heavy shop sign hangs from a boom sticking out horizontally from a building (see Fig. 14.68). The boom is hinged at the building and is supported by a diagonal wire, making an angle of 45° with the boom. The mass of the sign is 50 kg, and the boom and the wire are massless. What is the tension in the wire? What is the force with which the end of the boom pushes against the building?

FIGURE 14.68 Sign hanging from a boom.

91. Figure 14.69 shows cargo hanging from the loading boom of a ship. If the boom is inclined at an angle of 30° and the cargo has a mass of 2500 kg, what is the tension in the upper cable? What is the compressional force in the boom? Neglect the mass of the boom.

92. Repeat the calculation of Problem 91, but assume that the mass of the boom is 800 kg, and that this mass is uniformly distributed along the length of the boom.

93. A tractor pulls a trailer along a street (see Fig. 14.70). The rear wheels, which are connected to the engine by means of the axle, have a radius of 0.60 m. Draw a "free-body" diagram for one of the rear wheels; be sure to include the forces and

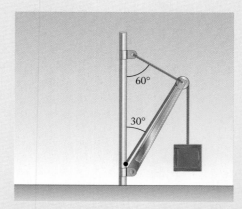

FIGURE 14.69 Cargo hanging from a boom.

FIGURE 14.70 Tractor pulling trailer.

the torque exerted by the axle on the wheel, but neglect the weight of the wheel. If the tractor is to provide a pull of 8000 N (a pull of 4000 N from each rear wheel), what torque must the axle exert on each rear wheel?

94. One end of a string is tied to a meterstick at the 80-cm mark, and the other end is tied to a hook in the ceiling. You push against the bottom edge of the meterstick at the 30-cm mark, so the stick is held horizontally (see Fig. 14.71). The mass of the meterstick is 0.24 kg. What is the magnitude of the force you must exert? What is the tension in the string?

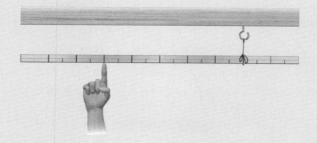

FIGURE 14.71 Meterstick tied to a hook.

95. A beam of steel hangs from a crane by means of cables attached to the upper corners of the beam making an angle of 60° with each other. The mass of the beam is M. Find the tensions in the cables and the compressional force in the beam.

96. Sheerlegs are sometimes used to suspend loads. They consist of two rigid beams leaning against each other, like the legs of the letter A (see Fig. 14.72). The load is suspended by a cable from the apex of the A. Suppose that a pair of sheerlegs, each at an angle of 30° with the vertical, are used to suspend an automobile engine of mass 400 kg. What is the compressional force in each leg? What are the horizontal and vertical forces that each leg exerts on the ground? Neglect the mass of the legs.

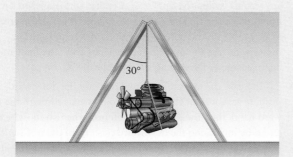

FIGURE 14.72 Sheerlegs supporting a load.

97. A 100-kg barrel is placed on a 30° ramp (see Fig. 14.73). What push, parallel to the ramp, must you exert against the middle of the barrel to keep it from rolling down? Assume that the friction between the barrel and the ramp prevents slipping of the barrel; that is, the barrel would roll without slipping if released.

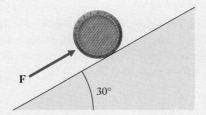

FIGURE 14.73 Barrel on a ramp.

98. Figure. 14.74 shows a pair of pliers and their dimensions. If you push against the handles of the pliers with a force of 200 N from each side, what is the force that the jaws of the pliers exert against each other?

99. To help his horses drag a heavy wagon up a hill, a teamster pushes forward at the top of one of the wheels (see Fig. 14.75). If he pushes with a force of 600 N, what forward force does he generate on the axle of the wagon? (Hint: The diameter of the wheel can be regarded as a lever pivoted at the ground.)

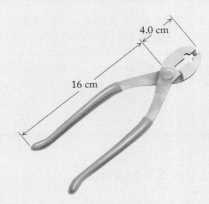

FIGURE 14.74 Pliers.

FIGURE 14.75 Teamster pushing on a wheel.

100. A flagpole points horizontally from a vertical wall. The pole is a uniform rod of mass M and length L. In addition to the pole mount at the wall end (which is hinged and exerts no torque), the pole is supported at its far end by a straight cable; the cable is attached to the wall a distance $L/2$ above the pole mount. What is the tension in the cable? What are the magnitude and direction of the force provided by the pole wall mount?

101. A wire stretches when subjected to a tension. This means that the wire can be regarded as a spring.

  (a) Express the effective spring constant in terms of the length of the wire, its radius, and its Young's modulus.

  (b) If a steel wire of length 2.0 m and radius 0.50 mm is to have the same spring constant as a steel wire of length 4.0 m, what must be the radius of the second wire?

102. If a steel rope and a nylon rope of equal lengths are to stretch by equal amounts when subjected to equal tensions, what must be the ratio of their diameters?

*103. A long rod of steel hangs straight down into a very deep mine shaft. For what length will the rod break off at the top because of its own weight? The density of mild steel is 7.8×10^3 kg/m³, and its tensile stress for breaking is 3.8×10^8 N/m².

104. A rope of length 12 m consists of an upper half of nylon of diameter 1.9 cm spliced to a lower half of steel of diameter 0.95 cm. How much will this rope stretch if a mass of 4000 kg is suspended from it? The Young's modulus for steel rope is 19×10^{10} N/m^2.

105. Suppose you drop an aluminum sphere of radius 10 cm into the ocean and it sinks to a depth of 5000 m, where the pressure is 5.7×10^7 N/m^2. Calculate by how much the diameter of this sphere will shrink.

Answers to Checkups

Checkup 14.1

1. If the bicyclist sits rigidly, the equilibrium is unstable: if tipped slightly, gravity will pull the bicycle and cyclist further over.

2. When you extend your legs while sitting on a swing, you are shifting your center of mass forward. To remain in equilibrium, the swing and your body will shift backward, and tilt, so as to keep your center of mass aligned below the point of support.

3. (a) Yes, the (vertical) support force is along the same line as the weight when holding the pole straight up (more precisely, it is slightly distributed around the edge of the pole). (b) No, the support force is provided by the more forward hand (which pushes up); an additional force from the rear hand pushes down, to balance the torques from the force of gravity and the support force.

4. (D) Neutral, stable, unstable. As our intuition might suggest, a cone on its side is in neutral equilibrium (after a small displacement, it remains on its side). A cone on its base is in stable equilibrium (after being tipped slightly, it will settle back on its base). Finally, a cone balanced on its apex is in unstable equilibrium (after being tipped slightly, the cone will fall over).

Checkup 14.2

1. When a ladder makes a large angle with the vertical, the weight of the ladder and the person climbing it exerts a large torque about the bottom, which can more easily overcome friction and make the ladder slip. When a ladder makes a small angle with the vertical, a person on the ladder can shift the center of mass to a point behind the bottom, causing the ladder to topple backward.

2. With the center of mass on the bottom, the box would have to be rotated 90° before toppling over. In that case, however, the box would then be on its side when it reaches the critical angle, where the center of mass is just above the support point.

3. For each side of the A, the force that one piece of lumber exerts on the other must exert a torque about the other's bottom that balances the torque due to the other's weight. Such a torque increases from zero when the pieces of lumber are vertical (when the tip of the A makes zero angle) to a maximum when the tip approaches 180°. Since the force exerted by one piece on the other acts with a smaller and smaller moment arm as the tip angle approaches 180°, the force must be very large as the tip angle approaches 180°.

4. (B) Increase the upward push and increase the downward push. If we consider the torques about an axis through the forward hand, then the downward pull from the fish must be balanced by increasing the downward push from the rear hand. The upward push of the forward hand must increase to balance those two increased downward forces.

Checkup 14.3

1. The arrangement shown in Fig. 14.23b has the larger ratio l/l', and thus has a greater mechanical advantage.

2. No. If, for example, the force F is not perpendicular to the lever, we must replace l by $l \sin \theta$, where θ is the angle between the force and the lever.

3. No. The pulley transmits tension to a different direction, independent of its size.

4. (C) 100 N. The weight of the rock is $w = mg = 100$ kg $\times$ 9.8 m/s^2 = 980 N $\approx$ 1000 N. The lever has a mechanical advantage of $l'/l \approx \frac{1}{10}$. So the force required to lift the rock is $F = (l'/l)F' = \frac{1}{10} \times 1000$ N = 100 N.

Checkup 14.4

1. The tension determines the *fractional* elongation [see Eq. (14.18)]; thus, for a piano wire of twice the length, the elongation will be twice as long, or 4.0 mm.

2. Yes, a cable can snap under its own weight (the downward weight below any point must be balanced by the upward tension at that point). Since the critical length for breaking is a condition of maximum tensile stress (a force per unit area), this depends on only the material and its mass density, not its area.

3. A material with a larger bulk modulus is stiffer, that is, its volume shrinks less in response to an applied pressure. We can rewrite Eq. (14.20) as $F/A = -B(\Delta V/V)$; thus, at constant pressure, a B that is larger by a given factor results in a fractional volume change that is smaller by the same factor. The volume of the copper sphere then shrinks by 0.005%.

4. (A) $\sqrt{2}$. Since the elongation is inversely proportional to the area of the elastic body [see Eq. (14.18)], if you want to decrease the elongation of a cable by a factor of 2, you must increase the cross-sectional area by a factor of 2; thus, you must increase the diameter by a factor of $\sqrt{2}$.

Oscillations, Waves, and Fluids

CONTENTS

CHAPTER 15   Oscillations

CHAPTER 16   Waves

CHAPTER 17   Sound

CHAPTER 18   Fluid Mechanics

At 1.1 times the speed of sound, this T-38 training jet generates shock waves, or sonic booms, in the surrounding air. The shock waves are made visible by Schlieren photography, a special technique that detects changes in the density of air.

Oscillations

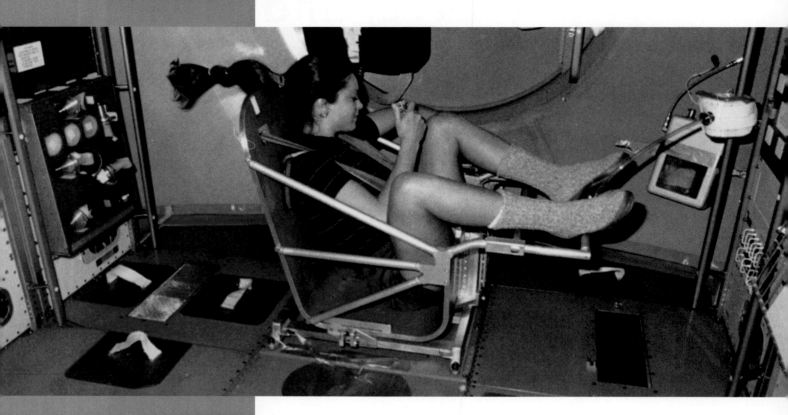

15.1 Simple Harmonic Motion

15.2 The Simple Harmonic Oscillator

15.3 Kinetic Energy and Potential Energy

15.4 The Simple Pendulum

15.5 Damped Oscillations and Forced Oscillations

CONCEPTS IN CONTEXT

Concepts
— *in* —
Context

The body-mass measurement device shown is used aboard the International Space Station for the daily measurement of the masses of the astronauts. The device consists of a spring coupled to a chair into which the astronaut is strapped. Pushed by the spring, the chair with the astronaut oscillates back and forth. We will see in this chapter that the frequency of oscillation of the mass–spring system depends on the mass, and therefore the frequency can serve as an indicator of the mass of the astronaut.

While learning about oscillating systems, we will consider such questions as:

❓ When the spring pushes and pulls the astronaut, what is the position of the astronaut as a function of time? The velocity of the astronaut? (Example 4, page 478)

? What is the total mechanical energy of the astronaut–spring system? What are the kinetic and potential energies as the spring begins to push? At later times? (Example 5, page 482)

? Good oscillators have low friction. How do we measure the quality of an oscillator? (Example 10, page 490)

The motion of a particle or of a system of particles is **periodic,** or **cyclic,** if it repeats again and again at regular intervals of time. The orbital motion of a planet around the Sun, the uniform rotational motion of a carousel or of a circular saw blade, the back-and-forth motion of a piston in an automobile engine or in a water pump, the swing-ing motion of a pendulum bob in a grandfather clock, and the vibration of a guitar string are examples of periodic motions. *If the periodic motion is a back-and-forth motion along a straight or curved line, it is called an* **oscillation.** Thus, the motion of the piston is an oscillation, and so are the motion of the pendulum and the motion of the indi-vidual particles of the guitar string.

In this chapter we will examine in some detail the motion of a mass oscillating back and forth under the push and pull exerted by an ideal, massless spring. The equa-tions that we will develop for the description of this mass–spring system are of great importance because analogous equations also occur in the description of all other oscil-lating systems. We will also examine some of these other oscillating systems, such as the pendulum.

15.1 SIMPLE HARMONIC MOTION

Online Concept Tutorial 16

Simple harmonic motion is a special kind of one-dimensional periodic motion. In any kind of one-dimensional periodic motion, the particle moves back and forth along a straight line, repeating the same motion again and again. In the special case of **simple harmonic motion,** *the particle's position can be expressed as a cosine or a sine func-tion of time.* As we will see later, the motion of a mass oscillating back and forth under the push and pull of a spring is simple harmonic (Fig. 15.1a), and so is the motion of a pendulum bob swinging back and forth (provided the amplitude of swing is small; see Fig. 15.1b), and so is the up-and-down motion of the blade of a saber saw (Fig. 15.1c). However, in this first section we will merely deal with the mathematical description of simple harmonic motion, and we will postpone until the next section the question of what causes the motion.

As a numerical example of simple harmonic motion, suppose that the tip of the blade in Fig. 15.1c moves up and down between $x = -0.8$ cm and $x = +0.8$ cm (where

(a)

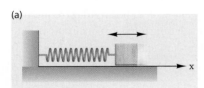

(b)

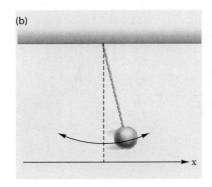

(c)

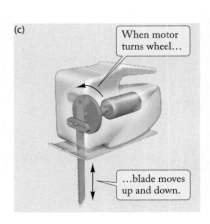

When motor turns wheel…

…blade moves up and down.

FIGURE 15.1 (a) The motion of a particle oscillating back and forth in response to the push and pull of a spring is simple harmonic. (b) The motion of a pendulum bob is approximately simple harmonic. (c) The motion of a saber saw blade is simple harmonic.

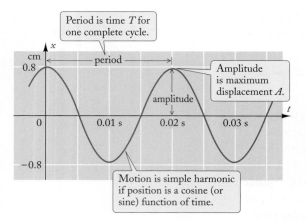

Period is time T for one complete cycle.

period

Amplitude is maximum displacement A.

amplitude

Motion is simple harmonic if position is a cosine (or sine) function of time.

FIGURE 15.2 Plot of position vs. time for a case of simple harmonic motion up and down along the x axis.

the x axis is assumed to be vertical); further suppose that the blade completes 50 up-and-down cycles each second. Figure 15.2 gives a plot of the position of the tip of the blade as a function of time. The plot in Fig. 15.2 has the mathematical form of a cosine function of the time t,

$$x = 0.8\cos(100\pi t) \tag{15.1}$$

where it is assumed that distance is measured in centimeters and time in seconds, and it is assumed that the "angle" $100\pi t$ in the cosine function is reckoned in radians. [The factor 100π multiplying t in Eq. (15.1) has been selected so as to obtain exactly 50 complete cycles each second, which is typical for saber saws; we will see below in Eq. (15.5) how the factor multiplying t in Eq. (15.1) is related to the period of the motion.]

Cosines and sines are called **harmonic functions**, which is why we call the motion *harmonic*. For the harmonic motion plotted in Fig. 15.2, at $t = 0$, the blade tip is at its maximum upward displacement [evaluating Eq. (15.1) at $t = 0$, we have $\cos 0 = 1$, so $x = 0.8$ cm] and is just starting to move; at $t = 0.005$ s, it passes through the midpoint [since $\cos(100\pi \times 0.005) = \cos(\pi/2) = 0$, Eq. (15.1) gives $x = 0$]; at $t = 0.010$ s, it reaches maximum downward displacement [$\cos(\pi) = -1$, so $x = -0.8$ cm]; at $t = 0.015$ s, it again passes through midpoint. Finally, at $t = 0.020$ s, the tip returns to its maximum upward displacement, exactly as at $t = 0$—it has completed one cycle of the motion and is ready to begin the next cycle. Thus, the **period** T, or the repeat time of the motion (the number of seconds for one complete cycle of the motion), is

$$T = 0.020 \text{ s} \tag{15.2}$$

and the **frequency** f of the motion, or the rate of repetition of the motion (the number of cycles per second), is

$$f = \frac{1}{T} = \frac{1}{0.020 \text{ s}} = 50/\text{s} \tag{15.3}$$

The points $x = 0.8$ cm and $x = -0.8$ cm, at which the x coordinate attains its maximum and minimum values, are the **turning points** of the motion; and the point $x = 0$ is the midpoint.

Equation (15.1) is a special example of simple harmonic motion. More generally, the motion of a particle is simple harmonic if the dependence of position on time has the form of a cosine or a sine function, such as

simple harmonic motion

$$x = A\cos(\omega t + \delta) \tag{15.4}$$

The quantities A, ω, and δ are constants. The quantity A is called the **amplitude** of the motion; it is simply the distance between the midpoint ($x = 0$) and either of the turning points ($x = +A$ or $x = -A$). The quantity ω is called the **angular frequency**; its value is related to the period T. To establish the relationship between ω and T, note that if we increase the time by T (from t to $t + T$), the argument of the cosine in Eq. (15.4) increases by ωT. For this to be one cycle of the cosine function, we must require $\omega T = 2\pi$. Thus, the repetition time of the motion, that is, the period T of the motion, is related to the angular frequency by

period and angular frequency

$$T = \frac{2\pi}{\omega} \qquad \text{or} \qquad \omega = \frac{2\pi}{T} \tag{15.5}$$

The repetition rate, or the frequency of the motion, is $1/T$, so we may write

$$f = \frac{1}{T} = \frac{\omega}{2\pi} \qquad \text{or} \qquad \omega = 2\pi f \qquad (15.6)$$

Note that the angular frequency ω and the frequency f differ by a factor of 2π, which corresponds to 2π radians $= 1$ cycle. The units of angular frequency are radians per second (radians/s). The units of frequency are cycles per second (cycles/s). Like the label *revolution* that we used in rev/s in rotational motion, the label *cycle* in cycle/s can be omitted in the course of a calculation, and so can the label *radian* in radian/s. But it is useful to retain these labels wherever there is a chance of confusion. The SI *unit of frequency is called the* **hertz** (Hz):

$$1 \text{ hertz} = 1 \text{ Hz} = 1 \text{ cycle/s} = 1/\text{s} \qquad (15.7)$$

For instance, in the example of the motion of the saber saw blade, the period of the motion is $T = 0.020$ s, the frequency is $f = 1/T = 1/(0.020 \text{ s}) = 50/\text{s} = 50$ Hz, and the angular frequency is

$$\omega = 2\pi f = 2\pi \times 50/\text{s} = 314 \text{ radians/s}$$

Here, in the last step of the calculation, the label *radians* has been inserted, so as to distinguish the angular frequency ω from the ordinary frequency f.

The argument $(\omega t + \delta)$ of the cosine function is called the **phase** of the oscillation, and the quantity δ is called the **phase constant**. This constant determines at what times the particle reaches the point of maximum displacement, when $\cos(\omega t + \delta) = 1$. One such instant is when

$$\omega t_{max} + \delta = 0$$

that is, when

$$t_{max} = -\delta/\omega \qquad (15.8)$$

Hence the particle reaches the point of maximum displacement at a time δ/ω before $t = 0$ (see Fig. 15.3). Of course, the particle also passes through this point at periodic intervals before and after this time. If the phase constant is zero ($\delta = 0$), then the maximum displacement occurs at $t = 0$.

Note that the preceding equations connecting *angular frequency,* period, and frequency are formally the same as the equations connecting *angular velocity,* period, and frequency of uniform rotational motion [see Eqs. (12.4) and (12.5)]. This coincidence

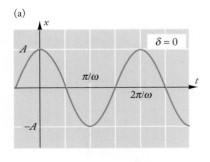

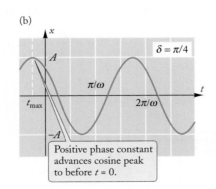

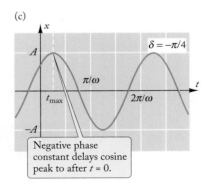

FIGURE 15.3 Examples of cosine functions $\cos(\omega t + \delta)$ for simple harmonic motion with different phase constants. (a) $\delta = 0$. The particle reaches maximum displacement at $t = 0$. (b) $\delta = \pi/4$ (or $45°$). The particle reaches maximum displacement before $t = 0$. (c) $\delta = -\pi/4$ (or $-45°$). The particle reaches maximum displacement after $t = 0$.

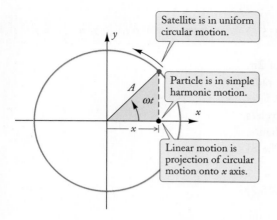

Satellite is in uniform circular motion.

Particle is in simple harmonic motion.

Linear motion is projection of circular motion onto x axis.

FIGURE 15.4 Particle oscillating along the x axis and satellite particle moving around reference circle. The particle and the satellite are always aligned vertically; that is, they have the same x coordinate.

arises from a special geometrical relationship between simple harmonic motion and uniform circular motion. Suppose that a particle moves with simple harmonic motion according to Eq. (15.4), with amplitude A and angular velocity ω; and consider a "satellite" particle that is constrained to move in uniform circular motion with angular velocity ω along a circle of radius A, centered on the midpoint of the harmonic motion, that is, centered on $x = 0$. Figure 15.4 shows this circle, called the **reference circle.** At time $t = 0$, both the particle and its satellite are on the x axis at $x = A$. After this time, the particle moves along the x axis, so its position is

$$x = A \cos(\omega t) \tag{15.9}$$

Meanwhile, the satellite moves around the circle, and its angular position is

$$\theta = \omega t$$

Now note that the x coordinate of the satellite is the adjacent side of the triangle shown in Fig. 15.4:

$$x_{\text{sat}} = A \cos \theta = A \cos(\omega t) \tag{15.10}$$

Comparing this with Eq. (15.9), we see that the x coordinate of the satellite always coincides with the x coordinate of the particle; that is, the particle and the satellite always have exactly the same x motion. This means that in Fig. 15.4 the satellite is always on that point of the reference circle directly above or directly below the particle.

This geometrical relationship between simple harmonic motion and uniform circular motion can be used to generate simple harmonic motion from uniform circular motion. Figure 15.5 shows a simple mechanism for accomplishing this by means of a slotted arm placed over a peg that is attached to a wheel in uniform circular motion. The slot is vertical, and the arm is constrained to move horizontally. The peg plays the role of "satellite," and the midpoint of the slot in the arm plays the role of "particle." The peg drags the arm left and right and makes it move with simple harmonic motion. A mechanism of this kind is used in electric saber saws and other devices to convert the rotational motion of an electric motor into the up-and-down motion of the saw blade or other moving component.

Finally, let us calculate the instantaneous velocity and instantaneous acceleration in simple harmonic motion. If the displacement is

$$x = A \cos(\omega t + \delta) \tag{15.11}$$

then differentiation of this displacement gives the velocity

$$v = \frac{dx}{dt} = -\omega A \sin(\omega t + \delta) \tag{15.12}$$

FIGURE 15.5 Rotating wheel with a peg driving a slotted arm back and forth.

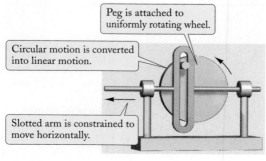

Peg is attached to uniformly rotating wheel.

Circular motion is converted into linear motion.

Slotted arm is constrained to move horizontally.

(a)   (b)

MATH HELP   DERIVATIVES OF TRIGONOMETRIC FUNCTIONS

Under the assumption that the argument of each trigonometric function is expressed in radians, the derivatives of the sine, cosine, and tangent are

$$\frac{d}{du}\sin bu = b\cos bu \qquad \frac{d}{du}\cos bu = -b\sin bu \qquad \frac{d}{du}\tan bu = b\sec^2 bu = \frac{b}{\cos^2 bu}$$

and differentiation of this velocity gives the acceleration

$$a = \frac{d^2x}{dt^2} = \frac{dv}{dt} = -\omega^2 A\cos(\omega t + \delta) \qquad (15.13)$$

Here we have used the standard formulas for the derivatives of the sine function and the cosine function (see Math Help: Derivatives of Trigonometric Functions). Bear in mind that the arguments of the sine and cosine functions in this chapter (and also the next) are always expressed in radians, as required for the validity of the standard formulas for derivatives.

As expected, the instantaneous velocity calculated from Eq. (15.12) is zero for $\omega t + \delta = 0$, when the particle is at the turning point. Furthermore, the instantaneous velocity attains a maximum magnitude of

$$v_{\text{max}} = \omega A \qquad (15.14)$$

maximum velocity

for $\omega t + \delta = \pi/2$, when the particle passes through the midpoint (note that the maximum magnitude of $\sin \omega t$ is 1).

Figure 15.6 shows a multiple-exposure photograph of the oscillations of a particle in simple harmonic motion. The picture illustrates the variations of speed in simple harmonic motion: the particle moves at low speed (smaller displacements between snapshots) near the turning points, and at high speed (larger displacements) near the midpoint.

The velocity (15.12) is a sine function, whereas the displacement (15.11) is a cosine function. When the cosine is at its maximum (say, $\cos 0 = 1$), the sine is small ($\sin 0 = 0$); when the cosine is small (say, $\cos \pi/2 = 0$), the sine is at its maximum

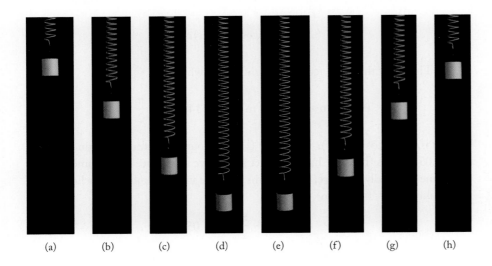

(a)   (b)   (c)   (d)   (e)   (f)   (g)   (h)

FIGURE 15.6 Sequence of snapshots at uniform time intervals of an oscillating mass on a spring (a-h). Note that the mass moves slowly at the extremes of its motion.

$(\sin \pi/2 = 1)$. Hence the displacement and the velocity are out of step—when one has a large magnitude, the other has a small magnitude, and vice versa. Figures 15.7a and b compare the velocity and the displacement for simple harmonic motion at different times. Graphically, the velocity is the slope of the position vs. time curve. When the position goes through a maximum or minimum, the slope is zero; when the position goes through zero, the magnitude of the slope is a maximum.

Comparison of Eqs. (15.11) and (15.13) shows that

$$\frac{d^2x}{dt^2} = -\omega^2 x \qquad (15.15)$$

Thus, the acceleration is always proportional to the displacement x, but is in the opposite direction; see Fig. 15.7c. This proportionality is a characteristic feature of simple harmonic motion, a fact that will be useful in the next section. Even when a phenomenon does not involve motion along a line (for example, rotational motion or the behavior of electric circuits), harmonic behavior occurs whenever the second derivative of a quantity is proportional to the negative of that quantity, as in Eq. (15.15). The sine and cosine functions (or a combination of them) are the *only* functions that have this property.

acceleration in simple harmonic motion

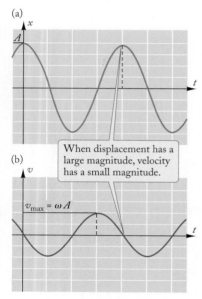

(a) x

When displacement has a large magnitude, velocity has a small magnitude.

(b) v

$v_{max} = \omega A$

(c) a

$a_{max} = \omega^2 A$

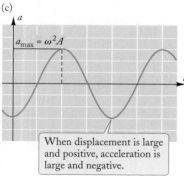

When displacement is large and positive, acceleration is large and negative.

FIGURE 15.7 (a) Position, (b) velocity, and (c) acceleration of a particle in simple harmonic motion as functions of time.

EXAMPLE 1   Consider the blade of a saber saw moving up and down in simple harmonic motion with a frequency of 50.0 Hz, or an angular frequency of 314 radians/s. Suppose that the amplitude of the motion is 1.20 cm and that at time $t = 0$, the tip of the blade is at $x = 0$ and its velocity is positive. What is the equation describing the position of the tip of the blade as a function of time? How long does the blade take to travel from $x = 0$ to $x = 0.60$ cm? To 1.20 cm?

SOLUTION: The position as function of time is given by Eq. (15.4):

$$x = A \cos(\omega t + \delta)$$

with $\omega = 314$ radians/s and $A = 0.0120$ m. Since $x = 0$ at $t = 0$, we must adopt a value of δ such that $\cos \delta = 0$. The smallest values of δ that satisfy this condition are $\delta = \pi/2$ and $\delta = -\pi/2$ (other possible values of δ differ from these by $\pm 2\pi$, $\pm 4\pi$, etc.). From Eq. (15.12), we see that to obtain a positive value of v at $t = 0$, we need a *negative* value of δ; that is, $\delta = -\pi/2$. So the equation describing the motion is

$$x = (0.0120 \text{ m}) \cos\left[(314/\text{s})t - \frac{\pi}{2}\right]$$

The tip of the blade reaches $x = 0.0060$ m when

$$0.0060 \text{ m} = (0.0120 \text{ m}) \cos\left[(314/\text{s}) t - \frac{\pi}{2}\right]$$

that is, when $\cos[(314/\text{s})t - \pi/2] = (\frac{1}{2})$. With our calculator we obtain $\cos^{-1}\frac{1}{2} = -1.05$ radians (here, we have to select a negative sign, since the argument of the cosine is initially negative, and remains negative until the motion reaches the full amplitude, $x = 0.0120$ m). So

$$(314/\text{s})t - \frac{\pi}{2} = -1.05$$

from which

$$t = \frac{-1.05 + (\pi/2)}{314/s} = 0.0017 \text{ s}$$

To find when the tip of the blade reaches $x = 0.0120$ m, we can use Eq. (15.8), which gives

$$t = -\frac{\delta}{\omega} = -\frac{(-\pi/2)}{314/s} = 0.0050 \text{ s}$$

COMMENT: Note that the time taken to reach a distance of one-half of the amplitude is not one-half of the time taken to reach the full amplitude, because the motion does not proceed at constant speed.

EXAMPLE 2   In an atomic-force microscope (AFM), a cantilever beam with a sharp tip (Fig. 15.8a) oscillates near a surface. We can map the topography of a surface (see Fig. 15.8b) by slowly moving the tip laterally as it oscillates vertically, much like a blind person tapping a cane on the ground. The AFM tip shown in Fig. 15.8a oscillates with a period of 3.0×10^{-6} s. The tip moves up and down with amplitude 9.0×10^{-8} m. What is the maximum vertical acceleration of the tip? Its maximum vertical velocity?

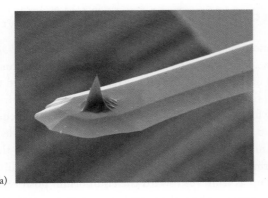

(a)

(b)

SOLUTION: As discussed above, the largest acceleration occurs at the point of maximum displacement. From Eq. (15.13) this maximum acceleration is [since the maximum value of $\cos(\omega t + \delta)$ is 1]

$$a_{max} = \omega^2 A \tag{15.16}$$

From Eq. (15.5) and the period $T = 3.0 \times 10^{-6}$ s, we obtain the angular frequency

$$\omega = \frac{2\pi}{T} = \frac{2\pi}{3.0 \times 10^{-6} \text{ s}} = 2.1 \times 10^6 \text{ radians/s}$$

Thus, with $A = 9.0 \times 10^{-8}$ m, the maximum acceleration is

$$a_{max} = \omega^2 A = (2.1 \times 10^6 \text{ radians/s})^2 \times 9.0 \times 10^{-8} \text{ m} = 4.0 \times 10^5 \text{ m/s}^2$$

This is more than 40 000 standard g's, an enormous acceleration.

The maximum velocity is, from Eq. (15.12),

$$v_{max} = \omega A = 2.1 \times 10^6 \text{ radians/s} \times 9.0 \times 10^{-8} \text{ m} = 0.19 \text{ m/s}$$

FIGURE 15.8 (a) Atomic-force microscope (AFM) cantilever and tip. (b) AFM image of the surface of a crystal, obtained by scanning the vibrating tip across the surface. The area shown is 2 μm $\times$ 2 μm. The ragged terraces are single atomic "steps."

✔ Checkup 15.1

QUESTION 1: Is the rotational motion of the Earth about its axis periodic motion? Oscillatory motion?

QUESTION 2: For a particle with simple harmonic motion, at what point of the motion does the velocity attain maximum magnitude? Minimum magnitude?

QUESTION 3: For a particle with simple harmonic motion, at what point of the motion does the acceleration attain maximum magnitude? Minimum magnitude?

QUESTION 4: Two particles execute simple harmonic motion with the same amplitude. One particle has twice the frequency of the other. Compare their maximum velocities and accelerations.

QUESTION 5: Are the x coordinates of the particle and the satellite particle in Fig. 15.4 always the same? The y coordinates? The velocities? The x components of the velocities? The accelerations? The x components of the accelerations?

QUESTION 6: Suppose that a particle with simple harmonic motion passes through the equilibrium point ($x = 0$) at $t = 0$. In this case, which of the following is a possible value of the phase constant δ in $x = A\cos(\omega t + \delta)$?

(A) 0   (B) $\pi/4$ (C) $\pi/2$ (D) $3\pi/4$ (E) π

15.2 THE SIMPLE HARMONIC OSCILLATOR

*The **simple harmonic oscillator** consists of a particle coupled to an ideal, massless spring that obeys Hooke's Law,* that is, a spring that provides a force proportional to the elongation or compression of the spring. One end of the spring is attached to the particle, and the other is held fixed (see Fig. 15.9). We will ignore gravity and friction, so the spring force is the only force acting on the particle. The system has an **equilibrium position** corresponding to the relaxed length of the spring. If the particle is initially at some distance from this equilibrium position (see Fig. 15.10), then the stretched spring supplies a restoring force that pulls the particle toward the equilibrium position. The particle speeds up as it moves toward the equilibrium position, and it overshoots the equilibrium position. Then, the particle begins to compress the spring and slows down, coming to rest at the other side of the equilibrium position, at a distance equal to its initial distance. The compressed spring then pushes the particle back toward the equilibrium position. The particle again speeds up, overshoots the equilibrium position, and so on. The result is that the particle oscillates back and forth about the equilibrium position—forever if there is no friction.

The great importance of the simple harmonic oscillator is that many physical systems are mathematically equivalent to simple harmonic oscillators; that is, these systems have an equation of motion of the same mathematical form as the simple harmonic oscillator. A pendulum, the balance wheel of a watch, a tuning fork, the air in an organ pipe, and the atoms in a diatomic molecule are systems of this kind; the restoring force and the inertia are of the same mathematical form in these systems as in the simple harmonic oscillator, and we can transcribe the general mathematical results directly from the latter to the former.

To obtain the equation of motion of the simple harmonic oscillator, we begin with Hooke's Law for the restoring force exerted by the spring on the particle [compare Eq. (6.11)]:

$$F = -kx \tag{15.17}$$

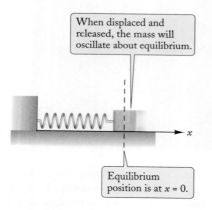

When displaced and released, the mass will oscillate about equilibrium.

Equilibrium position is at $x = 0$.

FIGURE 15.9 A mass attached to a spring slides back and forth on a frictionless surface. We regard the mass as a particle, whose position coincides with the center of the mass.

Here the displacement x is measured from the equilibrium position, which corresponds to $x = 0$. The constant k is the spring constant. Note that the force is negative if x is positive (stretched spring; see Fig. 15.10a); and the force is positive if the displacement is negative (compressed spring; see Fig. 15.10b).

With the force as given by Eq. (15.16), the equation of motion of the particle is

$$m\frac{d^2x}{dt^2} = -kx \qquad (15.18)$$

equation of motion for simple harmonic oscillator

This equation says that the acceleration of the particle is always proportional to the distance x, but is in the opposite direction. We now recall, from Eq. (15.15), that such a proportionality of acceleration and distance is characteristic of simple harmonic motion, and we therefore can immediately conclude that the motion of a particle coupled to a spring must be simple harmonic motion. By comparing Eqs. (15.18) and (15.15), we see that these equations become identical if

$$\omega^2 = \frac{k}{m}$$

and we therefore see that the angular frequency ω of the oscillation of the particle on a spring is

$$\omega = \sqrt{\frac{k}{m}} \qquad (15.19)$$

Consequently, the frequency and the period are

$$f = \frac{\omega}{2\pi} = \frac{1}{2\pi}\sqrt{\frac{k}{m}} \qquad (15.20)$$

angular frequency, frequency, and period for simple harmonic oscillator

and

$$T = \frac{1}{f} = 2\pi\sqrt{\frac{m}{k}} \qquad (15.21)$$

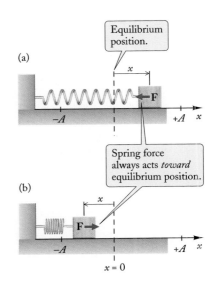

FIGURE 15.10 (a) Positive displacement of the particle; the force is negative. (b) Negative displacement of the particle; the force is positive.

With the value (15.19) for the angular frequency, the expression (15.4) for the position as a function of time becomes

$$x = A\cos\left(\sqrt{\frac{k}{m}}\,t + \delta\right) \qquad (15.22)$$

According to Eq. (15.20) the frequency of the motion of the simple harmonic oscillator depends *only* on the spring constant and on the mass. *The frequency of the oscillator is unaffected by the amplitude with which it has been set in motion*—if the oscillator has a frequency of, say, 2 Hz when oscillating with a small amplitude, then it also has a frequency of 2 Hz when oscillating with a large amplitude. This property of the oscillator is called **isochronism**.

Note that the period is long if the mass is large and the spring constant is small. This is as expected, since in each period the spring must accelerate and decelerate the mass, and a weak spring will give a large mass only little acceleration.

Spring scale oscillates about its shifted equilibrium.

FIGURE 15.11 A heavy book on a spring scale oscillates up and down.

EXAMPLE 3   When you place a heavy encyclopedia, of mass 8 kg, on a kitchen scale (a spring scale; see Fig. 15.11), you notice that before coming to equilibrium, the pointer of the scale oscillates back and forth around the equilibrium position a few times with a period of 0.4 s. What is the effective spring constant of the internal spring of the kitchen scale? (Neglect other masses in the scale.)

SOLUTION: The mass of 8 kg in conjunction with the internal spring of the scale forms a mass-and-spring system, to which we can apply Eq. (15.21). If we square both sides of this equation, we obtain

$$T^2 = 4\pi^2 \frac{m}{k}$$

which gives us

$$k = 4\pi^2 \frac{m}{T^2} \qquad (15.23)$$

With $m = 8$ kg and $T = 0.4$ s, this becomes

$$k = 4\pi^2 \times \frac{8 \text{ kg}}{(0.4 \text{ s})^2} = 2 \times 10^3 \text{ N/m}$$

COMMENT: In this example, there is not only the force of the spring acting on the mass, but also the force of gravity on the mass (the weight) and friction forces. The force of gravity determines where the spring will reach equilibrium, but this force has no direct effect on the frequency of oscillation around equilibrium. The friction forces cause the oscillations to stop after a few cycles, but only slightly reduce the frequency (see Section 15.5). For negligible friction, the frequency depends exclusively on the mass and the spring constant.

Concepts — in — Context

EXAMPLE 4   Suppose that the astronaut in the chapter photo has a mass of 58 kg, including the chair device to which she is attached. She and the chair move under the influence of the force of a spring with $k = 2.1 \times 10^3$ N/m. There are no other forces acting. Consider the motion to be along the x axis, with the equilibrium point at $x = 0$. Suppose that at $t = 0$, she is (instantaneously) at rest at $x = 0.20$ m. Where will she be at $t = 0.10$ s? At $t = 0.20$ s? What will her velocity be when she passes through the equilibrium point?

SOLUTION: Since the astronaut is initially at rest at $x = 0.20$ m, this must be one of the turning points of the motion; thus, the amplitude of the motion must be $A = 0.20$ m. Furthermore, since at $t = 0$ the astronaut is at the turning point, the phase constant $\delta = 0$ [see Eq. (15.8)]. Consequently, at time $t = 0.10$ s, the position of the astronaut will be

$$x = A \cos \omega t = 0.20 \text{ m} \times \cos(\omega \times 0.10 \text{ s})$$

To evaluate this, we need the angular frequency of the oscillation. By Eq. (15.19) this is

$$\omega = \sqrt{\frac{k}{m}} = \sqrt{\frac{2.1 \times 10^3 \text{ N/m}}{58 \text{ kg}}} = 6.0 \text{ radians/s}$$

With this value of ω,

$$x = 0.20 \text{ m} \times \cos(6.0 \text{ radians/s} \times 0.10 \text{ s})$$
$$= 0.20 \text{ m} \times \cos(0.60 \text{ radian}) = 0.20 \text{ m} \times 0.83 = 0.17 \text{ m}$$

Likewise, at time $t = 0.20$ s, the position will be

$$x = A \cos(\omega t) = 0.20 \text{ m} \times \cos(6.0 \text{ radians/s} \times 0.20 \text{ s})$$
$$= 0.20 \text{ m} \times \cos(1.2 \text{ radian}) = 0.20 \text{ m} \times 0.36 = 0.072 \text{ m}$$

The astronaut passes through the equilibrium point when $\omega t = \pi/2$ (which makes $\cos \omega t = 0$). To find her velocity when she passes through the equilibrium point, we take the derivative of x with respect to t, and then evaluate the resulting expression at $\omega t = \pi/2$:

$$v = \frac{dx}{dt} = \frac{d}{dt}(A \cos \omega t) = -\omega A \sin \omega t \qquad (15.24)$$
$$= -6.0 \text{ radians/s} \times 0.20 \text{ m} \times \sin(\pi/2) = -1.2 \text{ m/s}$$

Simple harmonic oscillators are used as the timekeeping element in modern watches. These watches use a quartz crystal as a spring-and-mass system. The crystal is elastic, with a high Young's modulus, and it therefore acts as a very stiff spring. The mass is not attached as a lump to the end of this spring, but it is uniformly distributed over the volume of the crystal (hence this spring–mass system is said to be "distributed," in contrast to a "lumped" system with separate springs and masses). The crystal is set into vibration by electric impulses, instead of mechanical pushes. The electric circuits attached to the crystal not only keep it vibrating, but also sense the frequency of vibration and control the display on the face of the clock.

The advantage of the quartz crystal as a timekeeping element is that the vibrations of the crystal are extremely stable, because any accelerations from bumping the watch are completely negligible compared with the immense accelerations of the oscillating masses in the crystal. Ordinary quartz clocks are accurate to within a few seconds per month; high-precision clocks are accurate to within 10^{-5} s per month.

 Checkup 15.2

QUESTION 1: For a particle with simple harmonic motion, at what point of the motion does the force on the particle attain maximum magnitude? Minimum magnitude?

QUESTION 2: Suppose we replace the particle in a simple harmonic oscillator by a particle of twice the mass. How does this alter the frequency of oscillation?

QUESTION 3: If we suddenly cut the spring of a simple harmonic oscillator when the particle is at the equilibrium point ($x = 0$), what is the subsequent motion of the particle? If we suddenly cut the spring when the particle is at maximum displacement ($x = A$)?

QUESTION 4: Suppose we replace the spring in a simple harmonic oscillator by a stronger spring, with twice the spring constant. What is the ratio of the new period of oscillation to the original period?

(A) 1/2   (B) 1/$\sqrt{2}$ (C) 1   (D) $\sqrt{2}$ (E) 2

15.3 KINETIC ENERGY AND POTENTIAL ENERGY

We know from Section 8.1 that the force exerted by a spring is a conservative force, for which we can construct a potential energy. With this potential energy, we can formulate a law of conservation of the mechanical energy: the sum of the kinetic energy and the potential energy is a constant; that is,

$$E = K + U = [\text{constant}] \tag{15.25}$$

In this section we will see how to calculate the kinetic energy and the potential energy of the simple harmonic oscillator at each instant of time, and we will verify explicitly that the sum of these energies is constant.

The kinetic energy of a moving particle is

$$K = \tfrac{1}{2}mv^2 \tag{15.26}$$

For simple harmonic motion, the speed is given by Eq. (15.12), and the kinetic energy becomes

$$\begin{aligned} K = \tfrac{1}{2}mv^2 &= \tfrac{1}{2}m[-\omega A \sin(\omega t + \delta)]^2 \\ &= \tfrac{1}{2}m\omega^2 A^2 \sin^2(\omega t + \delta) \end{aligned} \tag{15.27}$$

Since $m\omega^2 = k$ [see Eq. (15.18)], we can also write this as

$$K = \tfrac{1}{2}kA^2\sin^2(\omega t + \delta) \tag{15.28}$$

The potential energy associated with the force $F = -kx$ is [see Eq. (8.6)]

$$U = \tfrac{1}{2}kx^2 \tag{15.29}$$

For simple harmonic motion, with $x = A\cos(\omega t + \delta)$, this becomes

$$U = \tfrac{1}{2}kA^2\cos^2(\omega t + \delta) \tag{15.30}$$

The kinetic energy and the potential energy both depend on time. According to Eqs. (15.28) and (15.30), each oscillates between a minimum value of zero and a maximum value of $\tfrac{1}{2}kA^2$. Figure 15.12 plots the oscillations of the kinetic energy and the potential energy as functions of time; for simplicity, we set the phase constant at $\delta = 0$. At the initial time $t = 0$, the particle is at maximum distance from the equilibrium point and its instantaneous speed is zero; thus, the potential energy is at its maximum value, and the kinetic energy is zero. A quarter of a cycle later, the particle passes through the equilibrium point and attains its maximum speed; thus, the kinetic energy is at its maximum value and the potential energy is zero. Thus energy is traded back and forth between potential energy and kinetic energy.

Since the force $F = -kx$ is conservative, the total mechanical energy $E = K + U$ is a constant of the motion. To verify this conservation law for the energy explicitly, we take the sum of Eqs. (15.28) and (15.30),

$$\begin{aligned} E &= K + U \\ &= \tfrac{1}{2}kA^2\sin^2(\omega t + \delta) + \tfrac{1}{2}kA^2\cos^2(\omega t + \delta) \\ &= \tfrac{1}{2}kA^2[\sin^2(\omega t + \delta) + \cos^2(\omega t + \delta)] \end{aligned} \tag{15.31}$$

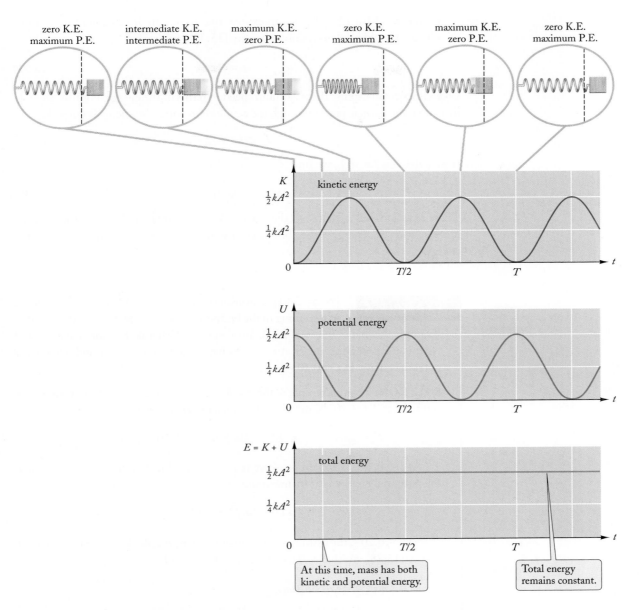

FIGURE 15.12 Kinetic energy and potential energy of a simple harmonic oscillator as a function of time.

We can simplify this expression if we use the trigonometric identity $\sin^2\theta + \cos^2\theta = 1$, which is valid for any angle θ. With this identity, we find that the right side of Eq. (15.31) is simply $\frac{1}{2}kA^2$:

$$E = \tfrac{1}{2}kA^2 \qquad\qquad (15.32)$$

energy of simple harmonic oscillator

This shows that *the energy of the motion is constant and is proportional to the square of the amplitude of oscillation.*

By means of Eq. (15.32), we can express the maximum displacement in terms of the energy. For this, we need only solve Eq. (15.32) for A:

$$x_{max} = A = \sqrt{2E/k} \tag{15.33}$$

Likewise, we can express the maximum speed in terms of the energy. For this, we note that when the particle passes through the equilibrium point, the energy is purely kinetic:

$$E = \tfrac{1}{2}mv_{max}^2 \tag{15.34}$$

If we solve this for v_{max}, we find

$$v_{max} = \sqrt{2E/m} \tag{15.35}$$

These equations tell us that both the maximum displacement and the maximum speed increase with the energy—they both increase in proportion to the square root of the energy.

Concepts —in— Context

EXAMPLE 5   For the 58-kg astronaut (with chair) moving under the influence of the spring in the body-mass measurement device described in Example 4, what is the total mechanical energy? What is the kinetic energy and what is the potential energy at $t = 0$? What is the kinetic energy and what is the potential energy at $t = 0.20$ s?

SOLUTION: From Example 4, the amplitude is $A = 0.20$ m and the spring constant is $k = 2.1 \times 10^3$ N/m. The total mechanical energy is

$$E = \tfrac{1}{2}kA^2 = \tfrac{1}{2} \times 2.1 \times 10^3\,\text{N/m} \times (0.20\,\text{m})^2 = 42\,\text{J}$$

At $t = 0$, the astronaut is at rest at $x = 0.20$ m. The kinetic energy is zero and the potential energy is at its maximum,

$$U = \tfrac{1}{2}kA^2 = 42\,\text{J}$$

At $t = 0.20$ s, the astronaut has nonzero speed, and the kinetic energy is given by Eq. (15.28). With $\delta = 0$ (see Example 4), we find

$$K = \tfrac{1}{2}mv^2 = \tfrac{1}{2}kA^2\sin^2(\omega t)$$
$$= \tfrac{1}{2} \times 2.1 \times 10^3\,\text{N/m} \times (0.20\,\text{m})^2$$
$$\times \sin^2(6.0\,\text{radians/s} \times 0.20\,\text{s})$$
$$= 36\,\text{J} \tag{15.36}$$

The potential energy is given by Eq. (15.30), again with $\delta = 0$:

$$U = \tfrac{1}{2}kA^2\cos^2(\omega t)$$
$$= \tfrac{1}{2} \times 2.1 \times 10^3\,\text{N/m} \times (0.20\,\text{m})^2$$
$$\times \cos^2(6.0\,\text{radians/s} \times 0.20\,\text{s})$$
$$= 6\,\text{J} \tag{15.37}$$

COMMENT: Note that the sum of the kinetic and potential energies is $K + U = 36\,\text{J} + 6\,\text{J} = 42\,\text{J}$, which agrees with our result for the total mechanical energy.

EXAMPLE 6 The hydrogen molecule (H_2) may be regarded as two particles joined by a spring (see Fig. 15.13). The center of the spring is the center of mass of the molecule. This point can be assumed to remain fixed, so this molecule consists of two identical simple harmonic oscillators vibrating in opposite directions. The spring constant for each of these oscillators is 1.13×10^3 N/m, and the mass of each hydrogen atom is 1.67×10^{-27} kg. Find the frequency of vibration in hertz. Suppose that the total vibrational energy of the molecule is 1.3×10^{-19} J. Find the corresponding amplitude of oscillation and the maximum speed.

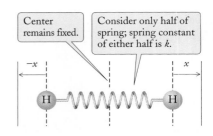

FIGURE 15.13 A hydrogen molecule, represented as two particles joined by a spring. The particles move symmetrically relative to the center of mass.

SOLUTION: The frequency is given by Eq. (15.20):

$$f = \frac{1}{2\pi}\sqrt{\frac{k}{m}} = \frac{1}{2\pi}\left(\frac{1.13 \times 10^3\,\text{N/m}}{1.67 \times 10^{-27}\,\text{kg}}\right)^{1/2} = 1.31 \times 10^{14}\,\text{Hz}$$

Thus molecular vibrational frequencies can be quite high, about a hundred thousand billion cycles per second.

Each atom has half the total energy of the molecule; thus, the energy per atom is

$$E = \tfrac{1}{2} \times 1.3 \times 10^{-19}\,\text{J} = 6.5 \times 10^{-20}\,\text{J}$$

According to Eqs. (15.33) and (15.35), the amplitude of oscillation and the maximum speed of each atom are then

$$x_\text{max} = \sqrt{\frac{2E}{k}} = \sqrt{\frac{2 \times 6.5 \times 10^{-20}\,\text{J}}{1.13 \times 10^3\,\text{N/m}}} = 1.1 \times 10^{-11}\,\text{m}$$

and

$$v_\text{max} = \sqrt{\frac{2E}{m}} = \sqrt{\frac{2 \times 6.5 \times 10^{-20}\,\text{J}}{1.67 \times 10^{-27}\,\text{kg}}} = 8.8 \times 10^3\,\text{m/s}$$

 Checkup 15.3

QUESTION 1: Two harmonic oscillators have equal masses and spring constants. One of them oscillates with twice the amplitude of the other. Compare the energies and compare the maximum speeds attained by the particles.

QUESTION 2: Two harmonic oscillators have equal spring constants and amplitudes of oscillation. One has twice the mass of the other. Compare the energies and the maximum speeds attained by the particles.

QUESTION 3: The period of a simple harmonic oscillator is 8.0 s. Suppose that at some time the energy is purely kinetic. At what later time will it be purely potential? At what later time again purely kinetic?

QUESTION 4: If the particle in a simple harmonic oscillator experiences a frictional force (say, air resistance), is the energy constant? Is the amplitude A constant?

QUESTION 5: The mass, frequency, and amplitude of one oscillator are each twice that of a second oscillator. What is the ratio of their stored energies, E_1/E_2?

(A) 2   (B) 4   (C) 8   (D) 16   (E) 32

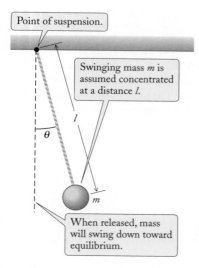

FIGURE 15.14 A pendulum swinging about a fixed suspension point. The angle θ is reckoned as positive if the deflection of the pendulum is toward the right, as in this figure.

15.4 THE SIMPLE PENDULUM

A **simple pendulum** consists of a bob (a mass) suspended by a string or a rod from some fixed point (see Fig. 15.14). The bob is assumed to behave like a particle of mass m, and the string is assumed to be massless. Gravity acting on the bob provides a restoring force. When in equilibrium, the pendulum hangs vertically, just like a plumb line. When released at some angle with the vertical, the pendulum will swing back and forth along an arc of circle (see Fig. 15.15). The motion is two-dimensional; however, the position of the pendulum can be completely described by a single parameter: the angle θ between the string and the vertical (see Fig. 15.14). We will reckon this angle as positive on the right side of the vertical, and as negative on the left side.

Since the bob and the string swing as a rigid unit, the motion can be regarded as rotation about a horizontal axis through the point of suspension, and the equation of motion is that of a rigid body [see Eq. (13.19)]:

$$I\alpha = \tau \tag{15.38}$$

Here the moment of inertia I and the torque τ are reckoned about the horizontal axis through the point of suspension, and α is the angular acceleration.

Figure 15.16 shows the "free-body" diagram for the string–bob system with all the external forces. These external forces are the weight $\mathbf{w}$ of magnitude $w = mg$ acting on the mass m and the suspension force $\mathbf{S}$ acting on the string at the point of support. The suspension force exerts no torque, since its point of application is on the axis of rotation (its moment arm is zero). The weight exerts a torque [see Eq. (13.3)]

$$\tau = -mgl \sin \theta \tag{15.39}$$

where l is the length of the pendulum, measured from the point of suspension to the center of the bob. The minus sign in Eq. (15.39) indicates that this is a restoring torque, which tends to pull the pendulum toward its equilibrium position.

The moment of inertia I of the string–bob system is simply that of a particle of mass m at a distance l from the axis of rotation:

$$I = ml^2$$

FIGURE 15.15 Stroboscopic photograph of a swinging pendulum. The pendulum moves slowly at the extremes of its motion.

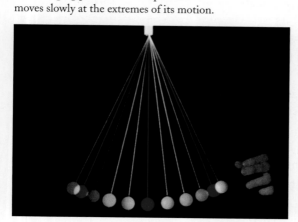

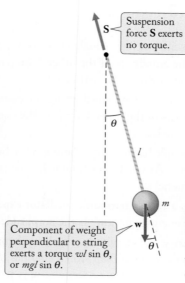

Component of weight perpendicular to string exerts a torque $wl \sin \theta$, or $mgl \sin \theta$.

FIGURE 15.16 "Free-body" diagram for the string–bob system. The torque exerted by the weight $\mathbf{w}$ has magnitude $wl \sin \theta$, or $mgl \sin \theta$.

Hence the equation of rotational motion (15.38) becomes

$$ml^2\alpha = -mgl\sin\theta \tag{15.40}$$

or

$$\alpha = -\frac{g}{l}\sin\theta \tag{15.41}$$

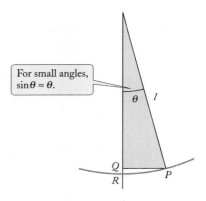

For small angles, $\sin\theta \approx \theta$.

We will solve this equation of motion only in the special case of small oscillations about the equilibrium position. If θ is small, we can make the approximation

$$\sin\theta \approx \theta \tag{15.42}$$

where the angle is measured in radians (see Math Help: Small-Angle Approximations for Sine, Cosine, and Tangent; and see Fig. 15.17).

With this approximation, the equation of motion becomes

$$\alpha = -\frac{g}{l}\theta \tag{15.43}$$

FIGURE 15.17 If the angle θ is small, the length of the straight line PQ is approximately the same as the length of the circular arc PR.

or, since the angular acceleration is $\alpha = d^2\theta/dt^2$,

$$\frac{d^2\theta}{dt^2} = -\frac{g}{l}\theta \tag{15.44}$$

This equation has the same mathematical form as Eq. (15.17). Comparing these two equations, we see that the angle θ replaces the distance x, the angular acceleration replaces the linear acceleration, l replaces m, and g replaces k. Hence the angular motion is simple harmonic. Making the appropriate replacements in Eq. (15.4), we find that the motion is described by the equation

$$\theta = A\cos(\omega t + \delta) \tag{15.45}$$

with an angular frequency [compare Eqs. (15.19) and (15.44)]

$$\tag{15.46}$$

$$\omega = \sqrt{\frac{g}{l}}$$

MATH HELP   **SMALL-ANGLE APPROXIMATIONS FOR SINE, COSINE, AND TANGENT**

With the assumption that an angle θ is expressed in radians and that this angle is small, the trigonometric functions have the simple approximations

$$\sin\theta \approx \theta$$
$$\cos\theta \approx 1 - \theta^2/2$$
$$\tan\theta \approx \theta$$

To understand how these approximations come about, consider the small angle θ shown in Fig. 15.17. The sine of this angle is $\sin\theta = PQ/l$. If θ is small, the length of the straight line PQ is approximately the same as the length of the curved circular arc PR (for small angles, the curved arc is almost a straight line). Thus, $\sin\theta \approx PR/l$. But the ratio PR/l is the definition of the angle θ expressed in radians, so $\sin\theta \approx \theta$. Similar arguments give the above approximations for the cosine and the tangent. These approximations are usually satisfactory if θ is less than about 0.2 radians, or about 10°.

The frequency and the period of the pendulum are then

$$f = \frac{\omega}{2\pi} = \frac{1}{2\pi}\sqrt{\frac{g}{l}}$$

and

$$T = \frac{1}{f} = \frac{2\pi}{\omega} = 2\pi\sqrt{\frac{l}{g}}$$

(15.47)

Note that these expressions for the frequency and the period depend only on the length of the pendulum and on the acceleration of gravity; they do not depend on the mass of the pendulum bob or on the amplitude of oscillation (but, of course, our calculation depends on the assumption that the angle θ, and thus the amplitude of motion, is small).

Like the simple harmonic oscillator, the pendulum has the property of isochronism—*its frequency is (approximately) independent of the amplitude with which it is swinging.* This property can be easily verified by swinging two pendulums of equal lengths side by side, with different amplitudes. The pendulums will continue to swing in step for a long while.

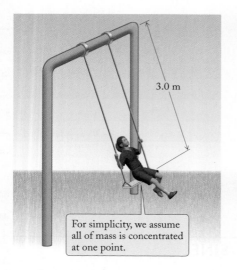

3.0 m

For simplicity, we assume all of mass is concentrated at one point.

FIGURE 15.18 Woman on a swing.

EXAMPLE 7   A woman sits in a swing of length 3.0 m (see Fig. 15.18). What is the period of oscillation of this swing?

SOLUTION: We can regard the swing as a pendulum of an approximate length 3.0 m. From Eq. (15.47) we then find

$$T = 2\pi\sqrt{\frac{l}{g}} = 2\pi\sqrt{\frac{3.0 \text{ m}}{9.81 \text{ m/s}^2}} = 3.5 \text{ s}$$

EXAMPLE 8   The "seconds" pendulum in a pendulum clock built for an astronomical observatory has a period of exactly 2.0 s, so each one-way motion of the pendulum takes exactly 1.0 s. What is the length of such a "seconds" pendulum at a place where the acceleration of gravity is $g = 9.81$ m/s^2? At a place where the acceleration of gravity is 9.79 m/s^2?

SOLUTION: If we square both sides of Eq. (15.47) and then solve for the length l, we find

$$l = \left(\frac{T}{2\pi}\right)^2 g$$

With $g = 9.81$ m/s^2 and the known period $T = 2.0$ s, this gives

$$l = \left(\frac{2.0 \text{ s}}{2\pi}\right)^2 \times 9.81 \text{ m/s}^2 = 0.994 \text{ m}$$

With $g = 9.79$ m/s^2, it gives

$$l = \left(\frac{2.0 \text{ s}}{2\pi}\right)^2 \times 9.79 \text{ m/s}^2 = 0.992 \text{ m}$$

FIGURE 15.19 This electromechanical clock, regulated by a pendulum, served as the U.S. frequency standard in the 1920s. Its master pendulum is enclosed in the canister at right.

The most familiar application of pendulums is the construction of pendulum clocks. Up to about 1950, the most accurate clocks were pendulum clocks of a special design, which were kept inside airtight flasks placed in deep cellars to protect them from disturbances caused by variations of the atmospheric pressure and temperature (see Fig. 15.19). The best of these high-precision pendulum clocks were accurate to within a few thousandths of a second per day. Later, such pendulums were superseded by quartz clocks (see Section 15.2) and then by atomic clocks (see Section 1.3).

Another important application of pendulums is the measurement of the acceleration of gravity g. For this purpose it is necessary only to time the swings of a pendulum of known length; the value of g can then be calculated from Eq. (15.47). The pendulums used for precise determinations of g usually consist of a solid bar swinging about a knife edge at one end, instead of a bob on a string. Such a pendulum consisting of a swinging rigid body is called a **physical pendulum**; its period is related to its size and shape.

EXAMPLE 9   A physical pendulum has a moment of inertia I about its point of suspension, and its center of mass is at a distance d from this point (see Fig. 15.20a). Find the period of this pendulum.

SOLUTION: Figure 15.20b shows the "free-body" diagram for the pendulum. The suspension force **S** has zero moment arm about the pivot, and so exerts no torque. The weight acts at the center of mass, at a distance of d from the point of suspension, and it exerts a torque [see Eq. (13.3)]

$$\tau = -mgd\sin\theta$$

Hence the equation of rotational motion (15.38) is

$$I\alpha = -mgd\sin\theta$$

where $\alpha = d^2\theta/dt^2$ is the angular acceleration for the rotational motion. With the usual small-angle approximation $\sin\theta \approx \theta$, this becomes

$$\frac{d^2\theta}{dt^2} = -\frac{mgd}{I}\theta$$

As in the case of the simple pendulum, we compare this with Eq. (15.17). Since the second time derivative of θ is proportional to the negative of θ, the motion will again be simple harmonic. Hence the angular frequency of oscillation is

(a)

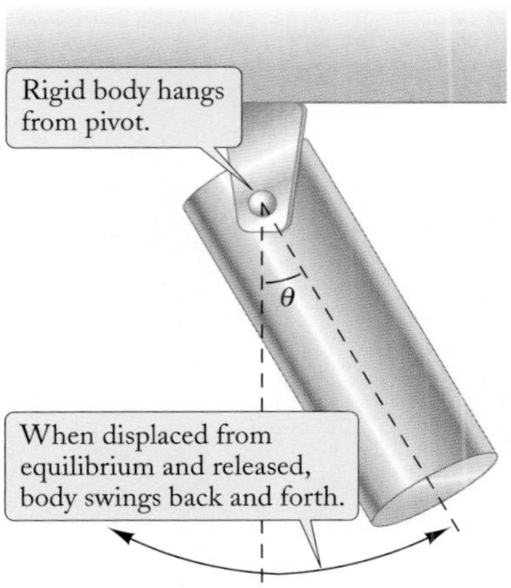

Rigid body hangs from pivot.

θ

When displaced from equilibrium and released, body swings back and forth.

(b)

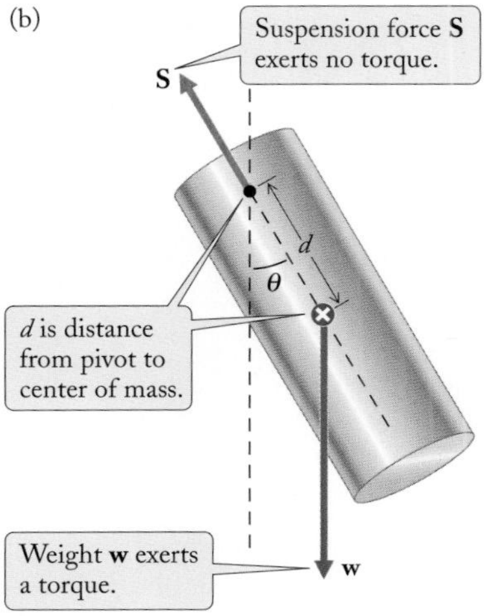

Suspension force **S** exerts no torque.

S

d

θ

d is distance from pivot to center of mass.

Weight **w** exerts a torque.   **w**

FIGURE 15.20 (a) A physical pendulum consisting of a rigid body swinging about a point of suspension. (b) "Free-body" diagram for the physical pendulum. The weight acts at the center of mass.

$$\omega = \sqrt{\frac{mgd}{I}} \tag{15.48}$$

and the period is

$$T = \frac{2\pi}{\omega} = 2\pi\sqrt{\frac{I}{mgd}} \tag{15.49}$$

COMMENT: Note that for a simple pendulum, the moment of inertia about the point of suspension is $I = ml^2$ and the distance of the center of mass from this point is $d = l$. Accordingly, Eq. (15.49) yields $T = 2\pi\sqrt{ml^2/mgl} = 2\pi\sqrt{l/g}$, which shows that the formula for the period of the simple pendulum is a special case of the general formula for the physical pendulum.

Finally, we must emphasize that the approximation contained in Eq. (15.43) is valid only for small angles. If the amplitude of oscillation of a pendulum is more than a few degrees—say, more than 10°—the approximation (15.43) begins to fail, and the motion of the pendulum begins to deviate from simple harmonic motion. At large amplitudes, the period of the pendulum depends on the amplitude—the larger the amplitude, the larger the period. For instance, a pendulum oscillating with an amplitude of 30° has a period 1.7% longer than the value given by Eq. (15.47).

 Checkup 15.4

QUESTION 1: If we shorten the string of a pendulum to half its original length, what is the alteration of the period? The frequency?

QUESTION 2: Two pendulums have equal lengths, but one has 3 times the mass of the other. If we want the energies of oscillation to be the same, how much larger must we make the amplitude of oscillation of the less massive pendulum?

QUESTION 3: A uniform metal rod of length l hangs from one end and oscillates with small amplitude. Such a rod, rotating about one end, has moment of inertia $I = \frac{1}{3}ml^2$ (Table 12.3). What is ω, the angular frequency of oscillation?

(A) $\sqrt{g/l}$ (B) $\sqrt{3g/2l}$ (C) $\sqrt{3g/l}$ (D) $\sqrt{6g/l}$

15.5 DAMPED OSCILLATIONS AND FORCED OSCILLATIONS

So far we have proceeded on the assumption that the only force acting on a simple harmonic oscillator or a pendulum is the restoring force $F = -kx$ or the restoring torque $\tau = -mgl\sin\theta$. However, in a real oscillator or a real pendulum, there is always some extra force caused by friction. For instance, if the pendulum starts its swinging motion with some initial amplitude, the friction against the air and against the point of support will gradually brake the pendulum, reducing its amplitude of oscillation. Although good oscillators have low friction, sometimes more friction is desirable for damping out unwanted oscillations, as with the kitchen scale of Example 3, so that a steady, equilibrium position can be attained.

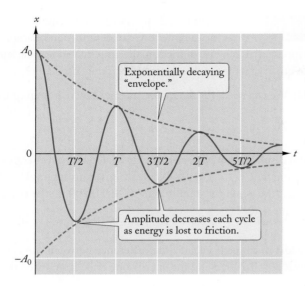

FIGURE 15.21 Plot of position vs. time for a particle with damped harmonic motion.

If the friction force is proportional to the velocity, the equation of motion becomes

$$m \frac{d^2x}{dt^2} = -kx - b \frac{dx}{dt} \qquad (15.50)$$

where b is called the friction constant, or the damping constant. Figure 15.21 is a plot of the position as a function of time for a harmonic oscillator with fairly strong friction. The amplitude of oscillation suffers a noticeable decrease from one cycle to the next. Such a gradually decreasing oscillation is called **damped harmonic motion**. The oscillation amplitude decreases exponentially with time, as indicated by the dashed line in Fig. 15.21. Increasing the friction shortens the time it takes for the amplitude to decrease, and slows the frequency of oscillation somewhat. If the damping is very large, a displaced "oscillator" merely moves back to its equilibrium position, without oscillating. In Section 32.4, we will examine the damped harmonic oscillator in detail.

Since the oscillator must do work against the friction, the mechanical energy gradually decreases. The energy loss per cycle is a constant fraction of the energy E that the oscillator has at the beginning of the cycle. If we represent the energy loss per cycle by ΔE, then ΔE is proportional to E:

$$\Delta E = \left(\frac{2\pi}{Q} \right) E \qquad (15.51)$$

Q of oscillator

Here, the constant of proportionality has been written in the somewhat complicated form $2\pi/Q$, which is the form usually adopted in engineering. The quantity Q is called the **quality factor** of the oscillator. In terms of the damping constant b,

$$Q = \frac{\sqrt{km}}{b} \qquad (15.52)$$

An oscillator with low friction has a high value of Q, and a small energy loss per cycle; an oscillator with high friction has a low value of Q, and a large energy loss per cycle. The value of Q roughly coincides with the number of cycles the oscillator completes before the oscillations damp away significantly. Mechanical oscillators of low friction, such as tuning forks or piano strings, have Q values of a few thousand; that is, they "ring" for a few thousand cycles before their oscillations fade noticeably.

EXAMPLE 10 The maximum displacement from equilibrium of the body-mass measurement device described in Examples 4 and 5 was 0.200 m. Suppose that, because of friction, the amplitude one cycle later is 0.185 m. What is the quality factor for this damped harmonic oscillator?

SOLUTION: We can solve for the quality factor Q by rearranging Eq. (15.51):

$$Q = 2\pi \frac{E}{\Delta E}$$

At maximum displacement, the total energy is all potential energy, so $E = \frac{1}{2}kA^2$. The spring constant $k = 2.1 \times 10^3$ N/m was given in Example 4. We found in Example 5 that the when the amplitude was $A = 0.200$ m, the energy stored was

$$E = \tfrac{1}{2}kA^2 = \tfrac{1}{2} \times 2.1 \times 10^3 \text{ N/m} \times (0.200 \text{ m})^2 = 42 \text{ J}$$

The energy lost during the cycle is the difference between the energy when the amplitude was $A = 0.200$ m and the energy one cycle later, when the amplitude is $A' = 0.185$ m:

$$\Delta E = \tfrac{1}{2}kA^2 - \tfrac{1}{2}kA'^2 = \tfrac{1}{2}k(A^2 - A'^2)$$

$$= \tfrac{1}{2} \times 2.1 \times 10^3 \text{ N/m} \times [(0.200 \text{ m})^2 - (0.185 \text{ m})^2] = 6.1 \text{ J}$$

Hence the quality factor is

$$Q = 2\pi \frac{E}{\Delta E} = 2\pi \times \frac{42 \text{ J}}{6.1 \text{ J}} = 43$$

To maintain the oscillations of a damped harmonic oscillator at a constant level, it is necessary to exert a periodic force on the oscillator, so the energy fed into the oscillator by this extra force compensates for the energy lost to friction. An extra force is also needed to start the oscillations of any oscillator, damped or not, by supplying the initial energy for the motion. Any such extra force exerted on an oscillator is called a **driving force**. A familiar example of a driving force is the "pumping" force that you must exert on a playground swing (a pendulum) to start it moving and to keep it moving at a constant amplitude. This is an example of a *periodic* driving force.

With the addition of a harmonic driving force of amplitude F_0 and angular frequency ω, the equation of motion (15.50) becomes

$$m\frac{d^2x}{dt^2} = -kx - b\frac{dx}{dt} + F_0 \cos \omega t \qquad (15.53)$$

If the frequency ω of the driving force coincides with the frequency ω_0 of the natural oscillations of the oscillator, then even a quite small driving force can gradually build up large amplitudes. Under these conditions the driving force steadily feeds energy into the oscillations, and the amplitude of these grows until the friction becomes so large that it inhibits further growth. The ultimate amplitude reached depends on the amount of friction; in an oscillator of low friction, or high Q, this ultimate amplitude can be extremely large. The buildup of a large amplitude by the action of a driving force in tune with the natural frequency of an oscillator is called **resonance**. Figure 15.22 shows the value of the final amplitude of oscillation attained as a function of the frequency of

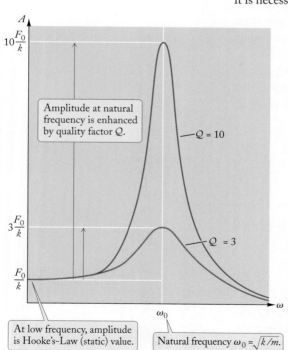

FIGURE 15.22 Amplitude of a forced damped harmonic oscillator as a function of the frequency of the oscillating force.

the harmonic driving force for two mass-and-spring systems with the same natural angular frequency $\omega_0 = \sqrt{k/m}$ but different values of Q, the quality factor. Notice that large amplitudes occur over a range of driving frequencies, and that some enhancement over the static Hooke's-Law displacement $x = -F_0/k$ occurs for any frequency of forced oscillation near or below the natural frequency ω_0. If the oscillator is forced precisely at resonance, the amplitude can be shown to take the value

$$A = \frac{F_0}{k} Q \qquad (15.54)$$

amplitude at resonance of damped, driven harmonic oscillator

This is simply the magnitude of the static displacement $x = -F_0/k$ multiplied by Q; thus the quality factor is equivalent to an amplitude enhancement factor for a system at resonance.

The phenomenon of resonance plays a crucial role in many pieces of industrial machinery—if one vibrating part of a machine is driven at resonance by a perturbing force originating from some other part, then the amplitude of oscillation can build up to violent levels and shake the machine apart. Such dangerous resonance effects can occur not only in moving pieces of machinery, but also in structures that are normally regarded as static. In a famous accident that took place in 1850 in Angers, France, the stomping of 487 soldiers marching over a suspension bridge excited a resonant swinging motion of the bridge; the motion quickly rose to a disastrous level and broke the bridge apart, causing the death of 226 of the soldiers (Fig. 15.23).

FIGURE 15.23 Resonance disaster: the collapse of the bridge at Angers, as illustrated in a contemporary newspaper.

 Checkup 15.5

QUESTION 1: Suppose that the driving force has a frequency half as large as the frequency of the oscillator. Would you expect a buildup of oscillations by resonance?

QUESTION 2: Suppose that the driving force has a frequency twice as large as the frequency of the oscillator. Would you expect a buildup of oscillations by resonance?

QUESTION 3: Suppose that a bell has a high Q (it continues to ring for a long time after you strike it). If you rest your hand against the bell after striking it, how does this alter the Q?

QUESTION 4: An oscillator begins with 1.00 J of mechanical energy. After 10 oscillations, the energy stored has dropped to 0.90 J. What is the approximate Q of the system?

   (A) 6.3   (B) 10   (C) 63   (D) 100   (E) 630

PHYSICS IN PRACTICE   CHAOS

The motions we examined in this chapter and in the preceding chapters are either periodic or else regular in some other sense. For instance, the motion of a simple harmonic oscillator and the motion of a planet around the Sun are periodic; and the motion of a particle under the influence of a constant force is highly regular, proceeding with constant acceleration. But there also exist mechanical systems with highly irregular motions, without periodicity, and with a pathological sensitivity to small changes in initial conditions, so a small change of the initial velocity or position quickly leads to very large changes in the motion. Such motions are called **chaotic**.

An example of a system with chaotic motion is the double-well oscillator, in which the attractive force $-kx$ of the simple harmonic oscillator is replaced by a sum of a repulsive force $+kx$ and an attractive force $-ax^3$. Such an oscillator can be constructed by clamping a leaf spring in a vertical position and attaching a fairly large mass to its top (see Fig. 1). The central position is an equilibrium position, but it is unstable—the leaf spring will flop sideways either to the left or to the right, attaining a bent equilibrium position, which is stable. If disturbed, it can oscillate about this left or right equilibrium position. The potential energy for this system has a minimum at the left equilibrium position, a minimum at the right equilibrium position, and a maximum at the vertical position in between; that is, the curve of potential energy has two wells and a hump in between. The equation of motion for this system cannot be solved exactly, but it can be solved numerically by a computer program that calculates derivatives by evaluating changes in the position in small time increments.

FIGURE 1 (a) Leaf spring clamped at the lower end with a mass attached at the upper end. (b) The leaf spring flops to the left (blue) or to the right (red), and it can oscillate about these bent equilibrium positions.

Figure 2 shows two numerical solutions for a double-well oscillator that includes both a frictional damping force and a periodic driving force (such as discussed in Section 15.5). Note that at first the oscillator moves erratically—it sometimes oscillates in the left well, then in the right well, then back again, etc. Also note that although the initial conditions for the two solutions barely differ at all and the motions are initially almost indistinguishable, they soon begin to differ drastically. Finally, one of the solutions settles down in the left well (blue), and the other solution settles down in the right well (red), and they then continue to oscillate in a steady manner about the left or the right equilibrium position with a frequency equal to that of the driving force. The steady modes of oscillations in the left and the right wells are called **attractors**, because the motion tends to settle into these modes.

The erratic motion that precedes the steady oscillations is an instance of chaos. How long the chaos lasts depends on the initial conditions and on the strengths of the driving force and the damping. For some values of these parameters, the chaos lasts forever. We can prepare plots such as those in Fig. 2 for a wide variety of initial conditions, and in each case examine whether the oscillator settles into the left or the right well. Figure 3 is a color-coded diagram that summarizes the results of 900×900 such calculations, with different initial positions and velocities. The initial positions are plotted horizontally and the initial velocities vertically; the color indicates where the oscillator settles: blue for the left well, and red for the right well. The solid blue and red zones indicate that for all initial conditions in these zones, the oscillator settles in the same final steady state of oscillation. But the other regions of the diagram, with fine striations of intermingled red and blue points, are characteristic of chaos. A very minor change in initial conditions takes us from a blue point to a red point (or vice versa), which means the final motion depends sensitively on small changes. The striations in Fig. 3 have a **fractal** character—if we examine any small patch in the striated zone at higher magnification, we find striations within striations within striations.

The chaotic behavior implies that although in principle the motion can be calculated from the initial conditions, in practice the motion is not predictable, except for a short time. Any small uncertainty in the initial conditions or any small uncertainty introduced by round-off errors in the numerical calculation will make it impossible to decide whether the initial conditions fall on a blue or a red dot—which means we can't decide whether the oscillator will ultimately settle into steady oscillation on the left or on the right.

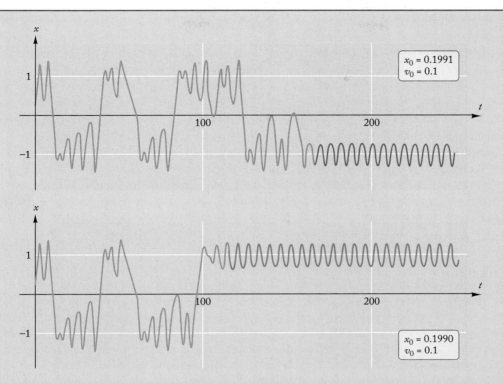

FIGURE 2   Motions for two slightly different initial conditions of the oscillator. At first the motion is chaotic (gray), but it ultimately settles into periodic oscillations about the left equilibrium position (blue) or the right equilibrium position (red). (For these plots, all the constants in the equation of motion were set equal to 1, except the damping constant and the strength of the driving force, which were set equal to 0.25.)

FIGURE 3   Plot of 900×900 initial positions (horizontal coordinate) and initial velocities (vertical coordinate). The color of each dot is blue if the oscillator ultimately settles into periodic oscillations about the left equilibrium position, red if about the right.

SUMMARY

MATH HELP Derivatives of trigonometric functions	(page 473)
MATH HELP Small-angle approximations for sine, cosine, and tangent	(page 485)
PHYSICS IN PRACTICE Chaos	(page 492)

SIMPLE HARMONIC MOTION
where A is the amplitude (the maximum displacement from $x = 0$); ω is the angular frequency, and δ is the phase constant.

$$x = A \cos(\omega t + \delta)$$

(15.4)

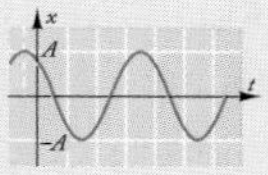

PERIOD   (time for one cycle)

$$T = 2\pi/\omega$$

(15.5)

FREQUENCY   (number of cycles per second)

$$f = 1/T = \omega/2\pi$$

(15.6)

PHASE CONSTANT AND TIME OF MAXIMUM DISPLACEMENT

$$\delta = -\omega t_{\text{max}}$$

(15.8)

MAXIMUM VELOCITY

$$v_{\text{max}} = \omega A$$

(15.14)

MAXIMUM ACCELERATION

$$a_{\text{max}} = \omega^2 A$$

(15.16)

EQUATION OF MOTION OF SIMPLE HARMONIC OSCILLATOR
where k is the spring constant.

$$m\frac{d^2x}{dt^2} = -kx$$

(15.18)

ANGULAR FREQUENCY AND PERIOD OF SIMPLE HARMONIC OSCILLATOR

$$\omega = \sqrt{\frac{k}{m}} \qquad T = 2\pi\sqrt{\frac{m}{k}}$$

(15.19, 15.21)

ENERGY OF SIMPLE HARMONIC OSCILLATOR

$$E = \tfrac{1}{2}kA^2 = \tfrac{1}{2}m\omega^2 A^2$$

(15.32)

ANGULAR FREQUENCY AND PERIOD OF SIMPLE PENDULUM

$$\omega = \sqrt{\frac{g}{l}} \qquad T = 2\pi\sqrt{\frac{l}{g}}$$

(15.46, 15.47)

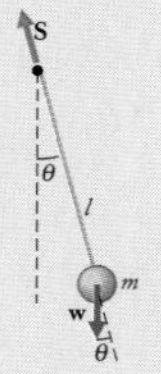

ANGULAR FREQUENCY AND PERIOD OF PHYSICAL PENDULUM
where I is the moment of inertia of the pendulum.

$$\omega = \sqrt{\frac{mgd}{I}} \qquad T = 2\pi\sqrt{\frac{I}{mgd}}$$

(15.48, 15.49)

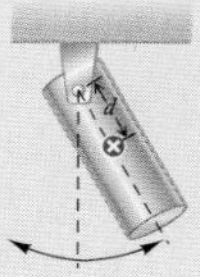

ENERGY LOSS PER CYCLE OF DAMPED OSCILLATOR   where Q is the quality factor.

$$\Delta E = -\frac{2\pi}{Q} E$$

(15.51)

AMPLITUDE AT RESONANCE OF DAMPED HARMONIC OSCILLATOR   where F_0 is the amplitude of a harmonic driving force.

$$A = \frac{F_0}{k} Q$$

(15.54)

QUESTIONS FOR DISCUSSION

1. Is the motion of the piston of an automobile engine simple harmonic motion? How does it differ from simple harmonic motion?

2. In our calculation of the frequency of the simple harmonic oscillator, we ignored the mass of the spring. Qualitatively, how does the mass of the spring affect the frequency?

3. A grandfather clock is regulated by a pendulum. If the clock is running late, how must we adjust the length of the pendulum?

4. Figure 15.24 shows the escapement of a pendulum clock, i.e., the linkage that permits the pendulum to control the rotation of the wheels of the clock. Explain how the wheel turns as the pendulum swings.

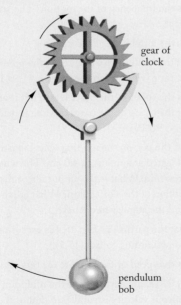

FIGURE 15.24 Escapement mechanism of a pendulum clock. At the instant shown, the tooth at the left has escaped from the left arm, and the tooth on the right is pushing against the right arm.

5. Would a pendulum clock keep good time on a ship?

6. Galileo claimed that the oscillators of a pendulum are isochronous, even for an amplitude of oscillation as large as 30°. What is your opinion of this claim?

7. Why would you expect a pendulum oscillating with an amplitude of nearly (but not quite) 180° to have a very long period?

8. Can a pendulum oscillate with an amplitude of more than 180°?

9. Figure 15.25 shows a "tilted pendulum" designed by Christiaan Huygens in the seventeenth century. When the pendulum is tilted, its period is longer than when the pendulum is vertical. Explain.

FIGURE 15.25 Huygens' tilted pendulum.

10. Most grandfather clocks have a lenticular pendulum bob which supposedly minimizes friction by "slicing" through the air. However, experience has shown that a cylindrical pendulum bob experiences less air friction. Can you suggest an explanation?

11. Galileo described an experiment to compare the acceleration of gravity of lead and of cork:

    I took two balls, one of lead and one of cork, the former being more than a hundred times as heavy as the latter, and suspended them from two equal thin strings, each four or five bracchia long. Pulling each ball aside from the vertical, I released them at the same instant, and they, falling along the circumferences of the circles having the strings as radii, passed through the vertical and returned along the same path. This free oscillation, repeated more than a hundred times, showed clearly that the heavy body kept time with the light body so well that neither in a hundred oscillations, nor in a thousand, will the former anticipate the latter by even an instant, so perfectly do they keep step.

    Since air friction affects the cork ball much more than the lead ball, do you think Galileo's results are credible?

Newton reported a more careful experiment that avoided the inequality of friction:

> I tried the thing in gold, silver, lead, glass, sand, common salt, wood, water, and wheat. I provided two equal wooden boxes. I filled the one with wood, and suspended an equal weight of gold (as exactly as I could) in the centre of oscillation of the other. The boxes, hung by equal threads of 11 feet, made a couple of pendulums perfectly equal in weight and figure . . . and, placing the one by the other, I observed them to play together forwards and backwards for a long while, with equal vibrations. . . . And by these experiments, in bodies of the same weight, one could have discovered a difference of matter less than the thousandth part of the whole.

Explain how Newton's experiment was better than Galileo's.

12. A simple pendulum hangs below a table, with its string through a small hole in the tabletop. Suppose you gradually pull the string while the pendulum is swinging. What happens to the frequency of oscillation? To the (angular) amplitude?

13. Shorter people have a shorter length of stride, but a higher rate of step when walking "naturally." Explain.

14. A girl sits on a swing whose ropes are 1.5 m long. Is this a simple pendulum or a physical pendulum?

15. A simple pendulum consists of a particle of mass m attached to a string of length l. A physical pendulum consists of a body of mass m attached to a string in such a way that the center of mass is at a distance l from the point of support. Which pendulum has the shorter period?

16. Suppose that the spring in the front-wheel suspension of an automobile has a natural frequency of oscillation equal to the frequency of rotation of the wheel at, say, 80 km/h. Why is this bad?

17. When marching soldiers are about to cross a bridge, they break step. Why?

PROBLEMS

15.1 Simple Harmonic Motion[†]

1. A particle moves as follows as a function of time:

$$x = 3.0 \cos(2.0t)$$

where distance is measured in meters and time in seconds.

(a) What is the amplitude of this simple harmonic motion? The frequency? The angular frequency? The period?

(b) At what time does the particle reach the midpoint, $x = 0$? The turning point?

2. A particle is performing simple harmonic motion along the x axis according to the equation

$$x = 0.6 \cos\left(\frac{\pi t}{2}\right)$$

where the distance is measured in meters and the time in seconds.

(a) Calculate the position x of the particle at $t = 0$, $t = 0.50$ s, and $t = 1.00$ s.

(b) Calculate the instantaneous velocity of the particle at these times.

(c) Calculate the instantaneous acceleration of the particle at these times.

3. A particle moves back and forth along the x axis between the points $x = 0.20$ m and $x = -0.20$ m. The period of the motion is 1.2 s, and it is simple harmonic. At the time $t = 0$, the particle is at $x = 0.20$ m and its velocity is zero.

(a) What is the frequency of the motion? The angular frequency?

(b) What is the amplitude of the motion?

(c) At what time will the particle reach the point $x = 0$? At what time will it reach the point $x = -0.10$ m?

(d) What is the speed of the particle when it is at $x = 0$? What is the speed of the particle when it reaches the point $x = -0.10$ m?

4. Suppose that the peg on the rotating wheel illustrated in Fig. 15.5 is located at a radius of 4.0 cm. The wheel turns at a rate of 600 rev/min. What is the amplitude of the simple harmonic motion of the slotted arm? What are the period, the frequency, and the angular frequency?

5. Consider that the particle in Fig. 15.4 is executing simple harmonic motion according to Eq. (15.1).

(a) What is the speed of the satellite for this case?

(b) At $t = 0.050$ s, the particle is at the midpoint and its instantaneous velocity is parallel to that of the satellite. What is the speed of the particle? How does it compare with the speed of the satellite?

6. A given point on a guitar string executes simple harmonic motion with a frequency of 440 Hz and an amplitude of 1.2 mm. What is the maximum speed of this motion? The maximum acceleration?

7. A piston in a windmill-driven water pump is in simple harmonic motion. The motion has an amplitude of 50 cm and the mass of the piston is 6.0 kg. Find the maximum net force on

[†]For help, see Online Concept Tutorial 16 at www.wwnorton.com/physics

the piston when it oscillates 80 times per minute. Find the maximum velocity.

8. A particle moves in simple harmonic motion according to $x = A \cos(\omega t + \delta)$. At $t = 0$, the particle is at $x = 0$ with initial velocity $v_0 > 0$. What is the phase constant δ?

9. The position of a body can be described by $x = A \cos(\omega t + \delta)$. The angular frequency ω, the initial position x_0, and the initial velocity v_0 are known. Find the amplitude A and the phase constant δ in terms of ω, x_0, and v_0.

10. The central part of a piano string oscillates at 261.7 Hz with an amplitude of 3.0 mm. What is the angular frequency of the motion? The period? What is the maximum velocity? What is the maximum acceleration?

11. In a modern nonlinear dynamics experiment, small beads (spheres) are vibrated on a plate; when the beads start to move, interesting patterns form (see Fig. 15.26). If the plate vibrates at 250 Hz, for what amplitude of motion will the beads start to lift off? (Hint: This will occur when the maximum acceleration of the plate equals $g = 9.81$ m/s^2.)

FIGURE 15.26 Oscillating beads.

12. A mass moves in a circle of radius 10 cm, centered on the origin in the x–y plane, with an angular velocity of $\pi/4$ radian/s. At $t = 0$, the mass is on the positive x axis. What are the x components of the position, velocity, and acceleration of the mass at $t = 1.0$ s? At $t = 2.0$ s?

13. A particle executes simple harmonic motion. Its displacement is given by $x = A \cos(\omega t + \delta)$, where as usual, the amplitude A is a positive constant. At $t = 0$, the particle is at the origin and moving in the positive x direction. What is the appropriate choice of the phase constant δ in this case?

*14. Experience shows that from one-third to one-half of the passengers in an airliner can be expected to suffer motion sickness if the airliner bounces up and down with a peak acceleration of $0.4\,g$ and a frequency of about 0.3 Hz. Assume that this up-and-down motion is simple harmonic. What is the amplitude of the motion?

*15. The frequency of a mass attached to a spring is 3.0 Hz. At time $t = 0$, the mass has an initial displacement of 0.20 m and an initial velocity of 4.0 m/s.

   (a) What is the position of the mass as a function of time?

   (b) When will the mass first reach a turning point? What will be its acceleration at that time?

15.2 The Simple Harmonic Oscillator

16. A man of mass 70 kg is bouncing up and down on a pogo stick (see Fig. 15.27). He finds that if he holds himself rigid and lets the stick do the bouncing (after getting it started), the period of the up-and-down motion is 0.70 s. What is the spring constant of the spring in the pogo stick? Assume that the bottom of the stick remains in touch with the floor and ignore the mass of the stick.

FIGURE 15.27 Man on pogo stick.

17. The cable described in Example 8 in Chapter 14 can be regarded as a spring. What is the effective spring constant of this spring? What is the frequency of oscillation when a mass of 7.1×10^3 kg is attached to the lower end of the cable and allowed to oscillate up and down? Neglect the mass of the cable in your calculation.

18. A simple harmonic oscillator consists of a mass sliding on a frictionless surface under the influence of a force exerted by a spring connected to the mass. The frequency of this harmonic oscillator is 8.0 Hz. If we connect a second, identical spring to the mass, parallel to the first spring, what will be the new frequency of oscillation?

19. The body of an automobile of mass 1100 kg is supported by four vertical springs attached to the axles of the wheels. In order to test the suspension, a man pushes down on the body of the automobile and then suddenly releases it. The body rocks up and down with a period of 0.75 s. What is the spring constant of each of the springs? Assume that all the springs are identical and that the compressional force on each spring

is the same; also assume that the shock absorbers of the automobile are completely worn out so that they do not affect the oscillation frequency.

20. Deuterium (D) is an isotope of hydrogen. The mass of the deuterium atom is 1.998 times larger than the mass of the hydrogen atom. Given that the frequency of vibration of the H_2 molecule is 1.31×10^{14} Hz (see Example 6), calculate the frequency of vibration of the D_2 molecule. Assume the "spring" connecting the atoms is the same in H_2 and D_2.

*21. Calculate the frequency of vibration of the HD molecule consisting of one atom of hydrogen and one of deuterium. See Problem 20 for necessary data. (Hint: The center of mass is stationary.)

22. A mass attached to a spring oscillates with an amplitude of 15 cm; the spring constant is $k = 20$ N/m. When the position is half the maximum value, the mass moves with velocity $v = 25$ cm/s. Determine the period of the motion. Find the value of the mass.

23. A mass of 150 g is attached to a spring of constant $k = 8.0$ N/m and oscillates without friction. The mass is displaced 20 cm from equilibrium and, at $t = 0$, is released from rest. If the position as a function of time is written $x = A \cos(\omega t + \delta)$, determine the values of A, ω, and δ. What is the maximum velocity of the mass? Its maximum acceleration?

24. The equilibrium position of the bottom end of a light, hanging spring shifts downward by 15 cm when a 200-g mass is hung from it. The mass is then displaced an additional 5.0 cm and released. What is the period of motion?

25. A *thickness monitor* is a laboratory instrument used to determine the thickness of a thin film that is deposited on the surface of a quartz crystal. We may treat the crystal as a spring-and-mass system with $k = 6.0 \times 10^5$ N/m and $m = 0.50$ g. What is the frequency of oscillation of this system? This frequency changes slightly as mass is added to the crystal. If the frequency decreases 0.010%, how much mass was deposited? If the area of the crystal is 2.0 cm^2 and the mass density of the film material is 7.5 g/cm^2, how thick was the deposited film?

*26. A thin metal rod is attached to the ceiling and a mass $M = 15$ kg is attached to the bottom of the rod. The rod is 2.0 m long and has a 9.0-mm^2 cross-sectional area. Regard the rod as a (stiff) spring. If the Young's modulus of the rod material is 22×10^{10} N/m^2, what is its spring constant (for small elongations and compressions)? If the mass is displaced vertically, what is its frequency of oscillation (in Hz)? Neglect the mass of the rod.

*27. A mass $m = 2.5$ kg hangs from the ceiling by a spring with $k = 90$ N/m. Initially, the spring is in its unstretched configuration and the mass is held at rest by your hand. If, at time $t = 0$, you release the mass, what will be its position as a function of time?

*28. The wheel of a sports car is suspended below the body of the car by a vertical spring with a spring constant 1.1×10^4 N/m. The mass of the wheel is 14 kg, and the diameter of the wheel is 61 cm.

(a) What is the frequency of up-and-down oscillations of the wheel? Regard the wheel as a mass on one end of a spring, and regard the body of the car as a fixed support for the other end of the spring.

(b) Suppose that the wheel is slightly out of round, having a bump on one side. As the wheel rolls on the street, it receives a periodic push each time the bump comes in contact with the street. At what speed of the translational motion of the car will the frequency of this push coincide with the natural frequency of the up-and-down oscillations of the wheel? What will happen to the car at this speed? (Note: This problem is not quite realistic because the elasticity of the tire also contributes a restoring force to the up-and-down motion of the wheel.)

*29. A mass m slides on a frictionless plane inclined at an angle θ with the horizontal. The mass is attached to a spring, parallel to the plane (Fig. 15.28); the spring constant is k. How much is the spring stretched at equilibrium? What is the frequency of the oscillations of the mass up and down on the plane?

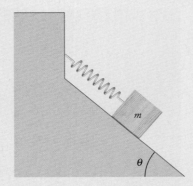

FIGURE 15.28 Mass sliding on a frictionless inclined plane.

**30. Two identical masses slide with one-dimensional motion on a frictionless plane under the influence of three identical springs attached as shown in Fig. 15.29. The magnitude of each mass is m, and the spring constant of each spring is k.

FIGURE 15.29 Two masses sliding on a frictionless plane.

(a) Suppose that at time $t = 0$, the masses are at their equilibrium positions and their instantaneous velocities are $v_1 = -v_2$. Find the position of each mass as a function of time. What is the frequency of the motion?

(b) Suppose that at time $t = 0$, the masses are at their equilibrium positions and their instantaneous velocities are $v_1 = v_2$. Find the position of each mass as a function of time. What is the frequency of the motion?

**31. A cart consists of a body and four wheels on frictionless axles. The body has a mass m. The wheels are uniform disks of mass M and radius R. The cart rolls, without slipping, back and forth on a horizontal plane under the influence of a spring attached to one end of the cart (Fig. 15.30). The spring constant is k. Taking into account the moment of inertia of the wheels, find a formula for the frequency of the back-and-forth motion of the cart.

FIGURE 15.30 A cart attached to a spring.

15.3 Kinetic Energy and Potential Energy

32. Suppose that a particle of mass 0.24 kg acted upon by a spring undergoes simple harmonic motion with the parameters given in Problem 3.

   (a) What is the total energy of this motion?

   (b) At what time is the kinetic energy zero? At what time is the potential energy zero?

   (c) At what time is the kinetic energy equal to the potential energy?

33. A mass of 8.0 kg is attached to a spring and oscillates with an amplitude of 0.25 m and a frequency of 0.60 Hz. What is the energy of the motion?

34. A simple harmonic oscillator consists of a mass of 2.0 kg sliding back and forth along a horizontal frictionless track while pushed and pulled by a spring with $k = 8.0 \times 10^2$ N/m. Suppose that when the mass is at the equilibrium point, it has an instantaneous speed of 3.0 m/s. What is the energy of this harmonic oscillator? What is the amplitude of oscillation?

35. A simple harmonic oscillator of mass 0.60 kg oscillates with a frequency of 3.0 Hz and an amplitude of 0.15 m. Suppose that, while the mass is instantaneously at rest at its turning point, we quickly attach another mass of 0.60 kg to it. How does this change the amplitude of the motion? The frequency? The energy? The maximum speed? The maximum acceleration?

*36. The separation between the equilibrium positions of the two atoms of a hydrogen molecule is 1.0×10^{-10} m. Using the data given in Example 6, calculate the value of the vibrational energy that corresponds to an amplitude of vibration of 0.5×10^{-10} m for each atom. Is it valid to treat the motion as small oscillation if the energy has this value?

37. A 500-g mass is connected to a spring and executes simple harmonic motion. The period of the motion is 1.5 s, and the total mechanical energy of the system is 0.50 J. Find the amplitude of motion.

38. A mass oscillates on a spring. At the points in a cycle when the kinetic energy is one-half of the potential energy, the displacement from equilibrium is 15 cm and the instantaneous velocity is ±25 cm/s. What is the period of the motion?

*39. One end of a horizontal spring of constant k is fixed and the other end is attached to a mass m on a frictionless surface. The spring is initially in its equilibrium position. At $t = 0$, a force F, constant thereafter, is applied in the direction of elongation of the spring. Sometime later, the mass has moved a distance d in the direction of the force. What is the kinetic energy at that time?

*40. A mass of 3.0 kg sliding along a frictionless floor at 2.0 m/s strikes and compresses a spring of constant $k = 300$ N/m. The spring stops the mass. How far does the mass travel while being slowed by the spring? How long does the mass take to stop?

*41. Two masses m_1 and m_2 are joined by a spring of spring constant k. Show that the frequency of vibration of these masses along the line connecting them is

$$\omega = \sqrt{\frac{k(m_1 + m_2)}{m_1 m_2}}$$

(Hint: The center of mass remains at rest.)

*42. Although it is usually a good approximation to neglect the mass of a spring, sometimes this mass must be taken into account. Suppose that a uniform spring has a relaxed length l and a mass m'; a mass m is attached to the end of the spring. The mass m' is uniformly distributed along the spring. Suppose that if the moving end of the spring has a speed v, all other points of the spring have speed directly proportional to their distance from the fixed end; for instance, a point midway between the moving and the fixed end has a speed $\frac{1}{2}v$.

   (a) Show that the kinetic energy in the spring is $\frac{1}{6}m'v^2$ and that the kinetic energy of the mass m and the spring is

   $$K = \tfrac{1}{2}mv^2 + \tfrac{1}{6}m'v^2 = \tfrac{1}{2}(m + \tfrac{1}{3}m')v^2$$

   Consequently, the effective mass of the combination is $m + \frac{1}{3}m'$.

   (b) Show that the frequency of oscillation is $\omega = \sqrt{k/(m + \tfrac{1}{3}m')}$.

   (c) Suppose that a spring has a mass of 0.05 kg. The frequency of oscillation of a 4.0-kg mass attached to this spring will then be somewhat smaller than calculated for a massless spring. How much smaller? Express your answer as a percentage of the value obtained for a massless spring.

15.4 The Simple Pendulum[†]

43. The longest pendulum in existence is a 27-m Foucault pendulum in Portland, Oregon. What is the period of this pendulum?

[†]For help, see Online Concept Tutorial 17 at www.wwnorton.com/physics

44. At a construction site, a bucket full of concrete hangs from a crane. You observe that the bucket slowly swings back and forth, 8.0 times per minute. What is the length of the cable from which the bucket hangs?

45. The elevator cage of a skyscraper hangs from a 300-m-long steel cable. The elevator cage is guided within the elevator shaft by railings. If we remove these railings and we let the elevator cage swing from side to side (with small amplitude), what is its period of oscillation?

46. On the Earth, a pendulum of length 0.994 m has a period of 2.00 s (compare Example 8). If we take this pendulum to the surface of Jupiter, where $g = 24.8$ m/s^2, what will be its period?

47. A mass suspended from a parachute descending at constant velocity can be regarded as a pendulum. What is the frequency of the pendulum oscillations of a human body suspended 7.0 m below a parachute?

48. A "seconds" pendulum is a pendulum that has a period of exactly 2.0 s; each one-way swing of the pendulum therefore takes exactly 1.0 s. What is the length of the seconds pendulum in Paris ($g = 9.809$ m/s^2), Buenos Aires ($g = 9.797$ m/s^2), and Washington, D.C. ($g = 9.801$ m/s^2)?

49. A grandfather clock controlled by a pendulum of length 0.9932 m keeps good time in New York ($g = 9.803$ m/s^2).

    (a) If we take this clock to Austin, Texas ($g = 9.793$ m/s^2), how many minutes per day will it fall behind?

    (b) In order to adjust the clock, by how many millimeters must we shorten the pendulum?

50. The pendulum of a grandfather clock has a length of 0.994 m. If the clock runs late by 1.0 minute per day, how much must you shorten the pendulum to make it run on time?

51. A small model of a 10-story construction crane used on a Hollywood movie set should appear realistic in motion. To make it look large, the mass hanging from the crane "cable" (actually, a rod) is constrained to oscillate with a period of 10 s. How long does this make the cable seem?

52. An astronaut lands on an asteroid and sets up a pendulum that has a period of 1.0 s on Earth. She finds that the pendulum has a period of 89 s on the asteroid. What is the local value of the acceleration due to gravity on the asteroid?

53. A circular painting is 2.00 m in diameter and has uniform thickness. It hangs on a wall, suspended by a nail 10 cm from the top edge. If it is pushed slightly, what is the period of small oscillations of the painting?

54. A hula hoop (a thin, uniform toy hoop) of radius 1.0 m hangs over a nail. If it is set to swinging with small amplitude, what is the period of motion?

*55. A **torsional oscillator** consists of a horizontal uniform disk of mass M and radius R attached at its center to the end of a massless vertical fiber. Some such oscillators can execute simple harmonic (twisting) motion with very large amplitudes (amplitudes greater than one rotation are possible). The restoring torque of the fiber is proportional to the angular rotation; that is, $\tau = -\kappa\theta$, where κ is called the torsional constant of the system.

(a) Find the angular frequency of oscillation in terms of M, R, and κ.

(b) If the disk is turned through an initial angle of θ_0 and released, what is the maximum rotational angular velocity of the subsequent motion?

(c) For what value of θ_0 do the answers to (a) and (b) have the same value?

*56. The balance wheel in a clock is a torsional oscillator with a period of 0.50 s. The restoring torque of the wheel spring is $\tau = -\kappa\theta$, where κ is the torsional constant. If the wheel is essentially a hoop, that is, all of its mass $m = 8.0$ g is concentrated at its radius $R = 1.0$ cm, what is the value of κ (in N·m/radian)?

*57. A pendulum hangs from an inclined wall (see Fig. 15.31). Suppose that this pendulum is released at an initial angle of 10° and it bounces off the wall elastically when it reaches an angle of −5°. What is the period of this pendulum?

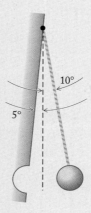

FIGURE 15.31 Pendulum hanging from an inclined wall.

*58. The pendulum of a pendulum clock consists of a rod of length 0.99 m with a bob of mass 0.40 kg. The pendulum bob swings back and forth along an arc of length 20 cm.

(a) What are the maximum velocity and the maximum acceleration of the pendulum bob along the arc?

(b) What is the force that the pendulum exerts on its support when it is at the midpoint of its swing? At the endpoint? Neglect the mass of the rod in your calculations.

*59. The pendulum of a regular clock consists of a mass of 120 g at the end of a (massless) wooden stick of length 44 cm.

(a) What is the total energy (kinetic plus potential) of this pendulum when oscillating with an amplitude of 4°?

(b) What is the speed of the mass when at its lowest point?

60. At the National Institute of Standards and Technology in Gaithersburg, Maryland, the value of the acceleration of gravity is 9.800 95 m/s^2. Suppose that at this location a very precise physical pendulum, designed for measurements of the acceleration of gravity, has a period of 2.103 56 s. If we take this pendulum to a new location at the U.S. Coast and Geodetic Survey, in nearby Washington, D.C., it has a period of 2.103 54 s. What is the value of the acceleration of gravity at this new location? What is the percentage change of the acceleration between the two locations?

61. Consider a meterstick swinging about a pivot through its upper end. What is the period of oscillation of this physical pendulum?

*62. A pendulum consists of a brass rod with a brass cylinder attached to the end (Fig. 15.32). The diameter of the rod is 1.00 cm and its length is 90.00 cm; the diameter of the cylinder is 6.00 cm and its length is 20.00 cm. What is the period of this pendulum?

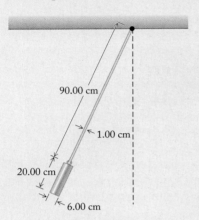

90.00 cm

1.00 cm

20.00 cm

6.00 cm

FIGURE 15.32
A physical pendulum.

*63. To test that the acceleration of gravity is the same for a piece of iron and a piece of brass, an experimenter takes a pendulum of length 1.800 m with an iron bob and another pendulum of the same length with a brass bob and starts them swinging in unison. After swinging for 12.00 min, the two pendulums are no more than one-quarter of a (one-way) swing out of step. What is the largest difference between the values of g for iron and brass consistent with these data? Express your answer as a fractional difference.

*64. Calculate the natural period of the swinging motion of a human leg. Treat the leg as a rigid physical pendulum with an axis at the hip joint. Pretend that the mass distribution of the leg can be approximated as two rods joined rigidly end to end. The upper rod (thigh) has a mass of 6.8 kg and a length of 43 cm; the lower rod (shin plus foot) has a mass of 4.1 kg and a length of 46 cm. Using a watch, measure the period of the natural swinging motion of *your* leg when you are standing on one leg and letting the other dangle freely. Alternatively, measure the period of the swinging motion of your leg when you walk at a normal rate (this approximates the natural swinging motion). Compare with the calculated number.

*65. A hole has been drilled through a meterstick at the 30-cm mark and the meterstick has been hung on a wall by a nail passing through this hole. If the meterstick is given a push so that it swings about the nail, what is the period of the motion?

*66. A physical pendulum has the shape of a disk of radius R. The pendulum swings about an axis perpendicular to the plane of the disk at a distance l from the center of the disk.

(a) Show that the frequency of the oscillations of this pendulum is

$$\omega = \sqrt{\frac{gl}{\frac{1}{2}R^2 + l^2}}$$

(b) For what value of l is the frequency a maximum?

*67. A physical pendulum consists of a massless rod of length $2l$ rotating about an axis through its center. A mass m_1 is attached at the lower end of the rod, and a smaller mass m_2 at the upper end (see Fig. 15.33). What is the period of this pendulum?

m_2

$2l$

m_1

FIGURE 15.33 A physical pendulum with two bobs.

*68. Suppose that a physical pendulum consists of a thin rigid rod of mass m suspended at one end. Suppose that this rod has an initial position $\theta = 20°$ and an initial angular velocity $\omega = 0$. Calculate the force **F** that the support exerts on the pendulum at this initial instant (give horizontal and vertical components).

*69. The door of a house is made of wood of uniform thickness. The door has a mass of 27 kg and measures 1.90 m × 0.91 m. The door is held shut by a torsional spring with $\kappa = 30$ N·m/radian arranged so that it exerts a torque of 54 N·m when the door is fully open (at right angles to the wall of the house). What angular speed does the door attain if it slams shut from the fully open position? What linear speed does the edge of the door attain?

**70. Galileo claimed to have verified experimentally that a pendulum oscillating with an amplitude as large as 30° has the same period as a pendulum of identical length oscillating with a much smaller amplitude. Suppose that you let two pendulums of length 1.5 m oscillate for 10 min. Initially, the pendulums oscillate in step. If the amplitude of one of them is 30° and the amplitude of the other is 5°, by what fraction of a (one-way) swing will the pendulums be out of step at the end of the 10-min interval? What can you conclude about Galileo's claim?

**71. A thin vertical rod of steel is clamped at its lower end. When you push the upper end to one side, bending the rod, the upper end moves (approximately) along an arc of circle[2] of radius R and the rod opposes your push with a restoring force $F = -\kappa\theta$, where θ is the angular displacement and κ is a constant. If you attach a mass m to the upper end, what will be the frequency of small oscillations? For what value of m does the rod become unstable; that is, for what value of m is $\omega = 0$? Treat the rod as massless in your calculations. (Hint: Think of the rod as an inverted pendulum of length R, with an extra restoring force $-\kappa\theta$.)

---

[2] The radius R of the approximating (osculating) circle is somewhat shorter than the length of the rod.

**72. According to a proposal described in Problem 83 of Chapter 1, very fast trains could travel from one city to another in straight subterranean tunnels (see Fig. 15.34). For the following calculations, assume that the density of the Earth is constant, so the acceleration of gravity as a function of the radial distance r from the center of the Earth is $g = (GM/R^3)r$.

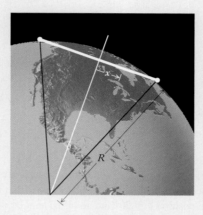

FIGURE 15.34 A straight tunnel connecting two points on the surface of the Earth.

(a) Show that the component of the acceleration of gravity along the track of the train is

$$g_x = -(GM/R^3)x$$

where x is measured from the midpoint of the track (see Fig. 15.34).

(b) Neglecting friction, show that the motion of the train along the track is simple harmonic motion with a period independent of the length of the track,

$$T = 2\pi\sqrt{\frac{R^3}{GM}}$$

(c) Starting from rest, how long would a train take to roll freely along its track from San Francisco to Washington, D.C.? What would be its maximum speed (at the midpoint)? Use the numbers you calculated in Problem 83 of Chapter 1 for the length and depth of the track.

**73. A physical pendulum consists of a long, thin cone suspended at its apex (Fig. 15.35). The height of the cone is l. What is the period of this pendulum?

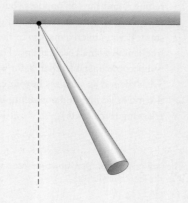

FIGURE 15.35
A long, thin cone.

**74. The net gravitational force on a particle placed midway between two equal spherical bodies is zero. However, if the particle is placed some distance away from this equilibrium point, then the gravitational force is not zero.

(a) Show that if the particle is at a distance x from the equilibrium point in a direction toward one of the bodies, then the force is approximately $4GMmx/r^3$, where M is the mass of each spherical body, m is the mass of the particle, and $2r$ is the distance between the spherical bodies. Assume $x \ll r$.

(b) Show that if the particle is at a distance x from the equilibrium point in a direction perpendicular to the line connecting the bodies, then the force is approximately $-2GMmx/r^3$, where the negative sign indicates that the direction of the force is toward the equilibrium point.

(c) What is the frequency of small oscillations of the mass m about the equilibrium point when moving in a direction perpendicular to the line connecting the bodies? Assume that the bodies remain stationary.

**75. The motion of a simple pendulum is given by

$$\theta = A\cos\left(\sqrt{\frac{g}{l}}\,t\right)$$

(a) Find the tension in the string of this pendulum; assume that $\theta \ll 1$. The mass of the suspended particle is m.

(b) The tension is a function of time. At what time is the tension maximum? What is the value of this maximum tension?

15.5 Damped Oscillations and Forced Oscillations

76. Roughly, what is the frequency of stomping of soldiers on the march? What must have been the resonant frequency of the bridge at Angers that broke when soldiers marched across it?

77. A pendulum of length 1.50 m is set swinging with an initial amplitude of 10°. After 12 min, friction has reduced the amplitude to 4°. What is the value of Q for this pendulum?

78. The pendulum of a grandfather clock has a length of 0.994 m and a mass of 1.2 kg.

(a) If the pendulum is set swinging, the friction of the air reduces its amplitude of oscillation by a factor of 2 in 13.0 min. What is the value of Q for this pendulum?

(b) If we want to keep this pendulum swinging at a constant amplitude of 8°, we must supply mechanical energy to it at a rate sufficient to make up for the frictional loss. What is the required mechanical power?

79. When a swing in motion is not being "pumped," the angular amplitude of oscillation decreases because of air and other friction. The motion of a 3.0-m-long swing decreases in amplitude from 12° to 10° after 5 complete cycles. What is the Q of the system? If the rider and seat are treated as a point mass with $m = 25$ kg, at what average rate is mechanical energy being dissipated?

80. A horizontal spring of constant k is attached to a mass m that slides on a slightly frictional floor. After the mass is displaced a distance A from equilibrium and released, the amplitude of oscillation decreases to $0.95A$ after 10 cycles. What is the Q of this system?

81. A harmonic force $F = F_0 \cos \omega t$, where $F_0 = 0.20$ N, is applied to a damped harmonic oscillator of spring constant $k = 15$ N/m and mass m, where $\omega = \sqrt{k/m}$. The amplitude of oscillation increases rapidly at first, and then settles to a constant value, $A = 40$ cm. What is the Q of the system? What would the amplitude be if the angular frequency of the force F had been much less than $\sqrt{k/m}$?

82. A microelectromechanical system (MEMS) consists of a microscopic silicon mechanical oscillator (see Fig. 15.36) with

FIGURE 15.36 A microelectromechanical system (MEMS) oscillator, the silicon membrane structure suspended above the faceted silicon trench.

a spring constant $k = 5 \times 10^{-3}$ N/m. When it oscillates in a vacuum-sealed device (to remove air friction), the Q of such an oscillator is large: $Q = 5 \times 10^6$. What amplitude of motion will the oscillator attain if an oscillating force of amplitude 1×10^{-18} N (near the current limits of force detection) is applied?

83. Using electron-beam lithography, engineers are attempting to fabricate nanoelectromechanical system (NEMS) oscillators with frequencies as high as 100 GHz (for communications and higher-speed computing). If the equivalent mass of such an oscillator is 1.0×10^{-18} g and a minimum amplitude of 0.10 nm is needed to detect an applied harmonic force of amplitude 1.0×10^{-10} N, what must the minimum Q of such an oscillator be?

*84. Consider the motion of the damped harmonic oscillator plotted in Fig. 15.21.

(a) According to this plot, what fraction of its amplitude does the oscillator lose in its first oscillation?

(b) What fraction of its energy does the oscillator lose in its first oscillation?

(c) According to Eq. (15.51), what is the value of Q for this oscillator?

*85. If you stand on one leg and let the other dangle freely back and forth starting at an initial amplitude of, say, 20° or 30°, the amplitude will decay to one-half of the initial amplitude after about four swings. Regarding the dangling leg as a damped oscillator, what value of Q can you deduce from this?

REVIEW PROBLEMS

86. A particle performs simple harmonic motion along the x axis with an amplitude of 0.20 m and a period of 0.80 s. At $t = 0$, the particle is at maximum distance from the origin; that is, $x = 0.20$ m.

(a) What is the equation that describes the position of the particle as a function of time?

(b) Calculate the position of the particle at $t = 0.10$ s, 0.20 s, 0.30 s, and 0.40 s.

87. In an electric saber saw, the rotational motion of the electric motor is converted into a back-and-forth motion of the saw blade by a mechanism similar to that shown in Fig. 15.5. Suppose the peg of the rotating wheel moves around a circle of diameter 3.0 cm at 4000 rev/min and thereby moves the slotted arm to which the saw blade is bolted. What are the amplitude and the frequency of the back-and-forth simple harmonic motion of the blade?

88. In response to a sound wave, the middle of your eardrum oscillates back and forth with a frequency of 4000 Hz and an amplitude of 1.0×10^{-5} m. What is the maximum speed of the eardrum?

89. Suppose that two particles are performing simple harmonic motion along the x axis with a period of 8.0 s. The first particle moves according to the equation

$$x = 0.30 \cos\left(\frac{\pi t}{4}\right)$$

and the second particle according to the equation

$$x' = 0.30 \sin\left(\frac{\pi t}{4}\right)$$

where the distance is measured in meters and the time in seconds.

(a) When does the first particle reach the midpoint? The turning point? Draw a diagram showing the particle and its satellite particle at these times.

(b) When does the second particle reach the midpoint? The turning point? Draw a diagram showing the particle and its satellite particle at these times.

(c) By some argument, establish that whenever the first particle passes through a point on the x axis, the second particle passes through this same point 2.0 s later.

90. A particle of 6.0 kg is executing simple harmonic motion along the x axis under the influence of a spring. The particle moves according to the equation

$$x = 0.20 \cos(3.0t)$$

where x is measured in meters and t in seconds.

(a) What is the frequency of the motion? What is the spring constant of the spring? What is the maximum speed of the motion?

(b) Suppose we replace the particle by a new particle of 2.0 kg (but we keep the same spring), and suppose we start the motion with the same amplitude of 0.20 m. What will be the new frequency of the motion? What will be the new maximum speed?

91. The motion of the piston in an automobile engine is approximately simple harmonic. Suppose that the piston travels back and forth over a distance of 8.50 cm and has a mass of 1.2 kg. What are its maximum acceleration and maximum speed if the engine is turning at its highest safe rate of 6000 rev/min? What is the maximum force on the piston?

92. A Small Mass Measurement Instrument (SMMI) was used in Skylab to measure the masses of biological samples, small animals, chemicals, and other such items used in life-sciences experiments while in orbit (see Fig. 15.37). The sample to be measured is strapped to a tray supported by leaf springs, and the mass is determined from the observed period of oscillation of the tray-and-mass. To calibrate this instrument, a test mass of 1.00 kg is first placed on the tray; the period of oscillation is then 1.08 s. Suppose that when the test mass is removed and an unknown sample is placed on the tray, the period becomes 1.78 s. What is the mass of the sample? Assume that the mass of the tray (and the straps) is 0.400 kg.

FIGURE 15.37 Small Mass Measurement Instrument.

93. A simple harmonic oscillator has a frequency of 1.5 Hz. What will happen to the frequency if we cut the spring in half and attach both halves to the mass so that both springs push jointly?

94. A physicist of 55 kg stands on a bathroom scale (a spring scale, with an internal spring). She observes that when she mounts the scale suddenly, the pointer of the scale first oscillates back and forth a few times with a frequency of 2.4 Hz.

(a) What value of the spring constant can she deduce from these data?

(b) If she then takes a child of 20 kg in her arms and again stands on the scale, what will be the new frequency of oscillation of the pointer?

95. Ropes used by mountain climbers are quite elastic, and they behave like springs. A rope of 10 m has a spring constant $k = 4.9 \times 10^3$ N/m. Suppose that a mountain climber of 80 kg hangs on this rope, which is stretched vertically down. What is the frequency of up-and-down oscillations of the mountain climber?

96. Consider a particle of mass m moving along the x axis under the influence of a spring of spring constant k. The equilibrium point is at $x = 0$, and the amplitude of the motion is A.

(a) At what point x is the kinetic energy of the particle equal to its potential energy?

(b) When the particle reaches the point $x = \frac{1}{2}A$, what fraction of its energy is potential, and what fraction is kinetic?

97. A simple harmonic oscillator consists of a mass of 3.0 kg sliding back and forth along a horizontal frictionless track while pushed and pulled by a spring with $k = 6.0 \times 10^2$ N/m. Suppose that initially the mass is released from rest at a distance of 0.25 m from the equilibrium point. What is the energy of this harmonic oscillator? What is the maximum speed it attains when passing through the equilibrium point?

98. A simple harmonic oscillator of mass 0.80 kg oscillates with a frequency of 2.0 Hz and an amplitude of 0.12 m. Suppose that, while the mass is instantaneously at rest at its turning point, we quickly shift the fixed end of the spring to a new fixed position, 0.12 m farther away from the mass. How does this change the amplitude of the motion? The frequency? The energy? The maximum speed? The maximum acceleration?

99. A pendulum has a length of 1.5 m. What is the period of this pendulum? If you wanted to construct a pendulum with exactly half this period, how long would it have to be?

100. An "interrupted" pendulum consists of a simple pendulum of length l that encounters a nail placed at a distance $\frac{3}{4}l$ below the point of support. If this pendulum is released from one side, it will begin to wrap around the nail as soon as it passes through the vertical position (Fig. 15.38). What is the period of this pendulum?

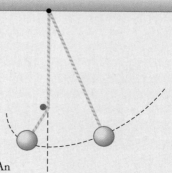

FIGURE 15.38 An "interrupted" pendulum.

101. A physical pendulum consists of a uniform spherical bob of mass M and radius R suspended from a massless string of length L (see Fig. 15.39). Taking into account the size of the bob, show that the period of small oscillations of this pendulum is

$$T = 2\pi\sqrt{\frac{\frac{2}{5}R^2 + (R + L)^2}{g(R + L)}}$$

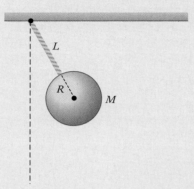

FIGURE 15.39 A physical pendulum with a large bob.

102. A uniform rod of length L is swinging about a pivot at a distance x from its center (see Fig. 15.40). Find the period of oscillation of this physical pendulum as a function of x. For what choice of x is the period shortest?

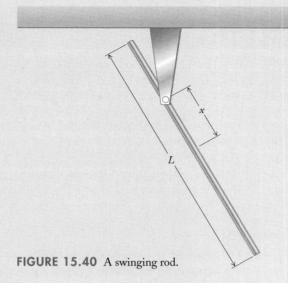

FIGURE 15.40 A swinging rod.

103. A swing of length 2.0 m hangs from a horizontal branch of a tree. With what frequency should you rock the branch to build up oscillations of the pendulum by resonance?

Answers to Checkups

Checkup 15.1

1. The Earth's rotational motion is periodic; it repeats with each daily cycle. It is not a back-and-forth motion along a line or arc, so it is not an oscillation.

2. The velocity attains its maximum magnitude at $x = 0$, that is, where the displacement is zero; the velocity attains its minimum magnitude at $x = \pm A$, that is, at the points of maximum displacement. This is because the displacement and velocity are 90° out of phase; if one is a cosine function, the other is a sine function [see Eqs. (15.11) and (15.12)].

3. The acceleration attains its maximum magnitude at $x = \pm A$, that is, at the point of maximum displacement from the origin; the acceleration attains its minimum magnitude at $x = 0$, that is, where the displacement is zero. This is because the displacement and acceleration are 180° out of phase; if one is a cosine function, the other is a negative cosine function [see Eqs. (15.11) and (15.13)].

4. If the maximum displacement is A, the maximum velocity is ωA [compare Eqs. (15.11) and (15.12)]. Thus, for the same amplitude, the particle with twice the frequency has twice the maximum velocity. Similarly, the maximum acceleration is $\omega^2 A$, so the particle with twice the frequency has 4 times the maximum acceleration.

5. As described in Section 15.1, the x coordinates of the particle and satellite are identical. Obviously, the y coordinates are not, since the particle is always at $y = 0$, while the satellite executes circular motion. The velocities are not the same, since the particle has zero y velocity, unlike the satellite. The x components of the velocities and accelerations are the same, since they are derived from the identical time dependence of the x coordinate. Since the particle is always at $y = 0$, the y components of the velocity and acceleration are not the same.

6. (C) $\pi/2$. If we insert $x = 0$ and $t = 0$ in $x = A\cos(\omega t + \delta)$ [Eq. (15.4)], then we see that $0 = \cos\delta$, which is true if $\delta = \pi/2$ or if $\delta = -\pi/2$. Of these two, only $\delta = \pi/2$ is listed.

Checkup 15.2

1. The force on the particle attains maximum magnitude at the extreme displacements (the turning points) of the motion, $x = \pm A$. The force on the particle attains the minimum magnitude of zero when the particle passes through the equilibrium point, $x = 0$.

2. Since the frequency is given by $\omega = \sqrt{k/m}$, doubling the mass decreases the frequency by a factor of $1/\sqrt{2}$.

3. If the spring is cut when the particle is at the equilibrium point, the particle will continue moving with the constant velocity it had there, $v_{max} = \omega A$. If we cut it when the particle is at $x = A$, where the particle is instantaneously at rest, the particle will remain at rest there.

4. (B) $1/\sqrt{2}$. A stronger spring causes oscillations with a higher frequency, and so a shorter period. The period varies inversely with the square root of the spring constant [Eq. (15.21)].

Checkup 15.3

1. Since energy is proportional to the square of the amplitude [Eq. (15.32)], the oscillator with twice the amplitude has 4 times the energy. Since the maximum speed is proportional to the amplitude, $v_{max} = \omega A$ [Eq. (15.34)], the oscillator with twice the amplitude also has twice the maximum speed.

2. Both oscillators have the same energy, since $E = \frac{1}{2}kA^2$. But the maximum speed is inversely proportional to the square root of the mass [Eq. (15.35)], so the particle with twice the mass has a smaller maximum speed by a factor of $1/\sqrt{2}$.

3. The energy is purely kinetic when the oscillator passes through equilibrium. The energy will be purely potential at maximum amplitude, which is one-quarter of a cycle later, or 2.0 s later. An oscillator passes through equilibrium twice each cycle (once in each direction), so the energy will be purely kinetic 4.0 s after the initial time, or another 2.0 s after the energy is purely potential.

4. Friction removes energy from the system, so the energy will decrease whenever the particle is moving, and will not remain constant. Since $E = \frac{1}{2}kA^2$, the amplitude A will also decrease each cycle due to friction.

5. (E) 32. The stored energy is $E = \frac{1}{2}kA^2$. But from Eq. (15.18), $k = m\omega^2$, so $E = \frac{1}{2}m\omega^2 A^2$; if each of m, ω, and A increases by a factor of 2, then the energy increases by a factor of $2^5 = 32$.

Checkup 15.4

1. The period of the simple pendulum is proportional to the square root of the length ($T = 2\pi\sqrt{l/g}$), so the period of the shorter pendulum will be decreased by a factor of $1/\sqrt{2}$. The frequency is the inverse of the period ($f = 1/T$), and so will increase by $\sqrt{2}$.

2. Two pendulums of the same length have the same angular frequency of oscillation, since $\omega = \sqrt{g/l}$. But the energy of a pendulum is $\frac{1}{2}m\omega^2 A^2$ (this formula is equally valid for the pendulum and the simple harmonic oscillater). Thus to have the same energy of oscillation, a mass 3 times smaller must move with an amplitude that is $\sqrt{3}$ times larger.

3. (B) $\sqrt{3g/2l}$. The angular frequency of such a physical pendulum is given by Eq. (15.48), $\omega = \sqrt{mgd/I}$. The distance d is measured from the point of suspension to the center of mass and thus is half of the length of the rod; that is, $d = l/2$. Inserting this value and the given moment of inertia yields
$$v = \sqrt{(mgl/2)/(ml^2/3)} = \sqrt{3g/2l}.$$

Checkup 15.5

1. Yes, at least some slight buildup always occurs at frequencies below resonance. Figure 15.22 indicates that forced oscillations far below the resonant frequency approach an amplitude $A \approx F_0/k$, the magnitude of the static-force spring displacement. This occurs because a slowly varying force allows the (faster) mass–spring system to follow the force over time.

2. No. In this case the slowly responding oscillator cannot follow the oscillating force; it is as if the response is averaged nearly equally over the positive and negative force contributions.

3. Your hand provides friction; you remove energy from the bell. When ΔE, the energy lost per cycle, increases, the Q must decrease.

4. (E) 630. We can solve Eq. (15.51) for Q and obtain $Q = 2\pi E/\Delta E$. Since 0.10 J is lost in 10 cycles, about 0.010 J is lost each cycle. Thus $Q = 2\pi(1.0\ \text{J})/(0.010\ \text{J}) = 200\pi \approx 630$.

Waves

Concepts — in — Context

CONCEPTS IN CONTEXT

At water parks, engineers have developed wave pools like this one, where water waves are generated by large alternating pumps at the deep end and propagate toward the shallow end. Design considerations include the following questions:

? How are the frequency and spacing of the waves related? (Example 2, page 511)

? What vertical acceleration does a swimmer feel as a wave passes by? (Example 3, page 513)

? The speed of a wave changes as it approaches the shallow end of the pool. Does its frequency change? Does its wavelength change? (Example 5, page 515)

16.1 Transverse and Longitudinal Wave Motion

16.2 Periodic Waves

16.3 The Superposition of Waves

16.4 Standing Waves

(a)

(b)

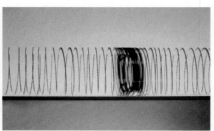

FIGURE 16.1 (a) A kink (a transverse deformation) traveling along a spring. (b) A compression (a longitudinal deformation) traveling along a spring.

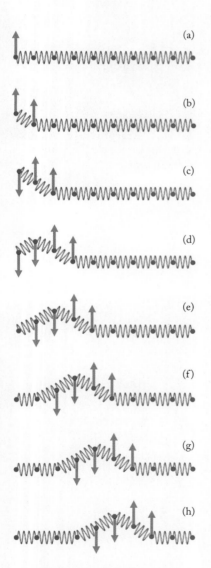

(a)

(b)

(c)

(d)

(e)

(f)

(g)

(h)

FIGURE 16.2 Particles joined by springs. A transverse disturbance propagates from left to right. The diagrams show snapshots at successive times. The particles move up and down.

A **wave** *is a vibrational, shaking motion in an elastic, deformable body.* The wave is initiated by some external force that acts on some part of the body and deforms it. The elastic restoring forces within the body communicate this initial disturbance from one part of the body to the next, adjacent part. The disturbance therefore gradually propagates along the elastic body. For instance, Fig. 16.1 shows a simple example of wave motion: a long spring, such as a Slinky, has been disturbed by a sudden up-and-down motion or a back-and-forth motion, which produced a kink (Fig. 16.1a) or a compressional deformation (Fig. 16.1b). The disturbance propagates along the spring as a wave pulse.

The elastic body in which the wave propagates is called the **medium.** Thus, a spring is the medium for the deformational waves illustrated in Figs. 16.1a and b, a stretched string is the medium for similar deformational "string" waves, water is the medium for water waves, air is the medium for sound waves, the crust of the Earth is the medium for seismic waves, and so on. When a wave propagates through a medium, *the particles in the medium vibrate back and forth, but the medium as a whole does not perform translational motion.* This is obvious in the case of a wave propagating on the spring, where we can see the spring vibrate and we know that the spring cannot travel anywhere, since it is held fixed at its ends. It is also obvious for a wave propagating on a stretched string, again held fixed at its ends. But the lack of motion of the medium as a whole is not so obvious for water waves—when watching ocean waves, we often gain the impression that the water travels with the wave, especially when the waves are large and when we see them crashing against a seawall or some other obstruction. But we can check that the water does not flow with the wave if we watch a chip of wood or some other flotsam on the water. Such a chip of wood only bobs up and down, and it rocks back and forth; it does not travel forward with the wave.

For the sake of simplicity, in this chapter we will concentrate on the motion of transverse waves on a stretched string. However, most of our mathematical results also apply to wave motion in other elastic bodies. In the next chapter we will examine some features of wave motion in air, that is, sound waves.

16.1 TRANSVERSE AND LONGITUDINAL WAVE MOTION

To gain some qualitative understanding of the mechanism of wave motion, consider a tightly stretched elastic string, such as a long rubber cord. The elastic string may be regarded as a row of particles connected by small, massless springs. If we shake one end of the string up and down with a flick of the wrist, a disturbance travels along the row of particles. Figure 16.2 shows in detail how such a traveling disturbance comes about. Initially, the particles are at their equilibrium positions, evenly spaced along the string. When we jerk the first particle upward, it will pull the second particle upward, and this will pull the third, and so on. If we then jerk the first particle back to its original position, it will pull the second particle back, and this will likewise pull the third, and so on. As the motion is transmitted from one particle to the next particle, the disturbance propagates along the row of particles. Such a disturbance, in which the particles move at right angles to the direction of propagation of the disturbance, is called a **transverse** wave pulse.

Alternatively, we can generate a disturbance by suddenly pushing the first particle toward the second and, soon after, pulling it back. Figure 16.3 shows how such a compressional disturbance propagates along the row of particles. This kind of disturbance, in which the particles move back and forth along the direction of propagation of the disturbance, is called a **longitudinal** wave pulse.

Note that although the wave pulse travels along the full length of the string, the particles do not—they merely move back and forth around their equilibrium positions. Also note that in the region of the wave pulse, the string has kinetic energy (due to the back-and-forth motion of the particles) and potential energy (due to the deformation of the springs between the particles). Hence, a wave pulse traveling along the string carries energy with it—the wave transports energy from one end of the string to the other.

Wave motion in air, water, or any other medium displays the same general features. The waves are propagating disturbances in the medium communicated by pushes and pulls from one particle to the next. The waves transport energy without transporting particles. The waves are longitudinal, transverse, or both. A sound wave in air is longitudinal; the air molecules move forward and backward, parallel to the direction of propagation of the wave. But a wave on the surface of the ocean is both longitudinal and transverse; the water molecules move up and down and, simultaneously, forward and backward—the net result of these simultaneous motions is that each water molecule traces out an elliptical path. Seismic waves in the body of the Earth can be either longitudinal (P waves) or transverse (S waves). These two kinds of seismic waves have different speeds, and their relative intensities depend on the characteristics of the earthquake that generated them.

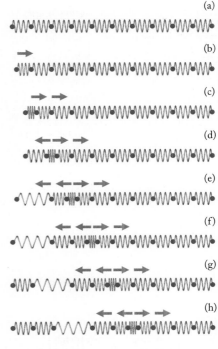

FIGURE 16.3 A longitudinal disturbance propagates from left to right. The particles move back and forth.

 ## Checkup 16.1

QUESTION 1: You shake a baseball bat back and forth. Does this produce a wave motion in the bat? You shake a bowl of Jell-O. Does this produce a wave motion in the Jell-O?

QUESTION 2: You stretch a rubber cord from the porch of a house to a tree in the garden. How can you initiate a transverse wave in this cord? A longitudinal wave?

QUESTION 3: When a guitar player plucks the string of her guitar, does she produce a transverse wave or a longitudinal wave?

QUESTION 4: You have a long rod of steel and a hammer. How must you hit the end of the rod to generate a longitudinal wave along the rod? A transverse wave?

QUESTION 5: An ocean wave travels from the coast of Africa to Florida. Does this wave carry water from Africa to Florida? Does it carry energy?

   (A) Yes; yes   (B) Yes; no   (C) No; no   (D) No; yes

16.2 PERIODIC WAVES

If we shake the end of a long string up and down and we continue shaking it steadily, we will generate a **periodic wave** on the string. Such a wave can be regarded as consisting of a steady succession of positive (upward) and negative (downward) wave pulses, which repeat at regular intervals. Figure 16.4a shows a periodic wave at one instant of time. The high points of the wave are called the **wave crests**, and the low points are called the **wave troughs**. *The distance from one crest to the next or from one*

periodic wave

wave crest and wave trough

wavelength

trough to the next is called the **wavelength**, designated by the symbol λ (the Greek letter *lambda*). The wavelength is the repeat distance of the wave pattern—a shift of the wave pattern by one wavelength to the right (or the left) reproduces the original wave pattern.

With the passing of time, the wave crests and wave troughs travel toward the right at a speed *v*. As the wave travels, the entire wave pattern shifts toward the right; that is, the wave pattern (but not the string) performs a rigid translational motion. Figures 16.4b–h show the wave at successive instants of time. These pictures span one **period** of the wave; that is, they span the interval *T* of time required for the wave pattern to travel exactly one wavelength to the right. The period is the repeat time of the wave pattern—after one period, each wave crest or wave trough will have traveled to the position previously occupied by the adjacent wave crest or wave trough, and the wave will have attained exactly the same configuration as it had at the initial time.

period

Since in one period, the wave travels a distance equal to one wavelength, the ratio of wavelength to period must equal the **wave speed**,

wave speed

$$\frac{\lambda}{T} = v \tag{16.1}$$

As in the case of simple harmonic motion, we define the frequency of the wave as the inverse of the period:

$$f = \frac{1}{T} \tag{16.2}$$

The frequency of the wave is simply the number of wave crests arriving at some point on the string per second. The unit for the frequency is cycles per second, or Hz (hertz). For example, if the period of the wave is 0.1 s, then in one second there will be 10 wave crests arriving at some point on the string, and consequently the frequency of the wave is 10 cycles per second, or 10 Hz.

In terms of the frequency, Eq. (16.1) becomes

$$\lambda f = v \tag{16.3}$$

This equation permits us to calculate the frequency from the wavelength, or the wavelength from the frequency, provided we know the speed of the wave.

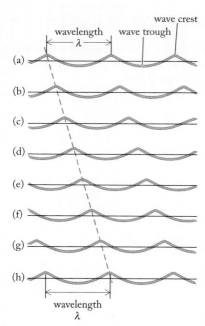

FIGURE 16.4 A periodic wave traveling to the right. The diagrams show snapshots of the wave at successive instants of time. The wave pattern (h) coincides with the wave pattern (a) because the wave has moved exactly one wavelength to the right.

EXAMPLE 1

A long clothesline is stretched horizontally between two trees. While shaking this clothesline up and down near one end at the rate of 4.0 cycles per second, you observe that the wavelength of the waves you generate is 1.0 m. What is the speed of these waves? If the distant end of the clothesline is 10 m away, how long does a wave pulse take to return to you?

SOLUTION: From Eq. (16.3),

$$v = \lambda f = 1.0 \text{ m} \times 4.0 \text{ Hz} = 4.0 \text{ m/s}$$

To return to you, the wave pulse has to complete a round-trip distance of *d* = 20 m. Hence, the time required is

$$t = \frac{d}{v} = \frac{20 \text{ m}}{4.0 \text{ m/s}} = 5.0 \text{ s}$$

EXAMPLE 2 At the deep end of a wave pool (see the chapter photo), the speed of water waves is 5.2 m/s. In order to avoid an excessively rough ride, the frequency of the waves is kept low, at 0.40 Hz. What is the period of such waves? What is the distance beween wave crests?

SOLUTION: The period is the inverse of the frequency [Eq. (16.2)]:

$$T = \frac{1}{f} = \frac{1}{0.40 \text{ Hz}} = 2.5 \text{ s}$$

The distance between crests is the wavelength; from Eq. (16.1), we know the wavelength is the product of the wave speed and the period:

$$\lambda = vT = 5.2 \text{ m/s} \times 2.5 \text{ s} = 13 \text{ m}$$

An important special case of a periodic wave is a **harmonic wave**. *This kind of wave has the shape of a harmonic function, that is, a sine curve or a cosine curve.* If we assume that a wave crest is at the origin at the initial time $t = 0$, the wavefunction is

$$y = A \cos kx \qquad \text{for the initial time } t = 0 \qquad (16.4)$$

harmonic wave

The constant A, which represents the height of the wave crests (and the depth of the wave troughs), is called the **wave amplitude**, and the constant k is called the **wave number**. Note that here, as in the preceding chapter, the argument of the cosine function is supposed to be expressed in radians.

wave amplitude

Figure 16.5 is a plot of the wavefunction (16.4). The wave crests (maxima) occur where $\cos kx = 1$, that is, at

$$kx = 0, \ 2\pi, \ 4\pi, \ 6\pi, \text{ etc.} \qquad \text{(maxima)} \qquad (16.5)$$

and the wave troughs (minima) occur where $\cos kx = -1$, or at

$$kx = \pi, \ 3\pi, \ 5\pi, \text{ etc.} \qquad \text{(minima)} \qquad (16.6)$$

From these equations we see that the distance from one crest to the next, or from one trough to the next, is $2\pi/k$. Thus the wavelength λ and the wave number k of a harmonic wave are related by

$$\lambda = \frac{2\pi}{k} \qquad \text{or} \qquad k = \frac{2\pi}{\lambda} \qquad (16.7)$$

wavelength and wave number

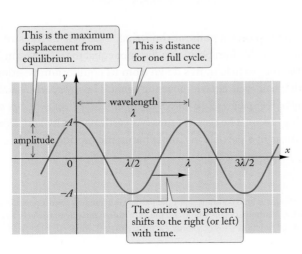

FIGURE 16.5 A harmonic wave.

Since the cosine function is periodic in intervals of 2π, we see that the relation $k = 2\pi/\lambda$ ensures that the value of the cosine function repeats whenever the position x in Eq. (16.4) changes by a distance equal to the wavelength λ.

At any later time, the harmonic wave will have traveled some distance to the right or to the left. This means that the wave pattern plotted in Fig. 16.5 shifts some distance to the right or to the left. If the speed of the wave is v, the wave pattern shifts a distance vt in a time t. The initial wavefunction (16.4) must then be replaced by a new wavefunction in which the value of the argument is shifted by a distance vt. Thus, we must replace x in Eq. (16.4) by $x - vt$ or by $x + vt$, for a wave that travels to the right or to the left, respectively. The wavefunction at time t is then

harmonic traveling wave

$$y = A\cos[k(x - vt)] \qquad \text{for a wave traveling in the positive } x \text{ direction} \qquad (16.8)$$

or

$$y = A\cos[k(x + vt)] \qquad \text{for a wave traveling in the negative } x \text{ direction} \qquad (16.9)$$

Note that according to Eq. (16.4) there is a wave crest at $kx = 0$, and that according to Eq. (16.8) the corresponding wave crest is at $k(x - vt) = 0$. Hence, at time t, this wave crest is at $x - vt = 0$, or at $x = vt$, as expected for a wave traveling in the positive x direction. In consequence of the *negative sign* in Eq. (16.8), as the time t increases, x must also increase to stay on the crest of the wave; the wave thus travels in the *positive x* direction.

For a harmonic wave it is customary to introduce the **angular frequency**

angular frequency

$$\omega = 2\pi f = \frac{2\pi}{T} = \frac{2\pi v}{\lambda} = kv \qquad (16.10)$$

In terms of the wavelength, period, wave number, and angular frequency, we can express the wavefunction (16.8) in the alternative forms

$$y = A\cos\left(2\pi\frac{x}{\lambda} - 2\pi\frac{t}{T}\right)$$

and

harmonic wavefunction

$$y = A\cos(kx - \omega t) \qquad (16.11)$$

When the wave passes a point of the string, a particle in the string at this point moves up from its equilibrium position a distance equal to the wave amplitude A; then, half a cycle later, the particle moves down from its equilibrium position a distance A; and then, another half cycle later, it moves up again. Thus, *the particle executes simple harmonic motion of a frequency and an amplitude equal to the frequency and the amplitude of the wave* (see Fig. 16.6). The particle thus has vertical velocity and acceleration of the same form as those of a particle in simple harmonic motion:[1]

---

[1] In this differentiation, x is held constant, since the particle remains at a fixed x position and only its y position changes with t. Mathematicians would use the *partial derivative* symbol $\partial y/\partial t$ to indicate that in the differentiation of the function $y = y(x, t)$ only t is differentiated, while x is held fixed.

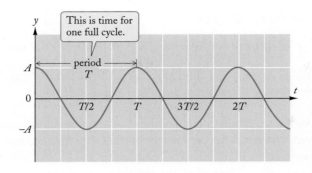

FIGURE 16.6 Each particle on the string executes simple harmonic motion with a period equal to the period of the wave, and an amplitude equal to the amplitude of the wave.

$$v_y = \frac{dy}{dt} = A\omega \sin(kx - \omega t) \tag{16.12}$$

$$a_y = \frac{dv_y}{dt} = -A\omega^2 \cos(kx - \omega t) \tag{16.13}$$

As mentioned previously, waves carry energy from one point to another. The kinetic energy of the particles involved in wave motion is proportional to the square of their velocity ($K = \frac{1}{2}mv^2$); combined with Eq. (16.12), this reveals a general feature of wave motion: *the energy stored in a wave is proportional to the square of the amplitude of the wave.*

Harmonic waves play a central role in the study of wave motion because, as we will see in the next section, any periodic wave of arbitrary shape can be regarded as a super-position, or sum, of several harmonic waves of suitably chosen amplitudes and wave-lengths. Thus, if we understand the motion of harmonic waves, we understand the motion of any kind of periodic wave. Hereafter, we will concentrate on harmonic waves.

Concepts
— in —
Context

EXAMPLE 3   The waves near the deep end of the wave pool in the chapter photo have an amplitude of 0.50 m. For the 0.40-Hz waves described in Example 2, what is the maximum vertical speed of a swimmer float-ing on the surface? What is the maximum vertical acceleration experienced by that swimmer?

SOLUTION: According to Eq. (16.11), for a swimmer at a given horizontal posi-tion x_0, the vertical displacement is

$$y = A \cos(kx_0 - \omega t)$$

and therefore the vertical speed is, as in Eq. (16.12),

$$v_y = \frac{dy}{dt} = \frac{d}{dt}[A \cos(kx_0 - \omega t)] = A\omega \sin(kx_0 - \omega t)$$

This has a maximum magnitude $A\omega$. Since $\omega = 2\pi f$, the maximum vertical speed is

$$v_{y,\text{max}} = A \times 2\pi f = 0.50 \text{ m} \times 2\pi \times 0.40 \text{ Hz} = 1.3 \text{ m/s}$$

Similarly, according to Eq. (16.13), the vertical acceleration at x_0 is

$$a_y = \frac{dv_y}{dt} = \frac{d}{dt}[A\omega \sin(kx_0 - \omega t)] = -A\omega^2 \cos(kx_0 - \omega t)$$

The maximum vertical acceleration is:

$$a_{y,\text{max}} = A\omega^2$$

Again using $\omega = 2\pi f$, this gives

$$a_{y,\text{max}} = A(2\pi f)^2 = 0.50 \text{ m} \times (2\pi \times 0.40 \text{ Hz})^2 = 3.2 \text{ m/s}^2$$

This acceleration is nearly one-third of g; it provides quite a thrilling experience.

The speed of a wave is determined by the characteristics of the medium. For a string, the relevant characteristics are the tension F in the string, its mass M, and its length L. By examining the implications of Newton's Second Law for a short deformed segment of string being accelerated by the tension forces acting on the ends of the segment, we can demonstrate that the speed of the wave on a string is

speed of wave on a string

$$v = \sqrt{\frac{F}{M/L}} \qquad\qquad (16.14)$$

To obtain this result, consider Fig. 16.7a, which shows a wave pulse on a string. We analyze the motion in a frame of reference moving to the right with the pulse; in that frame, the pulse is at rest, and the entire string moves to the left with the pulse speed v. Figure 16.7a shows one short segment ΔL in the pulse, which can be approximated as an arc of a circle with radius R, subtending a small angle $\Delta\theta = \Delta L/R$. The mass m of this segment is a fraction $\Delta L/L$ of the total mass M, $m = (\Delta L/L)M$; the centripetal acceleration is v^2/R. The radial components of the tension on either side of the segment provide the centripetal force; these sum to $F_{\text{net}} = 2F\sin(\Delta\theta/2) \approx F\Delta\theta = F\Delta L/R$ (see Fig. 16.7b). Newton's Second Law, $ma = F_{\text{net}}$, then implies

$$\frac{\Delta L}{L}M\frac{v^2}{R} = F\frac{\Delta L}{R} \qquad\qquad (16.15)$$

Solving Eq. (16.15) for v yields Eq. (16.14).

The characteristic of the string that enters into Eq. (16.14) is the ratio of mass to length (M/L), or the mass per unit length. If we use a given kind of string, of some given thickness, the mass per unit length will be the same, regardless of whether we use a short length of string or a long length in our experiments. Note that the speed of the wave is large if the tension is large and the mass per unit length is small (a thin string).

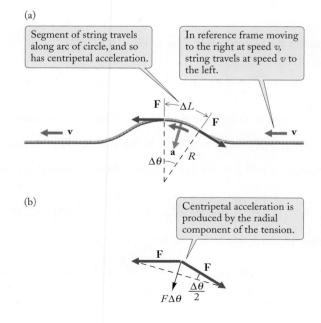

(a)

Segment of string travels along arc of circle, and so has centripetal acceleration.

In reference frame moving to the right at speed v, string travels at speed v to the left.

(b)

Centripetal acceleration is produced by the radial component of the tension.

FIGURE 16.7 (a) A wave pulse moving to the right, viewed in a frame of reference moving with the pulse, so the string moves to the left at a speed v. For a small segment ΔL, the centripetal acceleration is related to the radial component of the tension forces **F** by the geometry shown. (b) Vector sum of the tension forces **F**.

This is intuitively reasonable, since a large tension can accelerate a small mass very quickly, and therefore produces a quick back-and-forth motion of the mass elements in the string; that is, it produces a high frequency of the motion. According to Eq. (16.3), for a given wavelength, a high frequency implies a large speed of the wave.

Of course, Eq. (16.14) is valid not only for strings, but also for tightly stretched wires and ropes.

EXAMPLE 4   A long piece of piano wire of a mass of 3.9×10^{-3} kg per meter is under a tension of 1.0×10^3 N. What is the speed of transverse waves on this wire? What is the wavelength of a harmonic wave on this wire if its frequency is 262 Hz?

SOLUTION: If the mass per unit length is 3.9×10^{-3} kg per meter, the ratio M/L in Eq. (16.14) is 3.9×10^{-3} kg/m, and hence the speed of the wave is

$$v = \sqrt{\frac{F}{M/L}} = \sqrt{\frac{1.0 \times 10^3 \, \text{N}}{3.9 \times 10^{-3} \, \text{kg/m}}} = 5.1 \times 10^2 \, \text{m/s}$$

Consequently, with $f = 262$ Hz $= 262 \, \text{s}^{-1}$, the wavelength is

$$\lambda = \frac{v}{f} = \frac{510 \, \text{m/s}}{262 \, \text{s}^{-1}} = 1.9 \, \text{m}$$

Although a wave on a string is a rather special case of wave motion, the mathematical description of other kinds of waves is similar to that of waves on a string. The instantaneous configuration of the wave can always be described by a plot of the wave disturbance vs. position, such as the plot in Fig. 16.6, but the vertical axis of the plot must be adapted to the physical properties of the wave. For instance, to describe a sound wave in air, we can plot the pressure disturbance produced by the wave vs. the position.

Our Eq. (16.14) for the wave speed applies only to waves on a string (or a wire, or a rope). But this equation exhibits a general feature of wave propagation: in broad terms, this equation states that the speed of the wave depends on the restoring force and on the inertia of the elastic medium in which the wave is propagating. This is true for all kinds of waves. In all cases, some force within the medium opposes its deformation—tension tends to keep the string straight, the pressure within a gas tends to keep the density of the gas uniform, gravity tends to keep the surface of the sea smooth, and so on. But if something provides an initial disturbance, then the restoring force will cause it to propagate, as in Fig. 16.2, with a speed depending on the magnitude of the restoring force and on the amount of inertia or, equivalently, the amount of mass in the medium. In general, the speed will be large if the restoring force is large and the amount of mass in the medium is small.

EXAMPLE 5   Gravity provides the restoring force for the water waves in a wave pool. A somewhat complicated analysis of the motion of water, including the requirement that the vertical velocity is zero at the bottom of the pool, yields for the wave speed in shallow water the approximate formula

$$v = \sqrt{gD}$$

where g is the acceleration of gravity and D is the depth of the water. This formula is valid when the depth is much less than the wavelength. For the shallow wave pool described in Examples 2 and 3, the speed at the deep end was 5.2 m/s. Recall that pumps produced waves at a rate of 0.40 Hz, and the resulting wavelength at the deep end was 13 m. What is the speed at the shallow end, where $D = 0.50$ m? What are the frequency and wavelength of the wave there?

SOLUTION: Using the given expression, the speed at the shallow end is

$$v = \sqrt{gD} = \sqrt{9.81 \text{ m/s}^2 \times 0.50 \text{ m}} = 2.2 \text{ m/s}$$

The frequency of the wave is unchanged. This is so because at any fixed location, water elements are pushed by adjacent elements; ultimately, this process begins at the water-pump source, which sets the overall frequency. To confirm this, we can consider a counterexample: if the frequency were to decrease away from the source, oscillations would continually "pile up" somewhere, since more would be steadily produced at the source than would pass another location. This is an impossible consequence; thus, the frequency must be the same at different locations.

The wavelength is, by Eq. (16.3),

$$\lambda = \frac{v}{f} = \frac{2.2 \text{ m/s}}{0.40 \text{ s}^{-1}} = 5.5 \text{ m}$$

Thus the wavelength is shorter in a region where the speed is slower.

✔ Checkup 16.2

QUESTION 1: Are all periodic waves harmonic? Are all harmonic waves periodic?

QUESTION 2: Are the waves sketched in Fig. 16.4 harmonic?

QUESTION 3: Consider the piano wire described in Example 4. If we want to increase the wave speed by a factor of 2, by what factor must we increase the tension?

QUESTION 4: A steel wire stretched tightly across a room consists of two segments of the same material but of different diameters. If the first segment has a diameter half as large as the second segment, by what factor do the speeds of waves in the two segments differ?

QUESTION 5: A wave on a string has a wavelength of 30 cm and a frequency of 40 Hz. What is the frequency of a wave of wavelength 60 cm on this same string?

(A) 20 Hz   (B) 40 Hz   (C) 60 Hz   (D) 80 Hz   (E) 120 Hz

Online
Concept
Tutorial

16.3 THE SUPERPOSITION OF WAVES

Waves on a string and waves in other elastic bodies usually obey a **Superposition Principle:** *when two or more waves are present simultaneously in an elastic body, the resultant instantaneous displacement of a particle is the sum of the individual instantaneous displacements.* Such a superposition means that the waves do not interact; they have no effect on one another. Each wave propagates as though the other were not present, and the contribution that each makes to the displacement of a particle in the elastic body is as though the other were not present. For instance, if the sound waves from a violin and a flute reach us simultaneously, then each of these waves produces a displacement of the air molecules just as though it were acting alone, and the net displacement of the air molecules is simply the (vector) sum of these individual displacements.

For waves of low amplitude on a string and for sound waves of ordinary intensity in air, the Superposition Principle is very well satisfied. However, for waves of very large amplitude or intensity, the Superposition Principle fails. When a wave of very large amplitude is propagating on a string, it alters the tension of the string, and therefore affects the behavior of a second wave propagating on the same string. Likewise, a very intense sound wave (a shock wave, such as the loud bang from an explosion) produces significant alterations of the temperature and the pressure of the air, and therefore affects the behavior of a second wave propagating through this same region. However, we will not worry about such extreme conditions, and we will assume that the Superposition Principle is applicable.

As a first example of superposition, let us consider two waves propagating in the same direction with the same frequency and the same amplitude. If the wave crests and the wave troughs of the two waves coincide, the waves are said to be **in phase.** The superposition of these two waves yields a wave of twice the amplitude of the individual waves (see Fig. 16.8). *Such a reinforcement of one wave by another is called* **constructive interference.** If the wave crests of one wave match the wave troughs of the other, the waves are said to be **out of phase,** or to differ in phase by half a cycle. The superposition of these two waves yields a wave of zero amplitude (see Fig. 16.9). *Such a cancellation of one wave by another is called* **destructive interference.**[2]

If the two waves are out of phase but their amplitudes are not equal, then their cancellation will not be total; some portion of the wave that has the larger amplitude will be left over (see Fig. 16.10). Similarly, if two equal-amplitude waves are not exactly in phase or exactly out of phase, then their sum will have an amplitude somewhere between zero and twice the amplitude of either wave. In Chapter 35, we will examine interference of waves with the same frequency in more detail.

For a much different example of superposition, let us consider two waves of the same amplitude, but slightly different frequencies and, therefore, slightly different wavelengths. Figure 16.11 shows the two waves at one instant of time and their superposition. At $x = 0$, the waves are in phase, and they interfere constructively, giving a large net amplitude. But farther along the x axis, the difference in wavelengths gradually causes the waves to acquire a phase difference. At the point P, the waves are out of phase by half a cycle, and they interfere destructively, giving a net amplitude of zero. Beyond this point, the phase difference exceeds one half cycle. At the point R the phase difference has grown to one cycle; but since a phase difference of one cycle means that the crests of the two waves coincide, they interfere constructively, again giving a large amplitude, and so on. Thus, the superposition of the two waves displays regularly alternating regions of constructive and destructive interference, that is, alternating regions of large amplitude and small amplitude (see Fig. 16.11b).

---

[2] This cancellation raises a question: If the waves cancel, what happens to the energy they carry? To answer this question we must examine in detail how the two waves were brought together. For instance, if the two waves were initially propagating on two separate strings that merge into a single string at a junction or knot, then the cancellation of the waves beyond the junction is necessarily associated with a strong backward reflection of the two incident waves at the junction, and the waves reflected backward from the junction carry away the missing energy.

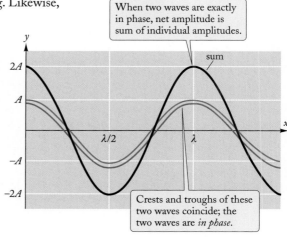

FIGURE 16.8 Constructive interference of two waves.

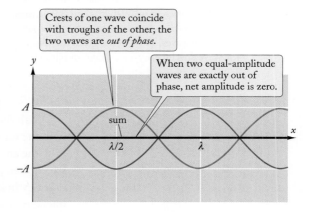

FIGURE 16.9 Destructive interference of two waves; the waves cancel everywhere.

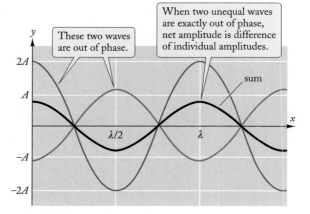

FIGURE 16.10 Destructive interference of two waves of different amplitudes. The sum is small, but not zero.

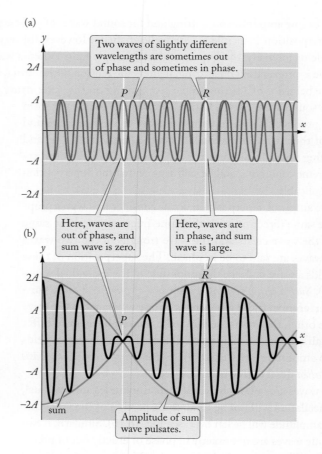

FIGURE 16.11 Superposition of two waves of slightly different wavelengths and frequencies. (a) The two waves before addition. (b) The sum of the two waves. The colored green line shows the wave envelope, or the average amplitude.

With the passing of time, the entire pattern in Fig. 16.11b moves toward the right with the wave velocity. This gives rise to the phenomenon of **beats.** At any given position, the amplitude of the wave pulsates—first the amplitude is large, then it becomes small, then again large, and so on. *The frequency with which the amplitude pulsates is called the* **beat frequency.** The beat frequency is simply the difference between the frequencies of the two waves:

beat frequency

$$f_{\text{beat}} = f_1 - f_2 \tag{16.16}$$

To establish this result, consider two traveling waves of the form (16.11), with different frequencies:

$$y_1 = A \cos(k_1 x - \omega_1 t)$$
$$y_2 = A \cos(k_2 x - \omega_2 t) \tag{16.17}$$

For simplicity, let us consider these wavefunctions at a single point, $x = 0$. The superposition of these waves is the sum

$$y = y_1 + y_2 = A[\cos(\omega_1 t) + \cos(\omega_2 t)] \tag{16.18}$$

where we have used $\cos(-\theta) = \cos\theta$. If we apply the trigonometric identity (see Appendix 3)

$$\cos\theta_1 + \cos\theta_2 = 2\cos\left(\frac{\theta_1 - \theta_2}{2}\right)\cos\left(\frac{\theta_1 + \theta_2}{2}\right) \tag{16.19}$$

we obtain

$$y = 2A \cos\left[\frac{(\omega_1 - \omega_2)}{2}t\right] \cos\left[\frac{(\omega_1 + \omega_2)}{2}t\right] \qquad (16.20)$$

The second cosine in Eq. (16.20) represents the rapid oscillations of the wave at the average angular frequency $(\omega_1 + \omega_2)/2$. The first cosine function in Eq. (16.20) represents the slow variation in the amplitude of that wave, producing the beats. Both the positive and negative parts of the slowly varying "envelope" in Fig. 16.11b produce large amplitudes; hence there are two beats per cycle of the function $\cos[(\omega_1 - \omega_2)t/2]$, and the angular beat frequency is $2 \times (\omega_1 - \omega_2)/2 = \omega_1 - \omega_2$. Thus the beat frequency is given by Eq. (16.16).

EXAMPLE 6   Suppose that two flutes generate sound waves of frequency 264 Hz and 262 Hz, respectively. What is the beat frequency?

SOLUTION: According to Eq. (16.16),

$$f_{\text{beat}} = f_1 - f_2 = 264 \text{ Hz} - 262 \text{ Hz} = 2 \text{ Hz}$$

Hence, a listener will hear a tone of average frequency 263 Hz, but with an amplitude pulsating 2 times per second.

Beats are a sensitive indication of small frequency differences, and they are very useful in the tuning of musical instruments. For example, when musicians want to bring two flutes in tune, they listen to the beats and, by trial and error, adjust one of the flutes so as to reduce the beat frequency; when the beats disappear entirely (zero beat frequency), the two flutes will be generating waves of exactly the same frequencies.

By the superposition of harmonic waves of different amplitudes and wavelengths, we can construct some rather complicated waveshapes. For example, Fig. 16.12 shows a periodic wave constructed by the superposition of three harmonic waves of wavelengths L, $L/3$, and $L/5$ whose amplitudes are in the ratio $1 : \frac{1}{3} : \frac{1}{5}$. It can be shown that any arbitrary periodic wave can be constructed by the superposition of a sufficiently large number of harmonic waves. If the desired periodic wave repeats each distance L, then the only harmonic waves needed in the sum are those with wavelengths equal to L divided by an integer. Since $f = v/\lambda$, the frequencies of the only harmonic waves needed are integer multiples of the frequency of the arbitrary periodic wave. This is called **Fourier's theorem.** As already mentioned in the preceding section, this theorem means that we can regard any arbitrary periodic wave as a sum of harmonic waves; and it means that the study of periodic waves is, in essence, the study of harmonic waves.

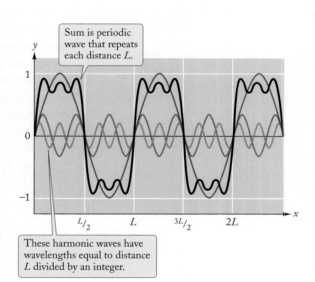

FIGURE 16.12 A wave constructed by superposition of three harmonic waves of wavelengths L (blue), $L/3$ (red), and $L/5$ (green). The harmonic waves are $y_{\text{blue}} = \sin(2\pi x/L)$, $y_{\text{red}} = \frac{1}{3}\sin[2\pi x/(L/3)]$, and $y_{\text{green}} = \frac{1}{5}\sin[2\pi x/(L/5)]$.

 ## Checkup 16.3

QUESTION 1: Consider a harmonic wave of small amplitude and short wavelength and a wave of much larger amplitude and much longer wavelength. What does the superposition of these two waves look like?

QUESTION 2: To check the tuning of a guitar, the lowest-pitch string with a finger on the fifth fret and the next string (open) are played simultaneously. If, for an out-of-tune guitar, the resulting tone pulsates 4 times per second and has a frequency of 107 Hz, at what frequencies are the two strings vibrating?

QUESTION 3: Two water waves of amplitudes 0.6 m and 0.8 m, respectively, arrive simultaneously at a buoy. What is the amplitude of the net wave if these two waves are in phase? If these two waves are out of phase?

(A) 0.2 m; 1.4 m   (B) 0.2 m; 0.2 m   (C) 1.0 m; 1.4 m
(D) 1.4 m; 0.2 m   (E) 1.4 m; 1.0 m

Online Concept Tutorial 18

16.4 STANDING WAVES

Next, we want to consider the superposition of two waves of the same amplitude and the same frequency, but of opposite directions of propagation. Figure 16.13a shows the two traveling waves and their sum at an initial instant of time. At this instant, the waves are in phase, and their sum is a wave of twice the amplitude of each. At a slightly later time, one wave has moved to the right and the other to the left (see Fig. 16.13b). At the points $P, Q, R, \ldots$ the waves were initially zero; now, one wave has a positive value at these points, and the other wave has an equally large negative value. Thus, the sum of the two waves still yields a result of zero at these points. The two waves will continue to cancel at these points at all times (see Figs. 16.13c, d, and e). These points at which the sum of the waves is zero are called **nodes**. They are one-half wavelength apart.

Midway between the nodes we find points at which the sum of the two waves is maximum (positive or negative). These points are called **antinodes**. Figure 16.13 shows that with the passing of time, the height of the wave crests at the antinodes oscillates, but the positions of these wave crests remain fixed. Thus, *the superposition of two waves traveling in opposite directions is a* **standing wave**. This means that the net wave travels neither to the right nor to the left; its wave crests remain at fixed positions while the entire wave increases and decreases in unison. The frequency of this pulsation of the standing wave is the same as the frequency of the two underlying traveling waves.

We can find the wavefunction for the standing wave by adding the wavefunctions of the individual traveling waves. The wave traveling to the right is

$$y_1 = A \cos(kx - \omega t) \tag{16.21}$$

and the wave traveling toward the left is

$$y_2 = A \cos(kx + \omega t) \tag{16.22}$$

The sum of these two waves is

$$y = y_1 + y_2 = A \cos(kx - \omega t) + A \cos(kx + \omega t) \tag{16.23}$$

To evaluate this sum, we again use the trigonometric identity (16.19)

$$\cos \theta_1 + \cos \theta_2 = 2 \cos[\tfrac{1}{2}(\theta_1 - \theta_2)] \cos[\tfrac{1}{2}(\theta_1 + \theta_2)] \tag{16.24}$$

which, with $\theta_1 = kx - \omega t$ and $\theta_2 = kx + \omega t$, gives us

$$y = 2\cos[\tfrac{1}{2}(kx - \omega t - kx - \omega t)] \cos[\tfrac{1}{2}(kx - \omega t + kx + \omega t)]$$
$$= 2A \cos(-\omega t) \cos(kx) = 2A \cos(\omega t) \cos(kx) \tag{16.25}$$

The time dependence in this wavefunction appears as an *overall factor* $\cos(\omega t)$, which shows that the entire wave indeed increases and decreases in unison. The dependence on x appears as an overall factor of $\cos(kx)$. This factor gives us the positions of the antinodes and nodes. The antinodes correspond to a maximum (positive or negative) value of $\cos(kx)$, which occurs at

$$kx = 0,\ \pi,\ 2\pi,\ 3\pi,\dots \qquad \text{for antinodes (maxima)} \qquad (16.26)$$

The nodes correspond to a zero value of $\cos(kx)$, which occurs at

$$kx = \pi/2,\ 3\pi/2,\ 5\pi/2,\dots \text{ for nodes (zeros)} \qquad (16.27)$$

With $k = 2\pi/\lambda$, the condition for the antinodes becomes

$$x = 0,\ \lambda/2,\ \lambda,\ 3\lambda/2,\dots \quad \text{for antinodes (maxima)} \qquad (16.28)$$

and the condition for the nodes becomes

$$x = \lambda/4,\ 3\lambda/4,\ 5\lambda/4,\dots \text{ for nodes (zeros)} \qquad (16.29)$$

As expected from the discussion of Fig. 16.13, the nodes are one-half wavelength apart, and the antinodes are midway between the nodes.

If this standing wave is a wave on a string, then each particle of this string executes simple harmonic motion. However, in contrast to the case of a traveling wave, where the amplitudes of the harmonic oscillations of all the particles are the same, the amplitudes of oscillation now depend on position: the amplitude is maximum at the antinodes, and it is minimum (zero) at the nodes.

So far, in our discussion of the waves on a string, we have assumed that the string is very long, and we have ignored the endpoints of the string. When a traveling wave arrives at an endpoint, something drastic will have to happen to it: the wave will either have to be absorbed at the endpoint or it will have to be reflected, with a reversal of its direction of propagation. If the endpoint is a fixed point (the string is attached to a rigid support), then the endpoint cannot absorb the energy of the wave, and the wave will be completely reflected. This results in the simultaneous presence of two waves of equal amplitudes and opposite directions of travel; that is, it results in a standing wave. Reflection will also occur if the endpoint is free to move but cannot absorb energy; in this case, the endpoint is an antinode. For the case of a fixed endpoint, the reflected wave is inverted, as illustrated by the behavior of a reflected pulse in Fig. 16.14a; this inversion ensures that the endpoint is a node. For the case of a free endpoint, the reflected wave is not inverted, as illustrated in Fig. 16.14b.

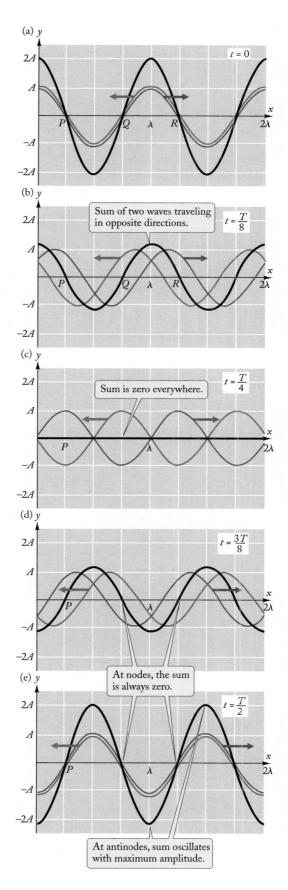

FIGURE 16.13 Superposition of two waves of the same amplitude but of opposite directions of propagation. (a) At $t = 0$ the waves are in phase. (b) At $t = \frac{1}{8}$ of a period $= \frac{1}{8}\,T$, one wave has moved $\frac{1}{8}$ of a wavelength to the right, and the other the same distance to the left. (c) At $t = \frac{1}{4}\,T$, one wave has moved $\frac{1}{4}$ of a wavelength to the right and the other the same distance to the left; the waves have therefore moved apart $\frac{1}{2}$ wavelength, and they are out of phase and they cancel everywhere. (d) At $t = \frac{3}{8}\,T$, the waves have moved $\frac{3}{8}$ of a wavelength to the right and the left, respectively. At $t = \frac{1}{2}\,T$, the waves are again in phase, and so on.

(a)

(b)

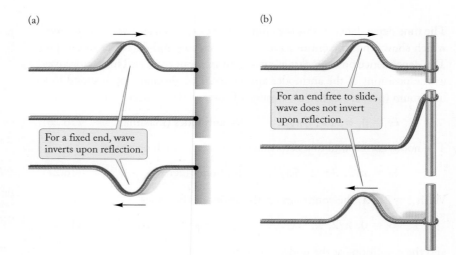

For a fixed end, wave inverts upon reflection.

For an end free to slide, wave does not invert upon reflection.

FIGURE 16.14 (a) A wave pulse approaches a fixed end of a string. The pulse is inverted upon reflection, as it must be for the superposition of the incoming and outgoing displacements to be zero at the endpoint. (b) A wave pulse approaching the end of a string that is free to slide vertically. The pulse does not invert upon reflection.

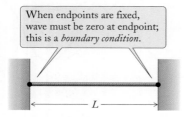

When endpoints are fixed, wave must be zero at endpoint; this is a *boundary condition.*

L

FIGURE 16.15 A tightly stretched string with fixed ends. The next figure shows possible standing waves on this string.

For a string with two fixed endpoints (see Fig. 16.15), the possible standing waves are subject to the restriction that the wave must be zero at each endpoint at all times. Such a restriction on what happens at the endpoints of a wave is called a **boundary condition.** Obviously, the boundary condition for our standing wave will be satisfied if the endpoints are nodes. Figure 16.16 shows possible standing waves on a string of some given length L; all these standing waves have nodes at the endpoints. *The possible standing-wave motions of the string shown in Fig. 16.16 are called the* **normal modes**. Figure 16.16a shows the **fundamental** mode; Fig. 16.16b, the **first overtone**; Fig. 16.16c, the **second overtone**. Figure 16.17 shows time-exposure photographs of a string with these modes.

(a) The fundamental mode

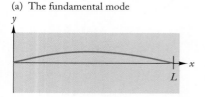

(b) The first overtone

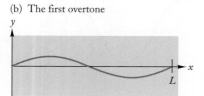

(c) The second overtone

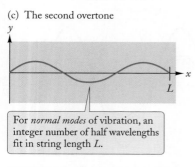

For *normal modes* of vibration, an integer number of half wavelengths fit in string length L.

FIGURE 16.16 Standing waves on a string. (a) The fundamental mode. (b) The first overtone. (c) The second overtone.

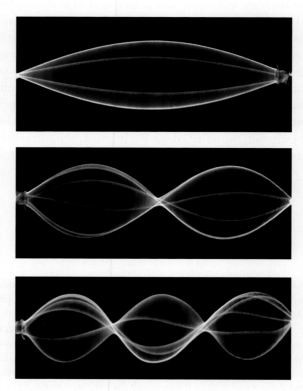

FIGURE 16.17 Stroboscopic photographs of standing waves on a string.

In all these modes, some integer number of half wavelengths exactly fits the length of the string. In the fundamental mode, one half wavelength fits the string; in the first overtone, two half wavelengths fit the string; in the second overtone, three half wavelengths fit the string; and so on. Thus, the normal modes occur when

$$L = \frac{\lambda}{2} \qquad L = 2\frac{\lambda}{2} \qquad L = 3\frac{\lambda}{2} \quad \cdots$$

and the wavelengths for the different modes are thus

$$\lambda_1 = 2L \qquad \lambda_2 = \frac{2L}{2} = L \qquad \lambda_3 = \frac{2L}{3} \quad \cdots \qquad (16.30)$$

wavelengths of normal modes of string

The frequencies of oscillation of the modes are related to the wavelengths in the usual way ($f = v/\lambda$, where v is the velocity):

$$f_1 = \frac{v}{2L} \qquad f_2 = 2\frac{v}{2L} \qquad f_3 = 3\frac{v}{2L} \quad \cdots \qquad (16.31)$$

The frequencies of these modes are called the normal frequencies, proper frequencies, or **eigenfrequencies** of the string. Note that all these frequencies are integer multiples of the **fundamental frequency**: the frequency of the first overtone is twice the frequency of the fundamental, the frequency of the second overtone is three times the frequency of the fundamental, and so on. The first overtone is also called the **second harmonic**; the second overtone is called the **third harmonic**; and so on.

eigenfrequencies

In general, any arbitrary motion of a freely vibrating string (with fixed endpoints) will be some superposition of several of the above normal modes. Which modes will be present in the superposition depends on how the motion is started. For instance, when a guitar player plucks a string on his guitar near the middle, he will excite the fundamental mode and also the second overtone and, to a lesser extent, some of the higher even-numbered overtones.

EXAMPLE 7   The low E string on the guitar vibrates with a frequency of 82.4 Hz when excited in its fundamental mode. What are the frequencies of the first, second, and third overtones of this string?

SOLUTION: According to Eq. (16.31), $f_2 = 2f_1$, $f_3 = 3f_1$, and $f_4 = 4f_1$. Hence the frequencies of the first, second, and third overtones are, respectively, $2 \times 82.4\ \text{Hz} = 165\ \text{Hz}$, $3 \times 82.4\ \text{Hz} = 247\ \text{Hz}$, and $4 \times 82.4\ \text{Hz} = 330\ \text{Hz}$.

The normal modes of vibration of a long, thin elastic rod or a beam fixed at both ends are mathematically similar to the normal modes of a string. However, such an elastic body can experience transverse deformations (like those of a string), compressional deformations, and rotational, or "torsional," deformations.

Recall from Chapter 15 that if an external driving force oscillates at the natural frequency of oscillation of a system, a large-amplitude oscillation can be built up, a phenomenon called **resonance**. Figure 16.18 shows a spectacular example of a torsional standing wave in the span of a bridge at Tacoma, Washington. This standing wave was excited by a wind blowing across the bridge, which generated a periodic succession of vortices, or regions of swirling motion in the air, with a frequency equal to

FIGURE 16.18 Standing wave on the deck of the Tacoma Narrows bridge, July 1, 1940. The bridge broke apart a short time after this picture was taken.

that of one of the normal modes of vibration of the span. Thus, the periodic generation of vortices was in resonance with the natural vibration of the bridge, and this gradually built up a large amplitude of vibration. The bridge vibrated for several hours, with increasing amplitude, and then broke apart.

✔ Checkup 16.4

QUESTION 1: Describe the third overtone and the fourth overtone for the wave on the string illustrated in Fig. 16.15.

QUESTION 2: Can the midpoint of a guitar string be an antinode? Can it be a node?

QUESTION 3: Is the number of nodes for a normal mode of a string fixed at both ends always larger than the number of antinodes? How much larger?

QUESTION 4: The lowest frequency on a guitar is obtained when the top string (the low E string) is played open, in its fundamental mode; the eighth harmonic of that frequency can be obtained on the bottom guitar string (the high E string) when it is constrained at its midpoint. Both strings have the same length. What is the ratio of the wave speed on the high E string to the wave speed on the low E string?

(A) 1   (B) 2   (C) 4   (D) 8   (E) 16

SUMMARY

WAVELENGTH, PERIOD, FREQUENCY, AND WAVE SPEED	$\dfrac{\lambda}{T} = \lambda f = v$	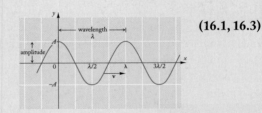	**(16.1, 16.3)**
WAVE NUMBER	$k = \dfrac{2\pi}{\lambda}$		**(16.7)**
ANGULAR FREQUENCY	$\omega = 2\pi f$		**(16.10)**
HARMONIC WAVE			
Wave traveling in the positive x direction:	$y = A\cos(kx - \omega t)$		**(16.8)**
Wave traveling in the negative x direction:	$y = A\cos(kx + \omega t)$		**(16.9)**
SPEED OF WAVE ON A STRING	$v = \sqrt{\dfrac{F}{M/L}}$		**(16.14)**

SUPERPOSITION PRINCIPLE FOR TWO OR MORE WAVES
The net instantaneous displacement is the sum of the
individual instantaneous displacements.

CONSTRUCTIVE INTERFERENCE   Waves meet crest to crest.

DESTRUCTIVE INTERFERENCE   Waves meet crest to trough.

BEAT FREQUENCY $f_{beat} = f_1 - f_2$ 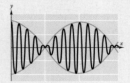   (16.16)

STANDING HARMONIC WAVE $y = A \cos(kx) \cos(\omega t)$ (16.25)

NODE   Point of zero oscillation.

ANTINODE   Point of maximum oscillation.

WAVELENGTHS OF NORMAL MODES OF STRING $\lambda_1 = 2L,\ \lambda_2 = L,\ \lambda_3 = 2L/3$, etc.   (16.30)
(of length L, fixed at both ends)

For *normal modes* of vibration, an integer number of half wavelengths fit in string length L.

EIGENFREQUENCIES $f_1 = v/2L$ fundamental mode (or first harmonic),   (16.31)
$f_2 = 2(v/2L)$ first overtone (or second harmonic),
$f_3 = 3(v/2L)$ second overtone (or third harmonic), etc.

QUESTIONS FOR DISCUSSION

1. You have a long, thin steel rod and a hammer. How must you hit the end of the rod to generate a longitudinal wave? A transverse wave?

2. Some people enjoy arranging long rows of dominoes on the floor, so the toppling of one domino triggers the toppling of all the others, by a chain reaction (see Fig. 16.19). The propagation of the disturbance along such a chain of dominoes has some of the properties of a wave pulse. In what way is it similar to a wave pulse? In what way is it different?

FIGURE 16.19
Toppling dominoes.

3. A wave pulse on a string transports energy. Does it also transport momentum? To answer this question, imagine a washer loosely encircling the string at some place; what happens to the washer when the wave pulse strikes it?

4. According to Eq. (16.14), the speed of a wave on a string increases by a factor of 2 if we increase the tension by a factor of 4. However, in the case of a rubber string, the speed increases by more than a factor of 2 if we increase the tension by a factor of 4. Why are rubber strings different?

5. A harmonic wave is traveling along a string. Where in this wave is the kinetic energy at maximum? The potential energy? The total energy?

6. Suppose that two strings of different densities are knotted together to make a single long string. If a wave pulse travels along the first string, what will happen to the wave pulse when it reaches the junction? (Hint: If the second string had the same density as the first string, the wave pulse would proceed

without interruption; if the second string were much denser than the first, the wave pulse would be totally reflected.)

7. Figure 16.13 shows a standing wave on a string. At time $t = T/4$, the amplitude of the wave is everywhere zero. Does this mean the wave has zero energy at this instant?

8. After an arrow has been shot from a bow, the bowstring will oscillate back and forth, forming a standing wave. Which of the overtones shown in Fig. 16.16 do you expect to be present?

9. In tuning a guitar or violin, by what means do you change the frequency of a string?

10. A mechanic can make a rough test of the tension on the spokes of a wire wheel (see Fig. 16.20) by striking the spokes with a wrench or a small hammer. A spoke under tension will ring, but a loose spoke will not. Explain.

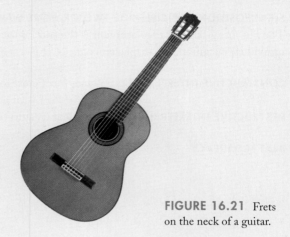

FIGURE 16.21 Frets on the neck of a guitar.

11. What is the purpose of the frets (see Fig. 16.21) on the neck of a guitar or a mandolin?

12. The strings of a guitar are made of wires of different thicknesses (the thickest strings are manufactured by wrapping copper or brass wire around a strand of steel). Why is it impractical to use wire of the same thickness for all the strings?

FIGURE 16.20 Wire wheels on an automobile.

PROBLEMS

16.2 Periodic Waves

1. The speed of light waves is 3.0×10^8 m/s. The wavelengths of light waves range from 4.0×10^{-7} m (violet) to 7.0×10^{-7} m (red). What is the range of frequencies of these waves?

2. An ocean wave has a wavelength of 120 m and a period of 8.77 s. Calculate the frequency, angular frequency, wave number, and speed of this wave.

3. Figure 16.22 is a record of a tsunami that struck the coast of Mexico. Approximately, what was the frequency of this wave? What was its wavelength in the open sea? Assume that the speed of the wave in the open sea was 740 km/h.

4. In deep water (where the depth is much larger than the wavelength), the speed of waves is given by the formula $v = \sqrt{g\lambda/2\pi}$. Calculate the speed of short water waves, with $\lambda = 1.0$ m. Calculate the speed of long waves with $\lambda = 300$ m.

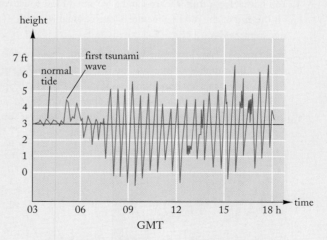

FIGURE 16.22 Height of a tsunami that struck the coast of Mexico.

5. In shallow water (where the depth is shorter than a wavelength), the speed of waves is given by the formula $v = \sqrt{gD}$, where D is the depth of the water.

   (a) Calculate the speed of water waves in a shallow pond with a depth of 2.0 m.

   (b) For ocean waves of extremely long wavelength, such as tidal waves, the oceanic basins can be treated as shallow ponds, since their depth is small compared with the wavelength. Calculate the speed of a tidal wave in the Pacific Ocean, where the mean depth is 4.3 km.

6. To determine the speed and the frequency of periodic waves on a lake, the owner of a motorboat first runs the boat in the direction of the waves, and he finds that when his boat keeps up with a wave crest the speed indicator shows 16 m/s. He then anchors the boat and finds that the waves make it bounce up and down 6.0 times per minute. What are the speed, frequency, and wavelength of the waves?

7. The speed of tidal waves in the Pacific is about 740 km/h.

   (a) How long does a tidal wave take to travel from Japan to California, a distance of 8000 km?

   (b) If the wavelength of the wave is 300 km, what is its frequency?

*8. In the open sea, a tsunami usually has an amplitude less than 30 cm and a wavelength longer than 80 km. Assume that the speed of the tsunami is 740 km/h. What are the maximum vertical velocity and acceleration that such a tsunami will give to a ship floating on the water? Will the crew of the ship notice the passing of the tsunami?

9. A transverse harmonic wave on a stretched string has an amplitude of 1.2 cm, a speed of 8.0 m/s, and a wavelength of 2.2 m.

   (a) What is the maximum transverse speed attained by a particle on the string? Does the particle attain this maximum speed when a wave crest passes the particle or at some other time?

   (b) What is the maximum transverse acceleration attained by a particle on the string? Does the particle attain this maximum acceleration when a wave crest passes the particle or at some other time?

10. When a periodic transverse wave travels along a clothesline, a ladybug sitting on the line experiences a maximum transverse velocity of 0.20 m/s and a maximum transverse acceleration of 4.0 m/s². Deduce the amplitude and the frequency of the wave.

11. A transverse wave travels along a stretched string with a wave speed of 14 m/s. A particle at a fixed location on this string oscillates up and down as follows as a function of time:

$$y = 0.020 \cos(9.0t)$$

where the displacement y is measured in meters and the time t is measured in seconds.

   (a) What is the amplitude of the wave?

   (b) What is the frequency of the wave?

   (c) What is the wavelength of the wave?

12. Suppose that the function $y = 6.0 \times 10^{-3} \cos(20x + 4.0t + \pi/3)$ describes a wave on a long string (distance is measured in meters and time in seconds).

   (a) What are the amplitude, wavelength, wave number, frequency, angular frequency, direction of propagation, and speed of this wave?

   (b) At what time does this wave have a maximum at $x = 0$?

13. A harmonic wave on a string has an amplitude of 2.0 cm, a wavelength of 1.2 m, and a velocity of 6.0 m/s in the positive x direction. At time $t = 0$, this wave has a crest at $x = 0$.

   (a) What are the period, frequency, angular frequency, and wave number of this wave?

   (b) What is the mathematical equation describing this wave as a function of x and t?

14. Ocean waves smash into a breakwater at the rate of 12 per minute. The wavelength of these waves is 39 m. What is their speed?

15. The velocity of sound in freshwater at 15°C is 1440 m/s, and at 30°C it is 1530 m/s. Suppose that a sound wave of frequency 440 Hz penetrates from a layer of water at 30°C into a layer of water at 15°C. What will be the change in the wavelength? Assume that the frequency remains unchanged.

16. A light wave of frequency 5.5×10^{14} Hz penetrates from air into water. What is its wavelength in air? In water? The speed of light is 3.0×10^8 m/s in air and 2.3×10^8 m/s in water; assume that the frequency remains the same.

17. The National Ocean Survey has deployed buoys off the Atlantic coast to measure ocean waves. Such a buoy detects waves by the vertical acceleration that it experiences as it is lifted and lowered by the waves. In order to calibrate the device that measures the acceleration, scientists placed the buoy on a Ferris wheel at an amusement park. The vertical acceleration (as a function of time) of a buoy riding on a Ferris wheel of radius 6.1 m rotating at 6.0 rev/min is used to simulate the vertical acceleration of a buoy riding on a wave. What is the maximum vertical acceleration? What is the wavelength of the corresponding wave? Assume that the waves are in deep water and that the buoy always rides on the surface of the wave. In deep water, the wave speed is given by $v = \sqrt{g\lambda/2\pi}$, where g is the acceleration of free fall and λ is the wavelength."

18. The wavefunction for a wave on a string is

$$y = A \cos(kx - \omega t + \delta)$$

where, as in Chapter 15, δ is a phase constant. If $A = 0.13$ m, $\omega = 20\pi$ radians/s, $k = 15\pi$ m^{-1}, and $\delta = \pi/4$, what is the speed of the wave? What is its wavelength? What is its frequency? What is the maximum transverse speed of a particle in the string? What is the vertical displacement of the string at $x = 0$ and $t = 0$?

19. Nine water wave crests and troughs pass a point in 15 s. If the horizontal distance between a crest and the nearest trough is 0.75 m, find the speed of the wave.

20. An astronaut wishes to measure gravitational acceleration. A 7.0-kg mass is suspended from a thin wire of length 2.0 m and

mass 6.0 g. A pulse travels the length of the wire in 33 ms. What is the local value of the acceleration due to gravity? (Neglect the mass of the wire when determining the tension.)

21. A copper wire is 100 m long and 0.50 mm in diameter. The wire is stretched to a tension of 75 N. How long does a wave pulse take to travel to the end of the wire?

22. A pendulum is made of a 2.0-kg mass hanging from a string of mass 5.0 g. For small amplitude, a complete pendulum oscillation takes 1.0 s. Find the speed of a transverse wave on the pendulum string.

23. After adjusting its tension to 150 N, an electrician taps a cable that is hanging between two utility poles. She notes that it takes 3.5 s for a pulse to travel to the next pole, 30 m away. What is the mass per unit length of the wire?

24. A piano wire of mass 35 g is 1.50 m long when unstretched, and has a cross-sectional area of 3.0 mm^2. The Young's modulus of the wire is 1.5×10^{11} N/m^2. If the wire is stretched 2.0 cm, find the speed of transverse waves on the wire.

25. A string has a length of 3.0 m and a mass of 12 g. If this string is subjected to a tension of 250 N, what is the speed of transverse waves?

26. A clothesline of length 10 m is stretched between a house and a tree. The clothesline is under a tension of 50 N, and it has a mass per unit length of 6.0×10^{-2} kg/m. How long does a wave pulse take to travel from the house to the tree and back?

27. A wire rope used to support a radio mast has a length of 20 m and a mass per unit length of 0.80 kg/m. When you give the wire rope a sharp blow at the lower end and generate a wave pulse, it takes 1.0 s for this wave pulse to travel to the upper end and to return. What is the tension in the wire rope?

28. A nylon rope of length 24 m is under a tension of 1.3×10^4 N. The total mass of this rope is 2.7 kg. If a wave pulse starts at one end of this rope, how long does it take to reach the other end?

*29. Ocean waves of wavelength 100 m have a speed of 6.2 m/s; ocean waves of wavelength 20 m have a speed of 2.8 m/s.[3] Suppose that a sudden storm at sea generates waves of all wavelengths. The long-wavelength waves travel fastest and reach the coast first. A fisherman standing on the coast first notices the arrival of 100-m waves; 10 hours later he notices the arrival of 20-m waves. How far is the storm from the coast?

*30. A motorboat is speeding at 12 m/s through a group of periodic ocean waves. When the motorboat travels in the same direction as the waves, it smashes into 6.5 waves per minute. When the motorboat is traveling in the direction opposite to the waves, it smashes into 30 waves per minute. Calculate the speed, frequency, and wavelength of the ocean waves.

*31. A string of mass per unit length μ is tied to a second string of mass per unit length μ'. A harmonic wave of speed v traveling along the first string reaches the junction and enters the second string. What will be the speed v' of this wave in the second string? Your answer should be a formula involving μ, μ', and v.

[3]These values are group velocities, or signal velocities.

*32. The maximum tensile stress that can be tolerated without breakage by steel is 5.0×10^8 N/m^2, and the density of steel is 7800 kg/m^3. If you apply a tension to a thin steel rod just barely less than the breaking tension, what is the speed of transverse waves on the rod? Does the answer depend on the diameter of the rod?

*33. A mass of 30 kg hangs from a string which, in turn, hangs from two other strings making an angle of 45° with each other (see Fig. 16.23). Each string has a length of 2.0 m, and the mass per unit length of each is 4.0×10^{-3} kg/m. How long does a wave pulse take to travel down from the upper end of this arrangement of strings to the mass?

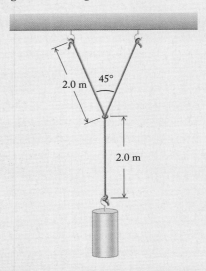

FIGURE 16.23 A mass hanging from an arrangement of strings.

*34. A steel wire of length 5.0 m and radius 0.30 mm is knotted to another steel wire of length 5.0 m and radius 0.10 mm. The wires are stretched with a tension of 150 N. How long does a transverse wave pulse take to travel the distance of 10 m from the beginning of the first wire to the end of the other? The density of steel is 7.8×10^3 kg/m^3.

**35. A long, uniform rope of length l hangs vertically. The only tension in the rope is that produced by its own weight. Show that, as a function of the distance z from the lower end of the rope, the speed of a transverse wave pulse of the rope is $\sqrt{gz}$. What is the time the wave pulse takes to travel from one end of the rope to the other?

**36. The speed of an ocean wave in shallow water is given by $v = \sqrt{gD}$, where D is the depth. A wave starts 50 m from shore, where the depth is 4.0 m. If the depth decreases linearly with distance as the wave approaches the shore, how long does it take for the wave to reach the shore?

**37. Suppose you take a loop of rope and make it rotate about its center at speed V. The centrifugal tendency of the segments of rope will then stretch it out along a circle of some radius R (see Fig. 16.24). What is the tension in the rope under these

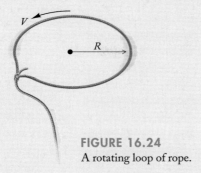

FIGURE 16.24
A rotating loop of rope.

conditions? Show that the speed of transverse waves on the rope (relative to the rope) coincides with the speed of rotation V.

**38. A flexible rope of length l and mass m hangs between two walls. The length of the rope is more than the distance between the walls (see Fig. 16.25), and the rope sags downward. At the ends, the rope makes an angle of α with the walls. At the middle, the rope approximately has the shape of an arc of a circle; the radius of the approximating (osculating) circle is R. What is the tension in the rope at its ends? What is the tension in the rope at its middle? What is the speed of transverse waves at the ends? At the middle?

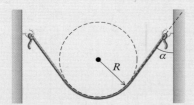

FIGURE 16.25 A rope hanging between two walls.

**39. The end of a long string of mass per unit length μ is knotted to the beginning of another long string of mass per unit length μ' (the tensions in these strings are equal). A harmonic wave travels along the first string toward the knot. The incident wave will be partially transmitted into the second string, and partially reflected. The frequencies of all these waves are the same. With the knot at $x = 0$, we can write the following expressions for the incident, reflected, and transmitted waves:

$$y_1 = A_{in} \cos(kx - \omega t)$$

$$y_2 = A_{refl} \cos(kx + \omega t)$$

$$y_3 = A_{trans} \cos(k'x - \omega t)$$

Show that

$$A_{refl} = \frac{k - k'}{k + k'} A_{in} = \frac{\sqrt{\mu} - \sqrt{\mu'}}{\sqrt{\mu} + \sqrt{\mu'}} A_{in}$$

$$A_{trans} = \frac{2k}{k + k'} A_{in} = \frac{2\sqrt{\mu}}{\sqrt{\mu} + \sqrt{\mu'}} A_{in}$$

[Hint: At $x = 0$, the displacement of the string must be continuous ($y_1 + y_2 = y_3$); if not, the string would break at the knot. Furthermore, the slope of the string must be continuous ($dy_1/dx + dy_2/dx = dy_3/dx$); if not, the string would have a kink and the (massless) knot would receive an infinite acceleration.]

16.3 The Superposition of Waves[†]

40. At one instant of time two transverse waves are traveling in the same direction along a stretched string. The instantaneous shapes of the wave are represented by

$$y = 0.020 \cos(4.0x) \quad \text{and} \quad y = 0.030 \cos(4.0x)$$

where the transverse displacements y and the position x are measured in meters.

(a) Are these waves in phase or out of phase?

(b) What are the amplitude and the wavelength of the net wave?

*41. At one instant of time two transverse waves are traveling in the same direction along a stretched string. The instantaneous shapes of the wave are represented by

$$y = 0.030 \cos(4.0x) \quad \text{and} \quad y = 0.030 \sin(4.0x)$$

where the transverse displacements y and the position x are measured in meters.

(a) What is the phase difference between these waves?

(b) Find the position x nearest the origin where the net wave has a wave crest. What is the amplitude of the net wave?

42. Three waves are traveling in the same direction; their individual amplitudes are 0.30 m, 0.50 m, and 0.80 m. What is the largest amplitude of the net wave that could occur? What is the smallest amplitude of the net wave that could occur, and how could that come about?

43. Two harmonic waves are described by

$$y_1 = A \cos(4x - 5t) \quad \text{and} \quad y_2 = 2A \cos(4x - 5t - \pi)$$

where $A = 6.0$ m, x is in meters, and t is in seconds. What are the amplitude, wavelength, and frequency of the superposition of these waves? At $x = 1.0$ m and $t = 1.0$ s, what is the net displacement?

44. At one point in space, two waves are described by

$$y_1 = A \cos(\omega_1 t) \quad \text{and} \quad y_2 = A \cos(\omega_2 t)$$

where $\omega_1 = 145$ radians/s and $\omega_2 = 152$ radians/s. When the two waves are superposed, how many beats are heard per second?

45. Two waves are described by

$$y_1 = A \cos(5x - 6t) \quad \text{and} \quad y_2 = A \cos(6x - 7t)$$

The two waves are superposed, and a snapshot of the resulting disturbance reveals a short-wavelength oscillation that gradually varies in amplitude over a longer length. What is the short

[†]For help, see Online Concept Tutorial 18 at www.wwnorton.com/physics

wavelength? What is the distance between the points where the amplitude of the short-wavelength oscillation goes to zero?

46. A periodic disturbance repeats every 2.0 m along a string (for example, a series of pulses). If we wish to describe this disturbance by a sum of harmonic waves, what are the only wavelengths that might be needed?

47. The wavefunctions for two transverse waves on a string are

$$y_1 = 0.030 \cos(6.0x - 18t + 1.5)$$

$$y_2 = 0.030 \cos(6.0x - 18t - 2.3)$$

where y and x are measured in meters and t in seconds.

(a) What is the amplitude of the sum of these waves?

(b) What is the transverse displacement of the string at $x = 0$ and $t = 0$?

*48. Consider the wavefunction $y = 3.0 \cos(5.0x - 8.0t) + 4.0 \sin(5.0x - 8.0t)$, which is a superposition of two wavefunctions expressed in some suitable units. Show that this wavefunction can be written in the form $y = A \cos(5.0x - 8.0t + \delta)$. What are the values of A and δ?

*49. A thin wire of length 1.0 m vibrates in a superposition of the fundamental mode and the second harmonic. The wavefunction is

$$y = 0.0060 \sin \pi x \cos 400\pi t - 0.0040 \sin 3\pi x \cos 1200\pi t$$

where y and x are measured in meters and t in seconds.

(a) What is the displacement at $x = 0.50$ m as a function of time?

(b) Plot this displacement as a function of time in the interval $0 \text{ s} \leq t \leq 0.0050$ s.

50. Two ocean waves with $\lambda = 100$ m, $f = 0.125$ Hz and $\lambda = 90$ m, $f = 0.132$ Hz arrive at a seawall simultaneously. What is the beat frequency of these waves?

51. A guitar player attempts to tune her instrument perfectly with the help of a tuning fork. If the guitar player sounds the tuning fork and a string on her guitar simultaneously, she perceives beats at a frequency of 4.0 per second. The tuning fork is known to have a frequency of 294.0 Hz. What fractional increase (or decrease) of the tension of the guitar string is required to bring the guitar in tune with the tuning fork? From the available information, can you tell whether an increase or decrease of tension is required?

*52. Figure 16.26 shows the height of the tide at Pakhoi. These tides can be regarded as a wave. The shape of the curve in

Fig. 16.26 indicates that the wave consists of two periodic waves of slightly different frequencies beating against each other. What are the frequencies of the two periodic waves? What are the periods? Which is caused by the Moon, which by the Sun?

16.4 Standing Waves[†]

53. The fundamental mode of the G string of a violin has a frequency of 196 Hz. What are the frequencies of the first, second, third, and fourth overtones?

54. Suppose that a vibrating mandolin string of length 0.34 m vibrates in a mode with five nodes (including the nodes at the ends) and four antinodes. What overtone is this? What is its wavelength?

55. A telegraph wire made of copper is stretched tightly between two telephone poles 50 m apart. The tension in the wire is 500 N, and the mass per unit length is 2.0×10^{-2} kg/m. What is the frequency of the fundamental mode? The first overtone?

56. A violin has four strings; all the strings have (approximately) equal tensions and lengths but they have different masses per unit length (kg/m), so that when excited in their fundamental modes they vibrate at different frequencies. The fundamental frequencies of the four strings are 196, 294, 440, and 659 Hz. What must be the ratios of the densities of the strings?

57. A car is being towed by means of a rope that has a mass per unit length of 0.080 kg/m. The length of the rope is 3.0 m, and the tension in the rope is 2.2×10^3 N. What is the eigenfrequency for a standing wave on this rope, in the fundamental mode?

58. You notice that a string fixed at both ends has a resonant frequency of 660 Hz, and also a resonant frequency of 440 Hz, but no resonant frequencies at any intermediate value. Identify what overtones these frequencies correspond to. Deduce the fundamental frequency of this string.

59. Two transverse harmonic waves are described by

$$y_1 = A \cos(\pi x - 3\pi t) \qquad \text{and} \qquad y_2 = A \cos(\pi x + 3\pi t)$$

where $A = 5.0$ m, x is in meters, and t is in seconds. What is the maximum amplitude of the superposition of these two waves at $x = 0.25$ m? What are the maximum transverse speed and acceleration at that point?

[†]For help, see Online Concept Tutorial 18 at www.wwnorton.com/physics

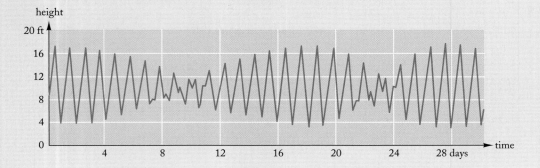

FIGURE 16.26 Height of the tide at Pakhoi as a function of time (because of exceptional local conditions, there is only one high and one low tide per day).

60. Several pulses of amplitude 3.0 cm are sent down a string that is attached to a pole at the far end. The pulses reflect there, and maintain their original amplitude. At the places where the forward and reflected pulses cross, what is the net amplitude if the string is fixed at the pole? If the string is free to slide up and down at the pole?

61. You pluck a guitar string upward near the center. The length of the string (fixed at both ends) is 65 cm. The wave speed is 70 m/s. How long does it take for the pulse you create to travel to the ends of the string and return to the center? When it returns, is the displacement now upward or downward? How many times per second does the string vibrate up and down?

62. If a human can hear up to 20 000 Hz, how many overtones of a low A ($f = 27.5$ Hz) can be heard by the human ear?

63. A 15-g string is 10 m long and is fixed at both ends. If the tension in the string is 40 N, what are the first five eigenfrequencies of the string?

64. The Bay of Fundy in Nova Scotia, Canada, is known for its extreme tides (Fig. 16.27); these are "resonance" tides, because the natural period of oscillation in the bay is about 12 hours, nearly matching the tidal period. Assume that a one-quarter-wavelength standing wave just fits the 250-km length of the bay, and that the wave speed is $v = \sqrt{gD}$, where D is the average depth of the bay. From this information, find the average depth of the Bay of Fundy.

65. The fundamental mode of the G string on a mandolin has a frequency of 196 Hz. The length of this string is 0.34 m, and its mass per unit length is 4.0×10^{-3} kg/m. What is the tension of this string?

66. Some automobiles are equipped with wire wheels (see Fig. 16.20). The spokes of these wheels are made of short segments of thick wire installed under large tension. Suppose that one of these wires is 9.0 cm long, 0.40 cm in diameter, and under a tension of 2200 N. The wire is made of steel; the density of steel is 7.8 g/cm^3. To check the tension, a mechanic gives the spoke a light blow with a wrench near its middle. With what frequency will the spoke ring? Assume that the frequency is that of the fundamental mode.

67. A light wave of wavelength 5.0×10^{-7} m strikes a mirror perpendicularly. The reflection of the wave by the mirror makes a standing wave with a node at the mirror. At what distance from the mirror is the nearest antinode? The nearest node?

68. A wave on the surface of the sea with a wavelength of 3.0 m and a period of 4.4 s strikes a seawall oriented perpendicularly to its path. The reflection of the wave by the seawall sets up a standing wave. For such a wave, there is an antinode at the seawall. How far from the seawall will there be nodes?

*69. The D string of a violin vibrates in its fundamental mode with a frequency of 294 Hz and an amplitude of 2.0 mm. What are the maximum velocity and the maximum acceleration of the midpoint of the string?

*70. The middle C string of a piano is supposed to vibrate at 261.6 Hz when excited in its fundamental mode. A piano tuner finds that in a piano that has a tension of 900 N on this string, the frequency of vibration is too low (flat) by 15.0 Hz. How much must he increase the tension of the string to achieve the correct frequency?

*71. The wire rope supporting the mast of a sailboat from the rear is under a large tension. The rope has a length of 9.0 m and a mass per unit length of 0.22 kg/m.

(a) If a sailor pushes on the rope sideways at its midpoint with a force of 150 N, he can deflect it by 7.0 cm. What is the tension in the rope?

(b) If the sailor now plucks the rope near its midpoint, the rope will vibrate back and forth like a guitar string. What is the frequency of the fundamental mode?

*72. Many men enjoy singing in shower stalls because their voice resonates in the cavity of the shower stall. Consider a shower stall measuring 1.0 m × 1.0 m × 2.5 m. What are the four lowest resonant frequencies of standing sound waves in such a shower stall? The speed of sound is 331 m/s.

*73. A piano wire of length 1.5 m fixed at its end vibrates in its second overtone. The frequency of vibration is 440 Hz, and the amplitude at the midpoint of the wire is 0.40 mm. Express this standing wave as a superposition of traveling waves. What are the amplitudes and speeds of the traveling waves?

FIGURE 16.27 Resonance tides in the Bay of Fundy.

*74. Two strings are tied together and stretched across a room. The strings have equal lengths of 2.0 m, and their masses are 1.0 g and 4.0 g, respectively. Find the frequencies of the standing-wave modes on these tied strings if the knot between the strings is a node.

*75. Consider a string with a mass density that depends on position. One end of the string is at $x = 0$, the other at $x = L$; and the mass density increases linearly from one end to the other, that is, [mass density] $= A + Bx$, where A and B are constants.

   (a) If the tension in the string is F, what is the speed of transverse waves as a function of x?

   (b) A wave on such a string does not have a well-defined wavelength. However, it can be described approximately as having a wavelength that depends on position. If the frequency of the wave is f, what is the wavelength $\lambda(x)$?

   (c) The condition for standing waves is then the usual condition: the number of wavelengths that fit within the length of the string must be 1/2, or 1, or 3/2, etc. Calculate the corresponding eigenfrequencies.

*76. (a) A long string is stretched along the x direction. Two transverse waves of equal wavelengths and equal frequencies travel simultaneously along this string. Suppose that one wave produces a displacement of the string in the y direction (see Fig. 16.28)

$$y = A \cos(kx - \omega t)$$

   and the other wave produces a displacement of the string in the z direction (see Fig. 16.28)

$$z = A' \cos(kx - \omega t)$$

   Show that the resulting motion of a particle on this string is back and forth along a line in the y–z plane. What is the angle of this line with respect to the y axis? The wave formed by a superposition of these y and z waves is called a wave of **linear polarization**, and the direction of the line of motion is called the direction of polarization.

   (b) Now suppose that the two waves have the same amplitude but the wave in the z direction is a quarter of a cycle out of phase with the wave in the y direction, so

$$y = A \cos(kx - \omega t)$$

   and

$$z = A \sin(kx - \omega t)$$

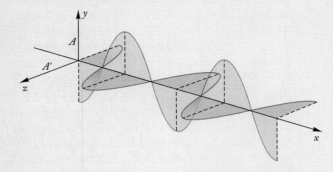

FIGURE 16.28 Displacements of a string in the y direction and the z direction.

Show that the resulting motion of a particle on the string is uniform circular motion. What are the radius, the frequency, the speed, and the centripetal acceleration of this circular motion? The wave formed by the superposition of these y and z waves is called a wave of **circular polarization.**

*77. Consider the superposition of two waves of the same frequency, opposite direction of propagation, and unequal amplitudes A_1 and A_2,

$$y_1 = A_1 \cos(kx - \omega t)$$
$$y_2 = A_2 \cos(kx + \omega t)$$

This superposition does not form a standing wave, but a wave with a modulated amplitude: the wave amplitude is large at some positions, and smaller at other positions. Show that the largest wave amplitude attained by the wave is $A_1 + A_2$. At what positions x is this large amplitude found? Show that the smallest wave amplitude attained by the wave is $|A_1 - A_2|$. At what positions x is this smallest amplitude found?

**78. A piano wire of length 0.18 m vibrates in its fundamental mode. The frequency of vibration is 494 Hz; the amplitude is 3.0×10^{-3} m. The mass per unit length of the wire is 2.2×10^{-3} kg/m. What is the energy of vibration of the entire wire? (Hint: Treat each small segment of the wire as a particle of mass dm with simple harmonic motion, and sum the energies of these simple harmonic motions.)

REVIEW PROBLEMS

79. A string is stretched along the x axis (horizontally), and it oscillates in the y direction (vertically). At one instant of time, a transverse traveling wave on this string is described by the mathematical formula

$$y = 0.030 \cos(1.2x)$$

where y and x are measured in meters.

   (a) What is the amplitude of the wave?

   (b) What is the wavelength of this wave?

   (c) Where are the first three wave crests and the first three wave troughs on the positive x axis nearest the origin?

*80. In the crust of the Earth, seismic waves of the P type have a speed of almost 5.0 km/s; waves of the S type have a speed of about 3.0 km/s. Suppose that after an earthquake, a seismometer placed at some distance first registers the arrival of P waves and 9.0 min later the arrival of S waves. What is the distance between the seismometer and the source of the waves?

81. A giant, freak wave encountered by a weather ship in the North Atlantic was 23.5 m high from trough to crest; its wavelength was 350 m and its period 15.0 s. Calculate the maximum vertical acceleration of the ship as the wave passed underneath; calculate the maximum vertical velocity. Assume that the motion of the ship was purely vertical.

*82. Many inhabitants of Tangshan, China, reported that during the catastrophic earthquake of July 28, 1976, they were thrown 2.0 m into the air as if hit by a "huge jolt from below."

(a) With what speed must a body be thrown upward to reach a height of 2.0 m?

(b) Assume that the vertical wave motion of the ground was simple harmonic with a frequency of 1.0 Hz. What amplitude of the vertical motion is required to generate a speed equal to that calculated in part (a)?

83. A harmonic transverse wave traveling on a tightly stretched wire has an amplitude of 0.020 m and a frequency of 100 Hz. What is the maximum speed attained by a particle on the string as this wave passes? What is the maximum acceleration? What is the time difference between the instant of maximum speed and the instant of maximum acceleration?

84. A passenger in an airplane flying over an anchored ship notices that ocean waves are smashing into the ship regularly at the rate of 10 per minute. He knows that ocean waves of this frequency have a speed of 9.4 m/s. He also notices that the length of the ship is the same as about three wavelengths. Deduce the length of the ship from this information.

85. While an anchored sailboat pulls on its anchor rope, the tension in the rope is 5.0×10^3 N. The anchor rope is nylon, of diameter 0.92 cm. The density of nylon is 1.1×10^3 kg/m^3. What is the mass per unit length for this rope? What is the speed of transverse waves on this rope?

86. Two strings are tied together and stretched across a room. The strings have equal lengths of 3.0 m, and their masses are 6.0 g and 9.0 g, respectively. If the tension in the strings is 200 N, what is the time a wave pulse takes to travel from the farthest end of one string to the farthest end of the other?

87. Three strings of identical material are tied together. They are under (different) tensions, and they make the angles indicated in Fig. 16.29.

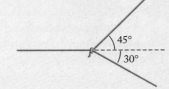

FIGURE 16.29 Three strings tied together.

(a) If the tension in the left string is T, what are the tensions in the other two?

(b) If the wave speed in the left string is 10 m/s, what are the speeds in the other two?

88. Consider two transverse harmonic waves of different wavelengths traveling in the same direction along a stretched string. At one instant of time, the shapes of the waves are given by

$$y = 0.012 \cos(3.0x) \qquad \text{and} \qquad y = 0.030 \cos(5.0x)$$

where the transverse displacements y and the position x are measured in meters.

(a) Is the superposition of these waves a harmonic wave? Is it a periodic wave?

(b) What is the wavelength of the net wave formed by the superposition of the two waves? What is the amplitude (the maximum transverse displacement) of the net wave?

89. Two cars, of identical make, have horns that generate sound waves of slightly different frequencies, 600 Hz and 612 Hz. What beat frequency do you hear if both of these cars are blowing their horns?

90. A standing wave on a string has the form

$$y = 0.020 \cos(15x) \cos(3.0t)$$

where distances are measured in meters and time in seconds.

(a) What are the amplitude and the frequency of this standing wave?

(b) What are the amplitudes, frequencies, and wavelengths of the two traveling waves whose superposition forms the standing wave?

(c) Where are the nodes and the antinodes of the standing wave?

(d) What are the maximum speed and the maximum acceleration of a particle on the string at one of the antinodes?

91. A uniform 20-m rope has a mass of 0.90 kg. The rope is hanging vertically from a support, so the only tension in the rope is that provided by its own weight.

(a) Find the speed of transverse waves in this rope, as a function of position along the rope. What is the speed at the top of the rope? At the midpoint? At the bottom?

(b) Find the time required for a wave pulse to travel from the top of the rope to the bottom.

92. An elevator of mass 2000 kg is hanging from a steel cable of length 60 m. The mass per unit length of this cable is 0.60 kg/m.

(a) What is the speed of transverse waves on this cable? (Neglect the mass of the cable when determining the tension.)

(b) What is the eigenfrequency of the fundamental mode on this cable? The first overtone? The second overtone?

93. A string of length L is fixed at one end and looped over a vertical frictionless rod at the other end (see Fig. 16.30). With

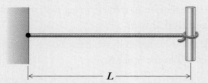

FIGURE 16.30 The left end of this string is fixed; the right end can slide up and down on the rod.

this arrangement the rod can maintain a tension in the string, but does not inhibit the up-and-down motion of the end of the string. Assume that the mass of the loop is negligible. What are the wavelengths of the standing waves of this string? What is the longest possible wavelength? (Hint: Since the loop is massless, the vertical forces that the tension in the string exerts on the loop must be zero at all times. This requires that the end of the string must be horizontal at all times.)

94. A mass of 15 kg hangs from two 6.0-m ropes, each of which leads downward at an angle of 30° with the vertical (Fig. 16.31). Each rope has a mass density $M/L = 0.012$ kg/m. Find the frequencies of those standing-wave modes for which the mass of 15 kg is at a node.

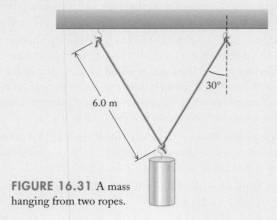

FIGURE 16.31 A mass hanging from two ropes.

Answers to Checkups

Checkup 16.1

1. The bat will remain essentially rigid under a small shaking motion, and so no appreciable wave motion is produced. The Jell-O (gelatin) is easily deformed, and so the shaking motion does produce a wave; the wiggling motion of the medium can be directly viewed.

2. A transverse wave can be generated by grabbing the cord and shaking it to the side and back (or up and down). A longitudinal wave can be generated by grabbing the cord and yanking it back and forth parallel to its length.

3. Such a wave on a string is transverse; except at the fixed endpoints, each point on the string moves up and down as it vibrates.

4. To generate a longitudinal wave, you must hit the end of the rod (causing a compression); to generate a transverse wave, you must hit the rod on its side (causing lateral motion).

5. (D) No; yes. The water wave does not carry the water from Africa to Florida; the particles merely oscillate about their equilibrium positions. The wave does carry energy from Africa to Florida; the motion of the arriving wave carries kinetic energy, and there is gravitational potential energy associated with the vertical displacements in water waves.

Checkup 16.2

1. All periodic waves are not harmonic; any shape of disturbance that repeats at regular intervals is periodic, but only a disturbance that takes the particular shape of a sine or cosine function is harmonic. However, all harmonic waves are periodic, since the sine and cosine functions repeat at regular intervals.

2. No. They are too pointy at the crests and flat in the troughs to be accurately described by a sine or a cosine function.

3. By Eq. (16.14), the speed is proportional to the square root of the tension; thus, if we want to increase the speed a factor of 2, we must increase the tension a factor of 4.

4. Since they are connected and stretched, they share the same tension. However, the first segment has a diameter half as large as the second; made of the same material, it must then have one-fourth of the cross-sectional area and thus one-fourth of the mass per unit length of the second. Since the speed is inversely proportional to the mass per unit length, the speed of waves in the first segment is twice that of the second.

5. (A) 20 Hz. Since it is the same string, the wave speed of the second wave must be the same as the first wave. Using Eq. (16.3), $v = \lambda f$, we see that if the wavelength is twice as long, the frequency must be half as large, or equal to 20 Hz.

Checkup 16.3

1. The superposition would look like the large-amplitude, long-wavelength wave, but with a quickly varying wiggle all along it due to the added small-amplitude, short-wavelength wave.

2. The superposition of the two waves vibrates at the average frequency, 107 Hz. The pulsating frequency of 4 Hz is the difference frequency (beats). So the two strings have frequencies of 105 Hz and 109 Hz.

3. (D) 1.4 m; 0.2 m. If the two waves are in phase, their amplitudes add, so that the net wave has amplitude 1.4 m. If the two waves are out of phase, their amplitudes subtract, giving a net wave of amplitude 0.2 m.

Checkup 16.4

1. For the third overtone, four half wavelengths fit in the string length L (two complete "cycles"). The fourth overtone has five half wavelengths in the string length L.

2. Yes; ordinarily, the midpoint of a guitar string is an antinode; the string primarily vibrates in its fundamental mode, similar to Fig. 16.16a. The midpoint (at the twelfth "fret") of the guitar string can also be a node; if constrained there (e.g., by holding a finger on it) while plucking the guitar string elsewhere, the string will vibrate primarily in its first overtone, similar to Fig. 16.16b.

3. Yes; if fixed at both ends, the number of nodes will always be one more than the number of antinodes. For example, the fundamental mode has two nodes (the ends) and one antinode (the center); each overtone adds one node and one antinode.

4. (C) 4. When the high E string is constrained, it has a node at its midpoint, and thus a half wavelength fits into half the length. The low E string vibrates in its fundamental mode, where one half wavelength fits into the full length. Since $v = \lambda f$, the high E string, with eight times the frequency and half the wavelength, has $8 \times \frac{1}{2} = 4$ times the speed.

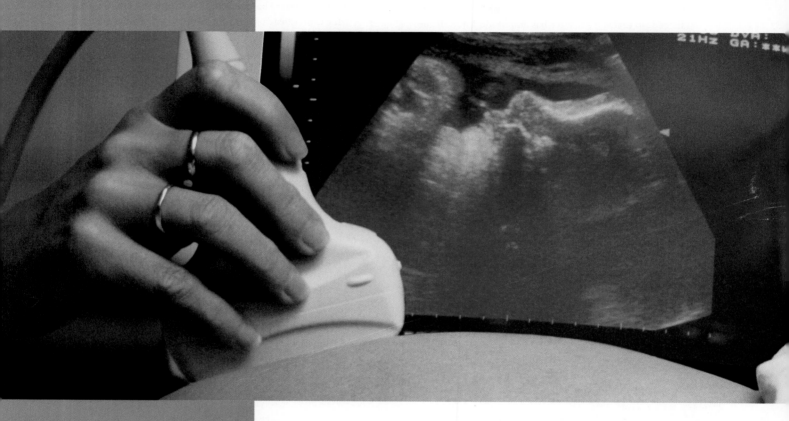

17.1 Sound Waves in Air

17.2 Intensity of Sound

17.3 The Speed of Sound; Standing Waves

17.4 The Doppler Effect

17.5 Diffraction

CONCEPTS IN CONTEXT

Concepts —in— Context

This ultrasound image of a fetus in the body of a pregnant woman was obtained through an echolocation technique similar to that of sonar and radar. To obtain an image, a probe sends a short pulse of ultrasound waves into the body, and then detects the echo returned by structures within the body; a computer analyzes the echoes and constructs the image. Such *sonography* is also widely used for examination of the heart.

As we learn about sound waves, we will consider such questions as:

? What are the frequencies of audible sound waves? Of ultrasound waves? (Section 17.1, page 539)

? How does ultrasound quantitatively determine the location of structures in the body? (Example 3, page 544)

? What determines the smallest size of a structure within the body that can be detected with ultrasound? (Example 8, page 555)

In the preceding chapter we dealt with waves propagating in one dimension, such as waves propagating on a string. These waves are confined; they can travel only in one direction (or the opposite direction) along the string. In this chapter we will deal with waves propagating in two or three dimensions. These waves spread out in all available directions, expanding as they move away from the source. For instance, Fig. 17.1a shows water waves spreading out in two dimensions on the surface of a pond from a pointlike disturbance caused by the impact of a pebble on the surface. And Fig. 17.2a shows sound waves spreading out in three dimensions from the earpiece of a telephone. For a simple diagrammatic representation of such two- or three-dimensional waves, we can use their **wave fronts**, that is, the locations of the wave crests at one instant of time. The wave fronts of the water waves are concentric circles (see Fig. 17.1b), and the wave fronts of the sound waves are concentric spherical shells (see Fig. 17.2b).

As time passes, the wave fronts spread outward, expanding as they move away from the source. This spreading of the waves is a characteristic feature of wave propagation in two or three dimensions. It implies that *the amplitude at a given wave front decreases as the wave front increases in size*. To understand this, consider the water waves in Fig. 17.1a. As a given circular wave front spreads outward, its circumference increases, and its energy is distributed along this larger circumference; thus, the amplitude at the wave front must decrease in accord with the decreased concentration of its energy. This decrease of the wave amplitude with distance is clearly visible in the water waves in Fig. 17.1a. Likewise, for the spherical sound wave spreading out from the telephone earpiece, the wave amplitude decreases with distance (but Fig. 17.2a fails to reveal this decrease, because the method used to make the sound waves visible is not sufficiently sensitive to the amplitude).

At a large distance from the source, the spherical wave fronts of a sound wave can be regarded as nearly flat, provided we concentrate our attention on a small region (see Fig. 17.3). Such *waves with flat, parallel wave fronts are called* **plane waves**.

(a)

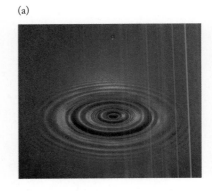

(b)

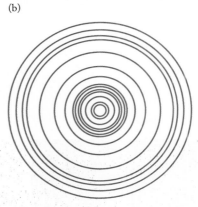

FIGURE 17.1 (a) Circular wave crests and wave troughs spreading on a water surface. (b) The wave fronts are concentric circles.

(a)

These sound waves are made visible by a moving bulb, synchronized with the waves.

telephone earpiece

(b)

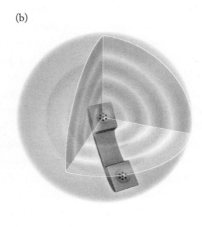

FIGURE 17.2 (a) Sound waves spreading out in air. The sound wave was made visible in this time-exposure photograph by means of a small electric lightbulb attached to a microphone that controlled the brightness of the bulb. The bulb and microphone were swept through the space in front of the telephone earpiece along arcs, as indicated by the fine pattern of ridges. The bulb and microphone were activated (gated) by an electronic circuit synchronized with the oscillations of the sound wave; this effectively produced a snapshot of the sound wave. (b) The wave fronts are concentric spherical shells.

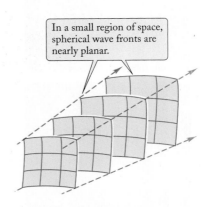

In a small region of space, spherical wave fronts are nearly planar.

FIGURE 17.3 Parallel wave fronts.

17.1 SOUND WAVES IN AIR

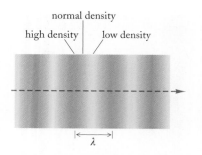

FIGURE 17.4 The sound wave consists of alternating zones of high and low density of air.

A sound wave in air consists of alternating zones of low density and high density (or, equivalently, zones of low pressure and of high pressure). Such zones of alternating density are generated by the vibrating diaphragm of a loudspeaker or the vibrating prong of a tuning fork, which exerts successive pushes on the air that is in contact with it. Figure 17.4 shows a sound wave in air at one instant of time. The alternating zones of low density and high density travel to the right, away from the source. However, although these density disturbances travel, the air as a whole does not travel—the air molecules merely oscillate back and forth.

The pushes of the loudspeaker or of the tuning fork on the air are longitudinal, and the sound wave itself is also **longitudinal**. The air molecules oscillate back and forth *along* the direction of propagation of the sound wave. The restoring force that drives these oscillations comes from the pressure of air. Wherever the density of molecules is higher than normal, the pressure also is higher than normal and pushes the molecules apart; wherever the density of molecules is lower than normal, the pressure also is lower than normal, and therefore the higher pressure of the adjacent regions pushes these molecules together. Thus, the pressure tends to keep the density uniform—it opposes the "deformation" of the air, and it gives air the elasticity required to permit the propagation of a wave.

Note that at the centers of the zones of high density in the wave, the molecules are instantaneously at rest while molecules from the right and the left have converged on them. At the centers of the zones of low density, the molecules also are instantaneously at rest, but molecules on the right and the left have moved away. Thus, *the zones of maximum and minimum density in the wave coincide with zones of zero displacement of the molecules. Conversely, the zones of zero density disturbance in the wave coincide with zones of maximum (positive or negative) displacement of the molecules.*

In Fig. 17.4, the density disturbances are shown much exaggerated. Even in an extremely intense sound wave, such as that produced by the engines of a jet airliner at takeoff, the displacements of the molecules are only about a tenth of a millimeter and the density enhancements only about 1%.

The frequency of the sound wave determines the pitch we hear; that is, it determines whether the tone is perceived as high or low by our ears (pitch is to sound what color is to light).

According to Fourier's theorem, mentioned in Section 16.3, a periodic sound wave of arbitrary shape can be regarded as a superposition of harmonic waves. The relative amplitudes of the harmonic waves in this superposition determine the perceived timbre, or quality, of the sound. Pure noise, or **white noise**, consists of a mixture of harmonic waves of all frequencies with equal strengths. White noise sounds like air rushing through a hole; to produce something like white noise, blow air out of your mouth, making a strong shushing sound (or turn up the volume on your TV set after selecting an inactive channel). In contrast, the musical tones emitted by a musical instrument consist of a mixture of just a few harmonic waves: the fundamental and its first few overtones. Figure 17.5 shows the waveforms emitted by a violin, a trumpet, and a clarinet when the musical note C is played on these instruments. In all cases the wave is periodic, repeating at the rate of 261.7 cycles per second; but the shapes of the waves and the amplitudes of the overtones are quite different in each case. It is because of this difference in the shapes of the waves that the ear can distinguish between diverse musical instruments.

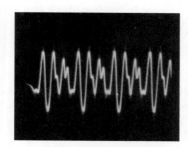

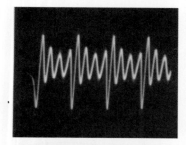

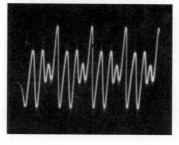

FIGURE 17.5 Waveforms emitted by (a) a violin, (b) a trumpet, and (c) a clarinet playing the note C. These pictures show the instantaneous wave amplitude as a function of time, displayed on an oscilloscope. The wave pattern repeats 261.7 times per second.

Table 17.1 lists the frequencies of the 12 notes of the chromatic musical scale. Other, nonchromatic scales are sometimes used, but we will not consider them here. The chromatic-scale frequencies are based on the system of equal temperament: successive frequencies in the scale differ by a factor of $2^{1/12} = 1.059$. The first entry in the table is middle C, with a frequency of 261.7 Hz (see Fig. 17.6). Any musical note not listed in the table can be obtained by multiplying or dividing the listed frequencies by a factor of 2, or 4, or 8, etc. Musical notes that differ by a factor of 2 in frequency are said to be separated by an **octave.** For example, C one octave above middle C has a frequency of 2×261.7 Hz $= 523.4$ Hz; C two octaves above middle C has a frequency of $2 \times 2 \times 261.7$ Hz $= 1046.8$ Hz, and so on. Incidentally: For a musician, the absolute values of these frequencies are not as important as the ratios of the frequencies. If an orchestra tunes its instruments so their middle C has a frequency of, say, 257 Hz, this will not do any noticeable harm to the music, provided that the frequencies of all the other notes are also decreased in proportion.

The ear performs the task of converting the mechanical oscillations of a sound wave into electric nerve impulses. Thus, it is similar to a microphone, which also converts the mechanical oscillations of sound into electric signals. However, the ear is unmatched in its ability to accommodate a wide range of intensities of sound—the intensities of the faintest and the loudest sounds acceptable to the ear differ by a factor of 10^{12}!

The range of frequencies audible to the human ear extends from 20 Hz to 20 000 Hz. These limits are somewhat variable; for instance, the ears of older people are less sensitive to high frequencies. Sound waves above 20 000 Hz are called **ultrasound**; some animals—dogs, cats, bats, and dolphins—can hear these frequencies.

Ultrasonic waves of very high frequency do not propagate well in air—they are rapidly dissipated and absorbed by air molecules. However, these waves propagate readily through liquids and solids, and this property has been exploited in applications of ultrasound. The use of ultrasound for images such as the chapter photo avoids the damage that X rays might do to the very sensitive tissues of the fetus. The sonography "cameras" that produce such images employ ultrasound waves of a frequency of about 10^6 Hz. Further development of this technique has led to the construction of acoustic microscopes. Most of these devices employ ultrasound waves of a frequency in excess of 10^9 Hz to produce highly magnified images of small samples of materials. The wavelength of sound waves of such extremely high frequency is about 10^{-6} m, roughly the same as the wavelength of ordinary light waves. The micrographs made by experimental acoustic microscopes compare favorably with micrographs made by ordinary optical microscopes (see Fig. 17.7).

TABLE 17.1	
THE CHROMATIC MUSICAL SCALE	
NOTE	**FREQUENCY**[a]
C	261.7 Hz
C#	277.2
D	293.7
D#	311.2
E	329.7
F	349.2
F#	370.0
G	392.0
G#	415.3
A	440.0
A#	466.2
B	493.9

[a] Based on a frequency of 440 Hz for A.

Concepts
— in —
Context

FIGURE 17.6 Middle C.

FIGURE 17.7 Acoustic micrograph of a portion of a transistor, at a magnification of 1000×. This picture was made with sound waves of 2.7×10^9 Hz.

✔ Checkup 17.1

QUESTION 1: In the picture of the sound wave shown in Fig. 17.2, where is the amplitude of the wave largest? Where smallest?

QUESTION 2: An old-fashioned speaking tube, for communication between different floors of a house, consists of a pipe that serves as a conduit of sound waves. For a sound wave traveling in such a tube, do the wave fronts expand in size? Does the amplitude of the wave decrease (apart from frictional losses)?

QUESTION 3: Consider a standing sound wave in air. At a density node, is the amplitude of the pressure oscillation maximum or minimum? Is the displacement of the air molecules maximum or minimum?

QUESTION 4: A pianist simultaneously strikes middle C on the piano and C one octave above that. Which note has the shorter wavelength? By what factor?

QUESTION 5: Apart from frictional losses, the amplitude of a wave pulse traveling on a string remains constant, but a wave pulse traveling on the surface of a pond or in air decreases with distance. What causes this difference?

QUESTION 6: At a density antinode, is the amplitude of the pressure oscillation maximum or minimum? Is the displacement of the air molecules maximum or minimum?

(A) Maximum; maximum   (B) Maximum; minimum
(C) Minimum; maximum   (D) Minimum; minimum

17.2 INTENSITY OF SOUND

A sound wave is intense and loud if it has a large amplitude. However, the amplitude of a sound wave is hard to measure directly, and it is more convenient to reckon the intensity of a sound wave by the energy it carries. The **intensity** I of a sound wave is defined as the energy per unit time transported by this wave per square meter of wave front, that is, the power transported by this wave per unit area:

intensity

$$I = \frac{[\text{energy/time}]}{[\text{area}]} = \frac{[\text{power}]}{[\text{area}]} = \frac{P}{A} \qquad (17.1)$$

Thus, to measure the intensity, we have to erect an area facing the wave, and we have to check how much energy the wave carries through this area per second (Fig. 17.8). Recall that the energy carried by a wave on a string was proportional to the square of the amplitude of the wave. Similarly, it can be shown that the intensity of a sound wave is proportional to the square of the pressure disturbance it produces in the air; equivalently, the intensity is proportional to the square of the density disturbance.

The unit of intensity is the watt per square meter (W/m^2). At a frequency of 1000 Hz, the minimum intensity audible to the human ear is about $2.5 \times 10^{-12} \text{ W/m}^2$. This intensity is called the threshold of hearing. There is no upper limit for the audible intensity of sound; however, an intensity above 1 W/m^2 produces a painful sensation in the ear.

Note that since the eardrum has an area of about $4 \times 10^{-5} \text{ m}^2$, the energy delivered per second by a sound wave of minimum intensity is only about $2.5 \times 10^{-12} \text{ J/m}^2 \times 4 \times 10^{-5} \text{ m}^2 \approx 10^{-16} \text{ J}$; this is a very small amount of energy, and it testifies to the extreme sensitivity of the ear.

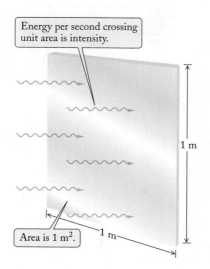

Energy per second crossing unit area is intensity.

1 m

Area is 1 m².

1 m

FIGURE 17.8 Area of 1 m² facing a wave. The intensity of the wave equals the energy incident on this area in 1 s.

The intensity of sound is often expressed on a logarithmic scale called the intensity level. The unit of intensity level is the **decibel** (dB); like the radian, this unit is a pure number, without any dimensions of m, s, or kg. The definition of intensity level is as follows: We take an intensity of 1.0×10^{-12} W/m^2 as our standard of intensity, which corresponds to 0 dB.[1] An intensity 10 times as large corresponds to 10 dB; an intensity 100 times as large corresponds to 20 dB; an intensity 1000 times as large corresponds to 30 dB; and so on. This scale of intensity level is intended to agree with our subjective perception of the loudness of sounds. We tend to underestimate increments in the intensity of sound—our ears perceive a sound of 100×10^{-12} W/m^2 as only twice as loud as a sound of 10×10^{-12} W/m^2.

Mathematically, the relationship between the intensity in W/m^2 and the intensity level in dB is given by a formula involving a logarithm:

$$[\text{intensity level in dB}] = (10\ \text{dB}) \times \log_{10}\left(\frac{[\text{intensity in W/m}^2]}{1.0 \times 10^{-12}\ \text{W/m}^2}\right) \qquad (17.2)$$

intensity level in decibels

EXAMPLE 1

Express the threshold of hearing (2.5×10^{-12} W/m^2) and the threshold of pain (1.0 W/m^2) in decibels.

SOLUTION: According to Eq. (17.2), we find that the intensity level corresponding to 2.5×10^{-12} W/m^2 is

$$(10\ \text{dB}) \times \log_{10}\left(\frac{2.5 \times 10^{-12}\ \text{W/m}^2}{1.0 \times 10^{-12}\ \text{W/m}^2}\right) = (10\ \text{dB}) \times \log_{10}(2.5)$$

With our calculator, we obtain $\log(2.5) = 0.40$ (on calculators, the base-10 logarithm "$\log_{10}$" usually appears simply as "log"), and hence the intensity level is $(10\ \text{dB}) \times 0.40 = 4.0$ dB.

Likewise, we find that the intensity level corresponding to 1.0 W/m^2 is

$$(10\ \text{dB}) \times \log_{10}\left(\frac{1.0\ \text{W/m}^2}{1.0 \times 10^{-12}\ \text{W/m}^2}\right) = (10\ \text{dB}) \times \log_{10}(10^{12})$$

Since $\log_{10}(10^{12}) = 12$, the intensity level is 120 dB.

Table 17.2 gives some examples of sounds of different intensities.

As a sound wave spreads out from its source, its intensity falls off because the area of the wave front grows larger, and therefore the wave energy per unit area grows smaller. Figure 17.9 helps to make this clear; it shows a spherical wave front at successive instants of time (such a spherical wave front is produced when a source emits waves uniformly in all directions). The wave front grows from an old radius r_1 to a new radius r_2; correspondingly, the area of the spherical wave front grows from $4\pi r_1^2$ to $4\pi r_2^2$. The total power carried by the wave front remains the same; hence, by the definition Eq. (17.1), the power per unit area, or the intensity, must be in inverse proportion to the area. If the intensity at r_1 is I_1 and the intensity at r_2 is I_2, then

$$I_1 \propto \frac{1}{4\pi r_1^2} \quad \text{and} \quad I_2 \propto \frac{1}{4\pi r_2^2} \qquad (17.3)$$

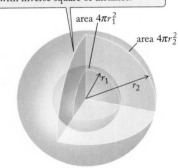

Same energy per second crosses increasingly larger areas as spherical wave spreads; thus, intensity decreases with inverse square of distance.

area $4\pi r_1^2$

area $4\pi r_2^2$

r_1

r_2

FIGURE 17.9 Concentric spherical wave fronts of a sound wave in air.

---

[1] More precisely, the standard of intensity, which corresponds to 0 dB, is 0.937×10^{-12} W/m^2. But it is conventional, and adequate for our purposes, to round this up to 1.0×10^{-12} W/m^2.

Taking the ratio of these two proportions, we obtain

$$\frac{I_2}{I_1} = \frac{r_1^2}{r_2^2} \tag{17.4}$$

This says that *the intensity of a spherical wave is inversely proportional to the square of the distance from the source.*

Although Fig. 17.9 shows the wave front spreading out uniformly in all directions, Eq. (17.4) remains valid if the wave is a beam, such as the beam emitted by a loudspeaker with a horn. Although the beam is aimed in some preferential direction, it spreads as it propagates, and the intensity falls off as the inverse square of the distance along any radial line within the beam.

TABLE 17.2	SOME SOUND INTENSITIES	
SOUND	**INTENSITY LEVEL**	**INTENSITY**
Rupture of eardrum	160 dB	1.0×10^4 W/m^2
Jet engine (at 30 m)	130	10
Threshold of pain	120	1.0
Rock music	115	0.30
Thunder (loud)	110	0.10
Subway train (New York City)	100	1.0×10^{-2}
Heavy street traffic	70	1.0×10^{-5}
Normal conversation	60	1.0×10^{-6}
Whisper	20	1.0×10^{-10}
Normal breathing	10	1.0×10^{-11}
Threshold of hearing	4	2.5×10^{-12}

EXAMPLE 2   At a distance of 30 m from a jet engine, the intensity of sound is 10 W/m^2, and the intensity level is 130 dB. What are the intensity and the intensity level at a distance of 180 m?

SOLUTION: According to Eq. (17.4), when we increase the distance from $r_1 = 30$ m to $r_2 = 180$ m (see Fig. 17.10), we decrease the intensity by a factor r_1^2/r_2^2. Hence the intensity will be

$$I_2 = \frac{r_1^2}{r_2^2}I_1 = \frac{(30 \text{ m})^2}{(180 \text{ m})^2}I_1 = 2.8 \times 10^{-2} \times I_1 = 2.8 \times 10^{-2} \times 10 \text{ W/m}^2$$

$$= 0.28 \text{ W/m}^2$$

By definition (17.2), the corresponding intensity level is

$$(10 \text{ dB}) \times \log_{10}\left(\frac{0.28 \text{ W/m}^2}{1.0 \times 10^{-12} \text{ W/m}^2}\right) = (10 \text{ dB}) \times \log_{10}(0.28 \times 10^{12})$$

$$= 114 \text{ dB}$$

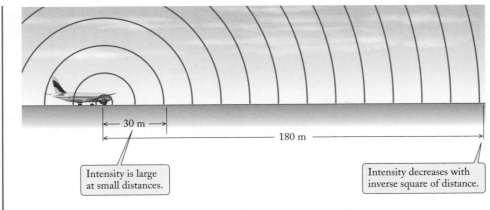

FIGURE 17.10 Sound wave spreading outward from a jet engine.

✔ Checkup 17.2

QUESTION 1: According to Table 17.2, by how many dB do the intensity levels of the noise of thunder and of a subway train differ? By what factor do the intensities differ?

QUESTION 2: If the intensity of sound at a distance of 1 m in front of a loudspeaker is 10^{-5} W/m², what is the intensity at a distance of 10 m? 100 m?

QUESTION 3: One source emits a spherical sound wave, uniformly in all directions. A second source emits a hemispherical sound wave, uniformly only over all the directions *in front of* the source. Both sources produce the same total power. What is the ratio of the intensity at some distance from the first source to the intensity at twice that distance in front of the second source?

(A) $\frac{1}{4}$ (B) $\frac{1}{2}$ (C) 1   (D) 2   (E) 4

17.3 THE SPEED OF SOUND; STANDING WAVES

As in the case of a wave on a string, the speed of sound in air depends on the restoring force and on the amount of mass. Since the restoring force is proportional to the pressure, and since the amount of mass is proportional to the density of the air, we expect the speed of sound to depend on the pressure and on the density. A somewhat involved calculation shows that the theoretical formula for the speed of sound in air is

$$v = \sqrt{1.40\frac{p_0}{\rho_0}} \tag{17.5}$$

where p_0 designates the air pressure and ρ_0 the mass density of the air. Note that this formula does display the expected dependence discussed in Section 16.2: the speed is the square root of an elastic property (the pressure, proportional to the force) divided by an inertial property (the density, proportional to the mass)—the speed is high if the pressure is large, and the speed is low if the density is large.

Under so-called standard conditions, the pressure of air is $p_0 = 1.01 \times 10^5$ N/m^2 and the density is $\rho_0 = 1.29$ kg/m^3 (at a temperature of 0°C); under these conditions Eq. (17.5) yields

$$v = \sqrt{1.40 \times \frac{1.01 \times 10^5 \text{ N/m}^2}{1.29 \text{ kg/m}^3}} = 331 \text{ m/s} \qquad (17.6)$$

The speed of sound in liquids and in solids is considerably higher than in air because, although the density is larger, the restoring force is much larger—liquids and solids offer much more opposition to compression than gases. Table 17.3 gives the values of the speed of sound in some materials.

Concepts — in — Context

EXAMPLE 3   In sonography (see the chapter photo), a probe pressed against the skin is used to send a sound pulse into the human body, and the same probe then detects the echo reflected by structures in the body. (a) Suppose that a pulse from the probe takes 1.6×10^{-4} s to travel to a bone and return to the probe. How far is the bone from the probe? (b) The probe must emit pulses that are short enough that the emitted pulse is complete before the reflected pulse returns. If we wish to detect a structure as close as 1.0 cm, what is the maximum pulse duration? The speed of sound in body tissues is about the same as that in water, 1500 m/s.

SOLUTION: (a) The round-trip distance traveled by the sound pulse is

$$x = v\,\Delta t = 1500 \text{ m/s} \times 1.6 \times 10^{-4} \text{ s} = 0.24 \text{ m} \qquad (17.7)$$

The distance to the bone is one-half as large as the round-trip distance, 0.12 m.

(b) For a structure at 1.0 cm, the round-trip distance is 0.020 m. Thus the pulse must be complete before a time

$$t = \frac{\Delta x}{v} = \frac{2.0 \times 10^{-2} \text{ m}}{1500 \text{ m/s}} = 13 \times 10^{-6} \text{ s}$$

The maximum pulse duration is 13 microseconds.

TABLE 17.3

THE SPEED OF SOUND IN SOME MATERIALS

MATERIAL	v
Air	
0°C, 1 atm	331 m/s
20°C, 1 atm	344
100°C, 1 atm	386
Helium gas, 0°C, 1 atm	965
Water (distilled)	1497
Water (sea)	1531
Aluminum	5104
Iron	5130
Glass	5000–6000
Granite	6000

A simple method for the measurement of the speed of sound in air takes advantage of standing waves in a tube open at one end and closed at the other (see Fig. 17.11). The *standing sound wave in the column of air in this tube must then have a displacement node at the closed end, since the motion of the air is restricted by the wall at this end. The wave must have a displacement antinode at the open end.* This second condition is not so obvious, but can be understood by first considering the pressure variations in the wave. The pressure excess, relative to normal atmospheric pressure, is large where the displacement of the molecules is small, and small where the displacement of the molecules is large (see the discussion of Fig. 17.4). Thus, the pressure antinodes are displacement nodes, and the pressure nodes are displacement antinodes. At the open end of our tube, the pressure must remain approximately constant, because the open end is accessible to the atmosphere, and hence any incipient decrease or increase of the pressure would immediately lead to an inflow or outflow of air from the surrounding atmosphere, canceling the pressure change. Thus, the atmosphere behaves as a "reservoir" of approximately constant pressure, and the pressure excess of the standing wave must have a node at the open end of the tube. This implies that the open end is a displacement antinode.

Closed end corresponds to a displacement node.

Open end corresponds to a displacement antinode.

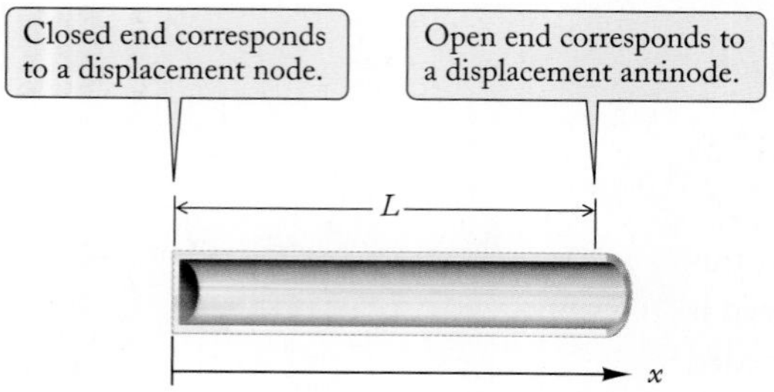

FIGURE 17.11 A tube open at one end and closed at the other.

With these boundary conditions for the closed end and the open end of the tube, the possible standing waves, or normal modes, are as shown in Fig. 17.12. If the length of the tube is L, then an odd number of quarter wavelengths must fit in the length L:

$$L = \frac{\lambda}{4}, \quad 3\frac{\lambda}{4}, \quad 5\frac{\lambda}{4}, \quad \ldots \tag{17.8}$$

so the wavelengths for these normal modes are

$$\lambda_1 = 4L, \quad \lambda_2 = \tfrac{4}{3}L, \quad \lambda_3 = \tfrac{4}{5}L, \quad \ldots \tag{17.9}$$

The eigenfrequencies of these normal modes are given by $f = v/\lambda$:

tube open at one end:
$$f_1 = \frac{v}{4L}, \quad f_2 = \frac{3v}{4L}, \quad f_3 = \frac{5v}{4L}, \quad \ldots \tag{17.10}$$

Note that the expressions (17.9) for the wavelengths of the normal modes of sound in a tube differ from the expressions (16.30) for the wavelengths of the normal modes of a string. This is due to the difference in boundary conditions: the tube has a node at one end and an antinode at the other end, whereas the string has nodes at both ends. For the tube closed at one end, we see from Eq. (17.10) that only odd-integer multiples (odd harmonics) of the fundamental frequency occur. In contrast, for a tube *open at both ends*, all integer harmonics occur. This is due to the boundary condition that the wave now has displacement antinodes (and pressure nodes) at *both* ends, and, like the string, requires that an integer number of half wavelengths must fit into the length L. The normal mode wavelengths are given by Eq. (16.30), and the eigenfrequencies are [see Eq. (16.31)]

tube open at both ends:
$$f_1 = \frac{v}{2L}, \quad f_2 = \frac{v}{L}, \quad f_3 = \frac{3v}{2L}, \quad \ldots \tag{17.11}$$

The speed of sound can be determined by means of Eq. (17.10) or Eq. (17.11) by measuring the resonant frequency of a tube of known length.

(a)   Fundamental mode

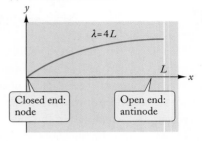

Closed end: node

Open end: antinode

(b)   First overtone

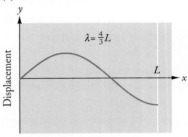

(c)   Second overtone

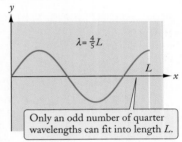

Only an odd number of quarter wavelengths can fit into length L.

FIGURE 17.12 Possible standing waves in tube open at one end. (a) The fundamental mode. (b) The first overtone. (c) The second overtone.

✔ Checkup 17.3

QUESTION 1: When you watch a thunderstorm, you see the lightning first, and you hear the thunder afterward. Why is the thunder delayed?

QUESTION 2: Consider a tube of length L *closed* at both ends. What is the frequency of the fundamental mode of a sound wave in this tube?

QUESTION 3: For a violin, does the wavelength of the emerging sound match the wavelength of the standing wave on its string? Does the frequency of the emerging sound match the frequency of the standing wave on the string?

QUESTION 4: You have two tubes of equal lengths. The first is open at one end and closed at the other; the second is open at both ends. What is the ratio of the fundamental frequency of sound waves in the first tube to that in the second tube?

(A) $\frac{1}{4}$ (B) $\frac{1}{2}$ (C) 1   (D) 2   (E) 4

PHYSICS IN PRACTICE   MUSICAL INSTRUMENTS

Standing waves play a crucial role in most musical instruments. Organs (Fig. 1), flutes, trumpets, trombones, and other wind instruments are essentially tubes open at their distant end with a blowhole or mouthpiece at the other end. Standing waves are excited within the tube by a stream of air blown across or into the blowhole or mouthpiece. In organs and flutes, the blowhole acts as an open end of the tube, and therefore the normal modes are those of a tube with *two* open ends. In trumpets and trombones, the lips of the player act approximately as a closed end, and the normal modes are those of a tube with one closed and one open end (however, the standing wave extends somewhat into the mouth cavity of the player, and the normal modes are quite complicated). The eigenfrequencies of the tube depend on its length. In many wind instruments—flutes, trumpets, French horns—the effective length of the tube can be varied by opening or closing valves, thereby changing the eigenfrequencies.

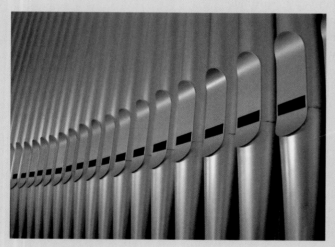

FIGURE 1 Organ pipes.

The excitation of vibrations in the tube of an organ or a flute by a steady stream of air blown across the blowhole arises from a rotational motion that develops behind the edges of the blowhole when the velocity of flow is high. As the air streams past the edge, it forms a vortex (see Fig. 2). This vortex soon breaks away from the edge and is replaced by another vortex, and another, and so on. The regular succession of vortices constitutes a vibration of the stream of air, and this excites standing waves in the tube of the organ by resonance.

The excitation of vibrations in the tube of a trumpet or trombone involves a different mechanism. These instruments have a cup-shaped mouthpiece, across which the player stretches his lips, which then behave somewhat like a pair of strings under tension, with a natural period of vibration. The vibration of the lips is triggered by the stream of air that the player blows out of his mouth. If the lips are initially close together and the gap between them is small, the pressure in the mouth builds up. This high pressure pushes the lips apart. But when the gap between them becomes wide, air rushes out and the pressure decreases. This permits the lips to snap back to their initial configuration. The gap between the lips therefore periodically widens and narrows, with a natural frequency that is determined by their tension. The periodic puffs of air produced by this vibration excite standing waves in the tube of the trumpet.

FIGURE 2 Vortices at the blowhole of an organ pipe.

Stringed instruments—violins, guitars, mandolins—use a resonant cavity to amplify and modify the sound produced by the string. The cavity is mechanically coupled to the string, and the vibrations of the latter excite resonant vibrations in the former. The resonant vibrations involve not only standing waves in the air in the cavity, but also standing waves in the solid material (wood) of the walls. Because the area of the body of, say, a violin is much larger than the area of its strings, the body pushes against much more air and radiates sound more efficiently than the strings (Fig. 3). Hence, most of the sound from a violin emerges from its body.

FIGURE 3 Vibrating violin.

17.4   THE DOPPLER EFFECT

Online
Concept
Tutorial

The speed of a sound wave in air is 331 m/s when measured in a reference frame at rest in the air. But when measured in a reference frame moving through the air, the speed of the sound wave will be larger or smaller, depending on the direction of motion of the reference frame. For example, if a train moving at 30 m/s approaches a stationary siren emitting sound waves (see Fig. 17.13), the speed of the sound waves relative to the train will be (331 + 30) m/s = 361 m/s. And if the train moves away from the siren, the speed of the sound waves relative to the train will be (331 − 30) m/s = 301 m/s.

The motion of the train affects not only the speed of the sound waves, but also their frequency. For instance, if the train approaches the siren, it runs head-on into the sound waves (see Fig. 17.13a) and hence encounters more wave fronts per second than if it were stationary; and if the train recedes from the siren, it runs with the sound waves (see Fig. 17.13b) and hence encounters fewer wave fronts per second. Consequently, *a receiver on the train will detect a higher frequency when approaching the siren, and a lower frequency when receding.* This frequency change caused by the motion of the receiver (or by motion of the source; see below) is called the **Doppler shift**.

To calculate the frequency shift, we note that in the reference frame of the air, we have the usual relation between the frequency, speed, and wavelength of the sound wave,

$$f = v/\lambda \tag{17.12}$$

and in the reference frame of the train, we have a corresponding relation

$$f' = v'/\lambda \tag{17.13}$$

The wavelengths in Eqs. (17.12) and (17.13) are exactly the same because the distance between the wave crests does not depend on the reference frame. Dividing Eq. (17.13) by Eq. (17.12), we obtain

$$\frac{f'}{f} = \frac{v'}{v} \tag{17.14}$$

We will designate by V_R the speed of the train acting as receiver of sound waves (R for receiver). In the reference frame of the train, the speed of sound is then $v' = v \pm V_R$, where the positive sign corresponds to motion of the train toward the source of sound

(a)   (b)

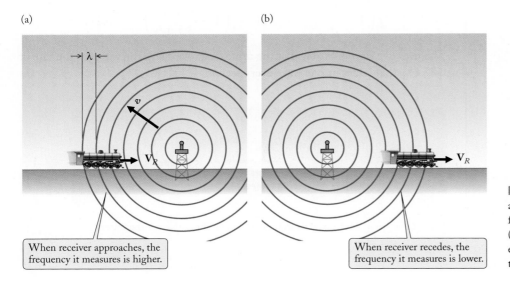

When receiver approaches, the frequency it measures is higher.

When receiver recedes, the frequency it measures is lower.

FIGURE 17.13 (a) Train approaching a siren. The train encounters *more* wave fronts per unit time than when stationary. (b) Train receding from a siren. The train encounters *fewer* wave fronts per unit time than when stationary.

and the negative sign to motion of the train away from the source of sound. With this expression for v', Eq. (17.14) yields

$$f' = f\left(1 \pm \frac{V_R}{v}\right)$$

$+$ for approaching receiver

$-$ for receding receiver

(17.15)

EXAMPLE 4   Suppose that a stationary siren emits a tone of frequency 440 Hz as the train moves away from it at 30.0 m/s. What is the frequency received on the train?

SOLUTION: From Eq. (17.14),

$$f' = f\left(1 - \frac{V_R}{v}\right) = 440\ \text{Hz} \times \left(1 - \frac{30.0\ \text{m/s}}{331\ \text{m/s}}\right) = 400\ \text{Hz}$$

Note that if the receiver is moving away from the source at a speed equal to the speed of sound ($V_R = v$), then the frequency f' is zero; this simply means that the receiver is moving exactly with the waves, and therefore no wave fronts catch up with it. If the receiver is moving away at a speed greater than the speed of sound ($V_R > v$), then Eq. (17.15) gives a negative frequency; this means that the receiver overruns the wave fronts from behind. A receiver speed equal to or larger than the speed of sound can be achieved only by mounting the receiver on a supersonic aircraft, an arrangement of no practical interest. However, Eq. (17.15) applies not only to sound waves, but also to water waves and other kinds of waves. For water waves it is not at all hard to arrange for a "receiver" with a speed V_R in excess of the speed v of the waves.

EXAMPLE 5   A motorboat (see Fig. 17.14) speeding at 6.0 m/s is moving in the same direction as a group of water waves of frequency 0.62 Hz and speed 2.5 m/s (relative to the water). What is the frequency with which the wave crests pound on the motorboat?

SOLUTION: We again use Eq. (17.14):

$$f' = f\left(1 - \frac{V_R}{v}\right) = 0.62\ \text{Hz} \times \left(1 - \frac{6.0\ \text{m/s}}{2.5\ \text{m/s}}\right) = -0.87\ \text{Hz}$$

The negative sign indicates that the motorboat overtakes the waves at the rate of 0.87 Hz, that is, about one wave every second.

FIGURE 17.14 Motorboat overtaking waves on the surface of the water.

A shift between the frequency emitted by the source of sound waves (or other waves) and the frequency detected by the receiver will also occur if the emitter is in motion relative to the air and the receiver is at rest. For example, if a train approaching a railroad crossing blows a whistle, the successive wave fronts emitted by the whistle are centered at intervals along the path of the whistle, and they will be crowded together in the forward direction and spread apart in the rearward direction (see Fig. 17.15). Consequently, *a stationary receiver will detect a higher frequency when the emitter of sound approaches, and a lower frequency when the emitter of sound recedes.* As the emitter passes by the stationary receiver, the detected frequency suddenly changes from high to low; that is, the pitch of the train's whistle suddenly drops. This also explains the sudden drop in pitch of the hum of a car engine that you hear when standing next to a road as the car passes by you.

We will designate by V_E the speed of the emitter of sound (E for emitter). To calculate the frequency change produced by the motion of the emitter, we begin by noting that in the time $1/f$ corresponding to one period, the train travels a distance $(1/f)V_E$ and hence the wavelength is shortened or lengthened from its normal value λ to a new value $\lambda' = \lambda \mp V_E/f$, where the negative sign corresponds to motion of the emitter toward the receiver and the positive sign to motion away from the receiver. The new frequency is therefore

$$f' = \frac{v}{\lambda'} = \frac{v}{\lambda \mp (V_E/f)} = \frac{v}{(v/f) \mp (V_E/f)} \qquad (17.16)$$

If we multiply both the numerator and the denominator of the right side of this equation by f/v, we obtain

$$f' = f\left[\frac{1}{1 \mp (V_E/v)}\right] \qquad \begin{array}{l} - \text{ for approaching emitter} \\[4pt] + \text{ for receding emitter} \end{array} \qquad (17.17)$$

By combining the formulas (17.15) and (17.17) we can obtain a general formula for the Doppler shift when both emitter and receiver are in motion. In this case, the

CHRISTIAN DOPPLER (1803–1853)
Austrian physicist. After Doppler discovered his formula for the frequency shift of sound, he recognized that light from a moving source should also be subject to a frequency shift, resulting in a change of color of the received light. However, unless the speed of the source is extremely large, such changes of the color of light are too small to be perceived by the eye.

Doppler shift, moving emitter

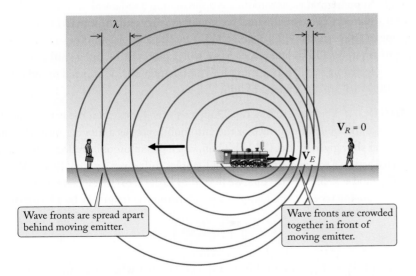

FIGURE 17.15 Train emitting sound waves while in motion. The wavelength ahead of the train is shorter and that behind the train is longer than when the train is stationary.

motion of the emitter causes a change of frequency by a factor of $1/[1 \mp (V_E/v)]$ and the motion of the receiver causes a further change of frequency by a factor of $[1 \pm (V_R/v)]$, leading to a final frequency

$$f' = f\left[\frac{1 \pm (V_R/v)}{1 \mp (V_E/v)}\right] \tag{17.18}$$

EXAMPLE 6   Suppose that the whistle of a train emits a tone of frequency 440 Hz as the train recedes from a stationary observer at 30.0 m/s. What frequency does the observer hear?

SOLUTION: According to Eq. (17.17), for a receding emitter,

$$f' = f\left[\frac{1}{1 + (V_E/v)}\right]$$

$$= 440 \text{ Hz} \times \frac{1}{1 + [(30.0 \text{ m/s})/(331 \text{ m/s})]} = 403 \text{ Hz}$$

COMMENT: If we compare the results of Examples 4 and 6, we see that motion of the emitter and motion of the receiver have nearly the same effect on the frequency—in both examples the frequency is decreased by about 10%. This symmetry of the Doppler shift has to do with the low speed of the motion. When the speed is low compared with the speed of sound, the effects of motion of the emitter and motion of the receiver are approximately the same; but when the speed is high, the effects of motion of the emitter and motion of the receiver are quite different.

EXAMPLE 7   Doppler radar units ("radar guns"), employed by police to measure the speeds of automobiles, consist of a transmitter of radar waves and a receiver. The transmitter sends a wave of frequency 8.00×10^9 Hz toward the target, and the receiver detects the reflected wave sent back by the target. Suppose that an automobile is approaching a radar unit at 100 km/h. What is the difference between the final received frequency and the initial transmitted frequency? The formulas (17.15) and (17.17) for the Doppler shift of sound waves are approximately valid for radar waves (we will explore the Doppler shift of radar and other electromagnetic waves more fully in Chapter 36). The speed of radar waves is 3.00×10^8 m/s, the speed of light.

SOLUTION: The wave suffers two Doppler shifts. First, when the wave is incident on the automobile, the surface of the automobile acts as a moving receiver, with a Doppler shift given by Eq. (17.15). Then, when the automobile reflects the wave, its surface acts as a moving emitter, with a Doppler shift given by Eq. (17.17). With $V_R = V_{auto}$ and $v = 3.00 \times 10^8$ m/s, we find from Eq. (17.14) that when the automobile acts as moving receiver, the Doppler shift increases the frequency by a factor of

$$\left(1 + \frac{V_{auto}}{v}\right)$$

And with $V_E = V_{auto}$, we find from Eq. (17.17) that when the automobile acts as moving emitter, the Doppler shift increases the frequency by an extra factor of

$$\frac{1}{1 - (V_{auto}/v)}$$

Multiplying the initial frequency by the product of these factors gives us the final frequency:

$$f' = f \times \left(1 + \frac{V_{\text{auto}}}{v}\right) \times \frac{1}{1 - (V_{\text{auto}}/v)} = f \times \frac{1 + (V_{\text{auto}}/v)}{1 - (V_{\text{auto}}/v)}$$

The difference between the final and the initial frequencies is

$$f' - f = f \times \left[\frac{1 + (V_{\text{auto}}/v)}{1 - (V_{\text{auto}}/v)} - 1\right] = f \times \frac{[1 + (V_{\text{auto}}/v)] - [1 - (V_{\text{auto}}/v)]}{1 - (V_{\text{auto}}/v)}$$

$$= f \times \frac{(2V_{\text{auto}}/v)}{1 - (V_{\text{auto}}/v)} \tag{17.19}$$

The speed of the automobile is

$$V_{\text{auto}} = 100 \text{ km/h} = 100 \text{ km/h} \times \frac{1000 \text{ m}}{1 \text{ km}} \times \frac{1 \text{ h}}{3600 \text{ s}} = 27.8 \text{ m/s}$$

Since V_{auto} is much, much smaller than v, we can here neglect the term V_{auto}/v in the denominator of Eq. (17.19), and we obtain the approximate result for the frequency shift

$$f' - f = f \times \frac{2V_{\text{auto}}}{v} = 8.00 \times 10^9 \text{ Hz} \times \frac{2 \times 27.8 \text{ m/s}}{3.00 \times 10^8 \text{ m/s}}$$

$$= 1.48 \times 10^3 \text{ Hz}$$

COMMENT: Although this is a very small shift (less than one part in a million), the electronics in a radar gun can accurately measure it by mixing the shifted frequency and the original frequency; this results in an easily detected difference frequency, as in the phenomenon of "beats" discussed in Chapter 16.

Finally, let us consider the case of an emitter, such as a fast aircraft, moving at a speed nearly equal to the speed of sound. If the aircraft emits sound of some frequency f, then Eq. (17.17) indicates that the frequency received at points just ahead of the aircraft is very large—in the limiting case of a speed V_E equal to the speed of sound, the frequency becomes infinite. This is because all the wave fronts are infinitely bunched together, and they all arrive at almost the same instant as the aircraft (see Fig. 17.16).

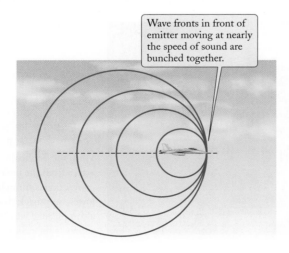

Wave fronts in front of emitter moving at nearly the speed of sound are bunched together.

FIGURE 17.16 A subsonic aircraft at a speed very close to the speed of sound emitting sound waves.

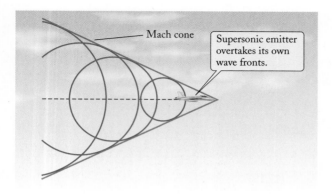

FIGURE 17.17 A supersonic aircraft emitting sound waves.

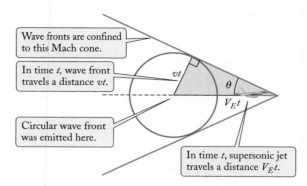

FIGURE 17.18 Mach cone. In a time t, the aircraft moves a distance $V_E t$ and the initial sound wave moves a distance vt.

If the speed of the aircraft exceeds the speed of sound, then the aircraft will overtake the wave fronts (see Fig. 17.17). In this case the sound is always confined to a conical region that has the aircraft at its apex and moves with the aircraft at the speed V_E; ahead of this region, the air has not yet been disturbed, although it will be disturbed when the aircraft and the cone move sufficiently far to the right. The cone is called the **Mach cone**.

The half-angle of the apex of the Mach cone is given by the formula

Mach cone

$$\sin \theta = v/V_E \qquad (17.20)$$

This can readily be seen from Fig. 17.18, which shows the aircraft at a time t and the wave front that was emitted by the aircraft at time zero. In the time t the sound travels a distance vt, and, simultaneously, the aircraft travels a distance $V_E t$. Thus, the radius of the wave front is vt. This radius is the opposite side of a right triangle of hypotenuse $V_E t$. Consequently,

$$\sin \theta = \frac{vt}{V_E t} \qquad (17.21)$$

which gives Eq. (17.20) if we cancel the factor t.

Any supersonic aircraft (or other body) will generate a Mach cone, regardless of whether or not it carries an artificial source of sound aboard (see Fig. 17.19). The motion of the body through the air creates a pressure disturbance that spreads out-

ERNST MACH (1838–1916)
Austrian philosopher and physicist. Mach obtained visual evidence for the Mach cone by photographing projectiles in flight. His book The Science of Mechanics *is a profound critical examination of the historical and logical foundations of mechanics.*

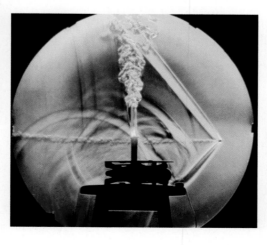

FIGURE 17.19 Schlieren photograph of a .22 caliber bullet passing a candle. With the Schlieren technique the shock waves from the bullet and turbulent convection column become visible. The lower sweptback shock wave reflects in a complicated way from the stool that supports the candle.

ward with the speed of sound and forms the cone. The cone trails behind the body much as a wake trails behind a ship. The sharp pressure disturbance at the surface of the cone is heard as a loud bang whenever the cone sweeps over the ear. This bang is called a **sonic boom**. For a large aircraft, such as the Concorde SST (Fig. 17.20), the noise level of the sonic boom reaches the pain threshold even if the aircraft is 20 km away. On a smaller scale, you can generate a sonic boom by cracking a whip. The crack of the whip occurs when the speed of the end of the whip through the air exceeds the speed of sound.

sonic boom

✔ Checkup 17.4

QUESTION 1: With what speed must a train approach you if you are to hear its whistle at twice the frequency it has when stationary?

QUESTION 2: With what speed must you approach a stationary whistle if you are to hear it at twice the frequency it has when you are stationary?

QUESTION 3: Can the Doppler-shifted frequency of a moving source of sound ever be zero?

QUESTION 4: A car races around a traffic circle while steadily blowing its horn. Describe the changes of pitch you hear if (a) you stand in the center of the circle or (b) you stand at the rim of the circle.

QUESTION 5: The frequency of the sound from the turbines of an approaching jet plane is twice the value of the frequency as the plane recedes. At what fraction of the speed of sound is the plane moving?

    (A) $\frac{1}{4}$ (B) $\frac{1}{3}$ (C) $1 - \sqrt{2}/2$ (D) $\frac{1}{2}$

FIGURE 17.20 Concorde SST.

17.5 DIFFRACTION

It is a characteristic feature of waves that they will deflect around the edges an obstacle placed in their path and penetrate into the "shadow" zone behind the obstacle. For example, Fig. 17.21 shows water waves striking a breakwater at the entrance of a harbor. The region directly behind the breakwater is out of the direct path of the waves, but nevertheless waves reach this region because each wave front spreads sideways once it has passed the entrance to the harbor. This lateral spreading of the wave fronts can be easily understood: The breakwater cuts a segment out of each wave front, and such a segment of wave front cannot just keep moving straight on as though nothing had happened—the end of the segment is a vertical wall of water where the breakwater has chopped off the wave front. The water at the ends will immediately begin to spill out sideways, producing a disturbance at the edge of the segment. This disturbance continues to spread out and gradually forms the curved wave fronts to the left and the right of the main beam of the wave.

Such a *deflection of waves at the edge of an obstacle is called* **diffraction**. We will perform a detailed analysis of diffraction phenomena in Chapter 35, when we examine the properties of light and other electromagnetic waves; here, we introduce a few of the main features of diffraction. It is a general rule for a wave passing through a gap that *the amount of diffraction increases with the ratio of wavelength to the size of the gap*. An increase of the wavelength (or a decrease of the size of the gap) makes the diffraction effects more pronounced; a decrease of the wavelength (or an increase of the size of the gap) makes

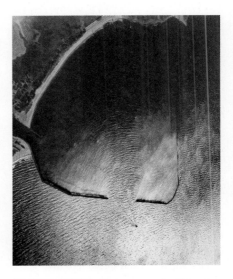

FIGURE 17.21 Ocean waves incident on a breakwater with an aperture.

(a) 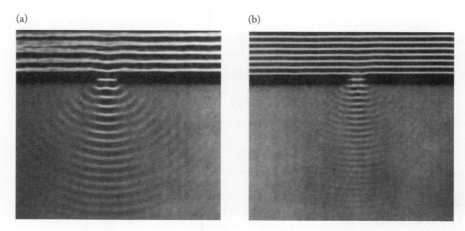 (b)

FIGURE 17.22 (a) Water waves of fairly long wavelength in a ripple tank exhibit strong diffraction when passing through an aperture. These waves spread out beyond the aperture in a fanlike pattern. (b) Water waves of shorter wavelength exhibit less diffraction.

the diffraction effects less pronounced. For instance, Fig. 17.22a shows diffraction of waves of relatively long wavelength by a small gap; the waves spread out very strongly, forming divergent, fanlike beams of concentric wave fronts. Figure 17.22b shows diffraction of waves of shorter wavelength; here the wave spreads out only slightly; most of the wave remains within a straight beam of nearly parallel wave fronts. Note that in Fig. 17.22b, the beam has a fairly well-defined edge—the region in the "shadow" of the barrier remains nearly undisturbed while the region facing the gap receives the full impact of the waves.

The fanlike beams of waves spreading out from the gap constitute a **diffraction pattern.** In Fig. 17.22a, the diffraction pattern consists of a central beam and two clearly recognizable secondary beams on each side. The beams are separated by nodal lines along which the wave amplitude is zero. Figures 17.23a and b show the diffraction patterns generated by a small island. Note that if the wavelength is large compared with the size of the island, then there exists no shadow zone; instead, the island merely produces some distortion of the waves.

Diffraction plays an important role in the propagation of sound. You can hear a person whose mouth is out of your direct line of sight—say, a person facing away from you or a person talking in an adjacent room—because the sound waves diffract through the open mouth or the open door and spread to fill the entire vicinity.

(a) 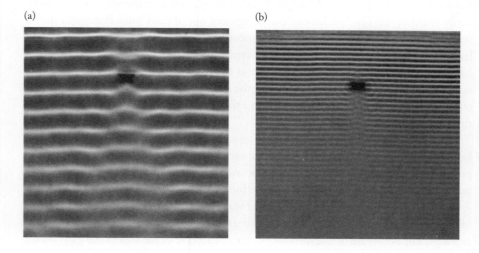 (b)

FIGURE 17.23 Diffraction by a small island. (a) Waves of long wavelength diffract around the island and spread into the shadow zone. (b) Waves of short wavelength do not spread into the shadow zone.

EXAMPLE 8   The minimum size of a structure that can be detected in an ultrasound image is limited by diffraction. With careful analysis of the echoes reflected from structures within the body, features as small as one-quarter of a wavelength can be imaged. Suppose that the probe uses ultrasound with a frequency of 2.0×10^6 Hz. What is the smallest feature that can be detected?

 Concepts —in— Context

SOLUTION: Diffraction effects are determined by the size of the features of interest relative to the wavelength. From Example 3, the speed of sound in body tissues is about 1500 m/s, so the wavelength of ultrasound waves of frequency 2.0×10^6 Hz is

$$\lambda = \frac{v}{f} = \frac{1500 \text{ m/s}}{2.0 \times 10^6 \text{ Hz}} = 7.5 \times 10^{-4} \text{ m}$$

If a feature as small as one-quarter of a wavelength can be detected, the size of such a structure is

$$\frac{\lambda}{4} = \frac{7.5 \times 10^{-4} \text{ m}}{4} = 1.9 \times 10^{-4} \text{ m}$$

Thus features as small as 0.2 mm can be detected.

✔ Checkup 17.5

QUESTION 1: If the water waves illustrated in Fig. 17.22 had a much longer wavelength, how would the distribution of waves beyond the barrier be different? If the waves had a much shorter wavelength?

QUESTION 2: Sound waves of frequency 1 kHz are traveling through a medium and strike an opening 1 m in width. The pattern of waves beyond the opening will be most spread out if the medium in which the sound waves are traveling is (hint: see Table 17.3.):

   (A) Air   (B) Helium gas   (C) Water

SUMMARY

PHYSICS IN PRACTICE   Musical instruments   **(page 546)**

SOUND WAVE PROPERTIES   Pressure (and density) antinodes correspond to displacement nodes (and vice versa).

CHROMATIC MUSIC SCALE   Successive notes are separated in frequency by a factor of $2^{1/12}$.

OCTAVE   Musical notes differing by a factor of 2 in frequency are separated by an octave.

INTENSITY LEVEL   (in dB) $(10 \text{ dB}) \times \log_{10}\left(\dfrac{[\text{intensity in W/m}^2]}{1.0 \times 10^{-12} \text{ W/m}^2} \right)$ **(17.2)**

INTENSITY   Transported energy per second (power) per unit area $I = \dfrac{P}{A}$ **(17.1)**

DECREASE OF INTENSITY OF SOUND WAVE WITH DISTANCE $I_2 = I_1 \dfrac{r_1^2}{r_2^2}$

> Same energy per second crosses increasingly larger areas as spherical wave spreads; thus, intensity decreases with the inverse square of distance.

area $4\pi r_1^2$
area $4\pi r_2^2$ **(17.4)**

SPEED OF SOUND IN AIR
(With pressure p_0 and density ρ_0) $v = \sqrt{1.40 \dfrac{p_0}{\rho_0}}$ **(17.5)**

SPEED OF SOUND IN AIR   (At 0°C and 1 atm)   331 m/s

STANDING WAVES IN TUBE OPEN AT ONE END
An odd number of quarter wavelengths fits in length L. $f_1 = \dfrac{v}{4L},\ f_2 = \dfrac{3v}{4L},\ f_3 = \dfrac{5v}{4L},\ \cdots$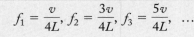

> Only an odd number of quarter wavelengths can fit into length L.   **(17.10)**

STANDING WAVES IN TUBE OPEN AT BOTH ENDS
An integer number of half wavelengths fits in length L. $f_1 = \dfrac{v}{2L},\ f_2 = \dfrac{v}{L},\ f_3 = \dfrac{3v}{2L},\ \cdots$ **(17.11)**

DOPPLER SHIFT   (v is sound speed; V_R is receiver speed; V_E is emitter speed)
Use upper sign when approaching ($f' > f$); lower sign when receding ($f' < f$).

Moving receiver, stationary emitter $f' = f\left(1 \pm \dfrac{V_R}{v}\right)$    **(17.15)**

Moving emitter, stationary receiver $f' = f\left[\dfrac{1}{1 \mp (V_E/v)}\right]$    **(17.17)**

MACH CONE   (For a supersonic emitter) $\sin\theta = \dfrac{v}{V_E}$ **(17.20)**

DIFFRACTION   The spreading of waves at an obstacle or gap. Diffraction increases with the ratio of the wavelength to the size of the obstacle or gap.

QUESTIONS FOR DISCUSSION

1. Could an astronaut be heard playing the violin while standing on the surface of the Moon?

2. A hobo can hear a very distant train by placing an ear against the rail. How does this help? (Hint: Ignoring frictional losses, how does the intensity of sound decrease with distance in air? In the rail?)

3. If you speak while standing in a corner with your face toward the wall, you will sometimes notice that your voice sounds unusually loud. Explain.

4. What happens to the frequency of musical notes if you play a $33\frac{1}{3}$-rpm record at 45 rpm?

5. Why does the wind whistle in the rigging of a ship or in the branches of a tree?

6. How does a flutist play different musical notes?

7. When inside a boat, you can often hear the engine noises of another boat much more loudly than when on deck. Can you guess why?

8. Many men like singing in the shower stall because the stall somehow enhances their voice. How does this happen? Would the effect be different for men and women?

9. According to a novel proposal for the reduction of engine noise inside aircraft cabins, loudspeakers installed along each side of the cabin are to cancel the noise by "antinoise," that is, sound waves of equal amplitude out of phase to the noise (see Fig. 17.24). The loudspeakers would be controlled by sensors and electronic circuits that detect the arriving engine noise and continuously adjust the amplitude and the phase of the required antinoise. Can such a noise cancellation system eliminate the noise throughout the cabin? In what part of the cabin would it be most effective? What happens to the energy in the arriving sound waves?

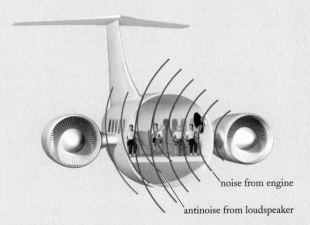

noise from engine

antinoise from loudspeaker

FIGURE 17.24 A system for the reduction of engine noise in an aircraft cabin.

10. The pipes that produce the lowest frequencies in a great organ are very long, usually 5.0 m. Why must they be so long? These pipes are also very thick. Why would a thin pipe give poor performance?

11. Some of the old European opera houses and concert halls renowned for their acoustic excellence have very irregular walls, heavily encrusted with an abundance of stucco ornamentation that reflects sound waves in almost all directions. How does the sound reaching a listener in such a hall differ from the sound reaching a listener in a modern concert hall with four flat, plain walls?

12. Electric guitars amplify the sound of the strings electronically. Do such guitars need a body?

13. Does the temperature of the air affect the pitch of a flute? A guitar?

14. The human auditory system is very sensitive to small differences between the arrival times of a sound signal at the right and left ears. Explain how this permits us to perceive the direction from which a sound signal arrives.

15. The depth finder (or "fish finder") on a boat sends a pulse of sound toward the bottom and measures the time an echo takes to return. The screen of the depth finder displays this echo time on a graph directly calibrated in distance units. Experienced operators can tell whether the bottom is clean rock or rock covered by a layer of mud, or whether a school of fish is swimming somewhere above the bottom. What echo times would you expect to see displayed on the screen of the depth finder in each of these instances?

16. The helmsman of a fast motorboat heading toward a cliff sounds his horn. A woman stands on the top of the cliff and listens. Compare the frequency of the horn, the frequency heard by the woman, and the frequency heard by the helmsman in the echo from the cliff. Which of these three is the highest frequency? Which is the lowest?

17. Two automobiles are speeding in opposite directions while sounding their horns. Describe the changes of pitch that each driver hears as they pass by one another.

18. A man is standing north of a woman while a strong wind is blowing from the south. If the man and the woman yell at each other, how does the wind affect the pitch of the voice of each as heard by the other?

19. A Concorde SST passing overhead at an altitude of 20 km produces a sonic boom with an intensity level of 120 dB lasting about half a second. How does this compare with some other loud noises? Would it be acceptable to let this aircraft make regular flights over populated areas?

20. Many people have reported seeing UFOs traveling through air noiselessly at speeds much greater than the speed of sound. If the UFO consisted of a solid impenetrable body, would you expect its motion to produce a sonic boom? What can you conclude from the absence of sonic booms?

21. When an ocean wave approaches a beach, its height increases. Why?

22. Occasionally ocean waves passing by a harbor entrance will excite very high standing waves ("seiches") within the harbor. Under what conditions will this happen?

23. Seismic waves of the S and P types have different speeds. Explain how a scientist at a seismometer station can take advantage of this difference in speed to determine the distance between his station and the point of origin of the waves.

24. The amplitude of an ocean wave initially decreases as the wave travels outward from its point of origin; but when the wave has traveled a quarter of the distance around the Earth, its amplitude *increases*. Explain how this comes about. (Hint: If the wave were to travel half the distance around the Earth, it would converge on a point, if no continents block its progress.)

25. Underground nuclear explosions generate seismic waves. How could you discriminate between the seismic waves received from such an explosion and the seismic waves from an earthquake? (Hint: Would you expect an explosion to produce mainly S waves or mainly P waves?)

26. Figure 17.25 shows a seismometer, an instrument used to detect and measure seismic waves. A vertical post is firmly set in the ground, and a large mass is suspended from it by a rigid

horizontal beam and a diagonal wire. The beam ends in a sharp point that rests against the post; the beam is therefore free to swing in the horizontal plane. Describe how the beam will swing if the ground moves and tilts the post. For what direction of motion of the post is this seismometer most sensitive?

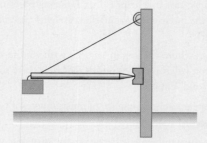

FIGURE 17.25 Seismometer.

27. If you are standing on the south side of a house, you can speak to a friend standing on the east side, out of sight around the corner. How do your sound waves reach into the shadow zone?

PROBLEMS[2]

17.1 Sound Waves in Air
17.2 Intensity of Sound

1. The range of frequencies audible to the human ear extends from 20 to 20000 Hz. What is the corresponding range of wavelengths?

2. The lowest musical note available on a piano is A, four octaves below that listed in Table 17.1; and the highest note available is C, four octaves above that listed in Table 17.1. What are the frequencies of these notes?

3. Both whales and elephants use low-frequency sound waves of a few hertz for communication. What is the wavelength of a whale song at 2.0 Hz (in water)? What is the wavelength of an elephant's rumble at 2.0 Hz (in air)?

4. Dogs can hear ultrasound of 40000 Hz; bats can hear ultrasound of 75000 Hz. What are the corresponding wavelengths?

5. Estimate the wavelength of the sound waves made visible in Fig. 17.2. Assume that the telephone handset is of standard size, like those found on pay telephones.

6. The lowest note that can be played on a guitar is E two octaves below the middle E listed in Table 17.1. This note is produced by the thickest string when vibrating in its fundamental mode. The length of the string is 0.62 m, and its mass per unit length is 5.4×10^{-3} kg/m. What must be the tension in this string?

7. Sound waves used for medical ultrasound scans of the soft tissues in the human body have frequencies in the range 0.80 to 15 MHz. What wavelengths correspond to these frequencies? The speed of sound in the soft tissues of the human body is 1500 m/s.

8. Designers of audio systems usually call sound waves of frequency below 800 Hz low frequency; from 800 Hz to 3500 Hz, middle frequency; and above 3500 Hz, high frequency. What are the wavelengths that correspond to these frequencies?

9. The maximum speed v_{max} acquired by particles of air when exposed to a sound wave of speed v is related to the intensity I by

$$I = \frac{v}{2} \rho_0 v_{max}^2$$

where ρ_0 is the density of air, 1.29 kg/m^3. Calculate the maximum speed acquired by the particles in a sound wave of intensity 1.0×10^4 W/m^2.

*10. A violin has four strings, each of them 0.326 m long. When vibrating in their fundamental modes, the four strings have frequencies of 196, 294, 440, and 659 Hz, respectively.

[2] In all the problems assume that the speed of sound in air is 331 m/s, unless otherwise stated.

(a) What is the wavelength of the standing wave on each string? What is the wavelength of the sound wave generated by the string?

(b) What are the frequency and the wavelength of the first overtone on each string? What is the corresponding wavelength of the sound wave generated by each string?

(c) According to Table 17.1, to what musical tones do the frequencies calculated above correspond?

*11. A mandolin has strings 34.0 cm long fixed at their ends. When the mandolin player plucks one of these strings, exciting its fundamental mode, this string produces the musical note D (293.7 Hz; see Table 17.1). In order to produce other notes of the musical scale, the player shortens the string by holding a portion of the string against one or another of several frets (small transverse metal bars) placed underneath the string. The player shortens the string by one fret to produce the note D#, by two frets to produce the note E, by three frets to produce the note F, etc. Calculate the correct spacing between the successive frets of the mandolin for one complete octave. Assume that the string always vibrates in its fundamental mode, and assume that the tension in the string is always the same.

12. In general, the numerical values of the intensity of sound in W/m^2 and of the intensity level in dB are different, but at one value of the intensity they are equal. What is this value?

13. The noise level in a quiet automobile is 50 dB. Find the sound intensity in W/m^2.

14. The intensity level of sound near a loud rock band is 120 dB. What is the intensity level of sound near two such rock bands playing together?

15. The highest frequency typically detectable by humans is 20 000 Hz. What note on the chromatic scale is closest to this frequency? How many octaves above the one listed in Table 17.1 is this frequency?

16. Assume that a particular loudspeaker emits sound waves equally in all directions; a total of 1.0 watt of power is in the sound waves. What is the intensity at a point 10 m from this source (in W/m^2)? What is the intensity level 20 m from this source (in dB)?

17. When 50 people are talking at once at a party, the intensity level is 70 dB. How much does the intensity level change when 25 people are talking?

18. A solo violinist generates an intensity level at the location of a listener of 60 dB. What is the intensity level there when 12 violinists play together?

19. A noisy machine produces an intensity level of 80 dB. What is the intensity level when two such machines operate at the same time?

20. The intensity level 50 m from an ambulance siren is 80 dB. What is the intensity level 1.0 m from the siren?

21. A sound source emits power equally in all directions. The intensity level 30 m from the source is 70 dB. What is the total sound power emitted by the source (in watts)?

*22. A loudspeaker receives 8.0 W of electric power from an audio amplifier and converts 3.0% of this power into sound waves. Assuming that the loudspeaker radiates the sound uniformly over a hemisphere (a vertical and horizontal angular spread of 180°), what will be the intensity and the intensity level at a distance of 10 m in front of the loudspeaker?

*23. An old-fashioned hearing trumpet has the shape of a flared funnel, with a diameter of 8.0 cm at its wide end and a diameter of 0.70 cm at its narrow end. Suppose that all of the sound energy that reaches the wide end is funneled into the narrow end. By what factor does this hearing trumpet increase the intensity of sound (measured in W/m^2)? By how many decibels does it increase the intensity level of sound?

17.3 The Speed of Sound; Standing Waves

24. Spectators at soccer matches often notice that they hear the sound of the impact of the ball on the player's foot (or head) sometime after seeing this impact. If a spectator notices that the delay time is about 0.50 s, how far is he from the player?

25. In the nineteenth century a signal gun was fired at noon at most harbors so that the navigators of the ships at anchor could set their chronometers. This method is somewhat inaccurate, because the sound signal takes some time to travel the distance from gun to ship. If this distance is 3.0 km, how long does the signal take to reach the ship? Can you suggest a better method for signaling noon?

26. In freshwater, sound travels at a speed of 1460 m/s. In air, sound travels at a speed of 331 m/s. Suppose that an explosive charge explodes on the surface of a lake. A woman with her head in the water hears the bang of the explosion and, lifting her head out of the water, hears the bang again 5.0 s later. How far is she from the site of the explosion?

27. While standing at some distance from a large stone cliff, you notice that if you clap your hands, an echo of the clap returns to you about 1.5 s later. What is the distance to the cliff?

28. The pitch of the vowels produced by the human voice is determined by the frequency of standing waves in several resonant cavities (larynx, pharynx, mouth, and nose). In an amusing demonstration experiment, a volunteer inhales helium gas and then speaks a few words. As long as his resonant cavities are filled with helium gas, the pitch of his voice will be much higher than normal. Given that the speed of sound in helium is about three times as large as in air, calculate the factor by which the eigenfrequencies of his resonant cavities will be higher than normal.

29. The Bay of Fundy (Nova Scotia) is about 250 km long. The speed of water waves of long wavelength in the bay is about 30 m/s.

(a) What are the frequency and the period of the fundamental mode of oscillation of the bay? Treat the bay as a long, narrow tube open at one end and closed at the other.

(b) The period of the tidal pull exerted by the Moon is about 12 h. Would you expect that the very large tidal oscillations (with heights of up to 15 m) observed in the Bay of Fundy are due to resonance?

30. You can estimate your distance from a bolt of lightning by counting the seconds between seeing the flash and hearing the thunder, and then dividing by 3 to obtain the distance in kilometers (or by 5 to obtain the distance in miles). Verify this rule.

31. The ultrasonic range finder on an automatic camera sends a pulse of sound to the target and determines the distance by the time an echo takes to return.

    (a) If the range finder is to determine a distance of 50 cm with an error no larger than ± 2 cm, how accurately (in seconds) must it measure the travel time?

    (b) If you aim this camera at an object placed beyond a sheet of glass (a window of a glass door), on what will the camera focus?

32. As described in Problem 31, the ultrasonic range finder on a camera sends a pulse of sound to the target and determines the distance by the time an echo takes to return. Suppose that, after waterproofing this camera somehow, you try to use it under water, in a swimming pool. If you aim the camera at a target 5.0 m away, what distance will the range finder indicate? The speed of sound in water is 1500 m/s.

33. The tube of a flute has a sliding joint that can be used to change the length, to tune the flute. Suppose that a flute has been tuned to perfect pitch while outdoors, where the temperature is 0°C and the speed of sound is 331 m/s. For this perfectly tuned flute, the frequency of middle C, which corresponds to the fundamental mode of the tube of the flute, is 261.7 Hz. Suppose that this flute is then taken indoors, where the temperature is 20°C and the speed of sound is 344 m/s. What will the frequency of the flute's middle C be now? To restore the flute to perfect pitch, how much must we increase the length of the tube of the flute? Express your answer as a percentage of the length.

34. The commonly accepted value for the speed of sound in dry air under standard conditions is 331.45 m/s. However, a scientist at the National Research Council of Canada recently discovered an error in the earliest determinations of the speed of sound, and he concluded that the correct value for the speed of sound is 331.29 m/s. What is the percent difference between the old and the new values? According to Eq. (17.5), what percent change of the pressure or of the density of the air will produce an equal change in the speed of sound?

35. Estimate the frequency of the sound waves made visible in Fig. 17.2.

36. A bat emits a pulse of ultrasound and detects an echo from a tree 0.20 s later. How far away is the tree?

37. A lightning flash is seen. Six seconds later, thunder is heard. How far away did the lightning occur?

38. Sound waves of wavelength 25.5 m travel in a piece of iron.

The vibrations of the iron generate sound waves in the air nearby. What is the wavelength of the sound in air?

39. A scuba diver clangs her hammer on an underwater pipeline. Another diver, with his hand on the pipeline, feels a vibration from the clang and, 1.5 seconds later, hears the clang through the water. How far along the pipeline are the two divers separated? The speed of sound in the metal pipe is 5100 m/s.

40. To measure the level of liquid helium, scientists often use a thin tube that permits oscillations of the gas in the tube above the liquid. When the bottom end of the tube is barely immersed in liquid (a closed end) and the top end is essentially open, a frequency f of oscillation is detected. When the bottom end of the tube is lifted out of the liquid, so that it is also open, what frequency of oscillations do you expect to occur?

41. With your ear on an iron railroad rail, you hear the sound of a distant train whistle through the iron. Eight seconds later, you hear the same whistle through the air. How far away is the train?

42. The musical note A (440 Hz) is played on a flute in air. What frequency would be heard if the air were replaced with helium gas?

*43. A rock is dropped into a deep well, and the sound of it hitting water is heard 4.62 s after the drop. Take the speed of sound to be 331 m/s. How far down the well is the water?

*44. The mass per unit length of a steel wire of diameter 1.3 mm is 0.010 kg/m, and the yield strength, or maximum tension that the wire can withstand, is 3.6×10^3 N. Is it possible to apply enough tension to the wire so that the speed of a transverse wave on this wire exceeds the speed of sound in the steel of the wire, 5000 m/s?

*45. In the eighteenth century, members of the French Academy organized the first careful measurement of the speed of sound in air. To compensate for wind speed, they adopted a reciprocal method. Cannons were fired alternately at Montmartre (in Paris) and Montlhéry, 29.0 km apart. Observers at each station measured the time delay between the muzzle flash seen at the *other* station and the arrival of the sound. Show that from the measurements of these travel times t_1 and t_2 of sound in both directions, the speed of sound in still air can be calculated as follows, independently of the speed of the wind blowing from one station to the other:

$$v = \frac{d}{2}\left(\frac{1}{t_1} + \frac{1}{t_2}\right)$$

where d is the distance between the stations. Evaluate numerically for the measured travel times of 87.4 s and 84.8 s.

*46. Because the human auditory system is very sensitive to small differences between the arrival times of a sound signal at the right and the left ear, we can perceive the direction from which a sound signal arrives to within about 5°. Suppose that a source of sound (a ringing bell) is 10 m in front of and 5° to the left of a listener. What is the difference in the arrival times of sound signals at the left and right ears? The separation between the ears is about 15 cm.

*47. Consider a tube of length L open at both ends. Show that the eigenfrequencies of standing sound waves in this tube are

$$f_n = n\frac{v}{2L} \qquad n = 1, 2, 3, \dots$$

Draw diagrams similar to those in Fig. 17.12 showing the displacement amplitude for each of the first four standing waves.

*48. The largest pipes in a great organ usually have a length of about 16 ft (4.8 m). These pipes are open at both ends so that a standing sound wave will have a displacement antinode at each end. What is the frequency of the fundamental mode of such a pipe?

*49. A flute can be regarded as a tube open at both ends. It will emit a musical note if the flutist excites a standing wave in the air column in the tube.

(a) The lowest musical note that can be played on a flute is C (261.7 Hz; see Table 17.1). What must be the length of the tube? Assume that the air column is vibrating in its fundamental mode (see Problem 47).

(b) In order to produce higher musical notes, the flutist opens valves arranged along the side of the tube. Since the holes in these valves are large, an open valve has the same effect as shortening the tube. The flutist opens one valve to play C#, two valves to play D, etc. Calculate the successive spacings between the valves of a flute for one complete octave. (The actual spacings used on flutes differ slightly from the results of this simple theoretical evaluation because the mouth cavity of the flutist also resonates and affects the frequency.)

50. The human ear canal is approximately 2.7 cm long. The canal can be regarded as a tube open at one end and closed at the other. What are the eigenfrequencies of standing waves in this tube? The ear is most sensitive at a frequency of about 3000 Hz. Would you expect that resonance plays a role in this?

*51. Consider a tube of length L closed at both ends. Show that the eigenfrequencies of standing sound waves in this tube are

$$f_n = n\frac{v}{2L} \qquad n = 1, 2, 3, \dots$$

Draw diagrams similar to those in Fig. 17.12 showing the displacement amplitude for each of the first four standing waves.

*52. If you stand in the vicinity of a picket fence and clap your hands, you will notice that the sound waves reflected by the fence and reaching your ear are strongly reinforced at a selected wavelength; that is, the picket fence seems to ring with a musical tone. This selective reinforcement of sound waves occurs whenever the wavelength is such that waves reflected by different boards in the fence arrive at your ear in phase, giving constructive interference. Suppose that the picket fence consists of boards separated by a distance d and you stand in line with this fence (see Fig. 17.26). Show that in this case the condition for constructive interference for waves traveling from you to the boards and back to you is that the wavelength be equal to $2d$.

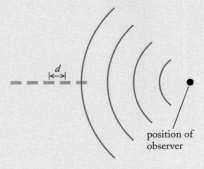

FIGURE 17.26 Picket-fence interference.

17.4 The Doppler Effect[†]

53. The horn of a stationary automobile emits a sound wave of 580 Hz. What frequency will you hear if you are driving toward this automobile at 80 km/h?

54. In an experiment performed shortly after Doppler proposed his theoretical formula for the frequency shift, several trumpeters were placed on a train and told to play a steady musical tone. As the train sped by, listeners standing on the side of the track judged the pitch of the tone received from the trumpeters. Suppose that the train had a speed of 60 km/h and that the trumpets on the train sounded the note of E (329.7 Hz; see Table 17.1). What was the frequency of the note perceived by a listener on the ground when the train was approaching? When the train was receding? Approximately to what musical notes do these Doppler-shifted frequencies correspond?

55. Ocean waves with a wavelength of 100 m have a period of 8.0 s. A motorboat, with a speed of 9.0 m/s, heads directly into the oncoming waves. What is the speed of the waves relative to the motorboat? With what frequency do wave crests hit the front of the motorboat?

56. The horn of an automobile emits a tone of frequency 520 Hz. What frequency will a pedestrian hear when the automobile is approaching at a speed of 85 km/h? Receding at the same speed?

57. An aircraft is flying at an altitude of 12000 m. On the ground, a sonic boom is heard 18 s after the jet passes directly overhead. What is the speed of the aircraft?

58. While watching a high-speed chase, a stationary observer measures the frequency of the siren of an approaching police car to be 497 Hz; after the car passes, the same observer measures 395 Hz. What is the speed of the police car? What is the frequency it emits?

59. A bat flies toward a wall and emits a pulse of ultrasound of frequency 50 kHz. The echo received by the bat is Doppler-shifted 800 Hz toward higher frequency. How fast is the bat flying?

[†]For help, see Online Concept Tutorial 19 at www.wwnorton.com/physics

60. An airplane emits a tone as it flies past an observation tower. The frequency that an observer measures as the airplane approaches is twice the frequency measured when the airplane recedes. What is the speed of the airplane?

61. The **Mach number** is the ratio of the speed of an aircraft (or body) to the speed of sound. When a jet flies at Mach 2, what angle does the shock wave make with the direction of travel of the jet?

62. What is the angle of the Mach cone of a meteor traveling at 15 km/s?

*63. In movies, a person who dives from a high cliff often screams all the way down to the water. Assume that such a person emits a constant frequency of 440 Hz. What frequency will an observer at the top of the cliff hear after 3.0 seconds? Take into account the travel time of the sound.

*64. A car horn sounds a pure tone as the car approaches a wall. A stationary listener behind the car hears an average frequency of 250 Hz, which pulsates (beats) at 12 Hz. What is the speed of the car?

*65. Two automobiles are driving on the same road in opposite directions. The speed of the first automobile is 90.0 km/h, and that of the second is 60.0 km/h. The horns of both automobiles emit tones of frequency 524 Hz. Calculate the frequency that the driver of each automobile hears coming from the other automobile. Assume that there is no wind blowing along the road.

*66. Repeat Problem 65 under the assumption that a wind of 40.0 km/h blows along the road in the same direction as that of the faster automobile.

*67. A train approaches a mountain at a speed of 75 km/h. The train's engine sounds a whistle that emits a frequency of 420 Hz. What will be the frequency of the echo that the engineer hears reflected off the mountain?

*68. Suppose that a moving train carries a source of sound and also a receiver of sound so that both have the same velocity relative to the air. Show that in this case the Doppler shift due to motion of the source cancels the Doppler shift due to motion of the receiver—the frequency detected by the receiver is the same as the frequency generated by the source.

*69. The whistle on a train generates a tone of 440 Hz as the train approaches a station at 30 m/s. A wind blows at 20 m/s in the same direction as the motion of the train. What is the frequency that an observer standing at the station will hear?

70. Figure 17.19 shows the shock wave of a bullet speeding through air. Measure the angle of the Mach cone and calculate the speed of the bullet.

71. A rifle bullet has a speed of 674 m/s. What is the half-angle of the Mach cone generated by this bullet?

72. According to tradition, Superman flies faster than a speeding bullet. The speed of a typical bullet is 700 m/s, so let us guess that the speed of Superman is 800 m/s. At this speed, what is the half-angle of the Mach cone that Superman produces?

*73. You may have noticed that at the instant a fast (but subsonic) jet aircraft passes directly over your head, the sound it makes seems to come from a point behind the aircraft.

(a) Show that the direction from which the sound seems to come makes an angle θ with the vertical such that $\sin\theta = V_E/v$, where V_E is the speed of the aircraft and v the speed of sound.

(b) If you hear the sound from an angle of 30° behind the aircraft, what is the speed of the aircraft?

17.5 Diffraction

74. Because of diffraction, it can be difficult to detect reflected sound from objects smaller even than one full wavelength of the sound used. Using this one-wavelength criterion, what is the size of the smallest insect a bat can echolocate when it emits ultrasound of frequency 40 kHz?

75. To circumvent diffraction effects, an ultrasonic imaging apparatus uses sound of a wavelength smaller than the size of most objects to be imaged. A particular medical apparatus operates at 5.0 MHz and can detect structures as small as one-half of a wavelength. What is the size of the smallest object that can be detected?

76. Ultrasonic microscopes are used to study features in materials as small as one-quarter of the wavelength of sound in the material; at smaller length scales, diffraction obscures such details. If ultrasound of frequency 2.0×10^9 Hz is used to examine aluminum, what size features can be detected?

REVIEW PROBLEMS

77. The speed of sound waves in water is 1500 m/s. Dolphins emit ultrasound waves at 1.0×10^6 Hz. What is the wavelength of these waves? What is the wavelength if these waves penetrate from water into air?

78. When vibrating in its lowest mode, the C string of the cello produces the musical note C that is two octaves below middle C. The length of this string is 0.68 m.

(a) What is the frequency of this note? What is the wavelength of the sound wave?

(b) What is the frequency of the standing wave on the string, and what is its wavelength?

79. The sound waves used for medical ultrasound scans of the human body typically have a frequency of 10 MHz and an amplitude of 8.0×10^{-8} m (this is the amplitude of the

simple harmonic motion of the particles in the tissues through which the wave passes). When subjected to such a sound wave, what is the maximum speed of the motion of a particle? What is the maximum acceleration of the particle? Express the acceleration in standard g's ($1g = 9.81$ m/s^2).

80. At a distance of 15 m from a pneumatic drill, the sound intensity is about 1.0×10^{-3} W/m^2. What is the intensity level in decibels?

81. In a screaming contest, a Japanese woman achieved 115 dB. How many such women would have to scream at you to bring you to the threshold of pain, 120 dB?

82. What is the energy incident per second on your eardrum (of diameter 7.0 mm) if exposed to a sound wave of 160 dB?

83. Suppose we turn the volume control of a loudspeaker up, and we increase the intensity of the sound reaching our ears by a factor of 2. What is the corresponding increase in intensity level, in dB?

84. Suppose that a whisper has an intensity level of 20 dB at a distance of 0.50 m from the speaker's mouth. At what distance will this whisper be below your threshold of hearing?

85. A bat can sense its distance from the wall of a cave (or whatever) by emitting a sharp ultrasonic pulse that is reflected by the wall. The bat can tell the distance from the time the echo takes to return.

    (a) If the bat is to determine the distance of a wall 10.0 m away with an error of less than ±0.5 m, how accurately must it sense the time interval between emission and return of the pulse?

    (b) Suppose that a bat flies into a cave filled with methane (swamp gas). By what factor will this gas distort the bat's perception of distances? The speed of sound in methane is 432 m/s.

86. In order to measure the depth of a ravine, a physicist standing on a bridge drops a stone and counts the seconds between the instant he releases the stone and the instant he hears it strike some rocks at the bottom. If this time interval is 6.0 s, how deep is the ravine? Take into account the travel time of the sound signal, but ignore air friction.

87. The Concorde SST had a cruising speed of 2160 km/h.

    (a) What was the half-angle of the Mach cone generated by this aircraft?

    (b) If the aircraft passed directly over your head at an altitude of 12 000 m, how long after this instant would the shock wave strike you?

88. The French TGV (*train à grande vitesse*; Fig. 17.27) attains a speed of 510 km/h. Suppose the train carries a whistle tuned to a frequency of 600 Hz. If it sounds this whistle, what is the frequency you hear standing alongside its track while the train is approaching? Receding?

FIGURE 17.27 The French TGV.

89. A man at the helm of a fast motorboat heading toward a cliff sounds his horn. The speed of the motorboat is 15 m/s, and the frequency of the sound emitted by the horn, when at rest, is 660 Hz. A woman stands on the top of the cliff and listens.

    (a) What is the frequency heard by the man?

    (b) What is the frequency heard by the woman?

    (c) What is the frequency heard by the man in the echo from the cliff?

Answers to Checkups

Checkup 17.1

1. The amplitude is largest near the earpiece of the telephone, where the sound waves have not spread appreciably; the amplitude is smallest far from the earpiece, where the energy in the wave has become distributed over a much larger area than near the earpiece.

2. No to both questions: because the sound wave is reflected from the inner wall of the tube, the wave fronts stay much the same size and maintain the same amplitude as they propagate along the tube.

3. At a density node, the pressure pattern also has a node; the density and pressure vary together. The displacement of the air molecules is maximum there.

4. The C that is an octave higher than the other has twice the frequency, and so a constant sound speed, $v = \lambda f$, requires that it has the shorter wavelength (by a factor of 2).

5. A wave pulse on the surface of a pond (or in air) is traveling in two (or three) dimensions, and so spreads out over a larger circumference (or a larger area), with a decreasing concentration of its energy. A wave pulse on a string travels in one dimension, and so does not spread.

6. (B) Maximum; minimum. At a density antinode, the pressure also experiences maximum amplitude oscillations. The air molecules experience minimum displacement there; thus a density and pressure antinode is a displacement node.

Checkup 17.2

1. From Table 17.2, the noise of thunder and of a subway train differ by 110 dB − 100 dB = 10 dB. The intensities differ by a factor of $0.1/(1.0 \times 10^{-2}) = 10$. Thus a change in intensity by a factor of 10 is equivalent to a change in intensity level by an increment of 10 dB.

2. The intensity decreases in proportion to the inverse of the square of the distance, so at 10 m it is 100 times smaller than at 1 m; thus it is $10^{-7}\,\text{W/m}^2$ there. At 100 m, it is $(100)^2 = 10^4$ times smaller than at 1 m; thus it is $10^{-9}\,\text{W/m}^2$ there.

3. (D) 2. Both sources emit the same power, but at any particular distance, that power is distributed over twice as much area for the first source, resulting in half the intensity. In both cases, the intensity will be inversely proportional to the square of the distance, and so will be another factor of 4 larger for the first source, resulting in an overall intensity ratio of $\frac{1}{2} \times 4 = 2$.

Checkup 17.3

1. The thunder is delayed compared with the lightning because the speed of sound is much less than the speed of light. The speed of sound is around 331 m/s, and the speed of light is roughly a million times faster; thus, the lightning flash reaches us almost instantly, and the delay of the thunderclap can be used to judge the listener's distance to the active cloud (about 3 seconds per kilometer, or 5 seconds per mile).

2. For a tube closed at both ends, each end must be a displacement node (a pressure antinode); two nodes (or antinodes) occur one-half wavelength apart. The fundamental mode thus contains a half wavelength in the length L, or $L = \lambda/2$. The fundamental frequency is then $f = v/\lambda = v/2L$.

3. The wavelengths do not match; the wavelength on the string is determined by the string length, while that in the air is determined by how far a disturbance propagates in one cycle (that is, it depends on the velocity of sound in air). The frequencies on the string and in the air do match; the oscillations of the string are mechanically transferred to the air, so the number of string oscillations per second matches the number of oscillations per second of the air.

4. (B) $\frac{1}{2}$. For the tube open at one end, a quarter wavelength must fit in the tube length; for the tube open at both ends, a half wavelength must fit. Thus the tube open at one end has twice the wavelength and, since $f = v/\lambda$, half the fundamental frequency of the tube open at both ends.

Checkup 17.4

1. For an emitter in motion, the increase in frequency is given by Eq. (17.17), where $f' = f/[1 - (V_E/v)]$ for an approaching emitter. Thus to hear twice the frequency, we require $f' = 2f$, which gives $\frac{1}{2} = 1 - (V_E/v)$, or $V_E = v/2$. Thus the train would have to approach at half the speed of sound!

2. For a receiver in motion, the increase in frequency is given by Eq. (17.15), where $f' = f[1 + (V_R/v)]$ for a receiver approaching the source of sound. Thus to hear twice the frequency, you must approach the whistle at the speed of sound, at $V_R = v$; this gives $f' = 2f$.

3. Not if the receiver remains stationary. This would require $V_E \to \infty$ for a receding emitter in Eq. (17.17).

4. (a) If you stand in the center of the circle, then there is no motion toward or away from you, and the pitch does not change. (b) If you stand at the rim of the circle, the pitch begins to increase when the car is diametrically opposite you (as the component of its velocity toward you increases from zero); the pitch continues to increase until the car is close to you (all of its velocity is toward you). The pitch then suddenly changes to its lowest value just after the car passes by. The pitch becomes less low until the car is again opposite you, when the pitch momentarily assumes its unshifted value.

5. (B) $\frac{1}{3}$. In both cases the emitter is moving, so the frequency change is given by Eq. (17.17). Since the frequency is twice as high when approaching, we have $1 + V_E/v = 2(1 - V_E/v)$, which implies $3V_E/v = 1$, or $V_E = \frac{1}{3}v$.

Checkup 17.5

1. If the waves had a much longer wavelength, diffraction effects would be greater, and the waves beyond the barrier would be even more spread out. If the waves had a much shorter wavelength, diffraction effects would be lessened, and the waves beyond the barrier would be less spread out, forming a more narrow "beam" of waves, similar to a shadow.

2. (C) Water. The waves are most spread out (diffraction effects are greatest) when the wavelength is longest. For waves of a given frequency, the wavelength is longest when the speed is largest ($v = \lambda f$). From Table 17.3, sound waves in water have a larger speed than in air or helium gas.

Fluid Mechanics

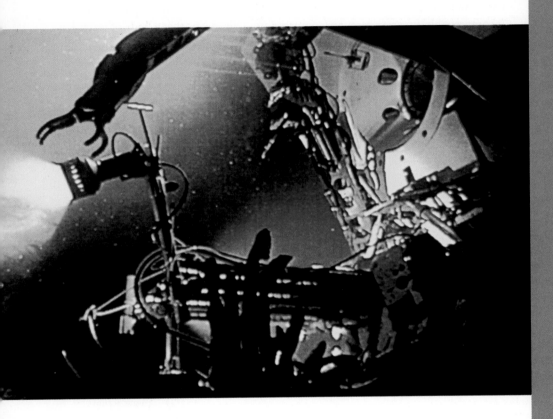

CONCEPTS IN CONTEXT

**Concepts
— in —
Context**

Deep-sea exploration demands submersibles of special design to withstand the immense underwater pressures and to provide maneuverability. The DSV (deep-sea vessel) *Alvin* was built in 1964. It carries a pilot and two passengers in a spherical cabin of thick titanium with several viewing ports. It was initially intended for dives of up to 1800 m, but it proved extremely successful in exploration and in salvage operations, and it is now certified for dives of up to 4500 m.

To appreciate the demands placed on such a submersible, we will consider these questions:

? What is the force that water pressure exerts on the viewing port of the vessel? (Example 4, page 574)

? What is the pressure at a depth of 4500 m? (Example 7, page 577)

? How does a submersible vessel change depth? (Section 18.5, page 580–581, and Example 11, page 582)

18.1 Density and Flow Velocity

18.2 Incompressible Steady Flow; Streamlines

18.3 Pressure

18.4 Pressure in a Static Fluid

18.5 Archimedes' Principle

18.6 Fluid Dynamics; Bernoulli's Equation

A **fluid** *is a system of particles loosely held together by their own cohesive forces or by the restraining forces exerted by the walls of a container.* Both liquids and gases are fluids—liquids are held together by their cohesive forces, and gases are held together by the restraining forces of a container (or, in the case of atmospheric air, by the weight of the atmosphere). In contrast to the particles in a rigid body, which are permanently locked into fixed positions, the particles in a fluid body are more or less free to wander about within the volume of the fluid body (see Fig. 18.1). A fluid will change its shape in response to external forces; for instance, a body of water or a body of air will change its shape in response to the forces exerted by gravity and by the container. The difference between these two kinds of fluids is that liquids are nearly incompressible, whereas gases are compressible. This means that a body of water has a constant volume independent of the container, whereas a body of air has a variable volume—the air always spreads out so as to entirely fill the container, and it can be made to expand or contract by increasing or decreasing the size of the container.

Although a fluid is a system of particles, the number of particles in, say, a cubic centimeter of water is so large that it is not feasible to describe the state of the fluid microscopically, in terms of the masses, positions, and velocities of all the individual particles in the system. Instead, we will describe the fluid in terms of its *density, velocity of flow,* and *pressure,* and we will see how the equations governing the statics and the dynamics of a fluid can be expressed in terms of these quantities.

In our discussion, we will neglect some other properties of fluids, such as the *viscosity* and the *surface tension.* Viscosity is an internal friction or stickiness within the fluid that offers resistance to its flow. For instance, honey is a fluid of high viscosity, whereas water is a fluid of fairly low viscosity. Surface tension is an elasticity of the exposed surface of fluid that tends to shrink this surface to a minimum area. Surface tension is responsible for the formation of drops of water in rain and for the beading of water splashed on a floor. Viscosity can be neglected when a fluid flows slowly, and surface tension can be ignored when the fluid has no exposed surface, as in the case of flow in completely filled pipes.

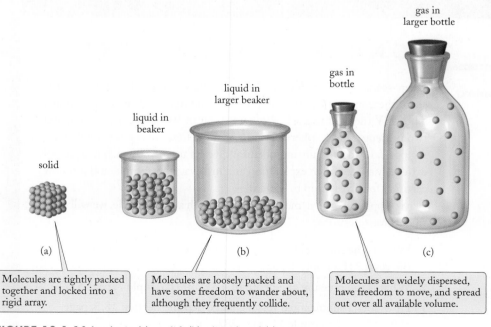

gas in
larger bottle

gas in
bottle

liquid in
larger beaker

liquid in
beaker

solid

(a)   (b)   (c)

Molecules are tightly packed together and locked into a rigid array.

Molecules are loosely packed and have some freedom to wander about, although they frequently collide.

Molecules are widely dispersed, have freedom to move, and spread out over all available volume.

FIGURE 18.1 Molecules in (a) a solid, (b) a liquid, and (c) a gas.

18.1 DENSITY AND FLOW VELOCITY

Online
Concept
Tutorial

Density, velocity of flow, and pressure give us a macroscopic description of the fluid—they tell us the average behavior of the particles in regions within the fluid, and they can be measured with large-scale instruments. For example, if our instruments indicate that the flow velocity of water in a fire hose is 4 m/s, this does not mean that all the individual water molecules have this velocity. The water molecules have a high-speed thermal motion of about 900 m/s; they move in short zigzags because they frequently collide with one another. This thermal motion of a water molecule is random; the motion is as likely to be in a direction opposite to the flow as along the flow (see Fig. 18.2). The flow of water molecules along the fire hose at 4 m/s represents a slow drift superimposed on the much faster random zigzag motion. However, on a macroscopic scale, we notice only the drift and not the random small-scale motion—we notice only the *average* motion of the water molecules.

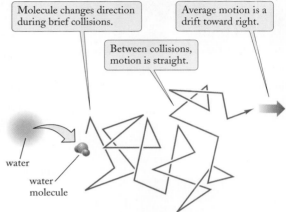

FIGURE 18.2 Motion of a molecule in water. The straight segments of the motion are typically 10^{-10} m long.

The **density** *is the amount of mass per unit volume.* Table 18.1 lists the densities of a few liquids and gases. The SI unit of density is the kilogram per cubic meter (kg/m^3). In the table, the density is designated by the customary symbol ρ, the Greek letter *rho*. Since density is the mass per unit volume, the total mass m in a volume V is the density times the volume:

$$m = \rho V \qquad (18.1)$$

The densities of gases depend on the temperature and the pressure (this dependence will be discussed in Chapter 19); unless otherwise noted, the values of the densities listed in Table 18.1 are for standard temperature and standard pressure (0°C and 1 atm). The densities of liquids depend only slightly on pressure, but they do depend appreciably on temperature. For instance, water has a maximum density at about 4°C (a few degrees above its freezing temperature), and lowest density at 100°C (when it begins to boil).

TABLE 18.1	**DENSITIES OF SOME FLUIDS**[a]		
LIQUID	ρ	**GAS**	ρ
Water		Air	
0°C	999.8 kg/m^3	0°C	1.29 kg/m^3
4°C	1000.0	20°C	1.20
20°C	998.2	Water vapor, 100°C	0.598
100°C	958.4	Hydrogen	0.0899
Seawater, 15°C	1025	Helium	0.178
Mercury	13600	Nitrogen	1.25
Sodium, liquid at 98°C	929	Oxygen	1.43
Texas crude oil, 15°C	875	Carbon dioxide	1.98
Gasoline, 15°C	739	Propane	2.01
Olive oil, 15°C	920		
Human blood, 37°C	1060		

[a] At 0°C and 1 atm, unless otherwise noted.

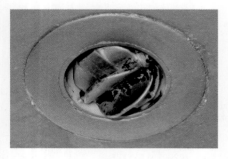

FIGURE 18.3 Paddle wheel of knot meter from a sailboat.

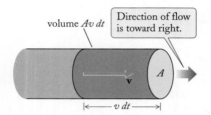

FIGURE 18.4 Flow of a fluid across an area A. In a time dt, the fluid within a distance $v\,dt$ reaches the area A, but fluid farther to the left does not.

The velocity of flow at some given point is the velocity of a small parcel of fluid passing this point. You can detect the flow velocity of the water at some point in a river by sprinkling bits of paper on the water. Quantitatively, the flow velocity can be measured by a small freely turning propeller or paddle wheel immersed in the fluid (Fig. 18.3). The motion of the fluid gives the paddle wheel a rotational speed directly proportional to the flow velocity, and the rotational speed indicator of the paddle wheel can be calibrated in terms of flow velocity.

From the magnitude v of the flow velocity of the fluid we can calculate the volume of fluid that flows across a unit area perpendicular to the direction of the velocity per unit time. Figure 18.4 shows a (stationary) area A perpendicular to the direction of flow and a volume of fluid about to cross this area. The fluid that crosses the area in a time dt is initially in a cylinder of base A and length $v\,dt$. The volume of fluid that crosses the area is therefore

$$dV = Av\,dt$$

and the volume that crosses per unit time is

$$\frac{dV}{dt} = Av \tag{18.2}$$

Here we assumed that the flow velocity v is constant over the area A. If the flow velocity in the fluid varies with position, then Eq. (18.2) is not valid for every area A, but only for a sufficiently small (infinitesimal) area A within which the flow velocity can be treated as constant.

EXAMPLE 1   The water in a fire hose of diameter 6.4 cm has a flow velocity of 4.0 m/s. At what rate does this hose deliver water? Give the answer in both cubic meters per second and kilograms per second.

SOLUTION: The radius of the hose is $r = 3.2$ cm $= 0.032$ m. The cross-sectional area of the hose is thus $A = \pi r^2 = \pi \times (0.032 \text{ m})^2 = 0.0032 \text{ m}^2$. Hence the rate of delivery is

$$\frac{dV}{dt} = Av = 0.0032 \text{ m}^2 \times 4.0 \text{ m/s} = 0.013 \text{ m}^3/\text{s} \tag{18.3}$$

This gives the answer in terms of cubic meters per second. To find the rate of delivery in terms of kilograms per second, we must multiply dV/dt by the (constant) density ρ [see Eq. (18.1)]:

$$\frac{dm}{dt} = \rho\frac{dV}{dt} = 1000 \text{ kg/m}^3 \times 0.013 \text{ m}^3/\text{s} = 13\,\text{kg/s} \tag{18.4}$$

 Checkup 18.1

QUESTION 1: According to Table 18.1, the mass of one cupful of mercury equals the mass of how many cupfuls of water?

QUESTION 2: What would be the rate at which the fire hose in Example 1 would deliver water if its diameter were half as large, that is, 3.2 cm instead of 6.4 cm?

QUESTION 3: The volume per second of water that flows through each of two pipes is the same. The flow velocity in the first pipe is one-quarter of that in the second pipe. What is the ratio of the radius of the first pipe to the radius of the second pipe?

(A) $\frac{1}{4}$ (B) $\frac{1}{2}$ (C) 1   (D) 2   (E) 4

18.2 INCOMPRESSIBLE STEADY FLOW; STREAMLINES

Online
Concept
Tutorial

In most of the examples in this chapter we will deal with **steady flow**, *for which the velocity at any given point of space remains constant in time.* Thus, in steady flow, each small parcel of fluid that starts at any given point follows exactly the same path as a small parcel that passes through the same point at an earlier (or later) time. For example, Fig. 18.5 shows velocity vectors for the steady flow of water around a cylindrical obstacle, say, the flow of the water of a broad river around a cylindrical piling placed in the middle. The water enters the picture in a broad stream from the left, and disappears in a similar broad stream toward the right.

For the steady flow of an incompressible fluid, such as water, the picture of velocity vectors can be replaced by an alternative graphical representation. Suppose we focus our attention on a small volume of water, say, 1 mm³ of water, and we observe the path of this 1 mm³ from the source to the sink. *The path traced out by the small volume of fluid is called a* **streamline**. Neighboring small volumes will trace out neighboring streamlines. In Fig. 18.6 we show the pattern of streamlines for the same steady flow of water that we already represented in Fig. 18.5 by means of velocity vectors. The streamlines on the far left (and far right) of Fig. 18.6 are evenly spaced to indicate the uniform and parallel flow in this region.

The steady flow of an incompressible fluid is often called **streamline flow**. Note that streamlines never cross. A crossing of two streamlines would imply that a small parcel of water moving along one of these streamlines has to penetrate through a small parcel of water moving along the other streamline. This is impossible—it would lead to disruption of both the small parcels and to destruction of the steadiness of flow. Because the streamlines for steady incompressible flow never cross, such flow is also called **laminar flow**, which refers to the layered arrangement of the streamlines.

If we know the velocity of flow throughout the fluid, we can trace out the motion of small parcels of fluid and therefore construct the streamlines. But the converse is also true—if we know the streamlines, we can reconstruct the velocity of flow. We can do this by means of the following rule:

> *The direction of the velocity at any one point is tangent to the streamline, and the magnitude of the velocity is proportional to the density of streamlines.*

The first part of this rule is self-evident, since the direction of motion of a small parcel of fluid is tangent to the streamline. To establish the second part, consider a bundle of streamlines forming a pipelike region, called a **stream tube**. Any fluid inside the stream tube will have to move along the tube; it cannot cross the surface of the tube because streamlines never cross. The tube therefore plays the same role as a pipe made of some impermeable material—it serves as a conduit for the fluid. If we consider a tube that is very narrow, so its cross-sectional area is very small, the velocity of flow will vary only along the length of the tube, and we can assume it will be the same

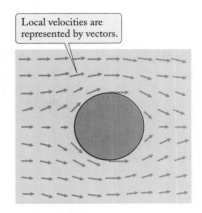

FIGURE 18.5 Velocity vectors for water flowing around a cylinder. The longest velocity vectors are found just above and just below the cylinder.

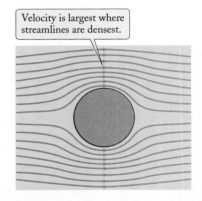

FIGURE 18.6 Streamlines for water flowing around a cylinder. The densest streamlines are found just above and just below the cylinder.

at all points on a given cross-sectional area. For instance, on the area A_1 (see Fig. 18.7) the velocity is v_1, and on the area A_2 the velocity is v_2. In a time dt, Eq. (18.2) implies that the fluid volume that enters across the area A_1 is $dV_1 = v_1 A_1\, dt$ and the fluid volume that leaves across the area A_2 is $dV_2 = v_2 A_2\, dt$. The amount of fluid that enters must match the amount that leaves, since, under steady conditions, fluid cannot accumulate in the segment of tube between A_1 and A_2. Hence $dV_1 = dV_2$, and

$$v_1 A_1\, dt = v_2 A_2\, dt$$

or, canceling the factor dt on both sides of the equation,

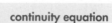 **continuity equation**

$$v_1 A_1 = v_2 A_2 \qquad (18.5)$$

This relation is called the **continuity equation**. It shows that along any stream tube, the speed of flow is inversely proportional to the cross-sectional area of the stream tube.

The density of streamlines inside the stream tube is the number of such lines divided by the cross-sectional area; since the number of streamlines entering A_1 is necessarily the same as that leaving A_2, the density of streamlines is inversely proportional to the cross-sectional area. This implies that the speed at any point in the fluid is directly proportional to the density of streamlines at that point. For example, in Fig. 18.6, the speed of the water is large at the top and bottom of the obstacle (large density of streamlines) and smaller to the left and right (smaller density of streamlines).

In experiments on fluid flow, the streamlines of a fluid can be made directly visible by several clever techniques. If the fluid is water, we can place grains of dye at diverse points within the volume of water; the dye will then be carried along by the flow, and it will mark the streamlines. The photograph in Fig. 18.8 shows a pattern of streamlines made visible by this technique. The water emerges from a pointlike source on the left and disappears into a pointlike sink on the right. The colored streamers were created by small grains of potassium permanganate dissolving in the water.

If the fluid is air, we can make the streamlines visible by releasing smoke from small jets at diverse points within the flow of air. The photograph in Fig. 18.9 shows fine trails of smoke marking the streamlines in air flowing past a scale model of the wing of an airplane in a wind tunnel. The experimental investigation of such streamline patterns plays an important role in airplane design. Incidentally: Under some conditions, the flow of air can be regarded as nearly incompressible, provided that the speed of flow is well below the speed of sound (331 m/s). Although the air will suffer some changes of density in its flow around obstacles, the changes are usually small enough to be neglected.

Finally, Fig. 18.10 shows an example of **turbulent flow**. In the region behind the wing, the streamers of smoke become twisted and chaotic. This is due to the generation of vortices, or swirls of air, in this region. As the vortices form, grow, break away, and disappear in quick succession, the velocity of flow fluctuates violently. The flow of the fluid becomes unsteady and irregular. The formation of vortices and the onset of turbulence have to do with viscosity in the fluid (see Problem 73). It is a general rule that vortices and turbulence will develop in a fluid of given viscosity whenever the velocity of flow, the length of the flow, or both exceed a certain limit. We can see the transition from steady flow to turbulent flow in the ascending smoke trail from a cigarette (see Fig. 18.11). The flow starts out steady, with smoke particles moving along well-defined streamlines; but at some height above the cigarette, where the length of the flow exceeds the critical limit, the flow becomes turbulent.

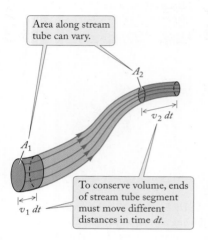

FIGURE 18.7 A stream tube.

Area along stream tube can vary.

A_2

$v_2\, dt$

A_1

$v_1\, dt$

To conserve volume, ends of stream tube segment must move different distances in time dt.

FIGURE 18.8 Streamlines in water flowing from a source (left) to a sink (right).

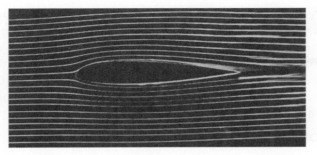

FIGURE 18.9 Fine trails of smoke indicate the streamlines in air flowing around the wing of an aircraft.

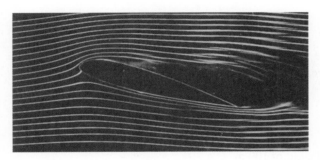

FIGURE 18.10 Here, the wing is in a partial stall, and the flow behind the wing has become turbulent.

FIGURE 18.11 Ascending smoke from a cigarette.

EXAMPLE 2   In the human circulatory system, the blood flows out of the heart via the aorta, which is connected to other arteries that branch out into a multitude of small capillaries (see Fig. 18.12). In the average adult, the aorta has a radius of 1.2 cm, and the speed of flow of the blood is 0.20 m/s. The radius of each capillary is about 3×10^{-6} m, and the number of open capillaries, under conditions of rest, is about 1×10^{10}. Calculate the speed of flow of the blood in the capillaries.

SOLUTION: The cross-sectional area of the aorta is

$$A_1 = \pi r_1^2 = \pi \times (0.012 \text{ m})^2 = 4.5 \times 10^{-4} \text{ m}^2$$

and the net cross-sectional area of all the capillaries is

$$A_2 = [\text{number of capillaries}] \times [\text{area of each}]$$

$$= 1 \times 10^{10} \times \pi r_2^2 = 1 \times 10^{10} \times \pi \times (3 \times 10^{-6} \text{ m})^2 = 3 \times 10^{-1} \text{ m}^2$$

From the continuity equation (18.5), with $v_1 = 0.20$ m/s, we then find that the speed of flow in the capillaries is

$$v_2 = \frac{A_1}{A_2} v_1 = \frac{4.5 \times 10^{-4} \text{ m}^2}{3 \times 10^{-1} \text{ m}^2} \times 0.20 \text{ m/s}$$

$$= 3 \times 10^{-4} \text{ m/s} = 0.3 \text{ mm/s}$$

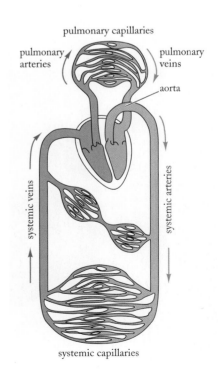

FIGURE 18.12 The human circulatory system.

FIGURE 18.13 The tall fountain at Fountain Hills, Arizona.

EXAMPLE 3   One of the world's tallest fountains (at Fountain Hills, Arizona; see Fig. 18.13) shoots water to a height of 170 m at the rate of 26 000 liters/min. If we ignore friction, the motion of small volumes of water is projectile motion. From this, it is easy to determine that the speed of flow must be 58 m/s at the base, and 37 m/s at a height of 100 m. Given these speeds, calculate the cross-sectional area of the water column at the base and at a height of 100 m.

SOLUTION: To find the cross-sectional area, we use Eq. (18.2),

$$A = \frac{1}{v}\frac{dV}{dt}$$

The rate of delivery of the fountain is (using 1 liter $= 1000$ cm$^3 = 10^{-3}$ m^3)

$$\frac{dV}{dt} = \frac{2.6 \times 10^4 \text{ liters}}{1 \text{ min}} = \frac{26 \text{ m}^3}{60 \text{ s}} = 0.43 \text{ m}^3/\text{s}$$

Therefore, at the base,

$$A = \frac{1}{58 \text{ m/s}} \times 0.43 \text{ m}^3/\text{s} = 0.0075 \text{ m}^2 = 75 \text{ cm}^2$$

and at a height of 100 m

$$A = \frac{1}{37 \text{ m/s}} \times 0.43 \text{ m}^3/\text{s} = 0.012 \text{ m}^2 = 120 \text{ cm}^2$$

Note that the water column is narrow at the base and increasingly widens toward the top.

FIGURE 18.14 Water flowing down from a faucet.

✓ Checkup 18.2

QUESTION 1: Is the continuity equation (18.5) valid for a compressible fluid, that is, for a fluid whose density changes as it flows?

QUESTION 2: Does steady flow mean that the velocity of a parcel of fluid remains constant?

QUESTION 3: Consider a gradually broadening river flowing out into the sea. Compare the densities of the streamlines before the mouth, at the mouth, and beyond the mouth. Where is the density of streamlines largest and where smallest?

QUESTION 4: According to the calculations in Example 2, is the density of streamlines larger in the aorta or in the capillaries?

QUESTION 5: When water flows vertically downward out of a faucet, the cross-sectional area of the stream of water gradually narrows (see Fig. 18.14). Explain.

QUESTION 6: Figure 18.6 shows the streamlines of water flowing around a cylinder. Where is the speed of flow largest?

(A) At top and bottom of cylinder   (B) At center of cylinder
(C) At left and right of cylinder

18.3 PRESSURE

The **pressure** within a fluid is defined as the force per unit area that a small volume of fluid exerts on an adjacent volume or on the adjacent wall of a container. Figure 18.15a shows two small adjacent cubical volumes of fluid that are within a larger volume of fluid surrounding them. The cube of fluid on the left presses against the cube on the right, and vice versa. Suppose that the magnitude of the perpendicular force between the two cubes is F and that the area of one face of one of the cubes is A; then the pressure p is defined as the magnitude F of the force divided by the area A:

$$p = \frac{F}{A} \tag{18.6}$$

According to this definition, pressure is simply the force per unit area. Note that, in contrast to the force, the pressure is a quantity *without direction,* that is, a scalar. We cannot associate a direction with the pressure, because each small volume exerts pressure forces in all directions perpendicular to its surface. For instance, each of the small cubes in Fig. 18.15a exerts pressure forces in all directions on the surrounding fluid (see Fig. 18.15b). We will see later that pressure can vary with position (for example, because of gravity); however, at any given location, the pressure is the same in all directions.

A simple mechanical device for the measurement of pressure consists of a hermetically sealed, evacuated cylindrical can with corrugated flexible bases (see Fig. 18.16). If this capsule is immersed in a fluid at high pressure, the bases will be compressed inward; if it is immersed in a fluid at low pressure, the bases will bulge outward. Thus, the deformation of the bases of the capsule serves as an indicator of the pressure, and a pointer linked to one base can be calibrated to read the pressure. This device is widely used in aneroid barometers for the measurement of atmospheric pressure.

In the SI system, the unit of pressure is the N/m², which has been given the name **pascal** (Pa),

$$1 \text{ Pa} = 1 \text{ N/m}^2 \tag{18.7}$$

Another unit in common use is the **atmosphere** (atm):

$$1 \text{ atm} = 1.01 \times 10^5 \text{ Pa} \tag{18.8}$$

(a)

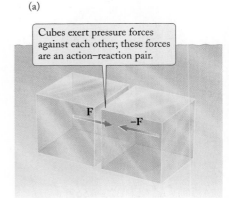

Cubes exert pressure forces against each other; these forces are an action–reaction pair.

F $-F$

(b)

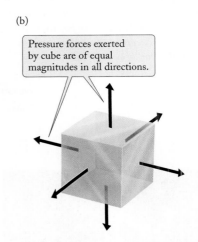

Pressure forces exerted by cube are of equal magnitudes in all directions.

FIGURE 18.15 (a) Adjacent small cubes of fluid exerting pressure forces on each other. (b) Each small cube exerts pressure forces in all directions on the surrounding fluid.

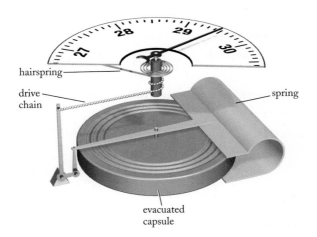

hairspring

drive chain

spring

evacuated capsule

FIGURE 18.16 Device for the measurement of atmospheric pressure consisting of a hermetically sealed evacuated cylindrical capsule made of thin sheet metal. A spring holds the flexible bases apart, and prevents the atmospheric pressure from collapsing the capsule. The upper flexible base of the capsule is linked to a pointer that indicates the atmospheric pressure.

BLAISE PASCAL (1623–1662) *French scientist. He made important contributions to mathematics and is regarded as the founder of modern probability theory. In physics, he performed experiments on atmospheric pressure and on the equilibrium of fluids.*

The common unit of pressure in the British system is the **pound-force per square inch** (lbf/in.^{2}, abbreviated psi); in psi, one atmosphere is

$$1\,\text{atm} = 14.7\,\text{psi} \tag{18.9}$$

This is the average value of the pressure of air at sea level. Note that this is quite a large pressure. For instance, the force that the atmospheric pressure of 1 atm exerts on the palm of your hand, of approximate area 0.006 m^2, is

$$F = Ap = 0.006\,\text{m}^2 \times 1.01 \times 10^5\,\text{Pa} = 600\,\text{N} \tag{18.10}$$

This is roughly the weight of 60 kg, but you do not notice this pressure force because an equal pressure force of opposite direction acts on the back of your hand, leaving the hand in equilibrium (you don't even notice that these opposed forces squeeze your hand, because the external pressure forces directed against your skin are compensated by the internal pressure forces exerted by your body fluids).

Table 18.2 gives diverse examples of values of pressure.

Concepts — *in* — Context

EXAMPLE 4   The viewing ports of the DSV *Alvin* have a diameter of 30 cm (see Fig. 18.17). What is the force that water pressure exerts on the outside of such a viewing port when the *Alvin* is at a depth of 4500 m below the surface of the ocean, where the water pressure is 4.5 × 10^7 Pa?

SOLUTION: The area of the porthole is $A = \pi r^2 = \pi \times (0.15\,\text{m})^2 = 0.071\,\text{m}^2$. According to Eq. (18.6), the force is then

$$F = A \times p = 0.071\,\text{m}^2 \times 4.5 \times 10^7\,\text{Pa} = 3.2 \times 10^6\,\text{N}$$

This is a weight of about 320 tons!

FIGURE 18.17 Viewing ports of the DSV *Alvin*.

TABLE 18.2	SOME PRESSURES
Core of neutron star	1×10^{38} Pa
Center of Sun	2×10^{16}
Highest sustained pressure achieved in laboratory	5×10^{11}
Center of Earth	4×10^{11}
Bottom of Pacific Ocean (5.5-km depth)	6×10^{7}
Water in core of nuclear reactor	1.6×10^{7}
Overpressure^a in automobile tire	2×10^{5}
Air at sea level	1.0×10^{5}
Overpressure at 7 km from 1-megaton explosion	3×10^{4}
Air in funnel of tornado	2×10^{4}
Overpressure in human heart	
Systolic	1.6×10^{4}
Diastolic	1.1×10^{4}
Lowest vacuum achieved in laboratory	10^{-14}

^{a} The *overpressure* is the amount of pressure in excess of normal atmospheric pressure.

Checkup 18.3

QUESTION 1: Consider a cube of solid concrete, 1 m × 1 m × 1 m, surrounded by air at 1 atm of pressure. What is the pressure force of air on its lower face? On its upper face? On its left face? On its right face?

QUESTION 2: A brick rests on a table. Does the atmospheric pressure force push this brick downward on the table?

QUESTION 3: A rubber balloon is filled with air and sealed tightly. Would you expect this balloon to change in size as the atmospheric pressure of air changes from one day to the next?

QUESTION 4: A closed, empty soda bottle, with a diameter of 10 cm at its base and 2 cm at its cap, is lying sideways on a desk in the air. What is the ratio of the value of the external pressure at its base to the value of the pressure at its cap?

(A) 25   (B) 5   (C) 1   (D) $\frac{1}{5}$ (E) $\frac{1}{25}$

18.4 PRESSURE IN A STATIC FLUID

A fluid is said to be in static equilibrium when the flow velocity is everywhere zero, that is, the fluid is at rest. An example of such a static fluid is the air in a closed room with no air currents. At first, we will neglect gravity and pretend that the only forces acting on the fluid are those exerted by the walls of the container. Under these conditions, the pressure at all points within the fluid must be the same. To see that this is so, consider two points 1 and 2, and imagine a long, thin parallelepiped of fluid with bases at these two points (see Fig. 18.18). The fluid outside the parallelepiped exerts pressure forces on the fluid inside the parallelepiped. The components of these forces along the long direction of the parallelepiped are entirely due to the forces on the bases at 1 and 2. If the parallelepiped of fluid is to remain static, these forces on the opposite bases must be equal in magnitude. Hence, the pressures at 1 and 2 must be equal. For example, the pressure of the air is the same at all points of a room—if the pressure is 1 atm in one corner of the room, it will be the same at any other point of the room.

The uniformity of pressure throughout a static fluid implies that if we apply a pressure to some part of the surface of a confined fluid by means of a piston or a weight pushing against the surface, then this pressure will be transmitted without change to all parts of the fluid. This rule for the transmission of pressure in a static fluid is called **Pascal's Principle**, and it finds widespread application in the design of hydraulic presses, jacks, and remote controls. Figure 18.19 is a schematic diagram illustrating the principle of a hydraulic press. The mechanism that generates a large force in such a press consists of two cylinders with pistons, one small and one large. The cylinders are filled with an incompressible fluid, and they are connected by a pipe. By pushing down on the small piston, we increase the pressure in the fluid; this increases the force on the large piston. Since the pressures on both pistons are the same, the forces on the pistons are in the ratio of the areas of their faces; thus, a small force on the piston in the small cylinder will generate a large force on the piston in the large cylinder. Figure 18.20 shows a hydraulic jack for an automobile. The pumping lever pushes on the small piston, and the hydraulic fluid communicates the

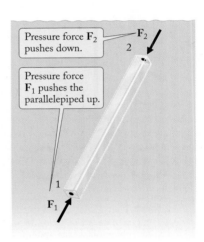

FIGURE 18.18 A long, thin parallelepiped of fluid within a static fluid.

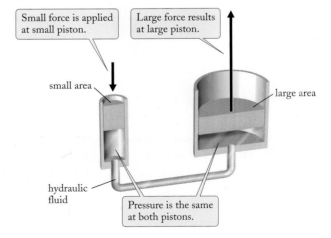

FIGURE 18.19 Mechanism of a hydraulic press (schematic).

FIGURE 18.20 Hydraulic car jack.

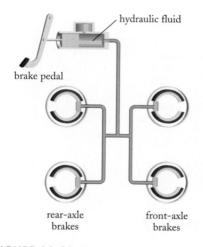

hydraulic fluid

brake pedal

rear-axle brakes   front-axle brakes

FIGURE 18.21 Hydraulic brake system.

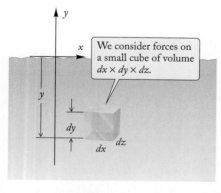

We consider forces on a small cube of volume $dx \times dy \times dz$.

FIGURE 18.22 A small cube of fluid at depth y below the surface of the fluid.

pressure in incompressible fluid

resultant pressure to the large piston, which lifts the automobile. The brake systems and other control systems on automobiles, trucks, and aircraft also employ such arrangements of cylinders connected by pipes filled with hydraulic fluid (see Fig. 18.21).

EXAMPLE 5   The diameter of the small piston in Fig. 18.19 is 1.5 cm and that of the large piston is 7.0 cm. If you exert a force of 300 N on the small piston, what force will this generate on the large piston?

SOLUTION: The pressure is the same at all points in a static fluid (neglecting gravity), so $p_1 = p_2$ requires $F_1/A_1 = F_2/A_2$. Thus the forces are in the ratio of the areas of the pistons, and these areas are in the ratio of the squares of the diameters:

$$F_2 = F_1 \frac{A_2}{A_1}$$

$$= 300 \text{ N} \times \frac{(7.0 \text{ cm})^2}{(1.5 \text{ cm})^2} = 6500 \text{ N}$$

Next, we want to take into account the effect of gravity on the pressure in a fluid. For a static fluid subjected to gravity, such as the water of a calm lake, the pressure force at any given depth must support the weight of the overlying mass of fluid; consequently, the pressure must increase with depth. To derive a formula for the dependence of pressure on depth, consider the condition for the equilibrium of a small cubical volume of fluid. Figure 18.22 shows a small cube at some depth y below the surface of the fluid. As always, the y coordinate is reckoned as positive in the upward direction; hence the depth of the cube corresponds to some negative value of y. The dimensions of the cube are $dx \times dy \times dz$, and hence its weight is

$$g\, dm = g\rho\, dx \times dy \times dz$$

where ρ is the mass density, or mass per unit volume, of the fluid. The weight of the cube must be balanced by the vertical forces contributed by the pressure. Suppose that the pressure at the top of the cube is p; the pressure at the bottom of the cube is then some larger value $p + dp$, where dp represents the change of pressure in the interval dy (note that if dy is negative, dp will be positive). Since the area of the top and bottom faces of the cube is $dx \times dz$, the difference between the vertical pressure forces on the top and bottom of the cube is

$$p \times dx \times dz - (p + dp) \times dx \times dz = -dp \times dx \times dz$$

For equilibrium, this pressure force must balance the weight:

$$-dp \times dx \times dz = g\rho\, dx \times dy \times dz$$

or

$$dp = -g\rho\, dy \qquad (18.11)$$

This formula gives us the small change in pressure for a small increase in depth. For an incompressible fluid, ρ is constant, and Eq. (18.11) shows that the small change in pressure is proportional to the small change in y. Hence the total change in pressure must be proportional to the total change in y; that is,

$$p - p_0 = -\rho g y \qquad \text{(incompressible fluid)} \qquad (18.12)$$

Here p_0 is the pressure at the surface of the liquid, where $y = 0$. Note that depths below the surface are reckoned as negative in Eq. (18.12), and therefore the right side of the equation is positive, which leads to a positive value of $p - p_0$, that is, an increase of pressure with depth.

EXAMPLE 6   What is the pressure at a depth of 10 m below the surface of a lake? Assume that the pressure of air at the surface of the lake is 1.0 atm.

SOLUTION: With $p_0 = 1 \text{ atm} = 1.01 \times 10^5 \text{ Pa}$, $\rho = 1000 \text{ kg/m}^3$, and $y = -10$ m, Eq. (18.12) gives

$$p = p_0 - \rho g y$$

$$= 1.01 \times 10^5 \text{ Pa} + 1000 \text{ kg/m}^3 \times 9.81 \text{ m/s}^2 \times 10 \text{ m} \qquad (18.13)$$

$$= 1.99 \times 10^5 \text{ Pa} = 2.0 \text{ atm}$$

Thus, the pressure is 1 atm at the surface of the lake and about 2 atm at a depth of 10 m; that is, the pressure increases by about 1 atm per 10 m of water, an easily remembered change with depth.

EXAMPLE 7   What is the pressure at a depth of 4500 m below the surface of the ocean, the maximum depth for which the DSV *Alvin* (see chapter photo) is certified? The density of seawater is 1025 kg/m^3.

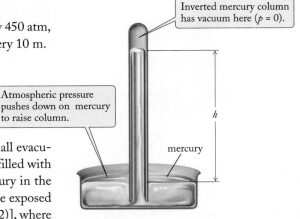

Concepts
— in —
Context

SOLUTION: The calculation proceeds as in the preceding example:

$$p = p_0 - \rho g y$$

$$= 1.01 \times 10^5 \text{ Pa} + 1025 \text{ kg/m}^3 \times 9.81 \text{ m/s}^2 \times 4500 \text{ m}$$

$$= 4.54 \times 10^7 \text{ Pa}$$

This is the value of the pressure used in Example 4. This is approximately 450 atm, in agreement with the expected pressure increase of about 1 atm for every 10 m.

Several simple instruments for the measurement of pressure make use of a column of liquid. Figure 18.23 shows a mercury **barometer** consisting of a tube of glass, about 1 m long, closed at the upper end and open at the lower end. The tube is filled with mercury, except for a small evacuated space at the top. The bottom of the tube is immersed in an open bowl filled with mercury. The atmospheric pressure acting on the exposed surface of mercury in the bowl prevents the mercury from flowing out of the tube. At the level of the exposed surface, the pressure exerted by the column of mercury is $\rho g h$ [see Eq. (18.12)], where $\rho = 1.36 \times 10^4 \text{ kg/m}^3$ is the density of mercury and h the height of the mercury column. For equilibrium, this pressure must match the atmospheric pressure:

Inverted mercury column has vacuum here ($p = 0$).

Atmospheric pressure pushes down on mercury to raise column.

h

mercury

FIGURE 18.23 A mercury barometer.

$$p_0 = \rho g h \qquad (18.14)$$

This equation permits a simple determination of the atmospheric pressure from a measurement of the height of the mercury column.

In view of the direct correspondence of the atmospheric pressure and the height of the mercury column, the pressure is often quoted in terms of this height, usually expressed in millimeters of mercury (abbreviated mm-Hg, where Hg is the symbol for the element mercury). The average value of the **atmospheric pressure** at sea level is 760 mm-Hg, which by definition is one atmosphere (atm). Hence,

$$1 \text{ atm} = 760 \text{ mm-Hg} = \rho \times g \times 0.760 \text{ m}$$

$$= 1.36 \times 10^4 \text{ kg/m}^3 \times 9.81 \text{ m/s}^2 \times 0.760 \text{ m} \qquad (18.15)$$

$$= 1.01 \times 10^5 \text{ N/m}^2 = 1.01 \times 10^5 \text{ Pa}$$

This value of the atmosphere has already been mentioned, in Eq. (18.8). The unit mm-Hg is also referred to as the **torr** (from Torricelli); that is, 1 torr = 1 mm-Hg. In British units, the atmospheric pressure is often quoted in inches of mercury; typical values are around 30 inches, since 760 mm × (1 inch)/(25.4 mm) = 29.9 inches.

Figure 18.24 shows an open-tube **manometer**, a device for the measurement of the pressure of a fluid, such as that contained in the tank shown on the left. The U-shaped tube contains mercury, or water, or oil. One side of the tube is in contact with the fluid in the tank; the other is in contact with the air. The fluid in the tank therefore presses down on one end of the mercury column and the air presses down on the other end. The difference h in the heights of the levels of mercury at the two ends gives the difference in the pressure at the two ends:

$$p - p_0 = \rho g h \qquad (18.16)$$

Hence, this kind of manometer indicates the amount of pressure in the tank in excess of the atmospheric pressure. This excess is called the **overpressure**, or **gauge pressure**. It is well to keep in mind that many pressure gauges used in engineering practice are calibrated in terms of the overpressure rather than absolute pressure. For instance, the pressure gauges used for automobile tires read overpressure.

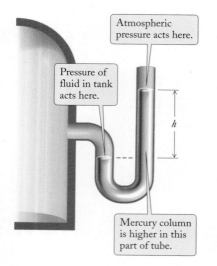

Atmospheric pressure acts here.

Pressure of fluid in tank acts here.

h

Mercury column is higher in this part of tube.

FIGURE 18.24 Open-tube manometer.

EVANGELISTA TORRICELLI (1608–1647) *Italian physicist, mathematician, and inventor. At age 33 he assisted Galileo, serving as both researcher and secretary. He was the first to understand air pressure in terms of the weight of the atmosphere and to explain the limitations of suction pumps. His accomplishments included contributions to the development of integral calculus and the invention of the barometer.*

EXAMPLE 8   What is the change in atmospheric pressure between the basement of a house and the attic, at a height of 10 m above the basement? Express the result in mm-Hg. Assume that the density of air has its standard value 1.29 kg/m^3.

SOLUTION: Although air is a compressible fluid, the change in its density is small if the change of altitude (and pressure) is small, as it is in the present example; we will examine the behavior for larger changes in the next chapter. Therefore, we can assume Eq. (18.12) is a good approximation:

$$p - p_0 = -\rho g y$$
$$= -1.29 \text{ kg/m}^3 \times 9.81 \text{ m/s}^2 \times 10 \text{ m} = -1.3 \times 10^2 \text{ Pa} \qquad (18.17)$$

Since $1.01 \times 10^5 \text{ Pa}$ equals 760 mm-Hg, $-1.3 \times 10^2 \text{ Pa}$ equals

$$-1.3 \times 10^2 \text{ Pa} \times \frac{760 \text{ mm-Hg}}{1.01 \times 10^5 \text{ Pa}} = -0.95 \text{ mm-Hg}$$

Hence the pressure change is indeed small; the pressure decreases by about 1 mm-Hg for a 10-m increase in height in air. This decrease in pressure can be detected by carrying an ordinary barometer from the basement to the attic of the house.

PHYSICS IN PRACTICE   THE SPHYGMOMANOMETER

Among the many practical applications of manometers is the **sphygmomanometer** used to measure cardiac blood pressure. This consists of a manometer connected to an air sac in the form of a cuff (see the figures). The air sac is wrapped around

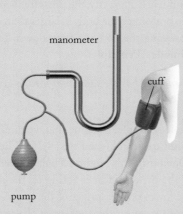

manometer

cuff

pump

FIGURE 1 A sphygmomanometer.

the upper arm of the patient, and is then inflated by means of a hand pump until the pressure of the sac against the arm collapses the brachial artery and cuts off the blood flow. The air is then slowly allowed to leak out of the sac. When the pressure drops to a value equal to the systolic (maximum) cardiac pressure, blood will intermittently squirt through the artery with each heartbeat. This initiation of intermittent blood flow can be readily detected by listening to the noise of the rushing blood with a stethoscope placed just below the cuff. As the pressure drops further, the intervals of intermittent blood flow become longer; and when the pressure has dropped to a value

equal to the minimum (diastolic) cardiac pressure, the blood flow becomes continuous. Thus, the onset of intermittent noises and the cessation of intermittent noises signal, respectively, the systolic and the diastolic cardiac pressures. Typical values of these pressures in healthy adults are 120 mm-Hg and 80 mm-Hg, usually reported in the abbreviated notation 120/80. Note that in order to obtain an accurate reading of the cardiac pressure, the cuff must be placed at the level of the heart; if it were placed lower (say, on a leg) or higher (say, on a raised arm), then there would be a pressure difference $\rho g h$ between the height of the heart and the height of the cuff. This pressure difference would amount to about 10 mm-Hg for every 10 cm of height difference—a significant discrepancy.

FIGURE 2

✔ Checkup 18.4

QUESTION 1: Pipes connect the faucets in a house to the water main that supplies the house. Is the pressure in the faucets on the first floor the same as the pressure on the second floor?

QUESTION 2: You carry an inflated rubber balloon up a mountain. Would you expect this balloon to change in size?

QUESTION 3: Does the increase of pressure with depth [Eq. (18.12)] contradict Pascal's Principle?

QUESTION 4: A scuba diver descends to 10 m below the surface of the sea. What is the pressure inside her muscle tissues when she is at the surface? When she reaches 10 m?

    (A) 0 atm, 0 atm   (B) 0 atm, 1 atm

    (C) 1 atm, 1 atm   (D) 1 atm, 2 atm

18.5 ARCHIMEDES' PRINCIPLE

If you try to push a beach ball below the surface of the water, you notice that the water exerts a strong upward push on the ball. You can barely force the ball under, and if you release it, the ball pops out of the water with violence. The upward force that water or some other fluid exerts on a partially or totally immersed body is called the **buoyant force**. This force results from the pressure difference between the bottom and the top of the body. In a fluid in equilibrium under the influence of gravity, the pressure increases with depth; hence the pressure of the fluid at the bottom of the body is larger than that at the top of the body, and there is more force pressing the body up than down. The magnitude of the buoyant force is given by **Archimedes' Principle:**

Archimedes' Principle

> *The buoyant force on an immersed body has the same magnitude as the weight of the fluid displaced by the body.*

The proof of this famous principle is simple. Imagine that we replace the immersed volume of the body by an equal volume of fluid (see Fig. 18.25). The volume of fluid will then be in static equilibrium. Obviously, this requires a balance between the weight of the fluid and the resultant of all the pressure forces acting on the surface enclosing this volume of fluid. But the pressure forces on the surface of the original immersed body are exactly the same as the pressure forces on the surface of the volume of fluid by which we have replaced it. Hence, the magnitude of the resultant of the pressure forces acting on the original body must equal the weight of the displaced fluid.

If a body of density less than that of water floats in water, equilibrium of the weight and of the buoyant force is achieved when the body is partially submerged—the weight of the body must match the weight of the water displaced by the submerged part of the body. Thus, the weight of a ship must match the weight of the water displaced by the submerged part of its hull. If we increase the weight of the ship, by loading more cargo, the hull will submerge more deeply until equilibrium is established (Fig. 18.26). If we increase the weight of the ship to such an extent that its average density exceeds that of water, equilibrium becomes impossible, and the ship sinks.

The human body has an average density slightly less than that of water; if inert, it floats with only the top of the head sticking out. But the equilibrium is quite delicate, and some people can make themselves sink by merely exhaling air, thereby reducing their chest volume and increasing their average density. Most species of bony fishes use a similar method to adjust their buoyancy; they have an internal swim bladder filled with gas,

(a)

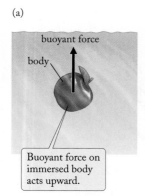

Buoyant force on immersed body acts upward.

(b)

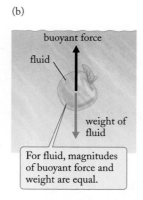

For fluid, magnitudes of buoyant force and weight are equal.

FIGURE 18.25 (a) A submerged body and the buoyant force that acts on this body. (b) "Free-body" diagram for a volume of fluid of the same shape as the body. For the volume of fluid, the buoyant force is balanced by the weight.

FIGURE 18.26 The markings along the hull of the ship indicate the immersion level.

and they preserve neutral buoyancy by making adjustments to the volume of this swim bladder. Submarines use much the same method to move down or up in the water. They have diving tanks ("ballast tanks") filled partially with water and partially with air. To dive, they pump air out of the tanks, into pressurized storage cylinders, and allow water to flood the tanks, thereby effectively increasing the mass and the density of the submarine. To surface, they blow high-pressure air into the tanks, and drive out the water.

ARCHIMEDES (287–212 B.C.) *Greek philosopher, mathematician, and physicist. He calculated an accurate value for π and obtained many other geometrical results, especially regarding the surface areas and volumes of curved bodies. He investigated the laws of the lever, the statics of fluids, and the static equilibrium of floating bodies. For these investigations, he relied on mathematical demonstrations, and he thereby initiated the mathematical analysis of physical phenomena, which is a cornerstone of theoretical physics.*

EXAMPLE 9   A chunk of ice floats in water (see Fig. 18.27). What percentage of the volume of ice will be above the level of the water? The density of ice is 917 kg/m³.

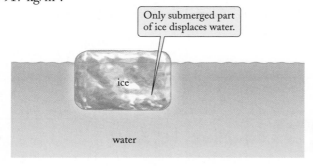

Only submerged part of ice displaces water.

ice

FIGURE 18.27
Ice floating in water.

water

SOLUTION: If the mass of the chunk of ice is, say, $M = 1000$ kg, it must displace an amount of water of the same weight, that is, 1.000 m³ of water. The volume of ice below the water level must then be 1.000 m³. For this mass of ice, the total volume is

$$V = \frac{M}{\rho} = \frac{1000 \text{ kg}}{917 \text{ kg/m}^3} = 1.091 \text{ m}^3$$

If the volume below water is 1.000 m³, the volume above water is 1.091 m³ − 1.000 m³ = 0.091 m³. The fraction of ice above the water level is therefore

$$\frac{\Delta V}{V} = \frac{0.091 \text{ m}^3}{1.091 \text{ m}^3} = 0.083$$

or 8.3%. Thus the expression "the tip of the iceberg" has come to mean a small part of the whole.

EXAMPLE 10   A hot-air balloon (Fig. 18.28) has a volume of 2.20×10^3 m³. What is the buoyant force that the surrounding cold air exerts on the balloon? Assume that the density of the surrounding air is 1.29 kg/m³.

SOLUTION: The mass M of the cold air displaced by the balloon volume V is

$$M = \rho V = 1.29 \text{ kg/m}^3 \times 2.20 \times 10^3 \text{ m}^3 = 2.84 \times 10^3 \text{ kg}$$

The weight of this air is $W = Mg$. By Archimedes' Principle, this weight gives us the magnitude of the buoyant force,

$$F = Mg = 2.84 \times 10^3 \text{ kg} \times 9.81 \text{ m/s}^2 = 2.79 \times 10^4 \text{ N}$$

If the balloon is to stay aloft, its weight (including the weight of the hot air inside it) must be less than or equal to the buoyant force of 2.79×10^4 N.

FIGURE 18.28 A hot-air balloon.

EXAMPLE 11   The deep-sea research diving vessel *Alvin* can operate at depths of 4500 m. With its ballast tanks empty, the vessel displaces a total volume of 16.8 m³ and has a total mass of 17 000 kg. What minimum volume must the ballast tanks have to permit the vessel to dive?

SOLUTION: To dive, the vessel must be slightly heavier than the weight that provides neutral buoyancy when fully submerged. The vessel displaces a mass of seawater equal to

$$m = \rho V = 1025 \text{ kg/m}^3 \times 16.8 \text{ m}^3 = 17\,200 \text{ kg}$$

The additional mass required to achieve neutral buoyancy is thus 17 200 kg − 17 000 kg = 200 kg. Accordingly, the ballast tanks must permit the entry of at least 200 kg of water into the vessel. They must therefore have a volume

$$V = \frac{m}{\rho} = \frac{200 \text{ kg}}{1025 \text{ kg/m}^3} = 0.20 \text{ m}^3$$

✔ Checkup 18.5

QUESTION 1: You completely immerse a sealed can of volume 1.0×10^{-3} m³ in water. What is the buoyant force on the can?

QUESTION 2: If a ship is holed below the water line and it fills with water, it will sink. Explain in terms of Archimedes' Principle.

QUESTION 3: Some people claim that it is easier to stay afloat in seawater than in freshwater. Is this claim justified?

QUESTION 4: A rock has a density of 5000 kg/m³. Will this rock float in water? Will it float in liquid mercury?

QUESTION 5: Suppose that in Example 9, the ice is floating in a glass of water that is filled to the brim, so that the ice above the water level sticks up above the rim of the glass. What happens when the ice melts?

  (A) Some water overflows.   (B) The water level drops below the rim.
  (C) The water level remains at the rim.

Online
Concept
Tutorial

18.6 FLUID DYNAMICS; BERNOULLI'S EQUATION

When moving air encounters an obstacle that slows down its motion, the air exerts an extra pressure on the obstacle. You can feel the push of this extra pressure if you stand in a strong wind, or if you put a hand out of the window of a speeding car. The pressure changes that occur when air flows around obstacles or when water or some other fluid flows through pipes of varying cross sections can be calculated in a simple way by exploiting the conservation theorem for the mechanical energy. In this section, we will formulate the conservation theorem for energy for the special case of steady flow of an incompressible fluid without viscosity. This conservation theorem is called *Bernoulli's equation.*

We know from Section 18.2 that steady incompressible flow can be described by streamlines. As in the derivation of the equation of continuity, we consider a bundle of streamlines forming a thin stream tube. The fluid flows inside this tube as though the surface of the tube were an impermeable pipe. Figure 18.29a shows a segment of this "pipe"; this segment contains some mass of fluid. The left end of the mass of fluid is at height y_1 and has area A_1; the right end is at height y_2 and has area A_2. Figure 18.29b shows the same mass of fluid at a slightly later time—the fluid has moved toward the right. During this movement, the external pressures at the left and at the right ends of the segment exert forces and do some work on the mass of fluid. By energy conservation, this work must equal the change of kinetic and potential energy. To express this mathematically, we begin by calculating the work done by pressure. As the left end of the mass of fluid moves through a distance Δl_1, the work done by the pressure is the force $A_1 p_1$ multiplied by the distance Δl_1:

$$W_1 = A_1 p_1 \Delta l_1 \tag{18.18}$$

Since the product $A_1 \Delta l_1$ is the volume ΔV vacated by the movement of the left end of the mass of fluid, we can also write this as

$$W_1 = p_1 \Delta V \tag{18.19}$$

Likewise, the work done by the pressure at the right end is

$$W_2 = -p_2 \Delta V \tag{18.20}$$

This is negative because the external force at the right end is opposite to the displacement. Note that the *same* volume ΔV appears in Eqs. (18.19) and (18.20)—the fluid is incompressible, and hence the volume vacated by the movement of the fluid at the left end must equal the volume newly occupied at the right end. The net work done by the pressure is then

$$\Delta W = W_1 + W_2 = p_1 \Delta V - p_2 \Delta V \tag{18.21}$$

The change in kinetic and potential energy is entirely due to the changes at the ends of the mass of fluid; everywhere else, the shift of the fluid merely replaces fluid of some kinetic energy and potential energy with fluid of exactly the same kinetic and potential energy. The change at the ends involves replacing a mass Δm of fluid, of speed v_1 at height y_1, by an equal mass Δm, of speed v_2 at height y_2. The corresponding change of kinetic and potential energy is

$$\Delta K + \Delta U = \tfrac{1}{2} \Delta m\, v_2^2 - \tfrac{1}{2} \Delta m\, v_1^2 + \Delta m\, g y_2 - \Delta m\, g y_1 \tag{18.22}$$

This change of mechanical energy must match the work done by the pressure:

$$\tfrac{1}{2} \Delta m\, v_2^2 - \tfrac{1}{2} \Delta m\, v_1^2 + \Delta m\, g y_2 - \Delta m\, g y_1 = p_1 \Delta V - p_2 \Delta V \tag{18.23}$$

If we divide both sides of this equation by ΔV and we move all terms with subscript 2 to the left side and all terms with subscript 1 to the right side, we obtain

$$\frac{1}{2} \frac{\Delta m}{\Delta V} v_2^2 + \frac{\Delta m}{\Delta V} g y_2 + p_2 = \frac{1}{2} \frac{\Delta m}{\Delta V} v_1^2 + \frac{\Delta m}{\Delta V} g y_1 + p_1 \tag{18.24}$$

or, since $\Delta m / \Delta V$ is the density ρ of the fluid,

$$\tfrac{1}{2} \rho v_2^2 + \rho g y_2 + p_2 = \tfrac{1}{2} \rho v_1^2 + \rho g y_1 + p_1 \tag{18.25}$$

From this we see that the quantity $\tfrac{1}{2}\rho v^2 + \rho g y + p$ has the same value at different locations along the streamline, which means it is a constant of the motion:

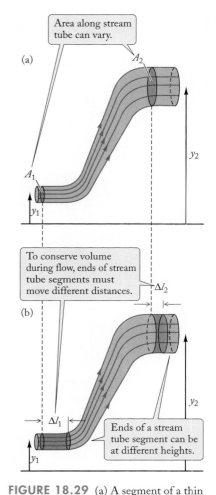

FIGURE 18.29 (a) A segment of a thin stream tube. The beginning and the end of the segment are marked by the dashed lines. (b) Motion of the fluid along the stream tube. The fluid enters the segment from the left and emerges on the right.

$$\tfrac{1}{2}\rho v^2 + \rho g y + p = \text{[constant]} \tag{18.26}$$

This is **Bernoulli's equation**. Note that $\tfrac{1}{2}\rho v^2$ is the density of kinetic energy (kinetic energy per unit volume) and $\rho g y$ is the density of potential energy (potential energy per unit volume). Hence Bernoulli's equation states that along any streamline, the sum of the density of kinetic energy, density of potential energy, and pressure is a constant. The kinetic and potential energy density terms correspond to the kinetic and potential energies in the Law of Conservation of Mechanical Energy for a particle [see Eq. (7.36)].

In the special case of a static fluid, with $v = 0$, Bernoulli's equation reduces to

$$p + \rho g y = \text{[constant]} \tag{18.27}$$

This is equivalent to Eq. (18.12) for the pressure in a static fluid.

According to Bernoulli's equation (18.26), in any region in which the term $\rho g y$ is constant or approximately constant, *the pressure along any given streamline must decrease wherever the speed increases.* Intuitively, we might expect that where the speed is large, the pressure is large; but energy conservation demands exactly the opposite! The relation between pressure and speed plays an important role in the design of wings for airplanes. Figure 18.30 shows an airfoil and the streamlines of air flowing around it, as seen in the reference frame of the airplane. The shape of the airfoil has been designed so that along its upper part the speed is large (high density of streamlines) and along its lower part the speed is small (low density of streamlines). Consider now one streamline passing just over the airfoil and one just under it. At a large distance to the right, the fluid in all streamlines has the same pressure and the same speed. Bernoulli's equation applied to each of the two streamlines therefore tells us that in the region just above the airfoil, the pressure is low, and in the region just below the airfoil, the pressure is high. This leads to a net upward force, or **lift**, on the airfoil—this lift force supports the airplane in flight. As Fig. 18.30 shows, at a large distance to the left, the streamlines have a slight downward trend; what has happened is that the air has given some upward momentum to the airfoil and, in return, acquired an equal amount of downward momentum. Ultimately, the downward flow of air presses against the ground and transmits the weight of the airplane to the ground.

To conclude this section, we will work out some examples in which Bernoulli's equation is applicable.

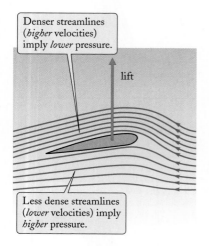

Denser streamlines (*higher* velocities) imply *lower* pressure.

lift

Less dense streamlines (*lower* velocities) imply *higher* pressure.

FIGURE 18.30 Flow of air around an airfoil.

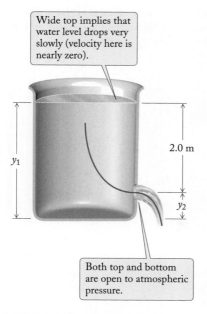

Wide top implies that water level drops very slowly (velocity here is nearly zero).

y_1

2.0 m

y_2

Both top and bottom are open to atmospheric pressure.

FIGURE 18.31 Streamline for a parcel of water flowing out of a tank.

EXAMPLE 12   A water tank has a (small) hole near its bottom at a depth of 2.0 m from the top surface (see Fig. 18.31). What is the speed of the stream of water emerging from the hole?

SOLUTION: Qualitatively, one of the streamlines for the water flowing out of the tank will look as shown in Fig. 18.31. Since the hole is small, the water level at the top of the tank drops only very slowly; we can therefore take $v_1 = 0$ in Eq. (18.25). Furthermore, the pressures at the top and in the emerging stream of water are the same; both are equal to the atmospheric pressure p_0. Thus, $p_1 = p_2 = p_0$. With this, Eq. (18.25) becomes

$$\tfrac{1}{2}\rho v_2^2 + \rho g y_2 + p_0 = \rho g y_1 + p_0 \tag{18.28}$$

We can cancel the terms p_0, and we can move the term $\rho g y_2$ to the right side of the equation:

$$\tfrac{1}{2}\rho v_2^2 = \rho g y_1 - \rho g y_2 \tag{18.29}$$

If we divide both sides by $\frac{1}{2}\rho$ and extract the square root of both sides, we find

$$v_2 = \sqrt{2g(y_1 - y_2)} \qquad (18.30)$$

With $y_1 - y_2 = 2.0$ m, the speed is

$$v_2 = \sqrt{2 \times 9.81 \text{ m/s}^2 \times 2.0 \text{ m}} = 6.3 \text{ m/s}$$

COMMENT: According to Eq. (18.30), the speed of the emerging water is exactly what it would be if the water were to fall freely through a height $y_1 - y_2$ [compare Eq. (2.29)]. This result is called **Torricelli's theorem**. It merely expresses conservation of energy: when a drop of water flows out at the bottom, the loss of potential energy of the water in the tank is equivalent to the removal of a drop of water from the top; the conversion of this potential energy into kinetic energy will give the drop the speed of free fall. We have already relied on this energy argument in Chapter 8, where we calculated the speed of water flowing out of a pipe in a hydroelectric storage plant.

EXAMPLE 13   The overpressure in a fire hose of diameter 6.4 cm is 3.5×10^5 Pa, and the speed of flow is 4.0 m/s. The (horizontal) fire hose ends in a metal tip of diameter 2.5 cm (see Fig. 18.32). What are the overpressure and the velocity of water in the tip?

SOLUTION: The magnitudes of the velocities in the hose and in the tip are related to the cross-sectional areas by the continuity equation [Eq. (18.5)],

$$v_2 = v_1 \frac{A_1}{A_2} \qquad (18.31)$$

The ratio of the cross-sectional areas is equal to the ratio of the squares of the diameters; hence the velocity in the tip is

$$v_2 = 4.0 \text{ m/s} \times \frac{(6.4 \text{ cm})^2}{(2.5 \text{ cm})^2} = 26 \text{ m/s}$$

The overpressure of the water in the tip can then be calculated from Eq. (18.25) with $y_1 = y_2 = 0$ for a horizontal hose:

$$\tfrac{1}{2}\rho v_2^2 + p_2 = \tfrac{1}{2}\rho v_1^2 + p_1$$

which yields

$$p_2 = p_1 + \tfrac{1}{2}\rho v_1^2 - \tfrac{1}{2}\rho v_2^2$$

$$= 3.5 \times 10^5 \text{ Pa} + \tfrac{1}{2} \times 1000 \text{ kg/m}^3 \times (4.0 \text{ m/s})^2$$

$$- \tfrac{1}{2} \times 1000 \text{ kg/m}^3 \times (26 \text{ m/s})^2$$

$$= 2 \times 10^4 \text{ Pa} \qquad (18.32)$$

EXAMPLE 14   The **Venturi flowmeter** is a simple device that measures the velocity of a fluid flowing in a pipe. It consists of a constriction in the pipe with a cross-sectional area A_2 that is smaller than the cross-sectional area A_1 of the pipe itself (see Fig. 18.33). Small holes in the constriction and in

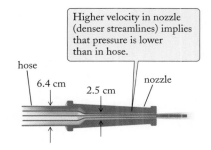

Higher velocity in nozzle (denser streamlines) implies that pressure is lower than in hose.

hose

6.4 cm   2.5 cm   nozzle

FIGURE 18.32 A fire-hose tip.

Fig. 24. Daniel Bernoulli.

DANIEL BERNOULLI (1700–1782)
Swiss physician, physicist, and mathematician. His great treatise Hydrodynamica *included the equation named after him. Several other members of the Bernoulli family made memorable contributions to mathematics and physics.*

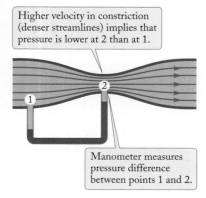

Higher velocity in constriction (denser streamlines) implies that pressure is lower at 2 than at 1.

Manometer measures pressure difference between points 1 and 2.

FIGURE 18.33 Venturi flowmeter.

the pipe permit the measurement of the pressures at these points by means of a manometer. Express the velocity of flow in terms of the pressure difference registered by the manometer.

SOLUTION: As in the preceding example, the magnitudes of the velocities at points 1 and 2 are related to the cross-sectional areas by the continuity equation,

$$v_2 = v_1 \frac{A_1}{A_2} \tag{18.33}$$

With $y_1 = y_2$, Eq. (18.25) then again gives us

$$\tfrac{1}{2}\rho v_2^2 + p_2 = \tfrac{1}{2}\rho v_1^2 + p_1$$

From this we obtain a pressure difference, in which we then substitute for v_2 from Eq. (18.33):

$$p_1 - p_2 = \tfrac{1}{2}\rho v_2^2 - \tfrac{1}{2}\rho v_1^2$$

$$= \frac{1}{2}\rho\left(v_1\frac{A_1}{A_2}\right)^2 - \frac{1}{2}\rho v_1^2 = \frac{1}{2}\rho v_1^2\left[\left(\frac{A_1}{A_2}\right)^2 - 1\right] \tag{18.34}$$

Taking the square root of both sides of this equation and solving for v_1, we find

$$v_1 = \sqrt{\frac{2(p_1 - p_2)}{\rho[(A_1/A_2)^2 - 1]}} \tag{18.35}$$

This says that the flow velocity is proportional to the square root of the pressure difference.

PROBLEM-SOLVING TECHNIQUES   BERNOULLI'S EQUATION

The application of Bernoulli's equation (18.26) to problems of fluid motion involves the same three steps we used in the application of energy conservation to problems of particle motion:

1 First evaluate the sum of the terms in Bernoulli's equation at one point of a streamline.

2 Then evaluate the sum of the terms in Bernoulli's equation at another point of the same streamline.

3 And then equate these two quantities.

When substituting the pressure into Bernoulli's equation, you can use either the absolute pressure (including atmospheric pressure) or the overpressure (excluding atmospheric pressure), but you must use the same kind of pressure on both sides of the equation. These two kinds of pressure differ by only

a constant, and adding or subtracting the same constant on both sides of the equation does not affect its validity. Also remember that the pressure in Bernoulli's equation must be expressed in SI units (N/m^2, or Pa), not atmospheres or mm-Hg.

If two points in a fluid system are in contact with the atmosphere, then they are at essentially equal pressure (as in Example 12. However, if both points are not in contact with the atmosphere, then the pressure difference depends on the flow velocity (as in Examples 13 and 14) and the depth (as in Examples 6 and 7).

If the fluid moves through a pipe or channel of varying cross section, then the other general equation governing the flow is the continuity equation $v_1 A_1 = v_2 A_2$, which relates the cross-sectional area to the flow velocity. Most problems involving steady incompressible flow can be solved by combining Bernoulli's equation and the continuity equation.

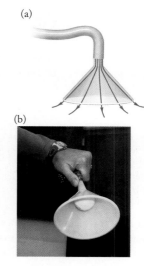

FIGURE 18.34 (a) Airflow through a funnel. (b) Levitation of a plastic ball.

 Checkup 18.6

QUESTION 1: A fire hose of constant diameter carries water from a fire truck to the top of a burning house. According to Bernoulli's equation, does the pressure difference between the ends of this hose depend on the speed of flow?

QUESTION 2: Consider water flowing downhill in a shallow, open channel. Which of the three terms in Bernoulli's equation are constant?

QUESTION 3: A lawn sprinkler is connected to a faucet by a horizontal hose of constant diameter. The water emerges from the nozzle of the sprinkler in an arc. Which of the terms in Bernoulli's equation are constant during the motion of the water in the hose? During the motion after the water emerges from the nozzle?

QUESTION 4: An air hose is connected to the narrow end of a funnel. Air from the hose flows into the narrow end and out of the wide end of the funnel (Fig. 18.34a), and this can levitate a body (Fig. 18.34b). Where is the air pressure the lowest?

(A) At the narrow end of the funnel   (B) At the wide end of the funnel
(C) Far from the funnel

SUMMARY

PHYSICS IN PRACTICE The sphygmomanometer		**(page 579)**

PROBLEM-SOLVING TECHNIQUES Bernoulli's equation		**(page 586)**

DENSITY, OR MASS PER UNIT VOLUME

$$\rho = \frac{m}{V}$$ **(18.1)**

CONTINUITY EQUATION

$$v_1 A_1 = v_2 A_2$$ **(18.5)**

PRESSURE IS FORCE PER UNIT AREA

$$p = \frac{F}{A}$$ **(18.6)**

UNIT OF PRESSURE

$$1 \text{ pascal} = 1 \text{ Pa} = 1 \text{ N/m}^2$$ **(18.7)**

ATMOSPHERIC PRESSURE

$$1 \text{ atm} = 760 \text{ mm-Hg} = 1.01 \times 10^5 \text{ Pa}$$ **(18.8)**

HYDROSTATIC PRESSURE (FOR INCOMPRESSIBLE FLUID)
For water, this is very close to 1 atm for every 10 m of depth.

$$p - p_0 = -\rho g y$$ **(18.12)**

ARCHIMEDES' PRINCIPLE   The upward buoyant force has the same magnitude as the weight of the displaced fluid.

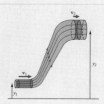

BERNOULLI'S EQUATION

$$\tfrac{1}{2}\rho v^2 + \rho g y + p = [\text{constant}]$$ **(18.26)**

QUESTIONS FOR DISCUSSION

1. The sheet of water of a waterfall is thick at the top and thin at the bottom (see Fig. 18.35). Explain.

FIGURE 18.35 Sheet of water on a smooth waterfall, narrowing at the bottom.

2. During construction, three lanes of a four-lane highway are closed. Suppose that bumper-to-bumper traffic on the four lanes funnels into the single open lane. If the cars on the four approaching lanes proceed at a crawl, say, 15 km/h, what will be their speed when they proceed along the single lane? (Hint: Think of the continuity equation.)

3. If you place a block of wood on the bottom of a swimming pool, why does the pressure of the water not keep it there?

4. Explain what holds a suction cup on a smooth surface.

5. Newton gave the following description of his classic experiment with a rotating bucket:

If a vessel, hung by a long cord, is so often turned about that the cord is strongly twisted, then filled with water, and held at rest together with the water; thereupon, by the sudden action of another force, it is whirled about the contrary way, and while the cord is untwisting itself, the vessel continues for some time in this motion; the surface of the water will at first be plain, as before the vessel began to move; but after that, the vessel, by gradually communicating its motion to the water, will make it begin sensibly to revolve, and recede by little and little from the middle, and ascend to the sides of the vessel, forming itself into a concave figure (as I have experienced), and the swifter the motion becomes, the higher will the water rise, till at last, performing its revolutions in the same times with the vessel, it becomes relatively at rest in it.

Explain why hydrostatic equilibrium requires that the rotating water be higher at the rim of the bucket than at the center.

6. When a physician measures the blood pressure of a patient, he places the cuff on the arm, at the same vertical level as the heart. What would happen if he were to place the cuff around the leg?

7. Scuba divers have survived short intervals of free swimming at depths of 430 m. Why does the pressure of the water at this depth not crush them?

8. Figure 18.36 shows a glass vessel with vertical tubes of different shapes. Explain why the water level in all the tubes is the same.

FIGURE 18.36 Glass vessel with vertical tubes.

9. In a celebrated experiment, Blaise Pascal attached a long metal funnel to a tight cask (Fig. 18.37). When he filled the cask by pouring water into this funnel, the cask burst. Explain.

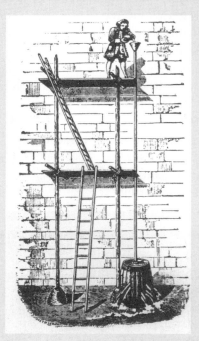

FIGURE 18.37 Blaise Pascal's experiment.

10. Face masks for scuba divers have two indentations at the bottom into which the diver can stick thumb and forefinger to pinch his nose shut. What is the purpose of this arrangement?

11. Figure 18.38 shows a grain silo held together by circumferential steel bands. Why has the farmer placed more bands near the bottom than near the top?

FIGURE 18.38 Grain silo.

12. Flooding on the low coasts of England and the Netherlands is most severe when the following three conditions are in coincidence: onshore wind, full or new moon, and low barometric pressure. Can you explain this?

13. Why do some men or women float better than others?

14. Will an ice cube float in a tub full of gasoline?

15. An ice cube floats in a glass full of water. Will the water level rise or fall when the ice cube melts?

16. The density of a solid body can be determined by first weighing the body in air and then weighing it again when it is immersed in water (Fig. 18.39). How can you deduce the density from these two measurements?

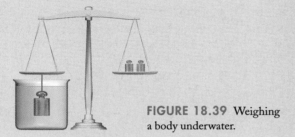

FIGURE 18.39 Weighing a body underwater.

17. A buoy floats on the water. Will the flotation level of the buoy change when the atmospheric pressure changes?

18. Will the water level in a canal lock rise or fall if a ship made of steel sinks in the lock? What if the ship is made of wood?

19. While training for the conditions of weightlessness they would encounter in an orbiting spacecraft, NASA astronauts were made to float submerged in a large water tank (Fig. 18.40). Small weights attached to their space suits gave the astronauts neutral buoyancy. To what extent does such simulated weightlessness imitate true weightlessness?

FIGURE 18.40 Astronauts floating underwater.

20. A girl standing in a subway car holds a helium balloon on a string. Which way will the balloon move when the car accelerates?

21. A slurry of wood chips is used in some tanning operations. What would happen to a man were he to fall into a vat filled with such a slurry?

22. The bathyscaphe *Trieste*, which set a record of 10917 m in a deep dive in the Marianas Trench in 1960, consists of a large tank filled with gasoline below which hangs a steel sphere for carrying the crew (Fig. 18.41). Can you guess the purpose of the tank of gasoline?

FIGURE 18.41 The bathyscaphe *Trieste*.

23. Gasoline vapor from, say, a small leak in the fuel system poses a very serious hazard in a motorboat, but only a minor hazard in an automobile. Explain. (Hint: Gasoline vapor is denser than air.)

24. Accidental release of carbon dioxide or of argon creates a much greater danger of suffocation than release of helium gas. Why?

25. How does a balloonist control the ascent and descent of a hot-air balloon?

26. When you release a bubble of air while underwater, the bubble grows in size as it ascends. Explain.

27. A "Cartesian diver" consists of a small inverted bottle floating inside a larger bottle whose mouth is covered by a rubber membrane (Fig. 18.42). By depressing the membrane, you can increase the water pressure in the large bottle. How does this affect the buoyancy of the small bottle?

FIGURE 18.42 A Cartesian diver.

28. To throw a curveball, a baseball pitcher gives the ball a spinning motion about a vertical axis. The air on the left and right sides of the ball will then be dragged along by the rotation and acquire slightly different speeds. Using Bernoulli's equation, explain how this creates a lateral deflecting force on the ball.

29. If you place a Ping-Pong ball in the jet of air from a hose aimed vertically upward, the Ping-Pong ball will be held in stable equilibrium within this jet. Explain this by means of Bernoulli's equation. (Hint: The speed of air is maximum at the center of the jet.)

PROBLEMS

18.1 Density and Flow Velocity
18.2 Incompressible Steady Flow; Streamlines[†]

1. The following table, taken from a firefighter's manual, lists the rate of flow of water (in liters per minute) through a fire hose of diameter 3.81 cm (1.5 in.) connected to a nozzle of given diameter; the listed rate of flow will maintain a pressure of 3.4 atm (50 lbf/in.^{2}) in the nozzle:

RATE OF FLOW FOR A 3.81-cm HOSE		
RATE OF FLOW	NOZZLE DIAMETER	NOZZLE PRESSURE
95 liters/min	0.95 cm	3.4 atm
190	1.27	3.4
284	1.59	3.4

For each case calculate the speed of flow (in meters per second) in the hose and in the nozzle.

2. The average rate of flow of blood in the human aorta is 92 cm^3/s. The radius of the aorta is 1.2 cm. From this calculate the average speed of the blood in the aorta.

3. What pump power is required to shoot 26 000 liters/min of water to a height of 170 m, as in the tallest fountain?

4. Using a garden hose, you take 4.0 minutes to fill a water barrel. The inner diameter of the garden hose is 2.5 cm, and the capacity of the barrel is 250 liters. What is the flow velocity of the water in the hose?

5. Water flows in a 10-cm-diameter pipe with flow velocity 6.0 m/s. If the water encounters a constricted region of the pipe with diameter 7.0 cm, what is the flow velocity in that region?

6. During a heavy storm, a steady stream of water flows down from a 5.0-cm-diameter hole in a gutter. The downward speed of the water exiting the hole is 0.50 m/s. What is the speed of the flow just before it hits the ground, 4.0 m below the hole? What is the diameter of the flow there?

7. A patient receives 0.50 liter of intravenous fluid in 30 minutes. The tubing carrying the fluid has an inner diameter of 2.0 mm; the needle at the end of the tube has an inner diameter of 0.20 mm. What is the speed of flow of the fluid in the tubing? In the needle?

8. Water flows though a pipe of a certain radius at a rate of 10 kg/s; propane flows through a pipe with a radius 10 times larger than that of the water pipe, also at a rate of 10 kg/s. What is the ratio of the flow velocities of these fluids, $v_{water}/v_{propane}$?

*9. A fountain shoots a stream of water vertically upward. Assume that the stream is inclined very slightly to one side so that the descending water does not interfere with the ascending water. The upward velocity at the base of the column of water is 15 m/s.

(a) How high will the water rise?

(b) The diameter of the column of water is 7.0 cm at the base. What is the diameter at the height of 5.0 m? At the height of 10 m?

18.3 Pressure

10. A metallic can is filled with a carbonated soft drink. The can is cylindrical, of radius 3.2 cm. The (over)pressure of the carbon dioxide in the liquid is 0.8 atm. What is the force acting on one of the circular bases of the can?

11. You can use a barometer as an altimeter. Suppose that when you carry the barometer up a hill, its reading decreases by 8.0 mm-Hg. What is the height of the hill? Assume the density of air is 1.29 kg/m^3.

12. Within the funnel of a tornado, the air pressure is much lower than normal—about 0.20 atm as compared with the normal value of 1.00 atm. Suppose that such a tornado suddenly envelops a house; the air pressure inside the house is 1.00 atm and the pressure outside suddenly drops to 0.20 atm. This will cause the house to burst explosively. What is the net outward pressure force on a 12 m × 3.0 m wall of this house? Is the house likely to suffer less damage if all the windows and doors are open?

13. What is the downward force that air pressure (1 atm) exerts on the upper surface of a sheet of paper ($8\frac{1}{2}$ in. × 11 in.) lying on a table? Why does this force not squash the paper against the table?

14. At a distance of 7.0 km from a 1-megaton nuclear explosion, the blast wave has an overpressure of 3.0×10^4 N/m^2. Calculate the force that this blast wave exerts on the front of a standing man; the frontal area of the man is 0.70 m^2. (The actual force on a man exposed to the blast wave is larger than the result of this simple calculation because the blast wave will be reflected by the man, and this leads to a substantial increase in pressure.)

15. The overpressure in the tires of a 1300-kg automobile is 2.4 atm. If each tire supports one-fourth the weight of the automobile, what must be the area of each tire in contact with the ground? Pretend the tires are completely flexible.

16. The shape of the wing of an airplane is carefully designed so that, when the wing moves through air, a pressure difference develops between the bottom surface of the wing and the top surface; this supports the weight of the airplane. A fully loaded DC-3 airplane has a mass of 10 900 kg. The (bottom) surface area of its wings is 92 m^2. What is the average pressure

difference between the top and the bottom surfaces when the airplane is in flight?

17. Pressure gauges used on automobile tires read the overpressure, that is, the amount of pressure in excess of atmospheric pressure. If a tire has an overpressure of 2.4×10^5 N/m^2 on a day when the barometric pressure is 0.95 atm, what will be the overpressure when the barometric pressure increases to 1.01 atm? Assume that the volume and the temperature of the tire remain constant.

18. Commercial jetliners have pressurized cabins enabling them to carry passengers at a cruising altitude of 10000 m. The air pressure at this altitude is 0.28 atm. If the air pressure inside the jetliner is 1.00 atm, what is the net outward force on a 1.0 m $\times$ 2.0 m door in the wall of the cabin?

19. In 1654 Otto von Guericke, the inventor of the air pump, gave a public demonstration of air pressure (see Fig. 5.44). He took two hollow hemispheres of copper whose rims fitted tightly together and evacuated them with his air pump. Two teams of 15 horses each, pulling in opposite directions, were unable to separate these hemispheres. If the evacuated sphere had a radius of 40 cm and the pressure inside was nearly zero, what force would each team of horses have had to exert to pull the hemispheres apart?

20. A long bar of iron has a diameter of 10 cm and a length of 2.0 m. If the mass of the bar is 123 kg and the bar stands on one end, what pressure does it exert on the surface supporting it?

21. A 50-kg boy hangs from the ceiling by a (completely evacuated) suction cup. What is the minimum area the suction cup must have?

22. A large, cylindrical, aboveground cistern of negligible mass holds 50000 liters of water and has a radius of 2.0 m. When full, what overpressure does it exert on the ground?

23. A 60-kg woman stands with her weight evenly distributed over the total area of her shoe bottoms, 300 cm^2. What pressure do the shoes exert on the floor? Then she balances on the bottom of one circular heel of diameter 1.0 cm. What pressure does the heel exert on the floor?

24. The baggage compartment of the DC-10 airliner is under the floor of the passenger compartment. Both compartments are pressurized at a normal pressure of 1.0 atm. In a disastrous accident near Orly, France, in 1974, a faulty lock permitted the baggage compartment door to pop open in flight, depressurizing this compartment. The normal pressure in the passenger compartment then caused the floor to collapse, jamming the control cables. At the time the airliner was flying at an altitude of 3800 m, where the air pressure is 0.64 atm. What is the net pressure force on a 1.0 m $\times$ 1.0 m square of the floor?

25. A pencil sharpener is held to the surface of a desk by means of a rubber suction cup measuring 6.0 cm $\times$ 6.0 cm. The air pressure under the suction cup is zero and the air pressure above the suction cup is 1.0 atm.

    (a) What is the magnitude of the pressure force pushing the cup against the table?

    (b) If the coefficient of static friction between the rubber and the table is 0.90, what is the maximum transverse force the suction cup can withstand?

18.4 Pressure in a Static Fluid

26. Porpoises dive to a depth of 520 m. What is the water pressure at this depth?

27. A tanker is full of oil of density 880 kg/m^3. The flat bottom of the hull is at a depth of 26 m below the surface of the surrounding water. Inside the hull, oil is stored with a depth of 30 m (Fig. 18.43). What is the pressure of the water on the bottom of the hull? The pressure of the oil? What is the vertical pressure force on 1.0 m^2 of bottom?

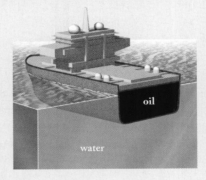

FIGURE 18.43 Cross section of an oil tanker.

28. (a) Calculate the mass of air in a column of base 1.00 m^2 extending from sea level to the top of the atmosphere. Assume that the pressure at sea level is 760 mm-Hg and that the value of the acceleration of gravity is 9.81 m/s^2, independent of height.

    (b) Multiply your result by the surface area of the Earth to find the total mass of the entire atmosphere.

29. Suppose that a zone of low atmospheric pressure (a "low") is at some place on the surface of the sea. The pressure at the center of the "low" is 64 mm-Hg less than the pressure at a large distance from the center. By how much will this cause the water level to rise at the center?

30. (a) Under normal conditions the human heart exerts a pressure of 120 mm-Hg on the arterial blood. What is the arterial blood pressure in the feet of a man standing upright? What is the blood pressure in the brain? The feet are 140 cm below the level of the heart; the brain is 40 cm above the level of the heart; the density of human blood is 1055 kg/m^3. Neglect the speed of flow of the blood; i.e., pretend it is at rest.

    (b) Under conditions of stress the human heart can exert a pressure of up to 190 mm-Hg. Suppose that an astronaut lands on the surface of a large planet where the acceleration of gravity is 61 m/s^2. Could the astronaut's heart maintain a positive blood pressure in his brain while he is standing upright? Could the astronaut survive?

31. A "suction" pump consists of a piston in a cylinder with a long pipe leading down into a well (Fig. 18.44). What is the maximum height to which such a pump can "suck" water?

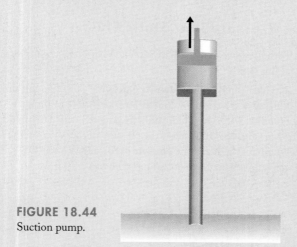

FIGURE 18.44
Suction pump.

32. On a day of high pressure, the barometer may read 790 mm-Hg; on a low-pressure day, it may drop to 730 mm-Hg. What are these pressures in units of atmospheres?

33. If a 2.0-m-high tank is filled with sand (with average density $\rho = 1.7$ g/cm^3), what is the overpressure at the bottom of the tank?

34. A hydraulic lift for cars has a small piston with radius 1.0 cm and a large piston with radius 12 cm. What pressure (in atm) must be exerted on the small piston for the large piston to lift a car of mass 1500 kg?

35. A 30-m-high dam holds back a full reservoir of water. What is the overpressure at the base of the dam?

36. A tube is bent so that both ends are upward; the tube contains some water. Gasoline is poured into one end of the tube as shown in Fig. 18.45; the region with gasoline is 20 cm high. How much higher is the gasoline–air surface compared with the water–air surface?

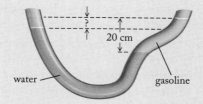

water   gasoline

FIGURE 18.45 A tube with two liquids.

37. As any diver who has ever "belly-flopped" knows, water is a very incompressible fluid. From Example 7, the pressure at 4500 m below the surface of the sea has the value 4.54×10^7 Pa. What is the percent volume change of water at this immense pressure? [Hint: The fractional volume change was given in Eq. (14.20); the value of the bulk modulus of water appears in Table 14.1.]

38. A swimming pool is 4.0 m in depth; a swimmer at this depth feels discomfort in the ear. Calculate the net force on a 0.50-cm-diameter eardrum, assuming the inner surface of the eardrum remains at 1 atm pressure.

*39. A 30-m-high dam holds back a full reservoir of water. The width of the dam is 70 m. What is the total force that the water exerts on the dam? (Hint: Sum the forces $dF = p\,dA$ on thin horizontal strips of area dA of the dam.)

*40. A man whirls a bucket full of water around a vertical circle at the rate of 0.70 rev/s. The surface of the water is at a radial distance of 1.0 m from the center.

(a) What is the pressure difference between the surface of the water and a point 1.0 cm below the surface when the bucket is at the lowest point of the circle? You may assume that a point on the surface and a point 1.0 cm below have practically the same centripetal acceleration.

(b) What if the bucket is at the highest point of the circle?

*41. A test tube filled with water is being spun around in an ultra-centrifuge with angular velocity ω. The test tube is lying along a radius, and the free surface of the water is at a radius r_0 (Fig. 18.46).

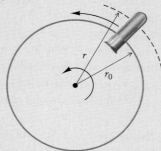

FIGURE 18.46 Test tube in an ultracentrifuge.

(a) Show that the pressure at radius r within the test tube is

$$p = \tfrac{1}{2}\rho\omega^2(r^2 - r_0^2)$$

where ρ is the density of the water. Ignore gravity and ignore atmospheric pressure.

(b) Suppose that $\omega = 3.8 \times 10^4$ radians/s and $r_0 = 10$ cm. What is the pressure at $r = 13$ cm?

*42. In a test centrifuge, a NASA scientist tolerated (suffered?) a sustained centripetal acceleration of 25 standard g's. Estimate the pressure difference between the front and back of his brain during the ordeal. Measure the relevant distance on your own head, and assume this distance was radial during the test.

*43. (a) Figure 18.47a shows a round conical flask filled with water of a depth h. The radius of the upper water surface is R_1, and that of the lower surface is R_2. What is the net force that the water exerts on the sides of the flask? On the bottom of the flask? What is the sum of these forces? Ignore atmospheric pressure.

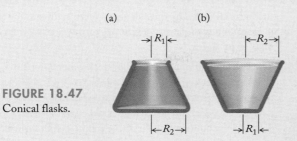

FIGURE 18.47 Conical flasks.

(b) Figure 18.47b shows another round conical flask, with the same radii. Answer the same questions for this flask.

18.5 Archimedes' Principle

44. What is the buoyant force on a human body of volume 7.4×10^{-2} m^3 when totally immersed in air? In water?

45. Icebergs commonly found floating in the North Atlantic (Fig. 18.48) are 30 m high (above the water) and 400 m × 400 m across. The density of ice is 920 kg/m^3, and the density of seawater is 1025 kg/m^3.

    (a) What is the total volume of such an iceberg (including the volume below the water)?

    (b) What is the total mass?

FIGURE 18.48 A large iceberg.

46. You can walk on water if you wear very large shoes shaped like boats. Calculate the length of the shoes that will support you; assume that each shoe is 30 cm × 30 cm in cross section.

47. A gasoline barrel, made of steel, has a mass of 20 kg when empty. The barrel is filled with 0.12 m^3 of gasoline with a density of 739 kg/m^3. Will the full barrel float in water? Neglect the volume of the steel.

48. A typical medium-size balloon used for scientific research (Fig. 18.49) is designed to attain an altitude of 40 km, at which altitude the helium in the balloon will have expanded to 570000 m^3. What is the buoyant force on the balloon under these conditions? The density of air at this altitude is 4.3×10^{-3} kg/m^3.

49. A spherical balloon of mass 4.0 g is filled with helium gas of density 0.18 kg/m^3. If the balloon can barely lift a mass of 150 g off the ground, what is its radius?

50. When submerged in water, what is the buoyant force on a Ping-Pong ball (radius 2.0 cm)? A basketball (radius 11.4 cm)? A beach ball (radius 50 cm)?

FIGURE 18.49 A research balloon being inflated with helium.

51. A balloon is filled with helium gas of density 0.18 kg/m^3; the air around it is at standard density. If released, what is the initial acceleration of the balloon? (For this simple estimate, neglect the mass of the balloon and the inertia of the surrounding air.)

52. An egg sinks to the bottom of 1.000 liter of freshwater. Table salt is slowly dissolved in the water until 83 g has been added; the egg then becomes neutrally buoyant. If the salt has negligible effect on the volume of the water, what is the average density of the egg?

53. You bet your friend that he cannot completely submerge your beach ball, which has a diameter of 60 cm and negligible mass. He must balance on the beach ball, not jump on it. What is the smallest mass that your friend can have and be successful?

*54. A supertanker has a mass of 220000 metric tons when empty and can carry up to 440000 metric tons of oil when fully loaded. Assume that the shape of its hull is approximately that of a rectangular parallelepiped 380 m long, 60 m wide, and 40 m high (Fig. 18.50).

FIGURE 18.50 Empty and fully loaded supertanker.

(a) What is the draft of the empty tanker, that is, how deep is the hull submerged in the water? Assume the density of seawater is 1025 kg/m^3.

(b) What is the draft of the fully loaded tanker?

*55. A supertanker has a draft (submerged depth) of 30 m when in seawater (density ρ = 1025 kg/m^3). What will be the draft of this tanker when it enters a river estuary with freshwater (density ρ = 1000 kg/m^3)? Assume that the sides of the ship are vertical.

*56. The Raven S-66A hot-air balloon has a volume of 4000 m^3 and a height of 27 m. Fully loaded, its mass is 1400 kg. If the density of the air outside the balloon is 1.29 kg/m^3, what must be the density of the air inside the balloon to achieve liftoff? What is the inside overpressure at the top of the balloon? (Hint: The bottom of the balloon is open, and therefore at the same pressure as the exterior air. Treat the air outside and the air inside the balloon as fluids of uniform densities.)

*57. A round log of wood of density 600 kg/m^3 floats in water. The diameter of the log is 30 cm. How high does the upper surface of the log protrude from the water?

*58. A child's rubber balloon of mass 2.5 g is filled with helium gas of density 0.33 kg/m^3. The balloon is spherical, with a radius of 12 cm. A long cotton string with a mass of 2.0 g per meter hangs from the bottom of the balloon. Initially, the string lies loosely on the floor, but when the balloon ascends, it pulls the string upward and straightens it out. At what height will the balloon stop ascending, having reached equilibrium with the hanging portion of the string? Assume that the surrounding air has the standard density 1.29 kg/m^3.

*59. When a force accelerates a body immersed in a fluid, some of the fluid must also be accelerated, since it must be pushed out of the way of the body and flow around it. Thus, the force must overcome not only the inertia of the body, but also the inertia of the fluid pushed out of the way. It can be shown that for a spherical body completely immersed in a nonviscous fluid, the extra inertia is that of a mass of fluid half as large as the fluid displaced by the body.

(a) From this, deduce that the downward acceleration of a spherical body of density ρ falling through a fluid of density ρ' is

$$a = \frac{\rho - \rho'}{\rho + \frac{1}{2}\rho'}\,g$$

(b) Find the upward acceleration of an empty bubble in the fluid.

(c) What value would you have found for the acceleration of an empty bubble if you had not taken into account the extra inertia of the displaced fluid?

*60. When a body is weighed in air on an analytical balance, a correction must be made for the buoyancy contributed by the air. Suppose that the weights used in the balance are made of brass, of density 8.7 × 10^3 kg/m^3, and that the density of air is 1.3 kg/m^3. By what percentage must you increase (or decrease)

the mass indicated by the balance in order to obtain the true mass of a body of density 1.0 × 10^3 kg/m^3? 5.0 × 10^3 kg/m^3? 10.0 × 10^3 kg/m^3?

*61. The bottom half of a tank is filled with water (ρ = 1.0 × 10^3 kg/m^3), and the top half is filled with oil (ρ = 8.5 × 10^2 kg/m^3). Suppose that a rectangular block of wood of mass 5.5 kg, 30 cm long, 20 cm wide, and 10 cm high is placed in this tank. How deep will the bottom of the block be submerged in the water?

**62. A hydrometer, the device used to determine the density of battery acid, wine, or other fluids, consists of a bulb with a long vertical stem (Fig. 18.51). The device floats in the liquid with the bulb submerged and the stem protruding above the surface. The density of the liquid is directly related to the length h of the protruding portion of the stem. Suppose that the bulb is a sphere of radius R and that the stem is a cylinder of radius R' and length l; the mass of both together is M. Derive a formula for the density of the liquid in terms of h, R, R', l, and M.

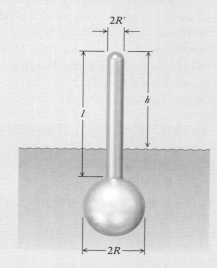

FIGURE 18.51 Hydrometer.

**63. A ship that displaces a weight of water equal to its own weight is in equilibrium with regard to vertical motion. If the ship were placed lower in the water, the buoyant force would exceed gravity and the ship would surge upward. If the ship were placed higher in the water, the buoyant force would be less than gravity and the ship would sink downward. Suppose that the sides of a ship are vertical above and below its normal waterline.

(a) Show that the frequency of small up-and-down oscillations of the ship is roughly $\omega = \sqrt{A\rho g/M}$, where A is the horizontal area bounded by the waterline, M is the mass of the ship, and ρ is the density of water. (This formula is only a rough approximation because, as the ship moves up and down, the water also has to move; hence the effective inertia is greater than M.)

(b) What is the frequency of such up-and-down oscillations for the fully loaded supertanker described in Problem 54?

**64. You drop a pencil, point down, into the water. If you release the pencil from a height of 4.0 cm (measured from the point to the water level), how far will it dive? Treat the pencil as a cylinder of radius 0.40 cm, length 19 cm, and mass 4.0 g. Ignore the frictional and inertial resistance of the water, but take into account the buoyant force.

18.6 Fluid Dynamics; Bernoulli's Equation[†]

65. A thin stream of water emerges vertically from a small hole on the side of a water pipe and ascends to a height of 1.2 m. What is the pressure inside the pipe? Assume the water inside the pipe is nearly static.

66. If you blow a thin stream of air with a speed of 7.0 m/s out of your mouth, what must be the overpressure in your mouth? Assume that the speed of the air in your mouth is (nearly) zero.

67. A pump has a horizontal intake pipe at a depth h below the surface of a lake. What is the maximum speed for steady flow of water into this pipe? (Hint: Inside the pipe the maximum speed of flow corresponds to zero pressure.)

68. The wind blows over the horizontal roof of a closed house at speed v. Find an expression for the difference in pressure inside and outside the house. For wind with speed $v = 20$ m/s, find the force on a square meter of the roof.

69. A small airplane has a total wing area of 8.0 m². If the air flowing above the wing has a speed of 130 m/s and the air below the wing a speed of 115 m/s, what is the net force on the wing?

70. A windmill has blades of length 20 m and extracts 5.0% of the kinetic energy of the wind that passes through the circle defined by the blade motion. If the wind blows at 5.0 m/s, how much power is extracted?

71. A certain fire extinguisher consists of a large sealed tank with a thin tube leading from the bottom of the tank to an exit nozzle; the air above the liquid in the tank is at an overpressure p_{tank} with respect to atmospheric pressure p_o. The exit nozzle is at the same level as the top of the liquid; both are a distance y above the bottom of the tank (see Fig. 18.52). The liquid has density ρ. Find an expression for the speed of the stream of liquid emerging from the nozzle.

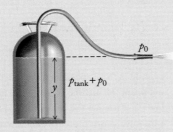

FIGURE 18.52 A fire extinguisher.

72. The friction experienced by real fluids is measured in terms of the **viscosity** η (the Greek letter *eta*), which for water at 20°C has the value $\eta = 1.0 \times 10^{-3}$ kg/m·s. Because of friction, the volume flow of fluid through a pipe requires a pressure difference Δp between the ends of the pipe. For a pipe of radius R and length L, the resulting volume flow is

$$\frac{dV}{dt} = \frac{\pi R^4}{8\eta L} \Delta p$$

Suppose that a 3.0-atm pressure difference is available. Calculate the volume flow rate for a 100-m length of pipe when the radius of the pipe is 1.0 cm (a garden hose) and when the radius is 20 cm (a water main).

73. As discussed in Section 18.2, laminar flow breaks down and the flow becomes turbulent when the velocity v of flow or the relevant dimension R of the flow becomes large. The transition to turbulence also depends on the viscosity η of the fluid (see Problem 72) and is related to a dimensionless quantity called the **Reynolds number** N. For flow through a tube of radius R, the Reynolds number is

$$N = \frac{2R\rho v}{\eta}$$

where ρ is the mass density of the fluid. Turbulence occurs when the Reynolds number gets large. Using $N = 2500$ as a threshold for turbulence, calculate the flow velocity at which turbulence will set in for water at 20°C when the tube radius is $R = 10$ cm.

*74. To fight a fire on the fourth floor of a building, firefighters want to use a hose of diameter 6.35 cm ($2\frac{1}{2}$ in.) to shoot 950 liters/min of water to a height of 12 m.

(a) With what minimum speed must the water leave the nozzle of the fire hose if it is to ascend 12 m?

(b) What pressure must the water have inside the fire hose? Ignore friction.

*75. Streams of water from fire hoses are sometimes used to disperse crowds. Suppose that a stream of water of diameter 2.5 cm emerging from a fire hose at a speed of 26 m/s impinges horizontally on a man. The collision of the water with the man is totally inelastic.

(a) What is the force that the stream of water exerts on the man; that is what is the rate at which the water delivers momentum to the man?

(b) What is the rate at which the water delivers energy?

*76. A **siphon** is an inverted U-shaped tube that is used to transfer liquid from a container at a high level to a container at a low level (Fig. 18.53).

(a) Using the lengths shown in Fig. 18.53 and the density ρ of the liquid, find a formula for the speed with which the liquid emerges from the lower end of the siphon. Assume that the containers are large, and that the lower end of the tube is *not* immersed in the container.

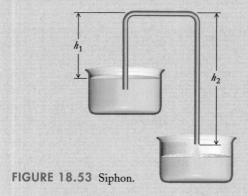

FIGURE 18.53 Siphon.

(b) Find the pressure at the highest point of the siphon.

(c) By setting this pressure equal to zero, find the maximum height h_1 with which the siphon can operate.

*77. The pump of a fire engine draws 1100 liters of water per minute from a pond with a water level 4.5 m below the pump and discharges this water into a fire hose of diameter 6.35 cm ($2\frac{1}{2}$ in.) at a pressure of 5.4 atm. In the absence of friction, what power (in hp) does this pump require?

*78. In the calculation of pressures in fire hoses, it is often necessary to take into account the frictional losses suffered by the water as it flows along the hose. Consider a horizontal hose of diameter 3.81 cm ($1\frac{1}{2}$ in.) and length 30 m carrying 380 liters/min of water. According to tables used by firefighters, a pressure of 6.1 atm at the upstream end of this hose will result in a pressure of only 3.4 atm at the downstream end; the loss of 2.7 atm is attributed to friction. At what rate (in hp) does friction remove energy from the water?

*79. A Venturi flowmeter in a water main of diameter 30 cm has a constriction of diameter 10 cm. Vertical pipes are connected to the water main and to the constriction (Fig. 18.54); these pipes are open at their upper ends, and the water level within them indicates the pressure at their lower ends. Suppose that the difference in the water levels in these two pipes is 3.0 m. What is the velocity of flow in the water main? What is the rate (in liters per second) at which water is delivered?

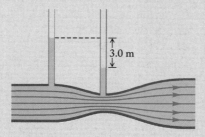

FIGURE 18.54 Venturi flowmeter.

*80. The **Pitot tube** is used for the measurement of the flow speeds of fluids, such as the flow speed of air past the fuselage of an airplane. It consists of a bent tube protruding into the airstream (Fig. 18.55) and another tube opening flush with the fuselage.

The pressure difference between the air in the two tubes can be measured with a manometer. Show that in terms of the pressures p_1 and p_2 in the two tubes, the speed of airflow is

$$v = \sqrt{\frac{2(p_2 - p_1)}{\rho}}$$

where ρ is the density of air. (Hint: p_1 is simply the static air pressure. To find p_2, consider a streamline reaching the opening of the bent tube; at this point, the velocity of the air is zero.)

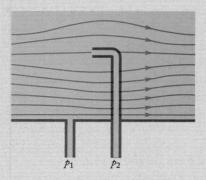

FIGURE 18.55 Pitot tube.

*81. The central front window of the cockpit of an airliner measures 30 cm × 30 cm. Estimate the force with which the air presses against this window when the airliner is flying at 900 km/h. [Hint: At the window, the air (almost) stops relative to the airliner.]

*82. A water tank filled to a height h has a small hole at the height z (see Fig. 18.56). Show that the stream of water emerging from this hole strikes the ground at a horizontal distance $2\sqrt{(h-z)z}$ from the base of the tank. What choice of z gives the largest horizontal distance? What is the largest horizontal distance?

FIGURE 18.56 Tank with small hole.

*83. Very high pressures can be generated (for a brief instant) by launching a projectile at high speed against a rigid target. According to one proposal, an electromagnetic launcher, or rail gun, would be used to give the projectile a speed of 15 km/s. Estimate the maximum pressure generated by the impact of such a projectile on a rigid target. Assume that the density of the projectile is 1.0×10^4 kg/m^3. (Hint: At extreme pressures, the solid projectile will flow like a liquid of approximately constant density; hence Bernoulli's equation is approximately valid.)

**84. A cylindrical tank has a base area A and a height l; it is initially full of water. The tank has a small hole of area A' at its bottom. Calculate how long it will take all the water to flow out of this hole.

**85. A rectangular opening on the side of a tank has a width l. The top of the opening is at a depth h_1 below the surface of the water, and the bottom is at a depth h_2 (Fig. 18.57). Show that the volume of water that emerges from the opening in a unit time is

$$\tfrac{2}{3}l\sqrt{2g}\,(h_2^{3/2} - h_1^{3/2})$$

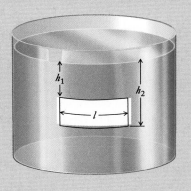

FIGURE 18.57 Tank with rectangular opening.

REVIEW PROBLEMS

86. (a) What is the arterial systolic pressure at your wrist if your arm is hanging straight down? Assume that the pressure is 120 mm-Hg at heart level, and that the wrist is 23 cm below this level.

   (b) What is the pressure if you raise your arm straight up? Assume that the wrist is then 78 cm above heart level.

87. Figure 18.58 shows a device for the measurement of venous pressure. A hypodermic needle is inserted into the vein, and saline solution from the syringe is pushed into the vertical manometer, until the hydrostatic pressure of the column of saline solution matches the pressure of the blood (there is then no blood flow or saline flow through the needle). If the height of the equilibrium column of saline is 103 mm, what is the blood pressure, in mm-Hg? The density of saline is the same as that of blood.

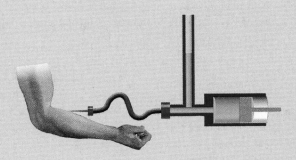

FIGURE 18.58 Device for the measurement of venous pressure.

88. A deep bowl has a layer of olive oil floating on top of a layer of water. The thickness of the layer of olive oil is 6.0 cm, and the thickness of the layer of water is 5.0 cm. The density of the olive oil is 918 kg/m^3. What is the pressure at the bottom of the bowl?

89. A diver attempts to breathe through a long snorkel, that is, a tube connecting his mouth to a float on the surface of the water (see Fig. 18.59). If the diver stays near the surface, he can breathe through this snorkel. But suppose the diver descends to a depth of 2.0 m. What is the water pressure on the outside of his chest? What is the air pressure on the inside of his chest? What is the net force on the front of his chest, of area 0.10 m^2? Can he breathe? (Hint: Could you breathe if two hefty football players were sitting on your chest?)

FIGURE 18.59 Diver with snorkel.

90. A submersible pump installed at the bottom of a well supplies water for a house. The pump is 30 m below the water level, and the house is 5 m above the water level. The house requires 20 liters of water per minute, at an overpressure of 3.0×10^5 N/m^2.

   (a) To supply this water, what power must the pump deliver in doing work against gravity?

   (b) What power must the pump deliver in doing work against the overpressure?

   (c) What is the total power required? Is a $\tfrac{1}{2}$-hp pump adequate for this purpose?

91. A large slick of very viscous oil dumped by a tanker floats on the surface of the sea. The density of the oil is 950 kg/m^3. If the slick is 10 cm thick at its center, how high is the center above the normal level of the sea?

92. A rectangular block of wood has a density of 600 kg/m^3. What fraction of this block will be submerged if the block floats in water?

93. A rectangular barge measures 5.0 m wide, 20 m long, and 3.0 m deep. Its mass is 5.0×10^4 kg when empty. What is the maximum load that this barge can carry, before the water swamps its deck?

94. A water trough is 1.0 m deep, 1.0 m wide, and 2.0 m long.

    (a) You slip a log of mass 200 kg and density 500 kg/m^3 into this trough. How much will the water level in the trough rise?

    (b) You slip a rock of mass 200 kg and density 2000 kg/m^3 into the trough. How much will the water level rise?

95. The mean density of the tissues in an average human body is 1.071×10^3 kg/m^3 (this value of the density excludes the volume of the lungs; that is, the air in the lungs is not counted as part of the body). What volume of air must a man of mass 80 kg take into his lungs if he wants to remain (barely) afloat in water?

96. A "gold" bracelet is made of a gold and copper alloy. The bracelet has a mass of 0.900 kg. To determine the amount of gold in the alloy, you weigh the bracelet when immersed in water. Its weight is then 0.820 N. What is the percentage of gold in the alloy? The density of gold is 19 300 kg/m^3, and the density of copper is 8960 kg/m^3.

97. In an experiment to measure the average density of the tissues of the human body, a man was first weighed in air, and then he was weighed while immersed in water. In air, his weight was 83.20×9.81 N; in water (after he expelled as much air from his lungs as he could, which caused him to sink), his weight was 4.30×9.81 N. The estimated residual volume of air in his lungs was 1.20 liters. What density do you deduce from these data?

98. An airplane has a wing of area (bottom area) 92 m^2. When in flight, the speed of the air along the bottom of the wing is 85 m/s, and along the top of the wing it is 95 m/s. Approximately what lift force does this wing provide? The density of air is 1.29 kg/m^3.

99. A tank full of water, 2.0 m deep, has a circular opening of radius 1.0 cm at its bottom (see Fig. 18.60). What is the rate (in m^3/s) at which water flows out of this opening?

2.0 m

FIGURE 18.60 Tank with circular opening.

100. Hurricanes with the lowest central barometric pressures have the highest wind speeds. Observations show that the square of the wind speed in a hurricane is roughly proportional to the difference between the barometric pressure outside the hurricane and the barometric pressure in the hurricane. Show that this proportionality is expected from Bernoulli's equation. (Hint: Consider a streamline of air that starts outside the hurricane and gradually spirals in toward the "eye.")

101. Suppose that a brisk wind of 18 m/s is blowing at your house from the north. The outside pressure is 1.0 atm.

    (a) If there were an open window on the north side, what would be the pressure inside your house? (Hint: Consider a streamline that enters the room and reaches a wall, where the speed of flow is zero.)

    (b) If, instead, there were an open window on the east side, would you expect the pressure to be larger or smaller than the atmospheric pressure?

Answers to Checkups

Checkup 18.1

1. From Table 18.1, we see that mercury has a density of 13 600 kg/m^3, that is, 13.6 times the density of water. Hence you need 13.6 cups of water to equal the mass of one cup of mercury.

2. For half the diameter, the cross-sectional area would be one-quarter as large ($A = \pi r^2$), so the rate would be $(1/4) \times$ 0.012 m$^2 \times 4.0$ m/s = 0.0032 m^3/s, or 3.2 kg/s.

3. (D) 2. The volume flow is given by Eq. (18.2), $dV/dt = Av$, so the product Av will be the same for each fluid. The pipe with one-quarter the flow velocity ratio thus has 4 times the area. Area is proportional to the square of the radius, so the first pipe has 2 times the radius of the second.

Checkup 18.2

1. No. If the fluid density can change, then matching the amounts of fluid that enter and leave a stream tube must include a density factor; this would give a more general continuity equation for compressible flow, $\rho_1 A_1 v_1 = \rho_2 A_2 v_2$.

2. No, a parcel of fluid may change velocity as it moves along to different locations. Steady flow means that as subsequent parcels of fluid pass a given point in space, they will each have the same velocity at that point.

3. The density of streamlines will be largest before the mouth (where the river is narrower), as required by the continuity equation (18.5); similarly, the density of streamlines will be

smallest out in the sea, beyond the mouth, where the flow has spread over a larger area.

4. From Example 2, the flow velocity is much larger in the aorta than in the capillaries; thus, the density of streamlines is larger in the aorta.

5. As the water falls, its velocity increases because of the acceleration of gravity. By the continuity equation (18.5), if the velocity increases, the cross-sectional area of the flow must decrease.

6. (A) At top and bottom of cylinder. The speed is the largest where the density of streamlines is greatest; from Fig. 18.6, thus is at the top and bottom of the cylinder.

Checkup 18.3

1. The pressure on each face from the air is one atmosphere; the value of the pressure force is this 1.01×10^5 N.

2. No. There is a large downward atmospheric pressure force on the upper face of the brick, but it is balanced by the upward atmospheric force on the lower surface of the brick, since there are many small pockets of air between the brick and the table.

3. Yes. If the balloon does not leak, it will expand on days of low atmospheric pressure and be compressed on high-pressure days.

4. (C) 1. The *pressure* is the same everywhere in any small region of the air, and so is the same at the base and at the cap of the bottle. (However, the *force* due to this pressure varies in proportion to the area, and so the force from the atmosphere on the base surface is 25 times the force on the bottle's cap.)

Checkup 18.4

1. No, the pressure is higher on the first floor, because of the weight of the water in the pipes going to the second floor. As we saw in Example 6, water pressure increases 1 atm for every 10 m of depth; for a typical distance between floors of about 3 m, this means the water pressure is about $\frac{1}{3}$ atm lower for each successive floor upward. In the upper floors of tall buildings, pumps are used to restore adequate water pressure.

2. Yes. As with any fluid, atmospheric air pressure decreases with height. As you climb the mountain, the decreased external pressure will allow the balloon to expand until the balloon's combined internal pressure and elastic forces reach equilibrium with the external pressure force.

3. No, Pascal's Principle deals with that part of the pressure not resulting from gravity. Gravity determines the pressure *difference* between different parts of a fluid that are at different heights [Eq. (18.12)]; Pascal's Principle states that a pressure applied at one point will change the pressure by the same amount throughout the fluid.

4. (D) 1 atm; 2 atm. The pressure in the muscle tissues balances the external pressure to maintain equilibrium. Thus, at the surface, the pressure in her muscle tissue is 1 atm; at a depth of 10 m, the pressure has increased to 2 atm, as in Example 6.

Checkup 18.5

1. By Archimedes' Principle, the buoyant force is equal to the weight of the fluid displaced. This is $W = Mg = \rho Vg$ $= 1000$ kg/m$^3 \times 1.0 \times 10^{-3}$ m$^3 \times 9.81$ m/s$^2 = 9.8$ N.

2. The ship still experiences the same buoyant force, but the water that gets in through the hole acts like extra cargo; if not limited, it increases the weight of the ship until the buoyant force is no longer sufficient to balance the weight.

3. Yes. Seawater is more dense than freshwater (by almost 3%; see Table 18.1), so for a given body weight in equilibrium, a smaller volume of fluid is displaced, and more of the body floats above the water level.

4. The rock will not float in water; since water has a density of 1000 kg/m^3, the buoyant force will be smaller than the weight of the rock even for total submersion. The rock will float in liquid mercury; because mercury has a much higher density of 13 600 kg/m^3, most of the rock will be above the mercury surface level.

5. (C) The water level remains at the rim. The weight of the ice cube equals the weight of the displaced water. When the ice melts to become water, this weight will thus precisely fill the volume previously displaced.

Checkup 18.6

1. No. Since the diameter is constant, the continuity equation (18.5) implies that the speed of flow does not change; according to Bernoulli's equation, the pressure difference between the ends depends on the difference in the square of the speeds, which is zero and thus independent of the speed of flow.

2. The channel is open to the atmosphere, so the pressure is the same at both ends; it is the only constant term. The height y will be different at the two ends (different potential-energy densities), and the speed will correspondingly change because of acceleration down the channel (different kinetic-energy densities).

3. Since the hose is horizontal, the height does not change, and the potential-energy term is constant. Since the diameter is constant, the speed does not change [continuity equation (18.5)], and so the kinetic-energy term is constant. Bernoulli's equation thus implies that the pressure is also constant: all three terms are constant in the hose. Outside the hose, the water is in projectile motion, but the pressure at every point is atmospheric pressure; thus $p = p_0$ is constant, but the kinetic- and potential-energy terms vary during the motion.

4. (A) At the narrow end of the funnel. Since the narrow end has the smaller area, the flow velocity is largest there [continuity equation (18.5)]. Bernoulli's equation tells us that where the velocity is largest, the pressure is lowest. This low pressure in an inverted funnel can be used to levitate an object, as in Fig. 18.34b.

3

Temperature, Heat, and Thermodynamics

CONTENTS

CHAPTER 19 The Ideal Gas
CHAPTER 20 Heat
CHAPTER 21 Thermodynamics

A liquid boils when bubbles of vapor form throughout its volume. The nature of the liquid and the atmospheric pressure determine the temperature required to form vapor. Pure water boils at 100° C at standard atmospheric pressure.

The Ideal Gas

19.1 The Ideal-Gas Law

19.2 The Temperature Scale

19.3 Kinetic Pressure

19.4 The Energy of an Ideal Gas

CONCEPTS IN CONTEXT

Concepts
— in —
Context

These spectacular hot-air balloons rely on buoyancy to fly: the warmed air in the balloon is less dense than the surrounding cold air.

To appreciate the properties of a gas that permit buoyancy, we will consider questions such as:

? How much air is in the balloon when cold? (Example 1, page 605)

? How much mass can the balloon lift when its air is heated to 50°C? (Example 6, page 612)

? How much energy (and fuel) does it take to heat the gas? (Example 9, page 618)

The particles of a gas—that is, the individual atoms or molecules of the gas—are well separated, and they fly about quite independently. The gas would disperse if it were not restrained by the forces exerted by the wall of the container. The gas molecules collide with each other, but these collisions do not alter the average speeds and average distribution of the gas molecules in the container. For most purposes, we can ignore these intermolecular collisions and regard the gas as a system of free particles. The motion of free gas particles is then very simple: they move with uniform velocity on straight lines, except when they collide with the walls of the container (Fig. 19.1). However, in spite of this simplicity of the motion of the gas particles, we cannot keep track of their motion in detail, because there are too many of them. For instance, one cubic centimeter of air contains about 2.7×10^{19} molecules. We have no way of ascertaining the initial position and velocity of each individual molecule; and even if we had, the calculation of the simultaneous motions of such an enormous number of molecules is far beyond the capabilities of even the fastest conceivable computer.

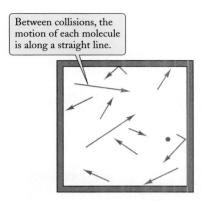

FIGURE 19.1 Random motion of the molecules of gas in a container.

In the absence of a microscopic description involving the individual positions and velocities of the molecules of the gas, we must be satisfied with a macroscopic description involving just a few parameters that characterize the average conditions in the volume of gas and that can be measured with large-scale laboratory instruments. We have already adopted such a macroscopic description for fluids in the preceding chapter. For a gas, the relevant macroscopic parameters characterizing the average conditions are *the mass, the number of moles, the volume, the density, the pressure, and the temperature.*

In this chapter we will study the macroscopic properties of gases, and we will see how these macroscopic properties are related to the average microscopic properties of the molecules of the gas.

19.1 THE IDEAL-GAS LAW

The pressure, volume, and temperature of a gas obey some simple laws. Before we state these laws, let us recall the definition of pressure from the preceding chapter. Imagine that the gas is divided into small adjacent cubical volumes. The pressure is the force that one of these cubes exerts on an adjacent cube, or on an adjacent wall, divided by the area of one face of the cube (see Fig. 18.15); that is, the pressure is the force per unit area:

$$p = \frac{F}{A} \tag{19.1}$$

pressure

As we know from Section 18.4, the pressure is the same throughout the entire volume of a container of gas. (Within a container of gas, gravity causes a small decrease in pressure from the bottom to the top, but this decrease in pressure can usually be neglected.)

Consider now a given amount of gas, say, n moles of gas. We saw in Chapter 1 that a mole of any chemical element (or chemical compound) is the amount of matter that contains exactly as many atoms (or molecules) as there are atoms in 12 g of carbon. The "atomic mass" of a chemical element (or the "molecular mass" of a compound) is the mass of 1 mole. Thus, according to the table of "atomic masses" (see Appendix 9), 1 mole of carbon has a mass of 12.0 g, 1 mole of oxygen molecules (O_2) has a mass of 32.0 g, 1 mole of nitrogen molecules (N_2) has a mass of 28.0 g, and so on. For a mixture, such as air (consisting of 76% nitrogen, 23% oxygen, and 1% argon by mass), the mass of 1 mole can be obtained by adding the masses of suitable fractions of moles of the constituents; 1 mole of air has a mass of 29.0 g.

Online
Concept
Tutorial
21

Suppose we place these n moles of gas in a container of volume V at a temperature T. The gas will then exert a pressure p. Experiments show that—to a good approximation—*the pressure p, the volume V, and the temperature T of the n moles of gas are related by the* **Ideal-Gas Law:**

Ideal-Gas Law

$$pV = nRT \qquad (19.2)$$

Here R is the **universal gas constant**, with the value

universal gas constant

$$R = 8.31 \text{ J/mole·K} \qquad (19.3)$$

From the Ideal-Gas Law, we can calculate one of the three quantities that characterize the state of the gas (pressure, temperature, volume) if the other two are known.

The temperature in Eq. (19.2) is measured on the **absolute temperature scale**, and the SI unit of temperature is the **kelvin,** abbreviated K. We have not previously given the definition of temperature because Eq. (19.2) plays a dual role: it is a law of physics and also serves for the definition of temperature. This is by now a familiar story—in Chapter 5 we already saw that Newton's Second Law is a law of physics and also serves as a definition of mass.

We will give the details concerning the definition of temperature in the next section. For now it will suffice to note that the freezing point of water corresponds to a temperature of of 273.15 K, and the boiling point of water corresponds to 373.15 K; hence, there is an interval of exactly 100 K between the freezing and the boiling points. The zero of temperature on the absolute scale is the absolute zero, $T = 0$ K. According to Eq. (19.2), the pressure of the gas vanishes at this point. Actually, the gas will liquefy or even solidify before the absolute zero of temperature can be reached; when this happens, Eq. (19.2) becomes inapplicable.

The Ideal-Gas Law is a simple relation between the macroscopic parameters that characterize a gas. At normal densities and pressures, real gases obey this law quite well; but if a real gas is compressed to an excessively high density, then its behavior will deviate from this law. We briefly examine the effects of such extreme conditions in Problem 61, but elsewhere in this chapter we will neglect any deviations from Eq. (19.2). An **ideal gas** is a gas that obeys Eq. (19.2) exactly. The ideal gas is a limiting case of a real gas when the density and the pressure of the latter tend to zero. The ideal gas may be thought of as consisting of atoms of infinitesimal size, exerting no forces on each other or on the walls of the container, except for instantaneous impact forces exerted during collisions.

The conditions of a temperature of 273 K and a pressure of 1 atm are called **standard temperature and pressure**, abbreviated **STP**. We can apply the Ideal-Gas Law to determine the volume of one mole of gas at STP. In SI units, a pressure of 1 atm is 1.01×10^5 N/m^2 [see Eq. (18.8)]. For one mole ($n = 1$ mole), the Ideal-Gas Law then gives us the volume

$$V = \frac{nRT}{p} = \frac{1 \text{ mole} \times 8.31 \text{ J/mole·K} \times 273 \text{ K}}{1.01 \times 10^5 \text{ N/m}^2}$$

$$= 2.24 \times 10^{-2} \text{ m}^3 = 22.4 \text{ liters}$$

In the last equality, we have used 1 liter = 1000 cm^3 = 10^{-3} m^3. Note that it makes no difference whether the gas in this calculation is air or something else—one mole of *any* gas at STP has a volume of 22.4 liters (see Fig. 19.2).

WILLIAM THOMSON, LORD KELVIN (1824–1907) *British physicist and engineer. Besides inventing the absolute temperature scale, he made many other contributions to the theory of heat. He was first to state the principle of dissipation of energy incorporated in the Second Law of Thermodynamics.*

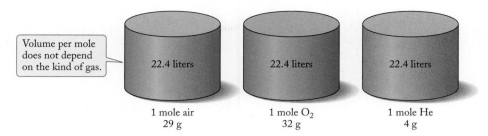

FIGURE 19.2 At a standard temperature of 273 K and pressure of 1 atm (STP conditions) a mole of any kind of gas occupies a volume of 22.4 liters.

EXAMPLE 1   The hot-air balloons pictured at the beginning of the chapter are large, with a typical volume of 2200 m³. How many moles of gas does such a balloon contain when cold (at STP)? What mass of air does the balloon contain? The mass of 1 mole of air is 29.0 g.

SOLUTION: We found just above that one mole of any gas occupies 22.4 liters at STP. The volume of the balloon is 2200 m³ = 2.20×10^6 liters. Thus the number of moles n in the balloon is the volume in liters times the molar density in moles per liter:

$$n = 2.20 \times 10^6 \text{ liters} \times \frac{1 \text{ mole}}{22.4 \text{ liters}} = 9.82 \times 10^4 \text{ moles}$$

The mass of this number of moles is

$$m = 29.0 \text{ g/mole} \times 9.82 \times 10^4 \text{ moles} = 2.85 \times 10^6 \text{ g} = 2.85 \times 10^3 \text{ kg}$$

The balloon contains almost three metric tons of air!

EXAMPLE 2   Suppose you heat 1.00 kg of water and convert it into steam at the boiling temperature of water, 373 K, and at normal atmospheric pressure, 1.00 atm. What is the volume of the steam?

SOLUTION: The molecular mass of water (H_2O) is the sum of the "atomic masses" of two hydrogen atoms and one oxygen atom; that is, 1.0 g + 1.0 g + 16.0 g = 18.0 g. Thus, the number of moles of steam in 1.00 kg is

$$n = \frac{1.00 \text{ kg}}{18.0 \text{ g/mole}} = \frac{1000 \text{ g}}{18.0 \text{ g/mole}} = 55.6 \text{ moles}$$

With this, and with p = 1.00 atm = 1.01×10^5 N/m², the Ideal-Gas Law tells us that the volume of the steam at 373 K is

$$V = \frac{nRT}{p} = \frac{55.6 \text{ moles} \times 8.31 \text{ J/mole·K} \times 373 \text{ K}}{1.01 \times 10^5 \text{ N/m}^2}$$

$$= 1.71 \text{ m}^3$$

Thus when we boil 1 liter of cold water, it becomes over 1700 liters of steam.

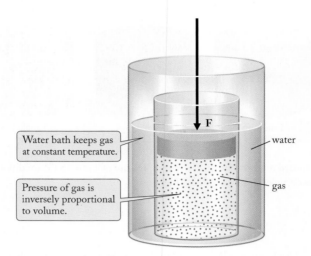

FIGURE 19.3 Compression of a gas at constant temperature by a force applied to the piston.

Water bath keeps gas at constant temperature.

water

Pressure of gas is inversely proportional to volume.

gas

The Ideal-Gas Law incorporates two other laws: the **Law of Boyle** and the **Law of Charles and Gay-Lussac**. The Law of Boyle asserts that if the temperature is held constant, then the product of pressure and volume must remain constant as a given amount of gas is compressed or expanded:

$$pV = [\text{constant}] \qquad \text{for } T = [\text{constant}] \tag{19.4}$$

This law can be tested experimentally by placing a sample of gas in a cylinder with a movable piston, surrounded by some substance at a fixed temperature, say, a water bath (see Fig. 19.3). By moving the piston in or out, we can vary the volume V and we can check that the pressure then varies in inverse proportion to the volume, as demanded by Eq. (19.4).

The Law of Charles and Gay-Lussac asserts that if the pressure is held constant, the ratio of volume to temperature remains constant as a given amount of gas is heated or cooled:

$$\frac{V}{T} = [\text{constant}] \qquad \text{for } p = [\text{constant}] \tag{19.5}$$

This law can be tested with a similar cylinder–piston arrangement with a fixed weight mounted on the piston and a heat source placed below the cylinder (Fig. 19.4). The weight mounted on the piston, in conjunction with the weight of the piston, subjects the gas to a fixed pressure. By increasing or decreasing the temperature of the gas we then cause the gas to expand or contract, and we can check that the volume is proportional to the temperature, as demanded by Eq. (19.5). Together, the experimental tests of Eqs. (19.4) and (19.5) amount to a test of the Ideal-Gas Law.

ROBERT BOYLE (1627–1691) *English experimental physicist. He invented a new air pump, with which he performed the experiments on gases that led to discovery of the law named after him.*

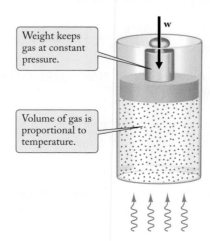

Weight keeps gas at constant pressure.

w

Volume of gas is proportional to temperature.

FIGURE 19.4 Expansion of a gas kept at constant pressure by a weight on the piston. The gas is being heated.

EXAMPLE 3 Early in the morning, at the beginning of a trip, the tires of an automobile are cold (280 K), and their air is at a pressure of 3.0 atm. Later in the day, after a long trip on hot pavements, the tires are hot (330 K). What is the pressure? Assume that the volume of the tires remains constant.

SOLUTION: At constant volume, the pressure is proportional to the temperature [see Eq. (19.2)]. Hence

$$p_2 = p_1 \times \frac{T_2}{T_1}$$

$$= 3.0 \text{ atm} \times \frac{330 \text{ K}}{280 \text{ K}} = 3.5 \text{ atm}$$

COMMENT: The pressure gauges for automobile tires are commonly calibrated to read **overpressure**, that is, the excess above atmospheric pressure (see also Section 18.4). Thus, a pressure gauge would read 2.0 atm in the morning, and 2.5 atm later in the day, if we assume that the atmospheric pressure remains constant at 1.0 atm (it rarely varies more than a few percent).

overpressure

The Ideal-Gas Law can also be written in terms of the number of molecules, instead of the number of moles. The number of molecules per mole is **Avogadro's number** N_A. As already mentioned in Chapter 1 [see Eq. (1.2)], the value of Avogadro's number is approximately

$$N_A = 6.02 \times 10^{23} \text{ molecules per mole} \tag{19.6}$$

Avogadro's number

Thus, if the number of moles is n, the number of molecules N is

$$N = N_A n \tag{19.7}$$

With this, the Ideal-Gas Law (19.2) becomes

$$pV = \frac{N}{N_A} RT \tag{19.8}$$

or

$$pV = NkT \tag{19.9}$$

Ideal-Gas Law in terms of number of molecules

where

$$k = \frac{R}{N_A} = \frac{8.31 \text{ J/mole·K}}{6.02 \times 10^{23}/\text{mole}} = 1.38 \times 10^{-23} \text{ J/K} \tag{19.10}$$

Boltzmann's constant

The constant k is called **Boltzmann's constant**. As we will see, this constant tends to make an appearance in equations relating macroscopic quantities (such as p or V) to microscopic quantities (such as the number N of molecules).

EXAMPLE 4 (a) What is the number of molecules in 1.00 cm^3 of air at a temperature of 273 K and a pressure of 1.00 atm? (As in Example 1, these conditions are *standard temperature and pressure*, or STP.) (b) What is the mass density of air at STP? Recall from above that the mass of 1 mole of air is 29.0 g.

LUDWIG BOLTZMANN (1844–1906)
Austrian theoretical physicist. He made crucial contributions in the kinetic theory of gases and in statistical mechanics.

SOLUTION: (a) With $p = 1.00 \text{ atm} = 1.01 \times 10^5 \text{ N/m}^2$ and $V = 1.00 \text{ cm}^3 = 1.00 \times 10^{-6} \text{ m}^3$, we can obtain the number of molecules N directly from the Ideal-Gas Law in the form (19.9):

$$N = \frac{pV}{kT} = \frac{1.01 \times 10^5 \text{ N/m}^2 \times 10^{-6} \text{ m}^3}{1.38 \times 10^{-23} \text{ J/K} \times 273 \text{ K}}$$

$$= 2.68 \times 10^{19} \text{ molecules}$$

Note that this result is valid for any kind of gas—the number of molecules in 1 cm^3 of any kind of gas under STP conditions is 2.68×10^{19}, or about 27 billion billion!

(b) The mass density is the mass per unit volume. We can find the mass of our 1.00-cm^3 volume by first obtaining the number of moles n from the number of molecules:

$$n = \frac{N}{N_A} = \frac{2.68 \times 10^{19} \text{ molecules}}{6.02 \times 10^{23} \text{ molecules/mole}} = 4.45 \times 10^{-5} \text{ moles}$$

Since the "molecular mass" is 29.0 grams per mole, the amount of mass in 1 cm^3 of air at STP is

$$4.45 \times 10^{-5} \text{ moles} \times 29.0 \text{ g/mole} = 1.29 \times 10^{-3} \text{ g} = 1.29 \times 10^{-6} \text{ kg}$$

Thus the density of air is $1.29 \times 10^{-3} \text{ g/cm}^3$, or 1.29 kg/m^3, in agreement with the value listed in Table 18.1.

PROBLEM-SOLVING TECHNIQUES   IDEAL-GAS LAW

When applying the Ideal-Gas Law to problems, care must be taken to use the correct units.

- The temperature T in the Ideal-Gas Law must be expressed in kelvins (K); the Ideal-Gas Law cannot be used in the forms (19.2) or (19.9) if the temperature is expressed in a different temperature scale, such as degrees celsius (°C) or degrees Fahrenheit (°F) (see Section 19.2).

- The pressure p in the Ideal-Gas Law must be the absolute pressure (not the overpressure), and it must be expressed in SI units (N/m², or pascals), not in atmospheres or mm-Hg.

 Checkup 19.1

QUESTION 1: If you heat the gas in a sealed jar from a temperature of 300 K to 600 K, by what factor does the pressure increase?

QUESTION 2: Suppose that the pressure of atmospheric air increases from 1.00 atm to 1.05 atm while the temperature remains constant. What happens to the density of air?

QUESTION 3: Suppose that the temperature of atmospheric air increases from 270 K to 290 K while the pressure remains constant. What happens to the density of air?

QUESTION 4: The air in the tires of an automobile is at an overpressure of 2 atm, but at the same temperature as the surrounding air. By what factor is the density of the air larger than the surrounding air?

QUESTION 5: A fixed amount of an ideal gas is in a variable-volume cylinder. The pressure is increased a factor of $\frac{3}{2}$, and the temperature is decreased by a factor of 2. The final volume occupied by the gas is what factor times the original volume?

(A) 3   (B) $\frac{4}{3}$ (C) 1   (D) $\frac{3}{4}$ (E) $\frac{1}{3}$

19.2   THE TEMPERATURE SCALE

As mentioned in the preceding section, the ideal gas can be used for the definition of temperature. For this purpose, we take a fixed amount of some gas, such as helium, and place it an airtight, nonexpanding container, such as a Pyrex glass bulb. According to Eq. (19.2), for a gas kept in such a constant volume, the pressure is directly proportional to the temperature. Thus, a simple measurement of pressure gives us the temperature.

To calibrate the scale of this thermometer, we must choose a standard reference temperature. The standard adopted in the SI system of units is the temperature of the **triple point of water**, that is, *the temperature at which water, ice, and water vapor coexist when placed in a closed vessel.* Figure 19.5 shows a triple-point cell used to achieve the standard temperature. This standard temperature has been assigned the value of 273.16 kelvins, or 273.16 K, which is ever so slightly above the freezing point of water. If the bulb of the gas thermometer is placed in thermal contact with this cell so that it attains a temperature of 273.16 K, it will read some pressure p_{triple}. If the bulb is then placed in thermal contact with some body at an unknown temperature T, it will read a pressure p that is greater or smaller than p_{triple} by some factor. The unknown temperature T is then greater or smaller than 273.16 K by this same factor; for instance, if the pressure p is half as large as p_{triple}, then $T = \frac{1}{2} \times 273.16$ K. *The temperature scale defined by this procedure is called the* **ideal-gas temperature scale**, *or the* **absolute temperature scale**.

When connecting a pressure gauge to the bulb of gas, we must take special precautions to ensure that the operation of the pressure gauge does not alter the volume available to the gas. Figure 19.6 shows a device designed for this purpose; this device is called a **constant-volume gas thermometer**. The pressure gauge used in this thermometer consists of a closed-tube manometer; one branch of the manometer is connected to the bulb of gas, and the other branch consists of a closed, evacuated tube. The difference h in the heights of the levels of mercury in these two branches is proportional to the pressure of the gas. The manometer is also connected to a mercury reservoir. During the operation of the thermometer, this reservoir must be raised or lowered so that the level of mercury in the left branch of the manometer tube always

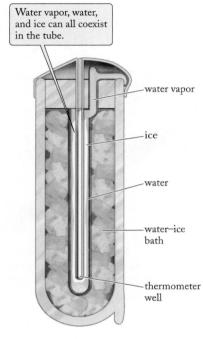

FIGURE 19.5 Triple-point cell of the National Institute of Standards and Technology. The inner tube (red) contains water, water vapor, and ice in equilibrium.

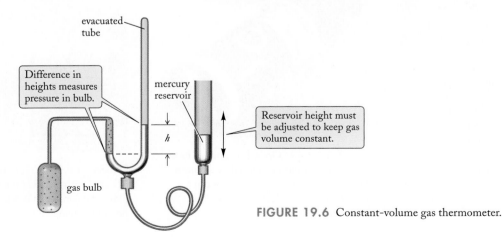

FIGURE 19.6 Constant-volume gas thermometer.

remains at a constant height; this keeps the gas in the bulb at a constant volume. The bulb of this thermometer may be put in contact with any body whose temperature we wish to measure, and the pressure registered by the manometer then gives us the absolute temperature.

Table 19.1 lists some examples of temperatures of diverse bodies.

For everyday and industrial use, the ideal-gas thermometer is somewhat inconvenient, and is often replaced by mercury-bulb thermometers, bimetallic strips, electrical-resistance thermometers, thermocouples, optical pyrometers, or color-strip thermometers (see Figs. 19.7–19.12). These must be calibrated in terms of the ideal-gas thermometer if they are to read absolute temperature.

FIGURE 19.7 Mercury-bulb thermometers. The thermal expansion of the mercury in the bulb causes it to rise in the thin capillary tube, indicating the temperature.

FIGURE 19.8 Bimetallic-strip thermometer. The thermometer contains a helix consisting of joined bands of different metals. With an increase of temperature, the bands expand by different amounts, which coils the helix and turns the pointer.

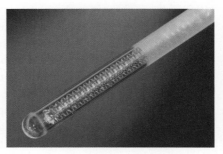

FIGURE 19.9 Platinum-resistance thermometer. The resistance that the fine coiled wire of platinum offers to an electric current serves as an indicator of the temperature (electrical resistance is discussed in Chapter 27).

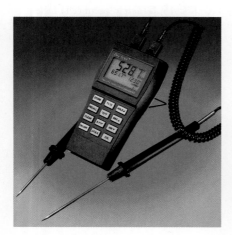

FIGURE 19.10 Thermocouple thermometer. A thermocouple consists of two wires of different metals, for example, one wire of platinum and one wire of platinum–rhodium alloy, joined at their ends. The other ends of the wires are at a reference temperature, usually 0°C; because of the temperature difference and the dissimilar materials, an electrical voltage develops between the other ends. Most digital thermometers use thermocouples.

FIGURE 19.11 Optical pyrometer. From the color distribution emitted by an incandescent material, the temperature can be determined (this effect is discussed in Chapter 37).

FIGURE 19.12 Color-strip thermometer. Material with different temperature-induced color transitions provides an indication of the temperature.

TABLE 19.1	SOME TEMPERATURES		
		KELVIN TEMPERATURE	CELSIUS TEMPERATURE
Interior of hottest stars		10^9 K	$10^{9\circ}$C
Center of H-bomb explosion		10^8	10^8
Highest temperature attained in plasma in laboratory		6×10^7	6×10^7
Center of Sun		1.5×10^7	1.5×10^7
Surface of Sun (a)		4.5×10^3	4.2×10^3
Center of Earth		4×10^3	3.7×10^3
Acetylene flame		2.9×10^3	2.6×10^3
Melting of iron (b)		1.8×10^3	1.5×10^3
Melting of lead		6.0×10^2	3.3×10^2
Boiling of water		373	100
Human body (c)		310	37
Surface of Earth (average)		287	14
Freezing of water		273	0
Liquefaction of nitrogen		77	−196
Liquefaction of hydrogen		20	−253
Liquefaction of helium		4.2	−269
Interstellar space		3	−270
Lowest temperature attained in laboratory		3×10^{-8}	≈ -273.15

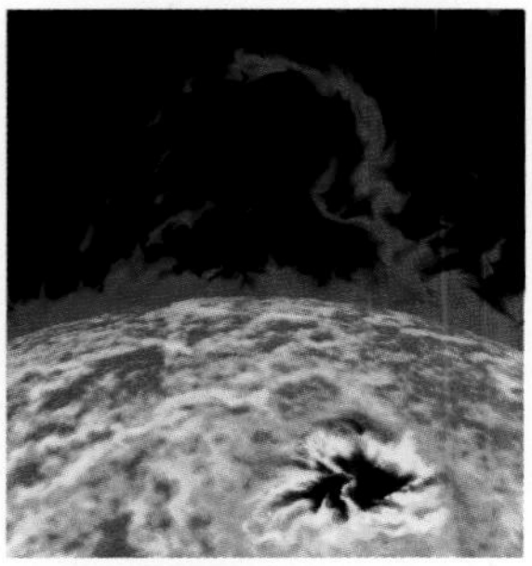

(a)

(b)

(c)

Although the absolute temperature scale is the only scale of fundamental significance, several other temperature scales are in practical use. The **Celsius scale** (formerly known as the **centigrade** scale) is shifted 273.15 K relative to the absolute scale:

$$T_C = T - 273.15°C \qquad (19.11)$$

Celsius scale

where the notation "°C" means "degrees Celsius." Note that on the Celsius scale, absolute zero is at −273.15°C. The triple point of water is then at 0.01°C, the freezing point at 0°C, and the boiling point at 100°C.

The **Fahrenheit scale** is shifted relative to the Celsius scale and, furthermore, uses degrees of smaller size, each degree Fahrenheit corresponding to $\frac{5}{9}$ degree Celsius:

$$T_F = \tfrac{9}{5}T_C + 32°F \qquad \text{or} \qquad T_C = \tfrac{5}{9}(T_F - 32°F) \qquad (19.12)$$

Fahrenheit scale

On this scale, the freezing point of water is at 32°F, the boiling point of water is at 212°F, and absolute zero is at −459.67°F. Figure 19.13 can be used for a rough conversion between the Fahrenheit and Celsius scales.

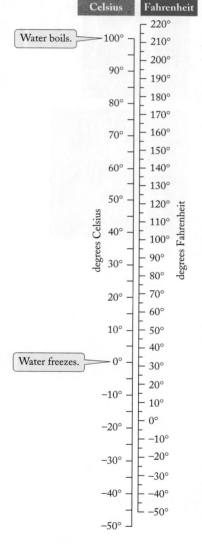

FIGURE 19.13 The correspondence between the Fahrenheit scale and the Celsius scale.

> **EXAMPLE 5**   The normal temperature of the human body is 98.6°F. Express this in °C.
>
> **SOLUTION:** According to Eq. (19.12)
>
> $$T_C = \tfrac{5}{9}(T_F - 32°F) = \tfrac{5}{9}(98.6°F - 32°F) = 37°C$$

Concepts —in— Context

> **EXAMPLE 6**   Hot-air balloons become buoyant when the gas inside is heated with a burner below the balloon. Recall from Example 1 that at STP, a typical balloon contained 2.85×10^3 kg of air. If the temperature of the gas is increased to 50°C (at constant pressure), what is the mass of the gas that remains in the balloon? What mass can be lifted by the balloon?
>
> **SOLUTION:** We must convert to absolute temperature to use the Ideal-Gas Law, so a temperature of 50°C is equivalent to an absolute temperature of
>
> $$T = T_C + 273.15\ \text{K} = 50 + 273.15\ \text{K}$$
> $$= 323\ \text{K}$$
>
> At constant pressure (and volume), the Ideal-Gas Law, $pV = nRT$, shows that the number of moles (and thus the mass) will be inversely proportional to temperature; hence the mass of air in the hot-air balloon is
>
> $$m_{323\,K} = m_{STP} \times \frac{273\ \text{K}}{323\ \text{K}} = 2.85 \times 10^3\ \text{kg} \times \frac{273\ \text{K}}{323\ \text{K}} = 2.41 \times 10^3\ \text{kg}$$
>
> The balloon will rise if its total mass (balloon, basket, equipment, and riders) is less than the difference between the displaced-air mass and the hot-air mass:
>
> $$2.85 \times 10^3\ \text{kg} - 2.41 \times 10^3\ \text{kg} = 4.4 \times 10^2\ \text{kg} = 440\ \text{kg}$$
>
> Thus the balloon can lift nearly one-half a metric ton.

✔ Checkup 19.2

QUESTION 1: Is a negative Fahrenheit temperature always a negative Celsius temperature? Is a negative Celsius temperature always a negative Fahrenheit temperature?

QUESTION 2: Three cups of water are at temperatures of 320 K, 20°C, and 90°F. Which is the hottest? Which the coldest?

QUESTION 3: Consider a cylindrical can filled with gas, closed off at the top by a piston. The weight of the piston then provides a constant force, which keeps the gas under a constant pressure. Explain how this device could be used as a "constant-pressure" gas thermometer.

QUESTION 4: To the nearest degree, at what temperature does the Celsius scale indicate the same numerical value of temperature as the Fahrenheit scale?

(A) −72°C   (B) −50°C   (C) −40°C   (D) −26°C

19.3 KINETIC PRESSURE

The pressure of a gas against the walls of its container is due to the impacts of the molecules on the walls. We will now calculate this pressure by considering the average motion of the molecules of gas. This permits us to understand how a macroscopic property, such as the pressure, emerges from the microscopic behavior of individual molecules of the gas. In our calculation, we will assume that the container is a cube of side L, that the gas molecules collide only with the walls but not with each other, and that the collisions are elastic. These assumptions are not required, but they simplify the calculations.

Figure 19.14 shows the container filled with gas molecules. The motion of each molecule can be resolved into x, y, and z components. Consider one molecule, and consider the component of its motion in the x direction. The component of the velocity in this direction is $\pm v_x$, and the magnitude of this velocity remains constant, since the collisions with the wall are elastic. The time that the molecule takes to move from the face of the cube at $x = L$ to $x = 0$ and back to $x = L$ is the distance divided by the speed:

$$t = \frac{2L}{v_x} \tag{19.13}$$

This is therefore the time between one collision with the face at $x = L$ and the next collision with the same face. When the molecule strikes the face, its x velocity is reversed from $+v_x$ to $-v_x$. Hence, during each collision at $x = L$, the x momentum of the molecule changes from $+mv_x$ to $-mv_x$, a net change of $-2mv_x$, where m is the mass of the molecule. Thus each collision transfers a momentum $+2mv_x$ to the face at $x = L$. The average rate at which the molecule transfers momentum to the face at $x = L$ is then the momentum transfer per collision divided by the time between collisions:

$$\frac{2mv_x}{t} = \frac{2mv_x}{2L/v_x} = \frac{mv_x^2}{L} \tag{19.14}$$

This average rate of momentum transfer equals the average force that the impacts of this one molecule exert on the wall. To find the total force exerted by the impacts of all the N molecules, we must multiply the force given in Eq. (19.14) by N; and to find the pressure, we must divide the total force by the area L^2 of the face of the cube. This leads to a pressure

$$p = \frac{N}{L^2}\frac{mv_x^2}{L} \tag{19.15}$$

or, in terms of the volume $V = L^3$ of the gas,

$$p = \frac{Nmv_x^2}{V} \tag{19.16}$$

In this calculation we made the implicit assumption that all the molecules have the same velocity. This is, of course, not true; the molecules of the gas have a distribution of velocities—some have high velocities, and some have low velocities. To account for this spread of velocities, we must replace the force in Eq. (19.14) due to one given molecule by the average over all the molecules. Consequently, we must replace v_x^2 by an average over all the molecules in the container. We will designate the average by an overbar, $\overline{v_x^2}$. Equation (19.16) then becomes

$$p = \frac{Nm\overline{v_x^2}}{V} \tag{19.17}$$

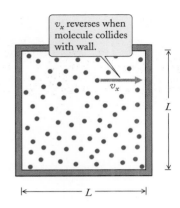

v_x reverses when molecule collides with wall.

FIGURE 19.14 Gas molecules in a container.

To proceed further, we note that on the average, molecules are just as likely to move in the x, y, or z direction. Hence, the average values of v_x^2, v_y^2, and v_z^2 are equal,

$$\overline{v_x^2} = \overline{v_y^2} = \overline{v_z^2} \tag{19.18}$$

The sum of the squares of the components of the velocity is the square of the magnitude of the velocity,

$$\overline{v_x^2} + \overline{v_y^2} + \overline{v_z^2} = \overline{v^2} \tag{19.19}$$

Since all three terms on the left side of this equation are equal, each of them must equal $\frac{1}{3}\overline{v^2}$. We can then write Eq. (19.17) as

$$p = \frac{Nm\overline{v^2}}{3V} \tag{19.20}$$

Let us now compare this result for the pressure with the Ideal-Gas Law [Eq. (19.9)], according to which the pressure is

$$p = \frac{NkT}{V} \tag{19.21}$$

The agreement between these two expression (19.20) and (19.21) for the pressure demands

$$\frac{m\overline{v^2}}{3} = kT \tag{19.22}$$

This shows that the average square of the molecular speed is proportional to the temperature. *The square root of $\overline{v^2}$ is called the* **root-mean-square speed**, or the **rms speed**, and it is usually designated by v_{rms}. If we divide both sides of Eq. (19.22) by $m/3$ and extract the square root of both sides, we find

root-mean-square (rms) speed

$$v_{\mathrm{rms}} = \sqrt{\overline{v^2}} = \sqrt{\frac{3kT}{m}} \tag{19.23}$$

This root-mean-square speed may be regarded as the typical speed of the molecules of the gas. Incidentally: There are other ways of calculating a typical speed; for example, we may want to calculate the average of all the molecular speeds, or the most probable of all the molecular speeds. The average and most probable speeds turn out to be somewhat less than the rms speed, but their calculation requires some further knowledge of the distribution of molecular speeds.

EXAMPLE 7   What is the root-mean-square speed of nitrogen molecules in air at 300 K? Of oxygen molecules?

SOLUTION: The "molecular mass" of N_2 molecules is 28.0 g, twice the "atomic mass" of nitrogen. Hence, the mass of one molecule is

$$m = \frac{28.0 \text{ g}}{N_A} = \frac{28.0 \times 10^{-3} \text{ kg}}{6.02 \times 10^{23}} = 4.65 \times 10^{-26} \text{ kg}$$

and Eq. (19.23) yields

$$v_{\mathrm{rms}} = \sqrt{\frac{3kT}{m}}$$

$$= \sqrt{\frac{3 \times 1.38 \times 10^{-23}\,\mathrm{J/K} \times 300\,\mathrm{K}}{4.65 \times 10^{-26}\,\mathrm{kg}}} = 517\,\mathrm{m/s} \qquad (19.24)$$

For O_2 molecules the "molecular mass" is 32.0 g, twice the "atomic mass" of oxygen. By a similar calculation we find that the mass of one oxygen molecule is 5.32×10^{-26} kg, and that v_{rms} is 483 m/s.

COMMENT: Note that the rms speed of nitrogen molecules is slightly larger than that of oxygen molecules. In general, Eq. (19.23) shows that the rms speed is inversely proportional to the square root of the mass of the molecule—at a given temperature, the molecules of lowest mass have the highest speeds.

The distribution of molecular speeds in a sample of gas at some given temperature can be deduced by means of kinetic theory; it is called the **Maxwell distribution** of molecular speeds. We will not attempt to deduce this distribution here, but only examine some of its qualitative features. The distribution of speeds in a sample of gas is described mathematically by a distribution function N_v, which specifies the number of molecules per unit speed interval. Thus, if dv is a small interval of speeds centered on a given speed v, then the number dN of molecules that have speeds in the interval dv is

$$dN = N_v\, dv$$

Keep in mind that when dealing with the distribution of speeds, or the distribution of any other kind of physical quantity that has a continuous range of variation, it is not reasonable to ask how many molecules have a speed v, since it is unlikely that any molecule has a speed *exactly* equal to v. The only reasonable question is how many molecules have speeds in some specified interval of speeds.

Figure 19.15 shows plots of the function N_v for the Maxwell distribution of speeds at two different temperatures. As we can see from these plots, the molecular speeds are spread over broad ranges. The distribution functions fade away as $v \to 0$ and as $v \to \infty$; thus, there are few molecules near zero speed, and few molecules of very large speed. The peaks of the distribution functions indicate the **most probable speed**. Comparing the peaks at the two different temperatures, we see that the most probable speed increases with temperature. For the Maxwell distribution, the most probable speed and the average speed are somewhat lower than the rms speed; the most probable speed and the average speed turn out to be approximately $0.82v_{\mathrm{rms}}$ and $0.92v_{\mathrm{rms}}$ respectively.

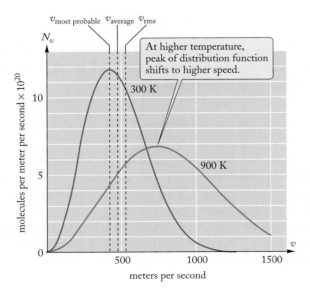

FIGURE 19.15 Maxwell distribution of speeds for one mole of N_2 gas molecules. The higher curve corresponds to a temperature of 300 K, the lower curve to 900 K. The most probable speed is indicated by the peak on the curve.

✔ Checkup 19.3

QUESTION 1: The pressure of a gas on the walls of its container increases with temperature. Is this increase due to a higher rate of impacts on the walls, or a higher amount of momentum transferred to the wall per impact, or both?

QUESTION 2: If we increase the temperature of a gas from 300 K to 600 K, by what factor do we increase the rms speed of its molecules?

QUESTION 3: Is the rms speed of water molecules in air higher or lower than that of oxygen molecules?

QUESTION 4: Consider hydrogen (H_2) gas at a temperature of 50 K and oxygen gas (O_2) at 200 K, both at the same pressure $p = 1$ atm. What is the ratio of the mass per unit volume of the hydrogen gas to that of the oxygen gas?

  (A) 64   (B) 4   (C) 1   (D) $\frac{1}{4}$ (E) $\frac{1}{64}$

19.4 THE ENERGY OF AN IDEAL GAS

Since a gas is a system of particles, its energy is the sum of the energies of all these particles. To calculate the energy of the gas, we begin with the kinetic energy of one molecule. According to Eq. (19.23), the average value of the square of the speed of a molecule in an ideal gas is

$$\overline{v^2} = \frac{3kT}{m}$$

Hence the average kinetic energy of a molecule of ideal gas is

$$\tfrac{1}{2}m\overline{v^2} = \tfrac{1}{2} \times 3kT = \tfrac{3}{2}kT \tag{19.25}$$

If we multiply this by the total number N of molecules, we obtain the total translational kinetic energy of all the molecules jointly. In an ideal gas, the molecules exert no forces on one another (they do not collide) and hence there is no intermolecular potential energy. The kinetic energy is then the total energy,

internal kinetic energy of monatomic gas

$$E = K = \tfrac{3}{2}NkT \tag{19.26}$$

This formula tells us how much energy is stored in the microscopic thermal motion of the gas. Since $Nk = nR$, where n is the number of moles [see Eqs. (19.7) and (19.10)], we can also write this formula for the energy as

$$E = \tfrac{3}{2}nRT \tag{19.27}$$

This energy is called the **thermal energy** of the gas. It is sometimes also called the **internal energy** of the gas, because it is stored (and hidden) in the microscopic motions in the interior of the gas, instead of being manifest in an overall macroscopic translational motion of the entire body of gas.

Note that we have assumed that the molecules behave as pointlike particles—each molecule has translational kinetic energy, but no internal energy. **Monatomic** gases, such as helium, argon, and krypton, consist of single atoms and behave in this way.

EXAMPLE 8   What is the thermal kinetic energy in 1.0 kg of helium gas at 0°C? How much extra energy must be supplied to this gas to increase its temperature to 60°C (at constant volume)?

SOLUTION: The "atomic mass" of helium is 4.0 g/mole; hence the number of moles in 1.0 kg of helium is

$$n = \frac{1.0 \text{ kg}}{4.0 \text{ g/mole}} = 250 \text{ moles}$$

According to Eq. (19.26), the energy at a temperature of 273 K is

$$E = \tfrac{3}{2}nRT = \tfrac{3}{2} \times 250 \text{ moles} \times 8.31 \text{ J/mole·K} \times 273 \text{ K}$$

$$= 8.5 \times 10^5 \text{ J}$$

Since thermal energy changes in proportion to absolute temperature, the extra energy needed to increase the temperature by 60°C, or by 60 K, is

$$\Delta E = \tfrac{3}{2}nR\,\Delta T = \tfrac{3}{2} \times 250 \text{ moles} \times 8.31 \text{ J/mole·K} \times 60 \text{ K}$$

$$= 1.9 \times 10^5 \text{ J}$$

Diatomic gases, which consist of two-atom molecules, such as N_2 and O_2, store an additional amount of energy in the internal motions of the atoms within each molecule. The molecules of these gases may be regarded as two pointlike particles rigidly connected together (a dumbbell; see Fig. 19.16). If such a molecule collides with another molecule or with the wall of the container, it will usually start rotating about its center of mass. We therefore expect that, on the average, an appreciable fraction of the energy of the gas will be in the form of this kind of rotational kinetic energy.

The molecule may rotate about either of the two axes through the center of mass perpendicular to the line joining the atoms (see Fig. 19.16). If the moments of inertia about these axes are I_1 and I_2, and if the corresponding angular velocities are ω_1 and ω_2, then the kinetic energy for these two rotations is [see Eq. (12.28)]

$$\tfrac{1}{2}I_1\omega_1^2 + \tfrac{1}{2}I_2\omega_2^2 \tag{19.28}$$

The average of this kinetic energy is

$$\tfrac{1}{2}I_1\overline{\omega_1^2} + \tfrac{1}{2}I_2\overline{\omega_2^2} \tag{19.29}$$

where, as in the preceding section, the overbars denote the average over all the molecules of the gas.

To discover the value of these average rotational energies, let us return to Eq. (19.22) and write it in terms of the x, y, and z components of velocity:

$$\tfrac{1}{2}m\overline{v_x^2} + \tfrac{1}{2}m\overline{v_y^2} + \tfrac{1}{2}m\overline{v_z^2} = \tfrac{3}{2}kT \tag{19.30}$$

We know that the average x, y, and z speeds are equal. Hence, Eq. (19.30) asserts that the kinetic energy for each component of the motion has the value $\tfrac{1}{2}kT$:

$$\tfrac{1}{2}m\overline{v_x^2} = \tfrac{1}{2}kT, \qquad \tfrac{1}{2}m\overline{v_y^2} = \tfrac{1}{2}kT, \qquad \text{and} \qquad \tfrac{1}{2}m\overline{v_z^2} = \tfrac{1}{2}kT \tag{19.31}$$

It turns out that this is true not only for the components of the translational motion, but also for rotational and other motion. The general result is known as the **equipartition**

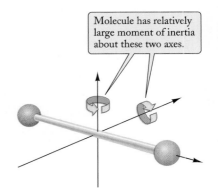

Molecule has relatively large moment of inertia about these two axes.

FIGURE 19.16 A diatomic molecule represented as two pointlike particles joined by a rod.

theorem. Applied to translational and rotational motion, the equipartition theorem states that

> *Each translational or rotational component of the random thermal motion of a molecule has an average kinetic energy of $\frac{1}{2}kT$.*

We will not prove this theorem, but we will make use of it.

According to this theorem, each of the terms in Eq. (19.29) has a value $\frac{1}{2}kT$. The total average rotational energy is then kT, and when we add this to the translational kinetic energy $\frac{3}{2}kT$, we obtain the total kinetic energy for one molecule:

$$kT + \tfrac{3}{2}kT = \tfrac{5}{2}kT \tag{19.32}$$

The energy of all the molecules of the diatomic gas taken together is then

$$E = \tfrac{5}{2}NkT \tag{19.33}$$

Note that in this calculation we have ignored the possibility of rotation about the longitudinal axis of the molecule. This means we have ignored the rotation of the atoms about an axis through them, just as we have ignored this kind of rotation of the atoms in a monatomic gas. The reason why we have ignored this rotational degree of freedom for monatomic and diatomic gases has to do with their very small moment of inertia for these rotations. The full analysis lies beyond the realm of classical physics; it lies in the realm of quantum physics. There it is established that only certain energies are allowed, and that for much smaller moments of inertia, the allowed energies are much higher. It turns out that the typical available thermal energy $\frac{1}{2}kT$ is not nearly enough to attain these higher energies, and thus their contribution is "frozen out" at normally attainable temperatures. If, however, the molecule in question is **polyatomic and non-linear** (like H_2O), then the third rotational degree of freedom would have the usual energy $\frac{1}{2}kT$ of random thermal motion, and the total rotational energy would be $\frac{3}{2}kT$, and the total energy would be $3kT$.

Similarly, we have ignored the vibrational motion of the atoms of the diatomic molecule. The interatomic forces do not really hold these atoms in a rigid embrace; rather, the forces act somewhat like springs (see Example 6 in Chapter 15), and they permit a restricted back-and-forth vibration of the atoms about their equilibrium positions. We have ignored the kinetic and potential energies associated with these vibrations because again the quantum-mechanically allowed energies are high. The vibration of atoms in a molecule can occur, however, if the temperature is rather high, 400°C or more. As we will see in Section 20.5, the energies calculated from Eqs. (19.27) and (19.33) actually agree quite well with experiments, provided we do not exceed this temperature limit.

Concepts
— in —
Context

EXAMPLE 9   We saw in Example 6 that a hot-air balloon achieves appreciable buoyancy when its air is heated from 273 K to 323 K. How much larger is the internal energy of the initial 9.82×10^4 moles of air when heated?

SOLUTION: Since air is 99% diatomic, it is a good approximation to treat it as a diatomic gas. Thus the energy change with temperature is given by Eq. (19.33):

$$\Delta E = \tfrac{5}{2}Nk\,\Delta T \tag{19.34}$$

We convert in the usual way from the number of molecules N to the number of moles n using [see Eqs. (19.7) and (19.10)]

$$Nk = nR \tag{19.35}$$

Substituting this and the given values into Eq. (19.34) gives

$$\Delta E = \tfrac{5}{2}nR\,\Delta T$$

$$= \tfrac{5}{2} \times 9.82 \times 10^4 \text{ moles} \times 8.31 \text{ J/mole·K} \times (323\,\text{K} - 273\,\text{K})$$

$$= 1.0 \times 10^8 \text{ J}$$

This is comparable to the energy released by the combustion of 1 gallon of gasoline, 1.3×10^8 J (see Table 8.1).

COMMENT: The balloonist must supply not only this internal energy, but also the work required to push some of the expanding gas out of the balloon against the pressure of the atmosphere. We will consider some aspects of gas expansion in the next chapter. Also, after initially heating the air, the balloonist must supply some energy thereafter to maintain the temperature, since heat is conducted away through the surface of the balloon.

 Checkup 19.4

QUESTION 1: If we increase the temperature of a gas from 300 K to 400 K (at constant volume), by what factor do we increase the thermal energy?

QUESTION 2: Why does a diatomic gas have a higher thermal energy per mole than a monatomic gas?

QUESTION 3: In Example 8 we assumed that the gas was heated at constant volume. If, instead, we were to heat the gas at constant pressure (in a container equipped with a piston), we would have to supply *more* energy to achieve the same increase of temperature. Why?

QUESTION 4: You have one mole of each of three gases at $T = 400$ K. Place them in order of increasing total thermal energy: (1) diatomic oxygen (O_2); (2) monatomic neon (Ne); (3) polyatomic (and nonlinear) water vapor (H_2O).

    (A) 1, 2, 3   (B) 1, 3, 2   (C) 2, 1, 3   (D) 2, 3, 1   (E) 3, 1, 2

SUMMARY

PROBLEM-SOLVING TECHNIQUES	Ideal-Gas Law	**(page 608)**
IDEAL-GAS LAW	$pV = nRT$ (n = number of moles)	**(19.2)**
	$pV = NkT$ (N = number of molecules)	**(19.9)**
UNIVERSAL GAS CONSTANT	$R = 8.31$ J/mole·K	**(19.3)**
BOLTZMANN'S CONSTANT k and R are related by Avogadro's number N_A.	$k = 1.38 \times 10^{-23}$ J/K $N_A k = R$	**(19.10)**
STANDARD TEMPERATURE AND PRESSURE (STP)	$T = 273$ K $p = 1$ atm $= 1.01 \times 10^5$ Pa	

TEMPERATURE SCALES

Absolute (kelvin or K)

Celsius (°C)

Fahrenheit (°F)

T

$T_C = T - 273.15$

$T_F = \frac{9}{5}T_C + 32$ **(19.12)**

ROOT-MEAN-SQUARE SPEED

$$v_{rms} = \sqrt{\overline{v^2}} = \sqrt{\frac{3kT}{m}}$$ **(19.23)**

INTERNAL KINETIC ENERGY OF IDEAL GAS

$E = \frac{3}{2}NkT$ (monatomic)   **(19.26)**

$E = \frac{5}{2}NkT$ (diatomic)   **(19.33)**

$E = 3NkT$ (nonlinear polyatomic)

EQUIPARTITION THEOREM   Each translational or rotational component of the random thermal motion of a molecule has an average kinetic energy of $\frac{1}{2}kT$.

QUESTIONS FOR DISCUSSION

1. Why do meteorologists usually measure the temperature in the shade rather than in the sun?

2. Why are there no negative temperatures on the absolute temperature scale?

3. The temperature of the ionized gas in the ionosphere of the Earth is about 2000 K, but the density of this gas is extremely low, only about 10^5 gas particles per cubic centimeter. If you were to place an ordinary mercury thermometer in the ionosphere, would it register 2000 K? Would it melt?

4. The temperature of intergalactic space is 3 K. How can empty space have temperature?

5. At the airport of La Paz, Bolivia, one of the highest in the world, pilots of aircraft find it preferable to take off early in the morning or late at night, when the air is very cold. Why?

6. If you release a rubber balloon filled with helium, it will rise to a height of a few thousand meters and then remain stationary. What determines the height reached? Is there an optimum pressure to which you should inflate the balloon to reach greatest height?

7. How can you use a barometer as an altimeter?

8. Explain why a real gas behaves like an ideal gas at low densities but not at high densities.

9. Helium and neon approach the behavior of an ideal gas more closely than do any other gases. Why would you expect this?

10. Ultrasound waves of extremely short wavelength cannot propagate in air. Why not?

11. Prove that it is impossible for all of the molecules in a gas to have the same speeds and to keep these speeds forever. (Hint: Consider an elastic collision between two molecules with the same speed. Will the speeds remain unchanged if the initial lines of motion are not parallel?)

12. If you increase the absolute temperature by a factor of 2, by what factor will you increase the average speed of the molecules of gas?

13. Air consists of a mixture of nitrogen (N_2), oxygen (O_2), and argon (Ar). Which of these molecules has the highest rms speed? The lowest?

14. Equipartition of energy applies not only to atoms and molecules, but also to macroscopic "particles" such as golf balls. If so, why do golf balls remain at rest on the ground instead of flying through the air like molecules?

15. If you open a bottle of perfume in one corner of a room, it takes a rather long time for the smell to reach the opposite corner (assuming that there are no air currents in the room). Explain why the smell spreads slowly, even though the typical speeds of perfume molecules are 300–400 m/s.

PROBLEMS

19.1 The Ideal-Gas Law[†]
19.2 The Temperature Scale[†]

1. Express the last six temperatures listed in Table 19.1 in terms of degrees Fahrenheit.

2. The hottest place on the Earth is Al-'Aziziyah, Libya, where the temperature has soared to 136.4°F. The coldest place is Vostok, Antarctica, where the temperature has plunged to −126.9°F. Express these temperatures in degrees Celsius and in kelvins.

3. A paper clip has a mass of 0.50 g. The paper clip is made of iron. How many atoms are in this paper clip?

4. What is the number of sodium and of chlorine atoms in one spoonful (10 g) of salt, NaCl?

5. Assume that air is 76% nitrogen and 24% oxygen by mass. What is the percent composition of air by number of molecules?

6. At an altitude of 160 km, the density of air is 1.5×10^{-9} kg/m³ and the temperature is approximately 500 K. What is the pressure?

7. How much does the frequency of middle C (see Table 17.1) played on the flute change when the air temperature drops from 20°C to −10°C? [Hint: The speed of sound in air is given by Eq. (17.5).]

8. In the Middle Ages, physicians applied suction cups to the skin, to draw out "bad humors." The cups produced "suction" by means of hot air. Suppose that a hot suction cup, at a temperature of 85°C, is applied to the skin and its rim makes an airtight seal against the skin. The cup initially contains air at 80°C and at atmospheric pressure, 1.0 atm. What will be the underpressure generated in the cup (that is, the difference between the pressure in the cup and atmospheric pressure) when the cup and the air trapped inside it cool from 85°C to 30°C?

9. In summer when the temperature is 30°C, the overpressure within an automobile tire is 2.2 atm. What will be the overpressure within this tire in winter when the temperature is 0°C? Assume that no air is added to the tire and that no air leaks from the tire; assume that the volume of the tire remains constant and that the atmospheric pressure remains at 1.0 atm.

10. What is the number of oxygen molecules in 1.00 cm³ of air at 273 K and 1.00 atm? Nitrogen molecules? What is the number of atoms?

11. A tank of a volume of 1.0 liter contains 1.0 g of nitrogen gas at 290 K. Another tank of equal volume at equal temperature contains 1.0 g of oxygen gas.

    (a) What is the pressure in each tank?

    (b) If we pump the oxygen gas into the nitrogen tank, what is the pressure produced by the mixture of the two gases? Assume that the temperature remains constant at 290 K.

12. Estimate the average distance between molecules of air at STP.

13. Repeat the calculation of Example 3 assuming that, because of the increase of pressure, the volume of the tire increases by 5%.

14. The storage tank of a small air compressor holds 0.30 m³ of air at a pressure of 5.0 atm and a temperature of 20°C. How many moles of air is this?

15. On a warm day, the outdoor temperature is 35°C and the indoor temperature in an air-conditioned house is 21°C. What is the difference between the densities of the air outdoors and indoors? Assume the pressure is 1.00 atm.

16. The lowest pressure attained in a "vacuum" in a laboratory on the Earth is 1.0×10^{-16} atm. Assuming a temperature of 20°C, what is the number of molecules per cubic centimeter in this vacuum?

17. Clouds of interstellar hydrogen gas have densities of up to 1.0×10^{10} atoms/m³ and temperatures of up to 1.0×10^4 K. What is the pressure in such a cloud?

18. The following table gives the pressure and density of the Earth's upper atmosphere as a function of altitude:

ALTITUDE	PRESSURE	DENSITY
20 000 m	5600 Pa	9.2×10^{-2} kg/m³
40 000	320	4.3×10^{-3}
60 000	28	3.8×10^{-4}
80 000	1.3	2.5×10^{-5}

Calculate the temperature at each altitude. The mean molecular mass for air is 29.0 g.

19. What is the density (in kilograms per cubic meter) of helium gas at 1.0 atm at the temperature of boiling helium liquid (see Table 19.1)?

20. The volume of an automobile tire is 2.5×10^{-2} m³. The pressure of the air in this tire is 3.0 atm and the temperature is 17°C. What is the mass of air? The mean molecular mass of air is 29.0 g.

21. When a bicycle tire is filled early in the morning at 22°C, the overpressure is 4.0 atm. What is the overpressure later in the day, when the temperature is 38°C? Assume that the atmospheric pressure of 1.0 atm is constant.

22. For gas storage, thick metal cylinders (Fig. 19.17) with an internal volume of 35 liters are used; typical cylinders can safely maintain a pressure of 180 atm. At 25°C, how many moles of gas can such a cylinder hold?

[†]For help, see Online Concept Tutorial 21 at www.wwnorton.com/physics

FIGURE 19.17
Gas cylinders.

23. Gas-storage cylinders that can maintain 400 atm pressure are commercially available. Treat nitrogen gas at such a high pressure approximately as an ideal gas. How does its mass density compare with that of liquid nitrogen? (Liquid nitrogen has a density of 800 kg/m³.)

24. A tube of argon gas at STP is sealed and placed in an oven at 850°C. What is the pressure in the gas at that temperature?

25. Carbon dioxide can be pressurized to 56 atm in the gaseous state at 25°C. What is the mass density of such high-pressure carbon dioxide vapor?

*26. Suppose you pour 10 g of water into a 1.0-liter jar and seal it tightly. You then place the jar into an oven and heat it to 500°C (a dangerous thing to do!). What will be the pressure of the vaporized water?

*27. A scuba diver releases an air bubble of diameter 1.0 cm at a depth of 15 m below the surface of a lake. What will be the diameter of this bubble when it reaches the surface? Assume that the temperature of the bubble remains constant.

*28. The helium atom has a volume of about 3.0×10^{-30} m³. What fraction of a volume of helium gas at STP is actually occupied by atoms?

29. A carbon dioxide (CO_2) fire extinguisher has an interior volume of 2.8×10^{-3} m³. The extinguisher has a mass of 5.9 kg when empty and a mass of 8.2 kg when fully loaded with CO_2. At a temperature of 20°C, what is the pressure of CO_2 in the extinguisher?

*30. (a) When you heat the air in a house, some air escapes because the pressure inside the house must remain the same as the pressure outside. Suppose you heat the air from 10°C to 30°C. What fraction of the mass of air originally inside will escape?

   (b) If the house were completely airtight, the pressure would have to increase as you heated the house. Suppose that the initial pressure inside the house is 1.00 atm. What is the final pressure? What force does this excess inside pressure exert on a window 1.0 m high and 1.0 m wide? Do you think the window can withstand this force?

*31. During the volcanic eruption of Mt. Pelée on the island of Martinique in 1902, a *nuée ardente* (burning cloud) of very hot gas and fine suspended ash rolled down the side of the volcano and killed the 30000 inhabitants of Saint-Pierre. The temper-

ature in the cloud has been estimated at 700°C. Treat this cloud as a gas of high molecular mass. What must have been this molecular mass to make this cloud as dense as, or denser than, the surrounding air (at 20°C)?

*32. A typical hot-air balloon has a volume of 2200 m³ and a mass of 730 kg (including balloon, gondola, four passengers, and a propane tank). Since the balloon is open at the bottom, the pressures of the internal and the external air are (approximately) equal. If the temperature of the external air is 20°C, what must be the minimum temperature of the internal air in the balloon to achieve liftoff? The density of the external air is 1.20 kg/m³.

*33. A research balloon (Fig. 19.18) ascends to an altitude of 40 km and floats in equilibrium. The pressure (outside and also inside the balloon) is 3.2×10^2 N/m², and the temperature is -13°C. The volume of the balloon is 8.5×10^5 m³, and it is filled with helium. What payload (including the mass of the fabric but excluding the helium) can this balloon carry? What was the volume of the balloon on the ground (at STP), before it was released?

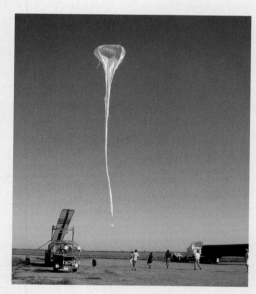

FIGURE 19.18 A research balloon just after launch.

*34. A sunken ship of steel is to be raised by making the upper part of the hull airtight and then pumping compressed air into it while letting the water escape through holes in the bottom. The mass of the ship is 50000 metric tons, and it is at a depth of 60 m. How much compressed air (in kilograms) must be pumped into the ship? The temperature of the air and the water is 15°C.

*35. A diving bell is a cylinder closed at the top and open at the bottom; when it is immersed in the water, any air initially in the cylinder remains trapped in the cylinder. Suppose that such a diving bell, 2.0 m high and 1.5 m across, is initially full of air and is immersed to a depth of 15 m measured from water level to water level (see Fig. 19.19).

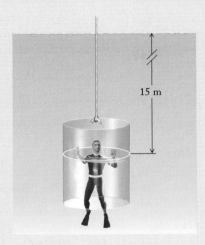

FIGURE 19.19 Submerged diving bell.

(a) How high will the water have risen within the diving bell?

(b) If compressed air is pumped into the bell, water will be expelled from the bell. How much air (in kilograms) must be pumped into the bell, and at what pressure, to get rid of all of the water? Assume that the temperature of the air is 15°C.

*36. At high altitudes, pilots and mountain climbers must breathe an enriched mixture containing more oxygen than the standard concentration of 23% found in ordinary air at sea level. At an altitude of 11 000 m, the atmospheric pressure is 0.24 atm. What oxygen concentration is required at this altitude if with each breath the same number of oxygen molecules is to enter the lungs as for ordinary air at sea level?

*37. Air is 75.54% nitrogen (N_2), 23.1% oxygen (O_2), and 1.3% argon (Ar) by mass. From this information and from the molecular masses of N_2, O_2, and Ar, deduce the mean molecular mass of air.

*38. (a) The gas at the center of the Sun is 38% hydrogen and 62% helium at a temperature of 1.50×10^7 K and a density of 1.48×10^5 kg/m^3. What is the pressure?

(b) The gas at a distance of 20% of the solar radius from the center of the Sun is 71% hydrogen and 29% helium at a temperature of 9.0×10^6 K and a density of 3.6×10^4 kg/m^3. What is the pressure?

*39. The pressure change with height given in Eq. (18.12), $p - p_0 = -\rho g y$, is not valid for a compressible fluid, like an ideal gas, except for small changes dp and dy in p and y. Show that at constant temperature, this and the Ideal-Gas Law lead to the relation

$$\frac{dp}{p} = -\frac{Mg}{RT} dy$$

where M is the mass of one mole of the gas.

**40. Show that if the temperature in the atmosphere is independent of altitude, then the pressure as a function of altitude y is

$$p = p_0 e^{-mgy/kT}$$

where m is the average mass per molecule of air. See Problem 39. (This formula is applicable only for altitudes less than about 2 km; higher up, the temperature depends on the altitude.)

19.3 Kinetic Pressure

41. What is the rms speed of molecules of water vapor in air at 0°C?

42. What is the rms speed of hydrogen ions on the surface of the Sun, where the temperature is 4.5×10^3 K? At the center of the Sun, where the temperature is 1.5×10^7 K?

43. In Example 7 we calculated the rms speed of nitrogen and of oxygen molecules in air at 0°C. If we want to increase these rms speeds by a factor of 2, what temperature do we need?

44. Consider separate samples of nitrogen gas and oxygen gas. The temperature of the nitrogen gas is 20°C. What must be the temperature of the oxygen gas if the rms speed of the oxygen molecules is to equal the rms speed of the nitrogen molecules?

45. Free neutrons in the core of a nuclear reactor have a temperature of 400 K. What are the rms speed and the average kinetic energy of such neutrons?

46. The fireball of a 1.0-megaton nuclear explosion attains a temperature of 7000 K. It contains ionized gas and free electrons. What are the rms speed and the average kinetic energy of free electrons in the fireball? What are the rms speed and the average kinetic energy of nitrogen ions?

47. In order to achieve a novel state of matter known as a **Bose–Einstein condensate**, a gas of rubidium atoms is cooled to $T = 5.0 \times 10^{-5}$ K. What is the rms speed of a rubidium atom at this temperature?

48. According to Eq. (17.5) the speed of sound in air is $\sqrt{1.4p/\rho}$.

(a) Show by means of the Ideal-Gas Law that this expression equals $\sqrt{1.4kT/m}$, where m is the average mass per molecule of air.

(b) Show that, in terms of the rms speed, the latter expression equals $\sqrt{1.4/3}v_{\text{rms}}$, or $0.68v_{\text{rms}}$.

(c) Calculate the speed of sound in air at 0°C, 10°C, 20°C, and 30°C.

49. What is the average kinetic energy of an oxygen molecule in air at STP? A nitrogen molecule?

50. What is the rms speed of a helium atom at 0°C? At −269°C?

51. At the top of the stratosphere, at an altitude of 30 km, the temperature is −38°C. What is the rms speed of an oxygen molecule at −38°C? Of an ozone (O_3) molecule? What are the average kinetic energies for these molecules?

52. The rms speed of nitrogen molecules in air at some temperature is 493 m/s. What is the rms speed of hydrogen molecules in air at the same temperature?

53. One method for the separation of the rare isotope ^{235}U (used in nuclear bombs and reactors) from the abundant isotope ^{238}U relies on diffusion through porous membranes. Both isotopes are first made into a gas of uranium hexafluoride (UF_6). The molecules of $^{235}UF_6$ have a higher rms speed and they

will diffuse faster through a porous membrane than the molecules of $^{238}\mathrm{UF}_6$. The "molecular masses" of $^{235}\mathrm{UF}_6$ and $^{238}\mathrm{UF}_6$ are 349 g and 352 g, respectively. What is the percent difference between their rms speeds at a given temperature?

54. Calculate the rms velocity of molecules of oxygen gas (O_2) and hydrogen gas (H_2) at room temperature; also, find the ratio of these two velocities.

55. By shaking containers of small grains or beads, some fluidlike behavior can be observed. If an evacuated container with a low density of small spheres of mass 1.0×10^{-9} g is shaken so the beads have an rms velocity of 2.0×10^{-2} m/s, what is the effective ideal-gas temperature of such a system?

56. Helium liquid at temperature 0.90 K (below its standard boiling point, achieved by **evaporative cooling**) is in equilibrium with 0.042 torr of helium gas pressure. What is the rms velocity of helium gas molecules at this temperature?

57. Using laser beams, physicists can cool a small amount of a gas of sodium atoms to extremely low temperature. Determine the rms speed of a sodium atom when such **laser cooling** results in a temperature of $T = 2.0 \times 10^{-4}$ K.

*58. For semiconductor fabrication, **ultrahigh vacuum** (UHV) is often needed to prevent surface contamination (Fig. 19.20). The time t for a surface to become appreciably contaminated by gas particles is inversely proportional to the particle flux $(N/V)v_{rms}/(2\sqrt{3})$ and to the surface area per particle a^2:

$$t = \frac{2\sqrt{3}}{(N/V)v_{rms}a^2}$$

where typically $a^2 = (3.0 \times 10^{-10}\ \mathrm{m})^2 = 9.0 \times 10^{-20}\ \mathrm{m}^2$. For room temperature, calculate this contamination time for hydrogen gas (a) at 1.0 atm, (b) at an "ordinary" vacuum pressure of 1.0×10^{-6} torr, and (c) at a UHV pressure of 1.0×10^{-11} torr.

FIGURE 19.20 An ultrahigh-vacuum (UHV) chamber.

59. Real gas molecules experience collisions with each other; the number of such collisions depends on the size of the gas molecule and the density N/V of the gas. The average distance the molecule travels before suffering one collision is called the **mean free path, denoted l. If we assume the molecules are

spheres of radius R_0, a collision occurs when two molecules come within a distance $2R_0$ of each other. By considering the volume swept out by a moving molecule, show that the mean free path is given by

$$l = \frac{1}{4\pi R_0^2(N/V)}$$

(Hint: For one collision, equate the effective cylindrical volume of radius $2R_0$ swept out by one molecule with the average volume per molecule.)

*60. The average distance an atom in a fluid travels before suffering a collision with another atom is called the **mean free path** l, given by (see Problem 59)

$$l = \frac{1}{4\pi R_0^2(N/V)}$$

where R_0 is the radius of the (spherical) atom.
(a) Calculate the mean free path of a helium atom in helium gas with $R_0 = 1.3 \times 10^{-10}$ m under STP conditions.
(b) On the average, how many collisions does the atom make per second?
(c) How many collisions do all the atoms in 1.0 cm^3 of helium gas make per second?

*61. The behavior of a real gas deviates from the Ideal-Gas Law, particularly at high density. There are two primary effects: the nonzero size of molecules decreases the available volume, and the long-range attractive forces of molecules decrease the momentum transfer to container walls. These effects are included in the **van der Waals equation of state**,

$$\left[p + A\left(\frac{N}{V}\right)^2\right](V - NV_m) = nRT$$

where A is a constant that depends on the attractive interaction, N and V are the total number of molecules and the total volume, and V_m is an effective volume of a molecule. Neglecting the pressure effect (set $A = 0$), calculate the percentage volume correction for helium atoms with $V_m = 3.7 \times 10^{-29}$ m^3 at room temperature for (a) $p = 1.0$ atm and (b) $p = 1000$ atm.

*62. In our calculation of the pressure on the walls of a box (see Section 19.3) we have ignored gravity. If we take gravity into account, the pressure on the bottom of the box will be greater than that at the top. Show that the pressure difference is $p - p_0 = (N/V)mgL$. (Hint: When a molecule falls from the top to the bottom, its speed increases according to $v^2 - v_0^2 = 2gL$.)

19.4 The Energy of an Ideal Gas

63. What is the thermal kinetic energy in 1.0 kg of oxygen gas at 20°C? What fraction of this energy is translational?

64. Assume that air consists of the diatomic gases O_2 and N_2. How much must we increase the thermal energy of 1.0 kg of air in order to increase its temperature by 1.0°C?

65. What is the thermal kinetic energy of 1.0 mole of helium gas at 300 K? How much does this kinetic energy increase if we increase the temperature by 20 K? If we increase the pressure by 3.0 atm (at fixed temperature)?

66. A 1.0-liter vessel contains a monatomic gas under STP conditions. If 50 J of energy is added to the gas (at constant volume), what is the new temperature of the gas? The new pressure?

67. A gas cylinder contains 30 liters of diatomic nitrogen gas at 273 K and a pressure of 140 atm. If the temperature is increased to 300 K, how much will the internal energy of the gas increase?

*68. Water vapor consists of nonlinear polyatomic molecules, so all three rotational modes are excited at ordinary temperatures.

What is the thermal kinetic energy of 1.0 mole of water vapor at 100°C?

*69. The two vibrational degrees of freedom (one kinetic, one potential) of diatomic hydrogen gas can be excited at very high temperatures (thousands of kelvins), resulting in seven total degrees of freedom. For one mole of hydrogen gas at such high temperatures, how much energy must be added to increase the temperature by 10 K?

*70. A container is divided into two equal compartments by a partition. One compartment is initially filled with helium at a temperature of 250 K; the other is filled with oxygen at a temperature of 310 K. Both gases are at the same pressure. If we remove the partition and allow the gases to mix, what will be their final temperature?

REVIEW PROBLEMS

71 What is the molecular mass of methanol, CH_3OH? What is the number of molecules in 1.0 kg of methanol?

72. A bicycle pump is a cylinder of diameter 2.5 cm and length 30 cm. Initially, when the piston is fully pulled out, the cylinder is filled with air at 25°C and 1.0 atm. If you slowly compress the air to half its initial volume, what force must you exert on the piston to hold it in the compressed position? Assume that the temperature of the air remains constant and that the valve is blocked, so no air escapes.

73. An airplane flies through air at a temperature of 5°C. The lift force generated by the flow of air over the wings is 1.2×10^3 N. What would be the lift force if the airplane were flying through air at a temperature of 35°C, other conditions remaining equal?

74. An oxygen cylinder for medical use contains oxygen at a pressure of 140 atm, at room temperature (20°C). The cylinder measures 20 cm in diameter and 110 cm in length. How many kilograms of oxygen does this cylinder contain? What will be the volume of the oxygen if it is allowed to expand slowly to normal atmospheric pressure, at room temperature?

75. The volume of air in the fully expanded human lungs is 5.0 liters. How many molecules are in the lungs? How many molecules of oxygen and how many of nitrogen? Assume that the air is 76% nitrogen and 24% oxygen by mass, at a temperature of 37°C and a pressure of 1.0 atm.

76. An inflatable life jacket, loosely inflated, provide a buoyant force of 50 N when completely immersed just below the surface of a lake. What buoyant force does this life jacket provide if you push it to a depth of 2.0 m below the surface? Assume that the initial pressure in the life jacket is 1.0 atm, and that the temperature of the air does not change when you push the life jacket down.

77. At the center of the Sun, the temperature is 1.5×10^7 K and the density is 1.5×10^5 kg/m^3.

   (a) Assume that the material in the Sun is a mixture of equal numbers of hydrogen ions and free electrons. Find the number of particles per unit volume, and find the pressure.

   (b) What would you have found for the pressure if you had assumed that the material of the Sun consists of hydrogen atoms (without free electrons)?

   (c) What if you had assumed that the material in the Sun consists of H_2 molecules?

78. In a mercury barometer (see Fig. 19.21), the space at the upper end of the tube is supposed to be evacuated. Suppose that by some mistake, a barometer has some small amount of air in this space. A barometer with this defect always under-reads the atmospheric pressure.

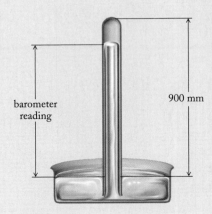

barometer reading

900 mm

FIGURE 19.21 A mercury barometer.

(a) Suppose that when the actual atmospheric pressure is 760 mm-Hg, the barometer reads 750 mm-Hg. What will this barometer read when the actual pressure increases to 780 mm-Hg? The length of the barometer tube is 900 mm (see Fig. 19.21), and the temperature of the air remains constant.

(b) Suppose that while the pressure increases as in part (a), the temperature decreases from 300 K to 270 K. What will the barometer read in this case?

79. In one method for the determination of the average density of the tissues of the human body, the subject is locked in a hermetic chamber of known volume V_c (see Fig. 19.22) containing an unknown volume V of air at an initial pressure p. Then, by means of a small piston, the volume of this chamber is reduced by an amount ΔV. This causes an increase of pressure Δp (at constant temperature). Show that the volume of *air* in the chamber is approximately given by

$$V = \frac{p}{\Delta p}\Delta V$$

and that therefore the volume of the body of the subject is given by

$$V_s = V_c - \frac{p}{\Delta p}\Delta V$$

[Hint: $pV = $ [constant], and hence $pV = (p + \Delta p)(V - \Delta V)$.]

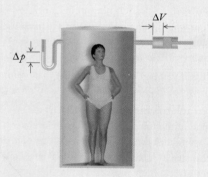

FIGURE 19.22 A volunteer locked in a hermetic chamber.

80. Estimate the number of impacts of air molecules per second on the palm of your hand (area 80 cm^2). Assume that the air is at 20°C and 1.0 atm, and assume that it consists entirely of nitrogen molecules.

81. What is the rms speed of molecules of water vapor in air at 0°C?

82. At a time of 100 000 years after the first instant of the Big Bang at the beginning of the Universe, the temperature was about 1.0×10^4 K, and the density of the hot gas filling the Universe was about 1.0×10^{-16} kg/m^3. By mass, this gas consisted of 75% H atoms (not molecules) and 25% He atoms.

(a) Calculate the number of atoms per unit volume.

(b) Calculate the pressure contributed by these atoms.

(c) Calculate the rms speed of the H atoms and of the He atoms.

83. A sample of gas has some initial pressure p, volume V, and temperature T. By what factor does the rms speed of its molecules increase or decrease if we do one of the following:

(a) Increase the temperature to $2T$?

(b) Decrease the volume to $V/3$?

(c) Increase the volume to $4V$ and simultaneously decrease the pressure to $p/2$?

(d) Decrease the temperature to $T/4$ and simultaneously decrease the pressure to $p/3$?

84. At the center of the Sun, the temperature is 1.5×10^7 K. What are the rms speed and the average kinetic energy of hydrogen ions? What are the rms speed and the average kinetic energy of helium ions?

85. Two moles of H_2 gas react with 1.0 mole of O_2 gas to form 2.0 moles of water vapor. If the initial temperature of the H_2 and O_2 gases is 300 K, what final temperature must we give to the water vapor, if its thermal kinetic energy is to equal the initial kinetic energy?

Answers to Checkups

Checkup 19.1

1. Since $pV = nRT$, the pressure is proportional to the temperature; when the temperature increases from 300 K to 600 K (a factor of 2), the pressure also increases a factor of 2.

2. Since $pV = nRT$, the pressure p is proportional to the density n/V. So when the pressure increases a factor of 1.05, the density also increases a factor of 1.05.

3. Since $pV = nRT$, or $p = (n/V)RT$, when the temperature increases at constant pressure, the density must decrease by the same factor, here by a factor of 270/290.

4. Since $pV = nRT$, the density (n/V) is proportional to the absolute pressure. For an *overpressure* of 2 atm, the absolute pressure is 3 atm, so the density is a factor of 3 larger inside the tire than in the surrounding air.

5. (E) $\frac{1}{3}$. Since the Ideal-Gas Law, $pV = nRT$, says that volume is directly proportional to temperature and inversely proportional to pressure, an increase in pressure and a decrease in temperature will each cause the volume to decrease by the corresponding factor, so overall it will decrease by $\frac{2}{3} \times \frac{1}{2} = \frac{1}{3}$.

Checkup 19.2

1. Since 0°C corresponds to 32°F, any negative Fahrenheit temperature will be far below this and will always correspond to a negative Celsius temperature. But temperatures slightly below 0°C will be negative on the Celsius scale and positive on the Fahrenheit scale, so the answer to the second question is no.

2. To determine which are the extreme temperatures, convert each to a common unit. In degrees Celsius, the first is $T_C = 320\,K - 273.15°C \approx 47°C$. From Fig. 19.13, we see that the last temperature is 90°F $\approx$ 32°C. Thus the first temperature (320 K) is the hottest, and the second temperature (20°C) is the coldest.

3. Since the thermometer is designed to maintain constant pressure, the volume of the gas would be proportional to temperature ($pV = nRT$). Thus the temperature could be determined by measuring the height of the piston, which determines the volume of the gas.

4. (C) $-40°C$. From Fig. 19.13, we see that the numerical values on the two scales coincide only at $-40°C$, a result that may also be obtained by equating the numerical values of T_C and T_F in Eq. (19.12).

Checkup 19.3

1. Both. For any given molecule, the root-mean-square speed will increase with temperature [Eq. (19.23)]. Thus it will make the round trip between impacts more quickly, increasing the rate of impacts in proportion to the speed. The transferred

momentum per impact, $2mv$, will also increase in proportion to the speed. Thus the two effects contribute equally.

2. The rms speed is proportional to the square root of the temperature [Eq. (19.23)], so an increase in T from 300 K to 600 K (a factor of 2) increases v_{rms} a factor of $\sqrt{2}$.

3. The "molecular mass" of oxygen molecules, O_2, is $2 \times 16 = 32$, which is larger than that of water, H_2O, which is $(2 \times 1) + 16 = 18$. From Eq. (19.23), $v_{rms} = \sqrt{3\,kT/m}$, we see that the lighter molecule, water, will have the higher rms speed.

4. (D) $\frac{1}{4}$. The Ideal-Gas Law, $p = (N/V)kT$, indicates that for the same pressure, the number density (N/V) will change inversely with temperature; thus N/V will be $200/50 = 4$ times larger for hydrogen gas. But the mass per unit volume is the number density times the mass per molecule; the latter is $m_{H_2}/m_{O_2} = 2/32 = 1/16$ times as large for hydrogen. Thus the mass per unit volume ratio will be $4 \times (1/16) = 1/4$.

Checkup 19.4

1. The thermal energy [see Eq. (19.27) or (19.33)] is proportional to temperature, so an increase to 400 K from 300 K increases the thermal energy a factor of 4/3.

2. The diatomic molecule has two rotational components of motion that can store energy at ordinary temperatures, in addition to the usual three translational components of motion.

3. As the gas expanded at constant pressure, it would apply a force to the moving piston, thus doing work.

4. (C) 2, 1, 3. The total energy is lowest for the monatomic gas (2), which has only translational kinetic energy, $\frac{3}{2}nRT$. With two rotational components of motion, the diatomic gas (1) has the next higher energy, $\frac{5}{2}nRT$. Finally, a nonlinear, polyatomic gas (3) has three rotational components of motion and thus the highest total energy, $3nRT$.

20.1 Heat as a Form of Energy Transfer

20.2 Thermal Expansion of Solids and Liquids

20.3 Thermal Conduction

20.4 Changes of State

20.5 The Specific Heat of a Gas

20.6 Adiabatic Expansion of a Gas

CONCEPTS IN CONTEXT

Concepts
— *in* —
Context

This image of the intensity of infrared energy emitted by a house is known as a *thermograph*; the colors indicate the different levels of heat loss through various parts of the building.

In this chapter, we will consider questions such as

? At what rate does thermal energy flow through a brick wall? (Example 6, page 639)

? How well does additional insulation reduce the flow of thermal energy through the wall? (Example 7, page 640)

? The flow of thermal energy through a window can be reduced by using two layers of glass with a layer of gas between them. What gas is best? (Checkup 20.5, Question 2, page 647)

In everyday language, heat is what makes things hot. When we place a kettle full of water on a stove, the water absorbs heat from the stove and becomes hot. But in the precise language of physics, what makes the water hot is **thermal energy**, *that is, the kinetic and potential energy of the random microscopic motions of molecules, atoms, ions, electrons, and other particles.* When the water is in contact with the hot stove, the atoms of the stove communicate some of their violent random microscopic motions to the water molecules. Thus, the thermal energy of the water molecules increases—they bounce around more violently than before. At the macroscopic level, such an increase of the energy or the random microscopic motions manifests itself as an increase of the temperature of the water.

In the language of physics, **heat** *is thermal energy transferred from a hotter body to a colder body.* The relationship of heat to thermal energy is analogous to the relationship of work to mechanical energy we studied in Chapter 7. Work done on a particle increases the mechanical energy of the particle. Thus, work is mechanical energy transferred by a force. Likewise, heat is thermal energy transferred by a temperature difference. This analogy between heat and work is not merely formal. In fact, heat can be regarded as microscopic work done by the particles in the hotter body on the particles in the colder body, and this microscopic work accomplishes the transfer of thermal energy. Although in a strict sense heat is a transfer of thermal energy, physicists sometimes use the word *heat* in a loose sense as a synonym for *thermal energy.* Thus, we speak of heat flow, heat storage, heat loss, etc., when there is a flow of thermal energy, storage of thermal energy, loss of thermal energy, etc.

The ambiguity in the usage of the word *heat* arises from historical roots. Until well into the nineteenth century, scientists did not have a clear understanding of the concept of energy, and they thought that heat was an invisible, weightless fluid, which they called "caloric." The first experiments to give conclusive evidence of the nature of heat were performed by Benjamin Thompson, Count Rumford, who showed that the mechanical energy lost in friction is converted into heat. You can verify such a frictional conversion of mechanical energy into heat by rubbing your hands against each other—a few seconds of rubbing produces a noticeable warming.

In this chapter, we will examine various effects caused by the application of heat in materials: increase of temperature, expansion of length or volume, thermal conduction, melting, and vaporization.

BENJAMIN THOMPSON, COUNT RUMFORD (1753–1814)
American–British scientist. On the basis of experimental observations that he collected while supervising the boring of cannon, Rumford argued against the prevailing view that heat is a substance, and he proposed that heat is nothing but the random microscopic motion of the particles within a body. **Robert von Mayer** *(1814–1878), German physician and physicist, calculated the mechanical equivalent of heat by comparing the work done on a gas during compression with the consequent increase of temperature. Finally,* **J. P. Joule** *(1818–1889) measured this quantity directly by means of his famous experiment.*

20.1 HEAT AS A FORM OF ENERGY TRANSFER

Online
Concept
Tutorial

We examined the connection between random microscopic motion and temperature in the preceding chapter, where we saw that the increase of the kinetic energy of the random microscopic motions of the molecules in a gas is directly proportional to the increase of temperature [see Eq. (19.26)]. In a liquid or a solid, the kinetic energy of the random microscopic motions also increases with the temperature. Furthermore, the atoms and molecules in a liquid or a solid have potential energies associated with the forces they exert on one another; these potential energies also increase with temperature. Thus, the microscopic view of thermal energy as kinetic and potential energy of the random motions of atoms and molecules agrees with the intuitive notion that absorption of thermal energy should lead to an increase of temperature.

Long before physicists recognized that heat is the transfer of kinetic and potential energy of the random microscopic motion of atoms, they had defined heat in terms of the temperature changes it produces in a body. A traditional, but non-SI, unit of heat is the **calorie** (cal), which was originally defined as the amount of heat needed to raise the temperature of 1 g of water by 1°C. The kilocalorie is 1000 cal,

$$1 \text{ kcal} = 1000 \text{ cal} \tag{20.1}$$

Incidentally: The "calories" marked on some packages of food in grocery stores are actually kilocalories, also called large calories. Sometimes this is made more explicit by use of a capital letter: 1 Cal = 1 kcal.

In the British system of units, the unit of heat is the **British thermal unit** (Btu), which is the heat needed to raise the temperature of 1 lb of water by 1°F. The relationship between this unit and the kcal is

$$1 \text{ Btu} = 0.252 \text{ kcal}$$

The heat necessary to raise the temperature of 1 kg *of a material by* 1°C *is called the* **specific heat capacity,** *or the* **specific heat,** usually designated by the symbol *c*. Thus, water has a specific heat of

$$c = 1.00 \frac{\text{kcal}}{\text{kg·°C}} \tag{20.2}$$

Table 20.1 lists the specific heats of some common substances. Note that water has a larger specific heat than all the other substances listed in Table 20.1. This means that, per kilogram, a temperature change in water requires more heat than an equal temperature change in these other substances. We might say that water has a large "thermal inertia"—it is capable of storing a large thermal energy with a small change of temperature. This makes water very useful for the storage and transport of thermal energy, for instance, in the heating system of a house (where water carries thermal energy from the boiler to the radiators) and in the cooling system of an automobile engine (where water carries thermal energy from the engine block to the radiator).

The specific heat of most substances varies slightly with temperature. For example, the specific heat of water varies by about 1% between 0°C and 100°C, reaching a minimum at 35°C.[1] Finally, the specific heat depends somewhat on the pressure to which the material is subjected during the heating. All the values listed in Table 20.1 were obtained at room temperature (20°C) and at a constant pressure of 1.0 atm.

The values in Table 20.1 give the amount of heat required to increase the temperature of 1 kg of a given substance by 1°C. For a mass *m* of this substance, the

TABLE 20.1   SOME SPECIFIC HEATS[a]

SUBSTANCE	c	
Aluminum	0.215 kcal/kg·°C	902 J/kg·°C
Brass	0.092	390
Copper	0.092	390
Iron, steel	0.106	445
Lead	0.031	130
Tin	0.054	230
Silver	0.056	240
Mercury	0.033	140
Water	1.000	4187
Seawater	0.93	3900
Ice, −10°C	0.530	2230
Ethyl alcohol	0.581	2430
Glycol	0.571	2390
Mineral oil	0.5	2000
Glass, thermometer	0.20	840
Marble	0.21	880
Granite	0.19	800

[a] At room temperature (20°C) and 1 atm, unless otherwise noted.

[1]This variation must be taken into account for precise definitions: a calorie is the heat needed to raise the temperature of 1 g of water from 14.5°C to 15.5°C; a British thermal unit is the heat necessary to raise the temperature of 1 lb of water from 63°F to 64°F.

amount of heat Q and the increase of temperature ΔT are related by

$$Q = mc\,\Delta T \tag{20.3}$$

heat and temperature change

This merely says that a large mass or a large temperature change requires more heat, in proportion to the mass or the temperature change.

EXAMPLE 1   You pour 0.10 kg of water at 20°C into an aluminum pot of 0.20 kg at the same temperature. How much heat must you supply to bring the water and the pot to a temperature of 100°C? (Neglect any heating of the environment.)

SOLUTION: The temperature change is $\Delta T = 80°C$. Hence the heat absorbed by the water is

$$Q_{water} = m_{water}\, c_{water}\, \Delta T = 0.10\text{ kg} \times 1.00\text{ kcal/(kg}\cdot°C) \times 80°C = 8.0\text{ kcal}$$

The specific heat of aluminum is $c_{Al} = 0.215$ kcal/(kg·°C) (see Table 20.1). Hence the heat absorbed by the aluminum is

$$Q_{Al} = m_{Al}\, c_{Al}\, \Delta T = 0.20\text{ kg} \times 0.215\text{ kcal/(kg}\cdot°C) \times 80°C = 3.4\text{ kcal}$$

The net heat absorbed by the water and the pot is then

$$Q_{total} = Q_{water} + Q_{Al} = 8.0\text{ kcal} + 3.4\text{ kcal} = 11.4\text{ kcal}$$

Since heat is a form of work, it can be transformed into macroscopic mechanical work and vice versa. The transformation of heat into work is accomplished by a steam engine, a steam turbine, or a similar machine; we will examine the theory of such heat engines in the next chapter. The transformation of work into heat requires no special machinery—any kind of friction will convert work into heat. Since heat is a form of energy transfer, the calorie is a unit of energy, and it must be possible to express it in joules. The conversion factor between these units is called the **mechanical equivalent of heat**.

The traditional method for the measurement of the mechanical equivalent of heat is Joule's experiment. A set of falling weights drives a paddle wheel that churns the water in a bucket (see Fig. 20.1). The bucket is surrounded by insulation, so no heat can escape from it. The friction inherent in the churning raises the temperature of the water in the bucket by a measurable amount, converting the initial gravitational potential energy of the falling weights into a measurable amount of heat. The best available experimental results for this conversion of mechanical energy into heat give

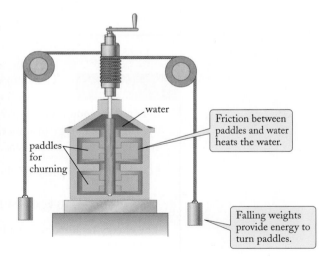

water

Friction between paddles and water heats the water.

paddles for churning

Falling weights provide energy to turn paddles.

FIGURE 20.1 Joule's apparatus.

$$1\text{ cal} = 4.187\text{ J} \tag{20.4}$$

mechanical equivalent of heat

for the mechanical equivalent of heat.

In the modern SI system of units, the calorie is taken as equal to 4.187 J by definition. This means that Joule's experiment is not needed anymore to find the mechanical equivalent of heat; instead, it is needed to determine the specific heat of water, which now must be regarded as a quantity to be measured experimentally. Hereafter we will mostly use joules to measure thermal energies and only occasionally revert to calories.

EXAMPLE 2   When an automobile is braking, the friction between the brake drums and the brake shoes converts translational kinetic energy into heat. If a 2000-kg automobile brakes from 25 m/s to 0 m/s, how much heat is generated in the brakes? If each of the four brake drums has a mass of 9.0 kg of iron of specific heat 450 J/(kg·°C), how much does the temperature of the brake drums rise? Assume that all the heat accumulates in the brake drums (there is not enough time for the heat to leak away into the air, and not much heat goes into the brake shoes).

SOLUTION: The initial kinetic energy of the automobile is

$$K = \tfrac{1}{2}\, mv^2 = \tfrac{1}{2} \times 2000 \text{ kg} \times (25 \text{m/s})^2 = 6.3 \times 10^5 \text{ J}$$

The brakes convert this kinetic energy into heat. Each of the four brake drums absorbs one-fourth of the total, or $Q = \tfrac{1}{4} \times 6.3 \times 10^5$ J. Since the mass of each brake drum is 9.0 kg, the temperature increase of each brake drum is, according to Eq. (20.3),

$$\Delta T = \frac{Q}{mc}$$

$$= \frac{\tfrac{1}{4} \times 6.3 \times 10^5 \text{ J}}{9.0 \text{ kg} \times 450 \text{ J/(kg·°C)}} = 39°\text{C} \qquad (20.5)$$

EXAMPLE 3   The complete metabolization of one apple supplies 110 kcal (110 Cal) of chemical energy. How high can you climb up a hill on this amount of energy? Assume that your muscles can completely convert the chemical energy into mechanical energy and that there is no frictional loss, and assume that your mass is 75 kg.

SOLUTION: In joules, the chemical energy is

$$110 \text{ kcal} \times 4.187 \times 10^3 \text{ J/kcal} = 4.6 \times 10^5 \text{ J}$$

The energy required to lift a body of mass m to a height y is $\Delta U = mgy$ [see Eq. (7.29)]. Hence

$$mgy = 4.6 \times 10^5 \text{ J}$$

and

$$y = \frac{4.6 \times 10^5 \text{ J}}{75 \text{ kg} \times 9.81 \text{ m/s}^2} = 6.3 \times 10^2 \text{ m} = 630 \text{ m}$$

COMMENT: In practice, the height you can climb with the energy supplied by one apple is much less, maybe 100 m. Your metabolic system fails to extract all of the available chemical energy, your muscles fail to convert all of the extracted chemical energy into mechanical energy, and finally there are frictional losses.

Checkup 20.1

QUESTION 1: You mix 1.0 kg of water at 80°C with 1.0 kg of water at 20°C. What is the final temperature?

QUESTION 2: The specific heat of iron is about one-ninth that of water. By how many degrees will the temperature of 1 kg of iron increase if you supply 1 kcal of heat to it?

QUESTION 3: Consider the following process: A laborer pushes a heavy crate over a rough but level floor at constant speed. Friction heats the bottom of the crate. Does the laborer do work on the crate? Does he transfer (or remove) mechanical energy to (or from) the crate? Does he transfer thermal energy to the crate? Does the floor do work on the crate? Does the crate do work on the floor?

QUESTION 4: As in Joule's experiment, suppose a weight of 1000 N slowly falls a distance of 4.187 m. The falling weight turns a paddle that churns 10 kg of water. Assuming ideal conditions, what is the change in water temperature?

(A) 0.010°C   (B) 0.10°C   (C) 1.0°C   (D) 10°C   (E) 100°C

20.2 THERMAL EXPANSION OF SOLIDS AND LIQUIDS

As we saw in the preceding chapter, if the pressure is held constant, the volume of a given amount of gas will increase with the temperature [see Eq. (19.4)]. Such an increase of volume with temperature also occurs for solids and liquids; this phenomenon is called **thermal expansion**. However, the thermal expansion of solids and of liquids is much less than that of gases. For example, if we raise the temperature of a piece of iron by 100°C, we will increase its volume by only 0.36%. During the expansion, the solid retains its shape, but all its dimensions increase in proportion. Figure 20.2a illustrates the expansion of a piece of metal; for the sake of clarity, the expansion has been exaggerated. An expanding liquid does not retain its shape; the liquid will merely fill more of the container that holds it. Figure 20.2b illustrates the thermal expansion of a liquid.

From a microscopic point of view, the thermal expansion of solids and liquids is due to the increase of thermal motion caused by the increase of temperature—in a solid the speed of the back-and-forth motions of the atoms about their equilibrium positions increases with temperature, and in a liquid the bouncing zigzag motions of the molecules increase with temperature. This increase of the random motions tends to push the atoms or molecules apart, and therefore leads to an increase of the volume of the solid or liquid.

thermal expansion

(a)   (b)

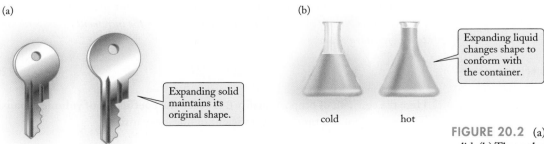

Expanding solid maintains its original shape.

Expanding liquid changes shape to conform with the container.

cold   hot

cold   hot

FIGURE 20.2   (a) Thermal expansion of a solid. (b) Thermal expansion of a liquid. The expansion of the flask has been neglected.

The thermal expansion of a solid can be best described mathematically by the increase in the linear dimensions of the solid (see Fig. 20.3). For most solids and for a broad range of temperatures near room temperature, *the increment ΔL in the length L is directly proportional to the increment of temperature and to the original length*:

linear expansion

$$\Delta L = \alpha L \Delta T \qquad (20.6)$$

The constant of proportionality α in this equation is called the **coefficient of linear expansion.** Table 20.2 lists the value of this coefficient for a few materials.

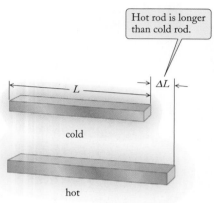

cold

hot

FIGURE 20.3 Increase of the length of a solid rod by thermal expansion. The initial length is L; the final length is $L + \Delta L$.

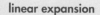
Hot rod is longer than cold rod.

TABLE 20.2	COEFFICIENTS OF EXPANSION^a			
SOLIDS	α **(linear)**		**LIQUIDS**	β **(volume)**
Lead	$29 \times 10^{-6}/°C$		Alcohol, ethyl (99%)	$1.01 \times 10^{-3}/°C$
Aluminum	24		Carbon tetrachloride	1.18
Brass	19		Ether	1.51
Copper	17		Gasoline	0.95
Iron, steel	≈ 12		Glycerine	0.49
Concrete	≈ 12		Olive oil	0.68
Glass	9.0		Mercury	0.182
Pyrex	3.6			
Quartz, fused	0.50			

^{a} At room temperature (20°C).

EXAMPLE 4   The highest tower in the world is the steel radio mast of Warsaw Radio in Poland, which has a height of 646 m. How much does its height increase between a cold winter day when the temperature is $-35°C$ and a hot summer day when the temperature is $+35°C$?

SOLUTION: The increment of temperature is $\Delta T = 70°C$. With a value of $\alpha = 12 \times 10^{-6}/°C$ for steel (see Table 20.2), we then find

$$\Delta L = \alpha L \Delta T = \frac{12 \times 10^{-6}}{°C} \times 646 \text{ m} \times 70°C = 0.54 \text{ m} \qquad (20.7)$$

Similarly, for temperatures near room temperature, for most materials *the increment in the volume of the solid is directly proportional to the increment of temperature and to the original volume*:

volume expansion

$$\Delta V = \beta V \Delta T \qquad (20.8)$$

Here the constant of proportionality β is called the **coefficient of volume expansion.** This coefficient is 3 times the coefficient of linear expansion:

$$\beta = 3\alpha \qquad (20.9)$$

To see how this relationship comes about, suppose that the solid has the shape of a cube of side L (see Fig. 20.4). The increment in the length of each side is ΔL [Eq. (20.6)], and treating this as a small (infinitesimal) quantity, the increment in the volume L^3 is

$$\Delta V = \Delta(L^3) = 3L^2 \Delta L$$

$$= 3L^2 \times \alpha L \Delta T \qquad (20.10)$$

$$= 3\alpha L^3 \Delta T = 3\alpha V \Delta T$$

Comparing this result for ΔV with Eq. (20.8), we see that, indeed, $\beta = 3\alpha$.

The increment in the volume of a liquid can be described by the same equation [Eq. (20.8)] as the increment in the volume of a solid. Table 20.2 gives values of coefficients of volume expansion for some liquids.

Water has not been included in this table because its behavior is quite peculiar: from 0°C to 4°C, the volume *decreases* with increasing temperature, but not uniformly; above 4°C the volume increases with increasing temperature. Figure 20.5 gives the volume of 1 kg of water for temperatures ranging from 0°C to 10°C. The strange behavior of the density of water at low temperatures can be traced to the crystal structure of ice. Water molecules have a rather angular shape that prevents a tight fit of these molecules; when they assemble in a solid, they adopt a very complicated crystal structure with large gaps. As a result, ice has a lower density than water—the density of ice is 917 kg/m³, and the volume of 1 kg of ice is 1091 cm³. At a temperature slightly above the freezing point, water is liquid, but some of the water molecules already have assembled themselves into microscopic (and short-lived) ice crystals; these microscopic ice crystals give the cold water an excess volume.

The maximum in the density of water at about 4°C has an important consequence for the ecology of lakes. In winter, the layer of water on the surface of the lake cools, becomes denser than the lower layer, and sinks to the bottom. This process continues until the temperature of the entire body of the lake reaches 4°C. Beyond this point, the cooling of the surface layer will make it *less dense* than the lower layers; thus, the surface layer stays in place, floating on the top of the lake. Ultimately, this surface layer freezes, becoming a solid sheet of ice while the body of the lake remains at 4°C. The sheet of ice inhibits the heat loss from the lake, especially if covered with an insulating blanket of snow. Besides, any further heat loss merely causes some thickening of the sheet of ice, without disturbing the deeper layers of water, which remain at a stable temperature of 4°C—fish and other aquatic life can survive the winter in this stable environment (Fig. 20.6).

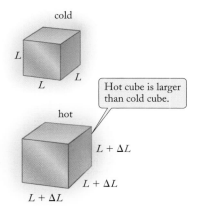

FIGURE 20.4 Thermal expansion of a solid cube. The initial volume is $L \times L \times L$; the final volume is $(L + \Delta L) \times (L + \Delta L) \times (L + \Delta L) = L^3 + 3L^2\Delta L + 3L(\Delta L)^2 + (\Delta L)^3 \approx L^3 + 3L^2\Delta L$.

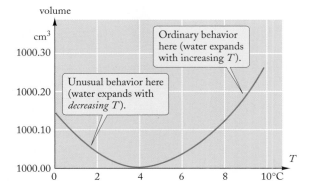

FIGURE 20.5 Volume of 1 kg of water as a function of temperature.

FIGURE 20.6 Fish under ice. Water is unusual: solid ice is *less* dense than liquid water and floats.

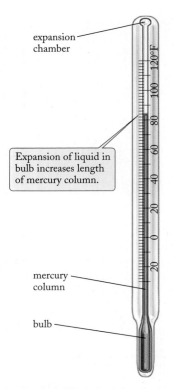

expansion
chamber

Expansion of liquid in
bulb increases length
of mercury column.

mercury
column

bulb

FIGURE 20.7 Mercury-bulb thermometer.

EXAMPLE 5   A glass vessel of volume 200 cm³ is filled to the rim with mercury (atomic symbol Hg). How much of the mercury will overflow the vessel if we raise the temperature by 30.0°C?

SOLUTION: The volume of mercury will increase by

$$\Delta V_{Hg} = \beta_{Hg}V\,\Delta T$$
$$= 0.182 \times 10^{-3}/°C \times 200\text{ cm}^3 \times 30.0°C = 1.09\text{ cm}^3 \tag{20.11}$$

The volume of the glass vessel will also increase, just as though all of the vessel were filled with glass (as in Fig. 20.2, the hole, or cavity, in the vessel expands as though it were completely filled with glass); hence

$$\Delta V_{glass} = \beta_{glass}V\Delta T = 3\alpha_{glass}V\Delta T \tag{20.12}$$
$$= 3 \times 9.0 \times 10^{-6}/°C \times 200\text{ cm}^3 \times 30.0°C = 0.16\text{ cm}^3$$

The difference

$$1.09\text{ cm}^3 - 0.16\text{ cm}^3 = 0.93\text{ cm}^3$$

is the volume of mercury that will overflow.

Ordinary thermometers and thermostats make use of thermal expansion to sense changes in temperature. The mercury-bulb thermometer (see Fig. 20.7) consists of a glass bulb filled with mercury connected to a thin capillary tube. Thermal expansion makes the mercury overflow into the capillary tube and increases the length of the mercury column; this length indicates the temperature. The bimetallic-strip thermometer (see Fig. 20.8) consists of two parallel strips of different metals—such as aluminum and iron—welded together and curled into a spiral. The differential thermal expansion increases the length of one side of the welded strip more than that of the other side. This causes the strip to curl up more tightly and rotates the upper end of the spiral relative to the lower end; a pointer attached to the upper end indicates the temperature.

PROBLEM-SOLVING TECHNIQUES

TEMPERATURE UNITS; THERMAL EXPANSION

- The temperature units in the tables of specific heats and coefficients of expansion in this chapter are degrees Celsius (°C). The temperature difference ΔT used in the calculations must therefore also be expressed in °C, not in °F. However, since the Celsius and the absolute temperature scales differ only by an additive constant (273.15°C), any temperature difference ΔT is the same in °C and in K, and these two units can be used interchangeably in any such temperature difference.

- If the linear dimensions of a solid body expand by, say, 0.001%, then the area of the body (or any part of the body)

expands by about 0.002%, and the volume of the body expands by about 0.003%; that is, the linear, area, and volume fractional expansions are in a ratio of 1:2:3.

- During the thermal expansion of a solid body with holes or cavities, such as the glass vessel in Example 5, the hole or cavity expands just as if it were filled with the same solid material—if the solid body expands by, say, 0.003% in volume, the hole or cavity likewise expands by 0.003% in volume.

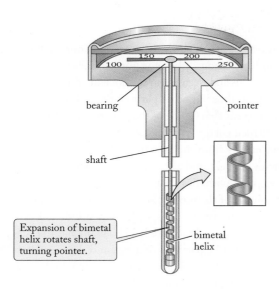

FIGURE 20.8 Bimetallic-strip thermometer.

FIGURE 20.9
Expansion joints in
deck of a bridge.

Thermal expansion must be taken into account in the design of long structures, such as bridges or railroad tracks. The decks of bridges usually have several expansion joints with gaps (see Fig. 20.9) that permit changes of length and prevent the bridge from buckling. Likewise, gaps are left between the segments of rail in a railroad track; but if the temperature changes exceed the expectations of the designers, the results can be disastrous (see Fig. 20.10).

Incidentally: Our ability to erect large buildings and other structures out of reinforced concrete hinges on the fortuitous coincidence of the coefficients of expansion of iron and concrete (see Table 20.2). Reinforced concrete consists of iron rods in a concrete matrix. If the coefficients of expansion for these two materials were appreciably different, the daily and seasonal temperature changes would cause the iron rods to move relative to the concrete—ultimately, the iron rods would work loose, and the reinforcement would come to an end.

✔ Checkup 20.2

FIGURE 20.10 Buckling of railroad rails due to heat.

QUESTION 1: Suppose that when we heat an aluminum rod, its length increases by 0.02%. What is the corresponding percent increase of the volume of the rod?

QUESTION 2: According to Fig. 20.5, what is the percent decrease of the volume of 1 kg of water between 0°C and 4°C?

QUESTION 3: An engineer proposes to build a bridge out of concrete reinforced with aluminum rods. What is wrong with this proposal?

QUESTION 4: Consider the thermal expansion of the key illustrated in Fig. 20.2a. During the thermal expansion, does the size of the hole in this key increase, decrease, or stay the same?

  (A) Increases   (B) Decreases   (C) Stays the same

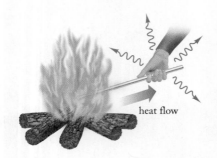

FIGURE 20.11 Heat flows from the hot end of the rod to the cold end.

20.3 THERMAL CONDUCTION

If you stick one end of an iron rod into a fire and hold the other end in your hand (see Fig. 20.11), you will feel the end in your hand gradually become warmer. This is an example of heat transfer by **conduction**. The atoms and electrons in the hot end of the rod have greater kinetic and potential energies than those in other parts of the rod. In random collisions, these energetic atoms and electrons share some of their energy with their less energetic neighbors; these, in turn, share their energy with their neighbors, and so on. The result is a gradual diffusion of thermal energy from the hot end to the cold end.

Most metals are excellent conductors of heat, and also excellent conductors of electricity. The high thermal and electric conductivities of a metal are due to an abundance of "free" electrons within the volume of the metal; these are electrons that have become detached from their atoms—they move at high speeds, they wander all over the volume of the metal with little hindrance, and they are held back only at the surface of the metal. The free electrons behave like particles of a gas, and the metal acts like a bottle holding this gas. Typically, a free electron will move past a few hundred atoms before it suffers a collision. Because the electrons move such fairly large distances between collisions, they quickly transport energy from one end of a metallic rod to the other. The motion of the free electrons transports the thermal energy much more efficiently than does the back-and-forth vibrational motion of the atoms.

heat flow

Quantitatively, we can describe the transport of heat by the **heat flow**, *or the heat current; this is the amount of heat that passes by some given place on the rod per unit time.* We will use the symbol $\Delta Q/\Delta t$ for heat flow. In the SI system, the unit of heat flow is the joule per second (J/s); however, in practice, the calorie per second (cal/s) and the British thermal unit per hour (Btu/h) are also used.

Consider a rod of cross-sectional area A and length Δx (see Fig. 20.12). Assume that the cold end of the rod is kept at a constant temperature T_1 and the hot end at a constant temperature T_2, so the difference of temperature between the ends is $\Delta T = T_2 - T_1$. If the ends are kept at these constant temperatures for a while, then the temperatures at all other points of the rod will settle to final steady values. Under such steady-state conditions, the heat flow of the rod is found to be directly proportional to the temperature difference ΔT and to the cross-sectional area A, and inversely proportional to the length Δx:

heat conduction

$$\frac{\Delta Q}{\Delta t} = k A \frac{\Delta T}{\Delta x} \qquad (20.13)$$

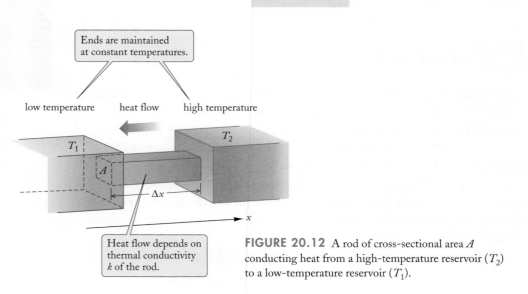

Ends are maintained at constant temperatures.

low temperature   heat flow   high temperature

Heat flow depends on thermal conductivity k of the rod.

FIGURE 20.12 A rod of cross-sectional area A conducting heat from a high-temperature reservoir (T_2) to a low-temperature reservoir (T_1).

TABLE 20.3	SOME THERMAL CONDUCTIVITIES[a]	
SUBSTANCE		k
Silver	102 cal/(s·m·°C)	427 J/(s·m·°C)
Copper	95	398
Aluminum	57	237
Iron, cast	11	46
Steel	11	46
Lead	8.3	35
Ice, 0°C	0.3	1.3
Snow, 0°C, compact	0.05	0.2
Glass, crown	0.25	1.0
Porcelain	0.25	1.0
Concrete	0.2	0.8
Brick	0.15	0.63
Wood (pine, across grain)	0.03	0.13
Fiberglass (batten)	0.010	0.042
Down	0.005	0.02
Styrofoam	0.002	0.008

[a] At room temperature (20°C), unless otherwise noted.

The direction of the heat flow is, of course, from the hot end of the rod toward the cold end. The constant of proportionality k in our equation is called the **thermal conductivity**. Table 20.3 lists values of k for some materials.[2]

heat conduction

EXAMPLE 6 A house is built of bricks, with walls 20 cm thick. The wall in one of the rooms of this house measures 5.0 m × 3.0 m (see Fig. 20.13). What is the heat flow through this wall if the inside temperature is 21°C and the outside temperature −18°C?

Concepts —in— **Context**

SOLUTION: The temperature change across the wall is $\Delta T = 39°C$, the thickness of the wall is $\Delta x = 20$ cm $= 0.20$ m, and the area of the wall is 5.0 m × 3.0 m $= 15$ m². Hence, with $k = 0.63$ J/(s·m·°C) from Table 20.3, Eq. (20.13) gives

$$\frac{\Delta Q}{\Delta t} = k A \frac{\Delta T}{\Delta x}$$

$$= 0.63 \text{ J/(s·m·°C)} \times 15 \text{ m}^2 \times \frac{39°C}{0.20 \text{ m}} \qquad (20.14)$$

$$= 1.8 \times 10^3 \text{ J/s}$$

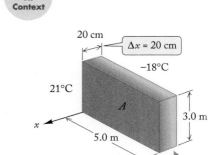

FIGURE 20.13 Heat flow through the brick wall of a house.

[2] The thermal conductivity must not be confused with the Boltzmann constant; both these quantities are designated with the same letter k, but they are not related.

Concepts
— in —
Context

EXAMPLE 7   To reduce the heat loss, the owner of the brick house described in Example 6 covers the brick wall with a 12-cm layer of fiberglass insulation (see Fig. 20.14). What is the heat loss now?

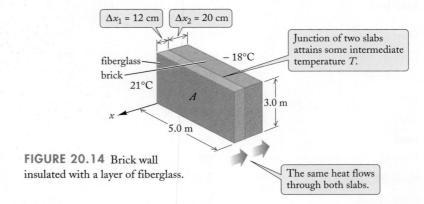

FIGURE 20.14 Brick wall insulated with a layer of fiberglass.

SOLUTION: The wall now consists of two layers. For steady-state conditions, heat does not accumulate anywhere; any heat that enters the first layer reaches the second, and hence the heat flow through each layer is the same. However, the temperature differences across the two layers are different, ΔT_1 for the brick layer and ΔT_2 for the fiberglass. The sum of these temperature differences must equal the net temperature change $\Delta T = 39°C$,

$$\Delta T_1 + \Delta T_2 = \Delta T \tag{20.15}$$

The heat flows through the brick and the fiberglass are, respectively,

$$\frac{\Delta Q_1}{\Delta t} = k_1 A \frac{\Delta T_1}{\Delta x_1} \quad \text{and} \quad \frac{\Delta Q_2}{\Delta t} = k_2 A \frac{\Delta T_2}{\Delta x_2}$$

Since, as discussed above, these heat flows must be equal,

$$k_1 A \frac{\Delta T_1}{\Delta x_1} = k_2 A \frac{\Delta T_2}{\Delta x_2} \tag{20.16}$$

Solving Eq. (20.16) for ΔT_2 and substituting into Eq. (20.15), we obtain

$$\Delta T_1 + \frac{k_1}{k_2} \frac{\Delta x_2}{\Delta x_1} \Delta T_1 = \Delta T$$

From this we find ΔT_1:

$$\Delta T_1 = \frac{\Delta T}{1 + (k_1 \Delta x_2)/(k_2 \Delta x_1)}$$

The heat flow is then

$$\frac{\Delta Q}{\Delta t} = k_1 A \frac{\Delta T_1}{\Delta x_1} = k_1 A \frac{\Delta T}{\Delta x_1 [1 + (k_1 \Delta x_2)/(k_2 \Delta x_1)]}$$

or more simply

two-layer heat conduction

$$\frac{\Delta Q}{\Delta t} = A \frac{\Delta T}{\Delta x_1/k_1 + \Delta x_2/k_2} \tag{20.17}$$

With $k_1 = 0.63$ J/(s·m·°C) and $k_2 = 0.042$ J/(s·m·°C), the heat flow is

$$\frac{\Delta Q}{\Delta t} = A \frac{\Delta T}{\Delta x_1/k_1 \ + \ \Delta x_2/k_2}$$

$$= 15 \ \text{m}^2 \times \frac{39°\text{C}}{[0.20 \ \text{m}/(0.63 \ \text{J/s·m·°C})] \ + \ [0.12 \ \text{m}/(0.042 \ \text{J/s·m·°C})]}$$

$$= 1.8 \times 10^2 \ \text{J/s}$$

Thus, the extra insulation reduces the heat loss of this wall by a factor of 10.

COMMENT: From the denominator of Eq. (20.17), we see that the heat flow is inversely proportional to the sum $\Delta x_1/k_1 + \Delta x_2/k_2$. In the description of home insulation, the quantity $\Delta x/k$ is commonly referred to as the **R value**, and it indicates the resistance that a layer offers to heat flow. Thus, for example, our 20-cm-thick brick wall has an R value of $(0.20 \ \text{m})/[0.63 \ \text{J}/(\text{s·m·°C})] = 0.32 \ \text{s·m}^2\text{·°C/J}$. In common engineering practice, a mixed system of units is often used, where R values are expressed in $\text{ft}^2\text{·°F·h/Btu}$; the conversion factor between these units is $1 \ \text{s·m}^2\text{·°C/J} = 5.54 \ \text{ft}^2\text{·°F·h/Btu}$.

Besides conduction, there are two other mechanisms of heat transfer: convection and radiation. *In* **convection***, the heat is stored in a moving fluid and is carried from one place to another by the motion of this fluid. In* **radiation***, the heat is carried from one place to another by electromagnetic waves*—for example, light waves, infrared waves, or radio waves. All three mechanisms of heat transfer are neatly illustrated by the operation of a hot-water heating system in a house. In this system, the heat is carried from the boiler to the radiators in the rooms by means of water flowing in pipes (convection); the heat then diffuses through the metallic walls of the radiators (conduction) and finally spreads from the surface of the radiators into the volume of the room (radiation and also convection of the air heated by direct contact with the radiators).

Radiation is the only mechanism of heat transfer that can carry heat through a vacuum; for instance, the heat of the Sun reaches the Earth by radiation. We will study thermal radiation in Chapter 37.

convection and radiation

Checkup 20.3

QUESTION 1: The walls of houses built in the Northeastern United States are commonly insulated with 6 in. (15 cm) of fiberglass insulation. If you wanted to achieve the same insulation with solid wood walls, what thickness of wood would you need?

QUESTION 2: A piece of ice at 0°C is in contact with a piece of steel. If heat is flowing from the ice into the steel, what can you say about the temperature of the steel?

QUESTION 3: Does each of the following changes increase or decrease the heat loss through the wall of the house calculated in Example 6: reduce wall thickness to 10 cm; increase wall area to 20 m²; reduce external temperature to −20°C; reduce internal temperature to 19°C?

QUESTION 4: The following devices are used to deliver heat to the human body: hair dryer, heat lamp, hot-water bottle. By which respective mechanism does each transfer heat?

(A) Conduction, convection, radiation   (B) Conduction, radiation, convection
(C) Convection, conduction, radiation   (D) Convection, radiation, conduction
(E) Radiation, convection, conduction

Online
Concept
Tutorial

20.4 CHANGES OF STATE

Heat absorbed by a body will not only increase the temperature, but it will also bring about a change of state from solid to liquid or from liquid to gas when the body reaches its melting point or its boiling point. At the melting temperature or the boiling temperature, the thermal motion of the atoms and molecules becomes so violent that the bonds holding them in the solid or liquid loosen or break. The loosening of the strong bonds in a solid transforms it into a liquid, and the breaking of the remaining weak bonds in a liquid transforms it into a gas.

While the body is melting or boiling, it absorbs some amount of heat without any increase of temperature. This heat represents the energy required to loosen and break the bonds that hold the atoms inside the solid or liquid. *The heat absorbed during the change of state is called the* **latent heat** *or the* **heat of transformation**, *and more specifically, the* **heat of fusion** *or the* **heat of vaporization**, for the change of state from solid to liquid or from liquid to gas, respectively. Table 20.4 lists the heats of fusion and vaporization for a few substances (at a pressure of 1 atm).

The quantities listed in Table 20.4 depend on the pressure. The decrease of the boiling point of water with a decrease of atmospheric pressure is a phenomenon familiar to people living at high altitude; for instance, in Denver, Colorado, at an altitude of 1600 m, the mean pressure is 0.83 atm, and the boiling point of water is 95°C.

heat of transformation

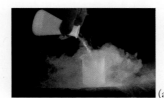

(a)

(b)

(c)

TABLE 20.4	**HEATS OF FUSION AND VAPORIZATION**[a]			
SUBSTANCE	MELTING POINT	HEAT OF FUSION	BOILING POINT	HEAT OF VAPORIZATION
Water	0°C	3.34×10^5 J/kg	100°C	2.26×10^6 J/kg
Nitrogen (a)	−210	2.6×10^4	−196	2.00×10^5
Oxygen	−218	1.4×10^4	−183	2.1×10^5
Helium	—	—	−269	2.06×10^4
Hydrogen	−259	6.3×10^4	−253	4.5×10^5
Aluminum	660	3.99×10^5	2467	1.1×10^7
Copper	1083	2.05×10^5	2567	5.2×10^6
Iron (b)	1535	2.7×10^5	2750	6.8×10^6
Lead	328	2.9×10^4	1740	8.5×10^5
Tin	232	5.9×10^4	2270	1.9×10^6
Silver	962	9.9×10^4	2212	2.4×10^6
Tungsten	3410	1.8×10^5	5660	4.9×10^6
Mercury	−39	1.1×10^4	357	2.92×10^5
Carbon dioxide[b] (c)	−79	—	—	5.8×10^5

[a] At a pressure of 1 atm.
[b] Undergoes direct vaporization (sublimation) from solid to gas.

EXAMPLE 8

How many ice cubes (at 0°C) must be added to a bowl containing 1.00 liter of boiling water at 100°C so that the resulting mixture reaches a temperature of 40°C? Assume that each ice cube has a mass of 20 g and that the bowl and the environment do not exchange heat with the water, and assume that the average specific heat of water is 4.2×10^3 J/(kg·°C).

SOLUTION: Since all of the heat released by the water is absorbed by the ice, the amount of heat released by the hot water during cooling must equal the amount of heat absorbed by the ice during melting and during the subsequent heating of the molten ice from 0°C to 40°C. Thus we can write an expression for each of these amounts of heat and equate them.

The heat released by the hot water during cooling from 100°C to 40°C is, according to Eq. (20.3),

$$Q = mc\,\Delta T = 1.00 \text{ kg} \times 4.2 \times 10^3 \text{ J/(kg·°C)} \times 60°C = 2.5 \times 10^5 \text{ J}$$

If the total mass of ice is m, then, from Table 20.4, the heat absorbed by this mass during melting is $m \times 3.34 \times 10^5$ J/kg, and the heat absorbed during the subsequent heating from 0°C to 40°C is $m \times 4.2 \times 10^3$ J/(kg·°C) $\times \Delta T = m \times 4.2 \times 10^3$ J/(kg·°C) $\times 40°C$. Hence the total heat absorbed is

$$Q = m \times 3.34 \times 10^5 \text{ J/kg} + m \times 4.2 \times 10^3 \text{ J/(kg·°C)} \times 40°C$$

$$= m \times (3.34 \times 10^5 \text{ J/kg} + 1.7 \times 10^5 \text{ J/kg}) = m \times (5.0 \times 10^5 \text{ J/kg})$$

These amounts of heat must be equal:

$$m \times (5.0 \times 10^5 \text{ J/kg}) = 2.5 \times 10^5 \text{ J}$$

Solving this for m, we find that the mass of ice required is

$$m = 0.50 \text{ kg}$$

Since each ice cube has a mass of 0.020 kg, this is

$$\frac{0.50 \text{ kg}}{0.020 \text{ kg/ice cube}} = 25 \text{ ice cubes}$$

 Checkup 20.4

QUESTION 1: Which of the materials listed in Table 20.4 are gases at a temperature of −200°C? Which are liquids? Which are solids?

QUESTION 2: Which of the materials listed in Table 20.4 releases the largest amount of heat (per kg) when it freezes?

QUESTION 3: For the substances listed in Table 20.4, is more heat required for the melting or for the vaporization of a given mass of substance?

QUESTION 4: Place the following in increasing order of the amount of heat required: (a) vaporizing 1.0 kg of water; (b) melting 1.0 kg of ice; and (c) heating 1.0 kg of water from 0°C to 100°C.

(A) a, b, c   (B) a, c, b   (C) b, a, c   (D) b, c, a   (E) c, b, a

20.5 THE SPECIFIC HEAT OF A GAS

If you heat a gas, the increase of temperature causes an increase of the pressure, and this tends to bring about an expansion of the gas. You can observe this expansion if you leave a loosely inflated beach ball or plastic bag in the sun; the beach ball soon becomes taut because the air warms and expands. The value of the specific heat of a gas depends on whether the container permits expansion during heating. If the container is perfectly rigid, the heating proceeds at constant volume (see Fig. 20.15). For gases it is customary to reckon the specific heat *per mole*, rather than per kilogram. The **molar specific heat at constant volume** is designated by C_V; it is the amount of heat needed to raise the temperature of 1 mole of gas by 1°C. The amount of heat Q required to increase the temperature of n moles by ΔT is proportional to n and to ΔT:

$$Q = nC_V \Delta T \tag{20.18}$$

This equation resembles Eq. (20.3), but the number of moles n appears instead of the mass m, because we now are reckoning the specific heat per mole.

If the container is fitted with a vertical piston whose weight presses down on the gas, the heating proceeds at a constant pressure determined by the weight of the piston and its area A (see Fig. 20.16). The **molar specific heat at constant pressure** is designated by C_p. For n moles of gas, the heat absorbed and the temperature increase are related by

$$Q = nC_p \Delta T \tag{20.19}$$

We expect C_p to be larger than C_V because, if we supply some amount of heat to the gas in Fig. 20.16, only part of this heat will go into a temperature increase of the gas; the rest will be converted into work as the expanding gas lifts the piston. Let us calculate the difference between the two heat capacities.

At constant volume, the gas does no work. Hence all the heat absorbed will go into the energy of the gas; if the gas absorbs a small amount of heat dQ, the energy increment dE must match dQ:

$$dQ = dE \tag{20.20}$$

or, according to Eq. (20.18),

$$nC_V \, dT = dE \tag{20.21}$$

At constant pressure, the gas does work against the moving piston. Suppose that the piston is displaced a small (infinitesimal) distance dx (see Fig. 20.17). The force of the gas on the piston is pA and the work done by the gas, $dW = F \, dx$, is

$$dW = pA \, dx \tag{20.22}$$

The product $A \, dx$ is simply the small change dV of the volume of the gas. Hence

$$dW = p \, dV \tag{20.23}$$

The heat absorbed by the gas must provide both the energy increase of the gas and the work done by the gas:

$$dQ = dE + dW = dE + p \, dV \tag{20.24}$$

or, according to Eq. (20.19),

$$nC_p \, dT = dE + p \, dV \tag{20.25}$$

In an ideal gas, the energy E depends on the temperature only [see Eqs. (19.26) and (19.33)]; consequently, if the temperature increment at constant pressure has the same

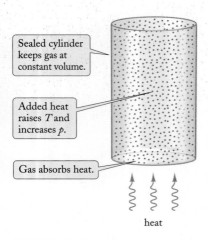

Sealed cylinder keeps gas at constant volume.

Added heat raises T and increases p.

Gas absorbs heat.

heat

FIGURE 20.15 A gas kept at constant volume while being heated.

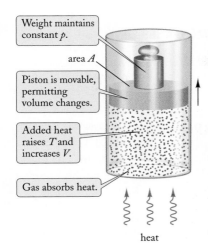

FIGURE 20.16 A gas kept at constant pressure while being heated. The pressure equals the atmospheric pressure plus the combined weight of the piston and the load divided by the area A.

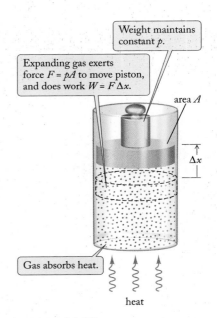

FIGURE 20.17 Expansion of a gas. During its expansion, the gas displaces the piston and performs work.

value as the temperature increment at constant volume, the increase dE of the energy must be the same. We can therefore insert the expression for dE given by Eq. (20.21) into Eq. (20.25) and obtain

$$nC_p \, dT = nC_V \, dT + p \, dV \tag{20.26}$$

From the Ideal-Gas Law [Eq. (19.2)], we find that, at constant pressure, the changes of volume and of temperature are related by

$$p \, dV = nR \, dT \tag{20.27}$$

With this, Eq. (20.26) becomes

$$nC_p \, dT = nC_V \, dT + nR \, dT \tag{20.28}$$

and, canceling the factors $n \, dT$ in this equation, we obtain the final result

$$C_p = C_V + R \tag{20.29}$$

molar specific heat relation

The numerical value of R is 8.31 J/K·mole, or 1.99 cal/K·mole. Hence, Eq. (20.29) shows that C_p is larger than C_V by about 8.3 J/K·mole—*it takes about 8.3 more joules to heat a mole of gas by 1°C at constant pressure than at constant volume.*

Note that although the above general argument did permit us to evaluate the difference between C_p and C_V, it does not permit us to find the individual values of C_p and C_V. For this, we must know something about the energy stored in the internal rotational or vibrational motions of the molecules. For a monatomic gas there is no such extra energy, and according to Eq. (19.27),

$$dE = \tfrac{3}{2}nR \, dT \tag{20.30}$$

By the definition of C_V [Eq. (20.21)], this leads to

$$C_V = \tfrac{3}{2}R \quad \text{(monatomic gas)} \tag{20.31}$$

and by Eq. (20.29)

$$C_p = C_V + R = \tfrac{3}{2}R + R = \tfrac{5}{2}R \quad \text{(monatomic gas)} \tag{20.32}$$

For a diatomic gas, the rotational energy of the molecules results in a larger value of E [see Eq. (19.33)], and consequently a larger value of C_V and C_p:

$$C_V = \tfrac{5}{2}R \quad \text{and} \quad C_p = \tfrac{7}{2}R \quad \text{(diatomic gas)} \tag{20.33}$$

TABLE 20.5	SPECIFIC HEATS OF SOME GASES[a]			
GAS	C_V	C_p	$C_p - C_V$	$\gamma = C_p/C_V$
Monatomic				
Helium (He)	12.5 J/mole·K	20.8 J/mole·K	8.3 J/mol·K	1.66
Argon (Ar)	12.5	20.9	8.4	1.67
Diatomic				
Nitrogen (N_2)	20.8	29.1	8.3	1.40
Oxygen (O_2)	20.8	29.1	8.3	1.40
Carbon monoxide (CO)	20.7	29.1	8.4	1.41
Polyatomic				
Ammonia (NH_3)	27.3	35.8	8.5	1.31
Methane (CH_4)	27.1	35.5	8.4	1.31

[a] At STP.

Table 20.5 lists the values of the specific heats of some gases. In the cases of monatomic and diatomic gases, these values are in reasonable agreement with Eqs. (20.31)–(20.33), although there are some minor deviations because the gases are not quite ideal gases. Note that in all cases the difference $C_p - C_V$ agrees quite well with Eq. (20.29).

EXAMPLE 9   During a sunny day, the sunlight warms the ground, which in turn warms the air in contact with it. How many joules must the ground supply to heat an initial volume of 1.00 m³of air from 0.0°C to 10.0°C? The atmospheric pressure is steady at 1.00 atm.

SOLUTION: Since the atmosphere surrounding the given amount of air provides a constant pressure, the relevant specific heat is C_p, the specific heat at constant pressure. Air is mostly N_2 and O_2, and Table 20.5 tells us that for N_2 and for O_2, $C_p = 29.1$ J/°C·mole. This must then also be the right value of C_p for any mixture of these two gases.

During the heating, the volume of air expands, but we can calculate the number of moles from the initial volume and temperature. For this calculation, we can either use the Ideal-Gas Law or, more simply, the known volume of 22.4 liters for one mole at STP (Example 1 in Chapter 19). By proportions, a volume of 1.00 m³, or 1000 liters, contains $n = (1000 \text{ liters})/(22.4 \text{ liters/mole}) = 44.6$ moles. Hence the amount of heat absorbed by the air is

$$Q = nC_p \, \Delta T = 44.6 \text{ moles} \times 29.1 \text{ J/°C·mole} \times 10.0°C = 1.30 \times 10^4 \text{ J}$$

(20.34)

This is a fairly small amount of heat (for comparison, if we wanted to heat a volume of 1 m³ of water by 10°C, we would need to supply about 4×10^7 J).

Checkup 20.5

QUESTION 1: Why is the specific heat of a gas at constant pressure larger than the specific heat at constant volume?

QUESTION 2: The flow of thermal energy through a window can be reduced by using two layers of glass with a layer of gas between them. Under appropriate conditions, the thermal conductivity of a gas is proportional to the specific heat of the gas and to the rms speed of the gas molecules. Under such conditions, which of the gases in Table 20.5 would best reduce the flow of thermal energy?

QUESTION 3: A rubber balloon and a sealed glass jar contain equal amounts of air. For equal increases of temperature, which requires more heat? Is the difference in the specific heats 8.31 J/°C·mole, or is it smaller?

QUESTION 4: Under what conditions might the specific heat of a gas exceed C_V by more than 8.31 J/°C·mole?

QUESTION 5: Heat is added to one mole of air at constant pressure, resulting in a temperature increase of 100°C. If the same amount of heat is instead added at constant volume, what is the temperature increase?

(A) 50°C   (B) 60°C   (C) 71°C   (D) 140°C   (E) 167°C

20.6 ADIABATIC EXPANSION OF A GAS

If an amount of gas at high pressure and temperature is placed in a container fitted with a piston, the gas will push the piston outward and do work on it. Such a process of expansion against a piston converts thermal energy into useful mechanical energy—the temperature of the gas decreases as it delivers work to the piston. This process is at the core of the operation of steam engines, automobile engines, and other heat engines.

In this section we will investigate the equation for the expansion of a gas. We will assume that the gas is thermally insulated (see Fig. 20.18), so it neither receives heat from its environment nor loses any. The temperature change of the gas is then entirely due to the work that the gas does on its environment. Such *a process occurring without the exchange of heat with the environment is called* **adiabatic**.

If the volume of gas increases by a small amount dV, the work done by the gas on the piston is [see Eq. (20.23)]

$$dW = p\,dV \tag{20.35}$$

The heat absorbed is zero; hence the change of energy of the gas is entirely due to the work done by the gas [see also Eq. (20.24)]:

$$dE = -dW = -p\,dV \quad \text{(adiabatic process)} \tag{20.36}$$

The change of energy can also be expressed in terms of the temperature change [see Eq. (20.21) and the discussion after Eq. (20.25)]:

$$dE = nC_V\,dT \tag{20.37}$$

Combining Eqs. (20.36) and (20.37), we find

$$nC_V\,dT = -p\,dV \tag{20.38}$$

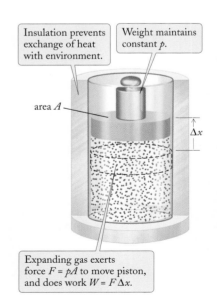

FIGURE 20.18 Expansion of a thermally insulated gas. The insulation prevents the escape of heat from the gas.

According to the Ideal-Gas Law, the pressure is $p = nRT/V$. Thus,

$$nC_V \, dT = -nRT \, \frac{dV}{V} \tag{20.39}$$

If we divide both sides of this equation by nTC_V, we obtain

temperature change in
adiabatic expansion

$$\frac{dT}{T} = -\frac{R}{C_V} \frac{dV}{V} \tag{20.40}$$

This shows that the fractional decrease of temperature (dT/T) is directly proportional to the fractional increase of volume (dV/V). For monatomic gases, $C_V = \frac{3}{2}R$ [see Eq. (20.31)]; hence $R/C_V = \frac{2}{3}$ and, consequently, $dT/T = -\frac{2}{3}dV/V$. This means that a 3% increase of volume of the gas will produce a 2% decrease of the absolute temperature.

The simple proportionality implied by Eq. (20.40) is valid only for *small* changes of volume and temperature; for large changes, we must integrate Eq. (20.40). We first define the quantity γ (the Greek letter *gamma*) to be the ratio of specific heats at constant pressure and constant volume:

$$\gamma = \frac{C_p}{C_V} \tag{20.41}$$

From Eqs. (20.31)–(20.33), we see that for a monatomic ideal gas, $\gamma = \frac{5}{3} \approx 1.67$ and for a diatomic ideal gas, $\gamma = \frac{7}{5} = 1.40$. Measured values of γ are given in Table 20.5; these are quite close to the ideal values. Since $R = C_p - C_V$ [Eq. (20.29)], we see that

$$\frac{R}{C_V} = \frac{C_p}{C_V} - 1 = \gamma - 1 \tag{20.42}$$

We substitute this form into Eq. (20.40) and integrate from an initial temperature T_1 and volume V_1 to a final temperature T_2 and volume V_2:

$$\int_{T_1}^{T_2} \frac{dT}{T} = -(\gamma - 1) \int_{V_1}^{V_2} \frac{dV}{V} \tag{20.43}$$

The integrals of $1/T$ and $1/V$ are the natural logarithms $\ln T$ and $\ln V$, respectively:

$$\ln\left(\frac{T_2}{T_1}\right) = -(\gamma - 1) \ln\left(\frac{V_2}{V_1}\right) \tag{20.44}$$

where we have used the property of logarithms $\ln A - \ln B = \ln(A/B)$. Taking the exponential of both sides, using $\exp(\ln x) = x$, and rearranging, we obtain

temperature and volume in
adiabatic process

$$T_1 V_1^{\gamma-1} = T_2 V_2^{\gamma-1} \tag{20.45}$$

Equation (20.45) can be used to relate any changes in temperature and volume during an adiabatic process. It is easy to see that if the volume increases, the temperature must drop.

The decrease of temperature of an expanding gas can be perceived quite readily when air is allowed to rush out of the valve of an automobile tire; this expanding air feels quite cool. This process is approximately adiabatic because the expanding air, although not insulated from its surroundings, expands so quickly that it does not have time to exchange heat with the surrounding atmospheric air. Conversely, the increase of temperature during the adiabatic compression of air can be perceived when operating a manual air pump. The compression of air in the barrel of the pump produces a noticeable warming of the pump.

Finally, we use the Ideal-Gas Law to eliminate T from Eq. (20.45); substituting $T = pV/nR$, we have

$$\frac{p_1 V_1}{nR} V_1^{\gamma-1} = \frac{p_2 V_2}{nR} V_2^{\gamma-1} \qquad (20.46)$$

or simply

$$p_1 V_1^{\gamma} = p_2 V_2^{\gamma} \qquad (20.47)$$

pressure and volume in adiabatic process

Intuitively, we expect the volume to decrease when the pressure increases. Equation (20.47) tells us quantitatively how pressure and volume vary in an adiabatic process.

EXAMPLE 10   In a diesel engine, the piston compresses the air–fuel mixture from an initial volume of 630 cm^3 to a final volume of 30.0 cm^3. The initial temperature of the mixture is 45°C. Assume that the compression occurs adiabatically. What is the final temperature? The value of γ for the air–fuel mixture is 1.37.

SOLUTION: For an adiabatic process, we use Eq. (20.45) to relate temperature and volume:

$$T_1 V_1^{\gamma-1} = T_2 V_2^{\gamma-1}$$

Solving for the final temperature T_2, we obtain

$$T_2 = T_1 \left(\frac{V_1}{V_2} \right)^{\gamma-1}$$

Inserting the initial temperature $T_1 = 45°C = 318$ K, the value $\gamma - 1 = 0.37$, and the given volumes, we find

$$T_2 = 318 \text{ K} \times \left(\frac{630 \text{ cm}^3}{30.0 \text{ cm}^3} \right)^{0.37} = 981 \text{ K} = 708°C$$

In a diesel engine, the high temperature resulting from nearly adiabatic compression triggers combustion without the need for the spark used for ignition in ordinary internal combustion engines.

 Checkup 20.6

QUESTION 1: When you switch on the heater in your room, the air warms and expands (and some air escapes from the room). Is this an adiabatic expansion?

QUESTION 2: The temperature on the tops of hills is usually lower than at the bottoms of valleys. In dry air, the temperature decrease with height is about 1°C per 100 m. This can be explained by examining the adiabatic behavior of parcels of air that drift upward (or downward). As a parcel drifts upward from the bottom of a valley to the top of a hill, what happens to its pressure, its volume, its temperature?

QUESTION 3: Three types of gases are initially at the same temperature. Each is adiabatically compressed to one-half of its original volume. Which type of gas attains the highest temperature?

      (A) Monatomic   (B) Diatomic   (C) Polyatomic

SUMMARY

PROBLEM-SOLVING TECHNIQUES Temperature units; thermal expansion		**(page 636)**

SPECIFIC HEAT OF WATER	$c = 1.00 \text{ kcal/kg·°C}$	**(20.2)**

HEAT Q FOR TEMPERATURE CHANGE ΔT OF A MASS m where c is the specific heat.	$Q = mc\,\Delta T$	**(20.3)**

MECHANICAL EQUIVALENT OF HEAT	$1.00 \text{ cal} = 4.187 \text{ J}$	**(20.4)**

THERMAL EXPANSION where α and β are the linear and volume coefficients of thermal expansion, respectively.	$\Delta L = \alpha L\,\Delta T$ $\Delta V = \beta V\,\Delta T$ $\beta = 3\alpha$	**(20.6)** **(20.8)** **(20.9)**

HEAT CONDUCTION   (energy per second) where k is the thermal conductivity, A is the area that the heat flows across, ΔT is the temperature difference, and Δx is the length.

$$\frac{\Delta Q}{\Delta t} = k A \frac{\Delta T}{\Delta x}$$

(20.13)

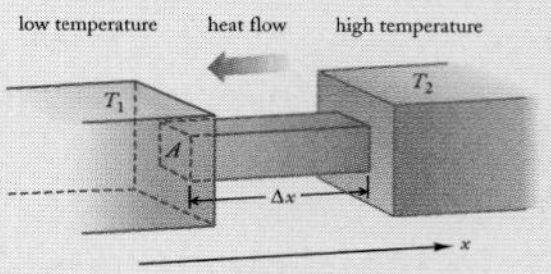

HEAT CONDUCTION THROUGH MULTIPLE SLABS

$$\frac{\Delta Q}{\Delta t} = A \frac{\Delta T}{(\Delta x_1/k_1) + (\Delta x_2/k_2) + \cdots}$$

(20.17)

CONVECTION   Transport of heat by a moving fluid.

RADIATION   Transport of heat by electromagnetic waves.

HEAT OF FUSION; HEAT OF VAPORIZATION   The heat required to change the state of a material from solid to liquid or from liquid to gas, respectively.

MOLAR SPECIFIC HEATS AT CONSTANT VOLUME	$C_V = \frac{3}{2}R$ (monatomic ideal gas)	**(20.31)**
(per mole)	$C_V = \frac{5}{2}R$ (diatomic ideal gas)	**(20.33)**

RELATION BETWEEN SPECIFIC HEATS OF A MOLE OF IDEAL GAS where C_p is the molar specific heat at constant pressure.	$C_p = C_V + R$	**(20.29)**

ADIABATIC EXPANSION OF GAS Using $\gamma = C_p/C_V$,	$\dfrac{dT}{T} = -\dfrac{R}{C_V}\dfrac{dV}{V}$		**(20.40)**
	$TV^{\gamma-1} = [\text{constant}]$		**(20.45)**
	$pV^{\gamma} = [\text{constant}]$		**(20.47)**

QUESTIONS FOR DISCUSSION

1. Can the body heat from a crowd of people produce a significant temperature increase in a room?

2. The expression "cold enough to freeze the balls off a brass monkey" originated aboard ships of the British Navy where cannonballs of lead were kept in brass racks ("monkeys"). Can you guess how the balls might fall off a "monkey" on a very cold day?

3. If the metal lid of a glass jar is stuck, it can usually be loosened by running hot water over the lid. Explain.

4. On hot days, bridges expand. How do bridge designers prevent this expansion from buckling the road?

5. At regular intervals, oil pipelines have lateral loops (shaped like a U; see Fig. 20.19). What is the purpose of these loops?

FIGURE 20.19 Loop in an oil pipeline.

6. When you heat soup in a metal pot, sometimes the soup rises at the rim of the pot and falls at the center. Explain.

7. A sheet of glass will crack if heated in one spot. Why?

8. When aluminum wiring is used in electrical circuits, special terminal connectors are required to hold the ends of the wires securely. If an ordinary brass screw were used to hold the end of an aluminum wire against a brass plate, what would be likely to happen during repeated heating and cooling of the circuit?

9. Suppose that a piece of metal and a piece of wood are at the same temperature. Why does the metal feel colder to the touch than the wood?

10. For lack of a better method, some nineteenth-century explorers in Africa and Asia measured altitude by sticking a thermometer into a pot of boiling water. Explain.

11. Can you guess why an alloy of two metals usually has a lower melting point than either pure metal?

12. In the cooling system of an automobile, how is the heat transferred from the combustion cylinder to the cooling water—by conduction, convection, or radiation? How is the heat transferred from the water in the engine to the water in the radiator? How is the heat transferred from the radiator to the air?

13. A fan installed near the ceiling of a room blows air down toward the floor. How does such a fan help to keep you cool in the summer and warm in the winter?

14. It is often said that an open fireplace sends more heat up the chimney than it delivers to the room. What is the mechanism

for the heat transport to the room? For the heat transport up the chimney?

15. A large fraction of the heat lost from a house escapes through the windows (Fig. 20.20). This heat is carried to the window-pane by convection—hot air at the top of the room descends along the windowpane, giving up its heat. Suppose that the windows are equipped with venetian blinds. In order to mini-mize the heat loss, should you close the blinds so that the slats are oriented down and away from the window or up and away?

FIGURE 20.20 Thermal photograph of a house. Bright regions indicate high heat loss.

16. Fiberglass insulation used in the walls of houses has a shiny layer of aluminum foil on one side. What is the purpose of this layer?

17. Is it possible to add heat to a system without changing its temperature? Give an example.

18. Why is boiling oil much more likely to cause severe burns on the skin than boiling water?

19. Would you expect the melting point of ice to increase or decrease with an increase of pressure?

20. A very cold ice cube, fresh out of the freezer, tends to stick to the skin of your fingers. Why?

21. What is likely to happen to the engine of an automobile if there is no antifreeze in the cooling system and the water freezes?

22. If an evacuated glass vessel, such as a TV tube, fractures and implodes, the fragments fly about with great violence. Where does the kinetic energy of these fragments come from?

23. When you boil water and convert it into water vapor, is the heat you supply equal to the change of the internal energy of the water?

24. A gas is in a cylinder fitted with a piston. Does it take more work to compress the gas at constant temperature or adiabati-cally?

25. According to the result of Section 11.2 [see Eqs. (11.13) and (11.14)], when a particle of small mass collides elastically with a body of very large mass, the particle gains kinetic energy if the body of large mass was approaching the particle before the collision. Using this result, explain how the collisions between the particles of gas and the moving piston lead to an increase of temperature during an adiabatic compression.

26. When a gas expands adiabatically, its temperature decreases. How could you take advantage of this effect to design a refrig-erator?

27. A sample of ideal gas is initially confined in a bottle at some given temperature. If we break the bottle and let the gas expand freely into an evacuated chamber of larger volume, will the temperature of the gas change?

PROBLEMS

20.1 Heat as a Form of Energy Transfer[†]

1. The immersible electric heating element in a coffeemaker converts 620 W of electric power into heat. How long does this coffeemaker take to heat 1.0 liter of water from 20°C to 100°C? Assume that no heat is lost to the environment.

2. The body heat released by children in a school makes a contri-bution toward heating the building. How many kilowatts of heat do 1000 children release? Assume that the daily food intake of each child has a chemical energy of 2000 kcal and that this food is burned at a steady rate throughout the day.

3. In 1847 Joule attempted to measure the frictional heating of water in a waterfall near Chamonix in the French Alps. If the water falls 120 m and all of its gravitational energy is con-verted into thermal energy, how much does the temperature of

the water increase? Actually, Joule found no increase of tem-perature because falling water cools by evaporation.

4. A nuclear power plant takes in $5.0 \times 10^6 \, \text{m}^3$ of cooling water per day from a river and exhausts 1200 megawatts of waste heat into this water. If the temperature of the inflowing water is 20°C, what is the temperature of the outflowing water?

5. Your metabolism extracts about 100 kcal of chemical energy from one apple. If you want to get rid of all this energy by jog-ging, how far must you jog? At a speed of 12 km/h, jogging requires about 750 kcal/h.

6. For basic subsistence a human body requires a diet with about 2000 kcal/day. Express this power in watts.

7. By turning a crank, you can do mechanical work at the steady rate of 0.15 hp. If the crank is connected to paddles churning 4.0 liters of water, how long must you churn the water to raise its temperature by 5.0°C?

[†]For help, see Online Concept Tutorial 22 at www.wwnorton.com/physics

8. A glass mug has a mass of 125 g when empty. It contains 180 g of coffee, and both the mug and the coffee are at 70°C. You add 15 g of cream at 5°C to the coffee. Assuming the mug–coffee system is isolated, find the common final temperature. Assume that coffee has the same specific heat as water and that the cream has a specific heat of 2900 J/kg·°C.

9. A cafeteria snack food label indicates that a "creme"-filled dessert contains 350 Calories (recall that 1 Calorie = 1000 cal = 1 kcal). How many steps would a person of mass 70 kg have to climb to do an amount of work equal to the energy contained in the snack food? Each step is 25 cm high.

10. A printing plate of mass 20 kg is made of lead. After casting, the solid lead cools to 90°C. It is then dropped into a 200-liter tank of water, initially at 25°C. What is the final equilibrium temperature of the lead and water?

11. Ten people can swim in a lap pool at the same time; each releases about 3.6×10^6 joules per hour. If the total volume of water in the pool is 600 m³, what temperature change occurs over a 12-h period, assuming that the pool is fully utilized?

12. An iron frying pan becomes dangerously hot above 400°C. After a 600-W stove burner is turned on full, how long will it take for a 5.0-kg frying pan (initially at 25°C) to become this hot? Assume for simplicity that all the heat goes to the frying pan.

13. An industrial polishing apparatus generates 300 W of heat due to friction. The heat is carried away by a water flow of 2.5 liters per minute. How much warmer is the water leaving the polishing station than the water entering it?

*14. You can warm the surfaces of your hands by rubbing one against the other. If the coefficient of friction between your hands is 0.60 and if you press your hands together with a force of 60 N while rubbing them back and forth at an average speed of 0.50 m/s, at what rate (in joules per second) do you generate heat on the surfaces of your hands?

*15. Problem 96 of Chapter 8 gives the relevant numbers for frictional losses in the Tennessee River. If all the frictional heat were absorbed by the water and if there were no heat loss by evaporation, how much would the water temperature rise per kilometer?

*16. The first quantitative determination of the mechanical equivalent of heat was made by Robert von Mayer, who compared available data on the amount of mechanical work needed to compress a gas and the amount of heat generated during the compression. From this comparison, Mayer deduced that the energy required for warming 1.00 kg of water by 1.00°C is equivalent to the potential energy released when a mass of 1.00 kg falls from a height of 365 m. By what percent does Mayer's result differ from the modern result given by Eq. (20.4)?

*17. On a hot summer day, the use of air conditioners raises the consumption of electric power in New York City to 22 400 megawatts. All of this electric power ultimately produces heat. Compare the heat produced in this way with the solar heat incident on the city. Assume that the incident flux of solar energy is 1.0 kW/m² and the area of the city is 850 km². Would you expect that the consumption of electric power significantly increases the ambient temperature?

*18. A simple gadget for heating water for showers consists of a black plastic bag holding 10 liters of water. When hung in the sun, the bag absorbs heat. On a clear, sunny day, the power delivered by sunlight per unit area facing the Sun is 1.0×10^3 W/m². The bag has an area of 0.10 m² facing the Sun. How long does it take for the water to warm from 20°C to 50°C? Assume that the bag loses no heat.

*19. In the heating system of a house, an oil furnace heats the water in a boiler, and the water is pumped into pipes connected to radiators. The hot water releases some of its heat in the radiators, and it then returns to the boiler, to be reheated. The furnace delivers 1.8×10^8 J of thermal energy per hour to the water, and the water leaves the boiler at 88°C and returns at 77°C. What rate of flow of the water (in m³/s) is required to achieve this?

*20. A solar collector consists of a flat plate that absorbs the heat of sunlight. A water pipe attached to the back of the plate carries away the absorbed heat (Fig. 20.21). Assume that the solar collector has an area of 4.0 m² facing the Sun and that the power per unit area delivered by sunlight is 1.0×10^3 W/m². What is the rate at which water must circulate through the pipe if the temperature of the water is to increase by 40°C as it passes through the collector?

*21. A fast-flowing stream of water in a horizontal channel strikes the bottom rim of an undershot waterwheel of radius 2.2 m (see Fig. 8.24). The water approaches the wheel with a speed of 5.0 m/s and leaves with a speed of 2.5 m/s; the amount of water passing through is 300 kg/s.

(a) At what rate does the water deliver angular momentum to the wheel? What torque does the water exert on the wheel?

(b) If the angular velocity of the wheel is 1.4 radians/s, what is the power delivered to the wheel?

(c) How much does the temperature of the water increase as it passes through the rim of the wheel?

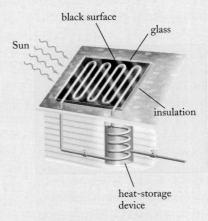

FIGURE 20.21 Collector of solar heat.

20.2 Thermal Expansion of Solids and Liquids

22. The tallest building in the United States is the Sears Tower in Chicago (Fig. 20.22), which is 443 m high. It is made of

concrete and steel. How much does its height change between a day when the temperature is 35°C and a day when the temperature is −29°C?

FIGURE 20.22 Sears Tower, Chicago.

23. The height of the Eiffel Tower is 321 m. What increase of temperature will lead to an increase of height by 10 cm?

24. Machinists use gauge blocks of steel as standards of length. A one-inch gauge block is supposed to have a length of 1 in., to within $\pm 10^{-6}$ in. In order to keep the length of the block within this tolerance, how precisely must the machinist control the temperature of the block?

25. A mechanic wants to place a sleeve (pipe) of copper around a rod of steel. At a temperature of 18°C the sleeve of copper has an inner diameter of 0.998 cm and the rod of steel has a diameter of 1.000 cm. To what temperature must the mechanic heat the copper to make it fit around the steel?

26. (a) Segments of steel railroad rails are laid end to end. In an old railroad, each segment is 18 m long. If they are originally laid at a temperature of −7°C, how much of a gap must be left between adjacent segments if they are to just barely touch at a temperature of 43°C?

    (b) In a modern railroad, each segment is continuously welded, typically 790 m long, for a smoother ride. A special expansion joint is used at each end. If no other allowance for expansion is provided, how much of a gap must be left between adjacent segments in this case?

27. A quartz photomask for silicon wafer fabrication must be positioned to within 1.0×10^{-7} m to match up with features from a previous quartz mask. If a wafer is 300 mm wide, what temperature change can be tolerated for accurate positioning across the entire wafer?

28. The outer walls of buildings include expansion joints (often filled with a soft caulking material). If the walls are made of concrete with expansion joints spaced every 10 m, how wide should the gaps be at low temperature to allow adequate expansion when the temperature rises 120°F?

29. Opticians immerse plastic eyeglass frames in a container of warm beads in order to expand the frames, permitting facile insertion of lenses. If the thermal expansion coefficient of such a plastic is 2.0×10^{-4}/°C and a lens is 0.75% larger than the opening into which it must be inserted, what minimum temperature increase of the plastic is required for easy insertion of the lens?

30. A particular experiment requires accurate positioning of a specimen at the end of a 1.5-m steel rod. In the experiment, the rod and sample are heated 200°C, while the surroundings stay at constant temperature. How far does the specimen move?

31. A 1.00-liter container made of glass is full to the brim with ethyl alcohol at −110°C (near its freezing point). How many cubic centimeters of alcohol overflow the container if the system is heated to +75°C (near the boiling point)?

32. An ordinary mercury thermometer consists of a glass bulb to which is attached a fine capillary tube. Given that the bulb has a volume of 0.20 cm³ and that the capillary tube has a diameter of 7.0×10^{-3} cm, how far will the mercury column rise up the capillary tube for a temperature increase of 10°C? Ignore the expansion of the glass and ignore the expansion of the mercury in the capillary tube.

*33. Suppose you heat a 1.0-kg cube of iron from 20°C to 80°C while it is surrounded by air at a pressure of 1.0 atm. How much work does the iron do against the atmospheric pressure while expanding? Compare this work with the heat absorbed by the iron. (The density of iron is 7.9×10^3 kg/m³.)

*34. When a solid expands, the increment of the area of one of its faces is directly proportional to the increment of temperature and to the original area. Show that the coefficient of proportionality for this expansion of area is 2 times the coefficient of linear expansion.

*35. A spring made of steel has a relaxed length of 0.316 m at a temperature of 20°C. By how much will the length of this spring increase if we heat it to 150°C? What compressional force must we apply to the hot spring to bring it back to its original length? The spring constant is 3.5×10^4 N/m.

*36. A wheel of metal has a moment of inertia I at some given temperature. Show that if the temperature increases by ΔT, the moment of inertia will increase by approximately $\Delta I = 2\alpha I \Delta T$.

*37. The pendulum (rod and bob) of a pendulum clock is made of brass.

    (a) What will be the fractional increment in the length of this pendulum if the temperature increases by 20°C? What will be the fractional increase in the period of the pendulum?

    (b) The pendulum clock keeps good time when its temperature is 15°C. How much time (in seconds per day) will the clock lose when its temperature is 35°C?

*38. (a) The density of gasoline is 730 kg/m³ when the temperature is 0°C. What will be the density of gasoline when the temperature is 30°C?

(b) The price of gasoline is 60 cents per liter. What is the price per kilogram at 0°C? What is the price per kilogram at 30°C? Is it better to buy cold gasoline or warm gasoline?

**39. In order to compensate for deviations caused by temperature changes, a pendulum clock built during the nineteenth century for an astronomical observatory uses a large cylindrical glass tube filled with mercury as a pendulum bob. This tube is held by a brass rod and bracket (Fig. 20.23); the combined length of the rod and bracket is l (measured from the point of suspension of the pendulum). Neglecting the mass of the brass and the glass and neglecting the expansion of the glass, show that the height of the mercury in the glass tube must be

$$h = \left(\frac{2\alpha_{\text{brass}}}{\beta_{\text{mercury}}} \right) l$$

if the center of mass of the mercury is to remain at a fixed distance from the point of suspension, regardless of temperature.

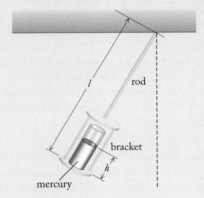

FIGURE 20.23 Pendulum with a temperature compensator.

20.3 Thermal Conduction

40. The walls of an igloo are made of compacted snow, 30 cm thick. What thickness of Styrofoam would provide the same insulation as the snow?

41. A pan of aluminum, filled with boiling water, sits on a hot plate. The bottom area of the pan is 300 cm², and the thickness of the aluminum is 0.10 cm. If the hot plate supplies 2000 W of heat to the bottom of the pan, what must be the temperature of the upper surface of the hot plate?

42. A rod of steel 0.70 cm in diameter is surrounded by a tight copper sleeve of inner diameter 0.70 cm and outer diameter 1.00 cm. What will be the heat flow along this compound rod if the temperature gradient along the rod is 50°C/cm? What fraction of the heat flows in the copper? What fraction in the steel?

43. A window in a room measures 1.0 m × 1.5 m. It consists of a single sheet of glass of thickness 2.5 mm. What is the heat

flow through this window if the temperature difference between the inside surface of the glass and the outside is 39°C? Compare the heat loss through the window with the heat loss through the wall calculated in Example 6.

44. The bottom of a teakettle consists of a layer of stainless steel 0.050 cm thick welded to a layer of copper 0.030 cm thick. The area of the bottom of the kettle is 300 cm². The copper sits in contact with a hot plate at a temperature of 101.2°C, and the steel is covered with boiling water at 100.0°C. What is the rate of heat transfer through the bottom of the kettle from the hot plate to the water?

45. According to Eq. (20.13), the heat flow through a rod or slab of cross-sectional area A can be expressed as

$$\frac{\Delta Q}{\Delta t} = \frac{A \, \Delta T}{R}$$

where $R = \Delta x/k$ is called the thermal resistance, or the **R value** (see also the comments in Example 7). Since the heat flow is inversely proportional to the R value, a good insulator has a high R value.

(a) What is the R value of a slab of fiberglass insulation, 10 cm thick?

(b) In the United States, R values of commercially available insulation are commonly expressed in units of ft²·°F·h/Btu. What is the R value of the 10-cm slab of fiberglass in these units?

46. An insulated coffee cup and lid contain coffee at 80°C. What is the initial heat flow per cm² of wall if the wall insulation is 1.0 cm thick and has a thermal conductivity of 0.080 J/s·m·°C? The ambient temperature is 22°C.

47. The water and the airspace in a tropical fish tank are maintained at 26°C by a heater when the temperature in the room is 18°C. If the walls, base, and lid of the fish tank are made of 3.0-mm-thick glass and the tank measures 80 cm × 50 cm × 30 cm, what average power (in watts) must be supplied by the heater?

48. What is the ratio of the heat flow though a glass door of thickness 3.0 mm to the heat flow through a wood door of thickness 25 mm? Each door has the same area and same temperature difference.

*49. Several slabs of different materials are piled one on top of another. All slabs have the same face area A, but their thicknesses and conductivities are Δx_i and k_i, respectively. Show that the heat flow through the pile of slabs is

$$\frac{\Delta Q}{\Delta t} = \frac{A \Delta T}{\displaystyle\sum_{i=1}^{n} \Delta x_i/k_i}$$

where ΔT is the temperature difference between the bottom of the first slab and the top of the last slab.

*50. A man has a skin area of 1.8 m²; his skin temperature is 34°C. On a cold winter day, the man wears a whole-body suit insulated with down. The temperature of the outside surface of his suit is −25°C. If the man can stand a heat loss of no more than 4.0×10^5 J/h, what is the minimum thickness of down required for his suit?

*51. The end of a rod of copper 0.50 cm in diameter is welded to a rod of silver of half the diameter. Each rod is 6.0 cm long. What is the heat flow along these rods if the free end of the copper rod is in contact with boiling water and the free end of the silver rod is in contact with ice? What is the temperature of the junction? Assume that there is no heat loss through the lateral surfaces of the rods.

*52. A ceiling skylight uses a "thermal window" of area 2.0 m². The window consists of two sheets of 3.2-mm-thick glass, separated by a 2.0-mm-thick layer of argon gas. Take the thermal conductivity of argon to be 0.017 J/s·m·°C. The outside temperature is 38°C, and the inside temperature is 25°C. For such a horizontal gas layer, heated from above, convection can be neglected. Assume an overcast day so that radiation can also be neglected. Find the thermal energy entering the skylight per day. (Hint: See the discussion in Example 7 or use the result of Problem 49.)

**53. A cable used to carry electric power consists of a metallic conductor of radius r_1 encased in an insulator of inner radius r_1 and outer radius r_2. The metallic conductor is at a temperature T_1, and the outer surface of the insulator is at a temperature T_2. Show that the radial heat flow across the insulator is given by

$$\frac{\Delta Q}{\Delta t} = \frac{2\pi k (T_2 - T_1) l}{\ln(r_2/r_1)}$$

where k is the thermal conductivity of the insulator and l is the length of the cable.

20.4 Changes of State[†]

54. The icebox on a sailboat measures 60 cm × 60 cm × 60 cm. The contents of this icebox are to be kept at a temperature of 0°C for 4 days by the gradual melting of a block of ice of 20 kg, while the temperature of the outside of the box is 30°C. What minimum thickness of the Styrofoam insulation is required for the walls of the icebox?

*55. On a cold winter day, the water of a shallow pond is covered with a layer of ice 6.0 cm thick. The temperature of the air is −20°C, and the temperature of the water is 0°C. What is the (instantaneous) rate of growth of the thickness of the ice (in centimeters per hour)? Assume that the windchill keeps the top surface of the ice at exactly the temperature of the air, and assume that there is no heat transfer through the bottom of the pond.

*56. Suppose that the pond described in the preceding problem has a layer of compacted snow 3.0 cm thick on top of the ice. What is the rate of growth of the thickness of the ice?

[†]For help, see Online Concept Tutorial 22 at www.wwnorton.com/physics

57. Thunderstorms obtain their energy by condensing the water vapor contained in humid air. Suppose that a thunderstorm succeeds in condensing *all* the water vapor in 10.0 km³ of air.

   (a) How much heat does this release? Assume the air is initially at 100% humidity and that each cubic meter of air at 100% humidity (at 20°C and 1.0 atm) contains 1.74×10^{-2} kg of water vapor. The heat of vaporization of water is 2.45×10^6 J/kg at 20°C.

   (b) The explosion of a nuclear bomb releases an energy of 8.0×10^{13} J. How many nuclear bombs does it take to make up the energy of one thunderstorm?

58. You place 1.0 kg of ice (at 0°C) in a pot and heat it until the ice melts and the water boils off, making steam. How much heat must you supply to achieve this?

59. Warm tea at 50°C is poured over ice to make iced tea. The specific heat of tea is the same as that of water. If 500 milliliters of tea are to be cooled to 5.0°C, how many grams of ice at 0.0°C are needed?

60. In magnets used for magnetic resonance imaging (MRI), both liquid nitrogen and liquid helium (which costs about 10 times more) are used; the magnet is precooled with liquid nitrogen to −196°C, and then cooled to −269°C with liquid helium. Considering only the latent heat of vaporization, how many liters of liquid nitrogen are needed to cool 20 kg of copper from 20°C to −196°C? How many liters of liquid helium would be needed to do the same job? (In reality, some cooling is also provided by the warming of the vaporized gas.)

61. Large central air conditioners for entire homes are typically rated in "tons"; this is the mass of ice that would have to melt in one day to remove the same amount of heat. If such a "ton" corresponds to 907 kg, how much heat in joules is removed per day by a 3.5-ton air-conditioning system? What is this rate of energy removal in watts?

*62. A liquid can be cooled and even frozen by pumping with a vacuum pump above the liquid; some of the liquid is vaporized, leaving behind a colder liquid, and, eventually, a solid. If 1.0 kg of liquid nitrogen at −196°C is pumped on, how much frozen nitrogen at −210°C is obtained? The specific heat of liquid nitrogen is 2.0×10^3 J/kg·°C. Assume for simplicity that the heats of fusion and vaporization do not vary significantly with temperature or pressure.

*63. The vaporization of liquid nitrogen is used in a popular demonstration to make ice cream. Assume that cream has a specific heat of 2900 J/kg·°C and has the same heat of fusion and melting point as water, and that ice cream has a specific heat of 2200 J/kg·°C. If two kilograms of cream are initially at 10°C, how much liquid nitrogen must be vaporized to cool, freeze, and further cool the cream to −10°C?

*64. The heat of vaporization of water at 100°C and 1.0 atm is 2.26×10^6 J/kg. How much of this energy is due to the work the water vapor does against atmospheric pressure? What would this work be at a pressure of 0.10 atm? At (nearly) zero pressure?

*65. During a rainstorm lasting 2 days, 7.6 cm of rain fell over an area of 2.6×10^3 km^2.

(a) What is the total mass of the rain (in kilograms)?

(b) Suppose that the heat of vaporization of water in the rain clouds is 2.43×10^6 J/kg. How many calories of heat are released during formation of the total mass of rain by condensation of the water vapor in these clouds?

(c) Suppose that the rain clouds are at a height of 1500 m above the ground. What is the gravitational potential energy of the total mass of rain before it falls? Express your answer in joules and in calories.

(d) Suppose that the raindrops hit the ground with a speed of 10 m/s. What is the total kinetic energy of all the raindrops taken together? Express your answer in joules and in calories. Why does your answer to part (c) not agree with this?

*66. If you pour 0.50 kg of molten lead at 328°C into 2.5 liters of water at 20°C, what will be the final temperature of the water and the lead? The specific heat of (solid) lead has an average value of 140 J/kg·°C over the relevant temperature range.

*67. Suppose you drop a cube of titanium of mass 0.25 kg into a Dewar flask (a thermos bottle) full of liquid nitrogen at −196°C. The initial temperature of the titanium is 20°C. How many kilograms of nitrogen will boil off as the titanium cools from 20°C to −196°C? The specific heat of titanium is 340 J/kg·°C.

*68. While jogging on a level road, your body generates heat at the rate of 750 kcal/h. Assume that evaporation of sweat removes 50% of this heat, and convection and radiation the remainder. The evaporation of 1.0 kg (or 1.0 liter) of sweat requires 580 kcal. How many kilograms of sweat do you evaporate per hour?

*69. The Mediterranean loses a large volume of water by evaporation. The loss is made good, in part, by currents flowing into the Mediterranean through the straits joining it to the Atlantic Ocean and the Black Sea. Calculate the rate of evaporation (in km^3/h) of the Mediterranean on a clear summer day from the following data: the area of the Mediterranean is 2.9×10^6 km^2, the power per unit area supplied by sunlight is 1.0×10^3 W/m^2, and the heat of vaporization of water is 2.43×10^6 J/kg (at a temperature of 21°C). Assume that all the heat of sunlight is used for evaporation.

20.5 The Specific Heat of a Gas

70. A TV tube of glass with zero pressure inside and atmospheric pressure outside suddenly cracks and implodes. The volume of the tube is 2.5×10^{-2} m^3. During the implosion, the atmosphere does work on the fragments of the tube and on the layer of air immediately adjacent to the tube. This amount of work represents the energy released in the implosion. Calculate this energy. If all of this energy is acquired by the fragments of the glass, what will be the mean speed of the fragments? The total mass of the glass is 2.0 kg.

71. The rear end of an air conditioner dumps 1.2×10^7 J/h of waste heat into the air outside a building. A fan assists in the removal of this heat. The fan draws in 15 m^3/min of air at a temperature of 30°C and ejects this air after it has absorbed the waste heat. With what temperature does the air emerge?

72. What are the specific heats C_V and C_p for air consisting of 75% nitrogen, 24% oxygen, and 1% argon (by mass) at STP? Use the values for C_V and C_p of nitrogen, oxygen, and argon listed in Table 20.5.

73. Table 20.5 gives the specific heat C_V per mole for several gases. Calculate the specific heat per kilogram for each gas. Which gas has the highest value of the specific heat per kilogram? The lowest?

74. If we heat 1.00 kg of hydrogen gas from 0.0°C to 50.0°C in a cylinder with a piston keeping the gas at a constant pressure of 1.00 atm, we must supply 7.08×10^5 J of thermal energy. How much work does the gas deliver to the cylinder during this process? How many joules of thermal energy must we supply to heat the same amount of gas from 0.0°C to 50.0°C in a container of constant volume?

75. The theoretical expression for the speed of sound in a gas is $\sqrt{\gamma p / \rho}$ [compare Eq. (17.5)]. Calculate the speed of sound in helium gas at STP. See Table 20.5 for $\gamma = C_p / C_V$.

76. A helium balloon consists of a large bag loosely filled with 600 kg of helium at an initial temperature of 10°C. While exposed to the heat of the Sun, the helium gradually warms to a temperature of 30°C. The heating proceeds at a constant pressure of 1.0 atm. How much heat does the helium absorb during this temperature change?

77. As in Example 9, consider an initial volume of 1.0 m^3 of air that is warmed by sunlight from 0°C to 10°C at a constant pressure of 1.0 atm. What is the change in volume of this air? During its expansion, how much work does the air do against the pressure of the surrounding atmosphere?

78. A quartz tube contains one mole of helium gas at 20°C. The gas is heated at constant volume to 300°C. How much thermal energy is transferred to the gas? If the same amount of gas were heated at constant pressure, how much energy would be required?

79. A gas mixture is made consisting of 1.00 mole of helium gas and 2.00 moles of oxygen gas. What is the molar specific heat of this mixture at constant volume? At constant pressure?

*80. An air conditioner removes heat from the air of a room at the rate of 8.0×10^6 J/h. The room measures 5.0 m × 5.0 m × 2.5 m, and the pressure is constant at 1.0 atm.

(a) If the initial temperature of the air in the room is 30.0°C, how long does it take the air conditioner to reduce the temperature of the air by 5.0°C? Pretend that the mass of the air in the room is constant.

(b) As the air in the room cools, it contracts slightly and draws in some extra air from the outside; hence the mass of air is not exactly constant. Repeat your calculation taking into account this increase of the mass of air. Assume that the extra air enters with an initial temperature of 30°C. Does the result of your second calculation differ appreciably from that of your first calculation?

*81. On a winter day you inhale cold air at a temperature of −30°C and at 0% humidity. The amount of air you inhale is 0.45 kg per hour. Inside your body you warm and humidify the air; you then exhale the air at a temperature of 37°C and 100% relative humidity. At a temperature of 37°C, each kilogram of air at 100% relative humidity contains 0.041 kg of water vapor. How many calories are carried out of your body by the air that passes through your lungs in one hour? Take into account both the heat needed to warm the air at constant pressure and the heat needed to vaporize the moisture that the exhaled air carries out of your body. The specific heat of air at constant pressure is 1.0×10^3 J/kg·°C; the heat of vaporization of water at 37°C is 2.42×10^6 J/kg.

20.6 Adiabatic Expansion of a Gas

82. (a) Show that for a small adiabatic expansion,

$$\frac{dp}{p} = -\left(1 + \frac{R}{C_V}\right)\frac{dV}{V}$$

Since $R = C_p - C_V$, this can also be written in the form

$$\frac{dp}{p} = -\frac{C_p}{C_V}\frac{dV}{V}$$

(b) If an ideal monatomic gas adiabatically expands by 3%, what is the percent decrease of pressure?

*83. Under normal conditions, the temperature of air in the atmosphere decreases with altitude; that is, the air temperature at the top of a mountain is lower than that at the foot of a mountain. This temperature difference is maintained by winds that move the air from one place to another. When a wind carries a parcel of air up the side of a mountain, the parcel expands adiabatically and cools; when the wind carries the parcel of air down, it is compressed adiabatically and warms. Calculate the temperature difference between the bottom and the top of a mountain 100 m high. The temperature at the bottom is 20°C. The pressure at the bottom is 1.00 atm, and the pressure at the top is 0.988 atm. For air, the specific heats are those of nitrogen and oxygen (see Table 20.5).

84. Argon gas is compressed without heat loss to one-tenth of its initial volume. If the initial temperature is 25°C, what is the final temperature?

85. A fire extinguisher is filled with 1.0 kg of nitrogen gas at a pressure of 1.2×10^6 N/m² and a temperature of 20°C.

(a) What is the volume of this gas?

(b) If the gas is allowed to escape adiabatically against atmospheric pressure, what will be the volume and temperature of the expanded gas?

86. Suppose that you have a sample of oxygen gas at a pressure of 300 atm and a temperature of −29°C. If you suddenly (adiabatically) let this gas expand to a final pressure of 1.00 atm, what will be the final temperature? (This method was used by Cailletet in 1877 to liquefy oxygen; oxygen liquefies at −183°C.)

87. Air in an automobile tire is at an overpressure of 2.0 atm and a temperature of 20°C. The pressure outside the tire is at 1.0 atm. If you let some air escape through the valve, what will be the final temperature of the escaping air? Assume that the air expands adiabatically.

*88. By means of a hand pump, you inflate an automobile tire from 0.0 atm to 2.4 atm overpressure. The volume of the tire remains constant at 0.10 m³. How much work must you do on the air with the pump? Assume that each stroke of the pump is an adiabatic process and that the air is initially at STP.

REVIEW PROBLEMS

89. In the cooling system of the engine of a boat, water is pumped from the outside through the engine and then returned to the outside. The engine produces 3.0 kW of waste heat, and the pump circulates 12 liters of water per minute through the engine. If all of the waste heat is carried away by the water, what is the increase in the temperature of the water?

90. A shallow pond has a depth of 50 cm. On a winter day, the initial temperature of the water is 6.0°C. If a snowstorm deposits a 50-cm layer of snow (which melts to form a 5.0-cm layer of water) on this pond, what will be the final temperature of the water? Assume that the initial temperature of the snow is 0.0°C, and that there is no heat exchange between the pond and the air or the ground.

91. A radiator, fed by hot water, is used to heat a room in a house. According to the manufacturer's specifications, the output of the radiator is 7.5×10^6 J/h when the inflowing water is at 88°C and the outflowing water at 77°C. What must be the rate of flow of the water through the radiator to achieve this output?

*92. The beam dump at the Stanford Linear Accelerator Center (SLAC) consists of a large tank with 12 m³ of water into which the accelerated electrons can be aimed when they are not wanted elsewhere (Fig. 20.24). The beam carries 3.0×10^{14} electrons/s; the kinetic energy per electron is 3.2×10^{-9} J. In the beam dump this energy is converted into heat.

(a) What is the rate of production of heat?

(b) If the water in the tank is stagnant and does not lose any heat to the environment, what is the rate of increase of temperature of the water?

(c) To prevent overheating, cooling water is pumped through the tank at the rate of 2.0 m³/min; this carries away the heat. If the temperature of the inflowing water is 20°C, what is the temperature of the outflowing water?

FIGURE 20.24 Beam dump at SLAC.

93. The largest ship is the supertanker *Seawise Giant*, with a length of 458 m and a beam of 69 m. By how many meters does this tanker expand in length and in width when it travels from the wintry North Atlantic (−20°C) to the hot Persian Gulf (+40°C)? By how many square meters does its deck area increase? Assume that the deck is approximately a rectangle, 458 m × 69 m.

94. The supertanker *Seawise Giant* (see also Problem 93) has an enclosed volume of 1.8×10^6 m³. By how many cubic meters does its volume expand when it travels from the wintry North Atlantic (−20°C) to the hot Persian Gulf (+40°C)?

*95. A Styrofoam box, used for the transportation of medical supplies, is filled with dry ice (carbon dioxide) at a temperature of −79°C. The box measures 30 cm × 30 cm × 40 cm, and its

walls are 4.0 cm thick. If the outside surface of the box is at a temperature of 20°C, what is the rate of loss of dry ice by vaporization?

*96. The wall of a house is to be built of a layer of wood, an adjacent layer of fiberglass insulation, and an adjacent layer of brick. The thickness of the wood is 1.2 cm, and that of the brick is 10 cm. The heat loss per square meter of wall is to be no more than 4.0×10^4 J/h when the temperatures inside and outside the house differ by 30°C. How thick must the fiberglass insulation be?

97. You dump a 0.30-kg chunk of dry ice (solid carbon dioxide) into a beaker containing a water–ice mixture. How much more ice will form in the beaker as the dry ice bubbles away, becoming carbon dioxide gas?

98. A blacksmith drops a 0.70-kg horseshoe of iron at a temperature of 1200°C into a bucket containing half a liter of water at an initial temperature of 30°C. How much of the water boils off? Assume that the bucket absorbs none of the heat.

99. How much heat must we supply to heat 1.0 mole of argon gas from 30°C to 100°C at constant volume? At constant pressure?

100. To discover whether an unknown gas is monatomic or diatomic, an experimenter takes a 2.0-liter sample of the gas at STP and heats this sample to 100°C at constant volume.

(a) The heat that the gas absorbs during this process is 180 J. Is the gas monatomic or diatomic? Assume it is an ideal gas.

(b) The experimenter weighs the sample and finds that its mass is 2.5 g. Can you tell what gas it is?

101. Suppose we heat 1.0 mole of oxygen gas at a constant pressure of 1.0 atm from 20°C to 80°C, and then cool it at a constant volume from 80°C back to 20°C.

(a) How much heat is absorbed by the gas during the first step?

(b) How much heat is released by the gas during the second step?

(c) What is the volume of the gas at the end of the first step? What is the pressure at the end of the second step?

(d) How much work does gas perform during the first step? During the second step?

Answers to Checkups

Checkup 20.1

1. Since the masses and heat capacities of the two entities being mixed are the same (and using the fact that the heat capacity of water is approximately independent of temperature), Eq. (20.3) and a common final temperature imply that the mixture will attain the average temperature 50°C.

2. Since 1 kcal would change the temperature of 1 kg of water by 1°C, Eq. (20.3) implies that for the same amount of heat and mass, but one-ninth the specific heat, the temperature change would be 9 times as large, or 9°C.

3. The laborer must apply a force during motion and so does work on the crate. This work transfers mechanical energy to the crate. The laborer does not transfer thermal energy to the crate. The floor does negative work on the crate because the floor exerts a frictional force in a direction opposite to the motion.

4. (B) 0.10°C. We equate the mechanical work done, $W = F \times \Delta y = 1000$ N $\times$ 4.187 m = 4187 J, with the heat of Eq. (20.3), $Q = mc\Delta T$: 4187 J = 10 kg $\times$ 4187 J/(kg·°C) $\times \Delta T$; and we obtain $\Delta T = 0.10$°C.

Checkup 20.2

1. The fractional volume expansion is 3 times the fractional linear expansion; thus, for a 0.02% linear expansion, the corresponding volume expansion is 0.06%.

2. From Fig. 20.5, between 0°C and 4°C the volume decreases from 1000.14 cm^3 to 1000.00 cm^3, a change of 0.14/1000, or 0.014%.

3. The coefficients of expansion of aluminum and concrete do not match (from Table 20.2, aluminum expands twice as much as concrete), so the rods would loosen from the concrete, or cause the concrete to crack.

4. (A) Increases. A hole or cavity in a material expands just as though it were filled with the same material as that which surrounds the hole or cavity.

Checkup 20.3

1. From Table 20.3, the thermal conductivity of wood is 3 times that of fiberglass. Thus, to provide the same insulation, the solid wood walls would have to be 3 times as thick, or 18 in. (45 cm).

2. Since heat always flows from the hotter body to the colder body, you can assert that the temperature of the steel is below 0°C.

3. Reducing the wall thickness will increase the heat loss [smaller Δx in Eq. (20.13)], increasing the wall area will also increase the heat loss [increasing A in Eq. (20.13)], reducing the external temperature to −20°C will increase the heat loss (larger ΔT), and reducing the internal temperature to 19°C will decrease the heat loss (smaller ΔT).

4. (D) Convection, radiation, conduction. The hair dryer blows warm air; heat transfer by fluid motion is convection. The heat lamp radiates light and infrared radiation. The hot-water bottle is used in contact with the body, transferring heat by conduction.

Checkup 20.4

1. The gases at −200°C are those materials with boiling points *below* −200°C; in Table 20.4, these are helium and hydrogen. The liquids are those materials with boiling points above −200°C and melting points below −200°C; nitrogen and oxygen satisfy these conditions. The solids are all the other materials; these all have melting points above −200°C.

2. The heat released when freezing refers to the heat of fusion; for the materials in Table 20.4, this is largest for aluminum.

3. For all of the materials in Table 20.4, the heat of vaporization is greater than the heat of fusion; thus, more heat is required for vaporization than for melting.

4. (D) b, c, a. From Table 20.4, vaporization of 1.0 kg requires 2.3×10^6 J and melting requires 3.3×10^5 J. From Eq. (20.4), heating from 0°C to 100°C will require approximately 4.2×10^5 J. Thus, in increasing order, they are melting, heating, and vaporization.

Checkup 20.5

1. C_p is larger than C_V because work is done when heating at constant pressure: to maintain constant pressure, the volume must increase when heating ($pV = nRT$). A force (pressure times area) applied during the displacement (expanding walls or moving piston) represents work done by the gas.

2. From Table 20.5, both helium and argon have the lowest value of the specific heat. Of these, argon atoms have the greater mass and thus the lower rms speed [see Eq. (19.23)], and so argon has the lower thermal conductivity (and is used in practice).

3. The gas in the balloon will require more heat, since it is both changing temperature (internal energy) and expanding against atmospheric pressure (work done). For an expanding balloon, the difference in specific heats will be smaller than 8.31 J/C·mole, since the elastic properties of the balloon imply that the gas is actually expanding against *increasing* pressure (so the volume change, and the work done, will be smaller than at constant pressure).

4. If the gas expands against a *decreasing* pressure, then the volume change (and thus the work done) will be greater than at constant pressure. This occurs, for example, with an air bubble in water, expanding while it ascends.

5. (D) 140°C. Air is essentially diatomic, so $C_p = \frac{7}{2}R$ and $C_V = \frac{5}{2}R$ [Eq. (20.33)]. Because C_V is smaller than C_p, the same added heat results in a temperature increase larger by a factor $C_p/C_V = \frac{7}{2}/\frac{5}{2} = \frac{7}{5} = 1.4$. Thus the temperature increase at constant volume is $1.4 \times 100°C = 140°C$.

Checkup 20.6

1. No. For an adiabatic process, no heat is transferred to or from the environment. For the heated room, both the heat supplied by the heater and the escaping air carrying heat to the environment violate adiabatic conditions.

2. In a fluid, pressure decreases with height. So as a parcel of air drifts upward, its pressure decreases. Under adiabatic conditions, the volume of the parcel must therefore increase [Eq. (20.47)], and the temperature must decrease [Eq. (20.45)].

3. (A) Monatomic. From Eq. (20.45), $T_2/T_1 = (V_1/V_2)^{\gamma-1}$. From Table 20.5, the monatomic gas has the largest value of γ, and so for any $V_1 > V_2$ (compression) the monatomic gas attains the highest temperature.

Thermodynamics

CONCEPTS IN CONTEXT

Concepts
— in —
Context

The operation of this steam-engine locomotive requires the transfer of heat in order to do work. As we learn about machines that generate mechanical energy, we can ask:

? How do we measure the efficiency of an engine, and how efficient is a typical steam engine? (Example 2, page 666)

? How efficient could an ideal steam engine be? (Example 4, page 671)

? How does a practical steam engine differ from an ideal engine in operation? (Section 21.2, page 671)

21.1 The First Law of Thermodynamics

21.2 Heat Engines; the Carnot Engine

21.3 The Second Law of Thermodynamics

21.4 Entropy

Thermodynamics *is the branch of physics that deals with the conversion of one form of energy into another, especially the conversion of heat into other forms of energy.* These conversions are governed by the two fundamental laws of thermodynamics. As we will see in this chapter, the first of these is essentially a general statement of the Law of Conservation of Energy, and the second is a statement about the maximum efficiency attainable in the conversion of heat into work.

Thermodynamics describes physical processes in terms of purely macroscopic parameters. Such a macroscopic, large-scale description is necessarily somewhat crude, since it overlooks the small-scale, microscopic phenomena, such as the collisions of molecules with the walls of the container we investigated in Section 19.3. However, in practical applications, a knowledge of the microscopic phenomena is often unnecessary. For instance, an engineer investigating the combustion of fuel in a rocket engine will find it satisfactory to deal with only such macroscopic quantities as temperature, pressure, density, and heat capacity, and ignore the microscopic behavior of the gases.

The development of steam engines for the industrial generation of (macroscopic) mechanical energy from heat motivated a careful examination of the theoretical principles underlying the operation of such engines, which led to the discovery of the Law of Conservation of Energy and to the recognition that heat is a form of energy transfer. Steam engines and other heat engines do not create energy; they merely convert thermal energy into mechanical energy, which can be used to perform useful work. For example, the steam engine of an old-fashioned locomotive converts thermal energy from the combustion of coal into mechanical energy, and the engine of an automobile converts thermal energy from the combustion of gasoline into mechanical energy.

Nineteenth-century engineers inaugurated the study of thermodynamics to discover what ultimate limitations the laws of physics impose on the operation of steam engines and other machines that generate mechanical energy. They soon established that perpetual motion machines, which earlier inventors had sought to build, are impossible. A **perpetual motion machine of the first kind** is a (hypothetical) device that supplies an endless output of work without any input of fuel or any other input of energy. Figure 21.1 shows a proposed design for such a machine. Weights are attached to the rim of a wheel by short, pivoted rods resting against pegs. With the rods in the position shown, there is an imbalance in the weight distribution causing a clockwise torque on the wheel; as the wheel turns, the rod coming to the top presumably flips over, maintaining the imbalance. This perpetual torque would not only keep the wheel turning, but would also continually deliver energy to the axle of the wheel. However, a detailed analysis demonstrates that the machine will not perform as intended—the wheel actually settles in a static equilibrium configuration such that the top rod just barely fails to flip over. The First Law of Thermodynamics, or the Law of Conservation of Energy, directly tells us of the failure of this machine: after one revolution of the wheel, the masses all return to their initial positions, their potential energy returns to its initial value, and they will not have delivered net energy to the motion of the wheel.

A **perpetual motion machine of the second kind** is a device that extracts thermal energy from some heat source, such as air or the water of the ocean, and converts it into mechanical energy. Such a device is not forbidden by conservation laws. The oceans are enormous reservoirs of thermal energy; if we could extract this thermal energy, a temperature drop of just 1°C of the oceans would supply the energy needs of the United States for the next 50 years. But, as we will see, the Second Law of Thermodynamics tells us that conversion of heat into work requires not only a heat

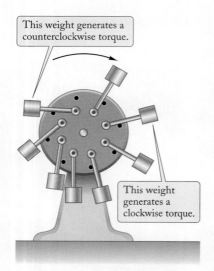

This weight generates a counterclockwise torque.

This weight generates a clockwise torque.

FIGURE 21.1 A hypothetical perpetual motion machine.

perpetual motion machine
of the first kind

perpetual motion machine
of the second kind

source, but also a heat sink. Heat flows out of a warm body only if there is a cooler body that can absorb heat. If we want heat to flow from the ocean into our machine, we must provide a low-temperature heat sink toward which the heat will tend to flow spontaneously. If no low-temperature sink is available, the extraction of heat from the ocean is impossible, and we cannot build a perpetual motion machine of the second kind.

21.1 THE FIRST LAW OF THERMODYNAMICS

Consider some amount of gas with a given initial volume V_1, pressure p_1, and temperature T_1. The gas is in a container fitted with a piston (see Fig. 21.2). Suppose we compress the gas to some smaller volume V_2 and also lower its temperature to some smaller value T_2; the pressure will then reach some new value p_2. (Such a compression and cooling process is of practical importance in the liquefaction of gases, say, oxygen or nitrogen—before the gas can be liquefied, it must be compressed and cooled.) Obviously, we can reach the new state V_2, p_2, and T_2 from the old state V_1, p_1, and T_1 in a variety of ways. For instance, we may first compress the gas and then cool it. Or else, we may first cool it and then compress it. Or we may go through small alternating steps of compressing and cooling. In order to compress the gas, we must do work on it [see Eq. (20.23)]; and in order to cool the gas, we must remove heat from it. The work done on or by the gas and the heat transferred from or to the gas result in a change of the internal energy of the gas. We can express this change of internal energy as

$$\Delta E = Q - W \tag{21.1}$$

where Q is the amount of heat transferred to the gas and W is the amount of work performed by the gas. Note the sign conventions in this equation: Q is positive if we add heat to the gas and negative if we remove heat; W is positive if the gas does work on us and negative if we do work on the gas (see Fig. 21.3).

The values of Q and W depend on the process. If we first compress the gas adiabatically and subsequently cool it at constant volume, then during the first step W is negative and Q zero; and during the second step W is zero and Q is negative. If we first cool the gas and then compress it, the values of W and Q will be quite different. Yet, it turns out that regardless of what sequence of operations we use to transform the gas from its initial state V_1, p_1, T_1 to its final state V_2, p_2, T_2, the net change ΔE in the internal energy is always the same: Q and W vary, but the sum of Q and $-W$ remains fixed. This is the **First Law of Thermodynamics**:

When several alternative processes involving heat and work are available to change a system from an initial state characterized by given values of the macroscopic parameters to a final state characterized by new values of the macroscopic parameters, the amounts Q of heat and W of work depend on the process. But the change in the internal energy of the system

$$\Delta E = Q - W \tag{21.2}$$

has a fixed value which does not depend on the process.

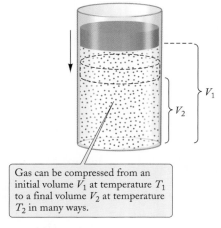

Gas can be compressed from an initial volume V_1 at temperature T_1 to a final volume V_2 at temperature T_2 in many ways.

FIGURE 21.2 Compression of a gas by a piston.

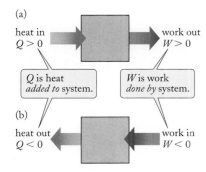

(a)

heat in
$Q > 0$ work out
$W > 0$

Q is heat *added to* system. W is work *done by* system.

(b)

heat out
$Q < 0$ work in
$W < 0$

FIGURE 21.3 (a) If the system receives heat from its surroundings and performs work on its surroundings, Q is positive and W is positive. (b) If the system delivers heat to its surroundings and the surroundings perform work on the system, Q is negative and W is negative.

First Law of Thermodynamics

Note that the First Law tells us that the energy is conserved—the change of internal energy of the system equals the input of heat minus the output of work, that is, it equals the input of microscopic work minus the output of macroscopic work. But the First Law tells us more than that. If we describe a system in terms of the detailed microscopic positions and velocities of all its constituent particles, then energy conservation is uncontestable—it is a theorem of mechanics. But if we describe a system in terms of nothing but macroscopic parameters, then it is not at all obvious that we have available enough information to determine the energy and to formulate a conservation law. The First Law of Thermodynamics tells us that a knowledge of the macroscopic parameters is indeed sufficient to determine the energy of the system.

free expansion

EXAMPLE 1   Some amount of gas at temperature T_1 is stored in a thermally insulated bottle. By means of a pipe with a valve, we connect this bottle to another insulated bottle which is evacuated (see Fig. 21.4). If we suddenly open the valve, the gas will rush from the first bottle into the second until the pressures are equalized. This is called a **free expansion** of the gas, because the gas expands without pushing against anything. What does the First Law say about the change of the internal energy of the gas in this process?

SOLUTION: The use of thermally insulated bottles makes the expansion adiabatic, so the expansion process neither adds nor removes heat from the gas, that is, $Q = 0$. Furthermore, the expansion process involves no work (the gas does not push against any moving piston), that is, $W = 0$. Consequently, Eq. (21.2) tells us that the internal energy of the gas does not change:

$$\Delta E = Q - W = 0 \tag{21.3}$$

We can take this conclusion a step further if we assume that the gas behaves like an ideal gas. If so, the internal energy E depends on the temperature only [E is directly proportional to T; see Eqs. (19.26) and (19.33)]. Since the energy does not change, we can then conclude that the temperature does not change. Thus, in the free expansion of the gas, the temperature remains constant.

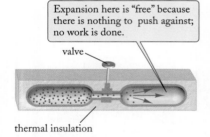

Expansion here is "free" because there is nothing to push against; no work is done.

valve

thermal insulation

FIGURE 21.4 Adiabatic free expansion of a gas.

✔ Checkup 21.1

QUESTION 1: Consider some fixed amount of gas sealed in a rubber balloon. For this system, what are the signs of Q and W if the balloon is heated by sunlight and expands?

QUESTION 2: Consider some fixed amount of gas sealed in a rigid glass jar. For this system, what are the signs of Q and W if the jar is heated?

QUESTION 3. An athlete pedals an exercise bicycle until he gets hot and begins to sweat. If we regard the athlete as the system, what are the respective signs of Q and W for this exercise process?

(A) Positive, positive
(B) Positive, negative
(C) Negative, positive
(D) Negative, negative

21.2 HEAT ENGINES; THE CARNOT ENGINE

Online
Concept
Tutorial

Steam engines and automobile engines convert heat into mechanical energy. The steam engine in Fig. 21.5 obtains heat from the combustion of coal or oil in a boiler; the automobile engine (Fig. 21.6) obtains heat from the (explosive) combustion of gasoline in its cylinders. If an engine is to convert heat continually into mechanical energy, it must operate cyclically. At the end of each cycle, it must return to its initial configuration, so it can repeat the process of conversion of heat into work over and over again. Steam engines and automobile engines are obviously cyclic—after one (or sometimes two) revolutions of their crankshaft or flywheel, they return to their initial configuration. These engines are not 100% efficient. The condenser of the steam engine and the radiator and exhaust of the automobile engine eject a substantial amount of heat into the environment; this waste heat represents lost energy.

Any device that converts heat into work by means of a cyclic process is called a **heat engine**. The engine absorbs heat from a heat reservoir at high temperature, converts this heat partially into work, and ejects the remainder as waste heat into a reservoir at low temperature. In this context, a **heat reservoir** is simply a body that remains at constant temperature, even when heat is removed from or added to it. In practice, the high-temperature heat reservoir is often a boiler whose temperature is kept constant by the controlled combustion of some fuel, and the low-temperature reservoir is usually a condenser in contact with a body of water or in contact with the atmosphere of the Earth, whose large volume permits it to absorb the waste heat without appreciable change of temperature.

Figure 21.7 is a flowchart for the energy, showing the heat Q_1 flowing into the engine from the high-temperature reservoir, the heat Q_2 (waste heat) flowing out of the engine into the low-temperature reservoir, and the work generated. The work generated by the engine is the difference between Q_1 and Q_2,

$$W = Q_1 - Q_2 \tag{21.4}$$

heat engine

heat reservoir

FIGURE 21.5 A steam engine.

FIGURE 21.6 An automobile engine.

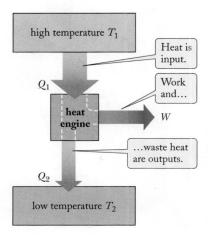

FIGURE 21.7 Flowchart for a heat engine. The center cube represents the heat engine.

The **efficiency** *of the engine is defined as the ratio of this work to the heat absorbed from the high-temperature reservoir:*

efficiency of heat engine

$$e = \frac{W}{Q_1} = \frac{Q_1 - Q_2}{Q_1} = 1 - \frac{Q_2}{Q_1} \tag{21.5}$$

This says that if $Q_2 = 0$ (no waste heat), then the efficiency would be $e = 1$, or 100%. If so, the engine would convert the high-temperature heat *totally* into work. As we will see later, this extreme efficiency is unattainable. Even under ideal conditions, the engine will produce some waste heat. It turns out that the efficiency of an ideal engine depends only on the temperatures of the heat reservoirs.[1]

EXAMPLE 2   The steam engine of a locomotive delivers 5.4×10^8 J of work per minute and receives 3.6×10^9 J of heat per minute from its boiler. What is the efficiency of this engine? How much heat is wasted per minute?

SOLUTION: From Eq. (21.5), the efficiency is the ratio of the work generated to the heat absorbed:

$$e = \frac{W}{Q_1} = \frac{5.4 \times 10^8 \text{ J}}{3.6 \times 10^9 \text{ J}} = 0.15 \tag{21.6}$$

Expressed in percent, this is 15%.

The wasted heat is the difference between the heat received and the work:

$$Q_2 = Q_1 - W = 3.6 \times 10^9 \text{ J} - 5.4 \times 10^8 \text{ J} = 3.1 \times 10^9 \text{ J} \tag{21.7}$$

EXAMPLE 3   During strenuous bicycling on a stationary bicycle (see Fig. 21.8), an athlete delivers 220 W of mechanical power to the pedals of the bicycle and, simultaneously, generates 760 W of waste heat. What overall efficiency is implied by these data?

SOLUTION: A power of 1 W is equal to 1 J per second. Thus, in one second, the mechanical work delivered by the athlete is 220 J and the waste heat generated is 760 J. Accordingly, the chemical energy flowing into the athlete's muscles must be $Q_1 = Q_2 + W = 760 \text{ J} + 220 \text{ J} = 980 \text{ J}$, which implies an efficiency of

$$e = \frac{W}{Q_1} = \frac{220 \text{ J}}{980 \text{ J}} = 0.22$$

Thus the efficiency of this process is 22%.

FIGURE 21.8 Athlete on a stationary bicycle.

[1] Note that according to the sign conventions summarized in Fig. 21.3, the heat flowing into the engine is supposed to be a positive quantity and the heat flowing out of the engine a negative quantity. But in discussions of the efficiency of engines, it is customary to treat both Q_1 and Q_2 as positive quantities, and an explicit minus sign therefore appears with Q_2 in Eq. (21.4); that is, $Q = Q_1$ for the absorbed heat, but $Q = -Q_2$ for the rejected heat.

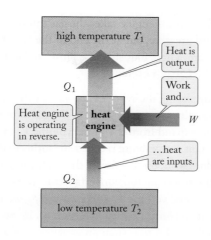

FIGURE 21.9 Flowchart for an engine operating in reverse. When the engine operates in the forward direction, it converts some amount of heat into work. When it operates in the reverse direction, it reconverts this amount of work into the original amount of heat.

We will now calculate the efficiency of an ideal heat engine that converts heat into work with maximum efficiency. As we will see in the next section, *for maximum efficiency*, the thermodynamic process within the engine should be **reversible**, *which means that the engine can, in principle, be operated in reverse and it then converts work into heat at the same rate as it converts heat into work when operating in the forward direction* (see Fig. 21.9).

The simplest kind of reversible engine is the **Carnot engine**, consisting of some amount of ideal gas enclosed in a cylinder with a piston (see Fig. 21.10). We can alternately place the cylinder in thermal contact with a high-temperature reservoir (where it absorbs heat) or a low-temperature reservoir (where it dumps waste heat). The gas delivers work when it pushes the piston outward, and the gas absorbs work when we push the piston inward. To achieve reversibility with this engine, the motion of the piston must be sufficiently slow, so that the gas is always in an equilibrium configuration. If we were to give the piston a sudden motion, a pressure disturbance would travel through the gas, and the motion of this pressure disturbance could not be reversed by giving the piston a sudden motion in the opposite direction—this would merely create a second pressure disturbance. Furthermore, the temperature of the gas must coincide with the temperature of the heat reservoir during contact. If the gas were at, say, lower temperature than that of the heat reservoir with which it was in contact, heat would rush from the reservoir into the gas, and this flow of heat could not be reversed by any manipulation of the piston. In practice, we cannot attain exact reversibility; but it is nevertheless worthwhile to consider the ideal Carnot engine with exact reversibility, because this tells us what is the best we can hope for when attempting to convert heat into work.

reversible process

Carnot engine

FIGURE 21.10 Carnot engine: a gas-filled cylinder with a piston.

p, V, and *T* can vary.

SADI CARNOT (1796–1832) *French engineer and physicist. In his book* On the Motive Power of Heat *he formulated the theory of the conversion of heat into work.*

The operation of the Carnot engine takes the gas through a sequence of four steps with varying volume and pressure, but at the end of the last step the gas returns to its initial volume and pressure. These four steps are illustrated in Fig. 21.11. The sequence of four steps is called the **Carnot cycle**:

Carnot cycle

1. We begin the cycle by placing the cylinder in contact with the high-temperature heat reservoir, which maintains the temperature of the gas at the constant value T_1. The gas is now allowed to expand from the initial volume V_1 to a new volume V_2. Such an expansion at constant temperature is called an **isothermal expansion**. During the expansion, the gas does work on the piston, that is, the engine absorbs heat Q_1 from the high-temperature reservoir and converts it into work.

isothermal expansion

2. When the gas has reached volume V_2 and pressure p_2, we remove it from the heat reservoir and allow it to continue the expansion on its own, in thermal isolation. We know from Section 20.6 that such an expansion of a thermally isolated gas, which neither receives heat from its surroundings nor loses any, is called an **adiabatic expansion**. During this expansion, the temperature of the gas decreases, as we saw in Section 20.6.

adiabatic expansion

3. When the temperature of the gas has decreased to the temperature T_2 of the low-temperature reservoir, we stop the piston and place the gas in contact with this low-temperature reservoir. The volume at this instant is V_3 and the pressure is p_3. We now begin to push the piston back toward its starting position, that is, we compress the gas isothermally. This means that the engine converts work to heat and ejects this heat Q_2 into the low-temperature reservoir.

4. When the gas has reached volume V_4 and pressure p_4, we remove it from the reservoir and continue to compress it adiabatically until the volume and the pressure return to their initial values. During this adiabatic compression, the temperature increases from T_2 to its initial value T_1.

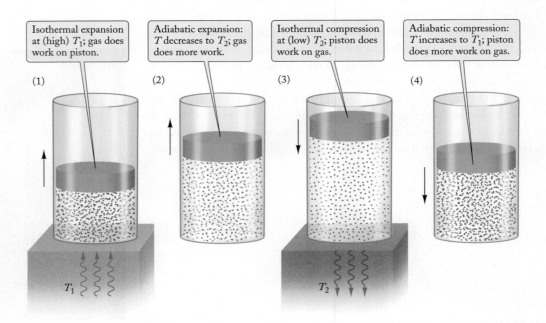

FIGURE 21.11 The Carnot cycle. The arrows indicate the displacements of the piston. (1) Isothermal expansion at temperature T_1 while in contact with a high-temperature heat reservoir. (2) Adiabatic expansion. (3) Isothermal compression at temperature T_2 while in contact with a low-temperature heat reservoir. (4) Adiabatic compression to the initial volume and pressure.

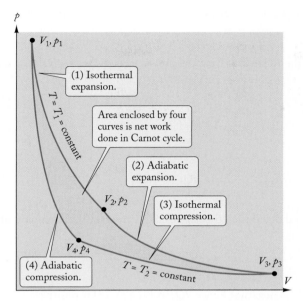

FIGURE 21.12 The Carnot cycle displayed on a p–V diagram.

To describe the operation of this engine mathematically, it is convenient to use a p–V diagram (see Fig. 21.12). Each point in this diagram represents a possible state of the gas; p and V can be read directly from the diagram, and T can then be calculated from the ideal-gas equation. Since $dW = p\, dV$ is the work done by the gas during a small increase of volume, the work done by the gas during a finite increase from volume V_1 to volume V_2 is $W = \int dW = \int_{V_1}^{V_2} p\, dV$; this integral is the area under the curve in the p–V diagram (see Fig. 21.13). When the volume decreases, work is instead done on the gas, that is, the work done by the gas is negative. Hence the net work done in the Carnot cycle is the area enclosed between the four curves representing the four steps of the cycle (see the green area in Fig. 21.12). However, in the following calculation of the efficiency we will not have occasion to use this interpretation of the area in the p–V diagram. Instead, we will directly evaluate the heat Q_1 that the engine absorbs from the high-temperature reservoir in step (1) and the heat Q_2 that it ejects into the low-temperature reservoir in step (3), and then calculate the efficiency from Eq. (21.5).

The calculations of Q_1 and Q_2 are somewhat difficult because the volume and the pressure vary in a somewhat complicated way along the curved lines in Fig. 21.12. To simplify the calculations, let us consider the small (infinitesimal) Carnot cycle shown in Fig. 21.14. In this cycle, the pressure and the volume vary by only small amounts during each of the four steps (1)–(4); these small changes of pressure and of volume are represented by the straight lines drawn in Fig. 21.14. We will designate the changes of volume for the four steps of the cycle by dV_{12}, dV_{23}, dV_{34}, and dV_{41}; the first two of these changes of volume are positive (expansion), and the last two are negative (compression). Since the net change of volume for the entire cycle is zero, the changes of volume satisfy the identity

$$dV_{12} + dV_{23} + dV_{34} + dV_{41} = 0 \qquad (21.8)$$

During the heat-absorption process (1), the temperature is constant; since the energy E of an ideal gas depends on the temperature only, we see that $dE = 0$ and, by Eq. (21.2), the amount of absorbed heat dQ_1 then equals the work dW done by the gas:

$$dQ_1 = dE + dW = dW \qquad (21.9)$$

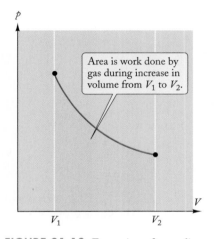

FIGURE 21.13 Expansion of a gas displayed on a p–V diagram. The work done by the gas is $W = \int dW = \int_{V_1}^{V_2} p\, dV$.

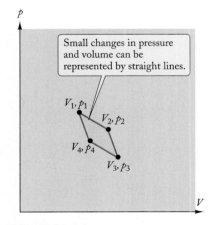

FIGURE 21.14 A small (infinitesimal) Carnot cycle, with small changes in temperature, volume, and pressure.

By Eq. (20.23), the work is the product of the pressure and the change of volume:

$$dW = p_1 dV_{12} \tag{21.10}$$

According to the Ideal-Gas Law, the pressure equals $p_1 = nRT_1/V_1$, and therefore

$$dW = nRT_1 \frac{dV_{12}}{V_1} \tag{21.11}$$

so

$$dQ_1 = nRT_1 \frac{dV_{12}}{V_1} \tag{21.12}$$

Likewise, we find that during the heat-ejection process (3), the amount of ejected heat is

$$dQ_2 = -nRT_2 \frac{dV_{34}}{V_1} \tag{21.13}$$

[Note that dV_{34} is negative; hence dQ_2 in Eq. (21.13) is positive.] For the calculation of the efficiency (21.5), we must evaluate the ratio dQ_2/dQ_1:

$$\frac{dQ_2}{dQ_1} = -\frac{nRT_2 dV_{34}}{nRT_1 dV_{12}} = -\frac{T_2}{T_1}\frac{dV_{34}}{dV_{12}} \tag{21.14}$$

To complete the evaluation of the right side of this equation, we need to know how the volume changes dV_{12} and dV_{34} are related. A relationship between these volume changes is given by Eq. (21.8), which contains not only the volume changes dV_{12} and dV_{34}, but also the volume changes dV_{23} and dV_{41}. The volume changes dV_{23} and dV_{41} correspond to the adiabatic processes (2) and (4), and we can exploit the results we obtained in Section 20.6 for the volume change in an adiabatic process. According to Eq. (20.40), the (small) volume change in an adiabatic process is directly proportional to the (small) temperature change. Since the temperature change in process (3) is simply the opposite of the temperature change in process (4), it follows that the volume changes dV_{23} and dV_{41} must also be opposites, that is, $dV_{23} = -dV_{41}$. We therefore see that the terms dV_{23} and dV_{41} in Eq. (21.8) cancel, leaving only

$$dV_{12} + dV_{34} = 0 \tag{21.15}$$

which implies that dV_{12} and dV_{34} are also opposites:

$$dV_{12} = -dV_{34} \tag{21.16}$$

With this result, the volume ratio on the right side of Eq. (21.14) is equal to -1, and the ratio of the amounts of heat is

$$\frac{dQ_2}{dQ_1} = \frac{T_2}{T_1} \tag{21.17}$$

This simply says that the amounts of heat are in direct proportion to the temperatures of the reservoirs.

With this simple expression for dQ_2/dQ_1, the formula (21.5) for the efficiency tells us that

efficiency of heat engine
in terms of absolute temperatures

$$e = 1 - \frac{dQ_2}{dQ_1} = 1 - \frac{T_2}{T_1} \tag{21.18}$$

This shows that *the efficiency depends only on the temperatures of the heat reservoirs.* Although our derivation of this formula for the efficiency was based on the small Carnot cycle shown in Fig. 21.14, it can be demonstrated that this formula is also valid for the large Carnot cycle shown in Fig. 21.12, because such a large cycle can be regarded as a combination of many small cycles (in the same way that any given area in a plane can be approximated by a large number of small rectangles).

Note that an efficiency of $e = 1$ (or 100%) can be achieved only if $T_2 = 0$, that is, if the low-temperature reservoir is at the absolute zero of temperature. Unfortunately, we have no such absolutely cold reservoir available anywhere.

EXAMPLE 4   The boiler of a locomotive steam engine produces steam at a temperature of 500°C. The engine exhausts its waste heat into the atmosphere, where the temperature is 20°C. As in Example 2, the actual efficiency of this steam engine is 0.15. In order to see how much improvement on this efficiency is possible, compare this with the efficiency of a Carnot engine operating between the same temperatures.

SOLUTION: Our analysis of efficiency is described in terms of absolute temperature, so the releevant temperatures of operation are

$$T_1 = (500 + 273)\text{K} = 773 \text{ K} \quad \text{and} \quad T_2 = (20 + 273)\text{K} = 293 \text{ K}$$

According to Eq. (21.18), the efficiency of a Carnot engine operating between these temperatures is

$$e = 1 - \frac{T_2}{T_1} = 1 - \frac{293 \text{ K}}{773 \text{ K}} = 0.62$$

or 62%; thus the maximum theoretical efficiency is here more than 4 times the typical efficiency.

The basic steps in the operation of a practical steam engine resemble the steps of the Carnot cycle. The **steam engine** has a piston that performs a cyclic motion of expansion and contraction. However, in contrast to the Carnot engine, which uses gas as working fluid, practical steam engines use gas and liquid (steam and water) as working fluid, and the cycle of a steam engine differs from the Carnot cycle. Figure 21.15 is a schematic diagram of the main parts of a simple steam engine with boiler, cylinder, and condenser. The boiler produces hot, high-pressure steam, which enters the cylinder, pushes against the piston, expands, and does work. The piston is linked to a crankshaft, by means of which the steam engine drives some external machinery, and does useful external work. Upon completion of the expansion, the low-pressure, spent steam is exhausted from the cylinder, and the piston returns to its initial position; these operations are controlled by the opening and closing of valves. Meanwhile, the spent steam is sent to a condenser, where an external coolant (air or flowing water) condenses the steam into liquid water. This liquid water is pumped back to the boiler. Each completed circulation of the fluid through the circuit can be regarded as one complete cycle of operation of the steam engine. Such simple steam engines have efficiencies of only 5–18%. Most modern steam engines employ a turbine wheel instead of the cylinder and piston; large engines of this kind achieve efficiencies of up to 40%.

For a description of the operation of an automobile engine see Physics in Practice: Efficiency of Automobiles on page 674.

steam engine

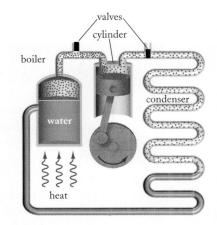

FIGURE 21.15 A steam engine (schematic diagram).

A Carnot engine can be operated in reverse, so the Carnot cycle begins with step (4) and ends with step (1). The Carnot engine then uses up work to transfer heat from the low-temperature reservoir to the high-temperature reservoir. This is the principle involved in the operation of **refrigerators**, air conditioners, and "heat pumps." The amount of work required to operate a Carnot engine in reverse can be calculated from Eq. (21.17); as stated above, this equation is valid not only for an infinitesimal Carnot cycle, but also for a finite Carnot cycle, for which the equation takes the form

$$\frac{Q_2}{Q_1} = \frac{T_2}{T_1} \tag{21.19}$$

refrigerator

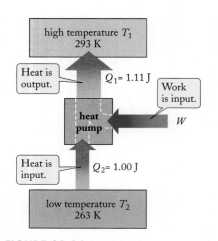

FIGURE 21.16 Flowchart for the operation of a heat pump.

| EXAMPLE 5 |

Suppose a homeowner uses a Carnot engine operating in reverse as a heat pump to extract heat from the outside air and inject it into his home. If the outside temperature is $-10°C$ and the inside temperature is $20°C$, what is the amount of work that must be supplied to pump 1.00 J of heat from the outside to the inside? How does this compare with the direct conversion of energy to heat, as in an electric heater?

SOLUTION: For a Carnot engine, the ratio of the heats exchanged at the low-temperature reservoir and at the high-temperature reservoir is given by Eq. (21.19):

$$\frac{Q_2}{Q_1} = \frac{T_2}{T_1} = \frac{-10 + 273 \text{ K}}{20 + 273 \text{ K}} = \frac{263 \text{ K}}{293 \text{ K}} = 0.898$$

If $Q_2 = 1.00$ J, then $Q_1 = Q_2/0.898 = 1.11$ J. The difference $Q_1 - Q_2$ represents the work that must be supplied (see the flowchart in Fig. 21.16); hence the work is 0.11 J.

COMMENTS: By the expenditure of 0.11 J of work, the heat pump delivers a total of 1.11 J of heat into the house. Thus a small amount of work can bring a large amount of heat into the home. This is obviously a much more economical heating method than the direct expenditure of 1.11 J of fuel or electric energy in a conventional furnace or electric heater. Heat pumps achieve their highest efficiencies when the temperatures T_1 and T_2 are nearly equal, and they are widely used in mild climates (see Fig. 21.17).

Practical refrigerators use a gas and liquid refrigerant (such as R15, an environmentally safe refrigerant that has replaced Freon) as working fluid, and their cycle differs from the Carnot cycle. The fluid refrigerants have a boiling point near room

PROBLEM-SOLVING TECHNIQUES   THERMODYNAMIC CALCULATIONS

- Be careful about units in thermodynamic calculations. If both heat and work appear in an equation, then you must express both in the same energy units. For instance, in the First Law of Thermodynamics, express both the heat and the work in joules.
- The temperature used in the formula (21.18) for the efficiency of a Carnot engine is the absolute temperature, in kelvins. The formula is not valid if you express the temperature in °C or °F.
- To keep track of the amounts of heat and work, it is helpful to draw a flowchart, such as the flowchart for the heat engine in Fig. 21.7 or the flowchart for the heat pump in Fig. 21.16.

FIGURE 21.17 A heat pump installed outside a home.

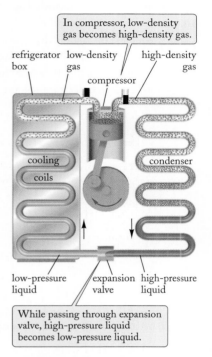

FIGURE 21.18 A refrigerator (schematic diagram).

temperature when at high pressure, but a boiling point below 0°C when at low pressure. Figure 21.18 is a schematic diagram of the parts of a practical refrigerator. Liquid refrigerant at low pressure enters the cooling coils in the refrigerator box and absorbs heat while evaporating into gas. This gas flows to the compressor, where its pressure and density are increased by the push of a piston. The high-pressure gas then circulates through the condenser coils, which are exposed to the atmospheric air (see Fig. 21.19). The gas loses its heat and condenses into liquid. This high-pressure liquid then passes through an expansion valve (a small orifice) where its pressure is reduced to match the low pressure in the cooling coils. This return of the fluid to the cooling coils completes the cycle.

An air conditioner employs a similar refrigeration cycle; and a "heat pump" is, in essence, an air conditioner turned around, so its cold end is outdoors and its warm end indoors.

The performance of a heat pump, an air conditioner, or a refrigerator is often rated using a **coefficient of performance** CP. A heat pump delivers heat to the high-temperature reservoir (the inside of a house), and so its coefficient of performance is defined as the ratio of the heat delivered to the work done: $CP = Q_1/W$. An air conditioner or refrigerator removes heat from a low-temperature reservoir; the coefficient of performance is then defined as the ratio of the heat removed to the work done: $CP = Q_2/W$. For practical devices, values of the coefficient of performance are usually greater than 1, often around 5.

coefficient of performance CP

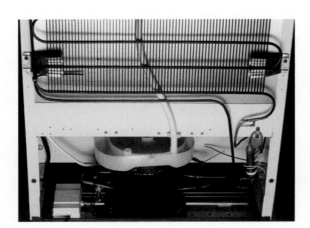

FIGURE 21.19 Condenser coils of a refrigerator.

PHYSICS IN PRACTICE   EFFICIENCY OF AUTOMOBILES

An automobile can be regarded as a thermodynamic engine that converts the chemical energy of fuel into mechanical energy, which is ultimately dissipated by friction. Figure 1 is a flowchart illustrating the successive energy conversions in an automobile propelled by a gasoline engine; Fig. 2 shows the piston's cycle: intake, compression, expansion, and exhaust.

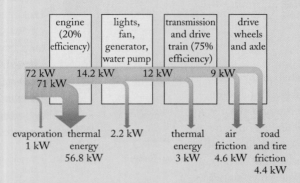

FIGURE 1 Energy conversions in an automobile. (Based on R.H. Romer, *An Introduction to Physics*.)

One liter of gasoline releases 3.4×10^7 J of chemical energy when burned with oxygen. While traveling at 65 km/h on a level road, the automobile described in this flowchart consumes 7.6 liters per hour, equivalent to a chemical energy of 2.6×10^8 J per hour, or 72 kW. The engine converts about 20% of this chemical energy into mechanical energy. It delivers most of this mechanical energy to the transmission and the axles of the automobile. The energy reaching the axles is in part fed into the body of the automobile (the axle pushes forward against its bearings, doing work on the body of the automo-

bile), and in part it is fed into the wheels. The energy delivered to the body is dissipated into heat by air friction, and the energy delivered to the wheels is dissipated into heat by rolling friction against the road. Of the 72 kW of chemical energy entering the engine, only 14 kW emerges as propulsive power, much of which does useful work against the external friction forces of air and road. Most of the energy is dissipated in the conversion process. Note that the largest waste occurs in the engine itself—in gasoline engines most of the chemical energy is ejected into the exhaust gases. Diesel engines achieve a somewhat higher efficiency, because they burn their fuel more completely.

However, in part, the inefficiency of these engines is a consequence of the basic laws of thermodynamics. The gasoline engine is a heat engine—it converts the heat released by the burning fuel into mechanical energy. The efficiency of an ideal heat engine is given by Eq. (21.18). In the gasoline engine, the heat source is the hot gas produced by the combustion, and the heat sink is the air surrounding the engine. The complete combustion of hydrocarbon fuel gives the residual gas a temperature of about 2400 K. The air surrounding the engine typically has a temperature of 300 K. Hence the maximum attainable efficiency is

$$e = 1 - \frac{T_2}{T_1} = 1 - \frac{300 \text{ K}}{2400 \text{ K}} = 0.88$$

Gasoline engines and other engines that burn hydrocarbon fuels do not come anywhere near this maximum efficiency of an ideal engine, because they do not burn the fuel completely, they eject exhaust gases at a temperature much higher than 300 K, and they suffer from internal friction.

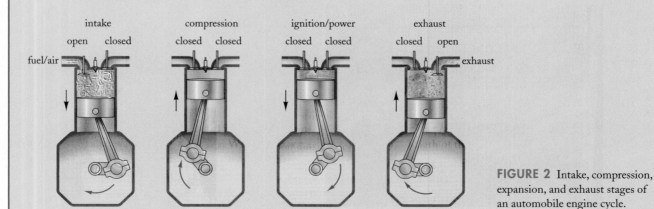

FIGURE 2 Intake, compression, expansion, and exhaust stages of an automobile engine cycle.

Checkup 21.2

QUESTION 1: What is the difference between an isothermal expansion and an adiabatic expansion?

QUESTION 2: Can the efficiency of an engine ever be larger than 1? Why or why not?

QUESTION 3: Is the operation of an automobile engine reversible?

QUESTION 4: Does the efficiency of a Carnot engine depend on the size of the cycle, that is, how far we let the piston in Fig. 21.11 travel within its cylinder?

QUESTION 5: Suppose we connect two Carnot engines together, so the work output of the first drives the second in reverse (see Figs. 21.7 and 21.9). What is the net effect?

    (A) Net heat is added.   (B) Net work is done.   (C) There is no net effect.

21.3 THE SECOND LAW OF THERMODYNAMICS

As we saw in the preceding section, a Carnot engine operating with an ideal gas as working fluid has a limited efficiency—it fails to convert all of the heat into work and instead produces some waste heat. The **Second Law of Thermodynamics** asserts that this is a limitation from which all heat engines suffer. This law may be stated in several ways. As formulated by Lord Kelvin and Max Planck, this law simply states,

> *An engine operating in a cycle cannot transform heat into work without some other effect on its surroundings.*

It is an immediate corollary of this law that *the efficiency of any heat engine operating between two heat reservoirs of high and low temperature is never greater than the efficiency of a Carnot engine; furthermore, the efficiency of any reversible engine equals the efficiency of a Carnot engine.* The proof of these statements, known as **Carnot's theorem**, is by contradiction. Imagine that some heat engine is more efficient than a Carnot engine, so this engine converts heat from a reservoir at a high temperature into work and ejects only a small amount of waste heat into a reservoir at low temperature. We can then use the work output of this engine to drive a Carnot engine in reverse, and thereby pump the waste heat from the low-temperature reservoir to the high-temperature reservoir. By hypothesis, the given engine is more efficient than the Carnot engine; hence only part of its work output will be needed to drive the reversed Carnot engine and return all of the waste heat to the high-temperature reservoir. The remainder of the output constitutes available work (see Fig. 21.20 for a flowchart). The net effect of the joint operation of both engines is the complete conversion of heat into work, without waste heat, in contradiction to the Second Law. We can avoid this contradiction only if the efficiency of any engine is never greater than that of a Carnot engine.

Second Law of Thermodynamics

Carnot's theorem

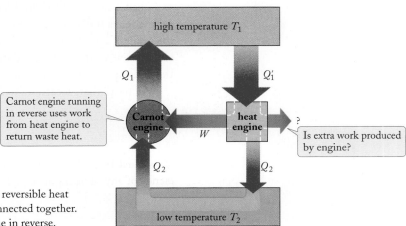

FIGURE 21.20 Flowchart for an arbitrary reversible heat engine (box) and a Carnot engine (circle) connected together. The arbitrary engine drives the Carnot engine in reverse.

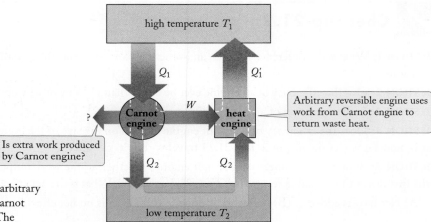

FIGURE 21.21 Flowchart for an arbitrary reversible heat engine (box) and a Carnot engine (circle) connected together. The Carnot engine drives the arbitrary engine in reverse.

To prove that the efficiency of any reversible engine equals that of a Carnot engine, we again consider the net effect of the joint operation of the two engines with the Carnot engine running in the forward direction and its work output driving the other reversible engine in the backward direction (see Fig. 21.21). By an argument similar to that given above, it now follows that the efficiency of the Carnot engine cannot be greater than that of the other engine. Thus, the efficiency of each engine can be no greater than that of the other; that is, they both must have exactly the same efficiency.

We recall that a perpetual motion machine of the second kind is an engine that takes heat energy from a reservoir and completely converts it into work. Thus, the Second Law asserts that no perpetual motion machine of the second kind can exist. Essentially, the operation of any heat engine hinges on the temperature difference between two heat reservoirs; heat tends to rush from the high-temperature reservoir to the low-temperature reservoir. By interposing a heat engine in the path of this rush of heat, we can force the heat to do useful work. The Second Law can be formulated in an alternative form which is based on this characteristic of the flow of heat. This formulation, due to Rudolf Clausius, states,

> *An engine operating in a cycle cannot transfer heat from a cold reservoir to a hot reservoir without some other effect on its surroundings.*

The Clausius and Kelvin–Planck formulations of the Second Law are equivalent—each implies the other. The proof is, again, by contradiction, and relies on examining the result of a joint operation of two engines: a Carnot engine, and a second engine that violates one of the two formulations of the Second Law. It is then easy to show that the joint operation violates the other formulation of the Second Law.

We recall that, according to Eq. (21.18), the efficiency of a Carnot engine is

$$e = 1 - \frac{T_2}{T_1} \qquad (21.20)$$

This must then also be the efficiency of any other reversible engine, since the efficiency of any such engine equals that of a Carnot engine.

Checkup 21.3

QUESTION 1: What happens to the efficiency of a Carnot engine if $T_2 = T_1$? What if $T_2 > T_1$?

QUESTION 2: A refrigerator transfers heat from a cold reservoir (the interior of the refrigerator) to a hot reservoir (the kitchen). Does this violate the Second Law of Thermodynamics?

QUESTION 3: What would happen if you were to try to operate a steam engine in a very hot environment, where the condenser is surrounded by air at about the same temperature as the boiler?

    (A) With a warm environment, heating is easier, so the efficiency increases.

    (B) Heating is easier, but heat flow decreases, so the efficiency is the same.

    (C) The cycle could not be completed, since the steam could not condense.

21.4 ENTROPY

In the operation of a Carnot engine, or any other reversible engine, the ratio of the heats absorbed and ejected at the high- and low-temperature reservoirs is, according to Eq. (21.19).

$$\frac{Q_2}{Q_1} = \frac{T_2}{T_1} \tag{21.21}$$

We can rewrite this equation in a suggestive way by multiplying both sides by $-Q_1/T_2$ and then moving all terms to the left side:

$$\frac{Q_1}{T_1} - \frac{Q_2}{T_2} = 0 \tag{21.22}$$

If, as in Section 20.1, we regard the heat change as positive when heat enters the system and negative when it leaves the system, then $\Delta Q_1 = Q_1$ for the heat exchange with the high-temperature reservoir and $\Delta Q_2 = -Q_2$ for the heat exchange with the low-temperature reservoir. Thus, Eq. (21.22) says that, for a Carnot cycle, the sum of all the heat exchanges ΔQ divided by their respective temperatures is zero:

$$\frac{\Delta Q_1}{T_1} + \frac{\Delta Q_2}{T_2} = 0 \tag{21.23}$$

This result can also be shown to be valid for any other reversible cycle, because any other reversible cycle can be approximated by a large number of small Carnot cycles. A general reversible cycle may have a large number of successive heat exchanges, each at some different temperature. For such a cycle, the generalized version of Eq. (21.23) is

$$\frac{dQ_1}{T_1} + \frac{dQ_2}{T_2} + \frac{dQ_3}{T_3} + \frac{dQ_4}{T_4} + \cdots = 0$$

If we proceed to the limit of a large number of small (infinitesimal) reversible heat exchanges, this sum becomes an integral,

Clausius' theorem

$$\int \frac{dQ}{T} = 0 \quad \text{(reversible cycle)} \qquad (21.24)$$

This result is called **Clausius' theorem**.

entropy

The importance of Clausius' theorem is that it permits us to define a new physical quantity called the **entropy**, or more precisely, the entropy difference. Given two states A and B of a system, the entropy difference ΔS between them is defined as the sum of the heat exchanges divided by the temperature for some reversible process that takes the system from the initial state A to the final state B:

$$\Delta S = \int_A^B \frac{dQ}{T} \quad \text{(reversible process)} \qquad (21.25)$$

Since there is usually a wide choice of different reversible processes that take the system from the given initial state to the given final state, the definition (21.25) would make no sense if these different processes gave different values to the sum (21.25). Clausius' theorem ensures that all these reversible processes lead to the same value for (21.25). To understand the connection between the definition of entropy and Clausius' theorem, consider the example of the ideal gas in the Carnot engine, for which different states are plotted in the p–V diagram in Fig. 21.12. If the initial state A corresponds to the point 1 in Fig. 21.12, and the final state B to the point 3, then one possible reversible path from 1 to 3 consists of the steps (1) and (2) of the Carnot cycle. The heat exchanged in step (1) is Q_1, and the heat exchanged in step (2) is zero; thus Eq. (21.25) yields

$$\Delta S = \frac{Q_1}{T_1} \qquad (21.26)$$

for the entropy change between points 1 and 3. But another reversible path from 1 to 3 consists of step (4) *in reverse* followed by step (3) *in reverse*—this also takes the system from point 1 to point 3. The heat exchanged in step (4) is zero, and the heat exchanged in step (3) is Q_2. Hence Eq. (21.25) now yields

$$\Delta S = \frac{Q_2}{T_2} \qquad (21.27)$$

But Eq. (21.22)—which is Clausius' theorem for the Carnot cycle—immediately tells us that the two expressions (21.26) and (21.27) for the entropy change are equal.

The SI unit of entropy is the unit of energy divided by the unit of temperature: joule per kelvin (J/K). An alternative unit is calories per kelvin (cal/K).

The entropy in thermodynamics plays a role somewhat analogous to that of the potential energy in mechanics. Just as the potential energy allows us to make some predictions about possible motions of a mechanical system, the entropy allows us to make some predictions about the possible behavior of a thermodynamic system. For instance, it can be shown from the Second Law of Thermodynamics that the entropy of a closed system—which is thermally and mechanically isolated from its surroundings—can never decrease. The entropy of such a system either remains constant (if only reversible processes are occurring in the system) or increases (if irreversible processes

RUDOLF CLAUSIUS (1822–1888)
German mathematical physicist. He was one of the creators of the science of thermodynamics. He contributed the concept of entropy, as well as the restatement of the Second Law of Thermodynamics.

are occurring). Thus an alternative formulation of the Second Law of Thermodynamics asserts that

The entropy of a closed system can never decrease.

The increase of entropy in closed systems is illustrated in the following examples.

EXAMPLE 6   A heat reservoir at a temperature of $T_1 = 400$ K is briefly put in thermal contact with a reservoir at $T_2 = 300$ K (see Fig. 21.22). If 1.00 J of heat flows from the hot reservoir to the cold reservoir, what is the change of the entropy of the system consisting of both reservoirs?

SOLUTION: Obviously, the heat flow from the hot to the cold reservoir is irreversible, and hence Eq. (21.25), which is restricted to reversible processes, is not directly applicable. To evaluate the entropy change by means of Eq. (21.25), we must first invent some process that reversibly takes the system from the initial state to the final state. We can imagine that 1.00 J flows from the hot reservoir into an auxiliary reservoir of a temperature just barely below 400 K, and that simultaneously 1.00 J flows from another auxiliary reservoir of a temperature just barely above 300 K into the cold reservoir. Then all processes are reversible, and Eq. (21.25) gives us the change of entropy:

$$\Delta S = \frac{\Delta Q_1}{T_1} + \frac{\Delta Q_2}{T_2} = \frac{-1.00 \text{ J}}{400 \text{ K}} + \frac{1.00 \text{ J}}{300 \text{ K}}$$

$$= 8.3 \times 10^{-4} \text{ J/K} \tag{21.28}$$

COMMENT: Note that when the magnitudes of the heat transfers are equal, as they are in this example, the contributions to the entropy are inversely proportional to the temperature; that is, "hot heat" contributes a smaller entropy change than "cold heat."

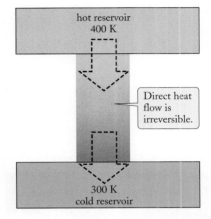

FIGURE 21.22 A hot reservoir (400 K) is in thermal contact with a colder reservoir (300 K). Heat flows through a conducting rod connecting the two reservoirs, from the hot reservoir into the colder reservoir.

EXAMPLE 7   A large stone, of mass 80 kg, slides down a hill of a vertical height of 100 m and is stopped by friction at the bottom. What is the increase of the entropy of the stone plus the surroundings? Assume that the temperature of the surroundings (hill and air) is 270 K.

SOLUTION: All of the initial mechanical energy of the stone is converted into heat:

$$Q = mgh = 80 \text{ kg} \times 9.81 \text{ m/s}^2 \times 100 \text{ m}$$

$$= 7.8 \times 10^4 \text{ J}$$

This heat is delivered to the surroundings at a temperature of 270 K.

Clearly, the process described is irreversible (the heat generated could not be completely reconverted into mechanical work). To calculate the entropy with Eq. (21.25), we must imagine a reversible process that brings the stone down the hill and delivers the heat into the surroundings. In principle, we can use an elevator to let the stone down slowly without friction and extract work while removing the potential energy; and afterward we can use a heat reservoir of a temperature barely above 270 K to supply reversibly the correct amount of heat (7.8×10^4 J) to the stone's surroundings. The first of these steps makes no contribution to the entropy, and the second makes a contribution

$$\Delta S = \frac{\Delta Q}{T} = \frac{7.8 \times 10^4 \, \text{J}}{270 \, \text{K}} = 290 \, \text{J/K} \tag{21.29}$$

From a microscopic point of view, the increase of entropy of a system is an increase of disorder. This is obvious in the preceding example of the conversion of mechanical energy into heat by friction. The translational motion of a macroscopic body is ordered energy—all the particles in the body move in the same direction with the same speed. Heat is disordered energy—the particles move in random directions with a mixture of speeds.

The increase of disorder also holds true in our other example of spontaneous flow of heat from a hot to a cold reservoir. When a given amount of heat is added to a reservoir of low temperature, it causes more disorder than when the same amount of heat is added to a reservoir of high temperature; in both reservoirs, the added heat generates extra random motion and extra disorder, but in the cold reservoir the percent increment of the random motions is larger than in the hot reservoir, and consequently the extra disorder is larger.

The connection between entropy and disorder can be given a precise mathematical meaning, but the details would require a discussion of statistical mechanics and information theory—and here we will not go into this. We merely assert that the Second Law can be reformulated to say,

Processes in a closed system always tend to increase the amount of disorder.

Human activities on the Earth, like any other processes in nature, cause an increase of disorder. Of course, *some* of our activities result in an increase of *order* in some portion of the system. For instance, when we extract dispersed bits of metal from ores and assemble them into a watch, we are obviously increasing the order of the bits of metal. However, we can do this only by simultaneously generating disorder somewhere else—the smelting of ores and the machining of metals demand an input of energy that is converted into waste heat, and increases the disorder of the environment. The net result is always an increase of disorder. All of our activities depend on a supply of highly ordered energy in the form of chemical or nuclear fuels or light from the Sun that can "soak up" disorder while becoming degraded into heat. We continually convert useful, ordered energy into useless, disordered energy.

We end this chapter with a brief statement of the **Third Law of Thermodynamics**. This law, as formulated by Walther Nernst, asserts that

The entropy of a system at absolute zero is zero.

We can understand this law in terms of the connection between entropy and disorder, mentioned above. As we lower the temperature of a system, we decrease the random thermal motions, and we decrease the disorder. According to classical theory, the random thermal motions cease completely as the temperature of the system approaches absolute zero. The system then tends to settle into a state of minimum disorder, that is, a state of minimum entropy.

WALTHER HERMANN NERNST
(1864–1941) *German physicist and chemist. He was a pioneer in physical chemistry and received in Nobel Prize in Chemistry in 1920 for his discovery of the Third Law of Thermodynamics.*

Third Law of Thermodynamics

Checkup 21.4

QUESTION 1: Two reservoirs are at temperatures of 300 K and 400 K, respectively. Each gains 100 J of heat from another reservoir. What is the entropy change in each?

QUESTION 2: Is the flowchart in Fig. 1 of Physics in Practice: Efficiency of Automobiles on page 674 consistent with the Second Law of Thermodynamics?

QUESTION 3: A pot full of hot water is placed in a cold room, and the pot gradually cools. Does the entropy of the water decrease? Does this violate the Second Law of Thermodynamics?

QUESTION 4: Which of the following processes results in an overall increase of entropy: (1) braking of an automobile, (2) gradual slowing of an automobile as it rolls uphill (without friction), (3) electric pump lifting water (without friction) from a well to a reservoir above, (4) firewood burning in a fireplace?

(A) 1 and 2   (B) 1 and 4   (C) 2 and 3   (D) 2 and 4   (E) 3 and 4

SUMMARY

PROBLEM-SOLVING TECHNIQUES Thermodynamic Calculations		**(page 672)**

PHYSICS IN PRACTICE Efficiency of Automobiles		**(page 674)**

FIRST LAW OF THERMODYNAMICS
where ΔE is the change in internal energy of the gas, Q is the heat added to the gas, and W is the work done by the gas.

$$\Delta E = Q - W \qquad (21.2)$$

EFFICIENCY OF HEAT ENGINE
where W is the work done by the engine, Q_1 is the heat absorbed from the high-temperature reservoir, and Q_2 is the waste heat.

$$e = \frac{W}{Q_1} = 1 - \frac{Q_2}{Q_1} \qquad (21.5)$$

CARNOT CYCLE   (1) Isothermal expansion at T_1; (2) adiabatic expansion to T_2; (3) isothermal compression at T_2; (4) adiabatic compression to T_1.

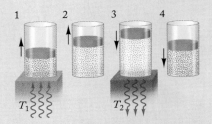

EFFICIENCY OF CARNOT ENGINE (OR ANY OTHER REVERSIBLE HEAT ENGINE)   Where T_1 and T_2 are absolute temperatures.

$$e = 1 - \frac{T_2}{T_1} \qquad (21.18)$$

CARNOT'S THEOREM   The efficiency of any engine cannot exceed that of a Carnot engine; the efficiency of any reversible engine equals that of a Carnot engine.

ENTROPY CHANGE (REVERSIBLE PROCESS)

$$\Delta S = \int_A^B \frac{dQ}{T} \tag{21.25}$$

SECOND LAW OF THERMODYNAMICS

- An engine operating in a cycle cannot transform heat into work without some other effect on its surroundings; or
- The entropy of a closed system can never decrease; or
- Processes in a closed system always tend to increase the amount of disorder.

THIRD LAW OF THERMODYNAMICS

The entropy is zero at absolute zero.

QUESTIONS FOR DISCUSSION

1. Figure 21.23 shows a perpetual motion machine designed by M. C. Escher. Exactly at what point is there a defect in this design?

FIGURE 21.23 Waterfall, lithograph by M. C. Escher.

2. An inventor proposes the following scheme for the propulsion of ships on the ocean without the input of energy: cover the hull of the ship with copper sheets and suspend an electrode of zinc in the water at some distance from the hull; since seawater is an electrolyte, the hull of the ship and the electrode will then act as the terminals of a battery, which can deliver energy to an electric motor propelling the ship. Will this scheme work? Does it violate the First Law of Thermodynamics?

3. The lowest-temperature heat sink available in nature is interstellar and intergalactic space, with a temperature of 3 K. Why don't we use this heat sink in the operation of a Carnot engine?

4. In some showrooms, salespeople demonstrate air conditioners by simply plugging them into an outlet without bothering to install them in a window or a wall. Does such an air conditioner cool the showroom or heat it?

5. Can mechanical energy be converted completely into heat? Give some examples.

6. If you leave the door of a refrigerator open, will it cool the kitchen?

7. Which of the following processes are irreversible, and which are reversible?

   (a) Burning a piece of paper.
   (b) Slow descent of an elevator attached to a perfectly balanced counterweight.
   (c) Breaking of a windowpane.
   (d) Exploding of a stick of dynamite.
   (e) Descent of a roller-coaster car from a hill, without friction.

8. If we measure the efficiency of a Carnot engine directly, we can use Eq. (21.20) to calculate the temperature of one of the heat reservoirs. What are the advantages and the disadvantages of such a thermodynamic determination of temperature?

9. In the absence of a heat sink, energy cannot be extracted from an ocean at uniform temperature. However, within the oceans there are small temperature differences between the warm water on the surface and the cooler water in the depths.

Design a Carnot engine that exploits this temperature difference.

10. Why are heat pumps used for heating houses in mild climates but not in very cold climates?

11. Does the Second Law of Thermodynamics forbid the spontaneous flow of heat between two bodies of equal temperature?

12. The inside of an automobile parked in the sun becomes much hotter than the surrounding air. Does this contradict the Second Law of Thermodynamics?

13. According to a story by George Gamow,[2] on one occasion Mr. Tompkins was drinking a cocktail when all of a sudden one small part at the surface of the liquid became hot and boiled with violence, releasing a cloud of steam, while the remainder of the liquid became cooler. Is this consistent with the First Law of Thermodynamics? With the Second Law?

14. A vessel is divided into two equal volumes by a partition. One of these volumes contains helium gas, and the other contains argon at the same temperature and pressure. If we remove the partition and allow the gases to mix, does the entropy of the system increase?

15. Does the motion of the planets around the Sun generate entropy?

16. Does static friction generate entropy?

17. Consider the process of emission of light by the surface of the Sun followed by absorption of this light by the surface of the Earth. Does this entail an increase in entropy?

18. Suppose that a box contains gas of extremely low density, say, only 50 molecules of gas altogether. The molecules move at random, and it is possible that once in a while all of the 50 molecules are simultaneously in the left half of the box, leaving the right half empty. At this instant the entropy of the gas is less than when the gas is more or less uniformly distributed throughout the box. Is this a violation of the Second Law of Thermodynamics? What are the implications for the range of validity of this law?

[2] G. Gamow, *Mr. Tompkins in Paperback.*

19. **Maxwell's demon** is a tiny hypothetical creature that can see individual molecules. The demon can make the heat flow from a cold body to a hot body as follows: Suppose that a box initially filled with gas at uniform temperature and pressure is divided into two equal volumes by a partition, equipped with a small door that is closed but can be opened by the demon (Fig. 21.24). Whenever a molecule of above-average speed approaches the door from the left, the demon quickly opens the door and lets it through. Whenever a molecule of below-average speed approaches from the right, the demon also lets it through. This selective action of the demon accumulates hot gas in the right volume, and cool gas in the left volume. Does this violate the Second Law of Thermodynamics? Do any of the activities of the demon involve an *increase* of entropy?

FIGURE 21.24 Maxwell's demon.

20. According to one cosmological model based on Einstein's theory of General Relativity, the Universe oscillates—it expands, then contracts, then expands, and then contracts, etc. If the Second Law of Thermodynamics is valid, can each cycle of oscillation be the same as the preceding cycle?

21. **Negentropy** is defined as the negative of the entropy ([negentropy] $= -S$). Explain the following statement: "In our everyday activities on the Earth, we do not consume energy, but we consume negentropy."

22. The amount of energy dissipated in the United States per year is 8×10^{19} J. Roughly, what is the entropy increase that results from this dissipation?

23. If you tidy up a messy room, you are producing a decrease of disorder. Does this violate the Second Law?

PROBLEMS

21.1 The First Law of Thermodynamics

1. We place a 1.00-mole sample of an ideal monatomic gas in a cylinder with a piston and we heat the gas so it expands and performs work against the piston. Suppose the temperature of the gas increases by 90°C while at the same time it performs 800 J of work. What is the change of the internal energy of the gas in this process? How much heat does the gas absorb during this process?

2. A sample of gas in a cylinder with a piston is in thermal contact with a heat reservoir at a temperature of 353 K. While keeping this gas at this constant temperature and at a constant pressure of 1.01×10^5 N/m^2, we permit the gas to expand by 1.50×10^{-5} m^3. How much heat does the gas absorb during this process?

3. A 4.0-liter sample of a monatomic ideal gas is initially at STP. We first heat this gas at constant volume, until its pressure is doubled. We then continue heating at constant pressure, and

allow the gas to expand until its volume is also doubled.

(a) What are the values of ΔE, Q, and W for the first step of this process?

(b) What are the values of ΔE, Q, and W for the second step of this process?

4. When pressurized air rushes out of the nozzle of a tire, the air must do work against the surrounding atmosphere while it expands. This means that the air suffers a loss of internal energy, and its temperature therefore drops (you can easily feel that the emerging air is quite cold). Suppose that each mole of emerging air does 1800 J of work, and suppose that the air rushes out so fast that it has no time to absorb heat from the surrounding atmosphere (adiabatic expansion). If the initial temperature of the air in the tire is 20°C, what is the temperature of the emerging air? Treat the air as an ideal diatomic gas.

5. In the cylinder of a diesel engine, the piston compresses the air–fuel mixture and does work on it. This work increases the internal energy of the mixture, and therefore heats it. The temperature attained by the compression is sufficient to ignite the mixture, without any need of a spark plug. How much work must the piston do on a parcel of gas, 0.0300 mole, to heat it from 40°C to 790°C? Assume that the compression is so fast that the gas loses no heat to the surroundings (adiabatic compression), and assume the gas behaves like an ideal diatomic gas.

6. A ball of lead of mass 0.25 kg drops from a height of 0.80 m, hits the floor, and remains there at rest. Assume that all the heat generated during the impact remains within the lead. What are the values of Q, W, and ΔE for the lead during this process? What is the increase of temperature of the lead?

7. A large closed bag of plastic contains 0.100 m³ of an unknown ideal gas at an initial temperature of 10°C and at the same pressure as the surrounding atmosphere, 1.0 atm. You place this bag in the sun and let the gas warm up to 38°C and expand to 0.110 m³. During this process, the gas absorbs 3500 J of heat. Assume the bag is large enough that the gas never strains against it, and therefore remains at a constant pressure of 1.0 atm.

(a) How many moles of gas are in the bag?

(b) What is the work done by the gas in the bag against the atmosphere during the expansion?

(c) What is the change in the internal energy of the gas in the bag?

(d) Is the gas a monatomic gas? A diatomic gas?

8. The vaporization of 1.00 kg of water converts 1.00 liter of liquid water at 100°C into a volume of water vapor at the same temperature. The water vapor occupies a larger volume than the water; hence, during vaporization, the water must do work on the surrounding atmosphere while it expands.

(a) How much work does 1.00 kg of water do while it expands into vapor, at a pressure of 1.00 atm? Given that the heat of vaporization of water is 2.26×10^6 J/kg, what is the change of internal energy of the water during vaporization?

(b) How much work does 1.00 kg of water do while it expands into vapor, at a pressure of 2.00 atm? Deduce the heat of vaporization of water at 2.00 atm (and 100°C). Assume that the change of the density of liquid water between 1.00 atm and 2.00 atm can be neglected in this calculation.

9. The melting of 1.00 kg of ice converts a volume of 1.091 liters of ice into 1.000 liter of liquid water at 0°C. The water occupies a smaller volume; hence, during melting, the surrounding atmosphere does work on the ice.

(a) How much work does the atmosphere do on 1.00 kg of ice while it melts at a pressure of 1.00 atm? Given that the heat of fusion of ice is 3.34×10^5 J/kg, what is the change of internal energy of the water during melting?

(b) How much work does the atmosphere do on 1.00 kg of ice while it melts at a pressure of 2.00 atm? Deduce the heat of fusion of ice at 2.00 atm (and 0°C). Assume that the change of the density of the ice and the water between 1.00 atm and 2.00 atm can be neglected in this calculation.

10. (a) For a small isothermal expansion, the work done by a gas is $dW = nRT \, dV/V$ [see Eq. (21.11)]. Integrate this to obtain the work done in a finite expansion from an initial volume V to a final volume V':

$$W = nRT \ln \frac{V'}{V}$$

(b) Suppose that 0.10 mole of helium gas absorbs 500 J of heat from a heat reservoir at 700 K. If the initial volume of the gas is 0.0020 m³, what is the final volume of the gas?

11. The volume of a gas is decreased by a constant external pressure of 0.25 atm, changing from 2.0 liters to 0.50 liter. During this compression, 75 J of energy flows out of the gas. Find the work done by the gas and its change in internal energy.

12. A piston of area 90 cm² slowly compresses a gas, which remains at a constant pressure of 15 atm. The final position of the piston is 12 cm from its initial position, and the internal energy of the gas drops by 15 J. Is heat removed from or added to the gas in this process? Find this amount of heat and the work done by the gas.

13. An initial volume of 1.00 liter of ethyl alcohol (ethanol) is heated from 5°C to 30°C at a pressure of 1.00 atm. Using the data in Tables 20.1 and 20.2, find the heat added to the ethanol and the work done by the ethanol in this process. The density of ethanol at 20° is 789 kg/m³.

14. In a hand pump, an initial volume of 80 cm³ of air at 1.0 atm and 300 K is quickly (adiabatically) compressed to a volume of 20 cm³. What is the change in internal energy just after compression? After a long time, the gas exchanges heat with its container and surroundings, and returns to 300 K, while remaining compressed to 20 cm³. What is the final internal energy of the gas? What is the heat transferred?

*15. An ideal gas is kept at 50°C throughout its expansion from 2.0 liters to 10 liters. If the quantity of gas is 10 moles, how much work is done by the gas? (Hint: Recall that when the pressure changes with volume, the work done is $W = \int p \, dV$.)

*16. 2.0 moles of an ideal gas at 100°C expands at constant temperature to a volume that is 10 times the initial volume. How much heat was added to the gas?

*17. Five moles of an ideal gas does 2.0×10^4 J of work as the gas expands at constant temperature to a final volume of 0.10 m^3 and a final pressure of 1.0 atm. Determine the temperature of the gas, the initial volume, and the initial pressure.

*18. A submerged scuba diver, 10 m below the surface of the water, blows a bubble of air from the valve of her tank. The final volume of the bubble is 4.0 cm^3, and the temperature of the air is that of the water, 15°C. Calculate how much heat the bubble absorbs as it forms, and how much work it does on the surrounding water. Assume that the temperature of the air in the bubble remains equal to the water temperature. The pressure in the tank is very high, so the initial volume of the air is negligible compared with the final volume at the bubble.

**19. At a pressure of 1.0 atm, the heat of vaporization of water is 2.26×10^6 J/kg; this is the heat required to convert 1.0 kg of water at 100°C into water vapor at the same temperature. Given that the specific heat of water is $c = 4187$ J/kg·°C and that of water vapor is (approximately) $c_p = 2010$ J/kg·°C, use the First Law of Thermodynamics to calculate the heat of vaporization of water at 20°C. [Hint: Instead of directly converting water at 20°C into water vapor, we can first heat the water to 100°C, then vaporize it, and then cool the vapor to 20°C (without condensation). The net work done against atmospheric pressure is the same during this indirect process as during direct vaporization (why?); hence, according to the First Law, the heat absorbed during each process is also the same.]

**20. Calculate the heat of vaporization of water at a temperature of 140°C, at a pressure of 1.0 atm. Use the data and the method of calculation described in Problem 19.

21.2 Heat Engines; the Carnot Engine[†]

21. A coal-burning power plant uses thermal energy at a rate of 850 megawatts and produces 300 megawatts of mechanical power for the generation of electricity. What is the efficiency of this power plant?

22. To produce 120 hp of mechanical power, an automobile engine requires a supply of heat of 4.40×10^5 J per second from combustion of the fuel. What is the efficiency of this engine?

23. While running up stairs at a (vertical) rate of 0.30 m/s, a man of 70 kg generates waste heat at a rate of 1300 J/s. What efficiency for the human body can you deduce from this?

24. Suppose that a heat engine takes 3.0×10^4 J of heat from the high-temperature reservoir to produce 2.0×10^4 J of mechanical work. What is the efficiency of this engine? How much waste heat does it produce?

[†]For help, see Online Concept Tutorial 23 at www.wwnorton.com/physics

25. A heat engine takes 2.0×10^7 J of heat from the high-temperature reservoir and ejects 8.0×10^6 J of heat into the low-temperature reservoir. How much work does it produce? What is its efficiency?

26. Electric motors convert electric energy into mechanical energy with an efficiency of 95%. If the electric current supplies a power of 3.0 kW to such an electric motor, what mechanical power will the motor produce? How much heat (in kW) will the motor produce?

27. An electric power plant consists of a coal-fired boiler that makes steam, a turbine, and an electric generator. The boiler delivers 90% of the heat of combustion of the coal to the steam; the turbine converts 50% of the heat of the steam into mechanical energy; and the electric generator converts 99% of this mechanical energy into electric energy. What is the overall efficiency of generation of electric power?

28. Although a sprinter running at a steady speed on level ground performs no external work (except for a small amount of work against air resistance), the sprinter performs a considerable amount of work to accelerate and decelerate his own limbs during each stride and to lift his limbs against gravity. The following table lists the mechanical power used by a sprinter for motion of limbs and body during a sprint (the power was calculated from data obtained from photographic analysis of the motion):

Acceleration of limbs	1.5 hp
Deceleration of limbs	0.67 hp
Work against gravity	0.1 hp
Speed changes of body	0.5 hp

What is the total mechanical power expended in these motions of the sprinter? According to a measurement of the oxygen consumption of the sprinter, his expenditure of chemical energy was 13 hp during the sprint. What is his efficiency of conversion of chemical energy into mechanical energy?

29. According to some naive speculations of the nineteenth century, the human body was supposed to be a heat engine in which the combustion of food produces body heat that is then somehow converted into mechanical work by the muscles. If this were true, what would be the maximum (ideal) efficiency of the human body? The temperature of the "hot" human body (the heat source) is 37°C, and the temperature of the heat sink is that of the surrounding air, about 20°C. How many kcal of food energy would you need to consume to climb up one flight of stairs, to a height of 3.0 m? (Actually, muscles convert chemical energy directly into work; they are not heat engines.)

30. The efficiency of striated muscle is typically 37%, that is, the muscle converts 37% of the chemical energy reaching it (in the form of glucose) into mechanical work, and converts the remainder into waste heat. If your biceps muscle performs mechanical work at the rate of 50 W, what is the rate at which it consumes chemical energy, and what is the rate at which it produces waste heat? Given that the oxidation of glucose

yields 3.7×10^3 kcal/kg, what is the rate at which this muscle consumes glucose?

31. Each of the two engines of a DC-3 airplane produces 1100 hp. The engines consume gasoline; the combustion of 1.0 kg of gasoline yields 4.4×10^7 J. If the efficiency of the engines is 20%, at what rate do the two engines consume gasoline?

32. A steam turbine in a power plant converts the thermal energy of hot steam into mechanical energy. The turbine consists of a high-pressure stage and a low-pressure stage operating in tandem. The steam first passes through the high-pressure stage and gives up some of its thermal energy. The waste heat from this stage is used to reheat the steam, which then enters the low-pressure stage, where it gives up some of its remaining thermal energy.

    (a) Draw a flowchart showing the two stages of the turbine and the heat and work inputs and outputs.

    (b) If the efficiency of the high-pressure stage is 40% and the efficiency of the low-pressure stage is 20%, what is the overall efficiency of the turbine?

33. In principle, nuclear reactions can achieve temperatures of the order of 10^{11} K. What is the efficiency of a Carnot engine taking in heat from such a nuclear reaction and exhausting waste heat at 300 K?

*34. A nuclear power plant generates 1000 megawatts of electric (or mechanical) power. If the efficiency of this plant is 33%, at what rate does the plant generate waste heat? If this waste heat is to be removed by passing water from a river through the plant, and if the water is to suffer a temperature increase of at most 8°C, how many cubic meters of water per second is required?

35. A Carnot engine operates between a high-temperature reservoir at 100°C and a low-temperature reservoir at 0°C. How much energy must the engine take from the high-temperature reservoir to produce 5.0×10^4 J of work? How much waste heat does it produce?

36. Consider a Carnot engine operating between heat reservoirs at 400 K and at 300 K. What is the efficiency of this Carnot engine? If we want to increase the efficiency by 10%, by how much must we increase the temperature of the high-temperature reservoir? Alternatively, by how much must we decrease the temperature of the low-temperature reservoir?

37. A heat engine absorbs 2500 J of heat from a hot reservoir and releases 1400 J of heat to a cold reservoir each second. Find the efficiency of the engine and its mechanical power output.

38. If a heat engine has an efficiency of 15%, does mechanical work at a rate of 12 kW, and releases 5.0×10^3 J of heat to a cold reservoir on each cycle, find the heat absorbed in each cycle and the time for one cycle.

39. A Texas home is cooled by a Carnot engine in reverse. The outside temperature is 104°F and the inside temperature is 75°F, and 5.0×10^6 J of heat must be removed each hour. What rate of work input does this Carnot air conditioner require?

40. When the hot reservoir of a Carnot engine is maintained at 350°C, the efficiency is 15%. If we keep the cold reservoir at constant temperature, to what temperature should the hot reservoir be increased to obtain an efficiency of 25%?

41. A Carnot heat pump brings heat from outside at 5°C into a house at 20°C. What is the ratio of the heat input to the house to the work input to the engine (the coefficient of performance) for this Carnot engine?

42. The Carnot engine can be used to determine an unknown temperature. Suppose that a Carnot engine operates between a heat reservoir at an unknown temperature and a heat sink consisting of a water and ice mixture at 0°C. If this Carnot engine is found to have an efficiency of $e = 0.322$, what temperature can you deduce for the heat reservoir? The temperature scale defined by means of a Carnot engine is called the **absolute thermodynamic temperature scale**.

43. The electric motor of your refrigerator uses 2.5×10^3 J of electric energy to remove 6.0×10^3 J of heat from the refrigerator compartment. How much heat does this refrigerator dump into your kitchen? What is its coefficient of performance?

44. Suppose we operate a Carnot engine in reverse to "pump" heat from a heat reservoir at 0°C into a heat reservoir at 100°C. How much mechanical work must we supply per joule of heat removed from the 0°C reservoir?

45. One mole of an ideal monatomic gas is enclosed in a cylinder with a piston and placed in contact with a heat reservoir of variable temperature. By suitable changes of volume and temperature, the gas is taken through the three-step cycle described by the p–V diagram illustrated in Fig. 21.25.

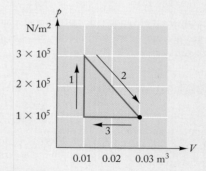

FIGURE 21.25 Three-step cycle.

    (a) In terms of the volumes and pressures given in Fig. 21.25, what is the net work done by the gas during this cycle?

    (b) What is the heat absorbed or ejected by the gas during steps 2 and 3?

    (c) What is the heat absorbed or ejected during step 1? [Hint: Use the results obtained in (a) and (b).]

    (d) What is the efficiency of this cycle?

46. One mole of an ideal monatomic gas is enclosed in a cylinder with a piston and placed in contact with a heat reservoir of

variable temperature. By suitable changes of volume and temperature, the gas is taken through the four-step cycle described by the p–V diagram illustrated in Fig. 21.26.

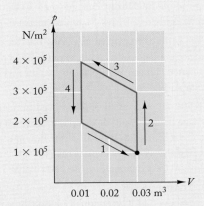

FIGURE 21.26 Four-step cycle.

(a) In terms of the volumes and pressures given in Fig. 21.26, what is the net work done by the gas during this cycle?

(b) What is the heat absorbed or ejected by the gas during steps 2 and 4?

(c) What is the heat absorbed or ejected during steps 1 and 3? (Hint: To find these amounts of heat, calculate W and ΔE for these steps.)

(d) What is the coefficient of performance of this refrigerator cycle?

47. A scheme for the extraction of energy from the oceans attempts to take advantage of the temperature difference between the upper and lower layers of ocean water. The temperature at the surface in tropical regions is about 25°C; the temperature at a depth of 300 m is about 5°C.

(a) What is the efficiency of a Carnot engine operating between these temperatures?

(b) If a power plant operating at the maximum theoretical efficiency generates 1.00 megawatt of mechanical power, at what rate does this power plant release waste heat?

(c) The power plant obtains the mechanical power and the waste heat from the surface water by cooling this water from 25°C to 5°C. At what rate must the power plant take in surface water?

48. In a nuclear power plant, the reactor produces steam at 520°C and the cooling tower eliminates waste heat into the atmosphere at 30°C. The power plant generates 500 megawatts of electric (or mechanical) power.

(a) If the efficiency is that of a Carnot engine, what is the rate of release of waste heat (in megawatts)?

(b) Actual efficiencies of nuclear power are about 33%. For this efficiency, what is the rate of release of heat?

49. An air conditioner removes 8.4×10^6 J/h of heat from a room at a temperature of 21°C and ejects this heat into the ambient air at a temperature of 27°C. This air conditioner requires 950 W of electric power.

(a) How much mechanical power would a Carnot engine, operating in reverse, require to remove this heat at the same rate?

(b) By what factor is the power required by the air conditioner larger than that required by the Carnot engine?

*50. An ice-making plant consists of a reversed Carnot engine extracting heat from a well-insulated box. The temperature in the icebox is −5°C and the temperature of the ambient air is 30°C. Water, of an initial temperature of 30°C, is placed in the icebox and allowed to freeze and to cool to −5°C. If the ice-making plant is to produce 10 000 kg of ice per day, what mechanical power is required by the Carnot engine?

*51. The boiler of a power plant supplies steam at 540°C to a turbine that generates mechanical power. The steam emerges from the turbine at 260°C and enters a steam engine that generates extra mechanical power. The steam is finally released into the atmosphere at a temperature of 38°C. Assume that the conversion of heat into work proceeds with the efficiency of a Carnot engine.

(a) What is the efficiency of the turbine? Of the steam engine?

(b) What is the net efficiency of both engines acting together? How does it compare with the efficiency of a single engine operating between 540°C and 38°C?

**52. One liter of water is initially at the same temperature as the surrounding air, 30°C. You wish to cool this water to 5°C by transferring heat from the water into the air. What is the minimum amount of work that you must supply in order to accomplish this? (Hint: Take a sequence of Carnot engines operating in reverse; use each engine to reduce the temperature by an infinitesimal amount.)

21.3 The Second Law of Thermodynamics
21.4 Entropy

53. What is the increase of entropy of 1.0 kg of water when it vaporizes at 100°C and 1.0 atm?

54. On a winter day heat leaks out of a house at the rate of 2.5×10^4 kcal/h. The temperature inside the house is 21°C and the temperature outside is −5°C. At what rate does this process produce entropy?

55. Consider the air conditioner described in Problem 49. Calculate the rate of increase of entropy contributed by the operation of this air conditioner.

56. Your body generates about 2000 kcal of heat per day. Estimate how much entropy you generate per day. Neglect the (small) amount of entropy that enters your body in the food you consume.

57. A steam engine operating between reservoirs at temperatures of 480°C and 27°C has an efficiency of 40%. The engine delivers 2000 hp of mechanical power. At what rate does this engine generate entropy?

58. Suppose that 1.0 kg of water freezes while at 0°C. What is the change of entropy of the water during this freezing process?

59. At a temperature of −79°C, solid carbon dioxide ("dry ice") transforms into a gas by sublimation (that is, direct vaporization from solid to gas). From the heat of transformation given in Table 20.4, calculate the increase of entropy per kilogram of carbon dioxide during sublimation.

60. Table 20.4 gives data for the melting and the vaporization of copper. What is the increase of entropy per kilogram of copper during melting? During vaporization?

61. Compare the increases of entropy during the melting of 1.0 kg of aluminum, iron, silver, and mercury (see Table 20.4 for data). Which of these has the largest change of entropy? Which has the smallest?

62. A parachutist of 80 kg descends at a constant speed of 5.0 m/s. What is the rate of increase of entropy of the parachute and the environment? The air temperature is 20°C.

63. For an automobile moving at a constant speed of 65 km/h on a level road, rolling friction, air friction, and friction in the drive train absorb a mechanical power of 12 kW. At what rate do these processes generate entropy? The temperature of the environment is 20°C.

64. An automobile of 2100 kg moving at 80 km/h brakes to a stop. In this process the kinetic energy of the automobile is first converted into thermal energy of the brake drums; this thermal energy later leaks away into the ambient air. Suppose that the temperature of the brake drums is 60°C when the automobile stops, and that the temperature of the air, and the final temperature of the brake drums, is 20°C.

    (a) How much entropy is generated by the conversion of mechanical energy into thermal energy of the brake drums?

    (b) How much extra entropy is generated as the heat leaks away into the air?

65. At Niagara Falls (Fig. 21.27), 5700 m³/s of water falls through a vertical distance of 50 m, dissipating all of its gravitational energy. Calculate the rate of increase of entropy contributed by this falling water. The temperature of the environment is 20°C.

FIGURE 21.27 Niagara Falls.

66. At Acapulco, a 70-kg high diver jumps 36 m from a cliff into the sea. If the ambient temperature is 295 K, what is the net change in entropy for this process?

67. In Example 6 of Chapter 20, 1.8×10^3 J/s of heat is transferred through the wall of a house when the inside tempera-

ture is 21°C and the outside temperature is −18°C. At what rate is entropy generated?

68. The melting of lead at 328°C requires 2.9×10^4 J/kg. Determine the entropy change when 20 kg of solid lead at 328°C is melted. Is this change an increase or a decrease?

*69. What is the entropy change during the free expansion of 1.00 mole of an ideal gas from an initial volume of 1.00 liter into an evacuated volume of 1.00 additional liter? (Hint: See Example 1.)

*70. A 0.500-kg piece of silver is removed from an annealing furnace at 950°C and dropped into a bucket containing 5.00 kg of water initially at 20°C. If the heat stays within the silver–water system, calculate the entropy change when a common final temperature has been reached.

*71. (a) Consider a material with a constant specific heat capacity c. Show that if we gradually supply heat to a mass m of this material, increasing its temperature from T_1 to T_2, the increase of entropy of the mass m is

$$S_2 - S_1 = mc \ln\left(\frac{T_2}{T_1}\right)$$

   (b) What is the increase of entropy in 1.0 kg of water that is heated from 20°C to 80°C?

*72. Consider the process of dissolving ice cubes in water as described in Example 8 of Chapter 20. What is the change of entropy of the system? (Hint: Use the formula derived in Problem 71.)

*73. You mix 1.0 liter of water at 20°C with 1.0 liter of water at 80°C. What is the increase of entropy?

*74. (a) Show that the increase of entropy of n moles of a gas heated at constant pressure from a temperature T_1 to a temperature T_2 is

$$\Delta S = nC_p \ln\frac{T_2}{T_1}$$

   (b) Show that the increase of entropy of a gas heated at constant volume from a temperature T_1 to a temperature T_2 is

$$\Delta S = nC_V \ln\frac{T_2}{T_1}$$

*75. Consider the free expansion (see Example 1) of a sample of n moles of an ideal gas from an initial volume V to a final volume V'. Show that the increase of entropy in this process is

$$\Delta S = nR \ln\frac{V'}{V}$$

[Hint: The initial pressure is p and the final pressure is $p \times V/V'$. Replace the (irreversible) process of free expansion by an expansion at constant pressure p from the volume V to the volume V', followed by a reduction of pressure at constant volume V' from the initial pressure p to the final pressure $p \times V/V'$. The entropy changes for each of these two processes are given by the formulas in Problem 74.]

REVIEW PROBLEMS

76. A cylinder with a piston contains one mole of helium gas. Suppose we heat the cylinder and allow the gas to expand, so it performs work on the piston, at constant pressure. If during this expansion, the gas absorbs 3000 J of heat and performs 2000 J of work on the piston, what is the change ΔE in the internal energy of the gas? What is the change of its temperature?

77. A cylinder with a piston contains one mole of air (regarded as an ideal diatomic gas). The cylinder is insulated, so no heat can escape from it. If we push the piston inward and compress the gas, the temperature of the gas will increase. How much does the temperature increase if we do 2500 J of work on the gas during the compression?

78. One mole of an ideal monatomic gas in a cylinder with a piston has an initial volume V_1 and pressure p_1. We allow this gas to expand to a final (larger) volume V_2 and a final (lower) pressure p_2. Assume that on a p–V diagram, the expansion process is represented by a straight line (see Fig. 21.28).

    (a) What is the work done by the gas?

    (b) What is the change of energy of the gas?

    (c) What is the heat absorbed by the gas?

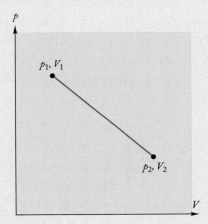

FIGURE 21.28 Expansion process.

79. 0.20 mole of an ideal gas is initially stored in a bottle of 0.0012 m³ at a temperature of 300 K. By means of a pipe with a stopcock, we connect this bottle to an insulated cylinder of 0.0010 m³ that is initially evacuated (see Fig. 21.29). We then open the stopcock, so the gas can rush from the first bottle into the cylinder in free expansion.

    (a) What are the initial and final pressures of the gas?

    (b) By pushing the piston of the cylinder inward, we now slowly compress the gas until all is expelled from the cylinder and its volume returns to the initial volume. During this compression, we keep the gas in contact with a heat reservoir at a temperature of 300 K. How much work must we do during this compression? How much heat is ejected into the heat reservoir?

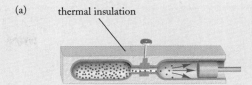

(a) thermal insulation

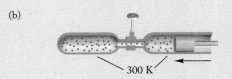

(b)

300 K

FIGURE 21.29 (a) Adiabatic free expansion. (b) Isothermal compression.

80. A tightly sealed plastic bag contains 4.0 liters of air at 0°C. The plastic bag is only loosely filled, and it exerts no compression on the air; the pressure of the air is therefore the pressure of the surrounding atmosphere, 1.0 atm. We place this plastic bag in the sun, and let it warm up to 60°C.

    (a) What is the new volume of the air? Assume that the plastic bag remains loosely filled with air, so the pressure remains 1.0 atm.

    (b) How much work did the air in the bag do on the surrounding atmosphere while it expanded? What is the change in the internal energy of the air? What is the amount of heat the air has absorbed? Assume that the air behaves like an ideal diatomic gas.

81. In an automobile proceeding at medium speed, the engine delivers 20 hp of mechanical power. The engine burns gasoline, which provides thermal energy at the rate of 6.3×10^4 J per second. What is the efficiency of the engine under these conditions? What is the rate at which the engine ejects waste heat?

82. In an experiment on the work efficiency of horses, a horse connected to an oxygen supply was made to do work on a treadmill (see Fig. 21.30). When the horse was delivering 869 watts of work to the treadmill, its rate of oxygen consumption was 10.8 liters per minute. The horse's metabolism yields 2.1×10^4 J of chemical energy per liter of oxygen consumed. Calculate the efficiency of conversion of chemical energy into external mechanical work.

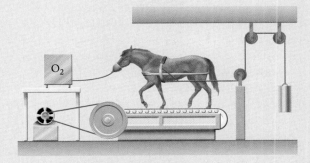

FIGURE 21.30 Horse doing work on a treadmill.

83. On a hot day a house is kept cool by an air conditioner. The outside temperature is 32°C and the inside temperature is 21°C. Heat leaks into the house at the rate of 3.8×10^7 J/h. If the air conditioner has the efficiency of a Carnot engine, what is the mechanical power that it requires to hold the inside temperature constant?

84. A Carnot engine operates between two heat reservoirs of temperature 500°C and 30°C, respectively.

   (a) What is the efficiency of this engine?

   (b) If the engine generates 1.5×10^3 J of work, how many calories of heat does it absorb from the hot reservoir? Eject into the cold reservoir?

85. One mole of an ideal monatomic gas is enclosed in a cylinder with a piston and placed in contact with a heat reservoir of variable temperature. By suitable changes of volume and temperature, the gas is taken through the four-step cycle described by the p–V diagram illustrated in Fig. 21.31.

   (a) Describe how you must vary the temperature of the heat reservoir and the volume of the gas to accomplish each of the four steps. Begin at the top left corner of the cycle and proceed clockwise.

   (b) From the volumes and pressures given in Fig. 21.31, calculate the work done by the gas during each step.

   (c) Calculate the heat absorbed or ejected by the gas during each step and verify that the net work done equals the net heat absorbed.

   (d) Calculate the efficiency of this cycle.

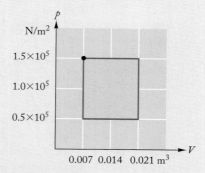

FIGURE 21.31 Four-step cycle.

86. A Carnot engine operating with an ideal gas uses a high-temperature reservoir at 600 K and a low-temperature reservoir at 200 K. In one cycle, the engine absorbs 1200 J from the high-temperature reservoir. What is the waste heat ejected into the low-temperature reservoir? What is the work produced by the engine during the cycle?

87. A geothermal power plant at Wairakei, New Zealand (Fig. 21.32), uses hot underground water at a temperature of 300°C as heat source and uses the atmosphere at a temperature of, say, 25°C as heat sink. What would be the efficiency of a Carnot engine operating between these temperatures? If this

FIGURE 21.32 Geothermal power plant at Wairakei, New Zealand.

Carnot engine produces 10 000 kW of mechanical power, what power does it eject as waste heat?

88. A refrigerator is operated by a Carnot engine. The temperature inside the refrigerator is 4°C and the temperature of the surrounding room is 20°C. How much mechanical work must the refrigerator's electric motor perform to remove 5.0×10^3 J of heat from the inside of the refrigerator and dump it into the room?

89. Compare the increases of entropy during the vaporization of 1.0 kg of nitrogen, oxygen, and hydrogen (see Table 20.4 for data). Which of these has the largest change of entropy? Which the smallest?

90. You dump 1.0 kg of molten lead (at 328°C) into 1.0 kg of crushed ice (at 0°C).

   (a) What is the final temperature of the mixture?

   (b) What is the decrease of entropy of the lead? (Hint: Use the formula derived in Problem 71.)

   (c) What is the increase of entropy of the water?

91. A Carnot engine operating with helium gas (approximately an ideal gas) uses a high-temperature reservoir at 700 K and a low-temperature reservoir at 300 K. In one cycle, the engine absorbs 800 J from the high-temperature reservoir.

   (a) What is the entropy change of the helium gas while in thermal contact with the high-temperature reservoir?

   (b) What is the waste heat ejected, and what is the entropy change of the helium gas while in thermal contact with the low-temperature reservoir?

   (c) What are the entropy changes during the adiabatic expansion and contraction?

Answers to Checkups

Checkup 21.1

1. The sunlight heats the balloon of gas, so Q, the amount of heat added, is positive. The balloon does work as it pushes against atmospheric pressure, so W, the work done by the gas, is positive.

2. If the jar is heated, Q, the amount of heat added to the system, is positive. The work done by the gas, W, is zero, since the volume of the gas does not change appreciably. (However, because the jar will experience a slight thermal expansion, W will actually have a very small positive value.)

3. (C) Negative, positive. The athlete loses heat to the environment, so Q, the heat added to the athlete, is negative. The athlete does work to move the bicycle (against friction), so W, the work done by the system, is positive.

Checkup 21.2

1. An isothermal expansion proceeds at constant temperature; an adiabatic expansion proceeds without loss or gain of heat.

2. No. An efficiency greater than 1 would mean more energy is coming out than going in, and this is not possible.

3. No. The burning of fuel cannot be reversed by performing mechanical work on the crankshaft.

4. No. The efficiency of a Carnot engine depends only on the two temperatures of operation.

5. (C) There is no net effect. Since the Carnot cycle is reversible, using the work output of the first to drive the second in reverse results in no net effect.

Checkup 21.3

1. At $T_2 = T_1$, the efficiency is zero; no useful work can be extracted. For $T_2 > T_1$, the heat flow will be opposite the usual direction; that is, the engine will absorb heat from the reservoir at T_2, so the efficiency is $e = 1 - T_1/T_2$.

2. No. The refrigerator has the effect of removing mechanical or electrical energy from the environment (work input to the refrigerator is required for operation).

3. (C) The cycle could not be completed, since the steam could not condense. The steam could not condense if the "condenser" was at very high temperature; also, with the two temperatures nearly equal, the efficiency would approach zero.

Checkup 21.4

1. The entropy changes are given by $\Delta Q/T$ and so are 100 J/300 K $= \frac{1}{3}$ J/K for the first reservoir and 100 J/400 K $= \frac{1}{4}$ J/K for the second.

2. Yes. The flowchart shows the release of waste heat, that is, only a partial conversion of heat into work, and so is consistent with the Second Law.

3. Yes, the entropy of the water decreases as its temperature is lowered (heat is removed, order increases). This does not violate the Second Law, since the entropy of the surroundings increases (nearby air absorbs heat).

4. (B) 1 and 4. The braking of an automobile converts mechanical work into heat; this process is not reversible, and so (1) creates an increase in entropy. Both the slowing of an automobile uphill and the pumping of water upward (without friction) are reversible, since the mechanical energy can be completely recovered by reversing the process; thus, (2) and (3) do not result in an increase in entropy. The burning of firewood is a conversion of chemical energy into heat, which is irreversible, and so (4) increases the entropy.

Appendix 1: Greek Alphabet

A	α	alpha	N	ν	nu
B	β	beta	Ξ	ξ	xi
Γ	γ	gamma	O	o	omicron
Δ	δ	delta	Π	π	pi
E	ϵ	epsilon	P	ρ	rho
Z	ζ	zeta	Σ	σ	sigma
H	η	eta	T	τ	tau
Θ	θ	theta	Y	υ	upsilon
I	ι	iota	Φ	ϕ	phi
K	κ	kappa	X	χ	chi
Λ	λ	lambda	Ψ	ψ	psi
M	μ	mu	Ω	ω	omega

Appendix 2: Mathematics Review

A 2.1 Symbols

$a = b$ means a equals b

$a \neq b$ means a is not equal to b

$a > b$ means a is greater than b

$a < b$ means a is less than b

$a \geq b$ means a is not less than b

$a \leq b$ means a is not greater than b

$a \propto b$ means a is proportional to b

$a \approx b$ means a is approximately equal to b

$a \gg b$ means a is much greater than b

$a \ll b$ means a is much less than b

$\pi = 3.141\ 59\ldots$

$e = 2.718\ 28\ldots$

A 2.2 Powers and Roots

For any number a, the *n*th *power* of the number is the number multiplied by itself n times. This is written as a^n, and n is called the **exponent**. Thus,

$$a^1 = a \quad a^2 = a \cdot a \quad a^3 = a \cdot a \cdot a \quad a^4 = a \cdot a \cdot a \cdot a \quad \text{etc.}$$

For instance,

$$3^2 = 3 \times 3 = 9 \quad 3^3 = 3 \times 3 \times 3 = 27 \quad 3^4 = 3 \times 3 \times 3 \times 3 = 81 \quad \text{etc.}$$

A negative exponent indicates that the number is to be divided n times into 1; thus

$$a^{-1} = \frac{1}{a} \quad a^{-2} = \frac{1}{a^2} \quad a^{-3} = \frac{1}{a^3} \quad \text{etc.}$$

A zero exponent yields 1, regardless of the value of a:

$$a^0 = 1$$

The rules for the combination of exponents in products, in ratios, and in powers of powers are

$$a^n \cdot a^m = a^{m+n}$$

$$\frac{a^n}{a^m} = a^{n-m}$$

$$(a^n)^m = a^{nm}$$

For instance, it is easy to verify that

$$3^2 \times 3^3 = 3^5$$

$$\frac{3^2}{3^3} = 3^{-1} = \frac{1}{3}$$

$$(3^2)^3 = 3^{2 \times 3} = 3^6$$

Note that for any two numbers a and b

$$(a \cdot b)^n = a^n \cdot b^n$$

For instance,

$$(2 \times 3)^3 = 2^3 \times 3^3$$

The nth root of a is a number such that its nth power equals a. The nth root is written $a^{1/n}$. The second root $a^{1/2}$ is usually called the square root, and designated by $\sqrt{a}$:

$$a^{1/2} = \sqrt{a}$$

As suggested by the notation $a^{1/n}$, roots are fractional powers, and they obey the usual rules for the combination of exponents:

$$(a^{1/n})^n = a^{n/n} = a$$

$$(a^{1/n})^m = a^{m/n}$$

A 2.3 Arithmetic in Scientific Notation

The scientific notation for numbers (see the first page of the Prelude) is quite handy for the multiplication and the division of very large or very small numbers, because we can deal with the decimal parts and the power-of-10 parts in the numbers separately. For example, to multiply 4×10^{10} by 5×10^{12}, we multiply 4 by 5 and 10^{10} by 10^{12}, as follows:

$$(4 \times 10^{10}) \times (5 \times 10^{12}) = (4 \times 5) \times (10^{10} \times 10^{12})$$

$$= 20 \times 10^{10+12} = 20 \times 10^{22} = 2 \times 10^{23}$$

To divide these numbers, we proceed likewise:

$$\frac{4 \times 10^{10}}{5 \times 10^{12}} = \frac{4}{5} \times \frac{10^{10}}{10^{12}} = 0.8 \times 10^{10-12} = 0.8 \times 10^{-2} = 8 \times 10^{-3}$$

When performing additions or subtractions of numbers in scientific notation, we must be careful to begin by expressing the numbers with the same power of 10. For example, the sum of 1.5×10^9 and 3×10^8 is

$$1.5 \times 10^9 + 3 \times 10^8 = 1.5 \times 10^9 + 0.3 \times 10^9 = 1.8 \times 10^9$$

A 2.4 Algebra

An equation is a mathematical statement that tells us that one quantity or a combination of quantities is equal to another quantity or combination. We often have to solve for one of the quantities in the equation in terms of the other quantities. For instance, we may have to solve the equation

$$x + a = b$$

for x in terms of a and b. Here a and b are numerical constants or mathematical expressions which are regarded as known, and x is regarded as unknown.

The rules of algebra instruct us how to manipulate equations and accomplish their solution. The three most important rules are:

1. Any equation remains valid if equal terms are added or subtracted from its left side and its right side.

This rule is useful for solving the equation $x + a = b$. We simply subtract a from both sides of this equation and find

$$x + a - a = b - a$$

that is,

$$x = b - a$$

To see how this works in a concrete numerical example, consider the equation

$$x + 7 = 5$$

Subtracting 7 from both sides, we obtain

$$x = 5 - 7$$

or

$$x = -2$$

Note that given an equation of the form $x + a = b$, we may want to solve for a in terms of x and b, if x is already known from some other information but a is a mathematical quantity that is not yet known. If so, we must subtract x from both sides of the equation, and we obtain

$$a = b - x$$

Most equations in physics contain several mathematical quantities which sometimes play the role of known quantities, sometimes the role of unknown quantities, depending on circumstances. Correspondingly, we will sometimes want to solve the equation for one quantity (such as x), sometimes for another (such as a).

2. Any equation remains valid if the left and the right sides are multiplied or divided by the same factor.

This rule is useful for solving

$$ax = b$$

We simply divide both sides by a, which yields

$$\frac{ax}{a} = \frac{b}{a}$$

or

$$x = \frac{b}{a}$$

Often it will be necessary to combine both of the above rules. For instance, to solve the equation

$$2x + 10 = 16$$

we begin by subtracting 10 from both sides, obtaining

$$2x = 16 - 10$$

or

$$2x = 6$$

and then we divide both sides by 2, with the result

$$x = \frac{6}{2}$$

or

$$x = 3$$

3. Any equation remains valid if both sides are raised to the same power.

This rule permits us to solve the equation

$$x^3 = b$$

Raising both sides to the power $\frac{1}{3}$, we find

$$(x^3)^{1/3} = b^{1/3}$$

or

$$x = b^{1/3}$$

As a final example, let us consider the equation

$$x = -\tfrac{1}{2}gt^2 + x_0$$

(as established in Chapter 2, this equation describes the vertical position of a particle that starts at a height x_0 and falls for a time t; but the meaning of the equation need not concern us here). Suppose that we want to solve for t in terms of the other quantities in the equation. This will require the use of all our rules of algebra. First, subtract x from both sides and then add $\tfrac{1}{2}gt^2$ to both sides. This leads to

$$0 = -\tfrac{1}{2}gt^2 + x_0 - x$$

and then to

$$\tfrac{1}{2}gt^2 = x_0 - x$$

Next, multiply both sides by 2 and divide both sides by g; this yields

$$t^2 = \frac{2}{g}(x_0 - x)$$

Finally, raise both sides to the power $\frac{1}{2}$, or, equivalently, extract the square root of both sides. This gives us the final result

$$t = \sqrt{\frac{2}{g}(x_0 - x)}$$

A 2.5 Equations with Two Unknowns

If we seek to solve for two unknowns simultaneously, then we need two independent equations containing these two unknowns. The solution of such simultaneous equations can be carried out by the method of elimination: begin by using one equation to solve for the first unknown in terms of the second, then use this result to eliminate the first unknown from the other equation. An example will help to make this clear. Consider the following two simultaneous equations with two unknowns x and y:

$$4x + 2y = 8$$
$$2x - y = -2$$

To solve the first equation for x in terms of y, subtract $2y$ from both sides and then divide both sides by 4:

$$x = \frac{8 - 2y}{4}$$

Next, substitute this expression for x into the second equation:

$$2 \times \frac{8 - 2y}{4} - y = -2$$

To simplify this equation, multiply both sides by 4:

$$2 \times (8 - 2y) - 4y = -8$$

and combine the two terms containing y:

$$16 - 8y = -8$$

This is an ordinary equation for the single unknown y, and it can be solved by the methods we discussed in the preceding section, with the result

$$y = 3$$

It then follows from the above expression for x that

$$x = \frac{8 - 2y}{4} = \frac{8 - 2 \times 3}{4} = \frac{2}{4} = \frac{1}{2}$$

A 2.6 The Quadratic Formula

The quadratic equation $ax^2 + bx + c = 0$ has two solutions:

$$x = \frac{-b \pm \sqrt{b^2 - 4ac}}{2a}$$

A 2.7 Logarithms and the Exponential Function

The **base-10 logarithm** of a (positive) number is the power to which 10 must be raised to obtain this number. Thus, from $10 = 10^1$ and $100 = 10^2$ and $1000 = 10^3$ and $10\,000 = 10^4$ we immediately deduce that

$$\log 10 = 1$$
$$\log 100 = 2$$
$$\log 1000 = 3$$
$$\log 10\,000 = 4, \text{ etc.}$$

Likewise

$$\log 1 = 0$$
$$\log 0.1 = -1$$
$$\log 0.01 = -2$$
$$\log 0.001 = -3, \text{ etc.}$$

Thus, the logarithm of a number between 1 and 10 is somewhere between 0 and 1, but to find the logarithm of such a number, we need the help of a computer program (many calculators have built-in computer programs that yield the value of the logarithm at the touch of a button). For some calculations, it is convenient to remember that $\log 2 = 0.301 \approx 0.3$ and $\log 5 = 0.699 \approx 0.7$.

The logarithm of the product of two numbers is the sum of the individual logarithms, and the logarithm of the ratio of two numbers is the difference of the individual logarithms. This rule makes it easy to find the logarithm of a number expressed in scientific notation. For example, the logarithm of 2×10^6 is

$$\log(2 \times 10^6) = \log 2 + \log 10^6 = 0.301 + 6 = 6.301$$

Note that the logarithm of any (positive) number smaller than 1 is negative. For example,

$$\log(5 \times 10^{-3}) = \log 5 + \log 10^{-3} = 0.699 - 3 = -2.301$$

The **exponential function** $\exp(x)$ is defined by the following infinite series:

$$\exp(x) = 1 + x + \frac{x^2}{2} + \frac{x^3}{3 \cdot 2} + \frac{x^4}{4 \cdot 3 \cdot 2} + \cdots$$

This function is equivalent to raising the constant $e = 2.718\,28 \ldots$ to the power x:

$$\exp(x) = e^x$$

The **natural logarithm** $\ln x$ is the inverse of the exponential function, so

$$x = e^{\ln x}$$

and

$$x = \ln(e^x)$$

Natural logarithms obey the usual rules for logarithms,

$$\ln(x \cdot y) = \ln x + \ln y$$

$$\ln\left(\frac{x}{y}\right) = \ln x - \ln y$$

$$\ln(x^a) = a \ln x$$

Note that

$$\ln e = 1$$

and

$$\ln 10 = 2.3026$$

If we designate the base-10 logarithm, or **common logarithm**, by log x, then the relationship between the two kinds of logarithm is as follows:

$$\ln x = \ln(10^{\log x}) = (\log x)(\ln 10) = 2.3026 \log x$$

Appendix 3: Geometry and Trigonometry Review

A3.1 Perimeters, Areas, and Volumes

[perimeter of a circle of radius r] $= 2\pi r$
[area of a circle of radius r] $= \pi r^2$
[area of a triangle of base b, altitude h] $= hb/2$
[surface area of a sphere of radius r] $= 4\pi r^2$
[volume of a sphere of radius r] $= 4\pi r^3/3$
[area of curved surface of a cylinder of radius r, height h] $= 2\pi rh$
[volume of a cylinder of radius r, height h] $= \pi r^2 h$

A3.2 Angles

The angle between two intersecting straight lines is defined as the fraction of a complete circle included between these lines (Fig. A3.1). To express the angle in **degrees**, we assign an angular magnitude of 360° to the complete circle; any arbitrary angle is then an appropriate fraction of 360°. To express the angle in **radians**, we assign an angular magnitude of 2π radians to the complete circle; any arbitrary angle is then an appropriate fraction of 2π. For example, the angle shown in Fig. A3.1 is $\frac{1}{12}$ of a complete circle, that is, 30°, or $\pi/6$ radian. In view of the definition of angle, the length of arc included between the two intersecting straight lines is proportional to the angle θ between these lines; if the angle is expressed in radians, then the constant of proportionality is simply the radius:

$$s = r\theta \tag{1}$$

Since 2π radians $= 360°$, it follows that

$$1 \text{ radian} = \frac{360°}{2\pi} = \frac{360°}{2 \times 3.141\,59} = 57.2958° \tag{2}$$

Each degree is divided into 60 minutes of arc (arcminutes), and each of these into 60 seconds of arc (arcseconds). In degrees, minutes of arc, and seconds of arc, the radian is

$$1 \text{ radian} = 57° \, 17' \, 44.8'' \tag{3}$$

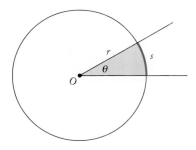

FIGURE A3.1 The angle θ in this diagram is $\theta = 30°$, or $\pi/6$ radian.

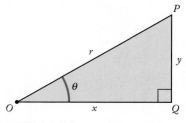

FIGURE A3.2 A right triangle.

A3.3 The Trigonometric Functions

The trigonometric functions of an angle are defined as ratios of the lengths of the sides of a right triangle erected on this angle. Figure A3.2 shows an acute angle θ and a right triangle, one of whose angles coincides with θ. The adjacent side OQ has a length x, the opposite side QP a length y, and the hypotenuse OP a length r. The **sine**, **cosine**, **tangent**, **cotangent**, **secant**, and **cosecant** of the angle θ are then defined as follows:

$$\text{sine} \qquad\qquad \sin\theta = y/r \qquad\qquad (4)$$

$$\text{cosine} \qquad\qquad \cos\theta = x/r \qquad\qquad (5)$$

$$\text{tangent} \qquad\qquad \tan\theta = y/x \qquad\qquad (6)$$

$$\text{cotangent} \qquad\qquad \cot\theta = x/y \qquad\qquad (7)$$

$$\text{secant} \qquad\qquad \sec\theta = r/x \qquad\qquad (8)$$

$$\text{cosecant} \qquad\qquad \csc\theta = r/y \qquad\qquad (9)$$

EXAMPLE 1   Find the sine, cosine, and tangent for angles of $0°$, $90°$, and $45°$.

SOLUTION: For an angle of $0°$, the opposite side is zero ($y = 0$), and the adjacent side coincides with the hypotenuse ($x = r$). Hence

$$\sin 0° = 0 \quad \cos 0° = 1 \quad \tan 0° = 0 \qquad\qquad (10)$$

For an angle of $90°$, the adjacent side is zero ($x = 0$), and the opposite side coincides with the hypotenuse ($y = r$). Hence

$$\sin 90° = 1 \quad \cos 90° = 0 \quad \tan 90° = \infty \qquad\qquad (11)$$

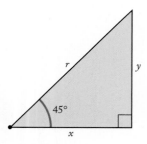

FIGURE A3.3 A right triangle with an angle of 45°.

Finally, for an angle of $45°$ (Fig. A3.3), the adjacent and the opposite sides have the same length ($x = y$) and the hypotenuse has a length of $\sqrt{2}$ times the length of either side ($r = \sqrt{2}x = \sqrt{2}y$). Hence

$$\sin 45° = \frac{1}{\sqrt{2}} \quad \cos 45° = \frac{1}{\sqrt{2}} \quad \tan 45° = 1 \qquad\qquad (12)$$

The definitions (4)–(9) are also valid for angles greater than $90°$, such as the angle shown in Fig. A3.4. In the general case, the quantities x and y must be interpreted as the rectangular coordinates of the point P. For any angle larger than $90°$, one or both of the coordinates x and y are negative. Hence some of the trigonometric functions will also be negative. For instance,

$$\sin 135° = \frac{1}{\sqrt{2}} \quad \cos 135° = -\frac{1}{\sqrt{2}} \quad \tan 135° = -1 \qquad\qquad (13)$$

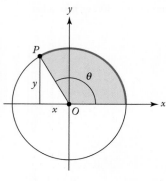

FIGURE A3.4 The angle θ in this diagram is larger than 90°.

Figure A3.5 shows plots of the sine, cosine, and tangent vs. θ.

(a)

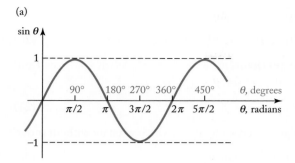

(b)

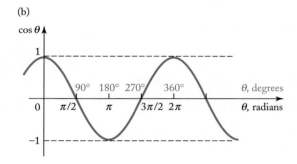

(c)

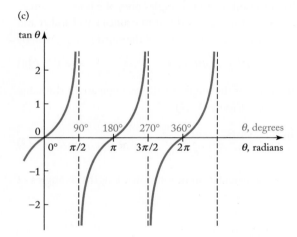

FIGURE A3.5 Plots of the sine, cosine, and tangent functions.

A3.4 Trigonometric Identities

From the definitions (4)–(9) we immediately find the following identities:

$$\tan \theta = \sin \theta / \cos \theta \tag{14}$$

$$\cot \theta = 1/\tan \theta \tag{15}$$

$$\sec \theta = 1/\cos \theta \tag{16}$$

$$\csc \theta = 1/\sin \theta \tag{17}$$

Figure A3.6 shows a right triangle with angles θ and $90° - \theta$. Since the adjacent side for the angle θ is the opposite side for the angle $90° - \theta$ and vice versa, we see that the trigonometric functions also obey the following identities:

$$\sin(90° - \theta) = \cos \theta \tag{18}$$

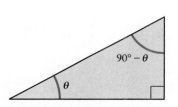

FIGURE A3.6 A right triangle with angles θ and $90° - \theta$.

$$\cos(90° - \theta) = \sin\theta \tag{19}$$

$$\tan(90° - \theta) = \cot\theta = 1/\tan\theta \tag{20}$$

According to the **Pythagorean theorem**, $x^2 + y^2 = r^2$. With $x = r\cos\theta$ and $y = r\sin\theta$, this becomes $r^2\cos^2\theta + r^2\sin^2\theta = r^2$, or

$$\cos^2\theta + \sin^2\theta = 1 \tag{21}$$

The following are a few other trigonometric identities, which we state without proof:

$$\sec^2\theta = 1 + \tan^2\theta \tag{22}$$

$$\csc^2\theta = 1 + \cot^2\theta \tag{23}$$

$$\sin 2\theta = 2\sin\theta\,\cos\theta \tag{24}$$

$$\cos 2\theta = 2\cos^2\theta - 1 \tag{25}$$

$$\sin(\alpha + \beta) = \sin\alpha\cos\beta + \cos\alpha\sin\beta \tag{26}$$

$$\cos(\alpha + \beta) = \cos\alpha\cos\beta - \sin\alpha\sin\beta \tag{27}$$

A3.5 The Laws of Cosines and Sines

In an arbitrary triangle the lengths of the sides and the angles obey the laws of cosines and of sines. The **law of cosines** states that if the lengths of two sides are A and B and the angle between them is γ (Figure A3.7), then the length of the third side is given by

$$C^2 = A^2 + B^2 - 2AB\cos\gamma \tag{28}$$

The **law of sines** states that the sines of the angles of the triangle are in the same ratio as the lengths of the opposite sides (Figure A3.7):

$$\frac{\sin\alpha}{A} = \frac{\sin\beta}{B} = \frac{\sin\gamma}{C} \tag{29}$$

Both of these laws are very useful in the calculation of unknown lengths or angles of a triangle.

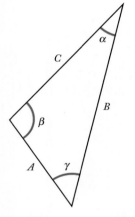

FIGURE A3.7 An arbitrary triangle.

Appendix 4: Calculus Review

A4.1 Derivatives

We saw in Section 2.3 that if the position of a particle is some function of time, say, $x = x(t)$, then the instantaneous velocity of the particle is the derivative of x with respect to t:

$$v = \frac{dx}{dt} \tag{1}$$

This derivative is defined by first looking at a small increment Δx that results from a small increment Δt, and then evaluating the ratio $\Delta x/\Delta t$, in the limit when both Δx and Δt tend toward zero. Thus

$$\frac{dx}{dt} = \lim_{\Delta t \to 0} \frac{\Delta x}{\Delta t} \tag{2}$$

Graphically, in a plot of position vs. time, the derivative dx/dt is the slope of the straight line tangent to the curved line at the time t (see Figure A4.1).

In general, if $f = f(u)$ is some given function of a variable u, the **derivative** of f with respect to u is defined by

$$\frac{df}{du} = \lim_{\Delta u \to 0} \frac{\Delta f}{\Delta u} \tag{3}$$

In a plot of f vs. u, this derivative is the slope of the straight line tangent to the curve representing $f(u)$.

Starting with the definition (3) we can find the derivative of any function (provided the function is sufficiently smooth so the derivative exists!). For example, consider the function $f(u) = u^2$. If we increase u to $u + \Delta u$, the function $f(u)$ increases to

$$f + \Delta f = (u + \Delta u)^2 \tag{4}$$

and therefore

$$\Delta f = (u + \Delta u)^2 - f = (u + \Delta u)^2 - u^2$$
$$= 2u \, \Delta u + (\Delta u)^2 \tag{5}$$

The derivative df/du is then

$$\frac{df}{du} = \lim_{\Delta u \to 0} \frac{\Delta f}{\Delta u} = \lim_{\Delta u \to 0} \frac{2u \, \Delta u + (\Delta u)^2}{\Delta u} \tag{6}$$

$$= \lim_{\Delta u \to 0} (2u) + \lim_{\Delta u \to 0} (\Delta u) \tag{7}$$

The second term on the right side vanishes in the limit $\Delta u \to 0$; the first term is simply $2u$. Hence

$$\frac{df}{du} = 2u \tag{8}$$

or

$$\frac{d}{du}(u^2) = 2u \tag{9}$$

This is one instance of the general rule for the differentiation of u^n:

$$\frac{d}{du}(u^n) = nu^{n-1} \tag{10}$$

This general rule is valid for any positive or negative number n, including zero. The proof of this rule can be constructed by an argument similar to that above. Table A4.1 lists the derivatives of the most common functions.

A4.2 Important Rules for Differentiation

1. Derivative of a constant times a function:

$$\frac{d}{du}(cf) = c\frac{df}{du} \tag{11}$$

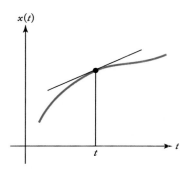

FIGURE A4.1 The derivative of $x(t)$ at t is the slope of the straight line tangent to the curve at t.

TABLE A4.1

SOME DERIVATIVES

$$\frac{d}{du}u^n = nu^{n-1}$$

$$\frac{d}{du}\ln u = \frac{1}{u}$$

$$\frac{d}{du}e^u = e^u$$

(In all the following formulas, u is in radian:)

$$\frac{d}{du}\sin u = \cos u$$

$$\frac{d}{du}\cos u = -\sin u$$

$$\frac{d}{du}\tan u = \sec^2 u$$

$$\frac{d}{du}\cot u = -\csc^2 u$$

$$\frac{d}{du}\sec u = \tan u \sec u$$

$$\frac{d}{du}\csc u = -\cot u \csc u$$

$$\frac{d}{du}\sin^{-1} u = 1/\sqrt{1 - u^2}$$

$$\frac{d}{du}\cos^{-1} u = -1/\sqrt{1 - u^2}$$

$$\frac{d}{du}\tan^{-1} u = \frac{1}{1 + u^2}$$

For instance,

$$\frac{d}{du}(6u^2) = 6\frac{d}{du}(u^2) = 6 \times 2u = 12u$$

2. Derivative of the sum of two functions:

$$\frac{d}{du}(f + g) = \frac{df}{du} + \frac{dg}{du} \qquad (12)$$

For instance,

$$\frac{d}{du}(6u^2 + u) = \frac{d}{du}(6u^2) + \frac{d}{du}(u) = 12u + 1$$

3. Derivative of the product of two functions:

$$\frac{d}{du}(f \times g) = g\frac{df}{du} + f\frac{dg}{du} \qquad (13)$$

For instance,

$$\frac{d}{du}(u^2 \sin u) = \sin u \frac{d}{du}u^2 + u^2 \frac{d}{du}\sin u$$

$$= \sin u \times 2u + u^2 \times \cos u$$

4. Chain rule for derivatives: If f is a function of g and g is a function of u, then

$$\frac{d}{du}f(g) = \frac{df}{dg}\frac{dg}{du} \qquad (14)$$

For instance, if $g = 2u$ and $f(g) = \sin g$, then

$$\frac{d}{du}\sin(2u) = \frac{d\sin(2u)}{d(2u)}\frac{d(2u)}{du}$$

$$= \cos(2u) \times 2$$

5. Partial derivatives: If f is a function of more than one variable, then the *partial derivative* of f with respect to one of the variables, say x, is denoted $\partial f/\partial x$, and is obtained by treating all the *other* variables as constants when differentiating. For instance, if $f = x^2y + y^2z$, then

$$\frac{\partial f}{\partial x} = 2xy, \quad \frac{\partial f}{\partial y} = x^2 + 2yz, \quad \text{and} \quad \frac{\partial f}{\partial z} = y^2$$

A4.3 Integrals

We have learned that if the position of a particle is known as a function of time, then we can find the instantaneous velocity by differentiation. What about the converse problem: if the instantaneous velocity is known as a function of time, how can we find the position? In Section 2.5 we learned how to deal with this problem in the special case of motion with constant acceleration. The velocity is then a fairly simple function of time [see Eq. (2.17)]

$$v = v_0 + at \qquad (15)$$

and the position deduced from this velocity is [see Eq. (2.22)]

$$x = x_0 + v_0 t + \tfrac{1}{2} a t^2 \tag{16}$$

where x_0 and v_0 are the initial position and velocity at the initial time $t_0 = 0$. Now we want to deal with the general case of a velocity that is an arbitrary function of time,

$$v = v(t) \tag{17}$$

Figure A4.2 shows what a plot of v vs. t might look like. At the initial time t_0, the particle has an initial position x_0 (for the sake of generality we now assume that $t_0 \neq 0$). We want to find the position at some later time t. For this purpose, let us divide the time interval $t - t_0$ into a large number of small time intervals, each of the duration Δt. The total number of intervals is N, so $t - t_0 = N \Delta t$. The first of these intervals lasts from t_0 to $t_0 + \Delta t$; the second from $t_0 + \Delta t$ to $t_0 + 2\Delta t$; etc.

In Figure A4.3 the beginnings and the ends of these intervals have been marked t_0, t_1, t_2, etc., with $t_1 = t_0 + \Delta t$, $t_2 = t_0 + 2\Delta t$, etc. If Δt is sufficiently small, then during the first time interval the velocity is approximately $v(t_0)$; during the second, $v(t_1)$; etc. This amounts to replacing the smooth function $v(t)$ by a series of steps (see Fig. A4.3). Thus, during the first time interval, the displacement of the particle is approximately $v(t_0) \, \Delta t$; during the second interval, $v(t_1) \, \Delta t$; etc. The net displacement of the particle during the entire interval $t - t_0$ is the sum of all these small displacements:

$$x(t) - x_0 \approx v(t_0)\Delta t + v(t_1)\Delta t + v(t_2)\Delta t + \cdots \tag{18}$$

Using the standard mathematical notation for summation, we can write this as

$$x(t) - x_0 \approx \sum_{i=0}^{N-1} v(t_i) \, \Delta t \tag{19}$$

We can give this sum the following graphical interpretation: since $v(t_i) \, \Delta t$ is the area of the rectangle of height $v(t_i)$ and width Δt, the sum is the net area of all the rectangles shown in Figure A4.3, i.e., it is approximately the area under the velocity curve. Note that if the velocity is negative, the area must be reckoned as negative!

Of course, Eq. (19) is only an approximation. To find the exact displacement of the particle we must let the step size Δt tend to zero (while the number of steps N tends to infinity). In this limit, the steplike horizontal and vertical line segments in Fig. A4.3 approach the smooth curve. Thus,

$$x(t) - x_0 = \lim_{\substack{\Delta t \to 0 \\ N \to \infty}} \sum_{i=0}^{N-1} v(t_i) \, \Delta t \tag{20}$$

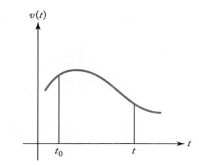

FIGURE A4.2 Plot of a function $v(t)$.

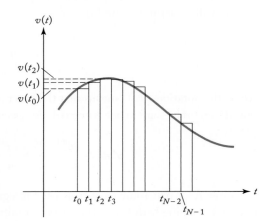

FIGURE A4.3 The interval $t - t_0$ has been divided into N equal intervals of duration Δt, so $t_1 = t_0 + \Delta t$, etc.

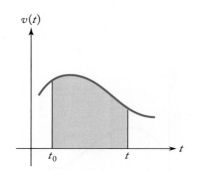

$v(t)$

t_0 t t

FIGURE A4.4 The area under the velocity curve.

In the notation of calculus, the right side of Eq. (20) is usually written in the following fashion:

$$x(t) - x_0 = \int_{t_0}^{t} v(t')\, dt' \tag{21}$$

The right side is called the **integral** of the function $v(t)$. The subscript and the superscript on the integration symbol $\int$ are called, respectively, the lower and the upper limit of integration; and t' is called the variable of integration (the prime on the variable of integration t' merely serves to distinguish that variable from the limit of integration t). Graphically, the integral is the exact area under the velocity curve between the limits t_0 and t in a plot of v vs. t (see Fig. A4.4). Areas below the t axis must be reckoned as negative.

In general, if $f(u)$ is a function of u, then the integral of this function is defined by a limiting procedure similar to that described above for the special case of the function $v(t)$. The integral over an interval from $u = a$ to $u = b$ is

$$\int_{a}^{b} f(u)\, du = \lim_{\substack{\Delta u \to 0 \\ N \to \infty}} \sum_{i=0}^{N-1} f(u_i)\, \Delta u \tag{22}$$

where $u_i = a + i\,\Delta u$. As in the case of the integral of $v(t)$, this integral can again be interpreted as an area: it is the area under the curve between the limits a and b in a plot of f vs. u.

For the explicit evaluation of integrals we can take advantage of the connection between integrals and antiderivatives. An **antiderivative** of a function $f(u)$ is simply a function $F(u)$ such that $dF/du = f$. For example, if $f(u) = u^n$ and $n \neq -1$, then an antiderivative of $f(u)$ is $F(u) = u^{n+1}/(n + 1)$. The fundamental theorem of calculus states that the integral of any function $f(u)$ can be expressed in terms of antiderivatives:

$$\int_{a}^{b} f(u)\, du = F(b) - F(a) \tag{23}$$

In essence, this means that integration is the inverse of differentiation. We will not prove this theorem here, but we remark that such an inverse relationship between integration and differentiation should not come as a surprise. We have already run across an obvious instance of such a relationship: we know that velocity is the derivative of the position, and we have seen above that the position is the integral of the velocity.

We will sometimes write Eq. (23) as

$$\int_{a}^{b} f(u)\, du = F(u)\Big|_{a}^{b} \tag{24}$$

where the notation $F(u)\big|_{a}^{b}$ means that the function $F(u)$ is to be evaluated at a and at b, and these values are to be subtracted. For example, if $n \neq -1$,

$$\int_{a}^{b} u^n\, du = \frac{u^{n+1}}{n + 1}\bigg|_{a}^{b} = \frac{b^{n+1}}{n + 1} - \frac{a^{n+1}}{n + 1} \tag{25}$$

Table A4.2 lists some frequently used integrals. In this table, the limits of integration belonging with Eq. (24) have been omitted for the sake of brevity.

TABLE A4.2	SOME INTEGRALS

$$\int u^n du = \frac{u^{n+1}}{n+1} \qquad \text{for } n \neq -1$$

$$\int \frac{1}{u} du = \ln u \qquad \text{for } u > 0$$

$$\int e^{ku} du = \frac{e^{ku}}{k}$$

$$\int \ln u \, du = u \ln u - u$$

$$\int \sin(ku) \, du = -\frac{1}{k} \cos(ku) \qquad \text{(where } ku \text{ is in radians)}$$

$$\int \cos(ku) \, du = \frac{1}{k} \sin(ku) \qquad \text{(where } ku \text{ is in radians)}$$

$$\int \frac{du}{1 + ku} = \frac{1}{k} \ln(1 + ku)$$

$$\int \frac{du}{\sqrt{k^2 - u^2}} = \sin^{-1}\left(\frac{u}{k}\right)$$

$$\int \frac{du}{\sqrt{u^2 \pm k^2}} = \ln\left(u + \sqrt{u^2 \pm k^2}\right)$$

$$\int \sqrt{k^2 - u^2} \, du = \frac{1}{2}\left[u\sqrt{k^2 - u^2} + k^2 \sin^{-1}\left(\frac{u}{k}\right)\right]$$

$$\int \frac{du}{k^2 + u^2} = \frac{1}{k} \tan^{-1}\left(\frac{u}{k}\right)$$

$$\int \frac{du}{u\sqrt{k^2 \pm u^2}} = -\frac{1}{k} \ln\left(\frac{k + \sqrt{k^2 \pm u^2}}{u}\right)$$

$$\int \frac{du}{(u^2 + k^2)^{3/2}} = \frac{u}{k^2\sqrt{u^2 + k^2}}$$

A4.4 Important Rules for Integration

1. Integral of a constant times a function:

$$\int_a^b cf(u) \, du = c\int_a^b f(u) \, du \qquad (26)$$

For instance,

$$\int_a^b 5u^2 \, du = 5\int_a^b u^2 \, du = 5\left(\frac{b^3}{3} - \frac{a^3}{3}\right) \qquad (27)$$

2. Integral of a sum of two functions:

$$\int_a^b [f(u) + g(u)] \, du = \int_a^b f(u) \, du + \int_a^b g(u) \, du \qquad (28)$$

For instance,

$$\int_a^b (5u^2 + u)\, du = \int_a^b 5u^2\, du + \int_a^b u\, du = 5\left(\frac{b^3}{3} - \frac{a^3}{3}\right) + \left(\frac{b^2}{2} - \frac{a^2}{2}\right) \quad (29)$$

3. Change of limits of integration:

$$\int_a^b f(u)\, du = \int_a^c f(u)\, du + \int_c^b f(u)\, du \quad (30)$$

$$\int_a^b f(u)\, du = -\int_b^a f(u)\, du \quad (31)$$

4. Change of variable of integration: If u is a function of v, then

$$\int_a^b f(u)\, du = \int_{v(a)}^{v(b)} f(u)\, \frac{du}{dv}\, dv \quad (32)$$

For instance, with $u = v^2$,

$$\int_a^b u^3\, du = \int_a^b v^6\, du = \int_{\sqrt{a}}^{\sqrt{b}} v^6 (2v)\, dv \quad (33)$$

Finally, let us apply these general results to some specific examples of integration of the velocity.

EXAMPLE 1   A particle with constant acceleration has the following velocity as a function of time [compare Eq. (15)]:

$$v(t) = v_0 + at$$

where v_0 is the velocity at $t = 0$.

By integration, find the position as a function of time.

SOLUTION: According to Eq. (21), with $t_0 = 0$,

$$x(t) - x_0 = \int_0^t v(t')\, dt' = \int_0^t (v_0 + at')\, dt'$$

Using rule 2 and rule 1, we find that this equals

$$x(t) - x_0 = \int_0^t v_0\, dt' + \int_0^t at'\, dt' = v_0 \int_0^t dt' + a \int_0^t t'\, dt' \quad (34)$$

The first entry listed in Table A4.2 gives $\int dt' = t'$ (for $n = 0$) and $\int t'\, dt' = t'^2/2$ (for $n = 1$). Thus,

$$x(t) - x_0 = v_0\, t' \Big|_0^t + \tfrac{1}{2}\, at'^2 \Big|_0^t$$

$$= v_0 t + \tfrac{1}{2}\, at^2 \quad (35)$$

This, of course, agrees with Eq. (16).

EXAMPLE 2 The instantaneous velocity of a projectile traveling through air is the following function of time:

$$v(t) = 655.9 - 61.14t + 3.26t^2$$

where $v(t)$ is measured in meters per second and t is measured in seconds. Assuming that $x = 0$ at $t = 0$, what is the position as a function of time? What is the position at $t = 3.0$ s?

SOLUTION: With $x_0 = 0$ and $t_0 = 0$, Eq. (21) becomes

$$x(t) = \int_0^t (655.9 - 61.14t' + 3.26t'^2)\, dt'$$

$$= 655.9 \int_0^t dt' - 61.14 \int_0^t t'\, dt' + 3.26 \int_0^t t'^2\, dt'$$

$$= 655.9(t') \Big|_0^t - 61.14(t'^2/2) \Big|_0^t + 3.26(t'^3/3) \Big|_0^t$$

$$= 655.9t - 61.14t^2/2 + 3.26t^3/3$$

When evaluated at $t = 3.0$ s, this yields

$$x(3.0) = 655.9 \times 3.0 - 61.14 \times (3.0)^2/2 + 3.26 \times (3.0)^3/3$$

$$= 1722 \text{ m}$$

EXAMPLE 3 The acceleration of a mass pushed back and forth by an elastic spring is

$$a(t) = B \cos \omega t \qquad (36)$$

where B and ω are constants. Find the position as a function of time. Assume $v = 0$ and $x = 0$ at $t = 0$.

SOLUTION: The calculation involves two steps: first we must integrate the acceleration to find the velocity, then we must integrate the velocity to find the position. For the first step we use an equation analogous to Eq. (21),

$$v(t) - v_0 = \int_{t_0}^t a(t')\, dt' \qquad (37)$$

This equation becomes obvious if we remember that the relationship between acceleration and velocity is analogous to that between velocity and position. With $v_0 = 0$ and $t_0 = 0$, we obtain from Eq. (29)

$$v(t) = \int_0^t B \cos \omega t'\, dt' = B\frac{1}{\omega} \sin \omega t' \Big|_0^t$$

$$= \frac{B}{\omega} \sin \omega t \qquad (38)$$

Next,

$$x(t) = \int_0^t v(t')\, dt' = \int_0^t \frac{B}{\omega} \sin \omega t'\, dt' = \frac{B}{\omega}\left(-\frac{1}{\omega} \cos \omega t' \right) \Big|_0^t$$

$$= -\frac{B}{\omega^2} \cos \omega t + \frac{B}{\omega^2} \qquad (39)$$

A4.5 The Taylor Series

Suppose that $f(u)$ is a smooth function of u in some neighborhood of a given point $u = a$, so the function has continuous derivatives of all orders. Then the value of the function at an arbitrary point near a can be expressed in terms of the following infinite series, where all the derivatives are evaluated at the point a:

$$f(u) = f(a) + \frac{df}{du}(u - a) + \frac{1}{2}\frac{d^2f}{du^2}(u - a)^2 + \frac{1}{3\cdot 2}\frac{d^3f}{du^3}(u - a)^3 + \cdots \quad (40)$$

This is called the **Taylor series** for the function $f(u)$ about the point a. The series converges, and is valid, provided u is sufficiently close to a. How close is "sufficiently close" depends on the function f and on the point a. Some functions, such as $\sin u$, $\cos u$, and e^u, are extremely well behaved, and their Taylor series converge for any choice of u and of a. The Taylor series gives us a convenient method for the approximate evaluation of a function.

EXAMPLE 4 Find the Taylor series for $\sin u$ about the point $u = 0$.

SOLUTION: The derivatives of $\sin u$ evaluated at $u = 0$ are

$$\frac{d}{du}\sin u = \cos u = 1$$

$$\frac{d^2}{du^2}\sin u = \frac{d}{du}\cos u = -\sin u = 0$$

$$\frac{d^3}{du^3}\sin u = \frac{d}{du}(-\sin u) = -\cos u = -1$$

$$\frac{d^4}{du^4}\sin u = \frac{d}{du}(-\cos u) = \sin u = 0, \quad \text{etc.}$$

Hence Eq. (32) gives

$$\sin u = 0 + 1 \times (u - 0) + \frac{1}{2} \times 0 \times (u - 0)^2 + \frac{1}{3\cdot 2} \times (-1) \times (u - 0)^3$$

$$+ \frac{1}{4\cdot 3\cdot 2} \times 0 \times (u - 0)^4 + \cdots$$

$$= u - \frac{1}{6}u^3 + \cdots$$

Note that for very small values of u, we can neglect all higher powers of u, so $\sin u \approx u$, which is an approximation often used in this book.

A4.6 Some Approximations

By constructing Taylor series, we can obtain the following useful approximations, all of which are valid for small values of u. It is often sufficient to keep just the first one or two terms on the right side.

$$\sqrt{1 + u} = 1 + \frac{1}{2}u - \frac{1}{8}u^2 + \frac{1}{16}u^3 + \cdots \quad (41)$$

$$\frac{1}{1 + u} = 1 - u + u^2 - u^3 + \cdots \tag{42}$$

$$\frac{1}{\sqrt{1 + u}} = 1 - \frac{1}{2}u + \frac{3}{8}u^2 - \frac{5}{16}u^3 + \cdots \tag{43}$$

$$\frac{1}{(1 + u)^n} = 1 - nu + \frac{n(n + 1)}{2}u^2 - \frac{n(n + 1)(n + 2)}{2 \cdot 3}u^3 + \cdots \tag{44}$$

$$e^u = 1 + u + \frac{1}{2}u^2 + \frac{1}{2 \cdot 3}u^3 + \cdots \tag{45}$$

$$\ln(1 + u) = u - \frac{1}{2}u^2 + \frac{1}{3}u^3 + \cdots \tag{46}$$

In all the following formulas, u is in radians:

$$\sin u = u - \frac{1}{6}u^3 + \frac{1}{120}u^5 + \cdots \tag{47}$$

$$\cos u = 1 - \frac{1}{2}u^2 + \frac{1}{24}u^4 - \frac{1}{720}u^6 + \cdots \tag{48}$$

$$\tan u = u + \frac{1}{3}u^3 + \frac{2}{15}u^5 + \cdots \tag{49}$$

$$\sin^{-1} u = u + \frac{1}{6}u^3 + \frac{3}{40}u^5 + \cdots \tag{50}$$

$$\tan^{-1} u = u - \frac{1}{3}u^3 + \frac{1}{5}u^5 + \cdots \tag{51}$$

Appendix 5: Propagating Uncertainties

Experimentalists carefully work to measure physical quantities and to determine the uncertainty in each quantity. We must often calculate a new result from a measured quantity or from several quantities; we must therefore understand the propagation of uncertainties through functions and formulas.

To keep things simple, we will make the assumption that the uncertainties in each quantity are symmetrically distributed about its measured value and that the various measured quantities are independent of each other. This is not always true. But by ignoring correlations and assuming symmetry, we can reduce all the necessary propagation of uncertainties to some simple formulas.

Suppose we have a measured quantity and its uncertainty, $x \pm \Delta x$, where Δx is a positive quantity and has the same units as x, and is also known as the *absolute uncertainty* in x. What, then, is the uncertainty of some function, $f(x)$, of this data? Under the assumption that the uncertainty is small, we can obtain the uncertainty from the first terms of the Taylor series expansion of f: $f(x + \Delta x) = f(x) + (df(x)/dx)\Delta x + \cdots$ From this we find the uncertainty $\Delta f = |f(x + \Delta x) - f(x)|$ in the function value $f(x)$ is

$$\Delta f = \left| \frac{df}{dx} \Delta x \right| \tag{1}$$

with the derivative evaluated at the point x. We can generalize this result to functions of several variables as follows: given the data $x \pm \Delta x$, $y \pm \Delta y$, . . ., the function $f(x, y, . . .)$ has the associated uncertainty

$$\Delta f = \left| \frac{\partial f}{\partial x} \Delta x \right| + \left| \frac{\partial f}{\partial y} \Delta y \right| + \cdots \tag{2}$$

where all the partial derivatives (see App. 4.2) are evaluated at the point $x, y, \ldots$. If we recall that we defined absolute uncertainties to be positive, we can write this as

$$\Delta f = \left| \frac{\partial f}{\partial x} \right| \Delta x + \left| \frac{\partial f}{\partial y} \right| \Delta y + \cdots \tag{3}$$

From this relationship, we can derive several simple results for uncertainty propagation.

EXAMPLE 1   **Addition and Subtraction.**

Given $f(x, y) = 3x + y - z + 5$, find Δf:

$$\Delta f = \left| \frac{\partial f}{\partial x} \right| \Delta x + \left| \frac{\partial f}{\partial y} \right| \Delta y + \left| \frac{\partial f}{\partial z} \right| \Delta z$$

$$= |3| \Delta x + |1| \Delta y + |-1| \Delta z$$

$$= 3\Delta x + \Delta y + \Delta z$$

Thus in addition or subtraction, the uncertainties add, and in multiplication by a constant, the uncertainty is multiplied by the same constant.

EXAMPLE 2   **Multiplication, Division, and Exponentiation.**

Given $f(x, y) = x^2 y/(5z)$, find Δf:

$$\Delta f = \left| \frac{\partial f}{\partial x} \right| \Delta x + \left| \frac{\partial f}{\partial y} \right| \Delta y + \left| \frac{\partial f}{\partial z} \right| \Delta z$$

$$= |2xy/(5z)| \Delta x + |x^2/(5z)| \Delta y + |-x^2 y/(5z^2)| \Delta z$$

Equivalently, for multiplication and division, we add *relative uncertainties* (e.g., $\Delta x/x$), and for exponentiation, we multiply the relative uncertainty by the magnitude of the exponent, to get the relative uncertainty of the product, quotient, or power.

EXAMPLE 3   **Numerical Application to Ohm's Law, $V = IR$.**

Given $V = 1.5 \pm 0.1$ Volt and $I = 0.50 \pm 0.02$ A, find R and ΔR:

Rearranging we find $R = V/I = (1.5 \text{ Volt})/(0.50 \text{ A}) = 3.0 \ \Omega$, and

$$\Delta R = \left| \frac{\partial R}{\partial V} \right| \Delta V + \left| \frac{\partial R}{\partial I} \right| \Delta I$$

$$= \left| \frac{1}{I} \right| \Delta V + \left| \frac{-V}{I^2} \right| \Delta I$$

$$= \left| \frac{1}{0.50 \text{ A}} \right| (0.1 \text{ Volt}) + \left| \frac{-1.5 \text{ Volt}}{(0.50 \text{ A})^2} \right| (0.02 \text{A})$$

$$= 0.2 \, \Omega + 0.12 \, \Omega = 0.4 \, \Omega$$

Note in the last step that unlike an ordinary calculation, we have rounded this final result up; uncertainties should always be rounded up, never down.

Appendix 6: The International System of Units (SI)

A6.1 Base Units

The SI system of units is the modern version of the metric system. The SI system recognizes seven fundamental, or base, units for length, mass, time, electric current, thermodynamic temperature, amount of substance, and luminous intensity.[b] The following definitions of the base units were adopted by the Conférence Générale des Poids et Mesures in the years indicated:

meter (m)   "The metre is the length of the path travelled by light in vacuum during a time interval of 1/299 792 458 of a second." (Adopted in 1983.)

kilogram (kg)   "The kilogram is . . . the mass of the international prototype of the kilogram." (Adopted in 1889 and in 1901.)

second (s)   "The second is the duration of 9 192 631 770 periods of the radiation corresponding to the transition between the two hyperfine levels of the ground state of the cesium-133 atom." (Adopted in 1967.)

ampere (A)   "The ampere is that constant current which, if maintained in two straight parallel conductors of infinite length, of negligible circular cross section, and placed one meter apart in vacuum, would produce between these conductors a force equal to 2×10^{-7} newton per meter of length." (Adopted in 1948.)

kelvin (K)   "The kelvin . . . is the fraction 1/273.16 of the thermodynamic temperature of the triple point of water." (Adopted in 1967.)

---

[b] At least two of the seven base units of the SI system are redundant. The mole is merely a certain number of atoms or molecules, in the same sense that a dozen is a number; there is no need to designate this number as a unit. The candela is equivalent to $\frac{1}{683}$ watt per steradian; it serves no purpose that is not served equally well by watt per steradian. Two other base units could be made redundant by adopting new definitions of the unit of temperature and of the unit of electric charge. Temperature could be measured in energy units because, according to the equipartition theorem, temperature is proportional to the energy per degree of freedom. Hence the kelvin could be defined as a derived unit, with $1 \text{ K} = \frac{1}{2} \times 1.38 \times 10^{-23}$ joule per degree of freedom. Electric charge could also be defined as a derived unit, to be measured with a suitable combination of the units of force and distance, as is done in the cgs system.

Furthermore, the definitions of the supplementary units—radian and steradian—are gratuitous. These definitions properly belong in the province of mathematics and there is no need to include them in a system of physical units.

TABLE A6.1	NAMES OF DERIVED UNITS		
QUANTITY	**DERIVED UNIT**	**NAME**	**SYMBOL**
frequency	$1/s$	hertz	Hz
force	$kg \cdot m/s^2$	newton	N
pressure	N/m^2	pascal	Pa
energy	$N \cdot m$	joule	J
power	J/s	watt	W
electric charge	$A \cdot s$	coulomb	C
electric potential	J/C	volt	V
electric capacitance	C/V	farad	F
electric resistance	V/A	ohm	Ω
conductance	A/V	siemen	S
magnetic flux	$V \cdot s$	weber	Wb
magnetic field	$V \cdot s/m^2$	tesla	T
inductance	$V \cdot s/A$	henry	H
temperature	K	degree Celsius	°C
luminous flux	$cd \cdot sr$	lumen	lm
illuminance	$cd \cdot sr/m^2$	lux	lx
radioactivity	$1/s$	becquerel	Bq
absorbed dose	J/kg	gray	Gy
dose equivalent	J/kg	sievert	Sv

TABLE A6.2	PREFIXES FOR UNITS	
FACTOR	**PREFIX**	**SYMBOL**
10^{24}	yotta	Y
10^{21}	zetta	Z
10^{18}	exa	E
10^{15}	peta	P
10^{12}	tera	T
10^{9}	giga	G
10^{6}	mega	M
10^{3}	kilo	k
10^{2}	hecto	h
10	deka	da
10^{-1}	deci	d
10^{-2}	centi	c
10^{-3}	milli	m
10^{-6}	micro	μ
10^{-9}	nano	n
10^{-12}	pico	p
10^{-15}	femto	f
10^{-18}	atto	a
10^{-21}	zepto	z
10^{-24}	yocto	y

mole   "The mole is the amount of substance of a system which contains as many elementary entities as there are atoms in 0.012 kilogram of carbon-12." (Adopted in 1967.)

candela (cd)   "The candela is the luminous intensity, in a given direction, of a source that emits monochromatic radiation of frequency 540×10^{12} Hz and that has a radiant intensity in that direction of $\frac{1}{683}$ watt per steradian." (Adopted in 1979.)

Besides these seven base units, the SI system also recognizes two supplementary units of angle and solid angle:

radian (rad)   "The radian is the plane angle between two radii of a circle which cut off on the circumference an arc equal in length to the radius."

steradian (sr)   "The steradian is the solid angle which, having its vertex in the center of a sphere, cuts off an area equal to that of a [flat] square with sides of length equal to the radius of the sphere."

A6.2 Derived Units

The derived units are formed out of products and ratios of the base units. Table A6.1 lists those derived units that have been glorified with special names. (Other derived units are listed in the tables of conversion factors in Appendix 8.)

A6.3 Prefixes

Multiples and submultiples of SI units are indicated by prefixes, such as the familiar *kilo*, *centi*, and *milli* used in *kilometer*, *centimeter*, and *millimeter*, etc. Table A6.2 lists all the accepted prefixes. Some enjoy more popularity than others; it is best to avoid the use of uncommon prefixes, such as *atto* and *exa*, since hardly anybody will recognize those.

Appendix 7: Best Values of Fundamental Constants

The values in the following table are the "2002 CODATA Recommended Values" by P. J. Mohr and B. N. Taylor Listed at the website physics.nist.gov/constants of the National Institute of Standards and Technology. The digits in parentheses are the one–standard deviation uncertainty in the last digits of the given value.

TABLE A7.1	BEST VALUES OF FUNDAMENTAL CONSTANTS			
QUANTITY	**SYMBOL**	**VALUE**	**UNITS**	**RELATIVE UNCERTAINTY (PARTS PER MILLION)**
UNIVERSAL CONSTANTS				
speed of light in vacuum	c	299 792 458	m·s^{-1}	(exact)
magnetic constant	μ_0	$4\pi \times 10^{-7}$	N·A^{-2}	
		$= 12.566\ 370\ 614 \ldots \times 10^{-7}$	N·A^{-2}	(exact)
electric constant $1/\mu_0 c^2$	ϵ_0	$8.854\ 187\ 817 \ldots \times 10^{-12}$	F·m^{-1}	(exact)
gravitational constant	G	$6.6742(10) \times 10^{-11}$	$\text{m}^3\text{·kg}^{-1}\text{·s}^{-2}$	1.5×10^{-4}
Planck constant	h	$6.626\ 0693(11) \times 10^{-34}$	J·s	1.7×10^{-7}
in eV·s	$\hbar$	$4.135\ 667\ 43(35) \times 10^{-15}$	eV·s	8.5×10^{-8}
$h/2\pi$		$1.054\ 571\ 68(18) \times 10^{-34}$	J·s	1.7×10^{-7}
in eV·s		$6.582\ 119\ 15(56) \times 10^{-16}$	eV·s	8.5×10^{-8}
ELECTROMAGNETIC CONSTANTS				
elementary charge	e	$1.602\ 176\ 53(14) \times 10^{-19}$	C	8.5×10^{-8}
magnetic flux quantum $h/2e$	Φ_0	$2.067\ 833\ 72(18) \times 10^{-15}$	Wb	8.5×10^{-8}
quantum	$2e^2/h$	$7.748\ 091\ 733(26) \times 10^{-5}$	S	3.3×10^{-9}
Josephson constant	$2e/h$	$483\ 597.879(41) \times 10^{9}$	Hz·V^{-1}	8.5×10^{-8}
Bohr magneton $e\hbar/2m_e$	μ_B	$927.400\ 949(80) \times 10^{-26}$	J·T^{-1}	8.6×10^{-8}
in eV·T^{-1}		$5.788\ 381\ 804(39) \times 10^{-5}$	eV·T^{-1}	6.7×10^{-9}
nuclear magneton $e\hbar/2m_p$	μ_N	$5.050\ 783\ 43(43) \times 10^{-27}$	J·T^{-1}	8.6×10^{-8}
in eV·T^{-1}		$3.152\ 451\ 259(21) \times 10^{-8}$	eV·T^{-1}	6.7×10^{-9}

(continued)

QUANTITY	SYMBOL	VALUE	UNITS	RELATIVE UNCERTAINTY (PARTS PER MILLION)
ATOMIC AND NUCLEAR CONSTANTS				
General				
fine-structure constant $e^2/4\pi\epsilon_0\hbar c$	α	$7.297\ 352\ 568(24) \times 10^{-3}$		3.3×10^{-9}
inverse fine-structure constant	α^{-1}	$137.035\ 999\ 11(46)$		3.3×10^{-9}
Rydberg constant $\alpha^2 m_e c/2h$	R_∞	$10\ 973\ 731.568\ 525(73)$	m^{-1}	6.6×10^{-12}
Bohr radius $4\pi\epsilon_0\hbar^2/m_e e^2$	a_0	$0.529\ 177\ 2108(18) \times 10^{-10}$	m	3.3×10^{-9}
Electron				
electron mass	m_e	$9.109\ 3826(16) \times 10^{-31}$	kg	1.7×10^{-7}
in u		$5.485\ 799\ 0945(24) \times 10^{-4}$	u	4.4×10^{-10}
energy equivalent in MeV	$m_e c^2$	$0.510\ 998\ 918(44)$	MeV	8.6×10^{-8}
electron-proton mass ratio	m_e/m_p	$5.446\ 170\ 2173(25) \times 10^{-4}$		4.6×10^{-10}
electron charge to mass quotient	$-e/m_e$	$-1.758\ 820\ 12(15) \times 10^{11}$	$C \cdot kg^{-1}$	8.6×10^{-8}
Compton wavelength $h/m_e c$	λ_C	$2.426\ 310\ 238(16) \times 10^{-12}$	m	6.7×10^{-9}
classical electron radius $\alpha^2 a_0$	r_e	$2.817\ 940\ 325(28) \times 10^{-15}$	m	1.0×10^{-8}
Thomson cross section $(8\pi/3)r_e^2$	σ_e	$0.665\ 245\ 837(13) \times 10^{-28}$	m^2	2.0×10^{-8}
electron magnetic moment	μ_e	$-928.476\ 412(80) \times 10^{-26}$	$J \cdot T^{-1}$	8.6×10^{-8}
to Bohr magneton ratio	μ_e/μ_B	$-1.001\ 159\ 652\ 1859(38)$		3.8×10^{-12}
to nuclear magneton ratio	μ_e/μ_N	$-1838.281\ 971\ 07(85)$		4.6×10^{-10}
electron magnetic moment anomaly $|\mu_e|/\mu_B - 1$	a_e	$1.159\ 652\ 1859(38) \times 10^{-3}$		3.2×10^{-9}
electron g-factor $-2(1 + a_e)$	g_e	$-2.002\ 319\ 304\ 3718(75)$		3.8×10^{-12}
Muon				
muon mass	m_μ	$1.883\ 531\ 40(33) \times 10^{-28}$	kg	1.7×10^{-7}
in u		$0.113\ 428\ 9264(30)$	u	2.6×10^{-8}
energy equivalent in MeV	$m_\mu c^2$	$105.658\ 3692(94)$	MeV	8.9×10^{-8}
muon-electron mass ratio	m_μ/m_e	$206.768\ 2838(54)$		2.6×10^{-8}
muon Compton wavelength $h/m_\mu c$	$\lambda_{C,\mu}$	$11.734\ 441\ 05(30) \times 10^{-15}$	m	2.5×10^{-8}
muon magnetic moment	μ_μ	$-4.490\ 447\ 99(40) \times 10^{-26}$	$J \cdot T^{-1}$	8.9×10^{-8}
to Bohr magneton ratio	μ_μ/μ_B	$-4.841\ 970\ 45(13) \times 10^{-3}$		2.6×10^{-8}
muon magnetic moment anomaly $|\mu_\mu|/(e\hbar/2m_\mu) - 1$	a_μ	$1.165\ 919\ 81(62) \times 10^{-3}$		5.3×10^{-7}
muon g-factor $-2(1 + a_\mu)$	g_μ	$-2.002\ 331\ 8396(12)$		6.2×10^{-10}
Proton				
proton mass	m_p	$1.672\ 621\ 71(29) \times 10^{-27}$	kg	1.7×10^{-7}
in u		$1.007\ 276\ 466\ 88(13)$	u	1.3×10^{-10}
energy equivalent in MeV	$m_p c^2$	$938.272\ 029(80)$	MeV	8.6×10^{-8}
proton-electron mass ratio	m_p/m_e	$1836.152\ 672\ 61(85)$		4.6×10^{-10}
proton-neutron mass ratio	m_p/m_n	$0.998\ 623\ 478\ 72(58)$		5.8×10^{-10}
proton charge to mass quotient	e/m_p	$9.578\ 833\ 76(82) \times 10^7$	$C \cdot kg^{-1}$	8.6×10^{-8}
proton Compton wavelength $h/m_p c$	$\lambda_{C,p}$	$1.321\ 409\ 8555(88) \times 10^{-15}$	m	6.7×10^{-9}
proton magnetic moment	μ_p	$1.410\ 606\ 71(12) \times 10^{-26}$	$J \cdot T^{-1}$	8.7×10^{-8}
to Bohr magneton ratio	μ_p/μ_B	$1.521\ 032\ 206(15) \times 10^{-3}$		1.0×10^{-8}
to nuclear magneton ratio	μ_p/μ_N	$2.792\ 847\ 351(28)$		1.0×10^{-8}
Neutron				
neutron mass	m_n	$1.674\ 927\ 28(29) \times 10^{-27}$	kg	1.7×10^{-7}
in u		$1.008\ 664\ 915\ 60(55)$	u	5.5×10^{-10}
energy equivalent in MeV	$m_n c^2$	$939.565\ 360(81)$	MeV	8.6×10^{-8}
neutron-electron mass ratio	m_n/m_e	$1838.683\ 6598(13)$		7.0×10^{-10}
neutron-proton mass ratio	m_n/m_p	$1.001\ 378\ 418\ 70(58)$		5.8×10^{-10}

(continued)

QUANTITY	SYMBOL	VALUE	UNITS	RELATIVE UNCERTAINTY (PARTS PER MILLION)
neutron Compton wavelength $h/m_{\mathrm{n}}c$	$\lambda_{\mathrm{C,n}}$	$1.319\,590\,9067(88) \times 10^{-15}$	m	6.7×10^{-9}
neutron magnetic moment	μ_{n}	$-0.966\,236\,45(24) \times 10^{-26}$	$\mathrm{J{\cdot}T^{-1}}$	2.5×10^{-7}
to Bohr magneton ratio	$\mu_{\mathrm{n}}/\mu_{\mathrm{B}}$	$-1.041\,875\,63(25) \times 10^{-3}$		2.4×10^{-7}
to nuclear magneton ratio	$\mu_{\mathrm{n}}/\mu_{\mathrm{N}}$	$-1.913\,042\,73(45)$		2.4×10^{-7}
Deuteron				
deuteron mass	m_{d}	$3.343\,583\,35(57) \times 10^{-27}$	kg	1.7×10^{-7}
in u		$2.013\,553\,212\,70(35)$	u	1.7×10^{-10}
energy equivalent in MeV	$m_{\mathrm{d}}c^2$	$1875.612\,82(16)$	MeV	8.6×10^{-8}
deuteron-electron mass ratio	$m_{\mathrm{d}}/m_{\mathrm{e}}$	$3670.482\,9652(18)$		4.8×10^{-10}
deuteron-proton mass ratio	$m_{\mathrm{d}}/m_{\mathrm{p}}$	$1.999\,007\,500\,82(41)$		2.0×10^{-10}
deuteron magnetic moment	μ_{d}	$0.433\,073\,482(38) \times 10^{-26}$	$\mathrm{J{\cdot}T^{-1}}$	8.7×10^{-8}
to Bohr magneton ratio	$\mu_{\mathrm{d}}/\mu_{\mathrm{B}}$	$0.466\,975\,4567(50) \times 10^{-3}$		1.1×10^{-8}
to nuclear magneton ratio	$\mu_{\mathrm{d}}/\mu_{\mathrm{N}}$	$0.857\,438\,2329(92)$		1.1×10^{-8}
Alpha Particle				
alpha particle mass	m_{α}	$6.644\,6565(11) \times 10^{-27}$	kg	1.7×10^{-7}
in u		$4.001\,506\,179\,149(56)$	u	1.4×10^{-11}
energy equivalent in MeV	$m_{\alpha}c^2$	$3727.379\,17(32)$	MeV	8.6×10^{-8}
alpha particle to electron mass ratio	$m_{\alpha}/m_{\mathrm{e}}$	$7294.299\,5363(32)$		4.4×10^{-10}
alpha particle to proton mass ratio	$m_{\alpha}/m_{\mathrm{p}}$	$3.972\,599\,689\,07(52)$		1.3×10^{-10}
PHYSICO-CHEMICAL CONSTANTS				
Avogadro constant	N_{A}	$6.022\,1415(10) \times 10^{23}$	$\mathrm{mole^{-1}}$	1.7×10^{-7}
atomic mass constant $m_{\mathrm{u}} = \frac{1}{12}\,m(^{12}\mathrm{C}) = 1\,\mathrm{u}$	m_{u}	$1.660\,538\,86(28) \times 10^{-27}$	kg	1.7×10^{-7}
energy equivalent in MeV	$m_{\mathrm{u}}c^2$	$931.494\,043(80)$	MeV	8.6×10^{-8}
Faraday constant $N_{\mathrm{A}}e$	F	$96\,485.3383(83)$	$\mathrm{C{\cdot}mole^{-1}}$	8.6×10^{-8}
molar gas constant	R	$8.314\,472\,(15)$	$\mathrm{J{\cdot}mole^{-1}{\cdot}K^{-1}}$	1.7×10^{-6}
Boltzmann constant R/N_{A}	k	$1.380\,6505(24) \times 10^{-23}$	$\mathrm{J{\cdot}K^{-1}}$	1.8×10^{-6}
in $\mathrm{eV{\cdot}K^{-1}}$		$8.617\,343(15) \times 10^{-5}$	$\mathrm{eV{\cdot}K^{-1}}$	1.8×10^{-6}
molar volume of ideal gas RT/p $T = 273.15$ K, $p = 101.325$ kPa	V_{m}	$22.413\,996(39) \times 10^{-3}$	$\mathrm{m^3{\cdot}mole^{-1}}$	1.7×10^{-6}
Loschmidt constant $N_{\mathrm{A}}/V_{\mathrm{m}}$	n_0	$2.686\,7773(47) \times 10^{25}$	$\mathrm{m^{-3}}$	1.8×10^{-6}
Stefan-Boltzmann constant $(\pi^2/60)k^4/\hbar^3 c^2$	σ	$5.670\,400(40) \times 10^{-8}$	$\mathrm{W{\cdot}m^{-2}{\cdot}K^{-4}}$	7.0×10^{-6}
Wien displacement law constant $b = \lambda_{\max}T$	b	$2.897\,7685(51) \times 10^{-3}$	$\mathrm{m{\cdot}K}$	1.7×10^{-6}

Appendix 8: Conversion Factors

The units for each quantity are listed alphabetically, except that the SI unit is always listed first. The numbers are based on "American National Standard; Metric Practice" published by the Institute of Electrical and Electronics Engineers, 1982.

Angle

1 radian $= 57.30° = 3.438 \times 10^3{}' = (1/2\pi)$ rev $= 2.063 \times 10^5{}''$

1 degree (°) $= 1.745 \times 10^{-2}$ radian $= 60' = 3600'' = \frac{1}{360}$ rev

1 minute of arc (') $= 2.909 \times 10^{-4}$ radian $= \frac{1}{60}° = 4.630 \times 10^{-5}$ rev $= 60''$

1 revolution (rev) $= 2\pi$ radians $= 360° = 2.160 \times 10^4{}' = 1.296 \times 10^6{}''$

1 second of arc ('') $= 4.848 \times 10^{-6}$ radian $= \frac{1}{3600}° = \frac{1}{60}{}' = 7.716 \times 10^{-7}$ rev

Length

1 meter (m) $= 1 \times 10^9$ nm $= 1 \times 10^{10}$ Å$= 6.685 \times 10^{-12}$ AU $= 100$ cm $=$ 1×10^{15} fm $= 3.281$ ft $= 39.37$ in. $= 1 \times 10^{-3}$ km $= 1.057 \times 10^{-16}$ light-year $= 1 \times 10^6\ \mu$m $= 5.400 \times 10^{-4}$ nmi $= 6.214 \times 10^{-4}$ mi $=$ 3.241×10^{-17} pc $= 1.094$ yd

1 angstrom (Å) $= 1 \times 10^{-10}$ m $= 1 \times 10^{-8}$ cm $= 1 \times 10^{-5}$ fm $=$ 3.281×10^{-10} ft $= 1 \times 10^{-4}\ \mu$m

1 astronomical unit (AU) $= 1.496 \times 10^{11}$ m $= 1.496 \times 10^{13}$ cm $=$ 1.496×10^8 km $= 1.581 \times 10^{-5}$ light-year $= 4.848 \times 10^{-6}$ pc

1 centimeter (cm) $= 0.01$ m $= 1 \times 10^8$ Å$= 1 \times 10^{13}$ fm $= 3.281 \times 10^{-2}$ ft $= 0.3937$ in. $= 1 \times 10^{-5}$ km $= 1.057 \times 10^{-18}$ light-year $= 1 \times 10^4\ \mu$m

1 fermi, or **femtometer** (fm) $= 1 \times 10^{-15}$ m $= 1 \times 10^{-13}$ cm $= 1 \times 10^5$ Å

1 foot (ft) $= 0.3048$ m $= 30.48$ cm $= 12$ in. $= 3.048 \times 10^5\ \mu$m $=$ 1.894×10^{-4} mi $= \frac{1}{3}$ yd

1 inch (in.) $= 2.540 \times 10^{-2}$ m $= 2.54$ cm $= \frac{1}{12}$ ft $= 2.54 \times 10^4\ \mu$m $= \frac{1}{36}$ yd

1 kilometer (km) $= 1 \times 10^3$ m $= 1 \times 10^5$ cm $= 3.281 \times 10^3$ ft $= 0.5400$ nmi $= 0.6214$ mi $= 1.094 \times 10^3$ yd

1 light-year $= 9.461 \times 10^{15}$ m $= 6.324 \times 10^4$ AU $= 9.461 \times 10^{17}$ cm $=$ 9.461×10^{12} km $= 5.879 \times 10^{12}$ mi $= 0.3066$ pc

1 micron, or **micrometer** (μm) $= 1 \times 10^{-6}$ m $= 1 \times 10^4$ Å$= 1 \times 10^{-4}$ cm $= 3.281 \times 10^{-6}$ ft $= 3.937 \times 10^{-5}$ in.

1 nautical mile (nmi) $= 1.852 \times 10^3$ m $= 1.852 \times 10^5$ cm $= 6.076 \times 10^3$ ft $= 1.852$ km $= 1.151$ mi

1 statute mile (mi) $= 1.609 \times 10^3$ m $= 1.609 \times 10^5$ cm $= 5280$ ft $=$ 1.609 km $= 0.8690$ nmi $= 1760$ yd

1 parsec (pc) $= 3.086 \times 10^{16}$ m $= 2.063 \times 10^5$ AU $= 3.086 \times 10^{18}$ cm $=$ 3.086×10^{13} km $= 3.262$ light-years

1 yard (yd) $= 0.9144$ m $= 91.44$ cm $= 3$ ft $= 36$ in. $= \frac{1}{1760}$ mi

Time

1 second (s) $= 1.157 \times 10^{-5}$ day $= \frac{1}{3600}$ h $= \frac{1}{60}$ min $=$ 1.161×10^{-5} sidereal day $= 3.169 \times 10^{-8}$ yr

1 day $= 8.640 \times 10^4$ s $= 24$ h $= 1440$ min $= 1.003$ sidereal days $=$ 2.738×10^{-3} yr

1 hour (h) $= 3600$ s $= \frac{1}{24}$ day $= 60$ min $= 1.141 \times 10^{-4}$ yr

1 minute (min) $= 60$ s $= 6.944 \times 10^{-4}$ day $= \frac{1}{60}$ h $= 1.901 \times 10^{-6}$ yr

1 sidereal day $= 8.616 \times 10^4$ s $= 0.9973$ day $= 23.93$ h $= 1.436 \times 10^3$ min
   $= 2.730 \times 10^{-3}$ yr

1 year (yr) $= 3.156 \times 10^7$ s $= 365.24$ days $= 8.766 \times 10^3$ h $=$
   5.259×10^5 min $= 366.24$ sidereal days

Mass

1 kilogram (kg) $= 6.024 \times 10^{26}$ u $= 5000$ carats $= 1.543 \times 10^4$ grains $=$
   1000 g $= 1 \times 10^{-3}$ t $= 35.27$ oz $= 2.205$ lb $= 1.102 \times 10^{-3}$ short
   ton $= 6.852 \times 10^{-2}$ slug

1 atomic mass unit (u) $= 1.6605 \times 10^{-27}$ kg $= 1.6605 \times 10^{-24}$ g

1 carat $= 2 \times 10^{-4}$ kg $= 0.2$ g $= 7.055 \times 10^{-3}$ oz $= 4.409 \times 10^{-4}$ lb

1 grain $= 6.480 \times 10^{-5}$ kg $= 6.480 \times 10^{-2}$ g $= 2.286 \times 10^{-3}$ oz $= \frac{1}{7000}$ lb

1 gram (g) $= 1 \times 10^{-3}$ kg $= 6.024 \times 10^{23}$ u $= 5$ carats $= 15.43$ grains $=$
   1×10^{-6} t $= 3.527 \times 10^{-2}$ oz $= 2.205 \times 10^{-3}$ lb $= 1.102 \times 10^{-6}$ short ton
   $= 6.852 \times 10^{-5}$ slug

1 metric ton, or **tonne** (t) $= 1 \times 10^3$ kg $= 1 \times 10^6$ g $= 2.205 \times 10^3$ lb $=$
   1.102 short tons $= 68.52$ slugs

1 ounce (oz) $= 2.835 \times 10^{-2}$ kg $= 141.7$ carats $= 437.5$ grains $= 28.35$ g $= \frac{1}{16}$ lb

1 pound (lb)^c $= 0.4536$ kg $= 453.6$ g $= 4.536 \times 10^{-4}$ t $= 16$ oz $=$
   $\frac{1}{2000}$ short ton $= 3.108 \times 10^{-2}$ slug

1 short ton $= 907.2$ kg $= 9.072 \times 10^5$ g $= 0.9072$ t $= 2000$ lb

1 slug $= 14.59$ kg $= 1.459 \times 10^4$ g $= 32.17$ lb

Area

1 square meter (m^2) $= 1 \times 10^4$ cm^2 $= 10.76$ ft^2 $= 1.550 \times 10^3$ in.^{2} $=$
   1×10^{-6} km^2 $= 3.861 \times 10^{-7}$ mi^2 $= 1.196$ yd^2

1 barn $= 1 \times 10^{-28}$ m^2 $= 1 \times 10^{-24}$ cm^2

1 square centimeter (cm^2) $= 1 \times 10^{-4}$ m^2 $= 1.076 \times 10^{-3}$ ft^2 $= 0.1550$ in.^{2}
   $= 1 \times 10^{-10}$ km^2 $= 3.861 \times 10^{-11}$ mi^2

1 square foot (ft^2) $= 9.290 \times 10^{-2}$ m^2 $= 929.0$ cm^2 $= 144$ in.^{2} $=$
   3.587×10^{-8} mi^2 $= \frac{1}{9}$ yd^2

1 square inch (in.^{2}) $= 6.452 \times 10^{-4}$ m^2 $= 6.452$ cm^2 $= \frac{1}{144}$ ft^2

1 square kilometer (km^2) $= 1 \times 10^6$ m^2 $= 1 \times 10^{10}$ cm^2
   $= 1.076 \times 10^7$ ft^2 $= 0.3861$ mi^2

1 square statute mile (mi^2) $= 2.590 \times 10^6$ m^2 $= 2.590 \times 10^{10}$ cm^2 $=$
   2.788×10^7 ft^2 $= 2.590$ km^2

1 square yard (yd^2) $= 0.8361$ m^2 $= 8.361 \times 10^3$ cm^2 $= 9$ ft^2 $= 1296$ in.^{2}

Volume

1 cubic meter (m^3) $= 1 \times 10^6$ cm^3 $= 35.31$ ft^3 $= 264.2$ gal $=$
   6.102×10^4 in.^{3} $= 1 \times 10^3$ liters $= 1.308$ yd^3

1 cubic centimeter (cm^3) $= 1 \times 10^{-6}$ m^3 $= 3.531 \times 10^{-5}$ ft^3 $=$
   2.642×10^{-4} gal $= 6.102 \times 10^{-2}$ in.^{3} $= 1 \times 10^{-3}$ liter

1 cubic foot (ft^3) $= 2.832 \times 10^{-2}$ m^3 $= 2.832 \times 10^4$ cm^3 $= 7.481$ gal $=$
   1728 in.^{3} $= 28.32$ liters $= \frac{1}{27}$ yd^3

^{c}This is the "avoirdupois" pound. The "troy" or "apothecary" pound is 0.3732 kg, or 0.8229 lb avoirdupois.

1 gallon (gal)^d = 3.785×10^{-3} m^3 = 0.1337 ft^3
1 cubic inch (in.^{3}) = 1.639×10^{-5} m^3 = 16.39 cm^3 = 5.787×10^{-4} ft^3
1 liter (l) = 1×10^{-3} m^3 = 1000 cm^3 = 3.531×10^{-2} ft^3 = 0.2642 gal
1 cubic yard (yd^3) = 0.7646 m^3 = 7.646×10^5 cm^3 = 27 ft^3 = 202.0 gal

Density

1 kilogram per cubic meter (kg/m^3) = 1×10^{-3} g/cm^3 =
6.243×10^{-2} lb/ft^3 = 8.345×10^{-3} lb/gal = 3.613×10^{-5} lb/in.^{3} =
8.428×10^{-4} short ton/yd^3 = 1.940×10^{-3} slug/ft^3
1 gram per cubic centimeter (g/cm^3) = 1×10^3 kg/m^3 = 62.43 lb/ft^3 =
8.345 lb/gal = 3.613×10^{-2} lb/in.^{3} = 0.8428 short ton/yd^3 = 1.940 slugs/ft^3
1 lb per cubic foot (lb/ft^3) = 16.02 kg/m^3 = 1.602×10^{-2} g/cm^3 =
0.1337 lb/gal = 1.350×10^{-2} short ton/yd^3 = 3.108×10^{-2} slug/ft^3
1 pound-per gallon (1 lb/gal) = 119.8 kg/m^3 = 7.481 lb/ft^3 = 0.2325 slug/ft^3
1 short ton per cubic yard (short ton/yd^3) = 1.187×10^3 kg/m^3 = 74.07 lb/ft^3
1 slug per cubic foot (slug/ft^3) = 515.4 kg/m^3 = 0.5154 g/cm^3 =
32.17 lb/ft^3 = 4.301 lb/gal

Speed

1 meter per second (m/s) = 100 cm/s = 3.281 ft/s = 3.600 km/h =
1.944 knots = 2.237 mi/h
1 centimeter per second (cm/s) = 0.01 m/s = 3.281×10^{-2} ft/s =
3.600×10^{-2} km/h = 1.944×10^{-2} knot = 2.237×10^{-2} mi/h
1 foot per second (ft/s) = 0.3048 m/s = 30.48 cm/s = 1.097 km/h =
0.5925 knot = 0.6818 mi/h
1 kilometer per hour (km/h) = 0.2778 m/s = 27.78 cm/s = 0.9113 ft/s
= 0.5400 knot = 0.6214 mi/h
1 knot, or **nautical mile per hour** = 0.5144 m/s = 51.44 cm/s =
1.688 ft/s = 1.852 km/h = 1.151 mi/h
1 mile per hour (mi/h) = 0.4470 m/s = 44.70 cm/s = 1.467 ft/s =
1.609 km/h = 0.8690 knot

Acceleration

1 meter per second squared (m/s^2) = 100 cm/s^2 = 3.281 ft/s^2 = 0.1020 g
1 centimeter per second squared (cm/s^2) = 0.01 m/s^2 =
3.281×10^{-2} ft/s^2 = 1.020×10^{-3} g
1 foot per second squared (ft/s^2) = 0.3048 m/s^2 = 30.48 cm/s^2 = 3.108×10^{-2} g
1 g = 9.807 m/s^2 = 980.7 cm/s^2 = 32.17 ft/s^2

Force

1 newton (N) = 1×10^5 dynes = 0.2248 lb-f = 1.124×10^{-4} short ton-force
1 dyne = 1×10^{-5} N = 2.248×10^{-6} lb-f = 1.124×10^{-9} short ton-force
1 pound-force (lb-f) = 4.448 N = 4.448×10^5 dynes = $\frac{1}{2000}$ = short ton-force
1 short ton-force = 8.896×10^3 N = 8.896×10^8 dynes = 2000 lb-f

^{d}This is the U.S. gallon; the U.K. and the Canadian gallon are 4.546×10^{-3} m^3, or 1.201 U.S. gallons.

Energy

1 joule (J) $= 9.478 \times 10^{-4}$ Btu $= 0.2388$ cal $= 1 \times 10^{7}$ ergs $=$
6.242×10^{18} eV $= 0.7376$ ft·lb-f $= 2.778 \times 10^{-7}$ kW·h

1 British thermal unit (Btu)[e] $= 1.055 \times 10^{3}$ J $= 252.0$ cal $=$
1.055×10^{10} ergs $= 778.2$ ft·lb-f $= 2.931 \times 10^{-4}$ kW·h

1 calorie (cal)[f] $= 4.187$ J $= 3.968 \times 10^{-3}$ Btu $= 4.187 \times 10^{7}$ ergs $=$
3.088 ft·lb-f $= 1 \times 10^{-3}$ kcal $= 1.163 \times 10^{-6}$ kW·h

1 erg $= 1 \times 10^{-7}$ J $= 9.478 \times 10^{-7}$ Btu $= 2.388 \times 10^{-8}$ cal $=$
6.242×10^{11} eV $= 7.376 \times 10^{-8}$ ft·lb-f $= 2.778 \times 10^{-14}$ kW·h

1 electron-volt (eV) $= 1.602 \times 10^{-19}$ J $= 1.602 \times 10^{-12}$ erg $=$
1.182×10^{-19} ft·lb-f

1 foot-pound-force (ft·lb-f) $= 1.356$ J $= 1.285 \times 10^{-3}$ Btu $= 0.3239$ cal $=$
1.356×10^{7} ergs $= 8.464 \times 10^{18}$ eV $= 3.766 \times 10^{-7}$ kW·h

1 kilocalorie (kcal), or **large calorie** (Cal) $= 4.187 \times 10^{3}$ J $= 1 \times 10^{3}$ cal

1 kilowatt-hour (kW·h) $= 3.600 \times 10^{6}$ J $= 3412$ Btu $= 8.598 \times 10^{5}$ cal $=$
3.6×10^{13} ergs $= 2.655 \times 10^{6}$ ft·lb-f

Power

1 watt (W) $= 3.412$ Btu/h $= 0.2388$ cal/s $= 1 \times 10^{7}$ ergs/s $=$
0.7376 ft·lb-f/s $= 1.341 \times 10^{-3}$ hp

1 British thermal unit per hour (Btu/h) $= 0.2931$ W $=$
7.000×10^{-2} cal/s $= 0.2162$ ft·lb-f/s $= 3.930 \times 10^{-4}$ hp

1 calorie per second (cal/s) $= 4.187$ W $= 14.29$ Btu/h $=$
4.187×10^{7} ergs/s $= 3.088$ ft·lb-f/s $= 5.615 \times 10^{-3}$ hp

1 erg per second (erg/s) $= 1 \times 10^{-7}$ W $= 2.388 \times 10^{-8}$ cal/s $=$
7.376×10^{-8} ft·lb-f/s $= 1.341 \times 10^{-10}$ hp

1 foot-pound-force per second (ft·lb-f/s) $= 1.356$ W $= 0.3238$ cal/s $=$
4.626 Btu/h $= 1.356 \times 10^{7}$ ergs/s $= 1.818 \times 10^{-3}$ hp

1 horsepower (hp)[g] $= 745.7$ W $= 2.544 \times 10^{3}$ Btu/h $= 178.1$ cal/s
$= 550$ ft·lb-f/s

1 kilowatt (kW) $= 1 \times 10^{3}$ W $= 3.412 \times 10^{3}$ Btu/h $= 238.8$ cal/s $=$
737.6 ft·lb-f/s $= 1.341$ hp

Pressure

1 newton per square meter (N/m^2), or **pascal** (Pa) $= 9.869 \times 10^{-6}$ atm $=$
1×10^{-5} bar $= 7.501 \times 10^{-3}$ mm-Hg $= 10$ dynes/cm^2 $= 2.953 \times 10^{-4}$ in.-Hg
$= 2.089 \times 10^{-2}$ lb-f/ft^2 $= 1.450 \times 10^{-4}$ lb-f/in.^{2} $= 7.501 \times 10^{-3}$ torr

1 atmosphere (atm) $= 1.013 \times 10^{5}$ N/m^2 $= 760.0$ mm-Hg $=$
1.013×10^{6} dynes/cm^2 $= 29.92$ in.-Hg $= 2.116 \times 10^{3}$ lb-f/ft^2
$= 14.70$ lb-f/in.^{2}

1 bar $= 1 \times 10^{5}$ N/m^2 $= 0.9869$ atm $= 750.1$ mm-Hg

1 dyne per square centimeter (dyne/cm^2) $= 0.1$ N/m^2 $=$
9.869×10^{-7} atm $= 7.501 \times 10^{-4}$ mm-Hg $= 2.089 \times 10^{-3}$ lb-f/ft^2 $=$
1.450×10^{-5} lb-f/in.^{2}

[e] This is the "International Table" Btu; there are several other Btus.
[f] This is the "International Table" calorie, which equals exactly 4.1868 J. There are several other calories; for instance, the thermochemical calorie, which equals 4.184 J.
[g] There are several other horsepowers; for instance, the metric horsepower, which equals 735.5 W.

1 inch of mercury (in.-Hg) $= 3.386 \times 10^3 \text{ N/m}^2 = 3.342 \times 10^{-2} \text{ atm} =$
25.40 mm-Hg $= 0.4912 \text{ lb-f/in.}^2$

1 pound-force per square inch (lb-f/in.^{2}, or psi) $= 6.895 \times 10^3 \text{ N/m}^2 =$
$6.805 \times 10^{-2} \text{ atm} = 6.895 \times 10^4 \text{ dynes/cm}^2 = 2.036 \text{ in.-Hg} =$
$7.031 \times 10^{-2} \text{ kp/cm}^2$

1 torr, or **millimeter of mercury** (mm-Hg) $= 1.333 \times 10^2 \text{ N/m}^2 = 1/760 \text{ atm} =$
$1.333 \times 10^{-3} \text{ bar} = 1.333 \times 10^3 \text{ dynes/cm}^2 = 0.03937 \text{ in.-Hg} = 0.01934 \text{ lb-f/in.}^2$

Electric Charge[h]

1 coulomb (C) $\Leftrightarrow 2.998 \times 10^9$ statcoulombs, or esu of charge $\Leftrightarrow$ 0.1 abcoulomb,
or emu of charge

Electric Current

1 ampere (A) $\Leftrightarrow 2.998 \times 10^9$ statamperes, or esu of current $\Leftrightarrow$ 0.1 abampere, or
emu of current

Electric Potential

1 volt (V) $\Leftrightarrow 3.336 \times 10^{-3}$ statvolt, or esu of potential $\Leftrightarrow 1 \times 10^8$ abvolts, or emu
of potential

Electric Field

1 volt per meter (V/m) $\Leftrightarrow 3.336 \times 10^{-5}$ statvolt/cm $\Leftrightarrow 1 \times 10^6$ abvolts/cm

Magnetic Field

1 tesla (T), or **weber per square meter** (Wb/m^2) $= 1 \times 10^4$ gauss

Electric Resistance

1 ohm (Ω) $\Leftrightarrow 1.113 \times 10^{-12}$ statohm, or esu of resistance $\Leftrightarrow 1 \times 10^9$ abohms, or
emu of resistance

Electric Resistivity

1 ohm-meter ($\Omega \cdot$m) $\Leftrightarrow 1.113 \times 10^{-10}$ statohm-cm $\Leftrightarrow 1 \times 10^{11}$ abohm-cm

Capacitance

1 farad (F) $\Leftrightarrow 8.988 \times 10^{11}$ statfarads, or esu of capacitance $\Leftrightarrow 1 \times 10^{-9}$ abfarad,
or emu of capacitance

Inductance

1 henry (H) $\Leftrightarrow 1.113 \times 10^{-12}$ stathenry, or esu of inductance $\Leftrightarrow 1 \times 10^9$ abhen-
rys, or emu of inductance

---

[h] The dimensions of the electric quantities in SI units, electrostatic units (esu), and electromagnetic units
(emu) are usually different; hence the relationships among most of these units are correspondences ($\Leftrightarrow$)
rather than equalities (=).

Appendix 9: The Periodic Table and Chemical Elements

TABLE A9.1	THE PERIODIC TABLE

Key
- ☐ Alkaline earth metals
- ☐ Alkali metals
- ☐ Transition metals
- ☐ Rare earths
- ☐ Nonmetals
- ☐ Halogens
- ☐ Noble gases
- ☐ Other metals

Group designation
Atomic number (of protons)
Symbol for element
Name
Atomic mass

Periods

IA / 1

VIIIA / 18

IIA / 2, **IIIB / 3**, **IVB / 4**, **VB / 5**, **VIB / 6**, **VIIB / 7**, **VIIIB (8, 9, 10)**, **IB / 11**, **IIB / 12**, **IIIA / 13**, **IVA / 14**, **VA / 15**, **VIA / 16**, **VIIA / 17**

Period 1
- 1 H — Hydrogen — 1.00794
- 2 He — Helium — 4.002602

Period 2
- 3 Li — Lithium — 6.941
- 4 Be — Beryllium — 9.012182
- 5 B — Boron — 10.811
- 6 C — Carbon — 12.0107
- 7 N — Nitrogen — 14.0067
- 8 O — Oxygen — 15.9994
- 9 F — Fluorine — 18.99840
- 10 Ne — Neon — 20.1797

Period 3
- 11 Na — Sodium — 22.98977
- 12 Mg — Magnesium — 24.3050
- 13 Al — Aluminum — 26.98154
- 14 Si — Silicon — 28.0855
- 15 P — Phosphorus — 30.97376
- 16 S — Sulfur — 32.065
- 17 Cl — Chlorine — 35.453
- 18 Ar — Argon — 39.948

Period 4
- 19 K — Potassium — 39.0983
- 20 Ca — Calcium — 40.078
- 21 Sc — Scandium — 44.955910
- 22 Ti — Titanium — 47.867
- 23 V — Vanadium — 50.9415
- 24 Cr — Chromium — 51.9961
- 25 Mn — Manganese — 54.938049
- 26 Fe — Iron — 55.845
- 27 Co — Cobalt — 58.93320
- 28 Ni — Nickel — 58.6934
- 29 Cu — Copper — 63.546
- 30 Zn — Zinc — 65.409
- 31 Ga — Gallium — 69.723
- 32 Ge — Germanium — 72.64
- 33 As — Arsenic — 74.92160
- 34 Se — Selenium — 78.96
- 35 Br — Bromine — 79.904
- 36 Kr — Krypton — 83.798

Period 5
- 37 Rb — Rubidium — 85.4678
- 38 Sr — Strontium — 87.62
- 39 Y — Yttrium — 88.90585
- 40 Zr — Zirconium — 91.224
- 41 Nb — Niobium — 92.90638
- 42 Mo — Molybdenum — 95.94
- 43 Tc — Technetium — 98.9072
- 44 Ru — Ruthenium — 101.07
- 45 Rh — Rhodium — 102.90550
- 46 Pd — Palladium — 106.42
- 47 Ag — Silver — 107.8682
- 48 Cd — Cadmium — 112.411
- 49 In — Indium — 114.818
- 50 Sn — Tin — 118.710
- 51 Sb — Antimony — 121.760
- 52 Te — Tellurium — 127.60
- 53 I — Iodine — 126.90447
- 54 Xe — Xenon — 131.293

Period 6
- 55 Cs — Cesium — 132.90545
- 56 Ba — Barium — 137.327
- 72 Hf — Hafnium — 178.49
- 73 Ta — Tantalum — 180.9479
- 74 W — Tungsten — 183.84
- 75 Re — Rhenium — 186.207
- 76 Os — Osmium — 190.23
- 77 Ir — Iridium — 192.217
- 78 Pt — Platinum — 195.078
- 79 Au — Gold — 196.96654
- 80 Hg — Mercury — 200.59
- 81 Tl — Thallium — 204.3833
- 82 Pb — Lead — 207.2
- 83 Bi — Bismuth — 208.98037
- 84 Po — Polonium — 208.9824
- 85 At — Astatine — 209.9871
- 86 Rn — Radon — 222.0176

Period 7
- 87 Fr — Francium — 223.0197
- 88 Ra — Radium — 226.0277
- 104 Rf — Rutherfordium — 261.1089
- 105 Db — Dubnium — 262.1144
- 106 Sg — Seaborgium — 263.118
- 107 Bh — Bohrium — 262.12
- 108 Hs — Hassium — 265.1306
- 109 Mt — Meitnerium — (268)
- 110 Ds — Darmstadtium — (271)
- 111 Uuu — Unununium — (272)
- 112 Uub — Un?nbium — (285)
- 114 Uuq — Ununquadium — (289)
- 116 Uuh — Ununhexium — (292)
- 118 — ---- — 0

Lanthanides
- 57 La — Lanthanum — 138.9055
- 58 Ce — Cerium — 140.116
- 59 Pr — Praseodymium — 140.90765
- 60 Nd — Neodymium — 144.24
- 61 Pm — Promethium — 144.9127
- 62 Sm — Samarium — 150.36
- 63 Eu — Europium — 151.964
- 64 Gd — Gadolinium — 157.25
- 65 Tb — Terbium — 158.92534
- 66 Dy — Dysprosium — 162.50
- 67 Ho — Holmium — 164.93032
- 68 Er — Erbium — 167.26
- 69 Tm — Thulium — 168.93421
- 70 Yb — Ytterbium — 173.04
- 71 Lu — Lutetium — 174.967

Actinides
- 89 Ac — Actinium — 227.0277
- 90 Th — Thorium — 232.0381
- 91 Pa — Protactinium — 231.0359
- 92 U — Uranium — 238.0289
- 93 Np — Neptunium — 237.0482
- 94 Pu — Plutonium — 244.0642
- 95 Am — Americium — 243.0614
- 96 Cm — Curium — 247.07003
- 97 Bk — Berkelium — 247.0703
- 98 Cf — Californium — 251.0796
- 99 Es — Einsteinium — 252.083
- 100 Fm — Fermium — 257.0951
- 101 Md — Mendelevium — 258.0984
- 102 No — Nobelium — 259.1011
- 103 Lr — Lawrencium — 262.110

TABLE A9.2   ATOMIC MASSES AND ATOMIC NUMBERS OF CHEMICAL ELEMENTS

Data were obtained from the National Institute for Standards and Technology; values are for the elements as they exist naturally on Earth or for the most stable isotope, with carbon-12 (the reference standard) having a mass of exactly 12 u. The estimated uncertainties in values between ± and ± 9 units in the last digit of an atomic mass are in parentheses after the atomic mass.
(Source: http://physics.nist.gov/PhysRefData/Compositions/index.html)

ELEMENT	SYMBOL	ATOMIC NUMBER	ATOMIC MASS (u)	ELEMENT	SYMBOL	ATOMIC NUMBER	ATOMIC MASS (u)
Actinium	Ac	89	227.027 7	Mercury	Hg	80	200.59 (2)
Aluminum	Al	13	26.981 538 (2)	Molybdenum	Mo	42	95.94 (1)
Americium	Am	95	243.061 4	Neodymium	Nd	60	144.24 (3)
Antimony	Sb	51	121.760 (1)	Neon	Ne	10	20.179 7 (6)
Argon	Ar	18	39.948 (1)	Neptunium	Np	93	237.048 2
Arsenic	As	33	74.921 60 (2)	Nickel	Ni	28	58.693 4 (2)
Astatine	At	85	209.987 1	Niobium	Nb	41	92.906 38 (2)
Barium	Ba	56	137.327 (7)	Nitrogen	N	7	14.006 7 (2)
Berkelium	Bk	97	247.070 3	Nobelium	No	102	259.101 1
Beryllium	Be	4	9.012 182 (3)	Osmium	Os	76	190.23 (3)
Bismuth	Bi	83	208.980 38 (2)	Oxygen	O	8	15.999 4 (3)
Bohrium	Bh	107	264.12	Palladium	Pd	46	106.42 (1)
Boron	B	5	10.811 (7)	Phosphorus	P	15	30.973 761 (2)
Bromine	Br	35	79.904 (1)	Platinum	Pt	78	195.078 (2)
Cadmium	Cd	48	112.411 (8)	Plutonium	Pu	94	244.064 2
Calcium	Ca	20	40.078 (4)	Polonium	Po	84	208.982 4
Californium	Cf	98	251.079 6	Potassium	K	19	39.098 3 (1)
Carbon	C	6	12.010 7 (8)	Praseodymium	Pr	59	140.907 65 (2)
Cerium	Ce	58	140.116 (1)	Promethium	Pm	61	144.912 7
Cesium	Cs	55	132.905 45 (2)	Protactinium	Pa	91	231.035 88 (2)
Chlorine	Cl	17	35.453 (9)	Radium	Ra	88	226.025 4
Chromium	Cr	24	51.996 1 (6)	Radon	Rn	86	222.017 6
Cobalt	Co	27	58.933 200 (9)	Rhenium	Re	75	186.207 (1)
Copper	Cu	29	63.546 (3)	Rhodium	Rh	45	102.905 50 (2)
Curium	Cm	96	247.070 3	Rubidium	Rb	37	85.467 8 (3)
Darmstadtium	Ds	110	271	Ruthenium	Ru	44	101.07 (2)
Dubnium	Db	105	262.114 4	Rutherfordium	Rf	104	261.108 9
Dysprosium	Dy	66	162.500 (1)	Samarium	Sm	62	150.36 (3)
Einsteinium	Es	99	252.083	Scandium	Sc	21	44.955 910 (8)
Erbium	Er	68	167.259 (3)	Seaborgium	Sg	106	263.118 6
Europium	Eu	63	151.964 (1)	Selenium	Se	34	78.96 (3)
Fermium	Fm	100	257.095 1	Silicon	Si	14	28.085 5 (3)
Fluorine	F	9	18.998 403 2 (5)	Silver	Ag	47	107.868 2 (2)
Francium	Fr	87	223.019 7	Sodium	Na	11	22.989 770 (2)
Gadolinium	Gd	64	157.25 (3)	Strontium	Sr	38	87.62 (1)
Gallium	Ga	31	69.723 (1)	Sulfur	S	16	32.065 (6)
Germanium	Ge	32	72.64 (1)	Tantalum	Ta	73	180.947 9 (1)
Gold	Au	79	196.966 55 (2)	Technetium	Tc	43	98.907 2
Hafnium	Hf	72	178.49 (2)	Tellurium	Te	52	127.60 (3)
Hassium	Hs	108	265.130 6	Terbium	Tb	65	158.925 34 (2)
Helium	He	2	4.002 602 (2)	Thallium	Tl	81	204.383 3 (2)
Holmium	Ho	67	164.930 32 (2)	Thorium	Th	90	232.038 1 (1)
Hydrogen	H	1	1.007 94 (7)	Thulium	Tm	69	168.934 21 (2)
Indium	In	49	114.818 (3)	Tin	Sn	50	118.710 (7)
Iodine	I	53	126.904 47 (3)	Titanium	Ti	22	47.867 (1)
Iridium	Ir	77	192.217 (3)	Tungsten	W	74	183.84 (1)
Iron	Fe	26	55.845 (2)	Ununbium	Uub	112	285
Krypton	Kr	36	83.798 (2)	Unununium	Uuu	111	272
Lanthanum	La	57	138.905 5 (2)	Ununquadium	Uuq	114	289
Lawrencium	Lr	103	262.110	Uranium	U	92	238.028 9 (1)
Lead	Pb	82	207.2 (1)	Vanadium	V	23	50.941 5 (1)
Lithium	Li	3	6.941 (2)	Xenon	Xe	54	131.293 (2)
Lutetium	Lu	71	174.967 (1)	Ytterbium	Yb	70	173.04 (3)
Magnesium	Mg	12	24.305 0 (6)	Yttrium	Y	39	88.905 85 (2)
Manganese	Mn	25	54.938 049 (9)	Zinc	Zn	30	65.409 (4)
Meitnerium	Mt	109	268	Zirconium	Zr	40	91.224 (2)
Mendelevium	Md	101	258.098 4				

Appendix 10: Formula Sheets

Chapters 1–21

$v = dx/dt$

$a = dv/dt = d^2x/dt^2$

$x = x_0 + v_0 t + \frac{1}{2}at^2$

$a(x - x_0) = \frac{1}{2}(v^2 - v_0^2)$

$A_x = A \cos \theta$

$A = \sqrt{A_x^2 + A_y^2 + A_z^2}$

$\mathbf{A} \cdot \mathbf{B} = AB \cos \phi$

$\qquad = A_x B_x + A_y B_y + A_z B_z$

$|\mathbf{A} \times \mathbf{B}| = AB \sin \phi$

$a = v^2/r$

$\mathbf{v}' = \mathbf{v} - \mathbf{V}_O$

$m\mathbf{a} = \mathbf{F}_{net}$

$w = mg$

$f_k = \mu_k N$

$f_s \leq \mu_s N$

$F = -kx$

$W = F_x \, \Delta x$

$W = \mathbf{F} \cdot \mathbf{s}$

$W = \int \mathbf{F} \cdot d\mathbf{s}$

$K = \frac{1}{2}mv^2$

$U = mgy$

$E = K + U = [\text{constant}]$

$U(x) = -\int_{x_0}^{x} F_x(x') \, dx'$

$F_x = -\dfrac{dU}{dx}$

$U = \frac{1}{2}kx^2$

$E = mc^2$

$P = dW/dt$

$P = \mathbf{F} \cdot \mathbf{v}$

$F = GMm/r^2$

$v^2 = GM_S/r$

$g = GM_E/R_E^2$

$U = -GMm/r$

$\mathbf{p} = m\mathbf{v}$

$\mathbf{r}_{CM} = \dfrac{1}{M}\displaystyle\int \mathbf{r} \, dm$

$\mathbf{I} = \displaystyle\int_0^{\Delta t} \mathbf{F} \, dt$

$v_1' = \dfrac{m_1 - m_2}{m_1 + m_2}v_1; \; v_2' = \dfrac{2m_1}{m_1 + m_2}v_1$

$\omega = d\phi/dt$

$\alpha = d\omega/dt = d^2\phi/dt^2$

$v = R\omega$

$K = \frac{1}{2}I\omega^2$

$I = \displaystyle\int R^2 \, dm$

$I_{CM} = MR^2 \text{ (hoop)}; \frac{1}{2}MR^2 \text{ (disk)};$

$\qquad \frac{2}{5}MR^2 \text{ (sphere)}; \frac{1}{12}ML^2 \text{ (rod)}.$

$I = I_{CM} + Md^2$

$\tau = FR \sin \theta$

$I\alpha = \tau$

$P = \tau\omega$

$L = I\omega$

$\mathbf{L} = \mathbf{r} \times \mathbf{p}$

$\dfrac{d\mathbf{L}}{dt} = \mathbf{r} \times \mathbf{F}$

$x = A \cos(\omega t + \delta)$

$T = 2\pi/\omega; \; f = 1/T = \omega/2\pi$

$m \, d^2x/dt^2 = -kx$

$\omega = \sqrt{k/m}$

$\omega = \sqrt{g/l}; \; T = 2\pi\sqrt{l/g}$

$\omega = \sqrt{mgd/I}$

$y = A \cos k(x \pm vt) = A \cos(kx \pm \omega t)$

$\lambda = 2\pi/k; \; f = v/\lambda; \; \omega = 2\pi f$

$v = \sqrt{F/(M/L)}$

$f_{beat} = f_1 - f_2$

$f' = f(1 \pm V_R/v)$

$f' = f/(1 \mp V_E/v)$

$\sin \theta = v/V_E$

$p - p_0 = -\rho gy$

$\frac{1}{2}\rho v^2 + \rho gy + p = [\text{constant}]$

$pV = NkT$

$T_C = T - 273.15$

$v_{rms} = \sqrt{3kT/m}$

$TV^{\gamma - 1} = [\text{constant}];$

$\quad pV^{\gamma} = [\text{constant}]; \; \gamma = C_p/C_V$

$\Delta E = Q - W$

$e = 1 - T_2/T_1$

$\Delta S = \displaystyle\int_A^B dQ/T$

$g = 9.81 \text{ m/s}^2$

$G = 6.67 \times 10^{-11} \text{ N·m}^2/\text{kg}^2$

$M_E = 5.98 \times 10^{24} \text{ kg}$

$R_E = 6.37 \times 10^6 \text{ m}$

$m_e = 9.11 \times 10^{-31} \text{ kg}$

$m_p = 1.67 \times 10^{-27} \text{ kg}$

$c = 3.00 \times 10^8 \text{ m/s}$

$N_A = 6.02 \times 10^{23}/\text{mole}$

$k = 1.38 \times 10^{-23} \text{ J/K}$

$1 \text{ cal} = 4.19 \text{ J}$

$$F = \frac{1}{4\pi\epsilon_0}\frac{qq'}{r^2}$$

$$E = \frac{1}{4\pi\epsilon_0}\frac{q'}{r^2}$$

$$E = \sigma/2\epsilon_0$$

$$p = lQ$$

$$\boldsymbol{\tau} = \mathbf{p} \times \mathbf{E}$$

$$U = -\mathbf{p} \cdot \mathbf{E}$$

$$\Phi_E = \int \mathbf{E} \cdot d\mathbf{A}$$

$$\oint \mathbf{E} \cdot d\mathbf{A} = \oint E_\perp\, dA = \frac{Q_{\text{inside}}}{\epsilon_0}$$

$$V = \frac{1}{4\pi\epsilon_0}\frac{q'}{r}$$

$$E_x = -\frac{\partial V}{\partial x}, \quad E_y = -\frac{\partial V}{\partial y}, \quad E_z = -\frac{\partial V}{\partial z}$$

$$U = \tfrac{1}{2}Q_1 V_1 + \tfrac{1}{2}Q_2 V_2 + \tfrac{1}{2}Q_3 V_3 + \cdots$$

$$u = \tfrac{1}{2}\epsilon_0 E^2$$

$$C = Q/\Delta V$$

$$C = \epsilon_0 A/d$$

$$E = E_{\text{free}}/\kappa$$

$$\oint \kappa E_\perp\, dA = \frac{Q_{\text{free, inside}}}{\epsilon_0}$$

$$\Delta V = -\int_{P_0}^{P} \mathbf{E} \cdot d\mathbf{s}$$

$$u = \tfrac{1}{2}\kappa\epsilon_0 E^2$$

$$I = \Delta V/R$$

$$R = \rho l/A$$

$$P = I\mathcal{E}\,;\ P = I\,\Delta V$$

$$F = \frac{\mu_0}{2\pi}\frac{qvI}{r}$$

$$\mathbf{F} = q\mathbf{v} \times \mathbf{B}$$

$$d\mathbf{B} = \frac{\mu_0}{4\pi}\frac{I\,d\mathbf{s} \times \mathbf{r}}{r^3}$$

$$\oint \mathbf{B} \cdot d\mathbf{s} = \oint B_\parallel\, ds = \mu_0 I$$

$$B = \mu_0 n I$$

$$r = \frac{p}{qB}$$

$$d\mathbf{F} = I\,d\mathbf{l} \times \mathbf{B}$$

$$\boldsymbol{\mu} = I \times [\text{area of loop}]$$

$$\boldsymbol{\tau} = \boldsymbol{\mu} \times \mathbf{B}$$

$$U = -\boldsymbol{\mu} \cdot \mathbf{B}$$

$$\mathcal{E} = vBl$$

$$\mathcal{E} = -\frac{d\Phi_B}{dt}$$

$$\Phi_B = \int \mathbf{B} \cdot d\mathbf{A}$$

$$\oint \mathbf{E} \cdot d\mathbf{s} = \oint E_\parallel\, ds = -\frac{d\Phi_B}{dt}$$

$$\Phi_B = LI$$

$$\mathcal{E} = -L\frac{dI}{dt}$$

$$U = \tfrac{1}{2}LI^2$$

$$u = \frac{1}{2\mu_0}B^2$$

$$\omega_0 = 1/\sqrt{LC}$$

$$Z = \sqrt{R^2 + \left(\omega L - \frac{1}{\omega C}\right)^2}$$

$$\mathcal{E}_2 = \mathcal{E}_1\frac{N_2}{N_1}$$

$$\oint \mathbf{B} \cdot d\mathbf{A} = \oint B_\perp\, dA = 0$$

$$\oint \mathbf{B} \cdot d\mathbf{s} = \oint B_\parallel\, ds = \mu_0 I + \mu_0\epsilon_0\frac{d\Phi_E}{dt}$$

$$B = E/c$$

$$\mathbf{S} = \frac{1}{\mu_0}\mathbf{E} \times \mathbf{B}$$

$$[\text{pressure}] = S/c$$

$$\frac{\partial^2 E}{\partial x^2} = \mu_0\epsilon_0\frac{\partial^2 E}{\partial t^2}$$

$$c = 1/\sqrt{\mu_0\epsilon_0}$$

$$v = c/n$$

$$n_1 \sin\theta_1 = n_2 \sin\theta_2$$

$$f = \pm\tfrac{1}{2}R$$

$$\frac{1}{s} + \frac{1}{s'} = \frac{1}{f}$$

Interference minima:

$$d \sin\theta = \tfrac{1}{2}\lambda, \tfrac{3}{2}\lambda, \tfrac{5}{2}\lambda, \ldots$$

Interference maxima:

$$d \sin\theta = 0, \lambda, 2\lambda, \ldots$$

Diffraction minima:

$$a \sin\theta = \lambda, 2\lambda, 3\lambda, \ldots$$

$$a \sin\theta = 1.22\lambda$$

$$f' = \sqrt{\frac{1 \mp v/c}{1 \pm v/c}}\, f$$

$$x' = \frac{x - Vt}{\sqrt{1 - V^2/c^2}}$$

$$y' = y$$

$$t' = \frac{t - Vx/c^2}{\sqrt{1 - V^2/c^2}}$$

$$\Delta t = \frac{\Delta t'}{\sqrt{1 - V^2/c^2}}$$

$$L = \sqrt{1 - V^2/c^2}\, L'$$

$$v'_x = \frac{v_x - V}{1 - v_x V/c^2}$$

$$\mathbf{p} = \frac{m\mathbf{v}}{\sqrt{1 - v^2/c^2}}\,;\ \ E = \frac{mc^2}{\sqrt{1 - v^2/c^2}}$$

$$E^2 = p^2 c^2 + m^2 c^4$$

$$E = hf$$

$$p = hf/c$$

$$\Delta y\, \Delta p_y \geq h/4\pi$$

$$L = n\hbar$$

$$E_n = -\frac{m_e e^4}{2(4\pi\epsilon_0)^2\hbar^2}\frac{1}{n^2} = -\frac{13.6\ \text{eV}}{n^2}$$

$$\lambda = h/p$$

$$\mu_{\text{spin}} = \frac{e\hbar}{2m_e}$$

$$E = \frac{J(J+1)\hbar^2}{2I}$$

$$R = (1.2 \times 10^{-15}\ \text{m}) \times A^{1/3}$$

$$n = n_0 e^{-t/\tau}\,;\ \ \tau = t_{1/2}/0.693$$

$$e = 1.60 \times 10^{-19}\ \text{C}$$

$$\epsilon_0 = 8.85 \times 10^{-12}\ \text{F/m}$$

$$\mu_0 = 1.26 \times 10^{-6}\ \text{H/m}$$

$$c = 3.00 \times 10^8\ \text{m/s}$$

$$h = 2\pi\hbar = 6.63 \times 10^{-34}\ \text{J·s}$$

$$m_e = 9.11 \times 10^{-31}\ \text{kg}$$

$$m_p = 1.67 \times 10^{-27}\ \text{kg}$$

Appendix 11: Answers to Odd-Numbered Problems and Review Problems

Chapter 1

1. 5.87 ft; 1.78 m (Assuming a height of 5 ft 10 in)
3. 48.7 m
5. 66 picas long and 51 picas wide
7. 12.7 mm; 6.35 mm; 3.18 mm; 1.59 mm; 0.794 mm; 0.397 mm
9. a) 1 mm; 3×10^6 m (Assuming grapefruit diameter = 0.1 m); b) 7 mm; 0.5 km (Assuming head diameter = 0.2 m)
11. 1 mm
13. 4.41 μm; 6.94 μm
15. 6.3×10^6 m
17. 1.4×10^{17} s
19. 7761 s
21. 23 h 56 min
23. 12 days
25. 3.7×10^7 beats/year
27. 0.25 min of arc; 0.463 km
29. 0.134 % in planets; 99.9% in sun
31. 0.021 % electrons; 99.98 % nucleus
33. 373.24 g
35. a) 8.4×10^{24} molecules; b) 4.3×10^{46} molecules; c) 1680 molecules
37. 28.95 g/mol
39. 6.9×10^8 m
41. 2.1×10^{22} m
43. a) 1 pc = 2.06×10^5 AU; b) 1 pc = 3.08×10^{16} m; c) 1 pc = 3.25 ly
45. 35.31 ft^3
47. 2.72 m
49. 8.9×10^3 kg/m^3; 5.6×10^2 lb/ft^3; 0.32 lb/in^3
51. 8.0 m^3/day
53. 10^8; 10^{13}
55. a) 7.4×10^2; b) 1.855×10^2; c) 8.47×10^{-3}
57. 6.0×10^7 metric tons/cm^3
59. 5.00×10^{-3} m^3/s; 5.00 kg/s
61. 7.1×10^{-15} m; 3.0×10^{-15} m
63. 354 m^2
65. 11°; 5.7°; 570 atoms
67. 359.76°; 1440.0°
69. 8.9 m; 9295 tons
71. 3.902×10^{-25} kg; 235.0 u
73. 2.8×10^{19} molecules
75. 0.125 mm
77. 88.5 km/h; 80.7 ft/s; 24.6 m/s
79. 3.81×10^9 s
81. yes, because the distance traveled while gliding = 18.7 km
83. a) 3840 km; b) 296 km; c) 0.315 or 1:3.2

Chapter 2

1. 0.3 s
3. 6.3×10^{-7} m/s; 5.4 cm/day
5. 32.5 km/h
7. 600 km/h
9. 14 km/h
11. 2.5×10^4 yr; 2.5×10^7 yr
13. 12.8 m/s; 46 km/h
15. 5.87 h; 150 h
17. 0.06 m
19. a) 14 s; 380 m; b) 72 m
21. 4.83 m/s
23. 2.0 m/s
25. a)

PLANET	ORBIT CIRC (km)	PERIOD (s)	SPEED	LOG SPEED	LOG CIRC
Mercury	3.64×10^8	7.61×10^6	47.8	1.68	8.56
Venus	6.79×10^8	1.94×10^7	35.0	1.54	8.83
Earth	9.42×10^8	3.16×10^7	29.8	1.47	8.97
Mars	1.43×10^9	5.93×10^7	24.1	1.38	9.16
Jupiter	4.89×10^9	3.76×10^8	13.0	1.11	9.69
Saturn	8.98×10^9	9.31×10^8	9.65	0.985	9.95
Uranus	1.80×10^{10}	2.65×10^9	6.79	0.832	10.26
Neptune	2.83×10^{10}	5.21×10^9	5.43	0.735	10.45
Pluto	3.71×10^{10}	7.83×10^9	4.74	0.676	10.57

b)

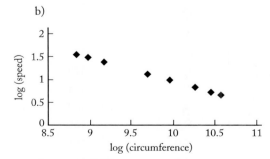

slope = -2.01

27. 20 m/s; 16.3 m/s

29. 1.2 m/s; 0.5 m/s

31. 12 m/s; 0 m/s

33. 0.67 m/s; 0.53 m/s

35. 32.4 m/s

37. 3.4×10^3 m/s^2

39. a)

t(s)	$\bar{a}$(m/s^2)	$\bar{a}$(in g)	
0	6.1	0.62	Method:
10	1.4	0.14	i) draw tangent to curve
20	0.83	0.085	ii) get slope of line by
30	0.56	0.057	counting squares
40	0.49	0.050	iii) to find Δv and Δt
			convert from km/h to m/s

b)

t(s)	$\bar{a}$(m/s^2)	$\bar{a}$(in g)
0	−0.74	−0.075
10	−0.44	−0.045
20	−0.44	−0.045
30	−0.31	−0.032
40	−0.22	−0.022

41. 0 s; 1 s; $x(0) = 0$ m; $x(1) = 1.2$ m

43. 1 m/s^2; 0.9 m/s^2; 1.3 m/s^2

45. at $t = 0$, $a = 0$; at $t = 2$ s, a $= -2.5$ m/s^2; at $t \rightarrow \infty$, $a \rightarrow 0$

47. a)

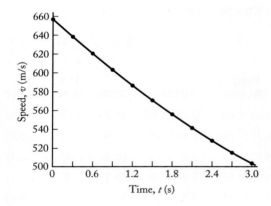

b)

TIME INTERVAL (s)	AVG SPEED (m/s)	DISTANCE TRAVELED (m)
0–0.3	647.5	194
0.3–0.6	628.5	189
0.6–0.9	611.5	183
0.9–1.2	596.0	179
1.2–1.5	579.5	174
1.5–1.8	564.0	169
1.8–2.1	549.5	165
2.1–2.4	535.0	161
2.4–2.7	521.0	156
2.7–3.0	508.0	152

c) 1722 ± 2 m

49. a)

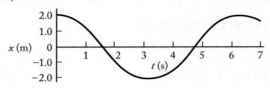

b) 1.6 s; 4.7 s; c) 0 s; 3.1 s; 6.3 s; $v(0) = v(3.1) = v(6.3) =$ 0 m/s; $a(0) = a(6.3) = -2$ m/s^2; $a(3.1) = 2$ m/s^2

51. 2.4 m/s^2

53. 6.36×10^7 s; 6.2×10^8 m/s

55. -350 m/s^2; will probably survive

57. -7.1 m/s^2; 3.8 s

59. 30 m/s; 300 m

61. 16 s

63.

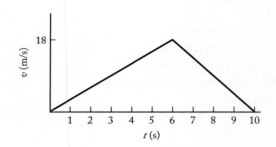

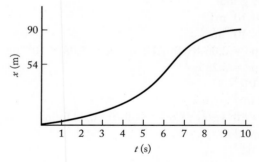

65. 32.9 m/s; 40.4 m/s

67. 0.875 m/s^2; 4.4 m/s

69.

v_0(km/h)	v_0(m/s)	$v_0 \Delta t$(m)	$-\dfrac{v_0^2}{2a}$ (m)	TOTAL STOPPING DISTANCE (m)
15	4.17	8.3	1.1	9.4
30	8.33	16.7	4.3	21.0
45	12.5	25.0	10	35.0
60	26.7	33.3	18	51.3
75	20.8	41.7	27	68.7
90	25.0	50.0	39	89.0

71. 15.5 m/s

73. $a = -ge^{-t/\tau}$

75. 66 m

77. 6.1 m/s

79. 44 m

81. 7.96 m or 3 floors; 22.1 m or 8 floors; 43.5 m or 15 floors

83. 2.8 s; 14 m/s up

85. 3.7 m above launching point; 8.6 m/s

87. 1.1 m/s; 5.5 m/s

89. 0.22%

91. 1.6×10^4 m/s^2

93. 1.9×10^3 m/s; 2.6×10^2 s

95. 80^2 m/s; 1.9 s

97. 14.9 m/s; 5.1 m/s

99. a) $n\sqrt{2h/g}$; b) $(3/4)h$ above the ground; c) $(2/3)h$

101. 18.3 m/s; 26.7 m/s; 33.3 m

105. 13.7 m

107. average speed = 1.3 m/s; average velocity = 0 m/s

109. 0.95 s; 28.8 m

111. 2.9 s

113. a) 4.3 m; b) 3.0 m/s; c) -6.0 m/s^2

115. 21.1 m/s

117. 33.1 m/s; 2.21×10^3 m/s^2

119. a) 8.10 m above ground; 11.1 m above ground; b) 9.8 m/s down; c) 0 m/s^2

Chapter 3

1. 11.8 km, 30° N of E

3. 11.2 km, 27.7° S of E

5. 612 m, 11.3° W of S

7. 436 km, 7.4° W of N

9. $\mathbf{B} = (-1.26 \text{ m})\mathbf{i} + (3.2 \text{ m})\mathbf{j}$

11. 13.6 nmi, 88° E of N

13. 1.88×10^4 km, 1.98×10^4 km

15. 6.07 mi, 78.3° W of S

17. 9.19 km N, 7.71 km W

19. 1.7 m

21. $(2\mathbf{i} + 5\mathbf{j})$ cm

23. $A_z = \pm 4.2$ units

25. a) $-3\mathbf{i} - 2\mathbf{j} - 2\mathbf{k}$ b) $-7\mathbf{i} - 4\mathbf{j} + 4\mathbf{k}$ c) $-16\mathbf{i} - 9\mathbf{j} + 11\mathbf{k}$

27. $x = -9.9$ m, $y = 9.9$ m

29. $(1/3)\mathbf{i} + (2/3)\mathbf{j} + (2/3)\mathbf{k}$

31. $c_1 = -8/7$, $c_2 = 9/7$

33. 4940 km

35. $(6/7)\mathbf{i} - (12/7)\mathbf{j} + (4/7)\mathbf{k}$

37. 9

39. $-8, 112°$

41. 56.1°

43. 45°

45. $(-3.9 \times 10^6 \text{ m})\mathbf{k}$

47. Because the vectors are nonzero, a zero result for the dot product means they must be perpendicular.

49. $B_x = -6.83, B_z = -4.5, C_z = 1.34$

51. $0.44\mathbf{i} - 0.22\mathbf{j} - 0.87\mathbf{k}$

53. $0.49\mathbf{i} + 0.81\mathbf{j} + 0.32\mathbf{k}$

55. 24

59. $-12\mathbf{i} - 14\mathbf{j} - 9\mathbf{k}$

61. $0.45\mathbf{i} - 0.59\mathbf{j} - 0.67\mathbf{k}$

65. Coordinate system rotated at $-26.6°$

67. 415 m, 29.8° W of N

69. $x = 1.0, y = 1.7$

71. $\mathbf{A} + \mathbf{B} = 5.4\mathbf{i} - 12.7\mathbf{j}$; $\mathbf{A} - \mathbf{B} = 5.4\mathbf{i} + 6.5\mathbf{j}$

73. 4.58

75. -304 m^2

77. 4.0, 5.0

Chapter 4

1. a) 7 km, 5 degrees E of N; b) 5.6 km/h, 5 degrees E of N; c) 8.24 km/h

3. 3.93 m

5. a) $2\mathbf{i} + (5 + 8t)\mathbf{j} - (2 + 6t)\mathbf{k}$; b) $8\mathbf{j} - 6\mathbf{k}$, magnitude = 10 m/s^2, direction = 37° below the y-axis in the y-z plane

7. 19.6 m at 90° below the direction of travel of the airplane at 2 s; 24.7 m at 83° below the direction of travel of the plane at 3 s

9. -13.3 km/h $\mathbf{i} - 123$ km/h $\mathbf{j}$

11. velocity = $(90\mathbf{i} - 15\mathbf{j})$ m/s, speed = 91 m/s; direction = 9.5° below the x-axis

13. a) $\mathbf{v} = (3t\mathbf{i} + 2t\mathbf{j})$ m/s; b) $\mathbf{r} = [(3t^2/2)]\mathbf{i} + t^2\mathbf{j}$ m

15. 2.4 m/s

17. 38 m/s

19. 65.8 m/s, 93.4 m/s

21. a) 7.25° b) 13 m

23. 1.74 sec, 14.9 m, 59.5 m

25. 3.13×10^3 m/s, 2.5×10^5 m/s, 452 sec

27. 64.8 m, 3.04 sec

29. 76°

31. 12 m/s, $\mathbf{r} = -21$ m $\mathbf{i} + 55$ m $\mathbf{j}$

33. 21 m/s

35. The lake surface is 34.3 m below the release point and the horizontal distance from release point is 68.8 m

37. Yes, puck passes 2.2 m above the goal, 0.391 sec

41. 63.4°

43. 9.29° and 80.5°

45. 5.19°, when angle off 0.03° in vertical direction arrow still hits bull's-eye (arrow hits 4.6 cm off center, which is still within 12 cm diameter), when angle off 0.03° in horizontal

direction arrow still hits bull's-eye (arrow hits 4.7 cm off center, which is still within 12 cm diameter)

47. $7.49°$
49. No, projectiles will never collide
51. 45.8 km
53. $\theta = \dfrac{1}{2}(d + \pi/2)$
55. $h = R(1 + \sin\theta) + \dfrac{u^2\cos^2\theta}{2g}$; the maximum possible height is 4.4 m
57. $\theta = 9.94°$, 17 km
59. 2.1 rev/min
61. 9.4 m/s^2
63. 3.95 m/s^2, $4 \times 10^5\,g$
65. 8.99×10^{13} m/s^2, $9.16 \times 10^{12}\,g$
67. 5.9×10^{-13} m/s^2
69. $a_{eq} = 3.39 \times 10^{-2}$ m/s^2, $a_{45} = 2.4 \times 10^{-2}$ m/s^2
71. $a_M = 0.0395$ m/s^2, $a_V = 0.0113$ m/s^2, $a_E = 0.00595$ m/s^2
73. 5.5 m/s, 2.5 m/s
75. 633 m/s $\mathbf{i}$ + 226 m/s $\mathbf{j}$
77. $V = 12$ m/s, $\theta = 83°$
79. 4.60 m/s
81. 60 cm/s at $34°$ above the horizontal
83. speed = 27 km/h, $\theta = 33°$
85. a) 50×10^3 m; b) 33×10^3 m, 67×10^3 m
87. 15.1 km/h at $15°$ E of N
89. $v_{rel} = \dfrac{2v_0}{(h^2/4v_0^2t^2 + 1)^{1/2}}$
91. 528 km/h at $8.5°$ N of W
93. a) vertical component = 62.1 km/h down, horizontal component = 232 km/h in direction of plane's travel; (b) 1.9 min
95. $v_E = 3.9$ km/h, $v_N = -1.4$ km/h (1.4 km/h south)
97. a) 13 m/s; b) $56.3°$
99. a) 25 km; b) 50 km; c) No
101. 26 m/s
103. 8.9 m/s^2
105. 10.8 sec

Chapter 5

1. 442 kg
3. 2.69×10^{-26} kg; 3.7×10^{25} atoms
5. 3.8 m/s^2; 6.2×10^3 N
7. 6.6×10^3 N; 12 times the weight
9. -1.2×10^3 N

11. -1.8×10^3 m/s^2; -1.3×10^5 N
13. 35 N
15. -4.2 m/s^2; -2.4 m/s^2; -1.0×10^3 N; -5.7×10^2 N
17. 0.063 m/s^2
19. -2.90×10^3 N; -1.82×10^3 N
21. $v = bx_0 \sin(bt)$; $a = b^2x_0 \cos(bt)$; $F = mb^2x_0 \cos(bt)$; $F = -mb^2(x - x_0)$
23. No, since the tension in the rope = 150 000 N > breaking tension
25. $36°$ south of east; 260 N
27. 3.7 m/s^2; $23.4°$ east of north
29. 4.7×10^{20} N; $25°$ clockwise from the Moon-Sun direction
31. 770 N in the positive x-direction
33. 2.6 kg; 34 N
35. 285 N on Mars; 1900 N on Jupiter
37. a) 9.9904×10^{-4}; b) no
39. 128 N in the upper cord; 29.4 N in the lower cord
41. $F = T_1 = (m_1 + m_2 + m_3)g$; $T_2 = (m_2 + m_3)g$; $T_3 = m_3g$
43. Mg; $Mg/2$
45. $T = \dfrac{\pi d^2 l\rho g}{4}$ at the upper end; $T = \dfrac{\pi d^2 l\rho g}{8}$ at the midpoint
47. 1.2 m/s^2; 36 N to the right; 36 N to the left
49. 600 N
51. 5.2×10^3 N
53. 165 N; $19.5°$ clockwise from positive x
57. 1.8×10^3 N upward
59. F in the first cable; $F\left(\dfrac{m_2 + m_3}{m_1 + m_2 + m_3}\right)$ in the second cable; $F\left(\dfrac{m_3}{m_1 + m_2 + m_3}\right)$ in the third cable; $a = \dfrac{F - (m_1 + m_2 + m_3)g}{m_1 + m_2 + m_3}$
61. 1.14×10^3 N or 265 lb; 820 N or 184 lb
63. 0.51 N
65. 1.9 m/s^2; 14 m/s
67. 64 m; 5.1 s
69. $\dfrac{mgR}{\sqrt{l(l + 2R)}}$

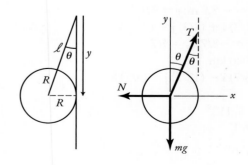

71. N = 680 N, F = 340 N

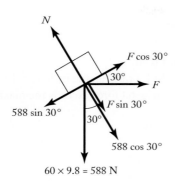

73. a) Incline forward; b) 22.3 m/s^2

75. $\pi/4$; $2\sqrt{\dfrac{\Delta x}{g}}$

77. $a = \dfrac{2g}{l}x$; $x(t) = x_0 e^{\sqrt{2g/l}\,t}$

79. 7.9 × 10^{-5} m/s^2; 0.14 m

81. a) $\mathbf{F}_{net} = -2\mathbf{i} + 3\mathbf{j} + 4\mathbf{k}$ N; b) $\mathbf{a} = -0.33\mathbf{i} + 0.50\mathbf{j} + 0.67\mathbf{k}$ m/s^2, 0.9 m/s^2

83. a)

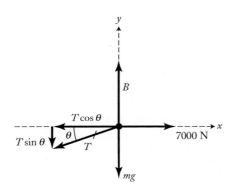

b) 7.1 × 10^3 N; c) 2.6 × 10^4 N

85. a)

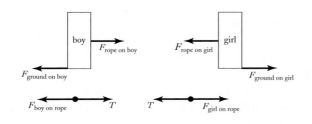

b) 250 N; c) 250 N; d) 250 N

87. a)

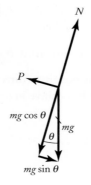

b) 1.8 × 10^4 N; c) same as b)

89. a) 590 N; b) 700 N; c) 590 N; d) 0 N

91. a) −0.98 m/s^2; b) 99 m; c) 50 km/h, same as speed when first decoupled

93. a) $a = \dfrac{m_2}{m_1 + m_2}g$;   b) $T = \dfrac{m_1 m_2}{m_1 + m_2}g$

Chapter 6

1. 5.7 × 10^3 people

3. 0.83

5. 1.6 × 10^2 m

7. 53 m

9. 3.4 m, so he will reach the plate; 1.9 m/s

11. 0.48

13. 2.1 × 10^3 N

15. 0.27

17. 2.0 × 10^2 N

19. 2.8 m/s

21. 1.9°

23. 0.78

25. 1.4 × 10^{-3} m/s (1.4 mm/s)

27. 39.5 m

29. a)

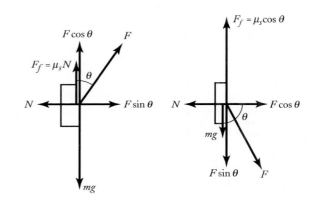

b) $F = \dfrac{mg}{\cos\theta + \mu_s \sin\theta}$; c) $\theta = \tan^{-1}\mu_s$;

$F = \dfrac{mg}{\sqrt{(1 + \mu_s^2)}}$; d) $\theta = \tan^{-1}\mu_s$, but now θ is the

angle between the force **F** and the positive x direction

31. The quantity $\cos\theta - \mu_k \sin\theta$ must be positive in order to find a solution for P

This gives the condition $\tan\theta > \dfrac{1}{\mu_k}$

33. 3.6 m/s^2

35. $T = \dfrac{\mu_k m_1 m_2 g \cos\theta}{(m_1 + m_2)\cos\phi - (m_1 + 2m_2)\mu_k \sin\phi}$

37. 23 N; 11 N

39. No, since k is not constant

41. 50 N/m

43. 8.8×10^2 N/m

45. 1.1×10^{-3} m

49. 4.4×10^2 N

51. 7.8×10^2 N at top; 8.1×10^2 N at bottom

53. $6.3°$

55. 0.13 m

57. 0.224 N

59. 6.9 m/s

61. 22 m/s

63. $68°$

65. $\sqrt{gl \sin\theta \tan\theta}$

67. 1.40×10^3 m

69. The equilibrium conditions when the balls are at maximum angular displacement is $\cos\theta_2/\cos\theta_1 = m_1/m_2$, and the condition when they are both vertical is $(m_1 - m_2)g = \dfrac{m_2 v_2^2 - m_1 v_1^2}{l}$. These conditions cannot both be satisfied, so the motion described is impossible

71. $T = \dfrac{mv^2}{2\pi r}$

73. $\phi = \tan^{-1}\left(\dfrac{\tan\theta}{1 - 4\pi^2 R/t^2 g}\right) - \theta$; at $\theta = 45°$, $\phi = 0.099°$

75. 0.89 m/s

77. 40 m; 3.2 s

79. a)

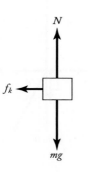

b) 3.9×10^2 **j** N; c) -3.1×10^2 **i** N.; d) -3.9×10^2 **j** N; -3.1×10^2 **i** N; e) a $= -7.85$**i** m/s^2; 31 m

81. 0.15 m

83. $a_1 = \dfrac{m_2 - \mu_k m_1}{m_1 + m_2} g$

85. 4.1 cm; 2.5 cm, 1.6 cm

87. a) 1.2×10^{-2} m; b) 1.7 m/s^2 up the incline; c) 1.0×10^{-2} m

89. Yes, since the centripetal force exceeds the maximum frictional force

Chapter 7

1. 1.5×10^3 J

3. 252 J

5. 3.7×10^3 J

7. 2.35×10^5 J; 357 J/s

9. 2.2×10^7 J by first tugboat; 1.0×10^7 J by second tugboat; 3.2×10^7 J total

11. 2.6×10^3 J

13. 7×10^4 J by gravity; -7×10^4 J by friction

15. $54°$

17. a) 1.3×10^4 J; b) 290 N; 1.3×10^4 J

19. a) 1.4×10^4 N; 8.3×10^3 N; b) 4.3×10^3 J; c) 2.2×10^4 J

21. a) 7.1×10^3 N; b) 2.2×10^5 J; 8.1×10^3 N

23. 6 J

25. -26 J

27. $3W_0$; $(2N + 1)W_0$

29. a) $k\left[y^2 + \dfrac{l^2}{2} - \dfrac{l}{2}\sqrt{(l/2)^2 + y^2}\right]$;

b) $P = 2ky\left[1 - \dfrac{(l/2)}{\sqrt{(l/2)^2 + y^2}}\right]$

31. 17 J

33. a) $x_{eq} = \sqrt[6]{\dfrac{A}{B}}$; b) $-\dfrac{B^2}{12A}$

35. 2.7×10^{33} J

37. 1.3×10^5 J; 5.8×10^3 J; 22

39. a) 4.0×10^5 J; b) 2.5×10^4 J; c) 1.2×10^6 J

41. 4.1×10^6 J

43. $K_{ball} = 46$ J; $K_{person} = 38$ J; they are of the same order of magnitude

45. 1.9 J; 0.44 m

47. 6.2×10^9 J

49. 196 m/s

51. 3.4×10^{-18} J

53. a) 80 J; b) -1.295×10^3 J; c) 1.375×10^3 J

55. 7.4×10^3 J; 1.6 % of the energy acquired by eating an apple

57. 8.2×10^6 m^3

59. 5.1 m

61. 99 m/s; 9.8×10^{10} J; 23 tons

63. 0.16

65. 79%

67. 7.7 m/s

69. 1.1×10^4 J

71. 53 N

73. 100 m/s

77. a) 24 m/s; b) 7.1 m; c) 26 m

79. 48.2°

81. 2.1×10^3 J

83. 1.69×10^5 J; 2.06×10^5 J

85. a) -8.8 m/s^2; 35.4 m; b) 1.06×10^4 N; 3.75×10^5 J

87. a) $U = 2.35 \times 10^7$ J; $K = 2.89 \times 10^6$ J; b) $U = 17.7 \times 10^6$ J; $K = 8.7 \times 10^6$ J; 120 m/s (or 430 km/h)

89. a) 150 J; b) 150 J; c) 122 m/s; d) 1520 m; e) 122 m/s

91. 4.1×10^4 J

93. a) 25.8 m/s; b) 10.3 m/s

Chapter 8

1. 0.076 m

5. $U(x) = \dfrac{Ax^4}{4}$, assuming that $x_0 = 0$; 5.6 m/s

7. $U(x) = x^2 + \dfrac{x^4}{4}$, assuming that $x_0 = 0$; 1.3 J; 8 J; 29.3 J

9. $F = -4x - 4x^3$

11. 64.5 m/s

13. a) bxy; b) $-bxy$

15. 2.61×10^6 J

17. $U(x) = \dfrac{A}{12x^{12}} - \dfrac{B}{6x^6}$

19. a) 13 kN; b) 13 kN. The force is independent of the rope length

21. $\mathbf{F} = a\dfrac{x\mathbf{i} + y\mathbf{j}}{(x^2 + y^2)^{3/2}}$

23. 1.89×10^5 N

25. a) -4.58 m; 14.9 m/s; b) -14.7 m; 21.7 m/s^2

27. a)

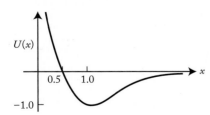

b) $x_{1,2} = (2 \pm \sqrt{2})(b/c)$; c) $x = (\sqrt{6} - 2)(b/c)$

29. $x = \pm \dfrac{mv^2}{2A}$

31. $x = \pm 1.0$ m; unbound for $E > 0$ J

33. a) 0.382 nm; b) -1.67×10^{-19} J; c) ± 0.34 nm, ± 0.89 nm

35. 6.3 eV/molecule

37.

VEHICLE	ENERGY PER MILE (J/mi)	ENERGY PER PASSENGER PER mi (J/ passenger-mi)
Motorcycle	1/60 gal/mi $\times$ 1.3 $\times$ 10^8 J/gal	
	= 2.2 $\times$ 10^6 J/mi	2.2 $\times$ 10^6
Snowmobile	1/12 $\times$ 1.3 $\times$ 10^8 = 1.1 $\times$ 10^7	1.1 $\times$ 10^7
Automobile	1/12 $\times$ 1.3 $\times$ 10^8 = 1.1 $\times$ 10^7	1.1 $\times$ 10^7/4 = 2.7 $\times$ 10^6
Bus	1/5 $\times$ 1.3 $\times$ 10^8 = 2.6 $\times$ 10^7	2.6 $\times$ 10^7/ 45 = 5.8 $\times$ 10^5
Jetliner	1/0.12 $\times$ 1.3 $\times$ 10^8 = 1.1 $\times$ 10^9	1.1 $\times$ 10^9/110 = 9.8 $\times$ 10^6
Concorde	1/0.1 $\times$ 1.3 $\times$ 10^8 = 1.3 $\times$ 10^9	1.3 $\times$ 10^9/ 360 = 3.6 $\times$ 10^6

Most efficient is the bus, least efficient is the snowmobile.

39. 183 m, assuming a mass of 70 kg for the climber

41. 1.05×10^4 kJ

43. Walking 1.7 kcal/kg; Slow running plus standing 2.8 kcal/kg; Fast running plus standing 2.8 kcal/kg

45. 1.88×10^9 eV

47. Thermal energy is 0.0001 % of mass energy

49. 511 keV; 939 MeV

51. 1.4×10^{-9} kg; .00000005% of the mass of the gasoline

53. 9.40×10^8 eV

55. 542 kcal

57. 18 kWh

59. 2.88×10^8 J

61. 1.5×10^{-3} hp; 23 kcal

63. a) 769 gal; b) 5.8 kW

65. 526 kWh/year; $79

67. 1100 W; 0 W

69. 0.61 hp

71. 746 J

73. 50°

75. -2.0 W

77. 4.24×10^5 W

79. 2500 km^2

81. a) 1.7×10^{10} J; b) 17 min; c) 3.73 km

83. 1.2×10^{-4} W

85. 195 m diameter

87. a) 487 hp; b) 2593 hp

89. 52 hp

91. 37%

93. a) 3.2×10^4 W; b) -784 W; c) -3.1×10^4 W

95. 2.3 W

97. 2.1×10^7 kW

99. 3.4 kW

101. a) 4.3×10^{-12} J; b) 6.39×10^{14} J; c) 6.1×10^{11} kg/s; d) 7.8×10^{10} years

103. a) 5 J; b) -4 J; no

105. a) 7.2×10^6 N; b) 0.414 m/s

107. 4.2×10^9 W

109. a) $P = -1.82 \times 10^6 + 3.63 \times 10^5\, t - 2.71 \times 10^4\, t^2 + 964\, t^3$; b) 9.757×10^6 J; 5.714×10^6 J; c) -1.35×10^6 W

111. 14 min

113. a) 1.6×10^7 kWh; b) 3.8×10^2 m^3/s

Chapter 9

1. 8.2 N

3. 3.46×10^8 m

5. $F_{Sun} = 0.41$ N, $F_{Moon} = 2.3 \times 10^{-3}$ N

7. $F_{Alpha} = 1.5 \times 10^{17}$ N, $F_{Earth} = 3.5 \times 10^{22}$ N

9. $a_J = 24.9$ m/s^2, $a_S = 10.5$ m/s^2, $a_U = 8.99$ m/s^2

11. 1×10^9 N

13. 2.54×10^{-10} N at $52°$

15. 2.76×10^{-4} g

17. $\Delta a_{Earth\text{-}Moon} = 2.21 \times 10^{-6}$ m/s^2, $\Delta a_{Earth\text{-}Moon}/g = 2.25 \times 10^{-7}$, $\Delta a_{Jupiter\text{-}Io} = 0.0123$ m/s^2, $\Delta a_{Jupiter\text{-}Io}/g = 0.00687$

19. 101 m/s

21. 3.08×10^3 m/s

23. 5.8×10^{15} sec $= 1.8 \times 10^8$ years, 3.1×10^5 m/s

25. $T_{Io} = 1.77$ days, $T_{Europa} = 3.55$ days, $T_{Ganymede} = 7.15$ days

27. 0.927 days

29. About 10 times

31. Same latitude $22.6°$ West, around Lincoln, Nebraska

33. $m_1/m_2 = 1.6$

35. 3.0×10^{10} m

37. a) 7.50×10^3 m/s, 8.32×10^3 m/s; b) 3.94×10^8 J, 4.85×10^8 J

39. 8.2×10^3 m/s

41. $T_S = 96.5$ min, $T_E = 115$ min

43. 5.33×10^{10} km, about 10 times Pluto's mean orbital distance from sun

45. 7.8×10^3 m/s, -1.4×10^{11} J

47. $\sqrt{2}$

49. $U = -1.04 \times 10^6$ J, $K = 5.2 \times 10^5$ J, $E = -5.2 \times 10^5$ J

51. 8.86 mm

53. 0.253

55. a) 1.11×10^4 m/s; b) 1.23×10^{11} J = 29 tons of TNT; c) 1.23×10^5 m/s^2

57. 2270 m/s, 1.11×10^4 m/s

59. elliptical

61. a) speed = 1680 m/s, time = 6510 sec = 109 min; b) this will give an elliptical orbit; c) this orbit will not be closed because the launch speed is greater than the escape speed

63. a) No, speed is less than that needed for circular orbit; b) 1.22×10^4 m/s

65. $E_S = -4.4 \times 10^9$ J, $E_E = -6.06 \times 10^8$ J

67. $v_{perigee} = 6.96 \times 10^4$ m/s, $v_{apogee} = 5.75 \times 10^3$ m/s

71. $h = 4.26 \times 10^6$ m, $v = 817$ m/s

73. a) 2.6×10^3 m/s; b) 2.8×10^3 m/s; c) 0.401 years; d) Venus moves $234.7°$, Earth moves $144°$

75. 1 rev/min

77. 1.6×10^{-9} N

79. 4.9 years

81. 1.90×10^{27} kg

83. a) $v_{rel} = 1.53 \times 10^4$ m/s; b) 3.91 g

85. height above earth = 9.89×10^6 m

87. a) -3.30×10^9 J; b) 3.30×10^9 J

Chapter 10

1. 9.0 kg·m/s; 3.2 kg·m/s.

3. 1.8×10^{29} kg·m/s; 4.3×10^7 kg·m/s; 3.8×10^4 kg·m/s; 95 kg·m/s; 2.0×10^{-24} kg·m/s

5. $(1.6 \times 10^{-25}\,\mathbf{i} + 7.7 \times 10^{-26}\,\mathbf{j})$ kg·m/s

7. a) $(9.7\mathbf{i} + 5.6\mathbf{j})$ kg·m/s; b) $(9.7\mathbf{i} + 0\mathbf{j})$ kg·m/s; c) $(9.7\mathbf{i} - 5.6\mathbf{j})$ kg·m/s

9. 9.81 kg·m/s down; 98.1 kg·m/s down

11. $-9.0\,\mathbf{i}$ m/s

13. $-2.2 \times 10^5\,\mathbf{j}$

15. $-2.0^{-2} \times 10^{-5}\,\mathbf{i}$ m/s

17. 8.26×10^3 m/s; 1.29×10^{-17} J

19. $-(1.3$ m/s$)\mathbf{i} + 0\mathbf{j}$

21. 66 N; 1.3×10^6 J

23. 150 N·s

25. $-(4.10 \times 10^3\mathbf{i} + 929\mathbf{j})$ m/s

27. 5.2 N

29. 5×10^{-11} kg/s; 5×10^{-5} N

31. $u\,\dfrac{m}{M}\displaystyle\sum_{k=1}^{n}\dfrac{1}{1 - km/M}$

33. 1.9 m from woman

35. 7.42×10^5 m; 0.107% of the sun's radius

37. $h/3$ along the height, away from the unequal side

39. 0.027 cm directly away from the 40 g piece

41. 0.23 nm from the hydrogen atom

43. $(950 \times 10^3, 180 \times 10^3, 820 \times 10^3)$ light-years

45. $(L/3, L/3, L/3)$

47. $(-0.061L, 0, 0)$

49. 950 m from the base

51. 31.5 J; 63 J

53. 9.0×10^7 J

55. CM is on the axis of symmetry, a distance $R/2$ away from either the base or the top of the hemisphere

57. 6.9×10^6 m/s in the direction of motion of the proton

59. 953 kg

61. 1.05 m

63. 1.6 m/s in the direction of motion of the bullet

65. $4D$ from the launch point

67. $(-17.4$ km/s, -17.4 km/s$)$

69. $K_{CM} = 3.8 \times 10^4$ J; $K_{TOT} = 6.38 \times 10^5$ J

71. 4.76×10^4 J; 3.6×10^5 J; 4.76×10^4 J

73. 3.9×10^3 J; 4.0×10^3 J

75. 0.955×10^{-3}

77. 1.1 m/s

79. $a_{boy} = 5.0$ m/s^2 toward girl; $a_{girl} = 6.7$ m/s^2 toward boy; 1.7 m from boy

81. 8 km/h

83. 0.0927 nm

85. 59 cm

87. If the stack of two books is at the top of the triangle, the CM is at a point halfway between the other two books and 0.43 m above the line connecting them

89. Halfway along the line joining the centers of the plates

Chapter 11

1. 1.13×10^3 kg·m/s, 2.3×10^4 N

3. a) 12 m/s, 7m/s, 3 m/s, 1 m/s, -1 m/s, 250 m/s^2, 200 m/s^2, 100 m/s^2, 100 m/s^2; b) 4.2×10^5 N, 3.4×10^5 N, 1.7×10^5 N, 1.7×10^5 N; c) 1.1×10^5 N·s

5. 12.6 kg·m/s, 4200 N

7. -1400 N

9. -1.8 kg·m/s, -1.35×10^4 N

11. 8.1 kg·m/s, 0.045 s

13. 18 kg·m/s

15. 7.5×10^{-2} s, 2.8×10^4 N

17. a) $v_{proj}{}' = -0.27$ m/s, $v_{targ}{}' = 0.53$ m/s; b) $K_{proj}{}' = 1.9 \times 10^{-2}$ J, $K_{targ} = 0$ J $K_{proj}{}' = 2 \times 10^{-3}$ J, $K_{targ}{}' = 1.7 \times 10^{-2}$ J

19. 39 m/s

21. 0.57 J

23. $0.22v$

25. $\dfrac{M}{7}$

27. $v_1{}' = 15$ m/s, $v_2{}' = 17$ m/s

29. Last ball has velocity $= v$ and other two balls have velocity $= 0$

31. a) mass m rises to $\dfrac{h}{9}$, mass $2m$ rises to $\dfrac{4h}{9}$; b) mass m rises to h, mass $2m$ stops and does not rise

33. 13.5 m/s

35. a) The 1400 kg mass has a velocity $= 1.3$ m/s and the 800 kg mass has a velocity $= 6.1$ m/s; b) t $= 0.98$ s $x = 1.7$ m

37. Yes

39. a) -7.5 m/s; b) -15 m/s; c) -15 m/s

41. -0.17 m/s, 0.18 m/s, 0.41 m/s, -0.34 m/s

43. 9.3 m/s

45. a) 9.8 m/s; b) 4.8×10^5 J; c) -130 m/s^2, 850 m/s^2

47. 4.0×10^{-13} J

49. a) 3.9×10^5 J, 3.9×10^5 J; b) 7.8×10^5 J, 3.9×10^5 J

51. 210 m/s

53. a) 440 m/s; b) -1200 J; c) 9.6 J; d) missing kinetic energy is energy that shows up as heat in bullet and block, compression/deformation, and noise

55. 620 m/s

57. 860 m/s

59. (a) 3 m/s $\hat{\mathbf{i}}$ (b) 79 m/s^2

61. 21 m/s

63. a) $\pi - 2\sin^{-1}\left(\dfrac{b}{2R}\right)$; b) $2mv\sin^{-1}\left(1 - \dfrac{b^2}{4R^2}\right)$

65. $v_1' = \dfrac{v}{2}(1 + \cos\theta)\,\mathbf{i} + \dfrac{v}{9}\sin\theta\,\mathbf{j}$; $v_2' = \dfrac{v}{2}(1 + \cos\theta)\,\mathbf{i} - \dfrac{v}{2}\sin\theta\,\mathbf{j}$ $\; v_1'\cdot v_2' = \dfrac{v^2}{4}(1 - (\sin^2\theta + \cos^2\theta)) = 0$ so $v_1' \perp v_2'$

67. 3.2 kg·m/s, 3200 N

69. $v_1' = 2.6$ km/h, $v_2' = 13$ km/h

71. 1.2×10^{-12} J

73. a) $v_1' = 0$ and $v_2' = 20$ m/s; b) The ball that had an initial velocity lands on ground next to fence and the ball with no initial velocity lands 11 meters away from the fence

75. 0.964, 0.036

77. a) 3.3×10^{-12} m/s; b) 2.4×10^7 tons of TNT; c) 8.1×10^{14} N

79. a) Ball height $h = \ell(1 - \cos\theta)$ where θ is the angle with the vertical; b) height $h = L/4\,(1 - \cos\theta)$ where $\theta = \cos^{-1}\left(\tfrac{3}{4} - \tfrac{1}{4}\cos\theta\right)$

80. a) 21 m/s and 11° west of north; b) 1.4×10^5 J

Chapter 12

1. 1.7×10^{-3} rad/s; 3.5×10^{-4} m/s

3. 81.5 rad/s; 13 rev/s

5. a) 0.52 m/s; 0.17 m/s; b) 1.8 m/s^2; 0.61 m/s^2

7. 22.0 rad/s; 1.28 m/s; 55.4 rad/s; 8.83 rev/s

9. 9.4 rad/s; 0.94 m/s; 1.9 rad/s^2

11. 88 rev/s for aluminum; 2.1 rev/s for steel

13. At $t = 0$, $\phi = 0$; $\omega = 0$; $\alpha = 40$ rad/s^2; at $t = 1.0$ s, $\phi = 15$ rad; $\omega = 25$ rad/s; $\alpha = 10$ rad/s^2; at $t = 2.0$ s, $\phi = 40$ rad; $\omega = 20$ rad/s; $\alpha = -20$ rad/s^2

15. 5.5×10^{-3} cm/s; 51 cm/s; 22 cm/s

17. 611 rad/s^2; 70 revolutions

19. 1.5×10^2 rad/s^2

21. -4.36×10^{-2} rad/s^2; 0.89 revolutions

23. -23 rad/s^2; 1.7 s

25. $-9.5 \times 10^{-22} \text{ rad/s}^2$

27. 10 rad/s; 8 rad

31. $3.8 \times 10^4 \text{ J}$

33. $1.21 \times 10^{-10} \text{ m}$

35. $6.50 \times 10^{-46} \text{ kg·m}^2$

37. 0.44 kg·m^2

39. 0.46 kg·m^2

41. $2.13 \times 10^{29} \text{ J}$

43. 0.96 kg·m^2; $1.5 \times 10^7 \text{ J}$

45. $-1.10 \times 10^{22} \text{ kg·m}^2$

47. $0.379 M_E R_E^2$

49. $\dfrac{ML^2 \sin^2 \theta}{12}$

51. $I = \dfrac{Ml^2}{3}$

53. $0.426 \, MR^2$

55. $\left(\dfrac{1}{2} - \dfrac{1}{10 + 30h/4R} \right) MR^2$

57. $\dfrac{2}{5} \dfrac{M(R_2^5 - R_1^5)}{R_2^3 - R_1^3}$

59. $I = \dfrac{3}{5} M_I^2$

61. a) 225 kg·m^2; b) $4.4 \times 10^3 \text{ J}$

63. $3.49 \times 10^5 \text{ J}$; 3.9%

65. $2 \times 10^{25} \text{ J/s}$; $1.05 \times 10^{15} \text{ s}$ or $3.3 \times 10^7 \text{ yr}$

67. $\dfrac{MR^2}{4}$

69. $\dfrac{Ml^2}{6}$

71. $\dfrac{MR^2}{4}$

73. $\dfrac{3}{10} \, mR^2$

75. (160 km/h) **i** at top; 0 km/h at bottom; (80 km/h) **i** − (80 km/h) **j** at front

77. 12 rad/s^2; 74.1 rad/s; 35 turns

79. 0.012 J; 0.37 m/s

81. $I_1 = 2 \times \dfrac{2}{5} MR^2 = \dfrac{4}{5} MR^2$;

$I_2 = 2 \left(\dfrac{2}{5} MR^2 + MR^2 \right) = \dfrac{14}{5} MR^2$

83. $\left(\dfrac{135\pi + 64}{24(9\pi + 4)} \right) MR^2 = 0.630 \, MR^2$

85. 4.4 m/s

Chapter 13

1. $(4610 \text{ N·m})R$, 613 kg, 940 kg

3. 310 N

5. 59 N·m

7. 130 hp, 176 N·m

9. 2900 N·m

11. 5.4 m/s, 7.7 m/s

13. 230 W

15. $1.9 \times 10^6 \text{ J}$

17. 19 J, 75 J

19. $5.6 \times 10^4 \text{ W}$, $1.1 \times 10^4 \text{ N·m}$

21. 4.6 rad/s

23. a) 1140 N·m; b) 2.1 m/s^2

25. 820 N

27. $2.7 \times 10^4 \text{ N·m}$

29. 9.7 m/s^2

31. Proof required.

33. 9.6 m/s^2

35. rolls to the right, $f = 2F/3$

37. 17 rev

39. 0.024 N·m, 0.16 N

41. Proof required.

43. 2.83 m/s^2, 2.94 m/s^2, 3.89 %

45. 1.6 m

47. $2.8 \times 10^{34} \text{ kg·m}^2/\text{s}$

49. $1.6 \text{ kg·m}^2/\text{s}$

51. $1.05 \times 10^{-34} \text{ kg·m}^2/\text{s}$, $2.11 \times 10^{-34} \text{ kg·m}^2/\text{s}$, $3.15 \times 10^{-34} \text{ kg·m}^2/\text{s}$

53. a) $1.8 \times 10^9 \text{ kg·m}^2/\text{s}$, upward; b) zero kg·m²/s

55. $7.9 \times 10^{22} \text{ rad/s}$, $7.9 \times 10^7 \text{ m/s}$, $2.1 \times 10^{-12} \text{ J}$, $1.5 \times 10^{-10} \text{ J}$

57. $0.051 \text{ kg·m}^2/\text{s}^2$

59. $5.6 \times 10^{41} \text{ kg·m}^2/\text{s}$, $3.14 \times 10^{43} \text{ kg·m}^2/\text{s}$, 1.8 %

61. 0.57 rad/s, 5.5 rad/s

63. 3.1 rev/s (or 19 rad/s)

65. $1.5 \times 10^{-19} \text{ rev/day}$

67. $v_1 = \sqrt{\dfrac{GM_E}{r_1}}$ and $v_2 = \sqrt{\dfrac{GM_E}{r_2}}$. The satellite closer to earth has the greater speed. $L_1 = m\sqrt{r_1 GM_E}$ and $L_2 = m\sqrt{r_2 GM_E}$. The satellite closer to earth has the smaller angular momentum.

69. $-4.3 \times 10^{-22} \text{ rad/s}^2$, $3.5 \times 10^{16} \text{ N·m}$, $2.6 \times 10^{12} \text{ W}$

71. $-7°$

73. a) 5.0 m/s; b) -0.009 rad/s; c) $-6860°$ or 19 rev

75. $\dfrac{m^2 v^2}{2g(m + \frac{1}{3}M)(m + \frac{1}{2}M)}$

77. $\dfrac{mv_0^3}{g\sqrt{32}}$, $\dfrac{mv_0^3}{g\sqrt{2}}$

79. 0.37 rad/s

81. The instantaneous change in angular momentum opposes the original direction of the angular momentum and makes the tilt worse

83. a) 110 kg·m^2/s; b) 34 N·m; c) 34 N·m

85. 1.6×10^8 kg·m^2/s, east or west, 1.2×10^4 N·m, 1.0×10^4 N

87. 3100 N·m

89. 3100 J

91. a) 76.2°; b) 290 N, -250 N

93. a) 4.10×10^{-5} kg·m^2; b) 0.12 rad/s^2; c) 4.8×10^{-6} N·m; d) 0.0023 J

95. 420 N

97. $8.50\mu_s$

99. a) 2.0×10^{-10}; b) 4.9×10^9; c) 4.9×10^9 rev/month or 1860 rev/s

101. 1.3×10^{14} kg·m^2/s, north

Chapter 14

1. 590 N

3. 5.5×10^6 N; 5.1×10^6 N

5. 5200 N

7. 8 cm; 18 cm

9. 5.88 kg

11. 3500 N; 6800 N

13. 51 N; 29 N

15. 1420 N; 2500 N

17. 30°

19. 7.65 m

21. Proof required.

23. $F = Mg \dfrac{\sqrt{R^2 - (R-h)^2}}{(R-h)}$; $F = Mg \dfrac{R}{\sqrt{R^2 - (R-h)^2}}$

25. a) 9×10^3 N; b) 2.6×10^3 N

27. $0.408\,Mg$

29. $T = Mg\left(\dfrac{L + R}{\sqrt{(L+R)^2 - R^2}}\right)$; $N = Mg\dfrac{R}{\sqrt{(L+R)^2 - R^2}}$

31. $0.62\,mg$

33. a); b) 26.6°; c) 56 J

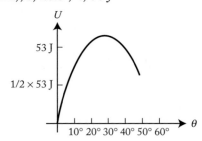

35. 1.04 L

37. 17.5 m/s

39. 7.2 m/s^2

41. a) 8.8 m/s^2; b) 5.4 m/s^2; c) 3.8 m/s^2

43. $2mg$; $\dfrac{mg}{\sqrt{3}}$; $\dfrac{mg}{\sqrt{3}}$

45. $\tan^{-1}(2\mu_s)$

47. $T_1 = \dfrac{\tau}{R}\left(\dfrac{1}{e^{\mu_s\pi} - 1}\right)$ and $T_2 = \dfrac{\tau}{R}\left(\dfrac{1}{1 - e^{-\mu_s\pi}}\right)$

49. 1580 N; 1340 N

51. 400 N

53. 240 N

55. 0.010 micrograms resolution; 0.20000 milligrams max load

57. 736 N

59. a) $\frac{1}{2}mg(R_1 - R_2)$; b) $F = \frac{1}{2}mg\dfrac{(R_1 - R_2)}{l}$; c) $\dfrac{F'}{F} = \dfrac{2l}{R_1 - R_2}$

61. 8.9

63. 393

65. 36

67. 2×10^{-3} m

69. 3.5×10^4 m

71. a) 4×10^{-3} m; b) 3.5×10^3 N

73. 0.057 m

75. 0.033%

77. 4.3×10^{-6} m

79. 0.52 cm

81. 1.5°

83. 3.96×10^8 N/m^2

85. 624 rad/s

87. 425 rad/s

89. 360 N

91. 1.23×10^4 N; 2.13×10^4 N

93. 2400 N·m

95. $0.577\,Mg$; $0.289\,Mg$

97. 490 N

99. 1200 N

101. 0.71 mm

103. 5.0×10^3 m

105. 4.85×10^{-5} N

Chapter 15

1. a) 3.0 m; 0.318 Hz; 2.0 rad/s; 3.14 s; b) $t_{\text{midpoint}} = 0.785$ s; $t_{\text{turn.point}} = 1.57$ s

3. a) 0.83 Hz; 5.2 rad/s; b) 0.20 m; c) 0.30 s; 0.40 s; d) 1.05 m/s; 0.91 m/s

5. a) 251 m/s; b) 251 m/s

7. 211 N; 4.2 m/s

9. $A = \sqrt{x_0^2 + \dfrac{v_0^2}{\omega^2}}$; $\delta = -\tan^{-1}\left(\dfrac{v_0}{\omega x_0}\right)$

11. 3.98×10^{-6} m

13. $\delta = \dfrac{3\pi}{2}$

15. a) $x = 0.292 \cos(6\pi t - 0.815)$; b) 0.043 s; -103 m/s^2

17. 2.8×10^6 N/m; 3.16 Hz

19. 1.9×10^4 N/m

21. 1.13×10^{14} Hz

23. 0.20 m; 7.3 rad/s; 0 rad; 1.46 m/s; 10.7 m/s^2

25. 5.51×10^3 Hz; 1.0×10^{-4} g; 6.7×10^{-6} cm

27. $x = 0.27 \cos 6t$, with the axis chosen so the initial position of the unstretched spring is at $x = 0.27$ m.

29. $\dfrac{mg \sin \theta}{k}$; $\dfrac{1}{2\pi}\sqrt{\dfrac{k}{m}}$

31. $f = \dfrac{1}{2\pi}\sqrt{\dfrac{k}{(m + 6M)}}$

33. 3.6 J

35. 2.12 Hz; E same; A same; 2.0 m/s; 26.7 m/s^2

37. 0.34 m

39. $Fd - \dfrac{kd^2}{2}$

43. 10.4 s

45. 34.8 s

47. 0.188 Hz

49. a) 0.73 min/day; b) 1.0 mm

51. 24.8 m

53. 3.0 s

55. a) $\sqrt{\dfrac{2\kappa}{MR^2}}$; b) $\theta_0\sqrt{\dfrac{2\kappa}{MR^2}}$; c) $\theta_0 = 1$ radian

57. $2.09 \sqrt{\dfrac{L}{g}}$

59. a) 1.26×10^{-3} J; b) 0.145 m/s

61. 1.64 s

63. 9.8×10^{-3} m/s^2

65. 1.6 s

67. $2\pi \sqrt{\dfrac{L}{g}\left(\dfrac{m_1 + m_2}{m_1 - m_2}\right)}$

69. 3.6 rad/s; 3.3 m/s

71. $m = \dfrac{\kappa}{g}$

73. $2\pi\sqrt{\dfrac{4L}{5g}}$

75. a) $mg\left[1 - \dfrac{A^2}{2} + \dfrac{3A^2}{2}\sin^2\left(\sqrt{\dfrac{g}{l}}\,t\right)\right]$;

b) $\dfrac{\pi}{2}\sqrt{\dfrac{l}{g}}$; $mg(1 + A^2)$

77. 1.0×10^3

79. 92; 0.32 W

81. 30

83. 395

85. 21

87. 1.5 cm; 66.7 Hz

89. a) midpoint at 2 s, 6 s, 10 . . .; turning point at 0 s, 4 s, 8 s, . . .;
b) midpoint at 0 s, 4 s, 8 s, . . .; turning point at 2 s, 6 s, 10 s, . . .

91. 26.7 m/s; 1.68×10^4 m/s^2; 2.0×10^4 N

93. 2.12 Hz

95. 1.25 Hz

97. 18.75 J; 3.54 m/s

99. 0.375 m

103. 0.35 Hz

Chapter 16

1. 4.3×10^{14} Hz to 7.5×10^{14} Hz (violet)

3. 2.08 cycles/hour, 356 km

5. a) 4.4 m/s; b) 205 m/s

7. a) 10.8 hour; b) 2.5 /hour

9. a) $v_{max} = 0.27$ m/s as it passes through equilibrium between crests; b) $|a_{max}| = 6.2$ m/s^2 at the wave crests.

11. a) 0.02 m; b) 1.4 Hz; c) 10 m

13. a) 0.2 sec, 5.0 Hz, 31 rad/sec, 5.2 m^{-1};
b) $y = 0.020 \cos (5.2 x - 31.4 t)$

15. wavelength decreases by 20 cm

17. $|a_{max}| = 2.41$ m/s^2; wavelength = 156 m

19. 0.45 m/s

21. 0.97 sec

23. 2.0 kg/m

25. 250 m/s

27. 1280 N

29. 184 km

31. $v = v\sqrt{\dfrac{\mu}{\mu}}$

33. 0.017 sec

35. $2\sqrt{l/g}$

37. $\dfrac{mV^2}{2\pi R}$; Proof required.

41. a) $\dfrac{\pi}{2}$; b) 0.20 m

43. 6.0 m, 1.57 m, 0.80 Hz; -3.2 m

45. 1.14 m, 6.28 m

47. a) 0.0194 m; b) −0.0179 m

49. a) $y = 0.0060 \cos(400\pi t) + 0.0040 \cos(1200\pi t)$; b)

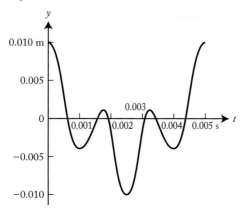

51. 0.028 is the fractional increase or decrease. We cannot tell which from the given information.

53. 392 Hz, 588 Hz, 784 Hz, 980 Hz

55. 1.58 Hz, 3.16 Hz

57. 28 Hz

59. 7.07 m, 66.6 m/s, 628 m/s^2

61. 9.3 ms, down, 54 Hz

63. 8.16 Hz, 16.3 Hz, 24.5 Hz, 32.7 Hz, 40.8 Hz

65. 71 N

67. 1.3×10^{-7} m, 2.5×10^{-7} m

69. 3.7 m/s, 6.8×10^3 m/s^2

71. a) 4.82×10^3 N; b) 8.2 Hz

73. $y = (0.20 \text{ mm}) \sin(2\pi x + 880\pi t) + (0.20 \text{ mm}) \sin(2\pi x - 880\pi t)$, $A = 0.20$ mm, $v = 440$ m/s

75. a) $v(x) = \sqrt{\dfrac{F}{A + Bx}}$; b) $\lambda(x) = \dfrac{1}{f}\sqrt{\dfrac{F}{A + Bx}}$;

    c) $f = \dfrac{n}{2L}\sqrt{\dfrac{F}{A + BL}}$

77. Large amplitude is at $x = \dfrac{n\pi}{k}$. Smallest amplitude is at $x = \dfrac{(2n + 1)\pi}{k}$ where n is an integer.

79. a) 0.030 m; b) 5.2 m; c) crests: 0 m, 5.2 m, 10.4 m, . . .; troughs: 2.6 m, 7.9 m, 13.5 m, . . .

81. 2.1 m/s^2, 4.9 m/s

83. 13 m/s, 7.9×10^3 m/s^2

85. 0.0731 kg/m, 261 m/s

87. a) $T_1 = \dfrac{\sqrt{2}}{1 + \sqrt{3}}\,T$, $T_2 = \dfrac{2}{1 + \sqrt{3}}\,T$;

    b) $v_1 = 7.2$ m/s, $v_2 = 8.6$ m/s

89. 12 Hz

91. a) $v(x) = \sqrt{g(L - x)}$, 14 m/s, 9.9 m/s, 0 m/s; b) 2.9 sec

93. $\lambda = 4L, \dfrac{4L}{3}, \dfrac{4L}{5}, \dfrac{4L}{7}, \ldots$, where $4L$ is the longest possible wavelength

Chapter 17

1. 17 m (20 Hz) to 1.7 cm (20 kHz)

3. 765 m; 166 m

5. about 9 cm

7. 1.9 mm and 0.10 mm

9. 6.8 m/s

11. D–D# 1.9 cm, D#–E 1.8 cm, E–F 1.7 cm, F–F# 1.6 cm, F#–G 1.5 cm, G–G# 1.5 cm, G#–A 1.3 cm, A–A# 1.3 cm, A#–B 1.2 cm, B–C 1.1 cm, C–C# 1.1 cm, C#–D 1.0 cm

13. 1.0×10^3 W/m^2

15. D#, 6 octaves above the one listed in Table 17.1

17. −3.0 dB

19. 83 dB

21. 0.11 W

23. 130 times (intensity measured in W/m^2), 21 dB

25. 9.1 sec

27. 249 m

29. a) $f = 3.0 \times 10^{-5}$ Hz; $T = 3.3 \times 10^4$ s (about 9 h); b) It's possible because the period of the first overtone is close to 1/4 of the tidal period.

31. a) 3.0×10^{-5} s; b) glass

33. 272 Hz, 3.9%

35. about 4000 Hz

37. 2.0 km

39. 3.3 km in sea water

41. 2.8 km

43. 92.4 m

45. 337 m/s

49. a) 0.632 m; b) C–C# 3.5 cm, C#–D 3.4 cm, D–D# 3.2 cm, D#–E 3.0 cm, E–F 2.8 cm, F–F# 2.7 cm, F#–G 2.5 cm, G–G# 2.4 cm, G#–A 2.2 cm, A–A# 2.1 cm, A#–B 2.0 cm, B–C 1.9 cm

51. $\lambda_n = \dfrac{2L}{n}$

    $f_n = \dfrac{nv}{2L}$, $n = 1, 2, 3, \ldots$

53. 619 Hz

55. 21.5 m/s, 0.215 Hz

57. 381 m/s

59. 2.63 m/s

61. 30°

63. 405 Hz

65. 594 Hz, 595 Hz

67. 476 Hz

69. 481 Hz

71. 29.4°

73. a) Proof required; b) 165 m/s

75. 0.15 mm

77. 1.5 mm, 0.33 mm

79. 5.0 m/s, 3.2×10^7g

81. 3 women

83. 3.0 dB

85. a) $3.0* 10^{-3}$ sec; b) The bat will think distances are 0.77 times the real distances.

87. a) 33.5°; b) 30.2 sec

89. a) 660 Hz; b) 691 Hz; c) 723 Hz

Chapter 18

1. In the hose: 1.39 m/s; 2.8 m/s; 4.2 m/s; In the nozzle: 22.3 m/s; 25.1 m/s; 23.9 m/s

3. 7.23×10^5 W

5. 12 m/s

7. 8.84 cm/s; 8.84 m/s

9. a) 11.5 m; b) 8.1 cm; 11.7 cm

11. 84 m

13. 1370 lbf = 6090 N

15. 132 cm^2

17. 2.34×10^5 Pa

19. 5.08×10^4 N

21. 48.6 cm^2

23. 2.0×10^4 Pa; 7.5×10^6 Pa

25. a) 360 N; b) 330 N

27. 3.56×10^5 Pa; 3.60×10^5 Pa; 4×10^3 N

29. 0.85 m

31. 10.3 m

33. 3.3×10^4 Pa

35. 2.94×10^5 Pa

37. 2.1%

39. 3.1×10^8 N

41. a) Proof required; b) 5.0×10^9 Pa

43. a) $F_b = \rho gh\pi\, R_2^2$; $F_\perp = \dfrac{\pi\rho gh(R_2 - R_1)(R_1 + 2R_2)}{3}$;

$$F_{total} = -\frac{\pi\rho gh\left(R_1^2 + R_1 R_2 + R_2^2\right)}{3};$$

b) $F_b = \rho gh\pi\, R_1^2$; $F_\perp = -\dfrac{\pi\rho gh(R_2 - R_1)(2R_1 + R_2)}{3}$;

$$F_{total} = -\frac{\pi\rho gh\left(R_1^2 + R_1 R_2 + R_2^2\right)}{3}$$

45. a) 4.7×10^7 m^3; b) 4.3×10^{10} kg

47. Yes

49. 0.32 m

51. 61 m/s^2

53. 113 kg

55. 31 m

57. 12.6 cm

59. a) Proof required; b) $-2g$; c) ∞

61. 4.4 cm

63. a) Proof required; b) 0.094 Hz

65. 1.13×10^5 Pa

67. $\sqrt{2gh + 2\dfrac{p_{atm}}{\rho}}$

69. 1.9×10^4 N

71. $v = \sqrt{\dfrac{2p_{tank}}{\rho}}$

73. 0.013 m/s

75. a) 332 N; b) Average rate = 4.3 kW; Peak rate = 8.6 kW

77. 12.4 hp

79. 2.7 m/s; 190 liters/s

81. 7.3×10^3 N

83. 1.12×10^{12} Pa

87. 8.06 mm-Hg

89. 1.21×10^5 Pa; 1.01×10^5 Pa; 2×10^3 N

91. 0.73 cm

93. 2.5×10^5 kg

95. 5.3×10^{-3} m^3

97. 1.07×10^3 kg/m^3

99. 2.0×10^{-3} m^3/s

101. The pressure inside increases by 209 Pa; Smaller

Chapter 19

1. 32°F, $-$ 380°F, $-$ 423.4°F, $-$ 452.2°F, $-$ 454°F, $-$ 459.67°F

3. 5.3×10^{21} atoms

5. 78% N$_2$, 22% O$_2$

7. The frequency decreases by 14 Hz

9. 1.9 atm

11. a) $p_{O2} = 7.5 \times 10^4$ Pa, $p_{N2} = 8.6 \times 10^4$ Pa; b) 1.6×10^5 Pa

13. 3.4 atm

15. $\rho_1/\rho_0 = 1.05$

17. 1.4×10^{-9} Pa

19. 12 kg/m^3

21. 4.3 atm

23. 500 kg/m^3

25. 100 kg/m^3

27. 1.3 cm

29. 4.5×10^7 Pa

31. 96.3 g

33. 3150 kg, 2.8×10^3 m^3

35. a) Water rises 1.2 m; b) 6.8 kg at 2.6×10^5 Pa

37. 29 g/mol

39. Differentiating $p - p_0 = -\rho gy$ yields $dp = -\rho gy$. Using the Ideal-Gas Law and $n = \dfrac{m}{M}$, $\rho = \dfrac{m}{V} = \dfrac{pM}{RT}$.

Substituting ρ into the differential equation: $dp = -\dfrac{pMg}{RT}\,dy$

or, $\dfrac{dp}{p} = -\dfrac{Mg}{RT}\,dy.$

41. 615 m/s
43. 1200 K
45. 4100 m/s, 1.4×10^{-20} J
47. 0.12 m/s
49. 5.65×10^{-21} J
51. For O_2, $v_{rms} = 428$ m/s, For O_3, $v_{rms}= 349$ m/s, For O_2, (translational) K = 4.87×10^{-21} J, for O_3, (translational) K = 4.85×10^{-21} J
53. 0.43%
55. 9.7×10^6 K
57. 0.47 m/s
59. Using the hint, the volume swept out per molecule with an effective radius $2R_0$ going a distance l is cylindrically shaped with volume $V/N = \pi\,(2R_0)^2\,l$. Solving for l yields the desired result.
61. a) for 1 atm: -0.091%; b) for 1000 atm: -91%
63. 1.9×10^5 J, translational 0.6, rotational 0.4
65. A 7% increase in kinetic energy by changing temperature, no change in the kinetic energy by changing the pressure.
67. 1.0×10^5 J
69. 291 J
71. 1.88×10^{25} molecules
73. 1.3×10^3 N
75. 1.1×10^{23} nitrogen molecules, 2.9×10^{22} oxygen molecules, 1.4×10^{23} total
77. a) 1.8×10^{32} particles/m^3 3.7×10^{16} Pa; b) 9.0×10^{31} particles/m^3, 1.9×10^{16} Pa; c) 4.5×10^{31} particles/m^3, 9.3×10^{15} Pa
79. From the Ideal-Gas Law at constant temperature $pV = p'V'$. So,

$$\dfrac{\Delta Z}{V} = \dfrac{V - V'}{V} = 1 - \dfrac{V'}{V} = 1 - \dfrac{p}{p'} = \dfrac{p' - p}{p'} = \dfrac{\Delta p}{p'}.$$

$$V = \Delta V\left(\dfrac{p'}{\Delta p}\right),$$

This can be rearranged to ―――――― where ΔV is the *decrease* in volume and Δp is the corresponding *increase* in pressure.

Furthermore, $\dfrac{p'}{\Delta p} = \dfrac{p + \Delta p}{\Delta p} = 1 + \dfrac{p}{\Delta p}$ and

$$V = \Delta V\left(1 + \dfrac{p}{\Delta p}\right) \approx \Delta V\left(\dfrac{p}{\Delta p}\right) \text{ for } \Delta p << p.$$

Using the specifics of the problem, $V_c = V_s + V$ or $V_s =$

$$V_c - V = V_c - \Delta V\left(\dfrac{p}{\Delta p}\right)$$

81. 615 m/s
83. a) $\sqrt{2}$; b) 1; c) 1; d) 0.5
85. 375 K

Chapter 20

1. 540 s
3. 0.28°C
5. 1.6 km
7. 750 s
9. 8500 steps
11. 0.17°C
13. 1.7°C
15. 1.7×10^{-4} °C/km
17. The heat produced from electric power 2.6% of the incident solar heat. This is enough to slightly increase the local temperature.
19. 1.1×10^{-3} m^3/s
21. a) 1.7×10^3 N·m; b) 2.3×10^3 W; c) 4.0×10^{-4} °C
23. 27°C
25. 136°C
27. 0.67°C
29. 38°C
31. 0.18 liter
33. 0.028 J of work done by iron, 2.7×10^7 J of heat absorbed by iron, amount of work is 1.0×10^{-6} times the heat absorbed
35. 4.9×10^{-4} m, 17 N
37. a) 3.8×10^{-4}, 1.9×10^{-4}; b) 16 s
39. Proof required.
41. 100.28°C
43. 23000 W, the rate through window is 13 times greater than the rate through the wall
45. a) 2.4 m^2·s·°C/J; b) 13.6 ft^2·h·°F/BTU
47. 4.2×10^3 W
49. The solution is a proof.
51. 11 W, 79°C
53. The solution is a proof.
55. 0.51 cm/h
57. a) 4.26×10^{14} J; b) 5.3 bombs
59. 270 g
61. 1.1×10^9 J, 1.2×10^4 W
63. 3.9 kg
65. a) 2.0×10^{11} kg; b) 1.1×10^{17} cal; c) 2.9×10^{15} J = 7.0×10^{14} cal; d) 1.0×10^{13} J = 2.4×10^{12} cal. The kinetic energy is smaller than the potential energy due to frictional losses with the air.
67. 0.092 kg
69. 4.3 km^3/h
71. 41°C

73.

GAS	C_V (J/(kG·K)
He	3.12×10^3
Ar	3.13×10^2
N_2	7.42×10^2
O_2	6.50×10^2
CO	7.39×10^2
NH_3	1.60×10^3
CH_4	1.69×10^3

The gas with the highest specific heat per kilogram is helium; and that with the lowest is argon.

75. 971 m/s

77. $\Delta V = 3.7 \times 10^{-2}$ m^3, $W = 3.7 \times 10^3$ J

79. $C_p = 26.3$ J/(mol·K), $C_V = 18.0$ J/(mol·K)

81. 17.9 kcal/h

83. 1 K

85. a) 0.072 m^3; b) 0.42 m^3, 145 K

87. 214 K

89. 3.6°C

91. 160 liters/h

93. 0.33 m, 0.050 m, 46 m^2

95. 2.3×10^{-5} kg/s

97. 0.52 kg

99. 880 J, 1500 J

101. a) 1700 J; b) 1200 J; c) 0.029 m^3, 7.8×10^4 N/m^2; d) 5.0×10^2 J, 0

Chapter 21

1. $Q = 1.9 \times 10^3$ J, $\Delta E = 1.1 \times 10^3$ J

3. a) $W = 0$, $Q = \Delta E = 610$ J; b) $W = 810$ J, $Q = 2.0 \times 10^3$ J, $\Delta E = 1.2 \times 10^3$ J

5. −470 J

7. a) 4.29 moles; b) $W = 1010$ J, $\Delta E = 2490$ J; c) 5/2, diatomic

9. a) $W = -9.19$ J, $\Delta E = 3.34 \times 10^5$ J; b) − 18.4 J, the heat of vaporization remains unchanged

11. $W = -37.9$ J, $\Delta E = -37.1$ J

13. 4.87×10^4 J, 2.56 J

15. 4.3×10^4 J

17. 0.014 m^3, 7.2×10^5 Pa

19. 2.43×10^6 J/kg

21. 35%

23. 14%

25. 60%, 1.2×10^7 J

27. 44.5%

29. 5.5%, $W = m_{you}$ (9.81 m/s^2) (3.0 m),

$$Q = \frac{m_{you}(9.81 \text{ m/s}^2)\,(3.0 \text{ m})}{0.055}\left(\frac{1 \text{ kcal}}{4187 \text{ J}}\right)$$

31. 8.2×10^6 J/s, 0.19 kg/s

33. 0.999 999 997

35. 1.4×10^5 J

37. 44%, 1100 W

39. 75 W

41. 19.5

43. 8.5×10^3 J, 3.4

45. a) 2.0×10^3 J; b) $Q_2 = 4.0 \times 10^4$ J (absorbed); $Q_3 = -2.3 \times 10^4$ J (ejected); c) $Q_1 = 3.0 \times 10^3$ J (absorbed); d) $e = 0.047$

47. a) 0.067; b) 1.39×10^7 W; c) 180 kg/s

49. a) 48 W; b) 20 times

51. a) $e_{turbine} = 0.34$, $e_{engine} = 0.42$; b) 0.62, the two efficiencies are the same

53. 9.5×10^3 J/K

55. 3.0 W/K

57. 12 400 W/K

59. 3×10^3 J/K

61. $\Delta S_{Al} = 430$ J/K, $\Delta S_{Fe} = 150$ J/K, $\Delta S_{Ag} = 80$ J/K, $\Delta S_{Hg} = 47$ J/K. The change in entropy seems to decrease with increasing atomic number. Largest is aluminum; smallest mercury.

63. 41 W/K

65. 9.5×10^6 W/K

67. 0.94 W/K

69. 5.8 J/K

71. a) Proof required; b $\Delta S = 780$ J/K

73. 37 J/K

75. Proof required.

77. 120 K

79. a) 4.16×10^5 Pa, 2.27×10^5 Pa; b) $W = \Delta Q = -3.0 \times 10^2$ J

81. 24%, 4.8×10^4 W

83. 4.0×10^2 W

85. a) Beginning with the point at the upper left, the gas undergoes an isobaric expansion in step 1 as the volume increases, followed by a isovolumetric reduction of pressure in step 2 as the temperature is reduced. The gas is then compressed isobarically in step 3 by reducing the volume, before an isovolumetric increase in pressure in step 4 by increasing the temperature.
b) $W_1 = 2100$ J, $W_2 = 0$ J, $W_3 = -700$ J, $W_4 = 0$ J; c) $Q_1 = 5260$ J, $Q_2 = -3160$ J, $Q_3 = -1750$ J, $Q_4 = 1050$ J; d) 22%

87. 48%, 1×10^7 W

89. $\Delta S_N = 2600$ J/K, $\Delta S_O = 2300$ J/K, $\Delta S_H = 22\,500$ J/K. Hydrogen is largest and oxygen smallest.

91. a) $\Delta S = 1.1$ J/K; b) $Q_2 = 340$ J, $\Delta S = 1.1$ J/K; c) $\Delta S = 0$ J/K

Photo credits

Part Openers: **Part Opener 1**: NASA/GRIN; **Part Opener 2**: L.Weinstein, NASA/Photo Researchers, Inc.; **Part Opener 3**: Alfred Pasieka/Photo Researchers, Inc.

Prelude:
10^0: John Markert/Junenoire Photography; 10^1: John Markert/Junenoire Photography; 10^2: John Markert/Junenoire Photography; 10^3: Jim Mairs; 10^4: Aerial Imagery Courtesy of GlobeXplorer.com; 10^5: Aerial Imagery Courtesy of GlobeXplorer.com; 10^6: Aerial Imagery Courtesy of GlobeXplorer.com; 10^7: Jacques Descloitres, MODIS Rapid Response Team, NASA/GSFC; 10^8: NASA/Corbis; p.5: 10^9: NASA/ JSC; 10^{21}: Dr. Fred Espenak/Photo Researchers, Inc.; 10^{22}: David Malin; 10^{23}: The Hubble Heritage Team AURA/STScl/NASA; 10^{24}: NASA/ESA/R. Thompson (University of Arizona); 10^{26}: Max Tegmark/SDSS Collaboration; 10^0: John Markert/Junenoire Photography; 10^{-1}: John Markert/Junenoire Photography; 10^{-2}: Professor Pietro M. Motta/Photo Researchers, Inc.; 10^{-3}: OMIKRON/Photo Researchers, Inc.; 10^{-4}: SPL/Photo Researchers, Inc.; 10^{-5}: G. Murti/Photo Researchers, Inc.; 10^{-6}: DOE/Science Source.; 10^{-7}: Kenneth Eward/Photo Researchers, Inc..; 10^{-8}: Eurelios/Phototake.

Chapter Opener 1: Roger Ressmeyer/Corbis; **fig. 1.5**: Photo by H. Mark Helfer/NIST; **fig. 1.7**: Photo by Barry Gardner; **fig. 1.8**: ©2004 Bruce Erik Steffine; **table 1.1a**: NASA/JSC; **table 1.1b**: Charles O Rear/Corbis; **table 1.1c**: U.S. Mint Handout/Reuters/Corbis; **table 1.1d**: Courtesy of Robert G. Milne, Plant Virus Institute, National Research Council, Turin, Italy; **fig. 1.9**: NIST; **fig. 1.10**: Robert Rathe/NIST; **fig. 1.11**: BIPM (International Bureau of Weights and Measures/Bureau International des Poids et Mesures), www.bipm.org; **table 1.7a**: NASA/JSC; **table 1.7b**: Gene Blevins/LA Daily News/Corbis; **table 1.7c**: Royalty-Free/Corbis; **table 1.7d**: Clouds Hill Imaging Ltd./Corbis; **fig. 1.13**: Reuters/Corbis; **fig. 1.15**: National Maritime Museum, UK; **fig. 1.19**: John Brecher/Corbis.

Chapter Opener 2: Reuters/Corbis; **table 2.1a**: Corbis; **table 2.1b**: Randy Wells/Corbis; **table 2.1c**: David Muench/Corbis; **fig. 2.19**: James Sugar/Black Star; **p.51**: The Granger Collection, New York; **fig. 2.22**: Tom Sanders/Photri/Microstock; **fig. 2.23**: Wally McNamee/Corbis; **fig. 2.26**: Taxi/Getty Images; **fig. 2.31**: NASA/JPL.

Chapter Opener 3: NOAA/Hurricane Research Division; **fig. 3.3**: Ron Watts/Corbis; **p.71**: (Pip) Galen Rowell/Corbis.

Chapter Opener 4: Dr. J. Alean (Stromboli); **p.99**: (Pip fig. 2) Photo Courtesy of Mark Wernet/NASA/Glenn Research Center, Cleveland Ohio; **fig. 4.7**: Krafft-Explorer/Photo

Researchers, Inc.; **fig. 4.9**: Richard Megna/Fundamental Photographs; **fig. 4.19**: Aero Graphics/Corbis; **fig. 4.15**: Richard Megna/Fundamental Photographs; **fig. 4.16**: Wally McNamee/Corbis; **fig. 4.22**: NASA/JSC; **fig. 4.27**: Roger Ressmeyer/Corbis; **fig. 4.34**: Fermilab; **fig. 4.35**: Robert Harrington, www.BobQat.com.

Chapter Opener 5: Bob Krist/Corbis; **fig. 5.1**: Courtesy of David Hammond; **p.131**: The Granger Collection, New York; **fig. 5.5**: Courtesy DYNCorp/NASA/JSC; **table 5.1a**: NASA/JSC; **table 5.1b**: Mark Bolton/Corbis; **fig. 5.6**: Bettmann/Corbis; **fig. 5.8**: TEK Image/Photo Researchers, Inc.; **fig. 5.14**: Tim Kiusalaas/Corbis; **fig. 5.15**: Jim Sugar/Corbis; **fig. 5.16**: Roger Ressmeyer/Corbis; **fig. 5.44**: Science Museum, London.

Chapter Opener 6: Michael Kim/Corbis; **p.174**: James L. Amos/Corbis; **fig. 6.17**: Kai Pfaffenbach/Reuters/Corbis; **p.189**: (Pip fig. 2) Photodisc Green/Getty Images; **p.189**: (Pip fig. 3) Courtesy Beckman Coulter, Inc.; **fig. 6.25**: Duomo/Corbis; **fig. 6.26**: George Hall/Corbis; **fig. 6.27**: October 2001 Physics Today (Volume 54, Number 10, p.39), Courtesy of John Yasaitis, Analog Devices, Inc.; **fig. 6.28**: Brian Erler/Corbis; **fig. 6.39**: Duomo/Corbis.

Chapter Opener 7: Lester Lefkowitz/Corbis; **p.207**: Hulton Deustch/Corbis; **fig. 7.17**: Jim Cummins/Corbis; **fig. 7.19**: Courtesy Klockit, Inc.; **p.221**: Bettmann/Corbis; **fig. 7.23**: David Cumming/Eye Ubiquitous/Corbis; **fig. 7.32**: Jim Cummins/Corbis; **fig. 7.34**: Lester Lefkowitz/Corbis; **fig. 7.37**: Image Bank/Getty Images.

Chapter Opener 8: Courtesy Blenheim-Gilboa Pumped Storage Power Project/New York Power Authority; **p.236**: The Granger Collection, New York; **p.242**: (Pip) Courtesy New York Power Authority; **fig. 8.11**: Paul A. Souder/Corbis; **p.248**: Corbis; **table 8.1a**: Courtesy International Dark Sky Association, Defense Meteorological Satellite Program (DMSP).**table 8.1b**: Royalty-Free/Corbis; **table 8.1c**: Mika/Zefa/Corbis; **p.254**: North Wind Picture Archives; **table 8.2a**: Earth Observatory/NASA; **table 8.2b**: Stone/Getty Images; **table 8.2c**: Taxi/Getty Images; **fig. 8.17**: Courtesy New York Power Authority; **fig. 8.18**: Corbis; **fig. 8.21**: Paul A. Souders/Corbis; **fig. 8.27**: National Archives.

Chapter Opener 9: Bettmann/Corbis; **fig. 9.5**: Science Museum and Society Picture Library; **p.277**: The Granger Collection, New York; **p.279**: The Granger Collection, New York; **fig. 9.8**: NASA/Corbis; **p.281**: (Pip) NASA; **table 9.1a**: JPL/NASA; **p.285**: The Granger Collection, New York; **fig. 9.15**: Bettmann/Corbis; **fig. 9.28**: NASA; **fig. 9.30**: ©2002 Calvin J. Hamilton; **fig. 9.36**: National Astronomy Observatiries/AP Images; **fig. 9.37**: Dennis di Cicco/Corbis; **fig. 9.40**: Photofest; **fig. 9.41**: JPL/NASA; **fig. 9.42**: Photographer: Mark Avino ©Copyright 1995 by Smithsonian Institution.

Chapter Opener 10: Reuters/Corbis; **fig. 10.6**: Richard Megna 1990 Fundamental Photographs; **fig. 10.13**: ©2005 Estate of Alexander Calder/Artist Rights Society (ARS), New York; **fig. 10.15**: Roger Ressmeyer/Corbis; **p.320**: (Pip fig. 1) Macduff Everton/Corbis; **p.320**: (Pip fig. 2) Right George D. Lepp/Corbis; **fig. 10.18**: Jason Reed/Reuters/Corbis; **fig. 10.22**: NASA/Roger Ressmeyer/Corbis; **fig. 10.25**: Philip James Corwin/Corbis.

Chapter Opener 11: Tom Wright/Corbis; **fig. 11.1**: Daimlerchrysler; **p.343**: (Pip: top and bottom) Robert Laberge/Allsport/Getty Images; **fig. 11.6**: Smithsonian Institute; **fig. 11.11**:

Charles D. Winters/Photo Researchers, Inc; **fig. 11.17**: Reuters/Corbis; **fig. 11.20**: Oxford University Press, UK; **fig. 11.21**: Charles & Josette Lenars/Corbis.

Chapter Opener 12: Courtesy Sandia National Laboratories; **fig. 12.1**: Courtesy Palm Press, Inc.; **fig. 12.3c**: Royalty-Free/Corbis; **fig. 12.3b**: Picture Arts/Corbis; **fig. 12.3d**: Hans C. Ohanian; **fig. 12.3d**: Richard T. Nowitz/Corbis; **fig. 12.3a**: Hans C. Ohanian; **fig. 12.20**: Corbis; **fig. 12.32**: Neil Rabinowitz/Corbis.

Chapter Opener 13: Don Harlan/Gravity Probe B; **fig. 13.6**: Courtesy of Gravity Probe B Photo Archive, Stanford University; **fig. 13.11**: Wally McNamee/Corbis; **fig. 13.14**: Nancy Ney/Corbis; **p.414**: (Pip fig. 2) Ron Keller, N.M. Museum of Space; **p.414**: (Pip fig. 1) Science & Society Picture Library; **fig. 13.25**: Lawrence Lucier/Getty Images; **fig. 13.34**: Reuters/Corbis; **fig. 13.37**: NASA/JSC.

Chapter Opener 14: Kroll Cranes A/S - Denmark; **fig. 14.5**: Corbis; **fig. 14.29**: Tom Pantages; **fig. 14.50**: Tim De Waele/Isosports/Corbis.

Chapter Opener 15: NASA/JSC; **fig. 15.6**: Courtesy of David Hammond; **fig. 15.8**: Courtesy of John Markert; **fig. 15.15**: Loren Winters/Visuals Unlimited; **fig. 15.19**: National Maritime Museum; **fig. 15.23**: Bibliotheque Nationale de France; **p.493**: (Pip fig. 3) Courtesy John Markert; **fig. 15.26**: P.B. Umbanhowar, F. Melo, and H. L. Swinney, "Localized excitations in a vertically vibrated granular layer," Nature 382, 793-796 (1996); **fig. 15.27**: Jim Craigmyle/Corbis; **fig. 15.36**: Courtesy of John Markert; **fig. 15.37**: NASA/JSC.

Chapter Opener 16: Vladimir Smolyakov/Stolichnaya Vechernyaya Gazeta /AP Images; **fig. 16.1**: Richard Megna Fundamental Photo; **fig. 16.17**: Richard Megna Fundamental Photo; **fig. 16.18**: Keystone/Getty Images; **fig. 16.19**: Royalty-Free/Getty Images; **fig. 16.20**: David Nock of British Car Specialists; **fig. 16.21**: Royalty-Free/Corbis; **fig. 16.27**: Francois Gohier/Photo Researchers, Inc.

Chapter Opener 17: Stone/Getty Images; **fig. 17.1**: Aaron Horowitz/Corbis; **fig. 17.2** *Engineering Applications of Lasers and Holography* by Winston E.Kock, Plenum Publishing Co., New York 1975; **fig. 17.5**: William B. Joyce; **fig. 17.7**: Courtesy of C.F. Quate and L. Lam Hansen Laboratory; **p.546**: (Pip fig. 1) www.777life.com/free photo stock; **p.546**: (Pip fig. 3) John Ross Buschert, Goshen College, IN; **p.549**: AIP Emilio Segre Visual Archives; **fig. 17.19**: Gary S. Settles/Photo Researchers, Inc.; **p.552**: The Granger Collection, New York; **fig. 17.20**: Museum of Flight/Corbis; **fig. 17.21**: John Shelton Photography; **fig. 17.22**: PSSC Physics 2nd edition 1965 DC Health & Company and Educational Development Center, Inc., Newtown, MA; **fig. 17.23**: PSSC Physics 2nd edition 1965 DC Health & Company and Educational Development Center, Inc., Newtown, MA; **fig. 17.27**: Tim Bird/Corbis.

Chapter Opener 18: Woods Hole Oceanographic Institution **fig. 18.3**: Hans C. Ohanian; **fig. 18.8**: AD Moore from Introduction to Electromagnetic; **fig. 18.9**: D.C. Hazen and R.F. Lehnert, Subsonic Aerodynamics Laboratory, Princeton; **fig. 18.10**: D.C. Hazen and R.F. Lehnert, Subsonic Aerodynamics Laboratory, Princeton; **fig. 18.11**: Royalty-Free/Corbis; **fig. 18.13**: Image Design by Nature; **fig. 18.14**: Paul Edmondson/Corbis; **fig. 18.17**: Tom Kleindinst, Woods Hole Oceanographic Institution; **p.574**: Stefano Bianchetti/Corbis; **fig. 18.20**: AC

Hydraulic; **p.579**: (Pip fig. 1) Anonia Reeve/Photo Researchers, Inc.; **fig. 18.26**: DW Stock Picture Library/S. Drossinos; **fig. 18.28**: Stan White Photography; **p.581**: Archivo Iconografico, S.A./Corbis; **p.585**: Bettmann/Corbis; **fig. 18.34**: Columbia University Physics Department; **fig. 18.37**: Principals of Physics, 1994; **fig. 18.38**: Peter Finger/Corbis; **fig. 18.40**: Roger Ressmeyer/Corbis; **fig. 18.41**: Bettmann/Corbis; **fig. 18.48**: Royalty-Free/Corbis; **fig. 18.49**: Roger Ressmeyer/Corbis; **fig. 18.50**: AFP/Hyundai Heavy Industries.

Chapter Opener 19: Vince Streano/Corbis; **p.604**: AIP Emilio Segre Visual Archives; **p.606**: Chemical Heritage Foundation Collection; **p.608**: Bettmann/Corbis; **fig. 19.7**: Royalty-Free/Corbis; **fig. 19.8**: Tom Pantages; **fig. 19.9**: Crown copyright 1999, Reproduced permission; **fig. 19.10**: Courtesy of Cole-Parmer Instrument Company; **fig. 19.12**: Liquid Crystal Resources, Glenview, IL; **table 19.1a**: Julian Baum/Photo Researchers, Inc.; **table 19.1b**: Visuals Unlimited; **table 19.1c**: Dr. Arthur Tucher/Photo Researchers, Inc.; **fig. 19.17**: Courtesy of Worthington Cylinders; **fig. 19.18**: Dr. Kimberly Strong, University of Toronto; **fig. 19.20**: National Institute of Standards and Technology.

Chapter Opener 20: Peter Arnold, Inc./Alamy; **p.629**: (bio) Burnstein Collection/Corbis; **fig. 20.1**: Bettmann/Corbis; **fig. 20.6**: Norbert Wu; **fig. 20.9**: Visuals Unimited; **fig. 20.10**: AP Images; **table 20.4a**: David Taylor/Corbis; **table 20.4b**: David Pollack/Corbis; **table 20.4c**: D.Winters/Photo Researchers, Inc.; **fig. 20.19**: Lowell Georgia/Corbis; **fig. 20.20**: Alfred Pasieka/Photo Researchers, Inc.; **fig. 20.22**; Private Collection; **fig. 20.24**: Stanford University.

Chapter Opener 21: The Image Bank/Getty Images; **fig. 21.5**: Science & Society Picture Library; **fig. 21.6**: Courtesy of BMW World; **fig. 21.8**: Jim Cummins/Corbis. **p.667**: Bridgeman Art Library; **fig. 21.17**: Tom Pantages; **fig. 21.19**: Tom Pantages; **p.678**: Bettmann/Corbis; **p.681**: Bettmann/Corbis; **fig. 21.23**: ©2005 The M.C. Escher Company - Holland. All rights reserved; **fig. 21.27**: Picture Arts/Corbis; **fig. 21.32**: Inga Spence/Visuals Unlimited.

Index

Page numbers in *italics* refer to biographies. Page numbers in **boldface** refer to figures. Page numbers followed by "*t*" refer to tables.

aberration:
  chromatic, 1144
  spherical, 1128, 1144
absolute acceleration, 132
absolute motion, 1217
absolute temperature scale, 604
absolute thermodynamic temperature scale, 609, 686
absolute zero, entropy at, 680
absorbed dose, 1375
Acapulco, divers, 688
accelerated charge:
  electric field of, 1075–79, **1076, 1077**
  radiation field of, 1075–76, **1076, 1077**
acceleration, 39–54
  absolute, 132
  angular, *see* angular acceleration
  average, 39, 39*t*, **40, 60**
  average, in three dimensions, 101–2
  average, in two dimensions, 96
  of center of mass, 323
  centripetal, 113–14, **113, 114,** 132, 184–90, **195,** 371, 372
  components of, 95–98, **101**
  as derivative of velocity, 41
  formulas for, 39, 41
  instantaneous, 40–41, **41**
  instantaneous, components of, 97
  instantaneous, in two dimensions, 96–97
  motion with constant, 42–49, **43, 63,** 102–4, **103, 104, 122**
  motion with variable, 54–56
  negative, 39
  positive, 39–40
  standard *g* as unit of, 52
  tangential, 371–72
  translational, 402
  vectors, 100–101
acceleration of free fall, 49–54, **64**
  universality of, 49, **49**
acceleration of gravity, 52–53, **64,** 274–75
  measurement of, 52–53
  variation of, with altitude, 274–75
accelerators:
  linear, 1413, 1415
  for particles, 1363, **1363,** 1398–99
  see CERN (LHC) Brookhaven Fermilab, SLAC, SSC
acceptor impurities, 1339
accidents, automobile, 339, **343,** 355
AC circuits, 1030–67

AC current, 1031
  hazards of, 913–14
acoustic micrograph, **539**
action and reaction, 144–51, **144, 145, 146**
action-at-a-distance, 274, 722
action-by-contact, 722, 723
action-reaction pairs, 144–51, **144, 145, 146, 149**
AC voltage, 1004
Adams, J. C., 272
addition law for velocity, Galilean, 1218
addition of vectors, 72–76, **72, 73, 74, 89**
  commutative law of, 74
  by components, 78–79
addition rule for velocities, 115–16, **117**
adiabatic equation, for gas, 649
adiabatic expansion, 668, **689**
adiabatic process, 647–49
air, composition by element and mass, 620, 623, 657
air bag, 343
air conditioner, 672, 673
airfoil, flow around, 570, 582–83
airplane:
  motion, pitch, roll, yaw, **366**
  propeller, **392**
air resistance, 49, 51, **61,** 181, **181**
  in projectile motion, 111
Al'Aziziyah, Libya, hottest temperature, 621
Alessandro, Conte Volta, *793*
Alpha Centauri, A and B, **xli,** 296, 1247, 1248, 1251
alpha decay, 1365–67
alpha-particle model, 823, **823**
alpha particles, 823, 1363–64
  scattering of, 1293–94
alpha rays, 1365
alternating current, 1031
  hazards of, 913–14
alternating emf, **1031, 1032, 1047,** *see* emf alternating
Alvin, DSV, 565, **565,** 574, **574,** 577, 582
ammeter, 905, **906,** 916. **916, 917**
Amontons, Guillaume, *174*
Ampère, André Marie, *940*
amperes, 697
Ampère's Law, 939–40, **944, 947**
  displacement current and, 1073–74
  electric flux and, 1073
  modified by Maxwell, **1071,** 1073, 1074, 1080, 1096, 1097, **1097**
amplitude:

of motion, 470
of wave, 511
Analytical Mechanics (LaGrange), *236*
analyzer, 1085, **1085**
Andromeda Galaxy, xliv, 1248
Angers, France, bridge collapse at, 491, **491**
angle:
  elevation, **109,** 111, **111**
  of incidence and of reflection, 1115–16, **1115, 1116**
angle in radians, 368
angular acceleration:
  average, 370
  constant, equations for, 374
  instantaneous, 370
  rotational motion with constant, 374–76
  time-dependent, 376–78
  torque and, 400
angular frequency, 470–71, **471**
  of simple harmonic oscillator, 477
  of wave, 512, 513–16
angular magnification:
  of magnifier, 1147, **1148**
  of microscope, 1149, **1150**
  of telescope, 1150, **1150**
angular momenta, some typical values, 407*t*
angular momentum, 284, 407*t*
  for circular orbit, 409
  in elliptical orbit, 291–92
  orbital, 409
  quantization of, 1322–23
  spin, 409
  torque and, 410–16
angular momentum, conservation of:
  in planetary motion, 284
  in rotational motion, 406–10
angular-momentum quantum number, 1296, 1322, 1324–26
angular momentum vector, 411, **411**
angular motion, 375
angular position, for time-dependent angular velocity, 376–77
angular resolution, of telescope, 1196–99
angular velocity, 369*t*, **471**
  average, 369
  instantaneous, 369
  for time-dependent angular acceleration, 376–77
annulus of sheet metal, **381**
antibaryons, 1404
antielectron, *see* positron

antihydrogen atom, 1425
antimatter-matter annihilation, 706, 1428
antimesons, 1404
antineutrino, 1368–70
antinodes and nodes, 520–21, **521**, 544–45, **544–45**
antiparticles, 706, 1403–4
antiquarks, 1413
aphelion, 282, 284, 409
  of planets, **xl**, 285
apogee:
  of artificial satellites, 286
  of moons, **xxxix**
  of planet, 286
Apollo 16, xxxix
Apollo astronaut, **295**
apparent depth, 1122, **1122**
apparent weight, 187–88, **187**
apple, chemical energy of, 632, 652
Archimedes, *581*
Archimedes' Principle, 580–82, 599
area, 13
areas, law of, 283–84
Arecibo radio telescope, 1198–99, **1198**
argon:
  compression, 658
  Lennard-Jones potential, 263
  monatomic kinetic energy, 616
  thermal window of, 656
artificial satellites, 271–72, 281, 286–87, 1344, **1344**, 1421
  apogee of, 286
  perigee of, 286
astigmatism, 1147
astrology, 295
astronaut, 581, 594, **602**, 605, 612, 618, 622
  weightlessness training, **589**
  see also Apollo; International Space Station; Skylab mission
astronomical unit (AU), 24
Atlas rocket, guidance system, **414**
atmosphere, 573
atmospheric electric field (on Earth), 785, 787, 804–05, 819, 823
atmospheric pressure, 577–78
  gauge, **573**
atom, 1397
  electron configuration of, 1328–32
  electron distribution in, 695
  nuclear model of, 1294, **1294**
  nucleus of, *see* nucleus
  quantum structure of, 1320–40
  stationary states of, 1299
  structure of, 695, 1287–95
atomic bomb (Hiroshima), 1251
atomic clock, Cesium, 9
atomic force microscope (AFM), 475, **475**, 1311, **1311**
atomic mass, 11–12
atomic mass unit, 11, 20, 1355
atomic number, 1356
atomic standard of mass, 11
atomic standard of time, 9
atomic states, quantum numbers of, 1328–32
atomic structure, 695, 1287–95
atom smashers, 1397
attractors, 492
Atwood's machine, 403, 421
automobile battery, 707, 890–91, **891, 892, 924**
automobiles:
  collisions, 339, **343**, 355
  crash tests of, 339, **340**, 355
  efficiency of, 674, **674**
  electric fields and, 805

energy conversions, **674**
engine cycle, **674**
impact speed, 343*t*
starter motor, **862**
automobile stopping distances, **45, 46**, 47, **47**
average acceleration, 39, 39*t*, **40, 60**
  formula for, 39
  in three dimensions, 101–2
  in two dimensions, 96
average angular acceleration, 370
average angular velocity, 369
average power, 253
average speed, 29–31, 30*t*
average velocity, 32–35, **33**, 101–2
  in two dimensions, 95
Avogadro's number, 11, 20, 607
axis of symetry, **380–82**, 382*t*

back emf, 1012
balance, 136–37
  beam, 136–37, **137**
  Cavendish torsion, 277
  Coulomb's, **698**, *700*
  spring, 136, **136, 151**
  watt, 11, **11**
ballistic curve, 111
ballistic pendulum, 349–50, **350**
balloons:
  hot air, 126, 581, 594, **602**, 605, 612, 618, 622
  Raven S-66A, **593**
  research, **622**
Balmer, Johann, 1291
Balmer series, 1291, 1291*t*
banked curve, 186–87, **187**
barometer, mercury, 577, **577**
baryon, 1403, 1404, 1404*t*, 1405, 1406, 1422
baryon number, 1406, 1407
  conservation law for, 1406
base units, 13
bathyscaphe, 589, **589**
battery:
  automobile, 707, 890–91, **891, 892**
  dry cell, 891, **891**
  internal resistance of, 895–96
  lead-acid, 707, 890–91, **891, 892**, 893
Bay of Fundy, 531, **531**, 559–60
beam balance, 136–37, **137**
beam dump, **659**
beat frequency, 518
beats, of a wave, 518
becquere (Bq), 1375
Becquerel, Antoine Henri, 1365, **1366**
Bellatrix, 1263
Bell Laboratories, 1421, **1421**
Bernoulli, Daniel, *585*
Bernoulli's equation, 582–85, 586, 587, 598, 599
beta decay, 1368–70
beta rays, 1365
betatron, 1023
Betelgeuse, **1263**
Bethe, Hans, *1384*
bicycle:
  rounding curve, **456**
  suspended, **431**
  upright, **433**
Big Bang, xliv, 626, 1420–21
Big European Bubble Chamber (BEBC), CERN, 1399
bimetallic strip thermometers, 610, **610**, 636, **637**
binary star system, 297
  resolution of telescope and, 1197

binding energy of nucleus, 1359–65
  curve of, 1361, 1377
binoculars, 1156, **1156**
Biot, Jean Baptiste, *950*
Biot-Savart Law, 948–50, **948, 949**
blackbody, spectral emittance of, 1259
blackbody radiation, 1255–58, 1259–61
black holes, 299
block-and-tackle, 443–44, **444**
blood pressure, 579
blood vessels, **xlvii**
blowhole, 546
blue, 1414–16
body-mass measurement device, 134, **134**, 468, **468**, 478, 482, 490
Bohr, Niels, *1295, 1296*, 1321
Bohr magneton, 976, 1348
Bohr radius, 1297
Bohr's postulates, 1296
boiling points, common substances, 642*t*
Boltzmann, Ludwig, *608*
Boltzmann's constant, 607
bomb, hydrogen, 1380
bomb calorimeter, 250, **250**
bonds, interatomic, 1333
bones as lever, 442, **442**
boom, sonic, 552–53, **552**
Born, Max, 1277, **1277**, 1302
Bose-Einstein condensate, 623
boson, 1403
bottom quark, 1415
boundary conditions, 522, **522**
bound charges, 838
bound orbit, 245
Boyle, Robert, *606*
Boyle's Law, 606
Brackett series, 1292
Brahe, Tycho, *285*
brake, hydraulic, 575–76
brake, power, 456
breeder reactor, 1383
Bremsstrahlung, 1090, 1273–74
Brewster's Law, 1124
bridge, 433, **433**
  thermal expansion and, 637, **637**
bridge collapse:
  at Angers, France, 491, **491**
  at Tacoma Narrows, 523–24, **524**
British system of units, 6–7, 12
British thermal unit (Btu), 630
Brookhaven AGS accelerator, 1250, 1252
Brown Mountain hydroelectric storage plant, 242–43, **242, 243**, 249, 257–58, **258**
bubble chamber, **986**, 1396, 1399–1400, **1400**, 1402, **1409**
bulk modulus, 447–48, 447*t*
bullet:
  impact on block, **350**
  measuring speed of, 356
bungee jumping, 246–47, **246, 247**
buoyant force, 580–81

cable, superconducting, 883
cable capacitance, 843
Cailletet, liquify oxygen, 658
calculus (review), A10–21
  antiderivative in, A14
  approximation of small values in, A18–19
  chain rule for derivatives in, A12
  derivatives in, 38, A10–12, A11*t*
  integral, A12–17, A15*t*
  integration rules in, A15–16
  partial derivatives in, A12

Taylor series in, A18
uncertainties, propagation of, A19–21,
    application to Ohm's Law of, A20–21
Calder, Alexander, mobile, 317
Caledonian Railway wheel set, **454**
California Speedway, **343**
calorie, 250, 630, 631
calorimeter, bomb, 250, **250**
camera, photographic, 1144–45, **1144**
camera, ultrasonic range finder in, 560
Canes Venatici, **1421**
capacitance, 811–16, 829–38
    of earth, 830
    of single conductor, 829
capacitance liquid-level sensor, 852, **852**
capacitive keyboard switch, **850**
capacitive reactance, 1036, **1036**
capacitor microphone, 833, **833**
capacitors, 811–16, 829–38
    circuit with, 1035–38, **1035**
    energy in, 844–47
    guard rings for, 848, **848**
    in loudspeakers, **1037**
    multiplate, 851, **851**, 852, **852**
    in parallel, 834
    parallel-plate, 831–32, 851, **851**, 852, **852**
    in series, 834–35
    two-conductor, **832**
    variable, **832**
Caph, spectrum of, 1288, **1290**
carbon:
    isotopes of, 1355–59, **1356**
    mass of, 1358
Carnot, Sadi, *667*
Carnot cycle, 668–69, 669, 671–73, 675–76
Carnot engine, 667–73
    efficiency of, 671–73
    Second Law of Thermodynamics and, 676
Carnot's theorem, 675–76
carrier, 1410
Cartesian diver, 589, **589**
cathode ray (TV) tube, 965, **966**
    schematic, **752**
Cavendish, Henry, *277*, 698
Cavendish torsion balance, 277, **277**
cavity radiation, 1257
ceiling fan, **367, 368, 371–72**
cell, triple-point, 609, **609**
cello, notes available on, 562
cells (of eye), rods, cones, xlvii
Celsius temperature scale, 611, **612**
Centaurus, xli
center of force, 240
center of mass, 313–23, **320**
    acceleration of, 323
    of continuous mass distribution, 316
    gravitational force acting on, 430–33
    motion of, 323–27
    velocity of, 323–24, 348
centrifugal compressor, 99
centrifugal force, 188–89, **189**
centrifuge, 114, **114**, 365, **365**, 373, **383**
centripetal acceleration, 113–14, **113, 114**, 132,
    184–90, **195**, 371, 372
    Newton's Second Law and, 185
centripetal force for circular motion, 185
centroid, 316
Cerenkov counters, 1399
CERN (Organisation Européenne pour la
    Recherche Nucléaire), 1225, 1238, 1398–99,
    1402, **1402**, 1411, 1425
Cesium atomic clock, 9
Cesium standard of time, 9

cgs system of units, 713
Chadwick, James, 1355
chain reaction, 1378–79, **1378**
Chamonix waterfall, 652
Champlain Canal, 334
changes of state, 642–43
chaos, 492–93
characteristic spectrum, 1274
characteristic time, 1016–17, **1016**
    of RC circuit, 909, **909**
characteristic X rays, 1303
charge, electric, 698–702, 729–30
    bound, 838–39
    bound, in dielectrics, 838–39
    on conductor/insulator, 708–11, **708**
    conservation of, 706–7
    of electron, 696–97, 698
    of electron, measurement of, 747
    of elementary particles, 1403t, 1404t, 1414
    by induction, 711, **711**
    of particles, 696–97, 698, **699**, 706, 741, **741**, 1414
    point, 699, **703, 716–17**, 722–28, **722–23, 725,
        748, 754, 760, 780–81, 811–12, 823**
    of proton, 696–97, 698
    quantization of, 706
    SI unit of, 972
    static equilibrium of, 774–75
    surface, on dielectric, 840–41
    see also electric charge
charge distribution, electric field of, 732, **782**
Charles' Law, 606
charm, of quarks, 1415
chemical elements, Periodic Table of, 1328–32, 1329t
chemical reactions, conservation of charge in, 707–8
Chernobyl, 1383
Chicago (Sears Tower), 653, **654**
chromatic aberration, 1144
chromatic musical scale, 539, **539**
chronometer, 21, **21**
circuit breaker, 903, **903**
circuit, electric:
    AC, 1030–67
    with ammeter, 906, **906, 916**
    with capacitor, 1035–38, **1035**
    DC, 1031, 1032
    frequency filter, 1037
    with inductor, 1038–41, **1038**
    LC, 1041–46, **1041, 1043–44**
    loop method for, 898
    multiloop, 897–900, **898, 899**, 919–20
    RC, 907–12, **907, 911, 912**
    with resistor, **876, 888, 889, 893–97**, 1031–35
    RL, 1015, **1015**, 1018, **1047, 1049**
    RLC, **1044, 1049–50**, 1063
    single-loop, 893–97, 917–918
    with voltmeter, 906, **906, 916**
circular aperture:
    diffraction by, 1196–1999, **1196**
    minimum in diffraction pattern of, 1196
circular motion:
    centripetal force for, 185
    translational speed in, 374
circular orbits, 278–82, **278**, 1321
    angular momentum for, 409
    energy for, 290–91
    in magnetic field, 966
circular polarization, 532, **532**
clarinet, sound wave emitted by, 538
classical electron radius, 1315
classical mechanics, quantum mechanics vs., 1287
Clausius, Rudolph, *678*
Clausius statement of Second Law of
    Thermodynamics, 676

Clausius' theorem, 678
clock:
    Cesium atomic, 9
    grandfather, 219, **219**
    pendulum, 487, **487**, 495, **495**
    synchronization of, 4, **5**, 133n, 1220–23
Coast and Geodetic Survey, U.S., 500
coaxial cable, **800**, 843, **843**
Cockroft-Walton accelerator, **1363**
coefficient of kinetic friction, 175–78, 175t
coefficient of linear thermal expansion, 633, 637
coefficient of performance (heat), 673
coefficient of restitution, 358
coefficient of static friction, 175t, 179–80
coefficient of volume thermal expansion, 634–35
coefficients of friction, 174–81, 175t
coherence of light, 1177
"cold resistance," 869
Collider Detector, Fermilab, 1399, **1399**
colliding beams, 1398–99
collisions, 338–64
    automobile, 339, 355
    impulsive forces and, 339–44
collisions, elastic, 342–47
    conservation of energy in, 344–45, 351–52, 353
    conservation of momentum in, 344–45, 351–52,
        353
    in one dimension, 344–47
    one-dimensional, speeds after, 345–47
    in three dimensions, 351–53
    in two dimensions, 351–53
collisions, inelastic, 348
    conservation of energy in, 351–52, 353
    conservation of momentum in, 351–52, 353
    in three dimensions, 351–53
    totally, 348
    in two dimensions, 351–53
color:
    of quarks, 1414–15
    of visible light, 1092
"color" force, 1405
color-strip thermometer, 610, **610**
Coma Berenices, **xliii**
combination principle, Rydberg-Ritz, 1293
comets, 291
    Hale-Bopp, 299, **299**
    Halley's, **291**, 298, **298**
    perihelion of, 291
    period of, 291
    Shoemaker-Levy, 299
communication satellites, 271–72, 281, 290–91
commutative law of vector addition, 74
compact disc, **367**, 1022, **1168**
compass needle, 927, **927**
Complimentary Principle, 1314
components, of vectors, 77–86, **78**, 95–98, **97, 99,
    101**
    formulas for, 77
compression, 446, 448–49
compressor, centrifugal, 99
Compton, Arthur Holly, 1269, **1269**
Compton effect, 1269–72
Compton wavelength, 1315
concave spherical mirror, 1128, **1130**
Concorde SST, **553**, 557, 563
    sonic boom of, 553
concrete, thermal expansion of, 637
condenser, 833
conducting strap, 714
conduction band, 1338
conduction of heat, 638–42
conductive suit, 805, **805**
conductivity, thermal, 638–41, 639t

conductors, 708, **708**, 709–10, 1337
  in electric field, 774–76, **804**
  insulators vs., 708–11, **708**, 871
  potential energy of, **803**, 812–13, **813**
conservation laws, 205
  for baryon number, 1406
  elementary particles and, 1406–7
  for mass number, 1406
conservation of angular momentum:
  in planetary motion, 284
  in rotational motion, 406–10
conservation of electric charge, 706–7
  in chemical reactions, 707–8
Conservation of Electric Charge, Law of, *710*
conservation of energy, 205, *207*, 223, 235–70,
    290, 797
  in analysis of motion, 223
  general law of, *248*, 249, 252, 662
  in inelastic collision, 351–52, 353
  law of, 790
  in one-dimensional elastic collision, 345
  in rotational motion, 397
  in simple harmonic motion, 483
  in two-dimensional elastic collision, 351–52, 353
  in two-dimensional inelastic collision, 351–52, 353
conservation of mass, 205, 252
conservation of mechanical energy, 238
  equation for, 239
  law of, 221–23, *221*, **222, 223**, 238
conservation of momentum, 307–12, **310**, 345, 348
  in elastic collisions, 344–45, 351–52, 353
  in fields, 723
  in inelastic collisions, 351–52, 353
  law of, 309
conservative electric field, 804–5
conservative force, 236–43, **238**
  gravity as, 288
  potential energy of, 236–43
constant, fine-structure, 1315
constant angular acceleration, equations for, 374
constant force, 205, 208
constant-volume gas thermometer, 609–10, **609**
Constitution, USS, 331
constructive and destructive interference, 517,
    517, 1169
  in Michelson interferometer, 1175–76
  for wave reflected by thin film, 1169–73
contact force, 142–43, **143**
"contact" forces, 697
continuity equation, 570
control rod, in nuclear reactor, 1381
convection, 641
conversion factors, 17–19, 20
conversion of units, 16–17, 18
convex spherical mirror, 1129
cooling, evaporative and laser, 624
Coordinated Universal Time (UTC), 9
coordinate grid, 3–4, **115**
coordinates, Galilean transformation of, 1218,
    1234, 1241
coordinates, origin of, 3, **3, 4, 44, 45, 46, 47**
Copernicus, Nicholas, *279*
copper wire gauges, 881
corner reflector, **1116**
Corona Borealis, 59
corona discharge, 710, **710**
corona wire, 709, **786**
cosecant, A8–10
cosine, 19, 473–74, 486, A8–10
  formula for derivatives of, 473
  law of cosines, A10
cosmic background radiation, 1421
Cosmological Principle, 1419

cosmology, 1416–23
cotangent, A8–10
Coulomb, Charles Augustin de, *700*
coulomb (C), 696–97, 972
Coulomb constant, 699
Coulomb potential, 794–95, **795**, 811
Coulomb's balance, **698**, *700*, 719
Coulomb's Law, 698–99, 702–04, 711–13, 762,
    790, 810, 1074
  in vector notation, 699
Crab Nebula, 935, **935**
crane (tower), *see* K-10000 tower crane
critical angle, for total internal reflection, 1123
cross product, 83–86, **84, 85,** 410–11, 743
  of unit vectors, 85
crossed fields, 987
crystals, atomic arrangement of, **xlix**
Curie, Marie Sklowdowska, *1373*
Curie, Pierre, *1373*
current balance, **971**
current, electric:
  AC, 1031
  AC, hazards of, 913–14, **913, 915**
  in capacitor circuit, 1035–38
  DC, 859–63, 862*t*, 887–925, 1031
  DC, hazards of, 913–14
  displacement, 1073–74
  generating magnetic field, 939–40, **940,** 948, **971**
  in inductor circuit, 1038–41
  "let-go," 913
  SI unit of, 972
  time-dependent, 907–12
current density, 869
current loop:
  potential energy of, 975–76
  right-hand rule for, 942
  torque on, 972–76
current resistor circuit, 1031–35, **1032**
curve:
  ballistic, 111
  banked, 186–87, **187**
  of binding energy, 1361, 1377
  of potential energy, 244–47, **244**
cutoff frequency, 1275
cutoff wavelength, 1274–75
cyclic motion, 469
cyclotron, 964, 967–68, **968,** 979–80
cyclotron emission, 1090
cyclotron frequency, 967
Cygnus X-1, 298
cylindrical symmetry, 767–68

da Costa, Ronaldo, 14
damped harmonic motion, 489, **1045**
damped oscillations, 488–91, **491**
damped oscillator:
  driving force on, 490
  harmonic, 1045, **1045**
  resonance of sympathetic oscillation of, **489,**
    490–91, **491**
dark energy, xlv, 1423
dark matter, 1423
daughter material, 1366
da Vinci, Leonardo, 174–75, *174*
Davisson, C. J., 1303
day:
  mean solar, 9
  sidereal, 294
  solar, 9
DC-3 airplane:
  efficiency, 686
  engines, 267
  take off speed, 266

DC-10 airliner, accident (Orly, France), 591
DC current, 887–925, 1031, 1032
  hazards of, 913–14
  instruments used in measurement of, 905–7
DC voltage, 1004
de Broglie, Louis Victor, Prince, 1302, **1303**
de Broglie wavelength, 1302, 1336
decay constant, 1373
decay of particles, 1365–70
decay rates of radioisotopes, 1374–76
deceleration, 40
decibel, 541–42, 542*t*
dees, 967–68
degree absolute, 604
Deimos, 715
density, 13, 316–17, 566
  of field lines, 739
  of fluid, 566–67, 567*t*
  of nucleus, 1359
depth finder, 557
depth of field, 1145
derivative, of the potential, 806–8
derivatives, rules for, 38
derived unit, 13–14, 20, A20–21, A21*t*
destructive and constructive interference, 517, 1169
  in Michelson interferometer, 1175–76
  for wave reflected by thin film, 1169–73
determinant, 86
deuterium, 498
dialectic strength, 841
diamagnets, 977
diatomic gas, energy of, 617–18
diatomic molecule, **244**
dielectric, 829, 838–47
  electric field in, 838–40
  energy density in, 844–45
  Gauss' Law in, 842
  linear, 838
dielectric constant, 840, 841*t*
dielectric slab, 838, 839, 840, **838–40,** 846, **846**
diffraction, 553–55, 1190, 1190–98
  at a breakwater, 553
  by circular aperture, 1196–99, **1196, 1197,**
    1209–10
  by a single slit, 1190–96, **1191–95,** 1207–09
  of sound waves, 554
  of water waves, 553, **1190**
diffraction pattern, 554, **554**
  of single slit, 1194
diffuse series, 1314
dimensional analysis, 16
dimensionless quantities, 17
dimensions, 16
diodes, 1340–42
dioxyribonucleic acid (DNA), **xlviii**
dip angle, 954
dip needle, 954
dipole, 742–44, 756
  potential energy of, 743
  torque on, 743
dipole moment, 743–44
  magnetic, 973
  permanent and induced, 887–925
Dirac, Paul Adrien Maurice, 1302, **1325, 1326**
direct current, 887–925, 1031, 1032
  hazards of, 913–14
Discoverer II satellite, 297
discus thrower, 109, **109**
dispersion, of light, 1125
displacement current, 1073–74, **1078**
displacement vector, 69, 70–72, **70, 71, 88, 96**
dog, hearing, 558
dominoes, toppling, **525**

donor impurities, semiconductors with, 1339
door (swinging), **367, 396**
Doppler, Christian, *549*
Doppler shift, 547–53, **547, 548,** 561
　of light, 1228–29
dot product:
　in definition of work, 208–9
　of vector components, 82–83, **82,** 86
　of vectors, 81–83, **81, 83,** 86, 208–9
double-well oscillator, 492, **493**
"doubling the angle on the bow," 89, **89**
down quark, 1413–14
drag forces, 180–81
drift velocity, of free-electron gas, 864
driving force, on damped oscillator, 490
dry cell battery, 891, **891**
dumbell, rotation, **411, 412**
dynamics, 29, 130–72
　fluid, 582–87
　of rigid body, 394–428
dynodes, 1267, 1269

Earth, **xxxix–xl,** 285*t,* 286, **286**
　angular momentum of, 409, 427
　capacitance of, 830
　coldest and hottest temperature, 621
　densities and pressures in upper atmosphere, 621*t*
　escape velocity from, 292
　mass distribution within, 390*t*
　moment of inertia of, 388, 389, 390–91
　perihelion of, 295
　polar ice cap melting, 425
　reference frame of, 132
　rotational motion of, **120**
　rotation of, 9, 132, 476
　translational motion of, **120**
earthquake, Tangshan, China, 533
echolocation, 536, **536**
Echo satellites, 1421
Eddington, Arthur, 1276
eddy currents, 1003
efficiency:
　of automobiles, 674
　of Carnot engine, 671–73
　of engines, 666–67
Eiffel Tower, 654
eigenfrequencies, 523, 545, 546
Einstein, Albert, 251, 394, 1217, *1217,* 1264–65, 1361
　photoelectric equation of, 1266–67
elastic body, 182, 445
　elongation of, 445–49
elastic collision, 342–47
　conservation of energy in, 344–45, 351–52, 353
　conservation of momentum in, 344–45, 351–52, 353
　in one dimension, 344–47
　speeds after one-dimensional, 345–47
　in three dimensions, 351–53
　in two dimensions, 351–53
elasticity of materials, 445–49
elastic moduli, some values, 447*t*
elastic potential energy, 236–37, 238
electrical conductivity, 882
electrical measurements, 905–7
electrical outlets, 876–77, 880
electric charge, 698–702, **722, 723, 725,** 729–30
　bound, 838
　bound vs. free, in dielectrics, 838
　conservation of, 706–7
　of electron, 696–97, 698
　of electron, measurement of, 747
　of elementary particles, 1403*t,* 1404*t,* 1414

of particles, 696–97, 698, 706, 1414
　point, 699
　of proton, 696–97, 698
　quantization of, 706
　SI unit of, 972
　static equilibrium of, 774–75
　surface, on dielectric, 840–41
Electric Charge, Law of Conservation of, *710*
electric circuit, *see* circuit, electric
electric constant, 699
electric dipole, 725, 742–44, 756
electric energy:
　in capacitors, 844–47
　of spherical charge distribution, 813
electric energy density in dielectric, 844–45
electric field, 721–55, **722,** 724*t,* **1078**
　of accelerated charge, 1075–79, **1077**
　atmosphere, 806–8
　calculation of, 810
　of charge distribution, 732, **731–32**
　conductors in, 774–76, **774, 776**
　conservative, 804–5
　definition of, 722–23
　in dielectric, 838–40
　electric dipole in, 742–44, **742–744**
　electric force and, 723, 724
　electric quadrupole, 749
　energy density of, 815
　as fifth state of matter, 722
　of flat sheet, 734–36, **734, 736, 750,** 759, **772**
　of harmonic traveling wave, 1098
　induced, 994
　motion in, 740–44
　of plane wave, 1079–80, **1079**
　of point charge, 723–24, **723, 725,** 728
　radial, **807**
　superposition of, 772–73, **773**
　of thundercloud, 725–28, **726–727**
　　charges, **748**
　in uniform wire, 860
electric field lines, 736–39
　made visible, with grass seeds, 859
　of point charge, 736–37, **737,** 739, 936
　sources and sinks of, **738,** 739–40, **763**
electric flux, 757–61, **757–60**
　Ampère's Law and, 1073
electric force, 695–99
　compared to gravitational force, 696, 699
　Coulomb's law for, 698–99
　electric field and, 723, 724
　in nucleus, 1355, 1359
　qualitative summary of, 695–97
　superposition of, 703, **703**
　and xerography, 709
electric fringing field, 848
electric generators, 892
electric ground, 803
electrical conductivity, 882
electricity:
　frictional, 695, 711
　Gauss' Law for, 1074
electric motor, 974, **974**
electric resistance thermometer, 610, **610**
electric shock, 913–14
electrolytes, 709
electromagnet, 946, **947**
electromagnetic flow meter, 996
electromagnetic force, 191, 695, 1405, 1406, 1406*t*
electromagnetic generator, 1000, **1000,** 1003–4, **1003**
　emf of, **1004**
electromagnetic induction, 993–1029
electromagnetic interactions, 1405, 1406, 1406*t*

electromagnetic launcher, **988**
electromagnetic radiation:
　kinds of, 1088–91
　wavelength and frequency bands of, 1090–91, **1091**
electromagnetic wave, 1070–1110, 1071, 1075–76, **1076, 1077**
　energy flux in, 1093–94
　generation of, 1088–91
　momentum of, 1094–96
　right-hand rule for, 1078–79, **1079**
　speed of, 1080
electromagnetic wave pulse, 1080–83, **1082**
electromagnetism, 1411
electromotive force, *see* emf (electromotive force)
electron:
　axis of spin, **976**
　charge of, 696, 698
　distribution of, in atoms, 695
　elliptical orbit of, 1321–22
　free, 708, 866–67
　free of gas, 711, 863–65
　magnetic moment of, 1325, 1355
　mass of, 137, 137*t,* 1356
　measurement of charge of, 748
　neutron and proton vs., 1356
　quantum behavior of, 1302–9, 1320–43
　spin of, 1324–25, 1328–29, 1355–56
　states of, 1327*t*
electron-attracting wire, 709
electron capture, 1390
electron configuration:
　of atoms, 1328–32
　of solids, 1337–39
electron field, 1409–10
electron-holography microscope, 694
electron microscopes, *see* microscope, electron
electrons, of neon, **xlix-1**
electron scanning microscope, 1310, **1310**
electron-volts (eV), 248, 796, 1302, 1359, 1397–98
electrostatic equilibrium, 774
electrostatic force, 695–97
electrostatic induction, 711
electrostatic potential, 790–98, **791, 792, 795, 800,** 810
　calculation of, 798–803, **799**
electrostatic precipitators, 735, **735**
electrostatic shielding, 804, **804**
electroweak force, 1411
elektron, 695
elementary particles, 1396–1430
　collisions between, 342–44
　conservation laws and, 1406–7
　electric charges of, 1403*t,* 1404*t,* 1414
　masses of, 1403*t,* 1404, 1404*t*
　spins of, 1403*t,* 1404*t*
elements, chemical:
　age of, 1421
　during Big Bang, 1421
　transmutation of, 1363
elephant, rumble (communication), 558
elevation angle, **109,** 111, **111**
elevator with counter weight, 154–55, **154, 155,** 157
ellipse, **xxxix,** 283
　major axis of, 282
　semimajor axis of, 282, 285, 291
elliptical orbits, 282–86, **282**
　angular momentum in, 291–92
　of electron, 1321–22
　energy in, 291–92
　vs. parabolic orbit for projectile, 287
elongation, 445–49, **445**

emf (electromotive force), 887–93, **889**, 917, 921
    alternating, 1004, 1032–33, 1035, 1046–53
    induced, 994, 998–1007
    motional, 993–97
    power delivered by, 901
    in primary and secondary circuits of trans-
        former, 1053–54
    sources of, 890–92
    steady, 1004
emf, alternating, 1004, **1031**, 1032–33, **1032**,
    1035, 1046–53, **1047**
    with capacitor, **1035–37**
    with inductor, **1038–39**
emission, stimulated, 1090
Empire State Building, 63, 227
energy, 204–23
    alternative units for, 248–49
    in capacitor, 844–47
    for circular orbit, 290–91
    conservation of mechanical, 221–23, *221, 222,*
        223, 238, 239
    dark, xlv, 1423
    of diatomic gas, 617–18
    electric, *see* electric energy
    in elliptical orbit, 291–92
    equivalent to atomic mass unit, 1361
    expenditure (U.S.), 1250
    gravitational potential, 218–23, **219, 220,** 238,
        288–93
    gravitational potential, of a body, 321
    of ideal gas, 616–19
    internal, 616
    kinetic, *see* kinetic energy
    law of conservation of, 790
    in LC circuits, 1042
    Lorentz transformations for, 1234–37
    magnetic, 1012
    magnetic, in inductor, 1013–15
    and mass, 1242–44
    mass and, 251–53
    mechanical, *see* mechanical energy
    momentum and, relativistic transformation for,
        1242–44
    in orbital motion, 288–93
    of photon, 1264–65
    of point charge, 796
    potential, *see* potential energy
    rate of dissipation of, 264
    rest-mass, 1242
    rotational, 1334
    of rotational motion of gas, 617–18
    sample values of some energies, 249*t*
    in simple harmonic motion, 480–83
    of stationary states of hydrogen, 1299
    of system of particles, 327–28
    thermal, 248, 616, 629
    threshold, 355
    total relativistic, 1243
    vibrational, 1333
    in wave, 1092–96
    zero-point, 1307
energy, conservation of, 205, *207,* 223, 235–70,
    290, 797
    in analysis of motion, 223
    general law of, *248, 249,* 252, 662
    in inelastic collision, 351–52, 353
    in one-dimensional elastic collision, 345
    in rotational motion, 397
    in simple harmonic motion, 483
    in three-dimensional elastic collision, 351–52,
        353
    in two-dimensional elastic collision, 351–52, 353
energy bands, 1337

energy banks, in solids, 1336–40
energy density:
    in dielectric, 844–45
    in electric field, 815
    in magnetic field, 1014
energy flux, in electromagnetic wave, 1093–94
energy level, 244
    of hydrogen, 1299
    in molecules, 1333–36
energy-level diagram, 1298–99, **1299,** 1334,
    1334–35, 1337, *1337*
energy quantization, of oscillator, 1259–60
energy quantum, 1258–63
energy-work theorem, 215, 236, 400
engine:
    automobile, **665,** 674
    Carnot, 667–73
    efficiency of, 666–67
    flowchart, **667**
    heat, 665
    steam, **665,** 671, **671**
enriched uranium, 1381
entropy, 678
    at absolute zero, 680
    change, in isothermal expansion of gas, 668
    disorder and, 680
    irreversible process, 679–80
    negative, 683
equation of motion, 151–53, **151,** 174
    integration of, 54–56
    of simple harmonic oscillator, 477
    of simple pendulum, 485
    see also Newton's Second Law
equilibrium:
    electrostatic, 774
    of fluid, 575
    of mass, 155
    neutral, 432, **432**
    static, *see* static equilibrium
    unstable, **432**
equilibrium point, 245, **245**
equilibrium position, 476–77
equipartition theorem, 617–18
equipotential surface, 808–9, **809**
escape velocity, 292
    from Earth, 292
    from Sun, 292
Escher, M. C., waterfall, **662**
ether, 1218–19, **1219**
ether wind, 1219
evaporative cooling, 624
evaporative loss, Mediterranean, 657
excited states, 1299
Exclusion Principle, 1321, 1328–32, 1337
expansion, free, of a gas, 664, 668
expansion, thermal:
    of concrete, 637
    linear, coefficient of, 634, 637
    of solids and liquids, 633–37, **633**
    of water, 635, **635**
expansion, volume, 634
expansion joints, bridge, **636**
expansion of railroad rails, 636
Explorer I, **286,** 298–99
Explorer III, **286**
Explorer X, 298
exponential function, 910
external field, 742
external forces, 311
eye:
    astigmatic, 1147, **1147**
    compound, 1202, **1202**
    insect, 1202, **1202**

    myopic and hyperopic, 1146
    nearsighted and farsighted, 1146, **1146**
    as optical instrument, 1145–47, **1145**
eye, components of:
    cone cells, **xlvii**
    iris, **xlvi**
    retina, **xlvii**
    rod cells, **xlvii**

Fabry-Perot interferometer, 1204
Fahrenheit temperature scale, 611, **612**
fallout, from nuclear accident, 1383
farad (F), 830
Faraday, Michael, 994, *998*
Faraday cage, 804–5, **804**
Faraday's constant, 715
Faraday's Law, 997–1001, **999,** 1005, 1006, 1008,
    1009, *1009,* 1071, **1071,** 1080, 1096–97, **1097**
farsightedness, 1146, **1146**
Fermi, Enrico, 1358, *1380*
fermi (fm), 1358
Fermilab, 125, 422
Fermi National Accelerator Laboratory
    (Fermilab), 1250, 1398–99, **1398,** 1415
    Collider Detector, 1399, **1399**
    Tevatron, 1398–99, **1398,** 1425
fermion, 1403
Ferris, George, 388, 421
Ferris wheel, **388,** 421, 527
ferromagnetic materials, 947
ferromagnets, 977
Feynman, Richard P., 1410, **1410**
Feynman diagram, 1410, **1410, 1427**
fibrillation, 913
field, 722
    depth of, 1145
    electric, *see* electric field
    electron, 1409–10
    as fifth state of matter, 722
    magnetic, *see* magnetic field
    quanta of, 1409–11, 1409*t*
    radiation, of accelerated charge, 1075–76, **1076,**
        1077
field lines, 736–39
    density of, 739
    of point charge, 736–37, **737,** 739
    sources and sinks of, 738–39
fifth state of matter, 722
filter, polarizing, 1084–86, **1084, 1086**
fine-structure constant, 1315
fire extinquisher, **595**
fire hose:
    pressure within, 596
    rate of water flow, 590*t*
First Law of Thermodynamics, 662–64
first overtone, 522, **522**
fission, 252, 1355, 1377–79
    in alpha-decay reaction, 1366, **1366**
    as a source of nuclear energy, **818,** 1242
Fizeau effect, 1250
flat sheet, electric field of, 734–36, 759
flip coil, 1019
flow:
    around an airfoil, 570, 582–83
    of free electrons, 866–67
    of heat, 638–39, **638**
    incompressible, 569
    laminar, 569
    methods for visualizing, 570
    in a nozzle, 586
    from source to sink, 570
    steady, 569
    streamline, 569

turbulent, 571, **571**
velocity of, 566, 568
around a wing, 571, 584
flowmeter, Venturi, 585–86, 596, **596**
fluid, 566
density of, 566–67, **567**
equilibrium of, 575
incompressible, 569
nuclear, 1360
static, 575–79, **575**
fluid dynamics, 582–87
fluid mechanics, 565–99
flute, 519, 546, 560, 561
flux:
electric, 757–61
magnetic, 997–1001, 999
f number, 1144, 1152
focal length, of spherical mirror, 1128
focal point:
of lenses, 1136
of mirror, 1128
foot, 6–7
force, 133, **133**, 135, 135t, **136**
buoyant, 580–81
calculated from potential energy, 241
center of, 240
centrifugal, 188–89, **189**
"color," 1405, 1414–15
conservative, 236–43, **238**, 288
conservative vs. nonconservative, 238–39
contact, 142–43, **143**, 697
as derivative of potential energy, 241
electric, see electric force
electromagnetic, 191, 695, 1405, 1406, 1406t
electromotive, see emf (electromotive force)
electrostatic, 695–97
electroweak, 1411
external, 311
fundamental, 191
gravitational, see gravitational force
impulsive, 339–44
internal and external, 311
inverse-square, 240
moment arm of, 400
motion with constant, 151–59
net, 138–40, **138**, **139**
normal, 143, **143**, **144**, **147**
nuclear, 1355
power delivered by, 255
between quarks, 1414–15
repulsive, **705**
restoring, 182–84, **183**
resultant, 138
of a spring, 182–84, **183**, **184**
"strong," 191, 1355, 1359–65, 1405, 1406,
1406t, 1414
torque and, 395–97
units of, 135–36, 141
"weak," 191, 1369, 1405, 1406, 1406t
work done by constant, 205, 208
work done by variable, 211–13, **211**, **212**, **213**,
214
force, magnetic, 695, 697, 927–31, **927**
due to magnetic field, 931
magnitude of, 933
on moving point charge, 928–30
right-hand rule for, 934, **935**, 936
vector, 933
on wire, 969–72
forced oscillations, 488, 490–91, 1046
Fornax constellation, xliv
Foucault pendulum, 499
four-wire measurement, 921

Fourier's theorem, 519, **519**, 538
fractal striations, 492, **493**
frames of reference, 3, 4, 20, 114, 115
in calculation of work, 208
of Earth, 132
freely falling, 142, **142**
inertial, 132, **132**, 133, 1219
for rotational motion, 366
Franklin, Benjamin, 710, **710**
Fraunhofer diffraction pattern, 1194
Fraunhofer Lines, 1290
"free-body" diagram, 146, **146**, **147**, **148**, **153**,
157, 158, **159**, 176, **180**, **181**, **184**, **186**, **187**,
188, **702**
for automobile tire, **403**
for backbone as lever, **442**
for box on truck, **440**
for box titled, **439**
for bridge, **433**
for ladder, **438**
for pulley, **402**, **443**
for string-bob system, **484**
for tower crane, **435–36**
free charges, 844
free-electron gas, 711
friction in, 863–65
free electrons, 708
flow of, 866–67
free expansion of a gas, 664, **689**
entropy change in, 668
free fall, 49–54, 51, **52**, **53**, **64**, 141, 142, **142**,
495–96
formulas for, 49, 50
universality of, 49, **49**
weightlessness in, 142, **142**
French Academy, speed of sound, 560
French horn, 546
freon, 672–73
frequency, 369–70
beat, 518
normal, 523
proper, 523
resonant, 1043
of simple harmonic motion, 470–71
threshold, 1266
of wave, 510–11
frequency bands, of electromagnetic radiation,
1090–91
frequency filter circuit, 1037
Fresnel, Augustin, 1191
Fresnel lens, 1156, **1156**
friction, 171–81
air, 49, 51, **61**, 181, **181**
coefficient of kinetic, 175–78, 175t
coefficients of, 174–81, 175t
of drag forces, 180–81
equation for kinetic, 175
equation for static, 179
in flow of free–electron gas, 863–65
heat produced by, 248
kinetic (sliding), 174–78, **175**, **176**, **177**, 190
loss of mechanical energy by, 238–39
microscopic and macroscopic area of contact
and, 174–75
as nonconservative force, 238–39
static, 178–80, **179**, **180**, 190
static, coefficient of, 175t, 179–80
of viscous forces, 181
frictional electricity, 695, 711
fringes, 1170
fringing field, 792, **792**, 848
fuel cells, 892–93, **892**
on Skylab, 892, **892**

fuel consumption, 263t
fuel rods, of nuclear reactor, 1381
function, oscillating, 1047, 1051
fundamental forces, 191
strength of, 191
fundamental frequency, 523
fundamental mode, 522, **522**
fuse, 903, **903**
fusion:
heat of, 642, 642t
nuclear, 1421

g, 40, 52–53, 189
measurement of, 52–53, 277–78
NASA centrifuge, 592
standard, 52–53
Gagarin, Yuri, 299
gain factor, 1343
galaxies, 1417–18
Milky Way, 1417, **1417**
Galilean addition law for velocity, 1218
Galilean coordinate transformations, 1218, 1234,
1241
Galilean telescope, 1142
Galilean transformation:
for momentum, 1239
for velocity, 1218
Galilean velocity transformation, 116
Galilei, Galileo, 51, 131, 495
claim on isochronous pendulum, 495, 501
experiment on free fall, 495
experiments on universality of free fall by, 495
isochronism of pendulum and, 501
pendulum experiments by, 495
tide theory of, 120
gallium arsenide (GaAs), 1287
Galvani, Luigi, 793
galvanometers, 975, **975**
gamma emission, 1369–70
gamma rays, xlix, 1090, 1365, 1421
gas:
adiabatic equation for, 649
diatomic, energy of, 617, **617**, **620**, 625
diatomic, molar heat, 645, 646t
distribution of molecular speeds in, 615
energy of ideal, 616–19
entropy change in isothermal expansion of, 668
of free electrons, 711, 863–65
free expansion of, 664, **664**, 668
ideal, see ideal gas
Law of Boyle for, 606
Law of Charles and Gay-Lussac for, 606
monatomic, energy of, 616–18, **620**
monatomic, molar heat, 645, 646t
polyatomic and non-linear, 618, **620**
polyatomic and non-linear, molar heat, 646t
root-mean-square speed of, 614–15
specific heat of, 644–47
gas constant, universal, 604
gas thermometer, constant-volume, 609–10, **609**
gauge, pressure, 578
gauge blocks, 6, **6**
Gauss, Karl Friedrich, 763
gauss (G), 934
Gaussian surface, 762, **765**, **767**, 769–71, 774–75,
779, **781**, **786**
Gauss' Law, 274, 757, 762–72, 790, **938**, 1422
in dielectrics, 842
for electricity, 1074
for magnetic field, 937, **937**, 940
for magnetism, 1074
Gay-Lussac's Law, 606
Gedankenexperiment, 287

Geiger counter, 782, 821
Geiger, H., 1294
Gell-Mann, Murray, 1413
General Relativity theory, 394
general wavefunction, 511
generator:
    electric, 892
    electromagnetic, 1003–4
    homopolar, 1004–5
geometric optics, 1112
geometry (review), A3–7
    angles, A3
    areas, perimeters, volumes, A3
geostationary orbit, 271–72, 281
geostationary satellite, **280**, 281, 290–91
geosynchronous orbit, 271–72
geothermal power plant (Waiakei, NZ), **690**
Germer, L., 1303
GeV, 1398–99, 1411
gimbals, 414
Glashow, Sheldon Lee, 1411
Global Positioning System (GPS), 25, 1216, **1216**, 1220, 1227, 1228
gluon, 1409, 1411
golf ball, club impact, **347**
Goudsmit, Samuel, 1325
gradient, of the potential, 806–10
grandfather clock, 219, **219**
grand unified theory (GUT), 1411
graphite, xlix
grating, 1184–85, **1184**
    diffraction, 1184, **1184**
    principal maximum of, 1184–85, **1184**
    reflection, 1185, 1207
    resolving power of, 1186
gravitation, 271–303
    law of universal, *131*, 272–76, 278
gravitational constant, 273
    measurement of, 277–78
gravitational force, 191, 272–76, 1405, 1406, 1406*t*
    acting on center of mass, 430–33
    compared to electric force, 696, 699
gravitational interactions, 1405, 1406, 1406*t*
gravitational potential energy, 218–23, **219**, **220**, 238, 288–93
    of a body, 321
graviton, 1409, 1411, 1415
gravity:
    action and reaction and, 146, **146**
    as conservative force, 288
    of Earth, *see* weight
    galaxies and, 1420
    work done by, 207, **207**
gravity, acceleration of, 52–53, **64**, 274–75
    measurement of, 52–53
    variation of, with altitude, 274–75
Gravity Probe B satellite, 394, **394**, 401, **401**, 406, 414
gray, 1375
Great Pyramid (Giza), 318, 334
Greek alphabet, A1
green, 1414–16
Greenwich time, 9
Griffiths-Joyner, Florence, **59**
ground, electric, 803, **803**
ground state, 1299
guard rings, for capacitors, 848, **848**
Guericke, Otto von, 161, 591
guitar, 523, **526**, 530, 531
    lowest note available, 558
gyrocompass, 414
gyroscope, 394, **394**, **401**, 406, 414, **414**
    precession of, 415–16

hadron, 1403
Hahn, Otto, 1377
Hale-Bopp comet, 299, **299**
half-life, 1372–73
Hall, Edwin Herbert, *981*
Hall coefficient, 990
Hall effect, 980–83, **981**, **983**
Hall microscope, **982**
Halley's comet, 298, **298**
Hall sensors, 982, **982**
Hall voltage, 981
hammer in free fall, **366**
harmonic function, 470
harmonic motion, damped, 489
    see also simple harmonic motion
harmonic oscillator, 477, 492, 1317
harmonic wave, 510–13, 1098
harmonic wavefunction, 513–16
Hawking radiation, 1281
hearing:
    dog, 558
    threshold of, 538
hearing trumpet, 559
heat, 248, 628–60
    as energy transfer, 662–63
    of fusion, 642, 642*t*
    latent, 642
    mechanical equivalent of, 631–32
    specific, *see* specific heat
    temperature changes and, 630–31
    transfer, by convection, 641
    transfer, by radiation, 641
    transfer; *see also* R value
    of transformation, 642
    of vaporization, 642, 642*t*
heat capacity, specific, 630
heat conduction, 638–42
    equation of, 638–39
    two-layer, 640, **640**
heat current, 638
heat engine, **655**, 665
heat flow, 638–39, **638**
    across electric insulator, 656
heat pump, 672, **672**, **673**
heat reservoir, 665
heating element, **916**
height, maximum, of projectiles, 109–11
Heisenberg, Werner, 1278, **1278**, 1302
Heisenberg's uncertainty relation, 1278
heliocentric system, *279*
helium:
    ion collision with O2, **352**
    and hydrogen, abundance of in universe, 1421
    monatomic kinetic energy, 616
helium liquid, 624
helium nuclei, 1366
Helmholtz, Hermann von, *248*
Helmholtz coil, 960, **960**
henry, 1010
Henry, Joseph, *1011*
hermetic chamber, **626**
Hertz, Heinrich, 1080, *1080*, 1265
hertz (Hz), 471
high-pass filter, 1063
high-voltage transmission line, power dissipated in, 904
Hobby-Eberly telescope, 1152, **1152**
"holes," 982
    in semiconductors, 1338–39, 1340–42
homopolar generator, 1004–5, **1005**
Hooke's law, 182–84, 445–46, 476
Hoover Dam, 993

horizontal velocity, 103–8, **103**, **106**, **107**, **108**
horse, work efficiency, 689, **689**
horsepower (hp), 254, *254*, **256**
house walls, heat flow in, 639–40, **652**, 659
Hubble, Edwin, 1416–17, 1418
Hubble's Law, 1418–19, 1420
Hubble Space Telescope, 1111, 1130, 1135, 1152, 1198, **1198**, *1198*, 1209
human:
    aorta, average blood flow, 590
    chemical energy conversion, 685–86
    circulatory system, 571, **571**
    ear, audible frequencies, 558
    heart pressure, 591
    muscle striation, 685
    power ed bicycling, 666, **666**
    powered flight (over English Channel), 264
    venous presssure, 597, **597**
    voice, pitch, 559
human body:
    lever-like motion of, 442, 451, 458–59
    mean tissue density, 598, **626**
    speed of sound in, 558
    temperature, 612
    ultrasound frequencies used in, 558, 562
hurricane, barometric pressure, 598
Huygens, Christiaan, *221*, 495
Huygens' Construction, 1113–14, **1113**
Huygens-Fresnel Principle, 1190–91
Huygens' tilted pendulum, 495, **495**
Huygens' wavelets, 1113, **1120**
Hyades Cluster, **xliii**
Hyatt Regency hotel, collapse of "skywalks" at, 451, **451**
hydraulic brake, 575–76, **576**
hydraulic car jack, **576**
hydraulic press, 575, **575**
hydroelectric pumped storage, *see* Brown Mountain
hydrogen:
    atom, stationary states of, 1298, 1300
    atomic mass, xlix
    electron orbit, 423
    energy levels of, 1300
    and helium, abundance of in universe, 1421
    interstellar gas, density and temperature, 621
    isotopes of, 1357
    quantum numbers for stationary states of, 1328
    spectral series of, 1291–93, 1299, 1324
hydrogen bomb, 1380
hydrogen molecule:
    diatomic gas, 625
    oscillations of, 483, **483**
    vibration frequency, 498
hydrogen spectrum, **1286**, 1291–92, 1291*t*
    produced by grating, **1184**
hydrometer, 594
hyperbola, 291
hyperbolic orbit, 291
hyperopic eye, 1146

ice:
    density of, 581, **581**
    density v. liquid water, **634**
iceberg, 593, **593**
ideal gas, 604
    energy of, 616–19
    kinetic theory and, 602–27
Ideal Gas Law, 603–8, 627, 646, 670
ideal-gas temperature scale, 609–12
ideal particle, 3, 20
ideal solenoid, 943–44, **943**

image:
  real, 1133, **1133**
  virtual, 1116, **1116, 1132, 1134, 1139–40**
image charges, 727, **727**
image orthicon, 1269
impact parameter, 1294, **1294**
impedance, for RLC circuit, 1050
impulse, 339–40
impulsive force, 339–44
impurities, acceptor, 1339
incidence, angle of, 1115–16, **1115**
inclined plane, **153, 157**
incompressible flow, 569
incompressible fluid, 569, 576
indeterminacy relation, 1278
index of refraction, 1118, 1119t, 1125–26, **1125**
induced dipole moments, 744
induced electric field, 994, 996, **1006**
induced emf, 994, **996,** 998–1007
induced magnetic field, 1071
  in capacitor, **1072,** 1082
inductance, 1008–12, **1009**
  mutual, 1009–10
  self-, 1011
induction:
  electromagnetic, 993–1029
  electrostatic, 711
  Faraday's Law of, 997–1001
induction furnace, 1024
induction microphone, 1000–1001, **1000**
inductive reactance, 1039–40, **1039**
inductor:
  circuit with, 1038–41, **1038**
  current in, 1038–41, **1041**
inelastic collision, 348
  conservation of energy in, 351–52, 353
  conservation of momentum in, 351–52, 353
  in three dimensions, 351–53
  totally, 348
  in two dimensions, 351–53
inertia:
  law of, 132, **132**
  moment of, *see* moment of inertia
inertial reference frames, 132, **132,** 133
infrared radiation, 1090
initial speed, 111
instantaneous acceleration, 40–41, **41**
  components of, 97
  in two dimensions, 96–97
instantaneous angular acceleration, 370
instantaneous angular velocity, 369
instantaneous configuration, **820, 822**
instantaneous power, 253
instantaneous velocity, 35–39, **36, 61**
  components of, 96
  as derivative, 38
  formulas for, 37, 38
  graphical method for, 37
  numerical method for, 37–39
  as slope, 35–37
  in two dimensions, 96, **96**
instantaneous velocity vector, 98, **98**
insulating material, *see* dielectric
insulators, 708
  conductors vs., 871
  electron configurations of, 1338
  resistivities of, 871
integrals, for work, 212–13
integrated circuit, 1345
integration, of equations of motion, 54–56
INTELSAT, 281, **281**
intensity:
  of sound waves, 538, 540–43, 542t, **543**

for multiple slit interference pattern, 1189, **1189**
for two-slit interference pattern, 1179, **1182, 1195**
interaction:
  "color," 1405
  electromagnetic, 1405, 1406, 1406t
  gravitational, 1405, 1406, 1406t
  "strong," 1405, 1406, 1406t
  "weak," 1369, 1405, 1406, 1406t
interatomic bonds, 1333
interference, 1169–86
  constructive and destructive, 517, 1169, 1169–75
  maxima and minima, for multiple slits, 1184, 1187–89
  maxima and minima for, 1179, **1179**
  from multiple slits, 1183–89, **1183,** 1206–07
  in thin films, 1169–73, **1173,** 1203–04
  two-slit, pattern for, 1177, **1177,** 1180, **1180**
  from two slits, 1177–83, **1178, 1181,** 1205–06
  principal and secondary maxima, 1183, **1183**
interferometer, Michelson, 1174–77, **1175**
interferometry, very-long-baseline, 1202
internal energy, 616
internal forces, 208, 311
internal kinetic energy, 348
internal resistance, of batteries, 895–96
International Bureau of Weights and Measures, 1175, **1175**
International Space Station, 389, 468, **1214**
international standard meter bar, 5–6, **5**
International System of Units (SI) 5, 14, 972
  see also system of units (SI)
International Thermonuclear Experimental Reactor, 1385
interstellar hydrogen gas, density and temperature, 621
invariance of speed of light, 1220–21
inverse Lorentz transformation equations, 1236, 1249
inverse-square force, 240
inverse-square law, 739
Io, 64, **64**
ion gun, 740–41, **740**
ionosphere, temperature and density, 620
ions, 697
  in electrolytes, 711
  in salt crystal, **749**
iris, of eye, **xlvi**
irreversible process, 678
  entropy change in, 679–80
isobars, 1389
isochronism, 477, 501
  of simple pendulum, limitations of, 486
isospin, 1407
isothermal compression of gas, **689**
isothermal expansion of gas, entropy change in, 668
isotopes, 1355–59
  of carbon, 1355–59, **1356**
  chart of, 1357t
  of hydrogen, 1357
  radioactive, 1372–76
  of uranium, 1362, 1383

J/Ψ meson, 1415
Jodrell Bank radio telescope, 1209
Jordan, P., 1302
Joule, James Prescott, *207, 629,* 652
joule (J), 206, 254
Joule heat, 902–5, **903,** 920–21
Joule's experiment and apparatus, 631–32, **631**
junction, **897**
Jupiter, 285t, 286
  moons Europa, Ganymede, Io, 296

K-10000 tower crane, 429, **429,** 435–37, **435–37,** 448, **448**
Keck Telescope, 1151
kelvin, 604
Kelvin, William Thomson, Lord, 458, *604*
Kelvin-Planck statement of Second Law of Thermodynamics, 675–76
Kelvin temperature scale, 604, 609–10
Kepler, Johannes, *285*
Kepler's Laws, 282–86
  of areas, 282–84
  First, 282
  limitations of, 288
  for motion of moons and satellites, 286–88
  Second, 282–84
  Third, 285–86
kilocalorie, 248–49, 250
kilogram, 5, 11, 13, 14
  multiples and submultiples of, 13t, 134
  standard, **11,** 134
kilometers per hour (km/h), 30
kilowatt-hours, 248, 250
kinematics, 29
kinetic energy, 214–17, **216,** 238
  equation for, 215
  of ideal monatomic gas, 616–18
  internal, 348
  relative examples of, 216t
  relativistic, 1240–41
  of rotation, 378–84
  in simple harmonic motion, 480–83
  of a system of particles, 327–28
kinetic friction, 174–78, **175, 176, 177,** 190
  coefficient of, 175–78, 175t
  equation for, 175
kinetic pressure, 613–16
kinetic theory, ideal gas and, 602–27
Kirchhoff, Gustav Robert, 894
Kirchhoff's current rule, 897–900
Kirchhoff's rule, 1031, 1032, 1038, 1044, 1046, 1054
Kirchoff's voltage rule, 894, 898, 900, 908
Knot meter, **568**
krypton, monatomic kinetic energy, 616

Lagrange, Joseph Louis, Comte, *236*
lamda particle, 1400, **1400**
laminar flow, 569
La Paz, Bolivia, airport, 620
Large Hadron Collider (CERN), 1250, 1425
Large Magellanic Cloud (galaxy), xliv
laser, 1090
  stabilized, 6, **6**
laser cooling, 624
Laser Interferometer Gravitational Observatory (LIGO), 1175
laser light, 1178
laser printers, 694, 709
laser range finder, **1104**
latent heat, 642
launch speed, 111
Lave, Max von, 1273
Lave spots, 1273
law of areas, 283–84
Law of Boyle, 606
Law of Charles and Gay-Lussac, 606
law of conservation of angular momentum, 407–9
Law of Conservation of Electric Charge, *710*
Law of Conservation of Energy, 662, 790
Law of Conservation of Energy, general, *248,* 249, 252
law of conservation of mechanical energy, 221–23, *221, ***222, 223,** 238, 584
law of conservation of momentum, 309

law of inertia, 132, **132**
Law of Malus, 1086, **1086–87**
law of radioactive decay, 1372–76
law of reflection, 1114–15
law of refraction, 1121, **1121**
law of universal gravitation, *131*, 272–76, 278
Lawrence, Ernest Orlando, 968
Lawrence Livermore National Lab (NIF), 828, 1385
laws of planetary motion, 282–86
Lawson's criterion, 1391
LC circuit, 1041–46
    energy in, 1042
    freely oscillating, 1041–46
    natural frequency of, 1042
lead-acid battery, 707, **707**, 890–91, **891**, 893
leaf spring oscillator, **492**
length, 5–8
    precision of measurement of, 6
    standard of, 5–6, **5, 6**
length contraction, 1230–32, **1230–31**, 1248–49
    visual appearance and, 1230–32
Lennard-Jones potential, 262, 263
lens:
    focal point of, 1135, 1136, **1135–36**
    magnification produced by, 1139
    objective, 1149–50
    ocular, 1149–50
    radius of curvature, **1136**
    thin, 1135–39, 1161–63
lens-maker's formula, 1135–36, 1162
Lenz' Law, 1001–2, **1001**, 1006
lepton, 1403, 1403*t*, 1406
lepton number, 1406, 1407, 1426
"let-go" current, 913
Leverrier, U. J. J., 272
levers, 441–45, **445**, 458–59
    human bones acting as, 442, **442**, 451, 458–59
    mechanical advantage and, 441, **441**, 444
Leyden jar, 856, **856**
Lichtenstein, R., 294
lift, 584
light:
    coherent, 1177, **1178**
    dispersion of, 1125
    Doppler shift of, 1228–29
    from laser, 1178
    polarized, **1125**, 1126–27
    pressure of, 1094–95
    quanta of, 1254–55
    reflection of, 1112, 1114–17, **1114–15**, 1156–57
    refraction of, 1112, 1117–27, **1119–22**, 1157–60
    spectral lines of, 1127
    spectrum of, 1126, 1184, 1291–93
    ultraviolet, 1090
    unpolarized, 1084, **1084**
    visible, 1090–91, **1092**
light, speed of, 6, 1218–20, 1247
    in air and water, 527
    invariance of, 1220–21
    in material medium, 1118
    measurement of, 1080
    universality of, 1220–21, 1238
light-emitting diode (LED), 1343–44
lightning, **710, 721, 805, 862**
    distance of, 560
lightning rod, 710–11, **711**
light waves, 1079–1110
    coherent, 1090, **1091**
    incoherent, 1090, **1091**
light-year, xlv, 24
linear dielectric, 838
linear magnification, 1141–42, **1142**

linear polarization, 532, **532**
lines of electric field, 736–39
    made visible, 859, **859**
liquid drop model of nucleus, 1360
liquids:
    bulk moduli for, 447–48
    thermal expansion of, 633–37, **633**
Lloyd's mirror, 1206
locomotive steam engine, **661**, 666, 671
Loki (volcano on Io), **64**
Long Island, xxxviii
longitudinal wave, **508**, 509, **509**
long wave, 1088
looping the loop, 187–88, **187**
loop method, for circuits, 898
Lorentz, Hendrik Antoon, *1235*
Lorentz force, 933
Lorentz transformations, 1232–38
    for momentum and energy, 1234–37
low-pass filter, 1063
Lyman series, 1292

Mach, Ernst, *552*
Mach cone, 552, **552**
Mach number, 562
macroscopic and microscopic parameters, 603
Magdeburg hemispheres, 161, **161**
magic nuclei, 1389
magnet, permanent, 977, **977, 1008–09**
magnetic constant, 929
magnetic dipole moment, 973
magnetic energy, 1012
    in inductor, 1013–15
magnetic field, 931–38, 935*t*
    at the center of a circular ring of current, **949**
    circular motion in, 965–69
    circular orbit in, 966
    closed path, **939**
    in Crab Nebula, **935**
    of the Earth, **942**
    energy density in, 1014
    Gauss' Law for, 937
    generated by current, 939–40, **940**, 948
    induced, in capacitor, 1072, **1072**
    made visible with iron filings, **937, 942**
    magnetic force and, 931
    MRI magnet, **935**
    near household wiring, **935**
    of plane wave, 1081–83, **1082, 1097**
    of point charge, 936
    of point charge, represented by field lines, 936–37
    principle of superposition for, 938
    in relation to current, **939**
    right-hand rule for, 932, **932–33, 935–36**
    of solenoid, 943–44
    of straight wire, 940–41, **941**
    in sunspot, **935**
magnetic field lines, **937, 938**
    of circular loop, **941, 942, 942**
    made visible with iron filings, 942, **942, 943**
    as a result of superposition, 953, **953**
magnetic flux, 997–1001, **997–98**, 999, **1000**
magnetic force, 695, 697, 927–31, **927**
    magnetic field and, 931*n*
    magnitude of, 933
    on moving point charge, 928–30
    in relation to current, **930, 931, 932**
    right-hand rule for, 934, **935, 936, 994**
    vector, **929**, 933
    on wire, 969–72, **969–71**
magnetic levitation, 1008–9
magnetic moment, 1355

of electron, 1325
    right-hand rule for, 974, **974**
    spin, 976
magnetic monopole, 953
magnetic permeability, 977
magnetic quantum number, 1321, 1328
magnetic recording media, 978–79, **978–79**
magnetic resonance imaging (MRI), 656, 977, **977, 1022**
magnetism, Gauss' Law for, 1074
magnetization, 977, 984, 989
magneto, 1019
magneton, 1348
    Bohr magneton, 976, 1348
    nuclear magneton, 1348
magnification, 1139
magnifier, 1147–49
major axis of ellipse, 282
Malus, Étienne, *1086*
Malus, law of, 1086
mandolin, 546
    frequencies, 530, 531, 558
manometer, 578, **578**
marker point, 29, **29**
Mars, **xl**, 285*t*, 286, **286**
Mars Climate Orbiter, 16, **16**
Marsden, E., 1294
mass, 11–13
    atomic, 11–12
    atomic standard of, 11
    of carbon, 1358
    center of, *see* center of mass
    conservation of, 205, 252
    definition of, 134–35
    of electron, 137, 137*t*, 1356
    of elementary particles, 1403*t*, 1404*t*
    and energy, 1242–44
    energy and, 251–53
    equilibrium of, 155
    molecular, 11
    moment of inertia of continuous distribution of, 379–80
    of neutron, 137, 137*t*, 1356
    of proton, 137, 137*t*, 1355
    relative examples of, 12*t*
    standard of, 134–35
    of universe, 1422–23, 1423*t*
    weight vs., 141
mass defect, 1362
mass-energy relation, 1242–44, 1361
mass number, 1356
    binding energy per nucleon vs., 1361
    conservation law for, 1406
mass spectrometer, 986, **986–87**, 1393
mathematics (review), A1–30
    algebra, A3–5
    arithmetic in scientific notation, A2–3
    equation with two unknowns, A5
    exponent function, A5–7
    logarithms, common (base–10) and natural, A5–7
    powers and roots, A1–2
    quadratic formula, A5
    symbols, A1
    see also calculus; geometry; trigonometry; uncertainties (propagation of)
matter, dark, 1423
matter-antimatter annihilation, 706
maximum height, of projectiles, 109–11
Maxwell, James Clerk, 1071, 1080, 1411
    formula for speed of light by, 1080
Maxwell-Ampère's Law, **1071**, 1073, 1074, 1080, 1096, 1097, **1097**

Maxwell distribution, 615, **615**
Maxwell's demon, 683
Maxwell's equations, 1071–72, 1074–75, 1112
Mayer, Robert von, *see* von Mayer, Robert
mean free path, 624, 881
mean lifetime, 1373
mean solar day, 9
mechanical advantage, 441, **441**, 443–44
mechanical energy, 238
    law of conservation of, 221–23, *221*, **222, 223,** 238
    loss of, by friction, 238–39
mechanical equivalent of heat, 631–32
mechanics, 29
    classical vs. quantum, 1287
    fluid, 565–99
Mediterranean, evaporative loss, 657
medium, wave propagation, 508
medium wave, 1088
Meissner effect, 989
Meitner, Lise, 1377, *1377*
melting points, common substances, 642*t*
Mercedes Benz automobile crash test, 340
Mercury, **xl,** 284, 285*t,* 286, **286**
mercury barometer, 577
mercury-bulb thermometers, 610, **610,** 636, **636**
merry-go-round, **367**
meson, 1403, 1404, 1405, 1405*t,* 1406, 1415
metal:
    resistivities, 870
    temperature dependence of resistivity of, 868–69
Meteor Crater (Arizona), **362**
meteoroid incidents, 301, (Tunguska, Siberia) 1426
meter, 5–6, 13–14
    cubic, 13
    cubic, multiples and submultiples of, 14*t*
    square, 13, **13**
    square, multiples and submultiples of, 14*t*
meters per second (m/s), 30
meters per second squared (m/s2), 39
metric system, 5
MeV, 1359, 1361–62, 1364, 1397, 1415
Michelson, Albert Abraham, *1175,* 1219, 1220
Michelson interferometer, 1174–77, **1175,** 1204–05
Michelson-Morley experiment, 1175–76, **1176,** 1219, 1220
Michelson-Morley inferometer, 1247
microelectromechanical system (MEMS), 503, **503**
microfarad, 830
micrograph, acoustic, **539**
microscope, 1149
    angular magnification of, 1149
    atomic-force (AFM), 475, **475,** 1311, **1311**
    electron-holographic, 694
    scanning capitance, 849
microscope, electron:
    scanning (SEM), **xlvii,** 1310, **1310**
    scanning tunneling (STM), **xlviii, xlix,** 1311, **1311,** 1320
    transmission scanning (TEM), **xlviii**
microscopic and macroscopic parameters, 603
microwaves, 1084, 1090, **1091**
Midas II satellite, 296
middle C, 539, **539**
Milky Way Galaxy, **xliii,** xliv, 1417, **1417**
Millikan, R. A., 1267
Millikan's experiment, 747
mirror:
    focal point of, 1128, **1128**
    image formed by, 1116–17
    magnification produced by, 1139

plane, 1116
    reflection by, 1116–17, **1117**
    spherical, 1128–35, **1128–35, 1155,** 1160–61
    in telescope, 1151
mirror equation, 1131, **1131**
mirror nuclei, 1388
mode, fundamental, 522
moderator, in nuclear reactor, 1381
modulation, of wave, 518
modulus, *see* bulk modulus; elastic moduli; shear modulus; Young's modulus
mole, 11, 20, 604–8
molecular mass, 11
molecular speeds:
    distribution of, in gas, 615
    Maxwell distribution, 615
    most probable speed, 615
    root mean square, 614
molecule, diatomic, **244**
molecules:
    energy levels in, 1333–36
    quantum structure of, 1321, 1333–36
    rotational energy of, 1334
    vibrational energy of, 1333
    water, 389, 743
moment, magnetic dipole, 973
moment arm, 400
moment of inertia:
    of continuous mass distribution, 379–80
    of Earth, 388, 389, 390–91
    of nitric acid molecule, 388
    of oxygen molecule, 388
    of system of particles, 378–84
    of water molecule, 389
momentum:
    angular, *see* angular momentum
    of an electromagnetic wave, 1094–96
    energy and, relativistic transformation for, 1242–44
    Galilean transformations for, 1239
    Lorentz transformations for, 1234–37
    of a photon, 1270
    rate of change of total, 312
    relativistic, 1239–40
    of a system of particles, 306–13, 324
momentum, conservation of, 307–12, **310,** 345, 348
    in elastic collisions, 344–45, 351–52, 353
    in fields, 723
    in inelastic collisions, 351–52, 353
    law of, 309
monatomic gas, kinetic energy of ideal, 616–18
Moon, xxxix
moons of Saturn, 296, 297*t*
Morley, E. W., **1175,** 1219, 1220
Moseley, Henry, *1332,* 1349
most probable speed, 615, **615**
motion:
    absolute, 1217
    along a straight line, 28–68, **32**
    amplitude of, 470
    angular, 375
    of a charge in magnetic field, **967**
    of center of mass, 323–27
    circular, translational speed in, 37
    with constant acceleration, 42–49, **43, 63**
    with constant acceleration, in three dimensions, 102–4, **103, 104, 122**
    with constant force, 151–59
    cyclic, 469
    energy conservation in analysis of, 223
    equation of, *see* equation of motion
    free-fall, 49–54, *51,* **52, 53, 64, 141,** 142, **142**
    harmonic, 489

Newton's Laws of, 130–72
    one-dimensional, 28–68, **32**
    parabolic, 108–9, **109**
    periodic, 469
    planetary, 282–86
    position vs. time in, 32–33, **33, 34,** 35, **35, 36,** 60
    of projectiles, 104–12, **122, 124**
    of projectiles, formulas for, 104
    as relative, 31, 115–18, 1217
    of rigid bodies, 366–67
    rotational, *see* rotational motion
    simple harmonic, *see* simple harmonic motion
    three-dimensional, 95
    with time-dependent angular acceleration, 376–78
    translational, 29, 95, **95, 120,** 366, 404
    two-dimensional, 94–129
    uniform circular, 112–15, **112, 113, 125,** 184–90
    in uniform electric field, 740–44
    with variable acceleration, 54–56
    wave, 508–9
motional emf, 993–97, **995, 996**
Mount Fuji, 334
Mt. Everest, descent, 127
Mt. Palomar telescope, 1151, **1151**
Mt. Pelée volcano, 622
multimeters, **905**
multiloop circuits, 897–900
multiplate capacitor, 851, **851,** 852, **852**
Multiple-Mirror Telescope, Mt. Hopkins, **1164**
multiwire chambers, 1401–2, **1402**
muon, 1403
muonic atom, 1313
musical instruments, standing waves, 546, **546**
mutual inductance, 1009–10
muzzle velocity, 287, 331
myopic eye, 1146

nanoelectromechanical system (NEMS), 503
NASA centrifuge, 592
NASA Spacecraft Center, 114, **114**
NASA weightlessness training, **589**
National Electrical Code, 882
National Ignition Facility (NIF), 828, **828,** 845, **1385**
National Institute of Standards and Technology, 9, 500, **609**
National Ocean Survey buoys, 527
National Research Council of Canada, speed of sound, correction, 560
navigation, vectors in, 71
near point, 1147
nearsightedness, 1146, **1146**
negative acceleration, 39
negative vectors, 75, **75**
negative velocity, 39
negentropy, 683
neon, 696
    atomic structure of, **xlix**
    nucleus of, **l–li**
Neptune, 285*t,* 286
    discovery of, 272
Nernst, Walther Hermann, *681*
net force, 138–40, **138, 139**
neutral equilibrium, 432, **432**
neutrino, 1368–69, 1403, 1423
neutron, 696, 1355, 1397
    beta-decay reaction and, 1368
    isotopes and, 1356
    mass of, **li,** 137, 137*t,* 1356
    nuclear binding energy and, 1359–62
    in nuclear reactor, 1381
    quark structure of, 1414

Newton, Isaac, 131, *131*, 287, 1358
  on action-at-a-distance, 722
  bucket experiment of, 588
  experiments on universality of free fall by, 496
  pendulum experiments by, 496
newton (N), 135, 141
newton-meter, 396
newton per coulomb, 724
Newton's cradle (pendulum), **355**
Newton's Laws:
  angular momentum and, 400
  First, 131–33, **131**
  of Motion, 130–72
  in rotational motion, 395
  Third, 144–51
  of universal gravitation, *131*, 272–76, 278
Newton's rings, 1173, **1173**
Newton's Second Law, 133–38, 171, 205, 214,
    400, 413, 514, 604
  centripetal acceleration and, 185
  empirical tests of, 135
  see also equation of motion
Newton's theorem, 274
New York City, **xxxvii–xxxviii**, 386
  electric power, 653
  Public Library, **xxxvii**
NGC 2997 (spiral galaxy), **xliv**
Niagara Falls, 266, **688**
nitric acid molecule:
  distance and angles between component
    atoms, 333
  moment of inertia of, 388
nitrogen molecule: diatomic kinetic energy, 617, **617**
nodes and antinodes, 520–21, **521**, 544–45
noise, white, 538
noise reduction, 557
nonuniform volume charge density, 783
normal force, 143, **143, 144**, 147
normal frequencies, 523
normalization condition, 1308
n-p-n junction transistor, 1342, **1342**
n-type semiconductor, 1338–42, **1341**, 1343–45
nuclear bomb, 1355
nuclear density, 1359
nuclear explosion, overpressure of blast wave, 590
nuclear fission, *see* fission
nuclear fluid, 1360
nuclear force, *see* "strong" force
nuclear fusion, 1421
nuclear magneton, 1348
nuclear model of atom, 1294, **1294**
nuclear power plant, 1382–83
nuclear radius, 1358
nuclear reactions, 1371
nuclear reactor, **1354**, 1381–82, **1381**
  containment shell of, 1382, **1382**
  control rods in, 1381
  fission, 1380–81
  fuel rods in, 1381
  heavy water, 358
  mass and, 1293
  moderator in, 1381
  neutrons in, 1381
  plutonium (Pu) produced in, 1383
  rods in, 1381
  water-moderated, 1383
nucleons, 1355, 1355*t*, 1359
  see also neutron; proton
nucleus, 1355
  binding energy of, 1359–65, **1359, 1360**
  density of, 1359
  electric force and, 1355, 1359
  of helium, 1366

liquid-drop model of, 1360
  magic, 1389
  mirror, 1388
  stable and unstable, 1360

Oak Ridge National Laboratory, 387
object, 1131
objective lens, 1149–50
observable universe, 1420
ocean, energy extraction, 687
ocean waves:
  amplitude of, 558
  diffraction of, 553
  see also seiche; tides; tsunami
octave, 539
ocular lens, 1149–50
Oerlikon Electrogyrobus, 391
Oersted, Hans Christian, *930*
Ohm, Georg Simon, *866*
ohm (Ω), 868
ohmmeter, 916, **916**
Ohm's Law, 866, 1034
oil pipeline, lateral loops, **651**
oil tanker:
  cross section, **591**
  see also supertanker
ommatidia, 1202
opera house acoustics, European, 557
optical fiber, 1124, **1124**
optical pyrometer, 610, **610**
optics:
  geometric, 1112
  geometric vs. wave, 1169
  wave, 1169
orbit:
  bound, 245
  circular, *see* circular orbits
  of comets, 290–91
  elliptical, 282–86, **282**, 287, 291–92, 1321–22
  geostationary, 271–72, 281
  geosynchronous, 271–72
  hyperbolic, 291
  parabolic, 287, 291
  period of, 279
  planetary, 279, **286**
  planetary, data on, 285–86, 285*t*
  synchronous, 271–72
  unbound, 245
orbital angular momentum, 409
orbital motion, energy in, 288–93
orbital quantum number, 1321, 1328
organ, pipe, 546, **546**, 561
Organisation Européenne pour la Recherche
    Nucléaire (CERN), 1225, 1238, 1398–99,
    1402, **1402**, 1411
origin of coordinates, 3, **3**, 4, **44, 45, 46, 47**
Orion, **1263**
orthicon, 1269
oscillating beads, **497**
oscillating emf, *see* emf, alternating
oscillating function, 1047, 1051
oscillating mass on spring, **473**
oscillations, 468–506
  forced, 488, 490–91, 1046
oscillator, **493**
  damped, *see* damped oscillator
  double-well, 492
  energy quantization of, 1259–60
  harmonic, 492
  leaf spring, **492**
  simple harmonic, 476–79, **481**
  torsional, 500
overpressure, 578

nuclear blast wave, 590
overtones, 522
Oxygen, liquify, 658
oxygen molecule:
  collision with He, **352**
  diatomic kinetic energy, 617, **617**
  moment of inertia of, 388

palladium, atoms, **xlix**
parabola, 291
parabolic motion, 108–9, **109**
parabolic orbit, 291
  vs. elliptical orbit for projectile, 287
parallel-axis theorem, 382–83
parallel-plate capacitor, 831–32, 851, **851**, 852, **852**
paramagnetism, 977
parbuckle, **459**
parent material, 1366
parity, 1407
parsec, 24
particle:
  accelerators for, 1363, **1363**, 1398–99
  alpha, 1293–94
  center of mass of system of, 313–23
  decays of, 1365–70
  electric charges of, 696–97, 698, 706, 1414
  elementary, *see* elementary particles
  ideal, 3
  kinetic energy of system of, 327–28
  moment of inertia of system of, 378–84
  momentum of system of, 306–13, 324
  in quantum mechanics, 1302–9
  scattering of, 1294
  stable, 1404
  system of, *see* system of particles
  unstable, 1404
  W, 1409–11
  W^-, 1409, 1411, 1415
  W^+, 1409, 1411, 1415
  wave vs., 1276–79
  Z, 1409–11
  Z^0, 1409, 1411, 1415
pascal, 573
Pascal, Blaise, *574*, 588
Pascal's Principle, 575, **575**, 599
Paschen series, 1292
Pauli, Wolfgang, *1325*, 1328–32, 1329*t*, 1368
pendulum, 476, 495, 496, **500**
  ballistic, 349–50
  Foucault, 499
  Huygens' tilted, 495
  physical, 487, 501
  see also Foucault pendulum; Huygens' tilted pen-
    dulum; Newton's cradle; simple pendulum
pendulum clocks, 487, 654–55, **655**
penetration depth, 989
Pentecost Island "land divers," 356
Penzias, Arno A., 1421
perigee:
  of artificial satellites, 286
  of comets, 291
  of Earth, 295
  of Moon, **xxxix**
perihelion, 282
  of planets, 286
perihelion, 282, 284, 409
  of comets, 291
  of Earth, 295
  of planets, **xl**, 285
period, 369
  of comets, 291
  of orbit, 279
  of planets, 285*t*

of simple harmonic motion, 470–71
of wave, 510, **510**
periodic motion, 469
Periodic Table of Chemical Elements, 1328–32,
    1329t
periodic waves, 509–16
permanent dipole moments, 743
permanent magnet, 977
permeability constant, 929
permittivity constant, 699
perpetual motion machine:
    of the first kind, 662
    hypothetical, **662**
    M. C. Escher, **662**
    of the second kind, 662
Perseus Cluster, **xliii**
phase, 471
phase constant, 471
phase Φ, RLC current of, 1050
phasor, 1047, **1047, 1048, 1050,** 1049, 1187, 1188,
    1187–88, 1194
Phobos, moon of Mars, **424**
photocopiers, 694
photoelectric effect, 1264–69
photoelectric equation, 1266
photographic camera, 1144–45, **1144, 1145**
photomultiplier, 1268
photomultiplier tube, 1266–67
photons, 1255, 1264–69
    in Compton effect, 1270–71
    emitted in atomic transition, 1300–1301
    energy of, 1264–65
    momentum of, 1270
    virtual, 1410
photovotaic arrays, **1214**
physical pendulum, 487, 501
piano:
    frequencies, 531
    notes available on, 558
picofarad, 830
pion, 1400, 1401, 1404, 1414
pitch, 366
Pitot tube, 596, **596**
planar symmetry, 770–72
Planck, Max, 675, 1259–61, **1260**
Planck's constant, 976, 1259, 1403
Planck's Law, 1259–60
plane mirror, 1116
planetary motion:
    conservation of angular momentum in, 284
    Kepler's laws of, 282–86
planetary orbit, 279
    aphelion of, 285
    data on, 285–86, 285t
    perihelion of, 285
    periodicity of, 492
    period of, 285t
plane wave, 537
    electric and magnetic fields of, 1081–83, **1082,
    1097**
plane wave pulse, electromagnetic, 1080–83, **1082**
plasma, 710
Pleiades Cluster, **xliii**
Plexiglas, electrical breakdown in, **841**
"plum-pudding" model, 1293, **1293**
Pluto, **xli,** 285t, 286
    xli, mass of, 294
plutonium (^{239}Pu), produced in nuclear reactor,
    1383
    in chain reaction, 1379
p–n junctions, 1340–42, **1341,** 1343–45, **1343**
point charge, 699
    electric field lines of, 736–37, **737,** 739, 936

electric field of, 723–24, 728
    energy of, 796
    magnetic field of, 936
    moving, magnetic force and, 928–30
    potential of, 794–99
Poisson spot, 1202, **1202**
polarization, 839–40, 1083, **1083,** 1124–25, **1125**
    circular and linear, 532, **532**
polarized light, 1126–27
polarized plane waves, electric and magnetic fields
    of, 1081–83, **1082**
polarizing filter, 1084–86, **1084, 1086**
Polaroid, 1085, **1086**
polyatomic and non-linear molecules, energy in,
    618, **620**
position, time vs., 32–33, **33, 34,** 35, **35, 36,** 60
position vector, 76–77
positive acceleration, 39–40
positive velocity, 39
positron, 1369
positronium, 1316
potential:
    derivatives of, 806–8
    electrostatic, 790–98, 798–803
    gradient of, 806–10
    of point charge, 794–99
    of spherical charge distribution, 799–800
potential difference, 793, **814**
potential energy, 218
    of argon (Lennard-Jones potential), 263
    of conductor, **803,** 812–13
    of conservative force, 236–43
    of current loop, 975–76
    curve of, 244–47, **244**
    of dipole, 743
    in double-well oscillator, 492
    elastic, 236–37, 238
    of force, equation for, 239
    force calculated from, 241
    gravitational, 218–23, **219, 220,** 238, 288–93
    gravitational, of a body, 321
    in simple harmonic motion, 480–83
    of a spring, 236–37, **237,** 241
    turning points and, 244–45
potentiometers, 872, 906
pound (lb), 12, 134
pound-force (lbf), 136
pound-force per square inch, 574
Powell, Asafa, **31**
power, 253–58
    average, 253
    delivered by force, 255
    delivered by source of emf, 901, **1053**
    delivered by torque, 397
    dissipated by resistor, 1033
    dissipated in high-voltage transmission line, 904
    instantaneous, 253
    sample values of somepowers, 257t
    time-average, 1033, **1033**
    transported by a wave, 516
power brake, 456
power cable, **444**
    copper contraction, 461
power factor, 1062
power line, **862**
power outlet, three prong, **1034**
Poynting, John Henry, 1093
Poynting flux, 1093, **1093**
Poynting vector, 1093–94
precession, 415
    of a gyroscope, 415–16
pressure, 448, 566, 567, 573–75, 574t, 590
    atmospheric, 577–78

barometric in hurricane, 598
blood, 579
gauge, 578
human vein, 597, **597**
in incompressible fluid, 576
kinetic, 613–16
standard temperature and, 604, 607–8
in a static fluid, 575–79
primary, of a transformer, 1053–54
Princeton Tokamak Test Reactor, **1385**
principal maximum, of grating, 1184–85
principal quantum number, 1321, 1328–29
principal rays, 1129
    of lens, 1137, **1137**
Principia Mathematica (Newton), *131,* **287**
principle of relativity, 1220–23
principle of superposition, 138, 703
    for magnetic field, 938
    for waves, 516–20, **518**
prism, **1123,** 1126, **1126, 1156**
probability interpolation of wave, 1277
problem-solving, guidelines for, 50
Procyon B, 1281
projectile:
    maximum height of, 109–11
    parabolic vs. elliptical orbit for, 287
    range of, 109, 111, **111,** 287
    time of flight of, 109–10
    trajectory of, **105, 106, 107,** 111, **111**
projectile motion, 104–12, **122, 124**
    with air resistance, 111
propagation of uncertainties, *see* uncertainties
proper frequencies, 523
proton, 1355
    charge of, 696, 698
    collision of, **361**
    isotopes and, 1356
    mass of, **li,** 137, 137t, 1356
    nuclear binding energy and, 1359–62
    quark structure of, 1414
Proxima Centauri, **xli,** 52
p-type semiconductor, 1338–42, **1341,** 1343–44
pulleys, 443–44, **458,** 459
    mechanical advantage and, 443–44
pulsar (neutron star), rotation, 391, 427
pupil, of eye, **xlvi**
p-V diagram, 668–69, 678
    four step, **687, 690**
    one step, **689**
    three step, **686**
pyrometer, optical, 610, **610**

Q, or quality factor, of oscillation, 489–91,
    502–3
quality factor (Q), 1045
quantization:
    of angular momentum, 1296, 1322–23
    of electric charge, 706
    of energy, 1260
quantum:
    of energy, 1260
    fields and, 1409–11, 1409t
    of light, 1254–55
quantum Hall effect, 982
quantum jump, 1296
quantum mechanics, 1287, 1302–9
quantum numbers, 1260
    angular-momentum, 1296, 1322, 1324–26
    of atomic states, 1328–32
    electron configuration and, 1328
    magnetic, 1328
    orbital, 1321, 1328
    principal, 1321, 1328

quantum numbers *(continued)*
  spin, 1324, 1328
  for stationary states of hydrogen, 1328
quantum structure, 1320–53
quark, 714, 1397, 1404*t*, 1412–16
  bottom, 1415
  charmed, 1415
  charm of, 1415
  color of, 1414–15
  down, 1413–14
  electric charges of, 706, 1414*t*
  force between, 1414–15
  strange, 1413–14
  top, 1415
  up, 1413–14
  up and down, within proton, **li**
quark structure of proton and neutron, 1414
Quito, Ecuador, 386

rad, 1375
radar "guns" (Doppler), 558
radians, position angle, 368
radiation, 641
  blackbody, 1255–58, 1259–61
  cavity, 1257
  electromagnetic, 1090–91
  electromagnetic, wavelength and frequency
    bands of, 1090–91, **1091**
  Hawking, 1281
  heat transfer by, 641
  infrared, 1090
  thermal, 1255, 1256–57, **1256**
radiation field, of accelerated charge, 1075–76,
  1076, 1077
radioactive dating, 1376, 1421
radioactive decay, law of, 1372–76
radioactive series, 1367–68
radioactivity, 1365–72
radio frequency (rf) choke, 1061
radioisotopes, decay rates of, 1374–76
radio station WWV, 9
radio telescope, 1186
  at Arecibo, 1198–99, *1198*
  at Jodrell Bank, 1209
  Very Large Array, 1186
radio waves, 1079–1110, 1089, **1094**
railroad tracks, thermal expansion and, 637, **637**
range, of projectiles, 109, 111, **111**, 287
rate of change, of momentum, 312
Rayleigh, John William Strutt, Lord, *1198*, 1258
Rayleigh's criterion, 1197, **1197**, 1209–10
rays, 1115, **1115**
  alpha, 1365
  beta, 1365
  gamma, 1090, 1365
  principal, 1129, **1130**, 1137, **1137**
rays, X, 1090, 1303
RC circuit, 907–12, **907, 911, 912**
reactance:
  capacitive, 1036
  inductive, 1039–40
reaction and action, 144–51, **144, 145, 146, 149**
reactor, nuclear, 1381–82, **1381**
  breeder, 1383
  containment shell of, 1382, **1382**
  control rods in, 1381
  fission, 1380–81
  fuel rods in, 1381
  moderator in, 1381
  neutrons in, 1381
  rods in, 1381
  water-moderated, 1383
real image, 1133, **1133**

recoil, 309–10
rectangular coordinates, 3, **3**
rectifier, 1340–42
red, 1414–16
red-shift, 1418
reference circle, 472
reference frames, 3, 4, 20, 114, 115
  in calculation of work, 208, **208**
  of Earth, 132
  freely falling, 142, **142**
  inertial, 132, **132**, 133
  for rotational motion, 366
reflection, 1112, 1114–17, **1114–15**, 1156–57
  angle of, 1115–16, **1115**
  critical angle for total internal, 1123
  law of, 1114–15
  by mirror, 1116–17
  polarization by, **1086**, 1125
  total internal, 1123
reflection grating, 1185
reflector, corner, 1116, **1116**
refraction, 1112, 1117–27, **1119–22, 1126,**
  1157–60
  index of, 1118, 1119*t*, 1125–26, **1125**
  law of, *see* Snell's Law
refrigerators, use of freon and, 672–73, **673**
relative biological effectiveness (RBE), 1375
relative measurement, 4
relativistic combination rule, 1237–38, **1238**
Relativistic Ion Collider, 1250
relativistic kinetic energy, 1240–41, **1240**, 1250
relativistic momentum, 1239–40, **1240**, 1250
relativistic total energy, 1243
relativistic transformation, for momentum and
  energy, 1242–44, 1251
relativity:
  of motion, 31, 115–18, 1217
  principle of, 1220–23, 1247
  of simultaneity, 1220–22
  special theory of, 1216–53
  of speed, 31, **31**
  of synchronization of clocks, 1222–23
Relativity, General, theory of, 394
rem, 1375
resistance, 866
  air, 49, 51, **61**, 111, 181, **181**
  in combination, 872–76
  of human body, 913, **913**
  internal, of batteries, 895–96
  of wires connecting resistors, 875–76
resistance thermometers, **610**, 870, **871**, 880
resistivity, 865, 866*t*
  of insulators, 871*t*
  of materials, 868–72
  of materials, temperature dependence of, 869
  of metals, 870
  of semiconductors, 871*t*
  of silicon, temperature dependence of, **869**
  temperature coefficient of, 880
  of tin, temperature dependence of, **871**
resistors, 872–76, **872**, 873*t*
  circuit with, 1031–35
  color code, 873, **873**
  infinite ladder, 884, **884**
  in parallel, 874, **875, 905**
  power dissipated by, 1033
  power dissipated in, 902, **1033**
  resistance of wires connecting, 875–76
  in series, **873**, 874, **875, 905**
resolution, angular, of telescope, 1196–99
resolving power, of grating, 1186
resonance, 523, 1404
  of bridge (Tacoma, WA), 523, **524**

of damped oscillator, 490–91, **491**
  in musical instruments, 546
  of oscillator, **1051, 1053**
  tides, 531
  width of, 1063
resonant frequency, 1043
rest-mass energy, 1242
restoring force, 182–84, **183**
resultant, of vectors, 72, **74**
resultant force, 138
retina, of eye, **xlvii**
reversible process, 667
Reynolds number, 595
rheostats, 872, **872**
Rigel, **1263**
right-hand rule, 84, **84, 85**, 743
  for a current loop, 942
  for electromagnetic waves, 1078–79, **1079**
  for magnetic field, 932, **932, 933**
  for magnetic force, 934, **935**, 936
  for magnetic moment, 974
  for solenoids, 944
right triangle, 19
rigid body:
  dynamics of, 394–428
  kinetic energy of rotation of, 378–84
  moment of inertia of, 378–84
  motion of, 366–67, **366–67**
  parallel-axis theorem for, 382–83
  rotation of, 365–93
  some moments of inertia for, 382*t*
  statics of, 430–33
  translational motion of, 366
RLC circuit, 1044–45, **1044, 1045**
  impedance Z of, 1050
  phase f of, 1050
RL circuit, 1015–18, **1015**, 1046
  without battery, **1017**
roll, 366
root-mean-square (rms) speed, of gas molecules,
  614–15
root-mean-square voltage, 1033
rotation:
  of the Earth, 9, 132, 476
  frequency of, 369–70
  kinetic energy of, 378–84
  period of, 369
  of rigid body, 365–93
rotational energy, of molecule, 1334
rotational motion:
  analogies to translational motion, 374*t*, 407*t*
  conservation of angular momentum in, 406–10
  conservation of energy in, 397
  with constant angular acceleration, 374–76
  of Earth, **120**
  equation of, 399–406
  about a fixed axis, 367–73
  of gas, 617–18
  Newton's laws in, 395
  reference frame for, 366
  torque and, 405
  work, energy, and power in, 395–99
roulette wheel, **367**
Rumford, Benjamin Thompson, Count, 629, *629*
Rutherford, Sir Ernest, 1287, 1293–95, *1294*,
  1355, 1363, 1365
Rutherford backscattering, 357
Rutherford scattering, 1293
Rutherford's apparatus, **1294**
R value, **640**, 641, 655
  three layers, 659
Rydberg constant, 1292, 1300–1301
Rydberg-Ritz combination principle, 1293

Sagittarius, constellation, xliii
Salam, Abdus, 1411
Sandia National Laboratory centrifuge, 365, **365**
satellites:
  artificial, 271–72, 281, 286–87, 1344, **1344**, 1421
  communication, 271–72, 281, **281**, 290–91
  geostationary, **280**, 281, 290–91
  Kepler's laws for motion of moons and, 286–88
  see also specific satellites
Saturn, 285*t*, 286
  moons of, 296, 297*t*
scalar, 72
scalar (dot) product, of vectors, 81–83, **81, 83**, 86,
  208–9
scale, chromatic musical, 539, **539**
scanning capitance microscope, 849
scanning electron micrograph, 704–05, **705**
scanning tunneling microscope, **1287**, 1311, **1311**,
  1320
scattering, of alpha particles, 1294
scattering cross section, 1315
Schlieren photograph (of bullet), **552**
Schrödinger, Erwin, 1302, 1303, *1304*
Schrödinger equation, 1303–6
Schwinger, Julian, *1410*
scientific notation, xxv, 14–15, A2–23
scissorjack, **460**
Sears Tower, 653, **654**
seat belt, 343
secant, A8–10
second, 5, 9–10, 13, 14
  multiples and submultiples of, 10*t*
secondary, of a transformer, 1053–54
second harmonic, 523
Second Law of Thermodynamics, 675–80
second overtone, 522, **522**
seiche, 558
seismic waves (P and S), 533, 558
seismometer, 558, **558**
selection rule, 1333
self-inductance, 1011, **1011**
semiconductors, 871, 1340–45
  with donor and acceptor impurities, 1339
  electron configurations of, 1338–39, **1339**
  "holes" in, 1338–39, 1340–42
  n-type, 1338–42, **1341**, 1343–45
  p-n junctions of, 1340–42, 1343–44
  p-type, 1338–42, **1341**, 1343–45
  resistivities of, 871*t*
  temperature dependence of resistivity of, 1338
semimajor axis of ellipse, 282, 285
  related to energy, 291
series limit, 1291
Sèvres, France, 5
shear, 445–46, 447*t*, 449
shear modulus, 447–48, 447*t*
sheerlegs, **463**
shells, 1330, 1330*t*
ship collison, *Andrea Doria* and *Stockholm*, 361
shock wave, of bullet, **552**
Shoemaker-Levy comet, 299
short wave, 1088
sidereal day, 294
sievert, 1375
significant figures, 14–15, 18
silicon:
  nanoparticle, 264
  wafer, 654
silicon solar cells, 893
silicon structures, micromachined, 195
silk, 711
silo, grain, **588**
silver-plating, 718, **718**

simple harmonic motion, 469–76
  conservation of energy in, 483
  frequency of, 470–71
  kinetic energy in, 480–83
  period of, 470–71
  phase of, 471
  potential energy in, 480–83
simple harmonic oscillator, 476–79
  angular frequency of, 477
  equation of motion of, 477
  as timekeeping element, 479
simple pendulum, 484–88, **484**
  equation of motion for, 485
  isochronism of, 486, 501
simultaneity, relative, 1220–22, **1222**
sine, 19, 473–74, 486, A8–10
  formula for derivatives of, 473
  law of sines, A10
single-loop circuits, 893–97
single slit:
  diffraction by, 1190–96
  diffraction pattern of, 1191
  minima in diffraction pattern of, 1191–92
sinks, of field lines, 738–39
siphon, 595, **596**
Sirius, star, **xliii**
SI units, *see* system of units (SI)
skater, figure, **408**
skiing, speed record, 267
skin, human, resistance of, 913
sky divers, **53**
Skylab mission, 134, **134**, 229, 295, **424**, 504
  body-mass measurement device on, 134, **134**
  fuel cell on, 892
sliding friction, 174–78, **175, 176, 177**, 190
slope, 17, **18**, 32–33
  instantaneous velocity as, 35–37
slug, 136*n*
small angle approximations, 485
Small Mass Measurement Instrument,
  504, **504**
Snell's Law, 1121, **1121**, 1124
sodium gas, 624
solar cells, 893, 1343–45
  silicon, 893
solar day, 9
solar heat collector, **653**
Solar panels, **1344**
Solar System, xli, 282
  data on, 285–87
solenoid, 943–46, **946**
  coaxial, **961**
  finite, 960, **960**
  magnetic field of, 943–44, **943–46, 1011**
  self-inductance of, 1012
solid:
  compression of, 446, 448
  elasticity of, 445–49
  elastic moduli for, 447*t*
  electron configuration of, 1337–39
  elongation of, 445–49, **445**
  energy banks in, 1336–40
  quantum structure of, 1321, 1336–45
  shear of, 445–46, 448
  thermal expansion of, 633–37, **633**
  ultimate tensile strength of, 447*t*
sonic boom, 552–53, **552**
sonography, 536, **536**, 544
sound:
  intensity level of, 542*t*
  speed of, 543–45, 544*t*, 559
  speed of (French Academy), 560
  speed of correction (NRC, Canada), 560

  speed of in air, 559–60
  speed of in freshwater, 527, 559
  speed of in human body, 558
  theoretical expression for speed in air, 543
  theoretical expression for speed in gas, 657
sound waves:
  in air, 538–39
  diffraction of, 554
  intensity of, 538, 540–43, **543**
  longitudinal, 538, **538**
  maximum speed in air, 558
  sources:
    of electromotive force, 890–92
    of field lines, 738–39
Soviet Union (toroid), 1027
Space Shuttle, 125, **142**, 280, **280**, 290–91, 426,
  1026
Space Telescope, 1209
Special Relativity theory, 1216–53
specific heat:
  of common substances, 630*t*
  at constant pressure, 644
  at constant volume, 644
  of a gas, 644–47, 646*t*
specific heat capacity, 630
spectra, color plate, 1288
spectral band, 1335
spectral emittance, 1257
  of blackbody, 1259
spectral lines, 1127, **1127**, 1286, 1287–88, **1289**,
  1290, **1290**
  splitting of, 1324–26
spectral series:
  diffuse of lithium, 1314
  of hydrogen, 1291–93, 1299, 1324
  principal of lithium, 1314
spectrum, 1288, **1290**
  of Caph, 1288
  of hydrogen, 1291–92, 1291*t*
  produced by grating, 1184, **1184**
  produced by prism, 1126
speed:
  average, 29–31, 30*t*
  of bullet, 356
  in circular motion, 371
  of electromagnetic wave, 1080
  initial, 111
  launch, 111
  of light, *see* speed of light
  of light in air and water, 527
  molecular, in gas, 615
  most probable, 615, **615**
  after one-dimensional elastic collision,
    345–47
  as relative, 31, **31**
  of sound, 543–45, 559, 560
  standard of, 6
  terminal, 53
  unit of, 30
  velocity vs., 33–34, 96
  of wave, 510
  of wave in Bay of Fundy, 559–60
  of waves, on a string, 513–16
speed of light, 6, 1218–20
  Fizeau effect, 1250
  invariance of, 1220–21, 1238
  in material medium, 1118
  measurement of, 1080
  universality of, 1220–21, 1238
spherical aberration, 1128, 1144
spherical charge distribution:
  electric energy of, 813
  potential of, 799–800

spherical mirrors, 1128–35
   concave, 1128, **1130**
   convex, 1129
   focal length of, 1128
sphygmomanometer, 579
spin:
   of electron, 1324–25, 1328–29, 1355–56
   of elementary particles, 1403*t*, 1404*t*
spin angular momentum, 409
spin magnetic moment, 976
spin quantum number, 1324, 1328
Spirit of America, 136, **136**
spokes of wire wheels, 389, 392, 426, **526**, 531
spring balance, 136, **136**, 151
spring constant, 183
springs, 476–79
   force of, 182–84, **183, 184**
   potential energy of, 236–37, **237,** 241
spring tides, 296
Sputnik I, **286,** 299, 301
Sputnik II, **286**
Sputnik III, **286**
SSC accelerator, 1252
stabilized laser, 6, **6**
stable equilibrium, 432, **432**
stable particle, 1404
standard *g,* 52–53
standard kilogram, **11,** 134
standard meter bar, international, 5–6, **5**
standard model, 1415
standard of length, 5–6, **5, 6**
standard of mass, 11, 134–35
standard of speed, 6
standard of time, 9
standard temperature and pressure (STP), 604, 607–8
standing wave, 520–24
   of bridge (Tacoma WA), 523, **524**
   mode and overtone, 522–23, **522, 545**
   in a tube, 545, **545**
Stanford Linear Accelerator (SLAC), **819,** 1250, 1411, 1413, 1415, 1427
   beam dump, 658, **659**
   beam pipe, **819**
Stanford Linear Collider (SLC), 1250
star, 55 Cancri system, 297, Cassiopeiae spectrum, **1290**
star, binary system:
   Cyguns, 298
   Kruger 60, 297
   PSR 1913+16, 297
states of aggregation (solid, liquid, gas), xxxvi
states of atoms, stationary, 1299
states of matter, fifth, 722
static equilibrium, 430–41
   condition for, 430
   of electric charge, 774–75
   examples of, 432–41, **432**
statics, 174
   of rigid body, 430–33
stationary state, 1296
   of hydrogen, 1298, 1300
steady emf, 1004
steady flow, 569
steam engines, 671, **671**
steel, maximum tensile stress and density, 528
steel rods, deformation, **449**
Stefan-Boltzmann Law, 1261–62
step-down transformer, 1054

step-up transformer, 1054
stimulated emission, 1090
stopping distances, automobile, **45, 46,** 47, **47**
stopping potential, 1266
STP (standard temperature and pressure), 604, 607–8
straight line, motion along, 28–68, **32**
strangeness, 1407
strange quark, 1413–14
Strassmann, Fritz, 1377
streamline flow, 569
streamlines, 569–71, **569, 570, 584, 585**
   velocity along, 584
stream tube, 569–70
stress, thermal, 461
stringed instruments, 546
   see also specific stringed instruments
string theory, li, 1416
"strong" force, 191, 1405, 1406, 1406*t,* 1414
   nuclear binding energy and, 1359–65, **1359, 1360**
   in nucleus, 1355
strong interactions, 1405, 1406, 1406*t*
strontium, radioactive decay of, 1372–76
sublimation, 642
suction pump, 592
Sun, **xl, 1254**
   age of, 1420
   escape velocity from, 292
   spectrum, **1290**
sunglasses, Polaroid, **1086**
superconducting cable, 883, **883**
Superconducting Super Collider, 1252, 1424
superconductivity, 858, 861, 870
superconductors, high-temperature, 870, **1008**
   commercial use, **1009**
superposition:
   of electric fields, 772–73, *773*
   of electric forces, 703
superposition principle, 138, 703, 729, 734
   for magnetic field, 938
   for waves, 516–20, **518**
supersonic aircraft, sonic boom and, 552–53, **552**
supertanker:
   empty and loaded, **593**
   Globtik Tokyo, 266
   Seawise Giant, 659
   see also oil tanker
surface tension, 566
symmetry, axis of, **380–82,** 382*t*
symmetry, in physics:
   cylindrical, 767–68
   planar, 770–72
synchronization of clocks, 4, **5,** 133*n,* 1220–23, **1221**
   relativity of, 1222–23, **1223**
synchronous (geostatry) orbit, 271–72
synchrotron, 1398
Syncom communications satellite, 271–72, 290–91
system of particles, 305–37
   center of mass in, 313–23
   energy of, 327–28
   kinetic energy of, 327–28
   moment of inertia of, 378–84
   momentum of, 306–13, 324
system of units (SI) 5, 14, 20
   base and derived, A21–23
   of current, 972
   of electric charge, 972
   for radioactive decay rate, 1374–75

Tacoma Narrows, 523–24, **524**

Tampa Bay, Fla., 99
tangent, 19, A8–10
tangent galvanometer, 954, **954**
tangential acceleration, 371–72
telescope, 1149–52
   angular magnification of, 1150, **1151**
   angular resolution of, 1196–99
   Arecibo radio-, 1198–99, **1198**
   Galilean, 1142
   Hobby-Eberly, 1152, **1152**
   Hubble Space Telescope, 1111, 1130, 1135, 1152, 1198, *1198,* 1209
   Jodrell Bank radio-, 1209
   Keck, 1151
   mirror, 1151
   Mt. Palomar, 1151, **1151**
   Multiple-Mirror Telescope, Mt. Hopkins, **1164**
   radio-, 1186, 1198–99, *1198,* 1209
   Space, 1209
   Very Large Array radio–, 1186, **1186**
television cameras, 1269
Telstar satellites, 1421
temperature:
   coefficient of resistivity, 880
   on Earth, hottest and coldest, 620
   some typical values, 611*t*
   standard pressure and, 604, 607–8
   at Sun's surface and center, 622, 625
temperature scales:
   absolute, 604
   absolute thermodynamic, 609
   Celsius, 611, **612**
   comparison of, 611, **612**
   Fahrenheit, 611, **612**
   ideal-gas, 609–12
   Kelvin, 604, 609–10
Tennessee River, frictional losses, 267, 653
tensile strength, of solids, 447*t*
tension, 149–50, **149, 151, 155**
terminal speed, 53
terminal velocity, 53
Tesla, Nikola, *934*
tesla (T), 934
Tethys, moon of Saturn, **297**
TeV, 1398
Tevatron, Fermilab, 1398–99, **1398,** 1425
thermal conductivity, 638–41, 639*t*
thermal energy, 248, 616, 629
thermal engine, efficiency of, 666–67
thermal expansion, **637**
   of concrete, 637
   linear, coefficient of, 634, 634*t, 637*
   of solids and liquids, 633–37, **633**
   volume, coefficient of, 634*t*
   of water, 635, **635**
thermal radiation, 1255, 1256–57, **1256**
thermal resistance (R value), **640,** 641, 655
thermal stress, 461
thermal units, 249
thermocouples, 610, **610**
thermodynamics, 661–91
   calculation techniques, 672
   First Law of, 662–64
   Second Law of, 675–80
   Third Law of, 680
thermodynamic temperature scale, absolute, 609
thermogram, **1256**
thermograph, **628,** 652
thermometer:
   bimetallic strip, 610, **610,** 636, **637**
   color-strip, 610, **610**
   constant-volume gas, 609–10, **609**
   electric resistance, 610, **610**

mercury-bulb, 610, **610**, 636, **636**
  resistance, **610**, 870, **871**, 880
  thermocouple, 610, **610**
thermonuclear bomb, **260**
thermos bottle, **1257**
thickness monitor, 498
thin films, interference in, 1169–73
thin-lens equation, 1138, **1138**, 1162
thin lenses, 1135–39
third harmonic, 523
Third Law of Thermodynamics, 680
Thomson, J.J., experiment, 990
Thomson model (of helium atom), **783**, 824
Thomson, Sir Joseph John, 1293, 1316
thread of a screw, 22
threshold energy, 355
threshold frequency, 1266
threshold of hearing, 538
thundercloud, electric field of, 725–28
tidal flow, 99
tides, 120, 530
  height at Bay of Fundy, 531, **531**
  height at Pakhoi, **530**
  resonance, 531
  spring, 296
tightrope walker, **418**
time:
  atomic standard of, 9
  Cesium standard of, 9
  Coordinated Universal, 9
  position vs., 32–33, **33**, **34**, 35, **35**, **36**, **60**
  standard of, 9
  unit of, 9–10
  velocity vs., **61**
time constant, 1373
  of RC circuit, 909
time-dependent angular acceleration, 376–78
time dilation, 1224–28, **1224–26**, 1247–48
time of flight, of projectiles, 109–10
time signals, 9
Titov, G.S. cosmonaut, 1247
TNT, 263
Tomonaga, Sin-Itiro, *1410*
top, 415
top quark, 1415
tornado, air pressure within, 590
toroid, 945, **945**
torque, 395–99
  angular acceleration and, 400
  angular momentum and, 410–16
  on a current loop, 972–76
  on dipole, 743
  equation of, 400
  as a function of angle, **975**
  power delivered by, 397
  rotational motion and, 405
  static equilibrium and, 430–41
torque vector, 410, **411**
torr, 578
Torricelli, Evangelista, *578*
Torricelli's theorem, 585
torsional oscillator, 500
torsion balance, 277, *277*
total internal reflection, 1123, **1123**
totally inelastic collision, 348
tower crane, *see* K–10000 tower crane
tracking chambers, 1401–2
traction apparatus, 149, **149**
train à grande vitesse (French TGV), 563, **563**
train box car collisions:
  elastic, **346**
  inelastic, **349**
trajectory of projectiles, **105**, **106**, **107**, 111, **111**

transfer of heat, 638–41, 641
transformation, heat of, 642
transformation equation:
  Galilean, 1218, 1234, 1239, 1241
  Lorentz, 1232–38, 1249–50
transformation of coordinates, 1218, **1233**, 1234, 1241
transformer, **862**, 1010, 1053–57, **1053–54, 1055, 1056**
  with center tap, 1063
  step-down, 1054
  step-up, 1054
  variac, 1064
transistors, 1340, 1342–43, **1343**
translational acceleration, 402
translational motion, 29, 95, **95**, 366, 404
  analogies to rotational motion, 374*t*, 407*t*
  of Earth, **120**
transmission line, power dissipated in, 904, **904**
transmutation of elements, 1363
transverse electric field, 1078
transverse magnetic field, **1078**
transverse radiation field of accelerated charge, 1076–77
transverse wave, 508, **508**
triangle, right, 19
Triangulum Galaxy, xliv
Trieste, 589, **589**
trigonometry (review), A8–10
  functions of (sine, cosine, tangent, secant, cosecant, cotangent), A8–10
  identities, A9–10
  laws of cosines and sines, A10
  of right triangle, 19, A8
  small angle approximation, 486
triode, 1340, 1342–43
triple-point cell, 609, **609**
triple point of water, 609, **609**
trombone, 546
trumpet, 546
  sound wave emitted by, 538
tsunami, height (Mexico), **526**, 527
tuning fork, 530
tunneling, 1311
turbulent flow, 571, **571**
turning points, 244–45
  of motion, 470
TV waves, 1088
twin paradox, 1226
two-conductor capacitor, **832**
two-slit interference, 1177–83, **1177**
  pattern for, 1180, **1180**

UA1 detector, CERN, 1402, **1402**
UFOs, 557
Uhlenbeck, George, 1325
ultimate tensile strength, 447*t*
ultrahigh vacuum, 624, **624**
ultrasound, **536**, 538, 539
  frequencies used in human body, 558, 562
ultraviolet catastrophe, 1259
ultraviolet light, 1090
unbound orbit, 245
uncertainties:
  application to Ohm's Law of, A20–21
  propagation of, A19–21
uncertainty relation, 1278
unified field theory, 1411
unified theory of weak and electromagnetic forces, 1411
uniform circular motion, 112–15, **112, 113, 125,** 184–90
Unionville, Md. rainfall, 332

unit, derived, *see* derived unit
unit of length, 5–8
units:
  consistency of, 16, 18, 20
  conversions of, 16–17, 18
  prefixes for, A21*t*
Units, International System of (SI), 5, 14
  of current, 972
  of electric charge, 972
units of force, 135–36, 141
unit vector, 79–80, **79**
  cross product of, 85
universal gas constant, 604
Universal Gravitation, Law of, *131*, 272–76, 278
universality of acceleration of free fall, 49, **49**
universality of free fall, 49, **49**
universality of speed of light, 1220–21, 1238
universe:
  contraction of, 1422
  expansion of, xlv, 1419–23
  mass of, 1422–23, 1423*t*
  observable, 1420
unpolarized light, 1084, **1084**
unstable equilibrium, 432, **432**
unstable particles, 1404
ununquadium, atomic mass of, xlix
up quark, 1413–14
uranium, 1355
  alpha decay of, 1365–67, **1366**
  enriched, 1381
  fission of, 1377–79, **1377**
  isotopes of, 1362
  as nuclear fuel, 1242
uranium isotopes (^{235}U, ^{238}U, UF_6), 623–24
Uranus, 272, 285*t*, 286
Ursa Major, **1417**, 1428

valence band, 1338
Van de Graaff generator, 793*t*
van der Waals equation of state, 624
Vanguard I, **286**
vaporization, 642, 642*t*
variable capacitor, **832**
variable force, 211–13, **211, 212, 213, 214**
Vasa (Swedish ship), 320
vector, magnetic force, 933
vector addition, 72–76, **72, 73, 74,** 89
  commutative law of, 74
  by components, 78–79
vector product, 83–86, **84, 85**
vectors, 69–93
  acceleration, 100–101
  addition of, *see* addition of vectors
  components of, 77–86, **78, 95–98, 97, 99, 101**
  cross product of, 83–86, **84, 85**
  definition of, 72
  displacement, 69, 70–72, **70, 71, 88, 96**
  dot (scalar) product of, 81–83, **81, 83,** 86, 208–9
  electric force, 703–05, 704, **704,** 736, **736**
  instantaneous velocity, 98, **98**
  multiplication of, 75, 81–86
  in navigation, 71
  negative, 75, **75**
  notation of, 71
  position, 76–77
  Poynting, 1093–94
  resultant of, 72, **74**
  subtraction of, 75
  three–dimensional, 79–81, **79, 80**
  unit, 79–86, **79**
  velocity, 98–100
vector triangle, 73

velocity:
   acceleration as derivative of, 41
   addition rule for, 115–16, **117**
   along streamlines, 584
   angular, 369*t*, 376–77, 471
   angular, average and instantaneous, 369
   average, 32–35, **33**
   average, in three dimensions, 101–2
   average, in two dimensions, 95
   of center of mass, 323–24, 348
   components of, 95–98, **197, 199**
   escape, 292
   of flow, 566, 568
   of galaxies, 1422
   Galilean addition law for, 1218
   horizontal, 103–8, **103, 106, 107, 108**
   instantaneous, *see* instantaneous velocity
   magnitude of, 96
   muzzle, 287, 331
   negative, 39
   positive, 39
   speed vs., 33–34, 96
   terminal, 53
   time vs., **61**
   transformation equations for, 1232–38, 1249–50
   vectors, 98–100, **955, 965**
Velocity Peak, Colorado, speed skiing, 267
velocity selector, 987
velocity transformation, Galilean, 1218
venous pressure, 597, **597**
Venturi flowmeter, 585–86, 596, **596**
Venus, 285*t*, 286, **286, x.**
Verne, Jules, 300
Very Large Array (VLA) radio telescope, 1186
very-long-baseline interferometry (VLBI), 1202
vibrational energy, of molecules, 1333
violin, 546, **546**
   frequencies, 530, 531, 558
   sound wave emitted by, 538
   vibration of standing wave, **546**
   wavefront emitted, **538**
Virgo Cluster (galaxies), xliv, **xlv**
virtual image, 1116, **1116, 1132, 1134, 1139–40.**
virtual photon, 1410
viscosity, 566, 595
viscous forces, 181, 196
visible light, 1090–91
   colors and wavelengths of, 1090–91, **1091**
volcanic bombs, 94, 100, **100, 105**
volt (V), 792
Volta, Alessandro, Conte, *793*
voltage, DC, 889–90, 1004
   AC, 1004
voltmeter, 905, **906**, 916, **916**, 917
volume, 13
   compression of, 446
volume expansion, 634
volume expansion coefficient, 634, 634*t*
Von Mayer, Robert, *629*, 653
vortex, 546, **546**, 571
Vostok, Antartica, coldest temperature, 621
Voyager spacecraft, 358

W⁺ particle, 1409, 1411, 1415
Wairakei, New Zealand (geothermal plant), **690**
Warsaw Radio tower, 634
water:
   thermal expansion of, 635, **635**
   triple point of, 609, **609**
   volume as function of temperature, **634**
waterfall:

Chamonix, 652
   hypothetical (M. C. Escher)
   see also Niagara Falls
water-moderated nuclear reactor, 1383
water molecule, 743
   distance and algle between component atoms, 332
   moment of inertia for, 388, 389
   motion, **566**
water park, **507**
waterwheel, 215–16, **216, 217**
   overshot, **267**
   undershot, **267**, 653
Watt, James, **254**
watt (W), 254, 540
watt balance, 11, **11**
wave, 507–64
   amplitude of, 511
   angular frequency of, 512, 513–16
   beats of, 518
   constructive and destructive interference for, 517
   crests, 509–10
   deep water, 526
   electromagnetic, *see* electromagnetic wave
   energy in, 1092–96, **1106–07**
   freak (North Atlantic), 533
   frequency of, 510–11
   harmonic, 510–13, **511**
   intensity of, **540**, 1093–94
   light and radio, 1079–1110
   long, 1088
   longitudinal, **508–9**, 509, 538, **538**
   measurement of momentum and position of, 1277–78
   medium, 1088
   modulation of, 518
   motion, 508–9
   number, 511
   ocean, 553
   ocean wavelength and speed, 528
   particle vs., 1276–79
   periodic, 509–16, **510**
   period of, 510, **510**, 513
   plane, 537, **537**, 1081–83, **1082, 1097**
   pool, **507**, 511, 515
   power transported by, 516
   probability interpolation and, 1277
   pulse, 508–9, **514, 522**
   radio, 1079–1110, **1089, 1094**
   shallow water, 515–16, 527, 528
   short, 1088
   sound, *see* sound waves
   speed, 510
   speed of, on a string, 513–16
   standing, 520–24, **524**
   standing, in a tube, 545, **545**
   superposition of, 516–20, **518**
   transverse, 508, **508**
   trough, 509–10, **510**
   TV, 1088
   water, 515–16, 526–28
wave equation, 1096–99, **1097**
wave fronts, 537, 541–42, **541**, 1113, **1113, 1115**
   circular, 537
   plane, 537
   spherical, 537
wavefunction, 511
   harmonic, **511**, 513–16
wavelength, 510–13, **510, 511**
   of visible light, 1090–91, **1091**

wavelength bands, of electromagnetic radiation, 1090–91, **1091**
wavelength shift, of photon, in Compton effect, 1270–71
wavelets, 1113, **1120**, 1276
wave mechanics, 1309, 1322
wave optics, 1169
wave pulse, electromagnetic, 1080–83, **1082**
waves, radio, 1089
wavicle, 1255, 1276
   measurement of, 1277–78
"weak" force, 191, 1369, 1405, 1406, 1406*t*
"weak" interactions, 1369, 1405, 1406, 1406*t*
Weber, Franz, speed skiing, 267
weber (Wb), 997
weight, 11, 141–42, **141**
   apparent, 187–88, **187**
   mass vs., 141
weightlessness, 142, **142**
   simulated, 589
Weinberg, Stephen, 1411
whale:
   breaching, 263, **264**
   song, 558
Wheatstone bridge, 906–7, **907**
wheel pottery, **409**
white noise, 538
Wien's Law, 1261
Wilson, Robert W., 1421
winches, 441, **441**, 458–59, **458**
   geared, **459**
wind instruments, 546
wing, flow around, 571, 584
wire:
   magnetic force on, 969–72
   straight, magnetic field of, 940–41, **941**
   uniform, electric field in, 860
work, 205–18
   calculation of, 218
   definition of, 205
   done by constant force, 205, 208
   done by gravity, **207**
   done by variable force, 211–13, **211, 212, 213, 214**
   dot product in definition of, 208–9
   frame of reference in calculation of, 208, **208**
   integrals for, 212–13
   internal, in muscles, 208
   in one dimension, 205–8, **206**
   in rotational motion, 395–99, **794**
   in three dimensions, 208–10, **209, 210**
   zero, 238–39
work-energy theorem, 215, 236, 400
work function, 1266
W⁻ particle, 1409, 1411, 1415
WWV (NIST radio station), 9, 23

xerographic copier, 786, **786**
xerography, 709, **709**
X-ray, scattering, **xlviii**
X rays, 1090, **1269**, 1273–75, **1273, 1274**, 1421
   characteristic, 1303

yaw, 366
Young, Thomas, 1169, 1170, *1170*, 1177
Young's modulus, 447–48, 447*t*, 460, 479

zero-point energy, 1307
zero work, and conservative force, 238–39
Z⁰ particle, 1409, 1411, 1415
Zweig, G., 1413